Aphids as Crop Pests

_____

Aphids as Crop Pests

Edited by

Helmut F. van Emden

Emeritus Professor of Horticulture, University of Reading, UK

and

Richard Harrington

Plant and Invertebrate Ecology Division, Rothamsted Research, UK

www.cabi.org

CABI is a trading name of CAB International

CABI Head Office	CABI North American Office
Nosworthy Way	875 Massachusetts Avenue
Wallingford	7th Floor
Oxfordshire OX10 8DE	Cambridge, MA 02139
UK	USA

Tel: +44 (0)1491 832111
Fax: +44 (0)1491 833508
E-mail: cabi@cabi.org
Website: www.cabi.org

Tel: +1 617 395 4056
Fax: +1 617 354 6875
E-mail: cabi-nao@cabi.org

A catalogue record for this book is available from the British Library, London, UK.

A catalogue record for this book is available from the Library of Congress, Washington, DC.

ISBN-13: 978 0 85199 819 0

Typeset by AMA DataSet Ltd, UK.
Printed and bound in the UK by Cromwell Press, Trowbridge.

The paper used for the text pages in this book is FSC certified. The FSC (Forest Stewardship Council) is an international network to promote responsible management of the world's forests.

In Memoriam

François Leclant

Emeritus Professor of Entomology, ENSA, Montpellier, France

who died while this book was in preparation

*He told me he particularly enjoyed writing his chapter,
since it was one of the few times he forgot his
terminal illness. His research was highly respected
internationally, he was a gifted teacher, and
a generous friend (HvE)*

Contents

_____

The colour plate section can be found following page 356

Colour plates sponsored by Syngenta and BASF
Many aphids show considerable colour variation; mostly only one colour variant is illustrated.

Preface

Following the publication in 1998 of *Thrips as Crop Pests,* edited by Trevor Lewis, CABI commissioned *Aphids as Crop Pests* as a second in the series. However, in prefacing this book, we wish to pay tribute to another predecessor, *Aphids: Their Biology, Natural Enemies and Control*, edited by Albert Minks and Paul Harrewijn and published in three volumes by Elsevier between 1987 and 1989. This comprehensive and impressive work has been the standard reference text for aphidologists for nearly 20 years, and we would not presume to make it redundant with the publication of *Aphids as Crop Pests*. Therefore, in planning the content of our book, we have concentrated on the applied aspects of aphidology and have not sought to match the comprehensive coverage of morphology, physiology and ecology found in *Aphids: Their Biology, Natural Enemies and Control*. As far as aphid biology is concerned, we have instead emphasized areas such as current taxonomic issues (Chapter 1), host selection and feeding (Chapter 4), nutrition and symbionts (Chapter 5, which includes full instructions for successful long-term rearing of *Myzus persicae* on synthetic diet), chemical ecology (Chapter 9) and reactions to stress (Chapter 11) where there have been major advances in the last fifteen years. We would especially direct the reader to Chapter 2 on Population Genetics; molecular markers were hardly available when Minks and Harrewijn's book was being put together. We have therefore given a large page allocation to this topic. In similar vein, molecular methods have enabled many substantial recent advances in our understanding of the mechanisms of aphid resistance to insecticides (Chapter 10). We also felt that life cycles and polymorphism (Chapter 3), Growth and Development (Chapter 6) and aphid movement (Chapter 7) were areas particularly relevant to aphids as pests.

We have been greatly encouraged by the readiness with which the potential authors we first contacted agreed to participate. Having recruited scientists we felt were at the forefront of their subject, it made sense to leave the selection of any co-authors to them. The result is that Chapter 12 (Population Dynamics) has been brought up to date with a strong modelling component, and that Chapter 8 on the natural enemies of aphids has involved four leading specialists – one for each of the major taxa. This led us to expand the page allocation for this chapter also.

The more identifiably applied part of the book starts with two chapters on the injury caused to crops by aphid feeding (Chapter 13) and by the transmission of plant viruses (Chapter 14). Both these chapters considerably update previous reviews on these topics.

Separate chapters then cover the four main building blocks of pest management. Chapter 15 on Chemical Control has much new material with the advent of noonicotinoid compounds, which are especially effective against aphids, and Chapter 16 (Cultural Control) highlights how far conservation biological control has come into practice. Chapter 17 covers Host Plant Resistance and Chapter 18 deals with Biological Control. This latter chapter describes practical interventions, and avoids making a stronger case for the method than is realistic. The involvement of a specialist on aphid pathogens enables these important tools to receive adequate treatment.

Chapter 19 on Monitoring and Forecasting is again very different from what could have been written on this topic in the 1980s, since advances in information technology have enabled huge databases to be feasible and have made possible rapid transmission of data and effective international networks.

Chapter 20 introduces the concepts and potential of IPM, pointing out that it originated in California with work on aphids. Specialists on aphid problems in selected cropping scenarios then review the status of IPM in a series of Case Studies (Chapters 21–30). We felt that the diversity of cropping-systems, pests and natural enemies, chemicals available etc. in different continents made a world-wide coverage for each crop too complex a proposition, and we encouraged authors to concentrate on the region in which they themselves worked. Imposing a template for these chapters would have clouded the very different issues that relate to aphids in these case studies, and we have taken the alternative approach towards consistency of ending each case study with an "Executive Summary", initially written by the editors on the basis of the text supplied by the chapter author(s) but then mutually agreed with them. The final Chapter on Decision Support Systems is another topic which has only developed strongly since the 1980s.

Inevitably, various problems forced a few of our original authors to withdraw and so we are especially grateful to those who only started on their chapters later and to the other authors for their patience when they saw these chapters delaying production of the book. We are also indebted to our authors for their invariably positive response to our comments and suggestions for changes; they all readily accepted that any large group of virtuoso performers needs a conductor!

The editors of a multi-author book face many problems, in particular that of repetition: take, for example, resistance to insecticides. Although our book has a chapter specifically on this topic, it also is a phenomenon of population genetics, it governs what chemicals can be used and so is relevant to the Chemical Control chapter and many of the IPM Case Studies. It would be tidy to eliminate duplication, but we are not so naïve as to think anyone is going to read our book from start to finish. Rather, each chapter needs to stand on its own as a comprehensive review. We have used cross references to other chapters as a partial solution, but have frequently left some discussion of a topic covered in another chapter where a cross reference alone would fail to make the point(s) the author had intended.

Another problem is the nomination of pesticides for aphid control, particularly in the IPM Case Studies. What products are permissible for use varies from country to country and changes almost weekly even within countries! Individual experts have therefore referred to products in relation to their own experience and environment. Mention of a compound should not be taken to mean it can be used elsewhere, nor necessarily after 2006.

The chapters are followed by a "Taxonomic Glossary". Here we have listed the Latin names of species mentioned in the book under the appropriate taxon, and then in alphabetical order together with the taxonomic authority and any generally accepted English common name(s). Such a glossary had to be compiled in order to get consistency between chapters in the citing of Latin names, and including it in the volume has meant we could avoid adding the taxonomic authority and Latin names of well-known crops in the individual chapters. In compiling the glossary we have encountered different practices in different taxa (e.g. insects, fungi, plants etc.) in how authorities and subspecies are cited. Thus entomologists tend to assume that a trinomial indicates a subspecies whereas botanists still insert "subsp." before the last name. We decided to follow the accepted practice for each taxon rather than impose uniformity. We are extremely grateful to the specialists who advised on current taxonomic usage, Dr Anne Baker (mites), Dr Roger Blackman (aphids), Mr Barry Bolton (ants), Dr Roland Fox (other fungi and bacteria), Dr Francis Gilbert (syrphids), Dr Simon Gowen (nematodes), Mr Paul Hillyard (spiders), Dr Stephen Jury and Mr Ronnie Rutherford (plants), Dr Ivo Kovar and Dr Mike Majerus (coccinellids), Dr Martin Luff (carabids), Professor Tim New (lacewings), Dr Judith Pell (entomopathogens) and Mr Nigel Wyatt (Diptera). Another specialist who provided valuable advice was Professor Roger Plumb in relation to the nomenclature of plant viruses.

Generous sponsorship from BASF and Syngenta has made it possible to include a section of colour plates of the aphids selected by the authors of Chapter 1 as the most serious aphid pests worldwide. The section also includes colour plates of representatives of important groups of natural enemies. We are extremely grateful for this encouragement and support from the agrochemical industry. The colour plates come from a variety of sources, and we greatly appreciate the permission of copyright holders to reproduce their images without paying a royalty. Many others (scientists and publishers) have also allowed us to use their material in the various chapters; their colour images have then been converted to grayscale. Our colleagues have without exception been most helpful in allowing us to use copyright material – in every case we have acknowledged the source in the legends to colour plates and other figures.

Part of the value of this book is in the comprehensive bibliographies accompanying each chapter. In order not to restrict authors in reviewing their topic, the list of references was not included in their page allocation. The work of editing the references was thus considerable and we are grateful to Ms Berit Pederson and Mr Greg Bentley for assistance with this chore.

We have both been researching on aphids for many years, and have become good friends in the process. We have enjoyed working together on this book and have both somehow found time to edit all chapters independently, though in each case one of us has taken the lead as "first editor". We will miss this collaboration now it has ended, but will feel rewarded if this book is appreciated by our aphidological colleagues and proves of value to them and those starting research on aphids for a good number of years to come.

Helmut van Emden
Richard Harrington

19th June 2006

List of Contributors

Caroline S. Awmack, *Department of Entomology, University of Wisconsin-Madison, Madison, WI 53706, USA* (Chapter 6)

Jeffrey S. Bale, *School of Biosciences, University of Birmingham, Edgbaston, Birmingham, B15 2TT, UK* (Chapter 11)

Sebastiano Barbagallo, *Dipartimento di Scienze e Tecnologie Fitosanitarie, University of Catania, Italy* (Chapters 29, 30)

Roger L. Blackman, *Department of Entomology, Natural History Museum, London, SW7 5BD, UK* (Chapter 1)

Jacques Brodeur, *Département de Phytologie, Université de Laval, Sainte-Foy, Québec, G1K 7P4, Canada* (Chapter 8)

John D. Burd, *Plant Science and Water Conservation Laboratory, USDA-ARS, Stillwater, OK 74075, USA* (Chapter 27)

Giuseppe Cocuzza, *Dipartimento di Scienze e Tecnologie Fitosanitarie, University of Catania, Italy* (Chapters 29, 30)

Rosemary H. Collier, *Warwick HRI, The University of Warwick, Wellesbourne, Warwick, CV35 9EF* (Chapter 21)

Piero Cravedi, *Istituto di Entomologia e Patologia Vegetale, University 'Cattolica Sacro Cuore', Piacenza, Italy* (Chapters 29, 30)

Jean-Philippe Deguine, *Centre de Coopération Internationale en Recherche Agronomique pour le Développement, Avenue Agropolis, 34398 Montpellier Cedex 5, France* (Chapter 23)

Gregor Devine, *Plant and Invertebrate Ecology Division, Rothamsted Research, Harpenden, Herts, AL5 2JQ, UK* (Chapter 10)

Alan L. Devonshire, *Biological Chemistry Division, Rothamsted Research, Harpenden, Herts, AL5 2JQ, UK* (Chapter 10)

Alan M. Dewar, *Entomology Research Group, Broom's Barn Research Station, Higham, Bury St. Edmunds, Suffolk, IP28 6NP, UK* Correspondance address: *Dewar Crop Protection Ltd, Drumlanrig, Great Saxham, Bury St Edmunds, Suffolk, IP29 5JR, UK* (Chapter 15)

Anthony F. G. Dixon, *School of Biological Sciences, University of East Anglia, Norwich, NR4 7TJ, UK* (Chapters 3, 12)

Angela E. Douglas, *Department of Biology, University of York, Heslington, York, YO10 5YW, UK* (Chapter 5)

Victor F. Eastop, *Department of Entomology, Natural History Museum, London, SW7 5BD, UK* (Chapter 1)

Helmut F. van Emden, *School of Biological Sciences, University of Reading, Whiteknights, Reading, Berks, RG6 6AJ, UK* (Chapters 5, 17, 20)

Stan Finch, *Warwick HRI, The University of Warwick, Wellesbourne, Warwick, CV35 9EF* (Chapter 21)

Stephen P. Foster, *Plant and Invertebrate Ecology Division, Rothamsted Research, Harpenden, Herts, AL5 2JQ, UK* (Chapter 10)

Bernd Freier, *Institute for Integrated Plant Protection, BBA, 14532 Kleinmachnow, Germany* (Chapter 25)

Robert T. Glinwood, *Department of Ecology, Swedish University of Agricultural Sciences, 750 07 Uppsala, Sweden* (Chapter 9)

Geoff M. Gurr, *School of Rural Management, Charles Sturt University, PO Box 883, Orange, New South Wales 2800, Australia* (Chapter 16)

Jim Hardie, *Division of Biology, Imperial College London, Silwood Park campus, Ascot, Berks, SL5 7PY, UK* (Chapter 4)

Richard Harrington, *Plant and Invertebrate Ecology Division, Rothamsted Research, Harpenden, Herts, AL5 2JQ, UK* (Chapter 19)

Maurice Hullé, *INRA, UMR BiO3P, 35653 Le Rheu, France* (Chapter 19)

Michael E. Irwin, *Department of Natural Resources and Environmental Sciences, University of Illinois, Urbana, IL 61801, USA* (Chapter 7)

Rufus Isaacs, *Department of Entomology, Michigan State University, East Lansing, MI 48824, USA* (Chapter 22)

Vojtěch Jarošík, *Department of Ecology, Charles University, 12844 Prague 2, Czech Republic* and *Institute of Botany, Czech Academy of Sciences, 25243 Pruhonice, Czech Republic* (Chapter 12)

Gail E. Kampmeier, *Section of Ecological Entomology, Illinois Natural History Survey, Urbana, IL 61801, USA* (Chapter 7)

Nikos I. Katis, *Aristotle University of Thessaloniki, Department of Agriculture, Laboratory of Plant Pathology, 541 24, Thessaloniki, Greece* (Chapter 14)

Pavel Kindlmann, *Faculty of Biological Sciences, University of South Bohemia and Institute of Landscape Ecology, Czech Academy of Sciences, 37005 České Budějovice, Czech Republic* (Chapter 12)

A. Michael Klüken, *Institute of Plant Protection and Plant Diseases, University of Hannover, 30419 Hannover, Germany* (Chapter 25)

Jonathan D. Knight, *Centre for Environmental Policy, Imperial College London, Silwood Park campus, Ascot, Berks, SL5 7PY, UK* (Chapter 31)

Shinkichi Komazaki, *Grape and Persimmon Research Station, National Institute of Fruit Tree Science, NARO, Hiroshima, Japan* (Chapters 29, 30)

Simon R. Leather, *Division of Biology, Imperial College London, Silwood Park campus, Ascot, Berks, SL5 7PY, UK* (Chapter 6)

François Leclant (the late), *Ecole Nationale Supérieure Agronomique, 34060 Montpellier Cedex 5, France* (Chapter 23)

Hugh D. Loxdale, *Plant and Invertebrate Ecology Division, Rothamsted Research, Harpenden, Herts, AL5 2JQ, UK* (Chapter 2)

Gugs Lushai, *Plant and Invertebrate Ecology Division, Rothamsted Research, Harpenden, Herts, AL5 2JQ, UK* (Chapter 2)

Manfred Mackauer, *Department of Biological Sciences, Simon Fraser University, Burnaby, B.C., V5A 1S6, Canada* (Chapter 8)

Gerald J. Michels, Jr., *Texas Agricultural Experiment Station, Bushland, TX 79012, USA* (Chapter 27)

Xinzhi Ni, *USDA-ARS Crop Genetics and Breeding Research Unit, Coastal Plain Experiment Station, Tifton, GA 31793, USA* (Chapter 13)

Judith K. Pell, *Plant and Invertebrate Ecology Division, Rothamsted Research, Harpenden, Herts, AL5 2JQ, UK* (Chapters 8, 18)

Jan Pettersson, *Department of Ecology, The Swedish University of Agricultural Sciences, 750 07 Uppsala, Sweden* (Chapter 4)

John A. Pickett, *Biological Chemistry Division, Rothamsted Research, Harpenden, Herts, AL5 2JQ, UK* (Chapter 9)

Manuel Plantegenest, *INRA, UMR BiO3P, 35653 Le Rheu, France* (Chapter 19)

Hans-Michael Poehling, *Institute of Plant Protection and Plant Diseases, University of Hannover, 30419 Hannover, Germany* (Chapter 25)

Katherine L. Ponder, *School of Biosciences, University of Birmingham, Edgbaston, Birmingham, B15 2TT, UK* (Chapter 11)

Glen Powell, *Division of Biology, Imperial College London, Wye campus, Ashford, Kent, TN25 5AH, UK* (Chapter 14)

Wilf Powell, *Plant and Invertebrate Ecology Division, Rothamsted Research, Harpenden, Herts, AL5 2JQ, UK* (Chapter 18)

Jeremy Pritchard, *School of Biosciences, University of Birmingham, Edgbaston, Birmingham, B15 2TT, UK* (Chapter 11)

Sharron S. Quisenberry, *College of Agriculture and Life Sciences, Virginia Tech, Blacksburg, VA 24061, USA* (Chapter 13)

Edward B. Radcliffe, *Department of Entomology, University of Minnesota, St. Paul, MN 55108, USA* (Chapter 26)

David W. Ragsdale, *Department of Entomology, University of Minnesota, St. Paul, MN 55108, USA* (Chapter 26)

Katherine A. Robinson, *National Centre for Advanced Bio-Protection Technologies, Lincoln University, Canterbury 7647, New Zealand* (Chapter 16)

Mark Stevens, *Broom's Barn Research Station, Higham, Bury St. Edmunds, Suffolk, IP28 6NP, UK* (Chapter 14)

Robert A. Surányi, *McLaughlin Gormley King Company, 8810 Tenth Ave. N, Minneapolis, MN 55427, USA* (Chapter 26)

G. Mark Tatchell, *Department of Biological Sciences, University of Warwick, Gibbet Hill Road, Coventry, CV4 7AL, UK* (Chapter 24)

Deborah J. Thackray, *Centre for Legumes in Mediterranean Agriculture, MO80, Faculty of Natural and Agricultural Sciences, University of Western Australia, 35 Stirling Highway, Crawley, WA 6009, Australia* and *Plant Pathology, Department of Agriculture and Food Western Australia, Locked Bag 4, Bentley Delivery Centre, WA 6983, Australia* (Chapter 31)

W. Fred Tjallingii, *Laboratory of Entomology, Wageningen University, PO Box 8031, Wageningen, The Netherlands* (Chapter 4)

John A. Tsitsipis, *University of Thessaly, Department of Agriculture, Crop Production and Rural Environment, Laboratory of Entomology and Agricultural Zoology, 384 46, Nea Ionia Magnissias, Greece* (Chapter 14)

Jason M. Tylianakis, *School of Biological Sciences, University of Canterbury, Private Bag 4800, Christchurch 8020, New Zealand* (Chapter 16)

Maurice Vaissayre, *Centre de Coopération Internationale en Recherche Agronomique pour le Développement, Avenue Agropolis, 34398 Montpellier Cedex 5, France* (Chapter 23)

Wolfgang Völkl, *Department of Animal Ecology, University of Bayreuth, 95440 Bayreuth, Germany* (Chapter 8)

Susan E. Webb, *Entomology and Nematology Department, University of Florida, Gainesville, FL 32611, USA* (Chapter 28)

Wolfgang W. Weisser, *Institute of Ecology, Friedrich-Schiller-University, 07743 Jena, Germany* (Chapter 7)

Iain S. Williams, *Science Directorate, DEFRA, Nobel House, Smith Square, London, SW1P 3JH, UK* (Chapter 3)

J. A. Trefor Woodford, *Scottish Crop Research Institute, Invergowrie, Dundee, DD2 5DA, UK* (Chapter 22)

Steve D. Wratten, *National Centre for Advanced Bio-Protection Technologies, Lincoln University, Canterbury 7647, New Zealand* (Chapter 16)

1 Taxonomic Issues

Roger L. Blackman and Victor F. Eastop

Department of Entomology, Natural History Museum, London, SW7 5BD, UK

Introduction

There are about 4700 species of Aphididae in the world (Remaudière and Remaudière, 1997). Of these, about 450 species have been recorded from crop plants (Blackman and Eastop, 2000), but only about 100 have successfully exploited the agricultural environment to the extent that they are of significant economic importance (Table 1.1). The agriculturally important species are mostly in the subfamily Aphidinae, not only because this is the largest subfamily, but also because it contains a very high proportion of the aphids that feed on herbaceous plants (Blackman and Eastop, 2006). Some quite large aphid subfamilies – the Calaphidinae and Lachninae, for example – are associated almost exclusively with woody plants, as are most of the smaller ones.

The Aphididae is one of three families of Aphidoidea, the other two being the Adelgidae, or conifer woolly aphids, and the Phylloxeridae, which are also nearly all associated with trees but include the notorious *Viteus* (= *Daktulosphaira*) *vitifoliae* (grape phylloxera) (Table 1.1). In the context of the order Hemiptera as a whole, the superfamily Aphidoidea is placed in the sub-order Sternorrhyncha, along with Coccoidea (scale insects and mealybugs), Aleyrodoidea (whiteflies) and Psylloidea (psyllids or jumping plant lice).

All these insects are phytophagous, and most of them are sap-sucking. Historically, the Sternorrhyncha have been grouped with the Auchenorrhycha (leafhoppers, cicadas, etc.) as Homoptera, but molecular work has provided strong support for the long-standing view, based on morphological and embryological evidence, that Sternorrhyncha and Auchenorrhyncha do not share a common ancestor, so this grouping is not phylogenetically sound. The general consensus now is that the primary division of the Hemiptera is into Sternorrhyncha and a sister group comprising Auchenorrhyncha plus Heteroptera (von Dohlen and Moran, 1995).

One major feature, their well-known cyclical parthenogenesis, sets aphids apart from other Hemiptera, and has influenced every aspect of their biology. The system of alternating one bisexual generation with a succession of parthenogenetic, all-female generations evolved in the ancestral line of all Aphidoidea, probably as far back as the Triassic. At first, the parthenogenetic females must have laid eggs like their sexual counterparts, as do the parthenogenetic females of present-day adelgids and phylloxerids. Then, in the line leading to all modern Aphididae, the parthenogenetic generations developed the further refinement of viviparity, which effectively 'telescoped' generations into one another, thereby greatly reducing the

©CAB International 2007. *Aphids as Crop Pests*
(eds H. van Emden and R. Harrington)

Table 1.1. Biological features and distribution of agriculturally important species among the major groups of aphids.

Taxon	Characteristic host plants	Geographical distribution	Host alternation?	No. of agriculturally important spp./total spp.	Representative genera
Adelgidae	Coniferae	Holarctic	Most	0/49	Adelges, Pineus
Phylloxeridae	Fagaceae, Juglandaceae	Holarctic (mostly nearctic)	Some	1/64	Phylloxera, Viteus
Aphididae:					
Eriosomatinae	Ulmaceae, Salicaceae, Anacardiaceae, roots of gymno- and angiosperms	Holarctic, oriental	Most	9/319	Eriosoma, Pemphigus
Hormaphidinae	Hamamelidaceae, Styracaceae, palms, bamboos	Mostly oriental	Many or most	5/176	Astegopteryx, Cerataphis, Ceratovacuna
Anoeciinae	Cornus, grass roots	Holarctic	Most	2/24	Anoecia
Calaphidinae	Deciduous trees	Holarctic	None	6/331	Chromaphis, Myzocallis, Therioaphis
Saltusaphidinae	Sedges	Holarctic	None	0/68	Iziphya, Subsaltusaphis, Thripsaphis
Chaitophorinae	Salicaceae, Acer, grasses	Holarctic	None	2/164	Atheroides, Sipha
Greenideinae	Dicotyledonous trees	Eastern palaearctic, oriental	None	2/151	Anomalosiphon, Greenidea
Pterocommatinae	Salicaceae	Holarctic	None	0/51	Pterocomma
Aphidinae:					
Aphidini	Many families including Rosaceae, Poaceae, Asteraceae	Holarctic, a few southern hemisphere	Some	20/738	Aphis, Rhopalosiphum, Schizaphis, Toxoptera
Macrosiphini			Some	60/2016	Acyrthosiphon, Brevicoryne, Macrosiphum, Myzus
Lachninae	Trees (dicotyledonous and coniferous)	Holarctic	None	4/356	Cinara, Lachnus, Trama
15 other subfamilies	Mostly dicotyledonous trees	Some are oriental and southern hemisphere	None	0/185	

generation time and enabling far more efficient exploitation of periods of rapid plant growth.

Aphids (Aphididae) can be recognized by a number of shared morphological characteristics that developed before the divergence into present-day subfamilies: e.g. siphunculi (secretory organs, but with their precise function still strangely enigmatic); five- or six-segmented antennae composed of two basal segments and a segmented flagellum with a terminal process; two-segmented tarsi with the first segment much shorter than the second; and a cauda, or tail, that is often used for flicking away droplets of honeydew from the anus. These features have been modified, reduced, or secondarily lost in some species, but are evident in most aphids that are pests of crop plants.

Of the present-day subfamilies of Aphididae, one in particular was successful at exploiting the rapid expansion of numbers and diversity of herbaceous flowering plants in the Tertiary period. This subfamily, the Aphidinae, with 2750+ extant species, is predominantly a northern temperate group, with life cycles closely tied to temperate seasonality and the phenologies of temperate plants. Originally on woody plants, they evolved a system of host alternation, migrating to completely unrelated herbaceous plants for the summer months, where their parthenogenetic generations could continue to utilize stages of rapid plant growth. However, today, only about 15% of Aphidinae host-alternate. Some of the other 85% live only on woody plants, but most of them, including some of the largest and most successful genera, have lost or given up the ancestral woody (primary) host, and now live all-year-round on herbaceous plants. Host alternation has, in fact, evolved independently in several other aphid subfamilies (von Dohlen and Moran, 2000), but the Aphidinae are the only subfamily to exploit numerous families and genera of flowering plants.

In the rest of this chapter, we will look at aphids from a taxonomist's viewpoint. We will highlight some of the problems of interpreting the observed variation within and between species, and discuss what the names we give to aphids really mean. Then, as examples, we will discuss 14 taxa that probably head the list of economically important aphid species.

Interpreting Variation in Aphids

In the year 2000, soybean crops in the USA and Australia were attacked for the first time by large numbers of an aphid closely resembling the well-known polyphagous species, *Aphis gossypii* (cotton or melon aphid), but obviously with a far greater affinity for soybean. Taxonomists identified the species as *Aphis glycines* (soybean aphid), previously known only from the Far East. It had been introduced probably a year or two earlier, building numbers and spreading until field entomologists and growers realized that they had something new on their hands. *Aphis glycines* is biologically quite different from *A. gossypii*; it is specific to soybean in summer and has host alternation, overwintering on *Frangula* spp. Identification immediately unlocked much crucial information about this species and its biology, as well as where to look for its natural enemies. In this case, the question 'What is it?' seems to have a fairly clear-cut answer; the soybean aphid is a relatively well-known and well-studied species in Eastern Asia. Although morphologically resembling *A. gossypii* and requiring specialist identification, it is clearly a distinct species. There are plenty of other cases, however, including other close relatives of *A. gossypii*, where the question of identity is not so easily answered. Several of these are discussed in the next section of this chapter. Sometimes, taxonomic difficulties arise as a result of founder effects and, even in the case of the soybean aphid, it may be necessary to bear in mind that the introduced population has been through a recent 'bottleneck', and can be expected to have less allelic diversity than East Asian populations.

Taxonomy and identification are a matter of interpreting observed variation. The first problem to be overcome is the effect of the environment on the phenotype, especially if the only available data are morphological. Most species are distinguished and described originally, often from one small sample only, using morphological criteria, and most identifications are based on keys that use morphological discriminants. As a group, aphids are renowned for the considerable extent to which the phenotype is influenced by environmental factors. Within any aphid species, there are a number of different forms (morphs) with discrete morphological differences, which may be triggered by specific environmental stimuli such as day length or crowding. It is well known, for example, that in most aphids, the parthenogenetic females can be winged or wingless, the differences being not just in the presence or absence of wings but involving every part of the body. Separate identification keys are therefore needed for each morph, but the structural distinction between the winged and wingless morphs is not always as clear as might be expected, as wingless individuals may occur with some tendency towards the characters of the winged morph. Likewise, intermediates between other morphs can occur; for example, between viviparous parthenogenetic females and oviparous sexual females.

The range of continuous morphological variation is also wider than in many other insect groups. Increases or decreases of size due to nutritional effects, for example, can accumulate over several generations, because the size of the mother can affect the size of her offspring. There may be large seasonal differences, with some species producing dwarf individuals when food quality is poor in midsummer. The parthenogenetic female that hatches from the overwintering egg – the fundatrix or stem mother – is morphologically different from the later generations of parthenogenetic females, and sometimes the particular features of the fundatrix are only gradually lost in subsequent generations. In some aphid groups – although not in most Aphidinae – the generations on the secondary host differ greatly in morphology from those on the primary host, and were originally described as different species and often placed in different genera. The temperature experienced during development can also have profound effects on morphology, involving not only general body size and pigmentation, but also having more subtle influences on the length relationships between body parts (allometry), which can play havoc with the morphometric ratios and functions often used to discriminate between closely related species (Blackman and Spence, 1994).

Thus, it is important to take into account the possible effects of season, host plant, and climate when examining samples of field-collected aphids. It is also necessary to bear in mind that a sample might consist of a single clone, especially if it comes from a warm temperate or subtropical region where the population has not gone through a sexual phase and is therefore more likely to be clonally structured. All the aphids in a clone are, discounting mutations, genetically identical, so a sample consisting of a single clone will give a misleading idea of the range of variation in the species. Sometimes 'abnormal' morphological features may occur, which may be due to mutation, but are often no more than character states near the extremes of a range of continuous variation, and part of the natural variation of all living organisms. In sexually reproducing organisms, any particular abnormal character state is likely to occur in only the occasional, rare individual within a population, and thus will be instantly recognizable as an extreme or rare condition. In aphids, it is possible to find a whole colony of individuals all with the same anomalous characters. Such a colony has sometimes been erroneously described as a new species.

The Taxonomy of Pest Aphids – What's in a Name?

The species that become pests are those that are best able to adapt to and exploit man-modified environments. Many of them belong to groups that were probably already speciating rapidly before human intervention, and the colonization of new geographical regions and/or new habitats are potent factors leading to further divergence and change. Several examples of this are discussed in the concluding part of this chapter.

Although it is clear that pest aphids are highly dynamic, rapidly evolving systems, there still seems to be a tendency, among agricultural entomologists in particular, to regard pest species identifications provided by taxonomists as names set in stone, and to react with alarm to the suggestion that certain common and well-known pest species may, in fact, be more accurately described as species complexes. The name has great importance, as it is the key to what we know about a species and its way of life, and to how we might expect it to behave. The name *Myzus persicae*, for example, identifies a set of populations that are closely related and share numerous attributes. But this must not be allowed to mask the heterogeneity that is also present, and which may in particular include populations that have diverged genetically to such an extent that they have evolved past the stage of being host races and achieved a degree of permanency, so that they can be regarded as incipient or sibling species (or subspecies) with particular attributes of their own.

The recognition of such divergent populations or taxa can add to our understanding of the ecology and evolution of a pest species complex, and increase the possibility of devising effective control measures. It is also important that they are named in some way, because without any consistent method of referring to them, new information is inaccessible, lost in the mass of literature about the species as a whole. Yet, both the recognition and the naming of such categories are fraught with difficulty. Discrimination of recently derived taxa may require sophisticated techniques, which are likely to be time-consuming and/or expensive. It is also hard to decide what to call them. Insect taxonomists are usually reluctant to provide formal names and descriptions and to give full species status to members of species complexes that cannot be distinguished readily by their morphology alone, even where consistent differences can be demonstrated in biology and/or host relationships. This is understandable, given that they probably will be asked to recognize and identify dead specimens of their 'new' species!

The subspecies category is the only intraspecific category recognized by the Zoological Code of Nomenclature, and therefore the only one that can be used formally to designate new taxa in animal taxonomy. It was developed mainly by vertebrate taxonomists, who define subspecies as geographically localized populations that are morphologically distinguishable. In aphid taxonomy, considerable use has been made of the subspecies category. Remaudière and Remaudière (1997) list 141 accepted subspecies names in the subfamily Aphidinae alone. Unfortunately, the subspecies designation has been used with an almost complete lack of consistency. In some cases, a single sample has been described as a subspecies, for no other reason than that it shows some deviation from the known range of variation of a species. This could be due simply to it comprising a single clone with certain anomalous features, or a colony that developed under unusual microclimatic conditions, or part of a continuous geographical cline of variation from which intermediate populations have not been sampled.

Aphid taxonomists studying groups of very closely related taxa ('species complexes') have also used subspecies in a completely different way to define populations that are morphologically very similar but that have been shown by field observation and/or experimental studies to differ in their life cycle or host-plant relationships (Müller, 1986). Such an approach recognizes that speciation processes in aphids may be very different from those of vertebrates, with changes in life cycle and/or host relations acting as the primary isolating mechanism and trigger for speciation, rather than spatial isolation (Guldemond and Mackenzie, 1994). Crucial evidence that such speciation processes may operate in aphids has been provided recently by Hawthorne and Via (2001). Thus, in rapidly speciating groups of aphids, one may expect to find incipient species that are at an early stage of divergence where they can be distinguished more easily by biological rather than by morphological properties.

Every speciation process is a unique event, or series of events, which may include transitional phases involving a wide variety of reproductive relationships between the incipient species. It is impractical to try to define such transitional phases except in the most flexible terms. Blackman (1995) suggested that the term 'semispecies' could convey this idea of 'species in the making', without implying anything about the nature of the speciation process or the degree of reproductive isolation. Semispecies have been used already for many years by both plant and animal taxonomists with experience of the problems of describing species complexes, including some of the best-studied ones such as those of the fruit fly, *Drosophila*. They have not, however, been used for aphids, or most other insect groups of agricultural importance, nor have they been accommodated into the Zoological Code of Nomenclature. On further reflection, it seems unrealistic to expect the wide acceptance and use in aphid taxonomy or agricultural entomology of the term semispecies and, even if it were accepted, that would still not resolve the problem of nomenclature.

Rakauskas (2004) has recently revived Müller's (1986) proposal for a broader use of the subspecies category to validate its application to aphid species complexes, and – of particular relevance to the present chapter – to meet the practical need of providing names that identify intraspecific categories in groups that include pest species. We support this idea and accommodate it in our discussion of the taxonomy of some of the major pest aphids that occupies the remainder of this chapter.

Blackman (1995) drew attention to the extensive misuse of the term 'species concept' in the literature on evolution and speciation, which has made it difficult for taxonomists and evolutionists working in different fields and on different groups of organisms to agree about the nature of species. The same difficulties have also impeded the use of the subspecies category. To emphasize the distinction between a concept and a working definition in the case of subspecies, we offer below (i) a description of the 'subspecies concept' that is applicable to all organisms irrespective of their modes of speciation, and (ii) a working definition for the subspecies category in aphids.

The subspecies as a concept. Subspecies can be conceived as a species in the making; populations or groups of populations that have diverged to the extent that they are recognizable entities which show certain consistent distinguishing properties, indicating an interruption or significant restriction of gene flow and making it reasonable to conclude that speciation is in progress, but has not reached the stage of irreversibility.

A working definition of an aphid subspecies. A group of populations that is recognizably part of an existing species yet maintains a consistent suite of properties distinguishing it from other populations within that species. The species should be sufficiently well known for it to be reasonably certain that the observed variation is discontinuous, and the consistency of this discontinuity should be demonstrated by samples from more than one time and place. The likely cause of the discontinuity should be identifiable, for example a difference in host-plant relationships, life cycle properties, or geographic location, and should not be of such a kind as to make it irreversible (e.g. permanent parthenogenesis). The description of a new subspecies should give a clear and accurate account of its morphological and biological properties in comparison with those of other populations within the species, and include the best possible morphological discriminants. One clone should be designated as the type, with slide-mounted specimens deposited in a national collection and other specimens of the same clone deep-frozen and/or preserved in absolute ethanol for future DNA studies.

The 14 Aphid Species of Most Agricultural Importance
(See also colour plates in this volume.)

There is a very large literature about all the major pest aphid species. Much of the recent work is readily accessible on the Internet, and summary accounts are provided by Blackman and Eastop (2000). The treatments that follow are therefore largely concerned with the taxonomic issues raised by these 14 species, and the two factors that have the greatest influence on intraspecific variation in aphids, the life cycle and the host plant. These factors are considered further in the next chapter from a population geneticist's viewpoint.

Acyrthosiphon pisum (pea aphid)

Acyrthosiphon pisum is a rather large, green or pink aphid with long, slender appendages, forming colonies on young growth and developing pods of many leguminous plants. Its host plants are mostly Fabaceae of the the tribes Genistae (*Cytisus, Genista, Sarothamnus, Spartium*), Trifoliae (*Medicago, Melilotus, Ononis, Trifolium, Trigonella*), Fabeae (*Lathyrus, Lens, Pisum, Vicia*), and Hedysareae (*Hippocrepis, Onobrychis*), and it also colonizes a few members of other tribes, e.g. *Lotus* (Loteae) and *Glycine* (Phaseolae). Many legumes, including some of economic importance (e.g. *Phaseolus*), and almost all other plants are not colonized, although under dry conditions it is sometimes found on *Capsella bursa-pastoris*. It is a vector of more than 30 virus diseases, including non-persistent viruses of beans, peas, beet, clover, cucurbits, *Narcissus*, and Brassicaceae, and the persistent viruses, *Pea enation mosaic virus* (PEMV) and *Bean leaf roll virus* (BLRV).

Originally a palaearctic species, *A. pisum* now has an almost worldwide distribution (see CIE Distribution Map 23, last revised 1982). In cold temperate regions, it is holocyclic, producing oviparae and males on various leguminous hosts. As in other members of the genus *Acyrthosiphon*, there is no true host alternation. The ancestral primary host was presumably a member of the Rosaceae, as in related genera, but this was probably lost long ago. Clones may produce either apterous or alate males, or both. At warmer latitudes, it overwinters without a sexual phase.

In Europe and Central Asia, *A. pisum* seems to be a complex of races and subspecies with different host ranges and preferences (Müller, 1980, 1985). Populations attacking peas (*Pisum sativum*) in Europe consist entirely of green genotypes and have some morphological differences from those colonizing other leguminous crops such as lucerne (*Medicago sativa*), which may be green or pink. They produce alate sexuparae and males in autumn, and go through a sexual phase on *Vicia*. Recent molecular work has shown that populations colonizing peas, lucerne, and red clover in France are genetically divergent and also differ in their symbionts (Simon *et al.*, 2003).

There seem to be several other speciation events currently in progress in Europe, with populations also in the process of diverging on *Lotus* and *Sarothamnus*. A form that lives on *Ononis* in Europe is morphologically recognizable and is currently regarded as a subspecies (*A. pisum ononis*), but could be a distinct species.

Populations introduced to other parts of the world must at least originally have been genetically depauperate, and hence their biology and host-plant relations may have diverged significantly from the species in Europe. The genotype(s) introduced to North America initially lacked the dominant red (pink) allele that is frequent in European populations, and were particularly well adapted to lucerne. They were named as a new species (Johnson, 1900), and this name (*destructor*) has been proposed as a subspecific name not only for the North American populations (Hille Ris Lambers, 1947), but also for the green, pea-adapted form in Europe from which it was assumed to have been derived (Müller, 1985; but see below). Variation within and between North American populations has been studied with respect to resistance in lucerne (Frazer, 1972) and pea (Cartier, 1963) varieties, life-history traits (MacKay *et al.*, 1993), photoperiodic responses (Smith and MacKay, 1990) and migratory tendencies (Lamb and MacKay, 1979). Harper *et al.* (1978) provided a bibliography of all the earlier work on *A. pisum*.

In ground-breaking work that has illuminated the evolutionary processes that may be going on in crop-colonizing aphids, Via (1991, 1999) demonstrated that populations of *A. pisum* on lucerne and red clover in north-eastern USA perform significantly better on their respective host plants, and are reproductively isolated from one another as a result of inherited differences in host selection by alatae, which lead to assortative mating. Hawthorne and Via (2001) have shown recently that there is close genetic linkage at several loci between the two key traits involved in host specialization, i.e. host selection by alatae, and subsequent performance (measured by fecundity) of populations on each host. This could lead to extremely rapid divergent selection of pea aphids on lucerne and clover.

The implications of this work are far-reaching, although much is still unclear. The origins of the populations studied by Via's group are unknown, but subsequent work in Europe by Simon *et al.* (2003) suggests that the observed differences may be long-standing and the result of separate introductions from the palaearctic of two genotypes with different host associations, as seems to have happened with some other pest aphids in North America (see under *Therioaphis trifolii*, *Schizaphis graminum*, and *Myzus persicae*). More detailed genetic comparisons of European and North American populations are needed to confirm this. However, regardless of their origins, Via's *A. pisum* populations seem to have a genetic system that can lead to rapid divergence and incipient speciation in crop environments. Add in the probability that the same processes of selection and reinforcement of correlations between key traits are occurring in large populations throughout the regions where these crops are grown and there seems to be a very potent mechanism for evolutionary change.

If the host-adapted forms of *A. pisum* demonstrated by Via's and Simon's groups are indeed 'incipient species', can they be classed as subspecies? It would obviously be undesirable to give names to locally divergent populations, even though the first stage in the speciation process can occur only at the level of the local population. To conform with the working definition of aphid subspecies above, it would be necessary to demonstrate that the observed differences were consistent in both time and space, and verify the existence of monophyletic host-adapted lineages. It is already apparent that the name used originally for the North American lucerne-feeding form (*destructor*) was applied wrongly to the pea-feeding populations in Europe, although it still may be valid for a lucerne-adapted clade if this is common to both continents.

Aphis craccivora (cowpea aphid)

This is a small, dark brown aphid with a shiny black dorsal shield. It occurs most commonly on legumes, but is much more polyphagous than *A. pisum*, with a wide range of hosts not only in the Fabaceae (e.g. *Arachis*, *Colutea*, *Glycine*, *Medicago*, *Melilotus*, *Trifolium*, *Vicia*), but also in many other plant families (altogether it attacks about 50 crops in 19 different plant families). It is a vector of about 30 plant virus diseases, including non-persistent viruses of beans, cardamom (*Elettaria cardamomum*), groundnuts, peas, beets, cucurbits and crucifers, and the persistent *Subterranean clover stunt virus*, *Peanut mottle virus*, and the complex of viruses causing groundnut rosette disease. Jones (1967) compared two clones from East and West Africa differing in ability to transmit strains of groundnut rosette and found that there were also differences in ability to colonize various host plants. Apart from this, there has been little work of any significance on intraspecific variation in *Aphis craccivora*.

Aphis craccivora now occurs in most parts of the world (CIE Distribution Map 99, revised 1983), but its origins are clearly in Europe, as the most polyphagous member of a group of closely related species (subgenus *Pergandeida*), most of which are specific to particular species of Fabaceae. None of this group of *Aphis* has host alternation. *Aphis craccivora* has a sexual phase on various Fabaceae in Central Europe (Germany), and sexual morphs have also been reported from India and Argentina, but through most of the world, reproduction seems to be exclusively parthenogenetic. It is particularly common as a pest in warmer climates, and it seems likely that the pest populations may have originated from the warmer part of its original distribution area in Southern Europe or the Middle East.

Aphis fabae (black bean aphid)

Aphis fabae is perhaps the most familiar aphid in Europe, due to its predilection for *Phaseolus* and *Vicia*, although it is probably just as important as a pest and virus vector of sugarbeet. Adult apterae in new colonies on young plant growth are matt black in life, most of the pigmentation of the body being internal, so that it is no longer present in cleared specimens prepared for the microscope. Individuals in older colonies and feeding on leaves tend to develop white wax markings.

Aphis fabae is a very polyphagous species, but the actual host range of the aphid that colonizes beans and sugarbeet is unclear, because it is a member of a bewildering complex of species, at least some of which also have wide host ranges. Many people have tried to sort out this complex. Stroyan (1984: 119–122) reviewed the taxonomy and host relations of the group as then understood. There has been significant work since then (e.g. Müller and Steiner, 1986; Thieme, 1987, 1988; Thieme and Dixon, 1996; Raymond et al., 2001), but a lot of questions still remain. To summarize the situation in Northern Europe as succinctly as possible, there are five closely related taxa, four of which go through their sexual phase on the spindle tree, Euonymus europaeus:

1. Aphis fabae sensu stricto (A. fabae ssp. fabae), which has Euonymus as its only primary host and migrates for the summer to a wide range of plants, including Vicia faba, sugarbeet, Chenopodium album, and poppies, but not Solanum nigrum.
2. Aphis solanella, which also has Euonymus as its only primary host and migrates to a wide range of plants including S. nigrum, the leaves of which it crumples and curls characteristically, but will not colonize beans, sugarbeet, Chenopodium album, or poppies. Aphis solanella, which has shorter hairs than A. fabae and a greater tolerance of high temperatures, has usually been classed as a subspecies of A. fabae, but Thieme and Dixon (2004) suggested that it should have full species status. It is a pest in its own right, especially on Solanaceae.
3. Aphis fabae cirsiiacanthoidis, which normally uses Euonymus as a primary host but can also go through its sexual phase on Viburnum opulus. This form is morphologically hardly distinguishable from A. fabae sensu stricto, but has Cirsium arvense as its most characteristic summer host. It can colonize certain other plants, but is not found on Vicia or Solanum.
4. Aphis euonymi, a brown aphid that stays on Euonymus all year round.
5. Aphis fabae mordvilkoi, again almost indistinguishable from A. fabae sensu stricto, but with sexual generations on Viburnum opulus or Philadelphus coronarius, and migrating for the summer mainly to various secondary hosts, but not colonizing Vicia, Solanum, or Cirsium, and often occurring on Arctium spp. and Tropaeolum majus.

However, the host ranges of all members of the group seem to overlap and some plants, such as Rumex obtusifolius, seem to be accepted by them all (Thieme, 1987, 1988).

To complicate the story still further, the primary host relationships of the members of the group are also not at all clear-cut. Aphis fabae sensu stricto and A. solanella are able to live on Viburnum and Philadelphus, the primary hosts of A. f. cirsiiacanthoidis and A. f. mordvilkoi, and may in autumn produce males on these plants, but not oviparae (Iglisch, 1968). Hybridization between any members of the group is therefore theoretically possible, and can be accomplished rather easily in the laboratory, even between slightly less closely related forms such as A. fabae and the brown species, A. euonymi (Müller, 1982). However, it is now clear that such results are misleading, and that strong prezygotic isolating mechanisms are likely to be operating to prevent hybridization and promote assortative mating in the field. Aphis fabae sensu stricto and A. solanella occurring together on Euonymus show differences in the diurnal patterns of pheromone release by oviparae and male responsiveness, and in choice experiments, males preferred the sex pheromones of conspecific females (Thieme and Dixon, 1996). Prezygotic isolating mechanisms have also been demonstrated in the laboratory between A. fabae sensu stricto and A. f. mordvilkoi, although these subspecies do not share the same primary host in nature (Raymond et al., 2001).

In spite of this evidence of substantial reproductive isolation between the members of the A. fabae complex, analysis of mitochondrial DNA and of a plasmid from their symbionts showed differences within, but not between, species (Raymond *et al.*, 2001). This is similar to the situation in *A. pisum* already discussed (where host/habitat selection rather than mate selection is the isolating factor), and suggests that such isolating mechanisms may evolve very rapidly, possibly by the process known as reinforcement (Coyne and Orr, 1989; Mackenzie and Guldemond, 1994).

On the evidence of its primary host and closest relatives, *A. fabae* must be of European origin. *Aphis fabae sensu stricto* occurs in Europe, Western Asia, Africa, and South America. It is a vector of more than 30 plant viruses, including non-persistent viruses of beans and peas, beets, crucifers, cucurbits, *Dahlia*, potato, tobacco, tomato, and tulip, and the persistent *Beet yellow net virus* (BYNV) and *Potato leaf roll virus* (PLRV). In warmer regions – the Mediterranean and the Middle East, the Indian subcontinent, and hotter parts of Africa and South America – it is replaced by *A. solanella*, reproducing parthenogenetically throughout the year on its secondary host plants, particularly Solanaceae, Asteraceae, and Polygonaceae. The *Tropaeolum*-feeding subspecies *A. f. mordvilkoi* has a more northerly, holarctic distribution, and uses *Viburnum trilobum* as its main primary host in Canada (Barber and Robinson, 1980; as *A. barbarae*).

Aphis gossypii (cotton or melon aphid)

If the multiplicity of populations that are lumped under the name *A. gossypii* are really all one species, then it is indeed a remarkable one, with greater diversity in terms of host relationships, life cycle, and geographical range than any other aphid. Small aphids that vary greatly in colour from pale yellow dwarfs at high temperatures, through dirty yellow-green to dark bluish-green or almost black at lower temperatures, occur on plants in numerous families, including nearly a hundred species of crop plants, throughout the world. Crops attacked include cotton, cucurbits, citrus, coffee, cocoa, aubergine, peppers, potato, okra, and many ornamental plants including chrysanthemums and *Hibiscus*. Populations on cotton and cucurbits can be particularly large and damaging. More than 50 plant viruses are transmitted, including non-persistent viruses of beans and peas, crucifers, celery, cowpea, cucurbits, *Dahlia*, lettuce, onion, pawpaw, peppers, soybean, strawberry, sweet potato, tobacco, and tulips, and the persistent *Cotton anthocyanosis virus*, *Lily symptomless virus*, PEMV, and lily rosette disease.

In all the warmer parts of the world, *A. gossypii* reproduces continuously by parthenogenesis. It is particularly abundant and widely distributed in the tropics, including many Pacific islands. During prolonged dry seasons in hot countries, small colonies may survive on a great variety of plants on which they are seldom seen during the growing season, including Poaceae. Deguine and Leclant (1997) provided a comprehensive account with an extensive bibliography.

Certain morphological features – short hairs on legs and antennae, and a cauda that is usually paler than the siphunculi and bears rather few hairs – make it easy to apply the name *A. gossypii* to aphids collected on crops or other non-indigenous plants anywhere in the world. This is, however, a considerable oversimplification of the taxonomic problem, as becomes evident, for example, when one compares accounts of *A. gossypii* in Europe and East Asia.

In Europe (Stroyan, 1984; Heie, 1986), *A. gossypii* is classed as a subspecies in the *Aphis frangulae* complex, a group of closely related and morphologically almost indistinguishable indigenous European species that use buckthorn (*Frangula alnus*) as their primary host. *Aphis gossypii* is regarded (that is, more or less defined) as the only member of the group that does not have a sexual phase on buckthorn, overwintering parthenogenetically in Northern Europe in protected situations such as glasshouses. One might conclude from this scenario that this worldwide pest originated in Europe as a permanently parthenogenetic, highly polyphagous and adaptable offshoot of the *A. frangulae* complex, and spread from there to all parts of the world as the classic 'general purpose genotype'.

However, such a conclusion is difficult or impossible to reconcile with accounts of what is purportedly the same species in Japan and China, where the parthenogenetic generations seem equally polyphagous, but there is an annual sexual phase. Overwintering as eggs occurs in East Asia on a variety of unrelated plants, including *Frangula* spp., but also *Hibiscus syriacus*, *Celastrus orbiculatus*, and *Rubia cordifolia* (Inaizumi, 1980; Zhang and Zhong, 1990). It is possible that some of these populations have diverged as a result of differential selection among these primary hosts; populations overwintering on *R. cordifolia* in Japan, for example, seem to be isolated from those on other primary hosts, and are possibly a separate taxon (Inaizumi, 1981). Earlier, Kring (1959) had demonstrated that populations in Connecticut, USA, also have a sexual phase, using *H. syriacus* and also *Catalpa bignonioides* as primary hosts.

So, did the worldwide pest called *A. gossypii* originate in Europe, East Asia, or North America? North America is unlikely to be the ancestral homeland, because there are no indigenous North American *Aphis* species of the group of species closely related to *A. gossypii* that use *Frangula* as primary hosts. In East Asia, there are indigenous species related to *A. gossypii* that have a sexual phase on *Frangula*, such as the soybean aphid *A. glycines*, but none of these seem quite so similar in morphology to *A. gossypii* as the European *A. frangulae* group. The answer to this enigma can only come from some extensive work encompassing the entire geographical and host-plant range of *A. gossypii* and including comparisons with related species in Europe and East Asia.

Vanlerberghe-Masutti and Chavigny (1998) provided some interesting pointers for further work, with a study of random amplified polymorphic DNA (RAPD) of 18 *A. gossypii* populations from Southern Europe, La Réunion, and Laos. They found certain RAPD bands that were fixed in populations collected from cucurbits and absent from those collected on non-cucurbit hosts, and the 18 populations therefore clustered into two groups according to host plant, irrespective of their geographical origins.

Particular host associations have been noted previously in *A. gossypii*, for example, in European glasshouses, where aphids from chrysanthemums will not colonize cucumber, and *vice versa* (Guldemond *et al.*, 1994). These glasshouse populations on both chrysanthemums and cucurbits are normally parthenogenetic, but can produce sexual morphs under certain conditions (Guldemond *et al.*, 1994 and Fuller *et al.*, 1999, respectively). The work of Vanlerberghe-Masutti and Chavigny is the first to show that forms of *A. gossypii* with particular host associations may be distributed over a wide area. One may hypothesize that two forms have become widely distributed, possibly with different geographic origins, both with rare sexual reproduction, one found particularly on Cucurbitaceae and Malvaceae (and causing a major pest problem on cotton), and the other specializing on Asteraceae. These may correspond to the two forms, one cucurbit-feeding and the other chrysanthemum-feeding, identified in European glasshouses. Further evidence of the existence of a distinct, widely distributed Asteraceae-feeding form of *A. gossypii* has come recently from a multivariate morphometric study (Margaritopoulos *et al.*, 2006). However, there were no consistent host-associated RAPD bands in Japanese *A. gossypii* populations studied by Komazaki and Osakabe (1998), where there is regular genetic recombination. The problems of identity and origin of *A. gossypii* and the taxonomic status of host-associated forms, thus still remain, and require investigation of the much more complex genetics of populations in places where *A. gossypii* and its relatives have an annual sexual phase.

Aphis spiraecola (green citrus aphid or spiraea aphid)

This is a small yellow or greenish-yellow aphid with black siphunculi and cauda, found in dense, ant-attended colonies, curling and distorting leaves near the stem apices of a wide range of plants, particularly those of shrubby habit. Its numerous hosts are in more than 20 plant families, especially Caprifoliaceae, Asteraceae, Rosaceae,

Rubiaceae, Rutaceae, and Apiaceae. Probably, its most important crop host is *Citrus*. Although not particularly efficient at transmitting viruses, very large populations occur in spring in some regions – the Middle East, for example – and can make it an important vector of *Citrus tristeza virus*. It also transmits *Cucumber mosaic virus* (CMV), *Plum pox virus*, an isolate of *Alfalfa mosaic virus* from *Viburnum*, *Water melon mosaic virus 2*, and *Zucchini yellow mosaic virus* (ZYMV).

Now almost worldwide, there seems little doubt that *A. spiraecola* is indigenous to East Asia. It has been in North America since at least 1907, and had reached Australia by 1926, New Zealand by 1931, Argentina by 1939, the Mediterranean region by about 1939, and Africa by 1961. Populations in most parts of the world are permanently parthenogenetic on secondary hosts, but in East Asia and North America, *A. spiraecola* has a sexual phase on *Spiraea*.

There is an extensive literature on *A. spiraecola*, particularly in relation to its economic importance on *Citrus* (it was referred to as *A. citricola* in the literature from 1975–1988 because of a misidentification). The most comprehensive accounts of *A. spiraecola* as a *Citrus* pest are by Barbagallo (1966) in Italy and Miller (1929) in Florida. On other plants, especially Rosaceae, *A. spiraecola* is often confused with *Aphis pomi* (green apple aphid). For example, Cottier (1953) wrote an account of *A. spiraecola* in New Zealand under the name *A. pomi*, and Singh and Rhomberg (1984) studied allozyme variation in populations nominally of *A. pomi* on apples in North America and found two forms, one of which was almost certainly *A. spiraecola*. *Aphis pomi* has a longer last rostral segment than *A. spiraecola*, more hairs on the cauda, and usually has lateral tubercles on abdominal segments 2–4 (see also Halbert and Voegtlin, 1992). There also is possible confusion of identity with *Aphis eugeniae* in East and South-east Asia and Australia that can occur on the same hosts; *A. eugeniae* can be recognized by the peg-like hairs on the hind tibia, and by the presence of a median sense peg between the pair of hairs on the first segment of the hind tarsus.

Compared with the problems raised by *A. fabae* and *A. gossypii*, the taxonomic status, origins, and identity of *A. spiraecola* seemed to be fairly clear. However, Komazaki *et al.* (1979) found that *A. spiraecola* in Japan was using *Citrus unshiu* as well as *Spiraea thunbergii* as a primary host, and experimental work demonstrated that there were genetically inherited differences in hatching time (Komazaki, 1983, 1986) and egg diapause (Komazaki, 1998), correlated with esterase differences (Komazaki, 1991), that seem to indicate a degree of genetic isolation and divergence between the populations on the two primary hosts. Presumably, the ancestral primary host was the rosaceous plant (*Spiraea*), and *Citrus unshiu* was acquired more recently as a primary host. This seems to be an example of incipient speciation, as in Via's populations of *A. pisum*, but again it is not yet clear whether it is more than a local phenomenon. There are no records of sexual generations on *Citrus* outside Japan, and it could be that they are unable to survive on other *Citrus* species. In Japan, alatae migrating from spring populations that had developed from overwintering eggs on *Citrus unshiu* seemed to be the most important source of infestations of *Citrus* groves at a time when the main migration from *Spiraea* had already taken place (Komazaki, 1983).

Diuraphis noxia (Russian wheat aphid)

This small, narrow-bodied, yellow-green aphid was little known outside southern Russia until the late 1970s. It then took only a little over 10 years to colonize the main wheat- and barley-growing areas of East Asia, South Africa, and both North and South America (see IIE Distribution Map 521, 1991). It is still expanding its range northward in Europe (Thieme *et al.*, 2001). Hughes and Maywald (1990) assessed the suitability of the Australian environment for *D. noxia*, although it still (as at 2006) has not reached Australia.

Diuraphis noxia feeds only on Poaceae, concentrating particularly on wheat and barley. It does best on late-sown crops on poor soils. It transmits *Barley yellow dwarf virus*, but its feeding also has a rapidly toxic effect on the plant, the leaves of which become rolled into tubes and desiccated, and infested ears become bent. In cold temperate regions of Europe and Asia, it has a sexual phase without host alternation on wheat and barley. In North America, it has been assumed generally to have no sexual phase. Kiriac *et al.* (1990) found some oviparae in Idaho and Oregon, and oviparae have appeared in glasshouse cultures (Puterka *et al.*, 1992) but no males have been found. Puterka *et al.* (1993) found some genetic variation in North American *D. noxia* using RAPD-PCR, although this was between, rather than within, populations, suggesting that it was generated only by rare mutational or recombinational events. More genetic variation was found in collections from the Middle East, Moldavia, Ukraine, and Kirghizia. Their data suggest a single source of spread of *D. noxia*, perhaps *via* Turkey to France, South Africa, and North America. An interesting addition to this story is that Mimeur (1942) collected *D. noxia* in North Africa in 1938, described it as a new species (*Cavahyalopterus graminearum*), and produced oviparae in culture, but no males. Thus, the absence of males may be a long-standing feature of the population, which has since become widely distributed.

There are no problems with the identity of *D. noxia*, although in Europe there is a very similar species, *Diuraphis muehlei*, which feeds specifically on *Phleum pratense*, turning the leaves yellow. This species has a shorter antennal terminal process than *D. noxia*; the ratio of the terminal process to the base of the last segment in apterae is 1.05–1.65 (*muehlei*), as opposed to 1.55–2.6 (*noxia*), and in alatae 1.2–1.9 (*muehlei*), as opposed to 1.8–2.7 (*noxia*).

There is an extensive older Russian literature on *D. noxia*, one of the most comprehensive studies being that of Grossheim (1914). The rapid spread and great economic importance of this aphid have resulted in more recent extensive studies; see Poprawski *et al.* (1992) for a bibliography, and general accounts by Pike and Allison (1991) and Hughes (1996). Berest (1980) studied the parasite and predator complex of *D. noxia*

in Ukraine, and Tanigoshi *et al.* (1995) described biological control measures. Chen and Hopper (1997) studied its population dynamics and the impact of natural enemies in southern France.

Lipaphis pseudobrassicae (mustard aphid, also known as the false cabbage aphid)

Lipaphis pseudobrassicae is a cosmopolitan pest of cruciferous crops. Apterae are small to medium-sized, yellowish, grey, or olive green, with a waxy bloom that, in humid conditions, becomes a dense coat of white wax. It can occur in large colonies on the undersides of leaves or in inflorescences of many species and genera of Brassicaceae, including *Barbarea*, *Brassica*, *Capsella*, *Erysimum*, *Iberis*, *Lepidium*, *Matthiola*, *Nasturtium*, *Raphanus*, *Rorippa*, *Sinapis*, *Sisymbrium*, and *Thlaspi*. Often, the leaves are curled and turn yellow. It is a vector of about 10 non-persistent viruses, including *Turnip mosaic virus* and *Cauliflower mosaic virus*. It occurs throughout the world (CIE Distribution Map 203, 1965), but particularly is a pest in warmer climates, reproducing throughout the year by continuous parthenogenesis.

The origin and identity of *L. pseudobrassicae* were long in doubt. In North America, it was at first confused with *Brevicoryne brassicae*, until Davis (1914) recognized it as distinct and named it *Aphis pseudobrassicae*. Because of its weakly clavate siphunculi, it was subsequently transferred by Takahashi (1923) to the genus *Rhopalosiphum*, and it was referred to in the economic literature as *Rhopalosiphum pseudobrassicae* (Davis) until 1964. Börner and Schilder (1932) recognized that *pseudobrassicae* should be placed in *Lipaphis*, a genus erected by Mordvilko (1928) for a palaearctic crucifer-feeding aphid, *erysimi*. *Lipaphis erysimi* is a holocyclic species with a 2n = 10 karyotype (Gut, 1976; Blackman and Eastop, 2000) that occurs commonly on wild crucifers in Northern and Central Europe, but is not usually found on *Brassica* crops (Müller, 1986; Heie, 1992). Hille Ris Lambers (1948) could not find characters to discriminate *pseudobrassicae* from *erysimi*, but nevertheless stopped short of making it a synonym. Others regarded it as a subspecies of *erysimi* (e.g. Eastop, 1958a; Müller, 1986). Despite these uncertainties, the name *erysimi* was used for the widely distributed crucifer pest from 1975–2000.

Although most *Lipaphis* populations throughout the world are continuously parthenogenetic, a holocycle does occur on cruciferous crops (*Brassica rapa, Raphanus sativus*) in western Honshu, Japan (Kawada and Murai, 1979). These aphids have 2n = 8, and thus differ in karyotype from holocyclic populations of *L. erysimi* in Northern Europe. Chen and Zhang (1985) also reported 2n = 8 for *Lipaphis* in China. In West Bengal, *Lipaphis* populations are economically important on field crops of mustard, *Brassica nigra*, and here the common karyotype is also 2n = 8 (Kar and Khuda-Bukhsh, 1991). Sexual morphs have been reported from northern India, but populations there are probably mostly anholocyclic. Most permanently parthenogenetic *Lipaphis* populations throughout the world have a 9-chromosome karyotype, probably derived from the 8-chromosome form by dissociation of one autosome to produce a small, unpaired element. In multivariate morphometric analysis, samples with 8 and 9 chromosomes group together and are separated from samples of European *L. erysimi* with 10 chromosomes (V.F. Eastop and R.L. Blackman, unpublished results), so it was concluded that *L. pseudobrassicae* should be re-introduced for the worldwide crucifer pest, and that it probably originated in Eastern Asia (Blackman and Eastop, 2000). Finding simple morphological discriminants for the two species is not easy. *Lipaphis pseudobrassicae* has relatively longer antennae and relatively shorter siphunculi, and the function '(length of antennal segment III + length of processus terminalis) + length of siphunculus' discriminates most specimens. The value of this function is more than 2.4 in 90% of apterae of *L. pseudobrassicae*, and less than 2.4 in 90% of apterae of *L. erysimi*. The equivalent discriminating value for alatae is 3.4.

Macrosiphum euphorbiae (potato aphid)

Macrosiphum euphorbiae is one of the few cosmopolitan aphid pests of field crops that are undoubtedly of North American origin. The earliest European record is from potato at Wye, Kent, England in 1917, after which it soon became common in Britain and spread to continental Europe (Eastop, 1958b). It is a medium-sized to large, spindle-shaped aphid, usually green but sometimes pink or magenta, the adult apterae often rather shiny in contrast to the immature stages, which have a light dusting of greyish-white wax. In northeastern USA, it has a sexual phase on *Rosa*, using both wild and cultivated species as primary hosts (Shands et al., 1972). In Europe, and probably elsewhere, *M. euphorbiae* is mainly anholocyclic, although sexual morphs are produced occasionally and the holocycle may sometimes occur (Möller, 1970). The pink form has become much more common in Europe in the past 10 years. On secondary hosts, *M. euphorbiae* is highly polyphagous, feeding on more than 200 plant species in more than 20 different plant families. It is a vector of more than 40 non-persistent and five persistent viruses including BYNV, PEMV, BLRV, *Sweet potato leaf-speckling virus*, ZYMV, and PLRV. It is an important pest of potato, but as a vector of PLRV under field conditions, it seems to be relatively unimportant in comparison with *Myzus persicae* (Robert, 1971; Woodford et al., 1995), although direct feeding by large numbers early in the season can cause 'false top roll'. There is a very large literature, but surprisingly little is known about intraspecific variation and specific aphid–host interactions of *M. euphorbiae*. In western North America there are several little-known, and even some undescribed, species closely related to and almost indistinguishable from *M. euphorbiae* (MacDougall, 1926 and V.F. Eastop, unpublished results). In Europe, there is a group of closely related species with more specific host associations (see Watson, 1982 and Heie, 1994); of these, *M. euphorbiae* is most easily confused with *Macrosiphum tinctum* (= *Macrosiphum epilobiellum*), which feeds only on *Epilobium* spp. It can be hybridized in the laboratory with *Macrosiphum stellariae*, which is usually found on *Stellaria holostea*, and can colonize other plants, but not potato (Möller, 1971).

Meier (1961) provided a general account of *M. euphorbiae* in Europe, Barlow (1962) studied its development on potato, and MacGillivray and Anderson (1964) studied the factors controlling sexual morph production in eastern Canada. Parasitoids and hyperparasitoids were studied in North America by Shands *et al.* (1965) and Sullivan and van den Bosch (1971).

Myzus persicae (peach–potato aphid)

Myzus persicae is an exceptional species in many respects; cosmopolitan, extremely polyphagous, highly efficient as a virus vector, and with a great range of genetically-based variability in properties such as colour, life cycle, host-plant relationships, and methods of resisting insecticides. Adult apterous parthenogenetic females of *M. persicae* are small to medium-sized, pale greenish-yellow, various shades of green, pink, red, or almost black (apart from the genetically-determined colour variation, any one genotype will be more deeply pigmented in cold conditions). Alatae have a shiny black dorsal abdominal patch, as in other members of the genus *Myzus*, and immature alatae are often red or pink, even of genotypes where the apterae are green. Immature males are always some shade of yellow or yellow-green.

The sexual phase of *M. persicae* occurs predominantly on *Prunus persica* (including v. *nectarina*), except in parts of north-eastern USA and eastern Canada, where *Prunus nigra* is the main primary host (Shands *et al.*, 1969). Host alternation occurs in the temperate regions of all continents, wherever peaches are available and the autumn temperatures are low enough to allow production of the sexual morphs (Blackman, 1974). Spring populations on peach become very dense, severely curling the leaves. In contrast to its extreme primary host specificity, the secondary hosts are in more than 40 different plant families. They include very many economically important plants,

on most of which the populations are highly dispersed and individuals are found feeding singly on the older leaves. The great economic importance of *M. persicae* is due to its efficiency as a virus vector. It has been shown to be able to transmit considerably more than 100 plant viruses, including the persistent viruses BLRV, *Beet western yellows virus*, *Beet mild yellowing virus*, BYNV, PEMV, PLRV, *Tobacco vein distorting virus*, *Tobacco yellow net virus*, and *Tobacco yellow vein virus*. The relationship with PLRV has received particular attention (e.g. Ponsen, 1972; Eskanderi *et al.*, 1979). *Myzus persicae* is also a very efficient vector of numerous non-persistent viruses; e.g. CMV and *Bean yellow mosaic virus* to lupins in Western Australia (Bwye *et al.*, 1997).

As its principal primary host is thought to originate from China, one would presume this to be the original homeland of *M. persicae*. This presumption is, however, not without its problems. First, one might expect to find its closest relatives in China. Yet the species that seem most closely related to *M. persicae*, including what many would regard as its sibling species, *Myzus certus*, and others with which it readily hybridizes in the laboratory such as *Myzus myosotidis*, are all European. It is difficult to see how this situation arose. There are no clues from biology, as all other species in the *Myzus* subgenus *Nectarosiphon* except *M. persicae* have lost their ancestral primary host and live all year round on their herbaceous host plants. A second problem concerns the relationship of *M. persicae* with PLRV, which seems to be intimate and therefore long-standing, but this is in conflict with their respective origins. It is possible, however, that potato leaf roll occurs in some unrecognizable or symptomless form in an Asian member of the Solanaceae.

As might be expected of such an adaptable and fast-evolving genome, biology and host relationships are likely to be changing faster than morphology, causing problems of identification and identity. Specimens of *M. certus* on slides are difficult enough to distinguish from those of *M. persicae*, although this species is clearly very different in its biology and host relations (living all year on Caryophyllaceae and Violaceae, and having its sexual phase on these plants, with apterous males). Two other taxa, *Myzus dianthicola* and *Myzus antirrhinii*, are even more like *M. persicae*, and individual slide-mounted specimens cannot be distinguished reliably from that species. These two are both permanently parthenogenetic as far as is known, and their karyotypes are structurally heterozygous (Blackman, 1980). *Myzus dianthicola* is found only on *Dianthus*, usually in glasshouses, where its consistently deep yellow-green colour and the leaf chlorosis that it causes distinguish it respectively from *M. certus* and *M. persicae*. *Myzus antirrhinii* may be almost as polyphagous as *M. persicae*, but has certain characteristic hosts such as *Antirrhinum* and *Buddleja* and a more consistent mid-green to dark green colour, and there are also differences in allozymes and at rDNA and microsatellite loci (Fenton *et al.*, 1998; Terradot *et al.*, 1999). Both *M. dianthicola* and *M. antirrhinii* are found in Europe and North America, the former also being found in New Zealand and the latter in Australia. Although probably of recent origin, they seem to be isolated from other members of the group by their obligate parthenogenesis, and are therefore best treated as discrete taxa (Blackman and Brown, 1991). The karyotype of *M. antirrhinii* is remarkably variable, and is of cytogenetic interest because fusions and dissociations of chromosomes have occurred in the absence of genetic recombination (Hales *et al.*, 2000).

Although complicating the practical identification of *M. persicae*, none of the forms discussed above have questionable taxonomic status. However, for many years it has been recognized that populations of *M. persicae* on tobacco (the 'tobacco aphid') are distinct from populations on other plants (de Jong, 1929; Brain, 1942; Müller, 1958; Takada, 1986). The aphid attacking commercial varieties of *Nicotiana tabacum* forms large, dense colonies at the growing points and on the youngest leaves, and seems able to avoid or tolerate the exudates of the glandular trichomes, which are not only sticky but contain repellent or toxic chemicals (Georgieva, 1998; Wang *et al.*, 2001).

Apterae on tobacco are predominantly pink/ red in colour, and have acquired resistance to insecticides far more slowly than those on other crops (Takada, 1979; Semtner *et al.*, 1990). Blackman (1987) demonstrated using multiple discriminant analysis that samples from tobacco in many parts of the world could be differentiated from *M. persicae* on other crops, indicating that populations on tobacco worldwide constituted a monophyletic lineage for which he proposed the name *Myzus nicotianae*. Most of the samples analysed by Blackman (1987) were from regions where populations are permanently parthenogenetic, but Margaritopoulos *et al.* (2000) found that holocyclic populations of tobacco aphids in Greece could also be discriminated from those collected from other crops, and from peach away from tobacco-growing regions. It has been suggested (Clements *et al.*, 2000a) that these morphological differences could be due to phenotypic plasticity. However, it is clear that the differences are genetically-based, as all the samples analysed were clones reared under controlled conditions on the same host plant. The clones originating from peach in tobacco-growing regions had never even seen a tobacco plant.

Genetic isolation between tobacco-adapted and non-tobacco-adapted forms cannot be complete, as the E_4 and FE_4 genes amplified in insecticide-resistant aphids are identical in the two forms (Field *et al.*, 1994). However, these genes apparently have taken many years to cross into tobacco aphids. For example, holocyclic populations of *M. persicae* on peach in Southern Europe have been resistant to organophosphates since about 1962, yet such resistance in tobacco aphids was first reported in holocyclic populations in northern Greece in the mid-1980s. This may be where introgression of these genes into tobacco-adapted genotypes occurred, selection then strongly favouring their spread to other populations. Absence of complete reproductive isolation between the two forms, perhaps in conjunction with a very recent origin of the tobacco-adapted form, may explain the failure to find consistent diagnostic genetic markers (Fenton *et al.*, 1998; Margaritopoulos *et al.*, 1998; Clements *et al.*, 2000a,b), or the divergence of gene sequence that one might normally expect to find between separate taxa (Clements *et al.*, 2000a). However, the degree of isolation must have been sufficient to preserve the integrity of the tobacco-adapted genome for at least 15–20 years, and it would be unwise to regard this form simply as synonymous with *M. persicae*, as suggested by Clements *et al.* (2000a,b), as this would lose important information. The nomenclatural problem can perhaps now be resolved, because the tobacco aphid conforms to the broader criteria for the subspecies category proposed by Müller (1986), and defined earlier in this chapter. We have therefore proposed (Eastop and Blackman, 2005) that the tobacco aphid should be called *M. persicae nicotianae*.

The literature on *M. persicae* (*sensu lato*) is immense. There have been extensive reviews of its ecology (van Emden *et al.*, 1969; Mackauer and Way, 1976), as well as discussions of migration and spatial dynamics (Taylor, 1977) and biological approaches to control (Blackman, 1976). This aphid has also been the subject of much laboratory research including, for example, studies of anatomy and function of the gut (Forbes, 1964), nutritional studies using host plants (e.g. van Emden, 1977) and artificial diets (e.g. Mittler, 1976), and photoperiodic responses (Takada, 1982). Genetic variation and the evolution of insecticide resistance in *M. persicae* are covered elsewhere in this book (Loxdale and Lushai, Chapter 2 and Foster *et al.*, Chapter 10).

Rhopalosiphum maidis (corn leaf aphid)

Apterae of *Rhopalosiphum maidis* are small to medium-sized, elongate oval, olive to bluish-green aphids with short antennae and dark legs, siphunculi, and cauda. They feed on young leaves of their host plants and particularly are a problem on *Zea mays*, *Sorghum bicolor*, and *Hordeum vulgare*, also colonizing many other grasses and cereals in more than 30 genera including *Avena*, *Secale*, *Triticum*, *Oryza*, and *Saccharum*, and also found occasionally on Cyperaceae and Typhaceae. *Rhopalosiphum maidis* is an important vector of the persistent *Barley yellow dwarf virus* (BYDV), *Millet red leaf virus* (MRLV), *Abaca mosaic virus* (AbaMV), *Sugarcane mosaic virus* (SCMV), and *Maize dwarf mosaic virus* (MDMV).

This is probably the most important aphid pest of cereals in tropical and warm temperate climates throughout the world, but the pest populations are all permanently parthenogenetic and cannot survive outdoors in regions with severe winter climates. Males occur sporadically, but the sexual phase so far has only been found in Pakistan (Remaudière and Naumann-Etienne, 1991) and seems to be tied to one species of *Prunus* native to that region, *Prunus cornuta*. The pest populations therefore presumably originated from this region, as one or more dispersals of permanently parthenogenetic genotypes.

Despite the apparent lack of sexual reproduction, pest populations of *R. maidis* show differences in host preference (e.g. Painter and Pathak, 1962), karyotype (Brown and Blackman, 1988), and rDNA (Lupoli *et al.*, 1990), although no mtDNA variation was detected by Simon *et al.* (1995), and there is conflicting data on whether there is any allozyme variation (Steiner *et al.*, 1985; Simon *et al.*, 1995). Host preference differences seem quite complex in *R. maidis*, and in some cases must involve many loci, especially when they result in a high degree of host species specificity, such as that shown by the 10-chromosome form that colonizes barley and eupanicoid grasses in the northern hemisphere (Blackman and Brown, 1991; Jauset *et al.*, 2000) but does not feed on maize or sorghum, which are colonized by aphids with 2n = 8. In eastern Australia, populations of *R. maidis* on eupanicoid grasses are characterized by a 9-chromosome karyotype and do not occur on maize and sorghum (de Barro, 1992). Karyotype variation is a common feature of permanently parthenogenetic aphids, but it is difficult to conceive how the complex genetic traits involved in selection of host species could have arisen in the absence of genetic recombination. This leads one to conclude that the host-related variation now observed in pest populations of *R. maidis* may be due to multiple origins from the sexually reproducing population in Asia, rather than to mutations within parthenogenetic lineages. Further analysis of variable regions of the nuclear DNA might help to resolve this problem. Absence of the primary host and sexual reproduction means, however, that in its genetic structure, and therefore in the way it is treated taxonomically, *R. maidis* must differ fundamentally from some of the other introduced pest aphids in North America such as *Acyrthosiphon pisum*, *Schizaphis graminum*, and *Therioaphis trifolii maculata*.

Rhopalosiphum padi (bird cherry–oat aphid)

Rhopalosiphum padi attacks all the major cereals and pasture grasses, and is probably the major pest of temperate cereal crops on a world scale. Apterae of *R. padi* on grasses and cereals are broadly oval, varying in colour from green mottled with yellowish-green to olive-green, dark olive or greenish-black, and often with rust-coloured patches around the bases of the siphunculi. For a grass-feeding species, it is relatively catholic in its tastes, for as well as feeding on numerous species of Poaceae, it can colonize many other monocotyledonous plants, and some dicotyledonous ones. It is a vector of BYDV (particularly strain BYDV-PAV) and of *Cereal yellow dwarf virus*–RPV, *Filaree red leaf virus*, *Maize leaf fleck virus*, and *Rice giallume virus*, as well as oat yellow leaf disease, AbaMV, *Onion yellow dwarf virus*, MDMV, and several other non-persistent viruses.

Now distributed worldwide, it is difficult to pin down its origins, as it has a sexual phase on *Prunus padus* (bird cherry) in Europe, and on *Prunus virginiana* (common choke-cherry) in North America, and seems equally at home on both. Halbert and Voegtlin (1998) argue for a North American origin of the genus *Rhopalosiphum* (and of BYDV), but *R. maidis* is clearly a palaearctic species (see above), as also is *Rhopalosiphum rufiabdominale* (rice root aphid), which has East Asian *Prunus* species as its primary hosts. The genus as a whole therefore has a holarctic distribution. *Rhopalosiphum padi* could be of nearctic origin, as there are several closely related North American species, but it has been in Europe at least since the time of Linnaeus (1758).

French populations of *R. padi* have been the subject of some interesting recent work (Delmotte *et al.*, 2001, 2003), which is discussed in some detail by Loxdale and Lushai, Chapter 2 this volume, but should be mentioned here as it has important relevance to the taxonomic treatment of parthenogenetic lineages. Using a combination of life cycle, mtDNA sequence, and microsatellite data, Delmotte *et al.* (2001) demonstrated the existence of an ancient monophyletic group of parthenogenetic lineages, with an mtDNA haplotype (I) quite distinct from that of all other lineages, which were of a second haplotype (II), the two haplotypes having possibly diverged for about 400,000 years. If they had been genetically isolated for that length of time, one might expect to be able to treat them as separate taxa. The microsatellite data, however, show that the nuclear genomes are not completely isolated. There is some gene flow from parthenogenetically to sexually reproducing lineages, and this is presumably mediated by the males that are still produced by the otherwise permanently parthenogenetic haplotype I lineages. Sequence analysis of two nuclear DNA markers (Delmotte *et al.*, 2003) then showed rather conclusively that the haplotype I lineages originated from one or more relatively recent hybridization events between European *R. padi* and another closely related species, possibly of Asian (or North American?) origin, which remains to be identified. This is the first clear evidence for hybrid origin of permanent parthenogenesis in an aphid, and it raises interesting questions about the taxonomic status of *R. padi*. These can only be resolved by further studies, which also need to take into account other apparently undescribed taxa in the *R. padi* group that have been introduced to Australia (Hales and Cowen, 1990) and New Zealand (Bulman *et al.*, 2005a,b).

Schizaphis graminum (greenbug)

Schizaphis graminum was the first introduced aphid to have a significant economic impact in the main winter wheat areas of North America. The small, yellowish- to bluish-green apterae with pale, dark-tipped siphunculi feed on the leaves of grasses and cereals, often causing yellowing and other phytotoxic effects. They restrict their feeding almost exclusively to Poaceae, but species in many genera are attacked, including *Agropyron*, *Avena*, *Bromus*, *Dactylis*, *Eleusine*, *Festuca*, *Hordeum*, *Lolium*, *Oryza*, *Panicum*, *Poa*, *Sorghum*, *Triticum*, and *Zea*. Several important viruses are vectored including BYDV (especially strain BYDV-SGV), MRLV, SCMV, and MDMV.

Schizaphis graminum is a palaearctic aphid, possibly of Middle Eastern or Central Asian origin, now distributed widely through Southern Europe, Asia, Africa, and North and South America. A problem with interpreting early records is that when *Metopolophium dirhodum* was first introduced to a region, it was sometimes misidentified as 'greenbug'. Records of *S. graminum* from Australia and the Philippines all seem to be referable to the Asian species, *Schizaphis hypersiphonata*. This feeds particularly on *Digitaria*, but can occur on other species of Poaceae, and is recorded on wheat in the Philippines, but it does not have the same phytotoxic effects as the greenbug. Records of *S. graminum* on grasses in Western Europe are now thought to apply to other species (Tambs-Lyche, 1959; Stroyan, 1960, 1984; Pettersson, 1971).

The history of the greenbug in North America is quite well documented, and is an interesting example of multiple introductions and their consequences. For many years, this story was confused, partly because of the description of a series of biotypes, but it has been made much clearer by DNA studies. The 'biotype concept' was useful to plant breeders, but the term 'biotype' referred only to host resistance-breaking traits, and was perhaps given greater significance than it deserved. These traits could be identified and studied when they first appeared, because aphids carrying them could be isolated and maintained in clonal cultures. A greenbug biotype was thus any genotype that had a particular resistance-breaking trait. In the absence of a sexual phase, this might be a single clone, but it is now clear that sexual morph production and genetic recombination occurs regularly in North American greenbug populations (Shufran *et al.*, 1991, 1997) so that there is great genetic diversity, and it follows that field populations may therefore have these traits portrayed in numerous genetic backgrounds.

The original introduction in about 1882 was of a highly virulent genetic stock, very damaging and phytotoxic to wheat and barley. In 1961, a very successful wheat variety (DS-28 A) carrying a gene for greenbug resistance was found to be susceptible to a 'new form' of the aphid, designated as biotype B to distinguish it from the original biotype A (Wood, 1961). Then, in about 1968, another 'new form' appeared, designated as biotype C, this one for the first time inflicting severe damage to cultivated sorghum. Since then, a series of new 'biotypes' (or more accurately, resistance-breaking traits) have been described.

Shufran *et al.* (2000) compared mitochondrial DNA sequences of clones of *S. graminum* representing all the available North American 'biotypes', and found that they fit into three distinct clades, the divergence of which pre-dates modern agriculture. These must therefore represent at least three separate introductions into North America. The form originally introduced ('biotype A') reproduced sexually in the northern States (Webster and Phillips, 1912), and is presumably represented by a clone started in 1958 (the 'NY isolate') and by clones with the resistance-breaking traits F and G, which cluster with it to form clade 2 (Shufran *et al.*, 2000). The second introduction (clade 3), with resistance-breaking trait B, has different probing behaviour and is more damaging to susceptible wheat and barley (Saxena and Chada, 1971). It reproduces possibly only parthenogenetically in North America, so this clade may represent a single parthenogenetic lineage. Shufran *et al.* (2000) found that it had a very similar mtDNA sequence to a sample from Germany, but it has subsequently emerged that the German sample originated from North America (T. Thieme, 2001, personal communication). The third introduction, originally recognized as the sorghum-adapted biotype C, seems to have been the source population for all the remaining resistance-breaking mutations (represented by biotypes C, E, K, I, and J), all of which have very similar mtDNA sequences (clade 1). This sorghum-adapted form, which has for many years now been the predominant form on wheat and sorghum in North America, differs morphologically from the others in several respects (V.F. Eastop, unpublished results), and may have come from further south in Europe or Asia, as suggested by its predilection for sorghum and by its ability to produce sexual morphs and overwinter eggs at more southerly latitudes in the USA.

These three clades have probably diverged sufficiently to provide a range of discriminant characters, using both multivariate morphometric and DNA diagnostics, and clades 1 and 2 have all the necessary attributes to warrant their description as subspecies according to the guidelines proposed earlier in this chapter. This would then provide a sounder basis for further work on the genetics and evolution of *S. graminum* in North America. It seems particularly important that the limitations of the 'biotype concept' are recognized. For example, the 'biotype-specific' patterns identified with RAPDs and analysis of the intergenic spacer region of the rDNA gene by Black *et al.* (1992) and Black (1993) presumably were applicable only to the clones studied. In populations that are undergoing periodic genetic recombination, such patterns cannot be expected to be associated consistently with particular resistance-breaking traits, unless they happen to be linked to the resistance-breaking loci.

Sitobion avenae (grain aphid)

The concept of the genus *Sitobion* has recently changed somewhat so that it now includes fewer species, all of which are Eurasian or African, and a greater proportion of which are grass feeders. The primitive life cycle of the genus probably involves host alternation from Rosaceae to grasses, as in the related genus *Metopolophium*. *Sitobion* are superficially very like *Macrosiphum*,

particularly in having similar siphunculi with polygonal reticulation, but this feature may have developed independently in these two genera. A closer relationship to *Metopolophium* is also supported by the karyotype, which is 2n = 16 or 18 in both *Sitobion* and *Metopolophium*, whereas *Macrosiphum* usually have 2n = 10.

The majority of *Sitobion* species no longer have host alternation, and *S. avenae* is one of these. Its apterae on grasses and cereals are medium-sized, broadly spindle-shaped, yellowish-green to dirty reddish-brown, sometimes rather shiny, with black antennae and siphunculi, and a pale cauda. It colonizes numerous species of Poaceae, including all the cereals and pasture grasses of temperate climates, and can also feed on many other monocotyledonous plants. As well as the direct damage caused to cereals by feeding on the developing ears, *S. avenae* is an efficient vector of BYDV (especially strains BYDV-PAV and BYDV-MAV) both within and between crops. Probably European in origin, *S. avenae* now occurs throughout the Mediterranean area, eastwards to India and Nepal, in northern and southern Africa, and in North and South America. Records from Eastern Asia are referable to *Sitobion miscanthi*, a closely related species that probably originated in the Far East and has spread to Australia, New Zealand, and several Pacific islands, including Hawaii.

Sitobion avenae produces sexual morphs and lays overwintering eggs on many species of Poaceae in Europe, but also continues reproducing parthenogenetically through the year wherever winters are mild enough. Studies of the genetic structure of populations in southern England (Sunnucks *et al.*, 1997a) and France (Haack *et al.*, 2000) have produced some unexpected results, described in more detail by Loxdale and Lushai (Chapter 2 this volume), which have certain taxonomic implications. This work has shown: (i) that based on analysis of microsatellite DNA, *S. avenae* sampled in southern England fell into three genotypic groups, apparently almost non-interbreeding, although there was evidence of high levels of genetic recombination within each group; (ii) two of these groups showed complete host specificity, even when collected at the same time and place, one of them occurring only on wheat and the other only on cocksfoot grass (*Dactylis glomerata*); (iii) the genotypic group found only on *Dactylis* had introgressed microsatellite alleles and mtDNA from the related species *Sitobion fragariae* (blackberry–cereal aphid), which colonizes *Dactylis* but rarely occurs on wheat; and (iv) microsatellite analysis of French (Brittany) populations of *S. avenae* showed that there were host-adapted genotypes on maize differing from those on adjacent wheat and barley, but also revealed the existence of some very common host-generalist clones that persisted parthenogenetically between years.

Host-related genetic divergence and potential incipient speciation thus seem to be going on in European *S. avenae*, in a way comparable to that found in *A. pisum*, but in the case of *S. avenae* there is an additional element, the introgression of alleles from a related species, which may also include genes influencing aphid–host interactions. Again, the question remains, are the observed divergences transient local population phenomena, or are they representative of longer-term evolutionary trends, and manifest over a wider geographical area?

Therioaphis trifolii (spotted alfalfa and yellow clover aphids)

Therioaphis trifolii is unusual among crop pest aphids in being a member of the subfamily Calaphidinae, most genera of which live on deciduous trees (Table 1.1). It has the characteristic features of this subfamily, such as a knobbed cauda and a bi-lobed anal plate. The members of the genus *Therioaphis* are all palaearctic, and all live on various species of Fabaceae. Apterae of *T. trifolii* are distinctive, pale yellow or greenish-white, rather shiny, with rows of dorsal tubercles, pigmented light or dark brown and bearing capitate hairs. It seems to be by far the most polyphagous species in the genus, as European populations can be found on numerous species of Fabaceae in the genera *Astragalus*, *Lotus*, *Medicago*, *Melilotus*, *Onobrychis*, *Ononis*, and *Trifolium*. However, to judge

from the host preferences shown by genotypes introduced to other parts of the world, there is also in this species a considerable amount of intraspecific genetic heterogeneity in the utilization of these host plants.

Therioaphis trifolii is a native of Europe, the Mediterranean area, and Southwest Asia, but in one form or another it now occurs as a pest of legumes in North and South America, South Africa, Japan, and Australia. The history of introductions of *T. trifolii* into North America has some remarkable parallels with that of *S. graminum* discussed above, and is an excellent example of the consequences of founder effects, worthy of inclusion in any textbook. The story has been told before (Blackman, 1981), so will only be summarized here. Two morphologically distinct forms of *T. trifolii* occur in North America, evidently due to separate introductions, about 70 years apart, of genotypes with different specific host associations: (i) the yellow clover aphid (YCA), which feeds almost exclusively on *Trifolium pratense*; and (ii) the spotted alfalfa aphid (SAA, also known as *T. trifolii maculata* – or in North America, simply as *T. maculata*), which feeds mainly on, and can be very injurious to, lucerne (alfalfa). Although originally introduced, respectively, to eastern and southwestern USA, these two forms now occur sympatrically and have annual sexual reproduction in the northern States, yet are effectively isolated by their association with different host plants, and thus function as separate species.

SAA has subsequently spread to South America, South Africa, Australia, New Zealand, and Japan (see Blackman and Eastop, 2000). In Australia, it now co-exists with a third form, the spotted clover aphid (SCA), which preferentially colonizes subclover, *Trifolium subterraneum* (subterranean clover). SCA can be distinguished from SAA morphologically and by RAPD-PCR, and the differences in mtDNA sequence between these two forms indicates that SCA is a separate, more recent immigrant to Australia (Sunnucks *et al.*, 1997b). It is possible that SCA is present also in South Africa, as populations there are reported to colonize *Trifolium*, as well as *Medicago*.

If these three forms are at all representative of the populations from which they originated, then there must be considerable partitioning of resources and subspecific structuring of *T. trifolii* on its different host plants in Europe and the Middle East. Hopefully, this will be the subject of future work. However, even on present evidence, there seems ample justification for giving SAA and SCA the status of subspecies. SAA can therefore become *T. trifolii maculata*, but SCA has yet to be formally named.

Conclusions

Studies of intraspecific variation and evolution of pest aphids have now reached a very interesting point, where enough light has been shed on some of the important pest species to show the way forward. Also, molecular tools and analytical methods are now available to cope with the complex population genetics of insects with variable life cycles. It has long been suspected that most major pest aphids show intraspecific partitioning of resources (Blackman, 1990), and there is now plenty of evidence of this among some of the best-studied species. It seems that speciation processes, where these involve assortative mating due to differential selection among potential host plants, could (i) progress very rapidly and (ii) take place in the face of gene flow, leading to situations where subspecies can be recognized by biological differences, yet show minimal genetic divergence of markers unlinked to host-related traits. If this is so, then some re-assessment may be needed of the taxonomic significance of genetic distance parameters calculated from molecular data.

This makes it all the more important not to lose sight of the continuing need for morphotaxonomy. The new knowledge always needs to be related to the old – with the 'old' mainly represented by museum specimens and a taxonomic literature based largely on morphology. We find it alarming that many population geneticists and others studying aphid variation are still not depositing voucher specimens in the major insect collections, especially when clonal reproduction makes it possible to have representative individuals of the actual genotypes preserved for morphological study.

Multivariate morphometrics – particularly multiple discriminant analysis and the use of canonical variates, with clones or samples from clearly defined populations as the groups in the analysis – has proved to be a very powerful technique for demonstrating genetic differences between closely related aphid taxa (Blackman, 1992), and even between different genotypes (Blackman and Spence, 1994), and is especially useful in conjunction with other methods (e.g. Blackman *et al.*, 1995; Blackman and de Boise, 2002). Subspecies that seem to be morphologically indistinguishable on the basis of single characters are very likely to be differentiated using a technique that compares the correlations between numerous characters. Developments in image analysis now make the acquisition of morphometric data far quicker and easier, and we hope that future work on aphid species complexes using molecular methods will also include parallel morphological studies.

References

Barbagallo, S. (1966) Contributo all conoscenza degli afidi degli agrumi. 1. *Aphis spiraecola* Patch. *Bollettino del Laboratorio di Entomologica Agraria 'Filippo Silvestri'* 24, 49–83.

Barber, R.P.A. and Robinson, A.G. (1980) Studies on two species of black aphids (Homoptera: Aphididae) on faba bean and nasturtium. *Canadian Entomologist* 112, 119–122.

Barlow, C.A. (1962) Development, survival and fecundity of the potato aphid, *Macrosiphum euphorbiae*, at constant temperatures. *Canadian Entomologist* 94, 667–671.

de Barro, P.J. (1992) Karyotypes of cereal aphids in South Australia with special reference to *Rhopalosiphum maidis* (Fitch) (Hemiptera: Aphididae). *Journal of the Australian Entomological Society* 31, 333–334.

Berest, Z.L. (1980) Parasites and predators of the aphids *Brachycolus noxius* and *Schizaphis graminum* in crops of barley and wheat in Nikalaev and Odessa regions. [In Russian.] *Vestnik Zoologii* 1980 (2), 80–81.

Black, W.C., IV (1993) Variation in the ribosomal RNA cistron among host-adapted races of an aphid (*Schizaphis graminum*). *Insect Molecular Biology* 2, 59–69.

Black, W.C., IV, DuTeau, N.M., Puterka, G.J., Nechols, J.R. and Pettorini, J.M. (1992) Use of random amplified polymorphic DNA polymerase chain reaction (RAPD-PCR) to detect DNA polymorphisms in aphids. *Bulletin of Entomological Research* 82, 151–160.

Blackman, R.L. (1974) Life-cycle variation of *Myzus persicae* (Sulz.) (Hom., Aphididae) in different parts of the world, in relation to genotype and environment. *Bulletin of Entomological Research* 63, 595–607.

Blackman, R.L. (1976) Biological approaches to the control of aphids. *Philosophical Transactions of the Royal Society of London B* 274, 473–488.

Blackman, R.L. (1980) Chromosome numbers in the Aphididae and their taxonomic significance. *Systematic Entomology* 5, 7–25.

Blackman, R.L. (1981) Aphid genetics and host plant resistance. *Bulletin OILB/ SROP* 4, 13–19.

Blackman, R.L. (1987) Morphological discrimination of a tobacco-feeding form from *Myzus persicae* (Sulzer) (Hemiptera: Aphididae), and a key to New World *Myzus* (*Nectarosiphon*) species. *Bulletin of Entomological Research* 77, 713–730.

Blackman, R.L. (1990) Specificity in aphid/plant genetic interactions, with particular attention to the role of the alate colonizer. In: Campbell, R.K. and Eikenbary, R.D. (eds) *Aphid–Plant Genotype Interactions*. Elsevier, Amsterdam, pp. 251–274.

Blackman, R.L. (1992) The use of ordination techniques to discriminate within pest aphid species complexes. In: Sorensen, J.T. and Foottit, R. (eds) *Ordination in the Study of Morphology, Evolution and Systematics*. Elsevier, Amsterdam, pp. 261–275.

Blackman, R.L. (1995) What's in a name? Species concepts and realities. *Bulletin of Entomological Research* 85, 1–4.

Blackman, R.L. and Brown, P.A. (1991) Morphometric variation within and between populations of *Rhopalosiphum maidis* with a discussion of the taxonomic treatment of permanently parthenogenetic aphids (Homoptera: Aphididae). *Entomologia Generalis* 16, 97–113.

Blackman, R.L. and de Boise, E. (2002) Morphometric correlates of karyotype and host plant in the genus *Euceraphis* (Hemiptera: Aphididae). *Systematic Entomology* 27, 323–335.

Blackman, R.L. and Eastop, V.F. (2000) *Aphids on the World's Crops: An Identification and Information Guide*, 2nd edn. Wiley, Chichester, 466 pp.

Blackman, R.L. and Eastop, V.F. (2006) *Aphids on the World's Herbaceous Plants and Shrubs*. Wiley, Chichester, 1439 pp.

Blackman, R.L. and Spence, J.M. (1994) The effects of temperature on aphid morphology, using a multivariate approach. *European Journal of Entomology* 91, 7–22.

Blackman, R.L., Watson, G.W. and Ready, P.D. (1995) The identity of the African pine woolly aphid: a multidisciplinary approach. *EPPO Bulletin* 25, 337–341.

Börner, C. and Schilder, F.A. (1932) Aphidoidea. In: *Sorauer's Handbuch der Pflanzenkrankheiten, Volume 5*, 4th edn. Paul Parey, Berlin, pp. 551–715.

Brain, C.K. (1942) The tobacco aphis. *Rhodesia Agricultural Journal* 39, 241–243.

Brown, P.A. and Blackman, R.L. (1988) Karyotype variation in the corn leaf aphid, *Rhopalosiphum maidis* species complex (Hemiptera: Aphididae), in relation to host plant and morphology. *Bulletin of Entomological Research* 78, 351–363.

Bulman, S.R., Stufkens, M.A.W., Nichol, D., Harcourt, S.J., Harrex, A.L. and Teulon, D.A.J. (2005a) *Rhopalosiphum* aphids in New Zealand. I. RAPD markers reveal limited variability in *Rhopalosiphum padi*. *New Zealand Journal of Zoology* 32, 29–36.

Bulman, S.R., Stufkens, M.A.W., Eastop, V.F. and Teulon, D.A.J. (2005b) *Rhopalosiphum* aphids in New Zealand. II. DNA sequences reveal two incompletely described species. *New Zealand Journal of Zoology* 32, 37–45.

Bwye, A.M., Proudlove, W., Berlandier, F.A. and Jones, R.A.C. (1997) Effects of applying insecticides to control aphid vectors and cucumber mosaic virus in narrow leafed lupins (*Lupinus angustifolius*). *Australian Journal of Experimental Agriculture* 37, 93–102.

Cartier, J.J. (1963) Varietal resistance of peas to pea aphid biotypes under field and greenhouse conditions. *Journal of Economic Entomology* 56, 205–213.

Chen, K. and Hopper, K.R. (1997) *Diuraphis noxia*. . . .population dynamics and impact of natural enemies in the Montpelier region of southern France. *Environmental Entomology* 26, 866–875.

Chen, X. and Zhang, G. (1985) The karyotype of aphids and its taxonomic significance. [In Chinese.] *Journal of the Graduate School, Academia Sinica* 2, 189–200.

Clements, K.M., Wiegmann, B.M., Sorenson, C.E., Smith, C.F., Neese, P.A. and Roe, R M. (2000a) Genetic variation in the *Myzus persicae* complex (Homoptera: Aphididae): evidence for a single species. *Annals of the Entomological Society of America* 93, 31–46.

Clements, K.M., Sorenson, C.E., Wiegmann, B.M., Neese, P.A. and Roe, R.M. (2000b) Genetic, biochemical and behavioral uniformity among populations of *Myzus nicotianae* and *Myzus persicae*. *Entomologia Experimentalis et Applicata* 95, 269–281.

Cottier, W. (1953) Aphids of New Zealand. *Bulletin of New Zealand Department of Scientific and Industrial Research* 106, 1–381.

Coyne, J.A. and Orr, H.A. (1989) Patterns of speciation in *Drosophila*. *Evolution* 43, 362–381.

Davis, J.J. (1914) New and little known species of Aphididae. *Canadian Entomologist* 46, 41–51, 71–87.

Deguine, J.P. and Leclant, F. (1997) Aphis gossypii *Glover (Hemiptera, Aphididae). Les déprédateurs du cotonnier en Afrique tropicale et dans le reste du monde, no 11.* CIRAD-CA, Montpellier, 114 pp.

Delmotte, F., Leterme, N., Bonhomme, J., Rispe, C. and Simon, J.-C. (2001) Multiple routes to asexuality in an aphid species. *Proceedings of the Royal Society of London B* 268, 2291–2299.

Delmotte, F., Sabater-Muñoz, B., Prunier-Leterme, N., Latorre, A., Sunnucks, P., Rispe, C. and Simon, J.-C. (2003) Phylogenetic evidence for hybrid origins of asexual lineages in an aphid species. *Evolution* 57, 1291–1303.

von Dohlen, C.D. and Moran, N.A. (1995) Molecular phylogeny of the Homoptera: a paraphyletic taxon. *Journal of Molecular Evolution* 41, 211–223.

von Dohlen, C.D. and Moran, N.A. (2000) Molecular data support a rapid radiation of aphids in the Cretaceous and multiple origins of host alternation. *Biological Journal of the Linnean Society* 71, 689–717.

Eastop, V.F. (1958a) *A Study of the Aphididae of East Africa.* HMSO, London, 126 pp.

Eastop, V.F. (1958b) The history of *Macrosiphum* in Europe. *Entomologist* 91, 198–201.

Eastop, V.F. and Blackman, R.L. (2005) Some new synonyms in Aphididae. *Zootaxa* 1089, 1–36.

van Emden, H.F. (1977) Failure of the aphid, *Myzus persicae*, to compensate for poor diet during early growth. *Physiological Entomology* 2, 43–48.

van Emden, H.F., Eastop, V.F., Hughes, R.D. and Way, M.J. (1969) The ecology of *Myzus persicae*. *Annual Review of Entomology* 14, 197–270.

Eskanderi, F., Sylvester, E.S. and Richardson, J. (1979) Evidence for lack of propagation of potato leaf roll virus in *Myzus persicae*. *Phytopathology* 68, 45–47.

Fenton, B., Woodford, J.A.T. and Malloch, G. (1998). Analysis of clonal diversity of the peach–potato aphid, *Myzus persicae* (Sulzer), in Scotland, UK and evidence for the existence of a predominant clone. *Molecular Ecology* 7, 1475–1487.

Field, L.M., Javed, N., Stribley, M.F. and Devonshire, A.L. (1994). The peach–potato aphid, *Myzus persicae*, and the tobacco aphid, *M. nicotianae*, have the same esterase-based mechanism of insecticide resistance. *Insect Molecular Biology* 3, 143–148.

Forbes, A.R. (1964) The morphology, history and fine structure of the gut of *Myzus persicae*. *Memoirs of the Entomological Society of Canada* 36, 1–74.

Frazer, B.D. (1972) Population dynamics and recognition of biotypes in the pea aphid (Homoptera: Aphididae). *Canadian Entomologist* 104, 1729–1733.

Fuller, S.J., Chavigny, P., Lapchin, L. and Vanleberghe-Massuti, F. (1999) Variation in clonal diversity in glasshouse infestations of the aphid, *Aphis gossypii* Glover, in southern France. *Molecular Ecology* 8, 1867–1877.

Georgieva, I.D. (1998) Possible relation between tobacco resistance to aphids (*Myzus nicotianae* Blackman) and phenolic compounds in glandular trichomes and leaf epidermis. *Annales du Tabac (Section 2)* 30, 3–9.

Grossheim, N.A. (1914) The barley aphid *Brachycolus noxius* Mordvilko. *Trudy Estestvenno-Istori eskago ear Tavri eskago Gubernskaya Zemstva Simferopol* 3, 35–73. [Abstract in *Review of Applied Entomology* 3 (1915), 307–308.]

Guldemond, J.A. and Mackenzie, A. (1994) Sympatric speciation in aphids I. Host race formation by escape from gene flow. In: Leather, S.R., Watt, A.D., Mills, N.J. and Walters, K.F.A. (eds) *Individuals, Populations and Patterns in Ecology.* Intercept, Andover, pp. 367–378.

Guldemond, J.A., Tigges, W.T. and de Vrijer, P.W.F. (1994) Host races of *Aphis gossypii* on cucumber and chrysanthemum. *Environmental Entomology* 23, 1235–1241.

Gut, J. (1976) Chromosome numbers of parthenogenetic females of fifty-five species of Aphididae new to cytology. *Genetica* 46, 279–285.

Haack, L., Simon, J.-C., Gauthier, J.-P., Plantegenest, M. and Dedryver, C.-A. (2000) Evidence for predominant clones in a cyclically parthenogenetic organism provided by combined demographic and genetic analyses. *Molecular Ecology* 9, 2055–2066.

Halbert, S.E. and Voegtlin, D.J. (1992) Morphological differentiation between *Aphis spiraecola* and *Aphis pomi* (Homoptera: Aphididae). *The Great Lakes Entomologist* 25, 1–8.

Halbert, S.E. and Voegtlin, D.J. (1998) Evidence for North American origin of *Rhopalosiphum* and barley yellow dwarf virus. In: Nieto Nafria, J.M. and Dixon, A.F.G. (eds) *Aphids in Natural and Managed Ecosystems*. Secretariado de Publicaciones, Universidad de León, León, pp. 351–356.

Hales, D.F. and Cowen, R. (1990) Genetic studies of *Rhopalosiphum* in Australia. *Acta Phytopathologica et Entomologica Hungarica* 25, 283–288.

Hales, D.H., Wilson, A.C.C., Spence, J.M. and Blackman, R.L. (2000) Confirmation that *Myzus antirrhinii* (Macchiati) (Hemiptera: Aphididae) occurs in Australia, using morphometrics, microsatellite typing and analysis of novel karyotypes by fluorescence *in situ* hybridisation. *Australian Journal of Entomology* 39, 123–129.

Harper, A.M., Miska, J.P., Manglitz, G.R., Irwin, B.J. and Armbrust, E.J. (1978) *The literature of arthropods associated with alfalfa. III. A bibliography of the pea aphid,* Acyrthosiphon pisum *(Harris) (Homoptera: Aphididae)*. Special Publication, Agricultural Experiment Station, College of Agriculture, University of Illinois at Urbana-Champaign, no. 50, 89 pp.

Hawthorne, D.J. and Via, S. (2001) Genetic linkage of ecological specialization and reproductive isolation in pea aphids. *Nature, London* 412, 904–907.

Heie, O.E. (1986) The Aphidoidea (Hemiptera) of Fennoscandia and Denmark. III. Family Aphididae: subfamily Pterocommatinae and tribe Aphidini of subfamily Aphidinae. *Fauna Entomologica Scandinavica* 17, 314 pp.

Heie, O.E. (1992) The Aphidoidea (Hemiptera) of Fennoscandia and Denmark. IV. Family Aphididae: Part 1 of tribe Macrosiphini of subfamily Aphidinae. *Fauna Entomologica Scandinavica* 25, 190 pp.

Heie, O.E. (1994) The Aphidoidea (Hemiptera) of Fennoscandia and Denmark. V. Family Aphididae: Part 2 of tribe Macrosiphini of subfamily Aphidinae. *Fauna Entomologica Scandinavica* 28, 242 pp.

Hille Ris Lambers, D. (1947) Contributions to a monograph of the Aphididae of Europe. III. *Temminckia* 7, 179–319.

Hille Ris Lambers, D. (1948) On Palestine aphids, with descriptions of new subgenera and new species (Hom. Aphid.). *Transactions of the Royal Entomological Society of London* 99, 269–289.

Hughes, R.D. (1996) *A synopsis of information on the Russian wheat aphid,* Diuraphis noxia *(Mordvilko) (revised edn)*. CSIRO, Australia, Division of Entomology Technical Paper, no.34, 97 pp.

Hughes, R.D. and Maywald, G.F. (1990) Forecasting the favourableness of the Australian environment for the Russian wheat aphid, *Diuraphis noxia* (Homoptera: Aphididae), and its potential impact on Australian wheat yields. *Bulletin of Entomological Research* 80, 165–175.

Iglisch, I. (1968) Über die Entstehung der Rassen der 'Schwarzen Blattlause' (*Aphis fabae* Scop. und verwandte Arten), über ihre phytopathologische Bedeutung und über die Aussichten für erfolgversprechende Bekämpfungsmassnahmen (Homoptera: Aphididae). *Mitteilungen aus der Biologischen Bundesanstalt für Land- und Forstwirtschaft Berlin-Dahlem* 131, 1–34.

Inaizumi, M. (1980) *Studies on the life cycle and polymorphism of* Aphis gossypii. Special Bulletin, College of Agriculture, Utsunomia University, no. 37, pp. 1–132.

Inaizumi, M. (1981) Life cycle of *Aphis gossypii* Glover (Homoptera, Aphididae) with special reference to biotype differentiation on various host plants. *Kontyû, Tokyo* 49, 219–240.

Jauset, A.M., Muñoz, M.P. and Pons, X. (2000) Karyotype occurrence and host plants of the corn leaf aphid (Homoptera: Aphididae) in a Mediterranean region. *Annals of the Entomological Society of America* 93, 1116–1122.

Johnson, W.G. (1900) The destructive green-pea louse. *Canadian Entomologist* 32, 52–60.

Jones, M.G. (1967) Observations on two races of the groundnut aphid, *Aphis craccivora*. *Entomologia Experimentalis et Applicata* 10, 31–38.

de Jong, J.K. (1929) *Enkele resultaten van het onderzoek naar de biologie van der tabaksluis,* Myzus persicae *Sulzer*. Bulletin Deli van het Proefstation te Medan, no. 28, 36 pp.

Kar, I. and Khuda-Bukhsh, A.R. (1991) Nucleolar organizer regions (NORs) in the chromosomes of an aphid *Lipaphis erysimi* Kalt. (Homoptera: Aphididae) with variable chromosome numbers. *Cytologia, Tokyo* 56, 83–86.

Kawada, K. and Murai, T. (1979) Apterous males and holocyclic reproduction of *Lipaphis erysimi* in Japan. *Entomologia Experimentalis et Applicata* 26, 343–345.

Kiriac, I., Gruber, F., Poprawski, T., Halbert, S. and Elberson, L. (1990) Occurrence of sexual morphs of Russian wheat aphid, *Diuraphis noxia* (Homoptera: Aphididae), in several locations in the Soviet Union and the northwestern United States. *Proceedings of the Entomological Society of Washington* 92, 544–547.

Komazaki, S. (1983) Overwintering of the spirea aphid, *Aphis citricola* van der Goot (Homoptera: Aphididae), on citrus and spirea plants. *Applied Entomology and Zoology* 18, 301–307.

Komazaki, S. (1986) The inheritance of egg hatching time of the spirea aphid, *Aphis citricola* van der Goot (Homoptera: Aphididae), on the two winter hosts. *Kontyû, Tokyo* 54, 48–53.

Komazaki, S. (1991) *Studies on the biology of the spirea aphid,* Aphis spiraecola *Patch, with special reference to biotypic differences.* [In Japanese.] Bulletin of the Fruit Tree Research Station, Extra no. 2, 60 pp.

Komazaki, S. (1998) Difference of egg diapause in two host races of the spirea aphid, *Aphis spiraecola*. *Entomologia Experimentalis et Applicata* 89, 201–205.

Komazaki, S. and Osakabe, Mh. (1998) Variation of Japanese *Aphis gossypii* clones in the life cycle, host suitability and insecticide susceptibility, and estimates of their genetic variability by DNA analysis. In: Nieto Nafria, J.M. and Dixon, A.F.G. (eds) *Aphids in Natural and Managed Ecosystems*. Secretariado de Publicaciones Universidad de León, León, pp. 83–89.

Komazaki, S., Sakagami, Y. and Korenaga, R. (1979) Overwintering of aphids on citrus trees. [In Japanese.] *Japanese Journal of Applied Entomology and Zoology* 23, 246–250.

Kring, J.B. (1959) The life cycle of the melon aphid *Aphis gossypii* Glover, an example of facultative migration. *Annals of the Entomological Society of America* 52, 284–286.

Lamb, R.J. and MacKay, P.A. (1979) Variability in migratory tendency within and among natural populations of the pea aphis, *Acyrthosiphon pisum*. *Oecologia* 39, 289–299.

Lupoli, R., Irwin, M.E. and Vossbrinck, C.R. (1990) A ribosomal DNA probe to distinguish populations of *Rhopalosiphum maidis* (Homoptera: Aphididae). *Annals of Applied Biology* 117, 3–8.

MacDougall, A.P. (1926) Some new species of *Macrosiphum* from British Columbia (Homoptera, Aphididae). *Pan-Pacific Entomologist* 2, 165–173.

MacGillivray, M.E. and Anderson, G.B. (1964) The effect of photoperiod and temperature on the production of gamic and agamic forms in *Macrosiphum euphorbiae* (Thomas). *Canadian Journal of Zoology* 42, 491–510.

Mackauer, M. and Way, M.J. (1976) *Myzus persicae* Sulz. an aphid of world importance. In: Delucchi, V.F. (ed.) *Studies in Biological Control*. Cambridge University Press, Cambridge, pp. 51–119.

MacKay, P.A., Lamb, R.J. and Smith, M.A.H. (1993) Variability in life history traits of the aphid *Acyrthosiphon pisum* (Harris), from sexual and asexual populations. *Oecologia* 94, 330–338.

Mackenzie, A. and Guldemond, J.A. (1994) Sympatric speciation in aphids II. Host race formation in the face of gene flow. In: Leather, S.R., Watt, A.D., Mills, N.J. and Walters, K.F.A. (eds) *Individuals, Populations and Patterns in Ecology*. Intercept, Andover, pp. 379–395.

Margaritopoulos, J.T., Mamuris, Z. and Tsitsipis, J.A. (1998) Attempted discrimination of *Myzus persicae* (Sulzer) and *Myzus nicotianae* Blackman (Homoptera: Aphididae) by Random Amplified Polymorphic DNA Polymerase Chain Reaction technique. *Annals of the Entomological Society of America* 91, 602–607.

Margaritopoulos, J.T., Tsitsipis, J.A., Zintzaras, E. and Blackman, R.L. (2000) Host-correlated morphological variation of *Myzus persicae* (Hemiptera: Aphididae) populations in Greece. *Bulletin of Entomological Research* 90, 233–244.

Margaritopoulos, J.T., Tzortzi, M., Zarpas, K.D., Tsitsipis, J.A. and Blackman, R.L. (2006) Morphological discrimination of *Aphis gossypii* (Hemiptera: Aphididae) populations feeding on Compositae. *Bulletin of Entomological Research* 96, 153–165.

Meier, W. (1961) Beiträge zur Kenntnis der grünstreifigen Kartoffelblattlaus, *Macrosiphum euphorbiae* (Thomas 1870) und verwandter Arten (Hemipt., Aphid.). *Mitteilungen der Schweizerischen Entomologischen Gesellschaft* 34, 127–186.

Miller, R.L. (1929) A contribution to the biology of and control of the green citrus aphid, *Aphis spiraecola* Patch. *Bulletin of Florida Agricultural Experimental Station* 203, 431–476.

Mimeur, J.M. (1942) Aphididae Nord-Africains. *Bulletin de la Société des Sciences naturelles du Maroc* (1941) 21, 67–70.

Mittler, T.E. (1976) Ascorbic acid and other chelating agents in the trace-mineral nutrition of the aphid *Myzus persicae* on artificial diets. *Entomologia Experimentalis et Applicata* 20, 81–98.

Möller, F.W. (1970) Die erste gelungene bisexuelle Fortpflanzung mit europäischen Herkünften von *Macrosiphum euphorbiae* (Thomas) (Homoptera: Aphididae). *Zoologischer Anzeiger* 184, 107–119.

Möller, F.W. (1971) *Macrosiphum stellariae* (Theobald) – eine bisher nicht von der grünstreifigen Kartoffelblattlaus *Macrosiphum euphorbiae* (Thomas) abgegrenzte Art. *Deutsche Entomologische Zeitschrift* 18, 207–215.

Mordvilko, A. (1928) Aphidodea. The plant lice. In: Filipjev, J.N. (ed.) *Keys to Insects of the European Part of the U.S.S.R.* [In Russian.] Novaya Derevnya, Moscow, pp.163–204.

Müller, F.P. (1958) Bionomische Rassen der grünen Pfirsichblattlaus *Myzus persicae* (Sulz.). *Archiv der Freunde der NaturGeschichte in Mecklenberg* 4, 200–233.

Müller, F.P. (1980) Wirstpflanzen, Generationenfolge und reproduktive Isolation infraspezifischer Formen von *Acyrthosiphon pisum*. *Entomologia Experimentalis et Applicata* 28, 145–157.

Müller, F.P. (1982) Das problem *Aphis fabae* (Homoptera: Aphididae). *Zeitschrift für Angewandte Entomologie* 94, 432–446.

Müller, F.P. (1985) Das Problem *Acyrthosiphon pisum* (Homoptera: Aphididae). *Zeitschrift für Angewandte Zoologie* 72, 317–334.

Müller, F.P. (1986) The rôle of subspecies in aphids for affairs of applied entomology. *Journal of Applied Entomology* 101, 295–303.

Müller, F.P. and Steiner, H. (1986) Beitrag zur vergleichenden Morphologie und Bionomie von *Aphis euonymi* F. *Deutsche Entomologische Zeitschrift* 33, 257–262.

Painter, R.H. and Pathak, M.D. (1962) The distinguishing features and significance of the four biotypes of the corn leaf aphid, *Rhopalosiphum maidis* (Fitch). *Proceedings of the 11th International Congress of Entomology, Vienna, 1960* 2, 110–115.

Pettersson, J. (1971) Studies on four grass inhabiting species of *Schizaphis*. III (a) host plants. *Swedish Journal of Agricultural Research* 1, 133–138.

Pike, K.S. and Allison, D. (1991) *Russian Wheat Aphid. Biology, Damage and Management.* Pacific Northwest Cooperative Extension Publication PNW371, 28 pp.

Ponsen, M.B. (1972) The site of potato leafroll virus multiplication in its vector, *Myzus persicae*. An anatomical study. *Mededlingen Landbouwhogeschool Wageningen* 72- 16, 1–147.

Poprawski, T.J., Underwood, N.L., Mercadier, G. and Gruber, F. (1992) *Diuraphis noxia (Kurdjumov) – A Bibliography on the Russian Wheat Aphid 1886–1992.* USDA Plant Soil and Nutrition Laboratory, Ithaca, 153 pp.

Puterka, G.J., Burd, J.D. and Burton, R.L. (1992) Biotypic variation in a worldwide collection of Russian wheat aphid (Homoptera: Aphididae). *Journal of Economic Entomology* 85, 1497–1506.

Puterka, G.J., Black, W.C., IV, Steiner, W.M. and Burton, R.L. (1993) Genetic variation and phylogenetic relationships among worldwide collections of the Russian wheat aphid, *Diuraphis noxia* (Mordvilko), inferred from allozyme and RAPD-PCR markers. *Heredity* 70, 604–618.

Rakauskas, R. (2004) What is the (aphid) subspecies? In: Simon, J.-C., Dedryver, C.A., Rispe, C. and Hullé, M. (eds) *Aphids in the New Millennium.* INRA, Paris, pp. 165–170.

Raymond, B., Searle, J.B. and Douglas, A.E. (2001) On the processes shaping reproductive isolation in aphids of the *Aphis fabae* (Scop.) complex (Aphididae: Homoptera). *Biological Journal of the Linnean Society* 74, 205–215.

Remaudière, G. and Naumann-Etienne, K. (1991) Découverte au Pakistan de l'hôte primaire de *Rhopalosiphum maidis* (Fitch) (Hom. Aphididae). *Compte rendu de l'académie d'agriculture de France* 77, 61–62.

Remaudière, G. and Remaudière, M. (1997) *Catalogue des Aphididae du Monde.* INRA, Paris, 473 pp.

Robert, Y. (1971) Epidémiologie de l'enroulement de la pomme de terre, capacité vectrice de stades et de formes des pucerons *Aulacorthum solani* (Kltb.), *Macrosiphum euphorbiae* (Thomas) et *Myzus persicae* (Sulz.). *Potato Research* 14, 130–139.

Saxena, P.N. and Chada, H.L. (1971) The greenbug, *Schizaphis graminum*. 1. Mouthparts and feeding habits. *Annals of the Entomological Society of America* 64, 897–904.

Semtner, P.J., Reed, T.D., Barnes, M.L. and Wilkinson, W.B., III (1990) Triazuron and other insecticides applied as foliar sprays for the control of the tobacco aphid on flue-cured tobacco. *Tobacco Science* 34, 39–43.

Shands, W.A., Simpson, G.W., Muesebeck, C.F.W. and Wave, H.E. (1965) *Parasites of Potato-infesting Aphids in North-eastern Maine.* Technical Bulletin of Maine Agricultural Experiment Station 719, 77 pp.

Shands, W.A., Simpson, G.W. and Wave, H.E. (1969) *Canada Plum,* Prunus nigra *Aiton, as a Primary Host of the Green Peach Aphid,* Myzus persicae *(Sulzer), in North-eastern Maine.* Technical Bulletin of University of Maine 39, 32 pp.

Shands, W.A., Simpson, G.W. and Wave, H.E. (1972) *Seasonal Population Trends and Productiveness of the Potato Aphid on Swamp Rose in North-eastern Maine.* Technical Bulletin of University of Maine 52, 35 pp.

Shufran, K.A., Black, W.C. (IV) and Margolies, D.C. (1991) DNA fingerprinting to study spatial and temporal distributions of an aphid, *Schizaphis graminum* (Homoptera: Aphididae). *Bulletin of Entomological Research* 81, 303–313.

Shufran, K.A., Peters, D.C. and Webster, J.A. (1997) Generation of clonal diversity by sexual reproduction in the greenbug, *Schizaphis graminum. Insect Molecular Biology* 6, 203–209.

Shufran, K.A., Burd, J.D., Anstead, J.A. and Lushai, G. (2000) Mitochondrial DNA sequence divergence among greenbug (Homoptera: Aphididae) biotypes: evidence for host-adapted races. *Insect Molecular Biology* 9, 179–184.

Simon, J.C., Hebert, P.D.N., Carillo, C. and de Melo, R. (1995). Lack of clonal variation among Canadian populations of the corn leaf aphid, *Rhopalosiphum maidis* Fitch (Homoptera: Aphididae). *Canadian Entomologist* 127, 623–629.

Simon, J.-C., Carré, S., Boutin, M., Prunier-Leterme, N., Sabater-Muñoz, B., Latorre, A. and Bournoville, R. (2003). Host-based divergence in populations of the pea aphid: insights from nuclear markers and the prevalence of facultative symbionts. *Proceedings of the Royal Society of London* B 270, 1703–1712.

Singh, R.S. and Rhomberg, L. (1984) Allozyme variation, population structure and sibling species in *Aphis pomi. Canadian Journal of Genetics and Cytology* 26, 364–373.

Smith, M.A.H. and MacKay, P.A. (1990) Latitudinal variation in the photoperiodic responses of populations of pea aphid. *Environmental Entomology* 19, 618–624.

Steiner, W.W.M., Voegtlin, D.J. and Irwin, M.E. (1985) Genetic differentiation and its bearing on migration in North American populations of the corn leaf aphid, *Rhopalosiphum maidis* (Fitch) (Homoptera: Aphididae). *Annals of the Entomological Society of America* 78, 518–525.

Stroyan, H.L.G. (1960) Three new subspecies of aphids from Iceland. *Entomologiske Meddelelser* 29, 250–265.

Stroyan, H.L.G. (1984) *Aphids – Pterocommatinae and Aphidinae (Aphidini). Handbooks for the Identification of British Insects, Volume 2, Part 6.* Royal Entomological Society, London, 232 pp.

Sullivan, D.J. and van den Bosch, R. (1971) Field ecology of the primary parasites and hyperparasites of the potato aphid, *Macrosiphum euphorbiae,* in the east San Francisco Bay area. *Annals of the Entomological Society of America* 64, 389–394.

Sunnucks, P., de Barro, P.J., Lushai, G., Maclean, N. and Hales, D. (1997a) Genetic structure of an aphid studied using microsatellites: cyclic parthenogenesis, differentiated lineages, and host specialisation. *Molecular Ecology* 6, 1059–1073.

Sunnucks, P., Driver, F., Brown, W.V., Carver, M., Hales, D.F. and Milne, W.M. (1997b) Biological and genetic characterization of morphologically similar *Therioaphis trifolii* (Hemiptera: Aphididae) with different host utilization. *Bulletin of Entomological Research* 87, 425–436.

Takada, H. (1979) Esterase variation in Japanese populations of *Myzus persicae* (Sulzer), with special reference to resistance to organophosphorus insecticides. *Applied Entomology and Zoology, Tokyo* 14, 245–255.

Takada, H. (1982) Influence of photoperiod and temperature on the production of sexual morphs in a green and a red form of *Myzus persicae* (Sulzer) (Homoptera, Aphididae). I. In the laboratory. II. Under natural conditions. *Kontyû, Tokyo* 50, 233–245, 353–364.

Takada, H. (1986) Genotypic composition and insecticide resistance of Japanese populations of *Myzus persicae* (Sulzer) (Hom., Aphididae). *Zeitschrift für Angewandte Entomologie* 102, 19–38.

Takahashi, R. (1923) *Aphididae of Formosa, Part 2.* Report of the Government Research Institute of the Department Agriculture, Formosa 4, pp. 1–173.

Tambs-Lyche, H. (1959) A new species of *Schizaphis* Börner attacking *Phleum pratense* in Norway. *Norsk Entomologisk Tidsskrift* 11, 1–2.

Tanigoshi, L.K., Pike, K.S., Miller, R.H., Miller, T.D. and Allison, D. (1995) Search for, and release of, parasitoids for the biological control of Russian wheat aphid in Washington State (USA). *Agriculture, Ecosystems and Environment* 52, 25–30.

Taylor, L.R. (1977) Migration and the spatial dynamics of an aphid, *Myzus persicae. Journal of Animal Ecology* 46, 411–423.

Terradot, L., Simon, J.-C., Leterme, N., Bourdin, D., Wilson, A.C.C., Gauthier, J.-P. and Robert, Y. (1999). Molecular characterization of clones of the *Myzus persicae* complex (Hemiptera: Aphididae) differing in their ability to transmit the potato leafroll luteovirus (PLRV). *Bulletin of Entomological Research* 89, 355–363.

Thieme, T. (1987) Members of the complex of *Aphis fabae* Scop and their host plants. In: Holman, J., Pelikan, J. and Dixon, A.F.G. (eds) *Population Structure, Genetics and Taxonomy of Aphids*. SPB Academic Publishing, The Hague, pp. 314–323.

Thieme, T. (1988) Zur Biologie von *Aphis fabae mordvilkoi* Börner and Janisch, 1922 (Hom., Aphididae). *Journal of Applied Entomology* 105, 510–515.

Thieme, T. and Dixon, A.F.G. (1996) Mate recognition in the *Aphis fabae* complex: daily rhythm of release and specificity of sex pheromones. *Entomologia Experimentalis et Applicata* 79, 85–89.

Thieme, T. and Dixon, A.F.G. (2004) The case for *Aphis solanella* being a good species. In: Simon, J.-C., Dedryver, C.A., Rispe, C. and Hullé, M. (eds) *Aphids in the New Millenium*. INRA, Paris, pp. 189–204.

Thieme, T., Heimbach, U. and Schliephake, E. (2001) Nachweis der 'Russischen Weizenlaus', *Diuraphis noxia* (Kurdjumov), in Deutschland. *Nachrichtenblatt für den Deutschen Pflanzenschutzdienst* 53, 35–40.

Vanlerberghe-Masutti, F. and Chavigny, P. (1998) Host-based genetic differentiation in the aphid *Aphis gossypii* Glover, evidenced from RAPD fingerprints. *Molecular Ecology* 7, 905–914.

Via, S. (1991) The genetic structure of host plant adaptation in a spatial patchwork: demographic variability among reciprocally transplanted pea aphid clones. *Evolution* 45, 827–852.

Via, S. (1999) Reproductive isolation between sympatric races of pea aphids. I. Gene flow restriction and habitat choice. *Evolution* 53, 1446–1457.

Wang, E., Wang, R., deParasis, J., Loughrin, J.H., Gan, S. and Wagner, G.J. (2001) Suppression of a P450 hydroxylase gene in plant trichome glands enhances natural-product-based aphid resistance. *Nature Biotechnology* 19, 371–374.

Watson, G.W. (1982) A biometric, electrophoretic and karyotypic analysis of British species of *Macrosiphum* (Homoptera: Aphididae). PhD thesis, University of London, London, UK.

Webster, F.M. and Phillips, W.J. (1912) *The Spring Grain Aphis or 'Greenbug'*. United States Department of Agriculture Bureau of Entomology Bulletin, no. 110, 153 pp.

Wood, E.A., Jr. (1961) Biological studies of a new greenbug biotype. *Journal of Economic Entomology* 54, 1171–1173.

Woodford, J.A.T., Jolly, C.A. and Aveyard, C.S. (1995) Biological factors influencing the transmission of potato leafroll virus by different aphid species. *Potato Research* 38, 133–141.

Zhang, G.-X. and Zhong, T.-S. (1990) Experimental studies on some aphid life cycle patterns and the hybridization of sibling species. In: Campbell, R.K. and Eikenbary, R.D. (eds) *Aphid–Plant Genotype Interactions*. Elsevier, Amsterdam, pp. 37–50.

2 Population Genetic Issues: The Unfolding Story Using Molecular Markers

Hugh D. Loxdale and Gugs Lushai

Plant and Invertebrate Ecology Division, Rothamsted Research, Harpenden, Herts, AL5 2JQ, UK

Introduction

Few insects can match the ecological flexibility of aphids with their range of environmentally and genetically determined phenotypes (Williams and Dixon, Chapter 3 this volume). They are therefore especially successful at exploiting diverse environmental conditions and host-plant resources.

In the past 25 years or so, molecular markers (protein markers, especially allozymes and, more recently, DNA markers) have revealed much new information about the biology of aphids. Some findings have been surprising, whilst others have only confirmed previous hypotheses, or have provided information of practical value for aphid control. Before discussing these markers and what they tell us, a short description of certain evolutionary and ecological aspects that make aphids so fascinating is appropriate.

Aphids are the 'Russian dolls' of the animal world, with the 'telescoping of generations' within an individual (Dixon, 1998) accelerating their reproduction (Harrington, 1994). Paradoxically, the evolutionary consequences of clonality are controversial (Lushai *et al.*, 2000; Gorokhova *et al.*, 2002; Lushai and Loxdale, 2002; Loxdale and Lushai, 2003a,b,c and references therein; Monro and Poore, 2004; Pineda-Krch and Lehtila, 2004; Van Doninck *et al.*, 2004; Morgan-Richards

and Trewick, 2005). In aphids, the reason for mass-producing genetically identical copies (assumed, but not proven; see below and Loxdale and Lushai, 2003a) of an asexual foundress is not clear, but sexual recombination alone may be insufficiently adaptive for these insects. Hence, a large 'evolutionary unit' (Janzen, 1977) seeks new habitats with seasonal environmental change, and can adapt to new microecological conditions (Mopper and Strauss, 1998; Lushai and Loxdale, 2004). This is an 'r' strategy (MacArthur and Wilson, 1963; Bruton, 1989), but a periodic reversal towards a 'K' type strategy is one beneficial evolutionary aspect of an overwintering egg stage. Other benefits include the purging of acquired genomic aberrations within annual parthenogenetic lines by selection, genetic drift, and recombination.

Some crop pest aphids have not freed themselves from a primary woody host on which cold-resistant eggs can overwinter (Blackman, 1980). However, the autumn search for a sometimes uncommon primary host presents huge risks for the migrants. In the case of *Rhopalosiphum padi* (bird cherry–oat aphid), only around 0.6% of winged colonizers is estimated to find the overwintering host, bird cherry (*Prunus padus*) (Ward *et al.*, 1998). This is another reason for producing large numbers of offspring. However, most aphids do not host

©CAB International 2007. *Aphids as Crop Pests*
(eds H. van Emden and R. Harrington)

alternate, but multiply exclusively on a single host, which may be either woody or herbaceous.

Another problem that aphids face is clonality itself. They may be considered 'cloning experts' (Blackman, 2000), but they are not necessarily very good at it. Thus, whilst the initial 'aim' may be to produce genetically identical offspring by a rapid process of mitotic (apomictic) reproduction (Blackman, 1980), because of the number of mutation-prone regions and processes at work within the genome (e.g. micro- and minisatellites, 'hotspots', transposons, etc.), exact identity is probably not maintained for long, if at all (Lushai *et al.*, 2000; Lushai and Loxdale, 2002; Loxdale and Lushai, 2003a). Furthermore, lineage longevity may decay due to genomic events such as reduction of chromosome telomere length, with the requirement for occasional recombination events to reset this (Loxdale and Lushai, 2003b; Lushai and Loxdale, 2007). In addition, so-called 'clones' are subject to the accumulation of random, genome-wide, mildly deleterious mutations (i.e. heterozygosity; Muller's ratchet – Muller, 1964; Normark and Moran, 2000), which may ultimately cause the demise of a particular clonal lineage by a process of 'mutational meltdown' (Lynch *et al.*, 1993; Lynch and Blanchard, 1998). Even so, such mechanisms have not yet been proven in aphids or in any other eukaryotic asexual lineages, except in yeasts (Zeyl *et al.*, 2001; see also Klekowski, 2003). Indeed, asexual lineages of some species like *Myzus ascalonicus* (shallot aphid) in which sexuals have never been found appear to persist for long periods of time (Blackman, 2000; but see also Blackman *et al.*, 2000 for other genetic/chromosomal consequences of long-term aphid parthenogenesis).

Rapid mutation could be advantageous by reducing competition between 'identical' siblings for resources and allowing mutant clones to arise and exploit novel ecological opportunities, including new plant host strains and species (Blackman, 1981). Nevertheless, asexual lineages, because of their non-recombinant (linked) genomes and linkage disequilibrium (Richardson *et al.*, 1986; Ridley, 1993), are unlikely ever to have as much adaptive genetic variation as lineages whose genomes go through recombination events, even if only very occasionally (Normark, 1999; Blackman, 2000).

The aforementioned examples of clonal studies include areas of aphid ecology and evolution that have recently been given close attention, and here molecular markers show great promise. Other aspects of aphid biology that have greatly benefited from the application of molecular markers include phylogenetics, life cycle strategies, host-plant adaptation and plant-mediated resistance, pathogenicity, insecticide resistance, dispersal and geographic range, and geographic colonization. The present synthesis describes the utility of genetic molecular markers in aphidology, especially during the past 10 years.

The greatest contribution of these markers has been to enable characterization of genetic individuals within a population. This is revolutionizing many aspects of the science by revealing patterns of genetic variation in relation to life cycle, ecology, demography, and climate. It is the recognition of genetic individuals, lineages, and the persistence of these, that we discuss first, after a brief introduction to the molecular techniques currently used. Thereafter, we discuss other aspects of aphid biology resolved using molecular markers, especially the demographic distribution of life cycle forms in relation to climate, host-plant preference and adaptation, migration and colonization events.

Application of Molecular Markers in Aphidology

The molecular techniques used in entomology have been reviewed in detail by Crampton *et al.* (1996), Loxdale and Lushai (1998), and Hoy (2003). Briefly, we describe these in relation to aphids as crop pests.

Detection of genetic variation

Initially, allozymes, the products of enzyme-coding alleles at specific loci (Richardson *et al.*, 1986), were used extensively in aphid studies, mainly for population genetics, clonal or biotype determinations, and higher-level taxonomic discriminations (Blackman *et al.*, 1989; Loxdale, 1994). Since about 1990, however, several DNA markers have been utilized, some of which involve the polymerase chain reaction (PCR) and sequencing. These techniques include:

1. Synthetic oligonucleotide probes of nuclear DNA restriction fragment length polymorphisms (RFLPs), e.g. (GATA)$_4$ (de Barro *et al.*, 1994a,b, 1995a);
2. Randomly amplified polymorphic DNA (RAPDs) (see also sequence character-ized amplified regions (SCARs), in which discriminatory RAPD bands are sequenced, from which specific homologous probes or flanking primers are produced, e.g. Simon *et al.*, 1999b);
3. Amplified fragment length poly-morphisms (AFLPs) (Forneck *et al.*, 2001);
4. Mitochondrial DNA (mtDNA) (Barrette *et al.*, 1994; Shufran *et al.*, 2000) and mito-chondrial partial 12S and 16S rDNA sequence analysis (Von Dohlen and Moran, 2000);
5. An ever increasing variety of nuclear markers based on amplification of specific sites of interest, e.g. intergenic spacer (IGS) and internal transcribed spacer (ITS) regions of the ribosomal genome (Shufran *et al.*, 1991; Shufran and Wilde, 1994; Fenton *et al.*, 1998a,b, 2003); the intron-bearing, nuclear protein-coding gene, elongation factor-1-α (EF-1α; Normark, 1999); and microsatellites (Sunnucks *et al.*, 1996; Vanlerberghe-Masutti and Chavigny, 1998; Simon *et al.*, 1999a), with primers now designed for a range of pest and non-pest species, e.g. *Sitobion* spp., *R. padi*, *Aphis gossypii* (cotton or melon aphid), *Macrosiphoniella tanacetaria* (tansy aphid), *Acyrthosiphon pisum* (pea aphid), and *Myzus persicae* (peach–potato aphid) (Yao *et al.*, 2003; Caillaud *et al.*, 2004a; Kurokawa *et al.*, 2004; see Wilson *et al.*, 2004 for an overview). Some of the microsatellite primers and nuclear sequences from sites adjacent to microsatellite repeats are known to cross-amplify to provide useful poly-morphisms in closely related taxa, e.g.

Sitobion avenae (grain aphid) and *Sitobion fragariae* (blackberry–cereal aphid); other *Sitobion* spp. and *Metopolophium dirhodum* (rose–grain aphid), and *Sitobion* spp. and *A. pisum* in the case of microsatellites (Llewellyn, 2000; Wilson *et al.*, 2004), and *Aphis pomi* (green apple aphid) and *Aphis spiraecola* (green citrus aphid) with nuclear sequences (Lushai *et al.*, 2004);
6. Molecular systems that speed up the detection of genetic variation, such as sin-gle nucleotide polymorphisms (SNPs – 'snips') (Stoneking, 2001); single-strand confirmation polymorphism (SSCP) markers of mitochondrial and nuclear DNA (Sunnucks *et al.*, 2000); and determination of DNA allele size in population genetic studies using MALDI-TOF technology (Kirpekar *et al.*, 1998; Kwon *et al.*, 2001); and
7. Molecular markers that involve screen-ing the functional genome, e.g. fluorescent *in situ* hybridization (FISH), which is an approach used in locating genes or DNA products to their sources on chromosomal bodies and detailing chromosomal rear-rangements (Blackman and Spence, 1996; Blackman *et al.*, 2000; see below); extensive chromosomal mapping performed by com-plex AFLP analysis (Hawthorne and Via, 2001; Braendle *et al.*, 2005b), and direct sequencing of large tracts of (and perhaps in the not too distant future, the entire) genome. This offers the ultimate resolution of the genetic identity of individuals, and hence of clones and clonal lineages (Caillaud *et al.*, 2004b).

The number of genotypes detected using DNA markers such as IGS has increased by an order of magnitude from the number found using allozymes (e.g. Brookes and Loxdale, 1987 *v.* Fenton *et al.*, 1998a; see also Delmotte *et al.*, 2002 for the case of allozymes *v.* microsatellite markers). This has ramifications in terms of clonality as sequencing of the genome, either of spe-cific regions or, more generally, for weakly deleterious mutations (Normark and Moran, 2000), will undoubtedly allow a more thor-ough understanding of the concepts that need to be re-addressed in genotype–phenotype associations.

Molecular phylogenetics and karyotypes

Issues concerning molecular phylogenetics and karyotypes are not discussed here (they are covered to some extent for pest species by Blackman and Eastop, Chapter 1 this volume), other than where they impinge on intraspecific variation and its consequences in ecology and evolution, for example in host-plant adaptation. Establishment of aphid species identity is crucial during population genetic studies, as failure to do so can lead to misinterpretation of data. Even before the widespread use of DNA markers, protein markers, especially allozymes, were used extensively (Eggers-Schumacher, 1987). These proved useful for taxonomic discriminations, either for characterizing full species (e.g. Loxdale and Brookes, 1989a,b; Blackman and Spence, 1992), so-called races or 'cryptic' species (Guldemond, 1990; Loxdale and Brookes, 1990; Fukatsu *et al.*, 2001), and even intraspecific differences, such as host-plant races, strains, or 'biotypes' of aphids (Eastop, 1973), e.g. Abid *et al.* (1989).

At one extreme, suitable DNA markers have allowed refinement of such discriminations, for example, mtDNA, RAPDs, and microsatellites in the case of *Sitobion* spp. (Figueroa *et al.*, 1999) and microsatellites with members of the *M. persicae* complex, e.g. *Myzus antirrhinii* (snapdragon aphid) from *M. persicae sensu stricto* (Hales *et al.*, 2000), to the point of distinguishing the morphs of individual clonal lineages and life cycle forms (see later). At the other extreme, such markers have facilitated gross phylogenetic differentiation, e.g. tribes, etc. (Von Dohlen and Moran, 2000). The understanding of important co-evolutionary associations of aphids with fungal entomopathogens (Tymon *et al.*, 2004) and, more especially, bacterial symbionts, e.g. *Buchnera* (Baumann *et al.*, 1995; Clark *et al.*, 2000; Abbot and Moran, 2002; Douglas *et al.*, 2003; Simon *et al.*, 2003), and the association of these with plant taxa (Von Dohlen and Moran, 2000) has been enhanced greatly by the application of molecular techniques.

Gross karyotypic analysis has revealed phylogenetic differences in chromosome number and type. Within species, chromosome polymorphisms, e.g. due to translocations and changes in ribosomal DNA (rDNA) regions, have been found to be associated with speciation events and/or the evolution of host races and strains (e.g. Blackman, 1980, 2000; Blackman *et al.*, 2000; Wilson *et al.*, 2003). In addition, molecular techniques have been used to determine the position of genes on chromosomes (e.g. rDNA on the X-chromosome of aphids using FISH – Blackman and Spence, 1996; Blackman *et al.*, 2000), as well as to reveal an apparent lack of recombination in this sex-determining chromosome in males and females using X-linked microsatellite loci (Hales *et al.*, 2002; but see Mandrioli *et al.*, 1999 in relation to the nucleolar organizer region, NOR); the genetic determinant system termed 'aphicarus' (api) for wing polymorphism in *A. pisum*, which is also on the X-chromosome (Caillaud *et al.*, 2002; Braendle *et al.*, 2005a,b); and the gene(s) controlling sexuality in relation to short photoperiods in *A. pisum* (Ramos *et al.*, 2003).

Clonal studies

Early work with allozymes

Even using allozymes, the presence could be assessed of a range of genotypes at some polymorphic loci, e.g. glutamate oxaloacetate transaminase (GOT) in *S. fragariae* or carboxylesterase (E4) in *M. persicae*, where two electrophoretic mobility variants and five quantitative variants were detected, respectively (Devonshire, 1989a,b; Loxdale and Brookes, 1990). These markers showed that species populations comprised a number of 'clones' or genotypes and were not necessarily monomorphic or homogeneous (Hodgson, 2001). Hence, by the late 1970s when the technology was first applied to aphids, clones or more accurately, asexual lineages, could be identified using even these low resolution but versatile genetic markers.

Use of high resolution DNA markers

Use of DNA markers, such as the first aphid 'genetic fingerprint' study using a Jeffrey's type probe (an RNA derivative 33.15 human minisatellite core sequence; Carvalho *et al.*, 1991), enhanced the resolving power for aphid ecology. More sensitive assessment of host-plant associations became possible (including those of biotypes) whilst showing that, within a population, many more genotypes existed than hitherto imagined. Thus, clonality *per se*, as a biological definition of genetic fidelity, was found to be closely related with, as well as a function of, the molecular marker type used and its ability to resolve variation within the sample population. Thus, it is ambiguous to use a few loci to determine clonality; multilocus approaches are preferable. The resolving power also depends on variability. For example, in *S. avenae* that alleles at 12 independent microsatellite loci were found to be identical in 18 individuals of one clone and 31 of another can be taken as good evidence of genotypic identity suggestive of descent from a single foundress (Haack *et al.*, 2000; Miller, 2000).

Defining clonality

Fenton *et al.* (2003) showed in *M. persicae* that asexual lineages derived from laboratory cultures and the field in Scotland in 1995, 1996, and 2001, which had a particular IGS genotype, generally also had consistent microsatellite genotypic profiles at four microsatellite loci tested. This showed that such multi-locus, multi-genotype lineages are probably the same clone, and thus derived from a common ancestor. Even so, some field collected material did show IGS variation, due to structural/mutational changes in these lineages, which is not surprising considering the vast natural populations sampled and the fact that IGS seems to evolve faster than microsatellite alleles (Fenton *et al.*, 2003, 2005; see also Shufran *et al.*, 2003).

In clonal lineages, where no recombination takes place (Hales *et al.*, 2002), loci are not re-assorted and are carried together in concert, leading to linkage disequilibrium. However, even though there is conformation of genotype, there may not necessarily be clonality *sensu stricto*, since other regions of the genome may differ due to mutations

of one form or another (Blackman *et al.*, 2000; Lushai *et al.*, 2000; Lushai and Loxdale, 2002). Using three microsatellite primer pairs, the 256 offspring of one parthenogenetic female *S. avenae*, representing five asexual generations reared on wheat (*Triticum aestivum*) seedlings, failed to show intraclonal differences (H.D. Loxdale and P. Moone, unpublished results). But here again, these genotypes are only proven stable at the loci and over the generation times tested and could, in theory, differ elsewhere in the genome. In contrast, intraclonal variation has been revealed in lineages founded from individual *S. avenae* virginoparae studied using RAPDs. Lineages were found to mutate within 14 generations to produce both *somatic* and, less commonly, *gametic* changes, as shown by new bands (Lushai *et al.*, 1998; see also de Barro *et al.*, 1994a for the case of (GATA)$_4$ markers). Similarly, using AFLP markers to test genetic variability within and among clonal lineages of *Viteus* (= *Daktulosphaira*) *vitifoliae* (grape phylloxera), the reported rate of intraclonal genetic change was even greater, with banding profile differences observed within five generations (Forneck *et al.*, 2001).

Intraclonal, intermorph differences

RAPD markers have also shown that clonal lines can differ in DNA fingerprint between morphs (winged, wingless, males and oviparae) (Lushai *et al.*, 1997) (Fig. 2.1). These intraclonal, intermorph differences may be due to variations in repetitive sequences between primer binding sites at low annealing temperatures, or to some unknown genetic phenomenon, e.g. related to transposonation effects. DNA methylation (i.e. epigenetic change) is probably not responsible for the observed differences, although the effects of secondary structure at low annealing temperatures differentially affecting primer binding cannot be ruled out.

Fenton *et al.* (1998b) have shown that individual *M. persicae* can have two different ITS1 haplotypes, one related to that of *Myzus certus* (black cherry aphid). The authors suggest that this is possibly the

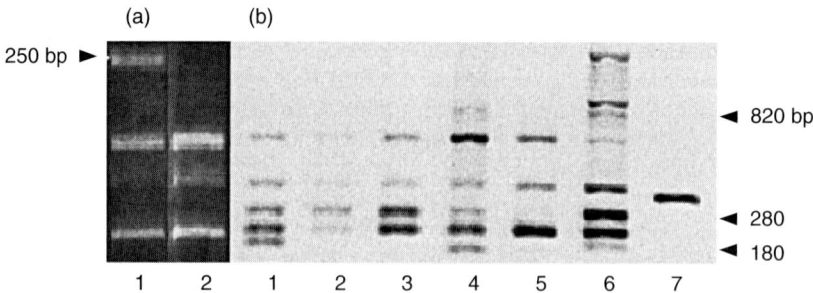

Fig. 2.1. DNA 'fingerprints' (RAPD-PCR) of cereal aphids showing *intra*clonal, *inter*morph profile differences. (a) Purified DNA of *Sitobion avenae*: lane 1 (L1), aptera; L2, alata. (b) crude DNA of *Rhopalosiphum padi* (inverse image): L1, aptera; L2, alata; L3, alata heads; L4, gynopara; L5, male; L6, ovipara; L7, water control. Apterous and alate virginoparae could be discriminated in *S. avenae* by the presence of a *ca.* 250 bp band in the former morph. With *R. padi* there was polymorphism across all morph types. The apterous and alate virginoparae were distinguished by a 180 bp band; both these morphs could be differentiated from the gynopara by an additional *ca.* 820 bp band in the latter. In the male, a 280 bp band, common to all the other morphs, was missing. The ovipara was distinguished from the gynopara by two further bands above 820 bp (from Lushai *et al.*, 1997 by permission of *Proceedings of the Royal Society of London B*).

result of recent divergence of the two species from a common ancestor, such that *M. persicae* contains an ancestral polymorphism, or that there is some degree of introgression between these closely related taxa. *M. persicae* also undergoes epigenetic changes within clonal lineages related to methylation of the DNA and switching on and off of the amplified E4 genes responsible for conferring resistance to insecticides (Hick *et al.*, 1996).

Clonal persistence

As mentioned earlier, it has been suggested that clones should be prone to the accumulation of mildly deleterious mutations leading to 'mutational meltdown'. Interestingly, the available (as yet, very limited) direct evidence from rearing does not support such an event in aphids. Thus, the clone of *Aphis fabae* (black bean aphid) established in 1949 at Imperial College is still going strong, as is the *M. persicae* clone reared on an artificial diet since 1976 at Reading University (van Emden, 1988), and other clones of *M. persicae* set up 20 years ago at Rothamsted to study insecticide resistance (S. Foster, personal communication).

Discriminating sexual and asexual lineages

The average number of microsatellite alleles per locus and heterozygosity of aphids in general have been shown recently to be similar to obligate, sexually reproducing insects (Miller, 2000). However, obligate asexual lines of *R. padi* have been shown to have half the number of microsatellite alleles per locus as their sexual counterparts (Delmotte *et al.*, 2001), and no multi-locus genotypes were shared between asexual and sexual lineages in samples collected from France (Delmotte *et al.*, 2002). According to Delmotte *et al.* (2002), each sexual lineage examined possessed a unique multi-locus genotype, as shown using a range of microsatellite markers. In sexual populations, five microsatellite loci were sufficient to identify between 94% and 100% of the genotypes in the sample, depending on the combination of loci (Fig. 2.2), showing that sexual genotypes differed greatly from each other. In asexual populations, around half of individuals were copies of genotypes and the number of genotypes did not increase significantly with sampling effort, whilst the mean number of genotypes increased slowly with the number of loci. A subset of asexual individuals

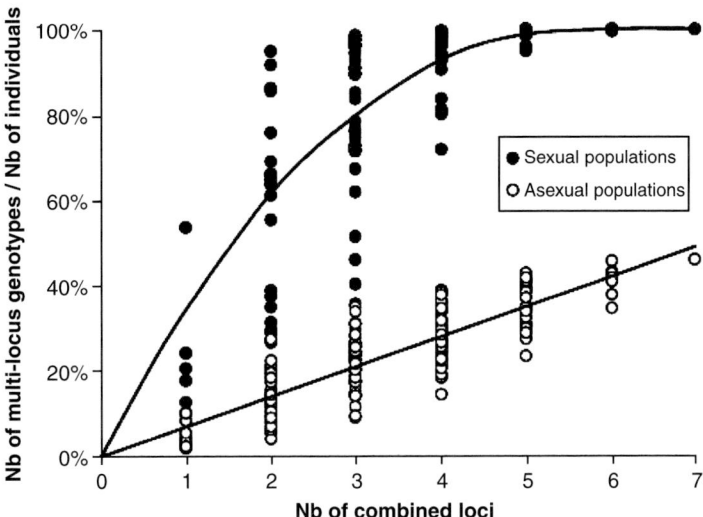

Fig. 2.2. Number (Nb) of genotypes discriminated (= (number of multi-locus genotypes/number of individuals sampled) expressed as a percentage) as a function of the number of loci combined to discriminate genotypes in sexual and asexual populations of *Rhopalosiphum padi* (from Delmotte *et al.,* 2002 by permission of *Molecular Ecology*).

belonging to the same seven-loci genotype was analysed at five additional microsatellite loci and still shared the same genotype. This suggests genotypic diversity to be largely resolved by the first seven loci tested. Lastly, the number of alleles in asexual populations was not very different with various combinations of a given number of loci, so that the proportion of asexual genotypes discriminated varied little.

Using both microsatellite and mtDNA markers, the very origin of asexual lineages has been revealed, for example by intraspecific, inter-lineage gene flow events in *R. padi* (Delmotte *et al.*, 2001), as well as by interspecific gene flow events, the result of multiple instances of introgression of this species with a closely related taxon at some time in the recent past (Delmotte *et al.*, 2003). Previously assumed ancient asexuals of the aphid tribe Tramini showed little or no heterozygosity when tested at the EF-1α locus, probably because rare sex has eliminated accumulated asexual variations (Normark, 1999).

Life cycle

That aphids show intraclonal variability is best illustrated by the fact that asexual lineages show alary polyphenisms triggered by crowding, nutritional cues, temperature, and photoperiod (Dixon, 1998). As host plants senesce and colonies become crowded, the propagation of winged, dispersive morphs increases (Lees, 1967), although generally these have lower fecundity than wingless morphs (Noda, 1960; Dixon, 1989, 1998; Williams and Dixon, Chapter 3 this volume). Some colonies produce individuals that are early dispersers or fly for longer periods than others (Kidd and Cleaver, 1984, 1986), and some gall-forming aphids of the families Pemphigidae and Hormaphididae produce soldier morphs that defend the colony and have very different morphology from the normal first instar (Aoki, 1977; Aoki and Akimoto, 1981; Stern and Foster, 1996; Foster, 2002). Thus, within a colony or population of aphids, different genotypes exist that can be categorized as life cycle types (asexual, etc.) or castes (e.g. soldiers). In addition, for any given life cycle form, levels of phenotypic plasticity may exist, which could have a genetic basis (e.g. Wool and Hales, 1997).

Demographic distribution of cereal aphid (Sitobion avenae and Rhopalosiphum padi) genotypes in relation to life cycle and climate

It has been shown for cereal aphids, notably *S. avenae*, that there is a demographic distribution of life cycle types in the UK in relation to latitude (Walters and Dewar, 1986; Newton and Dixon, 1988; Hand, 1989; Helden and Dixon, 2002). Anholocycly decreases in frequency and holocycly increases towards the north. This generally is assumed to be because the holocycle includes eggs, which have a survival advantage under cold conditions compared with aphids overwintering in the active stages. Yet even in colder climates within an aphid's range, some individuals may overwinter successfully as active stages, for example in thick tussocks of grass (Loxdale *et al.*, 1993; Powell and Bale, 2004), although, as found recently in the laboratory for *M. persicae*, obligate asexuals are no better adapted to low temperatures than the individuals from holocyclic clones (Vorburger, 2004). Life cycle synchrony in aphids is brought about by photoperiodism, which is related to latitude (Smith and Mackay, 1990; Mackay *et al.*, 1993).

STUDIES USING NUCLEAR MARKERS (MICROSATELLITES AND ALLOZYMES). Microsatellite markers have been used to study the relationship between reproductive mode and ecological parameters in *S. avenae* (Simon *et al.*, 1999a), revealing a strong geographic partitioning of life cycle types across a north–south cline. The species possesses a high level of allelic and genotypic diversity, a finding broadly similar to that using allozymes (Loxdale *et al.*, 1985). The fit of genotype numbers to Hardy-Weinberg expectations (indicative of random mating) measured by *F*-statistics (Raybould *et al.*, 2002; F_{IS} positive = heterozygote deficiency, equated with inbreeding; negative = heterozygote excess,

equated with outbreeding), along with assessment of linkage disequilibrium, showed isolated pools of genotype diversity to exist, which were then ranked in an ascending order from anholocyclic < androcyclic (males produced but no oviparae) < intermediate < holocyclic. Only holocyclic clones had positive F_{IS} for all loci (see Table 3 in Simon *et al.*, 1999a). The highest level of genetic recombination was associated with holocyclic forms (Simon *et al.*, 1999a), which proved the most closely related group, significantly different from the low relatedness exhibited in anholocyclic, androcyclic, and intermediate groups. Multivariate methods were used to plot the geographic distribution of life cycle types. This revealed a relationship between colour and life cycle connected with climatic and photoperiodic conditions prevalent at a given latitude. The widespread occurrence of some genotypes was taken as evidence of long-distance migration of *S. avenae* (Simon *et al.*, 1999a), supporting earlier claims by Loxdale *et al.* (1985) that the species is highly mobile, and subsequently reinforced by microsatellite studies of samples from suction traps (Llewellyn *et al.*, 2003). The relationship between climate and life cycle was further supported by a study showing that *S. avenae* populations from Romania, which generally has severe winters compared with France, comprised a high proportion of sexual forms with concomitant very high genotypic diversity and low linkage disequilibrium (Papura *et al.*, 2003).

Genetic variability in relation to life cycle strategy has also been investigated for *R. padi*. Simon and Hebert (1995) studied allozymes of Canadian populations and reported little polymorphism. Of that detectable (3 out of 51 loci tested), there was little geographic differentiation between populations from a single host category. However, gene frequency differences were observed between populations on the primary host, common chokecherry (*Prunus virginiana*) and secondary hosts (grasses and cereals, Poaceae), with heterozygosity substantially reduced on the latter. This suggested a dilution effect on the genotypic diversity from the primary host, probably due to immigrants arriving from further south. Alternatively, heterozygotes were disadvantaged on Poaceae, or were moving to different species sampled in the study (Simon and Hebert, 1995). Yet another suggestion was that interclonal competition produced dominant clones on the secondary hosts during a growing season, thereby reducing clonal diversity in the region (see also de Barro *et al.*, 1994b).

In another study involving *R. padi* (Simon *et al.*, 1996b), holocyclic populations were shown to be in Hardy-Weinberg equilibrium at individual allozyme loci in spring, with high genotypic diversity. This contrasted with anholocyclic lineages that showed the reverse, i.e. they had very low average heterozygosity, deviated from Hardy-Weinberg expectations, and often comprised a single genotype. When summer populations were analysed, populations in regions with cold winters (e.g. the north of France) comprised mainly holocyclic lineages, whilst areas with milder winters had predominantly anholocyclic clones, very much as found with *S. avenae* (Simon *et al.*, 1999a). The lowered level of heterozygosity observed here is related to a general trend to predominantly anholocyclic lineages rather than a dominant clone resulting from interclonal selection. This arises possibly from a transition to asexuality from sexual lineages, such that the commonest homozygotes tend to get fixed in populations. Alternatively, if such transition is frequent, loss of heterozygosity may be due to selection or to the mechanism(s) leading to asexuality itself. Lastly, it was suggested that homozygote excess might arise due to inbreeding (Simon *et al.*, 1996b). However, where sexual populations of this species occurred, Delmotte *et al.* (2002) found, using microsatellites, high allelic polymorphism and heterozygote deficits, ascribed to population subdivision, inbreeding, or selection. In contrast, asexual lineages showed less allelic polymorphism, with high heterozygosity at most loci tested. The authors suggested that this heterozygote excess, compared with the homozygosity found earlier using allozymes, is due possibly to allele sequence divergence during long-term asexuality or hybrid origin of asexual lineages, as briefly discussed below (see also Vorburger *et al.*, 2003b in the case of *M. persicae*).

STUDIES USING MITOCHONDRIAL MARKERS. *Rhopalosiphum padi* from a variety of locations (mainly Spain and France, but also Britain and the USA) has also been examined for mtDNA restriction site and length polymorphisms. Anholocyclic and androcyclic forms had an mtDNA haplotype I (hI) and plasmid haplotype I of the endosymbiotic bacterium, *Buchnera*. In contrast, holocyclic clones had other closely related haplotypes (hII, hIII, and hIV) and plasmid haplotype II (Martinez-Torres *et al.*, 1996). It was inferred from this that a sequence divergence had occurred 0.4 to 1.4. million years ago between the two maternally inherited molecules, respectively (see also Simon *et al.*, 1996a). This lack of gene flow between lineages may explain some of the aforementioned inter-host gene/genotype frequency differences. A discriminating RAPD marker was also identified that was diagnostic for the asexual and sexual lineages (Simon *et al.*, 1996a). The existence of two Spanish lineages with hII, but characterized as anholocyclic, was thought to result from nuclear gene flow between androcyclic and holocyclic lineages (Martinez-Torres *et al.*, 1996). Subsequently, Martinez-Torres *et al.* (1997) showed that these mitochondrial haplotype differences could be correlated with geographical location, rather as with colour and microsatellite markers in the *S. avenae* study of Simon *et al.* (1999a). Thus, a north–south cline was observed in France, with aphids on the secondary host (Poaceae) bearing high, almost fixed frequencies of mtDNA hI in the southwest, whilst aphids with hII, III, and IV (haplotypes combined) tended to be more common in the north (Martinez-Torres *et al.*, 1997).

STUDIES USING SCAR MARKERS. The discriminating RAPD marker, noted above, was subsequently used to produce a SCAR marker, a sequence-characterized, co-dominant marker useful for examining segregation patterns (Simon *et al.*, 1999b). High linkage between life cycle type and this marker was found, segregating populations with an association averaging 94% in field collected samples. The study also revealed further evidence for a unique and apparently ancient loss of sexuality in this aphid, and evidence that occasional gene flow nevertheless still occurred between the different life cycle types, facilitated by males produced by androcyclic lineages. This is a good example of 'sexual leakage' relating asexual and sexual lineages. Simon *et al.* (1999b) discuss the possibility of new asexual lineages arising by such a system, which ensures the longer-term persistence of asexuality, along with the ecological and evolutionary cost benefits of sex in aphids (Simon *et al.*, 2002).

COMBINED STUDIES USING NUCLEAR AND MITOCHONDRIAL MARKERS: THE ORIGINS OF ASEXUALITY. The origins of asexuality in aphids are explored in detail with *R. padi* by Delmotte *et al.* (2001). Mitochondrial and nuclear DNA (seven microsatellites) were combined with biological data obtained on cyclical and obligate forms of the aphid to investigate the frequency of transitions from sexuality to permanent asexuality (Fig. 2.3). Asexual lineages of the aphid were found to possess a subset of the microsatellite diversity found in sexually produced aphids (mean number of alleles for hI and hII mtDNA haplotypic genotypes was approximately 6 and 12, respectively). This was interpreted as representing derivation of asexual from sexual lineages. Interestingly, asexual lineages showed an increase in heterozygosity ($H_{obs.}$ 0.83) in mtDNA hI aphids compared with hII (H_{obs} 0.53), a value very similar to that observed for the mtDNA hII sexual lineages (Delmotte *et al.*, 2001). It is proposed that hI asexual lineages have undergone long-term asexuality, whilst hII asexuals, which are much less heterozygous for microsatellite polymorphisms, have a more recent origin.

Pair-wise genetic distance analysis of microsatellite alleles also revealed that hII type mtDNA asexual genotypes had a similar spread of heterozygotes with a mean closer to that of sexually produced hII genotypes compared with hI asexual forms. This again suggests a more recent sexual origin

for the hII compared to the hI genotypes. Cluster analysis (neighbourhood joining (NJ)) rooted using *Rhopalosiphum maidis* (corn leaf aphid) and *Rhopalosiphum insertum* (apple–grass aphid) showed that the asexual hI aphids form one clade and the hII, whether sexual or asexual, another. Lastly, a larger NJ tree based on shared microsatellite allele distance of *R. padi* collected from cereals and *P. padus* (the primary host) showed separation of asexual from sexual lineages. However, some hII asexual lines occurred in the 'sexual group' clade, emphasizing the possible recent, but different, origin of these forms from a sexual ancestor compared with the obligate asexuals (hI).

These findings suggest there are three independent origins of asexuality in *R. padi*, with two based on a complete loss of sex and repeated sexual leakage from asexual lineages to sexual ones. Delmotte *et al.* (2003) showed that many asexual lineages have a hybrid origin due to introgression of *R. padi* with an unknown sibling species, and are of 'modern origin', contradicting previous estimates that asexual *R. padi* lineages were of moderate longevity. By so doing, this study also contradicts the notion that such asexual lineages are of ancient origin, a popular theme amongst aphidologists; rather, a hybrid origin seems certain and may be the case in other aphid species also.

*Demographic distribution of peach–potato aphid (*M. persicae*) genotypes in relation to life cycle and climate*

Myzus persicae has been studied using microsatellites. Whilst several earlier allozyme studies had shown the species to be largely invariant for these true gene-coding nuclear markers (e.g. May and Holbrook, 1978; Brookes and Loxdale, 1987), significant variation was found by Sloane *et al.* (2001) at a number of microsatellite loci. These latter highly polymorphic 'neutral' markers have been used subsequently in various studies, notably by Fenton *et al.* (2003) for Scottish populations, as described earlier, by Wilson *et al.* (2002) for Australian populations (see 'Geographic colonization' below),

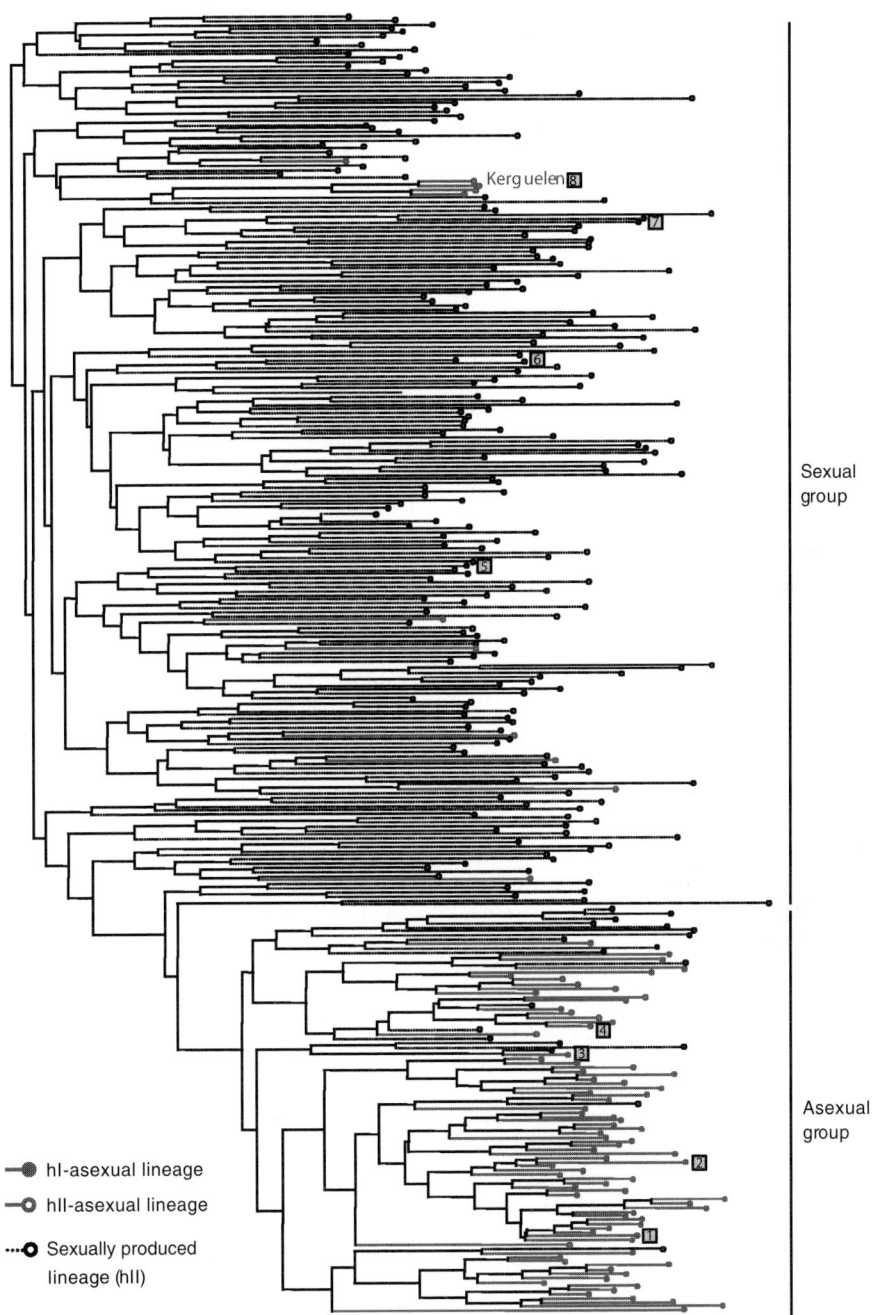

Fig. 2.3. Relationship between sexual and asexual lineages of *Rhopalosiphum padi*. Neighbour-joining tree based on allele shared distance (D_{AS}) calculated with seven microsatellite loci for 90 asexual and 224 sexually produced multi-locus genotypes of *R. padi*, collected on cereals and *Prunus padus*, respectively. Mitochondrial variation, scored as two different haplotypes (hI and hII) is also reported on the tree, the former haplotype being associated with asexual life cycle types, the latter with sexual types. Framed numbers in grey boxes refer to lineages for which the haplotype was sequenced (from Delmotte *et al.*, 2001 by permission of *Proceedings of the Royal Society of London B*).

and by Guillemaud *et al.* (2003) for French populations. In the latter study, three *M. persicae* microsatellites were employed, along with three cross-amplifying microsatellites from *Sitobion miscanthi* and one from *R. padi* (Guillemaud *et al.*, 2003; Wilson *et al.*, 2004). Aphids were sampled in regions of France with differing climates, either using 12.2 m high suction traps (Hullé, 1991) as representing both cyclically parthenogenetic and obligate asexual genotypes from the spring, summer, and autumn migrations, or from the primary host, peach (*Prunus persica*) in the south of the country, where the trees mainly grow and *M. persicae* populations are entirely holocyclic. In brief, it was shown that climate (as with *S. avenae*) and the availability of the primary host (as with *R. padi*) had a direct bearing on genotype, itself related to life cycle. Thus, aphids from peach had a maximum genetic diversity, no linkage disequilibrium and almost no clonal copies, and were similar to populations from the same host sampled in southeastern Australia (Wilson *et al.*, 2002, see below), whilst the autumn aerial sample collected at Colmar, a region in northeastern continental France with severe winters (see Guillemaud *et al.*, 2003 for supporting temperature data) and where males were most abundant, was closer to the sexual samples in terms of genotypic diversity and linkage disequilibrium. In contrast, populations from milder regions (i.e. Le Rheu in the northwest and Valence in the southeast), which were considered to be predominantly obligately asexual with many fewer males, showed frequent linkage disequilibrium and repeated genotypes. This last finding is thought to be the result of clonal amplification, a characteristic shaping aphid populations during the parthenogenetic stages of the life cycle and one observed in many species (Guillemaud *et al.*, 2003; Vorburger *et al.*, 2003b).

Suction-trapped aphids from Colmar were most similar in terms of genetic structure to those from peach and were deemed to be cyclically parthenogenetic. Population samples had an observed heterozygosity in the range 0.29–0.56, with most showing a deficiency of heterozygotes, including the

sexual and suction-trapped samples, perhaps because pooling of the populations by such sampling techniques leads to a distortion of true Hardy-Weinberg genotype proportions (the Wahlund effect; Ridley, 1993). Guillemaud *et al.* (2003) suggest that this general lack of difference is due to a limited accumulation of mutations in the obligate asexual compared with cyclical lineages and to the production of males by the latter forms perhaps being sufficient to prevent differences in heterozygosity between reproductive modes. The overall trend of the population diversity found in relation to climate is further support for Blackman's (1974) 'climate selection hypothesis'. As for the migratory abilities of this aphid species, putatively obligately parthenogenetic populations from aerial samples were significantly genetically differentiated at the national scale, with differentiation highest in autumn (see Table 3 of Guillemaud *et al.*, 2003). However, genetic differentiation of sexual samples was non-significant over small geographic scales (< 60 km), and low but significant at larger geographic scales (150–200 km). Guillemaud *et al.* (2003) conclude, as Wilson *et al.* (2002) did in their Australian study, that the data support the view that long-distance aerial movements are probably uncommon in this particular species.

In summary, these molecular-based data reveal the existence of north–south clines of life cycle types in several aphid species, as well as distributions related directly to both climate and primary host-plant availability, with anholocycly predominant in regions with milder winters and increasing holocycly in regions with colder winters, although sexual reproduction in host-alternating species is largely governed by the availability of the primary host. Holocyclic populations are genetically similar, probably a result of high recombination. Anholocyclic forms, including androcyclic and intermediate, are more diverse and appear to have distinct genetic pools which are, in some cases, quite ancient, excluding possibilities of more recent introgression events. In contrast, recent asexual lineages are thought to have evolved from holocyclic types.

Adaptation to host plants

Molecular markers have revealed an apparent genetic basis to host adaptation and host-plant resistance and that certain genotypes show host preference (de Barro *et al.*, 1995a,b; Lushai *et al.*, 2002). For example, populations of *S. avenae* can be recognized by their performance on different hosts. Different frequencies of life cycle characteristics (anholocycly, holocycly) and body colour have been found according to whether aphids were collected from barley (*Hordeum vulgare*), oats (*Avena sativa*), or wheat (Weber, 1985a), and some biological traits have now been matched with molecular evidence (de Barro *et al.*, 1995a,b, see later). Similar trends have been noted in *M. dirhodum* using enzyme markers (Weber, 1985b), although in the polyphagous *M. persicae* the data suggest a wider, continuously distributed variability in host-plant adaptation and a broad phenotypic plasticity (Weber, 1985c). However, with *M. persicae*, whilst enzymes and some kinds of high resolution DNA marker have failed to reveal host-related differences, others give a somewhat different story (see later). Before discussing this complex situation found in a species whose specific status is questionable, we consider cereal and other aphids where the situation appears clearer.

Host preference in cereal and other aphids

SITOBION SPECIES. Studies using RAPD markers revealed that *S. avenae* in southern Britain in the spring had significantly different genotypic structuring on wheat and cocksfoot (*Dactylis glomerata*) (de Barro *et al.*, 1995b). This structuring tended to break down as the season progressed, probably due to local inter-host movements. The existence of a host-based 'stratification' suggested by the data was supported by reciprocal host selection experiments demonstrating a genetic component in host performance (de Barro *et al.*, 1995c). Microsatellite analysis of these samples also supported these findings and revealed further details (Sunnucks *et al.*, 1997a). One 'wheat-specific' lineage had

alleles never found on cocksfoot, another group composed of interrelated genotypes was found on both host plants, whilst another bearing many *S. fragariae*-like alleles (apparently introgressed with the holocyclic *S. fragariae*) was found only on cocksfoot. The genotype with *S. fragariae* alleles also carried *S. fragariae*-like mtDNA in approximately 80% of cases, indicating 'asymmetrical hybridization' (Sunnucks *et al.*, 1997a).

Another microsatellite study (Haack *et al.*, 2000) revealed the existence of host-based differences in *S. avenae* collected from maize (*Zea mays*), wheat, and barley, as well as some more 'generalist' genotypes (Fig. 2.4). The generalists colonized all plant hosts sampled, were found over a large geographical area (France and England), and persisted asexually for several years. Again, the available evidence supports the notion of evolution of host-plant adaptation (see also Figueroa *et al.*, 2005 regarding *S. avenae* populations in Chile likewise showing a few (four) predominant genotypes, described as 'superclones', representing nearly 90% of the samples tested and persisting over large geographic regions; and Vorburger *et al.* (2003a) regarding two superclones of *M. persicae* in Australia, one constituting 24% and the other 17.4% of the collection).

Similarly, the use of a Latin square contingency distribution of four different hosts (wheat, barley, cocksfoot and Yorkshire fog – *Holcus lanatus*) revealed that the colony-founding, winged forms of *S. avenae* migrating into the crop early in the year preferentially select host species; again preference appears to have a genetic basis (Lushai *et al.*, 2002). In other studies on *Sitobion* spp., significant mean relative growth rate differences for four chromosomal races of *S. miscanthi* and five forms of *S.* sp. near *fragariae* (Hales *et al.*, 1990) were found with nearly all clones performing best on barley, followed by cocksfoot and rye (*Secale cereale*). These data suggested that parthenogenetic aphids evolve quickly, possibly in association with chromosomal rearrangements (Sunnucks *et al.*, 1998).

Vialatte *et al.* (2005), again using microsatellites, have examined patterns of

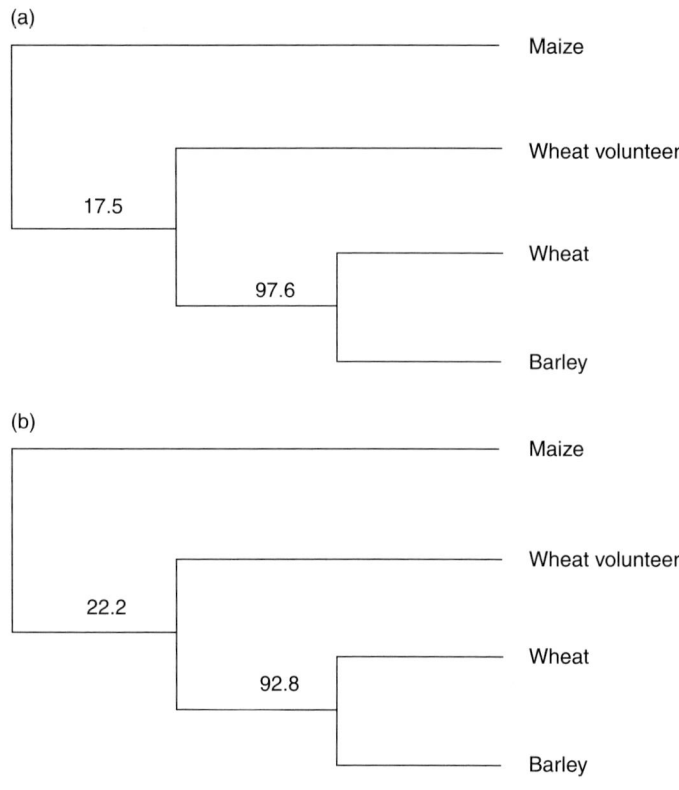

Fig. 2.4. Neighbour-joining trees derived from microsatellite data (five loci) relating individuals of *Sitobion avenae* grouped by host plant. These trees were constructed using: (a) Goldstein's distance; (b) − ln (shared allele proportion). Bootstrap values were computed from 1000 replications and are shown as percentages (from Haack *et al.*, 2000 by permission of *Molecular Ecology*).

host-based genetic differentiation in local populations of *S. avenae* in northwestern France. Aphids were collected from both cultivated fields and wild margins. Genetic differentiation was high between populations from wild Poaceae and those from cereal crops. Individuals or genotypes on cereals were genetically close, whereas those from wild Poaceae showed substantial genetic differentiation among populations. *Sitobion avenae* populations thus appear to exist as two main genetic groups within the agroecosystem, one largely restricted to cultivated cereals and comprising genetically related insects, the other associated with wild grasses and composed of more specialized genotypes.

GREENBUG. Numerous 'biotypes' (but see Blackman and Eastop, Chapter 1 this volume, on the use of this word) of *Schizaphis graminum* exist, adapted to particular cereal crops and cultivars (Puterka and Peters, 1990; Ono *et al.*, 1999), many distinguishable by

their ability to overcome different genes for host-plant resistance. These forms show clear differences in mtDNA RFLP patterns produced using a range of restriction enzymes and related to their divergence in the agricultural ecosystem (Powers *et al.*, 1989). However, some of these lineages appear to have diverged 0.3–0.6 million years ago (Powers *et al.*, 1989). More recent sequencing of part of the COI region of mtDNA has revealed the existence of three clades (Shufran *et al.*, 2000; Anstead *et al.*, 2002) (Fig. 2.5), also confirmed using RAPD markers (Shufran, 2003). These trends towards molecular divergence appear to be adaptations of *S. graminum* to host-plant groupings, and different from the cultivar resistance story, i.e. biotype formation (Shufran *et al.*, 2000; Anstead *et al.*, 2002, 2003). Such adaptation pre-dates agriculture and, therefore, cultivar management and breeding. The details deriving from this contentious issue are yet to be accepted and deployed in integrated pest management

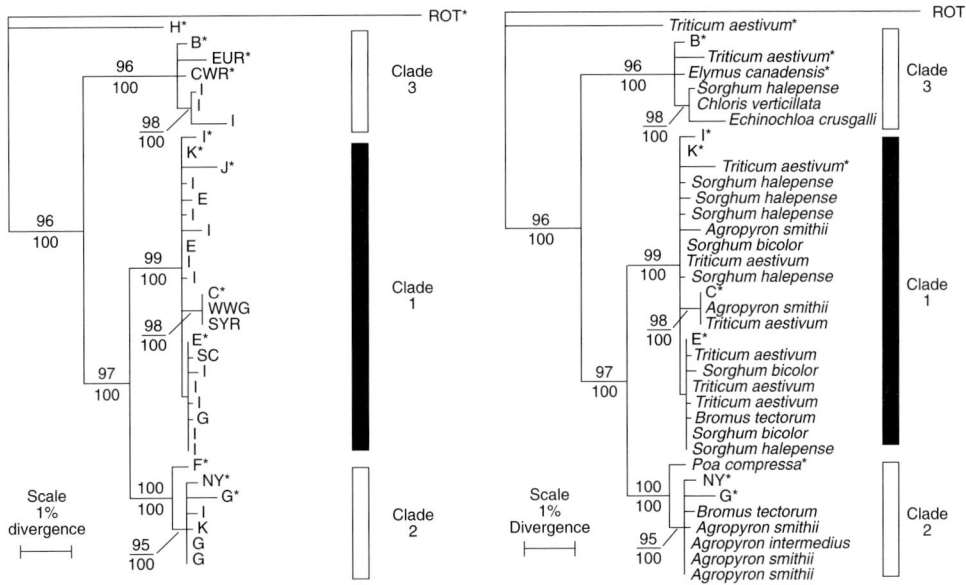

Fig. 2.5. Host association of biotypes of the greenbug, *Schizaphis graminum*, identified using mitochondrial COI gene sequencing. Maximum likelihood tree of *S. graminum* isolates by: left, biotype; right, host. For both distance/neighbour-joining and maximum parsimony analysis, 1000 bootstrap replications were performed. The percentage of replications supporting each branch is shown. The top value represents neighbour-joining, while the bottom value represents maximum parsimony. Biotypes are labelled C, E, F, G, H, I, J, and K. Clones: CWR = Canada wild rye; EUR = Europe; NY = New York; SC = South Carolina; SYR = Syria; WWG = intermediate wheatgrass; related aphid species, ROT = *Schizaphis rotundiventris* (from Anstead *et al.*, 2002, by permission of *the Bulletin of Entomological Research*).

(IPM) practices, but highlight the new information that molecular approaches can cast on long-held perspectives.

COTTON OR MELON APHID.   RAPDs have also been used to show that obligate asexual strains of *A. gossypii* from various geographical regions and infesting cucurbit and non-cucurbit hosts can be resolved into different host-based genotypes using clustering methods (Vanlerberghe-Masutti and Chavigny, 1998). In a subsequent microsatellite-based study on this species collected from glasshouses in southern France (Fuller *et al.*, 1999), 12 non-recombinant genotypes were detected, of which around 90% of individuals were one of three common genotypes. Here, no clear correlation between genotypic class and host-plant species was found. It was also shown that, as the spring and summer progressed, clonal diversity declined significantly, possibly due

to interclonal competition. Significant genetic substructuring was observed within glasshouse populations within a 500 m radius, probably relating to founder effects and restricted inter-population gene flow. The three main genotypes appeared in following years, along with previously unseen genotypes, suggesting that the aphid may have local overwintering refugia or disperse in from distant sources.

YELLOW CLOVER APHID.   Similarly host-based differences in Australian populations of *Therioaphis trifolii* (yellow clover aphid) have been discriminated using RAPDs and mtDNA markers (COI–II). Despite there being no differences in karyotype (2n = 16) or cuticular hydrocarbon profile, consistent differences were found in RAPD profile and mtDNA haplotype, as well as some morphological changes dependent upon host (lucerne – *Medicago sativa*, and subterranean clover – *Trifolium sub-terraneum*) (Sunnucks *et al.*, 1997b), a good indication of incipient speciation.

PEA APHID.   In *A. pisum*, AFLP-based studies showed that 'sympatric races' on lucerne and red clover (*Trifolium pratense*) are highly ecologically specialized and significantly isolated, including having specialized endosymbiotic bacteria (Simon *et al.*, 2003). Much of the restriction of gene flow is due to host preference by the winged colonizers (Via, 1999; Via *et al.*, 2000). Reciprocal host transfer experiments, including both aphid races, favoured the original host. Also, selection favoured host preference during colonization with heavy fitness consequences for inappropriate settling. Field studies supported these trends, with host-related differentiation declining as the season progressed, i.e. heavy fitness consequences of movement away from the original host. F_1 hybrids between the races also had significantly lower fitness than the parents on both hosts, leading to post-mating isolation between specialized races. Whilst hybrids had significant genetic variation, no 'generalist' genotypes were observed in field studies, supporting the view that these races have indeed evolved to the point of being effectively distinct evolutionary entities (see also Via and Hawthorne, 2002 and Del Campo *et al.*, 2003).

Allozyme and RAPD banding pattern differences have been detected in French clones of *A. pisum* differing in their aggressiveness to resistant cultivars of lucerne (Bournoville *et al.*, 2000). Interestingly, host-based variation has not been found in English and French samples of *A. pisum* following RFLP analysis of mtDNA and the DNA of plasmids of the symbiotic bacteria, *Buchnera* (Birkle and Douglas, 1999). Analysis of 29 parthenogenetic clones of three biotypes (adapted to pea – *Pisum sativum*, lucerne, and red clover) revealed only some length variation in the mitochondrial genomes tested, as well as limited variation in the plasmid DNA (see also Barrette *et al.*, 1994).

CABBAGE APHID.   In *Brevicoryne brassicae* (cabbage aphid), F_{ST} analysis of populations

in Mexico was performed following electrophoretic testing using seven polymorphic enzyme loci (out of 11 loci tested), and with aphids surveyed from two host plants (*Brassica rapa* ssp. *campestris* and *Brassica oleracea* var. *capitata*) within each of four localities. This study also revealed some degree of host-related population divergence, probably due to selection for host-preferring genotypes (Ruiz-Montoya *et al.*, 2003).

Karyotypic changes in the corn leaf aphid

Pertinent to the discussion of host races, aphids of different karyotype are known to have evolved and to have colonized different hosts. Thus, in the northern hemisphere, *R. maidis* has karyotypic forms specific to barley and *Sorghum* spp. (2n = 10) and maize (2n usually = 8; Brown and Blackman, 1988; Blackman and Eastop, 2000). Since males are extremely rare in this species and oviparae unknown (Blackman and Eastop, Chapter 1 this volume), significant gene flow is unlikely between these forms, so that their genetic isolation is more or less secured. Such an evolutionary trend reduces ecological flexibility (Brown and Blackman, 1988; Blackman and Eastop, 2000). Samples from clones of this aphid species tested from 12 sites around the world proved largely invariant for 32 allozyme and mtDNA markers, despite karyotypic differences associated with plant host as noted above (Simon *et al.*, 1995). This may be related to founder events and low recombination because of the lack of sexuals in this species.

The ongoing case of the peach–potato aphid – a 'good' species? Or is it?

Despite no obvious host-plant associations found using enzyme markers in *M. persicae*, a species notoriously invariant at most allozyme loci tested, both in Europe and the USA (May and Holbrook, 1978; Brookes and Loxdale, 1987), the aphid has been shown to display RAPD variation in Greek populations from peach and pepper (*Capsicum* spp.), compared with lineages from tobacco (*Nicotiana tabacum*), where significant divergence was found between holocyclic and anholocyclic populations from tobacco. The holocyclic clones had higher levels of heterozygosity compared with non-holocyclic clones (i.e. anholocyclic, androcyclic, and intermediate; Zitoudi *et al.*, 2001).

Margaritopoulos *et al.* (2005), using indoor choice and 'electrical penetration graph' recordings, showed that spring migrants of *Myzus persicae nicotianae* (tobacco aphid), which shares the same chromosome number as *M. persicae*, i.e. 2n = 12, yet is permanently parthenogenetic and discriminated from the parent species (from which it doubtless recently evolved) using allozyme (GOT) markers (Blackman, 1987; Blackman and Spence, 1992), preferentially chose tobacco, whereas equal proportions of *M. persicae* chose tobacco and pepper. In 'no-choice' host-plant tests, *M. persicae nicotianae* preferentially remained on tobacco, and fecundity was greater there. It appears that the aphids use chemical cues to select their host(s). Hence, winged colonizers are very important in host selection and in the maintenance of host specializations, perhaps leading to sympatric speciation, as suggested for other species (see Lushai *et al.*, 2002 and studies by Via and co-workers, cited below).

In contrast, research on *M. persicae* and *M. persicae nicotianae* by Clements *et al.* (2000) using RAPDs, mtDNA (COII), and nuclear EF-1α sequencing, failed to find either host-related or geographical differences in either red or green individuals taken from tobacco and non-tobacco hosts. Likewise, although Scottish populations of *M. persicae*, tested using a range of IGS markers, revealed a large number of clones (78) with one predominant clone, they failed to show clear host associations (Fenton *et al.*, 1998a).

Vorburger *et al.* (2003a), comparing the fitness of obligately and cyclically parthenogenetic genotypes of *M. persicae* on three unrelated host plants, found genetic variation for the relative performance on different hosts, but failed to find any difference in fitness between the two main life cycle forms, a finding contrary to the expectations of the GPG ('general-purpose genotype')

hypothesis (Vorburger *et al.*, 2003a; see also Vorburger's (2005) account of the lack of life-history trade-offs in this aphid, with clones distinguished using microsatellite markers and with heterozygosity increasing significantly with fitness).

CINARA AND PEMPHIGUS SPECIES. Ecological and molecular studies (involving mtDNA COI and nuclear EF-1α) on the conifer-feeding Cinara species have shown that host switching has been important in speciation of the genus in southwestern USA. This is reflected in the polyphyly of pinyon-feeding taxa. Furthermore, *Cinara* species sharing a common feeding site on different hosts are more closely related than those that feed at different sites on the same host. This suggests that feeding site fidelity plays a more important role in speciation than does host fidelity in general (Favret and Voegtlin, 2004). Similarly, ecological studies on two species of gall-forming aphids, *Pemphigus populitransversus* and *Pemphigus obesinymphae*, which both feed on eastern cottonwood (*Populus deltoides*), have suggested that host switching may be responsible for sympatric speciation. Perhaps this was related initially to allochronic changes resulting from 'a shift in timing of the life cycle in an ancestral population, correlated with an underlying phenological complexity in its host plant, which spurred divergence between the incipient species' (Abbot and Withgott, 2004).

The nature and role of genetic variability in the evolution of host preference

The lack of variation within populations shown using molecular markers is often related to founder events (May and Holbrook, 1978; Shufran *et al.*, 1991; Simon and Hebert, 1995; Nicol *et al.*, 1997; Wilson *et al.*, 1999; Figueroa *et al.*, 2005) or other bottlenecks such as scarcity or absence of primary hosts (aphids on sub-Antarctic islands are an extreme example – Hullé *et al.*, 2003), all of which are liable to reduce genetic variation (Delmotte *et al.*, 2001; Guillemaud *et al.*, 2003). This is especially so for allozyme variation (Brookes and

Loxdale, 1987; Loxdale and Brookes, 1988), these particular markers being relatively selective compared with more neutral DNA markers like microsatellites (Delmotte *et al.*, 2002, but see also Goldstein and Schlötterer, 1999 and Li *et al.*, 2002, who both cite evidence that, contrary to the perceived wisdom of DNA marker neutrality, some microsatellite regions are subject to selective constraints, including on length). Associated with these new aphid colonizations is the reduced predator pressure or 'enemy-free space' that allows new populations to thrive, e.g. *Diuraphis noxia* (Russian wheat aphid) in South Africa and North America following its introduction into these countries in the late 1970s and mid-1980s, respectively.

The possible reason that some DNA markers show host-based differences and others do not is that markers like RAPDs (which often contain microsatellite regions – Ender *et al.*, 1996) and microsatellite regions *per se* are integrated at random throughout the insect genome (Queller *et al.*, 1993; Goldstein and Schlötterer, 1999). In contrast, rDNA arrays are confined usually to the ends of both X-chromosomes in aphids (Blackman and Spence, 1996; but see Blackman *et al.*, 2000 for the case of the genus *Trama*, where rDNA occurs as single or multiple arrays on various chromosomes in several different species studied). Molecular selective pressures may be different in the two main cases. Microsatellites, for example, are sometimes carried by 'hitch-hiking' with genes under selection (Lanzaro *et al.*, 1998), whereas the rDNA regions are more isolated, except in the aforementioned *Trama* aphids.

That there are exceptions to the host-based trend may be because one of the limitations of RAPDs is that there is little clarity as to where these markers amplify, unless extensive mapping is performed. From another perspective, failure to find host differences in *M. persicae* using IGS markers (Fenton *et al.*, 1998a) may be because the Scottish populations sampled are not especially variable, perhaps due to their recent founding or, more probably, to winter-related selection against asexual forms, the primary host (peach) being rare in the

region sampled (Tatchell *et al.*, 1983; Foster *et al.*, Chapter 10 this volume). Use of other more widely genome-dispersed polymorphic markers such as microsatellites (Wilson *et al.*, 2002), AFLPs, and sequencing of diverse regions may clarify this.

To summarize, the concept that genetic variation leads to host-based differentiation goes through several levels of divergence. When asserting a host-based aphid speciation event, all the following should be present: (i) genetic variation – molecular and chromosomal, and sometimes associated with resistance-breaking genes, and perhaps phenotypic plasticity (e.g. Wool and Hales, 1997), which may also have a genetic basis; (ii) host differentiation (behavioural, including feeding and reinforcing mating preferences within races) and perhaps involving a restriction of gene flow between host-adapted forms; and (iii) morphological variation, not forgetting the occurrence of 'cryptic' species, that is genetically distinguishable entities possessing identical morphologies.

Insecticide resistance

Foster *et al.*, Chapter 10 this volume, discuss insecticide resistance in aphids, including the assaying for resistance genes and genotypes. Suffice here to say that rapid evolution of insecticide resistance has occurred through multiple mechanisms in several species, but often with associated fitness costs.

Pathogenicity

Studies involving molecular markers have shown differences in aphids in their ability to transmit plant viruses. Thus, a preliminary study by Terradot *et al.* (1999) examining allozymes, microsatellites, esterase-4 (E4), and karyotype demonstrated genetic differences amongst 27 clones of *M. persicae*, *M. persicae nicotianae*, and *M. antirrhinii* in their efficiencies in transmitting *Potato leaf roll virus* (PLRV). Previous work had shown some isolates, termed 'highly aphid transmissible (HAT)' of PLRV to be transmitted efficiently by all tested *M. persicae* clones. Other PLRV isolates, termed 'poorly aphid transmissible (PAT)', were transmitted inefficiently by most tested *M. persicae* clones, but efficiently by two clones (Bourdin *et al.*, 1998). According to Terradot *et al.* (1999), most of the 'old' *M. persicae* clones used in such studies are now known to be *M. antirrhinii*, and are poor vectors of the PAT-PLRV isolate. The authors conclude that the taxonomic separation of *M. persicae* from *M. antirrhinii* (Hales *et al.*, 2000) may correlate well with virus transmission efficiency. If so, the findings are pertinent with regard to understanding the relationship between aphid genotype and virus transmission.

Rhopalosiphum padi gynoparae and males returning to their primary host (*P. padus*) are unimportant in transmission of *Barley yellow dwarf virus* because they are unlikely to encounter crops. In contrast, overwintering asexuals of the same species potentially can transmit this virus. This contrast is clearly highly pertinent to assessing the role of life cycle forms in virus transmission. Thus, diagnostic SCAR marker band(s) mentioned earlier (Simon *et al.*, 1996a, 1999b) have considerable potential for tracking virus risk by discriminating individuals from holocyclic and anholocyclic clones.

Dispersal and geographic range

Prior to the introduction of molecular markers, much was known about aphid population structure and movement, especially through the pioneering studies by L.R. Taylor and C.G. Johnson at Rothamsted that evolved into the Europe-wide suction trap network (Harrington *et al.*, Chapter 19 this volume) and laboratory-based studies (e.g. Kennedy and Booth, 1963; Hardie, 1993), which allowed the assessment of finer scale physiological aspects (see Lushai and Loxdale, 2004).

The problems of tracking movement of individual aphids, especially of winged migrants, are principally related to their small size and the fact that even large populations are soon diluted by aerial currents, making recapture unlikely (Loxdale, 2001). Radioactive tracking has been tried, as have fluorescent dyes with *Bemisia tabaci* (sweet potato whitefly) (Byrne *et al.*, 1996), and whilst providing useful information at the field scale, dilution at greater scales is an insurmountable problem. Scanning radar has also been used to track aphids (Irwin and Thresh, 1988), but is not suitable for detecting individual movements. Surveying using vertical-looking radar (VLR) can identify individual insects but, currently, only those heavier than aphids (i.e. 2.0 mg) and flying at a height of at least 150 m (Chapman *et al.*, 2002; Osborne *et al.*, 2002).

Most approaches using molecular markers are indirect, relying on comparison of gene and genotype frequency changes from which gene flow is estimated and migration inferred (Loxdale and Lushai, 1998). Occasionally, direct deduction of movement is possible when a particular genotype, occurring at high frequency in a population, is found in another population from which it was previously absent, e.g. an insecticide-resistant aphid moving into an insecticide-susceptible population (Foster *et al.*, 1998). Even so, the possibility remains that the genotype had arrived earlier, but was not detected because of its low incidence (sampling effect).

It has often been assumed that genetic diversity should be related proportionally to spatial scale, i.e. the further populations are from one another, the less similar they should be, a phenomenon known as isolation by distance (IBD). However, the reality is usually more complex, and depends on flight behaviour of the species in question (see below). Population genetics studies using molecular markers can begin to reveal the extent of this.

Gall aphids

Gall aphids show a diversity of migratory ability, in a large part related to their unusual biology, often involving a significant amount of inbreeding. At the smallest spatial scale, colonies of single closed galls of *Ceratovacuna nekoashi* tested using RAPD markers appeared to be derived from a single fundatrix (Fukatsu and Ishikawa, 1994). Studies of other gall-forming aphids (e.g. *Melaphis rhois, Pemphigus spyrothecae, Pemphigus bursarius,* and *Baizongia pistaciae*), using a variety of molecular markers, showed that populations displayed a lack of gene flow as revealed by high F_{ST} and other statistical approaches, and tended to exist as genetically differentiated subpopulations or 'demes'. Thus, there appears to be little aerial movement by winged migrants between galls of these particular species, even over relatively small spatial scales, i.e. less than 20 km and, in the case of *B. pistaciae,* less than 0.15 km (Hebert *et al.,* 1991; Johnson, 2000; Miller, 2000, Johnson *et al.,* 2002; Miller *et al.,* 2003; Martinez *et al.,* 2005), except for the possibility of 'sneaky' immigrants of different genotype entering from nearby galls (Whitfield, 1998; Johnson *et al.,* 2002). On the other hand, some species of gall aphid, e.g. *P. obesinymphae,* do show higher levels of inter-gall movement and mixing (Abbot *et al.,* 2001), demonstrating that there are rarely general traits for a group.

In the case of *B. pistaciae* in Israel (Martinez *et al.,* 2005) infesting the primary woody host, *Pistacia palaestina,* in the Mediterranean maquis habitat, the secondary host (i.e. roots of various species of Poaceae) must be nearby to ensure a continuous genetic differentiation between demes during alternation of generations. In other studies of this and four other gall-forming species infesting *P. palaestina,* mature and old trees were shown to be infested more often than young ones, and with shrubs more often occupied than tree-like plants, although there was no difference in occupancy between isolated trees and those growing in clumps. Also, trees growing in the 'ecotone' (an area of transition between adjacent ecosystems) carried more galls than those growing in the maquis itself. The best patch for gall-inducing aphids was a small area comprising old *P. palaestina* trees standing in open landscape with nearby secondary hosts. Thus, a major determinant for the survival of these insects

in a metapopulation structure is a mosaic landscape (Martinez *et al.,* 2005; see also Inbar *et al.,* 2004).

Non-galling aphids

In colonies of some species, e.g. *Sitobion* spp. on wheat, *S. graminum* on sorghum and *Drepanosiphum platanoidis* (sycamore aphid) on single leaves of sycamore (*Acer pseudoplatanus*), both allozyme and DNA markers have shown that all comprised more than one, and often numerous, genotypes. In the case of *D. platanoidis,* due to flight within tree canopies, between nearby trees and, probably, longer-distance dispersal, genotypic diversity was rather similar at spatial scales ranging from within leaves to among trees, local clumps of trees, and trees more than 100 km apart (Wynne *et al.,* 1994). With *S. graminum,* IGS markers and assessment of genetic polymorphism using multivariate approaches revealed several genotypes, although diversity appeared unaffected by spatial scale from plant scale to county in Kansas, USA (Shufran *et al.,* 1991; Shufran and Wilde, 1994); as many aphid genotypes occurred on a plant as within a county! Such homogeneity in terms of genetic variability was thought to represent significant movement of individuals over hundreds of kilometres.

Loxdale *et al.* (1998) suggested that patterns of genetic variability produced by different species of aphid reflect migratory behaviour and that this is correlated with their respective peak abundance as observed in suction trap samples. The species that are common at peak abundance, and hence appear highly migratory, have homogeneous allele frequency patterns at both small and large geographic scales. Conversely, aphids that are relatively rare in the trap samples have heterogeneous patterns, even at small spatial scales (< 30 km). Aphid flight behaviour is influenced by host-seeking behaviour, initially attraction to the sky and then to green targets, which elicit orientation and landing (Hardie, 1993; Pettersson *et al.,* Chapter 4 this volume). It is known that gynoparae of *R. padi* fly higher on average than virginoparae, the former having to find the widely distributed and rather uncommon primary host, *P. padus* (Tatchell *et al.*, 1988).

Short-distance migrants

Sitobion fragariae, which has abundant local primary and secondary hosts (blackberry – *Rubus fruticosus* agg. and Poaceae, respectively), but is uncommon at peak abundance in suction traps compared with the closely related *S. avenae* (Woiwod *et al.*, 1988), also shows local patterns of genetic variation, more especially when on grasses. This suggests a restriction of gene flow between subpopulations of the aphid, very striking as seen even from the raw gene/genotype frequency data from

populations on the secondary host (Loxdale and Brookes, 1990).

Macrosiphoniella tanacetaria (tansy aphid) also shows a lack of short-range IBD, yet with restricted gene flow between local populations studied using microsatellite markers, with perturbations in local (meta)-population density from one year to another, indicative of selective sweeps resulting from drift/founder effects and local extinction events (Massonnet *et al.*, 2002; Massonnet and Weisser, 2004) (Fig. 2.6).

In *Phorodon humuli* (damson–hop aphid), the frequency of enzyme markers (elevated carboxylesterases associated with resistance to insecticides and phospho-gluconate dehydrogenase, 6-PGD) likewise revealed a restriction of gene flow, here

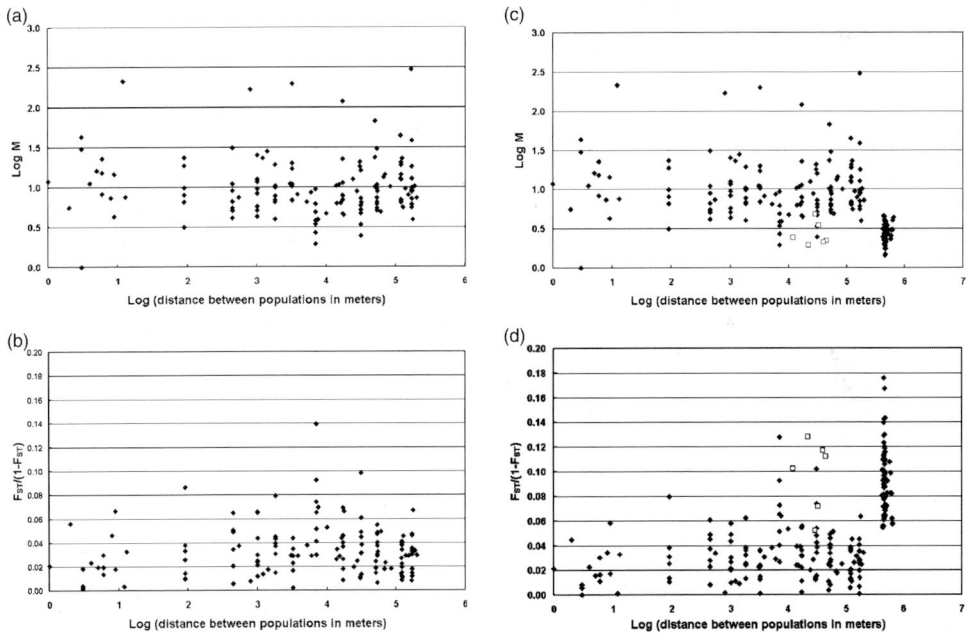

Fig. 2.6. Estimates of isolation by distance (IBD) using microsatellite analysis of metapopulation structure in colonies of *Macrosiphoniella tanacetaria*. Sampling of 17 populations from ramets (shoots) and genets (individual plants) was done along the Saale River in Germany ('Jena populations'), with distances between populations ranging from 1 m to 170 km. To test for genetic differentiation at a very large scale, four additional populations were tested from the Alsace region, France (see Massonet *et al.*, 2002 for collection details), an average of 470 km from the Jena populations. In (a) and (b), IBD was estimated by plotting the log-transformed geographic distances between the 17 German aphid populations *versus* measures of population differentiation: a, log-transformed $M\hat{}$ values, where $M\hat{} = (1/(F_{ST} - 1)/4$; b, $F_{ST}/(1 - F_{ST})$ values. In (c) and (d), the same parameters were plotted for the total of 21 aphid populations from Germany and France. The empty squares in (c) and (d) represent the six comparisons between the four French populations (from Massonnet and Weisser, 2004 by permission of *Heredity*).

between aphids from commercial (and hence intensively sprayed with insecticide) hop-growing regions (Hereford and Kent) and non-commercial (unsprayed) regions in southern Britain (Fig. 2.7), as well as a differential pattern of movement. Spring migrants moving from *Prunus* spp. (especially sloe – *Prunus spinosa* and plum – *Prunus domestica*) to the secondary host, hop (*Humulus lupulus*), appear to move further than autumn migrants on the return journey, although both ambits are probably generally restricted (i.e. < 20 km) for most individuals (Loxdale *et al.*, 1998). This agrees broadly with suction trap analyses, which also revealed a restricted ambit for the autumn migrants in the English hop-growing regions of Hereford and Kent (median approximately 15–20 km dispersal ambit for gynoparae and males; Taylor *et al.*, 1979).

Long-distance migrants

In contrast with the short-range, 'hedge-hopping' species mentioned above, species common at peak abundance in 12.2 m high suction traps, such as *S. avenae*, *R. padi*, and *D. platanoidis*, show homogeneous allele frequency patterns over large spatial scales (> 100 km) indicative of high gene flow/interpopulation movement (Loxdale *et al.*, 1985; Loxdale and Brookes, 1988; Wynne *et al.*, 1994; Delmotte *et al.*, 2002; Llewellyn *et al.*, 2003).

In the case of *S. avenae*, tests for microsatellite polymorphisms in aphids caught by suction trapping have shown that allele frequencies at four loci are generally similar for aphids captured along a north–south transect across Britain (Llewellyn, 2000; Llewellyn *et al.*, 2003). Clonal competition varied seasonally at one site (Rothamsted) in aphids sampled from winter wheat in June and again in July. In addition, in one sampling year (1997), there was a latitudinal cline of genotype frequencies, with the proportion of unique types rising northwards. This was interpreted as an increasing proportion of holocyclic clones, related to the incidence of overwintering as cold-hardy eggs (Llewellyn, 2000). This broadly confirms earlier direct observational and

breeding studies (Newton and Dixon, 1988; Hand, 1989; Helden and Dixon, 2002). By the following summer (1998), this clinal trend was not so obvious, perhaps because the previous winter was milder than that of 1996/1997, probably promoting asexual overwintering. Meanwhile, the fact that allele frequencies remained similar and homogeneous geographically for the selectively neutral markers is interpreted as representing high gene flow (equated with migration), at least in excess of the forces of drift and selection, which might otherwise be expected to perturb such apparent stability (but see Llewellyn *et al.*, 2004, who provide evidence for clonal selection at the field scale).

The finding of allelic homogeneity supports trends indicated by allozyme studies of the population genetics in the same species over large spatial scales in Europe (Britain and Spain), which also suggested high gene flow/aerial movements with little restriction due to geographical barriers, including mountains and the sea (Loxdale *et al.*, 1985).

Reimer (2005) has shown, using five polymorphic microsatellites, that *S. avenae* populations collected from wheat at two regions (Lower Saxony and Hesse) more than 100 km apart within central Germany, including a subset of 14 and 17 locations within regions, and another more distant site in northern Germany > 200 km away, displayed local variations in microsatellite allele and genotype structuring, thought to result from geographic factors. Overall, his results show that geographic factors enhance differences in regional population genetic structure and local distribution of genotypes, but that long-distance migration prevents the permanent isolation of populations between more distant areas (see also Figueroa *et al.*, 2005).

Other studies using RAPD markers when summer populations of *S. avenae* are established have shown rather little apparent movement over quite large distances (up to 60 km; de Barro *et al.*, 1995b). The discrepancies between these various studies are due possibly to the differences in the transect adopted and the spatial scale.

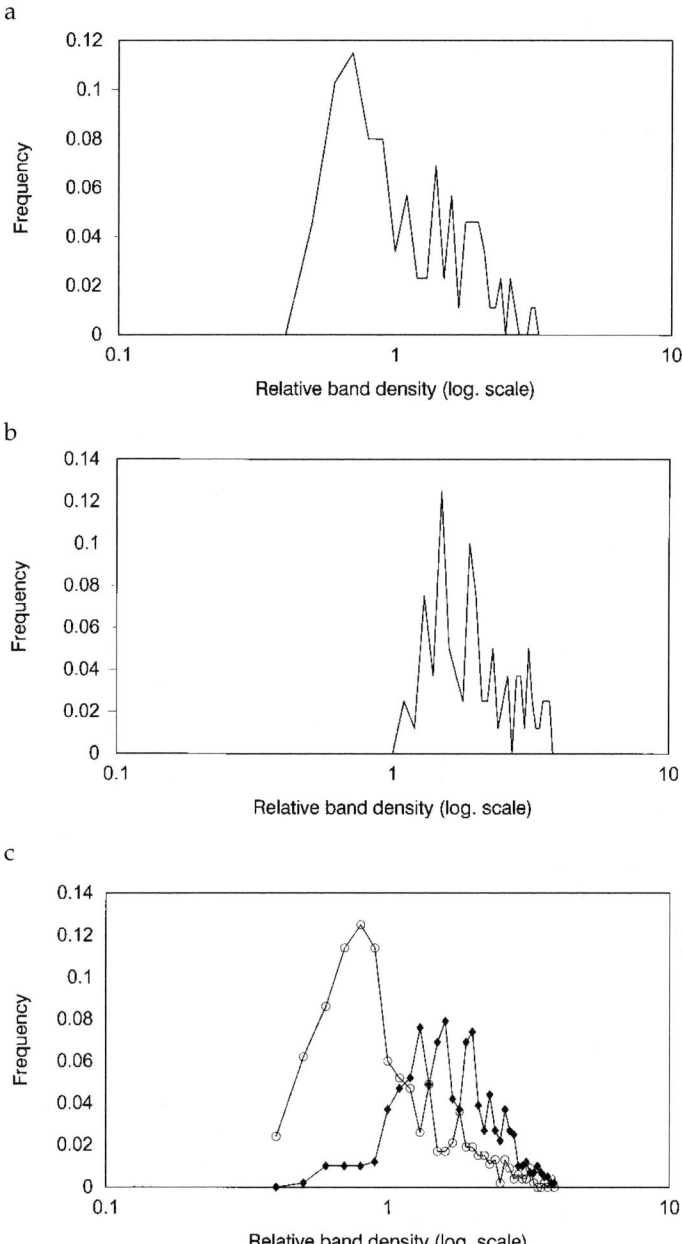

Fig. 2.7. Frequency of overall quantitative carboxylesterase (EST-4 to EST-7) variation in *Phorodon humuli* plotted against relative band density of variants. Samples collected in 1988 from: (a) wild hops (*Humulus lupulus*) in a non-commercial hop region from early June to mid-July; (b) *Prunus* species in a commercial hop growing region from late May to early July; (c) combined data from five non-commercial sites (aphids from wild hops; total $n = 434$, O), and five commercial hop growing sites (aphids from *Prunus* spp., total $n = 407$, ♦); n, sample size (from Loxdale *et al.*, 1998 by permission of *the Bulletin of Entomological Research*).

De Barro *et al.*'s (1995b) study was relatively local (Hampshire, southern Britain). Aphids were collected throughout the growing season (April to August), whereas Llewellyn *et al.* (2004) carried out their survey in the later part of the season (June and July) and along a much larger transect. These data sets may therefore highlight different dispersal events at local and wider geographic scales.

That *S. avenae* disperses over large distances in search of new hosts is further evidenced by the finding that the range of alleles at various microsatellite loci and their respective frequencies are similar between population samples collected in Britain and France (Llewellyn, 2000; J.-C. Simon, K.S. Llewellyn, L. Reimer and H.D. Loxdale, unpublished observations). Hence, the English Channel does not appear to be a significant barrier to gene flow, supporting the conclusions from Loxdale *et al.*'s (1985) study. In this respect, the behaviour of *S. avenae* seems fundamentally different to that of its congener, *S. fragariae*, and other aphids with apparently more restricted flight behaviours, including many gall aphids and the grape phylloxera, *V. vitifoliae* (Vorwerk *et al.*, 2006).

Other contributory evidence relating to migratory urge and flight duration

That patterns of allele frequency are related to flight behaviour also seems to be supported by the duration of the flight period of aphids in wind tunnel experiments, and an aphid's response to bright lights before it responds to green targets. Thus, short-distance flyers like *P. humuli* have shorter 'white light' flight times compared with longer-distance migrants, for example *R. padi* (Hardie, 1993; Hardie and Campbell, 1998). The molecular ecological data show that geographic scale is not necessarily proportional to diversity, and quite different patterns of movement are indicated by consideration of different relative spatial scales, e.g. a county (Hampshire) *v.* the length of Britain, or France in the case of *S. avenae*. In addition, because peak abundance and flight behaviour of given aphid species

appear very strongly correlated with dispersal, patterns of diversity clearly depend upon the species in question and should not be generalized. This being so, suitable control measures to combat pests must be considered case by case.

Geographic colonization

As shown, molecular markers have proved very useful for examining aphid movement at a range of spatial scales. A related aspect of this use is assessment of colonization events following an introduction. Good examples of this process may be found in Australia and New Zealand with the introduction of pest populations from distant geographic regions.

Sitobion *species*

In a study on *S. miscanthi* and *S.* sp. near *fragariae* (Hales *et al.*, 1990) in New Zealand, microsatellites, SSCP, and nuclear sequence markers as well as karyotype were used to examine population structure (Wilson *et al.*, 1999). In South-east Asia and Europe, the species are holocyclic, whereas in Australia they are completely anholocyclic. This has led to several important inferences about the introduction and establishment of these insects. The recent historical European colonization of Australia and New Zealand has also facilitated the assessment of the chronology and number of founder events thought to be involved.

In brief, *Sitobion* spp. populations in New Zealand appeared to comprise a subset of those found in Australia. Of these, some were identical to Australian genotypes, others shared alleles in common with Australian *S. miscanthi* (the majority), and one was unique. Cluster analysis revealed that the *Sitobion* on grasses in New Zealand had mixed affiliations with genotypes from both Australia and Taiwan. A heterogeneous genetic population structure was noted with high diversity in the north (< 39°S), declining in the south. In New Zealand, a single genotype of *S.* sp. near *fragariae* was predominant, virtually to the exclusion of others

(Wilson *et al.*, 1999). Genotypes were highly heterozygous (H$_{obs}$. > 0.5) at most loci studied, whilst there was extensive turnover of genotypes between 1996 and 1998, such that 36% were not sampled again in 1998 and 42% found in 1998 were new, suggesting clonal selective sweeps to be operative with some local annual persistence. Three new karyotypes were identified additional to those in Australia. This indicated asexual genotypes derived possibly from dispersing winged forms from Australia and Taiwan (although human agencies cannot be ruled out).

Other genotypes may have arisen from stepwise mutation of microsatellites and karyotypic change *in situ*. Low incidence of rare genotypes is also possible. Since strong geographic partitioning of genotypes/karyotypes was found predominantly over large geographical areas, interclonal selection appears to be important in the production and maintenance of the genetic diversity observed. From this, it is assumed that the original founders were of polyphyletic origin and then evolved by clonal mutation with the possible extinction of some 'missing' mutational intermediates (Wilson *et al.*, 1999).

Peach–potato aphid

In a separate study on *M. persicae* collected from the primary host, peach, in Bathhurst, New South Wales, Australia and from three sites in Tasmania, microsatellite and karyotypic analyses were performed in order to elucidate the pattern of colonization (Wilson *et al.*, 2002). The species was first recorded in Australia in 1893. The alleles and genotypes detected were found to represent a subset of European lineages as previously studied by Terradot *et al.* (1999). Common (worldwide) karyotypes were also found, including some highly insecticide-resistant aphids with the characteristic 1,3 autosomal translocation (Foster *et al.*, Chapter 10 this volume). Since the samples studied were highly heterozygous and indeed holocyclic, this supports the notion that a reasonable number of clones came into the continent. Interestingly, F_{ST} values were high, with significant substructuring at reasonably small spatial scales (< 50 km), again implying that this aphid is not very migratory compared with, say, *S. avenae*, and perhaps is more akin in its movements to less migratory species such as *S. fragariae* and *P. humuli* (Loxdale and Brookes, 1990; Loxdale *et al.*, 1998). Overall, the data suggest a polyphyletic origin for *M. persicae* in southeastern Australia. If so, the initial introduction of this important plant virus vector into Australia involved either a large diverse sample, which seems unlikely, or more probably, multiple and serial founder events over the past 100 years or so, mainly of European origin.

Rose–grain aphid

RAPD markers were used to genotype samples of *M. dirhodum* from a suction trap in New Zealand (Nicol *et al.*, 1997). Anholocyclic populations also appeared to comprise a subset of genotypes compared with Scottish samples, which are thought to be largely holocyclic. The aphid was first documented in New Zealand in about 1981. Probably, most such introductions are by human agency, *via* ships, aeroplanes, and the transport of commercial crops and plant stock, e.g. seed potatoes, as discussed by Loxdale *et al.* (1993).

Spotted alfalfa aphid

Although not detailed using molecular markers, one of the best examples of an asexual founder event that produced considerable adaptive variance in an agricultural context concerns clones of *Therioaphis trifolii maculata* (spotted alfalfa aphid) in North America, feeding on lucerne (*M. sativa*), and which became economically troublesome in the southwestern states in the mid-1950s (Dickson, 1962; Blackman and Eastop, 2000), although the original introduction (possibly from Europe) may have been earlier. This aphid displayed the ability to adapt to novel environments even without sexual recombination (Williams, 1975; Lushai *et al.*, 2003), although the species seemingly acquired the ability to produce viable sexuals and overwinter as eggs in about 1960, in the course of its spread to more northerly latitudes (Manglitz *et al.*, 1966, cited in Blackman and Eastop, 2000). This example also demonstrates the importance, relevant to control measures, of being able to track populations in terms of variation between individuals, more especially since the species first developed resistance to pesticides in the late 1950s in the USA (Stern and Reynolds, 1958) and subsequently elsewhere, e.g. it was introduced into Australia in 1977 and reported as resistant by 1979 (Holtkamp *et al.*, 1992; Blackman and Eastop, 2000; see also Sunnucks *et al.*, 1997b).

Long-distance migration is often claimed the probable source of such founder events, and this may certainly occur from time to time. Whatever the exact mode of transport of immigrants, assuming the aphids survive and are able to locate a suitable food plant on arrival and successfully reproduce on it, a new founder population may persist, thrive, and ultimately expand, both in terms of numbers and demography. Spread of newly colonized aphids has already been shown to follow classic radial patterns in areas lacking geographical barriers (Wellings, 1994). However, life cycle constraints may influence this establishment. For example, with a holocyclic colonizer, the species will not survive unless suitable overwintering primary hosts are present or a mutation arises that allows continuous parthenogenesis, in which case, environmental conditions must be suitable.

Conclusions

It is clear that a new age of aphidology has dawned with the employment of molecular techniques. Within the past decade of the 20th century and into the 21st, this has occurred using DNA technology, and a suite of markers (including sequencing) is now available which allows a much-enhanced resolution of ecological and evolutionary aspects of aphid biology, hitherto largely or completely impossible by other means. However, even the use of allozymes, despite

their undoubted selectivity (Nei and Graur, 1984), has been central to key discoveries from the 1970s onwards. These include the relationship between carboxylesterases and insecticide resistance, taxonomically discriminating polymorphic loci, detection of aphid genotypes with different plant resistance-breaking traits, cryptic species and genetic heterogeneity over different spatial and temporal scales, as well as detection of flight behaviours correlated with genotypic composition.

The employment of DNA markers has refined this understanding, especially and predominantly with respect to associations of genotype with host-plant species (e.g. Haack *et al.*, 2000; Lushai *et al.*, 2002) and life cycle strategy (e.g. Delmotte *et al.*, 2002; Guillemaud *et al.*, 2003; Vorwerk *et al.*, 2006). Even notions of clonality itself have changed in recent years as a result of these markers. Of special note is the discovery that asexual and sexual lineages can be discriminated and that these apparently have been genetically isolated. However, the story is never simple, as males from androcyclic lineages have been inferred to mate with females of other lineages, enabling gene flow between such lineages (Delmotte *et al.*, 2001). Some asexual lineages even appear to show a hybrid origin, as demonstrated in *R. padi* (Delmotte *et al.*, 2003). That clonal lineages can themselves be discriminated is also a novel discovery, although there are caveats related to intraclonal, intermorph variation found using some markers (e.g. RAPDs – Lushai *et al.*, 1997). The correlation of genetic variation with mobility and virus transmissibility should throw further light on these important aspects of virus epidemiology. The use of suction traps to collect airborne aphids with subsequent molecular testing has great potential for a whole raft of applications, including understanding the dynamics of insecticide resistance (Foster *et al.*, Chapter 10 this volume) and plant virus monitoring (Jones *et al.*, 1991).

How evolution and development relate to physiological aspects of egg diapause, wing formation, and other phenomena await the advent of population genomics (Black *et al.*, 2001). This will allow the move from mapping aphid genomes (Caillaud *et al.*, 2004b) to function (Hawthorne and Via, 2001; Braendle *et al.*, 2005a,b), and will ultimately have significant ramifications for understanding pest populations and development of strategies for control.

One might say that we are presently in the middle of a golden age of aphidology, the final outcome and achievements of which have been glimpsed, but not yet fully realized. Molecular advances, especially relating to direct sequencing of DNA and determination of DNA allele size in population genetic studies (e.g. MALDI-TOF technology – Kirpekar *et al.*, 1998; Kwon *et al.*, 2001), and single nucleotide polymorphisms (SNPs) (Stoneking, 2001), will quickly and efficiently reveal exciting aspects of the evolutionary relationships of aphids, not only with each other, but also pertaining to their host plants.

Acknowledgements

We thank Mark Taylor for help with aphid taxonomic matters; Drs Roger Blackman, Kate Llewellyn and James Anstead for their helpful comments on the manuscript, and the *Proceedings of the Royal Society Series B Biological Sciences*, *Molecular Ecology*, *Heredity* and the *Bulletin of Entomological Research* for permission to reproduce figures.

References

Abbot, P. and Moran, N.A. (2002) Extremely low levels of genetic polymorphism in endosymbionts (*Buchnera*) of aphids (*Pemphigus*). *Molecular Ecology* 11, 2649–2660.

Abbot, P. and Withgott, J.H. (2004) Phylogenetic and molecular evidence for allochronic speciation in gall-forming aphids (*Pemphigus*). *Evolution* 58, 539–553.

Abbot, P., Withgott, J.H. and Moran, N.A. (2001) Genetic conflict and conditional altruism in social aphid colonies. *Proceedings of the National Academy of Sciences of the United States of America* 98, 12068–12071.

Abid, H.S., Kindler, S.D., Jensen, S.G., Thomas-Compton, M.A. and Spomer, S.M. (1989) Isozyme characterization of sorghum aphid species and greenbug biotypes (Homoptera: Aphididae). *Annals of the Entomological Society of America* 82, 303–306.

Anstead, J.A., Burd, J.D. and Shufran, K.A. (2002) Mitochondrial DNA sequence divergence among *Schizaphis graminum* (Hemiptera: Aphididae) clones from cultivated and non-cultivated hosts: haplotype and host associations. *Bulletin of Entomological Research* 92, 17–24.

Anstead, J.A., Burd, J.D. and Shufran, K.A. (2003) Over-summering and biotypic diversity of *Schizaphis graminum* (Homoptera: Aphididae) populations on non-cultivated grass hosts. *Environmental Entomology* 32, 662–667.

Aoki, S. (1977) *Colophina clematis* (Homoptera, Pemphigidae), an aphid species with soldiers. *Kontyu* 45, 276–282.

Aoki, S. and Akimoto, S. (1981) Observations on *Pseudoregma alexanderi* (Homoptera, Pemphigidae), an aphid species producing pseudoscorpion-like soldiers on bamboos. *Kontyu* 49, 355–366.

Barrette, R.J., Crease, T.J., Hebert, P.D.N. and Via, S. (1994) Mitochondrial DNA diversity in the pea aphid *Acyrthosiphon pisum*. *Genome* 37, 858–865.

de Barro, P.J., Sherratt, T., Wratten, S.D. and Maclean, N. (1994a) DNA fingerprinting of cereal aphids using (GATA)$_4$. *European Journal of Entomology* 91, 109–114.

de Barro, P.J., Sherratt, T.N., Carvalho, G.R., Nicol, D., Iyengar, A. and Maclean, N. (1994b) An analysis of secondary spread by putative clones of *Sitobion avenae* within a Hampshire wheat field using the multilocus (GATA)$_4$ probe. *Insect Molecular Biology* 3, 253–260.

de Barro, P.J., Sherratt, T.N., Carvalho, G.R., Nicol, D., Iyengar, A. and Maclean, N. (1995a) Geographic and microgeographic genetic differentiation in two aphid species over southern England using the multilocus (GATA)$_4$ probe. *Molecular Ecology* 4, 375–382.

de Barro, P.J., Sherratt, T.N., Brookes, C.P., David, O. and Maclean, N. (1995b) Spatial and temporal variation in British field populations of the grain aphid *Sitobion avenae* (F.) (Hemiptera: Aphididae) studied using RAPD-PCR. *Proceedings of the Royal Society of London B* 262, 321–327.

de Barro, P.J., Sherratt, T.N., David, O. and Maclean, N. (1995c) An investigation of the differential performance of clones of the aphid *S. avenae* on two hosts. *Oecologia* 104, 379–385.

Baumann, P., Baumann, L., Lai, C.Y., Roubakhsh, D., Moran, N.A. and Clark, M.A. (1995) Genetics, physiology, and evolutionary relationships of the genus *Buchnera* – intracellular symbionts of aphids. *Annual Review of Microbiology* 49, 55–94.

Birkle, L.M. and Douglas, A.E. (1999) Low genetic diversity among pea aphid (*Acyrthosiphon pisum*) biotypes of different plant affiliation. *Heredity* 82, 605–612.

Black, W.C., Baer, C.F., Antolin, M.F. and DuTeau, N.M. (2001) Population genomics: genome-wide sampling of insect populations. *Annual Review of Entomology* 46, 441–469.

Blackman, R.L. (1974) Life cycle variation of *Myzus persicae* (Sulz.) (Hom., Aphididae) in different parts of the world, in relation to genotype and environment. *Bulletin of Entomological Research* 63, 595–607.

Blackman, R.L. (1980) Chromosomes and parthenogenesis in aphids. In: Blackman, R.L., Hewitt, G.M. and Ashburner, M. (eds) *Insect Cytogenetics. Proceedings of the Royal Entomological Society Symposium No. 10*. Blackwell, Oxford, pp. 133–148.

Blackman, R.L. (1981) Species, sex and parthenogenesis in aphids. In: Forey, P.L. (ed.) *The Evolving Biosphere*. Cambridge University Press, Cambridge, pp. 75–85.

Blackman, R.L. (1987) Morphological discrimination of a tobacco-feeding form from *Myzus persicae* (Sulzer) (Hemiptera: Aphididae), and a key to New World *Myzus (Nectarosiphon)* species. *Bulletin of Entomological Research* 77, 713–730.

Blackman, R.L. (2000) The cloning experts. *Antenna* 24, 206–214.

Blackman, R.L. and Eastop, V.F. (2000) *Aphids on the World's Crops: An Identification and Information Guide*, 2nd edn. Wiley, Chichester, 466 pp.

Blackman, R.L. and Spence, J.M. (1992) Electrophoretic distinction between the peach–potato aphid, *Myzus persicae*, and the tobacco aphid, *M. nicotianae* (Homoptera: Aphididae). *Bulletin of Entomological Research* 82, 161–165.

Blackman, R.L. and Spence, J.M. (1996) Ribosomal DNA is frequently concentrated on only one X chromosome in permanently apomictic aphids, but this does not inhibit male determination. *Chromosome Research* 4, 314–320.

Blackman, R.L., Brown, P.A., Furk, C., Seccombe, A.D. and Watson, G.W. (1989) Enzyme differences within species groups containing pest aphids. In: Loxdale, H.D. and den Hollander, J. (eds) *Electrophoretic Studies on Agricultural Pests. Systematics Association Special Volume No. 39*. Clarendon Press, Oxford, pp. 271–295.

Blackman, R.L., Spence, J.M. and Normark, B.B. (2000) High diversity of structurally heterozygous karyotypes and rDNA arrays in parthenogenetic aphids of the genus *Trama* (Aphididae: Lachninae). *Heredity* 84, 254–260.

Bourdin, D., Rouze, J., Tanguy, S. and Robert, Y. (1998) Variation among clones of *Myzus persicae* and *Myzus nicotianae* in the transmission of a poorly and a highly aphid-transmissible isolate of potato leafroll luteovirus (PLRV). *Plant Pathology* 47, 794–800.

Bournoville, R., Simon, J.-C., Badenhausser, I., Girousse, C., Guilloux, T. and Andre, S. (2000) Clones of pea aphid, *Acyrthosiphon pisum* (Hemiptera: Aphididae) distinguished using genetic markers, differ in their damaging effect on a resistant alfalfa cultivar. *Bulletin of Entomological Research* 90, 33–39.

Braendle, C., Friebe, I., Caillaud, M.C. and Stern, D.L. (2005a) Genetic variation for an aphid wing polyphenism is genetically linked to a naturally occurring wing polymorphism. *Proceedings of the Royal Society of London B* 272, 657–664.

Braendle, C., Caillaud, M.C. and Stern, D.L. (2005b) Genetic mapping of aphicarus – a sex-linked locus controlling a wing polymorphism in the pea aphid (*Acyrthosiphon pisum*). *Heredity* 94, 435–442.

Brookes, C.P. and Loxdale, H.D. (1987) Survey of enzyme variation in British populations of *Myzus persicae* (Sulzer) (Hemiptera: Aphididae) on crops and weed hosts. *Bulletin of Entomological Research* 77, 83–89.

Brown, P.A. and Blackman, R.L. (1988) Karyotype variation in the corn leaf aphid, *Rhopalosiphum maidis* (Fitch), species complex (Hemiptera: Aphididae) in relation to host-plant and morphology. *Bulletin of Entomological Research* 78, 351–363.

Bruton, M.N. (ed.) (1989) *Alternative Life-History Styles of Animals. Perspectives in Vertebrate Science, Volume 6*. Kluwer, Dordrecht, 624 pp.

Byrne, D.N., Rathman, R.J., Orum, T.V. and Palumbo, J.C. (1996) Localized migration and dispersal by the sweet potato whitefly, *Bemisia tabaci*. *Oecologia* 105, 320–328.

Caillaud, M.C., Boutin, M., Braendle, C. and Simon, J.-C. (2002) A sex-linked locus controls wing polymorphism in males of the pea aphid, *Acyrthosiphon pisum* (Harris). *Heredity* 89, 346–352.

Caillaud, M., Mondor-Genson, G., Levine-Wilkinson, S., Mieuzet, L., Frantz, A., Simon, J.-C.and D'Acier, A.C. (2004a) Microsatellite DNA markers for the pea aphid *Acyrthosiphon pisum*. *Molecular Ecology Notes* 4, 446–448.

Caillaud, M., Edwards, O., Field, L., Giblot-Ducray, D., Graye, S., Hawthorne, D., Hunter, W., Jander, G., Moran, N., Moya, A., Nakabachik, A., Robertson, H., Shufran, K., Simon, J.-C., Stern, D. and Tagu, D. (2004b) *Proposal to Sequence the Genome of the Pea Aphid* (Acyrthosiphon pisum). The International Aphid Genomics Consortium (IAGC) Steering Committee (www.princeton.edu/~dstern/Assets/ AphidGenomeWhitePaper.pdf).

Carvalho, G.R., Maclean, N., Wratten, S.D., Carter, R.E. and Thurston, J.P. (1991) Differentiation of aphid clones using DNA fingerprints from individual aphids. *Proceedings of the Royal Society of London B* 243, 109–114.

Chapman, J.W., Smith, A.D., Woiwod, I.P., Reynolds, D.R. and Riley, J.R. (2002) Development of vertical-looking radar technology for monitoring insect migration. *Computers and Electronics in Agriculture* 35, 95–110.

Clark, M.A., Moran, N.A., Baumann, P. and Wernegreen, J.J. (2000) Co-speciation between bacterial endosymbionts (*Buchnera*) and a recent radiation of aphids (*Uroleucon*). *Evolution* 54, 517–525.

Clements, K.M., Wiegmann, B.M., Sorenson, C.E., Smith, C.F., Neese, P.A. and Roe, R.M. (2000) Genetic variation in the *Myzus persicae* complex (Homoptera: Aphididae): evidence for a single species. *Annals of the Entomological Society of America* 93, 31–46.

Crampton, J.M., Beard, C.B. and Louis, C. (eds) (1996) *The Molecular Biology of Insect Disease Vectors*. Kluwer, Dordrecht, 604 pp.

Del Campo, M.L., Via, S. and Caillaud, M.C. (2003) Recognition of host-specific chemical stimulants in two sympatric host races of the pea aphid *Acyrthosiphon pisum*. *Ecological Entomology* 28, 405–412.

Delmotte, F., Leterme, N., Bonhomme, J., Rispe, C. and Simon, J.-C. (2001) Multiple routes to asexuality in an aphid. *Proceedings of the Royal Society of London B* 268, 2291–2299.

Delmotte, F., Leterme, N., Gauthier, J.P., Rispe, C. and Simon, J.-C. (2002) Genetic architecture of sexual and asexual populations of the aphid *Rhopalosiphum padi* based on allozyme and microsatellite markers. *Molecular Ecology* 11, 711–723.

Delmotte, F., Sabater-Munoz, B., Prunier-Leterme, N., Latorre, A., Sunnucks, P., Rispe, C. and Simon, J.-C. (2003) Phylogenetic evidence for hybrid origins of asexual lineages in an aphid species. *Evolution* 57, 1291–1303.

Devonshire, A.L. (1989a) The role of electrophoresis in the biochemical detection of insecticide resistance. In: Loxdale, H.D. and den Hollander, J. (eds) *Electrophoretic Studies on Agricultural Pests*. Clarendon Press, Oxford, pp. 363–374.

Devonshire, A.L. (1989b) Resistance of aphids to insecticides. In: Minks, A.K. and Harrewijn, P. (eds) *Aphids. Their Biology, Natural Enemies and Control, Volume 2C*. Elsevier, Amsterdam, pp. 123–139.

Dickson, R.C. (1962) Development of the spotted alfalfa aphid population in North America. *Proceedings of the 11th International Congress of Entomology, Vienna, 1960* 2, 26–28.

Dixon, A.F.G. (1989) Parthenogenetic reproduction and the rate of increase in aphids. In: Minks, A. and Harrewijn, P. (eds) *Aphids. Their Biology, Natural Enemies and Control, Volume 2A*. Elsevier, Amsterdam, pp. 269–287.

Dixon, A.F.G. (1998) *Aphid Ecology*, 2nd edn. Chapman and Hall, London, 300 pp.

Douglas, A.E., Darby, A.C., Birkle, L.M. and Walters, K.F.A. (2003) The ecological significance of symbiotic microorganisms in animals: perspectives from the microbiota of aphids. In: Hails, R.S., Beringer, J.E. and Godfray, H.C.J. (eds) *Genes in the Environment. Proceedings of the British Ecological Society Symposium No. 15*. Blackwell, Oxford, pp. 306–325.

Eastop, V.F. (1973) Biotypes of aphids. In: Lowe, A.D. (ed.) *Perspectives in Aphid Biology*. Entomological Society of New Zealand, Auckland, pp. 40–41.

Eggers-Schumacher, H.A. (1987) Enzyme electrophoresis in biosystematics and taxonomy of aphids. In: Holman, J., Pelikán, J., Dixon, A.F.G. and Weismann, L. (eds) *Population Structure, Genetics and Taxonomy of Aphids and Thysanoptera. Proceedings of an International Symposium, Smolenice, Czechoslovakia, September 1985*. SPB Academic Publishing, The Hague, pp. 63–70.

van Emden, H.F. (1988) The peach–potato aphid *Myzus persicae* (Sulzer) (Hemiptera: Aphididae) – more than a decade on a fully defined chemical diet. *Entomologist* 107, 4–10.

Ender, A., Schwenk, K., Städler, T., Streit, B. and Schierwater, B. (1996) RAPD identification of microsatellites in *Daphnia*. *Molecular Ecology* 5, 437–441.

Favret, C. and Voegtlin, D.J. (2004) Speciation by host-switching in pinyon *Cinara* (Insecta: Hemiptera: Aphididae). *Molecular Phylogenetics and Evolution* 32, 139–151.

Fenton, B., Woodford, J.A.T. and Malloch, G. (1998a) Analysis of clonal diversity of the peach–potato aphid, *Myzus persicae* (Sulzer), in Scotland, UK and evidence for the existence of a predominant clone. *Molecular Ecology* 7, 1475–1487.

Fenton, B., Malloch, G. and Germa, F. (1998b) A study of variation in rDNA ITS regions shows that two haplotypes coexist within a single aphid genome. *Genome* 41, 337–345.

Fenton, B., Malloch, G., Navajas, M., Hillier, J. and Birch, A.N.E. (2003) Clonal composition of the peach–potato aphid *Myzus persicae* (Homoptera: Aphididae) in France and Scotland: comparative analysis with IGS fingerprinting and microsatellite markers. *Annals of Applied Biology* 142, 255–267.

Fenton, B., Malloch, G., Woodford, J.A.T., Foster, S.P., Anstead, J., Denholm, I., King, L. and Pickup, J. (2005) The attack of the clones: tracking the movement of insecticide resistant peach–potato aphids *Myzus persicae* (Hemiptera: Aphididae). *Bulletin of Entomological Research* 95, 483–494.

Figueroa, C.C., Simon, J.-C., le Gallic, J.F. and Niemeyer, H.M. (1999) Molecular markers to differentiate two morphologically-close species of the genus *Sitobion*. *Entomologia Experimentalis et Applicata* 92, 217–225.

Figueroa, C.C., Simon, J.-C., le Gallic, J.F., Prunier-Leterme, N., Briones, L.M., Dedryver, C.A. and Niemeyer, H.M. (2005) Genetic structure and clonal diversity of an introduced pest in Chile, the cereal aphid *Sitobion avenae*. *Heredity* 95, 24–33.

Forneck, A., Walker, M.A. and Blaich, R. (2001) Ecological and genetic aspects of grape phylloxera *Daktulosphaira vitifoliae* Fitch (Hemiptera: Phylloxeridae) performance on rootstock hosts. *Bulletin of Entomological Research* 91, 445–451.

Foster, S.P., Denholm, I., Harling, Z.K., Moores, G.D. and Devonshire, A.L. (1998) Intensification of insecticide resistance in UK field populations of the peach–potato aphid, *Myzus persicae* (Hemiptera: Aphididae) in 1996. *Bulletin of Entomological Research* 88, 127–130.

Foster, W.A. (2002) Soldier aphids go cuckoo. *Trends in Ecology and Evolution* 17, 199–200.

Fukatsu, T. and Ishikawa, H. (1994) Differentiation of aphid clones by arbitrary primed polymerase chain reaction (AP-PCR) DNA fingerprinting. *Molecular Ecology* 3, 187–192.

Fukatsu, T., Shibao, H., Nikoh, N. and Aoki, S. (2001) Genetically distinct populations in an Asian soldier-producing aphid, *Pseudoregma bambucicola* (Homoptera: Aphididae), identified by DNA fingerprinting and molecular phylogenetic analysis. *Molecular Phylogenetics and Evolution* 18, 423–433.

Fuller, S.J., Chavigny, P., Lapchin, L. and Vanleberghe-Massuti, F. (1999) Variation in clonal diversity in glasshouse infestations of the aphid, *Aphis gossypii* Glover, in southern France. *Molecular Ecology* 8, 1867–1877.

Goldstein, D.B. and Schlötterer, C. (1999) *Microsatellites, Evolution and Applications*. Oxford University Press, Oxford, 352 pp.

Gorokhova, E., Dowling, T.E., Weider, L.J., Crease, T.J. and Elser, J.J. (2002) Functional and ecological significance of rDNA intergenic spacer variation in a clonal organism under divergent selection for production rate. *Proceedings of the Royal Society of London B* 269, 2373–2379.

Guillemaud, T., Mieuzet, L. and Simon, J.-C. (2003) Spatial and temporal genetic variability in French populations of the peach–potato aphid, *Myzus persicae*. *Heredity* 91, 143–152.

Guldemond, J.A. (1990) Evolutionary genetics of the aphid *Cryptomyzus*, with a preliminary analysis of the inheritance of host preference, reproductive performance and host alternation. *Entomologia Experimentalis et Applicata* 57, 65–76.

Haack, L., Simon, J.-C., Gauthier, J.-P., Plantegenest, M. and Dedryver, C.-A. (2000) Evidence for predominant clones in a cyclically parthenogenetic organism provided by combined demographic and genetic analyses. *Molecular Ecology* 9, 2055–2066.

Hales, D.F., Chapman, R.L., Lardner, R.M., Cowen, R. and Turak, E. (1990) Aphids of the genus *Sitobion* occurring on grasses in southern Australia. *Journal of the Australian Entomological Society* 29, 19–25.

Hales, D., Wilson, A.C.C., Spence, J.M. and Blackman, R.L. (2000) Confirmation that *Myzus antirrhinii* (Macchiati) (Hemiptera: Aphididae) occurs in Australia, using morphometrics, microsatellite typing and analysis of novel karyotypes by fluorescence *in situ* hybridisation. *Australian Journal of Entomology* 39, 123–129.

Hales, D.F., Wilson, A.C.C., Sloane, M.A., Simon, J.-C., Legallic, J.F. and Sunnucks, P. (2002) Lack of detectable genetic recombination on the X chromosome during the parthenogenetic production of female and male aphids. *Genetical Research* 79, 203–209.

Hand, S.C. (1989) The overwintering of cereal aphids on Gramineae in southern England, 1977–1980. *Annals of Applied Biology* 115, 17–29.

Hardie, J. (1993) Flight behaviour in migrating insects. *Journal of Agricultural Entomology* 10, 239–245.

Hardie, J. and Campbell, C.A.M. (1998) The flight behaviour of spring and autumn forms of the damson–hop aphid, *Phorodon humuli*, in the laboratory. In: Nieto Nafria, J.M. and Dixon, A.F.G. (eds) *Aphids in Natural and Managed Ecosystems*. Universidad de León, León, pp. 205–212.

Harrington, R. (1994) Aphid layer (letter). *Antenna* 18, 50.

Hawthorne, D.J. and Via, S. (2001) Genetic linkage of ecological specialization and reproductive isolation in pea aphids. *Nature, London* 412, 904–907.

Hebert, P.D.N., Finston, T.L. and Foottit, R. (1991) Patterns of genetic diversity in the sumac gall aphid, *Melaphis rhois*. *Genome* 34, 757–762.

Helden, A.J. and Dixon, A.F.G. (2002) Life-cycle variation in the aphid *Sitobion avenae*: costs and benefits of male production. *Ecological Entomology* 27, 692–701.

Hick, C.A., Field, L.M. and Devonshire, A.L. (1996) Changes in methylation of amplified esterase DNA during loss and reselection of insecticide resistance in peach–potato aphids, *Myzus persicae*. *Insect Biochemistry and Molecular Biology* 26, 41–47.

Hodgson, D.J. (2001) Monoclonal aphid colonies and the measurement of clonal fitness. *Ecological Entomology* 26, 444–448.

Holtkamp, R.H., Edge, V.E., Dominiak, B.C. and Walters, P.J. (1992) Insecticide resistance in *Therioaphis trifolii* f. *maculata* (Hemiptera, Aphididae) in Australia. *Journal of Economic Entomology* 85, 1576–1582.

Hoy, M.A. (2003) *Insect Molecular Genetics: An Introduction to Principles and Applications*, 2nd edn. Academic Press/Elsevier, San Diego, 560 pp.

Hullé, M. (1991) Agraphid, un réseau de surveillance des populations de pucerons: base de données associée et domaines d'application. *Annales de l'Association pour la Protection des Plantes* 2, 103–113.

Hullé, M., Pannetier, D., Simon, J.-C., Vernon, P. and Frenot, Y. (2003) Aphids of sub-Antarctic Iles Crozet and Kerguelen: species diversity, host range and spatial distribution. *Antarctic Science* 15, 203–209.

Inbar, M., Wink, M. and Wool, D. (2004) The evolution of host plant manipulation by insects: molecular and ecological evidence from gall-forming aphids on *Pistacia*. *Molecular Phylogenetics and Evolution* 32, 504–511.

Irwin, M.E. and Thresh, J.M. (1988) Long-range aerial dispersal of cereal aphids as virus vectors in North America. *Philosophical Transactions of the Royal Society of London B* 321, 421–446.

Janzen, D.H. (1977) What are dandelions and aphids? *American Naturalist* 111, 586–589.

Johnson, P.C.D. (2000) Genetic variation in the aphid *Pemphigus spyrothecae*. PhD thesis, University of Cambridge, Cambridge, United Kingdom.

Johnson, P.C.D., Whitfield, J.A., Foster, W.A. and Amos, W. (2002) Clonal mixing in the soldier-producing aphid *Pemphigus spyrothecae* (Hemiptera: Aphididae). *Molecular Ecology* 11, 1525–1531.

Jones, T.D., Buck, K.W. and Plumb, R.T. (1991) The detection of beet western yellows virus and beet mild yellowing virus in crop plants using the polymerase chain-reaction. *Journal of Virological Methods* 35, 287–296.

Kennedy, J.S. and Booth, C.O. (1963) Free flight of aphids in the laboratory. *Journal of Experimental Biology* 40, 67–85.

Kidd, N.A.C. and Cleaver, A.M. (1984) The relationship between pre-flight reproduction and migratory urge in alatae of *Aphis fabae* Scopoli (Hemiptera: Aphididae). *Bulletin of Entomological Research* 74, 517–527.

Kidd, N.A.C. and Cleaver, A.M. (1986) The control of migratory urge in *Aphis fabae* Scopoli (Hemiptera: Aphididae). *Bulletin of Entomological Research* 76, 77–87.

Kirpekar, F., Nordhoff, E., Larsen, L.K., Kristiansen, K., Roepstorff, P. and Hillenkamp, F. (1998) DNA sequence analysis by MALDI mass spectrometry. *Nucleic Acids Research* 26, 2554–2559.

Klekowski, E.J. (2003) Plant clonality, mutation, diplontic selection and mutational meltdown. *Biological Journal of the Linnean Society* 79, 61–67.

Kurokawa, T., Yao, I., Akimoto, S.I. and Hasegawa, E. (2004) Isolation of six microsatellite markers from the pea aphid, *Acyrthosiphon pisum* (Homoptera, Aphididae). *Molecular Ecology Notes* 4, 523–524.

Kwon, Y.-S., Tang, K., Cantor, C.R., Koster, H. and Kang, C. (2001) DNA sequencing and genotyping by transcriptional synthesis of chain-terminated RNA ladders and MALDI-TOF mass spectrometry. *Nucleic Acids Research* 29(3), 'Methods' (online), e11.

Lanzaro, G.C., Toure, Y.T., Carnahan, J., Zheng, L.B., Dolo, G., Traore, S., Petrarca, V., Vernick, K.D. and Taylor, C.E. (1998) Complexities in the genetic structure of *Anopheles gambiae* populations in west Africa as revealed by microsatellite DNA analysis. *Proceedings of the National Academy of Sciences of the United States of America* 95, 14260–14265.

Lees, A.D. (1967) The production of the apterous and alate forms in the aphid *Megoura viciae* Buckton, with special reference to the role of crowding. *Journal of Insect Physiology* 132, 289–318.

Li, Y.-C., Korol, A.B., Fahima, T., Beiles, A. and Nevo, E. (2002) Microsatellites: genomic distribution putative functions and mutational mechanisms: a review. *Molecular Ecology* 11, 2453–2465.

Llewellyn, K.S. (2000) Genetic structure and dispersal of cereal aphid populations. PhD thesis, University of Nottingham, Nottingham, United Kingdom.

Llewellyn, K.S., Loxdale, H.D., Harrington, R., Brookes, C.P., Clark, S.J. and Sunnucks, P. (2003) Migration and genetic structure of the grain aphid (*Sitobion avenae*) in Britain related to climate and clonal fluctuation as revealed using microsatellites. *Molecular Ecology* 12, 21–34.

Llewellyn, K.S., Loxdale, H.D., Harrington, R., Clark, S.J. and Sunnucks, P. (2004) Evidence for gene flow and local clonal selection in field populations of the grain aphid (*Sitobion avenae*) in Britain revealed using microsatellites. *Heredity* 93, 143–153.

Loxdale, H.D. (1994) Isozyme and protein profiles of insects of agricultural and horticultural importance. In: Hawksworth, D.L. (ed.) *The Identification and Characterization of Pest Organisms*. CAB International, Wallingford, pp. 337–375.

Loxdale, H.D. (2001) Tracking flying insects using molecular markers. *Antenna* 25, 242–250.

Loxdale, H.D. and Brookes, C.P. (1988) Electrophoretic study of enzymes from cereal aphid populations. V. Spatial and temporal genetic similarity of holocyclic populations of the bird-cherry–oat aphid, *Rhopalosiphum padi* (L.)(Hemiptera: Aphididae), in Britain. *Bulletin of Entomological Research* 78, 241–249.

Loxdale, H.D. and Brookes, C.P. (1989a) Separation of three *Rubus*-feeding species of aphid – *Sitobion fragariae* (Wlk.), *Macrosiphum funestum* (Macch.) and *Amphorophora rubi* (Kalt.) (Hemiptera: Aphididae) – by electrophoresis. *Annals of Applied Biology* 115, 399–404.

Loxdale, H.D. and Brookes, C.P. (1989b) Use of genetic markers (allozymes) to study the structure, overwintering and dynamics of pest aphid populations. In: Loxdale, H.D. and den Hollander, J. (eds) *Electrophoretic Studies on Agricultural Pests. Systematics Association Special Volume No. 39*. Clarendon Press, Oxford, pp. 231–270.

Loxdale, H.D. and Brookes, C.P. (1990) Genetic stability within and restricted migration (gene flow) between local populations of the blackberry–grain aphid *Sitobion fragariae* in southeast England. *Journal of Animal Ecology* 59, 495–512.

Loxdale, H.D. and Lushai, G. (1998) Molecular markers in entomology. *Bulletin of Entomological Research* 88, 577–600.

Loxdale, H.D. and Lushai, G. (2003a) Rapid changes in clonal lines: the death of a 'sacred cow'. *Biological Journal of the Linnean Society* 79, 3–16.

Loxdale, H.D. and Lushai, G. (2003b) Maintenance of aphid clonal lineages: images of immortality. *Infection, Genetics and Evolution* 3, 259–269.

Loxdale, H.D. and Lushai, G. (eds) (2003c) Intraclonal genetic variation: ecological and evolutionary aspects. *Biological Journal of the Linnean Society* 79, 1–208.

Loxdale, H.D., Tarr, I.J., Weber, C.P., Brookes, C.P., Digby, P.G.N. and Castañera, P. (1985) Electrophoretic study of enzymes from cereal aphid populations. III. Spatial and temporal genetic variation of populations of *Sitobion avenae* (F.) (Hemiptera: Aphididae). *Bulletin of Entomological Research* 75, 121–141.

Loxdale, H.D., Hardie, J., Halbert, S., Foottit, R., Kidd, N.A.C. and Carter, C.I. (1993) The relative importance of short- and long-range movement of flying aphids. *Biological Reviews* 68, 291–311.

Loxdale, H.D., Brookes, C.P., Wynne, I.R. and Clark, S.J. (1998) Genetic variability within and between English populations of the damson–hop aphid, *Phorodon humuli* (Hemiptera: Aphididae), with special reference to esterases associated with insecticide resistance. *Bulletin of Entomological Research* 88, 513–526.

Lushai, G. and Loxdale, H.D. (2002) The biological improbability of a clone (mini-review). *Genetical Research* 79, 1–9.

Lushai, G. and Loxdale, H.D. (2004) Tracking movement in small insect pests, with special reference to aphid populations. *International Journal of Pest Management* 50, 307–315.

Lushai, G. and Loxdale, H.D. (2007) The potential role of chromosome telomere resetting consequent upon sex in the population dynamics of aphids: an hypothesis. *Biological Journal of the Linnean Society* 90, 719–728.

Lushai, G., Loxdale, H.D., Brookes, C.P., von Mende, N., Harrington, R. and Hardie, J. (1997) Genotypic variation among different phenotypes within aphid clones. *Proceedings of the Royal Society of London B* 264, 725–730.

Lushai, G., de Barro, P.J., David, O., Sherratt, T.N. and Maclean, N. (1998) Genetic variation within a parthenogenetic lineage. *Insect Molecular Biology* 7, 337–344.

Lushai, G., Loxdale, H.D. and Maclean, N. (2000) Genetic diversity of clonal lineages. *The Journal of Reproduction and Development* 46 (supplement), 21–22.

Lushai, G., Markovitch, O. and Loxdale, H.D. (2002) Host-based genotype variation in insects revisited. *Bulletin of Entomological Research* 92, 159–164.

Lushai, G., Loxdale, H.D. and Allen, J.A. (2003) The dynamic clonal genome and its adaptive potential. *Biological Journal of the Linnean Society* 79, 193–208.

Lushai, G., Foottit, R., Maw, E. and Barette, R. (2004) Genetic variation in the green apple aphid, *Aphis pomi* De Geer (Aphididae, Homoptera) detected using microsatellite DNA flanking sequences. In: Simon, J.C., Dedryver, C.A., Rispe, C. and Hullé, M. (eds) *Aphids in a New Millennium*. INRA-Editions, Versailles, pp. 245–251.

Lynch, M. and Blanchard, J.L. (1998) Deleterious mutation accumulation in organelle genomes. *Genetica* 103, 29–39.

Lynch, M., Bürger, R., Butcher, D. and Gabriel, W. (1993) The mutational meltdown in asexual populations. *Journal of Heredity* 84, 339–344.

MacArthur, R.H. and Wilson, E.O. (1963) An equilibrium theory of insular zoogeography. *Evolution* 17, 373–387.

Mackay, P.A., Lamb, R.J. and Smith, M.A.H. (1993) Variability in life-history traits of the aphid, *Acyrthosiphon pisum* (Harris), from sexual and asexual populations. *Oecologia* 94, 330–338.

Mandrioli, M., Manicardi, G.C., Bizzaro, D. and Bianchi, U. (1999) NOR heteromorphism within a parthenogenetic lineage of the aphid, *Megoura viciae*. *Chromosome Research* 7, 157–162.

Manglitz, G.R., Calkins, C.O., Walstrom, R.J., Hintz, S.D., Kindler, S.D. and Peters, L.L. (1966) Holocyclic strain of spotted alfalfa aphid in Nebraska and adjacent states. *Journal of Economic Entomology* 59, 636–639.

Margaritopoulos, J.T., Tsourapas, C., Tzortzi, M., Kanavaki, O.M. and Tsitisipis, J.A. (2005) Host selection by winged colonizers within the *Myzus persicae* group: a contribution toward understanding host specialization. *Ecological Entomology* 30, 406–418.

Martinez, J.-J.I., Mokady, O. and Wool, D. (2005) Patch size and patch quality of gall-inducing aphids in a mosaic landscape in Israel. *Landscape Ecology* 20, 1013–1024.

Martinez-Torres, D., Simon, J.-C., Fereres, A. and Moya, A. (1996) Genetic variation in natural populations of the aphid *Rhopalosiphum padi* as revealed by maternally inherited markers. *Molecular Ecology* 5, 659–670.

Martínez-Torres, D., Moya, A., Hebert, P.D.N. and Simon, J.-C. (1997) Geographic distribution and seasonal variation of mitochondrial DNA haplotypes in the aphid *Rhopalosiphum padi* (Hemiptera: Aphididae). *Bulletin of Entomological Research* 87, 161–167.

Massonnet, B. and Weisser, W.W. (2004) Patterns of genetic differentiation between populations of the specialised herbivore *Macrosiphoniella tanacetaria* (Homoptera: Aphididae). *Heredity* 93, 577–584.

Massonnet, B., Simon, J.-C. and Weisser, W.W. (2002) Metapopulation structure of the specialised herbivore *Macrosiphoniella tanacetaria* (Homoptera, Aphididae). *Molecular Ecology* 11, 2511–2521.

May, B. and Holbrook, F.R. (1978) Absence of genetic variability in the green peach aphid, *Myzus persicae* (Hemiptera: Aphididae). *Annals of the Entomological Society of America* 71, 809–812.

Miller, N.J. (2000) Population structure and gene flow in a host-alternating aphid, *Pemphigus bursarius*. PhD thesis, University of Birmingham, Birmingham, United Kingdom.

Miller, N.J., Birley, A.J., Overall, A.D.J. and Tatchell, G.M. (2003) Population genetic structure of the lettuce root aphid, *Pemphigus bursarius* (L.), in relation to geographic distance, gene flow and host plant usage. *Heredity* 91, 217–223.

Monro, K. and Poore, A.G.B. (2004) Selection in modular organisms: is intraclonal variation in macroalgae evolutionarily important? *American Naturalist* 163, 564–578.

Mopper, S. and Strauss, S.Y. (eds) (1998) *Genetic Structure and Local Adaptation in Natural Insect Populations*. Chapman and Hall, New York, 449 pp.

Morgan-Richards, M. and Trewick, S.A. (2005) Hybrid origin of a parthenogenetic genus? *Molecular Ecology* 14, 2133–2142.

Muller, H.J. (1964) The relation of recombination to mutational advance. *Mutation Research* 1, 2–9.

Nei, M. and Graur, D. (1984) Extent of protein polymorphism and the neutral mutation theory. *Evolutionary Biology* 17, 73–118.

Newton, C. and Dixon, A.F.G. (1988) A preliminary study of variation and inheritance of life-history traits and the occurrence of hybrid vigour in *Sitobion avenae* (F.) (Hemiptera: Aphididae). *Bulletin of Entomological Research* 78, 75–83.

Nicol, D., Armstrong, K.F., Wratten, S.D., Cameron, C.M., Frampton, C. and Fenton, B. (1997) Genetic variation in an introduced aphid pest (*Metopolophium dirhodum*) in New Zealand and relation to individuals from Europe. *Molecular Ecology* 6, 255–265.

Noda, I. (1960) The emergence of winged viviparous females in aphids. VI. Difference in the rate of development between the winged and unwinged forms. *Japanese Journal of Ecology* 10, 97–102.

Normark, B.B. (1999) Evolution in a putatively ancient asexual aphid lineage: recombination and rapid karyotype change. *Evolution* 53, 1458–1469.

Normark, B.B. and Moran, N.A. (2000). Testing for the accumulation of deleterious mutations in asexual eukaryote genomes using molecular sequences. *Journal of Natural History* 34, 1719–1729.

Ono, M., Swanson, J.J., Field, L.M., Devonshire, A.L. and Seigfried, B.D. (1999) Amplification and methylation of an esterase gene associated with insecticide-resistance in greenbugs, *Schizaphis graminum* (Rondani) (Homoptera: Aphididae). *Insect Biochemistry and Molecular Biology* 29, 1065–1073.

Osborne, J.L., Loxdale, H.D. and Woiwod, I.P. (2002) Monitoring insect dispersal: methods and approaches. In: Bullock J.M., Kenward, R.E. and Hails, R.S. (eds) *Dispersal Ecology*. Blackwell, Oxford, pp. 24–49.

Papura, D., Simon, J.-C., Halkett, F., Delmotte, F., le Gallic, J.F. and Dedryver, C.A. (2003) Predominance of sexual reproduction in Romanian populations of the aphid *Sitobion avenae* inferred from phenotypic and genetic structure. *Heredity* 90, 397–404.

Pineda-Krch, M. and Lehtila, K. (2004) Costs and benefits of genetic heterogeneity within organisms. *Journal of Evolutionary Biology* 17, 1167–1177.

Powell, S.J. and Bale, J.S. (2004) Cold shock injury and ecological costs of rapid cold hardening in the grain aphid *Sitobion avenae* (Hemiptera: Aphididae). *Journal of Insect Physiology* 50, 277–284.

Powers, T.O., Jensen, S.G., Kindler, S.D., Stryker, C.J. and Sandall, L.J. (1989) Mitochondrial DNA divergence among greenbug (Homoptera: Aphididae) biotypes. *Annals of the Entomological Society of America* 82, 208–302.

Puterka, G.J. and Peters, D.C. (1990) Sexual reproduction and inheritance of virulence in the greenbug, *Schizaphis graminum* (Rondani). In: Campbell, R.K. and Eikenbary, R.D. (eds) *Aphid–Plant Genotype Interactions*. Elsevier, Amsterdam, pp. 289–318.

Queller, D.C., Strassmann, J.E. and Hughes, C. (1993) Microsatellites and kinship. *Trends in Ecology and Evolution* 8, 285–288.

Ramos, S., Moya, A. and Martinez-Torres, D. (2003) Identification of a gene over-expressed in aphids reared under short photoperiod. *Insect Biochemistry and Molecular Biology* 33, 289–298.

Raybould, A.F., Clarke, R.T., Bond, J.M., Welters, R.E. and Gliddon, C.J. (2002) Inferring patterns of dispersal from allele frequency data. In: Bullock J.M., Kenward, R.E. and Hails, R.S. (eds) *Dispersal Ecology*. Blackwell, Oxford, pp. 89–110.

Reimer, L. (2005) Clonal diversity and population genetic structure of the grain aphid *Sitobion avenae* (F.) in Central Europe. PhD thesis, University of Göttingen, Göttingen, Germany.

Richardson, B.J., Baverstock, P.R. and Adams, M. (1986) *Allozyme Electrophoresis. A Handbook for Animal Systematics and Population Studies*. Academic Press, London, 410 pp.

Ridley, M. (1993) *Evolution*. Blackwell, Cambridge, Massachussets, 670 pp.

Ruiz-Montoya, L., Nunez-Farfan, J. and Vargas, J. (2003) Host-associated genetic structure of Mexican populations of the cabbage aphid *Brevicoryne brassicae* L. (Homoptera: Aphididae). *Heredity* 91, 415–421.

Shufran, K.A. (2003) Polymerase chain reaction–restriction fragment length polymorphisms identify mtDNA haplotypes of greenbug (Hemiptera: Aphididae). *Journal of the Kansas Entomological Society* 76, 551–556.

Shufran, K.A. and Wilde, G.E. (1994) Clonal diversity in overwintering populations of *Schizaphis graminum* (Homoptera: Aphididae). *Bulletin of Entomological Research* 84, 105–114.

Shufran, K.A., Black, W.C., 4th and Margolies, D.C. (1991) DNA fingerprinting to study spatial and temporal distributions of an aphid, *Schizaphis graminum* (Homoptera: Aphididae). *Bulletin of Entomological Research* 81, 303–313.

Shufran, K.A., Burd, J.D. Anstead, J.A. and Lushai, G. (2000) Mitochondrial DNA sequence divergence among greenbug (Homoptera: Aphididae) biotypes: evidence for host-adapted races. *Insect Molecular Biology* 9, 179–184.

Shufran, K.A., Mayo, Z.B. and Crease, T.J. (2003) Genetic changes within an aphid clone: homogenization of rDNA intergenic spacers after insecticide selection. *Biological Journal of the Linnean Society* 79, 101–105.

Simon, J.-C. and Hebert, P.D.N. (1995) Patterns of genetic variation among Canadian populations of the bird cherry–oat aphid, *Rhopalosiphum padi* L. (Homoptera: Aphididae). *Heredity* 74, 346–353.

Simon, J.-C., Hebert, P.D.N., Carillo, C. and de Melo, R. (1995) Lack of clonal variation among Canadian populations of the corn leaf aphid, *Rhopalosiphum maidis* Fitch (Homoptera: Aphididae). *Canadian Entomologist* 127, 623–629.

Simon, J.-C., Martinez-Torres, D., Latorre, A., Moya, A. and Hebert, P.D.N. (1996a) Molecular characterization of cyclic and obligate parthenogens on the aphid *Rhopalosiphum padi* (L.). *Proceedings of the Royal Society of London B* 263, 481–486.

Simon, J.-C., Carrel, E., Hebert, P.D.N., Dedryver, C.A., Bonhomme, J. and le Gallic, J.-F. (1996b) Genetic diversity and mode of reproduction in French populations of the aphid *Rhopalosiphum padi*. *Heredity* 76, 305–313.

Simon, J.-C., Baumann, S., Sunnucks, P., Hebert, P.D.N., Pierre, J.S., le Gallic, J.F. and Dedryver, C.A. (1999a) Reproductive mode and population genetic structure of the cereal aphid *Sitobion avenae* studied using phenotypic and microsatellite markers. *Molecular Ecology* 8, 531–545.

Simon, J.-C., Leterme, N. and Latorre, A. (1999b) Molecular markers linked to breeding system differences in segregating and natural populations of the cereal aphid *Rhopalosiphum padi* L. *Molecular Ecology* 8, 965–973.

Simon, J.-C., Rispe, C. and Sunnucks, P. (2002) Ecology and evolution of sex in aphids. *Trends in Ecology and Evolution* 17, 34–39.

Simon, J.-C., Carre, S., Boutin, M., Prunier-Leterme, N., Sabater-Munoz, B., Latorre, A. and Bournoville, R. (2003) Host-based divergence in populations of the pea aphid: insights from nuclear markers and the prevalence of facultative symbionts. *Proceedings of the Royal Society of London B* 270, 1703–1712.

Sloane, M.A., Sunnucks, P., Wilson, A.C.C. and Hales, D.F. (2001) Microsatellite isolation, linkage group identification and determination of recombination frequency in the peach–potato aphid, *Myzus persicae* (Sulzer) (Hemiptera: Aphididae). *Genetical Research* 77, 251–260.

Smith, M.A.H. and Mackay, P.A. (1990). Latitudinal variation in the photoperiodic responses of populations of pea aphid (Homoptera, Aphididae). *Environmental Entomology* 19, 618–624.

Stern, D.L and Foster, W.A. (1996) The evolution of soldiers in aphids. *Biological Reviews* 71, 27–79.

Stern, V.M. and Reynolds, H.T. (1958) Resistance of the spotted alfalfa aphid to certain organophosphorus insecticides in southern California. *Journal of Economic Entomology* 51, 312–316.

Stoneking, M. (2001) Single nucleotide polymorphisms: from the evolutionary past. *Nature, London* 409, 821–822.

Sunnucks, P., England, P.E., Taylor, A.C. and Hales, D.F. (1996) Microsatellite and chromosome evolution of parthenogenetic *Sitobion* aphids in Australia. *Genetics* 144, 747–756.

Sunnucks, P., de Barro, P.J., Lushai, G., Maclean, N. and Hales, D. (1997a) Genetic structure of an aphid studied using microsatellites: cyclic parthenogenesis, differentiated lineages, and host specialisation. *Molecular Ecology* 6, 1059–1073.

Sunnucks, P., Driver, F., Brown, W.V., Carver, M., Hales, D.F. and. Milne, W.M. (1997b) Biological and genetic characterization of morphologically similar *Therioaphis trifolii* (Hemiptera: Aphididae) with different host utilization. *Bulletin of Entomological Research* 87, 425–436.

Sunnucks, P., Chisholm, D., Turak, E. and Hales, D.F. (1998) Evolution of an ecological trait in parthenogenetic *Sitobion* aphids. *Heredity* 81, 638–647.

Sunnucks, P., Wilson, A.C.C., Beheregaray, L.B., Zenger, K., French, J. and Taylor, A.C. (2000) SSCP is not so difficult: the application and utility of single-stranded conformation polymorphism in evolutionary biology and molecular ecology. *Molecular Ecology* 9, 1699–1710.

Tatchell, G.M., Parker, S.J. and Woiwod, I.P. (1983) *Synoptic Monitoring of Migrant Insect Pests in Great Britain and Western Europe IV. Host Plants and their Distribution for Pest Aphids in Great Britain*. Annual Report of Rothamsted Experimental Station, 1982. Part 2, pp. 45–159.

Tatchell, G.M., Plumb, R.T. and Carter, N. (1988) Migration of alate morphs of the bird cherry aphid (*Rhopalosiphum padi*) and implications for the epidemiology of barley yellow dwarf virus. *Annals of Applied Biology* 112, 1–11.

Taylor, L.R., Woiwod, I.P. and Taylor, R.A.J. (1979) The migratory ambit of the hop aphid and its significance in aphid population dynamics. *Journal of Animal Ecology* 48, 955–972.

Terradot, L., Simon, J.-C., Leterme, N., Bourdin, D., Wilson, A.C.C., Gauthier, J.-P. and Robert, Y. (1999) Molecular characterization of clones of the *Myzus persicae* complex (Hemiptera: Aphididae) differing in their ability to transmit the potato leafroll luteovirus (PLRV). *Bulletin of Entomological Research* 89, 355–363.

Tymon, A.M., Shah, P.A. and Pell, J.K. (2004) PCR-based molecular discrimination of *Pandora neoaphidis* isolates from related entomopathogenic fungi and development of species-specific diagnostic primers. *Mycological Research* 108, 419–433.

Van Doninck, K., Schön, I., Martens, K. and Backeljau, T. (2004) Clonal diversity in the ancient asexual ostracod *Darwinula stevensoni* assessed by RAPD-PCR. *Heredity* 93, 154–160.

Vanlerberghe-Masutti, F. and Chavigny, P. (1998) Host-based genetic differentiation in the aphid *Aphis gossypii* Glover, evidenced from RAPD fingerprints. *Molecular Ecology* 7, 905–914.

Via, S. (1999) Reproductive isolation between sympatric races of pea aphids. I. Gene flow restriction and habitat choice. *Evolution* 53, 1446–1457.

Via, S. and Hawthorne, D.J. (2002) The genetic architecture of ecological specialization: correlated gene effects on host use and habitat choice in pea aphids. *American Naturalist* 159 (supplement), 76–88.

Via, S., Bouck, A.C. and Skillman, S. (2000) Reproductive isolation between divergent races of pea aphids on two hosts. II. Selection against migrants and hybrids in the parental environments. *Evolution* 54, 1626–1637.

Vialatte, A., Dedryver, C.A., Simon, J.-C., Galman, M. and Plantegenest, M. (2005) Limited genetic exchanges between populations of an insect pest living on uncultivated and related cultivated host plants. *Proceedings of the Royal Society of London B* 272, 1075–1082.

Von Dohlen, C.D. and Moran, N.A. (2000) Molecular data support a rapid radiation of aphids in the Cretaceous and multiple origins of host alternation. *Biological Journal of the Linnean Society* 71, 689–717.

Vorburger, C. (2004) Cold tolerance in obligate and cyclical parthenogens of the peach–potato aphid, *Myzus persicae*. *Ecological Entomology* 29, 498–505.

Vorburger, C. (2005) Positive genetic correlations among major life-history traits related to ecological success in the aphid *Myzus persicae*. *Evolution* 59, 1006–1015.

Vorburger, C., Sunnucks, P. and Ward, S.A. (2003a) Explaining the coexistence of asexuals with their sexual progenitors: no evidence for general-purpose genotypes in obligate parthenogens of the peach–potato aphid, *Myzus persicae*. *Ecology Letters* 6, 1091–1098.

Vorburger, C., Lancaster, M. and Sunnucks, P. (2003b) Environmentally related patterns of reproductive modes in the aphid *Myzus persicae* and the predominance of two 'superclones' in Victoria, Australia. *Molecular Ecology* 12, 3493–3504.

Vorwerk, S., Blaich, R. and Forneck, S. (2006) Reproductive mode of grape phylloxera (*Daktulosphaira vitifoliae* Fitch) in Europe. *Genome* 49, 678–687.

Walters, K.F.A. and Dewar, A.M. (1986) Overwintering strategy and the timing of the spring migration of the cereal aphids *Sitobion avenae* and *Sitobion fragariae*. *Journal of Applied Ecology* 23, 905–915.

Ward, S.A., Leather, S.R., Pickup, J. and Harrington, R. (1998) Mortality during dispersal and the cost of host specificity in parasites: how many aphids find hosts? *Journal of Animal Ecology* 67, 763–773.

Weber, G. (1985a) On the ecological genetics of *Sitobion avenae* (F.) (Hemiptera, Aphididae). *Journal of Applied Ecology* 100, 100–110.

Weber, G. (1985b) On the ecological genetics of *Metopolophium dirhodum* (Walker) (Hemiptera, Aphididae). *Journal of Applied Ecology* 100, 451–458.

Weber, G. (1985c) Genetic variability in host plant adaptation of the green peach aphid, *Myzus persicae*. *Entomologia Experimentalis et Applicata* 38, 49–56.

Wellings, P.W. (1994) How variable are rates of colonization? *European Journal of Entomology* 91, 121–125.

Whitfield, J.A. (1998) Studies of the diversity and evolution of soldier aphids. PhD thesis, University of Cambridge, Cambridge, United Kingdom.

Williams, G.C. (1975) *Sex and Evolution*. Princeton University Press, Princeton, 193 pp.

Wilson, A.C.C., Sunnucks, P. and Hales, D.F. (1999) Microevolution, low clonal diversity and genetic affinities of parthenogenetic *Sitobion* aphids in New Zealand. *Molecular Ecology* 8, 1655–1666.

Wilson, A.C.C., Sunnucks, P., Blackman, R.L. and Hales, D.F. (2002) Microsatellite variation in cyclically parthenogenetic populations of *Myzus persicae* in southeastern Australia. *Heredity* 88, 258–266.

Wilson, A.C.C., Sunnucks, P. and Hales, D.F. (2003) Heritable genetic variation and potential for adaptive evolution in asexual aphids (Aphidoidea). *Biological Journal of the Linnean Society* 79, 115–135.

Wilson, A.C.C., Massonnet, B., Simon, J.-C., Prunier-Leterme, N., Dolatti, L., Llewellyn, K.S., Figueroa, C.C., Ramirez, C.C., Blackman, R.L., Estoup, A. and Sunnucks, P. (2004) Cross-species amplification of microsatellite loci in aphids: assessment and application. *Molecular Ecology Notes* 4, 104–109.

Woiwod, I.P., Tatchell, G.M., Dupuch, M.J., Macaulay, E.D.M., Parker, S.J., Riley, A.M. and Taylor, M.S. (1988) *Rothamsted Insect Survey: Nineteenth Annual Summary: Suction Traps 1987*. Annual Report of Rothamsted Experimental Station, 1987, Part 2, pp. 195–229.

Wool, D. and Hales, D.F. (1997) Phenotypic plasticity in Australian cotton aphid (Homoptera: Aphididae): host plant effects on morphological variation. *Annals of the Entomological Society of America* 90, 316–328.

Wynne, I.R., Howard, J.J., Loxdale, H.D. and Brookes, C.P. (1994) Population genetic structure during aestivation in the sycamore aphid *Drepanosiphum platanoidis* (Hemiptera: Drepanosiphidae). *European Journal of Entomology* 91, 375–383.

Yao, I., Akimoto, S.I. and Hasegawa, E. (2003) Isolation of microsatellite markers from the drepanosiphid aphid, *Tuberculatus quercicola*. *Molecular Ecology Notes* 3, 542–543.

Zeyl, C., Mizesko, M. and de Visser, J.A.G.M. (2001) Mutational meltdown in laboratory yeast populations. *Evolution* 55, 909–917.

Zitoudi, K., Margaritopoulos, J.T., Mamuris, Z. and Tsitsipis, J.A. (2001) Genetic variation in *Myzus persicae* populations associated with host-plant and life cycle category. *Entomologia Experimentalis et Applicata* 99, 303–311.

3   Life Cycles and Polymorphism

Iain S. Williams[1] and Anthony F.G. Dixon[2]

[1]Science Directorate, DEFRA, Nobel House, Smith Square, London, SW1P 3JH, UK;
[2]School of Biological Sciences, University of East Anglia, Norwich, NR4 7TJ, UK

Introduction

Aphids display a diverse range of relatively complicated life cycles. Each life cycle is divided into a number of stages, with each stage characterized by one or more specialist morphs. Each of these morphs has a specific function that is necessary for the completion of each stage of the life cycle. Typical aphid life cycles have morphs that specialize in reproduction, dispersal, and surviving severe or less favourable climatic or nutritional conditions. Not all morphs of pest species will infest crop plants. This chapter discusses how these life cycles and associated morphs influence the likelihood of aphids becoming crop pests and the importance of the different life cycles for applied entomologists.

The different aphid life cycles can have significant implications for the impact an aphid can have on crops. For example, the life cycle may determine whether a species is likely to encounter crops and the number of different crops likely to be encountered during a year. Also, the population density of an aphid species depends, in any one year, on factors such as temperature and natural enemies, but the relative importance of these factors depends on the aphid's life cycle. The morph of a particular aphid species infesting a crop also influences the degree of damage the aphid will inflict. For example,

wingless parthenogenetic morphs reproduce at a rate up to 70% greater than their winged counterparts (Noda, 1959; Dixon and Wratten, 1971), although obviously their capacity for long-range dispersal is removed. It is therefore important to understand the life cycles and morphs of aphids to determine if and how a species will damage a crop, and to assess and improve control measures.

This chapter describes first the major types of aphid life cycle and the different morphs associated with each stage of the cycle. Factors that induce the production of the morphs that cause damage to crops are then discussed. Finally, the significance of life cycle and polymorphism in determining the pest status of an aphid species is considered, together with implications for forecasting and controlling aphid populations. Examples come mainly from aphid groups, in particular the Aphidinae and Eriosomatinae, which contain the most important pests, although the life cycles of other aphid taxa are considered when they provide useful comparisons.

Types of Life Cycle

There are two major types of aphid life cycle based on how they utilize their host plants: host alternating (heteroecious) and non-host

©CAB International 2007. *Aphids as Crop Pests*
(eds H. van Emden and R. Harrington)

alternating (monoecious or autoecious). Host-alternating aphids live on one plant species in winter (primary host), migrate to an unrelated plant species (secondary host) in summer, and migrate back to the primary host in autumn. Eggs are produced on the primary host after males and sexual females have mated. Aphids that interrupt parthenogenesis with sexual reproduction in this way are termed holocyclic. Non-host-alternating aphids remain either on the same host species or migrate between closely related species throughout the year; in other words, they can produce eggs on the same group of host plant species that is fed on by all of the parthenogenetic generations. Some aphid species never produce an egg, and these are known as anholocyclic. Some species show both holocycly and anholocycly, but rarely both monoecy and heteroecy. These life cycle types are elaborated upon in this chapter, and their importance in determining the pest status of a species is discussed. There are significant crop pests in all the major categories.

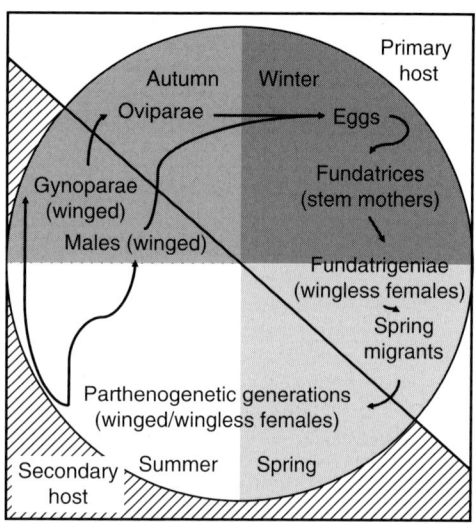

Fig. 3.1. A generalized life cycle of a host-alternating (heteroecious) aphid.

Host-alternating aphids

From an applied entomologist's point of view, host-alternating aphids are usually more important than non-host-alternating species, as the secondary host of the former is usually an herbaceous species, and frequently a crop plant. For this reason, and because of their unusual life cycle, these aphids have been particularly well studied, even though approximately only 10% of aphid species have a heteroecious life cycle. Despite its comparative rarity, host alternation is widespread across aphid groups, being found in four of the eight subfamilies of the Aphididae (the Aphidinae, Anoeciinae, Hormaphidinae, and Eriosomatinae) (Shaposhnikov, 1981), suggesting that this life cycle in aphids has evolved several times.

A generalized host-alternating life cycle for the subfamily Aphidinae is shown in Fig. 3.1. Typically, mating occurs in autumn on a primary host plant, usually a woody species. Eggs are laid and these overwinter. In spring, the eggs hatch and usually give rise to a sequence of two highly fecund,

apterous (wingless) morphs, the fundatrix, or stem mother, and the fundatrigenia. These produce spring migrants, which migrate to the secondary host, usually a herbaceous species, where they reproduce parthenogenetically through the summer. After several parthenogenetic generations, winged males and gynoparae are produced, and these migrate back to the primary host. The gynoparae produce sexual females (oviparae) on the primary host, which then mate with the males and produce the overwintering eggs. There are many variations on this general theme. For example, eggs of *Brachycaudus helichrysi* (leaf-curling plum aphid) hatch in autumn and the fundatrix overwinters. The number of generations between egg hatch and migration to the secondary host is variable in some species and may depend on the nutritional quality of the primary host.

The fundatrix usually has reduced sensory and dispersal abilities. It generally has shorter legs, cauda, siphunculi, and frequently, antennae with fewer antennal segments than other morphs. Depending on the species, the fundatrix can be winged, wingless, or brachypterous (possessing non-functional vestigial wings), and the fundatrices generally have more egg tubes (ovarioles) in their gonads (Dixon, 1975; Wellings *et al.*, 1980), enabling them to achieve higher rates of reproduction (Wellings *et al.*, 1980; Leather and Wellings, 1981). The bodies of fundatrices are more rounded than those of the other parthenogenetic morphs, and can be many times larger. Much of this morphology again is associated with achieving a high reproductive rate, which can, in extreme examples, be twenty times greater than that of later occurring parthenogenetic generations (Hille Ris Lambers, 1966). The fundatrigeniae have a similarly high reproductive rate and a similar morphology. These two morphs result in the production of large numbers of spring migrants. Of these spring migrants, it has been estimated that only 0.2–1% succeed in locating a host (Taylor, 1977; Ward *et al.*, 1998). The high reproduction on the primary host (by the fundatrix and fundatrigeniae), and later in the life cycle on the secondary hosts (by other parthenogenetic females), offset the potentially huge losses incurred during migration (Kundu and Dixon, 1995). The coexistence of abundant plants that are suitable for colonization at different times of year, and the ability of aphids to produce a number of highly reproductive generations in quick succession have resulted in aphids exploiting host alternation as a way of life (Kundu and Dixon, 1995).

During the summer on the secondary host, several parthenogenetic generations develop. It is this phase of the life cycle that usually causes damage to crops. Within these generations, there is frequently further polymorphism that is induced either by environmental factors or by direct genetic control, the latter depending on the generation. For example, in the Aphidinae, the first generation of individuals produced on the secondary host is usually wingless and often has shorter appendages and sometimes fewer antennal segments than individuals in the second and subsequent generations. These characteristics resemble those of the fundatrix, indicating a greater investment in reproduction at the expense of sensory and dispersal abilities. This is associated with a low predator pressure early in summer and, as natural enemy populations increase during the summer, the increasing investment in sensory and dispersal functions may afford some protection (Lang and Gsodl, 2001).

Aphids readily colonize new environments, such as newly emerging crops and annual plants, and rapidly increase in abundance, which predisposes them to become pests and results in them being difficult to control. Their ability to increase rapidly in abundance may also enable them to modify plant metabolism to improve host quality for themselves and subsequent generations (Prado and Tjallingii, 1997; Williams *et al.*, 1998). Such positive feedback is well known in galling aphids, but may also be a factor promoting the rapid population growth of non-galling aphids on crops.

Polymorphism in parthenogenetic wingless morphs is most marked in galling aphids. The fundatrices of these aphids often have very short appendages and a highly reduced sensory ability, and similar morphs are produced again after migration to the summer host. Enclosure within a gall makes it advantageous for aphids to reduce their dispersal and sensory ability to maximize reproduction as they are somewhat protected within the gall. Many of the species of Pemphiginae and Hormaphidinae produce soldiers, which are typically early instar nymphs and often reproductively sterile and morphologically and behaviourally specialized to defend the occupants of a gall. Surprisingly, these small and fragile soldiers, which are often present in large numbers, are very effective at protecting colonies against a wide range of insect natural enemies (Stern and Foster, 1996), and can even repair galls damaged by caterpillars (Kurosu *et al.*, 2003). Such species are, however, generally not very important as crop pests and thus will not be considered further here.

Gynoparae and males are usually winged and, when adult, differ in host-plant preference from their mothers and the spring migrants, preferring their primary rather than secondary host (Powell and Hardie, 2001; Tosh *et al.*, 2003). In the Aphidinae, the morphological differences between the winged parthenogenetic morphs and gynoparae are relatively small and generally restricted to the antennae (e.g. number of rhinaria).

The production of males and gynoparae often is temporally separated, which results in the offspring of the gynoparae, the oviparae, co-occurring in time with males. As with male aphids, oviparous females show little polymorphism and are either similar in size to the apterous parthenogenetic females produced in summer, or are smaller (Dixon *et al.*, 1998). The oviparae and males mate to produce one or more eggs. These may be laid on the host, or occasionally (e.g. in the Fordini), the oviparae die with a single egg inside their bodies. The eggs remain dormant and hatch when determined by genetic and environmental factors such as photoperiod and temperature (Via, 1992; Komazaki, 1995).

The evolutionary origins of host alternation have been the subject of much debate (Dixon, 1998). There are two main opposing hypotheses, the adaptive (or complementary) and the maladaptive (or fundatrix constraint) hypotheses. The adaptive hypothesis assumes that the significant losses incurred during migration (Taylor, 1977; Ward *et al.*, 1998) are outweighed by the increased fitness in summer compared to that if the aphid had remained on the primary host (Kundu and Dixon, 1995). The maladaptive hypothesis (Dixon, 1998) assumes that the fundatrix is so highly adapted to the primary host that the aphid has to return to the primary host (which is usually, but not always, phylogenetically older than the secondary host) as the fundatrix cannot survive on the secondary host. Whilst this debate is academic with respect to pest status, the idea central to both hypotheses is that the population growth rate on herbaceous plants in summer is significantly greater than it would be if the aphid remained on the primary host (Kundu and Dixon, 1995).

Non-host-alternating aphids

Many monoecious aphids live only on trees and are unlikely to be agricultural pests. Others, such as *Acyrthosiphon pisum* (pea aphid) and *Sitobion avenae* (grain aphid) live only on herbaceous hosts or grasses and therefore may be found on crops throughout all or part of the year. Many of the monoecious species that now live only on herbaceous plants have evolved from heteroecious

species that no longer utilize their primary host. A diagram of the monoecious life cycle is shown in Fig. 3.2. The same morphs generally are involved in the monoecious life cycle as are found in the heteroecious life cycle and their characteristics are largely as described above for heteroecious species. Here, only the particular characteristics of the morphs of monoecious species that differ from those of the corresponding morphs of heteroecious species are discussed.

The fundatrices of monoecious species are unlike those of most heteroecious species in being more similar to other morphs of the species. For example, the fundatrices of heteroecious species generally are more rotund and can have five times as many ovarioles as individuals of the following generation, whereas in most monoecious species, the number of ovarioles and their morphology may be only very slightly different between the two morphs. This high degree of specialization of heteroecious fundatrices is associated with the tendency to gall their primary host in spring. Furthermore, the spring migrants of monoecious species can also be produced over a greater time than the equivalent heteroecious generation, as the hosts of monoecious species are favourable for colonization for longer than

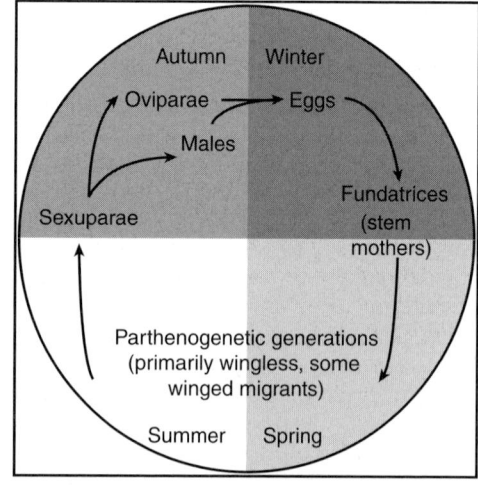

Fig. 3.2. A generalized life cycle of a non-host-alternating (monoecious) aphid from the sub-family Aphidinae.

the relatively ephemeral (either in terms of longevity or suitability for colonization) herbaceous hosts of heteroecious species.

The summer generations of monoecious species display similar characteristics to those of heteroecious species, producing sexual morphs in autumn that mate and produce overwintering eggs.

Two-year life cycles

Some aphids within the tribes Fordini and Hormaphidini do not have the annual life cycles described above. They exhibit biennial life cycles during which galls are produced on one of the host plants. Some of these galls remain closed for over a year before producing alatae that migrate to another host. Such life cycles are not discussed in detail here as these species are rarely of economic importance.

Holocyclic and anholocyclic life cycles

The life cycles described above, host alternation, non-host alternation, and two-year life cycles, are termed holocyclic; that is, they are 'complete' in that the aphids undergo sexual reproduction, produce eggs, and go through a parthenogenetic phase. However, aphids do not always complete such life cycles. Under specific circumstances, some can survive throughout the year reproducing parthenogenetically. This is termed anholocycly and can be particularly important in agricultural ecosystems. In these life cycles, the aphids survive the winter as mobile parthenogenetic females, although in temperate regions they frequently suffer exceptionally high mortality (Harrington and Cheng, 1984; Leather, 1993; Williams *et al.*, 2000).

Anholocyclic life cycles are found most frequently where the geographical range of an aphid is greater than that of its primary host plant. For example, the increase in the proportion of anholocyclic clones of *Rhopalosiphum padi* (bird cherry–oat aphid) in the UK from north to south is associated with a decline in the abundance of its primary host, bird cherry (*Prunus padus*). Another notable example is *Myzus persicae* (peach–potato aphid), a major pest of many broad-leaved crops in Northern Europe. The primary host of *M. persicae* is peach (*Prunus persica*), which is uncommon in the UK. As a result, *M. persicae* infrequently completes its full life cycle in the UK and overwinters on a large number of secondary hosts in the mobile stages (Williams *et al.*, 2000). Although mortality is considerable (Harrington and Cheng, 1984; Leather, 1993), enough adults survive the winter to colonize crops in spring in sufficient numbers for the species to become a significant pest.

Anholocycly is not confined to heteroecious species of aphids; approximately 3% of all aphid species are entirely anholocyclic throughout their range (Blackman, 1980). The summer and winter hosts of anholocyclic species are usually herbaceous.

For species to survive periods of adverse environmental conditions other than as eggs, specialist morphs may be produced (see below) or the adult females may adopt particular behavioural strategies (Dean, 1974; Dewar and Carter, 1984; McLeod, 1987). However, winter mortality of mobile morphs is generally much greater than that of eggs, as eggs are significantly more tolerant of low temperatures (Leather, 1993). The largest cause of mortality of eggs is frequently predation rather than low temperature.

There are clones of aphid species that have life cycles that are intermediate between anholocycly and holocycly. These clones produce only males or oviparae, but not both. For example, some clones of *M. persicae*, *S. avenae*, and *R. padi* that pass the winter in the mobile stages, produce males, but not oviparae. This is termed androcycly (Blackman, 1974; Simon *et al.*, 1991; Helden and Dixon, 2002).

Factors Determining the Production of Different Morphs

For an aphid either to progress from one stage of its life cycle to another or to escape environmental stress, it must respond to environmental or genetic cues either to produce offspring of a particular morph or for the early instar nymphs to develop into a specific morph.

Determination of sexual morphs

Males and mating females (oviparae) in temperate regions characteristically appear in autumn. In the case of host-alternating (heteroecious) aphids, however, asexual winged morphs are produced at this time on the secondary hosts and fly back to the primary host, where they produce either only oviparae or both oviparae and males. Those that produce only oviparae are called gynoparae and those that produce both sexes (Pemphiginae, Hormaphidinae, and Anoeciinae) are called sexuparae. That is, they appear in autumn and give rise to the sexual morphs, and therefore, although asexual, are included in this section.

The switch from parthenogenetic reproduction to the production of sexual morphs is determined by a number of environmental and genetic (intrinsic) factors that are similar for both heteroecious and monoecious species. As early as the mid-1800s, it was known that, given suitable conditions, aphids could continue indefinitely to reproduce parthogenetically. This led to the conclusion that environmental factors are involved in the production of sexual morphs.

Environmental determination

Marcovitch (1924) first demonstrated that photoperiod was important in the induction of sexual morphs, suggesting that a short photoperiod was associated with their production. However, as day length influences the growth of plants, which in turn affects the diet of aphids, it was not until the 1960s (Lees, 1961a; 1964) that the effect of photoperiod was unequivocally demonstrated to have a direct influence on aphids. Lees (1963) also demonstrated that the stimulus for the production of sexual morphs in the species he examined is actually the length of the night, rather than day length.

Temperature is also an important factor in the induction of sexual morphs. Sexual morphs are produced at shorter night lengths (longer day lengths) at lower temperatures, resulting in sexuals being produced earlier in the year when autumns are cool (Dixon and Glen, 1971). Lees (1963) calculated that for *Megoura viciae* (vetch aphid) a 5°C rise in temperature increased by 15 minutes, the length of the dark period required to induce sexual morph production.

The host plant itself can also play a role in the induction of sexual morphs. This has been demonstrated both for root-feeding aphids, such as *Pemphigus bursarius* (lettuce root aphid), which cannot be influenced directly by day length, and foliage feeders, such as *Aphis farinosa* and *Dysaphis devecta* (rosy leaf-curling aphid). These species produce sexual morphs when plant growth ceases, regardless of day length (Hille Ris Lambers, 1960; Forrest, 1970). Root-feeding aphids such as *Pemphigus betae* (sugarbeet root aphid) also produce sexual morphs in direct response to a decrease in temperature, regardless of day length (Moran *et al.*, 1993).

Intrinsic factors

When aphids emerge from eggs in the spring, the conditions experienced by the first few generations in terms of day length and temperature may be similar to those experienced by aphids in the autumn. However, an intrinsic 'clock' has been shown to delay the production of sexual morphs (Bonnemaison, 1951; Lees, 1961b). For example, the offspring of *M. viciae* fundatrices do not produce sexual morphs for 80–90 days, even when reared in conditions optimal for sexual morph production (Lees, 1960). In *Eucallipterus tiliae* (lime aphid) and *Drepanosiphum platanoidis* (sycamore aphid), the proportion of sexual morphs produced in each generation gradually increases; even when environmental conditions are kept constant.

The occurrence of males of some pest species in spring is relatively common in the UK, for example in the case of *Brevicoryne brassicae* (cabbage aphid) and *Metopolophium dirhodum* (rose–grain aphid) (Taylor *et al.*, 1998). This is contrary to what one would expect if an intrinsic clock prevented the early production of sexual morphs. As the occurrence of males in spring is more common after mild winters, this phenomenon is likely to be due to virginoparae from androcyclic clones surviving the winter rather than to virginoparae from holocyclic clones that passed the winter in the egg stage. As the intrinsic clocks of the mothers of spring males seem not to be operating, such clocks may only operate after the egg stage.

Determination of asexual winged morphs

The determination of winged morphs has long been known to be associated with crowding and the nutritional status of plants. The adaptive significance of this is clear, as under either of these conditions the production of winged morphs would presumably increase the chances of clonal survival as it facilitates escape from poor or depleted resources.

Crowding

Wadley (1923) was first to demonstrate experimentally that crowding was associated with the production of winged morphs. The stimulus appears to be physical contact between individual aphids (Johnson, 1965). A series of experiments in the 1960s showed that winged morphs could not only be induced by forcing individual aphids to come into contact with each other, but also by a range of physical stimuli including touching the abdomen with an inert object (Johnson, 1965; Lees, 1967; Auclair and Aroga, 1984).

However, not all aphid species respond in this manner, or at least do not have the same degree of sensitivity to such stimuli. Experiments with *M. persicae* show that this species does not always increase production of winged morphs in response to crowding (Awram, 1968; Williams *et al.*, 2000). Müller *et al.* (2001) postulated that this might be due to non-gregarious species such as *M. persicae* being less responsive to the crowding stimulus than the more gregarious species such as *A. fabae* and cereal aphids. This may be because, for less gregarious species, the crowding response is rarely present in the wild. Müller *et al.* (2001) surveyed 32 studies testing the influence of crowding on wing production in aphids. Experiments tended to show no increase in winged morph production when aphids were crowded in field-like conditions compared to isolated individuals. Evidence for aphids producing more winged morphs when crowded was strongest when aphids were reared in artificial conditions that ensured physical contact between individuals. This suggested that, whilst physical contact is a clear stimulus for winged morph production, more gregarious species are more responsive to such a stimulus.

Nutrition

The nutritional quality of a plant has also long been known to be associated with winged morph production (e.g. Wadley, 1923; Evans, 1938; Bonnemaison, 1951; Mittler and Dadd, 1966; Dixon and Glen, 1971). Artificial diets have been used to determine the effect of particular nutrients, such as folic acid (Raccah *et al.*, 1973), sugars (Raccah *et al.*, 1972), vitamins (Mittler and Kleinjan, 1970), amino acids (Mittler and Dadd, 1966; Dadd, 1968; Mittler and Sutherland, 1969; Leckstein and Llewellyn, 1973; Harrewijn, 1976), antibiotics (Mittler, 1971), and chemical 'tokens' or 'apterous promoting principles' (e.g. lithium) (Harrewijn, 1978) on the induction of winged morphs.

Whilst these studies show that certain dietary components are important for wing induction in parthenogenetic generations, it is the overall diet quality and/or crowding that is likely to determine their production. It is clear that when nutritional quality is low, production of winged morphs can be advantageous. This is supported by many studies (Mittler and Kleinjan, 1970; Sutherland and Mittler; 1971; Harrewijn, 1973, 1976; Mittler, 1972), although not by others (Mittler and Dadd, 1966; Dadd, 1968; Mittler and Sutherland, 1969; Schaefers and Judge, 1971; Harrewijn, 1973, 1976; Leckstein and Llewellyn, 1973; Schaefers and Montgomery, 1973; Williams *et al.*, 2000) (see Müller *et al.*, 2001 for a review of these studies).

The contradictory nature of these results may arise because the advantage of winged morph production is dependent on the predictability of future habitat quality. Such predictability is seen on an annual basis in trees, where the nutritional quality for aphids is high in spring, low in summer, and high again in autumn (Kundu and Dixon, 1995). This cycle of nutritional quality also occurs in individual leaves of annual plants where aphids have high growth and reproduction rates on very young and senescing leaves but not on mature leaves (Fig. 3.3) (Kennedy et al., 1950). The relative success of a strategy of 'sitting out' the period of poor nutritional quality and a strategy of producing winged morphs and moving depends partly on the risks associated with finding a better host. The latter depends on factors such as the abundance of the host plant, the degree of polyphagy of the aphid species, and the likely future quality of the current host.

Natural enemies, pathogens, and mutualists

The induction of winged morphs by natural enemies and pathogens is a relatively new area of research, but one that has significant implications in the context of aphids as crop pests (see Irwin et al., Chapter 7 this volume). The extent to which natural enemies induce or inhibit wing morph production would have implications for potential biological control strategies. Aphids spread plant pathogens, especially viruses, and

therefore any influence on morph induction would influence disease epidemiology.

Most species of aphid are attacked by parasitoids. Early aphid instars are more susceptible to parasitoid attack than later instars with, in the former case, the aphid being killed shortly before the parasitoid larva pupates. A few studies have assessed the influence of parasitoid attack on wing morph induction, the consensus of these studies being that parasitization reduces the production of winged morphs (Johnson, 1959; Liu and Hughes, 1984; Christiansen-Weniger and Hardie, 1998, 2000). The mechanism of reducing winged morph production is most likely to be through the composition of the calyx fluid released by the parasitoid into the haemolymph of the aphid during oviposition (Christiansen-Weniger and Hardie, 1998, 2000).

Whilst there is no evidence that aphids increase the production of winged morphs when attacked by parasitoids, there is evidence of them doing so when attacked by predatory beetles. This has been observed in A. pisum when attacked by ladybirds (Dixon and Agarwala, 1999; Weisser et al., 1999), although not when two other aphid species, A. fabae and M. viciae, were attacked by the same predatory species. This is likely to be due to these aphids having evolved other mechanisms that reduce their vulnerability to predators. For example, A. fabae is frequently ant attended (see below) and M. viciae is unpalatable to natural enemies.

If plant pathogens were to increase the proportion of winged morphs produced in an aphid population, then this would facilitate dispersal of the pathogens and clearly would have implications for their epidemiology. The relationship between virus infection in plants and winged morph production has been studied in many crops. The aphids, M. persicae, S. avenae, R. padi, and Aphis gossypii (cotton or melon aphid) have all been shown to increase their production of winged morphs when feeding on virus-infected plants (Macias and Mink, 1969; Gildow, 1980; Williams, 1995). However, it is unclear whether this is due to incidental changes in the nutritional quality of the phloem sap available to the aphids (Williams, 1995) or to

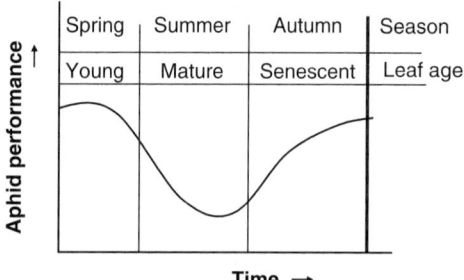

Fig. 3.3. The host quality of individual leaves of herbaceous plants during the summer and of trees for monoecious species (from Kennedy et al., 1950; Kundu and Dixon, 1995).

virus particles affecting the aphid's endocrine system.

The indirect role that mutualistic ants may have in reducing the production of winged morphs by mitigating the influence of ladybird attack (Dixon and Agarwala, 1999; Weisser *et al.*, 1999) has already been mentioned. Ant attendance may also directly reduce the production of winged morphs (El-Ziady and Kennedy, 1956; El-Ziady, 1960; Seibert, 1992) by improving the health of the colony by reducing fungal growth through the removal of honeydew (Way, 1963) and by stimulating feeding (El-Ziady, 1960). However, ant attendance in annual crops is rarely observed and therefore its significance in this context is likely to be minimal.

Other factors

Temperature certainly influences the production of sexual morphs as described above, but may also influence the morph of asexual generations, with higher temperatures tending to favour the production of wingless forms (e.g. Johnson, 1966; Lees, 1967; Schaefers and Judge, 1971; Liu, 1994). However, temperature is considered less important than either crowding or nutrition in determining a change in morph (Müller *et al.*, 2001).

Diapausing morphs

Diapausing morphs of aphids occur in summer (aestivation) and winter (hibernation), but this happens infrequently in pest species. As mentioned above, the nutritional quality of many woody hosts is poor for aphids during the summer months. Non-host-alternating species that remain on a woody host throughout the year have adopted numerous strategies to survive during the summer. For example, when the growth of its host slows down, *D. platanoidis* produces adults with poorly developed gonads and a large fat body to see them through the summer when food quality is poor and crowding often acute (Dixon, 1975). Other species produce more morphologically distinct aestivating morphs. For example, *Periphyllus testudinaceus* produces morphs that are flattened and covered with plates that protect them from natural enemy attack. Many of these morphs are much smaller than other stages of the life cycle and this may afford them some protection from natural enemy attack. The aestivating morphs of *Thelaxes dryophila* are sufficiently small to conceal themselves in the forks of leaf veins.

Hibernating morphs are termed hiemalis. These tend to have similar characteristics to aestivating morphs, such as large fat bodies and low reproductive capacity (Judge, 1967; Sutherland, 1969).

Eggs are particularly well adapted to survive periods of low temperature and clearly, once laid, they have no further nutritional requirements. For example, the eggs of *Rhopalosiphum insertum* (apple–grass aphid) can survive temperatures of –40°C (James and Luff, 1982). In addition, the termination of egg diapause and subsequent hatching are triggered by similar cues to those triggering the onset of plant growth in spring (Lushai *et al.*, 1996; Ro *et al.*, 1998). Thus, eggs are a very effective way of surviving winter in temperate regions.

The cues for the production of these diapausing morphs vary. For example, in *P. bursarius*, temperatures below 10°C stimulate the production of the hiemalis. A combination of food quality, crowding, and genetic factors produces the aestivating morphs in *D. platanoidis*.

Colour polymorphism

The colour of aphids can be due to pigmented internal tissues, pigmentation, or sclerotization of the cuticle or waxy exudates like those of *B. brassicae*. Colour either can be genetically fixed or environmentally triggered. Environmentally determined colour variation within a clone is usually reversible if conditions go back to their previous state.

Temperature and season are frequently associated with changes in aphid colour. For example, the highly colour-polymorphic *A. gossypii* is pale green at 25°C and dark green or dark brown at 12°C. Similarly, *D. platanoidis* develops dark pigmented dorsal bands on its abdomen when reared at low temperatures and red forms develop at high temperatures. Although this is tempered by genetic factors, as the offspring of fundatrices are green regardless of temperature, the influence of temperature becomes apparent only in later generations (Dixon, 1972a). Frequently, different morphs of the same species are also a different colour. For example, in *R. padi*, fundatrices are emerald green and fundatrigeniae are dark brown. The sexual morphs of many species are often a different colour from virginoparae.

Photoperiod influences the colour of *S. avenae*. The red form produces a larger proportion of green offspring when reared under short day lengths and low light intensities compared to the high proportion of red offspring produced under the reverse conditions (Markkula and Rautapää, 1967). Crowding can also induce colour change. The red form of *A. pisum*, for example, becomes pale green when reared under crowded conditions for a number of generations (Müller, 1961).

The nutritional quality of the diet can influence aphid colour. Green *M. persicae* produce pink offspring when reared on poor quality hosts (Williams *et al.*, 2000) and red *M. persicae* produce green offspring when crowded on poor diets (Ueda and Takada, 1977).

As little is known about the adaptive significance of different colour morphs, the implications for crop protection of environmentally determined colour polymorphism are not clear. However, when individuals of the same species can differ depending on environmental conditions, there are implications for the identification of pest species. Colour morphs can also be used to indicate stress in an aphid population, for example in *M. persicae* (Williams *et al.*, 2000), and therefore indicate when a population is likely to decline.

Colour within a species can be a genetic trait, as is well documented for *A. pisum*, *M. persicae*, and various *Macrosiphum* species. The adaptive significance of genotypic colour polymorphism is better understood than in the case of environmentally determined variation in colour, and may be useful for predicting the potential damage that may be caused by a colonizing aphid clone. For example, different coloured clones can have different reproduction rates, host preferences, feeding sites, tolerances of extreme temperatures, susceptibility to natural enemy attack, or other behavioural characteristics (Michaud and Mackauer, 1994; Battaglia *et al.*, 2000; Weisser and Braendle, 2001; Dromph *et al.*, 2002). However, why aphids often are not cryptically coloured is not clear, although tolerance of temperature extremes in pigmented aphids through their ability to absorb solar radiation and so increase their body temperature has been shown for some species (Markkula and Rautapää, 1967; Dixon, 1972b). Furthermore, brown forms of *S. avenae* are more protected against the damaging effects of ultraviolet radiation than are green forms (Thieme, 1997; Thieme and Heimbach, 1998). Colour has also been viewed as a form of defence from predators (Stroyan, 1949; Losey *et al.*, 1997).

The Importance of Different Morphs for Crop Protection

Pest status

The importance of an aphid as a pest depends partly on how it damages a crop. If the damage is due to direct feeding, the morphs with the greatest reproduction rate cause the most damage. Apterous virginoparae can have up to a 70% greater reproductive output than their winged counterparts (Dixon and Wratten, 1971). Even among the Drepanosiphinae, of which parthenogenetic generations are always winged (Heie, 1987), morphs with reduced wings (brachypterous), which are incapable of flight, are more fecund than those with fully developed wings (macropterous) (Dixon, 1972c).

Virus transmission

Virus transmission is considered in detail by Katis *et al.*, Chapter 14 this volume. The morphs most important for virus spread depend on the spatial scale being considered. Many viruses are brought into crops by winged aphids (primary infection), but the morph most important for spreading the virus within a crop (secondary spread) has been the subject of debate. Initially, winged aphids were considered to spread *Beet yellows virus* within fields of sugarbeet (Watson and Healy, 1953). However, more recent work has shown that wingless aphids walking from plant to plant are more likely to be the major source of within-field virus spread (Ribbands, 1964; Jepson and Green, 1983; Williams *et al.*, 2000). Molecular and immunological techniques that can detect virus particles in individual aphids (Harrington *et al.*, Chapter 19 this volume) are now used to identify the morphs important in virus epidemiology. Furthermore, the quantity of virus can be measured using PCR techniques.

Forecasting

Forecasting aphid outbreaks is discussed by Harrington *et al.*, Chapter 19 this volume. However, it is worth mentioning here that the life cycle and factors affecting morph determination of aphids can have a significant impact on the ability to forecast the timing and severity of aphid infestation of crops. For anholocyclic aphids, the population size in spring and early summer, and therefore potential severity of crop damage during particularly susceptible growth stages, can be predicted from winter temperatures (Watson *et al.*, 1975; Harrington *et al.*, 1989, 1990). However, this is not the case for species that are holocyclic in the region of concern. The forecasting of many aphid pests that are holocyclic, and therefore overwinter as eggs, is more complex as the winter mortality of such species is likely to be lower and certainly less dependent on temperature than that of anholocyclic species. However, the temperature in early spring, when the fundatrix and spring migrants are developing and reproducing, is more likely to be an important factor in determining the population size of such species in any particular summer. Several forecasts exist for holocyclic pest species, including *A. pisum*

(McVean *et al.*, 1999), *Phorodon humuli* (damson–hop aphid) (Jastrzebski and Solarska, 1998) and *A. fabae* (Thacker *et al.*, 1997).

The Importance of Polymorphism in Determining Pest Status

The above is an account of polymorphism from an applied entomologist's point of view. However, to what degree is polymorphism responsible for the pest status of aphids? By comparing the proportion of pest species in more or less polymorphic taxa, an attempt can be made to assess the role of polymorphism in determining the taxa's pest status.

Aphids can be compared with other insect taxa by determining the proportions of aphids and of other taxa that are pests. For this analysis, an aphid pest is defined as a species listed by Blackman and Eastop (2000). These range from extremely well known prolific pests such as *A. gossypii* to those where there is only one record of a species feeding on a crop. The list comprises 183 species. Remaudière and Remaudière (1997) estimated that there is a total of about 4700 aphid species. Therefore, about 3.9% of aphid species are pests (this excludes aphid vectors of non-persistent viruses). This proportion was compared to that for the Lepidoptera. Members of this insect order, like aphids, are predominantly herbivorous and thus have the potential to become crop pests. However, they are considerably less polymorphic than the aphids. It would be expected, therefore, that if polymorphism were the prime cause of aphids being such prolific pests there would be proportionally fewer pests among the Lepidoptera than among the aphids. The number of lepidopteran species that are found on crops is approximately 6000 (Zhang, 1994). This figure includes a few beneficial species, although the majority of species listed, taken from the Arthropod Name Index (ANI) database of CABI, are pests. There are approximately 150,000 species of Lepidoptera. Therefore, an estimation of the proportion of lepidopteran

Table 3.1. Proportion of pest species in aphid genera with predominantly monoecious or heteroecious life cycles.

Genus	No. of species (approximate)	No. of pest species	% pests	Predominant life cycle
Acyrthosiphon	100	12	12	Monoecious
Aphis	400	43	11	Heteroecious
Aulacorthum	50	8	16	Heteroecious
Brachycaudus	45	7	16	Heteroecious[a]
Dysaphis	100	16	16	Heteroecious
Greenidea	45	6	13	Monoecious
Illinoia	45	9	20	Monoecious
Macrosiphoniella	120	4	3	Monoecious
Macrosiphum	120	21	18	Monoecious[b]
Myzus	55	13	24	Heteroecious
Nasonovia	45	4	9	Monoecious[c]
Phylloxera	60	6	10	Monoecious
Pleotrichophorus	60	1	2	Monoecious
Prociphilus	50	2	4	Heteroecious
Schizaphis	40	5	13	Monoecious
Sitobion	75	17	23	Monoecious[c]
Uroleucon	180	9	5	Monoecious

[a]Mostly heteroecious, although some monoecious species are known.
[b]*Macrosiphum* species are generally monoecious, although some, including important pest species such as *Macrosiphum euphorbiae*, are heteroecious.
[c]Mostly monoecious, although some heteroecious species are known.

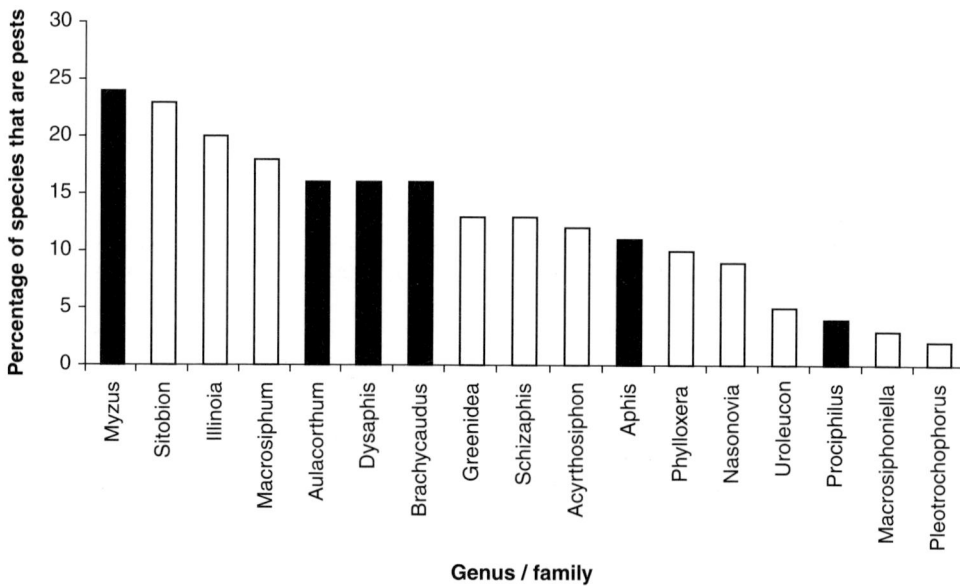

Fig. 3.4. The proportion of aphid species that are pests for aphid genera with more than 30 species, at least one of which is listed in Blackman and Eastop (2000). Heteroecious genera, solid bars; monoecious genera, white bars.

species that are pests is 4%. This is similar to the proportion of aphid species estimated to be pests.

To further consider the hypothesis that polymorphism predisposes insects to become pests, each aphid genus with more than 30 species with at least one species listed in Blackman and Eastop (2000) has been examined (Table 3.1). Host alternation was used as a measure of polymorphism, as heteroecious species are generally more polymorphic than monoecious species. To reject the null hypothesis, it would be expected that heteroecious genera would have a greater proportion of pest species than monoecious genera. Figure 3.4 shows that this is not the case, with 16% of heteroecious and 11% of monoecious species being pests ($\chi^2 = 3.75$; d.f. = 1; n.s.). This analysis suggests that polymorphism is not responsible for certain species of aphid being such significant pests.

Conclusions

Although it is widely accepted that the pest status of aphids is associated with their high degree of polymorphism, there appears to be no theoretical reason why this should be so. There is no evidence to suggest that there are proportionally more pest species of aphid amongst the more polymorphic aphid taxa than the less polymorphic taxa. Further, in comparison with other less polymorphic non-aphid taxa, there is no analysis that indicates that proportionally more aphids are pests.

Aphids can damage crops in a number of different ways; for example, by directly removing plant sap, through toxins in the saliva and indirectly by transmitting viruses. These types of damage can also occur at the same time caused by the same individual aphids. The more aphids present, the greater will be the direct damage. The most important factor in determining aphid abundance on a crop is the aphid's rate of increase. In terms of virus damage, aphid abundance is not as important as the ease with which they move from plant to plant. Therefore, the two features of aphids that are the likely determinants of their pest status are their prolific rate of increase and impressive powers of dispersal, both of which determine their abundance.

References

Auclair, J.L. and Aroga, R. (1984) Influence de l'effet de groupe et de la qualité de la plante-hôte sur le cycle évolutif de quatre biotypes du puceron du pois, *Acyrthosiphon pisum*. *Canadian Journal of Zoology* 62, 608–612.

Awram, W.J. (1968) Effects of crowding on wing morphogenesis in *Myzus persicae* Sulz. (Aphididae; Homoptera). *Quaestiones Entomologicae* 4, 3–29.

Battaglia, D., Poppy, G., Powell, W., Romano, A., Tranfaglia, A. and Pennacchio, F. (2000) Physical and chemical cues influencing the oviposition behaviour of *Aphidius ervi*. *Entomologia Experimentalis et Applicata* 94, 219–227.

Blackman, R.L. (1974) Life-cycle variation of *Myzus persicae* (Sulz.) (Hom., Aphididae) in different parts of the world, in relation to genotypes and environment. *Bulletin of Entomological Research* 63, 595–607.

Blackman, R.L. (1980) Chromosome numbers in the Aphididae and their taxonomic significance. *Systematic Entomology* 5, 7–25.

Blackman, R.L. and Eastop, V.F. (2000) *Aphids on the World's Crops: An Identification and Information Guide*, 2nd edn. Wiley, Chichester, 466 pp.

Bonnemaison, L. (1951) Contribution a l'étude des facteurs provoquant l'apparition des formes ailées et sexuées chez les Aphidinae. *Annales des Epiphyties* 2, 1–380.

Christiansen-Weniger, P. and Hardie, J. (1998) Wing development in parasitised male and female *Sitobion fragariae*. *Physiological Entomology* 23, 208–213.

Christiansen-Weniger, P. and Hardie, J. (2000) The influence of parasitism on wing development in male and female pea aphids. *Journal of Insect Physiology* 46, 861–867.

Dadd, R.H. (1968) Dietary amino acids and wing determination in *Myzus persicae*. *Annals of the Entomological Society of America* 82, 203–208.

Dean, G.J. (1974) The overwintering and abundance of cereal aphids. *Annals of Applied Biology* 76, 1–7.

Dewar, A.M. and Carter, N. (1984) Decision trees to assess the risk of cereal aphid (Hemiptera: Aphididae) outbreaks in summer in England. *Bulletin of Entomological Research* 74, 387–398.

Dixon, A.F.G. (1972a) Crowding and nutrition in the induction of macropterous alatae in *Drepanosiphum dixoni*. *Journal of Insect Physiology* 18, 459–464.

Dixon, A.F.G. (1972b) Control and significance of the seasonal development of colour forms in sycamore aphid, *Drepanosiphum platanoides* (Schr.). *Journal of Animal Ecology* 41, 689–697.

Dixon, A.F.G. (1972c) Fecundity of brachypterous and macropterous alatae in *Drepanosiphum dixoni* (Callaphididae, Aphididae). *Entomologia Experimentalis et Applicata* 15, 335–340.

Dixon, A.F.G. (1975) Seasonal changes in fat content, form, state of gonads and length of adult life in the sycamore aphid, *Drepanosiphum platanoides* (Schr.). *Transactions of the Royal Entomological Society of London* 127, 87–99.

Dixon, A.F.G. (1998) *Aphid Ecology*, 2nd edn. Chapman and Hall, London, 300 pp.

Dixon, A.F.G. and Agarwala, B.K. (1999) Ladybird-induced life-history changes in aphids. *Proceedings of the Royal Society of London B* 266, 1549–1553.

Dixon, A.F.G. and Glen, D.M. (1971) Morph determination in the bird cherry–oat aphid, *Rhopalosiphum padi* L. *Annals of Applied Biology* 68, 11–21.

Dixon, A.F.G. and Wratten, S.D. (1971) Laboratory studies on aggregation, size and fecundity in the black bean aphid, *Aphis fabae* Scop. *Bulletin of Entomological Research* 61, 97–111.

Dixon, A.F.G., Holman, J. and Thieme, T. (1998) Sex and size in aphids. In: Nieto Nafría, J.M. and Dixon, A.F.G. (eds) *Aphids in Natural and Managed Ecosystems*. Universidad de León, León, pp. 173–178.

Dromph, K.M., Pell, J.K. and Eilenberg, J. (2002) Influence of flight and colour morph on susceptibility of *Sitobion avenae* to infection by *Erynia neoaphis*. *Biocontrol Science and Technology* 12, 753–756.

El-Ziady, S. (1960) Further effects of *Lasius niger* L. on *Aphis fabae* Scopoli. *Proceedings of the Royal Entomological Society of London A* 35, 1–3.

El-Ziady, S. and Kennedy, J.S. (1956) Beneficial effects of the common garden ant, *Lasius niger* L., on the black bean aphid, *Aphis fabae* Scopoli. *Proceedings of the Royal Entomological Society London A* 31, 4–6.

Evans, A.C. (1938) The effect of the chemical composition of the plant on reproduction and production of winged forms in *Brevicoryne brassicae* L. *Annals of Applied Biology* 25, 558–572.

Forrest, W.A. (1970) The effects of maternal and larval experience on morph determination in *Dysaphis devecta*. *Journal of Insect Physiology* 15, 2179–2201.

Gildow, F.E. (1980) Increased production of alatae by aphids reared on oats infected by barley dwarf virus. *Annals of the Entomological Society of America* 73, 343–347.

Harrewijn, P. (1973) Function significance of indole alkylamines linked to nutritional factors in wing development of the aphid *Myzus persicae*. *Entomologia Experimentalis et Applicata* 16, 499–513.

Harrewijn, P. (1976) Host-plant factors in regulating wing production in *Myzus persicae*. *Symposium of Biology Hungary* 16, 79–83.

Harrewijn, P. (1978) The role of plant substances in polymorphism of the aphid *Myzus persicae*. *Entomologia Experimentalis et Applicata* 24, 198–214.

Harrington, R. and Cheng, X.N. (1984) Winter mortality, development and reproduction of a field population of *Myzus persicae* (Sulzer) (Hemiptera: Aphididae) in England. *Bulletin of Entomological Research* 74, 633–640.

Harrington, R., Dewar, A.M. and George, B. (1989) Forecasting the incidence of virus yellows in sugar beet in England. *Annals of Applied Biology* 114, 459–469.

Harrington, R., Tatchell, G.M. and Bale, J.S. (1990) Weather, life cycle strategy and spring populations of aphids. *Acta Phytopathologica et Entomologica Hungarica* 25, 423–432.

Heie, O. (1987) Palaeontology and phylogeny. In: Minks, A.K. and Harrewijn, P. (eds) *Aphids. Their Biology, Natural Enemies and Control, Volume 2A*. Elsevier, Amsterdam, pp. 367–391.

Helden, A.J. and Dixon, A.F.G. (2002) Life-cycle variation in the aphid *Sitobion avenae*: costs and benefits of male production. *Ecological Entomology* 27, 692–701.

Hille Ris Lambers, D. (1960) Some notes on morph determination in aphids. *Entomologie Berlin* 20, 110–113.

Hille Ris Lambers, D. (1966) Polymorphism in Aphididae. *Annual Review of Entomology* 11, 47–78.

James, B.D. and Luff, M.L. (1982) Cold-hardiness and development of eggs of *Rhopalosiphum insertum*. *Ecological Entomology* 7, 277–282.

Jastrzebski, A. and Solarska, E. (1998) The effect of weather conditions on spring migration of damson hop aphids. In: Nieto Nafría, J.M. and Dixon, A.F.G. (eds) *Aphids in Natural and Managed Ecosystems*. Universidad de León, León, pp. 579–584.

Jepson, P.C. and Green, R.E. (1983) Prospects for improving control strategies for sugar beet pests in England. *Advances in Applied Biology* 7, 175–250.

Johnson, B. (1959) Effect of parasitization by *Aphidius platensis* Brèthes on the developmental physiology of its host, *Aphid craccivora* Koch. *Entomologia Experimentalis et Applicata* 2, 82–99.

Johnson, B. (1965) Wing polymorphism in aphids II. Interaction between aphids. *Entomologia Experimentalis et Applicata* 8, 49–64.

Johnson, B. (1966) Wing polymorphism aphids III. The influence of the host plant. *Entomologia Experimentalis et Applicata* 9, 213–222.

Judge, F.D. (1967) Overwintering in *Pemphigus bursarius* (L.). *Nature, London* 216, 1041–1042.

Kennedy, J.S., Ibbotson, A. and Booth, C.O. (1950) The distribution of aphid infestation in relation to leaf age I. *Myzus persicae* (Sulz.) and *Aphis fabae* (Scop.) on spindle trees and sugar beet plants. *Annals of Applied Biology* 37, 651–679.

Komazaki, S. (1995) Collection and hatching of the aphid, *Aphis spiraecola* Patch, eggs in the laboratory. *Applied Entomology and Zoology* 30, 97–101.

Kundu, R. and Dixon, A.F.G. (1995) Evolution of complex life-cycles. *Journal of Animal Ecology* 64, 245–255.

Kurosu, U., Aoki, S. and Fukatsu, T. (2003) Self-sacrificing gall repair by aphid nymphs. *Proceedings of the Royal Society of London B* 270 Supplement 1, S12–S14.

Lang, A. and Gsodl, S. (2001) Prey vulnerability and active predator choice as determinants of prey selection: a carabid beetle and its aphid prey. *Journal of Applied Entomology* 125, 53–61.

Leather, S.R. (1993) Overwintering in six arable aphid pests: a review with particular reference to pest management. *Journal of Applied Entomology* 116, 217–233.

Leather, S.R. and Wellings, P.W. (1981) Ovariole number and fecundity in aphids. *Entomologia Experimentalis et Applicata* 30, 128–133.

Leckstein, P.M. and Llewellyn, M. (1973) Effect of dietary amino acids on the size and alary polymorphism of *Aphis fabae*. *Journal of Insect Physiology* 19, 973–980.

Lees, A.D. (1960) The role of photoperiod and temperature in the determination of parthenogenetic and sexual forms in the aphid *Megoura viciae* Buckton – II. The operation of the 'interval timer' in young clones. *Journal of Insect Physiology* 4, 154–175.

Lees, A.D. (1961a) Clonal polymorphism in aphids. In: Kennedy, J.S. (ed.) *Insect Polymorphism. Proceedings of the Royal Entomological Society Symposium No. 1*. Royal Entomological Society, London, pp. 261–280.

Lees, A.D. (1961b) Aphid clocks. *New Scientist* 355, 148–150.

Lees, A.D. (1963) The role of photoperiod and temperature in the determination of parthenogenetic and sexual forms in the aphid *Megoura viciae* Buckton – III. Further properties of the maternal switching mechanism in apterous aphids. *Journal of Insect Physiology* 9, 153–164.

Lees, A.D. (1964) The location of the photoperiodic receptors in the aphid *Megoura viciae* Buckton. *Journal of Experimental Biology* 41, 119–133.

Lees, A.D. (1967) The production of the apterous and alate forms in the aphid *Megoura viciae* Buckton. *Journal of Insect Physiology* 13, 289–318.

Liu, S.-S. (1994) Production of alatae in response to low temperature in aphids: a trait of seasonal adaptation. In: Danks, H.V. (ed.) *Insect Life-cycle Polymorphism*. Kluwer, Dordrecht, pp. 245–261.

Liu, S.-S. and Hughes, R.D. (1984) Effect of host age at parasitization by *Aphidius sonchi* on the development, survival, and reproduction of the sowthistle aphid, *Hyperomyzus lactucae*. *Entomologia Experimentalis et Applicata* 36, 239–246.

Losey, J.E., Ives, A.R., Harmon, J., Ballantyne, F. and Brown, C. (1997) A polymorphism maintained by opposite patterns of parasitism and predation. *Nature, London* 288, 269–272.

Lushai, G., Hardie, J. and Harrington, R. (1996) Diapause termination and egg hatch in the bird cherry aphid, *Rhopalosiphum padi*. *Entomologia Experimentalis et Applicata* 81, 113–115.

Macias, W. and Mink, G.I. (1969) Preference of green peach aphids for virus-infected sugarbeet leaves. *Journal of Economic Entomology* 62, 28–29.

Marcovitch, S. (1924) The migration of the Aphididae and the appearance of the sexual forms as affected by the relative length of daily light exposure. *Journal of Agricultural Research* 27, 513–533.

Markkula, M. and Rautapää, J. (1967) The effect of light and temperature on the colour of the English grain aphid *Macrosiphum avenae* (F.) (Hom., Aphididae). *Annales Entomologici Fennici* 33, 1–13.

McLeod, P. (1987) Effect of low temperature on *Myzus persicae* (Homoptera: Aphididae) on overwintering spinach. *Environmental Entomology* 16, 796–801.

McVean, R.I.K., Dixon, A.F.G. and Harrington, R. (1999) Causes of regional and yearly variation in pea aphid numbers in eastern England. *Journal of Applied Entomology* 123, 495–502.

Michaud, J.P. and Mackauer, M. (1994) The use of visual cues in host evaluation by Aphidiid wasps. 1. Comparison between three Aphidius parasitoids of the pea aphid. *Entomologia Experimentalis et Applicata* 70, 273–283.

Mittler, T.E. (1971) Some effects on the aphid *Myzus persicae* of ingesting antibiotics incorporated into artificial diets. *Journal of Insect Physiology* 17, 1333–1347.

Mittler, T.E. (1972) Aphid polymorphism as affected by diet. In: Lowe, A.D. (ed.) *Perspectives in Aphid Biology.* Entomological Society of New Zealand, Auckland, pp. 65–75.

Mittler, T.E. and Dadd, R.H. (1966) Food and wing determination in *Myzus persicae* (Homoptera: Aphididae). *Annals of the Entomological Society of America* 59, 1162–1166.

Mittler, T.E. and Kleinjan, J.E. (1970) Effect of artificial diet composition on wing-production by the aphid *Myzus persicae. Journal of Insect Physiology* 16, 833–850.

Mittler, T.E. and Sutherland, O.R.W. (1969) Dietary influences in aphid polymorphism. *Entomologia Experimentalis et Applicata* 12, 703–713.

Moran, N., Seminoff, J. and Johnstone, L. (1993) Induction of winged sexuparae in root-inhabiting colonies of the aphid *Pemphigus betae. Physiological Entomology* 18, 296–302.

Müller, C.B., Williams, I.S. and Hardie, J. (2001) The role of nutrition, crowding and interspecific interactions in the development of winged aphids. *Ecological Entomology* 26, 330–340.

Müller, F.D. (1961) Stabilität und Veränderlichkeit der Farbung bei Blattläusen. *Archiv der Freude der Naturgeschichte in Mecklenburg* 7, 228–239.

Noda, I. (1959) The emergence of winged viviparous female in aphid. VII. On the rareness of the production of the winged offsprings from the mothers of the same form. *Japanese Journal of Applied Entomology and Zoology* 3, 272–280.

Powell, G. and Hardie, J. (2001) The chemical ecology of aphid host alternation: how do return migrants find the primary host plant? *Applied Entomology and Zoology* 36, 259–267.

Prado, E. and Tjallingii, W.F. (1997) Effects of previous plant infestation on sieve element acceptance by two aphids. *Entomologia Experimentalis et Applicata* 82, 189–200.

Raccah, B., Tahori, A.S. and Applebaum, S.W. (1972) Effect of various sugars on development and wing production in the aphid *Myzus persicae. Israel Journal of Entomology* 7, 21–25.

Raccah, B., Applebaum, A.S. and Tahori, A.S. (1973) The role of folic acid in the appearance of alate forms in *Myzus persicae. Journal of Insect Physiology* 19, 1849–1855.

Remaudière, G. and Remaudière, M. (1997) *Catalogue of the World's Aphididae Homoptera Aphidoidea.* Institute National de la Researche Agronomique, Paris, 473 pp.

Ribbands, C.R. (1964) The spread of apterae of *Myzus persicae* (Sulz.) and of yellow viruses within a sugar beet crop. *Bulletin of Entomological Research* 54, 267–283.

Ro, T.H., Long, G.E. and Toba, H.H. (1998) Predicting phenology of green peach aphid (Homoptera: Aphididae) using degree-days. *Environmental Entomology* 27, 337–343.

Schaefers, G.A. and Judge, F.D. (1971) Effects of temperature, photoperiod, and host plant on alary polymorphism in the aphid, *Chaetosiphon fragaefolii. Journal of Insect Physiology* 17, 365–379.

Schaefers, G.A. and Montgomery, M.E. (1973) Influence of cytokinin (N6 benzyladenine) on development and alary polymorphism in strawberry aphid, *Chaetosiphon fragaefolii. Annals of the Entomological Society of America* 66, 1115–1119.

Seibert, T.F. (1992) Mutualistic interactions of the aphid *Lachnus allegheniensis* (Homoptera: Aphididae) and its tending ant *Formica obscuripes* (Hymenoptera: Formicidae). *Annals of the Entomological Society of America* 85, 173–178.

Shaposhnikov, G.C. (1981) *Populations and Species in Aphids and the Need for a Universal Species Concept.* Special Publication of the Research Branch, Agriculture, Canada, 61 pp.

Simon, J.C., Blackman, R.L. and le Gallic, J.F. (1991) Local variability in the life-cycle of the bird cherry–oat aphid, *Rhopalosiphum padi* (Homoptera, Aphididae) in Western France. *Bulletin of Entomological Research* 81, 315–322.

Stern, D.L. and Foster, W.A. (1996) The evolution of soldiers in aphids. *Biological Reviews* 71, 27–79.

Stroyan, H.L.G. (1949) The occurrence and dimorphism in Britain of *Metopeurum fuscoviride* nom. N. (*Pharalis tanaceti* auct. Nec L.) (Hemiptera, Aphidida). *Proceedings of the Royal Entomological Society of London A* 24, 79–82.

Sutherland, O.R.W. (1969) The role of crowding in the production of winged forms by two strains of the pea aphid *Acyrthosiphon pisum. Journal of Insect Physiology* 15, 1385–1410.

Sutherland, O.R.W. and Mittler, T.E. (1971) Influence of diet composition and crowding on wing production by the aphid *Myzus persicae*. *Journal of Insect Physiology* 17, 321–328.

Taylor, L.R. (1977) Migration and the spatial dynamics of an aphid, *Myzus persicae*. *Journal of Animal Ecology* 46, 411–423.

Taylor, M.S., Harrington, R. and Clark, S.J. (1998) Unseasonal male aphids. In: Nieto Nafria, J.M. and Dixon, A.F.G. (eds) *Aphids in Natural and Managed Ecosystems*. Universidad de León, León, pp. 287–294.

Thacker, J.I., Thieme, T. and Dixon, A.F.G. (1997) Forecasting of periodic fluctuations in annual abundance of the bean aphid: the role of density dependence and weather. *Journal of Applied Entomology* 121, 137–145.

Thieme, T. (1997) Adaptive significance of brown colouration in *Sitobion avenae*. *Bulletin IOBC/WPRS* 21(8), 7–13.

Thieme, T. and Heimbach, U. (1998) Einfluss von Getreidesorten auf Blattläuse. *Getreide* 4, 100–102.

Tosh, C.R., Powell, G. and Hardie, J. (2003) Decision making by generalist and specialist aphids with the same genotype. *Journal of Insect Physiology* 49, 659–669.

Ueda, N. and Takada, H. (1977) Differential relative abundance of green-yellow and red forms of *Myzus persicae* (Sulzer) (Homoptera: Aphididae) according to host plant and season. *Applied Entomology and Zoology* 12, 124–133.

Via, S. (1992) Inducing the sexual forms and hatching the eggs of pea aphids. *Entomologia Experimentalis et Applicata* 65, 119–127.

Wadley, F.M. (1923) Factors affecting the proportion of alate and apterous forms of aphids. *Annals of the Entomological Society of America* 16, 279–303.

Ward, S.A., Leather, S.R., Pickup, J. and Harrington, R. (1998) Mortality during dispersal and the costs of host-specificity in parasites: how many aphids find hosts? *Journal of Animal Ecology* 67, 763–773.

Watson, M.A. and Healy, M.J.R. (1953) The spread of beet yellows and beet mosaic virus in the sugar beet root crop II. The effect of aphid numbers on disease incidence. *Annals of Applied Biology* 40, 38–59.

Watson, M.A., Heathcote, G.D., Lauckner, F.B. and Sowray, P.A. (1975) The use of weather data and counts of aphids in the field to predict the incidence of yellowing viruses of sugar beet crops in England in relation to the use of insecticides. *Annals of Applied Biology* 40, 38–59.

Way, M.J. (1963) Mutualism between ants and honeydew-producing Homoptera. *Annual Review of Entomology* 8, 307–344.

Weisser, W.W. and Braendle, C. (2001) Body colour and genetic variation in winged morph production in the pea aphid. *Entomologia Experimentalis et Applicata,* 99, 217–223.

Weisser, W.W., Braendle, C. and Minoretti, N. (1999) Predator induced morphological shift in the pea aphid. *Proceedings of the Royal Society of London B* 266, 1175–1181.

Wellings, P.W., Leather, S.R. and Dixon, A.F.G. (1980) Seasonal variation in reproductive potential: a programmed feature of aphid life cycles. *Journal of Animal Ecology* 49, 975–985.

Williams, C.T. (1995) Effects of plant age and condition on the population dynamics of *Myzus persicae* (Sulz.) on sugar beet in field plots. *Bulletin of Entomological Research* 85, 557–567.

Williams, I.S., Dewar, A.M. and Dixon, A.F.G. (1998) The influence of size and duration of aphid infestation on host plant quality, and its effect on sugar beet yellowing virus epidemiology. *Entomologia Experimentalis et Applicata* 89, 25–33.

Williams, I.S., Dewar, A.M., Dixon, A.F.G. and Thornhill, W.A. (2000) Alate production of *Myzus persicae* on sugar beet – how likely is the evolution of sugar beet specific biotypes? *Journal of Applied Ecology* 37, 40–51.

Zhang, B.C. (1994) *Index of Economically Important Lepidoptera*, 1st edn. CAB International, Wallingford, Oxon, 600 pp.

4 Host-plant Selection and Feeding

Jan Pettersson[1], W. Fred Tjallingii[2] and Jim Hardie[3]

[1]Department of Ecology, The Swedish University of Agricultural Sciences, 750 07 Uppsala, Sweden; [2]Laboratory of Entomology, Wageningen University, PO Box 8031, Wageningen, The Netherlands; [3]Division of Biology, Imperial College London, Silwood Park campus, Ascot, Berks, SL5 7PY, UK

Introduction

Prominent traits in aphid ecology are rapid reproduction and an advanced adaptation to host-plant ecology/phenology, as well as to the physiology/biochemistry of the plant. To maximize survival and reproduction, it is necessary for aphids to have efficient mechanisms to locate and exploit the host-plant resource. In general, the dependence of an individual aphid on stimuli necessary for host-plant discrimination can be expected to reflect the host-plant range of the species. As only 5% of aphid species can be categorized as polyphagous (Blackman and Eastop, 1984) and 95% as monophagous or oligophagous, it might be expected that most aphid species are to some extent dependent on host-plant specific stimuli to distinguish between good host and poor/non-host plants. With few exceptions, aphids live in colonies and this lifestyle has advantages, as well as disadvantages. With regards to natural enemies, a dense colony favours contagious pathogens and some parasitoids, although predator pressure on individuals might be less. However, obviously aphids have evolved an ecological strategy relying on propagation and host-plant relations that outweighs these drawbacks (Cappucino, 1987, 1988). A developing aphid colony represents a heavy demand on an infested plant and may evoke plant responses, local as well as systemic (see Quisenberry and Ni, Chapter 13 this volume). Easily visible symptoms of these responses are necrotic spots, discoloration, disturbed growth, and galling. However, the aphid attack may also release less visible plant-stress responses and result in a gradual change of the host-plant quality. The interaction between plant responses to aphid attack and aphid dependence on high-quality food resources supports development of advanced sensory systems for host-plant discrimination and food-quality assessment by aphids. This review will follow the successive steps of host-plant orientation, finding, selection, and acceptance for an individual aphid, illustrating different mechanisms with examples from different species of Aphidoidea.

Orientation and Host-plant Finding (See also Irwin et al., Chapter 7 this volume)

Aphids are considered to be poor flyers in that they can only make headway at low wind speeds. However, they can remain airborne for many hours and thus can be transported considerable distances by air movements. This has been studied extensively in the applied context with suction traps (Wiktelius, 1984a; Wiktelius et al., 1990).

Visual responses

It has been known for over 50 years that aphids, like many phytophagous insects, will preferentially land on yellow (to the human eye) coloured surfaces (Moericke, 1955; Prokopy and Owens, 1983; Robert, 1987). The importance of visual stimuli for landing has been studied with different arrangements of coloured traps. These make it possible to estimate the potential infestation of a crop field by the airborne aphid population, in contrast to the passive (on the part of the aphid) suction trap catches that give an estimate of the total aerial population of aphids. Though most aphid species are responsive to yellow, there are some differences between species (Eastop, 1955; Moericke, 1955; Heathcote, 1957), and one species in particular, *Hyalopterus pruni* (the mealy plum aphid), has been shown to respond specifically to the hue of the host plant (Moericke, 1969). In a laboratory study, the maximum landing response was shown to unsaturated yellows that reflected some ultraviolet, resembling the light reflected from *Phragmites* reed, which is the summer host for this aphid species. Using monochromatic light in a vertical wind tunnel (see below), it could be shown that *Aphis fabae* (black bean aphid) and *Rhopalosiphum padi* (bird cherry–oat aphid), which are considerably more polyphagous aphid species, were most responsive to wavelengths in the green region of the spectrum, *c.* 550 nm (Hardie, 1989; Nottingham *et al.*, 1991a).

In wind tunnels with the airflow vertically downward and balanced against the rate of climb, it has been possible to study the flight behaviour of individual aphids with a known flight history (Kennedy and Booth, 1963; David and Hardie, 1988). Migratory behaviour (Irwin *et al.*, Chapter 7 this volume) is an initial period of flight during which the aphid remains unresponsive to those visual stimuli that will eventually lead to it landing on a plant. During migratory behaviour, the phototactic response of the insect to a bright, white, overhead light in an experimental chamber keeps the aphid flying in the centre of that chamber. These studies have shown that there are differences between morphs of *A. fabae*, *Phorodon humuli*

(damson–hop aphid) (Hardie and Campbell, 1998), and *R. padi* in the duration of migratory behaviour at the start of the maiden flight. Autumn migrants (gynoparae) of *A. fabae* and *R. padi* flew for 2–3 h before responding to plant-like stimuli (David and Hardie, 1988; Nottingham and Hardie, 1989; Nottingham *et al.*, 1991a). By contrast, summer virginoparae showed a short or negligible period of migratory behaviour. The readiness to land increased in both winged forms after starvation, and consequently the duration of the migratory period was reduced (Hardie and Schlumberger, 1996).

Detailed observations of landing strategies have received scant attention due to problems of observing such small insects. As aphid flight speed is limited, the wind conditions are restrictive. Wind tunnel and field experiments show that flying aphids can reach air speeds of 70–75 cm/s (Kennedy and Thomas, 1974; Hardie *et al.*, 1996), but in wind speeds greater than this, aphids cannot make headway. Nevertheless, aphid landing is usually an insect-guided rather than a passive event (see Irwin *et al.*, Chapter 7 this volume), as can be inferred from the effectiveness of yellow traps and in landing responses to plant material (Åhman *et al.*, 1985). Detailed analysis of the three-dimensional flight tracks of aphids in the laboratory, as well as in the field, has also shown that aphids can control ground speed and that they land into the wind (Hardie and Young, 1997; Storer *et al.*, 1999).

Olfactory responses

The importance of plant volatiles in host location by flying aphids seems to be more complex than the responses to wavelengths of light. Kennedy *et al.* (1963a,b) made observations of the landing/take-off behaviour of *Myzus persicae* (peach–potato aphid), *A. fabae*, and *Brevicoryne brassicae* (cabbage aphid) under field conditions. They observed that aphids appeared to land as readily on non-host plants as on neighbouring hosts, indicating that no discrimination occurred at a distance. They concluded that olfactory cues were not used prior to landing. However, results of later behavioural and physiological investigations (see below) made Kennedy (1986) change his opinion on aphid odour responses, and it is now accepted that plant odour can be important in host finding and discrimination, although the details of these mechanisms are not yet clear. Aphids certainly possess an array of olfactory organs on the antennae and it has been shown with electrophysiological methods that they respond to a variety of plant volatiles (Anderson and Bromley, 1987; Pickett *et al.*, 1992; Visser and Piron, 1997; Park and Hardie, 1998, 2004) (see Pickett and Glinwood, Chapter 9 this volume). The major olfactory organs present in all morphs, in adult and nymphal stages, are the primary rhinaria, one on each of the two last antennal segments. Electrophysiological recordings from these in adult insects showed that they contain receptors for common leaf volatiles (van Giessen *et al.*, 1994; Park and Hardie, 2002, 2004). Several studies have shown that walking apterae do respond to host-plant odours in olfactometers (e.g. Pettersson, 1970, 1973; Nottingham *et al.*, 1991b; Pettersson *et al.*, 1994). However, the precise role of these sensory cells in the aphid–plant interaction is not well understood. Host finding by these stages and forms may well be an unusual behaviour in the field, but it can be hypothesized that these sensory cells are receptors for stimuli contributing to host-plant finding and the initiation of probing.

Comparisons have been made between aphid species and morphs of electroantennogram (EAG) responses to plant volatiles. This electrophysiological technique records the response across all of the antennal olfactory receptors and Visser and Piron (1997) compared EAG responses from the cabbage specialist *B. brassicae* and the generalist *M. persicae* reared on the same host plant. They found different responses between species for 35 of the 80 plant volatiles tested. When EAGs from alate and apterous summer forms of *M. persicae* were compared, there was a difference for 8 out of 50 volatiles. Eleven of 54 test volatiles revealed different responses between winged morphs of two *M. persicae* clones and a similar number (6 out of 50 volatiles)

between apterous *M. persicae* reared on a host-plant or artificial diet. The EAG responses to fewer compounds differed between *M. persicae* clones and there were no differences between these data and those collected for alate and apterous morphs of a single clone of this species or between individuals reared on different host plants. The differences between morphs may be aphid species dependent, for in a comparison of EAG responses from the polyphagous alate summer virginoparae and the monophagous autumn gynoparae of *A. fabae*, no differences were found over 35 test volatiles, or between alate and apterous virginoparae (13 test volatiles) (Hardie *et al.*, 1994). However, a similar comparison in *R. padi*, using a more sensitive EAG technique, showed a significant difference in EAG responses between the winged virginoparae and the gynoparae. The former were more sensitive to benzaldehyde, which is a prominent volatile from bird cherry (*Prunus padus*), the winter host of *R. padi* (Park *et al.*, 2000), but only the gynoparae respond behaviourally to benzaldehyde in olfactometer studies (Pettersson, 1970; Park *et al.*, 2000).

Winged adults are the more mobile individuals in the aphid population and usually have a further group of olfactory receptors, the secondary olfactory rhinaria, predominantly on the third antennal segment. In most aphid species, the secondary rhinaria are fewer, less well-developed or absent on adult apterous individuals, which corresponds with the predicted role of apterae developing and reproducing close to where they are born. In contrast, winged aphids are more adapted for mobility and the colonization of new plants. However, the precise role of the secondary rhinaria in host selection is still not known. EAG studies indicate that they contribute little, if at all, to the overall EAG response to plant volatiles in winged females (van Giessen *et al.*, 1994; Park and Hardie, 2002), and more sensitive electrosensillogram studies (DC recordings of receptor potentials from individual placoid sensilla of the secondary rhinaria) revealed no response to 31 common plant volatiles (Park and Hardie, 2004).

As mentioned above, several olfacto-meter studies reveal that walking aphids show preferences for air streams containing host odour rather than pure air, and in some experiments it has also been found that non-host odours are avoided (e.g. Nottingham et al., 1991b; Hori, 1999). The response of flying aphids to odour sources has also been tested in wind tunnels. So far, too few species and morphs have been tested to permit general conclusions. Studies of flight behaviour of A. fabae and B. brassicae showed that non-host odours reduced landing frequencies, but host odours did not increase landing (Notting-ham and Hardie, 1993). In field experiments, flying individuals of Cavariella aegopodii (willow–carrot aphid) landed significantly more frequently in traps releasing volatiles typical for the summer host plants than in empty control traps (Chapman et al., 1981).

Plant Contact after Landing

After landing, aphid behaviour is affected both by plant morphology and chemistry. The optimal part of the plant for settling combines quantity and quality of food supply with physical properties such as protection from natural enemies and from weather factors. Aphids often show positive

geotaxis and negative phototaxis after land-ing, and prefer settling on the lower surface of the leaf (Fig. 4.1; Müller, 1984). Plant morphology sets the conditions for these movements, and mechanical obstacles such as hairiness and the structure of the epi-cuticular wax may constitute mechanical problems for walking aphids. The impor-tance of the structure of the wax in host-plant preferences has been shown (Klingauf et al., 1978; Powell et al., 1999). It has been particularly studied in Lipaphis pseudo-brassicae (mustard aphid) and B. brassicae on different plant cuticular wax genotypes of Brassica (Åhman, 1990). Here, it was found that the tarsal claws of L. pseudo-brassicae were too short to cope with the cuticular wax particles and, as a result, the aphids fell off the leaves. Physical entrap-ment of aphids by glandular trichomes on the plant's surface has been studied exten-sively (Lapointe and Tingey, 1986; Tingey, 1991). Such trichomes seem to produce repellent volatiles independent of the mechanical properties. The sticky mem-brane and volatile effects can temporarily be removed by washing, and aphids then seem to probe cultivars with trichomes as easily as those without them (Alvarez et al., 2006). This topic will be discussed further by van Emden (Chapter 17 this volume).

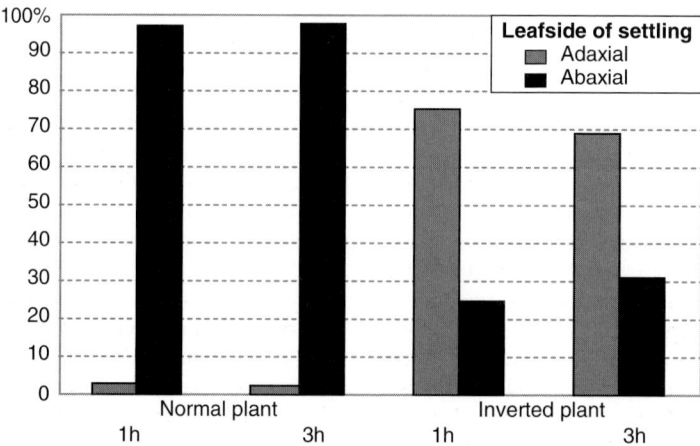

Fig. 4.1. Relation between leafside and settling by aphids. Although gravity and normal light direction caused most aphids to settle on the lower side of the inverted leaves, some aphids apparently remained, preferring the abaxial leafside, presumably on the basis of leaf properties. Three hours after release, this number tended to increase, but not significantly (after Müller, 1984).

Some aphid species show preferences for different parts of the host plant; preferences that change during the growth season. Thus, it is common to find individuals and small colonies of *R. padi* on a cereal plant at, or just beneath, the soil surface in early summer, but in the ears and on upper leaves later in the summer (Wiktelius *et al.*, 1990). Another example concerning *B. brassicae* on *Sinapis alba* is discussed below (Gabrys *et al.*, 1997).

Aphids would increase their time for reproduction if they had a rapid mechanism for rejection of unsuitable host plants. Such a response has been shown for *A. fabae*, where winged aphids respond to plant volatiles on landing and may take off from a suitable host plant if presented with a non-host odour immediately after landing. This response, however, seems to be switched off after some time has been spent on the plant (Storer *et al.*, 1996). Such behaviour clearly indicates a role for plant volatiles as part of a rapid host-plant screening mechanism in the field. There is an obvious risk of sensory adaptation of antennal receptors when aphids walk on a suitable host plant constantly exposed to positive volatile stimuli. This could mean a loss of part of the capacity for host-plant discrimination. It has been postulated that the antennal waving during walking may increase olfactory sensitivity by taking the olfactory receptors beyond the boundary layer of relatively still, odoriferous air at the plant surface and into regions of cleaner air (Hardie and Powell, 2000). During periods of stylet insertion into plant tissue, aphids tend to hold the antennae stationary and laid back over the thorax/abdomen (Hardie *et al.*, 1992). Such an action is open to the interpretation that internal plant chemical cues have taken over as more relevant to aphid behaviour than odours.

When the aphid has arrived on a potential new host plant, the role of contact chemoreceptors becomes important. Nevertheless, compared to other phytophagous insects, the number of external contact chemoreceptors on aphids appears low, and their sensitivity and role in host-plant selection is still unclear. Despite the early suggestion by Weber (1928) that the pegs on the labial tip could be contact chemoreceptors, all these 16 pegs and the other labial trichoid sensilla appear to be mechanoreceptors only (Wensler, 1977; Tjallingii, 1978). Aphids do, however, have candidate receptors on the tarsi (J. Hardie, unpublished results, Fig. 4.2). At least one trichoid sensillum, located at the tibial–tarsal junction, has four dendrites within the shaft, typical of a contact chemoreceptor. There is also ultra-structural evidence that the apical antennal hairs are contact chemoreceptors (Bromley *et al.*, 1980), and in *A. fabae* they have been shown to detect the antifeedant polygodial sprayed on the leaf surface (Powell *et al.*, 1995). As aphids walk over leaf surfaces and wave their antennae, the tips can touch the surface and detect non-volatile chemical cues associated with the cuticle. Surface contact by antennal tips has been demonstrated using fluorescent powder as a marker (Powell *et al.*, 1995). It has also been shown that non-volatile plant epicuticular lipids could be detected by aphids (Powell *et al.*, 1999), but whether antennal or tarsal chemoreceptors, or both, were responsible was not investigated. Prior to actual plant penetration, the secretion of watery saliva from the stylets on to the plant surface and its (possible) subsequent ingestion has been assumed to form a potential test phase that might explain the detection of soluble surface chemicals (Wensler, 1974). The subsequent ingestion and transfer to the pharyngeal gustatory organs (Wensler and Filshie, 1969) may offer a possible mechanism for tasting the leaf surface (Wensler, 1962).

For the most part, apterous aphids do not need to select a new host plant, but they are known to move both within and between plants (Ibbotson and Kennedy, 1950; Way and Banks, 1967; Mann *et al.*, 1995). In experiments with *A. fabae*, apterous adults and nymphs destined to be wingless adults left the host plant more readily than nymphs destined to be winged (Hardie, 1980). A feeding aphid is exposed to disturbance not only from other aphids in the colony but also from natural enemies, rain, wind, other herbivores, etc. After falling, some apterae of most aphid species succeed in regaining the plant and settling. Studies of the importance of colony density for the mobility of apterae of *R.*

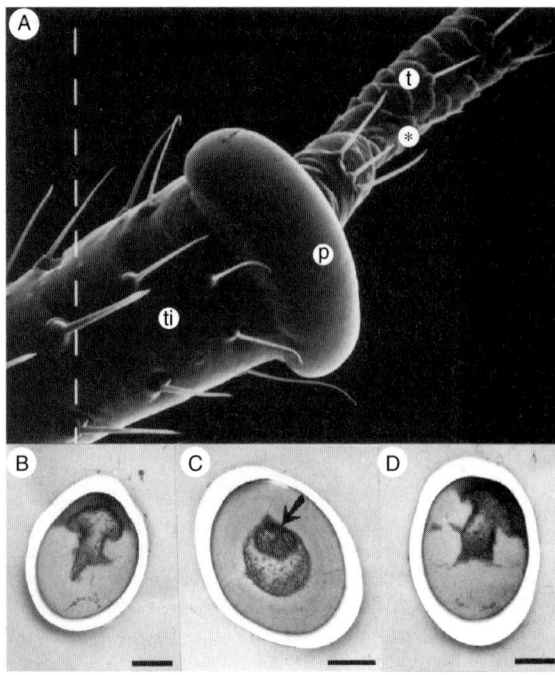

Fig. 4.2. (a), Scanning electron micrograph of part of the front leg of a wingless *Megoura viciae*, showing the lower part of the tibia (*ti*) with the everted pulvillus (*p*) and the tarsus (*t*). The first segment of the tarsus bears three trichoid sensilla, two longer outer hairs and a central shorter hair (asterisked). Scale bar = 10 µm (photo J. Hardie). (b), (c) and (d), Transmission electron micrographs showing transverse sections through the shafts of the outer hairs (b and d) and central hair (c). The central hair has four dendrites running along the shaft (c, arrowed) and is typical of a contact chemoreceptor, while the outer hairs do not and are mechanosensory in function. Scale bar = 0.5 µm (photo: J. Hardie).

padi showed that olfactory cues lowered the threshold for movement and also contributed to a spaced settling pattern (Pettersson *et al.*, 1995; Quiroz *et al.*, 1997). Aphid individuals walking on a plant receive and respond to both stimuli from the plant and from other aphids already settled on the plant, and to plant responses induced by their feeding. These plant responses may be either more or less favourable for the aphid individual that approaches an aphid group or the individual that is already settled on the plant. There is experimental evidence that feeding by a moderate colony density of *B. brassicae* causes changes in the food quality of the leaf that favour the settling and development of conspecifics (Way and Cammell, 1970). It may be hypothesized that there is an optimal size of the feeding aphid group initiating improved food quality promoted by plant responses to the stress caused by the feeding group (Way, 1973; cf. Miles, 1999 in relation to aphid enzyme systems).

Two aphid species using the same host plant will potentially create host-plant-mediated competition for plant resources, if the feeding-induced changes are aphid species-specific. Experiments with *Sitobion*

avenae (grain aphid) and *R. padi* lend support to the concept of such interspecific competition (Gianoli, 2000). There was a reciprocal negative effect on the performance of and host-plant acceptance by either species when fed on young wheat plants infested with the other aphid species. However, apterae of *S. avenae* preferred air containing the odour of feeding apterae of *R. padi* to clean air (Johansson *et al.*, 1997). In other olfactometer experiments, alate individuals of *L. pseudobrassicae* and *B. brassicae* avoided individuals of other species that had settled on a plant (Pettersson and Stephansson, 1991). Experimentally, it is difficult to distinguish between aphid and plant released/generated semiochemicals involved in the search for the optimal feeding place on a previously aphid-infested plant. It is still not known why some aphid species (such as *B. brassicae*) are indifferent to which species they share their feeding site with, while others (such as *L. pseudobrassicae*) are more discriminating. It may be hypothesized that this is related to sensitivity to specific dietary requirements.

Host alternation, i.e. seasonal movement between winter and summer host plants, occurs in a limited number (*c.* 10%) of aphid species. The winter host(s) is designated as the 'primary' and the summer host(s) as the 'secondary' host (Mordwilko, 1934; Dixon, 1998; Williams and Dixon, Chapter 3 this volume). The secondary host-plant range is usually the broader of the two, but there are exceptions to this, such as with *P. humuli*. This aphid uses several *Prunus* spp. as primary hosts, but only hops, *Humulus lupulus* (hop), as a secondary host (Blackman and Eastop, 1984). *Aphis fabae* uses *Euonymus europaeus* (spindle) as its main primary host and a broad spectrum of secondary host plants from numerous plant families (Thieme, 1987). Host-plant specificity of the migrating morphs varies between species. In *R. padi*, the gynoparae show an extreme preference in autumn for one specific winter host, *P. padus* (Fig. 4.3). Sloe (*Prunus spinosa*) sometimes grows alongside *P. padus* and it is accepted as a host plant by all morphs associated with the winter host, except for gynoparae migrating to *P. padus*. So, it appears that the exclusive key to host alternation in this species is the precise host-plant acceptance of the autumn migrants (Sandström and Pettersson, 2000).

Plant Penetration and Feeding

Plant penetration was reviewed in detail by Pollard (1973) and by Montllor (1991). In contrast to the plant-oriented activities of aphids prior to landing, probing is clearly the behavioural phase that is dominated by the role of non-volatile chemical and tactile information, used as cues for initiation, support, or termination of probing activities. Although often identified as 'feeding', i.e. ingestion, we consider probing as a synonym of plant penetration. Feeding is only one of the activities occurring during probing. Probing implies the involvement of the sensory system more than plant penetration does, and seems a more correct label. The first probes by an aphid after accessing a plant are often short and have been classified as 'test probes'. But apart from their shorter duration, these probes do not differ from later and longer probes. It is likely that some plant chemicals are sampled and tasted during all probes, so apart from mechanical stylet insertion and saliva excretion into plants (reviewed by Miles, 1999; Tjallingii, 2006), probing also involves gustatory monitoring. The 'pharyngeal' gustatory organ (Wensler and Filshie, 1969) may play the principal chemosensory role in host-plant selection for detecting non-volatile plant compounds once aphids have landed on a plant. This organ has a main epipharyngeal part with 14 papillae containing 60 taste cells, whereas the minor hypopharyngeal part contains about 30–40 taste cells. The organ in the posterior food channel thus contains about 100 gustatory cells in total.

Fig. 4.3. A colony of *Rhopalosiphum padi* on a young shoot of the winter host *Prunus padus*. The fundatrices can be seen as olive green, not wax-covered individuals, while their offspring are greyish, covered by wax (original colour photo courtesy of V. Ninkovic).

Box 4.1.   The electrical penetration graph (EPG).

In the EPG technique, an aphid (or another insect with piercing mouthparts) and a plant are made part of an electrical circuit by inserting a wire into the soil of a potted plant, and attaching a very thin wire to the insect. The circuit also incorporates an electrical resistor (Ri) and a voltage source (V), as illustrated below. As soon as the aphid stylets penetrate the plant, the circuit is completed and a fluctuating voltage, called the 'EPG signal', occurs at the measuring point which is then amplified and recorded. The voltage fluctuations appear in a number of distinct patters, referred to as 'waveforms'. The major waveforms that have been distinguished for aphids are given in Table 4.1 with their experimental correlations to probing activities and stylet tip positions in the plant.

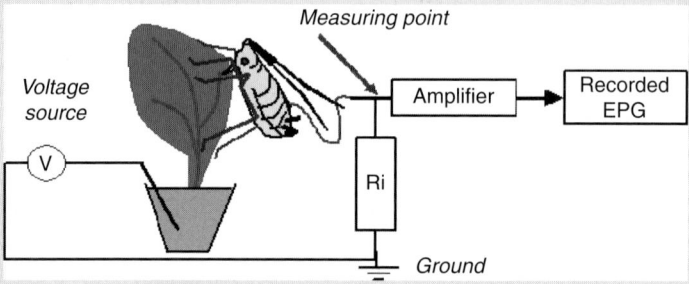

Electrical origin of the signal

The fluctuating voltages are caused by two different sources concurrently: 1) fluctuating electrical resistance of the aphid; and 2) voltages 'generated' in the insect–plant combination. These two sources cause signal components that are referred to as the resistance (R) components and the electromotive force (emf) components of the EPG, respectively, and they are superimposed at the measuring point. The R-components originate mainly from the activity of the valves in the stylet canals, the food and the salivary canal. The emf-components originate mainly from 'membrane potentials' of plant cells – when they are punctured by the stylets – and from 'streaming potentials' caused by the fluid movements in the two capillary stylet canals. Muscle and neural potentials in the insect are outside of the circuit, and therefore appear not to contribute to the EPG signal.

   Both, R and emf components include important biological information on the insect's activities and the stylet tip position in the plant tissue.

Measuring systems

The measuring system that was introduced by McLean and Kinsey (1964) used alternating current (AC) as a voltage source, and the signal was composed of modulations in voltage amplitude caused by resistance fluctuations in the insect, similar to signal processing in AM radio. This AC-system, however, appears to lose the emf-components during signal processing. Subsequently, the voltage source was replaced by a direct current (DC) source, and two DC-system variants were developed (Tjallingii, 1988). In one system, the input resistor (Ri) is very high ($>10^{12}\Omega$) so that resistance fluctuations of the insect become negligible and only the emf-components are recorded. The other DC-system has an input resistance of about the same value as the average electrical resistance of the aphid–plant combination, thus providing an optimal 1:1 ratio of resistance, which allows recording of both, the R- and emf-components. In summary, there are in fact 3 EPG systems: 1) the (regular) DC-system that records both signal components, 2) the 'emf-amplifier', the DC-system that only records the emf-components, and 3) the 'R-amplifier', the AC-system that only records the R-components. The EPG from the regular DC-system contains the widest range of biological information in the signal, and is therefore more complicated than the signals from the emf- and R-amplifiers. Some signal details from the regular DC-system are hidden (masked) in the EPG, and they become visible in the EPG from the R-amplifier (Jiang and Walker, 2001) or the emf-amplifier (Tjallingii, 1985, 1988, 2000) but in general, the regular DC-EPG contains the most complete and relevant biological information.

Pathway phase

Aphids are thought to suck up small samples of plant sap during the pathway phase of probing, i.e. during stylet penetration from the epidermal surface to the two target tissues, the phloem and the xylem. These samplings are of short duration and the sap presumably is transported very rapidly to the pharyngeal taste organ. Evidence of this comes from electrical penetration graph studies (EPG, see Box 4.1), which record waveforms reflecting the aphid's probing activities and stylet tip positions in plant tissue (Table 4.1). Despite earlier suggestions by Pollard (1973) and others, aphid stylets appear to penetrate exclusively intercellularly between the cellulose and hemi-cellulose fibres of the secondary cell walls, not through the middle lamella (Tjallingii and Hogen Esch, 1993). The role of salivary pectinases during tissue penetration seems unimportant (Cherqui and Tjallingii, 2000). *En route*, though, nearly every cell along the stylet track is also punctured intracellularly (Tjallingii and Hogen Esch, 1993). Probes with 50–100 or more such brief cell punctures are very common. The punctures are reflected in the EPG as the distinct potential drop (pd)

Table 4.1. Correlations of EPG waveforms in three behavioural phases and their corresponding aphid activities. Stylet pathway (path) includes all activities between the epidermis and a target tissue. During phloem and xylem phase, the stylet tips remain in one position. Waveform F represents a mechanical error, intrinsic of aphid and other homopteran stylet structures and function (from Tjallingii, 1987).

EPG waveform	Behavioural phase	Activity
A, B, C	Path	Intercellular penetration
Pd	Path	Intracellular puncture
E1	Phloem	Saliva secretion
E2	Phloem	Sap ingestion, passive
F	(Path)	Derailed stylet mechanics
G	Xylem	Sap ingestion, active

waveforms that typically last for only 5–10 s (Tjallingii, 1985), then the stylets are withdrawn and resume their intercellular path or are pulled out of the plant altogether. The pd waveform (Fig. 4.4) typically shows a pattern of 3 waveform phases (I–III); a brief period before the leading edge (I), a low intracellular voltage part (II), and a brief period after the second edge (III). The leading voltage drop reflects the transmembrane potential across the plasmalemma of the plant cell when the stylet tip passes through it. The second edge reflects the stylet tip withdrawal and return to the extracellular electrical potential. During phase II, the low intracellular voltage period, a waveform sequence of 3 sub-phases, is shown (Fig. 4.4 – II-1, II-2, and II-3; Powell *et al.*, 1995). The first sub-phase (II-1) represents watery saliva injection into the punctured cell, the second sub-phase is still unknown, and the third sub-phase represents ingestion of cell contents, as has been demonstrated by the sub-phase II-1 and II-3 relations to inoculation and acquisition of non-persistent plant viruses, respectively (Martin *et al.*, 1997).

During the short cell punctures, the maxillary stylets pierce the protoplast and subsequently most likely the tonoplast, the vacuole membrane. The vacuoles form the main storage sites for secondary metabolites of plants (potential allelochemicals). There is no reason to assume that the tonoplast, separating the protoplast and the vacuole compartments, will be injured during a cell puncture so as to mix the fluid contents. As the EPG shows, the membrane potential of the punctured cells remains intact, and also the organelles in micrographs of sectioned plant cells after stylet punctures are in good shape (Tjallingii and Hogen Esch, 1993; Fig. 4.5). Most allelochemicals are stored in the cell compartments in a non-toxic form, as amides or glycosides. In this chemical form, they are not toxic to the plant or plant-attacking organisms, but when mixed with enzymes occurring in different compartments, they are converted into toxins (Mathile, 1984). Aphids sample and taste the secondary metabolites directly from the cell compartments, in contrast to chewing insects that mostly encounter such metabolites in a toxic form after crushing

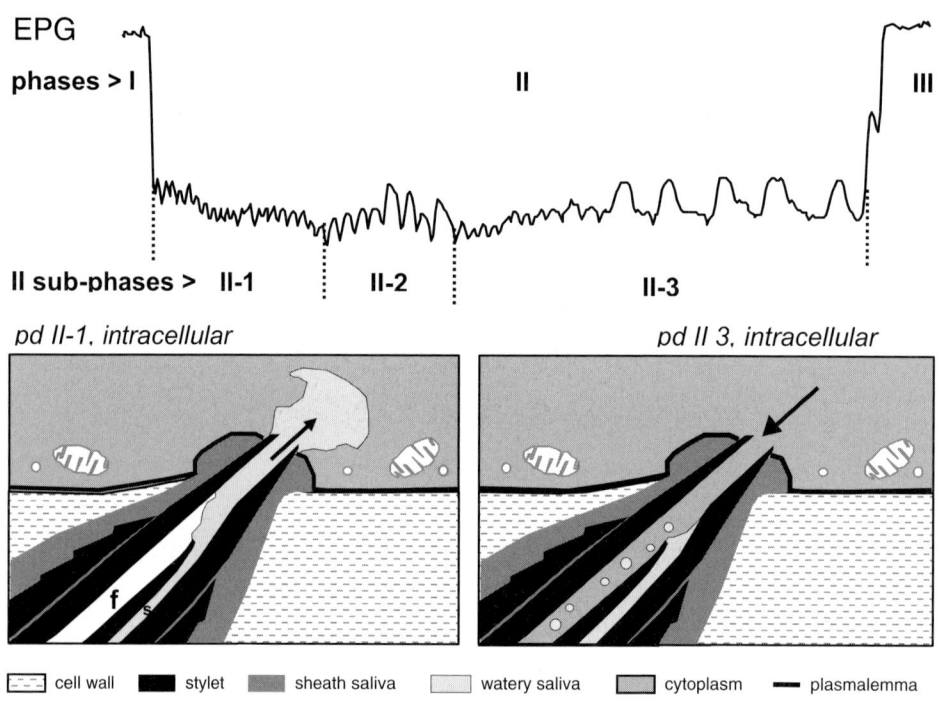

Fig. 4.4. Intracellular stylet puncture as recorded in an EPG (top) and its related stylet activities (diagrammatic representation) based on EM information as in Fig. 4.5 and EPG experiments on transmission of non-persistent viruses. The food (f) and the salivary canal (s) are fused near the tips of the maxillary stylets. Starting at the high intercellular voltage level, the leading downward edge in the trace represents the trans-membrane potential of the plant cell when the stylets are inserted. The later upward edge represents stylet withdrawal from the cell. At the II-1 (or II-2) sub-phase some watery saliva is injected, whereas at the II-3 sub-phase cytoplasm (sap) is sucked up from the cell contents. This sap sampling takes no longer than a few seconds (after Martin *et al.*, 1997).

the plant cells, thus mixing the allelochemicals with their specific enzymes – stored in separate plant compartments – that split off the linked amino acids or glucose molecules (Mullin, 1986). Thus, the delicate way that aphids penetrate plant tissues and cells avoids confrontation with an important part of the plant's chemical defence. For example, DIMBOA, a well-known allelochemical in some cereals that causes host-plant resistance to aphids (van Emden, Chapter 17 this volume), is stored as a glycoside and is not converted into the toxic aglucone. Nevertheless, it does exert an effect on probing aphids, demonstrating its importance as a 'token stimulus'. Electron micrographs of multiple-punctured plant cells indicate little cell damage is caused since the brief cell punctures show surviving

cells with intact mitochondria and chloroplasts (Fig. 4.5) even 1–2 h (the moment of tissue fixation) after the punctures occurred.

Thus, the 'safe' allelochemicals (as the plant stores them) will be ingested as soon as the stylets puncture the first cell. Generally, this is an epidermal cell, punctured as soon as 10–20 s after the start of a probe. Calculations based on ingestion rates of radioisotopes showed that the fluid is transported from the stylet tips to the pharyngeal organs in a fraction of a second (Spiller *et al.*, 1990; Tjallingii, 1995), a distance of about 1 mm. It is likely that these 'token stimuli' are used for host-plant selection (see below), but possibly additional cues from the punctured cells provide information for phloem finding.

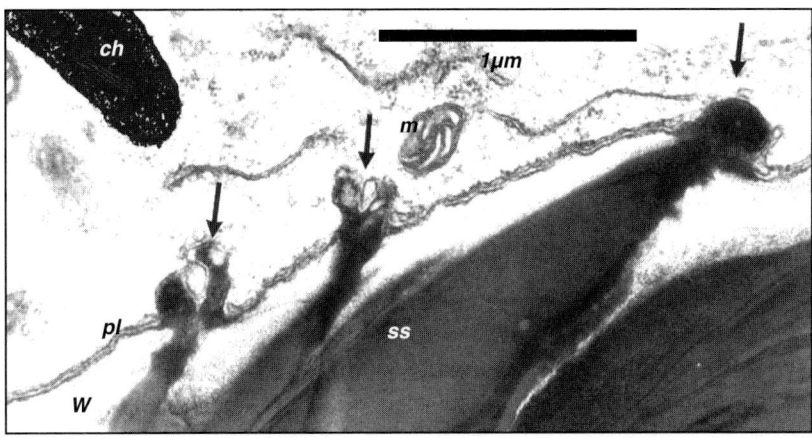

Fig. 4.5. Remnants of three intracellular punctures (arrows) with dark sheath saliva (ss) in the electron lucent material of the cell wall (w). Cell about 1.5 h after punctures (tissue fixation). Damage is restricted, as can be derived from the restored plasmalemma (pl) around the slightly invaded sheath material (especially clear in the most right puncture). Also, mitochondria (m) and chloroplasts (ch) seem intact and functional after a puncture (micrograph: W.F. Tjallingii and Th. Hogen Esch).

In this context, it is curious that, during plant penetration by whiteflies (Aleyrodidae), these brief cell punctures along the stylet route do not occur at all, or only once or twice in a small fraction of the probes. Whiteflies seem to select their host plants and to locate the phloem vessels as easily as aphids without brief intracellular punctures. Although many whiteflies are more polyphagous than most aphids, and may not need a host-selection cue as crucially as aphids do, they do need to find the phloem. Phloem finding, therefore, still forms one of the most intriguing problems of plant penetration by many Sternorrhyncha. All the suggestions made so far about the importance of gradients (of pH, sucrose, etc.) for phloem finding by Sternorrhyncha seem to have ignored the compartmental distribution of plant chemicals. Most gradients reported in the plant physiology literature imply that one cell differs in concentration to another cell. So in phloem cells, the pH and sucrose levels are high, whereas in epidermal, mesophyll, and vascular parenchyma cells, they are low. This seems to result in a situation where concentrations do not go up or down cell-by-cell, so that 'following' a gradient will be rather difficult. Even if a cell-by-cell gradient did exist – for example, due to symplastic phloem loading – then aphids might be able

to follow this gradient by their habit of puncturing cell-by-cell. However, this does not hold for whiteflies. It is likely that intercellular gradients may also exist. For example, due to apoplastic phloem loading, gradients of sugar and pH can be expected. However, the direction of the gradient will be opposite in source *versus* sink situations. So, how will the insects judge whether to move the stylets in a decreasing or increasing direction of the gradient?

The role of allelochemicals in host-plant selection by aphids is less studied and documented than in some other groups of insect herbivores. The best studied case is probably the role of glucosinolates, mainly occurring in Brassicaceae and used in host-plant recognition by *B. brassicae* (Wensler, 1962). Generalist aphids such as *M. persicae*, when on plants of the Brassicaceae, seem to tolerate the glucosinolates in water solutions up to certain concentrations (Nault and Styer, 1972). *Brevicoryne brassicae* greatly prefers the inflorescence stems of mustard plants to the leaves of *S. alba* (white mustard), whereas for *M. persicae*, it is the other way around. Although overall the leaves and these stem parts contained similar amounts of glucosinolates, especially sinalbin (Gabrys *et al.*, 1997), epidermal cells of the stem parts had a much higher concentration than the epidermal cells of the leaves. Moreover, EPGs of cabbage aphids on leaves showed many short (< 2 min) probes, a duration that is not sufficient to penetrate deeper than the epidermis. By contrast, on inflorescence stems, the very first probe usually lasted for much longer than 10 min and included phloem feeding. This strongly suggests that glucosinolate (sinalbin) information from these cells is used as a probing stimulus in *B. brassicae*, whereas in apterous *M. persicae*, the glucosinolates stimulated stylet withdrawal and walking. Glucosinolates (mainly sinigrin) have been added to artificial diets, or to water with the excised leaves of non-host plants without glucosinolates. This has always led to a better acceptance by *B. brassicae* of the diets or plant leaves than when no glucosinolates have been added (Gabrys and Tjallingii, 2002).

Whether the added glucosinolates remain in the intercellular fluids of xylem and the cell wall matrix, or they are also transferred to the intracellular compartments of the excised plant parts, has not been studied. So far, only intracellular fluid ingestion or sampling has been demonstrated, but intercellular fluid sampling by phloem-feeding insects cannot be excluded. Intracellular sampling has been inferred from experiments on the transmission of non-persistent plant viruses (Martin *et al.*, 1997). For the acquisition of these viruses, one brief puncture (5–10 s) of an epidermal cell during a single short probe into a virus-infected plant (Fig. 4.4, during subphase II-3) is enough to acquire the virus particles from the cell's plasma. After plant rejection and take-off, a subsequent landing on a healthy plant will only need one short probe with a cell puncture to inoculate the virus effectively. Many aphids transmit non-persistent viruses to non-host plants; for example, *Potato virus Y* is transmitted to potato by the cereal aphid *R. padi* (Sigvald, 1984; see also Katis *et al.*, Chapter 14 this volume and Radcliffe *et al.*, Chapter 26 this volume). Such transmission shows that the intracellular punctures apparently needed to gather information for plant rejection are sufficient for both acquisition and inoculation of non-persistent virus. Also, it indicates that the probing strategy of a short probe with a brief intracellular puncture is very

common in aphids. Additional evidence for this strategy comes from close-up video recordings in combination with EPG registrations (Hardie and Powell, 2000) of gynoparae of *A. fabae* on the summer host, broad bean (*Vicia faba*). It was shown that 75% of the gynoparae departed from the summer host within 5 min, and that the majority of these did so after a short probe during which brief cell punctures – including sap sampling – occurred that presumably caused the rejection.

When plant penetration continues, the stylets go deeper into successive tissues and these tissues may contain different cues that can stimulate or inhibit further probing before the phloem sieve elements are reached. In wheat, it has been demonstrated that DIMBOA prolongs pathway activities during aphid probing, presumably because this chemical is encountered in the sheath cells of the vascular bundle (Massardo *et al.*, 1994). Thus, the prolonged pathway phase delays the phloem phase initiation, but the phloem phase as such is hardly affected, at least as far as was possible to determine from the 8 h experiments (Givovich and Niemeyer, 1991; Givovich *et al.*, 1994).

Parturition in *A. fabae* is stimulated by components in the epidermis of the plant. A potent parturition stimulant has been isolated from spindle that induced larviposition in gynoparae but not in virginoparae of *A. fabae* (Powell and Hardie, 2001). Without this stimulant, there is little or no reproduction by gynoparae. It now seems that parturition, in general, is stimulated prior to the aphid reaching the phloem elements and feeding (Powell *et al.*, 2004). In the case of virginoparae, no correlation was found between the duration of ingestion from the phloem and the numbers of progeny produced by summer winged and wingless forms of *A. fabae* in 6 h EPG experiments. It has been shown that a significant majority of both forms of *A. fabae* reproduce prior to phloem feeding, although all show brief cell punctures (pd waveforms in the EPG) before reproducing (Tosh *et al.*, 2002). The interpretation is that the chemical cues identified in the peripheral plant cells are used by the aphids to identify suitable hosts on which to reproduce. Other aphid species may differ in the mechanisms and cues for such a crucial decision, although similar conclusions are being drawn from work with *Acyrthosiphon pisum* (pea aphid) (Caillaud and Via, 2000).

Phloem phase

Brief cell punctures occur in all tissues, including the sieve tube and companion cells, but it is at the level of the phloem that the final decision for host-plant acceptance is made in terms of feeding. Typical phloem activities always start with brief cell puncture activities (pd waveforms up to sub-phase II-3, the short intermediate level between the edge and the further lower signal level in Fig. 4.5, cf. also Fig. 4.4). During these initial brief puncture activities, the aphid seems to decide whether or not to switch to typical phloem phase activities that are characterized by the EPG waveforms E1 and E2 (Table 4.1; Fig. 4.5). In fact, many sieve elements may be reached and briefly punctured without leading to any phloem phase activity. In an EPG recording of such probes, the possible phloem contacts are indicated by an increased number of brief punctures, without phloem phase waveforms. It is not clear what makes the aphid switch from the brief cell puncture (pd) activities in a sieve element to phloem salivation (E1). In two well-documented cases of plant penetration by *A. fabae* on *V. faba*, a number (2 in one and 11 in the other case) of sieve elements showed EM evidence of punctures, while no phloem phase activities were detected in the EPG (Tjallingii and Hogen Esch, 1993). It was concluded that only brief cell punctures (pd waveforms) could have been involved here. Without exception, a phloem phase always starts with saliva excretion into the sieve element (E1 waveform, Fig. 4.6). On host plants, after about 1 min of phloem salivation, the second phloem phase activity often – but not always – follows, i.e. feeding with concurrent salivation (E2 waveform). Also here, it is unknown what cues are used by the aphid to switch or not from phloem salivation to feeding (E1 to E2). The phloem sap is under high pressure in the sieve element and phloem feeding is considered to

be a passive process (Mittler, 1958; Prado and Tjallingii, 2007), which presumably starts as soon as the pharyngeal (cibarial) valve opens. Consequently, this valve should be closed during phloem salivation to enable the saliva to be injected into the sieve element. On the other hand, the phloem sap pressure that causes passive feeding allows the continuously secreted saliva during E2 to be transported downstream only, i.e. into the food canal and foregut of the aphid (Prado and Tjallingii, 1994).

The composition of the E1 and E2 saliva is not known. It certainly can be considered as 'watery saliva' (Miles, 1999), and only speculation is possible with regard to the function. If a sieve element cell is severed, there are specific wound reactions that normally block the sieve element immediately (Knoblauch and van Bel, 1998). Phloem sap ingestion would be limited or impossible if such wound reactions occur upon puncturing by the stylets, and so it is assumed that aphids may avoid or suppress phloem wound responses. Salivation into the sieve element possibly plays an active role in such suppression, though many details remain uncertain. One type of phloem protein involved in wound responses is a crystal-like body in the sieve element – restricted to Fabaceae and called 'forisome' (Knoblauch and Peters, 2004) – that abruptly expands when calcium leaks into the cell. The expanded forisome blocks phloem transport. If the damage is weak or can be repaired, calcium is pumped out of the cell and the forisome will contract again and sieve element transport can be resumed. Remote damage, for example, a heat shock at some distance (burning the leaf tip), can also cause forisome expansion. Aphids that were phloem feeding (E2) at such a place of forisome expansion changed their behaviour to phloem salivation (E1), and after the heat shock and a time lapse (equal to what was needed for forisome contraction), phloem feeding was resumed (Will and van Bel, 2006). Some sieve element cues (such as pressure, free calcium, or phloem proteins) seem to be used by the aphids to switch backward and forward between E1 and E2 activities. Aphids not only need to prevent phloem protein reactions in the sieve tube, but also in

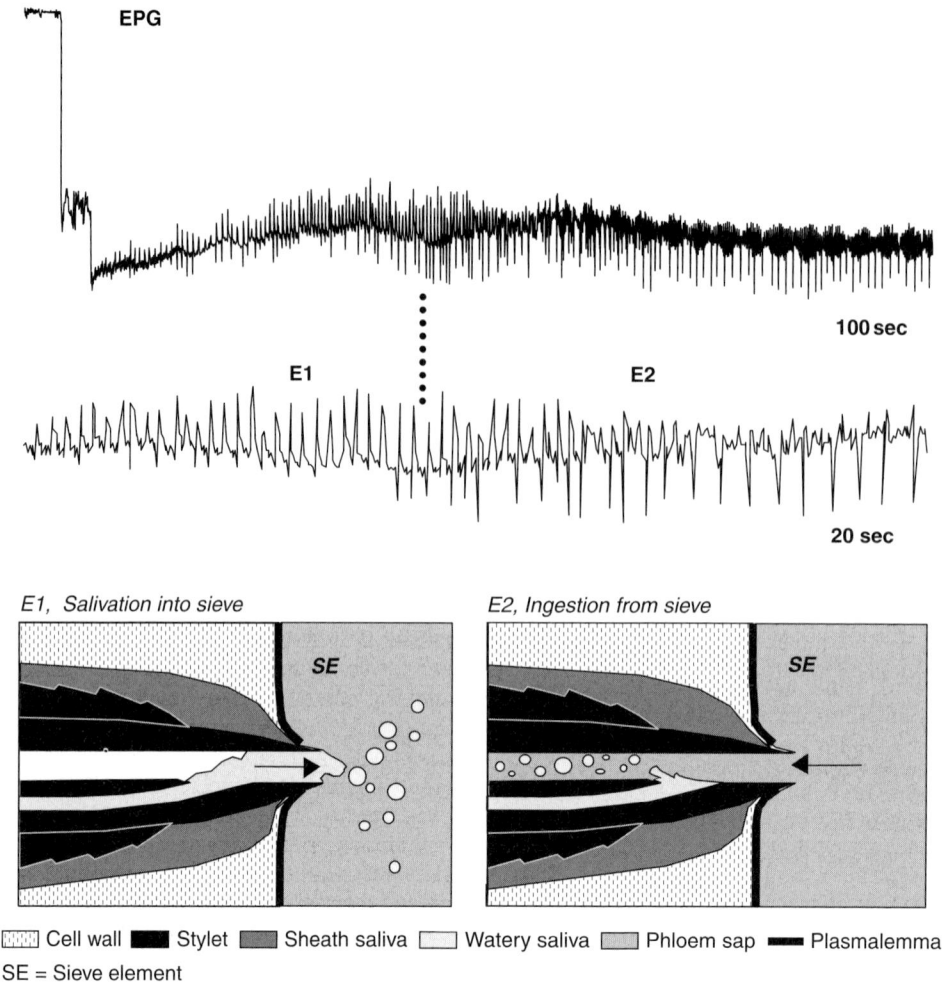

Fig. 4.6. Phloem phase waveforms in the EPG (top) and its related aphid activities. The fusion of the food and salivary canals near the tips of the maxillary stylets (cf. Fig. 4.2) allows injection of the E1 saliva only when the cibarial valve (in the head, not visible) is closed. When opened during E2, the high hydrostatic pressure forces the saliva into the food canal with the phloem sap (after Prado and Tjallingii, 1994).

the food canal of the stylets (Tjallingii and Hogen Esh, 1993). The way aphids prevent stylet canal blockage seems to be by the continuous addition of saliva to the ingested sap at the tip of the stylets during phloem feeding (E2 salivation; Fig. 4.6). Other information on phloem salivation/feeding switching derives from EPG studies on host-plant resistance (van Helden and Tjallingii, 1993; Klinger *et al.*, 1998; Kaloshian *et al.*, 2000). Data from these studies indicated that aphid resistance in some plants has a phloem-located mechanism. At least in some cases,

aphids did not switch from phloem salivation to feeding, or only occasionally did so. *Aphis gossypii* (melon aphid) showed phloem ingestion on resistant melon (*Cucumis melo* cv. TGR) only after extended phloem salivation (Garzo *et al.*, 2002), whereas *Macrosiphum euphorbiae* (potato aphid) showed a frequent alternation between the two phloem activities on tomato (*Lycopersicon esculentum* cv. 'Moneymaker') (W.F. Tjallingii, unpublished results). Similar to the heat shock situation, the aphids on the resistant plants were possibly confronted with a wound response situation that was unsuitable for phloem feeding. If the physico-chemical ability to suppress phloem wound responses is considered as 'compatible', these resistant plants may represent an incompatible situation. Resistance genes are often obtained from the germplasm of non-hosts or poor hosts. Certainly, such resistance mechanisms may also occur in other natural non-host plants. How general compatibility occurs in this sense of aphid–plant combinations and to what extent this may dictate host-plant specificity and host-plant range in aphids is unknown.

Although physico-chemical compatibility may or may not play an important role in host-plant acceptance, the gustatory and nutritional quality of the phloem sap will also be important during phloem feeding. In EPG studies on phloem-located resistance, it is not possible to distinguish between the physico-chemical, gustatory, or nutritional nature of the resistance since, irrespective of its cause, delayed or reduced phloem feeding (periods of waveform E2) will be the results. In traditional terminology, these phloem-located mechanisms of host resistance will be seen as antibiosis since reduced feeding (starvation) inevitably will affect aphid performance (survival, growth, and reproduction), the main trait of antibiosis. But such a resistance is based on behavioural 'phloem avoidance', and therefore should be defined as 'antixenosis' (cf. van Emden, Chapter 17 this volume, for definitions). Recently, it has been demonstrated that, within a group of potato-related *Solanum* species, there were several species with resistance to *M. persicae* (Alvarez *et al.*, 2006). On the basis of EPG waveform parameters, the resistance could be

qualified as surface-, epidermal-, mesophyll-, or phloem-located resistance, or a combination of these. In one *Solanum* species, however, there were no indications of resistance in any tissue location, but nevertheless aphid performance showed that the plant really was resistant, presumably on the basis of true chemical antibiosis: the presence of chemicals that do not alter feeding behaviour but are toxic to the insect.

Phloem feeding

In EPG studies, certain threshold durations (8, 10, or 15 min) have been used as criteria for 'sustained' or 'committed' phloem ingestion (Montllor *et al.*, 1983; Tjallingii, 1990). It has been demonstrated that when aphid ingestion has exceeded these thresholds, the insects have often continued ingesting phloem sap for a longer period of time. Committed phloem ingestion has been regarded as 'phloem acceptance', which might be considered as host-plant acceptance with respect to feeding. In this context, Kennedy and Booth's (1951) 'dual discrimination' theory still seems very relevant. This theory suggests that gustatory properties of allelochemicals (token stimuli) and the presence of nutrients are both important for host-plant selection and acceptance (sustained feeding). The time lag between landing of flying aphids, or plant access by walking aphids, and the final acceptance in the sense of committed phloem ingestion has been studied in several aphid species. It turns out that this time lag lasts hours, mostly including many separate probes with or without phloem phases (Fig. 4.7). For example, in *A.*

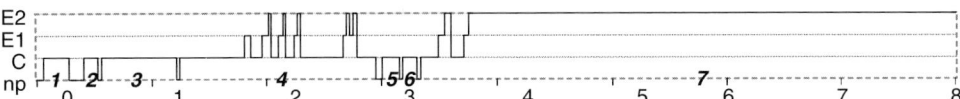

Fig. 4.7. Schematic representation of aphid probing activities during the first 8 h by *Brevicoryne brassicae* on the leaves of the aphid's host plant *Sinapis alba*. The activities are represented as levels on the Y axis, with time (h) on the X axis. Italic numbers indicate the successive probes. Some probes (*1, 2, 3, 5,* and *6*) show only pathway activities (C). The two other probes (*4* and *7*) include phloem phases, the first in probe *4* includes only phloem salivation (E1), all others also show phloem feeding with continuous salivation (E2). The last phloem phase in probe *4* switches from E1 to E2, back to E1 and then to E2 again. Only the last phloem phase has an E2 period of longer than 10 min, i.e. 'committed phloem ingestion' that is considered as phloem or host-plant acceptance, which in this aphid occurs after about 4 h.

fabae the average time between plant access and sustained phloem ingestion was found to be 5.2 h on *V. faba* leaves (Tjallingii, 1994). For other aphids on their host plants, it may take even longer. *Sitobion avenae* took an average of 6.3 h on barley but only 3.5 h on wheat (Tjallingii and Mayoral, 1992). This demonstrates that, within a range of host plants, the readiness of aphids to accept a food plant can be very different. It should be noted in the examples above that both aphids are polyphagous. A question that may arise is whether polyphagous species would show a longer time lag than monophagous species since they need to cope with many more different plant chemicals and morphological characteristics. There is no evidence for such a suggestion. The time lag for the monophagous *Drepanosiphum platanoidis* (sycamore aphid) on a leaf of sycamore (*Acer pseudoplantanus*) is about 4 h (W.F. Tjallingii, unpublished results), and comparisons of the generalist alate virginoparae and the specialist gynoparae of *A. fabae* show similar times (Tosh *et al.*, 2002). The times to sustained ingestion of phloem sap are quoted as mean values of 15–20 individuals, but the minimum time is often much shorter and can be as short as the time needed during a probe for sustained phloem ingestion to become identifiable (i.e. 10 min), in many species only some 20 min. Although the minimum time lag is often much shorter, the average values indicate that most aphids on their proper host plant need a long time, often including more than one probe, before any phloem activities and committed phloem ingestion is achieved. Under stable environmental conditions, such a period of phloem feeding may be sustained for hours, and even many days, during which the aphid continuously taps a single sieve element. In immature stages, therefore, the main reason for pulling out the stylets will presumably be moulting to the next instar.

Regulation of sap uptake is not well understood. On one hand, the pharyngeal valve might regulate (theoretically) the input of sap forced into the food canal (Fig. 4.6) by the hydrostatic pressure in the sieve elements. On the other, honeydew excretion data show constant droplet sizes as well as droplet intervals during long periods of EPG-recorded phloem ingestion (waveform E2). Therefore, it has been suggested that the control of phloem sap ingestion is by regulating the ingestion period rather than by regulating the ingestion flow (Tjallingii, 1995).

Xylem drinking

Aphids usually are considered to be phloem feeders, but they have also been found to ingest from xylem (Spiller *et al.*, 1990). Xylem ingestion is readily recognized in EPGs (as waveform G), and it has been recorded in increased amounts from aphids that were starved for long periods until given access to plants (Ramirez and Niemeyer, 2000). In contrast to phloem ingestion, xylem ingestion does not seem to contribute nutrition as such (at least not in aphids) but to compensate for water stress, and therefore should be considered as 'drinking' rather than feeding. Moreover, xylem drinking seems an active sucking process from the xylem elements with a low or often negative hydrostatic pressure, in contrast to the passive phloem feeding caused by the high pressure in the sieve element. Not much attention has been given to xylem ingestion in the literature for two reasons: (i) the vast majority of EPG studies have been done with non-starved, or only briefly starved apterous aphids, and (ii) if it occurs in aphids, the durations are very variable and seem to have no relationship to any of the treatments or plants studied. However, winged morphs of *A. fabae* frequently showed xylem ingestion (Powell and Hardie, 2002). A high percentage showed xylem ingestion on their appropriate host plants, winged virginoparae on broad bean (65%) and gynoparae on spindle (60%), respectively. They also ingested xylem on the non-preferred host, i.e. 85% of the summer forms (virginoparae) on spindle (the winter host) and 20% of autumn migrants (gynoparae) on broad bean. Apterous *A. fabae* never showed waveform G on broad bean. Possibly, alatae landing after a flight may need xylem ingestion to redress water balance more than walking apterous aphids. The absence of xylem ingestion by apterous *A. fabae* confirms earlier data (Prado and Tjallingii, 1997), but xylem ingestion by apterous *R. padi* was found to be rather common. Apterous *B. brassicae* (Gabrys *et al.*, 1997) and *A. pisum* showed xylem ingestion regularly, and *A. pisum* tended to show more xylem ingestion on non-hosts than on host plants (Tjallingii, 1986).

Leaving a Plant

Aphids leave a plant as a result of changes in the host plant and social cues related to population density. Free amino acids – the supposed main nutritional components for aphids – fluctuate not only with the age of the plant or plant part on which aphids feed, but there is also a diurnal periodicity in their concentrations (Douglas and van Emden, Chapter 5 this volume). Other components in the phloem sap show similar patterns of change. The concentrations of some amino acids may go up during the day and down at night, whereas those of other amino acids change in the opposite direction (Winter *et al.*, 1992). Sugars can also show diurnal changes in concentration. These changes in the phloem sap composition may cause aphids to stop feeding and to pull out their stylets. More aphids appeared to pull out their stylets and to start walking on their host plant during the night, as shown in honeydew collection experiments with *Nasonovia ribisnigri* (currant–lettuce aphid) on lettuce (*Lactuca sativa*) and *A. fabae* on beans (van Helden and Tjallingii, 1993; J. Hardie, unpublished results).

Tactile disturbance from other members in the colony increases with population density, and this has been shown to increase mobility of individuals in colonies of *D. platanoidis* (Dixon, 1998). In colonies of *R. padi*, increased mobility has also been attributed to volatile semiochemicals identified from dense colonies (Quiroz *et al.*, 1997). When the colony size exceeds a certain density threshold in terms of aphids per cm^2, the mobility of individuals increases. These chemical stimuli appear to act by lowering the response threshold for different sorts of disturbance from the environment.

Development of winged individuals plays a key role for aphid population dynamics. This is a seasonally cyclic occurrence in host-alternating species regulated by a short-day mechanism but can also be initiated directly by population density effects (crowding), interrupted feeding, and decreasing food quality in all species (Harrewijn, 1972; Dixon, 1998; Müller *et al.*, 2001). The close link between host-plant responses to feeding and the population biology of aphids can be illustrated with the seasonal migrations of two host-alternating species, *R. padi* and *A. fabae*.

The overwintering eggs of *R. padi* hatch on the winter host, bird cherry (*P. padus*), and colonies develop on the young leaves and shoots. Alate spring migrants are produced as a response to increased crowding in the first 2–3 generations (Wiktelius, 1984b; Wiktelius *et al.*, 1990) and leave the winter host for different grasses, which are the summer host plants. Aphid feeding on bird cherry leaves and shoots causes damage that induces production of methyl salicylate (see Pickett and Glinwood, Chapter 9 this volume), which acts to reduce aphids landing and interferes with cues keeping the colonies together. Thus, methyl salicylate acts as a key promoter for spring migrant take-off (Glinwood and Pettersson, 2000a,b). A similar behavioural effect of methyl salicylate has been shown for the hop aphid, *P. humuli*, when this species is leaving its winter host, *Prunus cerasifera* (Campbell *et al.*, 1990). Field experiments have shown that application of methyl salicylate to barley can give a significant reduction of aphid infestation (Ninkovic *et al.*, 2003). In the autumn, *R. padi* gynoparae and males migrate from grasses to leaves of bird cherry and settle in aggregates (Pettersson, 1993). Gynoparae give birth to oviparae, which develop to adults feeding on senescing leaves in the canopy of bird cherry. After mating, oviparae oviposit around the hibernating plant buds, but to stay too long on the leaves would involve an obvious risk of falling to the ground when the leaf is shed from the plant. In field experiments, oviparae were found to be capable of monitoring the leaf ageing process and left the leaves in due time to avoid falling with the leaf (Glinwood and Pettersson, 2000c).

The host-plant preference of the winged autumn gynoparae of *A. fabae* changes at the final moult, and all larval stages prefer to settle on the natal, summer host (Hardie and Glascodine, 1990). In this species, the readiness of the adult gynoparae to take off seems to be related to the extended period of initial migratory behaviour during which the insect will not land on a green plant-like visual target (David and Hardie, 1988; Nottingham and Hardie, 1989; see also above). A similar situation was also found for the gynoparae of *R. padi* moving to bird cherry (Nottingham *et al.*, 1991a). By contrast, spring migrants and alate summer morphs of both species do not show such a prolonged migratory flight phase.

Plant Responses and Predisposition to Aphid Feeding

The damage caused by aphid feeding may vary in how serious it is for the plant (see Quisenberry and Ni, Chapter 13 this volume). Mechanical damage is usually moderate, while the salivary secretions cause damage that in some cases is expressed as necrotic tissue and/or discoloration. Even moderate aphid numbers can have serious effects. An initial infestation of 15 *A. fabae* per plant killed broad bean plants with 4–5 true leaves in 1–2 weeks (Tjallingii, 2004). Typical symptoms were wilting of leaves and subsequent desiccation leaf by leaf, caused by a dried or necrotic area of the stem or petiole below each leaf. However, broad bean responds to attack by *A. pisum* in a different way. Here, leaf wilting coincides with a dried or necrotic area at the stem base in or just above the soil. Obviously, the plant response to aphid saliva can be very pronounced and also, to some extent, species-specific.

Aphids secrete two types of saliva. One forms 'the stylet sheath', which is proteinaceous, and the other is watery saliva carrying an active set of enzymes (see earlier). The stylet sheath supports stylet movements and actively absorbs or immobilizes potentially toxic phytochemicals, while the watery saliva contains important enzyme systems such as pectinases, cellulases, polyphenol oxidases, and peroxidases (Miles, 1999). There are similarities in the plant response to an aphid attack and that of a fungal pathogen, and the cascade of responses released in the plant is similar to the salicylic pathway described by Walling (2000). Thus, Forslund *et al.* (2000) studied the induction of pathogen-related (PR-) proteins following infestation by *R. padi* in two barley varieties, one a commercial variety susceptible to aphid infestation and one an aphid-resistant breeding line. One basic chitinase, two acidic β-1,3-glucanases, and two basic β-1,3-glucanases were induced by the aphid infestation, and the chitinase and the two acidic glucanases were more strongly induced in the resistant variety. This, together with the demonstrated effects of methyl salicylate as a plant resistance-inducing agent (Pettersson *et al.*, 1994; Shualev *et al.*, 1997; Ninkovic *et al.*, 2003), would indicate that the salicylic pathway may be important in aphid-induced plant resistance to aphids.

Plants usually coexist with other plants, and neighbouring plants may also influence host-plant acceptance by aphids for a specific individual plant. Studies of communication between attacked and unattacked plants have been reported. In laboratory experiments with cereal plants, Pettersson *et al.* (1996) showed that volatiles from a barley plant attacked by aphids could induce a change in neighbouring unattacked plants, making them significantly less acceptable to *R. padi*. The induced plant response was systemic, and non-preference (antixenosis) seems to play a significant role, although the active mechanism behind reduced aphid acceptance is less well known. It was observed that plants influenced by neighbouring attacked plants showed a temporal change in biomass allocation, favouring root growth, which could indicate a systemically mobilized barrier of resistance. This is in line with results of studies of aphid resistance in three aphid–plant combinations as soon as four days after infestation (E. Prado and W.F. Tjallingii, unpublished results). However, this resistance was only manifested in leaves that were not colonized. On the infested leaves, the aphids probed and settled as easily as on clean plants. The induced and systemically-spread resistance appeared to be phloem located, as was shown by EPG recording. It is hypothesized that the systemically spread resistance may be due to phloem phase salivation (watery saliva), whereas the resistance suppression in infested leaves may be due to pathway phase salivation inter- or intracellularly in mesophyll and parenchyma tissues.

Plants not attacked by herbivores may challenge other plants *via* semiochemicals (allelopathy *sensu* Molisch, 1937). Effects of communication between undamaged plants on aphid host-plant preferences have also been demonstrated in both field and laboratory (Fig. 4.8) experiments (Pettersson *et al.*, 2003). The results indicate that effects on plant–plant communication should be seen in a tritrophic perspective, i.e. they also affect the behaviour of natural enemies of aphids (allelobiosis *sensu* Pettersson *et al.*, 2003). The transmission of allelobiotically-active plant substances affecting the aphid ecosystem has been reviewed by Birkett *et al.* (2000), and some specific stress-related substances have been tested (Chamberlain *et al.*, 2001). Allelobiotically-active substances from couch grass (*Elytrigia repens*) and thistle (*Cirsium arvense*) cause a significant reduction of aphid acceptance of exposed barley plants (Glinwood *et al.*, 2003, 2004; see also Pickett and Glinwood, Chapter 9 this volume). Similar effects (Fig. 4.9) were obtained when certain genotypes of barley were paired and one was exposed to volatiles from another (Pettersson *et al.*, 1999; Ninkovic *et al.*, 2002).

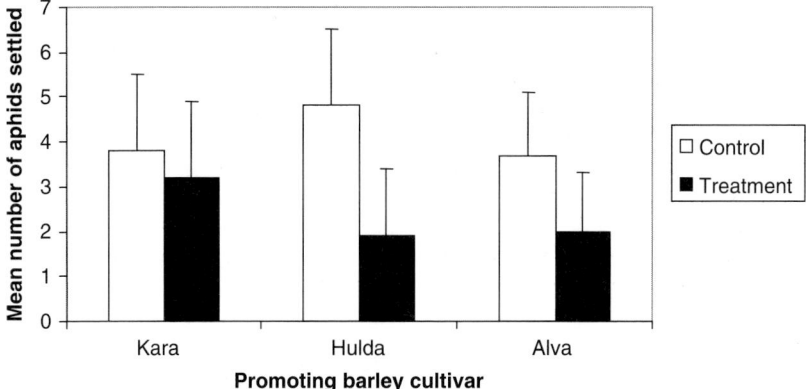

Fig. 4.8. Twin cage arrangements in which one plant in one cage is exposed to volatiles from another plant in a previous cage. The airflow through the two cages is provided *via* tubing connecting the second cage to a vacuum tank. The responses of the exposed plants can then be evaluated using different bioassays.

Fig. 4.9. Aphid settling on barley plants, cv. Kara after exposure to volatiles from cvs. Kara, Hulda, and Alva in twin cages (after Pettersson *et al.*, 1999).

Aphids and Host Plants – A Challenging Research Field

Visual and olfactory stimuli can act as specific and general landing cues to flying aphids. The primary rhinaria are the dominating antennal sensory organs and are known to be sensitive to plant odours. However, the impact of olfactory cues during the aphid landing and settling process is less well understood. It may vary between aphid species and, with reference to the limited host-plant range of the majority of the Aphidinae, avoidance of landing/settling on non-host plants may be of special importance. Several studies show synergistic interaction between host-plant odours and aphid pheromones such as alarm, aggregation, and sex pheromones. Thus, it seems that host-plant odours can reinforce the response to pheromone-directed mechanisms such as aggregation and finding a mate. Compared to other herbivorous insects, aphids show reduced contact chemoreceptors, but at least one sensory hair on the tarsus and the apical hairs on the antennae are likely to have this function. Further studies on the electrophysiology of these sensilla and the behavioural effects of the plant cuticle content of semiochemicals should fill in important gaps in knowledge of their role after the aphid lands on a plant and in aphid host-plant selection in general.

Aphids ultimately accept their food plant and their feeding site after the stylets have reached the phloem and sap ingestion has started, but gustatory discrimination of internal plant chemicals comes into action long before the phloem is reached. Aphids show a higher level of tissue specialization for feeding when compared with chewing insects. In each plant tissue along the stylet route, inter- and intracellular compartments are sampled and the chemical signals are tasted, most likely by a well-developed internal gustatory organ. Nevertheless, little is known about stimulating or deterrent properties of the allelochemicals, with a few exceptions, but any role in phloem finding has not been reported so far. On the other hand, it becomes more and more evident that aphids do affect the attacked plants, mainly due to the salivary secretions of probing and feeding aphids (Tjallingii, 2006). Along the stylet pathway, these secretions may cause induced effects, first locally, but later presumably also spreading systemically, whereas during the phloem phase, the saliva injected into sieve elements may be responsible more directly for induced systemic effects. Aphid probing and feeding has a systemic effect on the plant, mainly at the phloem level. However, induced effects on a specific site may favour individuals of the same species and inhibit feeding of competing aphid species. Therefore, host-plant reactions may actively mediate the competition between aphid species. A new wave of further studies on the whole organism and at the molecular level will contribute to general understanding.

A better mechanistic understanding of the adaptation process of aphids to new host plants would contribute to general aspects of aphid ecology and evolution. Special emphasis should be given to studies on the interaction between acceptance of new host plants and to the adaptability of active enzyme systems and the role of semiochemicals.

Experimental evidence for the effects of communication between plants (allelopathy) on herbivores has added a new dimension to aphid/plant ecology. In specific combinations, both attacked and unattacked plants may induce responses in neighbouring plants that affect aphid host-plant acceptance and support the habitat search of their natural enemies. This perspective can be seen as part of a tritrophic network with a general ecological relevance.

Conclusions

Host-plant selection by aphids is achieved by a sequence of responses from landscape to cellular scale. All responses represent behaviours with search-and-locate cycles in which sensory stimuli of different modalities play a role.

Though initially visual cues seem to play an important role, olfaction and taste increase in importance after plant contact.

Aphids are well equipped with chemosensory organs and the responses to volatile and non-volatile plant stimuli (olfaction and taste) are important but complex, and require further detailed study.

Plant penetration can be distinguished in three distinct phases; pathway, xylem, and phloem phases.

- The intercellular pathway is frequently interrupted by brief intracellular punctures.
- The xylem phase can be considered as drinking to relieve water stress.
- The phloem phase comprises the main feeding, but this is always preceded by sieve element salivation, presumably to suppress phloem wound responses.

During feeding, the composition of the phloem is continuously monitored and feeding can be ended on the basis of composition changes or signal molecules.

Changes in phloem sap can be caused by plant ageing, diurnal as well as seasonal changes (changed source–sink relations), but also by feeding of conspecific and non-specific insects (phloem feeders or not) and pathogens.

Plant–plant communication can change aphid host-plant preferences (allelobiosis). Though correlations of behavioural changes with plant infestations are convincing, their causal backgrounds and mechanisms need more research.

References

Åhman, I. (1990) Plant surface characteristics and movements of two *Brassica*-feeding aphids, *Lipaphis erysimi* and *Brevicoryne brassicae*. In: *Symposia Biologica Hungaria No. 39*. Publishing House of Hungarian Academy of Sciences, Budapest, pp. 119–125.

Åhman, I., Weibull, J. and Pettersson, J. (1985) The role of plant size and plant density for host finding in *Rhopalosiphum padi* (L.) (Hem.: Aphididae). *Swedish Journal of Agricultural Research* 15, 19–24.

Alvarez, A.E., Tjallingii, W.F., Garzo, E., Vleeshouwers, V., Dicke, M. and Vosman, B. (2006) Location of resistance factors in the leaves of potato and wild tuber-bearing *Solanum* species to the aphid *Myzus persicae*. *Entomologia Experimentalis et Applicata* (in press).

Anderson, M. and Bromley, A.K. (1987) Sensory system. In: Minks, A.K. and Harrewijn, P. (eds) *Aphids. Their Biology, Natural Enemies and Control, Volume 2A*. Elsevier, Amsterdam, pp. 153–162.

Birkett, M.A., Campbell, C.A.M., Chamberlain, K., Guerrieri, E., Hick, A.J., Martin, J.L., Matthes, M., Napier, J.A., Pettersson, J., Pickett, J.A., Poppy, G.M., Pow, E.M., Pye, B.J., Smart, L.E., Wadhams, G.H., Wadhams, L.J. and Woodcock, C.M. (2000) New roles for *cis*-jasmone as an insect semiochemical and in plant defense. *Proceedings of the National Academy of Sciences of the United States of America* 97, 9329–9334.

Blackman, R.L. and Eastop, V.F. (1984) *Aphids on the World's Crops: An Identification and Information Guide*. Wiley, Chichester, 466 pp.

Bromley, A.K., Dunn, J.A. and Anderson, M. (1980) Ultrastructure of the antennal sensilla of aphids II. Trichoid, chordotonal and campaniform sensilla. *Cellular and Tissue Research* 205, 493–511.

Caillaud, M.C. and Via, S. (2000) Specialized feeding behaviour influences both ecological specialization and assortative mating in sympatric host races of pea aphids. *American Naturalist* 6, 606–621.

Campbell, C.A.M., Dawson, G.W., Griffiths, D.C., Pettersson, J., Pickett, J.A., Wadhams, L.J. and Woodcock, C.M. (1990) Sex attractant pheromone of damson–hop aphid, *Phorodon humuli* (Homoptera, Aphididae). *Journal of Chemical Ecology* 16, 3455–3465.

Cappucino, N. (1987) Comparative population dynamics of two goldenrod aphids: spatial patterns and temporal constancy. *Ecology* 68, 1634–1646.

Cappucino, N. (1988) Spatial patterns of goldenrod aphids and the response of enemies to patch density. *Oecologia* 76, 607–610.

Chamberlain, K., Guerrieri, E., Pennacchio, F., Pettersson, J., Pickett, J.A., Poppy, G.M., Powell, W., Wadhams, L.J. and Woodcock, C.M. (2001) Can aphid-induced plant signals be transmitted aerially and through the rhizosphere? In: Dicke, M. and Bruin, J. (eds) *Chemical Information Transfer Between Wounded and Unwounded Plants. Biochemistry, Systematics and Ecology* 29, 1063–1074.

Chapman, R.F., Bernays, E.A. and Simpson, S.J. (1981) Attraction and repulsion of the aphid, *Cavariella aegopodii*, by plant odours. *Journal of Chemical Ecology* 7, 881–888.

Cherqui, A. and Tjallingii, W.F. (2000) Salivary proteins of aphids, a pilot study on identification, separation, and immunolocalisation. *Journal of Insect Physiology* 46, 1177–1186.

David, C.T. and Hardie, J. (1988) Visual responses of free-flying, summer and autumn forms of the black bean aphid, *Aphis fabae*. *Physiological Entomology* 13, 277–284.

Dixon, A.F.G. (1998) *Aphid Ecology*. Chapman and Hall, London, 300 pp.

Eastop, V.F. (1955) Selection of aphid species by different kinds of insect traps. *Nature, London* 176, 193.

Forslund, K., Pettersson, J., Bryngelsson, T. and Jonsson, L. (2000) Aphid infestation induces PR-proteins differently in barley susceptible or resistant to the bird cherry–oat aphid (*Rhopalosiphum padi* (L.)). *Physiologia Plantarum* 110, 496–502.

Gabrys, B. and Tjallingii, W.F. (2002) The role of sinigrin in host plant recognition by aphids during initial plant penetration. *Entomologia Experimentalis et Applicata* 104, 89–93.

Gabrys, B., Tjallingii, W.F. and van Beek, T.A. (1997) Analysis of EPG recorded probing by cabbage aphid on host plant parts with different glucosinolate contents. *Journal of Chemical Ecology* 23, 1661–1663.

Garzo, E., Soria, C., Gómez-Guiollamón, M.L. and Fereres, A. (2002) Feeding behaviour of *Aphis gossypii* on resistant accessions of different melon genotypes (*Cucumis melo*). *Phytoparasitica* 30, 129–140.

Gianoli, E. (2000) Competition in cereal aphids on wheat plants. *Environmental Entomology* 29, 213–219.

van Giessen, W.A., Fescemyer, H.W., Burrows, P.M., Peterson, J.K. and Barnett, O.W. (1994) Quantification of electroantennogram responses of the primary rhinaria of *Acyrthosiphon pisum* (Harris) to C4–C8 primary alcohols and aldehydes. *Journal of Chemical Ecology* 20, 909–912.

Givovitch, A. and Niemeyer, H.M. (1991) Hydroxamic acids affecting barley yellow dwarf virus transmission by the aphid *Rhopalosiphum padi*. *Entomologia Experimentalis et Applicata* 59, 79–85.

Givovich, A., Niemeyer, H., Sandström, J. and Pettersson, J. (1994) Presence of hydroxamic acid glucosides in wheat phloem sap, and its consequences for the performance of *Rhopalosiphum padi* (L.) (Homoptera: Aphididae). *Journal of Chemical Ecology* 20, 1923–1930.

Glinwood, R. and Pettersson, J. (2000a). Host choice and host leaving in *Rhopalosiphum padi* emigrants and repellency of aphid colonies on the winter host. *Bulletin of Entomological Research* 90, 57–61.

Glinwood, R. and Pettersson, J. (2000b) Change in response of *Rhopalosiphum padi* spring migrants to the repellent winter host component methylsalicylate. *Entomologia Experimentalis et Applicata* 94, 325–330.

Glinwood, R. and Pettersson, J. (2000c) Movement by oviparae of a host-alternating aphid: a response to leaf fall. *Oikos* 90, 43–49.

Glinwood, R., Pettersson, J., Ahmed, E., Ninkovic, V., Birkett, M. and Pickett, J. (2003) Change in acceptability of barley plants to aphids after exposure to allelochemicals from couch-grass (*Elytrigia repens*). *Journal of Chemical Ecology* 29, 261–274.

Glinwood, R., Pettersson, J., Ahmed, E. and Ninkovic, V. (2004) Barley exposed to aerial allelopathy from thistles (*Cirsium* spp.) becomes less acceptable to aphids. *Ecological Entomology* 29, 188–195.

Hardie, J. (1980) Behavioural differences between alate and apterous larvae of the black bean aphid, *Aphis fabae*: dispersal from the host plant. *Entomologia Experimentalis et Applicata* 28, 338–340.

Hardie, J. (1989) Spectral specificity for targeted flight in the black bean aphid, *Aphis fabae*. *Journal of Insect Physiology* 35, 619–626.

Hardie, J. and Campbell, C.A.M. (1998) The flight behaviour of spring and autumn forms of the damson–hop aphid, *Phorodon humuli*, in the laboratory. In: Nieto Nafría, J.M. and Dixon, A.F.G. (eds) *Aphids in Natural and Managed Ecosystems*. Universidad de León, León, pp. 205–212.

Hardie, J. and Glascodine, J. (1990) Polyphenism and host-plant preference in the black bean aphid, *Aphis fabae*. *Acta Phytopathologica et Entomologica Hungarica* 25, 323–330.

Hardie, J. and Powell, G. (2000) Close-up video combined with electronic monitoring of plant penetration and behavioural effects of an aphid (Homoptera: Aphididae) antifeedant. In: Walker, G.P. and Backus, E.A. (eds) *Principles and Applications of Electronic Monitoring and Other Techniques in the Study of Homopteran Feeding Behaviour*. Thomas Say Publications in Entomology, Entomological Society of America, Lanham, pp. 201–211.

Hardie, J. and Schlumberger, A. (1996) The early appearance of foraging flight associated with starvation in an aphid. *Entomologia Experimentalis et Applicata* 80, 73–75.

Hardie, J. and Young, S. (1997) Aphid flight-track analysis in three dimensions using video techniques. *Physiological Entomology* 22, 116–122.

Hardie, J., Holyoak, M., Taylor, N.J. and Griffiths, D.C. (1992) The combination of electronic monitoring and video-assisted observations of plant penetration by aphids and behavioural effects of polygodial. *Entomologia Experimentalis et Applicata* 63, 233–239.

Hardie, J., Visser, J.H. and Piron, P.G.M. (1994) Perception of volatiles associated with sex and food by different forms of the black bean aphid, *Aphis fabae*. *Physiological Entomology* 19, 278–284.

Hardie, J., Storer, J.R., Cook, F.J., Campbell, C.A.M., Wadhams, L.J., Lilley, R. and Peace, L. (1996) Sex pheromone and visual trap interactions in mate location strategies and aggregation by host-alternating aphid species in the field. *Physiological Entomology* 21, 97–106.

Harrewijn, P. (1972) Wing production by the aphid *Myzus persicae* related to nutritional factors in potato plants and artificial diets. In: Rodriguez, J. (ed.) *Insect and Mite Nutrition*. North-Holland, Amsterdam, pp. 575–588.

Heathcote, G.D. (1957) The comparison of yellow cylindrical, flat and water traps, and of Johnson suction traps, for sampling aphids. *Annals of Applied Biology* 45, 133–139.

van Helden, M. and Tjallingii, W.F. (1993) The resistance of lettuce (*Lactuca sativa* L.) to *Nasonovia ribisnigri*: the use of electrical penetration graphs to locate the resistance factor(s) in the plant. *Entomologia Experimentalis et Applicata* 66, 53–58.

Hori, M. (1999) Role of host-plant odours in the host-finding behaviours of aphids. *Applied Entomology and Zoology* 34, 293–298.

Ibbotson, A. and Kennedy, J.S. (1950) The distribution of aphid infestation in relation to leaf age. II. The progress of *Aphis fabae* Scop. infestations on sugar beet in pots. *Annals of Applied Biology* 37, 680–696.

Jiang, Y.X. and Walker, G.P. (2001) Pathway phase waveform characteristics correlated with length and rate of stylet advancement and partial stylet withdrawal in AC electrical penetration graphs of adult whiteflies. *Entomologia Experimentalis et Applicata* 101, 233–246.

Johansson, C., Pettersson, J. and Niemeyer, H.M. (1997) Odour recognition between apterae of the aphids *Sitobion avenae* and *Rhopalosiphum padi* on a wheat plant. *European Journal of Entomology* 94, 557–559.

Kaloshian, I., Kinsey, M.G., Williamson, V.M. and Ullman, D.E. (2000) Mi-mediated resistance against the potato aphid, *Macrosiphum euphorbiae* (Hemiptera: Aphididae) limits sieve element ingestion. *Environmental Entomology* 4, 690–695.

Kennedy, J.S. (1986) Some current issues in orientation to odour sources. In: Payne, T.L., Birch, M.C. and Kennedy, C.E.J. (eds) *Mechanisms in Insect Olfaction*. Clarendon Press, Oxford, pp. 11–25.

Kennedy, J.S. and Booth, C.O. (1951) Host alternation in *Aphis fabae* Scop. I. Feeding preference and fecundity in relation to the age and kind of leaves. *Annals of Applied Biology* 38, 25–64.

Kennedy, J.S. and Booth, C.O. (1963) Free flight of aphids in the laboratory. *Journal of Experimental Biology* 40, 67–85.

Kennedy, J.S. and Thomas, A.A.G. (1974) Behaviour of some low-flying aphids in wind. *Annals of Applied Biology* 76, 143–159.

Kennedy, J.S., Booth, C.O. and Kershaw, W.J.S. (1963a) Host finding by aphids in the field. I. Gynoparae of *Myzus persicae* (Sulzer). *Annals of Applied Biology* 47, 410–423.

Kennedy, J.S., Booth, C.O. and Kershaw, W.J.S. (1963b) Host finding by aphids in the field II. *Aphis fabae* Scop. gynoparae and *Brevicoryne brassicae* L.; with a re-appraisal of the role of host-finding behaviour in virus spread. *Annals of Applied Biology* 47, 424–444.

Klingauf, F., Nocker-Wenzel, K. and Rottger, U. (1978) Die Rolle periferer Pflanzenwachse für den Befall durch phytophage Insekten. *Zeitschrift für Pflanzenkrankheiten und Pflanzenschutz* 85, 227–237.

Klinger, J., Powell, G., Thompson, G.A. and Isaacs, R. (1998) Phloem specific resistance in *Cucumis melo* line AR5: effects on feeding behaviour and performance of *Aphis gossypii*. *Entomologia Experimentalis et Applicata* 86, 79–88.

Knoblauch, M. and Peters, W.S. (2004) Forisomes, a novel type of Ca^{2+}-dependent contractile protein motor. *Cell Motility and the Cytoskeleton* 58, 137–142.

Knoblauch, M. and van Bel, A.J.E. (1998) Sieve tubes in action. *Plant Cell* 10, 35–50.

Lapointe, S.L. and Tingey, W.M. (1986) Glandular trichomes of *Solanum neocardenasii* confer resistance to green peach aphid (Homoptera, Aphididae). *Journal of Economic Entomology* 79, 1264–1268.

Mann, J.A., Tatchell, G.M., Dupuch, M.J., Harrington, R., Clark, S.J. and McCartney, H.A. (1995) Movement of apterous *Sitobion avenae* (Homoptera, Aphididae) in response to leaf disturbances caused by wind and rain. *Annals of Applied Biology* 126, 417–427.

Martín, B., Collar, J.L., Tjallingii, W.F. and Fereres, A. (1997) Intracellular ingestion and salivation by aphids may cause acquisition and inoculation of non-persistently transmitted plant viruses. *Journal of General Virology* 78, 2701–2705.

Massardo, F., Zuñiga, G.E., Perez, L.M. and Corcuera, L.J. (1994) Effects of hydroxamic acids on electron transport and their cellular location in corn. *Phytochemistry* 35, 873–876.

Mathile, P. (1984) Das toxische Kompartiment der Pflanzenzelle. *Naturwissenschaften* 71, 18–24.

McLean, D.L. and Kinsey, M.G. (1964) A technique for electrically recording aphid feeding and salivation. *Nature, London* 205, 1130–1131.

Miles, P.W. (1999) Aphid saliva. *Biological Reviews of the Cambridge Philosophical Society* 74, 41–85.

Mittler, T.E. (1958) Studies on the feeding and nutrition of *Tuberolachnus salignus* (Gmelin) (Homoptera, Aphididae). II. The nitrogen and sugar composition of ingested phloem sap and excreted honeydew. *Journal of Experimental Biology* 35, 74–84.

Moericke, V. (1955) Über die Lebensgewohnheiten der geflügelten Blattläuse (Aphidina) unter besonderer Berücksichtigung des Verhaltens beim Landen. *Zeitschrift für Angewandte Entomologie* 37, 29–91.

Moericke, V. (1969) Host-plant specific colour behaviour by *Hyalopterus pruni* (Aphididae). *Entomologia Experimentalis et Applicata* 24, 409–420.

Molisch, H. (1937) *Der Einfluss einer Pflanze auf die Andere. Allelopathie.* Gustav Fischer, Jena, 106 pp.

Montllor, C.B. (1991) Influence of plant chemistry on aphid feeding behavior. In: Bernays, E.A. (ed.) *Insect–Plant Interactions, Volume 3.* CRC, Boca Raton, pp. 125–173.

Montllor, C.B., Campbell, B.C. and Mittler, T.E. (1983) Natural and induced differences in probing behavior of two biotypes of the green bug, *Schizaphis graminum*, in relation to resistance in sorghum. *Entomologia Experimentalis et Applicata* 34, 99–106.

Mordwilko, A.K. (1934) On the evolution of aphids. *Archiv für Naturgeschichte* 3, 1–60.

Müller, C., Williams, I. and Hardie, J. (2001) The role of nutrition, crowding and interspecific interactions in wing development in aphids. *Ecological Entomology* 26, 330–340.

Müller, F.P. (1984) Die Rolle des Lichtes bei dem Ansiedlungsverhalten zweier Rassen der Blattlaus *Acyrthosiphon pisum* (Harris). *Deutsche Entomologische Zeitschrift* 31, 201–214.

Mullin, C.A. (1986) Adaptive divergence of chewing and sucking arthropods to plant allelochemicals. In: Brattsten, L.B. and Ahmad, S. (eds) *Molecular Aspects of Insect–Plant Associations.* Plenum, New York, pp. 175–209.

Nault, L.R. and Styer, W.E. (1972) Effects of sinigrin on host selection by aphids. *Entomologia Experimentalis et Applicata* 15, 423–437.

Ninkovic, V., Olsson, U. and Pettersson, J. (2002) Field experiments with intraspecific plant communication and changes in host-plant acceptance of aphids. *Entomologia Experimentalis et Applicata* 102, 177–182.

Ninkovic, V., Ahmed, E., Glinwood, R. and Pettersson, J. (2003) Effects of two types of semiochemical on population development of the bird cherry–oat aphid *Rhopalosiphum padi* in a barley crop. *Agricultural and Forest Entomology* 5, 27–33.

Nottingham, S.F. and Hardie, J. (1989) Migratory and targeted flight in seasonal forms of the black bean aphid, *Aphis fabae. Physiological Entomology* 14, 451–458.

Nottingham, S.F. and Hardie, J. (1993) Flight behaviour of the black bean aphid, *Aphis fabae*, and the cabbage aphid, *Brevicoryne brassicae*, in host and non-host plant odour. *Physiological Entomology* 18, 389–394.

Nottingham, S.F., Hardie, J. and Tatchell, G.M. (1991a) Flight behaviour of the bird cherry aphid, *Rhopalosiphum padi. Physiological Entomology* 16, 223–229.

Nottingham, S.F., Hardie, J., Dawson, G.W., Hick, A.J., Pickett, J.A., Wadhams, L.J. and Woodcock, C.M. (1991b) Behavioural and electrophysiological responses of aphids to host- and non-host plant volatiles. *Journal of Chemical Ecology* 17, 1231–1242.

Park, K.C. and Hardie, J. (1998) An improved aphid electroantennogram. *Journal of Insect Physiology* 44, 919–928.

Park, K.C. and Hardie, J. (2002) Functional specialisation and polyphenism in aphid olfactory sensilla. *Journal of Insect Physiology* 48, 527–535.

Park, K.C. and Hardie, J. (2004) Electrophysiological characterisation of olfactory sensilla in the black bean aphid, *Aphis fabae. Journal of Insect Physiology* 50, 647–655.

Park, K.C., Elias, D., Donato, B. and Hardie, J. (2000) Electroantennogram and behavioural responses of different forms of the bird cherry–oat aphid, *Rhopalosiphum padi*, to sex pheromone and a plant volatile. *Journal of Insect Physiology* 46, 597–604.

Pettersson, J. (1970) Studies on *Rhopalosiphum padi* (L.): laboratory studies on olfactometric responses to the winter host *Prunus padus* L. *Lantbrukshögskolans Annaler* 36, 381–399.

Pettersson, J. (1973) Olfactory reactions of *Brevicoryne brassicae* (L.) (Hom.: Aph.). *Swedish Journal of Agricultural Research* 3, 95–103.

Pettersson, J. (1993) Odour stimuli affecting autumn migration of *Rhopalosiphum padi* (L.) (Hemiptera: Homoptera). *Annals of Applied Biology* 122, 417–425.

Pettersson, J. and Stephansson, D. (1991) Odour communication in two *Brassica*-feeding aphid species (Hom.:Aphidinea: Aphididae). *Entomologia Generalis* 16, 241–247.

Pettersson, J., Pickett, J.A., Pye, B.J., Quiroz, A., Smart, L.E., Wadhams, L.J. and Woodcock, C.M. (1994) Winter host component reduces colonization of summer hosts by the bird-cherry–oat aphid, *Rhopalosiphum padi* (L.) and other aphids in cereal fields. *Journal of Chemical Ecology* 20, 2565–2574.

Pettersson, J., Quiroz, A., Stephansson, D. and Niemeyer, H.M. (1995) Odour communication of *Rhopalosiphum padi* (L.) (Hom.: Aph.) on grasses. *Entomologia Experimentalis et Applicata* 76, 325–328.

Pettersson, J., Quiroz, A. and Fahad, A.E. (1996) Aphid antixenosis mediated by volatiles in cereals. *Acta Agriculturae Scandinavica* 46, 135–140.

Pettersson, J., Ninkovic, V. and Ahmed, F. (1999) Volatiles from different barley cultivars affect aphid acceptance of neighbouring plants. *Acta Agriculturae Scandinavica* 49, 152–157.

Pettersson, J., Ninkovic, V. and Glinwood, R. (2003) Plant activation of barley by intercropped conspecifics and weeds: allelobiosis. *Proceedings of the British Crop Protection Council International Congress on Crop Science and Technology, Glasgow, November 2003* 2, 1135–1144.

Pickett, J.A., Wadhams, L.J., Woodcock, C.M. and Hardie, J. (1992) The chemical ecology of aphids. *Annual Review of Entomology* 37, 67–90.

Pollard, D.G. (1973) Plant penetration by feeding aphids (Hemiptera, Aphidoidea): a review. *Bulletin of Entomological Research* 62, 631–714.

Powell, G. and Hardie, J. (2001) A potent morph-specific parturition stimulant in the overwintering host plant of the black bean aphid, *Aphis fabae*. *Physiological Entomology* 26, 194–201.

Powell, G. and Hardie, J. (2002) Xylem ingestion by winged aphids. *Entomologia Experimentalis et Applicata* 104, 103–108.

Powell, G., Pirone, T. and Hardie, J. (1995) Aphid stylet activity during potyvirus acquisition from plants and an *in vitro* system that correlates with subsequent transmission. *European Journal of Plant Pathology* 101, 411–420.

Powell, G., Maniar, S.P., Pickett, J.A. and Hardie, J. (1999) Aphid responses to non-host epicuticular lipids. *Entomologia Experimentalis et Applicata* 91, 115–123.

Powell, G., Tosh, C.R. and Hardie, J. (2004) Parturation by colonising aphids: no correlation with phloem ingestion. In: Simon, J.C., Dedryver, C.A., Rispe, C. and Hullé, M. (eds) *Aphids in a New Millennium*. INRA, Paris, pp. 485–489.

Prado, E. and Tjallingii, W.F. (1994) Aphid activities during sieve element punctures. *Entomologia Experimentalis et Applicata* 42, 157–165.

Prado, E. and Tjallingii, W.F. (1997) Effects of previous plant infestation on sieve element acceptance by two aphids. *Entomologia Experimentalis et Applicata* 82, 189–200.

Prado, E. and Tjallingii, W.F. (2007) Behavioural evidence for local reduction of aphid-induced resistance. *Journal of Insect Science* (in press).

Prokopy, R.J. and Owens, E.D. (1983) Visual detection of plants by herbivorous insects. *Annual Review of Entomology* 28, 337–364.

Quiroz, A., Pettersson, J., Pickett, J.A., Wadhams, L. and Niemeyer, H.M. (1997) Key compounds in a spacing pheromone in the bird cherry–oat aphid, *Rhopalosiphum padi* (L.) (Hemiptera, Aphididae). *Journal of Chemical Ecology* 23, 2599–2607.

Ramirez, C.C. and Niemeyer, H. (2000) The influence of previous experience and starvation on aphid feeding behavior. *Journal of Insect Behavior* 13, 699–709.

Robert, Y. (1987) Dispersion and migration. In: Minks, A.K. and Harrewijn, P. (eds) *Aphids. Their Biology, Natural Enemies and Control, Volume 2A*. Elsevier, Amsterdam, pp. 299–313.

Sandström, J. and Pettersson, J. (2000) Winter host specialization in a host-alternating aphid. *Journal of Insect Behavior* 13, 815–826.

Shualev, V., Silverman, P. and Raskin, I. (1997) Airborne signalling by methyl salicylate in plant pathogen resistance. *Nature, London* 385, 718–721.

Sigvald, R. (1984) The relative efficiency of some aphid species as vectors of PVY. *Potato Research* 27, 285–290.

Spiller, N.J., Koenders, L. and Tjallingii, W.F. (1990) Xylem ingestion by aphids – a strategy for maintaining water balance. *Entomologia Experimentalis et Applicata* 55, 101–104.

Storer, J.R., Powell, G. and Hardie, J. (1996) Settling responses of aphids in air permeated with non-host volatiles. *Entomologia Experimentalis et Applicata* 80, 76–78.

Storer, J.R., Young, S. and Hardie, J. (1999) Three-dimensional analysis of aphid landing behaviour in the laboratory and field. *Physiological Entomology* 24, 271–277.

Thieme, T. (1987) Members of the complex *Aphis fabae* Scop. and their host plants. In: Holman, J., Pelikan, J., Dixon, A.F.G. and Weissman, L. (eds) *Population Structure, Genetics and Taxonomy of Aphids and Thysanoptera*. SPB Academic Publishing, The Hague, pp. 314–323.

Tingey, W.M. (1991) Potato glandular trichomes – defensive activity against insect attack. *ACS Symposium Series, No. 449*, 126–136.

Tjallingii, W.F. (1978) Mechanoreceptors of the aphid labium. *Entomologia Experimentalis et Applicata* 24, 531–537.

Tjallingii, W.F. (1985) Membrane potentials as an indication for plant cell penetration by aphid stylets. *Entomologia Experimentalis et Applicata* 38, 187–193.

Tjallingii, W.F. (1986) Wire effects on aphids during electrical recording of stylet penetration. *Entomologia Experimentalis et Applicata* 40, 89–98.

Tjallingii, W.F. (1987) Stylet penetration activities by aphids: new correlations with electrical penetration graphs. In: Labeyrie, V., Fabres, G. and Lachaise, D. (eds) *Insect–Plants*. W. Junk, Dordrecht, pp. 301–306.

Tjallingii, W.F. (1988) Electrical recording of stylet penetration activities. In: Minks, A.K. and Harrewijn, P. (eds) *Aphids. Their Biology, Natural Enemies and Control, Volume 2B*. Elsevier, Amsterdam, pp. 95–108.

Tjallingii, W.F. (1990) Continuous recording of stylet penetration activities by aphids. In: Campbell, R.K and Eikenbary, R.D. (eds) *Aphid–Plant Genotype Interactions*. Elsevier, Amsterdam, pp. 89–99.

Tjallingii, W.F. (1994) Sieve element acceptance by aphids. *European Journal of Entomology* 91, 45–52.

Tjallingii, W.F. (1995) Regulation of phloem sap feeding by aphids. In: Chapman, R.F. and de Boer, G. (eds) *Regulatory Mechanisms in Insect Feeding*. Chapman and Hall, New York, pp. 190–209.

Tjallingii, W.F. (2000) Comparison of AC and DC systems for electronic monitoring of stylet penetration activities by homopterans. In: Walker, G.P. and Backus, E.A. (eds) *Principles and Applications of Electronic Monitoring and Other Techniques in the Study of Homopteran Feeding Behavior*. Thomas Say Publications in Entomology, Entomological Society of America, Lanham, pp. 41–69.

Tjallingii, W.F. (2004) What is a host plant? Aphid–plant interactions in three *Vicia faba* exploiting aphids. In: Simon, J.C., Dedryver, C.A., Rispe, C. and Hullé, M. (eds) *Aphids in a New Millennium*. INRA, Paris, pp. 513–519.

Tjallingii, W.F. (2006) Salivary secretions by aphids interacting with proteins of phloem wound responses. *Journal of Experimental Botany* 57, 739–745.

Tjallingii, W.F. and Hogen Esch, Th. (1993) Fine structure of the stylet route in plant tissues by some aphids. *Physiological Entomology* 18, 317–328.

Tjallingii, W.F. and Mayoral, A.M. (1992) Criteria for host-plant acceptance by aphids. In: Menken, S.B.J., Visser, J.H. and Harrewijn, P. (eds) *Proceedings of the 8th Symposium on Insect–Plant Relationships*. Kluwer, Dordrecht, pp. 280–282.

Tosh, C.R., Powell, G. and Hardie, J. (2002) Maternal reproductive decisions are independent of feeding in the black bean aphid, *Aphis fabae. Journal of Insect Physiology* 48, 619–629.

Visser, J.H. and Piron, P.G.M. (1997) Odour response profiles in aphids differentiating for species, clone, form and food. *Proceedings of the Section Experimental and Applied Entomology of the Netherlands Entomological Society* 8, 115–118.

Walling, L.L. (2000) The myriad plant responses to herbivores. *Journal of Plant Growth Regulation* 19, 195–216.

Way, M. (1973) Population structure in aphid colonies. In: Lowe, H.J.B. (ed.) *Perspectives in Aphid Biology*. Entomological Society of New Zealand, Auckland, pp. 76–84.

Way, M.J. and Banks, C.J. (1967) Intra-specific mechanisms in relation to the natural regulation of numbers of *Aphis fabae* Scop. *Annals of Applied Biology* 59, 189–205.

Way, M. and Cammell, M. (1970) Aggregation behaviour in relation to food utilization in aphids. In: Watson, A. (ed.) *Animal Populations in Relation to Their Food Resources. 10th Symposium of the British Ecological Society*. Blackwell, Oxford, pp. 229–247.

Weber, H. (1928) Skelett, Muskulatur, und Darm der schwarzen Blattlaus, *Aphis fabae* Scop. *Zoologica, Stuttgart* 76, 1–120.

Wensler, R.J.D. (1962) Mode of host selection by an aphid. *Nature, London* 195, 830–831.

Wensler, R.J.D. (1974) Sensory innervation monitoring movement and position in the mandibular stylets of the aphid *Brevicoryne brassicae* L. *Journal of Morphology* 143, 349–364.

Wensler, R.J.D. (1977) The fine structure of distal receptors on the labium of the aphid *Brevicoryne brassicae* L. (Homoptera). *Cell and Tissue Research* 181, 409–421.

Wensler, R.J.D. and Filshie, B.K. (1969) Gustatory sense organs in the food canal of aphids. *Journal of Morphology* 129, 473–492.

Wiktelius, S. (1984a) Long-range migration of aphids into Sweden. *International Journal of Biometeorology* 28, 185–200.

Wiktelius, S. (1984b) Studies on population development on the primary host and spring migration of *Rhopalosiphum padi* (L.) (Hom. Aphididae). *Zeitschrift für Angewandte Entomologie* 97, 217–222.

Wiktelius, S., Weibull, J. and Pettersson, J. (1990) Aphid host plant ecology: the bird cherry–oat aphid as a model. In: Campbell, R.K. and Eikenbary, R.D. (eds) *Aphid–Plant Genotype Interactions.* Elsevier, Amsterdam, pp. 21–36.

Will, T. and van Bel, A.J.E. (2006) Physical and chemical interactions between aphids and plants. *Journal of Experimental Botany* 57, 729–737.

Winter, H., Lohaus, G. and Heldt, H.W. (1992) Phloem transport of amino acids in relation to their cytosolic levels in barley leaves. *Plant Physiology* 99, 996–1004.

5 Nutrition and Symbiosis

Angela E. Douglas[1] and Helmut F. van Emden[2]

[1]Department of Biology, University of York, Heslington, York, YO10 5YW, UK;
[2]School of Biological Sciences, University of Reading, Whiteknights, Reading,
Berks, RG6 6AJ, UK

Introduction

A single question dominates the study of aphid nutrition: how does plant phloem sap, a nutritionally unbalanced diet of high C:N content, support the remarkably high growth and reproductive rates characteristic of many aphid species? The 'short answer' to this question has two elements. First, phloem sap offers a near continuous supply of small organic compounds, principally sugars and amino acids, that require the minimum of digestive processing, and consequently, the assimilation efficiency of aphids is exceptionally high. Second, the key nutritional inadequacy of phloem sap, the low essential amino acid content, is met by symbiotic bacteria in the aphid.

The 'long answer' to this opening question requires a mechanistic understanding of aphid nutrition and its interaction with the functions of the symbiotic microorganisms, and this is the topic of this chapter. Research on aphid nutrition has been dominated by carbon and nitrogen nutrition, and these topics are addressed in the third and fourth sections, respectively. The fifth section summarizes our current understanding of the mineral and micronutrient requirements of aphids. First, however, the key features of microbial symbiosis in aphids are reviewed.

Two technical advances have been vital in the development of research on aphid nutrition and symbiosis: first, chemically defined diets of a composition that is known precisely and can be manipulated readily; and second, aphids experimentally deprived of their symbiotic bacteria. Diet studies, often involving bacteria-free aphids, are now used routinely to investigate various aspects of aphid nutrition, e.g. identification of dietary requirements and analysis of nutrient utilization patterns. The value and limitations of such diet-based studies to explore aphid nutrition are assessed in the concluding section of this chapter.

Microbial Symbiosis in Aphids

Diversity of microorganisms

Aphids generally bear very few types of microorganisms, usually less than ten taxa. Nearly all aphids of the family Aphididae possess bacteria, known in the traditional literature as 'primary symbionts' and assigned by molecular criteria to a novel genus and species *Buchnera aphidicola* in the γ3-proteobacteria (*Escherichia coli* is also a member of this group) (Munson *et al.*, 1991). Many aphids additionally bear one to several other bacteria, known as 'secondary symbionts', or accessory bacteria, usually at a 10-fold lower density than *Buchnera*. The accessory bacteria in several species of Aphidinae have been studied, although none has been described formally. They include several different taxa in γ3-proteobacteria, known as *Serratia symbiotica*, *Hamiltonella defensa*, and *Regiella insecticola*, a member of α-proteobacteria known as S-type (= PAR), and a *Spiroplasma* (Chen *et al.*, 1996; Chen and Purcell, 1997; Fukatsu *et al.*, 2000, 2001; Darby *et al.*, 2001; Sandstrom *et al.*, 2001; Haynes *et al.*, 2003; Russell *et al.*, 2003; Moran *et al.*, 2005). *Buchnera* is coccoid and usually 2–4 μm in diameter, but 'giant' forms exceeding 10 μm occur occasionally. The accessory bacteria are small cocci (< 2 μm diam.) or rods of variable length (< 2–20 μm).

Aphids known to lack *Buchnera* are: some members of the subfamily Cerataphidini, which have pyrenomycete yeasts (Suh *et al.*, 2001); the Adelgidae, which have bacteria of unknown phylogenetic position 'resembling caraway seeds. . .[or]. . . blunt-ended tubes' (Buchner, 1966, p. 301); and the Phylloxeridae, which apparently have no symbiotic microorganisms (Buchner, 1966).

Location of symbiotic microorganisms

In general, parthenogenetic morphs of aphids bear *c*. 10^7 cells of *Buchnera* per mg fresh weight, occupying *c*. 8% of the aphid body volume (Baumann and Baumann, 1994; Humphreys and Douglas, 1997). Virtually all the *Buchnera* are located in insect cells called mycetocytes, or bacteriocytes (Fig. 5.1a), in the abdominal haemocoel of the insect. The bacteriocytes are large (up to 100 μm diam.), glistening silver or coloured with green, red, or brown pigment, varying between species, and often aggregated together as a coherent organ known as the mycetome, lying dorsal to the gut. The *Buchnera* cells are restricted to the cytoplasm (Fig. 5.1 b) and occupy *c*. 60% of the cytoplasmic volume (Whitehead and Douglas, 1993).

The tissue distribution of accessory bacteria is generally wider than that of *Buchnera*. Accessory bacteria (Fig. 5.1c) are very commonly located in 'sheath cells' surrounding mycetocytes and free in the haemolymph (Chen and Purcell, 1997; Fukatsu *et al.*, 2000). They also occasionally occupy mycetocytes; this has been interpreted as

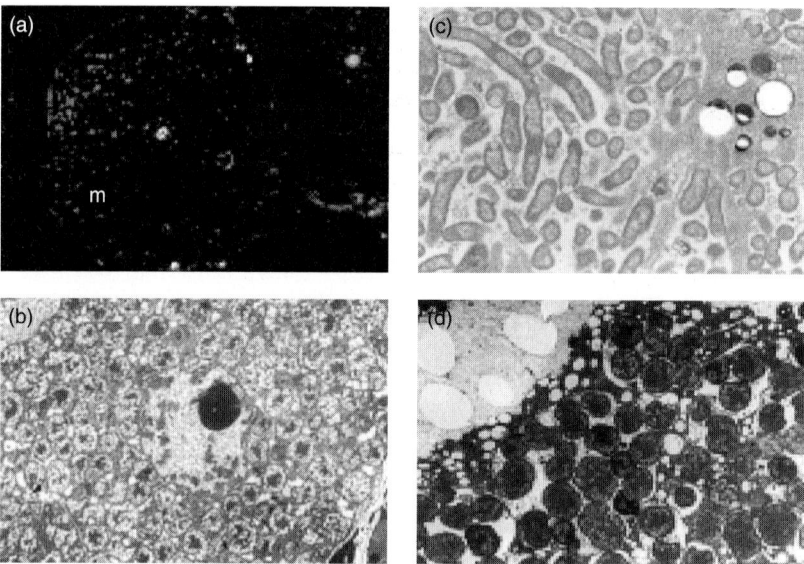

Fig. 5.1. The symbiotic bacteria associated with *Aphis fabae*. (a) *Buchnera* are restricted to mycetocytes (m) in haemocoel of aphid. (b) *Buchnera* are coccoid cells in cytoplasm of mycetocytes. (c) Rod-shaped accessory bacteria in aphid tissues. (d) *Buchnera* cells are transmitted to the cytoplasm of unfertilized eggs in the ovaries of oviparae. (Previously unpublished micrographs.)

usurpation of *Buchnera* from mycetocytes by accessory bacteria (Fukatsu *et al.*, 2000), but the possibility that cells bearing the accessory bacteria differentiate into mycetocytes has not been excluded.

The incidence of microorganisms in the gut lumen of aphids is variable. The alimentary tract of some aphid individuals is microbially sterile, others have taxa ingested with food and passing directly through the gut, and some bear a more or less resident microbial community, which may include species of the genera *Enterobacter*, *Erwinia*, and *Staphylococcus* (e.g. Grenier *et al.*, 1994; Harada *et al.*, 1997). *Buchnera* is not gutborne, and it is uncertain whether the tissue distribution of accessory bacteria includes the gut lumen (Darby and Douglas, 2003).

Acquisition of symbiotic microorganisms

Aphids acquire their complement of *Buchnera* exclusively from their mother. In other words, transmission of *Buchnera* is obligatorily vertical. The bacterial cells are released from mycetocytes abutting the germarium of the aphid ovaries and are transferred directly to the early embryo (blastoderm stage) in parthenogenetic morphs or cytoplasm of the unfertilized egg of oviparae (Fig. 5.1d) (Buchner, 1966; Hinde, 1971; Brough and Dixon, 1989). *Buchnera* probably have been transmitted faithfully from mother to offspring for more than 100 million years and possibly since the origin of the Aphididae, as indicated by the remarkable congruence between the phylogeny of aphids and *Buchnera* and the magnitude of sequence differences between rRNA genes of *Buchnera* in different aphid taxa (Moran *et al.*, 1993).

Aphids may acquire accessory bacteria from two sources: vertically from their mother, and horizontally. Vertical transmission of accessory bacteria is regular with efficiency of 98% (Darby and Douglas, 2003). The evidence for horizontal transmission is principally molecular. The accessory bacteria in different aphid species are often indistinguishable (e.g. *S. serratia* occurs in *Acyrthosiphon pisum* (pea aphid) and *Macrosiphum rosae* (rose aphid), but not in *Acyrthosiphon*

kondoi (blue alfalfa aphid) (Chen and Purcell, 1997; Chen *et al.*, 2000)), and some accessory bacteria are closely allied to forms in other arthropods (e.g. the 16 S rDNA sequence of S-type differs by < 2% from the sequence of *Rickettsia belli* in ticks (Chen *et al.*, 1996) and that of *H. defensa* is closely similar to a symbiotic bacterium in the whitefly *Bemisia tabaci* (sweet potato whitefly) (Darby *et al.*, 2001)). Horizontal transmission in nature *via* feeding and aborted parasitoid attack has been proposed (Darby *et al.*, 2001; Sandstrom *et al.*, 2001). Consistent with each mode of transmission, aphids acquire bacteria experimentally administered by both injection into the haemocoel and *via* feeding (Chen *et al.*, 2000; Fukatsu *et al.*, 2000, 2001; Darby and Douglas, 2003), but the rates of horizontal transmission in natural aphid populations are unknown. By contrast, experimental infections with *Buchnera*, either by injection or feeding, invariably have failed because the injected *Buchnera* cells are lysed in the aphid tissues.

Significance of symbiotic microorganisms to aphids

Treating the insects with antibiotics at a dose that has no apparent direct effect on aphid metabolism or behaviour, but eliminates the bacteria, reveals the crucial importance of *Buchnera* to aphids. These bacteria-free individuals are called 'aposymbiotic' aphids. When adult aphids are treated, reproductive output ceases within 1–2 days, reflecting the vital importance of *Buchnera* for embryo growth. Aphids treated at birth display normal development (although the final larval stadium may be prolonged) but grow slowly and, on reaching adulthood, are very small and produce no offspring, or a few offspring that usually die without growing or developing (Table 5.1). Aposymbiotic aphids display no known specific abnormalities, for example in feeding or embryogenesis (Wilkinson, 1998), and the role of *Buchnera* is predominantly, probably exclusively, nutritional.

The variable incidence of accessory bacteria in aphids indicates that the insect does not require these taxa. Two approaches

Table 5.1. Impact of antibiotics on the performance of apterous *Acyrthosiphon pisum*. Values of mean ± s.e. are shown (26 replicates for untreated aphids, 20 replicates for aphids treated with the antibiotic chlortetracycline) (data reproduced from Douglas, 1992).

Aphid	Larval development time (days)	Teneral weight (mg)	Number of offspring
Untreated	8.2 ± 0.12	3.06 ± 0.064	59 ± 3.1
Antibiotic-treated	10.0 ± 0.26	0.49 ± 0.017	0–5

have been adopted to investigate their significance to aphids: comparisons of the performance of – first, multiple clonal lines naturally bearing different complements of bacteria, and second, single aphid clones with their complement of accessory bacteria manipulated experimentally by injection. The results to date are fragmentary and not fully consistent. Accessory bacteria have been reported to depress, enhance, or have no detectable effect on aphids, varying between aphid and bacterial taxa and with environmental conditions, especially temperature and the plant of rearing (Chen *et al.*, 2000; Fukatsu *et al.*, 2001; Montllor *et al.*, 2002; Darby *et al.*, 2003; Russell and Moran, 2005). Further research is required to establish whether the accessory bacteria have a role in aphid nutrition.

Carbon Nutrition

Sugars

Phloem-mobile sugars are the principal source of carbon for aphid respiration and growth. Virtually all research has been conducted on aphid utilization of ingested sucrose, a disaccharide of glucose and fructose and the sole sugar in the phloem sap of many plants. Aphid utilization of galactose-based oligosaccharides (raffinose, stachyose, and higher homologues) that occur in the phloem sap of many Cucurbitaceae and Lamiaceae and of sugar alcohols (e.g. sorbitol, dulcitol, mannitol) in the sap of Celastraceae, Prunoideae and Rosaceae has not been studied.

Phloem sucrose poses two physiological 'problems' for aphids. First, sucrose is not assimilated directly across animal guts and its nutritional value depends on the hydrolysis to its constituent monosaccharides, glucose and fructose. Second, the sucrose concentration in phloem sap is very high, usually between 0.5 and > 1.0 M, and its osmotic pressure is considerably higher than that of aphid body fluids (Downing, 1978; Wilkinson *et al.*, 1997). The resultant osmotic gradient would be expected to cause uncontrolled flux of water from the aphid tissues to the gut lumen, and desiccation of the aphid.

Gut enzymes are central to the resolution of both these physiological problems. The aphid midgut has high activity of (i) sucrase, a α-glucosidase that hydrolyses sucrose to glucose and fructose, and (ii) transglycosidase, which synthesizes glucose-based oligosaccharides (3– >15 hexose units), so reducing the osmotic pressure per hexose unit in the gut lumen (Walters and Mullin, 1988; Ashford *et al.*, 2000; Cristofoletti *et al.*, 2003). The monosaccharides are assimilated across the gut wall into the haemolymph, and the oligosaccharides are voided *via* the honeydew (Fisher *et al.*, 1984; Rhodes *et al.*, 1997; Wilkinson *et al.*, 1997). It has been suggested that a single enzyme may mediate both sucrose hydrolysis and oligosaccharide synthesis by inserting a molecule of water and glucose, respectively, at the glucosidic bond (Ashford *et al.*, 2000). The enzyme activity is of aphid (and not microbial) origin as the activity is unaffected by experimental elimination of microorganisms with antibiotics (Wilkinson *et al.*, 1997). As in whitefly (Salvucci, 2000), the sucrase–transglucosidase activity may be mediated by multiple isozymes with slightly different kinetic properties.

The monosaccharides assimilated into the haemolymph are used as a respiratory substrate (Salvucci and Crafts-Brandner, 2000), and metabolized to trehalose (the principal blood sugar), lipid and, to a small extent, protein and nucleic acids (Rhodes *et al.*, 1996, 1997). Dietary sucrose is also a major source of the carbon diverted into the sugar alcohol mannitol by aphids subjected to thermal stress (Hendrix and Salvucci, 1998). The fructose moiety of sucrose is used preferentially (over the glucose moiety) as the substrate for both respiration and mannitol production (Hendrix and Salvucci, 1998; Ashford *et al.*, 2000), and glucose is the preferred substrate for lipid and trehalose synthesis (D.A. Ashford and A.E. Douglas, unpublished results).

Dietary sugars are important, not only as the principal source of carbon but also because they are the single, most important nutritional determinant of aphid feeding rate. The rate at which aphids ingest food varies inversely with the dietary concentration of sucrose (e.g. Mittler and Meikle, 1991; Douglas *et al.*, 2006) (Fig. 5.2). This response generally is interpreted as a compensatory feeding response, i.e. that aphids can compensate behaviourally, at least partially, for low dietary sucrose by feeding at higher rates. Two other factors may contribute to the low rate of food throughput at high sucrose concentrations. First, the viscosity of the diet increases with sucrose concentration, offering greater resistance to the flow of liquid through the stylets. Second, there is an osmotic requirement for complete hydrolysis and transglycosylation of ingested sugars before they are passed from the midgut to the hindgut; otherwise, the aphid would lose body water to the gut and tend to desiccate. Unfortunately, very little is known about the movement of water across aphid membranes, including the potential role of water channels in osmoregulation (Borgnia *et al.*, 1999).

Overlaying the feeding response illustrated in Fig. 5.2 is the role of sucrose as a phagostimulant, i.e. sustained feeding by aphids depends on the presence of dietary sucrose (Mittler and Dadd, 1964). As a result, aphids feed at a progressively increasing rate with declining sucrose concentration to a minimal concentration of 0.05–0.2 M (varying between species and with conditions), below which sustained feeding is not exhibited.

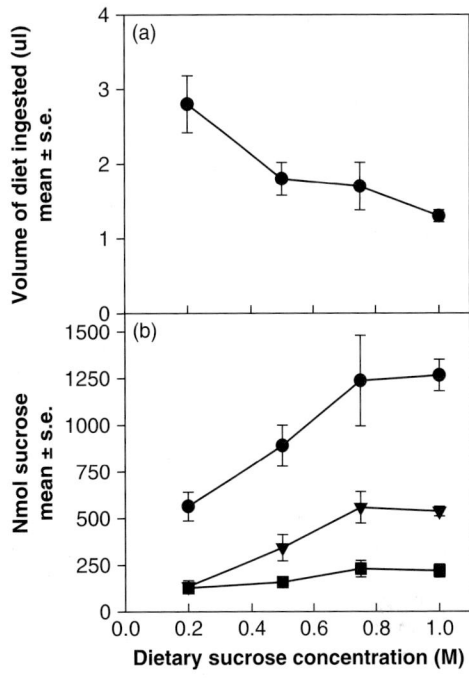

Fig. 5.2. Feeding response of *Acyrthosiphon pisum* to variation in concentration of dietary sucrose. Eight replicate final-instar apterous larvae were caged individually for 48 h to diets containing 0.2–1.0 M sucrose. (a) Volume of diet ingested. (b) Amount of sucrose ingested (circle) and recovered from aphid tissues (square) and honeydew (triangle) (previously unpublished data of W.A. Smith and A.E. Douglas; methodology as in Ashford *et al.*, 2000).

Research on the behavioural regulation of carbon acquisition by aphids has been conducted almost exclusively with aphids on chemically defined diets. The extent to which these data are relevant to aphids feeding on plants is far from clear. The key issues are:

- The sucrose concentration in plant phloem sap varies both with instantaneous irradiance and over the diurnal cycle (Geiger and Servaites, 1994), but the aphid behavioural and physiological responses to fluctuating sucrose concentrations are unknown.

- Phloem sucrose concentrations can often exceed 1.0 M (Ziegler, 1975), considerably higher than the concentrations usually used in diet studies.

- Phloem sap is under positive hydrostatic pressure, and aphids may not be able to regulate feeding rate from sieve elements as precisely as from diets of atmospheric pressure.

Lipids

Two lines of evidence indicate that aphid lipids are of endogenous origin: the natural diet of phloem sap contains very little lipid, and the chemically defined diets routinely used for maintaining aphids through multiple generations are lipid-free.

The composition of aphid lipids is most unusual for animals. The dominant lipids are triglycerides with acyl moieties of short chain length. For example, in *A. pisum*, myristic acid (C12) and hexanoic acid (C6) make up 90% of the fatty acids in the triglycerides, and the most abundant triglyceride is 1,3-dimyristoyl-2-hexanoyl glycerol (Rahbe *et al.*, 1994). The triglyceride content and composition of both *A. pisum* and *Macrosiphum euphorbiae* (potato aphid) (Rahbe *et al.*, 1994; Walters *et al.*, 1994) are not significantly modified by treatment with the antibiotic rifampicin (Fig. 5.3a), indicating that these compounds are synthesized by the aphid, and not the symbiotic microorganisms. The principal function of aphid triglycerides is as an energy source when aphids are not feeding (and so do not have access to sugars). The key periods of extended non-feeding are during moulting, flight, and for some species, the adult gynopara and male morphs. It has also been suggested (Walters *et al.*, 1994) that triglycerides are a crucial energy source for embryos but, to our knowledge, the relative importance of sugars and lipids for energy metabolism of maternal tissues and embryos in virginoparae has not been investigated systematically. The significance of the unusual fatty acid composition of aphid triglycerides is uncertain. The short acyl chain length ensures that the triglycerides are above their melting point, and this may be linked to one function of the haemolymph triglycerides as precursors of secretions from the siphunculi. When aphids are agitated, droplets rich in triglycerides of

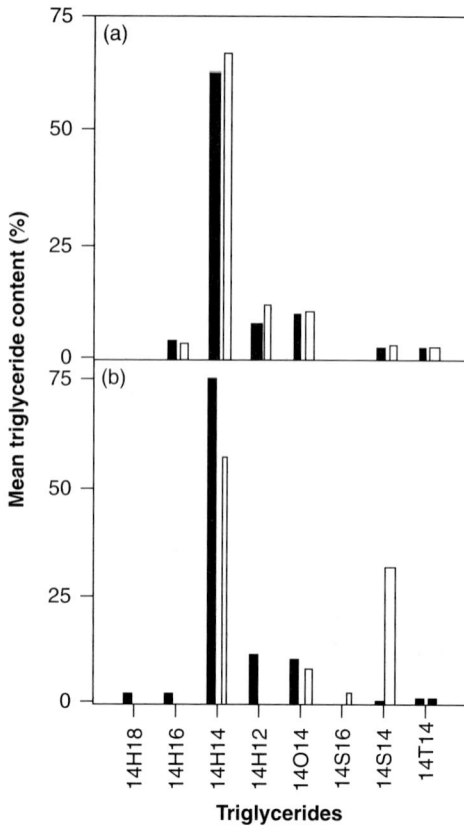

Fig. 5.3. Fatty acid composition of triglycerides in *Acyrthosiphon pisum*. (a) Aphids treated with rifampicin to eliminate the symbiotic bacteria (open bars) and untreated aphids (closed bars) reared on chemically defined diets. (b) Cornicle secretion (open bars) and fat body (closed bars) of plant-reared aphids. 14H18: 1-myristoyl-2-hexanoyl-3-stearoyl glycerol; 14H16: 1-myristoyl-2-hexanoyl-3-palmitoyl glycerol; 14H14: 1,3-dimyristoyl-2-hexanoyl glycerol; 14H12: 1-myristoyl-2-hexanoyl-3-lauroyl glycerol; 14O14: 1,3-dimyristoyl-2-octanoyl glycerol; 14S16: 1-myristoyl-2-sorboyl-3-palmitoyl glycerol; 14S14: 1,3-dimyristoyl-2-sorboyl glycerol; 14T14: 1,3-dimyristoyl-2-octatrienoyl glycerol (redrawn from Rahbe *et al.*, 1994).

composition similar (but not identical) to the haemolymph triglycerides (Fig. 5.3 b) are exuded from the siphuncles, and on release, these siphuncle secretions change into a sticky solid that deters potential predators. The triglycerides containing sorbic acid (C6:2) have also been implicated in defence against microbial antagonists (Brown, 1975) because sorbic acid is a potent fungistatic agent. However, the significance of this function is uncertain because, first, the fungistatic activity of sorbic acid is reduced greatly by esterification to glycerol, and second, the triglyceride content of the two key sites colonized by fungi, the cuticle and honeydew, is very low (Strong, 1963; Brey *et al.*, 1985).

The principal phospholipids in aphids are phosphatidylethanolamines and phosphatidylcholines; the levels of phosphatidylinositols are very low (Febvay *et al.*, 1992). Aphids are widely assumed to have a dietary requirement for choline and inositol (from which the phosphatidylcholines and phosphatidylinositols, respectively, are derived), but the possibility that aphids, like some other insects, can synthesize inositol from glucose has not been excluded definitively. The fatty acids in aphid phospholipids are generally of longer chain length than in the triglycerides. For example, in the phospholipids of *A. pisum*, the dominant fatty acid is linoleic acid (C18:2) (Ryan *et al.*, 1982). Radiotracer studies have confirmed that, as with the triglyceride fatty acids, the phospholipid fatty acids are synthesized endogenously by the aphid (de Renobles *et al.*, 1986).

Aphids require preformed sterols because, like other insects, they cannot synthesize these compounds *de novo*. Three lines of evidence clearly indicate that aphid sterol requirements are met from the phloem sap. First, phloem sap contains phytosterols (Forrest and Knights, 1972). It is anticipated that aphids are capable of de-alkylating these compounds to the principal animal sterol, cholesterol, from which the ecdysteroids and other steroidal compounds are derived. Second, the sole microorganisms associated with most aphids are bacteria and, in common with most bacteria, they lack the metabolic capability to synthesize sterols (Shigenobu *et al.*, 2000; Tamas *et al.*, 2002). Finally, aphids experimentally deprived of their bacteria do not display the characteristics of limitation by sterols, e.g. dysfunction and usually death at ecdysis (Wilkinson, 1998). In direct conflict with the expectation that

aphid sterols are of dietary origin is the ease with which various aphid species can be maintained over multiple generations on sterol-free diets. Direct analyses have excluded the possibility of inadvertent provision of sterol in the diet, *via* either chemical or fungal contamination (A.E. Douglas, unpublished results). In his comprehensive review of insect nutrition, Dadd (1985) decided to 'set. . .aside the vagaries and ambiguities of this aphid situation for future clarification'. Some 20 years later, we are still awaiting that clarification.

Nitrogen Nutrition

Amino acids and their sources

The chief nitrogenous compounds utilized by aphids are free amino acids derived from two sources: the diet of phloem sap, and the symbiotic bacteria in the aphids. These two sources are considered in turn.

Amino acids are the dominant nitrogenous compounds in the phloem sap of most plant taxa, and aphids assimilate them very efficiently. Additionally, aphids derive a supplementary supply of essential amino acids from the symbiotic bacteria in their tissues. The three core lines of evidence are that:

- Aphids do not have a dietary requirement for many essential amino acids (e.g. Dadd, 1985; Douglas *et al.*, 2001; Wilkinson and Douglas, 2003).
- The isolated bacteria are capable of essential amino acid synthesis (Sasaki and Ishikawa, 1995).
- Amino acids synthesized by the bacteria in aphid tissues are translocated to the aphid partner (Douglas, 1988; Douglas and Prosser, 1992).

It is widely accepted that amino acids are released from living bacterial cells and not by aphid-mediated lysis of bacteria, and there is some evidence that amino acids are made available to the insect as both free amino acids and in the form of protein (Douglas, 1998). The significance of *Buchnera* to the nitrogen nutrition of aphids has been confirmed spectacularly by the full genomic sequence of these bacteria. *Buchnera* has a much-reduced genome, just 0.6 kb, comprising a subset of the genes of *E. coli* (Shigenobu *et al.*, 2000). For most functional categories, *Buchnera* bears 20–50% of the genes that occur in *E. coli*. This pattern of gene loss indicates that *Buchnera* is incapable of independent existence and is metabolically and nutritionally dependent on its aphid host. Exceptionally, 48/51 (94%) of the genes involved in essential amino acid synthesis in *E. coli* have clear homologues in *Buchnera*, indicating that *Buchnera* are under strong selection pressure to retain the capacity to synthesize essential amino acids.

The fate of amino acids in aphids

Aphids and their symbiotic bacteria mediate considerable amino acid interconversions by transamination and other reactions. For example, dietary glutamic acid is metabolized by the aphid to proline, glutamine, aspartic acid, asparagine, and alanine, and additionally by *Buchnera* to lysine, threonine, and isoleucine in both *A. pisum* (Febvay *et al.*, 1995) and *Aphis fabae* (black bean aphid) (Douglas *et al.*, 2001). Protein synthesis is quantitatively a major fate for amino acids. More than 95% of the amino acid content of aphids is in the protein fraction, and protein accounts for 10–25% of aphid fresh weight.

Some amino acids have important functions other than in protein synthesis. Several amino acids are, or are precursors of, neurotransmitters and neurohormones, e.g. glutamic acid, dopamine (derived from tyrosine), serotonin (derived from tryptophan). Tyrosine is also a major precursor of cuticle synthesis. One consequence of the link between certain amino acids and cell-signalling interactions is that the biological ramifications of a shortfall in the supply of these compounds might be expected to extend beyond negative effects on aphid growth and reproductive rate. This issue has not been studied in detail but, as a possible example, 5-hydroxytryptamine (a derivative of dietary tryptophan) inhibits the production of alatiform *Myzus persicae* (peach–potato aphid) (Harrewijn, 1978).

Amino acids are also a vital constituent of the haemolymph of aphids. All 20 protein-amino acids are readily detectable in the haemolymph and, as in other insects, the haemolymph amino acids contribute to the osmoregulation of aphid body fluids. Haemolymph amino acids play a crucial role in nitrogen nutrition, because the haemolymph is the first destination of both dietary amino acids assimilated across the gut wall and *Buchnera*-derived amino acids released from the mycetocytes. Further research is required to quantify the flux of amino acids into the haemolymph and to establish the specificity and kinetic properties of the amino acid transporters on the gut and mycetocyte membranes. An indication that the flux is high, however, comes from evidence that the haemolymph amino acid pool has very high turnover. In addition to supporting protein synthesis and other functions (see above), haemolymph amino acids are consumed at a high rate in respiration. For example, in plant-reared *A. fabae*, haemolymph glutamic acid is metabolized to carbon dioxide at a rate of 1.2 nmol carbon/mg aphid weight/h (Wilkinson *et al.*, 2001b), and in diet-reared *A. pisum*, respiration accounts for 60% of the total glutamic acid assimilated (Febvay *et al.*, 1995). Perhaps amino acids are important to the energy metabolism of particular cell types, even though, in comparison with sucrose, they are minor respiratory substrates for aphids at the level of the whole organism (Rhodes *et al.*, 1996; Febvay *et al.*, 1999).

Ammonia is the principal nitrogenous product of amino acid catabolism. Aphid honeydew contains appreciable concentrations of ammonia, but not the alternative nitrogenous waste compounds, urea or uric acid (Sasaki *et al.*, 1990; Wilkinson and Douglas, 1995). It has been suggested that the symbiotic bacteria contribute to the overall nitrogen nutrition of aphids by nitrogen recycling, i.e. the bacteria assimilate aphid waste ammonia into essential amino acids, which are released back to the aphid. Two lines of evidence are consistent with a role in nitrogen recycling. First, preparations of isolated *Buchnera* take up ammonia *via* a high affinity transport system (Whitehead *et al.*, 1992). Second, the ammonia content of aphid honeydew is significantly elevated when the symbiotic bacteria are eliminated (Wilkinson and Douglas, 1995). As yet, however, definitive evidence, for example from [15] N-tracer experiments, is lacking.

Other dietary sources of nitrogen

Although research on the nitrogen nutrition of aphids has concentrated on the utilization of dietary amino acids, plant phloem sap does contain other nitrogenous compounds, including proteins (Thompson and Schulz, 1999) and polyamines (Antognoni *et al.*, 1998). In general, these compounds are at lower concentrations than free amino acids, but they may be biologically significant in certain plants or in relation to certain amino acids. For example, cucurbits contain high concentrations of protein (> 1 mg/ml), and phloem-mobile tripeptide glutathione (Rennenberg, 1982) may be a quantitatively important source of amino acid methionine.

The fate of peptides and proteins in the aphid gut is poorly understood. In the study by Rahbe *et al.* (1995) on *A. pisum*, various proteins added to the diet passed through the gut to the honeydew without chemical modification. Also, consistent with the interpretation that this species lacks substantial gut proteolytic function, protease activity was not detected in isolated guts. Aphids may, however, be able to hydrolyse ingested oligopeptides, as suggested by both aminopeptidase activity in the midgut of *A. pisum* (Rahbe *et al.*, 1995) and the limited capacity of dipeptide glycyl-methionine to support the dietary methionine requirements of one clone of *A. fabae* (Leckstein and Llewellyn, 1974).

Minerals and Micronutrients

Study of the mineral requirements of aphids has been dominated by the need for a convenient mineral mixture suitable for maintaining aphids on chemically defined diets, and there has been no systematic analysis of aphid requirements for inorganic ions. All the published diet formulations include relatively high concentrations of potassium and phosphate, reflecting their high concentrations in phloem sap, and of magnesium and sulphate: at micromolar concentrations, iron, copper, manganese, and zinc are included. It is recognized that aphid requirements for other metals may go undetected because these elements are present in the diet as impurities of other dietary constituents. In all probability, the inorganic requirements of aphids are comparable to those of other phytophagous insects.

Early studies (Dadd *et al.*, 1967) indicated that aphids have a dietary requirement for the seven core B vitamins: thiamine (B_1), riboflavin, nicotinic acid/nicotinamide, pyridoxine (B_6), pantothenic acid, folic acid, and biotin (B_{12}), and these compounds are included routinely in most chemically defined diet formulations. Some data, however, suggest that symbiotic bacteria may meet, partially or completely, the riboflavin requirements of aphids. In particular, Shigenobu *et al.* (2000) concluded that *Buchnera* has the genetic capability for riboflavin synthesis, and Nakabachi and Ishikawa (1999) reported that the performance of *A. pisum* is independent of dietary riboflavin.

The only other vitamin provided in published aphid diet formulations is ascorbic acid (vitamin C). This reflects its role as both a dietary essential and a chelator of dietary metal ions, especially Fe^{3+} (citric acid is also included in diets for the latter purpose) (Mittler, 1976). The lipophilic vitamins A, D, and E usually are not included in diets, suggesting that they are not dietary requirements for aphids. Aphids may, however, acquire these compounds, or precursors, from plants. For example, *A. fabae* accumulates carotenes, including β- carotene (provitamin A), from the host plant, broad bean (*Vicia faba*) (Brown, 1975).

Artificial Diet

History of artificial diets for aphids

There is considerable variation in the growth and reproductive performance of individual aphids, even within the progeny of the same mother. A major contribution to this variability is assumed to be variable nutrition. Such variation occurs between apparently identical plants and even between locations on the same leaf. Variance of the mean relative growth rate (MRGR) of *M. persicae* was significantly higher on replicate plants than on an artificial diet (Wojciechowicz-Zykto and van Emden, 1995). The possibility of rearing aphids on a nutritionally uniform substrate has therefore long been a goal for those working with aphids, and has been explored ever since Hamilton (1930) showed that aphids would feed on fluids through a natural membrane (onion epidermis). The real breakthrough came 30 years later, when Mittler and Dadd (1962) published a fully defined artificial diet for *M. persicae*, closely followed by Auclair and Cartier (1963) with a diet for *A. pisum*. Both pairs of workers used stretched Parafilm 'M' ™ as the membrane. The diet published by Dadd and Mittler (1966) has proved the most successful and widely used for several species of aphids, yet for many years the diet failed to sustain any species permanently, even for the originators. However, in February 1976, a culture of *M. persicae* was started on the diet at The University of Reading, and this culture has persisted without contact with plants for nearly 30 years. It is this successful diet system that is discussed here.

Recipe for the diet and practical procedures

The recipe for the diet is given in Table 5.2. The availability of nanopure water appears to have solved a major difficulty experienced over many years in transferring the technique from Reading to other laboratories. It is important that the ingredients are added in exactly the order given in the Table (beginning with those in the left-hand column), and that each ingredient is fully dissolved before the next is added. In practice, this involves setting up a magnetic stirrer for much of a day and alternating diet preparation with other work.

The diet is not sterilized at this stage, but is divided into convenient 25–30 ml aliquots, which are deep-frozen for up to 6 months until needed. As required, the aliquots are thawed and used to make large numbers of diet sachets, which can similarly be deep-frozen until individually needed.

At Reading, the diet tubes take the form of clear Perspex cylinders of 25 mm internal diameter and 25 mm long (Fig. 5.4a). It is during the preparation of these diet tubes that sterile procedures are required and the tubes, the thawed diet, and 50×35 mm squares of unstretched Parafilm 'M' ™ are placed in a laminar flow unit. A UV lamp in the roof of the unit is switched on for 40 min to sterilize the top surface of the Parafilm squares. There is a danger the Parafilm will deteriorate if exposed to more than 40 min of UV light. A metal filter holder incorporating a 0.45 µm bacterial filter above the needle (Fig. 5.4 b) is autoclaved, wrapped in aluminium foil and is also placed in the laminar flow unit.

A square of Parafilm is then stretched 2-ways from the underside (Fig. 5.4c), to keep the area that covers the opening of the tube sterile, and placed over one end of a diet tube. This forms one surface of the eventual 'sachet' of diet. Diet is poured into a syringe fitted with the autoclaved filter holder and needle, and a few drops of diet are pipetted through the bacterial filter on to the sterile surface of the Parafilm. The liquid is then covered with another stretched sheet of Parafilm, inverted so that the liquid is sandwiched between two sterile surfaces, and the projecting flaps of Parafilm pulled down on to the side of the tube. These flaps can be used to hold two diet tubes together at their open ends (Fig. 5.4d), and this is critical in the way sachets are replaced. The diet deteriorates quickly, especially since aphid stylets are not sterile, and the aphids have to be provided with new sachets every 2–3 days (2 whenever possible).

This replacement is done by fixing the open end of a new diet tube to the open end of the tube where the aphids are feeding and using the Parafilm to hold the tubes together. The upper Parafilm of the sachet with the aphids is then gently slit open with

Table 5.2. Composition of artificial diet for *Myzus persicae* – quantities (mg) of chemicals to be dissolved in 100 ml nanopure water (based on Dadd and Mittler, 1966). It is important that the chemicals are completely dissolved in the order shown, beginning with the left-hand column.

sucrose (15 g)	15000.0	L-serine	80.0
di-potassium hydrogen orthophosphate	750.0	L-threonine	140.0
magnesium sulphate	123.0	L-valine	80.0
L-tyrosine	40.0	ascorbic acid (vitamin C)	100.0
L-asparagine hydrate	550.0	aneurine hydrochloride (vitamin B)	2.5
L-aspartic acid	140.0	riboflavin	0.5
L-tryptophan	80.0	nicotinic acid	10.0
L-alanine	100.0	folic acid	0.5
L-arginine monohydrochloride	270.0	(+)-pantothenic acid (calcium salt)	5.0
L-cysteine hydrochloride, hydrate	40.0	inositol (meso) active	50.0
L-glutamic acid	140.0	choline chloride	50.0
L-glutamine	150.0	EDTA Fe(III)-Na chelate pure	1.5
L-glycine	80.0	EDTA Zn-Na$_2$ chelate pure	0.8
L-histidine	80.0	EDTA Mn-Na$_2$ chelate pure	0.8
L-isoleucine (allo free)	80.0	EDTA Cu-Na$_2$ chelate pure	0.4
L-leucine	80.0	pyridoxine hydrochloride (vitamin B$_6$)	2.5
L-lysine monohydrochloride	120.0	D-biotin, crystalline	0.1
L-methionine	40.0		
L-phenylalanine	40.0		
L-proline	80.0		

the end of a needle, and placed onto filter paper so that the old diet runs out through the slit and is absorbed on the paper. The aphids then rapidly climb the tubes to transfer to the new sachet, with no handling of the insects being necessary.

Relation between the diet and plant phloem sap

Dadd and Mittler's (1966) diet is based on Mittler's unpublished data (reported by Mittler, 1953, 1958) for willow (*Salix acutifolia*) of the composition of the phloem sap obtained from the cut ends of the inserted stylets of *Tuberolachnus salignus* (willow aphid). Since this work preceded the advent of column chromatography, the far less quantitative technique of paper chromatography was used for the amino acid analyses. In any case, developing a successful diet from this basis involved many changes as a result of experience; Dadd and Mittler (1966) tested and re-tested different concentrations

of individual components against many different backgrounds of a balance of the other constituents. The diet they published is very similar to Table 5.2, which is the diet used at Reading and differs only in the form in which some of the ingredients are purchased.

The diet is composed of 15% sucrose and 2.4% of amino acid compounds. Mittler (1958) reported percentages in willow phloem sap of around 8.3% sucrose and total nitrogen of between 0.03% and 0.13%, with an exceptional concentration of 0.2% during a brief period after winter when the buds were beginning to swell. An amino acid analysis (H.F. van Emden and M.A. Bashford, unpublished results) is also available for the phloem sap of glasshouse-grown castor oil plant (*Ricinus communis*), a shrub or small tree that is a host for *M. persicae*. The sap was obtained by the EDTA method (King and Zeevaart, 1974). Figure 5.5 plots the concentration in the artificial diet against the concentration in the castor bean plant phloem sap of those amino acid compounds for which data from both liquids were available.

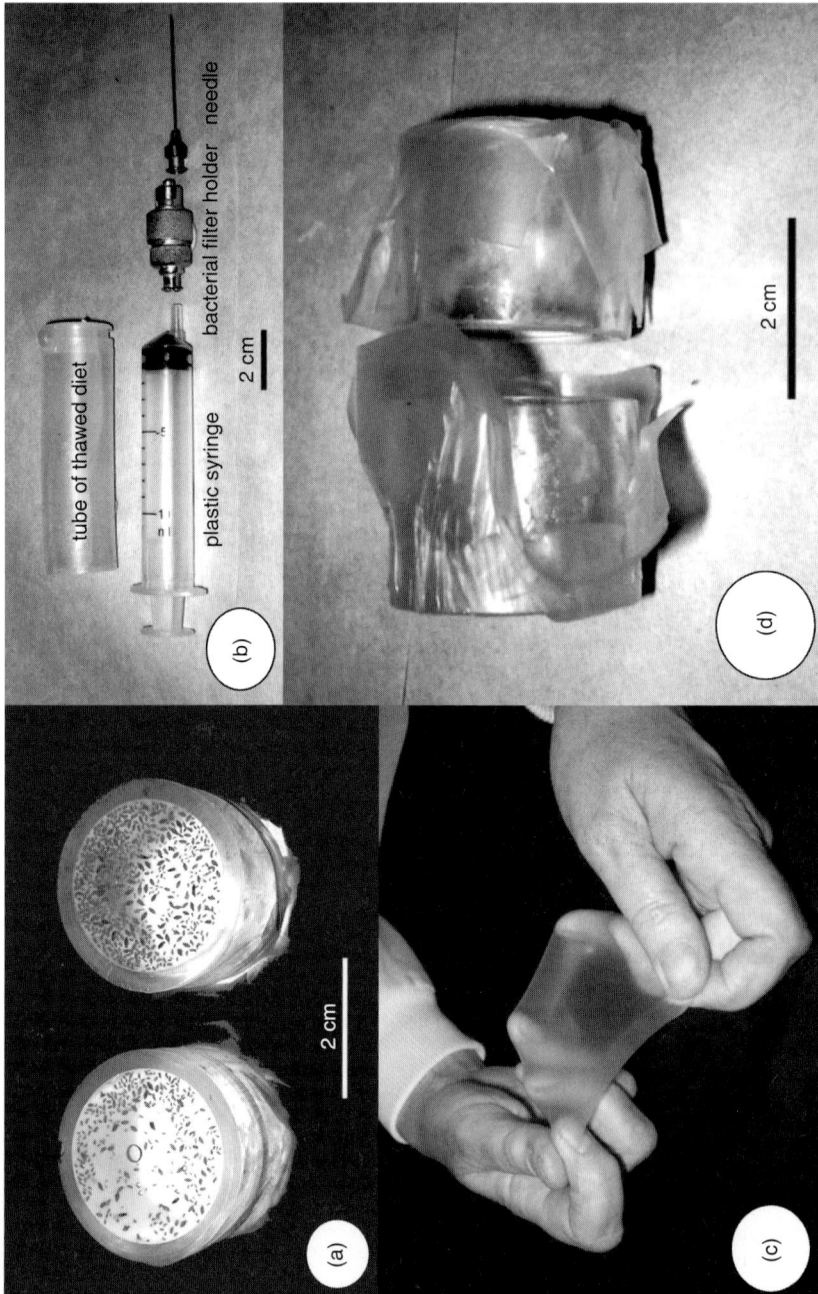

Fig. 5.4. The artificial diet technique. (a) *Myzus persicae* feeding on the diet. (b) Equipment for dispensing the diet. (c) Stretching the Parafilm so as to keep one surface sterile. (d) Old and new diet tubes, with free edges of Parafilm, which are used to hold them together in position.

Both Fig. 5.5 and Mittler's (1958) analyses point to the need for higher concentrations of both sucrose and most amino acids in the diet than in plant phloem sap; however, there is no consistent quantitative change in the ratio (contrast especially arginine and glutamic acid in Fig. 5.5).

That the concentration of most ingredients needs to be higher in the artificial diet than in phloem sap is not surprising, since uptake from the diet is much reduced compared with uptake from the phloem. The uptake from diet sachets by both *M. persicae* (Mittler, 1970) and *A. pisum* (Auclair, 1965) is 0.3–0.4 µl per aphid per day; this is only about one-eighth of the uptake from pea (*Pisum sativum*) plants reported for *A. pisum* by Auclair and Maltais (1961). It seems reasonable to assume that this reduced uptake from artificial diet is due to the lack of pressure compared with the considerable pressure of the phloem sap (Mittler, 1958). van Emden (1967) pressurized a 10% sucrose solution to 2 kg/cm^2, but could only extend the longevity of adult *A. fabae* by 20% to 4 days. One reason limiting the reduced uptake from diet is probably the high osmolality of the aphid haemolymph resulting from the constant high sugar concentration (see earlier under 'carbon nutrition'; also Cull and van

Emden, 1977), whereas sugar concentrations in the phloem fluctuate with ambient radiation, with a strong reduction at night. There may also be an element of gustatory 'habituation' to an unchanging food source. T.E. Mittler and H.F. van Emden (unpublished results) found that neonate *A. fabae* attained a mean weight of 147 mg if fed on a cycle of 17 h full diet: 7 h water for three days, but only 127 mg if the full diet was replaced with further full diet for the 7 h period. Moreover, the same workers found that *A. fabae* maintained a steady excretion rate of around 0.5 honeydew drops per aphid per hour, but that this rate rose to over 3 drops per hour (Fig. 5.6) for a few hours on full-strength diet after the aphids had been on half-strength diet for just 15 min (van Emden, 1996). Low uptake from diet therefore may well largely have a behavioural component.

Aphid performance on the diet

Even after nearly 30 years on the diet at Reading, *M. persicae* still performs badly compared with the same original genotype when plant-reared (van Emden, 1988; van Emden and Andrews, 1997). Generation times are extended and fecundity is reduced in

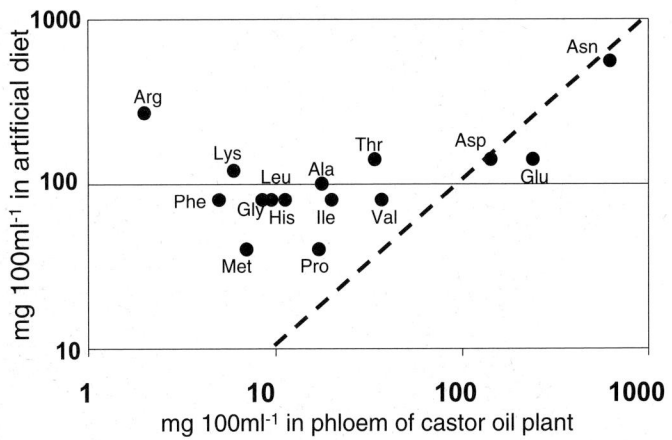

Fig. 5.5. Relationship between the concentration of individual amino acids in artificial diet and phloem sap of the castor oil plant (*Ricinus communis*); the dotted line is the line of equality in both substrates. Ala, alanine; Arg, argenine; Asn, asparagine; Asp, aspartic acid; Glu, glutamine; Gly, glycine; His, histidine; Ile, isoleucine; Leu, leucine; Lys, lysine; Met, methionine; Phe, phenylalanine; Pro, praline; Thr, threonine, Val, Valine.

diet-reared aphids (Fig. 5.7). The main contributor to extended generation times is the adult pre-reproductive period of several days characteristic of the diet-reared aphids. When monitored after 10 years, fecundity was only about one-third that on plants: after a further 10 years, fecundity and development rate had both improved a little, but fecundity was still down by more than 50%. Nutritionally, or through phagostimulation, the diet is clearly sub-optimal. Aphids returned to plants as neonates hardly improve their performance in the first generation (van Emden and Andrews, 1997; Fig. 5.7), but the MRGR of their offspring then rises to match that of individuals maintained on plants throughout.

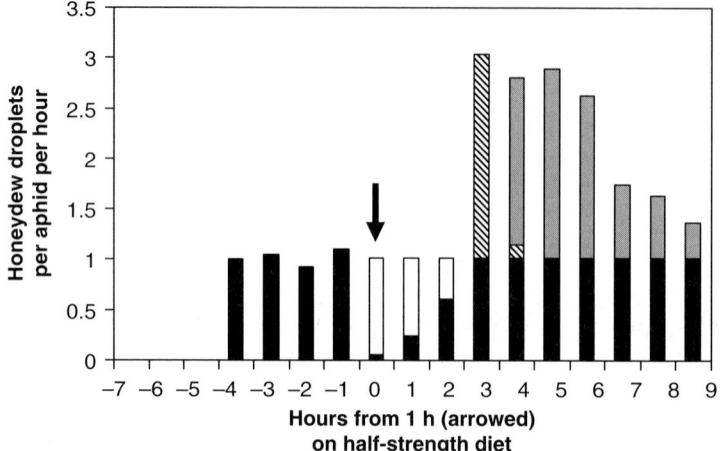

Fig. 5.6. Mean number of honeydew droplets excreted per hour by *Aphis fabae* on artificial diet before and after a 15 min period (arrowed) on half-strength diet. Black, baseline droplet excretion rate; white, reduction in rate following 1 h on half-strength diet; striped, amount of extra excretion to bring average back to baseline rate; grey, droplet excretion above average baseline rate.

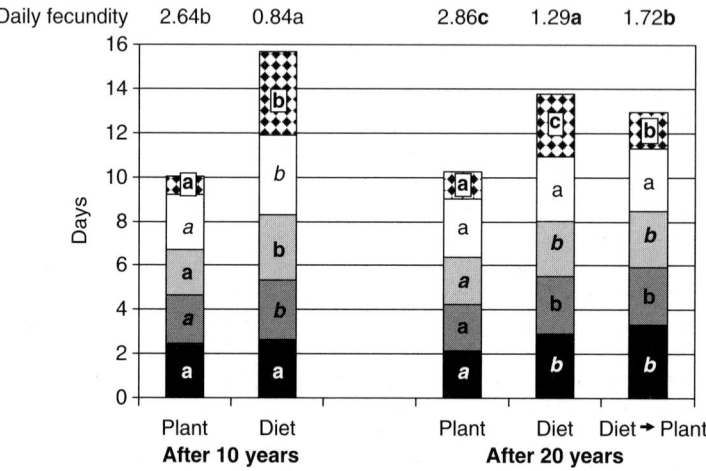

Fig. 5.7. Development of *Myzus persicae* on plants and artificial diet after 10 years on the diet (left) and (right) after 20 years, together with development of aphids born to mothers moved from diet to plants. The divisions of the columns (from bottom to top) are durations (days) of: 1st instar (black), 2nd instar (dark grey), 3rd instar (light grey), 4th instar (white), and the adult pre-reproductive period (chequered). Blocks within columns showing the same letter are not significantly different at p = 0.05.

How useful is the diet for studies on aphid nutrition?

The diet frequently has been used successfully for testing behaviour-modifying compounds such as attractants, repellents, and antifeedants against aphids (e.g. Corcuera *et al.*, 1985), although such short-term experiments usually can be conducted just as well with adult aphids and simple sucrose solutions in the sachets. Another successful use has been to introduce antibiotics into aphids to produce aposymbiotic individuals (Douglas, 1988; Prosser and Douglas, 1991).

However, what would seem an obvious use – to test cause and effect for apparent correlations between aphid performance and the concentration of nutrients, particularly amino acids, in plants – can in practice prove rather unsatisfactory. There seem to be three main reasons for this:

- It is not possible to increase the concentration of one nutrient without reducing another or changing the balance between compounds. In developing the diet for long-term aphid culture, the balances between ingredients seem to have been as critical to success as the concentration of individual compounds.
- The diet is not strictly a mimic of plant sap (as shown in Fig. 5.5). Moreover, Mittler (1972) pointed out that beneficial interactions, between for example amino acids and metal availability, which may occur in the homogeneous diet, may not occur in the plant.
- The laborious procedure of testing and re-testing employed by Dadd and Mittler in developing the diet ensured a mixture that maximized growth and reproduction of *M. persicae*. It is only to be expected therefore that any variation in the recipe will lower the performance of the aphid, and this has indeed proved to be the case (van Emden, 1996), with one exception. For the phagostimulant and essential amino acid methionine alone, doubling the concentration did significantly increase the growth rate of *M. persicae*, and halving it reduced it.

However, as considered earlier in this review, the diet has been used for nutritional studies with aposymbiotic aphids to determine the role of symbionts in nitrogen metabolism in aphids.

Conclusions and Future Prospects

The single most perplexing issue in aphid nutrition is the high rate at which the unbalanced diet of phloem sap is transformed into new aphid biomass (see Introduction). As this chapter summarizes, key factors are the speed and efficiency with which aphids assimilate, or otherwise dispose of, ingested nutrients. For example, aphids on many plants ingest a surfeit of sucrose. The sucrose is hydrolysed to completion in the gut lumen and, within 15–20 min of ingestion, the hydrolysis products are assimilated across the gut wall to the haemolymph, or transformed to oligosaccharides that pose no obvious osmotic hazard to the aphid, and voided *via* honeydew. In a comparable fashion, ingested amino acids are rapidly assimilated or voided. This efficiency is compatible with the aphid's absolute nutritional requirement for essential amino acids only because the symbiotic bacteria *Buchnera* provide the aphid with supplementary essential amino acids. The bacteria can be described as 'upgrading' the quality of the diet by converting non-essential amino acids (i.e. amino acids that the aphid can synthesize *de novo*) to essential amino acids (which the aphid cannot synthesize *de novo* and requires absolutely). The bacteria may additionally recycle ammonia, the chief nitrogenous waste product of aphid metabolism, but, contrary to some early reports, *Buchnera* has no capacity to promote aphid nitrogen nutrition by fixing atmospheric nitrogen (Shigenobu *et al.*, 2000).

Many issues relating to aphid nutrition and symbiosis remain to be resolved. In particular, we have no physiological information on aphid utilization of phloem sugars other than sucrose, even though many aphid species feed exclusively on plants with galactose-based oligosaccharides as the dominant phloem sugars. Similarly, the significance to aphid nutrition of phloem nitrogenous compounds other than free amino acids is obscure. Compounding these areas of ignorance is the dearth of information on the significance of accessory bacteria to aphids. There are indications in the recent literature that these bacteria may influence aphid host-plant range, thermal tolerance, and propensity to develop into alatiform (dispersal) morphs (Chen *et al.*, 2000; Wilkinson *et al.*, 2001a; Montllor *et al.*, 2002; Oliver *et al.*, 2003), but the physiological processes underlying these effects are, at present, unknown.

Finally, the main achievements in our understanding of aphid nutrition have come primarily through experiments conducted with aphids on chemically defined diets. However, we have only the most cursory understanding of the nutrition of plant-reared aphids, largely because of the technical difficulties with quantifying or manipulating phloem-mobile nutrients. With the advent of methodologies for routine sampling and analysis of single plant cells (e.g. Tomos and Sharrock, 2001), we are now well placed to address the nutrition of aphids on plants. There is every expectation that aphids display a suite of behavioural and physiological adaptations that optimize nutrient extraction from phloem sap, a diet of far greater biochemical complexity and variability than chemically defined diets.

Acknowledgements

We thank Dr M.E. Salvucci for helpful comments on the manuscript.

References

Antognoni, F., Fornale, S., Grimmer, C., Komor, E. and Bagni, N. (1998) Long-distance translocation of polyamines in phloem and xylem of *Ricinus communis* L. plants. *Planta* 204, 520–527.

Ashford, D.A., Smith, W.A. and Douglas, A.E. (2000) Living on a high sugar diet: the fate of sucrose ingested by a phloem-feeding insect, the pea aphid *Acyrthosiphon pisum*. *Journal of Insect Physiology* 46, 335–341.

Auclair, J.L. (1965) Feeding and nutrition of the pea aphid, *Acyrthosiphon pisum* (Homoptera: Aphididae), on chemically defined diets of various pH and nutrient levels. *Annals of the Entomological Society of America* 58, 855–875.

Auclair, J.L. and Cartier, J.J. (1963) Pea aphid: rearing on a chemically defined diet. *Science, New York* 142, 1068–1069.

Auclair, J.L. and Maltais, J.B. (1961) The nitrogen economy of the pea aphid, *Acyrthosiphon pisum* (Harris), on susceptible and resistant varieties of peas, *Pisum sativum* L. *Proceedings of the 11th International Congress of Entomology, Vienna, 1960* 1, 740–743.

Baumann, L. and Baumann, P. (1994) Growth kinetics of the endosymbiont *Buchnera aphidicola* in the aphid *Schizaphis graminum*. *Applied and Environmental Microbiology* 60, 3440–3443.

Borgnia, M., Nielsen, S., Engel, A. and Agre, P. (1999) Cellular and molecular biology of the aquaporin water channels. *Annual Reviews of Biochemistry* 68, 425–458.

Brey, P.T., Ohayon, H., Lesourd, M., Castex, H., Roucache, J. and Latge, J.P. (1985) Ultrastructure and chemical composition of the outer layers of the cuticle of the pea aphid *Acyrthosiphon pisum* (Harris). *Comparative Biochemistry and Physiology, Series A* 82, 401–411.

Brough, C.N. and Dixon, A.F.G. (1989) Ultrastructural features of egg development in oviparae of the vetch aphid, *Megoura vicia* Buckton. *Tissue and Cell* 22, 51–63.

Brown, K.S. (1975) The chemistry of aphids and scale insects. *Chemical Society Review* 4, 263–288.

Buchner, P. (1966) *Endosymbiosis of Animals with Plant Microorganisms*. Wiley, Chichester, 909 pp.

Chen, D.Q. and Purcell, A.H. (1997) Occurrence and transmission of facultative endosymbionts in aphids. *Current Microbiology* 34, 220–225.

Chen, D.Q., Campbell, B.C. and Purcell, A.H. (1996) A new *Rickettsia* from a herbivorous insect, the pea aphid *Acyrthosiphon pisum* (Harris). *Current Microbiology* 33, 123–128.

Chen, D.Q., Montllor, C.B. and Purcell, A.H. (2000) Fitness effects of two facultative endosymbiotic bacteria of the pea aphid *Acyrthosiphon pisum*, and the blue alfalfa aphid, *A. kondoi*. *Entomologia Experimentalis et Applicata* 95, 315–323.

Corcuera, L.J., Queirolo, C.B. and Argandona, V.H. (1985) Effects of 2-β-dscp-glucosyl-4-hydroxy-7-methoxy-1,4-benzoxazin-3-one on Schizaphis graminum Rondani (Insecta, Aphididae) feeding on artificial diets. *Experientia* 41, 514–516.

Cristofoletti, P.T., Ribeiro, A.F., Deraison, C., Rahbe, Y. and Terra, W.R. (2003) Midgut adaptation and digestive enzyme distribution in a phloem feeding insect, the pea aphid, *Acyrthosiphon pisum. Journal of Insect Physiology* 49, 11–24.

Cull, D.C. and van Emden, H.F. (1977) The effect on *Aphis fabae* of diet changes in their food quality. *Physiological Entomology* 2, 109–115.

Dadd, R.H. (1985) Nutrition: organisms. In: Kerkut, G.A. and Gilbert, L.I. (eds) *Comprehensive Insect Physiology, Biochemistry, and Pharmacology, Volume 4*. Pergamon, Oxford, pp. 313–390.

Dadd, R.H. and Mittler, T.E. (1966) Permanent culture of an aphid on a totally synthetic diet. *Experientia* 22, 832.

Dadd, R.H., Krieger, D.L. and Mittler, T.E. (1967) Studies on the artificial feeding of the aphid *Myzus persicae*. IV. Requirements for water-soluble vitamins and ascorbic acid. *Journal of Insect Physiology* 13, 249–272.

Darby, A.C. and Douglas, A.E. (2003) Elucidation of the transmission patterns of an insect-borne bacterium. *Applied and Environmental Microbiology* 69, 4403–4407.

Darby, A.C., Birkle, L.M., Turner, S.L. and Douglas, A.E. (2001) An aphid-borne bacterium allied to the secondary symbionts of whitefly. *FEMS Microbiology Ecology* 36, 43–50.

Darby, A.C., Tosh, C.R., Walters, K.F.A. and Douglas, A.E. (2003) The significance of a facultative bacterium to natural populations of the pea aphid *Acyrthosiphon pisum. Ecological Entomology* 28, 145–150.

Douglas, A.E. (1988) Sulphate utilisation in an aphid symbiosis. *Insect Biochemistry* 18, 599–605.

Douglas, A.E. (1992) Requirement of pea aphids (*Acyrthosiphon pisum*) for their symbiotic bacteria. *Entomologia Experimentalis et Applicata* 65, 195–198.

Douglas, A.E. (1998) Nutritional interactions in insect–microbial symbioses. *Annual Review of Entomology* 43, 17–37.

Douglas, A.E. and Prosser, W.A. (1992) Synthesis of the essential amino acid tryptophan in the pea aphid (*Acyrthosiphon pisum*) symbiosis. *Journal of Insect Physiology* 38, 565–568.

Douglas, A.E., Minto, L.B. and Wilkinson, T.L. (2001) Quantifying nutrient production by the microbial symbionts in a phytophagous insect. *Journal of Experimental Biology* 204, 349–358.

Douglas, A.E., Price, D.R.G., Minto, L.B., Jones, E., Pescod, K.V., Francois, C.L.M.J., Pritchard, J. and Boonham, N. (2006) Sweet problems: insect traits defining the limits to dietary sugar utilisation by the pea aphid, *Acyrthosiphon pisum. Journal of Experimental Biology* 209, 1395–1403.

Downing, N. (1978) Measurements of the osmotic concentrations of stylet sap, haemolymph and honeydew from an aphid under osmotic stress. *Journal of Experimental Biology* 77, 247–250.

van Emden, H.F. (1967) An increase in the longevity of adult *Aphis fabae* fed artificially through parafilm membrane on liquids under pressure. *Entomologia Experimentalis et Applicata* 10, 166–170.

van Emden, H.F. (1988) The peach–potato aphid *Myzus persicae* (Sulzer) (Hemiptera: Aphididae) – more than a decade on a fully-defined chemical diet. *Entomologist* 107, 4–10.

van Emden, H.F. (1996) Artificial diets for aphids – diet, debatable; artificial, without doubt. In: *Techniques in Plant–Insect Interactions and Biopesticides*. IFS, Stockholm, pp. 24–30.

van Emden, H.F. and Andrews, N. (1997) Twenty years of rearing the peach–potato aphid *Myzus persicae* (Sulzer) on a fully-defined chemical diet. *Entomologist* 116, 169–174.

Febvay, G., Pageaux, J.-F. and Bonnot, G. (1992) Lipid composition of the pea aphid, *Acyrthosiphon pisum* (Harris) (Homoptera: Aphididae), reared on host plant and on artificial media. *Archives of Insect Biochemistry and Physiology* 21, 103–118.

Febvay, G., Liadouze, I., Guillaud, J. and Bonnot, G. (1995) Analysis of energetic amino acid metabolism in *Acyrthosiphon pisum*: a multidimensional approach to amino acid metabolism in aphids. *Archives of Insect Biochemistry and Physiology* 29, 45–69.

Febvay, G., Rahbe, Y., Rynkiewicz, M., Guillaud, J. and Bonnot, G. (1999) Fate of dietary sucrose and neosynthesis of amino acids in the pea aphid, *Acyrthosiphon pisum*, reared on different diets. *Journal of Experimental Biology* 202, 2639–2652.

Fisher, D.B., Wright, J.P. and Mittler, T.E. (1984) Osmoregulation by the aphid *Myzus persicae*: a physiological role for honeydew oligosaccharides. *Journal of Insect Physiology* 30, 387–393.

Forrest, J.M.S. and Knights, B.A. (1972) Presence of phytosterols in the food of the aphid, *Myzus persicae*. *Journal of Insect Physiology* 18, 723–728.

Fukatsu, T., Nikoh, N., Kawai, R. and Koga, R. (2000) The secondary endosymbiotic bacterium of the pea aphid *Acyrthosiphon pisum* (Insecta: Homoptera). *Applied and Environmental Microbiology* 66, 2748–2758.

Fukatsu, T., Tsuchida, T., Nikoh, N. and Koga, R. (2001) *Spiroplasma* symbiont of the pea aphid *Acyrthosiphon pisum* (Insecta: Homoptera). *Applied and Environmental Microbiology* 67, 1284–1291.

Geiger, D.R. and Servaites, J.C. (1994) Diurnal regulation of photosynthetic carbon metabolism in C_3 plants. *Annual Review of Plant Physiology and Molecular Biology* 45, 235–256.

Grenier, A.M., Nardon, C. and Rahbe, Y. (1994) Observations on the micro-organisms occurring in the gut of the pea aphid *Acyrthosiphon pisum*. *Entomologia Experimentalis et Applicata* 70, 91–96.

Hamilton, M.A. (1930) Notes on the culturing of insects for virus work. *Annals of Applied Biology* 17, 487–492.

Harada, H., Oyaizu, H., Kosako, K. and Ishikawa, H. (1997) *Erwinia aphidicola*, a new species isolated from pea aphid. *Acyrthosiphon pisum*. *Journal of General and Applied Microbiology* 43, 349–354.

Harrewijn, P. (1978) The role of plant substances in polymorphism of the aphid *Myzus persicae*. *Entomologia Experimentalis et Applicata* 24, 198–214.

Haynes, S., Darby, A.C., Daniell, T.J., Webster, G., van Veen, F.J.F., Godfray, H.C.J., Prosser, J.I. and Douglas, A.E. (2003) Diversity of bacteria associated with natural aphid populations. *Applied and Environmental Microbiology* 69, 7216–7223.

Hendrix, D.L. and Salvucci, M.E. (1998) Polyol metabolism in homopterans at high temperatures: accumulation of mannitol in aphids (Aphididae: Homoptera) and sorbitol in whiteflies (Aleyrodidae: Homoptera). *Comparative Biochemistry and Physiology, Series A* 20, 487–494.

Hinde, R. (1971) The control of the mycetome symbiotes of the aphids *Brevicoryne brassicae*, *Myzus persicae* and *Macrosiphum rosae*. *Journal of Insect Physiology* 17, 1791–1800.

Humphreys, N. and Douglas, A.E. (1997) The partitioning of symbiotic bacteria between generations of an insect: a quantitative study of *Buchnera* in the pea aphid (*Acyrthosiphon pisum*) reared at different temperatures. *Applied and Environmental Microbiology* 63, 3294–3296.

King, R.W. and Zeevaart, J.A.D. (1974) Enhancement of phloem exudation from cut petioles by chelating agents. *Plant Physiology* 53, 96–103.

Leckstein, P.M. and Llewellyn, M. (1974) The role of amino acids in diet intake and selection and the utilization of dipeptides by *Aphis fabae*. *Journal of Insect Physiology* 20, 877–885.

Mittler, T.E. (1953) Amino-acids in phloem sap and their excretion by aphids. *Nature, London* 172, 207.

Mittler, T.E. (1958) Studies on the feeding and nutrition of *Tuberolachnus salignus* (Gmelin) (Homoptera, Aphididae). II. The nitrogen and sugar composition of ingested phloem sap and excreted honeydew. *Journal of Experimental Biology* 35, 74–84.

Mittler, T.E. (1970) Effects of dietary amino acids on the feeding rate of the aphid *Myzus persicae*. *Entomologia Experimentalis et Applicata* 13, 432–437.

Mittler, T.E. (1972) Interactions between dietary components. In: Rodriguez, J.G. (ed.) *Insect and Mite Nutrition*. North-Holland, Amsterdam, pp. 211–223.

Mittler, T.E. (1976) Ascorbic acid and other chelating agents in the trace-mineral nutrition of the aphid *Myzus persicae* on artificial diets. *Entomologia Experimentalis et Applicata* 20, 81–98.

Mittler, T.E. and Dadd, R.H. (1962) Artificial feeding and rearing of the aphid, *Myzus persicae* (Sulzer), on a completely defined synthetic diet. *Nature, London* 195, 404.

Mittler, T.E. and Dadd, R.H. (1964) Gustatory discrimination between liquids by the aphid *Myzus persicae* (Sulzer). *Entomologia Experimentalis et Applicata* 7, 315–328.

Mittler, T.E. and Meikle, T. (1991) Effects of dietary sucrose concentration on aphid honeydew carbohydrate levels and rates of excretion. *Entomologia Experimentalis et Applicata* 59, 1–7.

Montllor, C.B., Maxmen, A. and Purcell, A.H. (2002) Facultative bacterial endosymbionts benefit pea aphid *Acyrthosiphon pisum* under heat stress. *Ecological Entomology* 27, 189–195.

Moran, N.A., Munson, M.A., Baumann, P. and Ishikawa, H. (1993) A molecular clock in endosymbiotic bacteria is calibrated using the insect hosts. *Proceedings of the Royal Society of London B* 253, 167–171.

Moran, N.A., Russell, J.A., Koga, R. and Fukatsu, T. (2005) Evolutionary relationships of three new species of Enterobacteriaceae living as symbionts of aphids and other insects. *Applied and Environmental Microbiology* 71, 3302–3310.

Munson, M.A., Baumann, P. and Kinsey, M.G. (1991) *Buchnera* gen. nov. and *Buchnera aphidicola* sp. nov., a taxon consisting of the mycetocyte-associated, primary endosymbionts of aphids. *International Journal of Systematic Bacteriology* 41, 566–568.

Nakabachi, A. and Ishikawa, H. (1999) Provision of riboflavin to the host aphid, *Acyrthosiphon pisum*, by endosymbiotic bacteria, *Buchnera. Journal of Insect Physiology* 45, 1–6.

Oliver, K.M., Russell, J.A., Moran, N.A. and Hunter, M.S. (2003) Facultative bacterial symbionts in aphids confer resistance to parasitic wasps. *Proceedings of the National Academy of Sciences of the United States of America* 100, 1803–1807.

Prosser, W.A. and Douglas, A.E. (1991) The aposymbiotic aphid: an analysis of chlortetracycline-treated pea aphid, *Acyrthosiphon pisum. Journal of Insect Physiology* 37, 713–719.

Rahbe, Y., Delobel, B., Febvay, G. and Chantegrel, B. (1994) Aphid-specific triglycerides in symbiotic and aposymbiotic *Acyrthosiphon pisum. Insect Biochemistry and Molecular Biology* 24, 95–101.

Rahbe, Y., Sauvion, N., Febvay, G., Peumans, W.J. and Gatehouse, A.M.R. (1995) Toxicity of lectins and processing of ingested proteins in the pea aphid *Acyrthosiphon pisum. Entomologia Experimentalis et Applicata* 76, 143–155.

Rennenberg, H. (1982) Glutathione metabolism and possible biological roles in higher plants. *Phytochemistry* 21, 2771–2781.

de Renobles, M., Ryan, R.O., Heisler, C.R., McLean, D.L. and Blomquist, G.J. (1986) Linoleic acid biosynthesis in the pea aphid, *Acyrthosiphon pisum. Archives of Insect Biochemistry and Physiology* 3, 193–203.

Rhodes, J.D., Croghan, P.C. and Dixon, A.F.G. (1996) Uptake, excretion and respiration of sucrose and amino acids by the pea aphid *Acyrthosiphon pisum. Journal of Experimental Biology* 199, 1269–1276.

Rhodes, J.D., Croghan, P.C. and Dixon, A.F.G. (1997) Dietary sucrose and oligosaccharide synthesis in relation to osmoregulation in the pea aphid, *Acyrthosiphon pisum. Physiological Entomology* 22, 373–379.

Russell, J.A. and Moran, N.A. (2005) Costs and benefits of symbiont infection in aphids: variation among symbionts and across temperatures. *Proceedings of the Royal Society of London B* 273, 603–610.

Russell, J.A., Latorre, A., Sabater-Munoz, B., Moya, A. and Moran, N.A. (2003) Independent origins and horizontal transfer of bacterial symbionts of aphids. *Molecular Ecology* 12, 1061–1075.

Ryan, R.O., de Renobles, M., Dillwith, J.W., Heisler, C.R. and Blomquist, G.J. (1982) Biosynthesis of myristate in an aphid: involvement of a specific acylthioesterase. *Archives of Biochemistry and Biophysics* 213, 26.

Salvucci, M.E. (2000) Effect of the α-glucosidase inhibitor, bromoconduritol, on carbohydrate metabolism in the silverleaf whitefly, *Bemisia argentifolii. Archives of Insect Biochemistry and Physiology* 45, 117–128.

Salvucci, M.E. and Crafts-Brandner, S.J. (2000) Effects of temperature and dietary sucrose concentration on respiration in the silverleaf whitefly, *Bemisia argentifolii. Journal of Insect Physiology* 46, 1461–1467.

Sandstrom, J.P., Russell, J.A., White, J.P. and Moran, N.A. (2001) Independent origins and horizontal transfer of bacterial symbionts of aphids. *Molecular Ecology* 10, 217–228.

Sasaki, T. and Ishikawa, H. (1995) Production of essential amino acids from glutamate by mycetocyte symbionts of the pea aphid, *Acyrthosiphon pisum. Journal of Insect Physiology* 41, 41–46.

Sasaki, T., Aoki, T., Hayashi, H. and Ishikawa, H. (1990) Amino acid composition of the honeydew of symbiotic and aposymbiotic pea aphids *Acyrthosiphon pisum. Journal of Insect Physiology* 36, 35–40.

Shigenobu, S., Watanabe, H., Hattori, M., Sasaki, Y. and Ishikawa, H. (2000) Genome sequence of the endocellular bacterial symbiont of the aphids *Buchnera* sp. APS *Nature* 407, 81–86.

Strong, F.E. (1963) Studies on the lipids in some homopterous insects. *Hilgardia* 34, 43–61.

Suh, S.-O., Noda, H. and Blackwell, M. (2001) Insect symbiosis: derivation of yeast-like endosymbionts within an entomopathogenic filamentous lineage. *Molecular Biology and Evolution* 18, 995–1000.

Tamas, I., Klasson, L., Canback, B., Naslund, A.K., Eriksson, A.S., Wernegreen, J.J., Sandstrom, J.P., Moran, N.A. and Andersson, S.G.E. (2002) 50 million years of genomic stasis in endosymbiotic bacteria. *Science, New York* 296, 2376–2379.

Thompson, G.A. and Schulz, A. (1999) Macromolecular trafficking in the phloem. *Trends in Plant Sciences* 4, 354–360.

Tomos, A.D. and Sharrock, R.A. (2001) Cell sampling and analysis (SiCSA): metabolites measured at single cell resolution. *Journal of Experimental Botany* 52, 623–630.

Walters, F.S. and Mullin, C.A. (1988) Sucrose-dependent increase in oligosaccharide production and associated glycosidase activities in the potato aphid *Macrosiphum euphorbiae* (Thomas). *Archives of Insect Biochemistry and Physiology* 9, 35–46.

Walters, F.S., Mullin, C.A. and Gildow, F.E. (1994) Biosynthesis of sorbic acid in aphids: an investigation into symbiont involvement and potential relationship with aphid pigments. *Archives of Insect Biochemistry and Physiology* 26, 49–67.

Whitehead, L.F. and Douglas, A.E. (1993) Populations of symbiotic bacteria in the parthenogenetic pea aphid (*Acyrthosiphon pisum*) symbiosis. *Proceedings of the Royal Society of London B* 254, 29–32.

Whitehead, L.F., Wilkinson, T.L. and Douglas, A.E. (1992) Nitrogen recycling in the pea aphid (*Acyrthosiphon pisum*) symbiosis. *Proceedings of the Royal Society of London B* 250, 115–117.

Wilkinson, T.L. (1998) The elimination of intracellular microorganisms from insects: an analysis of antibiotic-treatment in the pea aphid (*Acyrthosiphon pisum*). *Comparative Biochemistry and Physiology, Series A* 119, 871–889.

Wilkinson, T.L. and Douglas, A.E. (1995) Why aphids lacking symbiotic bacteria have elevated levels of the amino acid glutamine. *Journal of Insect Physiology* 41, 921–927.

Wilkinson, T.L. and Douglas, A.E. (2003) Phloem amino acids and the host plant range of the polyphagous aphid, *Aphis fabae*. *Entomologia Experimentalis et Applicata* 106, 103–113.

Wilkinson, T.L., Ashford, D.A., Pritchard, J. and Douglas, A.E. (1997) Honeydew sugars and osmoregulation in the pea aphid *Acyrthosiphon pisum*. *Journal of Experimental Biology* 200, 2137–2143.

Wilkinson, T.L., Adams, D., Minto, L.B. and Douglas, A.E. (2001a) The impact of host plant on the abundance and function of symbiotic bacteria in an aphid. *Journal of Experimental Biology* 204, 3027–3038.

Wilkinson, T.L., Minto, L.B. and Douglas, A.E. (2001b) Amino acids as respiratory substrates in aphids: an analysis of *Aphis fabae* reared on plants and diets. *Physiological Entomology* 26, 225–228.

Wojciechowicz-Zytko, E. and van Emden, H.F. (1995) Are aphid mean relative growth rate and intrinsic rate of increase likely to show a correlation in plant resistance studies? *Journal of Applied Entomology* 119, 405–409.

Ziegler, W. (1975) Nature of transported substances. In: Zimmerman, M.H. and Milburn, J.A. (eds) *Encyclopedia of Plant Physiology – New Series. Volume 1*. Springer, Berlin, pp. 59–100.

6 Growth and Development

Caroline S. Awmack[1] and Simon R. Leather[2]

[1]*Department of Entomology, University of Wisconsin-Madison, Madison, WI 53706, USA;* [2]*Division of Biology, Imperial College London, Silwood Park Campus, Ascot, Berks, SL5 7PY, UK*

Introduction

The growth and developmental rates of individual aphids have been studied extensively since the early investigations of Davis (1915) because they can be reliable indicators of future population growth rates (Leather and Dixon, 1984; Acreman and Dixon, 1989). In this chapter, we discuss the methods used to measure aphid growth and development, the relationships between these measures of aphid performance, and the reliability of using the results of such experiments to predict the performance of field populations of pest aphids.

Individual aphids frequently have extremely high growth and developmental rates, allowing aphid populations to rapidly reach levels that are damaging to crop plants. Under optimal growth conditions, an individual aphid typically commences reproduction 7–10 days after it is born (Dixon, 1998). Such short development times are possible because newborn aphids contain the embryos of their first grand-daughters. This 'telescoping of generations' means that an individual aphid has already completed two-thirds of its development before it is born (Dixon, 1998). Growth and developmental rates can be used to predict future fecundity because somatic growth and reproductive development occur simultaneously in the developing nymph,

and are thus simultaneously affected by any change in the rearing environment.

Aphid growth and developmental rates are frequently measured in small-scale trials using single individuals, or small groups of similarly aged nymphs. In the first section of this chapter, the most commonly used measurements of aphid performance are described, with the relevant experimental techniques needed to investigate them. Factors that may affect the reliability of this approach are then discussed, with a particular emphasis on studies that have investigated the relationships between the various measures of individual performance, and between individual and population growth rates.

Definitions

Aphid growth and developmental rates have been used extensively to predict the performance of aphid populations on crop plants because they correlate well with potential fecundity, achieved fecundity, and the intrinsic rate of increase, r_m (Leather and Dixon, 1984; Dixon, 1990). Many aphid species show strong positive relationships between growth rates and potential fecundity (Lyth, 1985; Fereres *et al.*, 1989). Similarly, adult size is frequently correlated with both potential fecundity (Dixon and Dharma, 1980; Kempton *et al.*, 1980; Bintcliffe and Wratten, 1982; Llewellyn and Brown, 1985) and achieved fecundity (Leather and Dixon, 1984; Dixon, 1990; 1998).

'Growth' is defined here as an increase in aphid size; 'development' is used to imply increasing reproductive maturity; 'potential fecundity' is a measure of the reproductive potential of an individual aphid – for example, the number of mature embryos contained in an adult; while 'achieved fecundity' is the total number of progeny produced by an adult. Potential fecundity may be a useful measure of aphid performance, as potential and achieved fecundity frequently are strongly positively correlated (Dixon and Wratten, 1971; Dixon and Dharma, 1980; Frazer and Gill, 1981; Leather and Dixon, 1984). Developmental and reproductive rates can be combined in the intrinsic rate of increase, r_m (Birch, 1948; Wyatt and White, 1977), an estimate of future population growth rates based on the performance of individual aphids.

Uses of aphid growth and developmental rates

The use of aphid relative growth rates to measure performance was initially suggested by van Emden (1969) as a means of measuring the direct effects of plant nutrition on aphid performance, without the confounding effects of maternal experience. Dixon (1990) and Leather and Dixon (1984) later demonstrated that relative growth rates and the intrinsic rate of increase, r_m, (Wyatt and White, 1977) were strongly positively correlated (Fig. 6.1).

As a consequence, many authors have used mean relative growth rate (MRGR) or relative growth rate (RGR), defined below, when quick estimates of aphid performance are required; for example, to assess the resistance of plants to aphid attack (van Emden, 1969; Bintcliffe and Wratten, 1982; Givovich *et al.*, 1988; Leszczynski *et al.*, 1989; Wojciechowicz-Zytko and van Emden, 1995; Farid *et al.*, 1998; Telang *et al.*, 1999), the quality of different growth stages of the host plant (Leather and Dixon, 1981), the effects

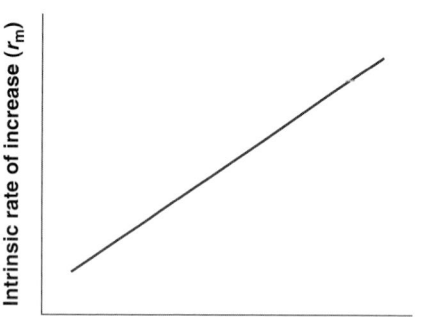

Fig. 6.1. Hypothetical relationship between aphid growth rates and the intrinsic rate of increase (r_m).

of temperature (Berg, 1984; Lamb *et al.*, 1987), the effects of air pollution (Dohmen, 1985; Holopainen *et al.*, 1997; Wu *et al.*, 1997), and the contribution of bacterial symbionts to aphid nutritional ecology (Adams and Douglas, 1997).

Similarly, the intrinsic rate of increase has been used to investigate the responses of individual aphids to changes in plant quality (Bintcliffe and Wratten, 1982; Lykouressis, 1984; Zuniga *et al.*, 1985; Leszczynski *et al.*, 1989), temperature (Landin and Wennergren, 1987; Liu and Yue, 2000), drought stress (Sumner *et al.*, 1986), atmospheric pollutants (Warrington *et al.*, 1987; Awmack *et al.*, 1997), host plant virus infection (Fereres *et al.*, 1989), and the sub-lethal effects of insecticide residues (Kerns and Gaylor, 1992).

Measurement of Aphid Growth and Developmental Rates

Growth rates

Most investigations of aphid growth rates have used living aphids to measure gain in fresh weight over a defined time period, weighing the same individuals at the beginning and the end of the experiment (e.g. Banks and Macaulay, 1964; Tsitsipis and Mittler, 1976; Frazer and Gill, 1981; Kindlmann *et al.*, 1992; Sunnucks *et al.*, 1998; Manninen *et al.*, 2000; Edwards, 2001), although some authors, such as Lamb (1992) have used dry weight and have therefore used different individuals for each measurement.

Other investigations of aphid growth rates have taken advantage of the allometric relationships between aphid skeletal structure and size (Reddy and Alfred, 1981), using hind tibia length (Campbell, 1983) or antennal length (Varty, 1964; Murdie, 1969a,b). However, since the ratios of the sizes of some aphid body parts may change as the developing aphid grows (Varty, 1964; Dixon, 1987), such allometric relationships should be used with caution. Aphid growth rates are usually measured using individual aphids or small groups of aphids, often the progeny of a single female. Young (typically newborn) aphids are removed from their host plants, weighed on a microbalance and returned to the host plants for the desired time period, after which they are removed and re-weighed.

Aphid growth rates are a function of birth weight (Dixon *et al.*, 1982; Carroll and Hoyt, 1986), as is weight loss when aphids are starved (e.g. Brough and Dixon, 1990). Since large aphids grow faster than small aphids (Dixon, 1998), measurements of aphid growth rates must correct for differences in initial weight. Both MRGR and RGR compensate for the increasing mass of the insect as it grows, since they are based on the logarithmic weight gain of the aphid.

The formula used to determine MRGR is based on an equation originally used by plant scientists (Radford, 1967):

MRGR $(\mu g/\mu g/day)$

$$= (\log W_2 - \log W_1)/t_2 - t_1 \qquad (6.1)$$

where W_1 = weight at the first weighing, W_2 = weight at the next weighing, $t_2 - t_1$ = the time (in days) between first (t_1) and second (t_2) weighing.

RGR is measured over the development time (D) of the aphid (i.e. from birth to the final moult but before the onset of reproduction), and therefore takes the effects of host-plant quality and maternal effects (such as ovariole number) into consideration:

RGR $(\mu g/\mu g/day) = (\log W_2 - \log W_1)/D$
$$\qquad (6.2)$$

Two major experimental drawbacks are associated with using MRGR or RGR to measure aphid performance. First, it is usually difficult to manipulate young aphids in the field. Second, a very accurate microbalance is required. Newborn aphids can weigh as little as 30 μg (Dixon, 1998), and hence small inaccuracies in the measurement of initial weight can have large effects on the final value of MRGR or RGR because of the logarithmic nature of insect growth. A simple solution is to measure groups of aphids and use the average initial and final weights to determine growth rates.

MRGR and RGR are, however, very simple ways to investigate treatment effects on aphid performance since they may be measured over as little as two days (van Emden and Bashford, 1969; Adams and van Emden, 1972). A second benefit of both MRGR and RGR is that they involve relatively little disturbance of the aphids (Adams and van Emden, 1972). The newly moulted adults produced after RGR is measured can be returned easily to the host plant to measure achieved fecundity and r_m, or dissected to investigate treatment effects on potential fecundity.

Developmental rates

Developmental rates are determined by recording the time period between particular events (for example, from birth to adult) and reporting the results as the reciprocal of the data (Berg, 1984; Carroll and Hoyt, 1986; Lamb *et al.*, 1987; Lamb, 1992; Cabrera *et al.*, 1995). Most aphids pass through four nymphal instars (Dixon, 1973) before moulting to the adult stage (although some species, such as *Rhopalosiphum nymphaeae*, may have five instars, depending on rearing temperature (Rohita and Penman, 1983; Ballou *et al.*, 1986). Instar duration may also be a useful measure of development, particularly in studies investigating treatment effects on the vulnerability of aphids to natural enemies that preferentially attack specific instars (Ives *et al.*, 1999; Chau and Mackauer, 2001).

Measurements of development are particularly useful in studies investigating treatment effects on aphids reared in field situations, as the aphids do not need to be removed from the host plant to be weighed. Developmental times or rates are particularly useful when predictions about treatment effects on future population growth rates are required, since they are an integral component of the intrinsic rate of increase, r_m.

The intrinsic rate of increase, r_m

The intrinsic rate of increase, r_m (Wyatt and White, 1977), relates the fecundity of an individual aphid to its development time:

$$r_m = (\ln Md \times c)/D \qquad (6.3)$$

where Md is the number of nymphs produced by the adult in the first D days of reproduction after the adult moult. The constant, c, has a value of 0.738 and is an approximation of the proportion of the total fecundity produced by a female in the first D days of reproduction. It is obvious from this equation that a small change in development time will have a greater effect on r_m than an increase in fecundity of a similar magnitude. Although r_m has limitations (Awmack *et al.*, 1997), it is a favoured way of estimating population growth as it is much easier than counting the thousands of aphids that are likely to occur in a real population.

Experimental Techniques

Aphid cages

Although the growth and development of individual aphids has been recorded under unrestricted field conditions (Cannon, 1984), most studies have used individual aphids reared in the laboratory or in controlled environments (van Emden, 1972; Dixon, 1998). Since newborn aphids are so small, it has become standard protocol to cage either an individual aphid or a group of aphids on the host plants.

The most commonly used aphid cage is the clip cage, originally designed by MacGillivray and Anderson (1957). Clip cages may range in size from those that cover a few cm^2 of the leaf to those that enclose whole leaves. As not all leaves used by aphids are flat, tubular cages based on gelatin capsules can also be used. These cages slip snugly over the entire leaf or stem, but can be opened easily without disturbing the occupants (Fisher, 1987). Alternatives to clip cages include gauze sleeves that can be used to confine aphids to specific parts of the plant (for example, stems). Whole plant cages, usually constructed of PVC tubes, open at the bottom and covered with gauze at the top, can be used to monitor colonies or aggregations of aphids that need free access to all or part of the plant (Markkula and Rautapaa, 1963).

Two clip cages, suitable for use on many crop species, are shown in Fig. 6.2. The Type I cage is the typical clip cage (MacGillivray and Anderson, 1957). The second cage (Type II) uses soft foam (Plastozote®, Watkins and Doncaster, UK) to minimize damage to the leaf surface and can also be supported with a standard plant stake, reducing strain on the leaf petiole, and can be opened without disturbing the aphids inside (Awmack, 1997; D. Huggett, personal communication).

Although some aphid species appear not to be adversely affected by regular removal from their host plants (Newton and Dixon, 1990b), disturbance should be kept to a minimum when measuring growth rates, because aphids may take several hours to locate a suitable phloem element and commence feeding (Tjallingii, 1995). Frequent disturbance may therefore lead to low estimates of aphid performance. The settling and feeding behaviour of aphids may also be affected by the treatment under investigation. Aphid settling and feeding behaviour may be affected by components of host-plant quality such as concentrations of plant defensive metabolites (Zehnder *et al.*, 2001), nutrient availability, and water stress (Ponder *et al.*, 2001), and by environmental factors such as elevated CO_2 atmospheres (Awmack *et al.*, 1996). Feeding and settling behaviour may also vary according to the prior experience of the aphids used (Ramirez and Niemeyer, 2000) and even between

Type
1

Type
2

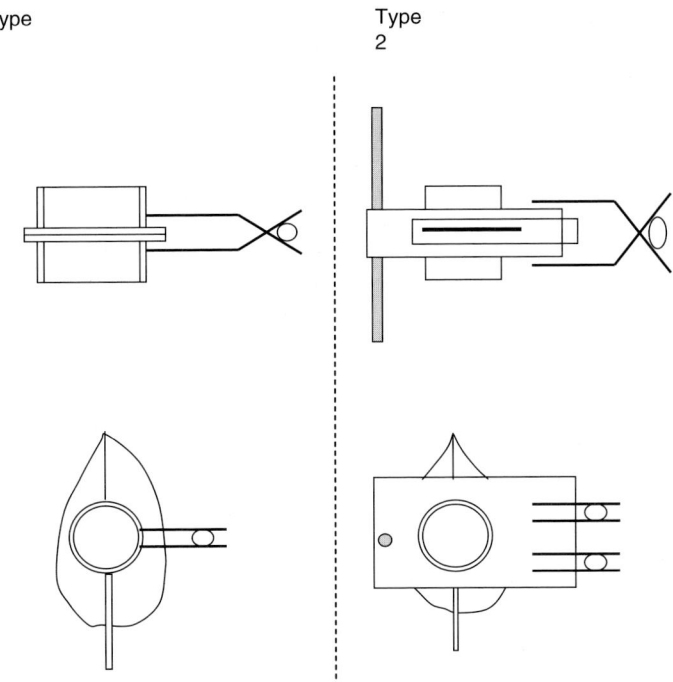

Fig. 6.2. Clip cages suitable for measuring aphid growth and development.

individuals within a species (Bernays and Funk, 2000).

Disadvantages of aphid cages

Many clip cage designs have been criticized because they interfere with leaf gas exchange and may damage the leaf surface, changing leaf quality in comparison with uncaged leaves (Crafts and Chu, 1999). The effects of larger cages on aphid performance have been documented from field experiments, as sunlight can be concentrated on to the plants leading to temperatures within the cage that are considerably higher than ambient external temperatures (Woodford, 1973). In this case, large, muslin-covered frames may be placed over the plants to minimize fluctuations in microclimate.

An equally serious concern associated with the use of cages to measure aphid growth and development is that aphid performance within cages may not reflect the performance of uncaged aphids. Figure 6.3 shows the adult weight, development time, and 7-day fecundity of *Rhopalosiphum padi* (bird cherry–oat aphid) reared without cages (control), in clip cages, and in PVC tubes on oat seedlings (*Avena sativa* cv. 'Aster'). The data clearly show that clip cages and PVC tubes had an adverse effect on the adult weight of the aphids, and a small but significant effect on development time. However, cages had no significant effects on 7-day fecundity (S.R. Leather, unpublished results). Adult weight and 7-day fecundity were positively correlated when the aphids were reared in clip cages ($r^2 = 0.218$, $P < 0.05$) and PVC tubes ($r^2 = 0.417$, $P < 0.01$), but not when they were reared on uncaged control plants ($r^2 = 0.005$). Cages therefore affected not only the performance of *R. padi*, but also the relationships between adult weight and potential or achieved fecundity. This example demonstrates clearly the problems inherent in comparing data collected using insect cages to data collected when the aphids are reared in more natural environments and able to select feeding sites.

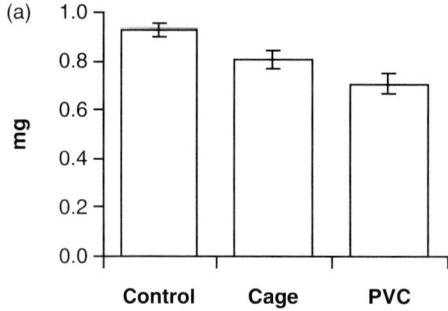

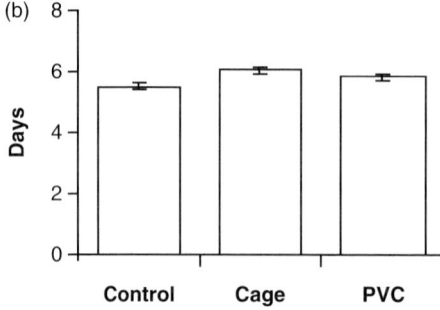

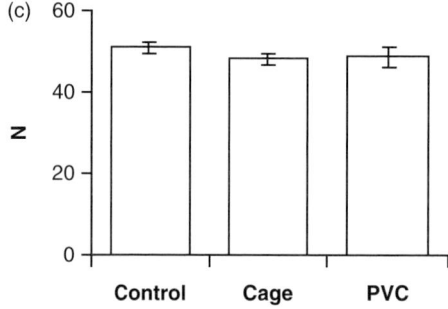

Fig. 6.3. Performance of the aphid *Rhopalosiphum padi* reared without cages, in clip cages, and in PVC tubes on oat seedlings (*Avena sativa* cv. 'Aster'). (a) Adult weight (mg) (F = 8.36, df = 2/45, P < 0.001). (b) Development time (days) (F = 6.10, df = 2/45, P < 0.001). (c) 7-day fecundity (F = 0.86, df = 2/45, P > 0.05). All data are presented as the means of 16 replicates and are shown with ± the standard error of the mean.

Factors Affecting Aphid Growth and Development

Aphid growth and developmental rates are affected by a wide range of both intrinsic and extrinsic factors such as diet quality

(e.g. Watt and Dixon, 1981; Gruber and Dixon, 1988; Tsai and Wang, 2001; Vacanneyt, 2001), plant growth stage (Zhou and Carter, 1992), the abiotic environment (Kenten, 1955; Dean, 1974; Walgenbach *et al.*, 1988; McVean and Dixon, 2001), maternal experience (Johnson, 1965; Dixon and Glen, 1971; Chambers, 1982; Kidd and Tozer, 1984), and maternal morph (Leather, 1989). In this section, some of the factors affecting the reliability of aphid growth rates as predictors of the performance of aphid populations are outlined, and some of the most common factors affecting the differences between measures of aphid growth and developmental rates and the performance of natural aphid populations are discussed.

A key consideration in these types of studies is that aphids used to measure growth and developmental rates in greenhouse-based investigations tend to have been raised at low densities at constant temperatures while natural aphid populations interact with a variable biotic and abiotic environment. Genetic variability may also contribute to the variability inherent in natural populations: in many studies, the aphids are derived from parthenogenetic lineages, produced by a single parthenogenetic female. Natural populations are rarely as uniform (except in the case of pests of greenhouse crops, which may have very low levels of genetic diversity (Rochat *et al.*, 1999). When apterous *Sitobion avenae* (grain aphid) were reared on oats (*A. sativa*), pink individuals developed more quickly than green individuals, highlighting the need to use multiple aphid genotypes (Araya *et al.*, 1996). Similarly, many aphid species such as *Myzus persicae* (peach–potato aphid), *Aphis craccivora* (cowpea aphid) (Edwards, 2001), *Sitobion miscanthi* and *Sitobion* near *fragariae* (Sunnucks *et al.*, 1998), *Acyrthosiphon pisum* (pea aphid) (Sandström, 1994), and *Phorodon humuli* (damson–hop aphid) (Lorriman and Llewellyn, 1983) show genetic variation in their growth and developmental rates, even when reared on hosts of similar quality.

Aphids used in laboratory-based studies frequently are confined to a specific part of the host plant (for example, in clip cages) and reared at very low densities in the absence of other pests and diseases. Many aphid species prefer specific host-plant parts, and experiments that cage aphids on the 'wrong' part of a plant do not give an accurate representation of the performance of natural aphid populations. Hopkins *et al.* (1998) showed that even though *M. persicae* prefers senescing leaves at the base of its host plants (three *Brassica* cultivars, differing in their glucosinolate content) while *Brevicoryne brassicae* (cabbage aphid) prefers young leaves, regardless of the cultivar of the host plant, aphid performance was unaffected by plant defensive chemistry. Similarly, Williams (1995) investigated the impacts of host plants of different ages, with and without *Beet yellows virus* infection, on *M. persicae*, and showed that performance was better on younger leaves than on old leaves, and that virus infection increased aphid performance. Table 6.1 shows the fecundity and development time of *S. avenae* when reared on different parts of oat and wheat plants (from Watt, 1979).

Factors Affecting the Reliability of Size × Fecundity Relationships

While aphid growth and developmental rates can often be used to predict future fecundity (and hence population growth rates), some treatments affect the reliability of these relationships (Leather, 1988; Awmack and Leather, 2002). Aphid size is determined predominantly by the quality of

the larval host plant (Banks and Macaulay, 1964; van Emden and Bashford, 1969; Leather and Dixon, 1982; Acreman and Dixon, 1989; Gange and Pryse, 1990; Caillaud *et al.*, 1994), although the quality of the adult's host plant is also important (Markkula and Rouka, 1970; Watt, 1979; Leather and Dixon, 1981; McLeod *et al.*, 1991) since aphids continue to mature offspring after the final adult moult. Some aphid species vary allocation of their resources between reproductive and somatic tissues, e.g. *Myzocallis boerneri* (Turkey oak aphid) (Sequeira and Dixon, 1996), *Drepanosiphum platanoidis* (sycamore aphid) (Douglas, 2000), *Megoura viciae* (vetch aphid) (Brough and Dixon, 1990), and *S. avenae* (Helden and Dixon, 1998). Thus, insect size may not necessarily be a reliable predictor of future fecundity (Leather, 1988). Maternal effects (i.e. the host plant on which the parent of the reproducing aphid was reared) may also affect aphid size × fecundity relationships (Leather, 1989; Messina, 1993).

While generally there are strong and positive correlations between aphid growth and developmental rates and the intrinsic rate of increase, r_m, it has been shown that this relationship varies with both the species of aphid under investigation and the growth stage of the host plant. Guldemond *et al.* (1998) demonstrated that the relationship between MRGR and r_m varied with both the species of aphid investigated (*Aphis gossypii* – cotton or melon aphid, or *M. persicae*) and the growth stage of the host plant (chrysanthemum, *Dendranthema* x *grandiflorum*). If experimental treatments affect such relationships (e.g. Kerns and Gaylor, 1992; Sarao and Singh, 1998), individual growth rates need not reflect population growth rates.

While individuals with high MRGRs also tend to have high RGRs, the two growth rates need not be the same for any individual aphid/treatment combination, and may not be the same throughout the entire nymphal development time. Some aphid species, such as *S. avenae*, have high nymphal growth rates during early instars (Newton and Dixon, 1990b), while growth rates often decrease in later instars as the

Table 6.1. Effects of rearing on different host plant parts of winter wheat (cv. 'Maris Huntsman') on the fecundity and development of the aphid *Sitobion avenae*.

	7-day fecundity	Time to adult moult (days)
Ears	41.7	8.5
Young leaves	26.5	8.4
Lower leaves	13.3	10.4
Flag leaves	17.6	9.8
Senescent leaves	14.5	9.0

aphid switches resources to embryo maturation (Kindlmann and Dixon, 1989, 1992; Newton and Dixon, 1990b). Instar duration frequently varies, with the fourth instar being significantly longer than the first three (Kieckhefer *et al.*, 1989; Araya *et al.*, 1996; Dixon, 1998), particularly in the case of individuals destined to be alate (Newton and Dixon, 1990a). Treatments may therefore not affect all aphid instars equally, and early aphid instars may also be more sensitive to changes in the quality of their host plants than later instars. When *Schizaphis graminum* (greenbug) was reared on maize (*Zea mays*) or sorghum (*Sorghum bicolor*) cultivars, the host plant significantly affected the development time of the first and second aphid instars, but not the third or fourth (McCauley *et al.*, 1990). Similarly, the development of first-instar *A. pisum* was slower on aphid-resistant red clover (*Trifolium pratense*) cultivars than on susceptible cultivars, but since later instars took less time to develop on resistant cultivars, the total developmental time was unaffected (Zeng *et al.*, 1993). Not all aphids show this variation in larval growth rates between instars; growth rates of *M. persicae* remained constant throughout larval development (van Emden, 1969). Since many other insect groups show similar declines in growth rates as they develop (Scriber and Slansky, 1981), *M. persicae* may be an exception, rather than the rule.

Changes in the reproductive rates of individual aphids may also affect the reliability of assumptions about the relationships between growth rates and adult size and fecundity, as not all aphids produce nymphs at a constant rate throughout their adult life. Zeng *et al.* (1993) showed that *A. pisum* produced more nymphs during the daytime than at night, demonstrating that measurements of fecundity must take place over at least 24 h. Many aphid species produce a rapid 'burst' of reproduction shortly after the final moult, and then show a reduced rate of reproduction for the remainder of their adult life (Dixon, 1998). Aphid reproductive strategies may also vary according to predictable changes in plant quality (Leather, 1987) or unpredictable environmental conditions such as starvation (Leather *et al.*, 1983; Ward *et al.*, 1983; Brough and Dixon, 1990; Kouame and Mackauer, 1992; Gruber and Dixon, 1988). The first-born nymphs may not be representative of the entire progeny (Dixon *et al.*, 1993) because birth order may affect subsequent performance. The first-born nymphs of *A. pisum* are smaller than those born on subsequent days (Murdie, 1969b), but after about 7 days of reproduction, nymphal weight begins to fall and by 14 days the nymphs born are smaller than those born in the first 2 days. Similarly, first-born *M. persicae* nymphs show greater cold hardiness than later born nymphs (Clough *et al.*, 1990).

While r_m includes both developmental and reproductive rates, and is often a more reliable measure of aphid performance than MRGR or RGR, it has several disadvantages. Measurements of r_m are very labour intensive, and the risk of losing replicates (and hence statistical power) increases with the duration of the experiment. r_m is also a less sensitive measure of small, but biologically significant, changes in aphid performance (Lykouressis, 1984). A further disadvantage of r_m is apparent when treatments affect adult longevity since it is inaccurate when a treatment affects longevity but not development time (Sumner *et al.*, 1986). The fundamental assumption underlying the r_m equation is that a reproducing female will produce 95% of her progeny in the first D days of reproduction (Wojciechowicz-Zytko and van Emden, 1995). If this assumption is not met (for example, if a female dies after only a few days of reproduction), values of r_m overestimate the contribution of individual aphids to the growth of the population. Population growth rates of the pea aphid *A. pisum* reared on *Vicia faba* (broad bean) and exposed to a neem-based insecticide depended on the age at which the aphids were exposed to the insecticide. When individual *A. pisum* were exposed to this insecticide from birth, the population growth rates were negative. However, when *A. pisum* were exposed as adults, there was no effect of this insecticide on r_m, highlighting the need to determine treatment effects on all aphid life stages (Stark and Wennergren, 1995).

Difference between Nymphs Destined to be Apterous and Alate

Measurements of aphid growth and developmental rates should also take into account differences between nymphs destined to be apterous adults and nymphs destined to be alate adults. Although alate aphids have a longer development time than apterae (Tsumuki *et al.*, 1990; Araya *et al.*, 1996), in some species (e.g. *R. padi*), alatae have greater longevity than apterae (Foster *et al.*, 1988) and may therefore make a greater contribution to population growth rates. However, nymphs that develop into alate adults allocate a smaller proportion of their resources to reproduction (Newton and Dixon, 1990b) since they must produce wings and the flight muscles needed to power them (Dixon, 1998). As a result, alate aphids typically have lower fecundity than apterae (Elliott *et al.*, 1988; Newton and Dixon, 1990b; Collins and Leather, 2001). Alate aphids may also have a lower ovariole number (Dixon and Dharma, 1980; Leather, 1987) than apterae, differ in their amino acid and carbohydrate metabolism (Tsumuki *et al.*, 1990), and have greater lipid reserves (Febvay *et al.*, 1992). These differences may therefore mean that alatae respond differently to experimental treatments such as starvation or environmental factors; e.g. Garsed *et al.* (1987) showed that the fecundity of alate *Aphis fabae* (black bean aphid) reared on *V. faba* increased as light levels increased, while that of apterae did not.

Alata production is stimulated by crowding (Watt and Dixon, 1981; Bergeson and Messina, 1997; Dixon, 1998; Williams and Dixon, Chapter 3 this volume), although the proportion of the population developing into alatae may vary among populations of the same aphid species on different host plant species (Bommarco and Ekbom, 1996). Alata production may also be stimulated on virus-infected host plants (Blua and Perring, 1992), and may either be stimulated (Liu and

Wu, 1994) or suppressed (Parish and Bale, 1990) by low temperatures, depending on the species of aphid involved. As apterous progeny of alatae may also have lower ovariole numbers than the progeny of apterae (Leather, 1987), the effects of experimental treatments on population growth rates may persist for more than one generation.

Treatments that affect reproductive rates but not development times or growth rates also affect the reliability of r_m. *S. avenae* reared on winter wheat (*Triticum aestivum*) at double-ambient CO_2 had greater fecundity than at ambient CO_2, but their development times were unaffected (Awmack *et al.*, 1996). In a similar experiment, when *Aulacorthum solani* (glasshouse and potato aphid) was reared on either broad bean or tansy (*Tanacetum vulgare*) at elevated CO_2, development time decreased and r_m increased on tansy but not on bean, while fecundity increased on bean but not tansy (Awmack *et al.*, 1997). In these examples, r_m is unlikely to be a reliable predictor of aphid population growth rates as aphid fitness parameters appear to become uncoupled on exposure to this novel environment. The relationships between adult size and r_m may also become uncoupled if the aphid is exposed to starvation and resorbs embryos to release essential nutrients (Ward and Dixon, 1982; Brough and Dixon, 1990). Thus, a fundamental assumption underlying the r_m equation is that experimental conditions remain constant throughout the life of the aphid.

Temperature

Aphid responses to temperature are similar to those of other insects. Most aphid species show a strong linear relationship between temperature and growth or development within a range of approximately 7 and 25°C (Campbell *et al.*, 1974; Frazer and Gill, 1981), followed by a decline at increasing temperatures. The exact shape and range of the curve depends on both the aphid species and the geographic origin of genotypes within the species (Auclair and Aroga, 1987; Lamb and Mackay, 1988; Akey and

Butler, 1989), and may also be genotype-specific within a population (Lamb *et al.*, 1987). Aphids reared at high temperature may also grow into small adults containing fewer embryos (Leather and Dixon, 1982; Collins and Leather, 2001), or have a poor ability to maintain embryo maturation after the adult moult (Carroll and Hoyt, 1986). High temperatures may also affect the slope of the relationship between adult weight and embryo number, making predictions of fecundity from adult size unreliable (Carroll and Hoyt, 1986).

Plant quality may also modify the impacts of temperature on aphid growth and development (Leather and Dixon, 1982; Acreman and Dixon, 1989), as may abiotic factors such as wind (Walters and Dixon, 1984). Other authors (Liu and Perng, 1987; Kieckhefer *et al.*, 1989; Xia *et al.*, 1999) have also demonstrated that aphid performance decreases when aphids are reared at fluctuating temperatures rather than the equivalent constant temperature. In contrast, Zhang *et al.* (1991) showed that the population growth rate of *R. padi* was greater at fluctuating temperatures than at constant temperatures.

Hodgson and Godfray, 1999; Awmack and Harrington, 2000; Busque and Schtozko, 2000; Awmack *et al.*, 2004), although in some cases, aphid density has no effect on population growth rates (e.g. Messina, 1993). Some plant species may also exhibit a hypersensitive response, which is only apparent above threshold aphid densities (Lyth, 1985; Belefant-Miller *et al.*, 1994). In contrast, many aphid species show enhanced performance when they are reared in groups, rather than singly (Way and Cammell, 1970; Dixon and Wratten, 1971) because they are able to divert nutrients from other plant tissues more effectively (Sandström *et al.*, 2000). Interspecific competition between aphid species exploiting the same host plant may also reduce population growth rates (Fisher, 1987; Moran and Whitham, 1990; Thirakhupt and Araya, 1992; Gianoli, 2000), as may the presence of other herbivores, either directly (Masters, 1995) or *via* apparent competition involving shared natural enemies (Muller and Godfray, 1997) and plant diseases (Coleman and Jones, 1988; Blua and Perring, 1992; Castle and Berger, 1993).

Population-scale factors

Although a detailed discussion of population-scale effects on aphid performance is beyond the scope of this chapter, population density also has significant effects on aphid performance. At high population densities, aphids frequently produce alatae and leave the host plant (e.g. Watt and Dixon, 1981). Crowding (and intraspecific competition) may also reduce the *per capita* rate of reproduction: as host-plant resources become limiting, the population growth rate decreases (e.g. Farid *et al.*, 1998;

Conclusions

Although it is tempting, and perhaps technically easier, to use the performance of individual aphids as an indicator of aphid performance, perhaps the most reliable way to predict aphid performance is to investigate population size. Comparisons of r_m with population size (a much simpler and more reliable measure of aphid population growth rates) also take crowding and sink induction into account (Larson and Whitham, 1991; Sandström *et al.*, 2000) and are much easier to use (Lykouressis, 1984).

References

Acreman, S.J. and Dixon, A.F.G. (1989) The effects of temperature and host quality on the rate of increase of the grain aphid (*Sitobion avenae*) on wheat. *Annals of Applied Biology* 115, 3–9.

Adams, D. and Douglas, A.E. (1997) How symbiotic bacteria influence plant utilisation by the polyphagous aphid, *Aphis fabae*. *Oecologia* 110, 528–532.

Adams, J.B. and van Emden, H.F. (1972) The biological properties of aphids and their host plant relationships. In: van Emden, H.F. (ed.) *Aphid Technology*. Academic Press, London, pp. 47–104.

Akey, D. and Butler, G. (1989) Developmental rates and fecundity of apterous *Aphis gossypii* (Homoptera, Aphididae) on seedlings of *Gossypium hirsutum*. *Southwestern Entomologist* 14, 295–301.

Araya, J., Cambron, S. and Ratcliffe, R. (1996) Development and reproduction of two color forms of English grain aphid (Homoptera: Aphididae). *Environmental Entomology* 25, 366–369.

Auclair, J. and Aroga, R. (1987) Influence de la temperature sur la croissance et la reproduction de quatre bio-types du puceron du pois, *Acyrthosiphon pisum* (Homoptera). *Annals de la Societé Entomologique de France (N.S.)* 23, 279–286.

Awmack, C.S. (1997) Aphid–plant interactions at ambient and elevated CO_2. PhD thesis, Imperial College, University of London, London, UK.

Awmack, C.S. and Harrington, R. (2000) Elevated CO_2 affects the interactions between aphid pests and host plant flowering. *Agricultural and Forest Entomology* 2, 57–61.

Awmack, C.S. and Leather, S.R. (2002) Host plant quality and fecundity in herbivorous insects. *Annual Review of Entomology* 47, 817–844.

Awmack, C.S., Harrington, R., Leather, S.R. and Lawton, J.H. (1996) The impacts of elevated CO_2 on aphid–plant interactions. In: Froud-Williams, R.J., Harrington, R., Hocking, T.J., Smith, H.G. and Thomas, J.H. (eds) *Implications of 'Global Environmental Change' for Crops in Europe. Aspects of Applied Biology, No. 45*, pp. 317–322.

Awmack, C.S., Harrington, R. and Leather, S.R. (1997) Host plant effects on the performance of the aphid *Aulacorthum solani* (Homoptera: Aphididae) at ambient and elevated CO_2. *Global Change Biology* 3, 545–549.

Awmack, C.S., Harrington, R. and Lindroth, R.L. (2004) Aphid individual performance may not predict population responses to elevated CO_2 or O_3. *Global Change Biology* 10, 1414–1423.

Ballou, J., Tsai, J. and Center, T. (1986) Effects of temperature on the development, natality, and longevity of *Rhopalosiphum nymphaeae* L. (Homoptera: Aphididae). *Environmental Entomology* 15, 1096–1099.

Banks, C.J. and Macaulay, E.D.M. (1964) The feeding growth and reproduction of *Aphis fabae* Scop. on *Vicia fabae* under experimental conditions. *Annals of Applied Biology* 53, 229–242.

Belefant-Miller, H., Porter, D., Pierce, M. and Mort, A. (1994) An early indicator of resistance in barley to Russian wheat aphid. *Plant Physiology Rockville* 105, 1289–1294.

Berg, G. (1984) The effect of temperature and host species on the population growth potential of the cowpea aphid, *Aphis craccivora* Koch (Homoptera: Aphididae). *Australian Journal of Zoology* 32, 345–352.

Bergeson, E. and Messina, F.J. (1997) Resource- versus enemy-mediated interactions between cereal aphids (Homoptera: Aphididae) on a common host plant. *Annals of the Entomological Society of America* 90, 425–432.

Bernays, E. and Funk, D. (2000) Electrical penetration graph analysis reveals population differentiation of host-plant probing behaviors within the aphid species *Uroleucon ambrosiae*. *Entomologia Experimentalis et Applicata* 97, 183–191.

Bintcliffe, E. and Wratten, S. (1982) Antibiotic resistance in potato cultures to the aphid *Myzus persicae*. *Annals of Applied Biology* 100, 383–392.

Birch, L.C. (1948) The intrinsic rate of natural increase of an insect population. *Journal of Animal Ecology* 17, 15–26.

Blua, M. and Perring, T. (1992) Alate production and population increase of aphid vectors on virus-infected host plants. *Oecologia* 92, 65–70.

Bommarco, R. and Ekbom, B. (1996) Variation in pea aphid population development in three different habitats. *Ecological Entomology* 21, 235–240.

Bosque, P.N.A. and Schtozko, D.J. (2000) Wheat genotype, early plant growth stage and infestation density effects on Russian wheat aphid (Homoptera: Aphididae) population increase and plant damage. *Journal of Entomological Science* 35, 22–38.

Brough, C.N. and Dixon, A.F.G. (1990) The effects of starvation on the development and reproductive potential of apterous virginoparae of vetch aphid *Megoura viciae*. *Entomologia Experimentalis et Applicata* 55, 41–46.

Cabrera, H.M., Argandona, V.H., Zuniga, G.E. and Corcuera, L.J. (1995) Effect of infestation by aphids on the water status of barley and insect development. *Phytochemistry* 40, 1083–1088.

Caillaud, C.M., Dedryver, C.A. and Simon, J.-C. (1994) Development and reproductive potential of the cereal aphid *Sitobion avenae* on resistant wheat lines (*Triticum monococcum*). *Annals of Applied Biology* 125, 219–232.

Campbell, A., Frazer, B., Gilbert, N., Gutierrez, A. and Mackauer, M. (1974) Temperature requirements of some aphids and their parasites. *Journal of Applied Ecology* 11, 431–438.

Campbell, C. (1983) Antibiosis in hop (*Humulus lupulus*) to the damson–hop aphid, *Phorodon humuli*. *Entomologia Experimentalis et Applicata* 33, 57–62.

Cannon, R.J.C. (1984) The development rate of *Metopolophium dirhodum* (Walker) (Hemiptera: Aphididae) on winter wheat. *Bulletin of Entomological Research* 74, 33–46.

Carroll, D. and Hoyt, S. (1986) Some effects of parental rearing conditions and age on progeny birth weight, growth, development, and reproduction in the apple aphid, *Aphis pomi* (Homoptera: Aphididae). *Environmental Entomology* 15, 614–619.

Castle, S.J. and Berger, P.H. (1993) Rates of growth and increase of *Myzus persicae* on virus-infested potatoes according to types of virus-vector relationship. *Entomologia Experimentalis et Applicata* 69, 51–60.

Chambers, R. (1982) Maternal experience of crowding and duration of aestivation in the sycamore aphid. *Oikos* 39, 100–102.

Chau, A. and Mackauer, M. (2001) Host-instar selection in the aphid parasitoid *Monoctonus paulensis* (Hymenoptera : Braconidae, Aphidiinae): assessing costs and benefits. *Canadian Entomologist* 133, 549–564.

Clough, M.S., Bale, J.S. and Harrington, R. (1990) Differential cold hardiness in adults and nymphs of the peach–potato aphid *Myzus persicae*. *Annals of Applied Biology* 116, 1–9.

Coleman, J.S. and Jones, C.G. (1988) Acute ozone stress on eastern cottonwood (*Populus deltoides* Bartr.) and the pest potential of the aphid, *Chaitophorus populicola* Thomas (Homoptera: Aphididae). *Environmental Entomology* 17, 207–212.

Collins, C.M. and Leather, S.R. (2001) Effect of temperature on fecundity and development of the giant willow aphid, *Tuberolachnus salignus* (Sternorrhyncha : Aphididae). *European Journal of Entomology* 98, 177–182.

Crafts, B.S. and Chu, C. (1999) Insect clip cages rapidly alter photosynthetic traits of leaves. *Crop Science* 39, 1896–1899.

Davis, J. (1915) The pea aphis in relation to forage crops. *US Department of Agriculture Bulletin No. 276*, 1–67.

Dean, G.J.W. (1974) Effect of temperature on the cereal aphids *Metopolophium dirhodum* (Wlk.), *Rhopalosiphum padi* and *Macrosiphum avenue* (F.) (Hem., Aphididae). *Bulletin of Entomological Research* 63, 401–409.

Dixon, A.F.G. (1973) *Biology of Aphids*. Edward Arnold, London, 58 pp.

Dixon, A.F.G. (1987) Seasonal development in aphids. In: Minks, A.K. and Harrewijn, P. (eds) *Aphids. Their Biology, Natural Enemies and Control, Volume 2A*. Elsevier, Amsterdam, pp. 315–320.

Dixon, A.F.G. (1990) Ecological interactions of aphids and their host plants. In: Campbell, R.K. and Eikenbary, R.D. (eds) *Aphid–Plant Genotype Interactions*. Elsevier, Amsterdam, pp. 7–19.

Dixon, A.F.G. (1998) *Aphid Ecology*, 2nd edn. Chapman and Hall, London, 300pp.

Dixon, A.F.G. and Dharma, T.R. (1980) Number of ovarioles and fecundity in the black bean aphid, *Aphis fabae*. *Entomologia Experimentalis et Applicata* 28, 1–14.

Dixon, A.F.G. and Glen, D.M. (1971) Morph determination in the bird cherry–oat aphid, *Rhopalosiphum padi* (L). *Annals of Applied Biology* 68, 11–21.

Dixon, A.F.G. and Wratten, S.D. (1971) Laboratory studies on aggregation, size and fecundity in the black bean aphid, *Aphis fabae* Scop. *Bulletin of Entomological Research* 61, 97–111.

Dixon, A.F.G., Chambers, R.J. and Dharma, T.R. (1982) Factors affecting size in aphids with particular reference to the black bean aphid, *Aphis fabae*. *Entomologia Experimentalis et Applicata* 32, 123–128.

Dixon, A.F.G., Kundu, R. and Kindlmann, P. (1993) Reproductive effort and maternal age in iteroparous insects using aphids as a model group. *Functional Ecology* 7, 267–272.

Dohmen, G. (1985) Secondary effects of air pollution: enhanced aphid growth. *Environmental Pollution* 39, 227–234.

Douglas, A.E. (2000) Reproductive diapause and the bacterial symbiosis in the sycamore aphid. *Ecological Entomology* 25, 256–261.

Edwards, O.R. (2001) Interspecific and intraspecific variation in the performance of three pest aphid species on five grain legume hosts. *Entomologia Experimentalis et Applicata* 100, 21–30.

Elliott, N., Kieckhefer, R. and Walgenbach, D. (1988) Effects of constant and fluctuating temperatures on developmental rates and demographic statistics for the corn leaf aphid (Homoptera: Aphididae). *Journal of Economic Entomology* 81, 1383–1389.

van Emden, H.F. (1969) Plant resistance to *Myzus persicae* induced by a plant regulator and measured by aphid relative growth rate. *Entomologia Experimentalis et Applicata* 12, 125–131.

van Emden, H.F. (ed.) (1972) *Aphid Technology*. Academic Press, London, 300 pp.

van Emden, H.F. and Bashford, M.A. (1969) A comparison of the reproduction of *Brevicoryne brassicae* and *Myzus persicae* in relation to soluble nitrogen concentration and leaf age (leaf position) in the Brussels sprout plant. *Entomologia Experimentalis et Applicata* 12, 351–364.

Farid, A., Quisenberry, S.S., Johnson, J.B. and Shafi, B. (1998) Impact of wheat resistance on Russian wheat aphid and a parasitoid. *Journal of Economic Entomology* 91, 334–339.

Febvay, G., Pageaux, J.F. and Bonnot, G. (1992) Lipid composition of the pea aphid, *Acyrthosiphon pisum* (Harris) (Homoptera: Aphididae), reared on host plant and on artificial media. *Archives of Insect Biochemistry and Physiology* 21, 103–118.

Fereres, A., Lister, R.M., Araya, J.E. and Foster, J.E. (1989) Development and reproduction of the English grain aphid (Homoptera: Aphididae) on wheat cultivars infected with barley yellow dwarf virus. *Environmental Entomology* 18, 288–293.

Fisher, M. (1987) The effect of previously infested spruce needles on the growth of the green spruce aphid, *Elatobium abietinum*, and the effect of the aphid on the amino acid balance of the host plant. *Annals of Applied Biology* 111, 33–41.

Foster, J., Stamenkovic, S. and Araya, J. (1988) Life cycle and reproduction of *Rhopalosiphum padi* (L.) (Homoptera: Aphididae) on wheat in the laboratory. *Journal of Entomological Science* 23, 216–222.

Frazer, B. and Gill, B. (1981) Age, fecundity, weight, and the intrinsic rate of increase of the lupine aphid, *Macrosiphum albifrons* (Homoptera: Aphididae). *Canadian Entomologist* 113, 739–745.

Gange, A. and Pryse, J. (1990) The roles of temperature and food quality in affecting the performance of the alder aphid, *Pterocallis alni*. *Entomologia Experimentalis et Applicata* 57, 9–16.

Garsed, S., Davey, H. and Galley, D. (1987) The effects of light and temperature on the growth of and balances of carbon, nitrogen and potassium between *Vicia faba* L. and *Aphis fabae* Scop. *New Phytologist* 107, 77–102.

Gianoli, E. (2000) Competition in cereal aphids (Homoptera: Aphididae) on wheat plants. *Environmental Entomology* 29, 213–219.

Givovich, A., Weibull, J. and Pettersson, J. (1988) Cowpea aphid performance and behaviour on two resistant cowpea lines. *Entomologia Experimentalis et Applicata* 49, 259–264.

Gruber, K. and Dixon, A.F.G. (1988) The effect of nutrient stress on development and reproduction in an aphid. *Entomologia Experimentalis et Applicata* 47, 23–30.

Guldemond, J.A., van den Brink, W.J. and den Belder, E. (1998) Methods of assessing population increase in aphids and the effect of growth stage of the host plant on population growth rates. *Entomologia Experimentalis et Applicata* 86, 163–173.

Helden, A.J. and Dixon, A.F.G. (1998) Generation specific life history traits of winged *Sitobion avenae*. *Entomologia Experimentalis et Applicata* 88, 163–167.

Hodgson, D.J. and Godfray, H.C.J. (1999) The consequences of clustering by *Aphis fabae* foundresses on spring migrant production. *Oecologia* 118, 446–452.

Holopainen, J.Q., Kainulainen, P. and Oksanen, J. (1997) Growth and reproduction of aphids and levels of free amino acids in Scots pine and Norway spruce in an open-air fumigation with ozone. *Global Change Biology* 3, 139–147.

Hopkins, R.J., Ekbom, B. and Henkow, L. (1998) Glucosinolate content and susceptibility for insect attack of three populations of *Sinapis alba*. *Journal of Chemical Ecology* 24, 1203–1216.

Ives, A.R., Schooler, S.S., Jagar, V.J., Knutson, S.E., Grbic, M. and Settle, W.H. (1999) Variability and parasitoid foraging efficiency: a case study of pea aphids and *Aphidius ervi*. *American Naturalist* 154, 652–673.

Johnson, B. (1965) Wing polymorphism in aphids II. Interaction between aphids. *Entomologia Experimentalis et Applicata* 8, 49–64.

Kempton, R., Lowe, H. and Bintcliffe, E. (1980) The relationship between fecundity and adult weight in *Myzus persicae*. *Journal of Animal Ecology* 49, 917–926.

Kenten, J. (1955) The effect of photoperiod and temperature on reproduction in *Acyrthosiphon pisum* (Harris) and on the forms produced. *Bulletin of Entomological Research* 46, 599–624.

Kerns, D. and Gaylor, M. (1992) Sublethal effects of insecticides on cotton aphid reproduction and color morph development. *Southwestern Entomologist* 17, 245–250.

Kidd, N. and Tozer, D. (1984) Host plant and crowding effects in the induction of alatae in the large pine aphid, *Cinara pinea*. *Entomologia Experimentalis et Applicata* 35, 37–42.

Kieckhefer, R., Elliott, N. and Walgenbach, D. (1989) Effects of constant and fluctuating temperatures on developmental rates and demographic statistics of the English grain aphid (Homoptera: Aphididae). *Annals of the Entomological Society of America* 82, 701–706.

Kindlmann, P. and Dixon, A.F.G. (1989) Developmental constraints in the evolution of reproductive strategies: telescoping of generations in parthenogenetic aphids. *Functional Ecology* 3, 531–538.

Kindlmann, P. and Dixon, A.F.G. (1992) Optimum body size: effects of food quality and temperature, when reproductive growth rate is restricted, with examples from aphids. *Journal of Evolutionary Biology* 5, 677–690.

Kindlmann, P., Dixon, A.F.G. and Gross, L. (1992) The relationship between individual and population growth rates in multicellular organisms. *Journal of Theoretical Biology* 57, 535–542.

Kouame, K.L. and Mackauer, M. (1992) Influence of starvation on development and reproduction in apterous virginoparae of the pea aphid, *Acyrthosiphon pisum* (Harris) (Homoptera: Aphididae). *Canadian Entomologist* 124, 87–95.

Lamb, R.J. (1992) Developmental rate of *Acyrthosiphon pisum* (Homoptera: Aphididae) at low temperatures: implications for estimating rate parameters for insects. *Environmental Entomology* 21, 10–19.

Lamb, R. and Mackay, P. (1988) Effects of temperature on developmental rate and adult weight of Australian populations of *Acyrthosiphon pisum* (Harris) (Homoptera: Aphididae). *Memoirs of the Entomological Society of Canada* 146, 49–56.

Lamb, R.J., Mackay, P.A. and Gerber, G.H. (1987) Are development and growth of pea aphids, *Acyrthosiphon pisum*, in North America adapted to local temperatures? *Oecologia* 72, 170–177.

Landin, J. and Wennergren, U. (1987) Temperature effects on population growth of mustard aphids. *Swedish Journal of Agricultural Research* 17, 13–18.

Larson, K.C. and Whitham, T.G. (1991) Manipulation of food resources by a gall-forming aphid: the physiology of sink–source interactions. *Oecologia* 88, 15–21.

Leather, S.R. (1987) Generation specific trends in aphid life history parameters. *Journal of Applied Entomology* 104, 278–284.

Leather, S.R. (1988) Size, reproductive potential and fecundity in insects: things aren't as simple as they seem. *Oikos* 51, 386–389.

Leather, S.R. (1989) Do alate aphids produce fitter offspring? The influence of maternal rearing history and morph on life-history parameters of *Rhopalosiphum padi* (L.). *Functional Ecology* 3, 237–244.

Leather, S.R. and Dixon, A.F.G. (1981) The effect of cereal growth stage and feeding site on the reproductive activity of the bird cherry–oat aphid, *Rhopalosiphum padi*. *Annals of Applied Biology* 97, 135–142.

Leather, S.R. and Dixon, A.F.G. (1982) Secondary host preferences and reproductive activity of the bird cherry–oat aphid, *Rhopalosiphum padi*. *Annals of Applied Biology* 101, 219–228.

Leather, S.R. and Dixon, A.F.G. (1984) Aphid growth and reproductive rates. *Entomologia Experimentalis et Applicata* 35, 137–140.

Leather, S.R., Ward, S.A. and Dixon, A.F.G. (1983) The effect of nutrient stress on life history parameters of the black bean aphid *Aphis fabae* Scop. *Oecologia* 57, 156–157.

Leszczynski, B., Wright, L. and Bakowski, T. (1989) Effect of secondary plant substances on winter wheat resistance to grain aphid. *Entomologia Experimentalis et Applicata* 52, 135–140.

Liu, S. and Wu, X. (1994) The influence of temperature on wing dimorphism in *Myzus persicae* and *Lipaphis erysimi*. *Acta Entomologica Sinica* 37, 292–297.

Liu, T. and Yue, B. (2000) Effects of constant temperatures on development, survival and reproduction of apterous *Lipaphis erysimi* (Homoptera: Aphididae) on cabbage. *Southwestern Entomologist* 25, 91–99.

Liu, Y. and Perng, J. (1987) Population growth and temperature-dependent effect of cotton aphid, *Aphis gossypii* Glover. *Chinese Journal of Entomology* 7, 95–112.

Llewellyn, M. and Brown, V. (1985) A general relationship between adult weight and the reproductive potential of aphids. *Journal of Animal Ecology* 54, 663–673.

Lorriman, F. and Llewellyn, M. (1983) The growth and reproduction of hop aphid (*Phorodon humuli*) biotypes resistant and susceptible to insecticides. *Acta Entomologica Bohemoslovaca* 80, 87–95.

Lykouressis, D. (1984) A comparative study of different aphid population parameters in assessing resistance in cereals. *Zeitschrift für Angewandte Entomologie* 97, 77–84.

Lyth, M. (1985) Hypersensitivity in apple to feeding by *Dysaphis plantaginea*: effects on aphid biology. *Annals of Applied Biology* 107, 155–162.

MacGillivray, M.E. and Anderson, G.B. (1957) Three useful insect cages. *Canadian Entomologist* 89, 43–46.

Manninen, A.M., Holopainen, T., Lyytikäinen-Saarenmaa, P. and Holopainen, J.K. (2000) The role of low-level ozone exposure and mycorrhizas in chemical quality and insect herbivore performance on Scots pine seedlings. *Global Change Biology* 6, 111–121.

Markkula, M. and Rautapaa, J. (1963) PVC rearing cages for aphid investigation. *Annales Agriculturae Fenniae* 2, 208–211.

Markkula, M. and Roukka, K. (1970) Resistance of plants to the pea aphid *Acyrthosiphon pisum* Harris (Hom., Aphididae) II. Fecundity on different red clover varieties. *Annales Agriculturae Fenniae* 9, 304–308.

Masters, G. (1995) The impact of root herbivory on aphid performance: field and laboratory evidence. *Acta Oecologica* 16, 135–142.

McCauley, G., Margolies, D., Collins, R. and Reese, J. (1990) Rearing history affects demography of greenbugs (Homoptera: Aphididae) on corn and grain sorghum. *Environmental Entomology* 19, 948–954.

McLeod, P., Morelock, T. and Goude, M. (1991) Preference, developmental time, adult longevity and fecundity of green peach aphid (Homoptera: Aphididae) on spinach. *Journal of Entomological Science* 26, 95–98.

McVean, R. and Dixon, A.F.G. (2001) The effect of plant drought-stress on populations of the pea aphid *Acyrthosiphon pisum*. *Ecological Entomology* 26, 440–443.

Messina, F. (1993) Effect of initial colony size on the per capita growth rate and alate production of the Russian wheat aphid (Homoptera: Aphididae). *Journal of the Kansas Entomological Society* 66, 365–371.

Moran, N.A. and Whitham, T.G. (1990) Interspecific competition between root-feeding and leaf-galling aphids mediated by host-plant resistance. *Ecology* 71, 1050–1058.

Muller, C.B. and Godfray, H.C.J. (1997) Apparent competition between two aphid species. *Journal of Animal Ecology* 66, 57–64.

Murdie, G. (1969a) The biological consequences of decreased size caused by crowding or rearing temperatures in apterae of the pea aphid, *Acyrthosiphon pisum* Harris. *Transactions of the Royal Entomological Society of London* 121, 443–455.

Murdie, G. (1969b) Some causes of size variation in the pea aphid, *Acyrthosiphon pisum* Harris. *Transactions of the Royal Entomological Society of London* 121, 423–442.

Newton, C. and Dixon, A. (1990a) Embryonic growth rate and birth weight of the offspring of apterous and alate aphids: a cost of dispersal. *Entomologia Experimentalis et Applicata* 55, 223–230.

Newton, C. and Dixon, A.F.G. (1990b) Pattern of growth in weight of alate and apterous nymphs of the English grain aphid, *Sitobion avenae*. *Entomologia Experimentalis et Applicata* 55, 231–238.

Parish, W. and Bale, J. (1990) Effects of short-term exposure to low temperature on wing development in the grain aphid *Sitobion avenae* (F.) (Hemiptera, Aphididae). *Journal of Applied Entomology* 109, 175–181.

Ponder, K., Pritchard, J., Harrington, R. and Bale, J. (2001) Feeding behaviour of the aphid *Rhopalosiphum padi* (Hemiptera: Aphididae) on nitrogen- and water-stressed barley (*Hordeum vulgare*) seedlings. *Bulletin of Entomological Research* 91, 125–130.

Radford, P.J. (1967) Growth analysis formulae – their use and abuse. *Crop Science* 7, 171–175.

Ramirez, C. and Niemeyer, H. (2000) The influence of previous experience and starvation on aphid feeding behavior. *Journal of Insect Behavior* 13, 699–709.

Reddy, M. and Alfred, J. (1981) Observations on the relationships between body length, breadth and weight of two pine aphids. *Entomon* 6, 307–309.

Rochat, J., Vanlerberghe, M.F., Chavigny, P., Boll, R. and Lapchin, L. (1999) Inter-strain competition and dispersal in aphids: evidence from a greenhouse study. *Ecological Entomology* 24, 450–464.

Rohita, B.H. and Penman, D.R. (1983) Effect of rearing temperature on the biology of bluegreen lucerne aphid, *Acyrthosiphon kondoi*. *New Zealand Journal of Zoology* 10, 299–308.

Sandström, J. (1994) High variation in host adaptation among clones of the pea aphid, *Acyrthosiphon pisum* on peas, *Pisum sativum*. *Entomologia Experimentalis et Applicata* 71, 245–256.

Sandström, J., Telang, A. and Moran, N. (2000) Nutritional enhancement of host plants by aphids: a comparison of three aphid species on grasses. *Journal of Insect Physiology* 46, 33–40.

Sarao, P. and Singh, G. (1998) Sublethal influence of insecticides on reproduction of mustard aphid, *Lipaphis erysimi* (Kaltenbach). *Journal of Insect Science* 11, 5–8.

Scriber, J.M. and Slansky, F. (1981) The nutritional ecology of immature insects. *Annual Review of Entomology* 26, 183–211.

Sequeira, R. and Dixon, A.F.G. (1996) Life history responses to host quality changes and competition in the Turkey-oak aphid, *Myzocallis boerneri* (Hemiptera: Sternorrhyncha: Callaphididae). *European Journal of Entomology* 93, 53–58.

Stark, J. and Wennergren, U. (1995) Can population effects of pesticides be predicted from demographic toxi-cological studies? *Journal of Economic Entomology* 88, 1089–1096.

Sumner, L., Dorschner, K., Ryan, J., Eikenbary, R., Johnson, R. and McNew, R. (1986) Reproduction of *Schizaphis graminum* (Homoptera: Aphididae) on resistant and susceptible wheat genotypes during sim-ulated drought stress induced with polyethylene glycol. *Environmental Entomology* 15, 756–762.

Sunnucks, P., Chisholm, D., Turak, E. and Hales, D. (1998) Evolution of an ecological trait in parthenogenetic *Sitobion* aphids. *Heredity* 81, 638–647.

Telang, A., Sandström, J., Dyreson, E. and Moran, N.A. (1999) Feeding damage by *Diuraphis noxia* results in a nutritionally enhanced phloem diet. *Entomologia Experimentalis et Applicata* 91, 406–412.

Thirakhupt, V. and Araya, J. (1992) Interactions between bird cherry–oat aphid (*Rhopalosiphum padi*) and English grain aphid (*Macrosiphum avenae*) (Homoptera: Aphididae) on Abe wheat. *Zeitschrift fur Pflanzenkrankheiten und Pflanzenschutz* 99, 201–208.

Tjallingii, W.F. (1995) Regulation of phloem sap feeding by aphids. In: Chapman, R.F. and de Boer, G. (eds) *Regulatory Mechanisms in Insect Feeding*. Chapman and Hall, New York, pp. 190–209.

Tsai, J. and Wang, J. (2001) Effects of host plants on biology and life table parameters of *Aphis spiraecola* (Homoptera : Aphididae). *Environmental Entomology* 30, 44–50.

Tsitsipis, J.A. and. Mittler, T.E. (1976) Development, growth, reproduction and survival of apterous virginoparae of *Aphis fabae* at different temperatures. *Entomologia Experimentalis et Applicata* 19, 1–10.

Tsumuki, H., Nagatsuka, H., Kawada, K. and Kanehisa, K. (1990) Comparison of nutrient reservation in apterous and alate pea aphids, *Acyrthosiphon pisum* (Harris): 1. Developmental time and sugar content. *Applied Entomology and Zoology* 25, 215–222.

Vancanneyt, G., Sanz, C., Farmaki, T., Paneque, M., Ortego, F., Castanera, P. and Sanchez-Serrano, J.J. (2001) Hydroperoxide lyase depletion in transgenic potato plants leads to an increase in aphid performance. *Proceedings of the National Academy of Sciences of the United States of America* 98, 8139–8144.

Varty, I. (1964) The morphology, life history and habits of *Betulaphis quadrituberculata* (Kalt.) on birch in New Brunswick (Homoptera: Callaphididae). *Canadian Entomologist* 96, 1172–1184.

Walgenbach, D., Elliott, N. and Kieckhefer, R. (1988) Constant and fluctuating temperature effects on devel-opmental rates and life table statistics of the greenbug (Homoptera: Aphididae). *Journal of Economic Entomology* 81, 501–507.

Walters, K.F.A. and Dixon, A.F.G. (1984) The effect of temperature and wind on the flight activity of cereal aphids. *Annals of Applied Biology* 104, 17–26.

Ward, S.A. and Dixon, A.F.G. (1982) Selective resorption of aphid embryos and habitat changes relative to lifespan. *Journal of Animal Ecology* 51, 859–864.

Ward, S.A., Wellings, P.W. and Dixon, A.F.G. (1983) The effect of reproductive investment on pre-reproduc-tive mortality in aphids. *Journal of Animal Ecology* 52, 305–314.

Warrington, S., Mansfield, T. and Whittaker, J. (1987) Effect of sulfur dioxide on the reproduction of pea aphids, *Acyrthosiphon pisum*, and the impact of sulfur dioxide and aphids on the growth and yield of peas. *Environmental Pollution* 48, 285–294.

Watt, A. D. (1979) The effect of cereal growth stages on the reproductive activity of *Sitobion avenae* and *Metopolophium dirhodum*. *Annals of Applied Biology* 91, 147–157.

Watt, A.D. and Dixon, A.F.G. (1981) The role of cereal growth stages and crowding in the induction of alatae in *Sitobion avenae* and its consequences for population growth. *Ecological Entomology* 6, 441–447.

Way, M.J. and Cammell, M. (1970) Aggregation behaviour in relation to food utilization by aphids. In: Watson, A. (ed.) *Animal Populations in Relation to their Food Source. Proceedings of the British Ecologi-cal Society Symposium No. 10*. Blackwell, Oxford, pp. 229–247.

Williams, C.T. (1995) Effects of plant age, leaf age and virus yellows infection on the population dynamics of *Myzus persicae* (Homoptera: Aphididae) on sugarbeet in field plots. *Bulletin of Entomological Research* 85, 557–567.

Wojciechowicz-Zytko, E. and van Emden, H.F. (1995) Are aphid mean relative growth rate and intrinsic rate of increase likely to show a correlation in plant resistance studies? *Journal of Applied Entomology* 119, 405–409.

Woodford, J.A.T. (1973) The climate within a large aphid-proof field cage. *Entomologia Experimentalis et Applicata* 16, 321.

Wu, K., Gong, P., Li, X., Wu, K.J., Gong, P. and Li, X. (1997) Effects of rape grown in SO_2-enriched atmo-spheres on performance of the aphid, *Myzus persicae* (Sulzer). *Entomologia Sinica* 4, 82–89.

Wyatt, I.J. and White, P.F. (1977) Simple estimation of intrinsic increase rates for aphids and tetranychid mites. *Journal of Applied Ecology* 14, 757–766.

Xia, J., van der Werf, W. and Rabbinge, R. (1999) Influence of temperature on bionomics of cotton aphid, *Aphis gossypii*, on cotton. *Entomologia Experimentalis et Applicata* 90, 25–35.

Zehnder, G., Nichols, A., Edwards, O. and Ridsdill-Smith, T. (2001) Electronically monitored cowpea aphid feeding behavior on resistant and susceptible lupins. *Entomologia Experimentalis et Applicata* 98, 259–269.

Zeng, F., Pederson, G., Ellsbury, M. and Davis, F. (1993) Demographic statistics for the pea aphid (Homoptera: Aphididae) on resistant and susceptible red clovers. *Journal of Economic Entomology* 86, 1852–1856.

Zhang, J., Zhang, G., He, F., Qu, G. and Yan, F. (1991) Studies on the experimental population dynamics of bird cherry–oat aphid, *Rhopalosiphum padi* (L.). *Sinzoologia* 8, 83–94.

Zhou, X. and Carter, N. (1992) Effects of temperature, feeding position and crop growth stage on the population dynamics of the rose grain aphid, *Metopolophium dirhodum* (Hemiptera: Aphididae). *Annals of Applied Biology* 121, 27–37.

Zuniga, G., Salgado, M. and Corcuera, L. (1985) Role of an indole alkaloid in the resistance of barley (*Hordeum vulgare*). *Phytochemistry (Oxford)* 24, 945–948.

7 Aphid Movement: Process and Consequences

Michael E. Irwin[1], Gail E. Kampmeier[2] and Wolfgang W. Weisser[3]

[1]Department of Natural Resources and Environmental Sciences, University of Illinois, Urbana, IL 61801, USA; [2]Section of Ecological Entomology, Illinois Natural History Survey, Urbana, IL 61801, USA; [3]Institute of Ecology, Friedrich-Schiller-University, 07743 Jena, Germany

Introduction

This chapter reviews the movement of agriculturally important aphids. It includes information on how different morphs and life stages redistribute themselves in response to intrinsic factors and extrinsic perturbations over time and through spatial scales that span walking behaviour on individual plants to aerial transport over very long distances. The chapter explores the economic consequences of aphid movement and weaves the multiple roles of movement into the tapestry of pest management, providing insight into ways of manipulating aphid movement and thereby mitigating the negative economic impacts resulting from it.

Pest Status

Some aphids are pests because they feed on the phloem of crop plants (see Quisenberry and Ni, Chapter 13 this volume). This feeding can result in yield decreases, sometimes of considerable proportions. The timing of aphid arrival on new hosts can influence plant health and yield. Rapid aphid reproduction early in the season can outstrip a plant's ability to recover from direct feeding pressure, whereas the same aphid pressure later in the season may have little effect on plant growth. Many of these same aphid species, and several others also, achieve pest status because they transmit plant-debilitating viruses that can, in turn, wreak havoc on a crop and its potential harvest.

Globally, nearly 5000 species of aphid have been described (Remaudière and Remaudière, 1997), although most are confined to the north and many to the south temperate regions. Individual species are usually restricted to one or a few plant species, and often move between primary (generally perennial) and secondary (mostly herbaceous) hosts (Dixon, 1988) at defined points in their life cycles. Most aphid species have, at most, a marginal impact on agriculture, with around 450 species (less than 10%) colonizing the world's food and fibre crops (see Blackman and Eastop, Chapter 1 this volume).

Blackman and Eastop (2000) consider only 18 (plus one subspecies) of the 450 aphid species found on the world's crops to be truly polyphagous, i.e. propagating on hosts from several plant families. Their definition of polyphagy excludes the traditional grain aphids (e.g. *Rhopalosiphum padi* – bird cherry–oat aphid, *Rhopalosiphum maidis* – corn leaf aphid, *Sitobion avenae* – grain aphid, *Schizaphis graminum* – greenbug, *Diuraphis noxia* – Russian wheat aphid, *Metopolophium dirhodum* – rose–grain aphid) that feed on multiple grass hosts, some of which (e.g. rice, maize, and wheat) are among our most widespread and important crops. Thus, aphids that directly impact agriculture comprise about 24 highly polyphagous and multi-grass feeders, or 0.5% of known aphid species, the majority of which are cosmopolitan and highly efficient *r*-strategists (Leather and Walters, 1984; Robert, 1987; see Kindlmann *et al.*, Chapter 12 this volume). A handful of additional aphid species are direct pests of specific crops, even though they have narrow host ranges (e.g. *Aphis glycines* – soybean aphid). Thus, worldwide, relatively few species are direct pests of agriculture. Many of these same multi-host aphid species are notable pests of crops and vectors of viruses in the tropics (Dixon *et al.*, 1987; Dixon, 1988).

Aphids are unquestionably important as crop pests in their own right. Their greatest impact on agriculture, however, is as vectors that transmit potentially crippling plant viruses (see Katis *et al.*, Chapter 14 this volume). Vector colonization and movement patterns influence the timing and pattern of virus epidemics, which, in turn, are linked to yield losses. Timings and patterns of virus epidemics differ substantially depending on transmission mode. Below, we explore aphid movement as it impacts the transmission of: (i) persistent and semi-persistent viruses, which, because of the considerable time required to acquire, process, and transmit the virus, are primarily vectored by the relatively few species of aphids that colonize the crop; and (ii) non-persistent viruses, which are acquired and inoculated by vectors during brief stylet probes (see Irwin and Nault, 1996; Nault, 1997; Katis *et al.*, Chapter 14 this volume, for discussions of modes of transmission).

Contrary to the aphid behaviours necessary for successful semi-persistently and persistently transmitted virus epidemics, lengthy aphid feeding and reproduction on the plant are counter to rapidly escalating epidemics of non-persistently transmitted virus diseases because these behaviours reduce aphid movement within a crop (Halbert *et al.*, 1981; Thomas, 1983; Gray and Lampert, 1986; Harrington *et al.*, 1987; Sigvald, 1987). Importantly, numerous aphid species well beyond those that are severe direct crop pests can be implicated in the spread of non-persistently transmitted viruses.

The Nature of Aphid Movement

Aphids often appear sedentary. As a general rule, aphids, even those that are winged as adults, spend a considerable period of their lives unable to fly. Wing muscles of alatae (winged morphs) of most host-alternating species begin to autolyse a few days after adults eclose (Haine, 1955; Dixon, 1988; Dixon *et al.*, 1993), confining the opportunity for flight to a narrow window shortly after the moult to the adult stage (Liquido and Irwin, 1986; Kobayashi and Ishikawa, 1993; Levin and Irwin, 1995). Furthermore, flight initiation is primarily, though not exclusively, restricted to daylight hours (Taylor, 1958; Johnson, 1969; Dixon, 1988; Isard and Irwin, 1993; Isard and Gage, 2001) and to instances when atmospheric conditions favour take-off (Jensen and Wallin, 1965; Walters and Dixon, 1984; Isard and Gage, 2001). Thus, when considering the entirety of an aphid's life cycle, the amount of time spent moving appears minimal, especially over spatial scales greater than a single plant. Nonetheless, aphids are predisposed to move in response to intrinsic stimuli (Kidd, 1977) and extrinsic perturbations (Bailey *et al.*, 1995; Mann *et al.*, 1995; Müller *et al.*, 2001; Smyrnioudis *et al.*, 2001; Zhang, 2002), and such movement can be of profound importance. Aphid movement, in particular migration (Loxdale *et al.*, 1993), may be perceived by some as minimal and inconsequential, but such a perception, we caution, seriously understates and grossly oversimplifies the ecological consequences of movement and ignores the profound economic impact aphids have, through the act of moving, on managed ecological systems.

Conceptual Framework for Aphid Movement

A conceptual framework best facilitates a pragmatic evaluation and assessment of aphid movement. Such a framework should take into account the forces that cause and the mechanisms that result in movement. Movement by different life stages and morphs should also be addressed. Finally, the scales of dispersal must be identified and linked with causal displacement forces and mechanisms. Through such a framework, a more realistic appraisal of movement can be gained, and the framework can serve as a conceptual model to link aphid movement more easily to resulting economic losses and pragmatic management solutions.

Modes of transport

An aphid moves from its point of origin (source) towards some other place (sink) by one of two types of transport mechanisms, either through 'inadvertent' (Fig. 7.1a) or 'intentional' (Fig. 7.1b) displacement. Inadvertent displacement is an involuntary act, leaving the aphid few options regarding its translocation; it is propelled by the force of impact, gravity, air currents, or a combination of these, or it can be transported by animals, farm machinery, automobiles, or aircraft. As a result, the aphid is involuntarily transported towards a sink. Inadvertent displacement by rising air currents might leave the aphid somewhat battered but, if it survives, it is often able to reproduce and transmit diseases (Zúñiga, 1985; Bailey *et al.*, 1995). The inadvertently displaced aphid may, during the act of being displaced, purposefully intervene by righting itself, walking, or flying, thus shifting the displacement mechanism from inadvertent to intentional.

Intentional displacement (Fig. 7.1b) is a voluntary act, being prompted by intrinsic forces or extrinsic perturbations. Intrinsic forces that result in aphid movement are pre-programmed, i.e. they are governed by the genetics of the organism. Extrinsic perturbations that result in aphid movement are governed by reactions to sensed stimuli in the environment. Movement can result from a combination of extrinsic and intrinsic forces. For instance, an extrinsic perturbation, e.g. the environmentally sensed attack by a natural enemy or altered chemical composition of a deteriorating host plant, can create an urge in the aphid to flee; this desire might well stem from an intrinsic, genetically wired response to the initial disturbance. It can be difficult, therefore, to ascertain whether the causal force was purely intrinsic, purely extrinsic, or a combination of both.

Intentional displacement manifests itself in either the intrinsically induced phenomenon known as migration (cf. Loxdale and Lushai, 1999), or in 'appetitive dispersal' (Wenks, 1981; cf. 'dispersal' of Loxdale and Lushai, 1999), the extrinsically or intrinsically induced dispersal mechanism that evokes a search response for specific resources (e.g. food, mate, oviposition target) or triggers an escape response (e.g. from natural enemies, chemical changes of senescing hosts) (*sensu* Loxdale and Lushai, 1999, concept of dispersal).

Scales of displacement

When an aphid is displaced or actively moves, it is translocated from some source to a sink, in not only a three-dimensional spatial context, but within a temporal context as well. The horizontal and temporal scales can best be visualized in terms of the relative magnitude of movement by aphids to adjacent plant parts, whole plants, fields, landscapes, eco-regions, continents, and transcontinental landmasses (Fig. 7.2). The movement scale that an aphid is able to achieve depends to a considerable extent on the height (vertical displacement) that an individual attains during the movement process. Although an aphid can be blown off its host plant and carried into the atmosphere, it rarely moves horizontally beyond the scale of neighbouring plants (Fig. 7.3). This is unless it is dislodged or otherwise propelled from a considerable height (tall tree, cliff face), is

(a)

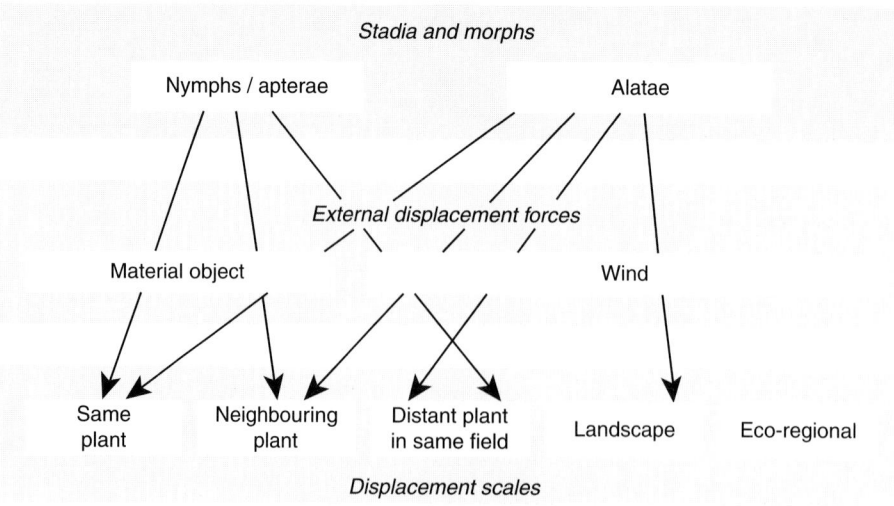

(b)

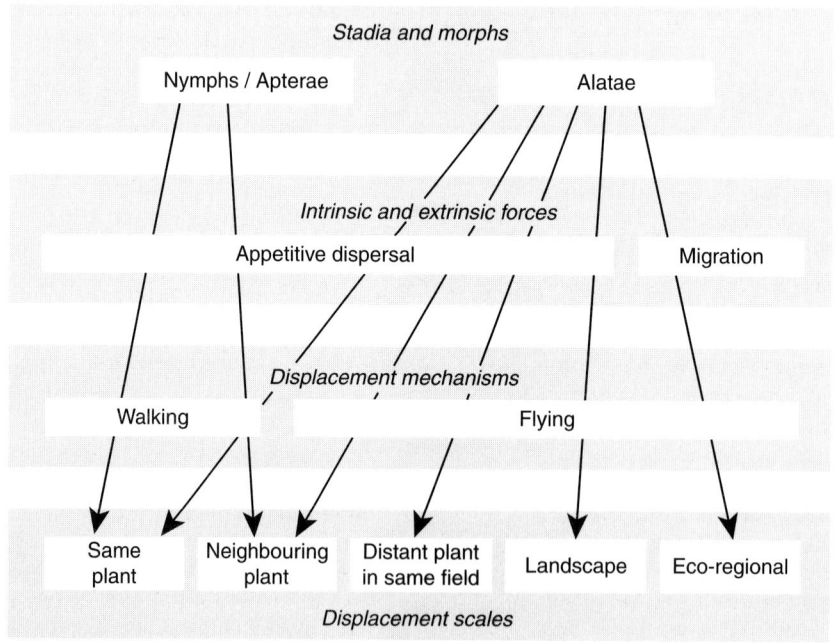

Fig. 7.1. (a) Movement patterns during inadvertent displacement. The schematic takes into account the stadia and morphs of aphids, the external displacement forces that cause movement, and displacement scales. (b) Schematic of movement patterns during intentional displacement. The schematic takes into account the stadia and morphs of aphids, intrinsic and extrinsic forces that cause movement, displacement mechanisms involved, and displacement scales.

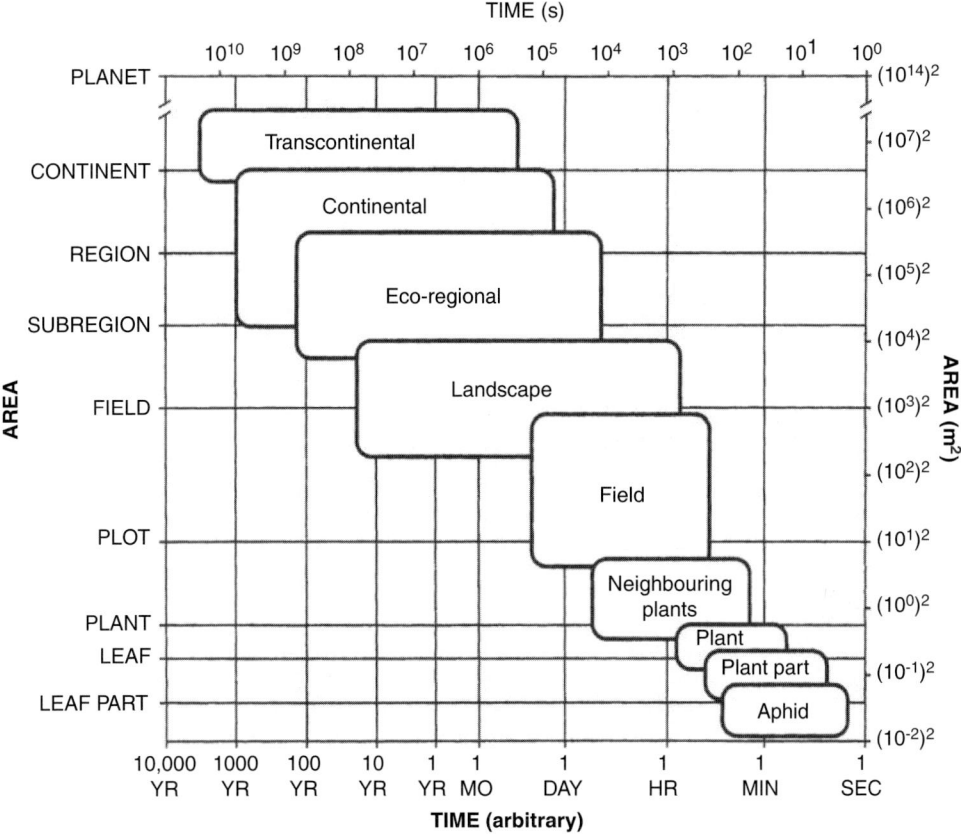

Fig. 7.2. Scaling temporal and spatial displacements that involve and affect aphid movement. Note that both the temporal and spatial scales are logarithmic.

caught up in a strong convective air current, or actively flies and thereby increases or maintains its height above ground (Isard and Gage, 2001). Individuals descending from dramatically different heights would, in fact, quite likely achieve different scales of movement (neighbouring plants *v.* field or at most landscape scale). However, the differences probably would not be great (Fig. 7.1a) unless alatae actively take flight, where the differences in scales of movement can be dramatic (Fig. 7.1b), depending on the inclination of the aphid, the altitude individuals attain, and the atmospheric motion systems active at the time.

The time it takes an aphid to reach a destination, and the duration of occupancy, provide the range of temporal backdrops for these scales. The duration of occupancy may be transient to alight, probe briefly, and decide to move on, or encompass colonization and maintenance at the sink. In a larger temporal scale, spanning seasons of hospitable and inhospitable climate or boom-or-bust resources, metapopulations, or generations of aphids, may also oscillate in space, moving, for example, into previously inhospitable territory after a harsh North American winter, and later southward as the day length decreases and resources deteriorate.

Aphid Life Stages, Morphs and Active Modes of Transport

Distinct aphid life stages, adult morphs, and the physiological condition and life cycle stage of aphids must be taken into

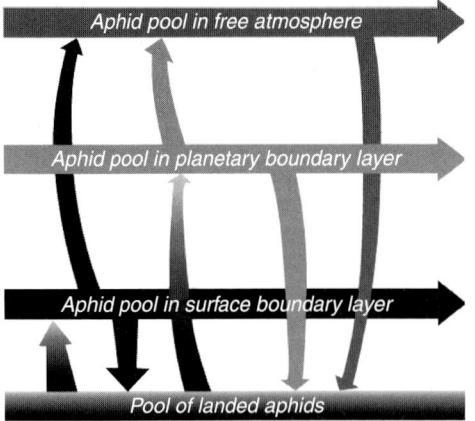

Fig. 7.3. Vertical movement patterns by pools of aphids in the atmosphere. Note that the atmosphere has been divided into three distinct layers; the free atmosphere, the planetary boundary layer, and the surface boundary layer. Most long-distance translational movement by aphids takes place in the planetary boundary layer.

account when formulating a movement framework. Distinct adult morphs constituting either apterae or alatae can appear at various stages in the life cycles of most host-alternating aphid species (see Williams and Dixon, Chapter 3 this volume). Innate genetic signalling causes some species to produce different proportions of alatae even under similar environmental conditions (Nakata, 1995).

Two kinds of life stages and morphs are considered in this chapter: flightless individuals (nymphs, apterae, brachypterae, and alatae that have lost the ability to use their wings through flight muscle autolysis) and flight-capable alatae (Bottenberg and Irwin, 1991). Thus, we seek to characterize the active mode of intentional displacement rather than the life stages and morphs that are moving. For flightless individuals, there is a single active mode: walking. For flight-capable individuals, there are potentially two such modes: walking and flying. These statements imply that all life stages and morphs walk the same and that all flight-capable morphs fly in the same manner – a faulty assumption. For example, according to Ferrar (1969), aphid nymphs are predisposed to greater interplant movement prior

to moulting, particularly during the last instar. Such movement can result in a less crowded and less plant-stressed environment for the next generation; it might also reduce predation and parasitism because the dispersed aphids would be scattered and less apparent to most natural enemies. A series of greenhouse experiments enabled Zhang (2002) to characterize the walking behaviour of wingless life stages and morphs of *R. padi*. When perturbed by modest wind gusts or when their host plants were brushed, fewer earlier nymphal instars than later instars or adult apterae tended to move from their hosts, although under these same perturbance conditions, similar proportions of late instar nymphs and adult apterae moved. A similar pattern was revealed concerning the distances aphids travelled; the more mature the aphid, the farther it tended to move. Furthermore, when testing movement patterns of non-alatoid and alatoid fourth-instar nymphs, no difference was detected in the proportion of individuals that moved or in the distances they walked.

Winged morphs also differ in movement behaviours. In a series of laboratory studies, Nottingham *et al.* (1991) found that three distinct alate morphs of *R. padi* responded to an intermittent resource target differently during their maiden flights. Autumn-produced gynoparae had a greater initial climb rate than both summer exules and spring emigrants. Gynoparae also flew for a longer period of time than summer migrants before responding to a resource target, while spring migrants responded after an intermediate time interval.

The two major modes of locomotion, walking and flying, translocate aphids from sources to sinks not only in different manners, but also potentially over considerably different spatial scales (Fig. 7.1b). Walking displaces the individual within a reasonably circumscribed and restricted setting, i.e. either the aphid moves from one part of a plant to another, or it walks from a plant to one that is in close proximity (usually within the bounds of a field). Aphid flight, in the sense of wing flapping to ascend or remain aloft, is confined to times when the phenological, physiological, and physical conditions favour their take-off and ascent. During such times, alatae ascend into the surface boundary layer, where they may be able to exert considerable control over their flight, and from which they may be able to descend during appetitive dispersal (Wenks, 1981) (Fig. 7.3).

The key to distinguishing appetitive dispersal from migration is not in the distances travelled, nor the mode of transport, but in the motivating force behind the movement (Irwin, 1999). Because the motivation for moving is difficult to deduce, discriminating between these modes can be problematic. Loxdale *et al.* (1993) suggested that, among movement modes of aphid species, migration is the exception and that movement largely conforms to the category of 'dispersal' (i.e. appetitive movement). Their hypothesis may appear accurate considering the large number of extant aphid species and the importance and amount of time spent in modes of appetitive dispersal by even the agriculturally important aphid species. However, the statement fails to bring to light that those aphid species that move great distances and, indeed, are known to migrate, are almost always those that belong to the small subset of aphids known to be agricultural pests.

Evidence supporting the hypothesis that migratory aphids often colonize crops can be found in several publications that list the species most often trapped aloft (Glick, 1939; Berry and Taylor, 1968; White, 1970, 1974; Irwin and Thresh, 1988). Because this book deals with aphids as pests of managed ecosystems, it seems fitting to treat both appetitive movement and migration as equally important categories of movement.

The Migratory Process

Aphid migration is predicated on innate behaviour and results in directed, undistracted movement (Kennedy, 1985; Dingle, 1996; Irwin, 1999), often culminating in dramatic displacements abetted by atmospheric motion systems (Irwin and Thresh, 1988; Gage *et al.*, 1999; Irwin, 1999; Irwin *et al.*, 2000; Isard and Gage, 2001). The concepts of

'long-distance' movement and aphid 'migration' are interwoven in the literature and often erroneously considered as synonymous (cf. Johnson, 1969; Robert, 1987 for reasons why these are not synonyms). Through a series of laboratory studies that determined when an aphid's flight trajectory could be altered by the presence of resource items (e.g. light of the same wavelength as a potential food source), Hardie (1993) experimentally disentangled the two terms and demonstrated the behavioural distinction between them. 'Long-distance' refers to an aphid that has moved far (from several to thousands of kilometres). This movement can be achieved in one of two ways. An aphid can actively seek a resource item during appetitive dispersal and encounter atmospheric conditions (e.g. thermals) that propel it upward, where it is then subject to horizontal translocation by atmospheric motion systems (i.e. inadvertent displacement over long distances). The other, more common, way is through the act of 'migration', during which an aphid actively flies towards an ultraviolet light source. It ignores all cues associated with resource items and climbs high into the atmosphere, where it is then positioned to move great distances, also by the prevailing atmospheric motion systems. Thus, migration is only one, albeit a more common, way in which an aphid can move long distances. Sometimes, the migratory urge can cease shortly after take-off, and this can result in rather short-distance dispersal events, demonstrating that not all migratory events end up in long-distance dispersal.

The migratory process is complex. When an alatoid nymph moults, becoming an adult alata, it may already be physiologically primed to migrate. Such an aphid, given the appropriate environmental setting, will actively ascend into the atmosphere, where it continues flying as the air currents aloft horizontally displace it. The process terminates when the aphid actively begins to descend and land. The migratory process, then, can be delineated by four phases: (i) events leading to take-off; (ii) take-off and ascent; (iii) horizontal translocation; and (iv) events leading to behaviour switching that culminates in descent and landing. Alighting behaviour, which is manifest in descent and landing events, occurs once the migratory process has been switched off and is treated under 'appetitive dispersal' below.

Events leading to take-off

Several factors govern which aphids are able to take flight and when they are able to do so. Migrants will take longer to develop, suffer delayed reproductive development, smaller gonads, and reduced fecundity as costs of flight (Dixon and Wratten, 1971; Dixon and Kindlmann, 1999; Müller *et al.*, 2001). They fuel their flight with sequestered energy reserves (Liquido and Irwin, 1986; Dixon *et al.*, 1993). Because aphids that migrate have a low probability of survival (i.e. few migrants find and colonize new hosts), the migratory strategy seems based on the notion that enough will encounter more favourable environments to colonize, reproduce, and make up for losses. When aphids are not physiologically primed to take off, they will not voluntarily do so (Dixon, 1977). Several environmental factors can be responsible for this physiological conditioning.

Host-plant physiology

Of these environmental factors, the phenological stage and quality of the host plants during the aphid's development are considered primary, but Müller *et al.* (2001) discovered in retrospective examination of the literature that not all species appear to react to the same stimuli. *Myzus persicae* (peach–potato aphid), in particular, was generally unaffected by host plant quality, and many of the experiments detailing this subject used *M. persicae* as a model. *Myzus persicae* was not alone in this assessment, however; other well-known pest species did not respond to host plant cues, including *Aphis fabae* (black bean aphid) and *Aphis gossypii* (cotton or melon aphid). Common pest species tested that did respond to deteriorating host-plant conditions included *Aphis*

craccivora (cowpea aphid), *Brevicoryne brassicae* (cabbage aphid), *Macrosiphum euphorbiae* (potato aphid), *R. padi*, and *S. avenae*. Clearly, it is important to know which aphid species react to deteriorating host conditions in targeted cropping systems.

Crowding

Explosive population growth results in severe crowding, which induces alata production and increases the chances of apterae purposefully moving or being jostled off their host plants (Robert, 1987). Although crowding nearly always appears to induce wing production, species are affected differently by the timing of the crowding event, with some species affected only by pre-natal crowding, some by post-natal crowding in the first two instars, and some by crowding at both stages (Müller *et al.*, 2001). Such priming can also pre-condition the eclosing alates for flight take-off (Walters and Dixon, 1984; Dixon, 1988).

Presence of natural enemies

Recent studies have shown that *Acyrthosiphon pisum* increases the proportion of winged morphs among its offspring when exposed to foraging predaceous ladybird beetle larvae and adults (Fig. 7.4) (Dixon and Agarwala, 1999; Weisser *et al.*, 1999), hover fly and lacewing larvae (Kunert and Weisser, 2003), or aphidiid parasitoids (Sloggett and Weisser, 2002). Similar results have been obtained with *Uroleucon nigrotuberculatum* (Mondor *et al.*, 2004) and the cotton aphid, *Aphis gossypii* (Mondor *et al.*, 2005). In contrast, neither *A. fabae* nor *Megoura viciae* (vetch aphid) increased alata production when placed in tubes or on plants previously infested with ladybird larvae and conspecifics (Dixon and Agarwala, 1999), or when exposed to foraging lacewing larvae (Kunert *et al.*, 2007). With respect to the mechanisms responsible for wing induction in the presence of natural enemies, an early study indicated that a chemical cue contained in ladybird larval tracks causes *A. pisum* to produce more alatae (Dixon and Agarwala, 1999). More recently,

Fig. 7.4. *Coccinella septempunctata* larva approaching colony of *Uroleucon tanaceti* on leaves of tansy, *Tanacetum vulgare* (original colour photo courtesy of H.H. Weisser).

aphid alarm pheromone has been shown to play a crucial role (Kunert *et al.*, 2005). Foraging predators or parasitoids cause movements in the aphid colony that are reinforced by the release of the alarm pheromone, (*E*)-ß-farnesene. The resulting increase in the number of physical contacts among individuals leads to the observed increase in winged morph production in the presence of natural enemies (e.g. the 'pseudo-crowding mechanism' (Weisser, 2001; Weisser and Sloggett, 2004; Kunert *et al.*, 2005).

Take-off and ascent

Physiological factors

Alate aphids that are in a migratory mode are physiologically primed to take flight towards an ultraviolet light source. Alatae have a relatively narrow window of time during which they can actively take off and ascend in migratory flight. This window opens within 12 h of the aphid moulting to the adult stage and begins to deteriorate about 2 days later, effectively closing 4 days after moulting to the adult stage (Liquido

and Irwin, 1986; Kobayashi and Ishikawa, 1993; Levin and Irwin, 1995).

The angle of aphid flight trajectories is physiologically regulated by the age of the aphid at take-off. Within a day of becoming adult, winged *R. padi* leave the canopy at a mean angle of 34° above horizontal; between 1 and 2 days old, the angle decreases to 24°; and by 2–3 days, the angle is only half what is was within the first 24 h (Isard and Irwin, 1996; Isard and Gage, 2001). Combined with the wind profile at the time of take-off, predictions may be made about the success of alatae of different ages and of their trajectories in enabling ascent into the planetary boundary layer (PBL) (also known as the mixing layer or atmospheric boundary layer, situated above the surface boundary layer and below the free atmosphere). This layer varies in depth with the time of day from as shallow as 0.3 km at night to as deep as 3 km during the day. If an aphid flies upward and reaches this layer, it can be translocated long distances along with the horizontal air currents of the stratum occupied by the aphid.

Many agriculturally important aphids are in a migratory mode during their maiden flights, at least for a short period of time (Hardie, 1993). The amount of time an aphid spends in purposeful, unobstructed, upwardly directed flight provides a measure of how high into the atmosphere it can ascend. Ascent rates of aphids in a migratory mode are initially rapid, e.g. 20–30 cm/s for *A. fabae* (Kennedy and Booth, 1963) and 22–67 cm/s for *S. graminum* (Halgren, 1970). Voluntary long-distance air transport is probably only achievable during an aphid's maiden flight, both because the rate of climb on subsequent flights is less, and because the aphid tends to shift sooner from an upward, UV-directed, migratory flight mode to a host-seeking, appetitive mode. This decreases the possibility of entering the PBL and drastically reduces the duration of flights subsequent to the initial one (Kennedy and Booth, 1963).

Atmospheric influences

The nature of the atmosphere directly influences whether an aphid will take off and ascend into the PBL. Two interactive atmospheric forces govern the ability of aphids to take flight and ascend into the PBL. Inertial forces sustain movement of air parcels (i.e. small, cohesive volumes of air) in a horizontal plane and permit aphids within them to move long distances, essentially riding the surrounding air currents. Buoyant forces, both positive and negative, allow air parcels, because of density differences, to move upwards or downwards in a vertical plane. Given the backdrop of a typical warm spring through autumn day, the interaction between these two forces throughout a diel cycle provides the windows of opportunity for aphids to ascend and dictates how high they will ascend into the atmosphere. Aphids will not voluntarily take off in the presence of strong winds. But in the absence of these, aphids are much more likely to ascend during times when the buoyant forces are neutral or positive with rising air parcels.

Unstable atmospheric conditions occur during daylight hours as the sun's energy heats the earth's surface. In turn, that energy warms the air nearest it, which becomes less dense than the air above and begins to rise. The cooler, denser air then descends (negative buoyancy), replacing the warmer, less dense air. This unstable atmospheric situation provides an excellent opportunity for aphids to take off and ascend into the PBL, where they can be moved far by inertial forces. The atmospheric mixing process continues throughout the daylight hours, driven by the solar heating of the earth, and results in an expansion in the depth of the PBL. As the air parcels rise, however, they are slowed by friction (another force acting on the system) as the vertically moving air parcels are dragged horizontally by inertial forces.

In the transition between unstable (daytime) and stable (night-time) atmospheric conditions, the ratio of buoyant to mechanical forces approaches zero, becoming neutral. This condition generally occurs in the early morning hours after sunrise, or for a shorter period around sunset. Under neutral conditions and when wind speeds are low, aphids primed for long-distance flight can ascend and reach the PBL. Their trajectory when escaping the surface boundary layer to the PBL is likely more shallow than that afforded during unstable atmospheric conditions because buoyancy neither aids (unstable conditions) nor hinders (stable conditions) their ascent (Isard and Gage, 2001).

Interactions of physiological and atmospheric influences

Aphids delay maiden and subsequent flights when wind speeds are high, usually preferring winds closer to their own flight speed of less than 1 m/s^2 (Haine, 1955; Robert, 1987; Kennedy, 1990; Bottenberg and Irwin, 1991); however, adverse wind conditions may delay but not prevent take-off (Walters and Dixon, 1984; Robert, 1987). Mixed cropping systems can form barriers that reduce wind speeds in the shorter crop (Castro *et al.*, 1991), allowing aphids to depart more readily than from canopies of uniform height (Bottenberg and Irwin, 1991). Both upper and lower temperature thresholds will inhibit aphid take-off, although these temperatures vary by species and season (Wiktelius, 1981).

With the exception of *S. graminum*, which can initiate flight at temperatures as high as 41–42°C (Berry, 1969; Dry and Taylor, 1970; Halgren, 1970), the lower threshold for most species varies between 13 and 16°C (Johnson and Taylor, 1957; Jensen and Wallin, 1965; Dry and Taylor, 1970; Walters and Dixon, 1984), while the upper threshold is generally around 31°C. Light intensity also affects the likelihood of aphid take-off, with few species ever initiating flight in the absence of light or during moonlight, which is up to 1 lux. According to Robert (1987), take-off usually occurs at intensities greater than 1000 lux, with no apparent upper limit (Berry, 1969) (sunlight ranges between 32,000 and 100,000 lux).

Horizontal translocation

Differences achieved in horizontal scales of displacement are related directly to the layering of the lower atmosphere (Fig. 7.3). The angle of ascent will determine whether the aphid will succeed in reaching a horizontal transport layer in the PBL. Alatae taking off under unstable atmospheric conditions, whether predisposed to long distance or local flight, are more likely to be transported longer distances than if they initiate flight in stable atmospheric conditions. When conditions are stable, aphids in neither appetitive nor migratory flight mode are able to escape the surface boundary layer and thus are unaided by positive atmospheric forces and cannot be moved long distances (Isard *et al.*, 1994; Isard and Gage, 2001). Once attaining the PBL, horizontal transport of an organism is largely governed by the stratification of the wind layers. Meteorological conditions favourable to long-distance transport within the PBL are seasonal in occurrence (Scott and Achtemeier, 1987; Isard and Gage, 2001), setting up migratory opportunities during specific periods of the year (usually spring, summer, and autumn) in association with certain types of meteorological events.

The most notable of these events is the 'low-level jet'. After dusk, when the earth's surface begins to cool, the PBL begins to stabilize and becomes compressed horizontally with air currents of different speeds and temperatures trapped and stratified within it. Under such circumstances, air currents that are warmer above than below (inversions), a common occurrence in the mid-latitudes, often result in and form what is commonly termed a 'low-level jet'. It is not known if aphids actively seek out low-level jets or if they are passively concentrated within them. Due to a lack of turbulence from below, these strong, horizontally directed, relatively warm strata of air retain the aphids trapped there after sundown. These aphids may then be transported great distances before dawn when the rising sun once more begins to heat the earth and mixing deprecates the integrity of the low-level jet; this is when the aphids trapped in those strata begin to descend (Isard and Gage, 2001).

Successful migratory flights, abetted by air currents in which they are embedded (Fig. 7.3), allow aphids to attain a more linear directionality, faster ground speed and, consequently, greater distance traversed than would be possible solely under their own power (Kennedy, 1985). In climates nearer the poles, when night temperatures become too low to sustain aphids in active flight overnight (i.e. flapping their wings and remaining aloft), the aerial transport of aphids seems limited to daytime movements. But in climates where night-time temperatures remain relatively warm, overnight transport is common and can result in displacements of hundreds, even thousands, of kilometres (Isard and Gage, 2001). Knowledge of the history of the wind speed and direction of these jets enables researchers to do back trajectory analyses to determine the probable source regions of the migrant populations (Scott and Achtemeier, 1987; Drake and Gatehouse, 1996).

Given the vagaries of weather and of physiological priming, one might presume that migration is a rather rare event. On the contrary, it can be a common, though seasonal, phenomenon, one that at times fills the sky with countless billions of aphids in long-distance transport (Grossheim, 1914; Dewar *et al.*, 1980; Irwin and Hendrie, 1985; Marking, 2002).

Switching off the migratory urge

At some point during its journey, although the mechanism is unknown, the aphid's behaviour switches and, in an effort to obtain specific resources, it will be attracted by cues other than ultraviolet light. When that occurs, the aphid ceases its migratory phase and enters an appetitive mode. The switch will occur at different times, depending on several factors, not the least of which is the alate morph's migratory state (see above). The timeliness of its success in descending and landing may also depend on meteorological conditions. Those aphids whose transport was abetted by favourable, low-level jet winds may be forced to remain flying until this horizontal transport layer breaks down in the early, post-dawn morning hours, when they can once again perceive landing sites and terminate their flight (Isard *et al.*, 1990; Isard and Gage, 2001). For those aphids in climates nearer the poles or elsewhere that cannot take advantage of low-level jets, the migratory urge may be satisfied by taking advantage of one or a series of rising thermals during the atmospherically unstable daylight hours. When the migratory phase has ceased, the aphids will begin the process of leaving the atmosphere in search of resources such as host plants, and all future flights appear to be appetitive in nature.

Appetitive Dispersal

Appetitive dispersal encompasses that set of behaviours other than migration that span the interval between leaving a host that is, for one reason or another, undesirable and seeking a host that is desirable. It also encompasses movement to seek a mate. By this definition, seeking an overwintering primary host as well as seeking a spring secondary host would not be considered acts of true 'migration' but rather included with those of appetitive dispersal. Winged adults produced on a primary host generally do not need to travel far to find suitable secondary hosts (MacKenzie and Dixon, 1991),

although Ward *et al.* (1998) estimated that the success rate of *R. padi* finding its primary host can be less than 1%. Our definition of appetitive dispersal includes the causal mechanisms for leaving and settling, as well as those of the dispersal event. Thus, appetitive dispersal must take into account both the stimuli that cause aphids to leave their feeding position on a plant and the cues that motivate them to cease their searching for a new feeding position or mate.

Stimuli that cause aphids to disperse in an appetitive manner

Extrinsic cues that stimulate aphids into an appetitive dispersal mode might be host-plant mediated or physical in origin, and could result from natural perturbations or crop management practices. Appetitive dispersal may even occur without any apparent external stimulus; instead, it might result from an innate strategy to make the best use of plant resources (Hodgson, 1991). Appetitive dispersal often occurs at the smallest scale herein considered, on a single plant, where an aphid might walk from one part of a plant to another to escape undesirable weather conditions, crowding, or natural enemies. However, this displacement mode includes plant-to-plant dispersal within fields, and even encompasses short duration, field-to-field movements *via* walking or flight. For appetitive flight, even if all other factors are favourable, including the physiological state of the aphid, the distance travelled will hinge on the atmospheric motion systems available at the time of take-off and ascent, horizontal transport, descent, and landing (Isard *et al.*, 1990; Gage *et al.*, 1999; Isard and Gage, 2001). It will depend ultimately on the motivation for moving in the first place.

Physical mechanisms

Aphids residing in agricultural fields may be inadvertently displaced. Mostly circumstantial evidence suggests that this sort of displacement is rather routine. Farm machinery passes through a field and brushes against an aphid or vibrates the plant on which the aphid resides (Ferrar, 1969; Schotzko and Knudsen, 1992; Bailey *et al.*, 1995). Raindrops pelt an aphid directly or hit the plant, jarring the aphid from the plant (Zúñiga, 1985), or gusts of air lift and propel an aphid from an exposed plant part (Schotzko and Knudsen, 1992; Bailey *et al.*, 1995; Mann *et al.*, 1995). Once dislodged, the aphid either falls to the ground, where it then dies or where it rights itself and walks to a new (or the original) host plant, or if an alata, begins to flap its wings. If in the proper physiological mode, it then ascends in active, controlled flight. Thus, if not incapacitated, an inadvertently displaced aphid invariably switches to an intentional displacement mode by seeking a host plant to recolonize or by fleeing a perceived threat. Intentional displacement, i.e. purposeful movement from one site to another, emanates from other environmental and agriculturally-related cues, including, but not limited to, light intensity, day length, temperature, humidity, crop canopy structure (Bottenberg and Irwin, 1992a,b), and plant density (Bottenberg and Irwin, 1992b).

Results of research into displacement responses to physical perturbations by nymphs, adult apterae, and alatae appear inconsistent. The contrasting results may be, in part, a reflection of the inherent differences in the aphid species being studied. Effects of natural forces such as rain and wind dislodge apterous aphids to some degree, as do acts of mechanical brushing or raking (Bailey *et al.*, 1995; Mann *et al.*, 1995; Zhang, 2002). Indeed, Bailey *et al.* (1995) found that under controlled conditions, wind proved to be the major physical disturbance that significantly altered dispersal patterns of *R. padi*. Conversely, in a study of apterous *S. avenae*, also conducted under controlled conditions, Mann *et al.* (1995) found precipitation was more important than wind in determining distances travelled.

Inconsistencies are not limited to the type of disturbance having the greatest effect; the relative magnitude of the influence also remains controversial. As an example, according to Mann *et al.* (1995), increased

wind duration extends the average distance apterous *S. avenae* walk; similarly, when these aphids were exposed to gentle gusts of air, they travelled less than when exposed to steady wind or strong gusts. However, while conducting a series of wind experiments, Zhang (2002) found that intensity and duration had little differential effect on movement of *R. padi*. Indeed, almost all aphid movement ceased until the wind had subsided, and the proportion that moved and the mean distances they travelled after wind cessation were influenced neither by wind intensity nor duration.

Regulated precipitation events provide another example of inconsistent results among studies. Mann *et al.* (1995) concluded that increased duration of a precipitation event correlated positively with an increase in the proportion of dislodged and displaced apterous *S. avenae*. Also, under heavy precipitation, the proportion of *S. avenae* dislodged and the distances they travelled were greater than when exposed to a drizzle or light rain. Under regulated precipitation events that were perhaps less extreme, Zhang (2002) found that apterous *R. padi* remained on the plants throughout the precipitation event, and both Bailey *et al.* (1995) and Zhang found that, once the precipitation event had ceased, similar numbers of apterae left the plants and, irrespective of the intensity or duration of the event, travelled about the same distances.

Zhang (2002) also found that when the host plants of apterous *R. padi* were vigorously stroked 5 or 10 times, or until all aphids were dislodged, the number of aphids relocating to neighbouring plants did not differ significantly among treatments and the minimum stroking threshold needed to result in an equal number of apterae relocating rapidly reached a plateau. Although intensity and duration of physical perturbations each influenced dispersal, neither was able to influence movement patterns once a disturbance threshold was reached (Zhang, 2002). Zhang's findings comparing brushing, wind, and precipitation events suggest that physical perturbations affect the numbers of *R. padi* that move, and the distances they travel, to similar degrees.

Host-plant influences

Biotic stimuli that cause aphids to move are frequently initiated by changes in host-plant quality (Hodgson, 1991; Bailey *et al.*, 1995; Zhang, 2002), including differences in plant structure and changed plant chemistry, the latter often attributed to plant stress (Haile, 2001). Direct, as well as indirect, factors can modify plant quality. Direct factors include relative host preferences, i.e. which plants and their cultivars are less preferred and thus more readily abandoned by an aphid (Ferrar, 1969; Zhang, 2002). Both chemical and physical attributes of plants can cause this exodus. Physical attributes include plant architecture (Hodgson and Elbakheit, 1985), hairiness (Gunasinghe *et al.*, 1988; Irwin and Kampmeier, 1989), leaf waxes (Bergman *et al.*, 1991; Powell *et al.*, 1999), and the availability of desirable feeding sites (Kobayashi and Ishikawa, 1993, 1994).

Aphids can be prone to abandon one plant species more readily than they do another, perhaps due to physical or chemical plant differences. Zhang (2002) found that late instar nymphs and adult apterae of *R. padi* more readily abandoned a wheat plant than either oats or barley, but that the numbers that relocated from three distinct cultivars of oats were similar. Likewise, aphids might more readily abandon their host plant during different times of the plant's life cycle. As plants mature and senesce, a mass exodus of aphids often occurs, presumably signalled by changes in plant chemistry (Parry, 1978; A'Brook, 1981).

Indirect factors that alter plant chemistry are brought about by plant stresses such as drought (Bailey *et al.*, 1995; Haile, 2001; Zhang, 2002), infection by plant pathogens (Fereres *et al.*, 1999; Olson, 2001), plant injury (Irwin and Thresh, 1990; Delaney and Macedo, 2001), and aphid crowding (Kidd, 1977; Hodgson, 1991; Bailey *et al.*, 1995; Zhang, 2002). Bailey *et al.* (1995) investigated impacts of drought and crowding on the movement of nymphal and adult apterous *R. padi* under greenhouse conditions and found that, although these indirect biotic factors tended to increase movement

of aphids, the differences between treatments were minimal and often insignificant. These results differed from those obtained by Zhang (2002), who found that the numbers of nymphs and apterous *R. padi* that relocated were considerably greater when perturbed by crowding, followed closely by drought, but that aphids travelled farther from the original host plants when these plants were subjected to drought rather than to crowding. Results of both Bailey *et al.* (1995) and Zhang (2002) indicate that host-plant-induced disturbances were generally not strongly differentiated from physical perturbations. Furthermore, their results suggest that the number of *R. padi* that moved and the distances they traversed after colonizing *Barley yellow dwarf virus*-infected hosts were similar to distances moved when colonizing hosts that are non-infected. This indicates that virus infection may not be a factor that stimulates this species to relocate or inhibits it from relocating.

Natural enemies and alarm pheromones

Aphids have evolved many behavioural defences in response to selection pressures exerted by predators and parasitoids (e.g. Klingauf, 1967; Gardner *et al.*, 1984; Dixon, 1988; Gerling *et al.*, 1990). Although waxing of natural enemies with fluids released from the siphunculi works well against small predators, the most effective defence reaction against both predators and parasitoids is to escape, either by walking from the feeding site or by dropping or flying from the host plant (e.g. Dixon, 1958; Klingauf, 1967; Stadler *et al.*, 1994; Weisser, 1995). Such escape reactions do not require direct contact with the predator, but may also be initiated in response to direct communication among conspecific aphids by means of aphid alarm pheromones (Phelan *et al.*, 1976; Calabrese and Sorensen, 1977; Montgomery and Nault, 1977a,b, 1978; Roitberg and Myers, 1978a; Wohlers, 1980, 1981; Clegg and Barlow, 1982; Gut and van Oosten, 1985; Nakamuta, 1991; Shah *et al.*, 1999; see also Pickett and Glinwood, Chapter 9 this volume). Aphids also respond to visual, tactile, and vibrational cues associated with the presence of natural enemies on a plant (Klingauf, 1967; Wohlers, 1980; McAllister and Roitberg, 1987; McAllister *et al.*, 1990; Losey and Denno, 1998b; Braendle and Weisser, 2001).

Dropping almost always leads to emigration from the host plant, except when dislodged aphids fall into leaf axils or onto leaves (McConnell and Kring, 1990). Aphid dislodgement by predators has been shown in many studies (e.g. Roitberg *et al.*, 1979; Brodsky and Barlow, 1986; McConnell and Kring, 1990; Sewell *et al.*, 1990; Minoretti and Weisser, 2000). In *A. pisum*, a certain proportion of aphids that walks away from the feeding site also walks from the host plant, and some individuals that at first escape by walking drop off the plant after a few steps (Minoretti and Weisser, 2000; Braendle and Weisser, 2001). The number of aphids emigrating from the host plant often exceeds the number being killed by the predator (Fig. 7.5) (Minoretti and Weisser, 2000). For *S. graminum*, the numbers dislodged for each individual consumed by adult *Coccinella septempunctata* (7-spot

ladybird) ranged from 1.8 to 7.2 in experiments over a wide range of aphid and predator densities (McConnell and Kring, 1990). Foraging parasitoid wasps also cause aphids to drop off or walk from the host plant (Tamaki *et al.*, 1970; Gowling and van Emden, 1994; Weber *et al.*, 1996). In addition, aphids parasitized by braconid wasps generally are more mobile (Weber *et al.*, 1996) and are more likely to drop from the plant when encountered by a predator (McAllister and Roitberg, 1987; McAllister *et al.*, 1990).

Several studies have quantified the effects of natural enemy action on aphid distribution at a larger spatial scale. In arenas with several host plants, the presence of predators resulted in significant dispersal and a more even distribution of aphids among plants (Niku, 1975; Roitberg *et al.*, 1979; Sewell *et al.*, 1990; Bailey *et al.*, 1995). In the field, aphids walking on soil between plants may be exposed to additional biotic and abiotic mortality factors (e.g. ground predators, high ground temperatures) as they search for a new host plant (Roitberg

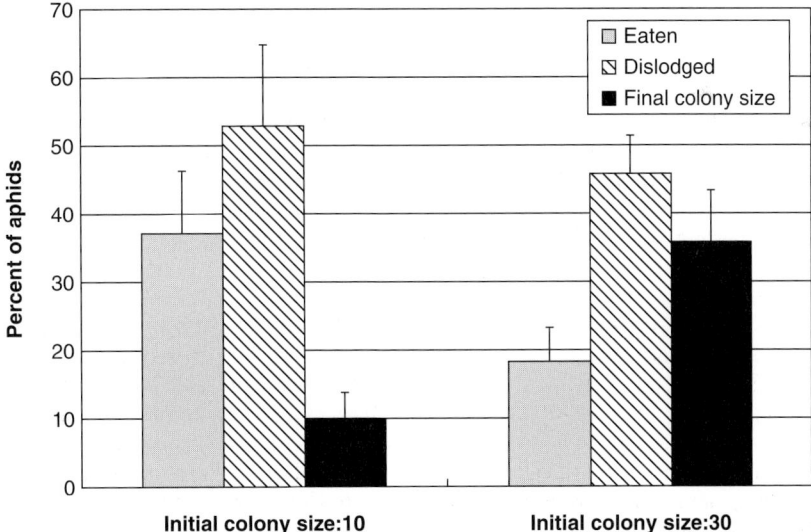

Fig. 7.5. Fate of aphids after a *Coccinella septempunctata* foraged on a broad bean (*Vicia faba*) plant. Initial colony sizes were 10 or 30 aphids when a ladybird with standardized hunger level (fed until satiation, then starved for 24 h) was introduced to the plant. Colonies consisted of a predetermined mixture of instars. The ladybird was free to leave the plant whenever it wanted and was observed continuously. Sample sizes were 6 (10-aphid colonies) and 8 (30-aphid colonies). Shown are means + SE (figure redrawn from Minoretti and Weisser, 2000).

and Myers, 1979; Losey and Denno, 1998a). Even though a large proportion of aphids walking on the ground may die before being able to settle on a new plant, available evidence suggests that predators affect both aphid redistribution and the spread of aphid-borne plant pathogens under field conditions (Niku, 1972; Roitberg and Myers, 1978b; Annan *et al.*, 1999). Annan *et al.* (1999) observed a positive correlation between the rate of within-field spread of *A. craccivora* and the incidence of natural enemies in cages placed in cowpea fields, but a number of other factors such as wind speed also influenced aphid dispersal. Available evidence therefore indicates that natural enemies play an important role in adult apterous and nymphal dispersal, but the relative importance of natural enemies compared with factors such as temperature, wind, rain, host-plant quality, or aphid population growth remains unclear. When estimating the effects of natural enemies on aphid populations in the field, both their positive effect of reducing aphid numbers and their effect on aphid dispersal and the subsequent spread of plant viruses need to be taken into consideration.

Response to multiple perturbations

When a chemical (e.g. a herbicide) that is not directly toxic to the aphid is applied to a field, the aphids that have colonized the crop are subjected to two independent types of perturbation, those related to the direct physical act of chemical application and those related to indirect changed host-plant chemistry as a consequence of chemical absorption or adsorption by the plant. Studies by Bailey *et al.* (1995) and Zhang (2002) determined that the application of such chemicals results in movement to the same degree and in the same manner as does the act of physically brushing colonized plants. This led Zhang (2002) to a series of interactive experiments that compared movement under single and multiple perturbation scenarios. Zhang found that the resultant movement was similar to the movement under the single factor that

independently resulted in the greatest movement. In other words, there was no interaction between perturbations; aphids simply moved in accordance with the most potent of those multiple stimuli.

Landing and alighting cues

Termination of aerial transport is usually directed by the aphid unless meteorological forces are controlling its displacement, as in the case of passively transported apterae or individuals trapped in columns or horizontal air currents. Many abiotic processes can account for undirected deposition of aphids, including the subsidence of convection currents, turbulence, and rain (Isard and Gage, 2001). Generally, however, if aphids do not actively flap their wings, e.g. when totally exhausted (Irwin and Hendrie, 1985), gravitational forces will take over and the aphids will descend. When aphids terminate migratory behaviours, losing their attraction to short wave length and UV light (Robert, 1987) and responding to cues that elicit what Hardie (1989) calls 'attack flight', they begin to act similarly to appetitive dispersers and are governed by cues of potential host plants. Favret and Voegtlin (2001) examined landing preference of alate aphids that were not native to the crop or natural habitat in which they landed. In studies that spanned 4 years in one location with row-cropped (soybean or maize) agriculture and restored prairie, more species and greater numbers of aphids landed in agricultural plots than in the nearby natural habitat. In a second location monitored for 3 years, again with similar agricultural plots but this time within a wooded area, migrant aphids once again landed more frequently in row-cropped agriculture each year. Clearly, cues to land are not random, and the preference for landing in agricultural fields has significant implications for the spread of non-persistently transmitted viruses.

Observations on numbers of aphids alighting in habitats lacking potential host plants suggest that discrimination of hosts from the air has less to do with chemical than with visual cues (Fereres *et al.*, 1999).

Aphids are attracted to shape, size, and colour of a landing target (Kennedy *et al.*, 1961; Fereres *et al.*, 1999), and seem especially attracted to edge effects (A'Brook, 1968; Schultz *et al.*, 1985). Landing aphids tend to respond to certain colours, particularly those reflecting strongly in the yellow-green wavelengths (520–580 nm) (Robert, 1987; Fereres *et al.*, 1999), and many species are sensitive to saturated yellow colours (Broadbent, 1948; Moericke, 1955; Kennedy *et al.*, 1961). The plants on which they alight may or may not be potential hosts; the discrimination appears to be sensed after landing and probing (Kring, 1972), or perhaps within centimetres of landing in a sniffing behaviour described by Kennedy *et al.* (1961).

The density of vegetation and its proportion in relation to bare ground are other factors that influence landing rates (A'Brook, 1968; Irwin and Goodman, 1981; Schultz *et al.*, 1985; Kring and Kring, 1991). For most pest species, *Aphis spiraecola* (green citrus aphid) being a notable exception (Halbert and Irwin, 1981), the amount of ground cover present is inversely proportional to aphid landing rates. Thus, aphids are often more attracted to young or widely spaced plantings (Kennedy *et al.*, 1961; Smith, 1976; Halbert and Irwin, 1981; Irwin and Kampmeier, 1989; Irwin *et al.*, 2000), or will more readily infest margins before the interior of fields. Barriers such as windbreaks, hedgerows, forested areas, and mixed cropping of plants of different heights (Kenny and Chapman, 1988; Irwin and Kampmeier, 1989; Bottenberg and Irwin, 1991; 1992a,b) can increase the number of aphids landing on the leeward side of such obstacles. This skewed distribution of landed aphids is hypothesized to result from both the behavioural attraction to these contrasts (Robert *et al.*, 1976) and the enhanced physical control over flight direction as the aphids approach the surface (Isard and Gage, 2001).

Incorporating Aphid Movement into Pest Management Strategies

The implications of aphid movement for the quality and quantity of harvestable products vary considerably among crops, with impacts resulting from interactions between the immigrating aphids and the crop on which they alight. Also of importance are the timing of the landing events relative to crop phenology, the aphid's reproductive potential, its abilities to transmit viruses to the crop, the residence time of the aphid on the plant and in the field, the types and impact of viruses it may transmit, and rates of ensuing virus epidemics. Aphid movement patterns and the potential implications of those patterns at multiple scales are requisite sets of information when considering control of aphids within an integrated pest management (IPM) programme. Where they are potential pests, either in their own right or as vectors of plant viruses, crop, cropping systems-based, landscape scale, area-wide, and eco-regional IPM packages should consider incorporating components to manage aphids and to track their movements.

Knowing the pest status of aphids that are found within agriculturally cropped settings is enormously valuable when considering how to manage them. Aphids entering a field and landing on a crop can be placed conveniently into one of four categories depending on the relationship between the aphid's colonizing potential and the way it is apt to damage the crop:

- **A transient non-vector** – a species of aphid that alights on a crop, cannot colonize it, and lacks the potential to transmit viruses to it.
- **A transient vector** – a species of aphid that alights on a crop, is incapable of colonizing it, but is able to transmit viruses that can infect it.
- **A colonizing non-vector** – a species of aphid that alights on a crop, can colonize it, but is incapable of transmitting viruses to it.
- **A colonizing vector** – a species of aphid that alights on a crop, can colonize it, and has the potential to transmit viruses to it.

These four categories of alighting aphid impact the crop in different ways and call for differing management practices.

Components of IPM programmes

Three functional components, fundamental, tactical, and strategic (see Irwin, 1999), delineate the stages necessary to develop and hone an IPM programme. These same components are essential for managing aphids that are potential pests.

Fundamentals

To manage aphids, no matter what the crop or cropping system or the scale of the endeavour, the identity of the species that alight and the biologies of those that can colonize the crop or transmit viruses to it are fundamental. Species identification is no easy task given that, in a number of agricultural regions worldwide, often more than 40 (Halbert et al., 1981) and as many as 80 (M.E. Irwin and G.E. Kampmeier, unpublished results) species of aphid can alight on plants within an annually cropped field in temperate climates in most growing seasons.

There are apt to be no more than four or five aphid species that colonize any given crop in a particular area of the world, and fewer than 35 species that are confirmed vectors of any of the viruses known to infect that crop. This renders the task of identification considerably less arduous because it narrows the potential list of species that need to be determined to a more manageable few dozen, which will likely differ by both crop and region. Identification guides, including pictorial keys, are available (e.g. Cottier, 1953; Medler and Ghosh, 1969; Taylor, 1984; Smith et al., 1992; Blackman and Eastop, 2000; Voegtlin et al., 2003), and can be of great help, especially to taxonomically trained personnel in the region for which the guide was written. However, many aphid identification guides base determinations partially on knowing the aphid's host plant, yet most aphids that fly into and land on crop plants in a field are not alighting on their hosts. As difficult as the identification process may be, it is an essential first step if the aphids that are coming into a field are grouped into one of the four agricultural pest status categories. Before the identified specimens can be categorized properly,

decisions need to be made concerning the economic implications of injury and damage potentially caused by the incoming aphids and the viruses they may be carrying. Blackman and Eastop (2000) can aid in determining whether any of the identified aphid species are potential colonizers or vectors of viruses capable of infecting the crop or crops.

The choice of sampling schemes can be key because they supply basic qualitative and quantitative information needed to execute successfully an IPM programme (see Harrington et al., Chapter 19 this volume). Sampling schemes for aphids most often reflect components of their population dynamics within the crop (e.g. Heathcote, 1957a,b; Heathcote et al., 1969; Irwin, 1980). Other schemes provide evidence of the aerial diversity and density of alate aphids at specific heights above the ground (Johnson, 1957a,b; Taylor, 1977; Isard et al., 1990, 1994), while still others document landing rates that point to aerial movements into and within fields (e.g. Irwin, 1980; Irwin and Ruesink, 1986; Raccah and Irwin, 1988). It is also important to consider where the sampling takes place. Visually monitoring primary hosts, for instance, can help predict the probability of significant outbreaks of aphids in small grains in England (Leather and Walters, 1984). Each of these sets of sampling or monitoring schemes provides complementary information about aspects of aphid biology and control.

Those schemes that emphasize population dynamics can help predict potential direct damage caused and measure the impact of persistently (and semi-persistently) transmitted viruses vectored by those species of aphids that colonize the crop. Those methods that monitor aerial densities, particularly where the sampling devices are situated at a considerable height above the ground (i.e. those positioned within the PBL) provide a measure of long-distance aerial transport. Those that emphasize alighting within the crop canopy provide information pertinent to the indirect damage aphids can cause as vectors of both non-persistently and persistently transmitted plant viruses, i.e. information useful to help forecast rates and timings of virus epidemics.

Suction trap networks, operated to monitor aerial densities of aphids within the PBL, are used widely in the UK, Europe (Taylor, 1977; Clark, *et al.*, 1992; Bommarco and Ekbom, 1995; Basky and Harrington, 2000) and parts of the USA (Blackmer and Bishop, 1991; Quinn *et al.*, 1991; Halbert *et al.*, 1992; Wright *et al.*, 1995), particularly with a view to helping forecast incoming infestations of aphids in fields of small grains, potatoes, and sugarbeet. Assuming the traps monitor within the PBL, reasonable data on numbers, sexes, morphs, and species determinations are all possible products of this type of sampling scheme. Within this context, these networks are capable of helping to determine when and in what numbers various species of aphids are moving within the atmosphere over fixed land targets. However, because the collecting orifices of these traps are at a static height and because the PBL is dynamic, shifting in height and depth with time of day, it is probable that the traps capture aphids within the PBL only some of the time. Also, because the insects are usually concentrated within specific parts of the PBL, some of the time the traps are sampling appropriate areas and some of the time they are not. Thus, it is difficult to characterize the density characteristics of the taxa sampled by such trapping networks.

Tactics

Tactics are practices that can help reduce threats posed by pests to crops. Those practices available for managing aphids usually can be placed into one of five groupings: chemical control, biological control, host-plant resistance, habitat manipulation, and legal control (note that genetic control is woven into the fabric of several of these groupings). These tactics are discussed below in relation to how they are affected by aphid movement.

Chemical control (see Dewar, Chapter 15 this volume) entails the introduction of chemically based compounds that kill or alter the behaviour of pests. Some juvenile hormones and ecdysone inhibitors can retard aphid development so that movement potential is restricted (Tamaki, 1973; Kuldova *et al.*, 1998) or flight is prevented in morphs that possess normally developed wings and flight muscles (White and Lamb, 1968). Although pesticides are not the best solution for the long-term control of aphids, intervention at critical points in the growing season may prove successful in stalling or reducing movement. In cropping systems where aphids are not the sole pests of interest, regular and frequent pesticide applications can cause pests, including aphids, to become resistant (ffrench-Constant and Devonshire, 1986; Field *et al.*, 1997; Foster *et al.*, Chapter 10 this volume), allowing pest populations to expand rapidly. Moreover, slower acting pesticides agitate aphids, causing them to move more readily, increasing the likelihood that they will spread non-persistently transmitted viruses to new plants before becoming incapacitated (Irwin and Kampmeier, 1989; Irwin, 1999).

Biological control (see Powell and Pell, Chapter 18 this volume) lowers populations of pests by introducing, augmenting, or inundatively releasing the pests' natural enemies, e.g. pathogens or their by-products, predators, or parasitoids. Classical biological control, e.g. using introduced parasitoids to control the invasive spotted alfalfa aphid, *Therioaphis trifolii maculata* (Schlinger *et al.*, 1959a,b), if at the source and not the sink, can successfully retard aphid build-up and therefore diminish their collective movement to a sink. This has the potential to reduce virus problems resulting from aphid movement. When successful, classical biological control is apt to span multiple cropping seasons, widening its influence over time.

Host-plant resistance (see van Emden, Chapter 17 this volume) is an effective measure for long-term control of pests, deploys crop cultivars that are resistant to colonizing aphids, deters aphids from landing or probing, or provides protection from the viruses they carry (Gunasinghe *et al.*, 1986; 1988; Martin *et al.*, 2003). In theory, the restlessness of aphids on resistant plants might be expected to increase the spread of non-persistent viruses, but this is not borne out in practice (discussed by van Emden, Chapter 17 this volume).

Habitat manipulation (see Wratten et al., Chapter 16 this volume) is perhaps the most versatile and potent of the tactics available for managing aphids and their movement. However, employing it requires considerable knowledge of the cropping system, including the aphids and viruses they transmit, and it is undoubtedly the most complex and difficult of the tactics to execute properly. Habitat manipulation may be key to making a crop less attractive or unapparent to both transient and colonizing aphids (Irwin et al., 2000). Species of aphids that are potentially damaging to crops often fly in large numbers and alight in fields during specific, predictable times of the season (A'Brook, 1981). Because aphids alight preferentially on isolated plants and, whenever possible, avoid alighting on closed or nearly closed canopies (A'Brook, 1968; Irwin and Kampmeier, 1989; Irwin et al., 2000), selecting planting dates and planting densities that provide the greatest canopy cover during peak flight periods can stave off colonizers and transients alike. Narrower row spacing can significantly reduce the time between sowing and canopy closure, rendering plants less attractive much sooner (Halbert and Irwin, 1981; Harvey et al., 1982). Reflective surfaces and coloured mulches can also help reduce aphid landings (Smith and Webb, 1969; Kring, 1972). Weeds and other non-crop plants that might harbour viruses or their vectors can be removed from within and near the field (Fereres et al., 1996). Locating newly planted fields upwind of and away from flight pathways of aphids emigrating from older, senescing crops that are apt to harbour them and the viruses they transmit is another way to avoid infestations and restrain virus epidemics. Exclusion covers can keep aphids from landing on and colonizing or spreading viruses within fields and, because of the added costs, are especially effective in fields planted to cash crops such as peppers (Avilla et al., 1997). Trap crops that are more sparsely planted or are sown later in the season than the main crop can help lure aphids, diverting them from a valued field. Moreover, on a landscape scale, multiple cropped systems may provide greater biodiversity, different micro-climates (Castro et al., 1991, 1992), non-virus hosts, and barriers that can influence movement in the canopy, resulting in altered landing and colonization rates (Bottenberg and Irwin, 1992b).

Legal control measures are usually enacted on a scale larger than that of a single field. When growers look beyond their own fields for sources of aphids and disease, communication through local or regional networks and cooperation among growers on an area-wide or regional basis provide an important tool. When government agencies or regional grower associations decide that measures beyond providing information and communication are necessary, policies that range from quarantine to mandated land- and water-use practices might be introduced. The foundation of the temporal and spatial aspects of these measures must be based firmly on knowledge of aphid movement. They are enacted for the common good of the growers of the area covered and for consumers and society at large. Although legal controls inevitably restrict individual freedoms, such measures are particularly applicable to highly mobile pest species, including agriculturally important aphids and the viruses they carry.

The strategic component

An IPM programme is ultimately founded on a strategy, the spatial integration and timely deployment of carefully selected, compatible, tactics. Many IPM programmes are primarily based on remedial measures. Under such programmes, active sampling plans are used to monitor densities of various pests, and when they exceed economic injury levels (i.e. pre-established thresholds), signal that some tactic or set of tactics should be implemented to reduce pest populations to levels below established thresholds. Of the five classes of tactics described above, only biological, chemical, and legal controls normally function in a remedial mode, and remedial IPM is often most appropriate for managing aphids that fall into either the 'colonizing non-vector' or 'colonizing vector' pest status category because this mode can lower population levels. In situations where an aphid species is potentially carrying viruses, population reductions cannot always occur before the targeted aphids are able to spread the virus.

Thus, practices to manage populations of aphids that transmit viruses into and within a field, by either transients or colonizers, should be anticipated and initiated before the crop is planted. Remedial IPM is generally ineffective for managing transient aphids.

Preventative measures attempt to contain potential pests before they become actual pests by anticipating and intervening to thwart their build-up. When protecting annual crops, preventative measures are almost always invoked before the crop is planted. Each of the five tactical categories can be executed in a preventative manner. Preventative IPM can alter an aphid's potential to move, the way it moves, its alighting behaviour, and its subsequent rate of increase.

Strategies need to be scaled to the movement potential of the target pest. If the focus is on colonizing non-vectors that move mainly within a field or between fields, a field or landscape scale circumscription is adequate. If, however, the aphid is migratory and capable of moving great distances, the IPM programme should be robustly large-scaled, even eco-regional in circumscription (Irwin, 1999).

The larger the programme circumscription, the more anticipatory it must be. Large-scale programmes should incorporate approaches that envisage the timing and abundance of pest influxes, including the use of predictive models that can forecast movement direction, distance, and timing of key aphid species. Much research has gone into developing simulation models of aphid movement (Taylor *et al.*, 1979; Stinner *et al.*, 1983; Ruesink and Irwin, 1986; Scott and Achtemeier, 1987; Kendall *et al.*, 1992). To date, however, few predictive models have been developed, and the success of forecasting aphid displacements is limited (Carter and Dewar, 1983; Dedryver and Pierre, 1987; Sigvald, 1992; Worner *et al.*, 1995; Masterman *et al.*, 1996; Harrington *et al.*, Chapter 19 this volume).

Outbreaks of aphid pests originating from afar (Hodson and Cook, 1960; Wallin *et al.*, 1967; Wallin and Loonan, 1971; Kieckhefer *et al.*, 1974; Zeyen *et al.*, 1987; Pedgley, 1993) can be unexpected and explosive. Yet, of the few forecasts that currently exist to alert growers to the timing and density of influxes of key aphid species into areas downwind of a source (EXAMINE, 2000; Manitoba Potato Industry, 2005; Scottish Agricultural Science Agency, 2005; AphidWatch, 2006; Rothamsted Insect Survey, 2006; Western Australia Department of Agriculture, 2006a,b), the IPM strategies being developed are grappling with ways to take full advantage of those forecasts. To better attune the management strategy, tactical activities to reduce populations or alter the timing of alata production need to be initiated at the source areas so that flight activities are suppressed and flight timings offset from particularly vulnerable stages of the crop being protected.

Predictive models depend on coupling atmospheric motion systems, especially those of the PBL, with the maiden take-off timings of alatae, thus linking source areas with sinks and vastly expanding the area under scrutiny. Such anticipatory models require greater knowledge of the population dynamics of aphids at their sources than at their sinks. This highlights the importance of research aimed at pinpointing source areas of aphids that might move into areas of concern.

The source area of an aphid in flight or alighting in a field is probably one of the most difficult pieces of the movement puzzle to decipher, let alone predict. Radar has been used to detect aphids in the atmosphere (Hendrie *et al.*, 1985; Irwin and Thresh, 1988). However, unlike the radar signature of larger insects such as locusts and moths, that of aphids is less reliable and the identity of the species flying cannot be ascertained without directly sampling them in the atmosphere. Aircraft-mounted sampling devices (Glick, 1939; Berry and Taylor, 1968; Isard *et al.*, 1990; Hollinger, *et al.*, 1991) can provide a quantitative measure (numbers per volume of air sampled) of aphids that can be identified, and some of the more sophisticated systems can provide data on the height at which they were flying (Hollinger *et al.*, 1991).

Mapping the meteorological features and where in the atmosphere aphids are located, is only part of the puzzle in determining possible sources of migrants. Calculating the flight energy reserves to match the amount of time spent in active flight from a source to a sink (Cockbain, 1961; Liquido and Irwin, 1986; Irwin and Thresh, 1988) provides information on the amount of time the aphid was airborne. This, coupled with the trajectory, provides a rough approximation of source area. Matching specific genetic markers of various species of aphid from a sink with population markers obtained at many potential sources is another, independent way of determining potential source areas (Steiner *et al.*, 1985; Voegtlin *et al.*, 1987; Loxdale and Lushai, Chapter 2 this volume). This complementary set of procedures provides a powerful tool for developing predictive models of source areas of immigrants.

Categories of aphids as agricultural pests

Unique aspects of managing each of the aphid pest categories mentioned earlier involve all three components of an IPM programme: fundamental, tactical, and strategic.

Transient non-vectors

On any given day, vast numbers of transient aphids can enter a field, land on a myriad of plants while moving through it, and make their way out of the field. These transients go undetected with sampling schemes that emphasize population dynamics and aerial densities. Aphids that are transient in a field and do not transmit viruses to that crop will not negatively impact the crop's potential harvest, even though they might be extremely numerous. This category often has the highest numbers of aphid species in the field at any given time. Therefore, their presence, when detected by IPM practitioners and crop managers, can spur growers to take remedial action unnecessarily, e.g.

apply pesticides. As a result of this action, production costs can increase, side effects can negatively impact human health, and other management practices can be severely disrupted. Aphids in this category, once identified and properly categorized, should be left alone. They are not true pests, at least of the crop being monitored.

Transient vectors

Transient vectors pose a serious threat to crops because, as they repeatedly alight, probe, and fly across a field, they are capable of rapidly spreading debilitating, non-persistently transmitted viruses. In many crops, potential vectors of these viruses are, for the most part, inadequately known, making the pest status categorization of those species problematic. Methods to test aphids for their potential to transmit non-persistent viruses are provided by Irwin and Ruesink (1986) and Raccah and Irwin (1988). Slightly modified versions of these same methods can also aid in determining vectoring potential of persistently and semi-persistently transmitted viruses.

Because transient aphids often go undetected, growers can be taken by surprise when non-persistent virus epidemics sweep through fields and across the landscape in a matter of days, devastating the crop and, if timed unfortunately, severely reducing its potential yield. By the time the epidemic has been detected, the damage will have already occurred.

Monitoring aphid movements into and within fields cannot provide information useful for curtailing non-persistent virus epidemics. Liquid-filled pan traps, especially those that are neutral with respect to crop canopy colour (Irwin, 1980; Irwin and Schultz, 1981), provide valuable data for quantitatively tracking the movement of aphids, by species, into and through a field (Irwin and Ruesink, 1986; Raccah and Irwin, 1988). These monitoring devices aid researchers in developing simulation models of aphid movement relative to chronological occurrence, crop phenology, crop canopy closure, and virus epidemics. Such models can also help in understanding yield reductions attributable to virus epidemics resulting from aphid movement into and within the crop. They can also provide information on potential impact of tactics employed to reduce the threat of non-persistently transmitted virus epidemics. Used in a predictive mode, these models can be of enormous help in deciding, for future seasons, which countermeasures to enact, where to focus those measures, and when to enact them.

Countermeasures should focus on a few aspects of the aphid or the viruses it can transmit to the crop under management. Theoretically, non-persistent virus epidemics can best be avoided by altering the aphids' time of appearance, controlling their frequencies of alighting, using virus-resistant genotypes, and by intercropping. Non-persistent virus epidemics can be managed through tactical measures enacted at either the aphids' source (i.e. where they mature) or at their sink. No matter where the measures are to be applied, they must be enacted in a preventative manner, before the crop is planted and well before the aphids can be detected by any sampling scheme placed at the sink.

Most pesticides are slow acting and unable to control transient aphids. The application at the sink of quick-acting contact pesticides such as synthetic pyrethroids, on the one hand might increase, like slow-acting pesticides, the spread of non-persistently transmitted viruses by increasing the likelihood that viruliferous vectors will move among plants, probe, and transmit the virus before dying. On the other hand, quick-acting pesticides, unlike those that are slow-acting, might help to curb epidemics because, although such pesticides do not inhibit acquisition or inoculation of such viruses, they can diminish the aphid's ability to accomplish both and to move to inoculate successive plants (Irwin *et al.*, 2000).

Biological control cannot effectively control transient aphids nor greatly reduce their ability to spread non-persistent viruses at a sink. Theoretically, introducing biological control agents or inundatively releasing natural enemies at the source, where the aphids mature, could effectively lower populations, thus reducing and retarding influxes of aphids at the sink.

Deploying plant genotypes resistant to aphids can influence the rapidity with which aphids move through a field by changing the rate of probing once the transient aphid has alighted on a leaf. Virus resistance can be even more effective in influencing the ability of the aphids to transmit viruses within a field. When a viruliferous aphid probes a virus-resistant plant or a plant that is not a host of the virus it may be carrying, the probability of the virus being transmitted is close to nil. Therefore, the probability of transmission the next time that aphid lands and probes, if on a non-infected, susceptible host of the virus, is greatly diminished. This concept is extendable to mixed plantings (e.g. intercropping), where the viruses that can infect one crop are not the same as those that infect the other, or mixtures of susceptible and resistant cultivars planted in a field. By so planting, the spread of non-persistent, aphid-borne viruses in that field can be constrained (Gunasinghe *et al.*, 1986; Martin *et al.*, 2003). Moreover, physical barriers to probing by viruliferous aphids, such as leaf hairiness, effectively reduce the probability of inoculation (Gunasinghe *et al.*, 1988).

Colonizing non-vectors

Although the pattern of settling may differ among colonizing species, when an immigrant encounters a suitable host, it will add to the infestation at its sink. Some species alight on a host plant in a field, begin vivipositing, and form a colony there. Others alight, produce a few nymphs, then fly to neighbouring plants, where they produce more nymphs, thus dispersing their offspring more widely and, at the same time, probing more host plants. Regardless of the settling pattern, colonizing aphids tend to spread from field margins to their interiors and from plant to plant along, rather than between rows (Bottenberg and Irwin, 1991). Once the initial colonies are formed, they expand, and subsequent spread within the field is often by plant-to-plant movements of nymphs, adult apterae, and even alatae.

Thus, aphid infestations often become randomly patchy within a field.

The colonizing non-vector category of aphids can be managed in much the same way as other potential insect pests. Because virus spread is not a concern, the aphids should be managed solely in accordance with the damage they might cause the crop and its potential yield. The propensity for movement of particular aphid species and morphs, the timing of predictable perturbations such as cultivation, and the likelihood of encountering hosts must all be taken into account when calculating the implications of movement on the management of colonizing aphids. Establishing economic injury levels, monitoring established aphid populations at the sink, and applying measures to limit infestations once populations exceed economic thresholds form appropriate, remedial responses to this category of pest.

Colonizing vectors

Aphids that can colonize a crop at a sink and are potential vectors of plant viruses pose a serious, ongoing threat to crops they colonize, often more so because they are capable of transmitting persistent and semi-persistent viruses than because they are a threat on their own. Timing of aphid movement relative to plant phenology must be taken into account when predicting the impact of viruliferous aphids. The distinction between the movement of already infective vectors and those that must first acquire the virus is critical for predicting the longer-term effects on the host. Early infections have a greater impact on yield parameters and the plant's ability to act as a source of inoculum for subsequent spread at the sink. Because aphids generally retain persistent viruses for long periods of time, a reasonable likelihood exists that aphids will transport them to a sink from quite distant sources. If, however, potential vectors are not yet infective when they arrive at a sink, these aphids must first settle on a virus-infected host, and their offspring must then move and colonize a non-infected host to spread the virus. Therefore, epidemics of persistent viruses often proceed at a slower rate than those of non-persistent viruses.

Unlike transient aphids that vector non-persistent viruses, colonizing vectors can be more successfully managed at the sink. Monitoring local primary or over-seasoning hosts to gauge developmental stages of the aphids; deploying liquid-filled pan traps to confirm landings; analysing data from a regional network of suction traps to determine what is currently flying; and periodically assessing population build-up of aphids and their natural enemies at both the source and sink all provide important information on which to base a management strategy. The tactics one might employ are many; a few are mentioned below.

Pesticides, even those that target aphids but are slow acting, function rapidly enough to disable an aphid before it can successfully move to and inoculate a neighbouring plant with a persistent or semi-persistent virus. Pesticides, therefore, can effectively reduce populations of colonizing aphids without the concomitant spread of persistent viruses. Even under these circumstances, pesticide use can have negative consequences. Beyond the added cost of chemicals and their application and of aphid genotypes becoming resistant to the chemicals, the sensitivity and susceptibility of natural enemies to some pesticides greatly impedes biological control efforts.

Minimizing row spacing and synchronizing planting dates at the sink to confront incoming colonizing vectors with as complete a canopy cover as possible will reduce the probability that the incoming aphids will alight in the field. Sparsely planted trap crops may also divert arriving aphids. For those aphids that do land, conservation and augmentation of natural enemies and the use of plant genotypes resistant to either the aphids or the viruses potentially being transmitted are also possible tactics that can be applied at the sink.

Conclusions

Given that aphids are sedentary for most of their lives, it seems ironic and somewhat counterintuitive that their ability to move is so pivotal in elevating them to the status of super pests. Even though only a small fraction of the extant species are pests, it is their ability to move over large spatial scales that allows aphids, through their vectoring capacity, to devastate crops over a very short time span. If aphids did not move much, or if they moved over only very small spatial scales, even those species that are excellent vectors would play a much-reduced role in spreading plant pathogenic viruses. That certain aphid species do range widely, transcending continental land masses and even crossing oceans, propels them, in an agricultural context, not only to super pest status, but also to the status of potential super vectors. To forecast the impact of aphids in an agricultural setting can be difficult and often requires knowledge of where they originated and the capacity to predict when they will arrive in the field. Thus, we believe that understanding aphid movement and how to modify it is key to shaping useful management strategies for aphids.

Acknowledgements

We thank members of the North Central Regional Committee (NCR-148) on Movement and Dispersal of Biota for their inspiration and encouragement in this important area of research. We also acknowledge the support, inspiration, and encouragement of the International Plant Virus Epidemiology Committee, and especially J. Michael Thresh (National Resources Institute, University of Greenwich, Chatham, UK), Alberto Fereres (Departamento de Protección Vegetal, Consejo Superior de Investigaciones Científicas, Madrid, Spain), and Benny Raccah (The Volcani Center, Bet Dagan, Israel). Colleagues, who through their collaboration contributed greatly to the conceptual nature of this paper, include Robert M. Goodman, The State University of New Jersey, William G. Ruesink, Illinois Natural History Survey, and Scott A. Isard, Pennsylvania State University. We are especially indebted to Isard for reviewing the meteorological portion of this chapter. Funding for the first two authors was provided by a number of institutions and agencies over the past 25 years. Among the many, we particularly acknowledge the University of Illinois at Urbana-Champaign, the Illinois Natural History Survey, the Illinois State Water Survey, the National Research Initiative, the United States Agency for International Development, the International Arid Lands Consortium, the International Soybean Program, the Consortium for International Crop Protection, and the Illinois Agricultural Experiment Station. W.W.W. was supported by the Deutsche Forschungsgemeinschaft, grant WE 2618/2-1.

References

A'Brook, J. (1968) The effect of plant spacing on the numbers of aphids trapped over the groundnut crop. *Annals of Applied Biology* 61, 289–294.

A'Brook, J. (1981) Forecasting the flight peaks of aphid vectors of cereal viruses. In: *Proceedings of the 3rd Conference on Virus Diseases of Gramineae in Europe, Rothamsted Experimental Station, Harpenden, 1980.* pp. 67–72.

Annan, I.B., Schaefers, G.A. and Saxena, K.N. (1999) Pattern and rate of within-field dispersal and bionomics of the cowpea aphid, *Aphis craccivora* (Aphididae), on selected cowpea cultivars. *Insect Science and its Application* 19, 1–16.

AphidWatch (2006) Welcome to Aphid Watch: Providers of Aphid Flight Records, Virus Forecasts and Other Aphid Information to New Zealand Farmers and Growers (www.aphidwatch.com).

Avilla, C., Collar, J.L., Duque, M., Perez, P. and Fereres, A. (1997) Impact of floating rowcovers on bell pepper yield and virus incidence. *HortScience* 32, 882–883.

Bailey, S.M., Irwin, M.E., Kampmeier, G.E., Eastman, C.E. and Hewings, A.D. (1995) Physical and biological perturbations: their effect on the movement of apterous *Rhopalosiphum padi* (Homoptera: Aphididae) and localized spread of barley yellow dwarf virus. *Environmental Entomology* 24, 24–33.

Basky, Z. and Harrington, R. (2000) Cereal aphid flight activity in Hungary and England compared by suction traps. *Anzeiger für Schädlingskunde* 73, 70–74.

Bergman, D.K., Dillwith, J.W., Zarrabi, A.A., Caddel, J.L. and Berberet, R.C. (1991) Epicuticular lipids of alfalfa relative to its susceptibility to spotted alfalfa aphids (Homoptera: Aphididae). *Environmental Entomology* 20, 781–785.

Berry, R.E. (1969) Effects of temperatures and light on takeoff of *Rhopalosiphum maidis* and *Schizaphis graminum* in the field (Homoptera: Aphididae). *Annals of the Entomological Society of America* 62, 1176–1184.

Berry, R.E. and Taylor, L.R. (1968) High-altitude migration of aphids in maritime and continental climates. *Journal of Animal Ecology* 37, 713–722.

Blackmer, J.L. and Bishop, G.W. (1991) Population dynamics of *Rhopalosiphum padi* (Homoptera: Aphididae) in corn in relation to barley yellow dwarf epidemiology in southwestern Idaho. *Environmental Entomology* 20, 166–173.

Blackman, R.L. and Eastop, V.F. (2000) *Aphids on the World's Crops: An Identification and Information Guide*, 2nd edn. Wiley, Chichester, 466 pp.

Bommarco, R. and Ekbom, B. (1995) Phenology and prediction of pea aphid infestations on peas. *International Journal of Pest Management* 41, 109–113.

Bottenberg, H. and Irwin, M.E. (1991) Influence of wind speed on residence time of *Uroleucon ambrosiae* alatae (Homoptera: Aphididae) on bean plants in bean monocultures and bean-maize mixtures. *Environmental Entomology* 20, 1375–1380.

Bottenberg, H. and Irwin, M.E. (1992a) Canopy structure in soybean monocultures and soybean–sorghum mixtures: impact on aphid (Homoptera: Aphididae) landing rates. *Environmental Entomology* 21, 542–548.

Bottenberg, H. and Irwin, M.E. (1992b) Flight and landing activity of *Rhopalosiphum maidis* (Homoptera: Aphididae) in bean monocultures and bean–corn mixtures. *Journal of Entomological Science* 27, 143–153.

Braendle, C. and Weisser, W.W. (2001) Variation in escape behavior of red and green clones of the pea aphid. *Journal of Insect Behaviour* 14, 497–508.

Broadbent, L. (1948) Aphid migration and the efficiency of the trapping method. *Annals of Applied Biology* 35, 379–394.

Brodsky, L.M. and Barlow, C.A. (1986) Escape responses of the pea aphid, *Acyrthosiphon pisum* (Harris) (Homoptera: Aphididae): influence of predator type and temperature. *Canadian Journal of Zoology* 64, 937–939.

Calabrese, E.J. and Sorensen, A.J. (1977) Dispersal and recolonization by *Myzus persicae* following aphid alarm pheromone exposure. *Annals of the Entomological Society of America* 71, 181–182.

Carter, N. and Dewar, A.M. (1983) Forecasting outbreaks of the grain aphid. In: *Proceedings of the 10th International Congress of Plant Protection, Brighton, November 1983* 1, 167.

Castro, V., Isard, S.A. and Irwin, M.E. (1991) The microclimate of maize and bean crops in tropical America: a comparison between monocultures and polycultures planted at high and low density. *Agricultural and Forest Meteorology* 57, 49–67.

Castro, V., Rivera, C., Isard, S.A., Gámez, R., Fletcher, J. and Irwin, M.E. (1992) The influence of weather and microclimate on *Dalbulus maidis* (Homoptera: Cicadellidae) flight activity and the incidence of diseases within maize and bean monocultures and bicultures in tropical America. *Annals of Applied Biology* 121, 469–482.

Clark, S.J., Tatchell, G.M., Perry, J.N. and Woiwod, I.P. (1992) Comparative phenologies of two migrant cereal aphid species. *Journal of Applied Ecology* 29, 571–580.

Clegg, J.M. and Barlow, C.A. (1982) Escape behaviour of the pea aphid *Acyrthosiphon pisum* (Harris) in response to alarm pheromone and vibration. *Canadian Journal of Zoology* 60, 2245–2252.

Cockbain, A.J. (1961) Fuel utilization and duration of tethered flight in *Aphis fabae* Scop. *Journal of Experimental Biology* 38, 163–174.

Cottier, W. (1953) Aphids of New Zealand. *New Zealand Department of Scientific and Industrial Research Bulletin No. 106*, 382 pp.

Dedryver, C.A. and Pierre, J.S. (1987) Estimation de la fonction de transfert entre les captures de pucerons des céréales au piège à succion et leurs populations sur les cultures par un module d'analyse de variance–covariance multivarie. In: Cavalloro, R. (ed.) *Aphid Migration and Forecasting 'Euraphid' Systems in European Community Countries*. CEC, Luxembourg, pp. 215–224.

Delaney, L.J. and Macedo, T.B. (2001) The impact of herbivory on plants: yield, fitness, and population dynamics. In: Peterson, R.K.D. and Higley, L.G. (eds) *Biotic Stress and Yield Loss*. CRC, Boca Raton, pp. 135–160.

Dewar, A.M., Woiwod, I. and de Janvry, E.C. (1980) Aerial migrations of the rose–grain aphid, *Metopolophium dirhodum* (Wlk.), over Europe in 1979. *Plant Pathology* 29, 101–109.

Dingle, H. (1996) *Migration. The Biology of Life on the Move.* Oxford University Press, Oxford, 474 pp.

Dixon, A.F.G. (1958) The escape response shown by certain aphids to the presence of the coccinellid *Adalia decempunctata* (L.). *Transactions of the Royal Entomological Society of London* 11, 319–334.

Dixon, A.F.G. (1977) Aphid ecology: life cycles, polymorphism and population regulation. *Annual Review of Ecology and Systematics* 8, 329–353.

Dixon, A.F.G. (1988) *Aphid Ecology*, 2nd edn. Chapman and Hall, London, 300 pp.

Dixon, A.F.G. and Agarwala, B.K. (1999) Ladybird-induced life-history changes in aphids. *Proceedings of the Royal Society London B* 266, 1549–1553.

Dixon, A.F.G. and Kindlmann, P. (1999) Cost of flight apparatus and optimum body size of aphid migrants. *Ecology* 80, 1678–1690.

Dixon, A.F.G. and Wratten, S.D. (1971) Laboratory studies on aggregation, size and fecundity in the black bean aphid, *Aphis fabae* Scop. *Bulletin of Entomological Research* 61, 97–111.

Dixon, A.F.G., Kindlmann, P., Leps, J. and Holman, J. (1987) Why there are so few species of aphids, especially in the tropics. *American Naturalist* 129, 580–592.

Dixon, A.F.G., Horth, S. and Kindlmann, P. (1993) Migration in insects: cost and strategies. *Journal of Animal Ecology* 62,182–190.

Drake, V.A. and Gatehouse, A.G. (1996) Population trajectories through space and time: a holistic approach to insect migration. In: Floyd, R.B., Sheppard, A.W. and de Barro, P.J. (eds) *Frontiers of Population Ecology*. CSIRO Publishing, Collingwood, Victoria, pp. 399–408.

Dry, W.W. and Taylor, L.R. (1970) Light and temperature thresholds for take-off by aphids. *Journal of Animal Ecology* 39, 493–504.

EXAMINE (2000) EXploitation of Aphid Monitoring systems IN Europe. Environment Project EVK2-1999-00151 (www.rothamsted.bbsrc.ac.uk/examine/).

Favret, C. and Voegtlin, D.J. (2001) Migratory aphid (Hemiptera: Aphididae) habitat selection in agricultural and adjacent natural habitats. *Environmental Entomology* 30, 371–379.

Fereres, A., Avilla, C., Collar, J.L., Duque, M. and Fernandez-Quintanilla, C. (1996) Impact of various yield-reducing agents on open-field sweet peppers. *Environmental Entomology* 25, 983–986.

Fereres, A., Kampmeier, G.E. and Irwin, M.E. (1999) Aphid attraction and preference for soybean and pepper plants infected with Potyviridae. *Annals of the Entomological Society of America* 92, 542–548.

Ferrar, P. (1969) Interplant movement of apterous aphids with special reference to *Myzus persicae* (Sulz.) (Hemiptera: Aphididae). *Bulletin of Entomological Research* 58, 653–660.

ffrench-Constant, R.H. and Devonshire, A.L. (1986) The effect of aphid immigration on the rate of selection of insecticide resistance in *Myzus persicae* by different classes of insecticides. In: *Crop Protection of Sugar Beet and Crop Protection and Quality of Potatoes. Aspects of Applied Biology, No. 13,* pp. 115–125.

Field, L.M., Anderson, A.P., Denholm, I., Foster, S.P., Harling, Z.K., Javed, N., Martinez Torres, D., Moores, G.D., Williamson, M.S. and Devonshire, A.L. (1997) Use of biochemical and DNA diagnostics for characterising multiple mechanisms of insecticide resistance in the peach–potato aphid, *Myzus persicae* (Sulzer). *Pesticide Science* 51, 283–289.

Gage, S.H., Isard, S.A. and Colunga, M. (1999) Ecological scaling of aerobiological dispersal processes. *Agricultural and Forest Meteorology* 97, 249–261.

Gardner, S.M., Ward, S.A. and Dixon, A.F.G. (1984) Limitation of superparasitism by *Aphidius rhopalosiphi*: a consequence of aphid defensive behaviour. *Ecological Entomology* 9, 149–155.

Gerling, D., Roitberg, B.D. and Mackauer, M. (1990) Instar specific defense of the pea aphid, *Acyrthosiphon pisum*: influence on oviposition success of the hymenopterous parasite *Aphelinus asychis*. *Journal of Insect Behavior* 3, 501–514.

Glick, P.A. (1939) The distribution of insects, spiders and mites in the air. *Technical Bulletin of the United States Department of Agriculture No. 673*, 1–151.

Gowling, G.R. and van Emden, H.F. (1994) Falling aphids enhance impact of biological control by parasitoids on partially aphid-resistant plant varieties. *Annals of Applied Biology* 125, 233–242.

Gray, S.M. and Lampert, E.P. (1986) Seasonal abundance of aphid-borne virus vectors (Homoptera: Aphididae) in flue-cured tobacco as determined by alighting and aerial interception traps. *Journal of Economic Entomology*, 79, 981–987.

Grossheim, N.A. (1914) The barley aphid, *Brachycolus noxius* Mordwilko [In Russian]. *Memoirs of the Natural History Museum of Zemstwo Province Tavaria* 3, 35–78 [Edited translation by Poprawski, T.J.,

Wraight, S.P. and Peresypkina, S. In: Morrison, W.P. (ed.) *Proceedings of the 5th Russian Wheat Aphid Conference, Fort Worth, January 1992*, pp. 1–2].

Gunasinghe, U.B., Irwin, M.E. and Bernard, R.L. (1986) Effect of a soybean genotype resistant to soybean mosaic virus on transmission-related behavior of aphid vectors. *Plant Disease* 70, 872–874.

Gunasinghe, U.B., Irwin, M.E. and Kampmeier, G.E. (1988) Soybean leaf pubescence affects aphid vector transmission and field spread of soybean mosaic virus. *Annals of Applied Biology* 112, 259–272.

Gut, J. and van Oosten, A.M. (1985) Functional significance of the alarm pheromone composition in various morphs of the green peach aphid, *Myzus persicae*. *Entomologia Experimentalis et Applicata* 37, 199–204.

Haile, F.J. (2001) Drought stress, insects, and yield loss. In: Peterson, R.K.D. and Higley, L.G. (eds) *Biotic Stress and Yield Loss*. CRC, Boca Raton, pp. 117–134.

Haine, E. (1955) Aphid take-off in controlled wind speeds. *Nature, London* 175, 474–475.

Halbert, S., Elberson, L. and Johnson, J. (1992) Suction trapping of Russian wheat aphid: what do the numbers mean? In: Morrison, W.P. (ed.) *Proceedings of the 5th Russian Wheat Aphid Conference, Fort Worth, January 1992*, pp. 282–297.

Halbert, S.E. and Irwin, M.E. (1981) Effect of soybean canopy closure on landing rates of aphids with implications for restricting spread of soybean mosaic virus. *Annals of Applied Biology* 98, 15–19.

Halbert, S.E., Irwin, M.E. and Goodman, R.M. (1981) Alate aphid (Homoptera: Aphididae) species and their relative importance as field vectors of soybean mosaic virus. *Annals of Applied Biology* 97, 1–9.

Halgren, L.A. (1970) Flight behavior of the greenbug, *Schizaphis graminum* (Homoptera: Aphididae), in the laboratory. *Annals of the Entomological Society of America* 63, 712–715.

Hardie, J. (1989) Spectral specificity for targeted flight in the black bean aphid, *Aphis fabae*. *Journal of Insect Physiology* 35, 619–626.

Hardie, J. (1993) Flight behavior in migrating insects. *Journal of Agricultural Entomology* 10, 239–245.

Harrington, R., Katis, N. and Gibson, R.W. (1987) Monitoring aphids to assess the spread of potato virus Y. In: Cavalloro, R. (ed.) *Aphid Migration and Forecasting 'Euraphid' Systems in European Community Countries*. CEC, Luxembourg, pp. 173–175.

Harvey, T.L., Hackerott, H.L. and Martin, T.J. (1982) Dispersal of alate biotype C greenbugs in Kansas. *Journal of Economic Entomology* 75, 36–39.

Heathcote, G.D. (1957a) The comparison of yellow cylindrical, flat water traps and of Johnson suction traps, for sampling aphids. *Annals of Applied Biology* 45, 133–139.

Heathcote, G.D. (1957b) The optimum size of sticky aphid traps. *Plant Pathology* 6, 104–107.

Heathcote, G.D., Palmer, J.M.P. and Taylor, L.R. (1969) Sampling for aphids by traps and by crop inspection. *Annals of Applied Biology* 63, 155–166.

Hendrie, L.K., Irwin, M.E., Liquido, N.J., Ruesink, W.G., Mueller, E.A., Voegtlin, D.J., Achtemeier, G.L., Steiner, W.M. and Scott, R.W. (1985) Conceptual approach to modeling aphid migration. In: Mackenzie, D.R., Barfield, C.S., Kennedy, G.G. and Berger, R.D. (eds) *The Movement and Dispersal of Agriculturally Important Biotic Agents, An International Conference on the Movement and Dispersal of Biotic Agents*. Claitor's Publishing Division, Baton Rouge, pp. 541–582.

Hodgson, C. (1991) Dispersal of apterous aphids (Homoptera: Aphididae) from their host plant and its significance. *Bulletin of Entomological Research* 81, 417–427.

Hodgson, C.J. and Elbakheit, I.B. (1985) Effect of colour and shape of 'target' hosts on the orientation of emigrating adult apterous *Myzus persicae* in the laboratory. *Entomologia Experimentalis et Applicata* 38, 267–272.

Hodson, A.C. and Cook, E.F. (1960) Long-range aerial transport of the harlequin bug and greenbug into Minnesota. *Journal of Economic Entomology* 53, 604–608.

Hollinger, S.E., Sivier, K.R., Irwin, M.E. and Isard, S.A. (1991) A helicopter-mounted isokinetic aerial insect sampler. *Journal of Economic Entomology* 84, 476–483.

Irwin, M.E. (1980) Sampling aphids in soybean fields. In: Kogan, M. and Herzog, D.C. (eds) *Sampling Methods in Soybean Entomology*. Springer, New York, pp. 239–259.

Irwin, M.E. (1999) Implications of movement in developing and deploying integrated pest management strategies. *Agricultural and Forest Meteorology* 97, 235–248.

Irwin, M.E. and Goodman, R.M. (1981) Ecology and control of soybean mosaic virus. In: Maramorosch, K. and Harris, K.F. (eds) *Plant Diseases and Vectors: Ecology and Epidemiology*. Academic Press, New York, pp. 181–200.

Irwin, M.E. and Hendrie, L.K. (1985) The aphids are coming. *Illinois Research* 27, 11.

Irwin, M.E. and Kampmeier, G.E. (1989) Vector behavior, environmental stimuli, and the dynamics of plant virus epidemics. In: Jeger, M.J. (ed.) *Spatial Components of Plant Disease Epidemics*. Prentice Hall, Englewood Cliffs, pp. 14–39.

Irwin, M.E. and Nault, L.R. (1996) Virus/vector control. In: Persley, G.J. (ed.) *Biotechnology and Integrated Pest Management. Biotechnology in Agriculture, No. 15.* CAB International, Wallingford, pp. 304–322.

Irwin, M.E. and Ruesink, W.G. (1986) Vector intensity: a product of propensity and activity. In: McLean, G.D., Garrett, R.G. and Ruesink, W.G. (eds) *Plant Virus Epidemics: Monitoring, Modelling, and Predicting Outbreaks.* Academic Press, Sydney, pp. 13–33.

Irwin, M.E. and Schultz, G.A. (1981) Soybean mosaic virus. *FAO Plant Protection Bulletin No. 29,* 41–55.

Irwin, M.E. and Thresh, J.M. (1988) Long-range aerial dispersal of cereal aphids as virus vectors in North America. *Proceedings of the Royal Society of London* B 321, 421–446.

Irwin, M.E. and Thresh, J.M. (1990) Epidemiology of barley yellow dwarf: a study in ecological complexity. *Annual Review of Phytopathology* 28, 393–424.

Irwin, M.E., Ruesink, W.G., Isard, S.A. and Kampmeier, G.E. (2000) Mitigating epidemics caused by non-persistently transmitted aphid-borne viruses: the role of the pliant environment. *Virus Research* 71, 185–211.

Isard, S.A. and Gage, S.H. (2001) *Flow of Life in the Atmosphere: An Airscape Approach to Understanding Invasive Organisms.* Michigan State University Press, East Lansing, 240 pp.

Isard, S.A. and Irwin, M.E. (1993) A strategy for studying the long-distance aerial movement of insects. *Journal of Agricultural Entomology* 10, 283–297.

Isard, S.A. and Irwin, M.E. (1996) Formulating and evaluating hypotheses on the ascent phase of aphid movement and dispersal. In: *Proceedings of the Twelfth Conference on Biometeorology and Aerobiology.* American Meteorological Society, Boston, pp. 430–433.

Isard, S.A., Irwin, M.E. and Hollinger, S.E. (1990) Vertical distribution of aphids (Homoptera: Aphididae) in the planetary boundary layer. *Environmental Entomology* 19, 1473–1484.

Isard, S.A., Irwin, M.E., Carter, M. and Holtzer, T.O. (1994) Temperature stratification and insect layer concentrations: a preliminary analysis of atmospheric measurements and concurrent aerial insect collections from northeastern Colorado and East Central Illinois. In: Preprint volume of the *21st Conference on Agricultural and Forest Meteorology and the 11th Conference on Biometeorology and Aerobiology, March 1994, San Diego.* The American Meteorological Society, Boston, pp. 407–410.

Jensen, R.E. and Wallin, J.R. (1965) *Weather and Aphids: A Review.* United States Department of Commerce Tech. Note, 5-AGMET-1, 19 pp.

Johnson, C.G. (1957a) The vertical distribution of aphids in the air and the temperature lapse rate. *Quarterly Journal of the Royal Meteorological Society* 83, 194–201.

Johnson, C.G. (1957b) The distribution of insects in the air and the empirical relation of density to height. *Journal of Animal Ecology* 26, 479–494.

Johnson, C.G. (1969) *Migration and Dispersal of Insects by Flight.* Methuen, London, 763 pp.

Johnson, C.G. and Taylor, L.R. (1957) Periodism and energy summation with special reference to flight rhythms in aphids. *Journal of Experimental Biology* 34, 209–221.

Kendall, D.A., Brain, P. and Chinn, N.E. (1992) A simulation model of the epidemiology of barley yellow dwarf virus in winter sown cereals and its application to forecasting. *Journal of Applied Ecology* 29, 414–426.

Kennedy, J.S. (1985) Migration, behavioral and ecological. In: Rankin, M.A. (ed.) *Migration: Mechanisms and Adaptive Significance. Contributions to Marine Science* 27 (supplement), pp. 5–26.

Kennedy, J.S. (1990) Behavioural post-inhibitory rebound in aphids taking flight after exposure to wind. *Animal Behaviour* 39, 1078–1088.

Kennedy, J.S. and Booth, C.O. (1963) Free flight of aphids in the laboratory. *Journal of Experimental Biology* 40, 67–85.

Kennedy, J.S., Booth, C.O. and Kershaw, W.J.S. (1961) Host finding by aphids in the field. III. Visual attraction. *Annals of Applied Biology* 49, 1–21.

Kenny, G.J. and Chapman, R.B. (1988) Effects of an intercrop on the insect pests, yield, and quality of cabbage. *New Zealand Journal of Experimental Agriculture* 16, 67–72.

Kidd, N.A.C. (1977) The influence of population density on the flight behavior of the lime aphid, *Eucallipterus tiliae. Entomologia Experimentalis et Applicata* 22, 251–261.

Kieckhefer, R.W., Lytle, W.F. and Spuhler, W. (1974) Spring movement of cereal aphids into South Dakota. *Environmental Entomology* 3, 347–350.

Klingauf, F. (1967) Abwehr- und Meidereaktionen von Blattläusen (Aphididae) bei Bedrohung durch Räuber und Parasiten. *Zeitschrift für Angewandte Entomologie* 59, 277–317.

Kobayashi, M. and Ishikawa, H. (1993) Breakdown of indirect flight muscles of alate aphids (*Acyrthosiphon pisum*) in relation to their flight, feeding and reproductive behavior. *Journal of Insect Physiology* 39, 549–554.

Kobayashi, M. and Ishikawa, H. (1994) Involvement of juvenile hormone and ubiquitin-dependent proteolysis in flight muscle breakdown of alate aphid (*Acyrthosiphon pisum*). *Journal of Insect Physiology* 40, 107–111.

Kring, J.B. (1972) Flight behavior of aphids. *Annual Review of Entomology* 17, 461–492.

Kring, J.B. and Kring, T.J. (1991) Aphid flight behavior. In: Peters, D.C., Webster, J.A. and Chlouber, C.S. (eds) *Aphid–Plant Interactions: Populations to Molecules*. Agricultural Experiment Station, Stillwater, pp. 203–214.

Kuldova, J., Hrdy, I. and Wimmer, Z. (1998) Response of the hop aphid, *Phorodon humuli* (Homoptera: Aphididae), to the application of juvenile hormone analogue in field trials. *Crop Protection* 17, 213–218.

Kunert, G. and Weisser, W.W. (2003) The interplay between density- and trait-mediated effects in predator–prey interactions: a study in aphid wing polymorphism. *Oecologia* 135, 304–312.

Kunert, G., Otto, S., Röse, U.S.R., Gershenzon, J. and Weisser, W.W. (2005) Alarm pheromone mediates production of winged dispersal morphs in aphids. *Ecology Letters* 8, 596–603.

Kunert, G., Schmook-Ortlepp, K., Reissmann, U., Creutzburg, S. and Weisser, W.W. (2007) The influence of natural enemies on wing induction in *Aphis fabae* and *Megoura viciae* (Hemiptera: Aphididae). *Bulletin of Entomological Research* (in press).

Leather, S.R. and Walters, K.F.A. (1984) Spring migration of cereal aphids. *Zeitschrift für Angewandte Entomologie* 97, 431–437.

Levin, D.M. and Irwin, M.E. (1995) Barley yellow dwarf luteovirus effects on tethered flight duration, wingbeat frequency, and age of maiden flight in *Rhopalosiphum padi* (Homoptera: Aphididae). *Environmental Entomology* 24, 306–312.

Liquido, N.J. and Irwin, M.E. (1986) Longevity, fecundity, change in degree of gravidity and lipid content with adult age, and lipid utilisation during tethered flight of alates of the corn leaf aphid, *Rhopalosiphum maidis*. *Annals of Applied Biology* 108, 449–459.

Losey, J.E. and Denno, R.F. (1998a) Interspecific variation in the escape responses of aphids: effect on risk of predation from foliar-foraging and ground-foraging predators. *Oecologia* 115, 245–252.

Losey, J.E. and Denno, R.F. (1998b) The escape response of pea aphids to foliar-foraging predators: factors affecting dropping behaviour. *Ecological Entomology* 23, 53–61.

Loxdale, H.D and Lushai, G. (1999) Slaves of the environment: the movement of herbivorous insects in relation to their ecology and genotype. *Philosophical Transactions: Biological Sciences* 354, 1479–1495.

Loxdale, H.D., Hardie, J., Halbert, S., Foottit, R., Kidd, N.A.C. and Carter, C.I. (1993) The relative importance of short- and long-range movement of flying aphids. *Biological Reviews of the Cambridge Philosophical Society* 68, 291–311.

Mackenzie, A. and Dixon, A.F.G. (1991) An ecological perspective of host alternation in aphids (Homoptera: Aphidinea: Aphididae). *Entomologia Generalis* 16, 265–284.

Manitoba Potato Industry (2005) Manitoba Potato News website (http://web2.gov.mb.ca/agriculture/potato/index.php).

Mann, J.A., Tatchell, G.M., Dupuch, M.J., Harrington, R., Clark, S.J. and McCartney, H.A. (1995) Movement of apterous *Sitobion avenae* (Homoptera: Aphididae) in response to leaf disturbances caused by wind and rain. *Annals of Applied Biology* 126, 417–427.

Marking, S. (2002) The soybean aphid blitz. *Soybean Digest, February 2002* (www.cornandsoybeandigest.com/ar/soybean_soybean_aphid_blitz_2/).

Martin, B., Rahbé, Y. and Fereres, A. (2003) Blockage of stylet tips as the mechanism of resistance to virus transmission by *Aphis gossypii* in melon lines bearing the Vat gene. *Annals of Applied Biology*, 142, 245–250.

Masterman, A.J., Foster, G.N., Holmes, S.J. and Harrington, R. (1996) The use of the Lamb daily weather types and the indices of progressiveness, southerliness and cyclonicity to investigate the autumn migration of *Rhopalosiphum padi*. *Journal of Applied Ecology* 33, 23–30.

McAllister, M.K. and Roitberg, B.D. (1987) Adaptive suicidal behaviour in pea aphids. *Nature, London* 328, 797–799.

McAllister, M.K., Roitberg, B.D. and Weldon, K.L. (1990) Adaptive suicide in pea aphids: decisions are cost sensitive. *Animal Behaviour* 40, 167–175.

McConnell, J.A. and Kring, T.J. (1990) Predation and dislodgement of *Schizaphis graminum* (Homoptera: Aphididae) by adult *Coccinella septempunctata* (Coleoptera: Coccinellidae). *Environmental Entomology* 19, 1798–1802.

Medler, J.T. and Ghosh, A.K. (1969) Keys to species of alate aphids collected by suction, wind, and yellow-pan water traps in the North Central States, Oklahoma and Texas. *University of Wisconsin Research Bulletin No. 277*, 99 pp.

Minoretti, N. and Weisser, W.W. (2000) The impact of individual ladybirds (*Coccinella septempunctata*, Coleoptera: Coccinellidae) on aphid colonies. *European Journal of Entomology* 97, 475–479.

Moericke, V. (1955) Über die Lebensgewohnheiten der geflügelten Blattläuse (Aphidina) unter besonderer Berücksichtigung des Verhaltens beim Landen. *Zeitschrift für Angewandte Entomologie* 37, 29–91.

Mondor, E.B., Tremblay, M.N. and Lindroth, R.L. (2004) Transgenerational phenotypic plasticity under future atmospheric conditions. *Ecology Letters* 7, 941–946.

Mondor, E.B., Rosenheim, J.A. and Addicott, J.F. (2005) Predator-induced phenotypic plasticity in the cotton aphid. *Oecologia* 142, 104–108.

Montgomery, M.E. and Nault, L.R. (1977a) Aphid alarm pheromone: dispersion of *Hyadaphis erysimi* and *Myzus persicae*. *Annals of the Entomological Society of America* 70, 1153–1156.

Montgomery, M.E. and Nault, L.R. (1977b) Comparative response of aphids to the alarm pheromone, (*E*)-beta-farnesene. *Entomologia Experimentalis et Applicata* 22, 236–242.

Montgomery, M.E. and Nault, L.R. (1978) Effects of age and wing polymorphism on the sensitivity of *Myzus persicae* to alarm pheromone. *Annals of the Entomological Society of America* 71, 788–790.

Müller, C.B., Williams, I.S. and Hardie, J. (2001) The role of nutrition, crowding and interspecific interactions in the development of winged aphids. *Ecological Entomology* 26, 330–340.

Nakamuta, K. (1991) Aphid alarm pheromone component, (*E*)-ß-farnesene, and local search by a predatory lady beetle, *Coccinella septempunctata bruckii* Mulsant (Coleoptera: Coccinellidae). *Applied Entomology and Zoology* 26, 1–7.

Nakata, T. (1995) Seasonal population prevalence of aphids with special reference to the production of alatoid nymphs in a potato field in Hokkaido, Japan. *Applied Entomology and Zoology* 30, 121–127.

Nault, L.R. (1997) Arthropod transmission of plant viruses: a new synthesis. *Annals of the Entomological Society of America* 90, 521–541.

Niku, B. (1972) Der Einfluss räuberischer Feinde auf die Ausbreitung von Erbsenläusen (*Acyrthosiphon pisum* Harr.) im Bestand. *Zeitschrift für Angewandte Entomologie* 70, 359–364.

Niku, B. (1975) Verhalten und Fruchtbarkeit ungeflügelter Erbsenläuse (*Acyrthosiphon pisum*) nach einer Fallreaktion. *Entomologia Experimentalis et Applicata* 18, 17–30.

Nottingham, S.F., Hardie, J. and Tatchell, G.M. (1991) Flight behaviour of the bird cherry aphid, *Rhopalosiphum padi*. *Physiological Entomology* 16, 223–229.

Olson, B.D. (2001) Crop disease and yield loss. In: Peterson, R.K.D. and Higley, L.G. (eds) *Biotic Stress and Yield Loss*. CRC, Boca Raton, pp. 185–203.

Parry, W.H. (1978) A reappraisal of flight regulation in the green spruce aphid, *Elatobium abietinum*. *Annals of Applied Biology* 89, 9–14.

Pedgley, D.E. (1993) Managing migratory insect pests – a review. *International Journal of Pest Management* 39, 3–12.

Phelan, P.L., Montgomery, M.E. and Nault, L.R. (1976) Orientation and locomotion of apterous aphids dislodged from their hosts by alarm pheromone. *Annals of the Entomological Society of America* 69, 1153–1156.

Powell, G., Maniar, S.P., Pickett, J.A. and Hardie, J. (1999) Aphid responses to non-host epicuticular lipids. *Entomologia Experimentalis et Applicata* 91, 115–123.

Quinn, M.A., Halbert, S.E. and Williams, L., III (1991) Spatial and temporal changes in aphid (Homoptera: Aphididae) species assemblages collected with suction traps in Idaho. *Journal of Economic Entomology* 84, 1710–1716.

Raccah, B. and Irwin, M.E. (1988) Techniques for studying aphid-borne virus epidemiology. In: Kranz, J. and Rotem, J. (eds) *Experimental Techniques in Plant Disease Epidemiology*. Springer, Berlin, pp. 209–222.

Remaudière, G. and Remaudière, M. (1997) *Catalogue des Aphididae du monde: Homoptera Aphidoidea*. Institut National de la Recherche Agronomique, Paris, 473 pp.

Robert, Y. (1987) Aphids and their environment. In: Minks, A.K. and Harrewijn, P. (eds) *Aphids. Their Biology, Natural Enemies, and Control, Volume 2A*. Elsevier, Amsterdam, pp. 299–313.

Robert, Y., Brunel, E., Malet, Ph. and Bautrais, P. (1976) Distribution spatiale de pucerons ailés et de diptères dans une parcelle de bocage, en fonction des modifications climatiques provoqueés par les haies. In: Missonnier, J. (ed.) *Les Bocages: Histoire, Écologie, Économie*. Edifat-Opida, Échauffour, pp. 427–435.

Roitberg, B.D. and Myers, J.H. (1978a) Adaptation of alarm pheromone responses of the pea aphid *Acyrthosiphon pisum* (Harris). *Canadian Journal of Zoology* 56, 103–108.

Roitberg, B.D. and Myers, J.H. (1978b) Effect of adult Coccinellidae on the spread of a plant virus by an aphid. *Journal of Applied Ecology* 15, 775–779.

Roitberg, B.D. and Myers, J.H. (1979) Behavioural and physiological adaptations of pea aphids (Homoptera: Aphididae) to high ground temperatures and predator disturbance. *Canadian Entomologist* 111, 515–519.

Roitberg, B.D., Myers, J.H. and Frazer, B. (1979) The influence of predators on the movement of apterous pea aphids between plants. *Journal of Animal Ecology* 48, 111–122.

Rothamsted Insect Survey (2006) Online at www.rothamsted.bbsrc.ac.uk/insect-survey/.

Ruesink, W.G. and Irwin, M.E. (1986) Soybean mosaic virus epidemiology: a model and some implications. In: McLean, G.D., Garrett, R.G. and Ruesink, W.G. (eds) *Plant Virus Epidemics: Monitoring, Modelling, and Predicting Outbreaks*. Academic Press, Sydney, pp. 295–313.

Schlinger, E.I., van den Bosch, R., Dietrick, E.J. Hagen, K.S. and Holloway, J.K. (1959a) The colonization and establishment of imported parasites of the spotted alfalfa aphid in California. *Journal of Economic Entomology* 52, 136–141.

Schlinger, E.I., van den Bosch, R., Dietrick, E.J. and Hall, I.M. (1959b) The role of imported parasites in the biological control of the spotted alfalfa aphid in southern California in 1957. *Journal of Economic Entomology* 52, 142–154.

Schotzko, D.J. and Knudsen, G.R. (1992) Use of geostatistics to evaluate a spatial simulation of Russian wheat aphid (Homoptera: Aphididae) movement behavior on preferred and nonpreferred hosts. *Environmental Entomology* 21, 1271–1282.

Schultz, G.A., Irwin, M.E. and Goodman, R.M. (1985) Relationship of aphid (Homoptera: Aphididae) landing rates to the field spread of soybean mosaic virus. *Journal of Economic Entomology* 78, 143–147.

Scott, R.W. and Achtemeier, G.L. (1987) Estimating pathways of migrating insects carried in atmospheric winds. *Environmental Entomology* 16, 1244–1254.

Scottish Agricultural Science Agency (2005) Aphid monitoring programme (www.sasa.gov.uk/seed_potatoes/aphids/aphmon.cfm).

Sewell, G.H., Storch, R.H., Manzer, F.E. and Forsythe, H.Y., Jr. (1990) The relationship between coccinellids and aphids in the spread of potato leafroll virus in a greenhouse. *American Potato Journal* 67, 865–868.

Shah, P.A., Pickett, J.A. and Vandenberg, J.D. (1999) Responses of Russian wheat aphid (Homoptera: Aphididae) to aphid alarm pheromone. *Environmental Entomology* 28, 983–985.

Sigvald, R. (1987) Aphid migration and the importance of some aphid species as vectors of potato virus Y (PVY) in Sweden. *Potato Research* 30, 267–283.

Sigvald, R. (1992) Progress in aphid forecasting systems. *Netherlands Journal of Plant Pathology* 98 (supplement 2), 55–62.

Sloggett, J.J. and Weisser, W.W. (2002) Parasitoids induce production of the dispersal morph of the pea aphid, *Acyrthosiphon pisum. Oikos* 98, 323–333.

Smith, C.F., Eckel, R.W. and Lampert, E. (1992) A key to many of the common alate aphids of North Carolina (Aphididae: Homoptera). *North Carolina Agricultural Research Service Technical Bulletin No. 299*, 92 pp.

Smith, F.F. and Webb, R.E. (1969) Repelling aphids by reflective surfaces, a new approach to the control of insect-transmitted viruses. In: Maramorosch, K. (ed.) *Viruses, Vectors, and Vegetation*. Interscience, New York, pp. 631–639.

Smith, J.G. (1976) Influence of crop background on aphids and other phytophagous insects on Brussels sprouts. *Annals of Applied Biology* 83, 1–13.

Smyrnioudis, I.N., Harrington, R., Clark, S.J. and Katis, N. (2001) The effect of natural enemies on the spread of barley yellow dwarf virus (BYDV) by *Rhopalosiphum padi* (Hemiptera: Aphididae). *Bulletin of Entomological Research* 91, 301–306.

Stadler, B., Weisser, W.W. and Houston, A.I. (1994) Defense reactions in aphids – the influence of state and future reproductive success. *Journal of Animal Ecology* 63, 419–430.

Steiner, W.W.M., Voegtlin, D.J. and Irwin M.E. (1985) Genetic differentiation and its bearing on migration in North American populations of the corn leaf aphid, *Rhopalosiphum maidis* (Fitch) (Homoptera: Aphididae). *Annals of the Entomological Society of America* 78, 518–525.

Stinner, R.E., Barfield, C.S., Stimac, J.L. and Dohse, L. (1983) Dispersal and movement of insect pests. *Annual Review of Entomology* 28, 319–335.

Tamaki, G. (1973) Insect developmental inhibitors: effect of reduction and delay caused by juvenile hormone mimics on the production of winged migrants of *Myzus persicae* (Hemiptera: Aphididae) on peach trees. *Canadian Entomologist* 105, 761–765.

Tamaki, G., Halfhill, J.E. and Hathaway, D.O. (1970) Dispersal and reduction of colonies of pea aphids by *Aphidius smithi* (Hymenoptera: Aphidiidae). *Annals of the Entomological Society of America* 63, 973–980.

Taylor, L.R. (1958) Aphid dispersal and diurnal periodicity. *Proceedings of the Linnean Society of London* 169, 67–73.

Taylor, L.R. (1977) Aphid forecasting and the Rothamsted Insect Survey. *Journal of the Royal Agricultural Society of England* 138, 75–97.

Taylor, L.R. (1984) *A Handbook for Aphid Identification,* 2nd edn. Rothamsted Experiment Station, Harpenden, 171 pp.

Taylor, L.R., Woiwod, I.P. and Taylor, R.A.J. (1979) The migratory ambit of the hop aphid and its significance in aphid population dynamics. *Journal of Animal Ecology* 48, 955–972.

Thomas, P.E. (1983) Sources and dissemination of potato viruses in the Columbia Basin of the northwestern USA. *Plant Disease* 67, 744–747.

Voegtlin, D.J., Steiner, W.W.M. and Irwin, M.E. (1987) Searching for the source of the annual spring migrants of *Rhopalosiphum maidis* (Homoptera: Aphididae) in North America. In: Holman, J., Pelikan, J., Dixon, A.F.G. and Weismann, L. (eds) *Population Structure, Genetics and Taxonomy of Aphids and Thysanoptera. Proceedings of International Symposia, Smolenice, Czechoslovakia, September 1985.* SPB Academic Publishing, Amsterdam, pp. 120–133.

Voegtlin, D., Villalobos, W., Sánchez, M.V., Saborío, G. and Rivera, C. (2003) Guía de los áfidos alados de Costa Rica. A guide to the winged aphids of Costa Rica. *Revista de Biología Tropical* 51 (Supplement 2), 228 pp.

Wallin, J.R. and Loonan, D.V. (1971) Low-level jet winds, aphid vectors, local weather, and barley yellow dwarf virus outbreaks. *Phytopathology* 61, 1068–1070.

Wallin, J.R., Peters, D. and Johnson, L.C. (1967) Low-level jet winds, early cereal aphid and barley yellow dwarf detection in Iowa. *Plant Disease Reporter* 51, 527–530.

Walters, K.F.A. and Dixon, A.F.G. (1984) The effect of temperature and wind on the flight activity of cereal aphids. *Annals of Applied Biology* 104, 17–26.

Ward, S.A., Leather, S.R., Pickup, J. and Harrington, R. (1998) Mortality during dispersal and the cost of host-specificity in parasites: how many aphids find hosts? *Journal of Animal Ecology* 67, 673–773.

Weber, C.A., Godfrey, L.D. and Mauk, P.A. (1996) Effects of parasitism by *Lysiphlebus testaceipes* (Hymenoptera: Aphidiidae) on transmission of beet yellows closterovirus by bean aphid (Homoptera: Aphididae). *Journal of Economic Entomology* 89, 1431–1437.

Weisser, W.W. (1995) Within-patch foraging behaviour of the aphid parasitoid *Aphidius funebris* – plant architecture, host behaviour, and individual variation. *Entomologia Experimentalis et Applicata* 76, 133–141.

Weisser, W.W. (2001) Predation and the evolution of dispersal. In: Woiwod, I.P., Reynolds, D.R. and Thomas, C.D. (eds) *Insect Movement: Mechanisms and Consequences. Proceedings of the Royal Entomological Society Symposium No. 20.* CAB International, Wallingford, pp. 261–280.

Weisser, W.W. and Sloggett, J.J. (2004) A general mechanism for predator- and parasitoid-induced dispersal in the pea aphid, *Acyrthosiphon pisum* (Harris). In: Simon, J.-C., Dedrvyer, C.-A., Rispe, C. and Hullé, M. (eds) *Aphids in a New Millennium.* INRA, Paris, pp. 79–85.

Weisser, W.W., Braendle, C. and Minoretti, N. (1999) Predator-induced morphological shift in the pea aphid. *Proceedings of the Royal Society of London B* 266, 1175–1182.

Wenks, P. (1981) Bionomics of adult blackflies. In: Laird, M. (ed.) *Blackflies: The Future for Biological Methods in Integrated Control.* Academic Press, New York, pp. 259–276.

Western Australia Department of Agriculture (2006a) Barley yellow dwarf virus and cereal aphid forecast (www.agric.wa.gov.au/pls/portal30/docs/FOLDER/IKMP/PW/PH/DIS/CER/bydv-forecast.htm).

Western Australia Department of Agriculture (2006b) CMV forecast for the current season (www.agric.wa.gov.au/pls/portal30/docs/FOLDER/IKMP/PW/PH/DIS/LP/cmv-cmvforecast.htm).

White, D.F. and Lamb, K.P. (1968) Effect of a synthetic juvenile hormone on adult cabbage aphids and their progeny. *Journal of Insect Physiology* 14, 395–402.

White, T.C.R. (1970) Airborne arthropods collected in South Australia with a drogue-net towed by a light aircraft. *Pacific Insects* 12, 251–259.

White, T.C.R. (1974) Semi-quantitative sampling of terrestrial arthropods occurring in the air over South Australia. *Pacific Insects* 16, 1–10.

Wiktelius, S. (1981) Diurnal flight periodicities and temperature thresholds for flight for different migrant forms of *Rhopalosiphum padi* L. (Hom., Aphididae). *Zeitschrift für Angewandte Entomologie* 92, 449–457.

Wohlers, P. (1980) Die Fluchtaktion der Erbenlaus *Acyrthosiphon pisum* ausgelöst durch Alarmpheromon und zusätzliche Reize. *Entomologia Experimentalis et Applicata,* 27, 156–168.

Wohlers, P. (1981) Effects of the alarm pheromone (*E*)-β-farnesene on dispersal behaviour of the pea aphid *Acyrthosiphon pisum. Entomologia Experimentalis et Applicata* 29, 117–124.

Worner, S.P., Tatchell, G.M. and Woiwod, I.P. (1995) Predicting spring migration of the damson–hop aphid *Phorodon humuli* (Homoptera: Aphididae) from historical records of host-plant flowering phenology and weather. *Journal of Applied Ecology* 32, 17–28.

Wright, L.C., Allison, D., Pike, K.S. and Cone, W.W. (1995) Seasonal occurrence of alate hop aphids (Homoptera: Aphididae) in Washington State. *Journal of Agricultural Entomology* 12, 9–20.

Zeyen, R.J., Stromberg, E.L. and Kuehnast, E.L. (1987) Long-range aphid transport hypothesis for maize dwarf mosaic virus: history and distribution in Minnesota, USA. *Annals of Applied Biology* 111, 325–336.

Zhang, H. (2002) Induced walking behavior of *Rhopalosiphum padi* L. (Homoptera: Aphididae). PhD thesis, University of Illinois, Urbana-Champaign, USA.

Zúñiga, E. (1985) Efecto de la lluvia en la abundancia de afidos y afidos momificados en trigo (Homoptera: Aphidae). *Revista Chilena de Entomología* 12, 205–208.

8 Predators, Parasitoids and Pathogens

Wolfgang Völkl[1], Manfred Mackauer[2], Judith K. Pell[3]
and Jacques Brodeur[4]

[1]*Department of Animal Ecology, University of Bayreuth, 95440 Bayreuth, Germany;*
[2]*Department of Biological Sciences, Simon Fraser University, Burnaby, B.C.,
V5A 1S6, Canada;* [3]*Plant and Invertebrate Ecology Division, Rothamsted Research,
Harpenden, Herts, AL5 2JQ, UK;* [4]*Département de Phytologie, Université
de Laval, Sainte-Foy, Québec, G1K 7P4, Canada*

Introduction

Aphids are found in most terrestrial habitats. They are commonly attacked by predators, parasitoids, and pathogens, often collectively termed Aphidophaga. Predators kill their prey by feeding on it. In some families of predatory insects (e.g. ladybirds – Coccinellidae), both the larvae and the adults are predaceous on aphids, whereas in other families (e.g. hover flies – Syrphidae, lacewings – Chrysopoidea, and midges – Itonididae), only the larvae are predaceous. Among insect parasitoids, all species in the braconid subfamily Aphidiinae and some genera in the family Aphelinidae develop as endoparasitoids of aphids, with one larva completing development in each host. At the end of larval development, the host is killed, and the parasitoid pupates within or below the hardened cuticle of its host (the mummy). The adult wasps are free-living; aphidiines feed on honeydew and extrafloral nectaries, whereas aphelinid females are predaceous and feed on the haemolymph of stung aphids. Some species of fungi are entomopathogenic, infecting aphids through the cuticle, eventually killing the host. The colour plates section in this volume illustrate life stages of the most important taxa of natural enemies.

Because of their importance in biological control, the taxonomy, systematics, and biology of natural enemies of aphids have been the subject of numerous studies, books, and reviews (e.g. Mackauer and Starý, 1967; Hagen and van den Bosch, 1968; Starý, 1970; Hodek, 1973; Canard *et al.*, 1984; Mackauer and Chow, 1986; Latgé and Papierok, 1988; Minks and Harrewijn, 1988; Gilbert, 1993; Majerus, 1994; Hodek and Honek, 1996; Dixon, 2000; Pell *et al.*, 2001).

In general, the impact of a natural enemy, and hence its potential contribution to a reduction in pest damage, depends on several factors. Although many aphid species are both widespread and relatively common, their density can vary over space and time. To contribute to control, a natural enemy must be effective in locating its target pest. A variety of sensory cues emanating from the aphid's host plant or the aphid itself are used to locate hosts or prey over a wide range of foraging distances. Once an aphid colony has been found, the potential voracity of a predator (or the potential fecundity of a parasitoid) plays an important role in determining the degree to which colonies will be exploited. Patterns of resource use are also influenced by population structure, searching behaviour, and degree of specialization.

Although specialized natural enemies are generally better than polyphagous ones at finding aphid colonies, and hence are considered more effective for control, polyphagous parasitoids and predators can also have a significant impact on aphid populations. For aphid pathogens, host location is largely passive; however, parameters affecting persistence, transmission, and dispersal influence the potential of pathogens to infect suitable hosts.

This chapter is organized in four sections, with the first two sections dealing respectively with predators and parasitoids. The focus is on foraging behaviour and on cues triggering successful search. Laboratory studies suggest that many natural enemies of aphids have the potential to decimate the host or prey population. Because of low searching success, however, they often fail to do so in the field, or fail to do so in a predictable manner. The third section on microbial agents focuses on parameters affecting the epizootiology, i.e. the potential of a pathogen to cause a population regulating epidemic or epizootic. We discuss also the relationship between host or prey specificity and resource utilization in terms of enemy effectiveness. In the fourth section, we consider mutualistic relationships and intraguild predation. Several recent studies have shown that ant–aphid mutualism and intraguild predation may limit the effectiveness of Aphidophaga as biocontrol agents in the field.

Predators

Coccinellidae (ladybird beetles)

General biology

Ladybirds (Coleoptera: Coccinellidae) have a cosmopolitan distribution, whose members are fungivorous, phytophagous, or entomophagous, the latter ones mainly developing on a variety of sternorrhynchan prey (Majerus, 1994; Hodek and Honek, 1996). Most aphidophagous coccinellids belong to the subfamilies Coccinellinae and Scymninae, but there are also aphidophagous species in other subfamilies. Ladybirds are probably the best-studied aphid predators, due to their considerable visibility and economic importance in a variety of crops. Both larvae and adults feed on the same type of prey species and occur in identical habitats (Majerus, 1994; Hodek and Honek, 1996; Dixon, 2000).

The lifetime fecundity varies greatly between species and may range from slightly more than 100 to more than 1500 eggs per female (Kawauchi, 1991). Eggs are usually laid in clusters (average: 11–30 eggs, depending on the species), but some species of the subfamily Scymninae may deposit eggs singly (for an overview, see Hodek and Honek, 1996). Developmental time varies greatly between species and is influenced by temperature, the amount of food consumed, and prey species (e.g. Obrycki and Orr, 1990; Majerus, 1994; Hodek and Honek, 1996). In *Cheilomenes sexmaculatus*, for example, larval development time varied between 5.4 and 12.9 days in relation to food regime (Ng, 1991). In *Adalia bipunctata* (2-spot ladybird), egg to adult development varied between 18.8 and 27.7 days, depending on prey species (Olszak, 1988).

In temperate regions, most species are either univoltine or bivoltine, while multivoltinism is more common in tropical regions (Hodek, 1973; Hodek and Honek, 1996). Most species in temperate regions hibernate as adults at selected sites, where extremely high ladybird numbers may be found. This enables them to respond very quickly and precisely to spatial variation in aphid availability in the following spring. Often, coccinellids hibernate in packed masses under and between rocks above the snow line on mountains near valleys with agriculture (Hagen, 1962). They reach these sites by orienting their flight towards the silhouette of upstanding objects (a behaviour termed 'hypsotaxis'). At such low temperatures, the beetles can maintain their metabolism at a very low level, and pathogens that have infected them or might do so cannot multiply.

Adult foraging and oviposition site selection

Coccinellids usually search their environment for plants with aphid prey randomly rather than systematically, as demonstrated by their frequent returns to the same sites on a plant (Banks, 1957; Dixon, 2000). In long-distance foraging, neither visual nor olfactory cues may be important for the location of an aphid colony (Hodek and Honek, 1996; Dixon, 2000, but see Ninkovic *et al.*, 2001 for contrasting evidence). At short-range foraging, however, there is a variety of cues that may influence foraging success. Upon arrival on an aphid-infested plant, plant structures may provide the first information. Adults search primarily along prominent structures such as twigs, pronounced leaf veins, or leaf edges, while trichomes or waxes on the surface may negatively influence coccinellid movements (e.g. Bänsch, 1964; Shah, 1982; Ferran and Deconchat, 1992; Vohland, 1996; Eigenbrode and Kabalo, 1999; White and Eigenbrode, 2000). Also, plant size and plant architecture may have a significant influence on adult foraging (e.g. Kareiva and Sahakian, 1990; Frazer and MacGregor, 1994). Kareiva (1990) showed that *Coccinella septempunctata* (7-spot ladybird) searched at a greater velocity on golden rod (*Solidago virgaurea*) than on the more structurally complex plants, broad bean (*Vicia faba*) and pea (*Pisum sativum*). On golden rod, they had greater foraging success and the greatest rate of aggregation. The influence of plant morphology was tested experimentally with two pea varieties, a normal one and a leafless one: beetles had greater foraging success on leafless plants than on plants with normal leaves (Kareiva and Perry, 1989; Kareiva, 1990; Kareiva and Sahakian, 1990; Messina and Hanks, 1998).

Adult ladybirds may forage in 'swarms' under certain circumstances. This unusual behaviour, which is not yet fully understood, may occur when many individuals emerge from pupae in an area where aphids have become scarce or are absent (Kareiva and Odell, 1987; Majerus, 1994).

Once foraging adults have encountered a prey item, or at least honeydew as an olfactory cue (Carter and Dixon, 1984a), they switch to an intensive 'area-restricted' search (Banks, 1957; Nakamuta, 1985), the duration of which depends on the adults' degree of satiation (Dixon, 1959; Carter and Dixon, 1984b). The timing of transition between extensive and this intensive search shows considerable individual variation, allowing a high degree of adaptation to short-term fluctuations in prey availability (Ferran *et al.*, 1994). During 'area-restricted' search, visual cues may play an important role in the successful capture of prey items.

When searching for oviposition sites, females respond to the amount and quality of the prey they encounter. Theoretically, ladybirds should prefer those aphid colonies for oviposition that contain a high proportion of nymphs and which are still growing in numbers (Kindlmann and Dixon, 1993; Dixon, 1997, 2000). In this way, adults ensure that aphids will not become scarce before the larvae complete their development. For example, *A. bipunctata* tends to lay eggs well before peak aphid abundance (Dixon, 2000), thus ensuring that larvae are very likely to feed in a growing aphid colony. In contrast, ladybirds seem to avoid laying eggs in colonies with a high risk of short-term extinction, such as colonies with an accumulation of honeydew (Johki *et al.*, 1988; Dixon, 2000). However, a certain threshold aphid density seems to be necessary to elicit adult oviposition behaviour (Wratten, 1973; Mills, 1979; Honek, 1980). In addition, the presence of potential competitors may negatively affect oviposition decisions (Dixon, 2000). Females of *A. bipunctata* responded to the presence of conspecific fourth-instar larvae by laying a reduced number of eggs, although conspecific egg batches or pupae did not have an influence on the number of eggs laid (Hemptinne and Dixon, 1991; Hemptinne *et al.*, 1992). Also, conspecific larval tracks reduce egg laying (Doumbia *et al.*, 1998). By this means, ladybirds may avoid cannibalism, which is a common feature in all aphidophagous species. Majerus (1994), Hodek and Honek (1996), and Dixon (2000) extensively discuss cannibalism in ladybirds.

Larval foraging and feeding

Larval foraging behaviour is influenced mainly by four factors that indirectly trigger movements. First, coccinellid larvae show a negative geotaxis and move to the top of host plants (Kesten, 1969; Ng, 1991). Thus, aphid colonies located at the tip of a host plant or at leaf edges are more likely to be found than colonies located elsewhere on the host plant. Second, as shown for adults, plant size and plant architecture have an enormous effect on foraging (Kareiva, 1990; Grevstad and Klepetka, 1992). In general, ladybird larvae forage more successfully in structurally simple environments, where encounters are much more likely than in a diversified environment, with characteristically greater densities of edge structures (e.g. leaf edges, exposed leaf veins) (Fig. 8.1). Third, the structure of the leaf surface is another important aspect of plant morphology. For example, long hairs, wax covers, or sticky glandular secretions may hinder larval movement; especially in first-instar larvae (Belcher and Thurston, 1982; Shah, 1982; Obrycki and Tauber, 1984; Heinz and Parrella, 1994; Eigenbrode *et al.*, 1998). Fourth, the number of encounters with prey has a significant impact on larval foraging. Since foraging activities are widely triggered by hunger (Dixon, 2000), foraging larvae alter their search tactics after encountering an aphid colony. They switch from an extensive search, characterized by rapid movement and few turns, to an intensive 'area-restricted' search, where they forage with a reduced speed and frequently turn (Carter and Dixon,

1982; Nakamuta, 1985; Ettifouri and Ferran, 1992; Ferran and Dixon, 1993).

If not influenced by plant characteristics, ladybird larvae forage at random (Hodek and Honek, 1996). In first-instar larvae, this random search increases mortality when prey density is low. First-instar larvae remain feeding on the empty egg-shell for up to a day before they disperse (Banks, 1957). It is essential that they feed within the next 24–36 h or they starve to death. Cannibalism prolongs the time interval before prey has to be found. Once a prey item is encountered, survival probability increases sharply, since the young larva will remain within or close to the aphid colony. Older larvae, like adults, aggregate close to aphid colonies because more individuals will immigrate than will leave. However, if too many larvae feed on a small colony, prey becomes scarce, and larvae leave the area when the rate of prey encounters reduces.

Prey specificity

Most aphidophagous ladybird species are able to develop on a variety of aphid prey. There are, however, size constraints. Small species consume small aphids, and early instar ladybird larvae prefer early larval stages of their prey species. Also, some species such as *Platynaspis luteorubra* or *Scymnus* spp. have specialized mouthparts that

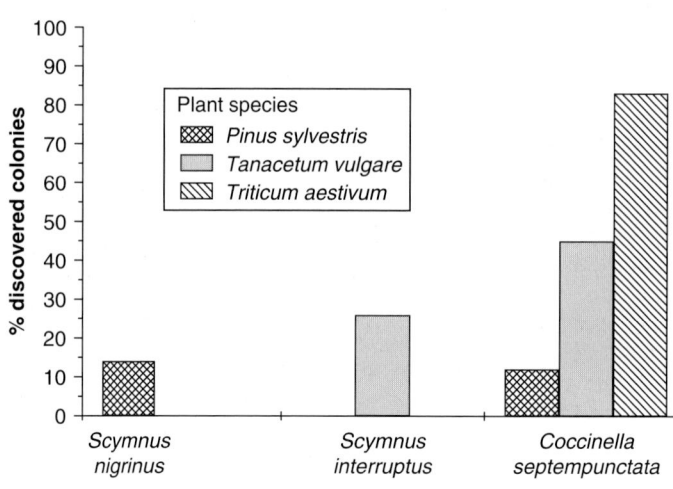

Fig. 8.1. Foraging success of fourth-instar larvae of three ladybird species on three different plant species with varying complexity and density of edge structures (redrawn from data in Völkl, 1990; Vohland, 1996; Völkl and Vohland, 1996).

may limit their prey size range (Ricci, 1979; Majerus, 1994; Hodek and Honek, 1996). Additionally, there may be ecological specializations arising from habitat specificity. For example, *Coccidula rufa* inhabits reeds and is, therefore, confined to feeding on aphids on those plant species. Various other ladybird species live mainly on pine and therefore consume pine aphids in nature, although they accept a wider range of prey in the laboratory (Majerus, 1994; Hodek and Honek, 1996; Sloggett and Majerus, 2000). Furthermore, several ladybird species need some 'essential prey species' (Hodek, 1973) to complete development with low mortality, while other 'accepted species' may serve as additional prey and an energy source, but they may not ensure complete ladybird development. Most aphid species of economic importance, however, seem to be in the category of 'essential prey'.

Some aphids may derive toxic secondary plant compounds (= allelochemicals) such as sambunigrin, quinolizidine alkaloids, or pyrolizidin alkaloids from the host plant, which make them unsuitable as prey even for polyphagous ladybirds (van Emden, 1995; Hodek and Honek, 1996). Such toxicity may interfere with biological control, as shown for *Toxoptera citricidus* (tropical citrus aphid) and *Macrosiphum albifrons* (lupin aphid), where for both species ladybirds are not economically important enemies. Not all plant allelochemicals, however, have a deleterious effect on ladybirds.

Some more recent studies have dealt with the effect of transgenic plants on ladybirds. Dogan *et al.* (1996) observed no negative effects if *Myzus persicae* (peach–potato aphid) reared on Bt-potato expressing a Cry 3A toxin (targeting the Colorado beetle) were fed to ladybirds. In contrast, some fitness parameters of *A. bipunctata* declined significantly if the beetles were fed with *M. persicae* reared on GNA potatoes (expressing snowdrop lectin – *Galanthus nivalis* agglutinin) (Birch *et al.*, 1999). However, a subsequent study showed there were no direct effects of GNA on *A. bipunctata*, but an indirect effect was caused by the fact that aphids reared on a GNA diet were a suboptimal food (Down *et al.*, 2000).

Ladybirds and aphid populations

Ladybirds are, due to their bright colour and their long residence times, the most conspicuous aphid predators in crops, and they have a long tradition in biological control. Since the famous success of the *Rodolia cardinalis* (vedalia beetle) in California (Caltagirone and Doutt, 1989), numerous, although frequently unsuccessful, attempts have been made to establish exotic ladybird populations (Gordon, 1985; Obrycki and Kring, 1998). However, it is worth mentioning that the arrival of exotic ladybirds may not necessarily be beneficial. At present, there is concern in the UK about the accidental introduction of *Harmonia axyridis* (harlequin ladybird) and the damage this might do to populations of indigenous ladybirds (Majerus and Roy, 2005).

There is controversy about how large is the impact of ladybirds in reducing aphid populations. On the one hand, many species of coccinellid share several characteristics of successful predators, such as high searching capacity, high voracity, appropriate food range, and the capability to develop on alternative food if aphids are scarce (Hodek and Honek, 1996). Thus, ladybirds are able to rapidly reduce high aphid densities, especially if aphid and ladybird peaks are coincident. On the other hand, it has been argued that a lack of synchronization and the restriction to one or two generations per year limit the efficiency of ladybirds in biological control (Hemptinne and Dixon, 1991; Kindlmann and Dixon, 1993; Dixon, 1997; Kindlmann *et al.*, Chapter 12 this volume). This shortcoming may be especially pronounced when aphid and coccinellid peak numbers do not coincide. Nevertheless, coccinellids are important predators of aphids, especially in cereals and in maize, where they contribute to a significant reduction in populations of economically important aphids. Also, some common and abundant ladybird species such as *C. septempunctata* or *A. bipunctata* are able to reduce the density of host-alternating pest aphid species on the primary host plants in hedges (Zwölfer *et al.*, 1984). Besides their impact on cereal aphid populations, ladybirds may play a role in reducing aphid populations on other crops such as potato or sugarbeet, in orchards, and even in tropical crops such as taro and banana.

Syrphidae (hover flies)

General biology

Hover flies (Diptera: Syrphidae) are one of the largest dipteran families. The larvae of about one third of the species, classified in the subfamily Syrphinae, are predators of Sternorrhyncha, usually aphids (Rotheray, 1989; Gilbert, 1993). Adults are active, diurnal flower visitors and feed on pollen and nectar (Rotheray, 1989; Gilbert, 1993).

Adult hover flies are characterized by a high fecundity. In *Episyrphus balteatus*, the average number of eggs per female during her lifetime ranges between 2000 and 4500 (Branquart and Hemptinne, 2000). Larvae hatch after 2–5 days and almost immediately start feeding on the aphids around them. Larvae feed by puncturing the aphid cuticle and sucking out the contents. There are three larval stages; the final instar pupates inside a puparium, usually located on the plant where the last prey was consumed. Developmental time varies between species and is influenced by prey species and temperature. In *E. balteatus*, it ranges between 6.8 and 13.8 days. The development of *Syrphus ribesii* takes on average slightly longer (between 9.6 and 11.0 days) but seems to be less variable (Sadeghi and Gilbert, 2000c).

Most species are univoltine in temperate regions, while multivoltinism is common in the tropics. Many species living in temperate regions enter diapause in the larval or pupal stage. *Episyrphus balteatus*, the most abundant species in Central Europe, hibernates in the adult stage either at specific sites in Central Europe or after migrating over long distances to Southern Europe (Krause and Poehling, 1996; Hart and Bale, 1998).

Oviposition site selection

Oviposition in hover flies is elicited by olfactory and visual cues. Females of *Eupeodes* *(Metasyrphus) corollae* and *E. balteatus* respond positively to stimuli originating from honeydew, and probably also to ones from aphid siphunculus secretion. Such stimuli may act both as long-distance kairomones and oviposition stimuli after the location of a plant with prey (Volk, 1964; Budenberg and Powell, 1992; Bargen *et al.*, 1998; Shonouda *et al.*, 1998; Sutherland *et al.*, 2001). Additionally, females of *E. corollae* respond to structural characters of plants, having a preference for vertical rather than horizontal surfaces and preferring darker to lighter strips (Sanders, 1983; Chambers, 1988). *Episyrphus balteatus* females also respond to leaf colour (Sutherland *et al.*, 2001). Eggs are often laid singly, either close to or within aphid colonies, although some species lay eggs in batches distant from the colony or even on uninfested plants (Chambers, 1988). In the latter case, young larvae may survive by cannibalizing conspecific eggs.

Aphid colony size has an important influence on the selection of the oviposition site. Since larvae have a rather limited dispersal ability and only forage occasionally between areas of the plant with aphids, the female's oviposition decision is of crucial importance to the offspring. Hover fly females seem to be able to adjust their egg number to aphid density, a behaviour that may be considered as adaptive since it secures both larval survival and optimizes the female's searching effort. Generally, the number of eggs deposited increases with aphid colony size (Dixon, 1959; Bargen *et al.*, 1998; Sutherland *et al.*, 2001). However, several authors have reported that females of many syrphid species prefer smaller aphid colonies, or aphid colonies with a high proportion of early aphid instars, for oviposition (e.g. Ito and Iwao, 1977; Kan and Sasakawa, 1986; Chambers, 1991; Hemptinne *et al.*, 1993). One explanation for this phenomenon may be that females avoid high aphid densities because colonies there are likely to be subject to an increased emigration of prey, rather than to continued colony growth, which would assure later hover fly larval survival (Kan, 1988a,b). Also, females of *Epistrophe nitidicollis* seem to prefer to lay fewer eggs in colonies of *Aphis fabae* (black bean aphid) on bean plants that are already predated by conspecific larvae (Hemptinne *et al.*, 1993). A similar pattern was found for *E. balteatus* on thistles (Fig. 8.2).

Ovipositing females also seem to discriminate between food types. Females of both *E. balteatus* and *S. ribesii* preferred *Acyrthosiphon pisum* (pea aphid) and *Macrosiphum rosae* (rose aphid) over a range of other aphid species, and *Microlophium carnosum* (nettle aphid) was least accepted by both species (Sadeghi and Gilbert, 2000a). These preferences remained unchanged by female age or by host deprivation (Sadeghi and Gilbert, 2000b).

Prey specificity

Field observations suggest that many hover fly species are apparently specialized to a range of prey species in the field, although they seem to be able to develop on a broader range of aphid species in the laboratory (e.g. *Dasysyrphus* spp., *Megasyrphus erraticus*, *Platycheirus parmatus* – see Goeldlin de Tiefenau, 1974; Laska, 1978). These species seem to be habitat specialists rather than prey specialists. There are, however, also truly specialized species, such as *Eupeodes (Metasyrphus) nielseni*, which consumes only a few aphid species on pine (Laska, 1978).

The most economically important syrphids accept a wide range of prey in the field (e.g. *E. balteatus*, *S. ribesii*, or *Eupeodes* spp. in cereals – Laska, 1978; Chambers, 1988; Gilbert and Owen, 1990; Tenhumberg and Poehling, 1995; Sadeghi and Gilbert, 2000a; or *Pseudodorus clavatus* in citrus – Michaud and Belliure, 2000). Even in polyphagous species, however, there may be an effect of different diet on developmental time and pupal weight, which in turn may influence adult fecundity (Růžička, 1976; Sadeghi and Gilbert, 2000a; Belliure and Michaud, 2001).

Larval foraging and feeding

The search of a syrphid larva is characterized by casting behaviour (Chambers, 1988), where the hind body remains attached to the substrate while the anterior end is extended forward and laterally, until a prey item is contacted. The current evidence concerning the importance of olfactory and gustatory cues for larval foraging behaviour is inconsistent. Bänsch (1964) and Chambers (1988) assumed that syrphid larvae recognized aphids only by tactile cues and that they did not use olfactory cues for prey location. In contrast, Bargen *et al.* (1998) found that first-instar larvae of *E. balteatus* exhibited a directed search over short distances and responded to volatile olfactory

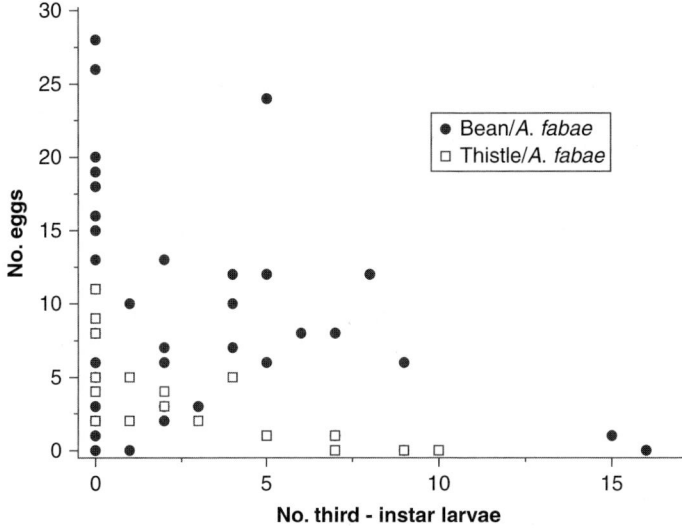

Fig. 8.2. Relationship between the number of eggs and third-instar larvae of hover flies: *Epistrophe nitidicollis* in *Aphis fabae* colonies on bean plants (*Vicia faba*) and *Episyrphus balteatus* in *A. fabae* colonies on creeping thistle (*Cirsium arvense*) (redrawn from data in Völkl, 1990; Hemptinne *et al.*, 1993).

cues from aphids, which were thought to be perceived by stretched papillae on the mouthparts (Gries, 1986). Such ability for rapid prey location would be especially important for first-instar larvae that had hatched from eggs which were not deposited directly in the aphid colony but somewhat distant on the same plant.

In the laboratory, hover fly larvae show great voracity. Single larvae of Syrphini may consume from 250 to 500 aphids, depending on aphid size and temperature (e.g. Chambers, 1988; Soleyman-Nezhadiyan and Laughlin, 1998). Larvae of Bacchini, which are considerably smaller, consume on average fewer prey (135–150 aphids – e.g. Chambers, 1988). In all studies, third-instar larvae consumed on average more than 80% of the aphids captured. Larvae of this age class were also most successful and efficient in capturing prey. When provided with a surplus number of aphids, hover fly larvae often kill more aphids than necessary for development; in this case, aphids are not consumed completely (Chambers, 1988). However, it is doubtful whether the voracity demonstrated in the laboratory is achieved in the field (Chambers, 1988, 1991).

Hover flies and aphid populations

Hover flies may be important antagonists of aphids, particularly in cereals (Ankersmit *et al.*, 1986; Chambers and Adams, 1986). The most striking effects of syrphids on aphid populations were observed when high levels of syrphid oviposition occurred early and large numbers of larvae hatched before aphid populations attained a rapid growth rate (Tenhumberg and Poehling, 1995). The effectiveness of syrphids, however, shows considerable variation between sites, and between years at the same sites (Poehling *et al.*, 1991). Besides climatic factors, the effectiveness of syrphids may be limited by immigration rates of adults into fields. Since adults are flower visitors, their foraging activity in crops generally can be enhanced by a continuous supply of flowers with easily available pollen sources such as Asteraceae and Umbelliferae in field margins (Ruppert and Molthan, 1991; Colley and

Luna, 2000; Morris and Li, 2000). Since cereal fields are usually characterized by a shortage of food for flower visitors, alternative agricultural practices that favour wild flowers (e.g. set-aside, herbicide-free buffer zones, conservation strips – van Emden, 2003; see also Wratten *et al.*, Chapter 16 this volume and Powell and Pell, Chapter 18 this volume) may lead to improved attraction of adult syrphids.

Besides their impact on cereal aphids, hover flies may play an important role in reducing aphid populations in citrus and apple orchards (e.g. Tracewski *et al.*, 1984; Michaud and Belliure, 2000), and even in tropical crops such as taro and banana (Stechmann and Völkl, 1990).

Chrysopidae and Hemerobiidae (lacewings)

General biology

Lacewings (Neuroptera: Chrysopidae and Hemerobiidae) are polyphagous predators feeding mainly on soft-bodied insects. Many species are frequently associated with aphids (e.g. Canard *et al.*, 1984; New, 1975, 1988). The members of two lacewing families (Chrysopidae and Hemerobiidae) are common predators of aphids; members of a third family (Coniopterygidae) have also been recorded using this type of prey, particularly on conifers.

The eggs of many lacewings have thin hyaline stalks (egg pedicels), although there are several species whose eggs do not possess such a structure. The pedicel protects both eggs and newly hatched larvae from cannibalism and from predation by other aphidophagous insects (Ruzicka, 1997). However, this protection is not always effective.

Fecundity in chrysopid species varies between 150 and 600 eggs per female and seems to be greater in hemerobiids, (between 600 and 1500) (New, 1975, 1988; Canard *et al.*, 1984; Chakrabarti *et al.*, 1991). Between 15 and 25°C, developmental time usually ranges from 6–12 days in hemerobiids and from 15–30 days in chrysopids (Canard *et al.*, 1984; New, 1988; Chakrabarti *et al.*, 1991; Liu and Chen, 2001; Michaud, 2001). In both families, there are three larval instars; pupation takes place in a globular cocoon, which is attached to the plant surface. In temperate regions, most lacewings seem to be multivoltine, although little is known about the biology of many species. Many lacewing species enter winter diapause (hibernation) either as adults (such as *Chrysoperla carnea* or *Micromus angulatus*) or as pupae or even prepupae (New, 1988).

Chrysopid and hemerobiid larvae are very active predators, as are most adults. Some adult chrysopids, such as *C. carnea*, feed on nectar, yeasts, pollen, and honeydew, which may attract them close to aphid colonies. Larval voracity depends on prey size and temperature. Chakrabarti *et al.* (1991) reported that larval *Cunctochrysa jubigensis* consumed an average of 23 *Brevicoryne brassicae* (cabbage aphid) daily at 23°C, while this rate decreased to 18 aphids per day at 20°C. However, the total number of aphids consumed during all larval instars was similar (245 ± 12 aphids *v.* 266 ± 14 aphids) since developmental time was extended at 20°C. Similarly, Michaud (2001) reported a relationship between food consumption and developmental time of larval *Chrysoperla plorabunda*, which consumed between 130 and 1650 *T. citricidus*, depending on aphid instar. *Chrysoperla carnea* larvae also responded with a varying voracity to a change in prey species, consuming on average more *Aphis gossypii* (cotton or melon aphid) (292) and *M. persicae* (273) than *Lipaphis pseudobrassicae* (mustard aphid) (146) (Liu and Chen, 2001). Larvae of the hemerobiid *Hemerobius pacificus* consumed an average of 41 *Therioaphis trifolii* (yellow clover aphid) per day (Neuenschwander *et al.*, 1975).

There are also some recent studies on the effect of transgenic plants on lacewings. Initial results suggested that *C. carnea* was sensitive to Cry 1Ab toxin (often included in Bt-plants) provided in artificial diet. However, no Bt-toxins have been detected in the phloem sap of transgenic maize plants, or in aphids feeding on transgenic maize. *Chrysoperla carnea* may, however, ingest Bt-toxins and be affected negatively when feeding on lepidopteran larvae as alternative prey on Bt-maize. A combined interaction of poor prey quality and Cry 1Ab toxin may then account for negative effects (Hilbeck *et al.*, 1998a,b; Dutton *et al.*, 2002).

Adult foraging and oviposition behaviour

Many chrysopid species are attracted to cues that are associated with the presence of aphids. In laboratory experiments, *C. carnea* responded positively to volatile breakdown products of the amino acid tryptophan, which is a common component of aphid honeydew (van Emden and Hagen, 1976), to aphid sex pheromone and to (*E*)-β-farnesene, the alarm pheromone of aphids (Zhu *et al.*, 1999). *Chrysopa cognata* showed a positive response to aphid sex pheromone but not to (*E*)-β-farnesene (Boo *et al.*, 1998). Plant volatiles, such as β-caryophyllene and 2-phenylethanol also may be attractive (Flint *et al.*, 1979; Zhu *et al.*, 1999). Alternatively, larvae of various *Chrysopa* species mark the substrate with an oviposition-deterring pheromone, which significantly reduces adult oviposition activity both intra- and interspecifically (Růžička, 1994, 1996, 1998).

After having selected suitable oviposition sites on a given plant, females of many neuropterans do not search for the direct vicinity of an aphid colony when ovipositing (Miermont and Canard, 1975; New, 1975; Coderre *et al.*, 1987; Fréchette and Coderre, 2000; Nakamura *et al.*, 2000). For example, although *Chrysopa oculata* laid eggs on maize plants only if there were aphids present, most eggs were found on leaves without aphid colonies. This means that newly hatched larvae have to seek intensively for prey (Coderre *et al.*, 1987). Similarly, random egg laying was observed in some other chrysopid species, such as *C. carnea*, whose adults feed on honeydew (Duelli, 1984, 1987). Also, the presence of alternative prey on an aphid-infested plant had no significant effect on egg distribution or on the distance between eggs and aphid prey (Fréchette and Coderre, 2000).

The first bottleneck in larval foraging is the interval between eclosion from the egg and the discovery of the first prey. During this period, many first-instar larvae may die of starvation and predation. Unfavourable plant structures such as dense trichomes may enhance dislodgement of small larvae from plants (Rosenheim *et al.*, 1999), thereby contributing to early larval mortality and making such plants less suitable for lacewing development.

Larval foraging

Larval foraging behaviour may be modified strongly by foliage density and plant architecture, which affects prey accessibility rather than influencing predator foraging (Canard and Duelli, 1984). Clark and Messina (1998) found that *C. carnea* was more effective in locating *Diuraphis noxia* (Russian wheat aphid) on Indian rice grass (*Oryzopis hymenoides*) (which is characterized by narrow linear leaves) than on crested wheat grass (*Agropyron desertorum*) with its flat, broad leaves (Fig. 8.3). This difference may explain the higher effectiveness of *C. carnea* in eliminating field-cage populations of *D. noxia* on rice grass in comparison to those on wheat grass (Messina *et al.*, 1995, 1997). Similarly, alternative prey deflected foraging larvae of *C. plorabunda* away from *D. noxia* when it was feeding on sites that were easily accessible (Bergeson and Messina, 1998). Larvae of *C. plorabunda* were also more effective against *D. noxia* on resistant wheat plants than on susceptible ones, suggesting that even a modest reduction in aphid population growth may produce synergistic effects for pest reduction by natural enemies (van Emden and Wearing, 1965; van Emden, 1986; Messina and Sørenson, 2001).

Lacewings and aphid populations

The role of Neuroptera in reducing aphid populations depends heavily on local conditions. Chrysopids have the potential to reduce aphid numbers significantly (New, 1988), but there are only a few studies dealing with their effectiveness on aphid populations. Chrysopids also have a considerable potential for manipulation, which makes them particularly successful in closed environments such as greenhouses. For field conditions, there are reports that releases of mass-produced Chrysopidae successfully reduced aphid populations (e.g. in apple orchards – Hagley, 1989), while only a slight or no additional effect of releases was found in other studies (Campbell, 1990; Grasswitz and Burts, 1995). Thus, further studies are needed to determine the conditions under which lacewings might act as biocontrol agents against aphid key pests.

Cecidomyiidae (predatory midges)

General biology

Within the dipteran family Cecidomyiidae, there are at least five predatory species of the genera *Aphidoletes* and *Monobremia*

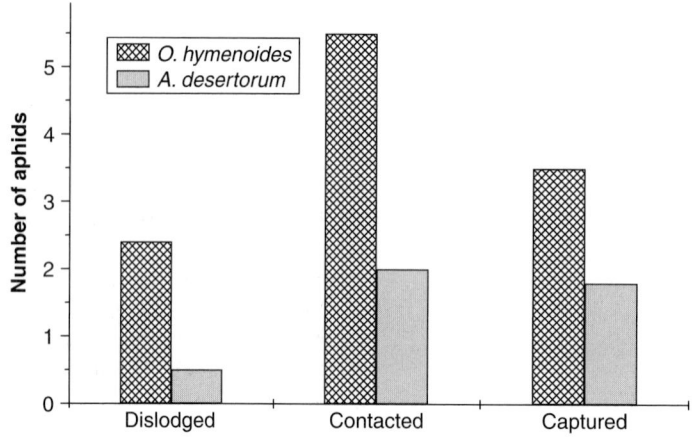

Fig. 8.3. Average number of aphids dislodged, contacted, or captured by *Chrysoperla carnea* on seedlings of crested wheat grass (*Agropyron desertorum*) and Indian rice grass (*Oryzopis hymenoides*) (redrawn after Clark and Messina, 1998).

whose larvae prey exclusively on aphids (Harris, 1973; Nijveldt, 1988), while the adults feed on nectar or honeydew. The best-known species is *Aphidoletes aphidimyza*, which is commonly used in biological control programmes (Markkula and Tiitanen, 1985; van Schelt and Mulder, 2000). Adult flies are nocturnal (Nijveldt, 1988; Kulp *et al.*, 1989). The average number of eggs laid per female ranges between 50 and 150; fecundity depends both on larval and adult nutrition of the midge and on the geographic origin (Havelka and Růžička, 1984; Havelka and Zemek, 1999). Larvae hatch after 2–4 days and almost immediately start sucking aphids that are feeding around them. There are three larval stages; the final instar pupates in soil. Developmental time varies with temperature and nutritional condition, ranging between approximately 30 and 55 days from egg to adult (Harris, 1973; Nijveld, 1988; Havelka and Zemek, 1999). In Central Europe, there are two or three generations per year (Harris, 1973). *Aphidoletes aphidimyza* hibernates in the pupal stage in the soil. Diapause induction in autumn is triggered by shorter day length and low temperature (Gilkeson and Hill, 1986).

Oviposition behaviour

Females lay single eggs or small clusters of eggs on foliage, usually within or close to the aphid colony. Females are able to discriminate among plant species or varieties and especially between infested and uninfested plants (e.g. El Titi, 1972; Mansour, 1975, 1976). The latter ability is crucial for *A. aphidimyza* survival, because neonate larvae cannot detect prey unless they are very close to them and will die from starvation if they are more than 63 mm distant from food (Wilbert, 1973). Furthermore, ovipositing females respond to morphological host plant characters such as leaf pubescence (Lucas and Brodeur, 1999). Evidence for a female response to varying aphid density is inconsistent. Usually, there is a positive correlation between clutch size and aphid density (El Titi, 1972, 1974; Wilbert, 1973; Nijveldt, 1988; Lucas and Brodeur, 1999). Additionally, Růžička and Havelka (1998) found that larvae of *A. aphidimyza* secrete a species-specific, oviposition-deterring pheromone so that females respond by laying significantly fewer eggs in colonies already attacked (Fig. 8.4). This finding may explain the reluctance of

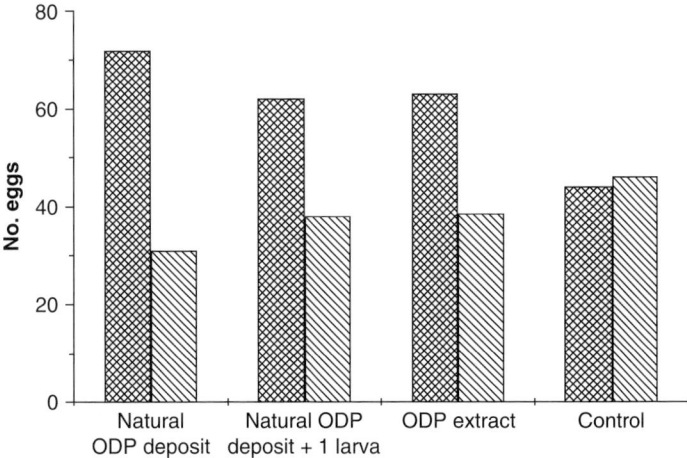

Fig. 8.4. Oviposition activity measured as the mean number of eggs laid per female of *Aphidoletes aphidimyza* on aphid-infested broad bean plants with (shaded bars) and without (cross-hatched bars) oviposition deterring pheromone (ODP). The ODP treatments were either the natural pheromone deposit left on the plant by 10 third-instar larvae during 4 h, or a water extract from deposits left by 50 larvae on glass over 4 h. Clean plants were used as the control (redrawn after Růžička and Havelka, 1998).

A. aphidimyza females to oviposit in large aphid colonies that are already colonized by conspecifics (Hafez, 1961; Havelka, 1982).

Prey specificity and larval feeding

Larvae of *A. aphidimyza* feed on a variety of aphids, mainly on herbaceous plants or deciduous trees (Harris, 1973; Nijveldt, 1988; Kulp *et al.*, 1989). The number of aphids killed during the larval stage varies considerably. One larva needs only a minimum of seven adult aphids to complete the life cycle, yet it may consume as many as 80, depending on aphid size but also on environmental conditions such as humidity (e.g. Barnes, 1929; George, 1957; Uygun, 1971; Nijveldt, 1988; Kulp *et al.*, 1989). Thus, larvae may kill many more aphids than they need for their development (Uygun, 1971; Havelka, 1982; Nijveldt, 1988), a fact that enhances their success in biological control programmes, especially in glasshouse systems.

Aphidoletes aphidimyza is a multivoltine species and its density in the field tends to increase during the vegetation period, larvae being most abundant from mid to late summer in areas with soil conditions favouring pupation. Furthermore, *A. aphidimyza* is rare in agricultural ecosystems where cultivation methods like ploughing in late autumn or early spring do not favour successful hibernation.

Other arthropod predators

Predatory bugs

ANTHOCORIDAE (FLOWER BUGS). Within the family Anthocoridae (Hemiptera: Heteroptera), members of the genera *Anthocoris* and *Orius* are important predators of aphids, especially on woody host plants. Anthocorids, or flower bugs, are mainly diurnal. Females lay their eggs singly just beneath the epidermis of the prey's host plant. Mean egg numbers of 50 per female are common, but there may be up to 200 (Hodgson and Aveling, 1988; Lattin, 1999). There are five nymphal stages. Their developmental time is markedly influenced by temperature, ranging from between 22 days at 20°C and

53 days at 14°C in *Anthocoris nemorum*. Other species, such as *Anthocoris nemoralis* and *Orius* spp., may develop even faster (Hodgson and Aveling, 1988). In *A. nemorum*, the number of generations per year depends on climate; the species is univoltine in Scotland, bivoltine in England, and has at least three generations in France. Anthocorid bugs usually hibernate as adults in leaf litter, under bark, or in hollow stems (Parker, 1975).

As with many other aphid predators, females of *A. nemorum* and *A. nemoralis* may use herbivore-induced volatiles for prey location (Dwumfour, 1992; Scutareanu *et al.*, 1997). This ability may be enhanced by associative learning (Drukker *et al.*, 2000) and may also lead to a concentration of ovipositions on prey-infested plants. During development, *Anthocoris* nymphs may consume between 60 and 240 aphids, depending on aphid size and temperature (Hodgson and Aveling, 1988).

Anthocorid bugs are very mobile and may respond rapidly to changes in prey population and aggregate in areas with high prey densities, especially when the prey is feeding in sheltered areas, like *Schizoneura ulmi* in the rolled leaves of Dutch elm (*Ulmus glabra*) or *Metopeurum fuscoviride* on the inflorescences of tansy (*Tanacetum vulgare*). This may often lead to a significant correlation between anthocorid numbers and prey populations (Hodgson and Aveling, 1988). In such situations, both nymphal and adult anthocorid bugs may contribute to a significant reduction in aphid numbers and help to increase biological control, especially in orchards.

OTHER BUGS. Nabid and mirid bugs (Heteroptera: Nabidae and Miridae) are also aphid predators. These, like anthocorids, benefit from a diversified landscape (Müller and Godfray, 1999) and may have some impact on cereal aphid populations.

Predatory bugs are considered as 'true' top predators that may interfere with other biocontrol agents. Therefore, they may disrupt biocontrol through intraguild predation (see later – 'Intraguild interactions, mutualistic ants and aphid symbionts').

Chamaemyiidae (aphid flies)

This is a small dipteran family of the Brachycera whose larvae feed exclusively on aphids (Sunderland, 1988). The adult flies are diurnal. Females lay single eggs close to or within an aphid colony, even colonies within folded leaves (Gaimari and Turner, 1997). Hatched first instars subsequently search actively for aphid colonies, where they usually attack aphids that have walked over them (Gaimari and Turner, 1997). Second- and third-instar larvae search more actively for prey. Puparia are usually formed within the aphid colony. Larvae and puparia can be distinguished from syrphid larvae and puparia by the two distal siphons. Most chamaemyiid flies seem to be specialized to particular habitats (e.g. conifer forests) and/or to a few aphid species (Tracewski *et al.*, 1984; Sunderland, 1988; Mizuno *et al.*, 1997). Aphid flies were found to be abundant predators, especially in New England apple orchards (Tracewski *et al.*, 1984) and in Florida citrus orchards (Michaud and Belliure, 2000), where they may play an important role in aphid reduction. *Leucopis* is perhaps the commonest genus.

Polyphagous predators

Aphids are attacked by a variety of polyphagous predators such as carabid beetles (Coleoptera: Carabidae), wolf spiders (Araneae: Lycosidae), and to a lesser extent, rove beetles (Coleoptera: Staphylinidae). These polyphagous predators are abundant predators of the ground layer (= epigeal predators) in many temperate agroecosystems. They consume a wide variety of crop pests including aphids (see Snyder and Wise, 1999 for further references), and there is growing evidence that suggests that these taxa can contribute significantly to a reduction of aphid densities (e.g. Sunderland, 1988; Holland and Thomas, 1997; Sunderland and Samu, 2000). Immigrating carabids may aggregate around aphid colonies (Monsrud and Toft, 1999), and their exclusion with barriers led to elevated densities of cereal aphids (Edwards *et al.*, 1979; Chiverton, 1986). Within the vegetation layer, web-making linyphiid spiders may also cause a high mortality

among aphids, especially in cereals and in orchards (e.g. Sunderland *et al.*, 1986; Wyss *et al.*, 1995; Samu *et al.*, 1996). The abundance and efficiency of generalist predators can be increased significantly by a diversification of cropping systems (Wratten and van Emden, 1995; Sunderland and Samu, 2000). Because of the ease of quantifying epigeal predators with pitfall traps, these beneficials have been used, perhaps more than their importance otherwise warrants (Madsen *et al.*, 2004), to study the effects of agricultural practices, insecticides, and habitat diversification on beneficial insects 'in general'.

The European earwig (Dermaptera, Forficulidae: *Forficula auricularia*) is common and sometimes an important predator of *Eriosoma lanigerum* (woolly apple aphid) in orchards. Earwigs may also reduce aphid numbers in cereal fields (Sunderland, 1988). The occurrence and efficacy of earwigs is highly dependent on the availability of a sufficient number of shelters (where this nocturnal species can hide during the day) within or in close proximity to the crop.

Parasitoids

General biology

Aphids (Aphidoidea or 'true' aphids) are commonly attacked by hymenopteran parasitoids. With some 50 described genera and over 600 species, the subfamily Aphidiinae (Hymenoptera: Braconidae) comprises the largest number of species of aphid parasitoids (Mackauer and Starý, 1967). No parasitoids are recorded for any species in the oviparous superfamily Phylloxeroidea, which suggests that aphid parasitization evolved after the division between Phylloxeroidea and Aphidoidea (Mackauer, 1965; Mackauer *et al.*, 1996). Among the Aphelinidae, all species of the genus *Aphelinus* and related genera (Starý, 1988a) and several species of *Encarsia* (Evans *et al.*, 1995) use aphids as hosts. In addition, several species of gall midges (Diptera: Cecidomyiidae) are parasitic on aphids; one species, *Endaphis gregaria*, is gregarious (Mackauer and Foottit, 1979). Except for some host records, only scant information is available on the biology of the aphid-parasitic *Encarsia* and *Endaphis* species.

Aphidiinae

All species in this subfamily are solitary endoparasitoids (Starý, 1970, 1988b; Mackauer and Chow, 1986). Females normally deposit a single egg in an aphid, although super-parasitism may occur when unparasitized hosts are scarce or not available (Mackauer, 1990). Supernumerary larvae are eliminated by contest in the early first instar, or by physiological suppression in later instars, so that only one larva per host completes development (Chow and Mackauer, 1984, 1986; Mackauer, 1990). After eclosion from the egg, the larva feeds first on the aphid's haemolymph (Couchman and King, 1977), but later feeds destructively on other tissues, thereby killing the host (Polaszek, 1986). The mature larva spins a cocoon either inside (e.g. species of *Ephedrus*, *Aphidius*, *Lysiphlebus*, *Pauesia*, and *Trioxys*) or below (species of *Praon* and *Dyscritulus*) the mummy.

Except for some thelytokous species, in which males are rare or absent (Nemec and Starý, 1985), the majority of aphidiine wasps are bisexual. Females can determine offspring sex by controlling sperm release, with unfertilized and fertilized eggs developing into sons and daughters, respectively (i.e. arrhenotoky) (Cook, 1993). Because daughters (as opposed to sons) may gain relatively more in fitness from increased size, mothers are expected to deposit fertilized eggs selectively in large hosts assumed to have higher quality than small hosts (Cloutier *et al.*, 1991). Most females collected in the field are mated (Mackauer, 1976), with average offspring sex ratios showing a moderate female bias (Singh and Pandey, 1997). Both partial sib-mating and local mate competition on the natal patch and off-patch matings among non-sibs are common features of aphidiine biology (Mackauer and Völkl, 2002).

Females mature eggs throughout their reproductive life. Eggs are relatively small and nutrient-poor (Le Ralec, 1991). A female may store between 200 and 400 mature eggs in her ovaries; however, eggs that are not laid cannot be resorbed (Völkl and Mackauer, 1990; Le Ralec, 1991). Lifetime fecundity is potentially very high, ranging between 300 and over 1800 eggs per female under optimal laboratory conditions when sufficient hosts are available (Kambhampati and Mackauer, 1989; Hågvar and Hofsvang, 1991a), but fecundity is probably much lower in the field where adults may not survive for more than 2–3 days (Mackauer, 1983). Adults feed on aphid honeydew and extrafloral nectaries. The majority of species have several generations per year. Exceptions are *Monoctonia pistaciaecola* and *Pseudopauesia prunicola*, which apparently are obligatorily monovoltine (Halme, 1986; Starý, 1988b). In temperate climates, diapause is induced by a variety of abiotic (e.g. temperature, photoperiod) and biotic (e.g. host and/or host–plant) signals (Polgár and Hardie, 2000). Although parasitoids typically have higher temperature requirements than their hosts, and hence appear later in the spring (Messenger, 1970; Campbell *et al.*, 1974), the combination of high fecundity, short generation time, and female-biased sex ratio results in a high intrinsic rate of increase (r_m), ranging between 0.2 and 0.6 females per female per day (Force and Messenger, 1964; Cohen and Mackauer, 1987; Kambhampati and Mackauer, 1989; Völkl, 1990).

Aphelinus *and related species*

These parasitoids are also solitary. Unlike aphidiine parasitoids, *Aphelinus* females must feed on host haemolymph for egg maturation, often using low-quality hosts for feeding and high-quality hosts for oviposition (Boyle and Barrows, 1978; Bai and Mackauer, 1990; Takada and Tokumaru, 1996). Females have almost perfect host discrimination and rarely superparasitize (Bai and Mackauer, 1990). When few hosts are available, the relatively large and nutrient-rich eggs are resorbed, thereby prolonging the female's lifespan (Michel, 1967; Mackauer, 1982). The mature larva pupates inside the host mummy, which is bluish-black and loosely attached to the substrate.

Aphelinid wasps are relatively small, measuring only 1–2 mm in length, which limits the female's ability to handle large aphids (Gerling et al., 1990). Mated females deposit fertilized and unfertilized eggs differentially in large and small hosts, respectively (Mueller et al., 1992; Asante and Danthanarayana, 1993); this pattern of offspring sex allocation is independent of the mother's body size (Honek et al., 1998). Under optimal laboratory conditions, aphelinid wasps may survive for 3–4 weeks and produce between 200 and 800 progeny (Force and Messenger, 1964; Mackauer, 1982; Honek et al., 1996). However, the number of hosts killed by feeding and oviposition can be much lower, averaging only three aphids per day over 27 days in Aphelinus flavus (Hamilton, 1973).

Foraging behaviour and host finding

Although all parasitoids cause host mortality, potential impact on the host population is determined by the parasitoid's reproductive strategy. For example, the female's between- and within-patch foraging behaviour determines the number and kind of hosts parasitized in each aphid colony. Therefore, the identification of patterns and processes of foraging behaviour and of the cues involved in host finding and acceptance may help in the planning and implementation of effective biological control programmes.

Between-patch foraging behaviour

Aphidiine wasps use a variety of host-plant or host-borne cues to locate an aphid colony (Mackauer et al., 1996). The relative importance of these cues can vary considerably depending on the spatial scale (Bell, 1990; Völkl, 2000). When foraging between habitats or between host plants (i.e. over relatively long distances so that hosts cannot be found by walking), females often respond to olfactory cues by directed movement. Volatile host-plant compounds serving as semiochemicals can provide the parasitoid

with first information about the potential presence of aphids (Read et al., 1970; Wickremasinghe and van Emden, 1992; Hou et al., 1997; Micha et al., 2000). For example, Vaughn et al. (1996) showed that females of Diaeretiella rapae, a parasitoid of B. brassicae, responded to volatiles of the aphid's host plant in the Brassicaceae. Similarly, Pauesia picta distinguished between a host plant (Scots pine – Pinus sylvestris) and a non-host plant (silver birch – Betula pendula) in choice experiments, independent of the presence of its aphid host, which are species of Cinara (Völkl, 2000).

A more reliable source than plant-borne cues are aphid-borne ones, which, however, might be less detectable over a long distance. Siphuncular secretions (Grasswitz and Paine, 1992; Battaglia et al., 1993; Guerrieri et al., 1993), alarm pheromone (Micha and Wyss, 1996; Foster et al., 2005), and sex pheromones (Hardie et al., 1994; Glinwood et al., 1999) have all been shown as attractive to aphidiine parasitoids. Herbivore-induced plant volatiles can also be used for long-range orientation (Du et al., 1998; Mölck et al., 2000). These cues are particularly important for generalists that parasitize aphids feeding on a variety of plants. Wind tunnel and olfactometer experiments have shown that the searching behaviour of Aphidius ervi was influenced by synomones released by broad bean (V. faba) infested with A. pisum (Du et al., 1996, 1998; Powell et al., 1998; Guerrieri et al., 1999). In uninfested bean plants, synomone induction may be mediated by root contact with an infested plant (Guerreri et al., 2002). The directed flights of Aphidius rosae within a rose bush towards shoots infested with M. rosae is guided by both aphid-borne cues and herbivore-induced volatiles (Völkl, 1994a). However, a parasitoid's response to plant volatiles apparently depends on prior experience (Guerrieri et al., 1997; Rodriguez et al., 2002); naive females of A. ervi showed no directed response towards plants infested with pea aphids in a semi-natural setting (Schwörer and Völkl, 2001). Random search may also be involved in host location, at least in some species (van Roermund and van Lenteren, 1995; van Roermund et al.,

1997). Sheehan and Shelton (1989) reported that foraging *D. rapae* did not respond to increased host-plant density and hence a higher concentration of plant volatiles.

At the intermediate foraging scale, which corresponds roughly to Ayal's (1987) 'elementary unit of foraging', a potential host can be found within walking distance or short-range flight, but outside the distance allowing visual host recognition by the parasitoid (Völkl, 1994a). Aphid parasitoids often search systematically (Li *et al.*, 1992) and use gustatory and olfactory cues for rapid host location. Aphid honeydew is an important contact kairomone for both parasitoids (Bouchard and Cloutier, 1984; Budenberg, 1990; Hågvar and Hofsvang, 1991b) and hyperparasitoids (Buitenhuis *et al.*, 2004, 2005). Parasitoids remain longer on a honeydew-contaminated plant, which increases the probability of host location (Ayal, 1987; McGregor and Mackauer, 1989; Cloutier and Baudouin, 1990; Budenberg and Powell, 1992). Aphid alarm pheromone and secretions from the siphuncles can also influence searching at the intermediate scale (Grasswitz and Paine, 1992; Battaglia *et al.*, 1993; Foster *et al.*, 2005). Some aphidiine species apparently use the presence of honeydew-collecting ants as indirect information in order to find potential hosts. For example, females of *Pauesia pini*, a parasitoid of *Cinara piceicola* feeding on Norway spruce (*Picea abies*), searched spruce seedlings more intensively after a contact with foraging *Formica polyctena* (a red wood ant) than after a contact with honeydew in the absence of ants (Völkl, 2000). Aphidiine wasps, such as *Lysiphlebus* species, exploiting aphids associated with trophobiotic ants, may respond similarly and possibly even use the ants' communication system (e.g. trail pheromones) to locate their hosts (Völkl, 1992, 1997).

Within-patch foraging behaviour

For short-range detection, aphid parasitoids use distance-restricted visual and gustatory information to locate and identify suitable hosts. Visual cues, including aphid colour, shape, size, and movement, can be evaluated from a short distance without physical contact (Michaud and Mackauer, 1994, 1995; Battaglia

et al., 1995; Chau and Mackauer, 2000). Contact chemosensory cues (i.e. gustatory cues) located in the aphid's cuticle may induce various behavioural responses in aphidiine wasps (Mackauer *et al.*, 1996) and hence can play a role in the recognition of aphids and non-aphid insects (Michaud, 1995). In laboratory tests, females of *Aphidius colemani*, *Aphidius picipes*, and *Aphelinus abdominalis* responded to aphids at a distance of 3–4 mm, independent of aphid size; but they were apparently unable to discriminate between host and non-host aphids at that distance (Le Ralec *et al.*, 2005).

Having successfully located a potential host, the female uses antennation and ovipositor probing to determine if the host is suitable for her offspring. Although females of most aphidiine species oviposit in a broad range of hosts differing in size, including aphid embryos (Mackauer and Kambhampati, 1988), females generally show a preference for particular larval instars and morphs; this preference appears to be correlated with the host's nutritional quality. In several studies, Sequeira and Mackauer (1987, 1992a,b, 1994) showed that quality is determined by both species-specific and individual-specific attributes; the latter include aphid size and age at parasitization, attributes that determine an aphid's potential growth until death. Host quality may differ for female and male progeny (Mackauer, 1996). Host quality is an intrinsic property of the host that, presumably, can be evaluated according to absolute criteria, whereas host value is a relative and dynamic property, which varies with parasitoid state variables such as age, egg load, and prior experience, as well as with the female's mortality risks (Weisser *et al.*, 1994; Mackauer *et al.*, 1996). For example, after several ovipositions and hence a low egg load, a female may place a low value on a high-quality host (and reject it); by contrast, it may place a high value on a low-quality host (and accept it) if hosts of a higher quality are not available (Michaud and Mackauer, 1995).

A female can gain in terms of fitness of offspring by attacking a high-quality, large aphid; however, compared with small aphids, large aphids are able to defend themselves better against an attack and can escape

(Gerling *et al.*, 1990). To counter escape by the host, females of *Monoctonus paulensis* hold an aphid with the forelegs for oviposition (Chau and Mackauer, 2000, 2001a). Although offspring developing in large hosts gain in body size and hence have higher fecundity, females selectively attack the smallest available aphids, which need less time for handling and are less likely to escape (Chau and Mackauer, 2001b).

Because only one offspring per host survives, females generally reject already parasitized aphids when unparasitized counterparts are available (Mackauer, 1990). Parasitized aphids are marked with a contact pheromone (Chow and Mackauer, 1986; Hofsvang, 1988). Discrimination between self-parasitized, conspecific-parasitized, and heterospecific-parasitized aphids (Bai and Mackauer, 1990; McBrien and Mackauer, 1990, 1991; Völkl and Mackauer, 1990) suggests that marking pheromones vary among parasitoid species, and possibly also among conspecific females.

Male foraging behaviour

Few studies have examined foraging behaviour and mate finding by males. Females of aphidiine wasps release a sex pheromone to attract males (McNeil and Brodeur, 1995; Nazzi *et al.*, 1996; Cloutier *et al.*, 2000; Marchand and McNeil, 2000). In contrast to females, which mate only once, males can copulate several times, apparently throughout their life (Singh and Pandey, 1997). The virtual absence of virgin females (which can produce only sons) in field populations of aphidiine wasps suggests that mate finding is very effective (Mackauer, 1976; Mackauer and Völkl, 2002).

Virgin females of *Aphelinus asychis* produce a trail sex pheromone, which induces searching behaviour in males (Fauvergue *et al.*, 1995).

Patterns of resource use

Life-table studies indicate that most species of aphid parasitoids can, at least in theory, overwhelm their host population (Messenger,

1970; Mackauer, 1983) and hence should make effective biological control agents. In practice, however, average rates of parasitism in the field are often less than 5%, which is insufficient for control in most cases (Hughes, 1989). A number of factors, both extrinsic and intrinsic, can account for low impact (Mackauer and Völkl, 1993). Extrinsic factors, which include environmental conditions and the risk of female mortality (e.g. from predation), are discussed first.

Adverse weather conditions such as wind and rain cause a reduction in foraging activity, chiefly by preventing females from dispersing in search of hosts (Fink and Völkl, 1995; Weisser *et al.*, 1997; Schwörer and Völkl, 2001). Although a wasp's mean residence time per patch is longer during adverse weather conditions than during optimal conditions, the total number of ovipositions declines, which could explain why many parasitoids have low reproductive success in the field despite their high potential fecundity (Fink and Völkl, 1995; Weisser *et al.*, 1997). Moreover, parasitoids may change their foraging behaviour in response to perceived adult mortality risks (Rosenheim, 1998). For example, Völkl and Kroupa (1997) observed that females of *Pauesia silvestris*, a parasitoid of *Cinara* species on Scots pine (*P. sylvestris*), foraged more thoroughly on the bark (as opposed to needles) in the absence of honeydew-collecting ants. However, in the presence of ants, which may capture and kill parasitoids, females searched more intensively on needles, where they were safe from such attacks.

Three general patterns of resource use can be distinguished: host resources are exploited well below the female's actual egg load; the degree of resource exploitation varies with the female's egg load; and the degree of resource exploitation varies with the host type. These patterns are not mutually exclusive, but overlap to some degree.

Low resource utilization

Most aphidiine and *Aphelinus* species parasitize a relatively small number of aphids in each colony, regardless of colony size and egg load (Mackauer and Völkl, 1993) (Table 8.1).

Females of *A. rosae* laid on average only 2.8 eggs in each colony of rose aphids comprising 10–40 aphids, although the ovaries of each female contained at least 200 mature eggs (Fig. 8.5). A short residence time combined with a low number (< 5) of ovipositions, independent of egg load and host availability, was also observed in *Aphidius*

funebris parasitizing the aphid *Uroleucon jaceae* on *Centaurea* (Weisser, 1995), in *A. ervi* parasitizing *A. pisum* on broad bean (Schwörer and Völkl, 2001), and in *Pauesia unilachni* parasitizing *Schizolachnus pineti* (Lachnidae) on Scots pine (Völkl and Kraus, 1996). In *Binodoxys angelicae*, patch residence time and oviposition activity varied

Table 8.1. Host plant–aphid–parasitoid associations characterized by low resource utilization of the parasitoid throughout the year.

Host plant	Host aphid	Parasitoid	Reference
Brassica spp.	*Brevicoryne brassicae*	*Diaeretiella rapae*	Paetzold and Vater, 1967
Acer pseudoplatanus	*Drepanosiphum platanoidis*	*Trioxys cirsii*	Hamilton, 1974
Acer pseudoplatanus	*Drepanosiphum platanoidis*	*Falciconus pseudoplatani*	Hamilton, 1974
Cereals	Various species	Various species	Dean *et al.*, 1981; Höller *et al.*, 1993
Cedrus atlantica	*Cinara (Cedrobium) laportei*	*Pauesia cedrobii*	Fabre and Rabasse, 1987
Lupinus spp.	*Macrosiphum albifrons*	*Ephedrus californicus*	Cohen and Mackauer, 1987
Centaurea spp.	*Uroleucon* spp.	*Aphidius funebris*	Mackauer and Völkl, 1993
Cirsium arvense	*Aphis fabae*	*Binodoxys angelicae*	Mackauer and Völkl, 1993
Pinus spp.	*Eulachnus* spp.	*Diaeretus leucopterus*	Murphy and Völkl, 1996
Pinus resinosa	*Schizolachnus piniradiatae*	*Pauesia californica*	Sharma and Laviolette, 1968
Rosa spp.	*Sitobion fragariae*	*Aphidius rosae*	Zwölfer and Völkl, 1997

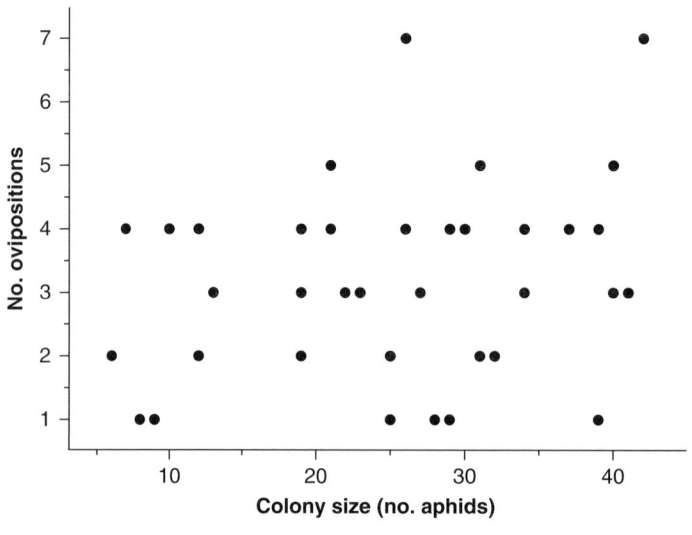

Fig. 8.5. Oviposition activity of *Aphidius rosae* in relation to the size of *Sitobion fragariae* colonies on *Rosa* sp. (redrawn after Völkl, 1994a).

with the number of *A. fabae* per colony on thistles (*Cirsium arvense*). Females were most successful attacking aphids in small and medium-sized colonies, while they abandoned large colonies after a short time and few ovipositions (Mackauer and Völkl, 1993).

Resource utilization varies with egg load

This pattern is typically found in *Lysiphlebus* species. Völkl (1994b) examined the behaviour of *Lysiphlebus cardui* foraging for *A. fabae cirsiiacanthoides* on *C. arvense*. Foraging behaviour was influenced by trophobiotic ants. In the presence of ants, females generally remained in a discovered aphid colony for long periods, often more than 1 day, until they had depleted the supply of 200–300 mature eggs in their ovaries or had parasitized all available aphids (Fig. 8.6). However, females laid significantly fewer eggs and dispersed earlier if aphid colonies were not attended by ants. The aphids' defence behaviour was reduced in the presence of guarding ants, which favoured increased oviposition by the parasitoid. Average rates of parasitism in ant-attended colonies increased with the season and could reach 100% due to the parasitoids' increasing colonization success (Weisser and Völkl, 1997; Weisser, 2000).

Similar patterns of resource exploitation were observed in *Lysiphlebus hirticornis* parasitizing *M. fuscoviride* on tansy (Fig. 8.7 – Mackauer and Völkl, 1993; Weisser, 2000), *Lysiphlebus fabarum* parasitizing *A. fabae* (Starý, 1970; Völkl and Stechmann, 1998), and *Lysiphlebus testaceipes* parasitizing *Aphis* spp. (Starý *et al.*, 1988; Völkl, 1997).

Resource utilization varies with the host type

In *P. pini*, patterns of host utilization differed markedly with the generation of its main host, *C. piceicola*, feeding on Norway spruce (Fig. 8.8). Average patch times were longest, and females laid the highest number of eggs when searching for fundatrices and fundatrigeniae in early spring. Oviposition numbers were constrained only by the number of hosts or eggs available, whichever was lower (Völkl and Novak, 1997; W. Völkl, unpublished results). By contrast, females laid fewer eggs (independent of colony size) when foraging for viviparous aphids or oviparae later in the season. Rates of parasitism and the number of mummies per colony were lowest during midsummer when *P. pini* exploited alternate hosts (e.g. *Cinara pinea*) on pine. This seasonal pattern of parasitization was correlated with variation in host size (Fig. 8.9). Differences

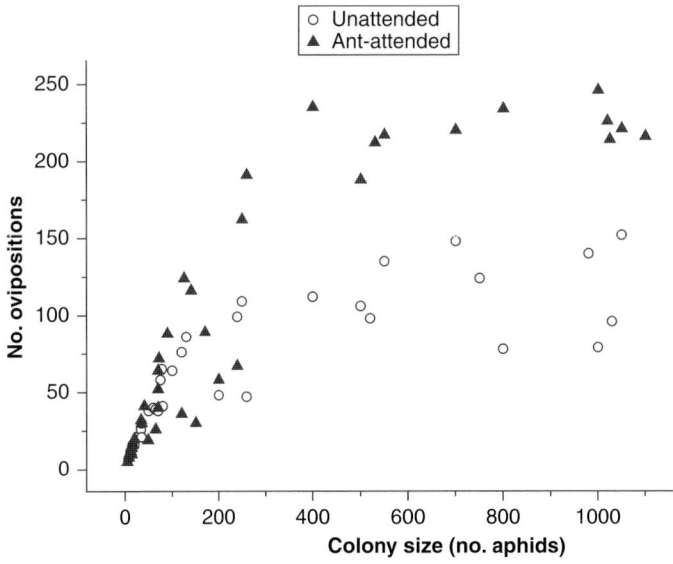

Fig. 8.6. Oviposition activity of *Lysiphlebus cardui* in relation to the size of *Aphis fabae* colonies on *Cirsium arvense* in the presence and absence of honeydew-collecting ants (redrawn after Völkl, 1994b).

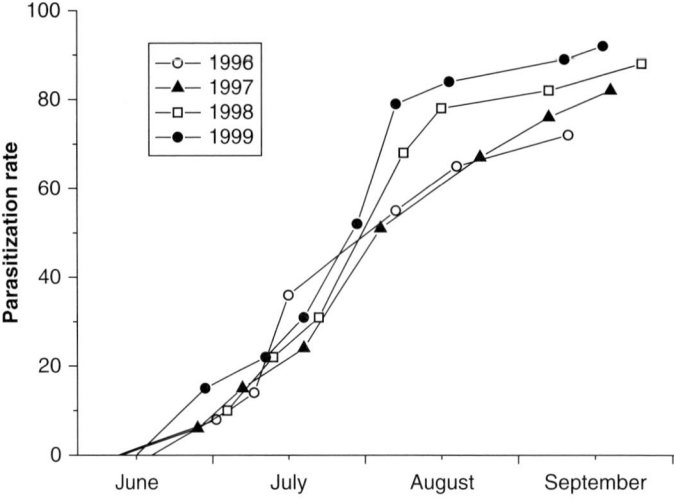

Fig. 8.7. Percentage parasitism by *Lysiphlebus hirticornis* in ant-attended colonies of *Metopeurum fuscoviride* feeding on *Tanacetum vulgare* in ruderal areas near Bayreuth, Germany (redrawn after Fischer *et al.*, 2001; W. Völkl and M. Mackauer, unpublished results).

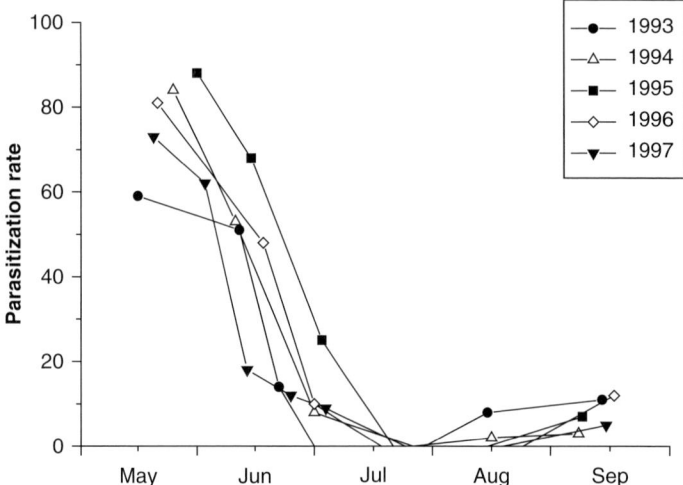

Fig. 8.8. Percentage parasitism by *Pauesia pini* of *Cinara piceicola* feeding on *Picea abies* in the Fichtelgebirge, Germany (W. Völkl, unpublished results).

between years were due mainly to variations in environmental conditions. Low temperatures in spring favoured *P. pini*, while high temperatures led to decreased parasitism rates. This result can be explained by differences between the activity patterns of *P. pini* and worker wood ants (*F. polyctena*) that protected the aphids. Parasitoid females were very active at low temperatures and showed increased oviposition rates, while the activity of the worker ants was drastically reduced. A similar pattern of parasitism was observed in *Euaphidius*

cingulatus, a parasitoid of aphids in the genus *Pterocomma* on willows (Völkl, 1997).

Aphid Pathogens

Although viruses have been recorded infecting aphids (e.g. Darcy *et al.*, 1981; Laubscher and von Wechmar, 1992; van den Heuvel *et al.*, 1997), the most significant microbial natural enemies of aphids in the field all belong to the 'true' fungi. Most entomopathogenic species are found in the divisions

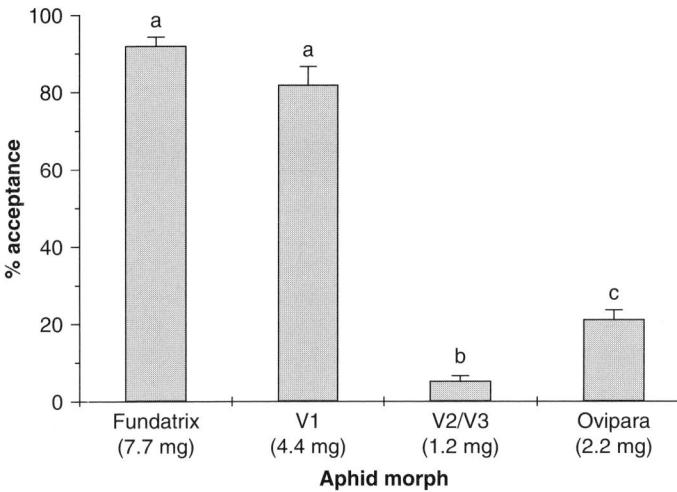

Fig. 8.9. Percentage acceptance by *Pauesia pini* of different morphs of *Cinara piceicola*. The average fresh weight is indicated below each morph. V1 = virginoparae, first generation (= fundatrigeniae), V2/V3 = virginoparae, second and third generation since fundatrix. Columns sharing the same letter are not significantly different at $p = 0.05$ (W. Völkl, unpublished results).

Zygomycota and Ascomycota. The most common fungal pathogens that contribute commonly to the regulation of aphid populations are from the Zygomycota, and in particular, the order Entomophthorales. There are six species of entomophthoralean fungi commonly recorded from pest and non-pest aphids worldwide: *Conidiobolus obscurus, Entomophthora planchoniana, Neozygites fresenii, Pandora neoaphidis, Zoophthora phalloides,* and *Zoophthora radicans* (Wilding and Brady, 1984). When disease epizootics develop, they can have a significant impact on aphid populations, with the potential to reduce them to practically zero at a local scale; hence, the potential of entomopathogenic fungi as biological control agents (e.g. Wraight *et al.*, 1992; McLeod *et al.*, 1998; Hemmati, 1999; Nielsen *et al.*, 2001a). Recent studies on the biology and ecology of these organisms have focused largely on *P. neoaphidis, N. fresenii,* and *Z. radicans.* Some aphid-pathogenic Entomophthorales are impossible or extremely difficult to culture *in vitro*, and most are difficult to produce on a large scale, and so microbial control using these species has often focused on inoculative augmentation and conservation (Pell *et al.*, 2001; Powell and Pell, Chapter 18 this volume).

Outside the Entomophthorales, *Lecanicillium longisporum* (in *L. lecanii sensu lato*) is the only species recorded causing occasional epizootics in aphid populations in geographically diverse regions and in a number of different aphid species (e.g. Ozino *et al.*, 1988; Feng *et al.*, 1990; Sanchez-Pena, 1993; Kish *et al.*, 1994; Milner, 1997; Hatting *et al.*, 1999; Mathew *et al.*, 1999). *Metarhizium anisopliae* has also been recorded causing epizootics in *Pemphigus bursarius* (lettuce root aphid) in salt marshes in the UK (Foster, 1975). Other species that have been recorded rarely and never causing epizootics include *Beauveria bassiana* and *Paecilomyces* spp. (Feng *et al.*, 1990; Kish *et al.*, 1994). Although these mitosporic fungi are not common natural enemies under field conditions, they are readily mass-produced and can be suspended in aqueous media and sprayed. For this reason, they have been commercialized as inundative microbial control agents of insects, including aphids (Shah and Goettel, 1999; Inglis *et al.*, 2001; Powell and Pell, Chapter 18 this volume).

Biology of entomophthoralean fungi

The infective propagule of the fungus is the spore or conidium. Conidia are responsible for infection during the season when aphids are active. On the death of the aphid, fungi break out from within the cadaver, particularly through the intersegmental membranes (Butt *et al.*, 1990). Vast numbers of primary conidia are then actively discharged by hydrostatic pressure. Many thousands of conidia can be produced from a single aphid, with numbers depending on aphid biomass (Steinkraus *et al.*, 1993; Hemmati *et al.*, 2001b). In comparison with mitosporic fungi, these conidia are large – 18–35 × 10–15 μm for *P. neoaphidis* (Wilding and Brady, 1984).

Conidia are covered with pre-formed mucous which aids attachment, and if they land on an aphid host, they have the potential to germinate rapidly and infect by direct penetration through the cuticle. Once within the aphid, they grow initially as protoplasts (e.g. *P. neoaphidis*), which lack cell walls. Other species, such as *Z. radicans*, grow as unicellular hyphal bodies or coenocytic hyphae and do not have a protoplastic stage (Butt *et al.*, 1981, 1990; Kobayashi *et al.*, 1984; Wraight *et al.*, 1990). Host death is probably caused by starvation of the host when the fungus has consumed all the nutrients within the aphid; it occurs after approximately 4 days at 20°C. Time to death varies with temperature and, to a certain extent, with the initial inoculum dose (Hemmati, 1999). Cadavers are fixed in place on the plant by fungal rhizoids, which emerge through the ventral surface of the cadaver (*P. neoaphidis*, *Zoophthora* spp.) or through the proboscis (*C. obscurus*). These ensure that the fungus remains in the environment of transmission.

If primary conidia land on non-host surfaces, they can produce another conidium (a secondary conidium), which, in *P. neoaphidis*, is also actively discharged, increasing the ability of the fungus to reach and infect new aphid hosts. In *Zoophthora* spp., higher order conidia are not always actively discharged, but rather produced on fine capillaries or capilliconidiophores. The capilliconidium is borne on the top of the conidiophore some distance above the surface and remains attached until a host dislodges it. Both primary and secondary conidia of these fungi are infective (Glare *et al.*, 1985a,b; Wraight *et al.*, 1990; Pell *et al.*, 1993). In *N. fresenii*, primary conidia are not infective, but serve only as a method of dispersal. It is the secondary conidia, also produced on capilliconidiophores, which are infective.

Most entomophthoralean species attacking aphids also produce resting spores at the end of the season as a mechanism of survival during periods when aphids are not present or at low population densities (Bitton *et al.*, 1979; Wilding and Brady, 1984). These spores can be produced sexually (zygospores) or asexually (azygospores) and are resistant to abiotic stress, having a thick double wall. Dead, infected aphids containing resting spores remain on foliage and bark or fall to the ground, where the spores can remain dormant for many years. Resting spores germinate throughout the period of time that aphids are present in the field by producing forcibly ejected germ conidia or capilliconidia (the latter only in *N. fresenii*), either of which are infective. Although resting spores are known for most of the species infecting aphids, they have not been recorded for *P. neoaphidis*. The mechanism by which *P. neoaphidis* overwinters is uncertain, but it has been suggested that the fungus may survive as specialized spherical hyphal bodies inside mycosed insect cadavers on above-ground plant substrates where relative humidity is low (Feng *et al.*, 1992), or in the soil as dormant 'loricoconidia' (Nielsen *et al.*, 2003). In the laboratory, survival as dried hyphal bodies certainly can be prolonged at low relative humidities (Wilding, 1973). Specialized hyphal bodies within winter collected aphid cadavers have also been observed for *E. planchoniana* (Keller, 1987). Where aphids are always present, even through the winter, it is possible that *P. neoaphidis* survives through cycles of re-infection during this period (Feng *et al.*, 1991).

Epizootiology of entomophthoralean fungi

Epizootiology is 'the science of causes and forms of the mass phenomena of disease at all levels of intensity in a host population' (Fuxa and Tanada, 1987). Although the basic fungal life cycle described above is relatively simple, the epizootiology of fungi within populations is regulated by factors relating to the host population (density, susceptibility, distribution), the fungus population itself (density, virulence, distribution), and the environment (temperature and humidity), which are all linked to transmission and dispersal (Fuxa and Tanada, 1987). Both aphid and fungus populations are discontinuous in the environment, and therefore mechanisms that encourage infection and dispersal within and between host populations are essential for fungal survival.

The most important limiting abiotic factor is low relative humidity. Aphid pathogenic entomophthoralean fungi (as other fungi) cannot germinate, infect, or sporulate if the ambient relative humidity falls below 90–93% (Wilding, 1969; Millstein *et al.*, 1982, 1983; Milner and Bourne, 1983; Hemmati *et al.*, 2001a). *Pandora neoaphidis*-infected cadavers in contact with liquid water produce more conidia than those in a saturated environment (Wilding, 1969). Ekbom and Pickering (1990) showed a positive correlation between fungus-induced aphid mortality (largely *N. fresenii* and *E. planchoniana*) and the sum of hours of leaf wetness prior to sampling. Several authors have also reported that high infection levels were associated with high rainfall at critical periods of aphid and fungus establishment (Nielson and Barnes, 1961; Dean and Wilding, 1973; Hemmati *et al.*, 2001a). Ambient humidity may play a significant role in the initiation of an epizootic, assuming that suitable aphid hosts are present, but once an epizootic is initiated, the importance of humidity becomes secondary as, during the night at least, conditions within the crop are usually favourable (Hemmati *et al.*, 2001a). Both *P. neoaphidis* and *Z. radicans* have evolved to exploit this window of opportunity during the night when humidity is high and ultraviolet radiation limited. Using an internal clock set by the time of dawn, the fungus kills the aphid at the end of the photophase, ensuring that sporulation and infection will occur during the night (Milner *et al.*, 1984). *Neozygites fresenii* is also known to sporulate most profusely during the night, which suggests that a similar mechanism may be operating (Steinkraus *et al.*, 1996, 1999). The importance of a humid microclimate can also be seen in studies where irrigation has been used to enhance infection levels or where the canopy was thought to increase local humidity (Powell *et al.*, 1986a; Wilding *et al.*, 1986; Pickering *et al.*, 1989).

Temperature affects the number of conidia produced by an aphid cadaver and the rate at which sporulation and infection (time to death) occur; these in turn affect the rate of epizootic establishment. At suboptimal temperatures, the number of conidia produced from a cadaver by *Z. radicans* and *P. neoaphidis* is reduced, and the time to infect and kill extended (Milner and Lutton, 1983; Leite *et al.*, 1996; Dromph *et al.*, 1997; Hemmati *et al.*, 2001b). Of all the entomophthoralean species infecting aphids, only *N. fresenii* is adapted to function in hot weather (Steinkraus *et al.*, 1991; Keller, 1997), making it the most important aphid pathogen in tropical regions (Steinkraus *et al.*, 1995).

As conidia are actively discharged, they escape the boundary layer and enter the air stream, which facilitates local and distant dispersal. Conidia of *N. fresenii* and *P. neoaphidis* have been detected in the air above and some distance from crops infested with diseased aphids (Hemmati, 1999; Steinkraus *et al.*, 1999; Hemmati *et al.*, 2001a). This is a very important mechanism for dispersal between patchily distributed hosts and highly effective when large numbers of hosts are present. However, it is not targeted and relies entirely on the wind direction, and so many conidia are likely to be lost from the transmission cycle. A more targeted mechanism of dispersal occurs via co-occurring insect natural enemies, particularly predators. *Coccinella septempunctata*, a ladybird predator of aphids, can vector disease between *P. neoaphidis*-infected and uninfected aphid colonies in laboratory and small-scale field trials (Pell *et al.*, 1997; Roy *et al.*, 2001). The importance of this in the field, and particularly in respect to host location by the fungus when host numbers are low, requires further investigation.

The host range of entomophthoralean fungi is relatively limited; *P. neoaphidis*, *E. planchoniana*, and *N. fresenii* only infect aphids (Nielsen *et al.*, 2001a; Pell *et al.*, 2001). *Z. radicans* has been recorded from numerous different insect orders; however, isolates are often more infective to the host species from which they were isolated than to other species (e.g. Pell *et al.*, 1993). Within a taxon of insect hosts, such as aphids, it can be essential for survival of the fungus that it is able to infect a number of different species. This is particularly true for *P. neoaphidis*, which does not produce resting spores. *P. neoaphidis* has been recorded from numerous different pest and non-pest aphid species, and it has been suggested that aphids on wild plants, in addition to crop aphids, can act as alternate hosts for this fungus (Keller and Suter, 1980; Powell *et al.*, 1986a; Steenberg and Eilenberg, 1995; Shah and Pell, 2003). Studies comparing the susceptibility of seven pest aphid species to different isolates of *P. neoaphidis* suggest that some aphid species, particularly *Rhopalosiphum padi* (bird cherry–oat aphid), are resistant (Shah *et al.*, 2004). This confirms field observations by Dedryver (1983), although others have recorded *R. padi* infection by *P. neoaphidis* and other entomophthoralean species in the field, sometimes at high levels (e.g. Steenberg and Eilenberg, 1995; Basky and Hopper, 2000). The aphid species most susceptible to a wide range of *P. neoaphidis* isolates is *A. pisum* (Shah *et al.*, 2004), although great variability in susceptibility has been demonstrated amongst different clones of this aphid (Milner, 1982; Ferrari *et al.*, 2001; Ferrari and Godfray, 2003) and amongst different morphs of a single clone (Milner, 1982, 1985; Lizen *et al.*, 1985). There is also huge variation amongst the isolates in infectivity to different aphid species (Shah *et al.*, 2004). The epizootiology of different isolates will therefore depend on a number of virulence parameters associated with the isolate and susceptibility parameters of the host, which could interact in complex ways. Molecular techniques under development will help to elucidate these interactions at a field scale (Rohel *et al.*, 1997; Sierotzki *et al.*, 2000; Nielsen *et al.*, 2001b; Tymon *et al.*, 2004).

Intraguild Interactions, Mutualistic Ants and Aphid Symbionts

Intraguild predation and competition

Individual species of natural enemies can have a significant effect on aphid populations; however, species do not exist in isolation but generally are part of larger complexes within guilds (Frazer *et al.*, 1981; Gutierrez *et al.*, 1984; Dennis, 1991; Booij and Noorlander, 1992). If there is only minimal interference between different natural enemies, the total impact on the aphid population can be greater than if each species is acting alone (Chang, 1996). Alternatively, if interference is significant, aphid populations can increase in spite of natural enemy pressure (Hochberg and Lawton, 1990).

Many aphid predators prey, occasionally or regularly, on other predators when they compete for the same prey within an aphid colony, which results in intraguild predation (Rosenheim *et al.*, 1995; Lucas, 2005). Predator type and predator size are two important factors determining the outcome of intraguild predation. The relatively larger individuals generally kill the relatively smaller ones (Lucas *et al.*, 1998; Hindayana *et al.*, 2001). Smaller individuals are also likely to be eaten by larger predators in the case of cannibalism between conspecifics and in interspecific predation among predators belonging to the same family (e.g. larvae of ladybird beetles differing in size – Völkl and Vohland, 1996) or to different families (Agarwala and Yasuda, 2001). For example, Lucas *et al.* (1998) showed that interactions between larvae of the lacewing *Chrysoperla rufilabris* and the ladybird *Coleomegilla maculata* (12-spot ladybird) resulted in reciprocal predation, i.e. both predators were at risk of being eaten by a competitor. Hindayana *et al.* (2001) found that larvae of the hover fly *E. balteatus* were susceptible to predation by larvae of the lacewing *C. carnea* and the ladybird *C. septempunctata*, while *E. balteatus* pupae were consumed only by lacewing larvae. Lucas *et al.* (1998) suggested that in interactions between members of the same guild, specialist species were more likely than generalists to become prey. In contrast, competition between larvae of the predatory midge *A. aphidimyza* and other predators was always asymmetrical in that gall midge larvae generally served as prey. This species is not adapted to kill prey other than aphids due to its highly specific feeding requirements (Lucas *et al.*, 1998; Hindayana *et al.*, 2001).

Aphid parasitoids may experience mortality due to intraguild predation (Brodeur and Rosenheim, 2000). Working in cotton, Colfer and Rosenheim (2000) showed that adult coccinellid beetles consumed a considerable proportion of aphid mummies, although beetles preferred living aphids as prey. A large percentage of mummies of *L. hirticornis* on tansy was destroyed by the anthocorid bug *A. nemorum* (Fig. 8.10), which is the most abundant predator in this system (Fischer *et al.*, 2001); ant-attendance had a significant influence on the percentage of destroyed mummies (see below). Meyhöfer and Hindayana (2000) reported that mummies of *L. fabarum* on broad beans were at a high risk of predation, especially in colonies not protected by ants. Foraging parasitoid females, however, did not avoid colonies in which aphidophagous predators were present or displayed any effective defences in direct confrontations with adult *C. septempunctata* and larvae of *E. balteatus* (Meyhöfer and Klug, 2002). Avoidance behaviour may vary with the interval between successive foraging bouts by potential competitors. Nakashima *et al.* (2004) demonstrated that *A. ervi* females avoided leaves visited by larvae and adults of *C. septempunctata* during the previous 24 h, but not if the interval was greater.

Fungal pathogens with a broad host range capable of infecting predators and parasitoids can also be involved in intraguild interactions. This applies in particular to some mitosporic fungi developed as myco-insecticides for use against aphids, although perhaps not to the commonest fungal pathogens of aphids, which have extremely narrow host ranges (see above). For example, *B. bassiana*, *Paecilomyces fumosoroseus*, and *L. longisporum* can infect aphid predators and parasitoids, but often only at high doses and/or under laboratory conditions favouring the fungus and/or stressing the host (Feng *et al.*, 1994; James and Lighthart, 1994; Lacey *et al.*, 1997; Askary and Brodeur, 1999; Yeo, 2000; Pell and Vandenberg, 2002). In other laboratory experiments, however, different isolates from the same species of mitosporic fungi showed no or only limited infectivity to insect natural enemies of aphids (Yeo, 2000). Physiological susceptibility under laboratory conditions often represents the 'worst case scenario' for infection of beneficial insects, which may differ from ecological susceptibility in the field. Beneficial arthropods and mitosporic myco-insecticides can, therefore, coexist even when the potential of some isolates and species for infecting beneficial organisms has been demonstrated in the laboratory (James *et al.*, 1995, 1998).

Within the guild of aphid natural enemies, arthropod predators can be intraguild predators of fungi. Parasitoids and fungi can also compete for the same host resources. Furthermore, both predators and parasitoids can affect fungal dispersal and transmission

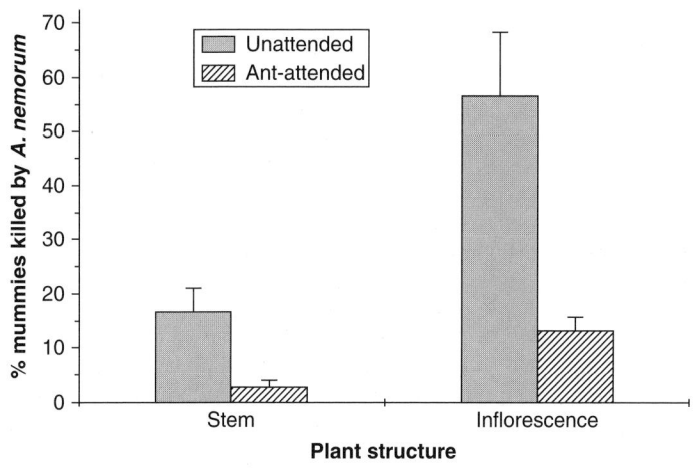

Fig. 8.10. Percentage of mummies of *Lysiphlebus hirticornis* preyed on by *Anthocoris nemorum* on tansy (*Tanacetum vulgare*) in relation to plant structure and ant attendance (W. Völkl and M. Fischer, unpublished results).

(Roy and Pell, 2000). For example, the ladybird *C. septempunctata* and the carabid beetle *Pterostichus madidus* both feed on aphid cadavers sporulating with *P. neoaphidis*, in contrast to larvae of the lacewing *C. carnea*, and the hover fly *E. balteatus*, which do not consume *P. neoaphidis*-infected aphids (Roy *et al.*, 2001). Predation by carabids is unlikely to have a significant impact on fungal dynamics, however, because these beetles feed at soil level on cadavers that are already lost from the environment where fungal transmission occurs. In the studies by Roy *et al.* (2001), *P. neoaphidis*-infected aphid cadavers were less palatable to *C. septempunctata* larvae than uninfected ones. Hungry larvae did attempt to consume cadavers, but never consumed them entirely. These findings agree with the results of previous studies using adults (Pell *et al.*, 1997) and studies on the foraging behaviour of another ladybird, *Hippodamia convergens* (convergent ladybird), preying on *D. noxia* infected with *P. fumosoroseus* (Pell and Vandenberg, 2002). Although spore production from partially consumed *P. neoaphidis*-infected aphids was significantly reduced, this had no effect on subsequent transmission. Rates of transmission were significantly increased in the presence of foraging ladybird beetles (Roy *et al.*, 2001), which suggests that predators, in fact, may enhance the transmission and dispersal of entomopathogenic fungi rather than having a negative impact as intraguild predators (Roy *et al.*, 2001; Pell and Vandenberg, 2002).

Interactions between parasitoids and pathogens are generally asymmetrical in favour of the pathogen (Hochberg and Lawton, 1990). The pathogen invariably kills the host faster than the parasitoid can complete its development, although the final outcome of such competitive interactions is dependent on the relative timing of parasitism and infection (Kim *et al.*, 2005). Development of the parasitoid *Aphidius nigripes* was impeded by the fungus *L. longisporum*, but many parasitoids still emerged from aphids infected 4 days after parasitization (Askary and Brodeur, 1999). A similar 'priority effect' was observed in interactions between *Aphidius rhopalosiphi* and *P. neoaphidis* in

the laboratory; however, competition was minimized in the field because parasitoids and fungi were favoured by different environmental conditions (Powell *et al.*, 1986b). Interactions between pathogens and parasitoids can also be affected by plant resistance. Fuentes-Contreras *et al.* (1998) reported that aphids and parasitoids developed more slowly on aphid-resistant wheat cultivars, but the time required for an aphid to be killed by a fungus remained the same. Some parasitoids avoid infected hosts (Brobyn *et al.*, 1988). There is some evidence that the combination of parasitoids and fungal pathogens can have an additive effect with respect to aphid control (Hagen and van den Bosch, 1968), i.e. that parasitoids are not negatively affected by entomopathogenic fungi. Recent evidence from field experiments indicates that aphid control can be enhanced by the combination of different natural enemies. For example, the combination of the parasitoid *A. asychis* and the fungus *P. fumosoroseus* resulted in better control of *D. noxia* than by an application of a single antagonist; percentage parasitism and parasitoid emergence were not affected by the fungus (Mesquita *et al.*, 1997).

The effects of mutualistic interactions with ants on predation and parasitism

Many aphid species have evolved mutualistic relationships with honeydew-collecting ants (Zwölfer, 1958; Way, 1963; Buckley, 1987). Both partners may derive benefits from this association. The ants gain access to an important source of nutrients and, by attacking all intruders into an aphid colony as potential competitors for the carbohydrate source, act as effective guards against the aphids' natural enemies. Several studies show that adults and larvae of coccinellid species were more abundant in areas where ants were naturally absent or were excluded experimentally (Banks, 1962; Way, 1963; Bradley, 1973; Jiggins *et al.*, 1993; Sloggett and Majerus, 2000). There is less information about the influence of ants on the density of other predator species, especially syrphid and chrysopid larvae. Müller and

Godfray (1999) and Fischer *et al.* (2001) demonstrated a general decrease of predator density in ant-attended colonies, which in turn survived for a longer time. On tansy, the predacious bug *A. nemorum* was significantly less abundant in ant-attended colonies of *M. fuscoviride* than in unattended ones, where the presence of predators significantly reduced colony persistence (Fischer *et al.*, 2001). Bishop and Bristow (2001) found that the density of lacewings and salticid spiders in aphid colonies on conifers was reduced when *Formica exsectoides* (Allegheny mound ant) was present, while mirid bugs were hardly affected.

A number of predators have evolved various morphological and/or behavioural adaptations to gain access to ant-attended resources, however. The ladybirds *P. luteorubra* and *Coccinella magnifica*, for example, are protected by chemical camouflage (Völkl, 1995; Sloggett and Majerus, 2000). Larvae of *Scymnus nigrinus* benefit from a dorsal wax cover that smears the ants' mouthparts (Völkl and Vohland, 1996). Similarly, the larvae of some chrysopid species cover themselves with waxes obtained from their prey to escape ant aggression (Eisner *et al.*, 1978; Mason *et al.*, 1991; Milbrath *et al.*, 1993). In all cases, the predators benefit from their ability to exploit ant-attended aphid colonies where there is a reduced risk of parasitism or intraguild predation.

Ant-attendance does not only lead to a decreased predation risk but also may reduce the risk of parasitism. Völkl (1997) reported that, in Central Europe, 14 of 40 parasitoid species attacking ant-attended aphids were generally treated aggressively by honeydew-collecting ant workers when approaching an aphid colony. Therefore, resource utilization by these parasitoid species was largely restricted to unattended aphid colonies. For example, *Periphyllus* spp. on *Acer* spp. and *Symydobius oblongus* on *B. pendula* were heavily tended by ants and were significantly less parasitized by *Trioxys falcatus* and *Trioxys betulae*, respectively (Völkl, 1997). Also, *A. fabae* benefited from the presence of ants through a reduced parasitism by *B. angelicae* (Völkl, 1992). This finding was consistent across different host plants,

although some evidence suggests that large aphids can be guarded less effectively than small counterparts. Foraging parasitoids seem to be more effective when aphid colonies are feeding on plant structures with a high structural diversity (e.g. leaves and flowers of *Chenopodium* spp.) than on structures with a low structural diversity (e.g. stems of the thistle *C. arvense*) (Mackauer and Völkl, 1993; Völkl, 1997).

Some parasitoid species have evolved mechanisms to avoid or escape ant aggression. Species of the genus *Pauesia*, which attack *Cinara* spp. feeding on conifers, can escape attacking ants by their quick and agile movements, although they are recognized by ant workers. Furthermore, wasps seem to learn during aggressive interactions, thereby enhancing their parasitization success in subsequent attempts (Völkl, 2001). An even more specific adaptation is chemical mimicry of aphids, which results in foraging parasitoids not being recognized by ant workers (Starý, 1986; Völkl and Mackauer, 1993). For example, evidence suggests that the parasitoid *L. cardui* mimics the epicuticular hydrocarbon profile of its host, *A. fabae cirsiiacanthoidis,* to confuse honeydew-collecting ant workers (Liepert and Dettner, 1993). Because females can oviposit in ant-attended aphid colonies without being attacked, *L. cardui* has evolved a strategy resulting in the parasitism of a high proportion of available hosts (Weisser *et al.*, 1994). Ant-attended colonies were parasitized even more heavily, and found earlier, than unattended ones (Völkl, 1992). Similar observations were reported for two closely related species, *L. hirticornis* and *L. fabarum* (Starý, 1970; Mackauer and Völkl, 1993; Völkl and Stechmann, 1998), which also employ chemical mimicry (Liepert, 1996). Moreover, aphid parasitoids that can exploit ant-attended resources suffer significantly less from predation (Fig. 8.10) and hyperparasitism than parasitoids developing in unattended colonies (Mackauer and Völkl, 1993; Sullivan and Völkl, 1999; Kaneko, 2002, 2003).

Ant–parasitoid and ant–predator interactions can affect the outcome of biological control attempts. For example, the polyphagous parasitoid *A. colemani* was introduced in the South Pacific for the biological control of *Pentalonia nigronervosa* (banana aphid) (Völkl *et al.*, 1990). Small aphid colonies are found mainly in concealed areas of the banana plant such as the space between pseudo-stem and leaf sheaths. *Aphidius colemani* females foraged intensively in these areas but were heavily attacked by honeydew-collecting ants, which prevented parasitism and thus caused *A. colemani* to be ineffective for control (Stechmann *et al.*, 1996). Moreover, the abundance of predators was reduced significantly by ant-attendance (Stechmann and Völkl, 1990). In contrast, the success of *L. fabarum* parasitizing *A. fabae* on sugarbeet (*Beta vulgaris*) increased significantly in the presence of ants, while unattended aphid colonies were hardly parasitized (Fig. 8.11). Thus, a significant reduction of *A. fabae* densities by *L. fabarum* was found only in field margins within the foraging range of honeydew-collecting ants (Völkl and Stechmann, 1998).

Aphid symbionts and resistance to parasitism

Not all parasitized aphids are also susceptible to parasitism, however. Griffiths (1961) observed that *Monoctonus crepidis* oviposited in various aphid species feeding on lettuce (*Lactuca sativa*) but offspring only developed in *Nasonovia ribisnigri*. Even within the same aphid species, considerable variation can exist for resistance to a specific parasitoid. For example, Henter and Via (1995) found variation in susceptibility to *A. ervi* among clones of the pea aphid collected from a single population. Eggs deposited in resistant aphids failed to develop. Although lines of *A. ervi* varied in virulence (Henter, 1995), there was no evidence that the aphids responded to selection by becoming more resistant to the parasitoid over the course of a season (Henter and Via, 1995). Clones of pea aphids resistant to *A. ervi* were also resistant to *Aphidius eadyi* and the pathogen *P. neoaphidis* (Ferrari *et al.*, 2001). Non-susceptibility in pea aphids was correlated with the presence of facultative accessory bacteria (secondary symbionts), which, in turn, varied with the aphids' host-plant use (Ferrari *et al.*, 2004). Although Ferrari *et al.* (2001) found no relationship between resistance and fecundity in pea aphids, Gwynn *et al.* (2005) suggested that non-susceptibility may be costly in terms of other fitness attributes. They reported that resistance to *A. ervi* was negatively correlated with fecundity and positively correlated with off-plant survival time and body size, which latter is unexpected (Murdie, 1969; Dixon, 1987) and indicates that the actual mechanisms underlying non-susceptibility may be complex and involve other factors in addition to the kind and diversity of bacterial symbionts.

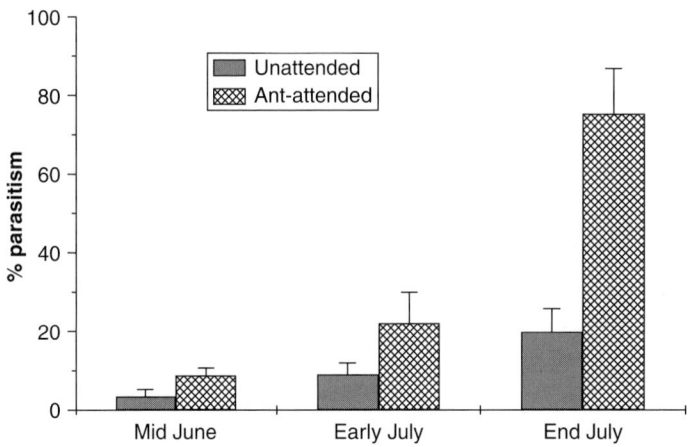

Fig. 8.11. Percentage parasitism by *Lysiphlebus fabarum* in the presence and absence of honeydew-collecting ants (redrawn after Völkl and Stechmann, 1998).

Conclusions

Aphids are attacked by a wide range of natural enemies, many of which have a high fecundity and, at least in theory, should be capable of reducing aphid populations below the economic threshold. However, Aphidophaga often do not achieve their expected potential in the field (see also both Powell and Pell, Chapter 18 and the IPM case studies in this volume). Many predators are not very effective in locating aphid prey, and specific habitat requirements (such as the availability of a pollen or nectar supply as food for adults) additionally reduce their activity in the vicinity of aphid colonies. Parasitoids seem to be very successful foragers within their habitat, but their species-specific oviposition strategies often prevent a thorough exploitation of an aphid colony, which leads to their departure before all available hosts are parasitized. Entomophthoralean fungi have effective transmission and dispersal mechanisms but can be limited by their requirement for high humidity. Intraguild predation additionally may lower the impact of Aphidophaga.

Despite these shortcomings, Aphidophaga play an important role in the reduction of aphid numbers in crops. Although an extension of biological control measures in the field and a reduction of pesticide use is a promising strategy for the future (Powell and Pell, Chapter 18 this volume), the practice of biological control currently does not reflect all the available information on parasitoids, predators, and pathogens. In particular, attempts at controlling exotic aphid pests often rely exclusively on the introduction of agents from the country of origin, with little regard for the integration of different control measures (Michaud, 2002). Moreover, agents are selected for release solely on the basis of life-history attributes such as fecundity and longevity, so long as they are specific to the target pest. Evidence from field experiments that intraguild interactions play a significant role in aphid biological control is not given sufficient attention. For example, several recent studies point to unexpected consequences of parasitoid introductions to control *A. pisum* in North America, including non-adaptive changes in parasitoid virulence (Hufbauer, 2002a,b) and non-target effects of agricultural practices on exotic and native parasitoids competing for the same host resource (Schellhorn *et al.*, 2002). Clearly, opportunities for the biological control of aphids exist but will require a system-based approach to minimize interference between different aphidophagous agents.

References

Agarwala, B. and Yasuda, H. (2001) Larval interactions in aphidophagous predators: effectiveness of wax cover as defence of *Scymnus* larvae against predation from syrphids. *Entomologia Experimentalis et Applicata* 100, 101–107.

Ankersmit, G.W., Dijman, H., Keuning, N.J., Mertens, H., Sins, A. and Tacoma, H.M. (1986) *Episyrphus balteatus* as a predator of the aphid *Sitobion avenae* on winter wheat. *Entomologia Experimentalis et Applicata* 42, 271–277.

Asante, S.K. and Danthanarayana, W. (1993) Sex ratio in natural populations of *Aphelinus mali* (Hym., Aphelinidae) in relation to host size and host density. *Entomophaga* 38, 391–403.

Askary, H. and Brodeur, J. (1999) Susceptibility of larval stages of the aphid parasitoid *Aphidius nigripes* to the entomopathogenic fungus *Verticillium lecanii*. *Journal of Invertebrate Pathology* 73, 129–132.

Ayal, Y. (1987) The foraging strategy of *Diaeretiella rapae*. I. The concept of the elementary unit of foraging. *Journal of Animal Ecology* 56, 1057–1068.

Bai, B. and Mackauer, M. (1990) Oviposition and host-feeding patterns in *Aphelinus asychis* (Hymenoptera: Aphelinidae) at different aphid densities. *Ecological Entomology* 15, 9–16.

Banks, C.J. (1957) The behaviour of individual coccinellid larvae on plants. *British Journal of Animal Behaviour* 5, 12–24.

Banks, C.J. (1962) Effects of the ant, *Lasius niger* (L.), on insects preying on small populations of *Aphis fabae* Scop. on bean plants. *Annals of Applied Biology* 50, 669–679.

Bänsch, R. (1964) Vergleichende Untersuchungen zur Biologie und zum Beutefangverhalten aphidivorer Coccinelliden, Chrysopiden und Syrphiden. *Zoologische Jahrbücher für Systematik* 91, 271–340.

Bargen, H., Sudhoff, K. and Poehling, H.M. (1998) Prey finding by larvae and adult females of *Episyrphus balteatus*. *Entomologia Experimentalis et Applicata* 87, 245–254.

Barnes, H.F. (1929) Gall midges (Dipt., Cecidomyiidae) as enemies of aphids. *Bulletin of Entomological Research* 20, 433–442.

Basky, Z. and Hopper, K.R. (2000) Impact of plant density and natural enemy exclosure on abundance of *Diuraphis noxia* (Kurdjumov) and *Rhopalosiphum padi* (L.) (Hom., Aphididae) in Hungary. *Journal of Applied Entomology* 124, 99–103.

Battaglia, D., Pennacchio, F., Marincola, G. and Tranfaglia, A. (1993). Cornicle secretion of *Acyrthosiphum pisum* (Homoptera: Aphididae) as a contact kairomone for the parasitoid *Aphidius ervi* (Hymenoptera: Braconidae). *European Journal of Entomology* 90, 423–428.

Battaglia, D., Pennacchio, F., Romano, A. and Tranfaglia, A. (1995) The role of physical cues in the regulation of host recognition and acceptance behavior of *Aphidius ervi* Haliday (Hymenoptera: Braconidae). *Journal of Insect Behavior* 8, 739–750.

Belcher, D.W. and Thurston, R. (1982) Inhibition of movement of larvae of the convergent lady beetle by leaf trichomes of tobacco. *Environmental Entomology* 11, 91–94.

Bell, W. (1990) Searching behaviour patterns in insects. *Annual Review of Entomology* 35, 447–467.

Belliure, B. and Michaud, J.P. (2001) Biology and behaviour of *Pseudodorus clavatus* (Diptera: Syrphidae), an important predator of citrus aphids. *Annals of the Entomological Society of America* 94, 91–96.

Bergeson, E. and Messina, F.J. (1998) Effects of co-occurring aphids on the susceptibility of the Russian wheat aphid to lacewing predators. *Entomologia Experimentalis et Applicata* 87, 103–108.

Birch, A.N.E., Geoghegan, I.E., Majerus, M.E.N., McNicol, J.W., Hackett, C.A., Gatehouse, A.M.R. and Gatehouse, J.A. (1999) Tri-trophic interactions involving pest aphids, predatory 2-spot ladybirds and transgenic potatoes expressing snowdrop lectin for aphid resistance. *Molecular Breeding* 5, 75–83.

Bishop, D.B. and Bristow, C.M. (2001) Effect of Allegheny mound ant (Hymenoptera: Formicidae) presence on homopteran and predator populations in Michigan jack pine forests. *Annals of the Entomological Society of America* 94, 33–40.

Bitton, S., Kenneth, R.G. and Ben-Ze'ev, I. (1979) Zygospore overwintering and sporulative germination in *Triplosporium fresenii* (Entomophthoraceae) attacking *Aphis spiraecola* on citrus in Israel. *Journal of Invertebrate Pathology* 34, 295–302.

Boo, K.-S., Chung, I.B., Han, K.S., Pickett, J.A. and Wadhams, L.J. (1998) Responses of the lacewing *Chrysopa carnea* to pheromones of its aphid prey. *Journal of Chemical Ecology* 24, 631–643.

Booij, C.J.H. and Noorlander, J. (1992) Farming systems and insect predators. *Agriculture, Ecosystems and Environment* 40, 125–135.

Bouchard, Y. and Cloutier, C. (1984) Honeydew as a source of host-searching kairomones for the aphid parasitoid *Aphidius nigripes* (Hymenoptera, Aphidiidae). *Canadian Journal of Zoology* 62, 1513–1520.

Boyle, H. and Barrows, E.M. (1978) Oviposition and host feeding behavior of *Aphelinus asychis* (Hymenoptera: Chalcidoidea: Aphelinidae) on *Schizaphis graminum* (Homoptera: Aphididae) and some reactions of aphids to this parasite. *Proceedings of the Entomological Society of Washington* 80, 441–455.

Bradley, G.A. (1973) Effect of *Formica obsuripes* (Hymenoptera: Formicidae) on the predator–prey relationship between *Hyperaspis congressis* (Coleoptera: Coccinellidae) and *Toumeyella numismaticum* (Homoptera: Coccidae). *Canadian Entomologist* 105, 1113–1118.

Branquart, E. and Hemptinne, J.L. (2000) Development of ovaries, allometry and reproductive traits and fecundity of *Episyrphus balteatus* (Diptera: Syrphidae). *European Journal of Entomology* 97, 165–170.

Brobyn, P.J., Clark, S.J. and Wilding, N. (1988) The effect of fungus infection of *Metopolophium dirhodum* (Hom.: Aphididae) on the oviposition behaviour of the aphid parasitoid *Aphidius rhopalosiphi* (Hym.: Aphidiidae). *Entomophaga* 33, 333–338.

Brodeur, J. and Rosenheim, J.A. (2000) Intraguild interactions in aphid parasitoids. *Entomologia Experimentalis et Applicata* 97, 93–108.

Buckley, R. (1987) Interactions involving plants, Homoptera, and ants. *Annual Review of Ecology and Systematics* 18, 111–135.

Budenberg, W.J. (1990) Honeydew as a contact kairomone for aphid parasitoids. *Entomologia Experimentalis et Applicata* 55, 139–148.

Budenberg, W.J. and Powell, W. (1992) The role of honeydew as an oviposition stimulant for two species of syrphids. *Entomologia Experimentalis et Applicata* 64, 57–61.

Buitenhuis, R., McNeil, J.N., Boivin, G. and Brodeur, J. (2004) The role of honeydew in host searching of aphid hyperparasitoids. *Journal of Chemical Ecology* 30, 273–285.

Buitenhuis, R., Vet, L.E.M., Boivin, G. and Brodeur, J. (2005) Foraging behaviour at the fourth trophic level: a comparative study of host location in aphid hyperparasitoids. *Entomologia Experimentalis et Applicata* 114, 107–117.

Butt, T.M., Beckett, A. and Wilding, N. (1981) Protoplasts in the *in vivo* lifecycle of *Erynia neoaphidis*. *Journal of General Microbiology* 127, 417–421.

Butt, T.M., Beckett, A. and Wilding, N. (1990) A histological study of the invasive and developmental process of the aphid pathogen *Erynia neoaphidis* (Zygomycetes: Entomophthorales) in the pea aphid *Acyrthosiphon pisum*. *Canadian Journal of Botany* 68, 2153–2163.

Caltagirone, L.E. and Doutt, R.L. (1989) The history of the vedalia beetle importation to California and its impact on the development of biological control. *Annual Review of Entomology* 34, 1–16.

Campbell, A., Frazer, B.D., Gilbert, N., Gutierrez, A.P. and Mackauer, M. (1974) Temperature requirements of some aphids and their parasites. *Journal of Applied Ecology* 11, 431–438.

Campbell, C.A.M. (1990) Introduced predators for biological control of hop pests. *Proceedings of the British Crop Protection Conference, Pests and Diseases, Brighton, November 1990* 1, 333–336.

Canard, M. and Duelli, P. (1984) Predatory behaviour of larvae and cannibalism. In: Canard, M., Semeria, Y. and New, T.R. (eds) *Biology of Chrysopidae. Series Entomologica, Volume 27.* W. Junk, The Hague, pp. 92–100.

Canard, M., Semeria, Y. and New, T.R. (eds) (1984) *Biology of Chrysopidae. Series Entomologica, Volume 27.* W. Junk, The Hague, 294 pp.

Carter, M.C. and Dixon, A.F.G. (1982) Habitat quality and the foraging behaviour of coccinellid larvae. *Journal of Animal Ecology* 51, 865–878.

Carter, M.C. and Dixon, A.F.G. (1984a) Honeydew: an arrestant stimulus for coccinellids. *Ecological Entomology* 9, 383–387.

Carter, M.C. and Dixon, A.F.G. (1984b) Foraging behaviour of coccinellid larvae: duration of intensive search. *Entomologia Experimentalis et Applicata* 36, 133–136.

Chakrabarti, S., Debnath, N. and Ghosh, D. (1991) Developmental rate, larval voracity and oviposition of *Cunctochrysa jubigensis* (Neuroptera: Chrysopidae), an aphidophagous predator in the Western Himalaya. In: Polgár, L., Chambers, R., Dixon, A.F.G. and Hodek, I. (eds) *Behaviour and Impact of Aphidophaga.* SPB Academic Publishing, The Hague, pp. 107–113.

Chambers, R.J. (1988) Syrphidae. In: Minks, A.K. and Harrewijn, P. (eds) *Aphids. Their Biology, Natural Enemies and Control, Volume 2B.* Elsevier, Amsterdam, pp. 259–270.

Chambers, R.J. (1991) Oviposition by aphidophagous hoverflies (Diptera: Syrphidae) in relation to aphid density and distribution in winter wheat. In: Polgár, L.A., Chambers, R., Dixon, A.F.G. and Hodek, I. (eds) *Behaviour and Impact of Aphidophaga.* SPB Academic Publishing, The Hague, pp. 115–121.

Chambers, R.J. and Adams, T.H.L. (1986) Quantification of the impact of hoverflies (Diptera: Syrphidae) on cereal aphids in winter wheat: an analysis of field populations. *Journal of Applied Ecology* 23, 895–904.

Chang, G.C. (1996) Comparison of single versus multiple species of generalist predators for biological control. *Environmental Entomology* 25, 207–212.

Chau, A. and Mackauer, M. (2000) Host-instar selection in the aphid parasitoid *Monoctonus paulensis* (Hymenoptera, Braconidae, Aphidiinae): a preference for small pea aphids. *European Journal of Entomology* 97, 347–353.

Chau, A. and Mackauer, M. (2001a) Adaptive self-superparasitism in a solitary parasitoid wasp: the influence of clutch size on offspring size. *Functional Ecology* 15, 335–343.

Chau, A. and Mackauer, M. (2001b) Host-instar selection in the aphid parasitoid *Monoctonus paulensis* (Hymenoptera, Braconidae, Aphidiinae): assessing costs and benefits. *Canadian Entomologist* 133, 549–564.

Chiverton, P.A. (1986) Predator density manipulation and its effect on populations of *Rhopalosiphum padi* (Hom.: Aphididae) in spring barley. *Annals of Applied Biology* 109, 49–60.

Chow, F.J. and Mackauer, M. (1984) Inter- and intraspecific larval competition in *Aphidius smithi* and *Praon pequodorum* (Hymenoptera: Aphidiidae). *Canadian Entomologist* 116, 1097–1107.

Chow, F.J. and Mackauer, M. (1986) Host discrimination and larval competition in the aphid parasite *Ephedrus californicus*. *Entomologia Experimentalis et Applicata* 41, 243–254.

Clark, T.L. and Messina, F.J. (1998) Foraging behavior of lacewing larvae (Neuroptera: Chrysopidae) on plants with divergent architectures. *Journal of Insect Behavior* 11, 303–317.

Cloutier, C. and Baudouin, F. (1990) Searching behaviour of the aphid parasitoid *Aphidius nigripes* (Hymenoptera: Aphidiidae) foraging on potato plants. *Environmental Entomology* 19, 222–228.

Cloutier, C., Lévesque, A., Eaves, D.M. and Mackauer, M. (1991) Maternal adjustment of sex ratio in response to host size in the aphid parasitoid *Ephedrus californicus*. *Canadian Journal of Zoology* 69, 1489–1495.

Cloutier, C., Duperron, M., Tertuliano, M. and McNeil, J.N. (2000) Host instar, body size and fitness in the koinobiotic parasitoid *Aphidius nigripes*. *Entomologia Experimentalis et Applicata* 97, 29–40.

Coderre, D., Provencher, L. and Tourneur, J.-C. (1987) Oviposition and niche partitioning in aphidophagous insects on maize. *Canadian Entomologist* 119, 195–203.

Cohen, M.B. and Mackauer, M. (1987) Intrinsic rate of increase and temperature coefficients of the aphid parasite *Ephedrus californicus* Baker (Hymenoptera: Aphidiidae). *Canadian Entomologist* 119, 231–237.

Colfer, R.G. and Rosenheim, J.A. (2000) Predation on immature parasitoids and its influence on aphid population suppression. *Oecologia* 126, 292–304.

Colley, M.R. and Luna, J.M. (2000) Relative attractiveness of potential beneficial insectary plants to aphidophagous hoverflies (Diptera: Syrphidae). *Environmental Entomology* 29, 1054–1059.

Cook, J.M. (1993) Sex determination in the Hymenoptera: a review of models and evidence. *Heredity* 71, 421–435.

Couchman, J.P. and King, P.E. (1977) Morphology of the larval stages of *Diaeretiella rapae* (M'Intosh) (Hymenoptera: Aphidiidae). *International Journal of Insect Morphology and Embryology* 6, 127–136.

Darcy, C.J., Burnett, P.A. and Hewings, A.D. (1981) Detection, biological effects and transmission of a virus of the aphid *Rhopalosiphum padi*. *Virology* 114, 268–272.

Dean, G.J. and Wilding, N. (1973) Infection of cereal aphids by the fungus *Entomophthora*. *Annals of Applied Biology* 74, 133–138.

Dean, G.J., Jones, M.G. and Powell, W. (1981) The relative abundance of the hymenopterous parasites attacking *Metopolophium dirhodum* (Walker) and *Macrosiphum avenae* (F.) (Hem., Aphidiidae). *Bulletin of Entomological Research* 71, 307–315.

Dedryver, C.A. (1983) Field pathogenesis of three species of Entomophthorales of cereal aphids in western France. In: Cavalloro, R. (ed.) *Aphid Antagonists*. Balkema, Rotterdam, pp. 11–19.

Dennis, P. (1991) The temporal and spatial distribution of arthropod predators of the aphids *Rhopalosiphum padi* W. and *Sitobion avenae* F. in cereals next to field margin habitats. *Norwegian Journal of Agricultural Sciences* 5, 79–88.

Dixon, A.F.G. (1959) An experimental study of the searching behaviour of the predatory coccinellid beetle *Adalia decempunctata* L. *Journal of Animal Ecology* 28, 259–281.

Dixon, A.F.G. (1987) Parthenogenetic reproduction and the rate of incease in aphids. In: Minks, A.K. and Harrewijn, P. (eds) *Aphids. Their Biology, Natural Enemies and Control, Volume 2A*. Elsevier, Amsterdam, pp. 269–287.

Dixon, A.F.G. (1997) Patch quality and fitness in predatory ladybirds. *Ecological Studies* 130, 205–223.

Dixon, A.F.G. (2000) *Insect Predator–Prey Dynamics: Ladybird Beetles and Biological Control*. Cambridge University Press, Cambridge, 257 pp.

Dogan, E.B., Berry, R.E., Reed, G.L. and Rossignol, P.A. (1996) Biological parameters of convergent lady beetle (Coleoptera: Coccinellidae) feeding on aphids (Homoptera: Aphidiidae) on transgenic potatoes. *Journal of Economic Entomology* 89, 1105–1108.

Doumbia, M., Hemptinne, J.L. and Dixon, A.F.G. (1998) Assessment of patch quality by ladybirds: role of larval tracks. *Oecologia* 113, 197–202.

Down, R.E., Ford, L., Woodhouse, S.D., Raemaekers, R.J.M., Leitch, B., Gatehouse, J.A. and Gatehouse, A.M.R. (2000) Snowdrop lectin (GNA) has no acute toxic effect on a beneficial insect predator, the 2-spot ladybird (*Adalia bipunctata* L.). *Journal of Insect Physiology* 46, 379–391.

van Driesche, R. (ed.) *Proceedings of the International Symposium on Biological Control of Arthropods, Honolulu, January 2002*. USDA Forest Service, Morgantown, West Virginia, pp. 199–208.

Dromph, K., Pell, J.K. and Eilenberg, J. (1997) Sporulation of *Erynia neoaphidis* from alate *Sitobion avenae*. *Insect Pathogens and Insect Parasitic Nematodes, Bulletin IOBC/WPRS* 21, 91–94.

Drukker, B., Bruin, J. and Sabelis, M.W. (2000) Anthocorid predators learn to associate herbivore-induced plant volatiles with presence or absence of prey. *Physiological Entomology* 25, 260–265.

Du, Y.-J., Poppy, G.M. and Powell, W. (1996) Relative importance of semiochemicals from first and second trophic levels in host foraging behavior of *Aphidius ervi*. *Journal of Chemical Ecology* 22, 1591–1605.

Du, Y.-J., Poppy, G.M., Powell, W.J., Pickett, J.A., Wadhams, L. and Woodcock, C.L. (1998) Identification of semiochemicals released during aphid feeding that attract the parasitoid *Aphidius ervi*. *Journal of Chemical Ecology* 24, 1355–1368.

Duelli, P. (1984) Oviposition. In: Canard, M., Semeria, Y. and New, T.R. (eds) *Biology of Chrysopidae. Series Entomologica, Volume 27*. W. Junk, The Hague, pp. 129–133.

Duelli, P. (1987) The influence of food on the oviposition-site selection in a predatory and in a honeydew-feeding lacewing species (Planipennia: Chrysopidae). *Neuroptera International* 4, 205–210.

Dutton, A., Klein, H., Romeis, J. and Bigler, F. (2002) Uptake of Bt-toxin by herbivores feeding on transgenic maize and consequences for the predator *Chrysoperla carnea*. *Ecologica Entomology* 27, 441–447.

Dwumfour, E.F. (1992) Volatile substances evoking orientation in the predatory flower bug *Anthocoris nemorum* (Heteroptera: Anthocoridae). *Bulletin of Entomological Research* 82, 465–469.

Edwards, C.A., Sunderland, K.D. and George, K.S. (1979) Studies on polyphagous predators of cereal aphids. *Journal of Applied Ecology* 16, 811–823.

Eigenbrode, S.D. and Kabalo, N.N. (1999) Effects of *Brassica oleracea* waxblooms on predation and attachment by *Hippodamia convergens*. *Entomologia Experimentalis et Applicata* 91, 125–130.

Eigenbrode, S.D., White, C., Rohde, M. and Simon, C.J. (1998) Behavior and effectiveness of adult *Hippodamia convergens* (Coleoptera: Coccinellidae) as a predator of *Acyrthosiphon pisum* on a glossy-wax mutant of *Pisum sativum*. *Environmental Entomology* 27, 902–909.

Eisner, T., Hicks, K., Eisner, M. and Robson, D.S. (1978) 'Wolf-in-sheep's-clothing' strategy of a predaceous insect larva. *Science, New York* 199, 790–794.

Ekbom, B.S. and Pickering, J. (1990) Pathogenic fungal dynamics in a fall population of the blackmargined aphid (*Monellia caryella*). *Entomologia Experimentalis et Applicata* 57, 29–37.

El Titi, A. (1972) Einfluss von Beutetierdichte und Morphologie der Wirtspflanze auf die Eiablage von *Aphidoletes aphidimyza* (Rond.) (Diptera: Cecidomyiidae). *Zeitschrift für Angewandte Entomologie* 72, 400–415.

El Titi, A. (1974) Zur Auslösung der Eiablage bei der aphidophagen Gallmücke *Aphidoletes aphidimyza* (Rond.) (Diptera: Cecidomyiidae). *Entomologia Experimentalis et Applicata* 17, 9–21.

van Emden, H.F. (1986) The interaction of plant resistance and natural enemies: effects on populations of sucking insects. In: Boethel, D.J. and Eikenbary, R.D. (eds) *Interactions of Plant Resistance and Parasitoids and Predators of Insects*. Ellis Horwood, Chichester, pp. 138–150.

van Emden, H.F. (1995) Host plant–aphidophaga interactions. *Agriculture, Ecosystems and Environment* 52, 3–11.

van Emden, H.F. (2003) Conservation biological control: from theory to practice. In: van Emden, H.F. and Hagen, K.S. (1976) Olfactory reactions of the green lacewing, *Chrysopa carnea*, to tryptophan and certain breakdown products. *Environmental Entomology* 5, 469–473.

van Emden, H.F. and Wearing, C.H. (1965) The role of the aphid host plant in delaying economic damage levels in crops. *Annals of Applied Biology* 56, 323–324.

Ettifouri, M. and Ferran, A. (1992) Influence d'une alimentation préalable et du jeune sur l'apparition de la recherche intensive des proies chez *Semiadalia undecimnotata*. *Entomologia Experimentalis et Applicata* 65, 101–111.

Evans, G.A., Polaszek, A. and Bennett, F.D. (1995) The taxonomy of the *Encarsia flavoscutellum* species-group (Hymenoptera: Aphelinidae), parasitoids of Hormaphididae (Homoptera, Aphidoidea). *Oriental Insects* 29, 33–45.

Fabre, J.P. and Rabasse, J.M. (1987) Introduction dans le sud-est de la France d'un parasite, *Pauesia cedrobii* (Hymenoptera: Aphidiidae) du puceron, *Cedrobium laportei* (Homoptera: Lachnidae) du cèdre de l'atlas: *Cedrus atlantica*. *Entomophaga* 32, 127–141.

Fauvergue, X., Hopper, K.H. and Antolin, M.F. (1995) Mate finding via a trail sex pheromone by a parasitoid wasp. *Proceedings of the National Academy of Sciences of the United States of America* 92, 900–904.

Feng, M.G., Johnson, J.B. and Kish, L.P. (1990) Survey of entomopathogenic fungi naturally infecting cereal aphids (Homoptera: Aphididae) of irrigated grain crops in southwestern Idaho. *Environmental Entomology* 19, 1534–1542.

Feng, M.G., Johnson, J.B. and Halbert, S.E. (1991) Natural control of cereal aphids (Homoptera: Aphididae) by entomopathogenic fungi (Zygomycetes: Entomophthorales) and parasitoids (Hymenoptera: Braconidae and Encyrtidae) on irrigated spring wheat in southwestern Idaho. *Environmental Entomology* 20, 1699–1710.

Feng, M.G., Nowierski, R.M., Klein, R.E., Scharen, A.L. and Sands, D.C. (1992) Spherical hyphal bodies of *Pandora neoaphidis* (Remaudière and Hennebert) Humber (Zygomycetes: Entomophthorales) on *Acyrthosiphon pisum* Harris (Homoptera: Aphididae): a potential overwintering form. *Pan-Pacific Entomologist* 68, 100–104.

Feng, M.G., Poprawski, T.J. and Khachatourians, G.G. (1994) Production, formulation and application of the entomopathogenic fungus *Beauveria bassiana* for insect control: current status. *Biocontrol Science and Technology* 4, 3–34.

Ferran, A. and Deconchat, R. (1992) Exploration of wheat leaves by *Coccinella septempunctata* L. (Coleoptera, Coccinellidae) larvae. *Journal of Insect Behavior* 5, 147–159.

Ferran, A. and Dixon, A.F.G. (1993) Foraging behaviour of ladybird larvae (Coleoptera: Coccinellidae). *European Journal of Entomology* 90, 383–402.

Ferran, A., Ettifouri, M., Clement, P. and Bell, W.J. (1994) Sources of variability in the transition from extensive to intensive search in coccinellid predators (Coleoptera: Coccinellidae). *Journal of Insect Behavior* 7, 633–647.

Ferrari, J. and Godfray, H.C.J. (2003) Resistance to a fungal pathogen and host plant specialization in the pea aphid. *Ecology Letters* 6, 111–118.

Ferrari, J., Müller, C.B., Kraaijeveld, A.R. and Godfray, H.C.J. (2001) Clonal variation and covariation in aphid resistance to parasitoids and a pathogen. *Evolution* 55, 1805–1814.

Ferrari, J., Darby, A.C., Daniell, T.J., Godfray, H.C.J. and Douglas, A.E. (2004) Linking the bacterial community in pea aphids with host-plant use and natural enemy resistance. *Ecological Entomology* 29, 60–65.

Fink, U. and Völkl, W. (1995) The effect of abiotic factors on foraging and oviposition success of the aphid parasitoid, *Aphidius rosae. Oecologia* 103, 371–378.

Fischer, M.K., Hoffmann, K.H. and Völkl, W. (2001) Competition for mutualists in an ant–homopteran interaction mediated by hierarchies of ant-attendance. *Oikos* 92, 531–541.

Flint, H.M., Slater, S.S. and Walters, S. (1979) Caryophyllene: an attractant for the green lacewing. *Environmental Entomology* 8, 1123–1125.

Force, D.C. and Messenger, P.S. (1964) Fecundity, reproductive rate and innate capacity of increase of three parasites of *Therioaphis maculata* (Buckton) reared at various constant temperatures. *Ecology* 45, 706–715.

Foster, S.P., Denholm, I., Thompson, R., Poppy, G.M. and Powell, W. (2005) Reduced response of insecticide-resistant aphids and attraction of parasitoids to aphid alarm pheromone; a potential fitness trade-off. *Bulletin of Entomological Research* 95, 37–46.

Foster, W.A. (1975) Life history and population biology of an intertidal aphid, *Pemphigus trehernei* Foster. *Transactions of the Royal Entomological Society of London* 127, 192–207.

Frazer, B.D. and MacGregor, R.R. (1994) Searching behaviour of adult female Coccinellidae (Coleoptera) on stem and leaf models. *Canadian Entomologist* 126, 389–399.

Frazer, B.D., Gilbert, N., Nealis, V. and Raworth, D.A. (1981) Control of aphid density by a complex of predators. *Canadian Entomologist* 113, 1035–1041.

Fréchette, B. and Coderre, D. (2000) Oviposition strategy of the green lacewing *Chrysoperla rufilabris* (Neuroptera: Chrysopidae) in response to extraguild prey availability. *European Journal of Entomology* 97, 507–510.

Fuentes-Contreras, E., Pell, J.K. and Niemeyer, H.M. (1998) Influence of plant resistance at the third trophic level: interactions between parasitoids and entomopathogenic fungi of cereal aphids. *Oecologia* 117, 426–432.

Fuxa, J.R. and Tanada, Y. (1987) *Epizootiology of Insect Diseases*. Wiley, New York, 555 pp.

Gaimari, S.D. and Turner, W.J. (1997) Behavioural observations on the adults and larvae of *Leucopis ninae* and *L. gaimarii* (Diptera: Chamaemyiidae), predators of the Russian wheat aphid, *Diuraphis noxia* (Homoptera: Aphididae). *Journal of the Kansas Entomological Society* 70, 153–159.

George, K.S. (1957) Preliminary investigations on the biology and ecology of the parasites and predators of *Brevicoryne brassicae* (L.). *Bulletin of Entomological Research* 48, 619–629.

Gerling, D., Roitberg, B.D. and Mackauer, M. (1990) Instar-specific defense of the pea aphid, *Acyrthosiphon pisum*: influence on oviposition success of the parasite *Aphelinus asychis* (Hymenoptera: Aphelinidae). *Journal of Insect Behavior* 3, 501–514.

Gilbert, F.S. (1993) *Hoverflies*. Naturalists' Handbooks No. 5, 2nd edn. Richmond Press, Slough, 72 pp.

Gilbert, F. and Owen, J. (1990) Size, shape, competition, and community structure in hoverflies (Diptera, Syrphidae). *Journal of Animal Ecology* 59, 21–39.

Gilkeson, L.A. and Hill, S.B. (1986) Diapause prevention in *Aphidoletes aphidimyza* (Diptera: Cecidomyiidae) by low-intensity light. *Environmental Entomology* 15, 1067–1069.

Glare, T.R., Chilvers, G.A. and Milner, R.J. (1985a) A simple method for inoculating aphids with capilliconidia. *Transactions of the British Mycological Society* 85, 353–354.

Glare, T.R., Chilvers, G.A. and Milner, R.J. (1985b) Capilliconidia as infective spores in *Zoophthora phalloides* (Entomophthorales). *Transactions of the British Mycological Society* 85, 463–470.

Glinwood, R.T., Du, Y.-J. and Powell, W. (1999) Responses to aphid sex pheromones by the pea aphid parasitoids *Aphidius ervi* and *Aphidius eadyi*. *Entomologia Experimentalis et Applicata* 92, 227–232.

Goeldlin de Tiefenau, P. (1974) Contribution to the systematics and ecology of the Syrphidae (Diptera) of western Switzerland. *Mitteilungen der Schweizerischen Entomologischen Gesellschaft* 47, 151–252.

Gordon, R.D. (1985) The Coccinellidae (Coleoptera) of America north of Mexico. *Journal of the New York Entomological Society* 93, 1–912.

Grasswitz, T.R. and Burts, E.C. (1995) Effect of native natural enemies and augmentative releases of *Chrysoperla rufilabris* Burmeister and *Aphidoletes aphidimyza* (Rondani) on the population dynamics of the green apple aphid, *Aphis pomi* DeGeer. *International Journal of Pest Management* 41, 176–183.

Grasswitz, T.R. and Paine, T.D. (1992) Kairomonal effect of an aphid cornicle secretion on *Lysiphlebus testaceipes* (Cresson) (Hymenoptera: Aphidiidae). *Journal of Insect Behavior* 5, 447–457.

Grevstad, F.S. and Klepetka, B.W. (1992) The influence of plant architecture on the foraging efficiencies of a suite of ladybird beetles feeding on aphids. *Oecologia* 92, 399–404.

Gries, G. (1986) Zum Beutefangverhalten der Schwebfliegenlarve *Syrphus balteatus* Deg. (Diptera, Syrphidae). *Journal of Applied Entomology* 102, 309–313.

Griffiths, D.C. (1961) The development of *Monoctonus paludum* Marshall (Hym., Braconidae) in *Nasonovia ribisnigri* on lettuce, and immunity reactions in other lettuce aphids. *Bulletin of Entomological Research* 52, 147–163.

Guerrieri, E., Pennacchio, F. and Tremblay, E. (1993) Flight behaviour of the aphid parasitoid *Aphidius ervi* (Hymenoptera: Braconidae) in response to plant and host volatiles. *European Journal of Entomology* 90, 415–421.

Guerrieri, E., Pennacchio, F. and Tremblay, E. (1997) Effect of adult experience on in-flight orientation to plant and plant-host complex volatiles in *Aphidius ervi* Haliday (Hymenoptera: Braconidae). *Biological Control* 10, 159–165.

Guerrieri, E., Poppy, G.M., Powell, W., Tremblay, E. and Pennacchio, F. (1999) Induction and systemic release of herbivore-induced synomones regulating in-flight orientation of *Aphidius ervi* (Hymenoptera: Braconidae). *Journal of Chemical Ecology* 25, 1247–1261.

Guerrieri, E., Poppy, G.M., Powell, W., Rao, R. and Pennacchio, F. (2002) Plant-to-plant communication mediating in-flight orientation of *Aphidius ervi*. *Journal of Chemical Ecology* 28, 1703–1715.

Gutierrez, A.P., Baumgaertner, J.U. and Summers, C.G. (1984) Multitrophic models of predator prey energetics. III. A case study in an alfalfa ecosystem. *Canadian Entomologist* 116, 950–963.

Gwynn, D.M., Callaghan, A., Gorham, J., Walters, K.F.A. and Fellowes, M.D.E. (2005) Resistance is costly: trade-offs between immunity, fecundity and survival in the pea aphid. *Proceedings of the Royal Society B* 272, 1803–1808.

Hafez, M. (1961) Seasonal fluctuations of population density of the cabbage aphid *Brevicoryne brassicae* (L.) in the Netherlands, and the role of its parasite *Aphidius (Diaeretiella) rapae* (Curtis). *Tijdschrift voor Plantenziekten* 67, 445–548.

Hagen, K.S. (1962) Biology and ecology of predaceous Coccinellidae. *Annual Review of Entomology* 7, 289–326.

Hagen, K.S. and van den Bosch, R. (1968) Impact of pathogens, parasites and predators of aphids. *Annual Review of Entomology* 13, 325–384.

Hagley, E.A.C. (1989) Release of *Chrysoperla carnea* Stephens (Neuroptera: Chrysopidae) for control of the green apple aphid, *Aphis pomi* DeGeer (Homoptera: Aphididae). *Canadian Entomologist* 121, 309–314.

Hågvar, E.B. and Hofsvang, T. (1991a) Aphid parasitoids (Hymenoptera, Aphidiidae): biology, host selection and use in biological control. *Biocontrol News and Information* 12, 13–41.

Hågvar, E.B. and Hofsvang, T. (1991b) Effect of honeydew and hosts on plant colonization by the aphid parasitoid *Ephedrus cerasicola*. *Entomophaga* 34, 495–501.

Halme, J. (1986) Aphidiidae (Hymenoptera) of Finland. II. *Pseudopauesia prunicola* gen. n., sp. n., a new parasitoid of *Rhopalosiphum padi* (L.) (Homoptera: Aphididae). *Annales Entomologici Fennici* 52, 20–27.

Hamilton, P.A. (1973) The biology of *Aphelinus flavus* (Hym. Aphelinidae), a parasite of the sycamore aphid *Drepanosiphum platanoides* (Hemipt. Aphididae). *Entomophaga* 18, 449–462.

Hamilton, P.A. (1974) The biology of *Monoctonus pseudoplatani, Trioxys cirsii* and *Dyscritulus planiceps*, with notes on their effectiveness as parasites of the sycamore aphid, *Drepanosiphum platanoides*. *Annales de la Societé Entomologique Française, N.S.* 10, 821–840.

Hardie, J., Hick, A.J., Höller, C., Mann, J., Merritt, L., Nottingham, S.F., Powell, W., Wadhams, L.J., Witthinrich, J. and Wright, A.F. (1994) The responses of *Praon* spp. parasitoids to aphid sex pheromone components in the field. *Entomologia Experimentalis et Applicata* 71, 95–99.

Harris, K.M. (1973) Aphidophagous Cecidomyiidae (Diptera): taxonomy, biology and assessment of field populations. *Bulletin of Entomological Research* 63, 305–325.

Hart, A.J. and Bale, J.S. (1998) Factors affecting the freeze tolerance of the hoverfly *Syrphus ribesii* (Diptera: Syrphidae). *Journal of Insect Physiology* 44, 21–29.

Hatting, J.L., Humber, R.A., Poprawski, T.J. and Miller, R.M. (1999) A survey of fungal pathogens of aphids from South Africa with special reference to cereal aphids. *Biological Control* 16, 1–12.

Havelka, J. (1982) Biology and ecology of the predacious gall midge *Aphidoletes aphidimyza* (Rond.) (Diptera, Cecidomyiidae). Its laboratory mass rearing and use against aphids in glasshouses. PhD thesis, All-Union Plant Protection Institute, Leningrad, USSR.

Havelka, J. and Růžička, Z. (1984) Selection of aphid species by ovipositing females and effect of larval food on the development and fecundity in *Aphidoletes aphidimyza* (Rondani) (Diptera: Cecidomyiidae). *Zeitschrift für Angewandte Entomologie* 98, 432–437.

Havelka, J. and Zemek, R. (1999) Life-table parameters and oviposition dynamics of various populations of the predacious gallmidge *Aphidoletes aphidimyza*. *Entomologia Experimentalis et Applicata* 91, 481–484.

Heinz, K.M. and Parrella, M.P. (1994) Poinsettia (*Euphorbia pulcherrima* Willd. ex Koltz) cultivar-mediated differences in the performance of five natural enemies of *Bemisia argentifolii* Bellows and Perring n.sp. (Homoptera: Aleyrodidae). *Biological Control* 4, 305–308.

Hemmati, F. (1999) Aerial dispersal of the entomopathogenic fungus *Erynia neoaphidis*. PhD thesis. The University of Reading, Reading, UK.

Hemmati, F., Pell, J.K., McCartney, H.A. and Deadman, M.L. (2001a) Airborne concentrations of conidia of *Erynia neoaphidis* above cereal fields. *Mycological Research* 105, 485–489.

Hemmati, F., Pell, J.K., McCartney, H.A., Clark, S.J. and Deadman, M.L. (2001b) Active discharge in the aphid pathogenic fungus *Erynia neoaphidis*. *Mycological Research* 105, 715–722.

Hemptinne, J.L. and Dixon, A.F.G. (1991) Why have ladybirds been generally so ineffective in biological control? In: Polgár, L.A., Chambers, R., Dixon, A.F.G. and Hodek, I. (eds) *Behaviour and Impact of Aphidophaga*. SPB Academic Publishing, The Hague, pp. 149–157.

Hemptinne, J.L., Dixon, A.F.G. and Coffin, J. (1992) Attack strategy of ladybird beetles (Coccinellidae): factors shaping the numerical response. *Oecologia* 90, 238–245.

Hemptinne, J.L., Dixon, A.F.G., Doucet, J.L. and Petersen, J.E. (1993) Optimal foraging by hoverflies (Diptera: Syrphidae) and ladybirds (Coleoptera: Coccinellidae): mechanisms. *European Journal of Entomology* 90, 451–455.

Henter, H. (1995) The potential for coevolution in a host-parasitoid system. II. Genetic variation within a population of wasps in the ability to parasitize an aphid host. *Evolution* 49, 439–445.

Henter, H. and Via, S. (1995) The potential for coevolution in a host-parasitoid system. I. Genetic variation within an aphid population in susceptibility to a parasitic wasp. *Evolution* 49, 427–438.

van den Heuvel, J.F.J.M., Hummelen, H., Verbeek, M., Dullemans, A.M. and van der Wilk, F. (1997) Characteristics of *Acyrthosiphon pisum* virus, a newly identified virus infecting the pea aphid. *Journal of Invertebrate Pathology* 70, 169–176.

Hilbeck, A., Baumgartner, M., Fried, M.P. and Bigler, F. (1998a) Effects of transgenic *Bacillus thuringiensis* corn-fed prey on mortality and development time of immature *Chrysoperla carnea* (Neuroptera: Chrysopidae). *Environmental Entomology* 27, 480–487.

Hilbeck, A., Moar, W.J., Pusztai-Carey, M., Filippini, A. and Bigler, F. (1998b) Toxicity of *Bacillus thuringiensis* Cry1Ab toxin to the predator *Chrysoperla carnea* (Neuroptera: Chrysopidae). *Environmental Entomology* 27, 1255–1263.

Hindayana, D., Meyhöfer, R., Scholz, D. and Poehling, H.M. (2001) Intraguild predation among the hover fly *Episyrphus balteatus* de Geer (Diptera: Syrphidae) and other aphidophagous predators. *Biological Control* 20, 236–246.

Hochberg, M.E. and Lawton, J.H. (1990) Competition between kingdoms. *Trends in Evolution and Ecology* 5, 367–371.

Hodek, I. (1973) *Biology of Coccinellidae*. W. Junk, The Hague, 260 pp.

Hodek, I. and Honek, A. (1996) *Ecology of Coccinellidae*. Kluwer, Dordrecht, 464 pp.

Hodgson, C. and Aveling, C. (1988) Anthocoridae. In: Minks, A.K. and Harrewijn, P. (eds) *Aphids. Their Biology, Natural Enemies and Control, Volume 2B*. Elsevier, Amsterdam, pp. 279–292.

Hofsvang, T. (1988) Mechanisms of host discrimination and intraspecific competition in the aphid parasitoid *Ephedrus cerasicola*. *Entomologia Experimentalis et Applicata* 48, 233–239.

Holland, J.M. and Thomas, J.R. (1997) Quantifying the impact of polyphagous invertebrate predators in controlling cereal aphids and in preventing wheat yield and quality reductions. *Annals of Applied Biology* 131, 375–397.

Höller, C., Borgemeister, C., Haardt, H. and Powell, W. (1993) The relationship between primary parasitoids and hyperparasitoids of cereal aphids: an analysis of field data. *Journal of Animal Ecology* 62, 12–21.

Honek, A. (1980) Population density of aphids at the time of settling and ovariole maturation in *Coccinella septempunctata* (Col., Coccinellidae). *Entomophaga* 25, 427–430.

Honek, A., Jarosik, V., Lapchin, L. and Rabasse, J.-M. (1996) The effect of parasitism by *Aphelinus abdominalis* and drought on the walking movement of aphids. *Entomologia Experimentalis et Applicata* 87, 191–200.

Honek, A., Jarosik, V., Lapchin, L. and Rabasse, J.-M. (1998) Host choice and offspring sex allocation in the aphid parasitoid *Aphelinus abdominalis* (Hyemnoptera: Aphelinidae). *Journal of Agricultural Entomology* 15, 209–221.

Hou, Z.Y., Chen, X., Zhang, Y., Guo, B.Q. and Yan, F.S. (1997) EAG and orientation tests on the parasitoid *Lysiphlebia japonica* (Hym., Aphidiidae) to volatile chemicals extracted from host plants of cotton aphid *Aphis gosssypii* (Hom., Aphididae). *Journal of Applied Entomology* 121, 495–500.

Hufbauer, R.A. (2002a) Evidence for nonadaptive evolution in parasitoid virulence following a biological control introduction. *Ecological Applications* 12, 66–78.

Hufbauer, R.A. (2002b) Aphid population dynamics: does resistance to parasitism influence population size? *Ecological Entomology* 27, 25–32.

Hughes, R.D. (1989) Biological control in the open field. In: Minks, A.K. and Harrewijn, P. (eds) *Aphids. Their Biology, Natural Enemies and Control, Volume 2C*. Elsevier, Amsterdam, pp. 167–198.

Inglis, G.D., Goettel, M.S., Butt, T.M. and Strasser, H. (2001) Use of hyphomycetous fungi for managing insect pests. In: Butt, T.M., Jackson, C. and Magan, N. (eds) *Fungi as Biocontrol Agents: Progress, Problems and Potential*. CAB International, Wallingford, pp. 23–70.

Ito, K. and Iwao, S. (1977) Oviposition behaviour of the syrphid, *Episyrphus balteatus*, in relation to aphid density on the plant. *Japanese Journal of Applied Entomology and Zoology* 21, 130–134.

James, R.R. and Lighthart, B. (1994) Susceptibility of the convergent lady beetle (Coleoptera: Coccinellidae) to four entomogenous fungi. *Environmental Entomology* 23, 190–192.

James, R.R., Schaffer, B.T., Croft, B. and Lighthart, B. (1995) Field evaluation of *Beauveria bassiana*: its persistence and effects on the pea aphid and a non-target coccinellid in alfalfa. *Biocontrol Science and Technology* 5, 425–437.

James, R.R., Croft, B.A., Shaffer, B.T. and Lighthart, B. (1998) Impact of temperature and humidity on host–pathogen interactions between *Beauveria bassiana* and a coccinellid. *Environmental Entomology* 27, 1506–1513.

Jiggins, C., Majerus, M.E.N. and Gough, U. (1993) Ant defence of colonies of *Aphis fabae* Scopoli (Hemiptera: Aphididae), against predation by ladybirds. *British Journal of Entomology and Natural History* 6, 129–137.

Johki, Y., Obata, S. and Matsui, M. (1988) Distribution and behaviour of five species of aphidophagous ladybirds (Coleoptera) around aphid colonies. In: Niemczyk, E. and Dixon, A.F.G. (eds) *Ecology and Effectiveness of Aphidophaga*. SPB Academic Publishing, The Hague, pp. 35–38.

Kambhampati, S. and Mackauer, M. (1989) Multivariate assessment of inter- and intraspecific variation in performance criteria of several pea aphid parasites (Hymenoptera: Aphidiidae). *Annals of the Entomological Society of America* 82, 314–324.

Kan, E. (1988a) Assessment of aphid colonies by hoverflies. I. Maple aphids and *Episyrphus balteatus* (DeGeer) (Diptera: Syrphidae). *Journal of Ethology* 6, 39–48.

Kan, E. (1988b) Assessment of aphid colonies by hoverflies. II. Pea aphids and 3 syrphid species: *Betasyrphus serarius* (Wiedemann), *Metasyrphus frequens* Matsumura and *Syrphus vitripennis* (Meigen) (Diptera: Syrphidae). *Journal of Ethology* 6, 135–142.

Kan, E. and Sasakawa, M. (1986) Assessment of the maple aphid colony by the hoverfly *Episyrphus balteatus* (DeGeer) (Diptera: Syrphidae). *Journal of Ethology* 4, 121–127.

Kaneko, S. (2002) Aphid-attending ants increase the number of emerging adults of the aphid's primary parasitoids and hyperparasitoids by repelling intraguild predators. *Entomological Science* 5, 131–146.

Kaneko, S. (2003) Different impacts of two aphid-attending ants with different aggressiveness on the number of emerging adults of the aphid's primary parasitoids and hyperparasitoids. *Ecological Research* 18, 199–213.

Kareiva, P. (1990) The spatial dimensions in pest–enemy interactions. In: Mackauer, M., Ehler, L.E. and Roland, J. (eds) *Critical Issues in Biological Control*. Intercept, Andover, pp. 213–227.

Kareiva, P. and Odell, G. (1987) Swarms of predators exhibit 'preytaxis' if individual predators use area-restricted search. *American Naturalist* 130, 233–270.

Kareiva, P. and Perry, R. (1989) Leaf overlap and the ability of ladybirds to search among plants. *Ecological Entomology* 14, 127–129.

Kareiva, P. and Sahakian, R. (1990) Tritrophic effects of a single architectural mutation in pea plants. *Nature, London* 345, 433–434.

Kawauchi, S. (1991) Selection for highly prolific females in three aphidophagous coccinellids. In: Polgár, L.A., Chambers, R., Dixon, A.F.G. and Hodek, I. (eds) *Behaviour and Impact of Aphidophaga*. SPB Academic Publishing, The Hague, pp. 177–181.

Keller, S. (1987) Observations on the overwintering of *Entomophthora planchoniana*. *Journal of Invertebrate Pathology* 50, 333–335.

Keller, S. (1997) The genus *Neozygites* (Zygomycetes, Entomophthorales) with special reference to species found in tropical regions. *Sydowia* 49, 118–146.

Keller, S. and Suter, H. (1980) Epizootiologische Untersuchungen über das Entomophthora-Auftreten bei feldbaulich wichtigen Blattlausarten. *Acta Oecologica, Oecologia Applicata* 1, 63–81.

Kesten, U. (1969) Zur Morphologie und Biologie von *Anatis ocellata* (L.) (Coleoptera, Coccinellidae). *Zeitschrift für Angewandte Entomologie* 63, 412–445.

Kim, J.J., Kim, K.C. and Roberts, D.W. (2005) Impact of the entomopathogenic fungus *Verticillium lecanii* on development of an aphid parasitoid, *Aphidius colemani*. *Journal of Invertebrate Pathology* 88, 254–256.

Kindlmann, P. and Dixon, A.F.G. (1993) Optimal foraging in ladybirds (Coleoptera: Coccinelllidae) and its consequences for their use in biological control. *European Journal of Entomology* 90, 443–450.

Kish, L.P., Majchrowicz, I. and Biever, K.D. (1994) Prevalence of natural fungal mortality of green peach aphid (Homoptera: Aphididae) on potatoes and non-solanaceous hosts in Washington and Idaho. *Environmental Entomology* 23, 1326–1330.

Kobayashi, Y., Mogami, K. and Aoki, J. (1984) Ultrastructural studies on the hyphal growth of *Erynia neoaphidis* in the green peach aphid *Myzus persicae*. *Transactions of the Mycological Society of Japan* 25, 425–434.

Krause, U. and Poehling, H.M. (1996) Overwintering, oviposition and population dynamics of hoverflies (Diptera: Syrphidae) in Northern Germany in relation to small- and large-scale landscape structure. *Acta Jutlandica* 71, 157–169.

Kulp, D., Fortmann, M., Hommes, M. and Plate, H.-P. (1989) Die räuberische Gallmücke *Aphidoletes aphidimyza* (R.) (Dipt.: Cec.). Ein bedeutender Blattlausprädator. Nachschlagewerk zur Systematik, Verbreitung, Biologie, Zucht und Anwendung. *Mitteilungen Biologische Bundesanstalt für Land- und Forstwirtschaft* 250, 1–126.

Lacey, L.A., Mesquita, L.M., Mercadier, G., Debire, R., Kazmer, D.J. and Leclant, F. (1997) Acute and sublethal activity of the entomopathogenic fungus *Paecilomyces fumosoroseus* (Deuteromycotina: Hyphomycetes) on adult *Aphelinus asychis* (Hymenoptera: Aphelinidae). *Environmental Entomology* 26, 1452–1460.

Laska, P. (1978) Current knowledge of feeding specialization of different species of aphidophagous larvae of Syrphidae. *Annales des Zoologie-Écologie Animale* 10, 395–397.

Latgé, J.P. and Papierok, B. (1988) Aphid pathogens. In: Minks, A.K. and Harrewijn, P. (eds) *Aphids. Their Biology, Natural Enemies and Control, Volume 2B*. Elsevier, Amsterdam, pp. 323–335.

Lattin, J.D. (1999) Bionomics of the Anthocoridae. *Annual Review of Entomology* 44, 207–231.

Laubscher, J.M. and von Wechmar, M.B. (1992) Influence of aphid lethal paralysis virus and *Rhopalosiphum padi* virus on aphid biology at different temperatures. *Journal of Invertebrate Pathology* 60, 134–140.

Leite, L.G., Alves, S.B., Wraight, S.P., Galaini-Wraight, S., Roberts, D.W. and Magalhães (1996) Conidiogênese de *Zoophthora radicans* (Brefeld) Batko sobre *Empoasca kraemeri* (Ross and Moore) à diferentes temperaturas. *Arquivos Instituto Biológico, São Paulo* 63, 47–53.

Le Ralec, A. (1991) Les Hyménoptères parasitoides: adaptions de l'appareil reproducteur femelle. Morphologie et ultrastructure de l'ovaire, de l'oeuf et de l'ovipositeur. Dissertation, Université de Rennes I, Rennes, France.

Le Ralec, A., Curty, C. and Wajnberg, E. (2005) Inter-specific variation in the reactive distance of different aphid–parasitoid associations: analysis from automatic tracking of the walking path. *Applied Entomology and Zoology* 40, 413–420.

Li, C., Roitberg, B.D. and Mackauer, M. (1992) The search pattern of a parasitoid wasp, *Aphelinus asychis*. *Oikos* 65, 207–212.

Liepert, C. (1996) Chemische Mimikry bei Blattlausparasitoiden der Gattung *Lysiphlebus* (Hymenoptera, Aphidiidae). *Bayreuther Forum für Ökologie* 39, 1–141.

Liepert, C. and Dettner, K. (1993) Recognition of aphid parasitoids by honeydew-collecting ants: the role of cuticular lipids in a chemical mimicry system. *Journal of Chemical Ecology* 19, 2143–2153.

Liu, T.X. and Chen, T.Y. (2001) Effects of three aphid species (Homoptera: Aphididae) on development, survival and predation of *Chrysoperla carnea* (Neuroptera: Chrysopidae). *Applied Entomology and Zoology* 36, 361–366.

Lizen, E., Latteur, G. and Oger, R. (1985) Sensibilite à l'infection par l'Entomophthorale *Erynia neoaphidis* Remaud. et Henn. du puceron *Acyrthosiphon pisum* (Harris) selon sa forme, son stade et son age. *Parasitica* 41, 163–170.

Lucas, E. (2005) Intraguild predation among aphidophagous predators. *European Journal of Entomology* 102, 351–363.

Lucas, E. and Brodeur, J. (1999) Oviposition site selection by the predatory midge *Aphidoletes aphidimyza* (Diptera: Cecidomyiidae). *Environmental Entomology* 28, 622–627.

Lucas, E., Coderre, D. and Brodeur, J. (1998) Intraguild predation among aphid predators: characterization and influence of extraguild prey. *Ecology* 79, 1084–1092.

Mackauer, M. (1965) Parasitological data as an aid in aphid classification. *Canadian Entomologist* 97, 1016–1024.

Mackauer, M. (1976) The sex ratio in field populations of some aphid parasites. *Annals of the Entomological Society of America* 69, 453–456.

Mackauer, M. (1982) Fecundity and host utilization of the aphid parasitoid *Aphelinus semiflavus* (Hymenoptera: Aphelinidae) at two host densities. *Canadian Entomologist* 114, 721–726.

Mackauer, M. (1983) Quantitative assessment of *Aphidius smithi* (Hymenoptera, Aphidiidae): fecundity, intrinsic rate of increase and functional response. *Canadian Entomologist* 115, 399–415.

Mackauer, M. (1986) Growth and developmental interactions in some aphids and their hymenopterous parasites. *Journal of Insect Physiology* 32, 275–280.

Mackauer, M. (1990) Host discrimination and larval competition in solitary endoparasitoids. In: Mackauer, M. (1996) Sexual size dimorphism in solitary parasitoid wasps: influence of host quality. *Oikos* 76, 265–272.

Mackauer, M. and Chow, F.J. (1986) Parasites and parasite impact on aphid populations. In: McLean, G.D., Garret, R.G. and Ruesink, W.G. (eds) *Plant Virus Epidemics: Monitoring, Modelling and Predicting Outbreaks*. Academia Press, Sydney, pp. 95–117.

Mackauer, M. and Foottit, R. (1979) A gall midge, *Endaphis* sp. (Diptera: Cecidomyiidae), as a gregarious aphid parasite. *Canadian Entomologist* 111, 615–620.

Mackauer, M. and Kambhampati, S. (1988) Parasitism of aphid embryos by *Aphidius smithi*: some effects of extremely small host size. *Entomologia Experimentalis et Applicata* 49, 167–173.

Mackauer, M. and Starý, P. (1967) *World Aphidiidae*. Le François, Paris, 195 pp.

Mackauer, M. and Völkl, W. (1993) Regulation of aphid populations by aphidiid wasps: does aphidiid foraging behaviour or hyperparasitism limit impact? *Oecologia* 94, 339–350.

Mackauer, M. and Völkl, W. (2002) Brood-size and sex-ratio variation in field populations of three species of solitary aphid parasitoids (Hymenoptera: Braconidae, Aphidiinae). *Oecologia* 131, 296–305.

Mackauer, M., Michaud, J.P. and Völkl, W. (1996) Host choice by aphidiid parasitoids (Hymenoptera: Aphidiidae): host recognition, host quality and host value. *Canadian Entomologist* 128, 959–980.

Madsen, M., Terkiildsen, S. and Toft, S. (2004) Microscosm studies on control of aphids by generalist arthropod predators: effects of alternative prey. *Biocontrol* 49, 483–504.

Majerus, M.E.N. (1994) *Ladybirds*. Harper Collins, London, 367 pp.

Majerus, M.E.N. and Roy, H.E. (2005) Scientific opportunities presented by the arrival of the harlequin ladybird, *Harmonia axyridis*, in Britain. *Antenna* 29, 196–208.

Mansour, M.H. (1975) The role of plants as a factor affecting oviposition by *Aphidoletes aphidimyza* (Dipt: Cecidomyiidae). *Entomologia Experimentalis et Applicata* 18, 173–179.

Mansour, M.H. (1976) Some factors influencing egg laying and site of oviposition by *Aphidoletes aphidimyza* (Dipt: Cecidomyiidae). *Entomophaga* 21, 281–288.

Marchand, D. and McNeil, J.N. (2000) Effects of wind speed and atmospheric pressure on mate searching behavior in the aphid parasitoid *Aphidius nigripes* (Hymenoptera, Aphidiidae). *Journal of Insect Behavior* 13, 187–199.

Markkula, M. and Tiitanen, K. (1985) Biology of the midge *Aphidoletes* and its potential for biological control. In: Hussey, N.W. and Scopes, N. (eds) *Biological Pest Control: The Glasshouse Experience*. Cornell University Press, Ithaca, pp. 74–81.

Mason, R.T., Fales, H.M., Eisner, M. and Eisner, T. (1991) Wax of a whitefly and its utilization by a chrysopid larva. *Naturwissenschaften* 78, 28–30.

Mathew, M.J., Venugopal, M.N. and Saju, K.A. (1999) First record of some entomogenous fungi on cardamon aphid, *Pentalonia nigronervosa* f. *caladii* van der Goot (Homoptera: Aphididae). *Insect Environment* 4, 147–148.

McBrien, H. and Mackauer, M. (1990) Heterospecific larval competition and host discrimination in two species of aphid parasitoids: *Aphidius smithi* and *Aphidius ervi*. *Entomologia Experimentalis et Applicata* 56, 145–153.

McBrien, H. and Mackauer, M. (1991) Decision to superparasitize based on larval survival: competition between aphid parasitoids *Aphidius smithi* and *Aphidius ervi*. *Entomologia Experimentalis et Applicata* 59, 145–150.

McGregor, R. and Mackauer, M. (1989) Toxicity of carbaryl to the pea-aphid parasite *Aphidius smithi*: influence of behaviour on pesticide uptake. *Crop Protection* 8, 193–196.

McLeod, P.J., Steinkraus, D.C., Correll, J.C. and Morelock, T.E. (1998) Prevalence of *Erynia neoaphidis* (Entomophthorales: Entomophthoraceae) infections of green peach aphid (Homoptera: Aphididae) on spinach in the Arkansas River Valley. *Environmental Entomology* 27, 796–800.

McNeil, J.N. and Brodeur, J. (1995) Pheromone-mediated mating in the aphid parasitoid, *Aphidius nigripes* (Hymenoptera, Aphidiidae). *Journal of Chemical Ecology* 21, 959–972.

Mesquita, A.L.M., Lacey, L.A. and Leclant, F. (1997) Individual and combined effects of the fungus *Paecilomyces fumosoroseus* and parasitoid *Aphelinus asychis* Walker (Hym.: Aphelinidae) on confined populations of the Russian wheat aphid, *Diuraphis noxia* (Mordvilko) (Hom.: Aphididae) under field conditions. *Journal of Applied Entomology* 121, 155–163.

Messenger, P.S. (1970) Bioclimatic inputs to biological control and pest management programs. In: Rabb, R.L. and Guthrie, F.E. (eds) *Concepts of Pest Management*. North Carolina State University, Raleigh, pp. 84–102.

Messina, F.J. and Hanks, J.B. (1998) Host plant alters the shape of the functional response of an aphid predator (Coleoptera: Coccinellidae). *Environmental Entomology* 27, 1196–1202.

Messina, F.J. and Sørenson, S.M. (2001) Effectiveness of lacewing larvae in reducing Russian wheat aphid populations on susceptible and resistant wheat. *Biological Control* 21, 19–26.

Messina, F.J., Jones, T.A. and Nielson, D.C. (1995) Host plant affects the interaction between the Russian wheat aphid and a generalist predator. *Journal of the Kansas Entomological Society* 68, 313–319.

Messina, F.J., Jones, T.A. and Nielson, D.C. (1997) Host plant affects the efficacy of two predators attacking Russian wheat aphids (Homoptera: Aphididae). *Environmental Entomology* 26, 1398–1404.

Meyhöfer, R. and Hindayana, D. (2000) Effects of intraguild predation on aphid parasitoid survival. *Entomologia Experimentalis et Applicata* 97, 109–114.

Meyhöfer, R. and Klug, T. (2002) Intraguild predation on the aphid parasitoid *Lysiphlebus fabarum* (Marshall) (Hymenoptera: Aphidiidae): mortality risks and behavioral decisions made under the threats of predation. *Biological Control* 25, 239–248.

Micha, S.G. and Wyss, U. (1996) Aphid alarm pheromone (E)-β-farnesene: a host-finding kairomone for the aphid primary parasitoid *Aphidius uzbekistanicus* (Hymenoptera: Aphidiinae). *Chemoecology* 7, 132–139.

Micha, S.G., Kistenmacher, S., Mölck, G. and Wyss, U. (2000) Tritrophic interactions between cereals, aphids and parasitoids: discrimination of different plant-host complexes by *Aphidius rhopalosiphi* (Hymenoptera: Aphidiidae). *European Journal of Entomology* 97, 539–543.

Michaud, J.P. (1995) Static and dynamic criteria in host evaluation by aphid parasitoids (Hymenoptera: Aphidiidae). PhD thesis, Simon Fraser University, Burnaby, B.C., Canada.

Michaud, J.P. (2001) Evaluation of green lacewings, *Chrysoperla plorabunda* (Fitch) (Neuropt., Chrysopidae), for augmentative release against *Toxoptera citricida* (Hom., Aphididae) in citrus. *Journal of Applied Entomology* 125, 383–388.

Michaud, J.P. (2002) Classical biological control: a critical review of recent programs against citrus pests in Florida. *Annals of the Entomological Society of America* 94, 531–540.

Michaud, J.P. and Belliure, B. (2000) Consequences of foundress aggregation in the brown citrus aphid *Toxoptera citricida*. *Ecological Entomology* 25, 307–314.

Michaud, J.P. and Mackauer, M. (1994) The use of visual cues in host evaluation by aphidiid wasps. I. Comparison between three *Aphidius* parasitoids of the pea aphid. *Entomologia Experimentalis et Applicata* 70, 273–283.

Michaud, J.P. and Mackauer, M. (1995) The use of visual cues in host evaluation by aphidiid wasps. II. Comparison between *Ephedrus californicus*, *Monoctonus paulensis* and *Praon pequodorum*. *Entomologia Experimentalis et Applicata* 74, 267–275.

Michel, M.F. (1967) Importance écologique du comportement prédateur d'*Aphelinus asychis* Walker (Hym. Aphelinidae), endoparasite de pucerons (Hom. Aphididae). *Comptes Rendues Hebdomadaires des Séances de l'Académie des Sciences, Paris* 264, 936–939.

Miermont, Y. and Canard, M. (1975) Biologie du prédateur aphidiphage *Eumicromus angulatus* (Neur.: Hemerobiidae): études au laboratoire et observation dans le sud-ouest de la France. *Entomophaga* 20, 179–191.

Milbrath, L.R., Tauber, M.J. and Tauber, C.A. (1993) Prey specifity in *Chrysopa*: an interspecific comparison of larval feeding and defensive behaviour. *Ecology* 74, 1384–1393.

Mills, N.J. (1979) *Adalia bipunctata* (L.) as a generalist predator of aphids. PhD thesis, University of East Anglia, Norwich, UK.

Milner, R.J. (1982) On the occurrence of pea aphids, *Acyrthosiphon pisum*, resistant to isolates of the fungal pathogen *Erynia neoaphidis*. *Entomologia Experimentalis et Applicata* 32, 23–27.

Milner, R.J. (1985) Distribution in time and space of resistance to the pathogenic fungus *Erynia neoaphidis* in the pea aphid *Acyrthosiphon pisum*. *Entomologia Experimentalis et Applicata* 37, 235–240.

Milner, R.J. (1997) Prospects for biopesticides for aphid control. *Entomophaga* 42, 227–239.

Milner, R.J. and Bourne, J. (1983) Influence of temperature and duration of leaf wetness on infection of *Acyrthosiphon pisum*. *Annals of Applied Biology* 102, 19–27.

Milner, R.J. and Lutton, G.G. (1983) Effect of temperature on *Zoophthora radicans* (Brefeld) Batko: an introduced microbial control agent of the spotted alfalfa aphid, *Therioaphis trifolii* (Monell) f. *maculata*. *Journal of the Australian Entomological Society* 22, 167–173.

Milner, R.J., Holdom, D.G. and Glare, T.R. (1984) Diurnal patterns of mortality in aphids infected by entomophthoralean fungi. *Entomologia Experimentalis et Applicata* 36, 37–42.

Millstein, J.A., Brown, G.C. and Nordin, G.L. (1982) Microclimatic humidity influence on conidial discharge in *Erynia* sp. (Entomophthorales: Entomophthoraceae), an entomopathogenic fungus of the alfalfa weevil (Coleoptera: Curculionidae). *Environmental Entomology* 11, 1166–1169.

Millstein, J.A., Brown, G.C. and Nordin, G.L. (1983) Microclimatic moisture and conidial production in *Erynia* sp. (Entomophthorales: Entomophthoraceae): *in vivo* moisture balance and conidiation phenology. *Environmental Entomology* 12, 1339–1343.

Minks, A.K. and Harrewijn, P. (eds) (1988) *Aphids. Their Biology, Natural Enemies and Control, Volume 2B.* Elsevier, Amsterdam, 364 pp.

Mizuno, M., Itioka, T., Tatematsu, Y. and Ito, Y. (1997) Food utilization of aphidophagous hoverfly larvae (Diptera: Syrphidae, Chamaemyiidae) on herbaceous plants in an urban habitat. *Ecological Research* 12, 239–248.

Mölck, G., Micha, S.G. and Wyss, U. (2000) Attraction to odour of infested plants and learning behaviour in the aphid parasitoid *Aphelinus abdominalis*. *Journal of Plant Diseases and Protection* 106, 557–567.

Monsrud, C. and Toft, S. (1999) The aggregative numerical response of polyphagous predators to aphids in cereal fields: attraction to what? *Annals of Applied Biology* 134, 265–270.

Morris, M.C. and Li, F.Y. (2000) Coriander (*Coriandrum sativum*) 'companion plants' can attract hoverflies, and may reduce pest infestation in cabbages. *New Zealand Journal of Crop and Horticulture Science* 28, 213–217.

Mueller, T.E., Blommers, L.H.M. and Mols, P.J.M. (1992) Woolly apple aphid (*Eriosoma lanigerum* Hausm., Hom., Aphididae) parasitism by *Aphelinus mali* Hal. (Hym., Aphelinidae) in relation to host stage and host colony size, shape and location. *Journal of Applied Entomology* 114, 143–154.

Müller, C. and Godfray, H.C.J. (1999) Predators and mutualists influence the exclusion of aphid species from natural communities. *Oecologia* 119, 120–125.

Murdie, G. (1969) The biological consequences of decreased size caused by crowding or rearing temperatures in apterae of the pea aphid, *Acyrthosiphon pisum* Harris. *Transactions of the Royal Entomological Society of London* 121, 443–455.

Murphy, S.T. and Völkl, W. (1996) Population dynamics and foraging behaviour of *Diaeretus leucopterus* Haliday (Hymenoptera: Braconidae), and its potential for the biological control of pine damaging *Eulachnus* spp. (Homoptera: Aphididae). *Bulletin of Entomological Research* 86, 397–405.

Nakamura, M., Nemoto, H. and Amano, H. (2000) Ovipositional characteristics of lacewings, *Chrysoperla carnea* (Stephens) and *Chrysopa pallens* (Rambur) (Neuroptera: Chrysopidae) in the field. *Japanese Journal of Applied Entomology and Zoology* 44, 17–26.

Nakamuta, K. (1985) Mechanism of the switchover from extensive to area-concentrated search behaviour of the ladybird beetle *Coccinella septempunctata bruckii*. *Journal of Insect Physiology* 31, 849–856.

Nakashima, Y., Birkett, M.A., Pye, B.J., Pickett, J.A. and Powell, G. (2004) The role of semiochemicals in the avoidance of the seven-spot ladybird, *Coccinella septempunctata*, by the aphid parasitoid, *Aphidius ervi*. *Journal of Chemical Ecology* 30, 1103–1116.

Nazzi, F., Powell, W., Wadhams, L.J. and Woodcock, C.M. (1996) Sex pheromone of the aphid parasitoid *Praon volucre* (Hymenoptera, Braconidae). *Journal of Chemical Ecology* 22, 1169–1175.

Nemec, V. and Starý, P. (1985) Population diversity in deuterotokous *Lysiphlebus* species, parasitoids of aphids (Hymenoptera, Aphidiidae). *Acta Entomologica Bohemoslovaca* 82, 170–174.

Neuenschwander, P., Hagen, K.S. and Smith, R.F. (1975) Predation on aphids in California's alfalfa fields. *Hilgardia* 43, 53–78.

New, T.R. (1975) The biology of the Chrysopidae and Hemerobiidae (Neuroptera), with reference to their usage in biological control. *Transactions of the Royal Entomological Society of London* 127, 115–140.

New, T.R. (1988) *Neuroptera*. In: Minks, A.K. and Harrewijn, P. (eds) *Aphids. Their Biology, Natural Enemies and Control, Volume 2B*. Elsevier, Amsterdam, pp. 249–258.

Ng, S.M. (1991) Voracity, development and growth of larvae of *Menochilus sexmaculatus* (Coleoptera: Coccinellidae) fed on *Aphis spiraecola*. In: Polgár, L.A., Chambers, R., Dixon, A.F.G. and Hodek, I. (eds) *Behaviour and Impact of Aphidophaga*. SPB Academic Publishing, The Hague, pp. 199–206.

Nielsen, C., Eilenberg, J., Harding, S., Oddsdottir, E. and Haldórsson, G. (2001a) Geographical distribution and host range of Entomophthorales infecting the green spruce aphid *Elatobium abietinum* Walker in Iceland. *Journal of Invertebrate Pathology* 78, 72–80.

Nielsen, C., Sommer, C., Eilenberg, J., Hansen, K.S. and Humber, R.A. (2001b) Characterization of aphid pathogenic species in the genus *Pandora* by PCR techniques and digital image analysis. *Mycologia* 93, 864–874.

Nielsen, C., Hajek, A.E., Humber, R.A., Bresciani, J. and Eilenberg, J. (2003) Soil as an environment for winter survival of aphid-pathogenic Entomophthorales. *Biological Control* 28, 92–100.

Nielson, M.W. and Barnes, O.L. (1961) Population studies of the spotted alfalfa aphid in Arizona in relation to temperature and rainfall. *Annals of the Entomological Society of America* 54, 441–448.

Nijveldt, W. (1988) Cecidomyiidae. In: Minks, A.K. and Harrewijn, P. (eds) *Aphids. Their Biology, Natural Enemies and Control, Volume 2B*. Elsevier, Amsterdam, pp. 271–277.

Ninkovic, V., Al-Abassi, S. and Petterson, J. (2001) The influence of aphid-induced plant volatiles on ladybird beetle searching behavior. *Biological Control* 21, 191–195.

Obrycki, J.J. and Kring, T.J. (1998) Predaceous Coccinellidae in biological control. *Annual Review of Entomology* 43, 295–321.

Obrycki, J.J. and Orr, C.J. (1990) Suitability of three prey species for Nearctic populations of *Coccinella septempunctata*, *Hippodamia variegata* and *Propylaea quattuordecimpunctata*. *Journal of Economic Entomology* 83, 1292–1297.

Obrycki, J.J. and Tauber, M.J. (1984) Natural enemy activity on glandular pubescent potato plants in the greenhouse: an unreliable predictor of effects in the field. *Environmental Entomology* 13, 679–683.

Olszak, R. (1988) Voracity and development of three species of Coccinellidae, preying upon different species of aphids. In: Niemczyk, E. and Dixon, A.F.G. (eds) *Ecology and Effectiveness of Aphidophaga*. SPB Academic Publishing, The Hague, pp. 47–53.

Ozino, O.I., Arzone, A. and Alma, A. (1988) Entomogenous fungi of *Sitobion avenae* (F.) in Piedmont cereal crops. *Redia* 71, 173–183.

Paetzold, G. and Vater, D. (1967) Populationsdynamische Untersuchungen an den Parasiten und Hyperparasiten von *Brevicoryne brassicae* (L.) (Homoptera, Aphididae). *Acta Entomologica Bohemoslovaca* 64, 83–90.

Parker, N.J.B. (1975) An investigation of reproductive diapause in two British populations of *Anthocoris nemorum* (Hemiptera: Anthocoridae). *Journal of Entomology A* 49, 173–178.

Pell, J.K. and Vandenberg, J.D. (2002) Interactions among *Diuraphis noxia*, the fungal pathogen *Paecilomyces fumosoroseus* and the coccinellid *Hippodamia convergens*. *Biocontrol Science and Technology* 12, 217–224.

Pell, J.K., Wilding, N., Player, A.L. and Clark, S.J. (1993) Selection of an isolate of *Zoophthora radicans* (Zygomycetes: Entomophthorales) for biocontrol of the diamondback moth *Plutella xylostella* (Lepidoptera: Yponomeutidae). *Journal of Invertebrate Pathology* 61, 75–80.

Pell, J.K., Pluke, R., Clark, S.J., Kenward, M.G. and Alderson, P.G. (1997) Interactions between two aphid natural enemies, the entomopathogenic fungus, *Erynia neoaphidis* and the predatory beetle, *Coccinella septempunctata*. *Journal of Invertebrate Pathology* 69, 261–268.

Pell, J.K., Eilenberg, J., Hajek, A.E. and Steinkraus, D.S. (2001) Biology, ecology and pest management potential of Entomophthorales. In: Butt, T.M., Jackson, C. and Magan, N. (eds) *Fungi as Biocontrol Agents: Progress, Problems and Potential*. CAB International, Wallingford, pp. 71–154.

Pickering, J., Dutcher, J.D. and Ekbom, B.S. (1989) An epizootic caused by *Erynia neoaphidis* and *E. radicans* (Zygomycetes: Entomophthoraceae) on *Acyrthosiphon pisum* (Homoptera: Aphidae) on legumes under overhead irrigation. *Journal of Applied Entomology* 107, 331–333.

Poehling, H.M., Tenhumberg, B. and Groeger, U. (1991) Different pattern of cereal aphid population dynamics in northern and southern areas of West Germany. *Bulletin IOBC/WPRS* 14, 1–12.

Polaszek, A. (1986) The effects of two species of hymenopterous parasitoids on the reproductive system of the pea aphid, *Acyrthosiphon pisum*. *Entomologia Experimentalis et Applicata* 40, 285–292.

Polgár, L.A. and Hardie, J. (2000) Diapause induction in aphid parasitoids. *Entomologia Experimentalis et Applicata* 97, 21–27.

Powell, W., Dean, G.J. and Wilding, N. (1986a) The influence of weeds on aphid-specific natural enemies in winter wheat. *Crop Protection* 5, 182–189.

Powell, W., Wilding, N., Brobyn, P.J. and Clark, S.J. (1986b) Interference between parasitoids (Hym.: Aphidiidae) and fungi (Entomophthorales) attacking cereal aphids. *Entomophaga* 31, 293–302.

Powell, W., Pennacchio, F., Poppy, G.M. and Tremblay, E. (1998) Strategies involved in the location of hosts by the parasitoid *Aphidius ervi* Haliday (Hymenoptera: Braconidae: Aphidiinae). *Biological Control* 11, 104–112.

Read, D.P., Feeny, P.P. and Root, R.B. (1970) Habitat selection by the aphid parasite *Diaeretiella rapae* and the hyperparasite *Charips brassicae*. *Canadian Entomologist* 102, 1567–1578.

Ricci, C. (1979) L'apparato boccale pungente succhiante della larva di *Platynaspis luteorubra* Goeze (Col., Coccinellidae). *Bollettino del Laboratorio di Entomologia Agraria 'Filippo Silvestri', Portici* 36, 179–198.

Rodriguez, L.C., Fuentes-Contreras, E. and Niemeyer, H.M. (2002) Effect of innate preferences, conditioning and adult experience on the attraction of *Aphidius ervi* (Hymenoptera: Braconidae) toward plant volatiles. *European Journal of Entomology* 99, 285–288.

van Roermund, H.J.W. and van Lenteren, J.C. (1995) Residence times of the whitefly parasitoid *Encarsia formosa* Gahan (Hym.: Aphelinidae) on tomato leaflets. *Journal of Applied Entomology* 119, 465–471.

van Roermund, H.J.W., van Lenteren, J.C. and Rabbinge, R. (1997) Analysis of the foraging behaviour of the whitefly parasitoid *Encarsia formosa* on a plant: a simulation study. *Biocontrol Science and Technology* 7, 131–151.

Rohel, E., Couteaudier, Y., Papierok, B., Cavelier, N. and Dedryver, C.A. (1997) Ribosomal internal transcribed spacer size variation correlated with RAPD-PCR pattern polymorphisms in the entomopathogenic fungus *Erynia neoaphidis* and some closely related species. *Mycological Research* 101, 573–579.

Rosenheim, J.A. (1998) Higher-order predators and the regulation of insect herbivore populations. *Annual Review of Entomology* 43, 421–447.

Rosenheim, J.A., Kaya, H.K., Ehler, L.E., Marois, J.J. and Jaffee, B.A. (1995) Intraguild predation among biological control agents: theory and evidence. *Biological Control* 5, 303–335.

Rosenheim, J.A., Limburg, D.D. and Colfer, R.G. (1999) Impact of generalist predators on a biological control agent, *Chrysoperla carnea*: direct observations. *Ecological Applications* 9, 409–417.

Rotheray, G.E. (1989) *Aphid Predators. Cambridge Naturalists' Handbooks, No. 7.* Cambridge University Press, Cambridge, 86 pp.

Roy, H.E. and Pell, J.K. (2000) Interactions between entomopathogenic fungi and other natural enemies: implications for biological control. *Biocontrol Science and Technology* 10, 737–752.

Roy, H.E., Pell, J.K. and Alderson, P.G. (2001) Targeted dispersal of the aphid pathogenic fungus *Erynia neoaphidis* by the aphid predator *Coccinella septempunctata*. *Biocontrol Science and Technology* 11, 99–110.

Ruppert, V. and Molthan, J. (1991) Augmentation of aphid antagonists by field margins rich in flowering plants. In: Polgár, L.A., Chambers, R., Dixon, A.F.G. and Hodek, I. (eds) *Behaviour and Impact of Aphidophaga*. SPB Academic Publishing, The Hague, pp. 243–247.

Růžička, Z. (1976) Prey selection by larvae of *Metasyrphus corollae* (Diptera, Syrphidae). *Acta Entomologica Bohemoslovaca* 73, 305–311.

Růžička, Z. (1994) Oviposition-deterring pheromone in *Chrysopa oculata* (Neuroptera: Chrysopidae). *European Journal of Entomology* 91, 361–366.

Růžička, Z. (1996) Oviposition-deterring pheromone in chrysopids: intra and interspecific effects. *European Journal of Entomology* 93, 161–166.

Růžička, Z. (1997) Protective role of the egg stalk in Chrysopidae (Neuroptera). *European Journal of Entomology* 94, 111–114.

Růžička, Z. (1998) Further evidence of oviposition-deterring allomone in chrysopids (Neuroptera: Chrysopidae). *European Journal of Entomology* 95, 35–39.

Růžička, Z. and Havelka, J. (1998) Effects of oviposition deterring pheromone and allomones in *Aphidoletes aphidimyza* (Diptera: Cecidomyiidae). *European Journal of Entomology* 95, 211–216.

Sadeghi, H. and Gilbert, F. (2000a) Oviposition preferences by aphidophagous hoverflies. *Ecological Entomology* 25, 91–100.

Sadeghi, H. and Gilbert, F. (2000b) The effect of egg load and host deprivation on oviposition behaviour in aphidophagous hoverflies. *Ecological Entomology* 25, 101–108.

Sadeghi, H. and Gilbert, F. (2000c) Aphid siutability and its relationship to oviposition preference in predatory hoverflies. *Journal of Animal Ecology* 69, 771–784.

Samu, F., Sunderland, K.D., Topping, C.J. and Fenlon, J.S. (1996) A spider population in flux: selection and abandonment of artificial web-sites and the importance of intraspecific interactions in *Lepthyphantes tenuis* (Araneae: Linyphiidae). *Oecologia* 106, 228–239.

Sanchez-Pena, S.R. (1993) Entomogenous fungi associated with the cotton aphid in the Texas high plains. *Southwestern Entomologist* 18, 69–71.

Sanders, W. (1983) Das Suchverhalten von *Syrphus corollae* and seine Abhängigkeit von optischen Reizen. *Zeitschrift für Angwandte Zoologie* 70, 235–247.

Schellhorn, N.A., Kuhman, T.R., Olson, A.C. and Ives, A.R. (2002) Competition between native and introduced parasitoids of aphids: nontarget effects and biological control. *Ecology* 83, 2745–2757.

van Schelt, J. and Mulder, S. (2000) Improved methods of testing and release of *Aphidoletes aphidimyza* (Diptera: Cecidomyiidae) for aphid control in glasshouses. *European Journal of Entomology* 97, 511–515.

Schwörer, U. and Völkl, W. (2001) The foraging behavior of *Aphidius ervi* (Haliday) (Hymenoptera: Braconidae: Aphidiinae) at different spatial scales: resource utilization and suboptimal weather conditions. *Biological Control* 21, 111–119.

Scutareanu, P., Drukker, B., Bruin, J., Posthumus, M.A. and Sabelis, M.W. (1997) Volatiles from *Psylla*-infested pear trees and their possible involvement in attraction of anthocorid predators. *Journal of Chemical Ecology* 23, 2241–2260.

Sequeira, R. and Mackauer, M. (1987) Host instar preference of the aphid parasite *Praon pequodorum* Viereck (Hymenoptera: Aphidiidae). *Entomologia Generalis* 12, 259–265.

Sequeira, R. and Mackauer, M. (1992a) Nutritional ecology of an insect host parasitoid association: the pea aphid–*Aphidius ervi* system. *Ecology* 73, 183–189.

Sequeira, R. and Mackauer, M. (1992b) Covariance of adult size and development time in the parasitoid wasp *Aphidius ervi* in relation to the size of its host, *Acyrthosiphon pisum*. *Evolutionary Ecology* 6, 34–44.

Sequeira, R. and Mackauer, M. (1994) Variation of selected life-history parameters of the parasitoid wasp, *Aphidius ervi*: influence of host developmental stage. *Entomologia Experimentalis et Applicata* 71, 15–22.

Shah, M.A. (1982) The influence of plant surfaces on the searching behaviour of coccinellid larvae. *Entomologia Experimentalis et Applicata* 31, 377–380.

Shah, P.A. and Goettel, M.S. (1999) *Directory of Microbial Control Products and Services*. Microbial Control Division, Society for Invertebrate Pathology, Gainesville, 31 pp.

Shah, P.A. and Pell, J.K. (2003) Managed field margins as refugia for *Pandora neoaphidis*. In: *Proceedings of the 36th Annual Meeting of the Society for Invertebrate Pathology, Burlington, Vermont, July 2003*. Society for Invertebrate Pathology, Gainesville, p. 28.

Shah, P.A., Clark, S.J. and Pell, J.K. (2004) Determination of *Pandora neoaphidis* host range using a tiered evaluation process. *Biological Control* 29, 90–99.

Sharma, M.L. and Laviolette, R. (1968) Fluctuations des populations de *Schizolachnus piniradiatae* (Davidson) (Aphididae-Homoptera) dans l'Est en 1967. *Annals of the Entomological Society of Quebec* 13, 89–97.

Sheehan, W.E. and Shelton, A.M. (1989) Parasitoid response to concentration of herbivore food plants: finding and leaving plants. *Ecology* 70, 993–998.

Shonouda, M.L., Bombosch, S., Shalaby, A.M. and Osman, I. (1998) Biological and chemical characterization of a kairomone excreted by the bean aphids, *Aphis fabae* Scop. (Hom. Aphididae), and its effect on the predator *Metasyrphus corollae* Fabr. II. Behavioural response of the predator *M. corollae* to the aphid kairomone. *Journal of Applied Entomology* 122, 25–28.

Sierotzki, H., Camestral, F., Shah, P.A., Tuor, U. and Aebi, M. (2000) Biological characteristics of selected *Erynia neoaphidis* isolates. *Mycological Research* 104, 213–219.

Singh, R. and Pandey, S. (1997) Offspring sex ratio in Aphidiinae (Hymenoptera: Braconidae): a review and bibliography. *Journal of Aphidology* 11, 61–82.

Sloggett, J.J. and Majerus, M.E.N. (2000) Aphid-mediated coexistence of ladybirds (Coleoptera: Coccinellidae) and the wood ant *Formica rufa*: seasonal effects, interspecific variability and the evolution of a coccinellid myrmecophily. *Oikos* I89, 345–349.

Snyder, W.E. and Wise, D.H. (1999) Predator interference and the establishment of generalist predator populations for biocontrol. *Biological Control* 15, 283–292.

Soleyman-Nezhadiyan, E. and Laughlin, R. (1998) Voracity of larvae, rate of development in eggs, larvae and pupae, and flight seasons of adults of the hoverflies *Melangyna viridiceps* Macquart and *Simosyrphus grandicornis* Macquart (Diptera: Syrhidae). *Australian Journal of Entomology* 37, 243–248.

Starý, P. (1970) *Biology of Aphid Parasites, with Respect to Integrated Control. Seria Entomologica, No. 6.* W. Junk, The Hague, 643 pp.

Starý, P. (1986) Creeping thistle, *Cirsium arvense*, as a reservoir of aphid parasitoids (Hymenoptera, Aphidiidae) in agroecosystems. *Acta Entomologica Bohemoslovaca* 83, 425–431.

Starý, P. (1988a) *Aphelinidae.* In: Minks, A.K. and Harrewijn, P. (eds) *Aphids. Their Biology, Natural Enemies and Control, Volume 2B.* Elsevier, Amsterdam, pp. 185–188.

Starý, P. (1988b) *Aphidiidae.* In: Minks, A.K. and Harrewijn, P. (eds) *Aphids. Their Biology, Natural Enemies and Control, Volume 2B.* Elsevier, Amsterdam, pp. 171–184.

Starý, P., Lyon, J.P. and Leclant, F. (1988) Biocontrol of aphids by the introduced *Lysiphlebus testaceipes* (Cress.) (Hym., Aphidiidae) in Mediterranean France. *Zeitschrift für Angewandte Entomologie* 105, 74–87.

Stechmann, D.H. and Völkl, W. (1990) A preliminary survey of aphidophagous insects of Tonga, with regards to the biological control of the banana aphid. *Journal of Applied Entomology* 110, 408–415.

Stechmann, D.H., Völkl, W. and Starý, P. (1996) Ants as a critical factor in the biological control of the banana aphid *Pentalonia nigronervosa* in Oceania. *Journal of Applied Entomology* 120, 119–123.

Steenberg, T. and Eilenberg, J. (1995) Natural occurrence of entomopathogenic fungi on aphids at an agricultural field site. *Czech Mycology* 48, 89–96.

Steinkraus, D.C., Kring, T.J. and Tugwell, N.P. (1991) *Neozygites fresenii* in *Aphis gossypii* on cotton. *Southwestern Entomologist* 16, 118–122.

Steinkraus, D.C., Boys, G.O. and Slaymaker, P.H. (1993) Culture, storage, and incubation period of *Neozygites fresenii* (Entomophthorales: Neozygitaceae), a pathogen of the cotton aphid. *Southwestern Entomologist* 18, 197–202.

Steinkraus, D.C., Hollingsworth, R.G. and Slaymaker, P.H. (1995) Prevalence of *Neozygites fresenii* (Entomophthorales: Neozygitaceae) on cotton aphids (Homoptera: Aphididae) in Arkansas cotton. *Environmental Entomology* 24, 465–474.

Steinkraus, D.C., Hollingsworth, R.G. and Boys, G.O. (1996) Aerial spores of *Neozygites fresenii* (Entomophthorales: Neozygitaceae): density, periodicity, and potential role in cotton aphid (Homoptera, Aphididae) epizootics. *Environmental Entomology* 25, 48–57.

Steinkraus, D.C., Howard, M.N., Hollingsworth, R.G. and Boys, G.O. (1999) Infection of sentinel cotton aphids (Homoptera: Aphididae) by aerial conidia of *Neozygites fresenii* (Entomophthorales: Neozygitaceae). *Biological Control* 14, 181–185.

Sullivan, D.J. and Völkl, W. (1999) Hyperparasitism: multitrophic ecology and behavior. *Annual Review of Entomology* 44, 291–315.

Sunderland, K.D. (1988) Carabidae and other invertebrates. In: Minks, A.K. and Harrewijn, P. (eds) *Aphids. Their Biology, Natural Enemies and Control, Volume 2B.* Elsevier, Amsterdam, pp. 293–310.

Sunderland, K.D. and Samu, F. (2000) Effects of agricultural diversity on the abundance, distribution, and pest control potential of spiders: a review. *Entomologia Experimentalis et Applicata* 95, 1–13.

Sunderland, K.D., Fraser, A.M. and Dixon, A.F.G. (1986) Field and laboratory studies on money spiders (Linyphiidae) as predators of cereal aphids. *Journal of Applied Ecology* 23, 433–447.

Sutherland, J.P., Sullivan, M.S. and Poppy, G.M. (2001) Oviposition behaviour and host colony size discrimination in *Episyrphus balteatus* (Diptera: Syrphidae). *Bulletin of Entomological Research* 91, 411–417.

Takada, H. and Tokumaru, S. (1996) Observations on oviposition and host-feeding behaviour of *Aphelinus gossypii* Timberlake (Hymenoptera: Aphelinidae). *Applied Entomology and Zoology* 31, 263–270.

Tenhumberg, B. and Poehling, H.M. (1995) Syrphids as natural enemies of cereal aphids in Germany: aspects of their biology and efficacy in different years and regions. *Agriculture, Ecosystems and Environment* 52, 39–43.

Tracewski, K.T., Johnson, P.C. and Eaton, A.T. (1984) Relative densities of predacious Diptera (Cecidomyiidae, Chamaemyiidae, Syrphidae) and their prey in New Hampshire apple orchards. *Protection Ecology* 6, 199–207.

Tymon, A.M., Shah, P.A. and Pell, J.K. (2004) PCR-based molecular discrimination of *Pandora neoaphidis* isolates from related entomopathogenic fungi and development of species-specific diagnostic primers. *Mycological Research* 108, 1–15.

Uygun, N. (1971) Zum Einfluss der Nahrungsmenge auf Fruchtbarkeit und Lebensdauer von *Aphidoletes aphidimyza* (Rondani) (Diptera: Itoniidae). *Zeitschrift für Angewandte Entomologie* 69, 234–258.

Vaughn, T., Antolin, M. and Bjostad, M.B. (1996) Behavioral and physiological responses of *Diaeretiella rapae*. *Entomologia Experimentalis et Applicata* 78, 187–196.

Vohland, K. (1996) The influence of plant structure on searching behaviour and resource exploitation in the ladybird, *Scymnus nigrinus* (Coleoptera: Coccinellidae). *European Journal of Entomology* 93, 151–160.

Volk, S. (1964) Untersuchungen zur Eiablage von *Syrphus corollae*. *Zeitschrift für Angewandte Entomologie* 54, 365–386.

Völkl, W. (1990) Fortpflanzungsstrategien von Blattlausparasitoiden (Hymenoptera, Aphidiidae): Konsequenzen ihrer Interaktionen mit Wirten und Ameisen. PhD thesis, University of Bayreuth, Bayreuth, Germany.

Völkl, W. (1992) Aphids or their parasitoids: who actually benefits from ant-attendance? *Journal of Animal Ecology* 61, 273–281.

Völkl, W. (1994a) Searching at different spatial scales: the foraging behaviour of the aphid parasitoid *Aphidius rosae* in rose bushes. *Oecologia* 100, 177–183.

Völkl, W. (1994b) The effect of ant-attendance on the foraging behaviour of the aphid parasitoid *Lysiphlebus cardui*. *Oikos* 70, 149–155.

Völkl, W. (1995) Behavioural and morphological adaptations of the coccinellid *Platynaspis luteorubra* for exploiting ant-attended resources. *Journal of Insect Behavior* 8, 653–670.

Völkl, W. (1997) Interactions between ants and aphid parasitoids: patterns and consequences for resource utilization. *Ecological Studies* 130, 225–240.

Völkl, W. (2000) Foraging behaviour and sequential multisensory orientation in the aphid parasitoid, *Pauesia picta* (Hymenoptera, Aphidiidae) at different spatial scales. *Journal of Applied Entomology* 124, 307–314.

Völkl, W. (2001) Parasitoid learning during interactions with ants: how to deal with an aggressive antagonist. *Behavioural Ecology and Sociobiology* 49, 135–144.

Völkl, W. and Kraus, W. (1996) Foraging behaviour and resource utilization of the aphid parasitoid *Pauesia unilachni*: adaptation to host distribution and mortality risks. *Entomologia Experimentalis et Applicata* 79, 101–108.

Völkl, W. and Kroupa, A.S. (1997) Effects of adult mortality risks on parasitoid foraging tactics. *Animal Behaviour* 53, 349–359.

Völkl, W. and Mackauer, M. (1990) Age-specific pattern of host discrimination by the aphid parasitoid *Ephedrus californicus* Baker (Hymenoptera: Aphidiidae). *Canadian Entomologist* 122, 349–361.

Völkl, W. and Mackauer, M. (1993) Interactions between ants and parasitoid wasps foraging for *Aphis fabae* spp. *cirsiiacanthoidis* on thistles. *Journal of Insect Behavior* 6, 301–312.

Völkl, W. and Novak, H. (1997) Foraging behaviour and resource utilization of the aphid parasitoid, *Pauesia pini* on spruce: influence of host species and ant attendance. *European Journal of Entomology* 94, 211–220.

Völkl, W. and Stechmann, D. (1998) Parasitism of the black bean aphid (*Aphis fabae*) by *Lysiphlebus fabarum* (Hymenoptera, Aphidiidae): the influence of host plant and habitat. *Journal of Applied Entomology* 122, 201–206.

Völkl, W. and Vohland, K. (1996) Wax covers in larvae of two *Scymnus* species: do they enhance coccinellid larval survival? *Oecologia* 107, 498–503.

Völkl, W., Stechmann, D.H. and Starý, P. (1990) Suitability of five species of Aphidiidae (Hymenoptera) for the biological control of the banana aphid *Pentalonia nigronervosa* Coq. (Homoptera, Aphididae) in the South Pacific. *Tropical Pest Management* 36, 249–257.

Way, M.J. (1963) Mutualism between ants and honeydew-producing homoptera. *Annual Review of Entomology* 8, 307–344.

Weisser, W.W. (1995) Within-patch foraging behaviour of the aphid parasitoid *Aphidius funebris*: plant architecture, host behaviour and individual variation. *Entomologia Experimentalis et Applicata* 76, 133–141.

Weisser, W.W. (2000) Metapopulation dynamics in an aphid-parasitoid system. *Entomologia Experimentalis et Applicata* 97, 83–92.

Weisser, W.W. and Völkl, W. (1997) Dispersal in the aphid parasitoid, *Lysiphlebus cardui*. *Journal of Applied Entomology* 121, 23–28.

Weisser, W.W., Houston, A.I. and Völkl, W. (1994) Foraging strategies of solitary parasitoids: the trade-off between female and offspring mortality. *Evolutionary Ecology* 8, 587–597.

Weisser, W.W., Völkl, W. and Hassell M.P. (1997) The importance of adverse weather conditions for behaviour and population ecology of an aphid parasitoid. *Journal of Animal Ecology* 66, 386–400.

White, C. and Eigenbrode, S.D. (2000) Leaf surface waxbloom in *Pisum sativum* influences predation and intra-guild interactions involving two predator species. *Oecologia* 124, 252–259.

Wickremasinghe, M.G.V. and van Emden, H.F. (1992) Reactions of adult female parasitoids, particularly *Aphidius rhopalosiphi*, to volatile chemical cues from the host plants of their prey. *Physiological Entomology* 17, 297–304.

Wilbert, H. (1973) Zur Suchfähigkeit der Eilarven von *Aphidoletes aphidimyza* (Diptera: Cecidomyiidae). *Entomologia Experimentalis et Applicata* 16, 514–524.

Wilding, N. (1969) Effect of humidity on the sporulation of *Entomophthora aphidis* and *E. thaxteriana*. *Transactions of the British Mycological Society* 53, 126–130.

Wilding, N. (1973) The survival of *Entomophthora* spp. in mummified aphids at different temperatures and humidities. *Journal of Invertebrate Pathology* 21, 309–311.

Wilding, N. and Brady, B.L. (1984) *Descriptions of Pathogenic Fungi and Bacteria. Set 82, Nos. 812, 814, 815, 817, 820.* CMI, Kew, UK.

Wilding, N., Mardell, S.K. and Brobyn, P.J. (1986) Introducing *Erynia neoaphidis* into a field population of *Aphis fabae*: form of the inoculum and effect of irrigation. *Annals of Applied Biology* 108, 373–385.

Wraight, S.P., Butt, T.M., Galaini-Wraight, S., Allee, L.L., Soper, R.S. and Roberts, D.W. (1990) Germination and infection processes of the entomophthoralean fungus *Erynia radicans* on the potato leafhopper *Empoasca fabae*. *Journal of Invertebrate Pathology* 56, 157–174.

Wraight, S.P., Poprawski, T.J., Meyer, W.L. and Peairs, F.B. (1992) Natural enemies of Russian wheat aphid (Homoptera: Aphididae) and associated cereal aphid species in spring-planted wheat and barley in Colorado. *Environmental Entomology* 22, 1383–1391.

Wratten, S.D. (1973) The effectiveness of the coccinellid beetle, *Adalia bipunctata* (L.), as a predator of the lime aphid, *Eucallipterus tiliae* (L.). *Journal of Animal Ecology* 42, 785–802.

Wratten, S.D. and van Emden, H.F. (1995) Habitat management for enhanced activity of natural enemies of insect pests. In: Glen, D.M. and Greaves, M.P. (eds) *Ecology of Integrated Farming Systems*. Wiley, Chichester, pp. 117–145.

Wyss, E., Niggli, U. and Nentwig, W. (1995) The impact of spiders on aphid populations in a strip-managed orchard. *Journal of Applied Entomology* 119, 473–478.

Yeo, H. (2000) Mycoinsecticides for aphid management: a biorational approach. PhD thesis, University of Nottingham, Nottingham, UK.

Zhu, J.W., Cosse, A.A., Obrycki, J.J., Boo, K.-S. and Baker, T.C. (1999) Olfactory reactions of the twelve-spotted lady beetle *Coleomegilla maculata* and the green lacewing *Chrysoperla carnea* to semiochemicals released from their prey and host plant: electroantennogram and behavioral responses. *Journal of Chemical Ecology* 25, 1163–1177.

Zwölfer, H. (1958) Zur Systematik, Biologie und Ökologie unterirdisch lebender Aphiden (Hom., Aphidoidea) (Anoeciinae, Tetraneurini, Pemphigini und Fordinae). *Zeitschrift für Angewandte Entomologie* 43, 1–52.

Zwölfer, H. and Völkl, W. (1997) Der Einfluß des Verhaltens adulter Insekten auf Ressourcen-Nutzung und Populationsdynamik: ein Drei-Komponenten-Modell der Populationsdichte-Steuerung. *Entomologia Generalis* 21, 129–144.

Zwölfer, H., Bauer, G., Heusinger, G. and Stechmann, D.H. (1984) Die tierökologische Bedeutung und Bewertung von Hecken. *Berichte der ANL* 2 (supplement), 1–155.

9 Chemical Ecology

John A. Pickett[1] and Robert T. Glinwood[2]

[1]*Biological Chemistry Division, Rothamsted Research, Harpenden, Herts, AL5 2JQ, UK;* [2]*Department of Ecology, Swedish University of Agricultural Sciences, 750 07 Uppsala, Sweden*

Introduction

The ecology of aphids is, like that of most insects, highly dependent upon signals. Signals from host and non-host plants convey information that is vital for selecting feeding, larviposition, and mating sites. Signals from aphids themselves are important in attracting a mate, aggregating with conspecifics, avoiding competition, and sensing, or giving warning of, threats. Chemical signals (semiochemicals) are relatively efficient to produce, specific, easy to disperse into the environment and, not least, easy to detect. Aphid life cycles are characterized by complex interactions, and those species that alternate between hosts and have a sexual phase are faced with considerable challenges such as locating the correct winter (primary) host, finding mates, leaving the winter host in the spring, and successfully colonizing the summer (secondary) host. Therefore, it is no surprise that aphids make extensive use of semiochemicals, both in gathering information from their environment and in signalling to each other. Parasitoids and predators have also evolved responses to some of these semiochemicals. It is has been 15 years since the last major review of aphid chemical ecology (Pickett *et al.*, 1992). In this time, significant advances have been made in studies on systemic production of aphid-induced volatiles by plants,

volatiles that sometimes have a composition specific to an aphid species. There have also been advances in respect of the role of aphid sex pheromone-related chemistry beyond parasitoid chemical ecology and into predator behaviour.

There are around 4000 species of aphid. However, for two reasons, this review will necessarily focus on a very small number, principally in the subfamily Aphidinae. First, the great majority of information that exists on aphids as a group has been obtained through the study of species from the temperate regions of the northern hemisphere (Dixon, 1998), and secondly, studies have naturally tended to focus, not exclusively but in large measure, on those species that have attained pest status due to their interference with human activities. This focus has been particularly accentuated in relation to aphid chemical ecology, since the principles and techniques of chemical ecology have so often been directed towards more effective and sustainable management of pests. Arguably the most studied aphids in terms of their chemical ecology have been *Aphis fabae* (black bean aphid) and *Rhopalosiphum padi* (bird cherry–oat aphid), both from the Aphidinae and both host-alternating species that regularly manifest themselves as pests in the northern hemisphere. In the case of *R. padi*, the importance of chemical signals in the complex life cycle of a host-alternating aphid is readily apparent, and is summarized in Fig. 9.1. We now also see the first application of molecular biological techniques to aphid chemical ecology and, in particular, in the study of olfactory mechanisms and alarm pheromone biosynthesis (Field *et al.*, 2000). Already, the first aphid gene involved in processing secondary plant metabolites (Jones *et al.*, 2001; Pontoppidan *et al.*, 2001) has been identified (Jones *et al.*, 2002).

In this review, we examine the many varied interactions in aphid ecology in which semiochemicals play important roles. Taking a trophic level approach, we present examples of chemically mediated interactions between aphids, between aphids and plants, and between aphids and their natural enemies.

Methods

Since this is a review of aphid chemical ecology, the 'Methods' section will be brief, but may be useful in highlighting some recent developments that have played an important generic role in the advances described in subsequent sections. It may also be of value to those workers contemplating new studies on aphid chemical ecology.

Bioassay techniques have continued to employ the Pettersson olfactometer (Pettersson, 1970a) and the linear track olfactometer with or without the modification (Hardie *et al.*, 1994a) allowing continuous release of test compounds. Approaches to obtaining volatile samples involve entrainment of air above the natural system by absorption onto porous polymers. Use of the solid-phase microextraction (SPME) technique has become widespread and is extremely convenient. However, SPME does not provide a single consistent sample for multiple analysis, suffers from some other disadvantages, and has been the subject of comparative quantitative investigation, which confirms the value of entrainment systems giving samples as solutions (Agelopoulos and Pickett, 1998).

When studying semiochemical release from intact plants, or plants colonized by particular herbivores such as individual aphid species, any cutting of petioles or damage to roots can alter the volatile profile. Entraining from semi-open vessels placed over the plant material can solve this problem. To avoid pressure or other damage at hermetic seals, a loose seal is established in the enclosing vessel and air is sampled from the positive pressure that can be created within (Dicke *et al.*, 1990; Turlings *et al.*, 1991; Röse *et al.*, 1996; Agelopoulos *et al.*, 1999).

Electrophysiological recordings coupled to capillary gas chromatography (GC) are important approaches for initial identification of volatile semiochemicals and can comprise electroantennogram (EAG) (Birkett *et al.*, 2000) as well as single cell (neuron) recordings (SCR) (Pickett *et al.*, 1992). However, tentative identification by GC coupled mass spectrometry (MS) can remain problematic as the threshold detection level for MS, or at least the level for obtaining useful spectra, can be higher than for GC-SCR, or even the less sensitive technique of GC-EAG. Where GC is directly coupled to behavioural measurement, for example, wing fanning by aphid parasitoids (Nazzi *et al.*, 1996), the response threshold can be even lower than for GC-SCR. Increased sample size can cause loss of GC resolution, but sometimes biological solutions to the problem can be found; for example, if a plant compound is involved, related cultivars or species that may contain more of the active compound can be investigated (L.J. Wadhams, personal communication).

Non-volatile semiochemicals can also mediate important aspects of chemical ecology (Powell and Hardie, 2001; Powell *et al.*, 2006; see also Pettersson *et al.*, Chapter 4 this volume). Techniques for studying these materials have advanced with liquid chromatography (LC) coupled MS using reverse phase columns and electrospray ionization in both positive and negative ion modes (Takemura *et al.*, 2002, 2006). High-performance LC can also be used to produce fractions and pure samples of the putative non-volatile semiochemicals for analysis by

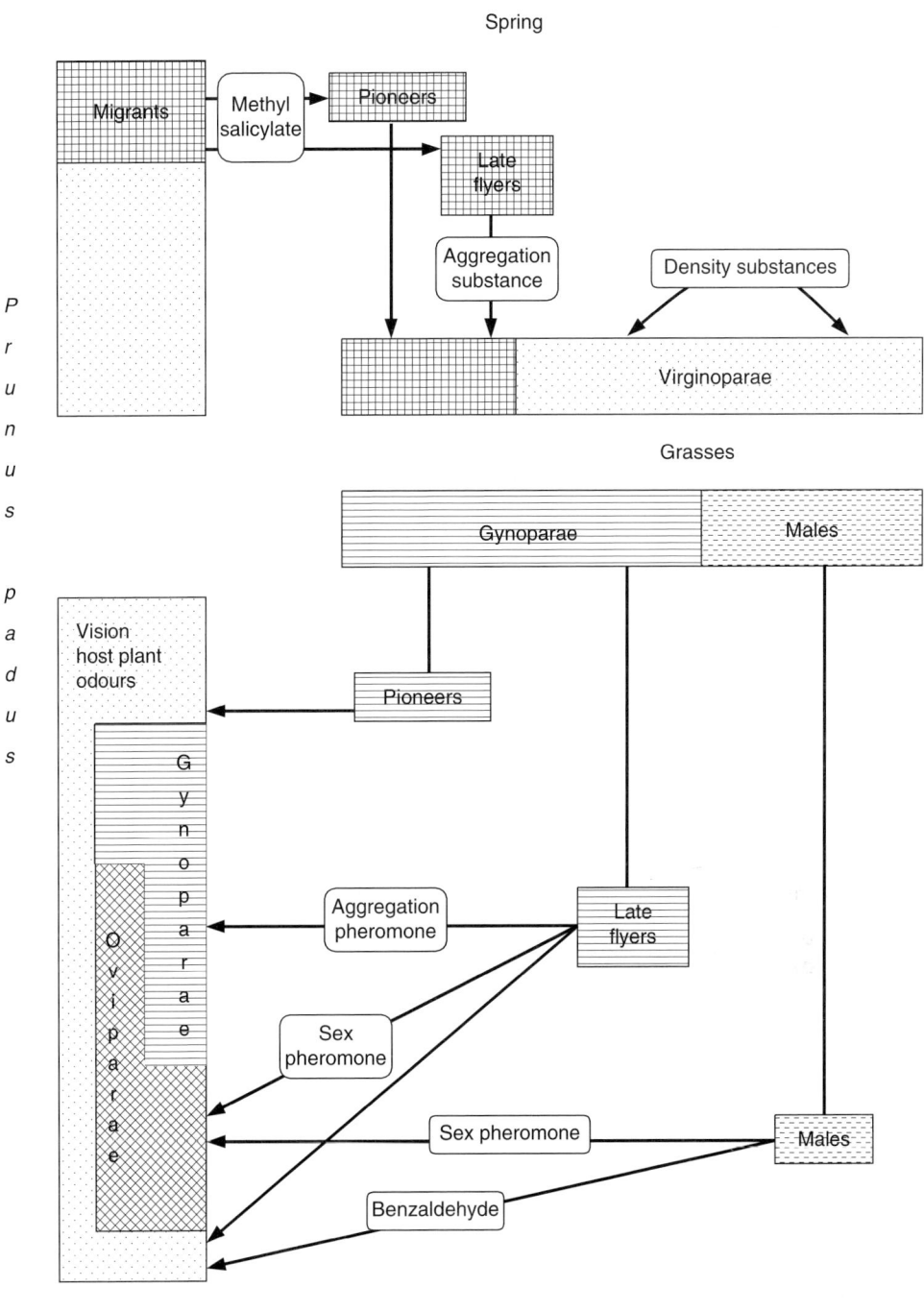

Fig. 9.1. Chemical ecology of host alternation in *Rhopalosiphum padi* (bird cherry–oat aphid). The diagram summarizes the interactions in which chemical signals are important for the aphid as it migrates from its winter host bird cherry *Prunus padus*, to grasses and cereals, and back to *P. padus* in the autumn to undergo the sexual phase (based on Pettersson, 1994; Pettersson *et al.*, 1994, 1995; Hardie *et al.*, 1996; Glinwood and Pettersson, 2000a; Park *et al.*, 2000).

nuclear magnetic resonance spectroscopy using a 500 MHz Fourier Transform instrument (Takemura *et al.*, 2002).

Interactions between Aphids

Sex pheromones

Aphid sex pheromones were unequivocally demonstrated by Pettersson (1970b), and the first was chemically characterized in 1987 (Dawson *et al.*, 1987a). The subject was reviewed in 1992 (Pickett *et al.*, 1992) and later by Hardie *et al.* (1999). The sex pheromones, released by the sexual females, excite males and increase mating success (Fig. 9.2). The pheromones are produced in glandular epidermal cells lying beneath scent plaques on the tibiae of the hind legs of the sexual females and released through porous cuticles above the plaques. During pheromone release, the female engages in typical 'calling' behaviour, with the hind legs raised (Fig. 9.3). The olfactory receptors for sex pheromones are placoid sensilla, in the

secondary rhinaria on the antennae of male aphids. The secondary rhinaria are mainly on the third, fourth, and sometimes fifth antennal segments, and are highly sensitive to sex pheromone components (see also Pettersson *et al.*, Chapter 4 this volume). There are separate receptor cells for individual pheromone components, which usually have different response amplitudes, both cells being necessarily activated to trigger a behavioural response.

Sex pheromones have been chemically characterized from a number of species, all of which are in the subfamily Aphidinae. GC-SCR has played a major role in these studies, followed by GC-MS and confirmation of structures by synthesis and spectroscopy (Hardie *et al.*, 1999; Boo *et al.*, 2000). The pheromones usually comprise (4a*S*, 7*S*, 7a*R*)-nepetalactone (**1**, Fig. 9.4) and (1*R*, 4a*S*, 7*S*, 7a*R*)-nepetalactol (**2**, Fig. 9.4), which are monoterpenoids in the cyclopentanoid or iridoid series. However, *Phorodon humuli* (damson–hop aphid) employs neither compound **1** nor **2**, but a mixture of the two diastereoisomeric (1*S*)- and (1*R*,4a*R*,7*S*,

Fig. 9.2. Mating in the aphid *Megoura viciae*. The darker, slightly thinner male (top) is attracted by sex pheromone released by the ovipara, the sexual female aphid (original colour photo courtesy Rothamsted Research).

Fig. 9.3. The typical 'calling' position of a sexual female (ovipara) of *Megoura viciae*, with the hind legs raised. Sex pheromone components are produced in glandular epidermal cells lying beneath scent plaques on the tibiae of the hind legs (original colour photo courtesy Rothamsted Research).

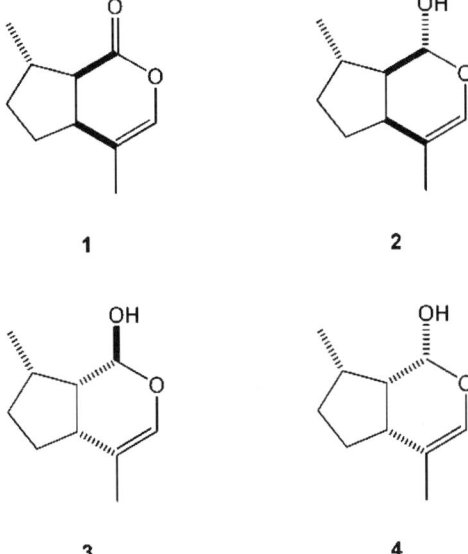

Fig. 9.4. Chemical structures for aphid sex pheromone components (4a*S*,7*S*,7a*R*)-nepetalactone (**1**), (1*R*,4a*S*,7*S*,7a*R*)-nepetalactol (**2**), (1*S*,4a*R*,7*S*,7a*S*)-nepetalactol (**3**) and (1*R*,4a*R*,7*S*,7a*S*)-nepetalactol (**4**).

Nepeta cataria (Lamiaceae), and can be chemically reduced selectively to the nepetalactol **2**. This now provides a commercial (Botanix Ltd) route to these compounds in which specific cultivars of *N. cataria* are grown as arable crops, harvested wet, and then directly subjected to steam distillation to yield the essential oil with a very high proportion of **1**. Further conversion by hydride reduction to the nepetalactol **2** can also be achieved on a commercial scale. Formulation (AgriSense-BCS Ltd) is as a polymeric rope where the length determines the overall release rate or field dose. The nepetalactone giving the nepetalactols **3** and **4** is also found in the related *Nepeta racemosa* (Birkett and Pickett, 2003).

The biosynthesis of nepetalactones in plants has received considerable attention and may be achieved by a similar route in aphids. The isoprenoid precursor is geraniol and/or the (*Z*)-isomer nerol, or possibly citronellol, and a detailed analytical study of the aphid sex pheromone from various species has revealed that citronellol (with undetermined stereochemistry) is a common, but minor, component of volatiles associated with sex pheromone release. It has shown no electrophysiological or behavioural activity, at least in connection with mate location, and may therefore be the precursor for the cyclopentanoid biosynthesis of aphid sex pheromones (Dawson *et al.*, 1996).

7a*S*)-nepetalactols (**3** and **4**, Fig. 9.4) (Campbell *et al.*, 2003), although EAG responses to **1** have been recorded in males of this species (Pope *et al.*, 2004).

The nepetalactone **1** has long been familiar as a compound from the catmint plant

Most aphids examined so far employ a limited range of pheromone components, but there are differences in relative and absolute compositions. Only *P. humuli* produces **3** and **4**, while the other species produce blends of **1** and **2** or the individual components (Hardie *et al.*, 1999; Boo *et al.*, 2000; Goldansaz *et al.*, 2004). However, there does not seem to be sufficient variability to allow species-specificity even merely within the Aphidinae. Indeed, a number of species use the same single component **1** alone, for example, *Sitobion avenae* (grain aphid), *Sitobion fragariae* (blackberry–cereal aphid), and *Brevicoryne brassicae* (cabbage aphid) (Hardie *et al.*, 1999). Laboratory males are more responsive to the pheromone blend closest to that released by the conspecific sexual females than to other blends. Addition of **2** inhibited the olfactometric responses of males of *S. avenae* (Lilley *et al.*, 1995) and *S. fragariae* (Lilley and Hardie, 1996) to their sex pheromone component **1**. Field trap-catch data show a remarkable specificity. Water traps releasing only **1**, recorded males of 21 aphid species, including mainly *S. fragariae* (Hardie *et al.*, 1992). Even the addition of minimal quantities of **1** reduced catches of *R. padi*, which employs only **2** (Hardie *et al.*, 1997), to levels similar to unbaited traps. In traps releasing **1**, **2**, or a 1 : 1 blend of **1** and **2**, 66% of male *Aphis* spp. and 85% of male *Dysaphis* spp., including *Dysaphis plantaginea* (rosy apple aphid), were captured by the blend, while 88% of *Rhopalomyzus lonicerae* were found in traps releasing **1**. *Myzus cerasi* (cherry blackfly) was also captured in traps releasing only **1** (Hardie *et al.*, 1994b; J. Hardie, R. Harrington and L.J. Wadhams, unpublished results). The sex pheromone for *D. plantaginea* comprises compounds **1** and **2**, with the latter the most abundant (S.Y. Dewhirst, personal communication). Trap catches appear to be associated with pheromonally mediated anemotaxis, although this was, before the availability of synthetic pheromone, considered to be an unlikely possibility. Indeed, it has been shown that *P. humuli* can fly towards a source of its pheromone against wind speeds of up to 0.7 m/s, and can detect pheromone 3–4 m

downwind (Hardie *et al.*, 1996). The sex pheromones have potential for use in population monitoring and control of pest aphids, and there are a number of practical development programmes under way, but the main agricultural use under development is in the attraction of parasitoids (see 'Natural enemies' below). One particular opportunity for direct aphid control is the use of pheromone in traps for disseminating an entomopathogen from the trap into the aphid population (Hartfield *et al.*, 2001).

There is now clear evidence for an interaction between sex pheromone and host-plant volatiles in some aphid species (Hardie *et al.*, 1999; Powell and Hardie, 2001). More *P. humuli* males were caught in traps releasing volatiles extracted from winter host plants, *Prunus* spp., with **3/4** than with **3/4** alone. Similar results were found with *R. padi* when combining release of volatile leaf extracts of bird cherry (*Prunus padus*), with **2**, which was at least partially accounted for by the presence of benzaldehyde in the leaf volatiles. Placing pheromone traps in the winter host rather than the non-host tree *Malus sylvestris* also enhanced catches of male *R. padi* (J.R. Storer and J. Hardie, unpublished results). However, this area of aphid chemical ecology requires further investigation (Hardie *et al.*, 1999). It should also be mentioned that the various semiochemicals can interact with visual cues, the latter also affecting the behaviour of male aphids. For example, surface colour is crucial and catches of males of three species in water traps were dramatically affected by trap colour (Hardie *et al.*, 1996).

Alarm pheromones

The asexual forms, and most often the wingless females, of many aphids release an alarm pheromone when disturbed. Nearby aphids exhibit a variety of behaviours, ranging from removal of mouthparts from the plant and moving away, to running, dropping off the plant, and even attacking the predator. This aspect of aphid chemical ecology has been reviewed in depth (Hardie *et al.*, 1999), and so will be dealt with briefly for completeness, highlighting work subsequent to that review.

Variation in response to alarm pheromone is seen both intra- and interspecifically and relates to the relative risks of predation and costs of escape. A lack of response from the early instars may arise from the risk of predation to these larvae being lower than the risk involved in ceasing to feed. Winged adults, on the other hand, are more responsive to alarm pheromone, perhaps because they move more readily off the natal host. However, with *Ceratovacuna lanigera* (sugar cane woolly aphid), it is the first-instar nymphs that show attack behaviour in response to alarm pheromone, while older nymphs and adults merely disperse (Arakaki, 1989). In this species, however, the first-instar nymphs have long frontal horns and are able to mount an effective defence against predators. Exposure to alarm pheromone can lead to an increase in the production of winged morphs in an aphid colony (Kunert *et al.*, 2005). Since this effect was seen when groups of aphids, but not isolated individuals, were exposed to the pheromone, it seems likely that the exposure causes a 'pseudo crowding' effect, with increased physical contacts triggering the shift in morph production.

Considerable variation exists among aphid species in their sensitivity to alarm pheromones in both the speed and the form of the response, variation that often can be related to the ecology of the species. For example, some aphids, particularly those tended by ants, respond by walking or 'waggling' their abdomens rather than falling off the plant (H.F. van Emden, personal communication). Response to alarm pheromone is affected by many additional factors, and in some cases may be dependent upon physical cues, or other semiochemicals, as discussed below (Hardie *et al.*, 1999).

In terms of variation within species, susceptibility to insecticides has also been found to correlate with alarm pheromone responses, discovered originally in 1983 (Dawson *et al.*, 1983). Insecticide-resistant *Myzus persicae* (peach–potato aphid) responded to alarm pheromone more slowly and in lower numbers than insecticide-susceptible forms (Foster *et al.*, 1999). This was true for a wide range of aphid clones carrying different combinations of metabolic (carboxylesterase) and target site (kdr) resistance mechanisms (Foster *et al.*, 2005; Foster *et al.*, Chapter 10 this volume). This may be due to pleiotropic physiological effects associated with resistance affecting mobility, or to impairment of nerve function (Foster *et al.*, 1999).

Aphids that have dropped from a plant may either recolonize it or search for another host plant, but in both cases face a greatly increased risk of predation (Griffiths *et al.*, 1985; Sunderland *et al.*, 1986) or starvation. Fallen aphids exposed to alarm pheromone often appear behaviourally reluctant to regain their host plant (Klingauf, 1976).

Droplets secreted from the siphunculi, on attack or mechanical interference, contain the minor, rapidly vaporizing fraction that is the alarm pheromone. A waxy fraction, consisting mainly of triglycerides, that crystallizes on contact with foreign particles outside of the aphid's body is also produced. The waxy component appears to function as a sticky irritant to predators and parasitoids.

The main component of the alarm pheromone of many aphids is the sesquiterpene hydrocarbon (E)-β-farnesene (**5**, Fig. 9.5) (Bowers *et al.*, 1972; Edwards *et al.*, 1973; Wientjens *et al.*, 1973; Pickett and Griffiths, 1980). Other components may also be present. For example, the alarm pheromone of *Megoura viciae* (vetch aphid) contains the monoterpenes (–)-α-pinene, (–)-β-pinene, (Z,E)-α-farnesene, and (E,E)-α-farnesene, in addition to **5**, and these can synergize the activity of **5**. There is a high degree of cross-activity of both natural alarm pheromone and **5** among species within the Aphidinae and Chaitophorinae, typical of insect alarm pheromones in general. The main component of the alarm pheromone of *Therioaphis trifolii maculata* (spotted alfalfa aphid) and *Therioaphis riehmi* (sweet clover aphid) (Calaphidinae), however, is the cyclic sesquiterpene (–)-(10S)germacrene-A (**6**, Fig. 9.5). Both **5** and **6** are volatile, highly labile compounds and produce a short-lived, though for some species dramatic, alarm signal.

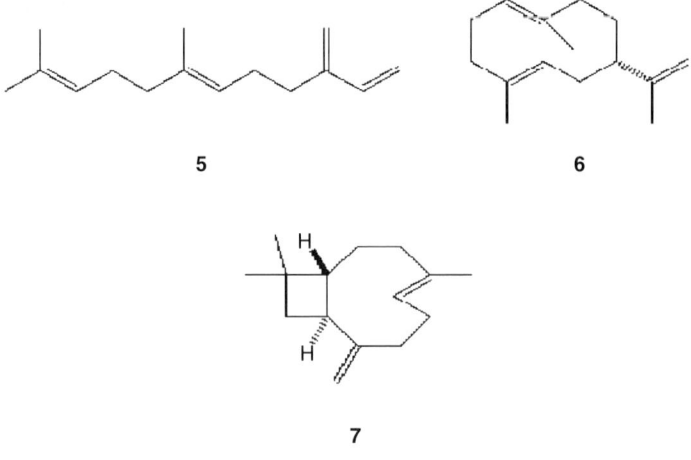

5   **6**

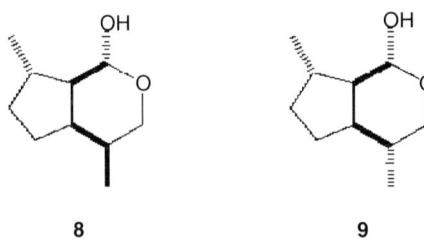

7

OH   OH

8   **9**

Fig. 9.5.   Chemical
structures for other aphid
related semiochemicals
(E)-β-farnesene (**5**),
(–)-(10S)germacrene-A (**6**),
(–)-β-caryophyllene (**7**),
(1R,4S,4aR,7S,7aR)-dihydro
nepetalactol (**8**), and
(1R,4R,4aR,7S,7aR)-dihydr
onepetalactol (**9**).

The alarm pheromone is detected by the primary rhinaria on the fifth and sixth antennal segments, which can be seen from recordings from single olfactory cells, for example with *M. persicae* (Dawson *et al.*, 1987b).

Although the biosynthesis of (E)-β-farnesene has not been studied in aphids, the gene for an enzyme from peppermint (*Mentha × piperita*) that effects conversion of farnesyl pyrophosphate to this sesquiterpene has been cloned and functionally expressed in *Escherichia coli* (Crock *et al.*, 1997). However, many other sesquiterpenes are also produced that could cause inhibition of the alarm response (see below).

Sesquiterpene hydrocarbons, including **5**, are commonly produced by plants. Indeed, compound **5** in the volatiles produced by leaf hairs of the aphid-resistant wild potato *Solanum berthaultii* caused alarm behaviour in wingless adult *M. persicae* (Gibson and Pickett, 1983). However, the sesquiterpene (–)-β-caryophyllene (**7**, Fig 9.5), together with

other sesquiterpenes commonly accompanying **5** in plants, was found to inhibit the alarm activity (Dawson *et al.*, 1984). This enables the aphid to distinguish whether **5** emanates from alarm pheromone or from a plant source. Olfactory cells in aphids have been found that are sensitive to either **5** or **7**, but not to both, and the two cell types are located in rhinaria on different antennal segments (Campbell *et al.*, 1993). For *Lipaphis pseudobrassicae* (mustard aphid), there is synergy between **5** and the generally toxic defence compounds such as organic isothiocyanates, which are produced from glucosinolates apparently sequestered from host plants in the Brassicaceae (Dawson *et al.*, 1986). The aphid enzyme responsible for the initial hydrolytic elimination of glucose from the glucosinolate has been characterized (Jones *et al.*, 2001; Pontoppiddan *et al.*, 2001), the associated gene cloned, and the sequence, although similar, shown to confer a structural difference from the corresponding plant thioglucosidase or myrosinase (Jones *et al.*, 2002). Hydrolysis of glucosinolates by the aphid myrosinase proceeds by general acid base catalysis using two glutamate residues as proton donor and nucleophile. Plant myrosinases differ from other members of this family in that one of the glutamate residues has been replaced by a glutamine (Burmeister *et al.*, 1997) and ascorbic acid substitutes for the second proton donor (Burmeister *et al.*, 2000). Thus, unexpectedly, aphid myrosinase acts mechanistically similarly to the *O*-glucosidases in having two glutamate residues; the absence of a proton donor, as in plant myrosinases, is not a prerequisite for the hydrolysis of glucosinolates, as was once thought (Burmeister *et al.*, 1997). The crystal structure at 1.1 Å resolution of the myrosinase of *B. brassicae* has been obtained from milligram amounts of the pure recombinant myrosinase obtained by over-expression in *E. coli*. However, the only residue specific for the aphid myrosinase in proximity of the glycosidic linkage is Tyr180, which may have a catalytic role. The aglycone binding site differs strongly from plant myrosinase, whereas due to the presence of Trp424 in the glucose binding site, this part of the active site is more similar to plant β-*O*-glucosidases, as plant myrosinases carry a phenylalanine residue at this position (Husebye *et al.*, 2005).

An antibody has been raised to the aphid myrosinase and its localization in the insect determined by immunocytochemistry and electron microscopy (Bridges *et al.*, 2002). The enzyme was found to be located in muscle of the head and the thorax and is present as regular crystal-like structures. This represents an apparently unique organizational arrangement, where myrosinase is held in the sarcoplasm of the muscle, while glucosinolates are most likely present in the haemolymph. Remarkably, this arrangement is similar to the brassicaceous plants that the insect feeds on, where myrosinase is compartmentalized into special myrosin cells and glucosinolates appear to be held in separate cells. In both cases, the spatial arrangement suggests a defence mechanism, with the enzyme and substrate remaining separated until tissue damage brings them together.

The use of (*E*)-β-farnesene in aphid control strategies has been hindered by its instability. However, there is renewed interest in exploiting plant essential oils rich in the substance, such as that of *Hemizygia petiolata* (Lamiaceae), which contains > 70% (*E*)-β-farnesene. Although laboratory studies suggest that the presence of minor components (+)-bicyclogermacrene and (–)-germacrene D may inhibit the alarm response of certain aphid species, a slow release formulation reduced *Acyrthosiphon pisum* (pea aphid) numbers in the field, giving a promising route for the deployment of (*E*)-β-farnesene in pest aphid management (Bruce *et al.*, 2005).

Social interactions: aggregation, density regulation, and avoidance

Aphids live in colonies that often reach a high density of individuals. A number of factors explain the existence of this habit and the selective forces that maintain it, and the capacity for rapid reproduction combined with low mobility can be seen as either reasons for, or adaptations to, colonial living. One benefit of aggregation is increased protection from natural enemies (Turchin and Kareiva, 1989), although aphids forming heterospecific aggregations could satisfy this. Therefore, it is possible that conspecific aggregation improves the food quality of the host plant (Way and Cammell, 1970; Way, 1973; see Douglas and van Emden, Chapter 5 this volume). Aphid colonies can grow to such a size that their sustainability on the plant is in no way assured (Way and Banks, 1967), and consequently there should be a point at which the advantage switches from recruitment of further individuals to the plant to their deterrence from it. Aphids of different species can have deleterious effects on each other when feeding on the same plant (Chongrattanmeteekul *et al.*, 1991). Therefore, control over social interactions both within and between species is important for aphid survival, and it is no surprise that semiochemicals play a role in maintaining such interactions. Current knowledge is biased towards aphid species that are considered to be economically important, and aggregation and avoidance interactions may be especially important in host-alternating species.

Aggregation

Gynoparae of *R. padi* show an aggregated distribution of settling on the winter host *P. padus* that cannot be explained by differences in microclimate or leaf quality (Pettersson, 1993). Gynoparae were attracted to the odour of other gynoparae in an olfactometer (Pettersson, 1993) and to water traps baited with gynoparae (Pettersson, 1994). Gynoparae were also attracted to the odour of *P. padus* leaves, but there was no increased response to leaves infested with gynoparae. Thus, aggregation of colonizing aphids on the winter host is mediated by an aggregation pheromone. Although it is unclear whether gynoparae of *R. padi* feed on the winter host, it is likely that one benefit of aggregation is the subsequent aggregated distribution of oviparae, which should maximize the attraction of males to the sex pheromone.

Aggregation of *R. padi* gynoparae is further enhanced by attraction of late flyers to the sex pheromone nepetalactol (**2**, Fig. 9.4), produced by oviparae already on the winter host (Fig. 9.1). Nepetalactol-baited traps captured gynoparae in the field (Hardie *et al.*, 1996), but since the traps caught only a relatively small proportion of the aerial population, it was suggested that the response of gynoparae to nepetalactol is weaker than that of males. This was later confirmed by EAG and behavioural studies (Park *et al.*, 2000). Since gynoparae of *S. fragariae* and *P. humuli* are also attracted to sex pheromones, this aggregation mechanism may be common among host-alternating species (Hardie *et al.*, 1996; Lösel *et al.*, 1996).

Semiochemicals are also involved in causing aphids to aggregate to their summer host plants. In olfactometer tests, spring migrants of *P. humuli* were more attracted by spring migrants feeding on a leaf of their summer host, hop (*Humulus lupulus*), than to an uninfested leaf (Campbell *et al.*, 1993). Entrainment of hops, followed by GC-MS, identified three compounds released by

the plant on feeding: (*E*)-2-hexenal, β-caryophyllene, and methyl salicylate, all of which were shown by GC-SCR to elicit responses from separate olfactory receptors on the aphid antenna. Interestingly, addition of methyl salicylate caused the other two compounds to lose attractiveness in the olfactometer, and it was suggested that methyl salicylate might be produced by the plant in response to high densities of feeding aphids, effectively preventing recruitment of further migrants to the plant. Analogous to the above results for *P. humuli*, *R. padi* spring migrants are more attracted to odour produced by spring migrants feeding on oat leaves than to uninfested leaves (Pettersson, 1994), again suggesting the presence of an aggregation substance for the migrating morph. In this case, the origin of the substance or substances has been difficult to identify. The migrant–oat complex is highly attractive to the migrants in the olfactometer. Migrants alone, however, are not attractive when removed from the leaf and presented immediately in the olfactometer, and when a previously attractive migrant–oat complex was separated and the odours of both the aphids and the previously infested leaf presented simultaneously, they did not elicit attraction (R. Glinwood, unpublished results). The rapid loss of attractiveness suggests that aggregation is mediated by an aphid pheromone, production of which is disrupted by removing the aphid from the plant, rather than by an aphid-induced plant volatile.

Aggregation pheromones in non-migratory morphs have also been reported. Alatae of two brassica-feeding aphids, *B. brassicae* and *L. pseudobrassicae*, emit an odour that attracts conspecific alatae, but not conspecific apterae (Pettersson and Stephansson, 1991). Apterae of *Aphis craccivora* (cowpea aphid) were attracted to the odours emitted by both apterae and alatae in small groups (≤ 10 individuals), and alatae were also attracted to small groups of apterae (Pettersson *et al.*, 1998). In both these studies, it appears that the aggregation signals are also involved in density regulation and avoidance interactions (see following section). Arrestment of alatae by odour from settled alatae was shown in *A. fabae* (Kay, 1976). In the examples reviewed here, early colonizers can be considered as pioneers that act as 'beacons' for later arriving individuals (Pettersson, 1994).

Density regulation and avoidance

Based on observations of the increase in walking activity of *R. padi* apterae when colony density passes a critical threshold, Pettersson *et al.* (1995) investigated the presence of a semiochemical-based, density-regulating mechanism. In olfactometer tests, colonies of *R. padi* apterae on oat leaves became repellent to other apterae when the density threshold was exceeded. Air entrainment, followed by GC-MS, identified three compounds associated with high, but not lower, densities of *R. padi* apterae feeding on wheat; 6-methyl-5-hepten-2-one, 6-methyl-5-hepten-2-ol, and 2-tridecanone, with the enantiomeric composition of 6-methyl-5-hepten-2-ol determined as a 1 : 3 ratio of (+) and (−) (Quiroz *et al.*, 1997). In olfactometer tests, all the compounds were repellent, the strongest effect being obtained with the ratio of compounds present in air entrainment samples. However, when the enantiomers of 6-methyl-5-hepten-2-ol were presented either individually or in the incorrect ratio, they were behaviourally inactive (Quiroz and Niemeyer, 1998). Chemicals that influence aphid behaviour, especially those that are repellent, are of great interest as components of integrated pest management strategies. Trials are under way using the *R. padi* density related substances applied to cereal crops in the form of slow release wax pellets, and initial results are encouraging, especially when used in combination with other behaviour modifying chemicals such as methyl salicylate (Ninkovic *et al.*, 2003).

A similar density-regulating mechanism to that of *R. padi* may operate in *A. craccivora*. Odours from groups of apterae repelled both apterae and alatae in an olfactometer when a critical group size of 20 or more individuals was surpassed (Pettersson *et al.*, 1998). The aggregation pheromones emitted by alatae of the brassica-feeding aphids *B. brassicae* and *L. pseudobrassicae*, described earlier, also appear to facilitate interspecies avoidance, with alatae of both species repelled by the odour of alatae of the other species (Pettersson and Stephansson, 1991).

Although, in some cases, semiochemicals that influence aphid social interactions are pheromonal, in other cases their origins are unclear. In functional terms, it makes little difference whether the aphid of the plant produces the substance; the message is the same. Indeed, induction of plant chemicals by aphids and their ability to detect them has certain advantages, such as avoidance of the costs of pheromone production and the conveyance of additional information about the condition of the plant. It is interesting that species that share a common host plant have the ability to detect each other using odour alone (Pettersson and Stephansson, 1991; Johansson *et al.*, 1997).

Interaction with Plants

Host-plant semiochemicals

Where host-plant semiochemicals have been identified, their perception has been found generally to be associated with highly specific cells, usually in the primary rhinaria of the fifth and sixth antennal segments.

Semiochemicals of the primary host of host-alternating aphids

The use of chemicals by return migrants searching for their primary hosts in the autumn has been reviewed recently (Powell and Hardie, 2001), providing evidence of semiochemical use by *A. fabae*, *P. humuli*, *R. padi*, and *S. fragariae*, among others.

In the autumn, the orientation of *R. padi* to its primary host *P. padus* is largely dependent upon aphid-produced signals, as described earlier, although the role of the host chemical, benzaldehyde, has also been demonstrated (Pettersson, 1970a). In the spring, the challenge facing the migrating morphs is to leave *P. padus* and, rather than recolonizing it, to successfully locate a suitable summer host. In the search for substances that could be used for control of the summer generations, Pettersson *et al.* (1994) identified methyl salicylate as a volatile chemical occurring in the primary, but not the summer, host. The substance was repellent to *R. padi* spring migrants in olfactometer tests, and also switched off the response to their aggregation cue (the odour of migrant-infested cereals). Pettersson *et al.* (1994) hypothesized that methyl salicylate could be used by aphids as a host discrimination cue, and also pointed out the potential role of the substance in plant defence signalling and as a promoter for induced resistance (Shulaev *et al.*, 1997).

Methyl salicylate may also function as a host-leaving stimulus. Using air entrainments *in situ* on intact *P. padus* trees in the field, Glinwood and Pettersson (2000a) showed that volatiles from foliage infested with developing spring populations of *R. padi* were repellent to spring migrants in olfactometer tests, whereas volatiles from uninfested foliage were behaviourally neutral. Subsequent collection and identification of volatiles from intact uninfested and infested foliage on *P. padus* saplings revealed that methyl salicylate is released in response to aphid infestation (Fig. 9.6). Thus, it appears that methyl salicylate is a damage-induced plant signal that is used as a behavioural cue by the aphid that induces its production. Methyl salicylate has been deployed successfully in integrated control of cereal aphids (Pettersson *et al.*, 1994), although there is evidence that the response of *R. padi* to this substance is dynamic and varies with the physiological state and/or age of the aphid (Glinwood and Pettersson, 2000b). It may be used most successfully in combination with other behaviour modifying chemicals such as the density-regulating substances described earlier (Ninkovic *et al.*, 2003).

Aphis fabae on spindle does not seem to parallel the *R. padi–P. padus* system in

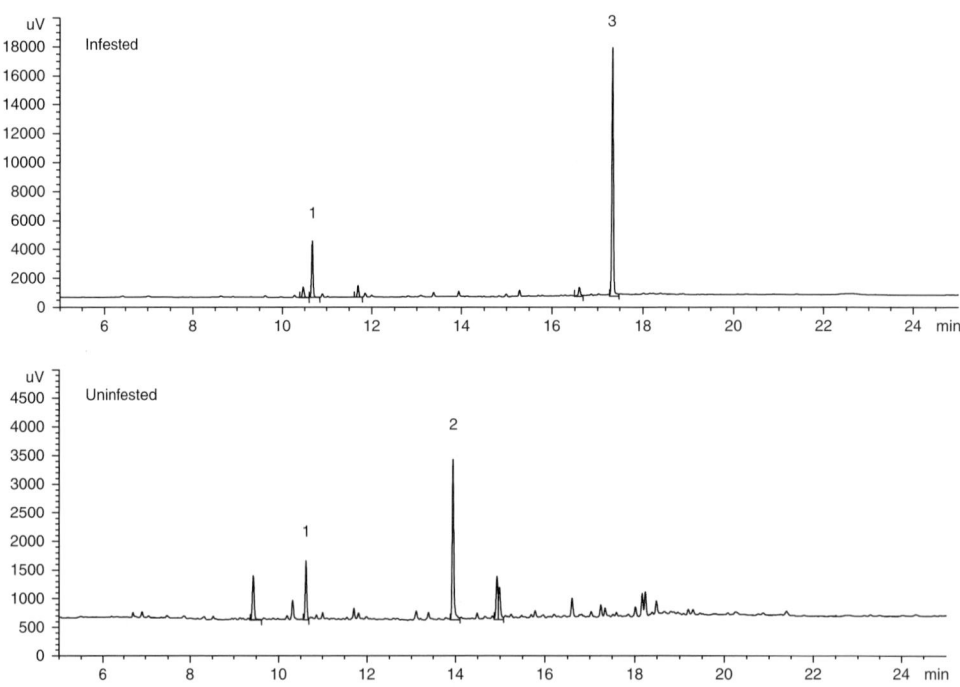

Fig. 9.6. GC of volatiles collected from *Prunus padus* that was either infested or uninfested with spring generations of *Rhopalosiphum padi* (bird cherry–oat aphid). Major peaks identified by GC-MS as (1) (Z)-3-hexenyl-acetate, (2) benzaldehyde, (3) methyl salicylate.

terms of the importance of chemical signals. Evidence of attraction of gynoparae to spindle volatiles is contradictory (Powell and Hardie, 2001), and there is no evidence of induction of volatiles by feeding spring generations analogous to the induction of methyl salicylate in the *R. padi–P. padus* system (R. Glinwood, unpublished results). These two aphid–plant systems therefore provide an interesting arena for comparative studies.

Semiochemicals of the secondary host of host-alternating aphids

A substantial part of the existing knowledge in this area was reviewed extensively by Pickett *et al.* (1992) and therefore will not be dealt with here. While the ability of aphids to perceive chemical cues (Visser and Piron, 1995; Park and Hardie, 2003, 2004) and the importance of these cues in plant finding is undoubted, eventual host acceptance is expected to depend upon the interaction of chemical, visual, and nutritional cues available to the aphid (Hori, 1999; Powell *et al.*, 2006; see also Pettersson *et al.*, Chapter 4 this volume). It should be noted that volatiles from host plants can also have negative effects on aphids, for example, by directly reducing fecundity (Hildebrand *et al.*, 1993) or by inducing responses in neighbouring plants (Ninkovic *et al.*, 2002; Glinwood *et al.*, 2003, 2004).

Non-host-plant semiochemicals

The study of semiochemicals used by aphids to avoid unsuitable hosts, although also reviewed in 1992 (Pickett *et al.*, 1992), has since provided important new discoveries. In 1994, there was evidence that, besides taxonomically unsuitable hosts being avoided by aphid migrants (Nottingham and Hardie, 1993; Hardie *et al.*, 1994a), in the spring, the primary host became a non-host (Pettersson *et al.*, 1994) and that the associated repellency was perceived by specific olfactory cells, again in the primary rhinaria, that responded to non-host semiochemicals. Two of these, (1*R*,5*S*)-myrtenal from the *A. fabae* non-host Lamiaceae and methyl salicylate

from *P. padus*, the primary host for *R. padi*, not only caused repellency but at lower levels also interfered with attraction to host plants (Hardie *et al.*, 1994a; Pettersson *et al.*, 1994). Fieldwork with methyl salicylate showed that the compound reduced populations of several species of cereal aphids in addition to *R. padi*, including *S. avenae* (Pettersson *et al.*, 1994; Ninkovic *et al.*, 2003). The effect on aphid species other than *R. padi* can be seen as a result of this substance acting as a promoter for plant-induced resistance. Plants under stress from herbivory also produced methyl salicylate. The compound presumably is derived from salicylate, which was considered a product of the inducible phenylalanine ammonia lyase pathway, although an alternative involvement of isochorismate synthase has been suggested (Wildermuth *et al.*, 2001).

During studies on the semiochemical basis for host alternation by *Nasonovia ribisnigri* (currant–lettuce aphid), a number of compounds from the primary host *Ribes nigrum* (Saxifragaceae) were identified that interfered with the attractancy of the secondary host lettuce (*Lactuca sativa*) (Asteraceae). One, *cis*-jasmone, was found to repel other aphids and also to attract *Coccinella septempunctata* (7-spot ladybird) and the parasitoid *Aphidius ervi* (Birkett *et al.*, 2000). This positive effect on the higher trophic level was also observed for methyl salicylate. Since methyl salicylate had, furthermore, been shown directly to induce defence against pathogens by intact plants (Shulaev *et al.*, 1997), the possible role of *cis*-jasmone as a plant signal was investigated. *cis*-Jasmone, although commonly induced during herbivory, was widely considered and originally expected by ourselves to be semiochemically inactive, except when released from flowers as a pollination attractant. However, surprisingly, when released into air above intact broad bean (*Vicia faba*) plants, *cis*-jasmone caused induction of a defence response resulting in the attraction of parasitoids, which persisted long after the *cis*-jasmone could no longer be detected (Birkett *et al.*, 2000). Although *cis*-jasmone is considered to be a biosynthetic product from jasmonic acid (Koch *et al.*, 1997), this compound and its volatile methyl ester do not give such a persistent effect in comparative studies also using *V. faba* (Birkett *et al.*, 2000). Since then, an effect of *cis*-jasmone on cereals, which causes repellency of aphids, has been exploited in the field (Bruce *et al.*, 2003). Because of the availability of microarrays for searching gene expression, studies were directed to *Arabidopsis thaliana*, and genes possibly associated with aphid defence have been identified that are specifically induced by *cis*-jasmone (Matthes *et al.*, 2003). However, in terms of plant–plant interactions, the plant produced insufficient *cis*-jasmone during aphid feeding to account solely for induction of plant defence against aphids. For crop protection, though, the persistence of the *cis*-jasmone induction effect, with the initial success in the field and the prospects of exploiting the developing molecular mechanism by which *cis*-jasmone acts, is promising for the future (Birkett *et al.*, 2001; Chamberlain *et al.*, 2001).

Aphid effects on the plant

The physical and chemical association between a feeding aphid and its host plant is relatively intimate, and aphids cause changes in the chemistry of the plants on which they feed. These changes are reviewed thoroughly in Petterson *et al.*, Chapter 4 this volume and Quisenberry and Ni, Chapter 13 this volume. There is evidence that barley plants infested with *R. padi* produce volatile signals that induce neighbouring plants to become less acceptable to the aphid (Pettersson *et al.*, 1996). The study and identification of such aphid-induced signals is likely to be an area of great interest to chemical ecologists.

Interaction with Natural Enemies

A disadvantage of employing semiochemicals in intraspecific communication is that they can betray the often-sophisticated crypsis that an insect has attained. An individual may be coloured to blend into the foliage, or concealed on a lower leaf surface or inside a rolled leaf, but when it releases chemicals into the environment it will reveal its presence to any natural enemy that has evolved the ability to detect those signals. Of course, insects leave further clues from which searching predators and parasitoids can elicit information, such as waste products and other excretions. Given the importance of chemical signals to aphids, it is no surprise that there are several examples of their exploitation by natural enemies. Recently, major developments have been made in the study of the use by aphid parasitoids of damage-induced chemicals produced by aphid-attacked plants. These will be dealt with separately to chemicals of direct aphid origin.

Responses of natural enemies to aphid-produced chemicals

Parasitoids

Initial studies on the use of kairomones as host location cues by aphid parasitoids focused on aphid honeydew. It is an obvious target for interest since it is relatively apparent and is produced as a necessity by all aphids. Indeed, it is strongly indicative of the presence of aphids, and parasitoids respond to it with intensified searching behaviours (Gardner and Dixon, 1985; Cloutier and Bauduin, 1990) and by increasing their residence time in honeydew-contaminated patches (Hågvar and Hofsvang, 1989; Budenberg *et al.*, 1992). These experimental studies all strongly suggest that honeydew is detected on contact by the parasitoid, although there are reports of parasitoid attraction to honeydew odour in an olfactometer. Such attraction was shown by Bouchard and Cloutier (1985) for *Aphidius nigripes* and by Wickremasinghe and van Emden (1992) for *A. ervi*, *Aphidius rhopalosiphi*, *Aphelinus flavus*, *Lysiphlebus fabarum*, *Trioxys* sp. and *Praon* sp. It has also been shown that *A. rhopalosophi* has olfactory attraction to indole-3-acetaldehyde, a volatile breakdown product of tryptophan in the honeydew of aphids (van Emden and Hagen, 1976).

The first evidence that parasitoids were attracted to aphid sex pheromones came during field experiments designed to observe the response of field-flying male aphids to synthetic sex pheromone (Hardie *et al.*, 1991). Besides capturing male aphids, large numbers of parasitoids of the genus *Praon* were recovered from water traps baited with nepetalactone (**1**, Fig. 9.4). All the wasps were female, and 98.5% were found in pheromone-baited traps, as opposed to unbaited controls. Since clear traps and synthetic pheromone were used, all other visual and olfactory cues were ruled out, clearly demonstrating that parasitoids were attracted by the pheromone.

When the experiments were repeated during the following autumn in winter cereal fields, attraction to nepetalactone was again demonstrated (Powell *et al.*, 1993; Hardie *et al.*, 1994c). This time, 89% of the captured wasps were *Praon volucre*, with over 99% occurring in the pheromone-baited traps. The trials were replicated at sites in southwestern, central and northern England, and in northern Germany, and the response to sex pheromone was apparent at all sites that had parasitoid populations. In several of the trials, nepetalactol-baited traps were included, but caught significantly fewer parasitoids than those baited with nepetalactone.

Attraction to aphid sex pheromones in laboratory assays has been demonstrated for *P. volucre* (Lilley *et al.*, 1994), *Aphidius eadyi* and *A. ervi* (Glinwood *et al.*, 1999a), *Aphidius matricariae* (Isaacs, 1994), *A. rhopalosiphi* and *Ephedrus plagiator* (Glinwood, 1998). *Diaeretiella rapae* was attracted to traps baited with nepetalactone in the field (Gabrys *et al.*, 1997). Electroantennogram responses have been shown in *Lysiphlebia japonica* (Hou and Yan, 1995).

There are opportunities for applied use of aphid sex pheromones in integrated management of aphid pests because aphid parasitoids are important members of the aphid natural enemy complex, yet their impact is often diminished by poor synchrony between their arrival in crops and aphid population development (Carter *et al.*, 1980; Wratten and Powell, 1991). Aphid sex pheromone deployed in field margins in the autumn (Powell *et al.*, 1993) or in the crop in the spring (Glinwood, 1998) might reduce this asynchrony. Attempts to increase parasitization in aphid colonies in the field using sex pheromones have been successful. When potted cereal plants were artificially infested with *S. avenae* and exposed in the field, the presence of nepetalactone significantly increased parasitism of *S. avenae* by *Praon* spp. (Lilley *et al.*, 1994; Glinwood *et al.*, 1998) and *A. rhopalosiphi* (Glinwood *et al.*, 1998). In a field trial (Glinwood, 1998), the deployment of nepetalactone lures in spring wheat significantly increased the numbers of parasitized aphids in treated plots (Fig. 9.7). Although the impact of parasitization was not sufficient to control the aphid population, levels of parasitization in pheromone plots were clearly shifted to earlier in aphid population development, indicating that this strategy may indeed enhance the synchrony discussed previously.

Initially, it was concluded that the parasitoid response to aphid sex pheromones was either induced by environmental stimuli (i.e. the changes in temperature and photoperiod which occur in the autumn) or developed in response to parasitoids learning to associate suitable hosts with the presence of pheromone by attacking sexual aphids when they appeared in the autumn (Hardie *et al.*, 1994c). However, parasitoids have since been shown to respond to sex pheromones in laboratory bioassays at temperatures and photoperiods that do not match those which exist in the field in autumn, thus ruling out the environmental induction theory. Furthermore, since test insects were prevented from having contact with either sexual aphids or sex pheromones, the response does not appear to be learned *via* association. Therefore, the parasitoid response to aphid sex pheromones appears to be innate.

Both *P. volucre* and *A. ervi* were attracted equally by plant-extracted and fully synthetic (4a*S*,7*S*,7a*R*)-nepetalactone (**1**, Fig. 9.4) in wind tunnel bioassays (Glinwood *et al.*, 1999b).

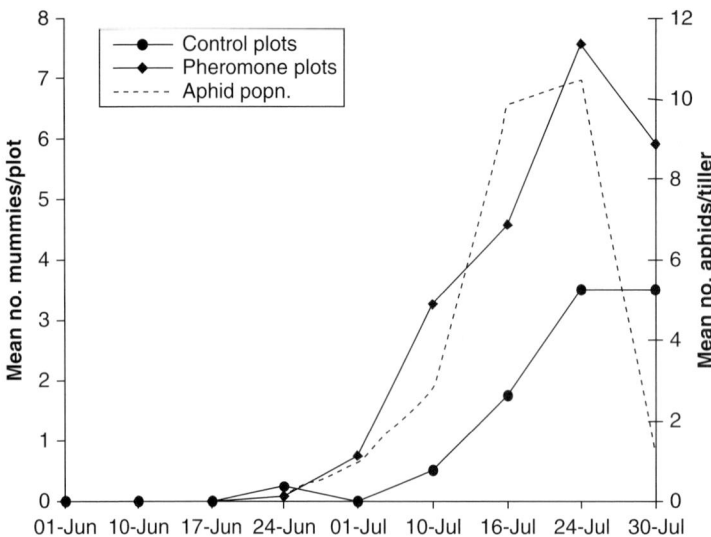

Fig. 9.7. Effect of aphid sex pheromone on parasitization levels of cereal aphids in plots of spring wheat. Parasitization increased earlier in plots treated with pheromone, and in greater synchrony with the increase in the aphid population (from Glinwood, 1998).

However, in field tests with baited aphid-infested plants, the synthetic 7R isomer was unattractive, and even rendered the 7S isomer unattractive when combined in a 50% blend with it (Glinwood et al., 1999b). This indicated that enantiomeric purity of the pheromone is more important for parasitoid response than is the presence of plant-derived contaminants. This is in contrast to the response of male aphids (Hardie et al., 1997), which raises interesting questions about the evolution of the response of the two different insects to this signal.

The parasitoid *Aphidius uzbekistanicus* was attracted to the aphid alarm pheromone (E)-β-farnesene (**5**, Fig. 9.5) in a simple Y-tube olfactometer bioassay (Micha and Wyss, 1996). Detailed measurements of the foraging response of the parasitoid *D. rapae* to **5** showed increased movement and tighter turning angles, behaviours likely to increase the probability of host encounters (Foster et al., 2005).

The searching behaviour of aphid parasitoids seems to be influenced by cues present in the aphid mummy that are experienced by the adult wasp upon emergence. Several reports have shown that this experience can prime parasitoids to preferentially search for and oviposit in a particular aphid species (van Emden et al., 1996; Storeck

et al., 2000; Blande et al., 2004). This underlines the subtlety of chemical interactions between parasitoids and their aphid hosts, about which more surely remains to be discovered.

Predators

Predators generally have been considered as less sophisticated users of host location kairomones than parasitoids. One reason may be that a parasitoid's reproductive success is more critically dependent upon finding suitable hosts for oviposition, and therefore suitable for its offspring, than that of a predator whose search is usually directed towards a food resource for that individual. However, in the case of predators such as coccinellids, syrphids, and chrysopids, the location of aphid colonies by adults is also very important for the survival of their offspring. Thus, they may be considered as similar to parasitoids in this respect.

The responses of several ladybird species to aphid siphunculus secretions and alarm pheromone ((E)-β-farnesene, **5**, Fig. 9.5) have been investigated, and there appears to be some variation in response between species and between larval and adult stages. In olfactometer tests, *Adalia bipunctata* (2-spot ladybird) larvae were attracted to the odour of aphids under attack by conspecific larvae, whereas there was no response to larvae or aphids alone (Hemptinne *et al.*, 2000). The hypothesis was that the larval response is adaptive since it allows them to share the aphid prey of larvae that have hunted successfully. Adults of *A. bipunctata* have also been shown to respond to **5** in an olfactometer (Francis *et al.*, 2004). Adults of *Harmonia axyridis* (harlequin ladybird) did not show any response to siphunculus droplets produced by *A. pisum* (Mondor and Roitberg, 2000). Responses of larvae were not tested in this study. Electroantennograms in response to **5** have been recorded from *Coleomegilla maculata* (12-spot ladybird) (Zhu *et al.*, 1999). Behavioural studies were not reported, but the authors claimed to have obtained behavioural responses from two further species, *Hippodamia convergens* and *H. axyridis*, in preliminary studies subsequently confirmed to involve **5** for attraction of *H. convergens* (Acar *et al.*, 2001).

The most comprehensive study of ladybird response to **5** (Al Abassi *et al.*, 2000) demonstrates both electrophysiological and behavioural responses in *C. septempunctata* (7-spot ladybird). Not only was attraction to **5** demonstrated, but it was also shown that increasing concentrations of the plant-derived alarm pheromone inhibitor (–)-β-caryophyllene (**7**, Fig. 9.5), analogous to the system that operates inhibition of response in aphids (Dawson *et al.*, 1984), inhibited ladybird response. The receptor cells for **5** and **7** in *C. septempunctata* are paired and are frequently found in close proximity, with larger amplitude consistently recorded for the cell specifically responding to **5**. The co-location of these cells may provide the mechanism by which different behavioural responses are exhibited when the inhibitor is present together with the attractant (Al Abassi *et al.*, 2000). Although **5** did not influence the fine-scale searching behaviours of *C. septempunctata* such as turning rate and 'meander' (Nakamuta, 1991), it is likely that it plays some role as a host location kairomone in this and other ladybird species.

The behavioural responses of ladybirds to aphid sex pheromones either have not been studied or studies have not been reported due to a negative outcome. Electroantennogram responses to the nepetalactone and nepetalactol sex pheromone components have, however, been shown in *C. maculata* (Zhu *et al.*, 1999). The same authors also obtained EAG responses to both of these sex pheromone components from *Chrysoperla carnea* (green lacewing). Another lacewing, *Chrysopa cognata*, showed EAG responses to both nepetalactone and nepetalactol, as well as positive behavioural responses in an olfactometer and attraction to pheromone-baited traps in the field (Boo *et al.*, 1998). *Chrysopa oculata* showed EAG responses to sex pheromone components, and the (1*R*, 4a*S*,7*S*, 7a*R*)-nepetalactol attracted adults to field traps (Zhu *et al.*, 2005). Interestingly, the same study also demonstrated field attraction of *C. carnea* females to 2-phenylethanol, a component of its prey aphid's host plant. *Chrysoperla carnea* showed an EAG response to **5** (Zhu *et al.*, 1999), whereas *C. cognata* did not (Boo *et al.*, 1998).

Earlier literature revealed that the lacewing *Chrysopa pallens* was attracted to the vine *Actinidia polygama*, a plant that generates two dihydronepetalactols previously named neomatatabiol and isoneomatatabiol (Hyeon *et al.*, 1968), structurally related to nepetalactone and nepetalactol but incompletely characterized. The enantiomerically pure diastereoisomers (1*R*,4 *S*,4a*R*,7*S*,7a*R*)-(**8**, Fig. 9.5) and (1*R*,4*R*,4a*R*,7*S*,7a*R*)-dihydronepetalactol (**9**, Fig. 9.5) were synthesized (Hooper *et al.*, 2002) diastereoselectively from (4a*S*,7*S*,7a*R*)-nepetalactone (**1**). The stereochemistry of the compounds was determined by NMR spectroscopy and X-ray crystallography, and the compounds were shown to be identical to those previously referenced as neomatatabiol and isoneomatatabiol, respectively. (1*R*,4*S*,4a*R*,7*S*,7a*R*)-dihydronepetalactol (**8**) was found to catch significant numbers of three species of lacewing in the field: in Korea, *C. cognata* and, in the UK, *Nineta vittata* and most notably, *Peyerimhoffina gracilis*. All species caught in significant numbers were found more frequently in traps releasing (1*R*,4*S*,4a*R*,7*S*, 7a*R*)-(**8**) rather than (1*R*,4*R*,4a*R*,7*S*,7a*R*)-dihydronepetalactol (**9**), while more *C. cognata*,

Chrysopa formosa, and *Chrysopa phyllochroma* were found in traps releasing the nepetalactol (Hooper *et al.*, 2002; Boo *et al.*, 2003). The capture of *P. gracilis* with (1*R*,4 *S*,4a*R*,7*S*,7a*R*)-dihydronepetalactol (**8**) was of particular interest as this lacewing has been recorded only recently in the UK; indeed, its discovery was made possible by attraction to the aphid sex pheromone components (Donato *et al.*, 2001).

The efficacy of certain sprayed yeast hydrolysates in attracting lacewings into lucerne crops in California was ascribed to the fact that they contain tryptophan, an amino acid present in aphid honeydew (Hagen *et al.*, 1971). van Emden and Hagen (1976) found that, for *C. carnea*, this attraction stemmed from indole acetaldehyde, a volatile breakdown product of tryptophan (see analogous results with parasitoids above). Interestingly, later work (Hagen, 1986) showed that the response of the lacewing was not triggered in the absence of a synomone from the lucerne plant.

Hover fly responses to aphid semiochemicals are notably absent from the literature, and again it may be that they have been sought but not found. Non-volatile kairomones present in extracts of the siphunculus secretion of *A. fabae* influenced the searching behaviour of adult females of *Eupeodes (Metasyrphus) corollae* (Shonouda *et al.*, 1998). Effects included stimulation of oviposition and arrestment on bean plants treated with the extract. Olfactory responses to aphid honeydew have been shown in the aphidophagous gall midge, *Aphidoletes aphidimyza* (Choi *et al.*, 2004).

There is one report of attraction of non-aphid specialist predators to an aphid semiochemical. In olfactometer tests, adults of the ground beetles *Pterostichus melanarius* and *Harpalus rufipes* (Coleoptera: Carabidae) responded positively to (*E*)-β-farnesene (**5**) (Kielty *et al.*, 1996).

Responses of natural enemies to aphid-induced plant signals

Herbivore-induced chemicals from infested plants provide natural enemies with information that is both easily detectable and a highly reliable indicator of the presence of hosts or prey (Vet and Dicke, 1992). These signals are often considered to be plant SOS signals, i.e. synomones that enhance recruitment of the attacking herbivores' enemies. However, in most cases, the consequences of production of these signals for plant fitness are unknown.

Parasitoids

In wind tunnel bioassays, *A. ervi* was attracted equally to the plant–host complex (*V. faba–A. pisum*) and to aphid-damaged plants from which aphids had been removed (Du *et al.*, 1996). The parasitoid was able to discriminate between bean plants previously infested with *A. pisum* and a non-host aphid, *A. fabae*, even after the plants had been washed to remove traces of aphid honeydew and exuviae. Thus, the plant produces aphid species-specific volatile cues that can be used by female parasitoids in host location. The plant response that leads to the production of these volatiles is now known to be systemic, the upper leaves becoming attractive to parasitoids 48–72 h after aphid infestation of basal leaves (Guerrieri *et al.*, 1999). Air entrainment of the bean plants, followed by GC-MS, identified several volatiles that were produced at higher levels by plants infested with *A. pisum*, including 6-methyl-5-hepten-2-one, (*E*)-β-ocimene, (*Z*)-3-hexen-1-ol, (*Z*)-3-hexenyl acetate, linalool, and (*E*)-β-farnesene from the plant (Du *et al.*, 1998). All the chemicals were behaviourally active, with 6-methyl-5-hepten-2-one eliciting significantly more upwind flights by *A. ervi* in a wind tunnel than the other substances. Recent work on the transfer of damage-induced messenger substances between plants has demonstrated that root exudates of *A. pisum*-infested bean plants contain compounds that are systemically translocated and, when taken up by an intact plant, elicit the release of the same volatiles that are induced by aphid feeding (Chamberlain *et al.*, 2001). These volatiles are also attractive to the parasitoid *A. ervi*.

Predators

Studies on predator attraction to aphid-induced plant signals are in their infancy compared with those on parasitoids. Adult female ladybirds (*A. bipunctata*) were more attracted to the odour of *V. faba* infested with *A. pisum* than to uninfested *V. faba* (Raymond *et al.*, 2000). However, when aphids and aphid products (exuviae and honeydew) were removed from the plant, the attraction was lost. Therefore, attraction of ladybirds to honeydew and alarm pheromone cannot be ruled out in this study. There is stronger evidence for attraction to aphid-induced signals in *C. septempunctata* (Ninkovic *et al.*, 2001). Olfactometer responses to solvent extracts from air entrainments of barley and the aphid *R. padi* showed that attraction to a host-damaged plant from which aphids had been removed was as strong as attraction to the plant–host complex. There was no response to the undamaged plant or to aphids alone, and it is likely that the aphids in this case were undisturbed and thus not producing alarm pheromone (J. Pettersson, personal communication).

It is clear from laboratory experiments that a broad range of aphid parasitoids and predators has evolved behavioural responses to both aphid kairomones and aphid-induced plant signals. What is not as clear, however, is the importance of such responses for the success of natural enemies in their natural habitats and, in the case of plant signals, whether there are genuine fitness benefits for the emitting plants. It is also unclear how useful these substances will prove as components of integrated aphid control strategies. Nevertheless, the few attempts to use aphid kairomones to concentrate natural enemies on plants have been encouraging (Lilley *et al.*, 1994; Glinwood *et al.*, 1998; Shonouda *et al.*, 1998).

Conclusions

The intimate relationship that aphids have with their host plants has resulted in the evolution of fascinating chemical ecological interactions. The range of molecular structures of the semiochemicals involved is restricted by the limited array of plant secondary metabolites with appropriate physicochemical characteristics, and has given rise to diverse activities from relatively simple compounds such as methyl salicylate, 6-methyl-5-hepten-2-one and nepetalactone, which appear in various roles in this review. This parsimony in usage of semiochemicals in aphid chemical ecology bodes well for their applied use in pest management through restricting the number of compounds required for registration. However, it should be borne in mind that there are still areas, for example, involving the chemical ecology of aphid parasitoids and predators, where new chemistries will be identified.

It appears from recent work that the main problem in developing semiochemicals for aphid pest management, which at least for arable crops is cheap delivery, is now being solved. Thus, the discovery that genes relating to production of aphid semiochemicals by plants can be 'switched on' by benign chemical signals means that the plants themselves can generate target semiochemicals. Already, volatile signals from damaged plants that have this potential have been identified, but further work is needed, both in this area and on signals active in the rhizosphere. Signalling by undamaged plants is also an area of increasing interest (Pettersson *et al.*, Chapter 4 this volume). These studies will be enhanced greatly by a deeper understanding of the molecular biology of aphid chemosensory receptors, and of the genetics underpinning plant secondary metabolism, particularly those pathways induced by external signals. Currently, the latter studies rely heavily on *A. thaliana*, but as genomic information from other plants more relevant to pest aphid chemical ecology becomes available, more rapid advances will be made.

Finally, it must be emphasized that determination of a complete aphid genome is now seen to be essential for a full understanding and exploitation of aphid chemical ecology.

References

Acar, E.B., Medina, J.C., Lee, M.L. and Booth, G.M. (2001) Olfactory behavior of convergent lady beetles (Coleoptera: Coccinellidae) to alarm pheromone of green peach aphid (Hemiptera: Aphididae). *Canadian Entomologist* 133, 389–397.

Agelopoulos, N.G. and Pickett, J.A. (1998) Headspace analysis in chemical ecology: effects of different sampling methods on ratios of volatile compounds present in headspace samples. *Journal of Chemical Ecology* 24, 1161–1172.

Agelopoulos, N.G., Hooper, A.M., Maniar, S.P., Pickett, J.A. and Wadhams, L.J. (1999) A novel approach for isolation of volatile chemicals released by individual leaves of a plant *in situ*. *Journal of Chemical Ecology* 25, 1411–1425.

Al Abassi, S., Birkett, M.A., Pettersson, J., Pickett, J.A., Wadhams, L.J. and Woodcock, C.M. (2000) Response of the seven-spot ladybird to an aphid alarm pheromone and an alarm pheromone inhibitor is mediated by paired olfactory cells. *Journal of Chemical Ecology* 26, 1765–1771.

Arakaki, N. (1989) Alarm pheromone eliciting attack and escape responses in the sugar cane woolly aphid, *Ceratovacuna lanigera* (Homoptera, Pemphigidae). *Journal of Ethology* 7, 83–90.

Birkett, M.A. and Pickett, J.A. (2003) Aphid sex pheromones: from discovery to commercial production. (Molecules of interest review.) *Phytochemistry* 62, 651–656.

Birkett, M.A., Campbell, C.A.M., Chamberlain, K., Guerrieri, E., Hick, A.J., Martin, J.L., Matthes, M., Napier, J.A., Pettersson, J., Pickett, J.A., Poppy, G.M., Pow, E.M., Pye, B.J., Smart, L.E., Wadhams, G.H., Wadhams, L.J. and Woodcock, C.M. (2000) New roles for *cis*-jasmone as an insect semiochemical and in plant defense. *Proceedings of the National Academy of Sciences of the United States of America* 97, 9329–9334.

Birkett, M.A., Chamberlain, K., Hooper, A.M. and Pickett, J.A. (2001) Does allelopathy offer real promise for practical weed management and for explaining rhizosphere interactions involving higher plants? *Plant and Soil* 232, 31–39.

Blande, J.D., Pickett, J.A. and Poppy, G.M. (2004) Attack rate and success of the parasitoid *Diaeretiella rapae* on specialist and generalist feeding aphids. *Journal of Chemical Ecology* 30, 1781–1795.

Boo, K.S., Chung, I.B., Han, K.S., Pickett, J.A. and Wadhams, L.J. (1998) Responses of the lacewing *Chrysopa carnea* to pheromones of its aphid prey. *Journal of Chemical Ecology* 24, 631–643.

Boo, K.S., Choi, M.Y., Chung, I.B., Eastop, V.F., Pickett, J.A., Wadhams, L.J. and Woodcock, C.M. (2000) Sex pheromone of the peach aphid, *Tuberocephalus momonis*, and optimal blends for trapping males and females in the field. *Journal of Chemical Ecology* 26, 601–609.

Boo, K.S., Kang, S.S., Park, J.H., Pickett, J.A. and Wadhams, L.J. (2003) Field trapping of *Chrysopa cognata* (Neuroptera: Chrysopidae) with aphid sex pheromone components in Korea. *Journal of Asia-Pacific Entomology* 6, 29–36.

Bouchard, Y. and Cloutier, C. (1985) Role of olfaction in host finding by aphid parasitoid *Aphidius nigripes* (Hymenoptera: Aphidiidae). *Journal of Chemical Ecology* 11, 801–808.

Bowers, W.S., Nault, L.R., Webb, R.E. and Dutky, S.R. (1972) Aphid alarm pheromone: isolation, identification, synthesis. *Science, New York* 177, 1121–1122.

Bridges, M., Jones, A.M.E., Bones, A.M., Hodgson, C., Cole, R., Bartlet, E., Wallsgrove, R., Karapapa, V.K., Watts, N. and Rossiter, J.T. (2002) Spatial organisation of the glucosinolate-myrosinase system in brassica specialist aphids is similar to that of the host plant. *Proceedings of the Royal Society of London B* 269, 187–191.

Bruce, T.J.A., Martin, J.L., Pickett, J.A., Pye, B.J., Smart, L.E. and Wadhams, L.J. (2003) *cis*-Jasmone treatment induces resistance in wheat plants against the grain aphid, *Sitobion avenae* (Fabricius) (Homoptera: Aphididae). *Pest Management Science* 59, 1031–1036.

Bruce, T.J.A., Birkett, M.A., Blande, J., Hooper, A.M., Martin, J.L., Khambay, B., Prosser, I., Smart, L.E. and Wadhams, L.J. (2005) Response of economically important aphids to components of *Hemizygia petiolata* essential oil. *Pest Management Science* 61, 1115–1121.

Budenberg, W.J., Powell, W. and Clark, S.J. (1992) The influence of aphids and honeydew on the leaving rate of searching aphid parasitoids from wheat plants. *Entomologia Expermentalis et Applicata* 63, 259–264.

Burmeister, W.P., Cottaz, S., Driguez, H., Iori, R., Palmieri, S. and Henrissat, B. (1997) The crystal structures of *Sinapis alba* myrosinase and a covalent glycosyl-enzyme intermediate provide insights into the substrate recognition and active-site machinery of an *S*-glycosidase. *Structure* 5, 663–675.

Burmeister, W.P., Cottaz, S., Rollin, P., Vasella, A. and Henrissat, B. (2000) High resolution x-ray crystallography shows that ascorbate is a cofactor for myrosinase and substitutes for the function of the catalytic base. *Journal of Biological Chemistry* 275, 39385–39393.

Campbell, C.A.M., Pettersson, J., Pickett, J.A., Wadhams, L.J. and Woodcock, C.M. (1993) Spring migration of damson–hop aphid, *Phorodon humuli* (Homoptera: Aphididae), and summer host-plant derived semiochemicals released on feeding. *Journal of Chemical Ecology* 19, 1569–1576.

Campbell, C.A.M., Cook, F.J., Pickett, J.A., Pope, T.W., Wadhams, L.J. and Woodcock, C.M. (2003) Responses of the aphids *Phorodon humuli* and *Rhopalosiphum padi* to sex pheromone stereochemistry in the field. *Journal of Chemical Ecology* 29, 2225–2234.

Carter, N., McLean, I.F.G., Watt, A.D. and Dixon, A.F.G. (1980) Cereal aphids – a case study and review. *Applied Biology* 5, 271–348.

Chamberlain, K., Guerrieri, E., Pennacchio, F., Pettersson, J., Pickett, J.A., Poppy, G.M., Powell, W., Wadhams, L.J. and Woodcock, C.M. (2001) Can aphid-induced plant signals be transmitted aerially and through the rhizosphere? *Biochemical Systematics and Ecology* 29, 1063–1074.

Choi, M.Y., Roitberg, B.D., Shani, A., Raworth, D.A. and Lee, G.H. (2004) Olfactory response by the aphidophagous gall midge, *Aphidoletes aphidimyza* to honeydew from green peach aphid, *Myzus persicae*. *Entomologia Experimentalis et Applicata* 111, 37–45.

Chongrattanmeteekul, W., Foster, J., Shukle, R.H. and Araya, J.E. (1991) Feeding behavior of *Rhopalosiphum padi* (L.) and *Sitobion avenae* (F.) (Homoptera: Aphididae) on wheat as affected by conspecific and interspecific interactions. *Journal of Applied Entomology* 111, 361–364.

Cloutier, C. and Bauduin, F. (1990) Searching behaviour of the aphid parasitoid *Aphidius nigripes* (Hymenoptera: Aphidiidae) foraging on potato plants. *Environmental Entomology* 19, 222–228.

Crock, J., Wildung, M. and Croteau, R. (1997) Isolation and bacterial expression of a sesquiterpene synthase cDNA clone from peppermint (*Mentha × piperita*, L.) that produces the aphid alarm pheromone, (*E*)-β-farnesene. *Proceedings of the National Academy of Sciences of the United States of America* 94, 12833–12838.

Dawson, G.W., Griffiths, D.C., Pickett, J.A. and Woodcock, C.M. (1983) Decreased response to alarm pheromone by insecticide resistant aphids. *Naturwissenschaften* 70, 254–255.

Dawson, G.W., Griffiths, D.C., Pickett, J.A., Smith, M.C. and Woodcock, C.M. (1984) Natural inhibition of the aphid alarm pheromone. *Entomologia Experimentalis et Applicata* 36, 197–199.

Dawson, G.W., Griffiths, D.C., Pickett, J.A., Wadhams, L.J. and Woodcock, C.M. (1986) Plant compounds that synergise activity of the aphid alarm pheromone. *Proceedings of the Brighton Crop Protection Conference, Pests and Diseases, November 1986* 3, 829–834.

Dawson, G.W., Griffiths, D.C., Janes, N.F., Mudd, A., Pickett, J.A., Wadhams, L.J. and Woodcock, C.M. (1987a) Identification of an aphid sex pheromone. *Nature, London* 325, 614–616.

Dawson, G.W., Griffiths, D.C., Pickett, J.A., Wadhams, L.J. and Woodcock, C.M. (1987b) Plant-derived synergists of alarm pheromone from turnip aphid, *Lipaphis (Hyadaphis) erysimi* (Homoptera, Aphididae). *Journal of Chemical Ecology* 13, 1663–1671.

Dawson, G.W., Pickett, J.A. and Smiley, D.W.M. (1996) The aphid sex pheromone cyclopentanoids: synthesis in the elucidation of structure and biosynthetic pathways. *Bioorganic and Medicinal Chemistry* 4, 351–361.

Dicke, M., van Beek, T.A., Posthumus, M.A., Ben Dom, N., van Bokhoven, H. and de Groot, A.E. (1990) Isolation and identification of volatile kairomone that affects acarine predator–prey interactions. Involvement of host plant in its production. *Journal of Chemical Ecology* 16, 381–396.

Dixon, A.F.G. (1998) *Aphid Ecology*, 2nd edn. Chapman and Hall, London, 300 pp.

Donato, B., Brooks, S.J., Pickett, J.A. and Hardie, J. (2001) *Peyerimhoffina gracilis* (Schneider) (Neuroptera: Chrysopidae): a green lacewing new to Britain. *Entomologist's Record* 113, 131–135.

Du, Y.-J., Poppy, G.M. and Powell, W. (1996) The relative importance of semiochemicals from the first and second trophic level in the host foraging behaviour of *Aphidius ervi*. *Journal of Chemical Ecology* 22, 1591–1605.

Du, Y.-J., Poppy, G.M., Powell, W., Pickett, J.A., Wadhams, L.J. and Woodcock, C.M. (1998) Identification of semiochemicals released during aphid feeding that attract parasitoid *Aphidius ervi*. *Journal of Chemical Ecology* 24, 1355–1368.

Edwards, L.J., Siddall, J.B., Dunham, L.L., Uden, P. and Kislow, C.J. (1973) *Trans*-β-farnesene, alarm pheromone of the green peach aphid, *Myzus persicae* (Sulzer). *Nature, London* 241, 126–127.

van Emden, H.F. and Hagen, K.S. (1976) Olfactory reactions of the green lacewing, *Chrysopa carnea*, to tryptophan and certain breakdown products. *Environmental Entomology* 5, 469–473.

van Emden, H.F., Sponagl, B., Wagner, E., Baker, I., Ganguly, S and Douloumpaka, S. (1996) Hopkins' 'host selection principle', another nail in its coffin. *Physiological Entomology* 21, 325–328.

Field, L.M., Pickett, J.A. and Wadhams, L.J. (2000) Molecular studies in insect olfaction. *Insect Molecular Biology* 9, 545–551.

Foster, S.P., Woodcock, C.M., Williamson, M.S., Devonshire, A.L., Denholm, I. and Thompson, R. (1999) Reduced alarm response by peach–potato aphids, *Myzus persicae* (Hemiptera: Aphididae), with knock-down resistance to insecticides (*kdr*) may impose a fitness cost through increased vulnerability to natural enemies. *Bulletin of Entomological Research* 89, 133–138.

Foster, S.P., Denholm, I., Thompson, R., Poppy, G.M. and Powell, W. (2005) Reduced response of insecticide-resistant aphids and attraction of parasitoids to aphid alarm pheromone; a potential fitness trade-off. *Bulletin of Entomological Research* 95, 37–46.

Francis, F., Lognay, G. and Haubruge, E. (2004) Olfactory responses to aphid and host plant volatile releases: (*E*)-beta-farnesene, an effective kairomone for the predator *Adalia bipunctata*. *Journal of Chemical Ecology* 30, 741–755.

Gabrys, B.J., Gadomski, H.J., Klukowski, Z., Pickett, J.A., Sobota, G.T., Wadhams, L.J. and Woodcock, C.M. (1997) Sex pheromone of cabbage aphid *Brevicoryne brassicae*: identification and field trapping of male aphids and parasitoids. *Journal of Chemical Ecology* 23, 1881–1890.

Gardner, S.M. and Dixon, A.F.G. (1985) Plant structure and the foraging success of *Aphidius rhopalosiphi* (Hymenoptera: Aphidiidae). *Ecological Entomology* 10, 171–179.

Gibson, R.W. and Pickett, J.A. (1983) Wild potato repels aphids by release of aphid alarm pheromone. *Nature, London* 302, 608–609.

Glinwood, R.T. (1998) Responses of aphid parasitoids to aphid sex pheromones: laboratory and field studies. PhD thesis, University of Nottingham, UK.

Glinwood, R.T. and Pettersson, J. (2000a) Host choice and host leaving in *Rhopalosiphum padi* (L.) emigrants and repellency of aphid colonies on the winter host. *Bulletin of Entomological Research* 90, 57–61.

Glinwood, R.T. and Pettersson, J. (2000b) Change in response of *Rhopalosiphum padi* spring migrants to the repellent winter host component methyl salicylate. *Entomologia Experimentalis et Applicata* 94, 325–330.

Glinwood, R.T., Powell, W. and Tripathi, C.P.M. (1998) Increased parasitization of aphids on trap plants alongside vials releasing synthetic aphid sex pheromone and effective range of the pheromone. *Biocontrol Science and Technology* 8, 607–614.

Glinwood, R.T., Du, Y.-J. and Powell, W. (1999a) Pea aphid parasitoids (Hymenoptera: Aphidiinae) respond to aphid sex pheromones. *Entomologia Experimentalis et Applicata* 92, 227–232.

Glinwood, R.T., Du, Y.-J., Smiley, D.W.M. and Powell, W. (1999b) Comparative responses of parasitoids to synthetic and plant-extracted nepetalactone component of aphid sex pheromones. *Journal of Chemical Ecology* 25, 1481–1488.

Glinwood, R.T., Pettersson, J., Ninkovic, V., Ahmed, E., Birkett, M. and Pickett, J.A. (2003) Change in acceptability of barley plants to aphids after exposure to allelochemicals from couch-grass (*Elytrigia repens*). *Journal of Chemical Ecology* 29, 259–272.

Glinwood, R., Ninkovic, V., Ahmed, E. and Pettersson, J. (2004) Barley exposed to aerial allelopathy from thistles (*Cirsium* spp.) becomes less acceptable to aphids. *Ecological Entomology* 29, 188–195.

Goldansaz, S.H., Dewhirst, S., Birkett, M.A., Hooper, A.M., Smiley, D.W.M., Pickett, J.A., Wadhams, L. and McNeil, J.N. (2004) Identification of two sex pheromone components of the potato aphid, *Macrosiphum euphorbiae* (Thomas). *Journal of Chemical Ecology* 30, 819–834.

Griffiths, E., Wratten, S.D. and Vickerman, G.P. (1985) Foraging by the carabid *Agonum dorsale* in the field. *Ecological Entomology* 10, 181–189.

Guerrieri, E., Poppy, G.M., Powell, W., Tremblay, E. and Pennacchio, F. (1999) Induction and systemic release of herbivore-induced plant volatiles mediating in-flight orientation of *Aphidius ervi*. *Journal of Chemical Ecology* 25, 1247–1261.

Hagen, K.S. (1986) Ecosystem analysis: plant cultivars (HPR), entomophagous species and food supplements. In: Boethel, D.J. and Eikenbary, R.D. (eds) *Interactions of Plant Resistance and Parasitoids and Predators of Insects*. Wiley, Chichester, pp. 151–197.

Hagen, K.S., Sawall, E.F., Jr. and Tassan, R.L. (1971) The use of food sprays to increase effectiveness of entomophagous insects. In: *Proceedings of the Tall Timbers Conference on the Ecological Animal Control by Habitat Management* 2, pp. 59–81.

Hågvar, E.B. and Hofsvang, T. (1989) The effect of honeydew on plant colonization by the aphid parasitoid *Ephedrus cerasicola*. *Entomophaga* 34, 495–501.

Hardie, J., Nottingham, S.F., Powell, W. and Wadhams, L.J. (1991) Synthetic aphid sex pheromone lures female aphid parasitoids. *Entomologia Experimentalis et Applicata* 61, 97–99.

Hardie, J., Nottingham, S.F., Dawson, G.W., Harrington, E., Pickett, J.A. and Wadhams, L.J. (1992) Attraction of field-flying aphid males to synthetic sex pheromone. *Chemoecology* 3, 113–117.

Hardie, J., Isaacs, R., Pickett, J.A., Wadhams, L.J. and Woodcock, C.M. (1994a) Methyl salicylate and (–)-(1*R*,5*S*)-myrtenal are plant-derived repellents for black bean aphid, *Aphis fabae* Scop. (Homoptera: Aphididae). *Journal of Chemical Ecology* 20, 2847–2855.

Hardie, J., Storer, J.R., Nottingham, S.F., Peace, L., Harrington, R., Merritt, L.A., Wadhams, L.J. and Wood, D.K. (1994b) The interaction of sex pheromone and plant volatiles for field attraction of male bird-cherry aphid, *Rhopalosiphum padi*. *Proceedings of the British Crop Protection Conference, Pests and Diseases, Brighton, November 1994* 3, 1223–1230.

Hardie, J., Hick, A.J., Höller, C., Mann, J., Merritt, L., Nottingham, S.F., Powell, W., Wadhams, L.J., Witthinrich, J. and Wright, A.F. (1994c) The responses of *Praon* spp. parasitoids to aphid sex pheromone components in the field. *Entomologia Experimentalis et Applicata* 71, 95–99.

Hardie, J., Storer, J.R., Nottingham, S.F., Cook, F.J., Campbell, C.A.M., Wadhams, L.J., Lilley, R. and Peace, L. (1996) Sex pheromone and visual trap interactions in mate location strategies and aggregation by host-alternating aphids in the field. *Physiological Entomology* 21, 97–106.

Hardie, J., Peace, L., Pickett, J.A., Smiley, D.W.M., Storer, R.J. and Wadhams, L.J. (1997) Sex pheromone stereochemistry and purity affect field catches of male aphids. *Journal of Chemical Ecology* 23, 2547–2554.

Hardie, J., Pickett, J.A., Pow, E.M. and Smiley, D.W.M. (1999) Aphids. In: Hardie, J. and Minks, A.K. (eds) *Pheromones of Non-Lepidopteran Insects Associated with Agricultural Plants*. CAB International, Wallingford, pp. 227–249.

Hartfield, C.M., Campbell, C.A.M., Hardie, J., Pickett, J.A. and Wadhams, L.J. (2001) Pheromone traps for the dissemination of an entomopathogen by the damson–hop aphid *Phorodon humuli*. *Biocontrol Science and Technology* 11, 401–410.

Hemptinne, J.L., Gaudin, M., Dixon, A.F.G. and Lognay, G. (2000) Social feeding in ladybirds: adaptive significance and mechanism. *Chemoecology* 10, 149–152.

Hildebrand, D.F., Brown, G.C., Jackson, D.M. and Hamilton-Kemp, T.R. (1993) Effects of some leaf-emitted volatile compounds on aphid population increase. *Journal of Chemical Ecology* 19, 1875–1887.

Hooper, A.M., Donato, B., Woodcock, C.M., Park, J.H., Paul, R.L., Boo, K.S., Hardie, J. and Pickett, J.A. (2002) Characterization of (1*R*,4S,4a*R*,7S,7a*R*)-dihydronepetalactol as a semiochemical for lacewings, including *Chrysopa* spp. and *Peyerimhoffina gracilis*. *Journal of Chemical Ecology* 28, 849–864.

Hori, M. (1999) Antifeeding, settling inhibitory and toxic activities of labiate essential oils against the green peach aphid, *Myzus persicae* (Sulzer) (Homoptera: Aphididae). *Applied Entomology and Zoology* 34, 113–118.

Hou, Z. and Yan, F. (1995) Electroantennogram responses of *Lysiphlebia japonica* Ashmead (Hymenoptera: Aphidiidae) to some cotton plant volatiles and cotton aphid pheromones. *Entomologia Sinica* 2, 253–264.

Husebye, H., Arzt, S., Burmeister, W.P., Härtel, F.V., Brandt, A., Rossiter, J.T. and Bones, A.M. (2005) Crystal structure at 1.1 Å resolution of an insect myrosinase from *Brevicoryne brassicae* shows its close relationship to β-glucosidases. *Insect Biochemistry and Molecular Biology* 35, 1311–1320.

Hyeon, S.B., Isoe, S. and Sakan, T. (1968) The structure of neomatatabiol, the potent attractant for *Chrysopa* from *Actinidia polygama*. *Tetrahedron Letters* 51, 5325–5326.

Isaacs, R. (1994) Studies on the chemical ecology of the black bean aphid, *Aphis fabae*. PhD thesis, University of London, UK.

Johansson, C., Pettersson, J. and Niemeyer, H.M. (1997) Interspecific recognition through odours by aphids (Sternorryncha: Aphididae) feeding on wheat plants. *European Journal of Entomology* 94, 557–559.

Jones, A.M.E., Bridges, M., Bones, A.M., Cole, R. and Rossiter, J.T. (2001) Purification and characterisation of a non-plant myrosinase from the cabbage aphid *Brevicoryne brassicae* (L.). *Insect Biochemistry and Molecular Biology* 31, 1–5.

Jones, A.M.E., Winge, P., Bones, A.M., Cole, R. and Rossiter, J.T. (2002) Characterisation and evolution of a myrosinase from the cabbage aphid *Brevicoryne brassicae*. *Insect Biochemistry and Molecular Biology* 32, 275–284.

Kay, R.H. (1976) Behavioural components of pheromonal aggregation in *Aphis fabae* Scopoli. *Physiological Entomology* 1, 249–254.

Kielty, J.P., Allen Williams, L.J., Underwood, N. and Eastwood, E.A. (1996) Behavioral responses of three species of ground beetle (Coleoptera: Carabidae) to olfactory cues associated with prey and habitat. *Journal of Insect Behavior* 9, 237–250.

Klingauf, F. (1976) Die Bedeuting der 'Stimmung' im Leben phytophager Insekten am Beispiel des Wirtswahl-Verhaltens von Blattläusen. *Zeitschrift für Angewandte Entomologie* 82, 200–209.

Koch, T., Bandemer, K. and Boland, W. (1997) Biosynthesis of *cis*-jasmone: a pathway for the inactivation and the disposal of the plant stress hormone jasmonic acid to the gas phase? *Helvetica Chemica Acta* 80, 838–850.

Kunert, G., Otto, S., Röse, U.S.R., Gershenzon, J. and Weisser, W.W. (2005) Alarm pheromone mediates production of winged dispersal morphs in aphids. *Ecology Letters* 8, 596–603.

Lilley, R. and Hardie, J. (1996) Cereal aphid responses to sex pheromones and host-plant odours in the laboratory. *Physiological Entomology* 21, 304–308.

Lilley, R., Hardie, J. and Wadhams, L.J. (1994) Field manipulation of *Praon* populations using semiochemicals. *Norwegian Journal of Agricultural Science* 16 (supplement), 221–226.

Lilley, R., Hardie, J., Pickett, J.A. and Wadhams, L.J. (1995) The sex pheromone of the English grain aphid *Sitobion avenae*. *Chemoecology* 5/6, 43–46.

Lösel, P.M., Lindemann, M., Scherkenbeck, J., Maier, J., Engelhard, B., Campbell, C.A.M., Hardie, J., Pickett, J.A., Wadhams, L.J., Elbert, A. and Theilking, G. (1996) The potential of semiochemicals for control of *Phorodon humuli* (Homoptera: Aphididae). *Pesticide Science* 48, 293–303.

Matthes, M., Napier, J.A., Pickett, J.A. and Woodcock, C.M. (2003) New chemical signals in plant protection against herbivores and weeds. *Proceedings of the BCPC International Congress – Crop Science and Technology, Glasgow, November 2003* 2, 1227–1236.

Micha, S.G. and Wyss, U. (1996) Aphid alarm pheromone (*E*)-beta-farnesene: a host finding kairomone for the aphid primary parasitoid *Aphidius uzbekistanicus* (Hymenoptera: Aphidiinae) *Chemoecology* 7, 132–139.

Mondor, E.B. and Roitberg, B.D. (2000) Has the attraction of predatory coccinellids to cornicle droplets constrained aphid alarm signalling behaviour? *Journal of Insect Behavior* 13, 321–329.

Nakamuta, K. (1991) Aphid alarm pheromone component, (*E*)-beta-farnesene, and local search by a predatory lady beetle, *Coccinella septempunctata brukii* Mulsant (Coleoptera: Coccinellidae). *Applied Entomology and Zoology* 26, 1–7.

Nazzi, F., Powell, W., Wadhams, L.J. and Woodcock, C.M. (1996) Sex pheromone of aphid parasitoid *Praon volucre* (Hymenoptera, Braconidae). *Journal of Chemical Ecology* 22, 1169–1175.

Ninkovic, V., Al Abassi, S. and Pettersson, J. (2001) The influence of aphid-produced plant volatiles on ladybird beetle searching behavior. *Biological Control* 21, 191–195.

Ninkovic, V., Olsson, U. and Pettersson, J. (2002) Mixing barley cultivars affects aphid host plant acceptance in field experiments. *Entomologia Experimentalis et Applicata* 102, 177–182.

Ninkovic, V., Ahmed, E., Glinwood, R. and Pettersson, J. (2003) Effects of two types of semiochemical on population development of the bird cherry oat aphid *Rhopalosiphum padi* in a barley crop. *Agricultural and Forest Entomology* 5, 27–33.

Nottingham, S.F. and Hardie, J. (1993) Flight behaviour of the black bean aphid, *Aphis fabae*, and the cabbage aphid, *Brevicoryne brassicae*, in host and non-host plant odour. *Physiological Entomology* 18, 389–394.

Park, K.C. and Hardie, J. (2003) Electroantennogram responses of aphid nymphs to plant volatiles. *Physiological Entomology* 28, 215–220.

Park, K.C. and Hardie, J. (2004) Electrophysiological characterisation of olfactory sensilla in the black bean aphid, *Aphis fabae*. *Journal of Insect Physiology* 50, 647–655.

Park, K.C., Elias, D., Donato, B. and Hardie, J. (2000) Electroantennogram and behavioural responses of different forms of the bird cherry–oat aphid, *Rhopalosiphum padi*, to sex pheromone and a plant volatile. *Journal of Insect Physiology* 46, 597–604.

Pettersson, J. (1970a) Studies on *Rhopalosiphum padi* (L.) I. Laboratory studies on olfactometric responses to the winter host *Prunus padus* L. *Lantbrukshögskolans Annaler* 36, 381–389.

Pettersson, J. (1970b) An aphid sex attractant. I. Biological studies. *Entomologica Scandinavica* 1, 63–73.

Pettersson, J. (1993) Odour stimuli affecting autumn migration of *Rhopalosiphum padi* (L.) (Hemiptera: Homoptera). *Annals of Applied Biology* 122, 417–425.

Pettersson, J. (1994) The bird cherry–oat aphid, *Rhopalosiphum padi* (Hom.: Aph.), and odours. In: Leather, S.R., Wyatt, A., Kidd, N.A.C. and Walters, K.F.A. (eds) *Individuals, Populations and Patterns in Ecology*. Intercept, Andover, pp. 3–12.

Pettersson, J. and Stephansson, D. (1991) Odour communication in two brassica feeding aphid species (Homoptera: Aphidinae: Aphididae). *Entomologia Generalis* 16, 241–247.

Pettersson, J., Pickett, J.A., Pye, B.J., Quiroz, A., Smart, L.E., Wadhams, L.J. and Woodcock, C.M. (1994) Winter host component reduces colonisation by bird cherry–oat aphid, *Rhopalosiphum padi* (L.) (Homoptera, Aphididae), and other aphids in cereal fields. *Journal of Chemical Ecology* 20, 2565–2574.

Pettersson, J., Quiroz, A., Stephansson, D. and Niemeyer, H.M. (1995) Odour communication of *Rhopalosiphum padi* on grasses. *Entomologia Experimentalis et Applicata* 76, 325–328.

Pettersson, J., Quiroz, A. and Fahad, A.E. (1996) Aphid antixenosis mediated by volatiles in cereals. *Acta Agriculturae Scandinavica, Section B, Soil and Plant Science* 46, 135–140.

Pettersson, J., Karunaratne, S., Ahmed, E. and Kumar, V. (1998) The cowpea aphid, *Aphis craccivora*, host plant odours and pheromones. *Entomologia Experimentalis et Applicata* 88, 177–184.

Pickett, J.A. and Griffiths, D.C. (1980) Composition of aphid alarm pheromones. *Journal of Chemical Ecology* 6, 349–360.

Pickett, J.A., Wadhams, L.J., Woodcock, C.M. and Hardie, J. (1992) The chemical ecology of aphids. *Annual Review of Entomology* 37, 67–90.

Pontoppidan, B., Ekbom, B., Eriksson, S. and Meijer, J. (2001) Purification and characterization of myrosinase from the cabbage aphid (*Brevicoryne brassicae*), a brassica herbivore. *European Journal of Biochemistry* 268, 1041–1048.

Pope, T.W., Campbell, C.A.M., Hardie, J. and Wadhams, L.J. (2004) Electroantennogram responses of the three migratory forms of the damson–hop aphid, *Phorodon humuli*, to aphid pheromones and plant volatiles. *Journal of Insect Physiology* 50, 1083–1092.

Powell, G. and Hardie, J. (2001) The chemical ecology of aphid host alternation: how do return migrants find the primary host plant? *Applied Entomology and Zoology* 36, 259–267.

Powell, G., Tosh, C.R. and Hardie, J. (2006) Host-plant selection by aphids: behavioral, evolutionary, and applied perspectives. *Annual Review of Entomology* 51, 309–330.

Powell, W., Hardie, J., Hick, A.J., Höller, C., Mann, J., Merritt, L., Nottingham, S.F., Wadhams, L.J., Witthinrich, J. and Wright, A.F. (1993) Responses of the parasitoid *Praon volucre* (Hymenoptera: Braconidae) to aphid sex pheromone lures in cereal fields in autumn: implications for parasitoid manipulation. *European Journal of Entomology* 90, 435–438.

Quiroz, A. and Niemeyer, H.M. (1998) Activity of enantiomers of sulcatol on apterae of *Rhopalosiphum padi*. *Journal of Chemical Ecology* 24, 361–371.

Quiroz, A., Pettersson, J., Pickett, J.A., Wadhams, L.J. and Niemeyer, H.M. (1997) Semiochemicals mediating spacing behavior of bird cherry–oat aphid, *Rhopalosiphum padi* feeding on cereals. *Journal of Chemical Ecology* 23, 2599–2607.

Raymond, B., Darby, A.C. and Douglas, A.E. (2000) The olfactory responses of coccinellids to aphids on plants. *Entomologia Experimentalis et Applicata* 95, 113–117.

Röse, U.S.R., Manukian, A., Heath, R.R. and Tumlinson, J.H. (1996) Volatile semiochemicals released from undamaged cotton leaves: a systemic response of living plants to caterpillar damage. *Plant Physiology* 111, 487–495.

Shonouda, M.L., Bombosch, S., Shalaby, A.M. and Osman, S.I. (1998) Biological and chemical characterization of a kairomone excreted by the bean aphid, *Aphis fabae* Scop. (Hom., Aphididae), and its effect on the predator *Metasyrphus corollae* Fabr. II. Behavioural response of the predator *M. corollae* to the aphid kairomone. *Journal of Applied Entomology* 122, 25–28.

Shulaev, V., Silverman, P. and Raskin, I. (1997) Airborne signalling by methyl salicylate in plant pathogen resistance. *Nature, London* 385, 718–721.

Storeck, A., Poppy, G.M., van Emden, H.F. and Powell, W. (2000) The role of plant chemical cues in determining host preferences in the generalist parasitoid *Aphidius colemani*. *Entomologia Expermentalis et Applicata* 97, 41–46.

Sunderland, K.D., Fraser A.M. and Dixon, A.F.G. (1986) Field and laboratory studies on money spiders (Lynyphiidae) as predators of aphids. *Journal of Applied Ecology* 23, 433–447.

Takemura, M., Nishida, R., Mori, N. and Kuwahara, Y. (2002) Acylated flavonol glycosides as probing stimulants of a bean aphid, *Megoura crassicauda*, from *Vicia angustifolia*. *Phytochemistry* 61, 135–140.

Takemura, M., Kuwahara, Y. and Nishida, R. (2006) Feeding responses of an oligophagous bean aphid, *Megoura crassicauda*, to primary and secondary substances in *Vicia angustifolia*. *Entomologia Experimentalis et Applicata* 121, 51–57.

Turchin, P. and Kareiva, P. (1989) Aggregation in *Aphis varians*: an effective strategy for reducing predation risk. *Ecology* 70, 1008–1016.

Turlings, T.C.J., Tumlinson, J.H., Heath, R.R., Proveaux, A.T. and Doolittle, R.E. (1991) Isolation and identification of allelochemicals that attract the larval parasitoid, *Cotesia marginiventris* (Cresson), to the microhabitat of one of its hosts. *Journal of Chemical Ecology* 17, 2235–2251.

Vet, L.E.M. and Dicke, M. (1992) Ecology of infochemical use by natural enemies in a tritrophic context. *Annual Review of Entomology* 37, 141–172.

Visser, J.H. and Piron, P.G.M. (1995) Olfactory antennal responses to plant volatiles in apterous virginoparae of the vetch aphid *Megoura viciae*. *Entomologia Experimentalis et Applicata* 77, 37–46.

Way, M. (1973) Population structure in aphid colonies. In: Lowe, A.D. (ed.) *Perspectives in Aphid Biology*. Entomological Society of New Zealand, Auckland, pp. 76–84.

Way, M.J. and Banks, C.J. (1967) Intra-specific mechanisms in relation to the natural regulation of numbers of *Aphis fabae* Scop. *Annals of Applied Biology* 59, 189–205.

Way, M. and Cammell, M. (1970) Aggregation behaviour in relation to food utilization in aphids. In: Watson, A. (ed.) *Animal Populations in Relation to Their Food Resources. 10th Symposium of the British Ecological Society, 1969.* Blackwell, Oxford, pp. 229–247.

Wickremasinghe, M.G.V. and van Emden, H.F. (1992) Reactions of adult female parasitoids, particularly *Aphidius rhopalosiphi*, to volatile chemical cues from the host plants of their aphid prey. *Physiological Entomology* 17, 297–304.

Wientjens, W.H.J.M., Lawijk, A.C. and van der Marel, T. (1973) Alarm pheromone of grain aphids. *Experientia* 29, 658–660.

Wildermuth, M.C., Dewdney, J., Wu, G. and Ausubel, F. (2001) Isochorismate synthase is required to synthesize salicylic acid for plant defence. *Nature, London* 414, 562–565.

Wratten, S.D. and Powell, W. (1991) Cereal aphids and their natural enemies. In: Firbank, L.G., Carter, N., Darbyshire, J.F. and Potts, G.R. (eds) *The Ecology of Temperate Cereal Fields. Symposium of the British Ecological Society No. 32.* Blackwells, Oxford, pp. 233–257.

Zhu, J.-W., Crosse, A.A., Obrycki, J.J., Boo, K.-S. and Baker, T.C. (1999) Olfactory reactions of twelve-spotted lady beetle, *Coleomegilla maculata* and the green lacewing, *Chrysoperla carnea* to semiochemicals released from their prey and host plant: electroantennogram and behavioral responses. *Journal of Chemical Ecology* 25, 1163–1177.

Zhu, J., Obrycki, J.J., Ochieng, S.A., Baker, T.C., Pickett, J.A. and Smiley, D. (2005) Attraction of two lacewing species to volatiles produced by host plants and aphid prey. *Naturwissenschaften* 92, 277–281.

10 Insecticide Resistance

Stephen P. Foster[1], Gregor Devine[1] and Alan L. Devonshire[2]

[1]*Plant and Invertebrate Ecology Division and* [2]*Biological Chemistry Division,
Rothamsted Research, Harpenden, Herts, AL5 2JQ, UK*

Introduction

Insecticide resistance is an example of a dynamic evolutionary process in which chance mutations conferring protection against insecticides are selected in treated populations. Over the past 20 years, rapid advances have been made in the characterization and understanding of such adaptations. These have provided valuable insights into the origin and nature of selection and micro-evolution in agricultural environments.

In practical terms, the evolution of insecticide resistance has undoubtedly contributed to overall increases in the application of chemicals to crops. About 500,000 t of insecticide are now applied each year in the USA alone, with significant implications for both human health and the environment. Despite this, resistant insects such as aphids continue to affect our agricultural productivity. As a result, the phenomenon imposes a huge economic burden upon much of the world, e.g. an estimated annual cost of US$ 1.4 billion in crop and forest productivity in the USA (Pimentel *et al.*, 1992). Moreover, it is proving impossible to combat resistance by embarking on a chemical arms race. The development of a new insecticide takes 8 to 10 years at a cost of US$ 20–40 million, and the rate of discovery of new insecticidal molecules, unaffected by current resistance mechanisms, appears to be on the wane. Only by monitoring, characterizing, and predicting the appearance and spread of resistance can we hope to use existing chemical tools in a sustainable manner.

Of the thousands of aphid species that exist globally, only a few have been reported as having developed insecticide resistance. However, some of these are ranked among the most problematic pests worldwide (Table 10.1). Some species are highly polyphagous, while others are virtually confined to a single crop type. They damage agricultural and horticultural crops through direct feeding, the transmission of plant viruses, and by their impact on the aesthetic value of crops (by depositing honeydew or simply by being present). This has resulted in them being subjected, historically, to intense selection by aphicides, which has led to the evolution of a variety of resistance mechanisms. This chapter focuses on their diagnosis and characterization, their impact on insecticide efficacy and, where known, the dynamics of resistance in aphid populations and the factors driving them. Emphasis is placed on developments since publication of the last major reviews of this topic (Devonshire and Rice, 1988; Devonshire, 1989).

©CAB International 2007. *Aphids as Crop Pests*
(eds H. van Emden and R. Harrington)

Table 10.1. Insecticide resistant aphid species (resistant species of primary importance shown in bold).

Species	Resistance first recorded	Compounds resisted[a]	Regions
Myzus persicae	Anthon, 1955	BPU, C, Nic, OC, OP, Py	Europe Asia USA S. America Australia
Aphis gossypii	Ghong *et al.*, 1964	C, OC,OP, Py	Europe Asia USA Africa Australia
Schizaphis graminum	Peters *et al.*, 1975	OP	Asia USA Africa
Phorodon humuli	Dicker, 1965	C, OP, Py	Europe
Nasonovia ribisnigri	Rufingier *et al.*, 1997	C, OC, OP	Europe
Dysaphis plantaginea	Wiesmann, 1955	C, OP, Py	Europe
Macrosiphum euphorbiae	Foster *et al.*, 2002d	C, OP, Py	Europe
Aphis craccivora	Smirnova and Ivanova, 1974	Nic, OC, OP, Py	Asia
Aphis fabae	Boness and Unterstenhofer, 1974	C, OP	Europe
Aphis nasturtii	Laska, 1981	C	Europe
Lipaphis pseudobrassicae	Tang *et al.*, 1988	OP, Py	Asia
Rhopalosiphum padi	Anonymous, 1974	OP	Europe Asia
Therioaphis trifolii maculata	Stern and Reynolds, 1958	C, OP	USA Australia

[a]BPU: benzoylphenyl ureas; C: carbamates; Nic: nicotine/neonicotinoids; OC: organochlorines; OP: organophosphates; Py: pyrethroids

Diagnosis of Resistance in Aphids

Although many laboratory bioassay methods have been developed for detecting and characterizing resistance, most of these are limited to defining phenotypes and provide little or no information on the underlying genes or mechanisms. Nonetheless, bioassays remain the indispensable mainstay of most large-scale resistance monitoring programmes and are essential prior to the development of alternative diagnostic techniques based on knowledge of the mechanism of resistance.

The phenotypic expression of resistance is assessed in small-scale laboratory bioassays by exposing aphids to topical, foliar, or systemic applications of insecticides (Devonshire and Rice, 1988). The response of potentially insecticide-resistant forms is, whenever possible, compared with known insecticide-susceptible standards. The only way of being certain that a population is truly susceptible is to find one that previously has not been exposed to any insecticidal selection pressure. Such populations (of pest species) may be extremely rare and therefore, in the majority of instances, susceptibility is a relative rather than an absolute concept. Moreover, even the natural, unselected variability between different populations of a species may result in differing responses to insecticides. A further subjective distinction may be made, therefore, between 'resistance' as a selected mutation conferring substantial protection against pesticide(s) and 'tolerance' that can exist in the absence of selection pressure.

In addition to quantifying resistance *per se* in bioassays, it is important to establish whether resistance quantified in this manner is of practical importance in the field. One way to do this is to apply insecticides in a more realistic manner under field and pseudo-field conditions, e.g. using field cages (ffrench-Constant *et al.*, 1987) and 'field simulators' (Foster *et al.*, 2002b). These larger-scale approaches ensure that bioassays are correlated with potential field control problems. By allowing the expression of 'natural' insect behaviour under realistic field conditions (e.g. on whole plants whose architecture can provide insecticide-free refugia), the actual exposure of the insect to an insecticide is more closely mimicked.

Relevance of Bioassays to Field Control

Field experiments and field-simulator experiments have been used to study the relative performance of various established insecticides applied at recommended field rates against *Myzus persicae* (peach–potato aphid) carrying different combinations of various resistance mechanisms (Dewar *et al.*, 1998; Foster and Devonshire, 1999; Foster *et al.*, 2002b) and low-level resistance to imidacloprid (Foster *et al.*, 2003a). These studies exploited well-defined clones that can be identified with the help of biochemical and molecular characterization. Results of the two approaches proved comparable and showed that the effectiveness of different insecticides is very much dependent on the resistance mechanisms present. Experiments have also shown the consequences of different treatment regimes for the build-up of metabolic (esterase)-based resistance in open field and caged populations (ffrench-Constant *et al.*, 1987, 1988a,b). As resistance

mechanisms accumulate, more and more products become resisted, leading to a progressive depletion in chemical control options.

Field-simulator experiments have explored the implications of low-level variation in susceptibility to neonicotinoids for the performance of imidacloprid under exposure conditions likely to be encountered in the field (Foster *et al.*, 2003a). While even imidacloprid-tolerant forms were well controlled by imidacloprid applied as a soil drench at the recommended field rate, reduced rates led to substantial survival and rates of population increase. This implies the existence of selection favouring tolerant forms when imidacloprid is present at concentrations lower than those expected at the time that correct treatments are applied. Such conditions can arise from the natural decay of the insecticide with time, or the application of reduced rates, either to cut costs or to target pests other than aphids. In Northern Europe, the latter situation may arise from the increasing cultivation of oilseed rape, using seed treated with one-sixth the dose of imidacloprid recommended for aphid control on sugarbeet in order to manage early infestations of *Psylloides chrysocephala* (cabbage stem flea beetle). Besides the decreased kill of these tolerant variants, other sub-lethal advantages in the form of reduced inhibition of feeding and of higher reproduction have been recorded under exposure to extremely low doses of neonicotinoids (Devine *et al.*, 1996). Such studies demonstrate how variation in susceptibility may be the first evolutionary step towards resistance capable of compromising neonicotinoids when they are applied at full field rates.

Biochemistry and Molecular Basis of Resistance

The information provided by the use of small-scale and field-scale bioassay techniques is essential. However, attention is being focused increasingly on the development of more incisive diagnostics that not only offer greater precision and throughput, but also identify the specific mechanism(s) present and even the genotype of the resistant insect. This is exemplified by research on *M. persicae*, where significant advances have led to rapid and precise methods for the detection of different resistance mechanisms in individual insects.

Resistance mechanisms in *Myzus persicae* (peach–potato aphid)

Several approaches have been adopted for identifying the various resistance mechanisms, including electrophoretic or immunological detection of resistance-causing enzymes, kinetic and end-point assays for quantifying the activity of enzymes or their inhibition by insecticides, and DNA-based diagnostics for mutant resistance alleles. One of the advantages of using an aphid species such as *M. persicae* as a model insect is that it is easy to maintain in pure-breeding, clonal lines in the laboratory. This facilitates a more incisive analysis of resistance than when dealing with obligate holocyclic species that are therefore predominantly sexual and genetically heterogeneous.

In Northern Europe, *M. persicae* possesses at least three coexisting resistance mechanisms: (i) an overproduced carboxylesterase conferring resistance to organophosphates (OPs) and carbamates and some resistance to pyrethroids; (ii) an altered acetylcholinesterase (AChE) conferring resistance to certain carbamates; and (iii) target site (kdr) resistance to pyrethroids. These mechanisms collectively confer strong resistance to virtually all available aphicides (Table 10.1) and are illustrated in Box 10.1. It is now possible to diagnose all three mechanisms in individual aphids using an immunoassay for the overproduced esterase, a kinetic microplate assay for the mutant AChE, and a molecular diagnostic for the *kdr* allele. The combined use of these techniques against field populations provides information on the dynamic occurrence of these mechanisms and is used to provide up-to-date advice to growers on potential control problems with particular insecticides (Foster *et al.*, 1998, 2002c; Foster, 2000). It has also produced valuable

insights into factors influencing the development of resistance and its impact on aphid management strategies.

Metabolic mechanisms

Over the past 40 years, the majority of commercial aphicides in use has belonged to the organophosphorus (OP), carbamate, and pyrethroid classes. All of these have one chemical feature in common; they are esters. Such bonds are particularly prone to cleavage and, consequently, resistance to these compounds is commonly mediated by the enhanced hydrolysis of ester bonds. Hydrolytic enzymes can often be measured readily in the laboratory by the use of chromogenic compounds in conjunction with ester substrates such as naphthyl-acetate and butyrate. An association between resistance in bioassays and enhanced 1-naphthyl acetate esterase activity was established in *M. persicae* over 30 years ago (Needham and Sawicki, 1970). This discovery proved central to our understanding of this type of resistance in this and other species. Devonshire (1977) confirmed that the enzyme hydrolysing the model substrate was the same as that responsible for resistance. This specific enzyme shows a very high affinity for binding the insecticidal esters. It is present in such large molar amounts that it can also sequester a substantial proportion of a toxic dose, even without hydrolytic cleavage. Quantification of the esterase with 1-naphthyl acetate has relied on measurements in crude homogenates (Devonshire, 1975) by staining the enzyme after resolution from other esterases using polyacrylamide gel electrophoresis (PAGE) (Devonshire, 1975), or by trapping the enzyme selectively in microplates with a specific antiserum (Devonshire *et al.*, 1986). Both techniques provide increased discrimination by allowing the relevant enzyme to be measured in the absence of contamination by the other esterases present in crude homogenates. These overproduced carboxylesterases (E4 or FE4) sequester or degrade insecticide esters before they reach their target sites in the nervous system (Devonshire, 1977; Devonshire and Moores, 1982) and confer strong resistance to

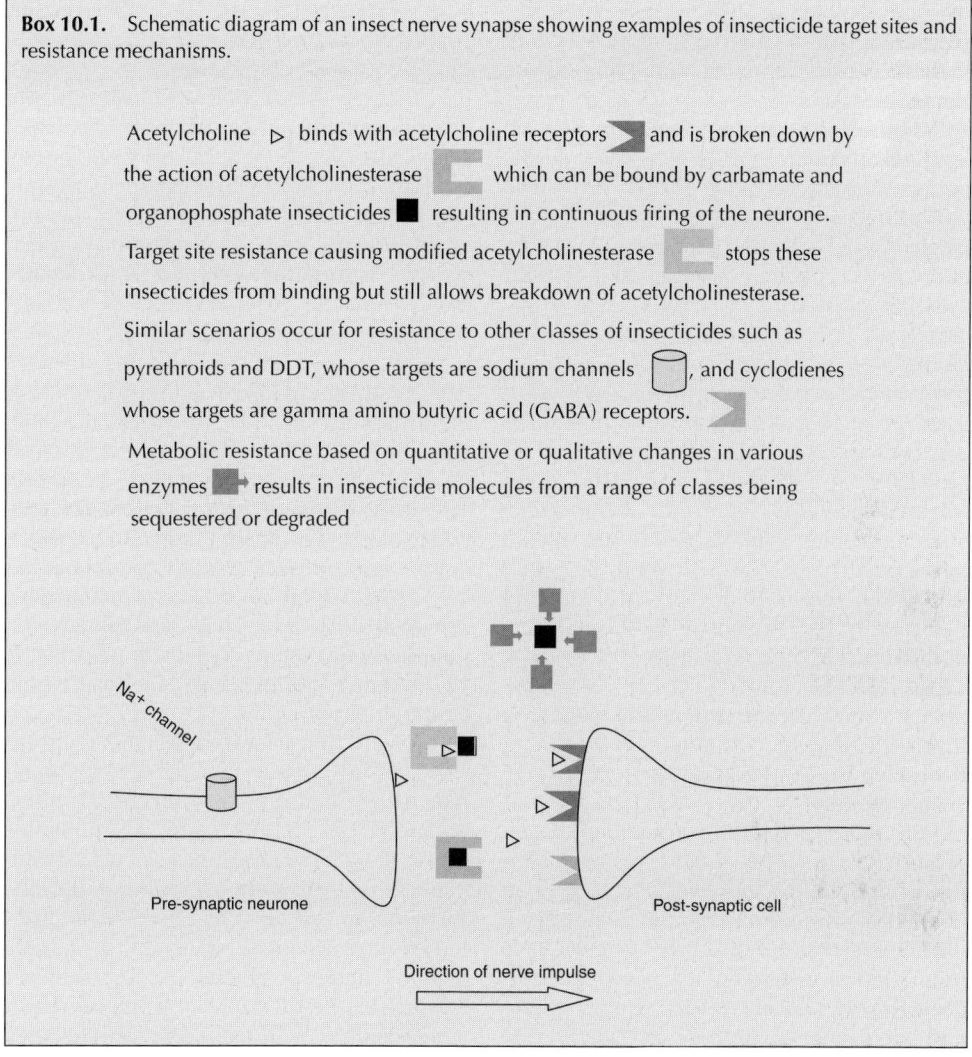

Box 10.1. Schematic diagram of an insect nerve synapse showing examples of insecticide target sites and resistance mechanisms.

Acetylcholine ▷ binds with acetylcholine receptors ▶ and is broken down by the action of acetylcholinesterase ⌐ which can be bound by carbamate and organophosphate insecticides ■ resulting in continuous firing of the neurone. Target site resistance causing modified acetylcholinesterase ⌐ stops these insecticides from binding but still allows breakdown of acetylcholinesterase. Similar scenarios occur for resistance to other classes of insecticides such as pyrethroids and DDT, whose targets are sodium channels ▯, and cyclodienes whose targets are gamma amino butyric acid (GABA) receptors. ▶

Metabolic resistance based on quantitative or qualitative changes in various enzymes ■▶ results in insecticide molecules from a range of classes being sequestered or degraded

Na+ channel

Pre-synaptic neurone

Post-synaptic cell

Direction of nerve impulse

OPs and lower resistance to carbamates and pyrethroids (Devonshire and Moores, 1982). Depending on their esterase content and response in laboratory bioassays, individual aphids can be classified broadly into one of four groups; S (susceptible), R_1 (moderately resistant), R_2 (highly resistant), or R_3 (extremely resistant) (Devonshire *et al.*, 1986). Furthermore, there is a close association between the form of esterase produced and the karyotype of the aphid; E4-overproducing aphids have an autosomal 1,3 translocation, which appears to confer some reproductive isolation through androcycly, whereas FE4 over-producers are of normal karyotype (Blackman *et al.*, 1978). The immunoassay is especially powerful as it allows as many as 1000 individuals to be characterized by one person in a day, so enabling detailed insight into *M. persicae* population structure and how it can change in response to different treatment regimes.

Detailed analyses of the underlying molecular genetic changes have implicated gene amplification as the cause of this overproduction, i.e. the acquisition of multiple copies of the esterase structural gene (Devonshire and Sawicki, 1979; Field *et al.*, 1988, 1999). These studies have provided a paradigm for the development of resistance by gene amplification, with many subsequent studies of insecticide resistance in a wide variety of species drawing on the *M. persicae* model.

The esterase mechanism has become well established worldwide, with forms of *M. persicae* carrying greater resistance appearing to predominate only when they are under regular selection pressure. This probably stems primarily from their apparent reduced fitness under certain conditions, discussed below, together with a genetic change that can prevent amplified E4 genes from being expressed (this has not been seen for amplified FE4 genes), a phenomenon associated with methylation levels of the encoding DNA (Field *et al.*, 1989). Such 'revertant' aphids retain the capacity to switch on their esterase genes again when subsequently exposed to insecticide selection (Hick *et al.*, 1996) and so constitute potential 'hidden resistance' when analysing populations by bioassay or esterase assays. Their identification requires a DNA diagnostic, which at present is not suited to high throughput (Field *et al.*, 1996), so that the incidence of such forms in the field currently remains unclear.

In many areas of commercial peach production, spraying on the primary host can compound the problems of selection on the summer host. For example, major problems of resistance in Southern Europe have long been associated with peach-growing areas (Devonshire *et al.*, 1983, 1999; Cervato and Cravedi, 1995). In these areas, holocycly is common and the sexual forms typically overproduce the FE4, rather than the E4, form of esterase.

Target site mechanisms

Mutant forms of AChE, showing reduced inhibition by OPs and carbamates, have been demonstrated in several aphid species. These compounds exert their toxicity by inhibiting AChE, thereby impairing the transmission of nerve impulses across the cholinergic synapses. Biochemical and molecular analyses of insecticide-insensitive AChE have shown that pests may possess several different mutant forms of this enzyme

with contrasting insensitivity profiles, thereby conferring distinct patterns of resistance to these two insecticide classes.

For many years, increased esterase production was the only mechanism identified in aphids; even though others were known to play a primary role in other species, especially target site insensitivity to insecticides. AChE and the voltage-dependent sodium channel, the targets for OP/carbamate and pyrethroid insecticides, respectively, have been shown to contribute to resistance in aphids. Insensitive AChE in aphids was first identified in *Aphis gossypii* (cotton or melon aphid) in 1984 (Silver *et al.*, 1995) and subsequently in *M. persicae* (Moores *et al.*, 1994a) and *Schizaphis graminum* (greenbug) (Zhu and Gao, 1999). This mechanism is genetically dominant, so heterozygotes show resistance. Such genotypes can be distinguished in a microplate assay (Moores *et al.*, 1994b). In parthenogenetic populations, such as those common in the UK, the heterozygous genotype persists within clonal lineages.

As with the esterase mechanism, insensitive AChE (also termed MACE, or **M**odified **A**cetyl**C**holin**E**sterase) can be identified in individual aphids using a microplate assay. Since the *M. persicae* esterase immunoassay requires only 0.02 of an aphid, the AChE assay can be done on the same individual homogenates, so enabling a more detailed analysis of changes in genotype frequencies resulting from aphicide treatments both in field trials and regionally in aphid populations. This target site mechanism confers strong resistance specifically to the dimethyl-carbamates, pirimicarb and triazamate (Moores *et al.*, 1994a; Foster and Devonshire, 1999; Foster *et al.*, 2002b). MACE in *M. persicae* was discovered initially in Southern Europe, and subsequently Japan and South America. Recently, it has shown an apparent northward expansion in its European distribution, being found for the first time on UK crops in 1996 (Foster *et al.*, 1998). In the UK, MACE continues to cause sporadic control problems as a result of the protection it confers to the insecticides, pirimicarb and triazamate. The molecular basis of this resistance has now been identified in *M. persicae* (Nabeshima *et al.*, 2003; Andrews *et al.*, 2004) and *A. gossypii* (Li and Han, 2004). It involves point mutations that alter amino acids close to the catalytic centre of the enzyme. The mutation in *M. persicae* changes a serine to phenylalanine, which is the residue normally found at this position in most AChEs. This raises the possibility that the success of pirimicarb arose from its ability to inhibit the enzyme having serine at this position, which so far appears to be unique to aphids.

A further target site mechanism, termed knockdown resistance (kdr), has been identified as the main cause of resistance to pyrethroid insecticides (Martinez-Torres *et al.*, 1999). Although detected only recently in aphids, this target site resistance, conferred by changes in a voltage-gated sodium channel protein, has been found in a number of *M. persicae* clones established in the laboratory over the past 20 years, indicating that it has long been present in this species. The close linkage of kdr with elevated E4 (but not FE4) esterases (which also contribute to pyrethroid resistance), probably as a consequence of permanent parthenogenesis in the former clones carrying the A1,3 chromosomal translocation (Blackman *et al.*, 1978), appears to have masked its presence until its recent discovery (Devonshire *et al.*, 1998).

Initial recognition of the kdr mechanism in aphids was dependent on a molecular biological approach in which a highly conserved point mutation in the sodium channel gene was identified based on studies in kdr houseflies and other pest species with the same mechanism (Martinez-Torres *et al.*, 1997, 1999). Molecular diagnostic methods continue to be essential in understanding the role of this mechanism in resistant aphids (Guillemaud *et al.*, 2003), including characterization of its enhanced form, super-kdr. The availability of molecularly defined clones has enabled the development of a sensitive, discriminating dose bioassay using DDT (to which kdr confers cross-resistance, but which is unaffected by the esterase and MACE mechanisms (Martinez-Torres *et al.*, 1999)) and deltamethrin (Anstead *et al.*, 2004). Intriguingly, *M. persicae* clones showing heterozygous kdr genotypes appear to be highly resistant to a range of pyrethroids in aphid bioassays, despite the prediction from other species that they should show susceptible phenotypes. Furthermore, the addition of synergists in these bioassays has not revealed any evidence of additional metabolic-type resistance to pyrethroids that could be linked genetically with kdr. The reasons underlying the phenomenon are currently being investigated. The kdr mechanism was found at high frequency in populations in France reproducing sexually on peaches. It was homozygous in 20–56% of the aphids sampled and heterozygous in most of the remainder, regardless of whether the orchards were sprayed with pyrethroids during the sampling period (Guillemaud *et al.*, 2003). However, the frequency increased overall by the end of the winter periods, presumably due to pyrethroid usage. Sequencing of the intron regions within the sodium channel gene in *M. persicae* collected from different geographical areas suggests that the kdr and super-kdr target site mutations have multiple independent origins (Anstead *et al.*, 2005).

Although endosulfan is the only cyclodiene used for aphid control, molecular biological analysis has shown an association in *M. persicae* between resistance to this compound and the presence of a point mutation for resistance to dieldrin (*rdl*) in the gamma-aminobutyric acid (GABA) receptor found in a broad range of other species. Increased resistance levels (26-fold) to endosulfan were correlated with the replacement of a single amino acid (alanine 302) with glycine (Anthony *et al.*, 1998). Other authors have also documented pronounced endosulfan resistance (Unruh *et al.*, 1996; Kerns *et al.*, 1998). Molecular diagnosis of French populations on peaches (Guillemaud *et al.*, 2003) showed that the frequency of the *rdl* mechanism decreased over winter, presumed to be due to a fitness cost (as found in other insect species).

The biochemical and molecular genetic diagnostics described above have provided an unparalleled understanding of the incidence and impact of the different resistance mechanisms, especially in *M. persicae*.

In particular, it is clear that, as a consequence of prolonged parthenogenesis in field populations, combinations of mechanisms can become 'locked together' in clones so that they are co-selected, even when the insecticide used is unaffected by some of the mechanisms present (Devonshire *et al.*, 1998).

Resistance mechanisms in other aphid species

Metabolic mechanisms

APHIS GOSSYPII (COTTON OR MELON APHID). Several studies on *A. gossypii* carried out in the USA (O'Brien *et al.*, 1992) and Japan (Saito, 1989; Hosoda *et al.*, 1992) have found a link between esterase activity and OP resistance. It has been reported that OP-resistant clones can have 15–35 times more carboxylesterase activity than susceptible clones and that particular esterase banding patterns are closely associated with high total enzyme titres and moderate OP resistance (Suzuki and Hama, 1998). Resistance has not, however, been ascribed to a single enzyme but to a group of different isoenzymes. If carboxylesterases confer resistance to OPs in *A. gossypii*, it is clearly through a very different mechanism to that in *M. persicae*, where usually only one or the other of two very closely related isoenzymes (initially arising from a gene duplication) becomes amplified. Furthermore, OP resistance in *A. gossypii* cannot always be ascribed to carboxylesterase-based mechanisms. An association of OP resistance with high oxidase titre has been inferred (Sun *et al.*, 1970) and Delorme *et al.* (1997), using synergists, questioned whether oxidases or esterases played a role in OP resistance. Han *et al.* (1998) also concluded that esterases play a minor role in OP resistance in *A. gossypii* clones from around the world, and that most resistance is due to an altered target site.

PHORODON HUMULI (DAMSON–HOP APHID). The underlying mechanisms of resistance in *P. humuli* are not well documented, apart, perhaps, from an association between esterases and OP resistance. In *P. humuli*

collected from wild and cultivated hop plants in Switzerland, a positive correlation was observed between esterase activity and OP resistance (Beck and Buchi, 1980). In the UK, clones of OP-resistant *P. humuli* had high esterase isoenzyme activity in a complex of bands that was almost absent in susceptible populations (Lewis and Madge, 1984). The antiserum to carboxylesterase E4 from *M. persicae*, which is a closely related species, cross-reacted with homogenates of *P. humuli* (Devonshire *et al.*, 1986), showing that increased esterase activity arises from the overproduction of protein(s) homologous to E4. Others have also described methods for measuring esterase activity in *P. humuli*, and correlated these with OP resistance (Wachendorff and Zoebelein, 1988). Although the association of esterases with OP and carbamate resistance is established, Buchi (1981) found that pyrethroid resistance in some populations of both *P. humuli* and *M. persicae* did not correlate with high carboxylesterase activity. This is now supported by the recent discovery of the kdr pyrethroid resistance mechanism in *M. persicae* (Martinez-Torres *et al.*, 1997), but not so far in *P. humuli*. This species is now controlled in many hop gardens throughout Europe by stem application of imidacloprid to the vines, so far with no reported resistance problems (Vostrel, 1995; Elbert *et al.*, 1996).

SCHIZAPHIS GRAMINUM (GREENBUG). Between 1988 and 1991, a rise in the incidence of *S. graminum* control failures was noted in several areas of Kansas, USA. Clonal lines established from insects collected from areas of control failure exhibited 20- to 30-fold resistance to OP insecticides such as parathion (Sloderbeck *et al.*, 1991). As in *M. persicae*, these resistant forms displayed enhancement of general esterases based on PAGE and staining with 1-naphthyl acetate. Carbofuran and chlorpyrifos, however, remained effective.

In subsequent investigations, two different patterns of esterase isozymes with different genetic bases were detected in resistant forms (e.g. Siegfried and Ono, 1995). These have been categorized variously as Type I/Type II patterns or R1/R2 patterns. We will use the former categorization. Type I forms exhibit a single esterase band that is either absent in susceptible insects or expressed at levels below the limits of detection. Type II forms exhibit a different pattern of enhanced esterase isozymes (Ono *et al.*, 1994a). After testing over 10,000 individuals, Shufran *et al.* (1996) found that none displayed both Type I and Type II esterase polymorphisms. In addition to marked differences in electrophoretic mobility, the two resistant forms show striking differences in general esterase activity when measured spectrophotometrically from *S. graminum* homogenates. Both forms are active against a series of 1-naphthol esters, but Type II forms display 15-fold higher levels of activity than Type I forms. Type I forms appear to result from the enhanced production of an esterase enzyme that cross-reacts with an antiserum to E4 from *M. persicae* (Siegfried *et al.*, 1997). This similarity between the two species has been reinforced by comparison of the esterase gene sequences and the demonstration that, as in *M. persicae*, the genes are amplified and also methylated (Ono *et al.*, 1999). For both esterase forms, activity towards parathion is similar, but strong inhibition by the active metabolite paraoxon has been noted. This suggests that the mechanism of resistance does not involve true enzymatic hydrolysis, but rather that these enzymes have a sequestrating role (Ono *et al.*, 1994b). Zhu and He (2000) also suggest that Type II esterases cannot efficiently utilize paraoxon and chlorpyrifos-oxon as their substrates and that elevated esterases contribute to OP resistance by sequestration.

Partially purified enzymes from Type I-resistant and -susceptible *S. graminum* strains exhibited similar elution profiles by ion exchange chromatography, although the activity peak from the resistant strains was greater in both height and total area. Kinetic analysis of the resulting activity indicated that the Km of the esterase was identical for the resistant and susceptible strains. Resistance is therefore associated with overproduction of isozymes already present in the susceptible strain rather than the presence of enzymes with altered properties (Siegfried and Zera, 1994).

DNA from Type I forms and a clone derived by crossing Type I and Type II displayed a restriction fragment pattern different from that of Type II and susceptible clones, and an increased hybridization signal compared to susceptible insects. This suggested that the mechanism behind Type I esterase activity may be gene amplification, but that Type II elevated esterase activity results from a different genetic mechanism (Rider *et al.*, 1998). Shufran *et al.* (1997a) found that resistance patterns to a variety of OP and carbamate compounds in Type II and Type I strains differ.

NASONOVIA RIBISNIGRI (CURRANT–LETTUCE APHID). In some areas of Europe, a large proportion of the *N. ribisnigri* population occurs on lettuce during the summer months. The crop is subject to heavy insecticide use and, as a result, this aphid is under strong selection pressure. Resistance was first detected in strains collected from southern France and Spain. These showed resistance factors ranging from 12-fold to acephate (an OP) and up to 660-fold to the cyclodiene, endosulfan, in lettuce leaf discbioassays (Rufingier *et al.*, 1997). Maximum resistance to the pyrethroid, deltamethrin, and the carbamate, pirimicarb, was 28-fold and 19-fold, respectively. It was speculated that both enhanced detoxification and altered target site(s) were involved.

In the late 1990s, following increasing reports of reduced insecticide success in the field, leaf dip-bioassays were used to compare the activity of several insecticides against suspected resistant *N. ribisnigri* strains from UK lettuce (Barber *et al.*, 1999). In comparison to a susceptible strain, these showed widespread low-level resistance to pirimicarb (7- to 11-fold), and very low (2- to 4-fold) resistance to pyrethroids and OPs. In some *N. ribisnigri* strains, resistance was associated with a strong esterase band disclosed by staining with naphthyl acetate after PAGE. This might imply a similar over-expression of esterases analogous to that causing resistance in *M. persicae*, but the molecular basis of increased esterase activity remains to be investigated.

Target site mechanisms

APHIS GOSSYPII (COTTON OR MELON APHID). I n *vitro* studies of OP resistance in *A. gossypii* have shown that it is commonly linked to the presence of insensitive forms of AChE (Gubran *et al.*, 1992; Delorme *et al.*, 1997). However, the earliest example of this type of resistance mechanism in any aphid species was very specifically to the carbamate pirimicarb (Gubran *et al.*, 1992; Silver *et al.*, 1995), a compound introduced in 1969. This form of resistance then became common in the 1970s and proved to be particularly problematic when pirimicarb became a mainstay of many integrated aphid control programmes due to its general low toxicity to beneficial insects (Furk *et al.*, 1980). The esterase-banding pattern typical of OP resistance is not associated with pirimicarb resistance (Takada and Murakami, 1988; Saito and Hama, 2000).

Mutated AChE in some strains of *A. gossypii* from around the world (e.g. Gubran *et al.*, 1992) is very specific, protecting against pirimicarb, but not the monomethyl or oxime carbamates or OPs (Silver *et al.*, 1995), whilst other forms of the enzyme also confer OP resistance (Moores *et al.*, 1996; Han *et al.*, 1998). Indeed, there is evidence of slight hypersensitivity to bendiocarb conferred by the pirimicarb-insensitive enzyme (Villatte *et al.*, 1999), although it is not sufficient for exploitation in an anti-resistance strategy owing to fitness costs negating practical benefits.

Pyrethroid resistance is also common in *A. gossypii* from around the world, e.g. in Israel (Ishaaya and Mendelson, 1987), China (Tang, 1992; Guilin *et al.*, 1997), Japan (Saito *et al.*, 1995), and the USA (O'Brien *et al.*, 1992). In the cotton-growing areas of the Sudan, resistance to all the major insecticide groups occurs concurrently (Gubran *et al.*, 1992). Some of these resistant forms have been found to have super-kdr-type mutations, in the absence of kdr mutations, in their sodium channel genes (X. Yang and M. Williamson, personal communication), suggesting this to be the underlying mechanism of pyrethroid resistance. Resistance to this class of insecticide can be extremely high

(29,000-fold), and it can be dramatically affected by the host plant used for the bioassay. Wang *et al.* (2001) reported greater pyrethroid resistance on cotton compared to cucumber. However, the low level of imidacloprid resistance (*c.* 5-fold) they found in one *A. gossypii* strain was only barely affected by the plant host in the bioassay.

SCHIZAPHIS GRAMINUM (GREENBUG). Analyses of *S. graminum* cDNA have resulted in the identification of two AChE DNA fragments. One is closely related to the AChEs found in several insect species, mutant forms of which are now believed to mediate OP and carbamate resistance in a range of insects. The second fragment resembles a sequence more related to nematode, human, and other vertebrate AChEs (Gao and Zhu, 2001, 2002). In studies on OP-resistant clones of *S. graminum*, Zhu *et al.* (2000) examined AChE activity and found it to be slightly less sensitive (2.5-fold) to inhibition by various OP compounds. This is quite different to the very strong insensitivity seen in *M. persicae* and *A. gossypii*. These authors suggested that alterations to AChE are responsible for resistance to OP insecticides. Northern blot analysis (which examines expression levels) did not show over-expression of the AChE gene and so it is thought that a qualitative, rather than quantitative, change in *S. graminum* mediates resistance.

NASONOVIA RIBISNIGRI (CURRANT-LETTUCE APHID). Rufingier *et al.* (1999) showed that insensitive AChE in *N. ribisnigri* conferred resistance to pirimicarb and propoxur, but was not responsible for methomyl, acephate, or paraoxon resistance. Glutathione transferases were inferred to contribute to endosulfan resistance, based on synergism studies with DEF and enhanced activity with the model substrate DCNB. However, there was no evidence for the altered target site (GABA receptor) commonly found in other cyclodiene-resistant insects (ffrench-Constant *et al.*, 1996), including *M. persicae* (Anthony *et al.*, 1998). An altered AChE did not appear to contribute to pirimicarb resistance in UK *N. ribisnigri* strains (Barber *et al.*, 1999).

Neonicotinoid resistance

In addition to the well-characterized resistance to OPs, carbamates, and pyrethroids, *M. persicae* has developed resistance to a number of other chemical classes, the mechanisms of which are, to date, not fully elucidated. Of those chemical classes, neonicotinoids are among the most important compounds to have been registered, and are now used widely as both foliar and systemic applications for aphid control.

Resistance (< 20-fold) to imidacloprid has been measured in *M. persicae* clones collected from the UK, mainland Europe, USA, Zimbabwe, and Japan (Devine *et al.*, 1996; Nauen *et al.*, 1996; Kerns *et al.*, 1998; Foster *et al.*, 2003a; see Nauen and Denholm, 2005 for review of resistance to neonicotinoids in a variety of insect pests). However, to date this has not led to any reported difficulties controlling *M. persicae* with these compounds when they are applied at recommended field rates for aphids. Nonetheless, the significant variation in worldwide response to imidacloprid suggests that this species may have taken the first evolutionary step towards developing more potent resistance under intense selection from imidacloprid and other neonicotinoids.

Mechanism(s) underpinning this variation are still unclear. Cross-resistance to imidacloprid, acetamiprid, and nitenpyram in *M. persicae* (Foster *et al.*, 2003b) implies a phenomenon influencing responses to neonicotinoid insecticides as a whole, as well as to nicotine (Devine *et al.*, 1996) and cartap (Nauen *et al.*, 1996), both of which also act at the nicotinic acetylcholine receptor. However, work on a Japanese imidacloprid-resistant strain of *M. persicae* showed no detectable difference in the binding of tritiated imidacloprid to this receptor, indicating that target site insensitivity was probably not a contributing factor in this strain (Nauen *et al.*, 1996). An explanation linking imidacloprid resistance to

naturally acquired nicotine tolerance in the tobacco-feeding subspecies *M. persicae nicotianae* (see Blackman and Eastop, Chapter 1 this volume) seems plausible, but such tolerance occurs in countries such as the UK where tobacco is not produced commercially. It is possible that long-distance movement from tobacco-growing countries, by aphid migration or, more likely, on imported plant material (Cannon *et al.*, 1999; Smith, 1999; Vierbergen, 2001), perhaps coupled with genetic introgression between *M. persicae sensu stricto* and *M. p. nicotianae*, as demonstrated for amplified esterase genes (Field *et al.*, 1994), has led to the trait being selected in some areas of the world and then becoming more widespread. Its appearance in Chile in the past decade seems to be as a single, widespread anholocyclic clone of *M. p. nicotianae* carrying R_1 resistance and lacking the kdr genotype (Fuentes-Contreras *et al.*, 2004). Studies of resistance mechanisms in *M. p. nicotianae* (Wolff *et al.*, 1994) have served simply to corroborate earlier studies on *M. persicae sensu stricto*. Besides widespread aphid movement, tolerance of nicotine extending to neonicotinoid insecticides could have been selected locally by use of nicotine as a fumigant or foliar spray in insect pest management strategies.

Low-level resistance to neonicotinoid insecticides, whatever the cause, is unrelated to the occurrence of the esterase, MACE, and kdr mechanisms of insecticide resistance in *M. persicae*. This is reassuring for growers in countries like the UK, where all three major mechanisms exist and have had serious consequences for aphid control in the recent past (Foster *et al.*, 2000). There are therefore opportunities to incorporate neonicotinoids into resistance management strategies for *M. persicae*, following principles advocated for neonicotinoid use generally. However, such strategies must concede that while neonicotinoids may not be greatly compromised by the variation in response that currently exists between *M. persicae* clones, excessive or indiscriminate use of these compounds will invariably risk selecting for novel resistance genes, or ones that enhance the potency of current mechanisms.

Factors Affecting the Dynamics of Insecticide Resistance in the Field

Selection pressures

One of the most intensive studies of resistance dynamics in any pest species involves UK *M. persicae* populations. Thirty years ago, R_2 and R_3 esterase forms were documented only in glasshouses. Since then, there has been a gradual general increase in their frequency in the field, but annual monitoring of aphids on a range of crops and in aerial trap samples (Muggleton *et al.*, 1996; Foster *et al.*, 2002c) has shown that R_1 and S aphids still tend to predominate. Periodically, however, esterase resistance can escalate, as in 1996 when samples taken from a range of crops contained high proportions of R_2 and R_3 aphids (Foster *et al.*, 2002c). Furthermore, many of these individuals also carried the MACE and kdr mechanisms; a combination that led to significant control failures on potatoes and Brussels sprouts (Foster *et al.*, 1998). The reasons for this rapid and dramatic upsurge in resistance remain unclear, but probably include early-season

applications of pyrethroids aimed at the unusually large influx of the noctuid *Autographa gamma* (silver Y moth) into the UK in that year. This may have inadvertently killed many aphid predators and parasitoids, allowing *M. persicae* to proliferate in the favourable spring and summer conditions. In certain areas, growers were then forced to apply regularly a variety of insecticides, which probably selected heavily for the esterase, MACE, and kdr mechanisms. However, in subsequent years up to 2000, field samples taken from various crops showed a progressive decline in the frequencies of high esterase and MACE forms (Fig. 10.1) (Foster *et al.*, 2002c). This apparent instability most probably reflected selection gradually acting through associated fitness costs (Crow, 1957 – see below), coupled with the response of growers to up-to-date advice on resistance management. In more recent years, high esterase and, particularly, MACE forms have become more common in UK field samples.

In the UK at least, fitness costs appear sufficient to maintain a fluctuating polymorphism of the susceptible- and resistant-esterase forms despite prolonged and often

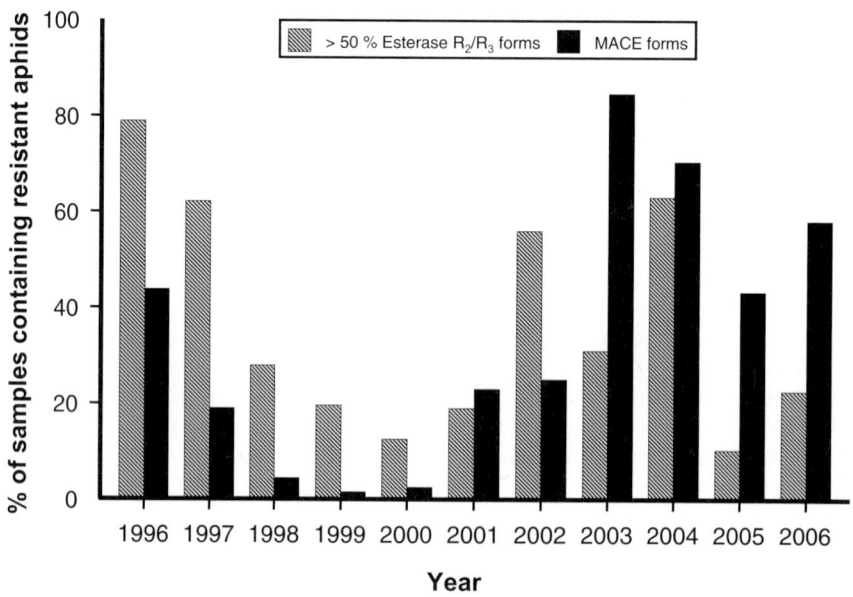

Fig. 10.1. Frequency of *Myzus persicae* samples, collected from UK field crops between 1996 and 2006, which contained greater than 50% high (R_2 and R_3) esterase aphids or at least one MACE aphid.

intense selection by insecticides. The prevalence of anholocycly in UK populations of *M. persicae* and the partial reproductive isolation (through androcycly) of esterase-R_2 and -R_3 aphids producing the E4 form of the carboxylesterase reinforces any existing associations between different resistance mechanisms. As a result, they can 'hitch-hike' together from generation to generation benefiting or suffering from any fitness advantages or costs that each may confer.

Ecological factors

Given the breadth and diversity of resistance mechanisms reported to date, it can confidently be assumed that no insecticide, however novel or unconventional in its effect, is immune to the appearance of genes conferring resistance. The probability of these achieving detectable frequencies depends instead on a suite of ecological and genetic factors and how these interact with pesticide usage patterns (reviewed by McKenzie, 1996; Denholm *et al.*, 1998). As a consequence, the same product can face very different resistance risks with different pest species, and even within the same species in different cropping systems.

The prevalence of resistance to the established insecticide groups (OPs, carbamates, and pyrethroids) among aphid pests of UK agriculture and horticulture provides a case in point. The three aphid species that historically have proved most problematic are *P. humuli*, *A. gossypii*, and *M. persicae*. Their capacity to evolve resistance can, to an extent, be rationalized in hindsight in terms of differing life histories and the crop systems they inhabit. During the summer, *P. humuli* is virtually restricted to wild and cultivated hops, on the latter of which insecticides remain the predominant means of aphid control. If insecticide resistance in UK populations of *P. humuli* on wild hops and on primary hosts were low (Lewis and Madge, 1984; Furk and Baxter, 1988), immigration of susceptible insects into commercial crops might help to ameliorate selection for resistance. This clearly has not been the case in the UK, nor in Czechoslovakia,

where *P. humuli* populations from untreated 'wild' hops were found also to be resistant to insecticides (Hrdy, 1975; Hrdy *et al.*, 1986). The very different profiles of resistance to OPs, pyrethroids, and carbamates exhibited by different *P. humuli* populations indicate that resistance originates in distinct geographical loci and then spreads throughout the extended complexes of hop fields and wild hosts. Loxdale *et al.* (1998) investigated the genetic heterogeneity of the esterases associated with insecticide resistance in *P. humuli* populations collected from primary and secondary hosts and concluded that migration, even within a 30 km region, was restricted. *Phorodon humuli*, therefore, tends to remain relatively localized and rapid dispersal over wide areas is unlikely in the UK. In this species, untreated 'refugia' appear, at least in practice, to exert a negligible effect on the development of resistance.

Aphis gossypii is restricted to glasshouses in the UK, where selection pressures are generally far more intense than in the open field (Denholm *et al.*, 1998). *Myzus persicae* is also a glasshouse pest but, in addition, infests a range of field crops including potatoes, brassicas, lettuce, and sugarbeet, all treated with insecticides for aphid control. The overall effect of exposing *M. persicae* to the same range of chemicals on glasshouse and field crops has been to select for a number of mechanisms that collectively threaten all available insecticides except the more novel agents pymetrozine and imidacloprid (Foster *et al.*, 2002a, 2003a).

Just as one can speculate on the factors that have encouraged the spread and development of resistance in some species, it is also possible to suggest why resistance has not occurred in other species. Good examples are the cereal aphids, *Sitobion avenae* (grain aphid), *Metopolophium dirhodum* (rose–grain aphid), and *Rhopalosiphum padi* (bird cherry–oat aphid). Despite a long history of control with insecticides, these species have tended to retain susceptibility in laboratory tests (e.g. Furk *et al.*, 1983; Stribley *et al.*, 1983), although in the latter species, there are sparse and geographically diverse reports of resistance disclosed by bioassay (Anonymous, 1974; Wei *et al.*, 1988; Sekun *et al.*, 1990). However, no reductions in field performance have been reported. Compared with *P. humuli*, *A. gossypii*, and *M. persicae*, these cereal aphids appear to present a contrasting situation in which migration between treated crops and extensive areas of wild grass hosts apparently has precluded any increase in the frequency of resistance genes. The failure of these species to develop resistance probably reflects a lower intensity of insecticide selection, presumably as a result of the ecology and life cycles (many cereal aphids are holocyclic for example). A large proportion of cereal aphids that exist on untreated crops and grasses will not be exposed to selection, and the extremely large numbers of these species trapped from aerial populations (Dewar *et al.*, 1984) suggest that gene flow between populations is extremely high and should serve to dilute any incipient resistance. Other reasons for the apparent lack of resistance in some species might include the absence of population bottlenecks on crops treated with insecticides, or conceivably lower mutation rates or greater fitness costs associated with resistance genes. In some areas of Europe, a large proportion of the *N. ribisnigri* population occurs on lettuce and is heavily selected by insecticides. However, the presence of resistance was only confirmed to have developed very recently (Rufingier *et al.*, 1997; Barber *et al.*, 1999). In this instance, it may be that the proportion of the population that exists in other localities on other untreated hosts might help ameliorate the selection pressure. In contrast, *P. humuli* is largely restricted to a single species of heavily sprayed crop grown in disjunct localities in the UK (hops). Resistance in this species was confirmed at least two decades earlier than in *N. ribisnigri*.

Despite this speculation, it is clear that resistance risks are not easily predicted. Other species illustrate that the association between ecological factors and the development of resistance is not readily reconciled. For example, the ecology of *Macrosiphum euphorbiae* (potato aphid) has much in common with that of *M. persicae*. Both inhabit glasshouses and a range of field crops and yet low resistance in the former, first confirmed in 1998, appears to have taken at least 30 years longer to evolve (Foster *et al.*, 2002d). *Schizaphis graminum* is widely recognized in the USA as resisting a range of insecticides. Intriguingly, *S. graminum* is a pest with a wide distribution and yet reports of resistance are rare outside the USA. Resistance in this species to a number of compounds has been noted in Egypt (El-Ghareeb, 1994a,b) and China (Tang *et al.*, 1988) but has yet to be explicitly reported from the Russian federation, Canada, or South America. The frequencies of esterase-based resistance mechanisms in *S. graminum* are thought to be related to insecticide use patterns and to potential fitness costs associated with each mechanism (Shufran *et al.*, 1997b). It is probable, therefore, that this species is under higher selection pressure in the USA because of the vast areas of sprayed crops involved, coupled with the facts that it is the most damaging of the cereal aphids and best adapted to aphid-resistant plant cultivars.

Pleiotropic effects of resistance

One of the tenets of resistance management is that resistant forms are less fit than the 'normal' susceptible genotype in the absence of insecticides; otherwise, the former would probably be present at greater frequencies prior to selection by insecticides. Significant selection pressure from synthetic insecticides dates back only 50 years, but the intensity of usage in some contexts already has imposed extremely strong selection. Nevertheless, susceptible forms persist and often increase in proportion within a population when insecticide selection pressures are relaxed, for example over winter months when many field crops are either unavailable or untreated. However, the fitness of resistant forms in the absence of insecticides has been studied in only a limited number of aphid species, sometimes with conflicting results, as discussed below. Furthermore, detecting what are likely to be subtle fitness deficits can pose substantial hurdles in some species. Some of the best evidence of pleiotropic effects of resistance genes comes from recent work on *M. persicae* where field and laboratory studies suggest the existence of adverse selection in the form of poor winter survival, maladaptive behaviour, and reduced reproductive fitness imposed primarily during times of stress.

Reduced overwintering ability

Monitoring of UK *M. persicae* populations in the late 1980s showed an apparent fall in the frequencies of high (R_2 and R_3) esterase forms during the winter months (Furk *et al.*, 1990), possibly due to counteracting selection in the absence of insecticides. This view was supported by lower levels of high esterase resistance in winged aphids caught during the spring/summer migrations of 1993 compared with aphids caught in the previous autumn. Subsequent winter field experiments using augmented *M. persicae* clones showed that aphids carrying higher levels of esterase resistance suffered greater mortality than their lower esterase counterparts during colder, wetter, and windier weather conditions (Foster *et al.*, 1996). This apparent fitness cost may underlie the dynamics of esterase resistance and an associated mechanism, MACE, seen in *M. persicae* collected from UK field crops over a number of years (Fig. 10.1).

Maladaptive behaviour

Movement of aphids from ageing leaves to younger leaves is an important aspect of aphid fitness as individuals remaining at the time of leaf abscission risk starvation during the period of locating another host plant (Harrington and Taylor, 1990). This danger increases dramatically under cold and wet conditions when movement can be severely restricted. Aphids are therefore under strong selection pressure to recognize the cues associated with leaf senescence and to respond accordingly. Studies of this important aspect of behaviour using UK *M. persicae* clones, performed at low temperatures (around 5°C) in the laboratory and the field, have shown that the rate of aphid movement is inversely

associated with esterase resistance level (Foster *et al.*, 1996, 1997), i.e. aphids with greater esterase resistance tended to move at slower rates from deteriorating leaves. As a result, these forms appear to run greater risks of becoming separated from their host plants after leaf fall.

Another component of aphid fitness is response to the alarm pheromone, (*E*)-β–farnesene. Any reduction in response could be highly maladaptive as the dispersal of aphids through this stimulus is considered to be an important behavioural adaptation for avoiding attack by natural enemies (Pickett *et al.*, 1992). Response studies in bioassays on *M. persicae* clones with different combinations of kdr and esterase resistance have shown reduced alarm response associated with both mechanisms (Foster *et al.*, 1999, 2003b). Both heterozygous and homozygous kdr aphids were far less responsive than aphids without kdr. There was also a significant inverse correlation between esterase resistance level and response of non-kdr aphids. Reduced pheromone response has also been recorded in OP-resistant leaf rollers (El-Sayed *et al.*, 2001).

Reduced reproductive success

Development time and fecundity have been measured in the laboratory using various *M. persicae* clones reared on excised Chinese cabbage leaves (Foster *et al.*, 2000). Esterase-R_3 aphids showed a very reduced reproductive fitness, measured by intrinsic rate of increase (Wyatt and White, 1977), compared with the significantly higher rates of aphids carrying lower levels of esterase resistance both at optimal (21°C) and suboptimal (5.5°C) temperatures. The trend was repeated for esterase-R_3 and susceptible forms held in clip cages on Chinese cabbage plants. It would appear therefore that reduced reproductive success is restricted to *M. persicae* capable of producing the highest (R_3) levels of esterase.

Studies on biological differences and fitness costs associated with resistance have also been conducted in *P. humuli* and *S. graminum*, but far less exhaustively than in *M. persicae* and never under suboptimal conditions. The results of these studies provide

conflicting conclusions on the possible fitness costs of resistance. In *P. humuli*, one
study found no consistent differences between
rates of reproduction, or any correlation
between their reproductive capacity and
overall body size between OP-resistant and
susceptible clones (Hampson and Madge,
1986, 1987). Another found adults from a
resistant strain to be larger and to have higher
reproductive rates and greater overall fecundity than susceptible adults (Lorriman and
Llewellyn, 1983).

In *S. graminum*, the reproductive
potential of two phenotypes harbouring different esterase forms was compared to that
of susceptible strains. Several clones of each
were manipulated into their sexual cycle
and inbred or cross-bred to determine their
reproductive capacity (Rider and Wilde, 1998;
Stone *et al.*, 2000). Crosses with susceptible
females averaged the highest number of eggs,
followed by Type I esterase females, and
then by Type II esterase females. They suggested insecticide resistance was therefore
associated with reduced reproductive success in certain clones. Shufran *et al.* (1997c)
showed that a Type 1 strain had a lower
intrinsic rate of increase than either susceptible or Type II resistant strains and
produced fewer offspring on a resistant sorghum hybrid. The Type II strain produced
fewer offspring on a susceptible hybrid
than the insecticide-susceptible strain. More
offspring were produced by the Type I
strain on the susceptible hybrid than on
the resistant hybrid. Stone *et al.* (2000) compared differences and similarities in pre-
reproductive time, progeny production, and
intrinsic rate of increase between one susceptible and three resistant strains. They
found that interactions of sorghum hybrid
with different *S. graminum* strains did not
significantly influence any of the parameters
measured, but that the Type I strain exhibited a significantly longer pre-reproductive
period than the other strains. In addition,
the Type I strain had a significantly lower
intrinsic rate of increase than the Type II or
susceptible strains. Thus, it seems that
under certain conditions, resistance can be
associated with differences in life history
parameters.

Aphids in which Resistance Poses a Minor or Potential Threat

There are a number of aphid species that
appear just sporadically in the resistance
literature, and for which the phenomenon
seems limited to a particular region or crop,
or to a particular time. It is unclear whether
these represent adaptations that evolved in
highly unusual circumstances that have not
re-arisen, or whether they are indicative of a
real resistance risk which, in an ever-changing
agricultural environment, might one day
prove to be of real economic concern. Examples of these include:

MACROSIPHUM EUPHORBIAE (POTATO APHID).   This
is a highly polyphagous species occurring
on potatoes, lettuce, sugarbeet, and ornamentals. However, unlike *M. persicae*,
with which it shares the same host range,
this species had never been documented as
posing a threat to the efficacy of insecticides.
Despite this, many *M. euphorbiae* clones
isolated from field potatoes in England
and Wales since 1998 have produced cross-
reactions in an immunoassay designed to
measure levels of E4/FE4 carboxylesterase
resistance in *M. persicae* (Foster *et al.*, 2002d).
Mean immunoassay values of each *M.
euphorbiae* clone fell within the S–R_2 range
used to categorize individual *M. persicae*.
These values were closely correlated with
resistance of the *M. euphorbiae* clones to
pirimicarb and lambda-cyhalothrin measured in laboratory leaf dip-bioassays. This
suggests that *M. euphorbiae* has developed
a metabolic resistance mechanism based on
overproduction of an esterase analogous to
that in *M. persicae*. Furthermore, levels of
esterase resistance appeared to be sufficient
to confer selective advantages in the field,
as a substantial majority of the aphid samples taken from crops recently treated with
insecticides contained high esterase forms
compared with samples that had not received
treatment.

DYSAPHIS PLANTAGINEA (ROSY APPLE APHID).
After resistance to organophosphates was first
reported by Wiesmann (1955), there followed a long period in which resistance in

D. plantaginea went unrecorded. However, it is now well established that its pest status in French, Swiss, and German orchards is due partly to the loss of efficacy of some aphicides. Hohn *et al.* (1995, 1996) noted slight resistance to pirimicarb, and Delorme *et al.* (1997, 1999) proposed that resistance to OPs, carbamates, and pyrethroids in this species was correlated with both a modified AChE and an amplified carboxylesterase. Schaub *et al.* (2001) provided evidence that this species also was becoming tolerant to neonicotinoids.

LIPAPHIS PSEUDOBRASSICAE (MUSTARD APHID OR TURNIP APHID). There are some reports of resistance in *L. pseudobrassicae* from India and China. One Indian population was 24-fold resistant to endosulfan and others have shown *c.* 5-fold resistance. Some of these were also 3- to 6-fold resistant to some OPs (Udeaan and Narang, 1988). Other populations have shown increases of 10- to 30-fold in resistance to some pyrethroids between 1980 and 1988 (Dhingra and Singh, 1988). Aphids from vegetables in Beijing developed 600-fold resistance to fenvalerate following selection in the laboratory, with cross-resistance to deltamethrin and cypermethrin, but little (5- to 7-fold) resistance to the fluorine-containing pyrethroids, cyhalothrin, flucythrinate, and fluvalinate (Rui *et al.*, 1996).

APHIS NASTURTII (BUCKTHORN–POTATO APHID). Poor control using pirimicarb was first reported by Laska (1981). However, only one report of inferred resistance to chemical control has appeared since for this species on potatoes in Belgium and northern France (Duvauchelle *et al.*, 1997a,b).

APHIS FABAE (BLACK BEAN APHID). This species was reported to show OP resistance of up to 11-fold as long ago as 1974 (Boness and Unterstenhofer, 1974), but none was found in the UK in the 1980s (Stribley *et al.*, 1983). More recently, Ioannidis (2000) reported that *A. fabae* from sugarbeet in Greece showed 50-fold resistance to methamidophos, 8-fold resistance to pirimicarb, and 7-fold resistance to imidacloprid, the latter being of the same order as that found in *M. persicae*.

APHIS CRACCIVORA (COWPEA APHID). Resistance to OPs was first reported by Smirnova and Ivanova (1974) in the Asian region of the USSR. Broad-spectrum resistance to OPs, nicotine, pyrethroids, and organochlorine insecticides has also been found in *A. craccivora* populations on various leguminous and non-leguminous hosts in India (Dhingra, 1993, 1994).

APHIS SPIRAECOLA (GREEN CITRUS APHID). Resistance has been recorded to pirimicarb, phosphamidon, and omethoate in populations from heavily sprayed apples in Korea. This was associated with increased esterase activity and a very marked insensitivity (200- to 300-fold) of their AChE to inhibition by the first two compounds and modest insensitivity (6-fold) to the latter (Song *et al.*, 1995).

THERIOAPHIS TRIFOLII MACULATA (SPOTTED ALFALFA APHID). Problems in controlling *T. t. maculata* caused by OP and carbamate resistance in the late 1970s in Australia (Walters and Forrester, 1979) appear to have been largely circumvented by the increased adoption of aphid-resistant lucerne cultivars (Holtkamp *et al.*, 1992). Nevertheless, insecticide use continued into the 1980s with chlorpyrifos as the recommended compound. Resistance to pirimicarb (750-fold) has also been very high, with moderate resistance (10- to 40-fold) to some OPs but little resistance to endosulfan or fenvalerate. Intriguingly, control problems with chlorpyrifos in 1990 were associated with low-level resistance to this compound in laboratory bioassays (Holtkamp *et al.*, 1992).

Conclusions

In many respects, the continuing battle against resistance can be equated with co-evolution and is a clear illustration of how such processes generate biological diversity. Over the past twenty years, researchers have made great progress in monitoring and characterizing pesticide resistance and in understanding some of the genetic, ecological, and operational factors that affect the speed of its development. Nowhere has this been more obvious than in the study of aphids. In doing so, researchers have provided valuable insights into evolutionary biology and genetics; on the origin and nature of adaptations and in understanding genetic responses to changes in the environment. Regardless of the progress that has been made, it is clear that the 'arms race' between insect evolution and human ingenuity will continue to present major challenges. In order to meet these, it will be necessary to gain a greater understanding of the processes that mediate the development of resistance. In particular, there is a need for empirical research on the mechanisms conferring resistance to more novel chemical groups, and an understanding of the breadth of resistance that these mechanisms confer. Additionally, there is a paucity of information on the ecological factors that mediate resistance development – the role and manipulation of untreated refugia for susceptible forms, and the value of dispersal and migration. Armed with such information, pest management specialists will be better able to develop models that realistically assess 'resistance risk' and thereby strengthen future risk-mitigation tactics.

Acknowledgements

We are grateful to Ian Denholm, Lin Field, and Graham Moores for useful discussion and comments. Rothamsted Research receives grant-aided support from the Biotechnology and Biological Sciences Research Council of the UK.

References

Anonymous (1974) cited in Georgiou, G.P. and Lagunes-Tejeda (1991) *The Occurrence of Resistance to Pesticides in Arthropods: An Index of Cases Reported through 1989*. FAO, Italy, 318pp.

Andrews, M.C., Callaghan, A., Field, L.M., Williamson, M.S. and Moores, G.D. (2004) Identification of mutations conferring insecticide-insensitive AChE in the cotton–melon aphid, *Aphis gossypii* Glover. *Insect Molecular Biology* 13, 555–561.

Anstead, A., Williamson, S. and Denholm, I. (2004) High-throughput detection of knockdown resistance in *Myzus persicae* using allelic discriminating quantitative PCR. *Insect Biochemistry and Molecular Biology* 34, 871–877.

Anstead, A., Williamson, S. and Denholm, I. (2005) Evidence for multiple origins of identical insecticide resistance mutations in the aphid, *Myzus persicae*. *Insect Biochemistry and Molecular Biology* 35, 249–256.

Anthon, E.W. (1955) Evidence for green peach aphid resistance to organophosphorous insecticides. *Journal of Economic Entomology* 48, 56–57.

Anthony, N., Unruh, T., Ganser, D. and ffrench-Constant, R. (1998) Duplication of the Rdl GABA receptor subunit gene in an insecticide-resistant aphid, *Myzus persicae*. *Molecular and General Genetics* 260, 165–175.

Barber, M.D., Moores, G.D., Tatchell, G.M., Vice, W.E. and Denholm, I. (1999) Insecticide resistance in the currant–lettuce aphid, *Nasonovia ribisnigri* (Hemiptera: Aphididae) in the UK. *Bulletin of Entomological Research* 89, 17–23.

Beck, A.K. and Buchi, R. (1980) Esterase assay for the detection of resistance to insecticides in the hop aphid, *Phorodon humuli* Schrk. *Zeitschrift für Angewandte Entomologie* 89, 113–121.

Blackman, R.L., Takada, H. and Kawakami, K. (1978) Chromosomal rearrangement involved in insecticide resistance of *Myzus persicae*. *Nature, London* 271, 450–452.

Boness, M. and Unterstenhofer, G. (1974) Insecticide resistance in aphids. *Zeitschrift für Angewandte Entomologie* 77, 1–19.

Buchi, R. (1981) Evidence that resistance against pyrethroids in aphids *Myzus persicae* and *Phorodon humuli* is not correlated with high carboxylesterase activity. *Zeitschrift für Pflanzenkrankheiten und Pflanzenschutz* 88, 631–634.

Cannon, R.J.C., Pemberton, A.W. and Bartlett, P.W. (1999) Appropriate measures for the eradication of unlisted pests. *Bulletin OEPP* 29, 29–36.

Cervato, P. and Cravedi, P. (1995) Resistance to insecticides in *Myzus persicae* (Sulzer) in Italian peach orchards. *Bollettino di Zoologia Agraria e di Bachicoltura II* 27, 191–199.

Crow, J.F. (1957) Genetics of insect resistance to chemicals. *Annual Review of Entomology* 2, 227–246.

Delorme, R., Auge, D., Bethenod, M.T. and Villatte, F. (1997) Insecticide resistance in a strain of *Aphis gossypii* from Southern France. *Pesticide Science* 49, 90–96.

Delorme, R., Ayala, V., Touton, P., Auge, D. and Vergnet, C. (1999) The rosy apple aphid (*Dysaphis plantaginea*): insecticide resistance mechanisms. *Annales de la 5e ANPP Conférence Internationale sur les Ravageurs en Agriculture, Montpellier, December 1999* 1, 89–96.

Denholm, I., Horowitz, A.R., Cahill, M. and Ishaaya, I. (1998) Management of resistance to novel insecticides. In: Ishaaya, I. and Degheele, D. (eds) *Insecticides with Novel Modes of Action: Mechanism and Application*. Springer, Berlin, pp. 260–282.

Devine, G.J., Harling, Z.K., Scarr, A.W. and Devonshire, A.L. (1996) Lethal and sublethal effects of imidacloprid on nicotine-tolerant *Myzus nicotianae* and *Myzus persicae*. *Pesticide Science* 48, 57–62.

Devonshire, A.L. (1975) Studies of the carboxylesterases of *Myzus persicae* resistant and susceptible to organophosphorus insecticides. *Proceedings of 8th Insecticides and Fungicides Conference, Brighton, November 1975* 1, 67–73.

Devonshire, A.L. (1977) The properties of a carboxylesterase from the peach–potato aphid, *Myzus persicae* (Sulz.), and its role in conferring insecticide resistance. *Biochemical Journal* 167, 675–683.

Devonshire, A.L. (1989) Resistance of aphids to insecticides. In: Minks, A.K. and Harrewijn, P. (eds) *Aphids. Their Biology, Natural Enemies and Control, Volume 2C*. Elsevier, Amsterdam, pp. 123–139.

Devonshire, A.L. and Moores, G.D. (1982) A carboxylesterase with broad substrate specificity causes organophosphorus, carbamate and pyrethroid resistance in peach–potato aphids (*Myzus persicae*). *Pesticide Biochemistry and Physiology* 18, 235–246.

Devonshire, A.L. and Rice, A.D. (1988) Aphid bioassay techniques. In: Minks, A.K. and Harrewijn, P. (eds) *Aphids. Their Biology, Natural Enemies and Control, Volume 2B*. Elsevier, Amsterdam, pp. 119–128.

Devonshire, A.L. and Sawicki, R.M. (1979) Insecticide resistant *Myzus persicae* as an example of evolution by gene duplication. *Nature, London* 280, 140–141.

Devonshire, A.L., Moores, G.D. and Chiang, C. (1983) The biochemistry of insecticide resistance in the peach–potato aphid (*Myzus persicae*). In: *Pesticide Chemistry: Human Welfare and the Environment. Proceedings of the 5th International Congress of Pesticide Chemistry, Kyoto, August 1982, Volume 3*. Pergamon, Oxford, pp. 191–196.

Devonshire, A.L., Moores, G.D. and ffrench-Constant, R.H. (1986) Detection of insecticide resistance by immunological estimation of carboxylesterase activity in *Myzus persicae* (Sulzer) and cross-reaction of the antiserum with *Phorodon humuli* (Schrank) (Hemiptera: Aphididae). *Bulletin of Entomological Research* 76, 97–107.

Devonshire, A.L., Field, L.M., Foster, S.P., Moores, G.D., Williamson, M.S. and Blackman, R.L. (1998) The evolution of insecticide resistance in the peach–potato aphid, *Myzus persicae*. *Philosophical Transactions of the Royal Society of London B* 353, 1677–1684.

Devonshire, A.L., Foster, S.P. and Denholm, I. (1999) Insecticide resistance in the peach–potato aphid, *Myzus persicae*. In: *Proceedings of the ENMARIA Symposium: Combating Insecticide Resistance, Thessaloniki, May 1999*. AgroTypos SA, Athens, pp. 79–85.

Dewar, A.M., Tatchell, G.M. and Turl, L.A.D. (1984) A comparison of cereal-aphid migrations over Britain in the summers of 1979 and 1982. *Crop Protection* 3, 379–389.

Dewar, A.M., Haylock, L.A., Campbell, J., Harling, Z.K., Foster, S.P. and Devonshire, A.L. (1998) Control on sugarbeet of *Myzus persicae* with different insecticide-resistance mechanisms. In: Dale, M.F.B., Dewar, A.M., Fisher, S.J., Haydock, P.P.J., Jaggard, K.W., May, M.J., Smith, H.G., Storey, R.M.J. and Wiltshire, J.J.J. (eds) *Protection and Production of Sugar Beet and Potatoes. Aspects of Applied Biology, No. 52*. pp. 407–414.

Dhingra, S. (1993) Development of resistance in the bean aphid, *Aphis craccivora* Koch., to various synthetic pyrethroids with special reference to change in susceptibility of some important aphid species during the last one and a quarter decades. *Journal of Entomological Research* 17, 247–250.

Dhingra, S. (1994) Development of resistance in the bean aphid, *Aphis craccivora* Koch., to various insecticides used for nearly a quarter century. *Journal of Entomological Research* 18, 105–108.

Dhingra, S. and Singh, D.S. (1988) Impact of formulation on the level of resistance in mustard aphid, *Lipaphis erysimi* Kalt., to synthetic pyrethroids. *Journal of Entomological Research* 12, 56–60.

Dicker, G.H.L. (1965) cited in Georgiou, G.P. and Lagunes-Tejeda (1991) *The Occurrence of Resistance to Pesticides in Arthropods: An Index of Cases Reported Through 1989*. FAO, Italy, 318pp.

Duvauchelle, S., Dubois, L. and Nguyen, N. (1997a) Aphids and viruses on ware potatoes in northern France particularly in 1995 and 1996. *Mededelingen Faculteit Landbouwkundige en Toegepaste Biologische Wetenschappen Universiteit Gent* 62, 545–546.

Duvauchelle, S., Trouve, C., Delorme, R., Dubois, L., Chudzicki, A.M. and Ducatillon, C. (1997b) Recent problems for the control of aphids in potato crops in France and Belgium. *Annales de la 4e ANPP Conférence Internationale sur les Ravageurs en Agriculture, Montpellier, January 1997* 3, 895–902.

Elbert, A., Nauen, R., Cahill, M., Devonshire, A.L., Scarr, A.W., Sone, S. and Steffens, R. (1996) Resistance management with chloronicotinyl insecticides using imidacloprid as an example. *Pflanzenschutz Nachrichten* 49, 5–54.

El-Ghareeb, A.M. (1994a) Insecticide resistance of the greenbug aphid *Schizaphis graminum* (Rondani): insecticidal potency, development of resistance and acetylcholinesterase inhibition in field populations. *Assiut Journal of Agricultural Sciences* 25, 61–86.

El-Ghareeb, A.M. (1994b) Insecticide resistance of the greenbug aphid *Schizaphis graminum* (Rondani) (Homoptera: Aphididae). Resistance mechanism in two resistant field populations. *Assiut Journal of Agricultural Sciences* 25, 87–112.

El-Sayed, A.M., Fraser, H.M. and Trimble, R.M. (2001) Modification of the sex-pheromone communication system associated with organophosphorus-insecticide resistance in the obliquebanded leafroller (Lepidoptera: Tortricidae). *Canadian Entomologist* 133, 867–881.

ffrench-Constant, R.H., Devonshire, A.L. and Clark, S.J. (1987) Differential rate of selection for resistance by carbamate, organophosphorus and combined pyrethroid insecticides in *Myzus persicae* (Sulzer) (Hemiptera: Aphididae). *Bulletin of Entomological Research* 77, 227–238.

ffrench-Constant, R.H., Clark, S.J. and Devonshire, A.L. (1988a) Effect of repeated applications of insecticides to potatoes on numbers of *Myzus persicae* (Sulzer) (Hemiptera: Aphididae) and on the frequencies of insecticide resistant variants. *Crop Protection* 7, 55–61.

ffrench-Constant, R.H., Harrington, R. and Devonshire, A.L. (1988b) Effect of decline of insecticide residues on selection for insecticide resistance in *Myzus persicae* (Sulzer) (Hemiptera: Aphididae). *Bulletin of Entomological Research* 78, 19–29.

ffrench-Constant, R.H., Anthony, N.M. and Andreev, D. (1996) Single versus multiple origins of insecticide resistance: inferences from the cyclodiene resistance gene *Rdl*. In: Brown, T.M. (ed.) *Molecular Genetics and Evolution of Pesticide Resistance. ACS Symposium Series No. 645.* American Chemical Society, Washington D.C., pp.106–116.

Field, L.M., Devonshire, A.L. and Forde, B.G. (1988) Molecular evidence that insecticide resistance in peach–potato aphids (*Myzus persicae* Sulz.) results from amplification of an esterase gene. *Biochemical Journal* 251, 309–315.

Field, L.M., Devonshire, A.L., ffrench-Constant, R.H. and Forde, B.G. (1989) Changes in DNA methylation are associated with loss of insecticide resistance in the peach–potato aphid *Myzus persicae* (Sulz.). *Federation of European Biochemical Societies Letters* 243, 323–327.

Field, L.M., Javed, N., Stribley, M.F. and Devonshire, A.L. (1994) The peach–potato aphid *Myzus persicae* and the tobacco aphid *Myzus nicotianae* have the same esterase-based mechanisms of insecticide resistance. *Insect Molecular Biology* 3, 143–148.

Field, L.M., Crick, S.E. and Devonshire, A.L. (1996) Polymerase chain reaction-based identification of insecticide resistance genes and DNA methylation in the aphid *Myzus persicae* (Sulzer). *Insect Molecular Biology* 5, 197–202.

Field, L.M., Blackman, R.L., Tyler-Smith, C. and Devonshire, A.L. (1999) Relationship between amount of esterase and gene copy number in insecticide-resistant *Myzus persicae* (Sulzer). *Biochemical Journal* 339, 737–742.

Foster, S.P. (2000) Knock-down resistance (kdr) to pyrethroids in peach–potato aphids (*Myzus persicae*) in the UK: a cloud with a silver lining? *Proceedings of the British Crop Protection Council Conference, Pests and Diseases, Brighton, November 2000* 1, 465–472.

Foster, S.P. and Devonshire, A.L. (1999) Field simulator study of insecticide resistance conferred by esterase-, MACE- and kdr-based mechanisms in the peach–potato aphid, *Myzus persicae* (Sulzer). *Pesticide Science* 55, 810–814.

Foster, S.P., Harrington, R., Devonshire, A.L., Denholm, I., Devine, G.J., Kenward, M.G. and Bale, J.S. (1996) Comparative survival of insecticide-susceptible and resistant peach–potato aphids, *Myzus persicae* (Sulzer) (Hemiptera: Aphididae), in low temperature field trials. *Bulletin of Entomological Research* 86, 17–27.

Foster, S.P., Harrington, R., Devonshire, A.L., Denholm, I., Clark, S.J. and Mugglestone, M.A. (1997) Evidence for a possible fitness trade-off between insecticide resistance and the low temperature movement that is essential for survival of UK populations of *Myzus persicae* (Hemiptera: Aphididae). *Bulletin of Entomological Research* 87, 573–579.

Foster, S.P., Denholm, I., Harling, Z.K., Moores, G.D. and Devonshire, A.L. (1998) Intensification of insecticide resistance in UK field populations of the peach–potato aphid, *Myzus persicae* (Hemiptera: Aphididae) in 1996. *Bulletin of Entomological Research* 88, 127–130.

Foster, S.P., Woodcock, C.M., Williamson, M.S., Devonshire, A.L., Denholm, I. and Thompson, R. (1999) Reduced alarm response for peach–potato aphids (*Myzus persicae*) with knock-down resistance to insecticides (*kdr*) may impose a fitness cost through increased vulnerability to natural enemies. *Bulletin of Entomological Research* 89, 133–138.

Foster, S.P., Denholm, I. and Devonshire, A.L. (2000) The ups and downs of insecticide resistance in peach–potato aphids (*Myzus persicae*) in the UK. *Crop Protection* 19, 873–879.

Foster, S.P., Denholm, I. and Thompson, R. (2002a) Bioassay and field-simulator studies of the efficacy of pymetrozine against peach–potato aphids, *Myzus persicae* (Hemiptera: Aphididae), possessing different mechanisms of insecticide resistance. *Pest Management Science* 58, 805–810.

Foster, S.P., Denholm, I. and Devonshire, A.L. (2002b) Field-simulator studies of insecticide resistance to dimethyl-carbamates and pyrethroids conferred by metabolic- and target site-based mechanisms in peach–potato aphids, *Myzus persicae* (Hemiptera: Aphididae). *Pest Management Science* 58, 811–816.

Foster, S.P., Harrington, R., Dewar, A.M., Denholm, I. and Devonshire, A.L. (2002c) Temporal and spatial dynamics of insecticide resistance in *Myzus persicae* (Hemiptera: Aphididae). *Pest Management Science* 58, 895–907.

Foster, S.P., Hackett, B., Mason, N., Moores, G.D., Cox, D.M., Campbell, J. and Denholm, I. (2002d) Resistance to carbamate, organophosphate and pyrethroid insecticides in the potato aphid (*Macrosiphum euphorbiae*). *Proceedings of the British Crop Protection Council Conference, Pests and Diseases, Brighton, November 2002* 2, 811–816.

Foster, S.P., Denholm, I. and Thompson, R. (2003a) Variation in response to neonicotinoid insecticides in peach–potato aphids, *Myzus persicae* (Hemiptera: Aphididae). *Pest Management Science* 59, 166–173.

Foster, S.P., Young, S., Williamson, M.S., Duce, I., Denholm, I. and Devine, G.J. (2003b) Analogous pleiotropic effects of insecticide resistance genotypes in peach–potato aphids and houseflies. *Heredity* 91, 98–106.

Fuentes-Contreras, E., Figueroa, C.C., Reyes, M., Briones, L.M. and Niemeyer, H.M. (2004) Genetic diversity and insecticide resistance of *Myzus persicae* (Hemiptera: Aphididae) populations from tobacco in Chile: evidence for the existence of a single predominant clone. *Bulletin of Entomological Research* 94, 11–18.

Furk, C. and Baxter, C. (1988) Monitoring for pyrethroid resistance in the damson–hop aphid, *Phorodon humuli*. *Bulletin IOBC/WPRS* 11, 10–21.

Furk, C., Powell, D.F. and Heyd, S. (1980) Pirimicarb resistance in the melon and cotton aphid, *Aphis gossypii* Glover. *Plant Pathology* 29, 191–196.

Furk, C., Cotten, J. and Gould, H.J. (1983) Monitoring for insecticide resistance in aphid pests of field crops in England and Wales. *Proceedings of the Brighton Crop Protection Conference, Pests and Diseases, November 1983* 2, 637.

Furk, C., Hines, C.M., Smith, C.D.J. and Devonshire, A.L. (1990) Seasonal variation of susceptible and resistant variants of *Myzus persicae*. *Proceedings of the Brighton Crop Protection Conference, Pests and Diseases, November 1990* 3, 1207–1212.

Gao, J.R. and Zhu, K.Y. (2001) An acetylcholinesterase purified from the greenbug (*Schizaphis graminum*) with some unique enzymological and pharmacological characteristics. *Insect Biochemistry and Molecular Biology* 31, 1095–1104.

Gao, J.R. and Zhu, K.Y. (2002) Biochemical and molecular analyses of acetylcholinesterase conferring organophosphate resistance in the greenbug, *Schizaphis graminum* (Homoptera: Aphididae). *Abstracts of Papers of the American Chemical Society* 223, 009–AGRO.

Ghong, K., Zhang, G. and Zhai, G. (1964) Resistance of cotton aphids to demeton. *Journal of Entomology* 13, 1–9.

Gubran, E.M.E., Delorme, R., Auge, D. and Moreau, J.P. (1992) Insecticide resistance in cotton aphid *Aphis gossypii* (Glov.) in the Sudan Gezira. *Pesticide Science* 35, 101–107.

Guilin, C., Runzxi, L., Mingjiang, H., Kefu, X. and Shiju, J. (1997) Comparison of the cotton aphid resistance level between Xinjiang and Shandong populations. *Resistant Pest Management News Letter* 9, 10–12.

Guillemaud, T., Brun, A., Anthony, N., Sauge, M.-H., Boll, R., Delorme, R., Fournier, D., Lapchin, L. and Vanlerberghe-Masutti, F. (2003) Incidence of insecticide resistance alleles in sexually-reproducing populations of the peach–potato aphid *Myzus persicae* (Hemiptera: Aphididae) from southern France. *Bulletin of Entomological Research* 93, 289–297.

Hampson, M.J. and Madge, D.S. (1986) Morphometric variation between clones of the damson–hop aphid, *Phorodon humuli* (Schrank) (Hemiptera: Aphididae). *Agriculture, Ecosystems and the Environment* 16, 255–264.

Hampson, M.J. and Madge, D.S. (1987) Reproduction rates of insecticide-resistant and susceptible strains of the damson–hop aphid, *Phorodon humuli* (Hemiptera, Aphididae). *Acta Entomologica Bohemoslovaca* 84, 181–184.

Han, Z., Moores, G.D., Denholm, I. and Devonshire, A.L. (1998) Association between biochemical markers and insecticide resistance in the cotton aphid, *Aphis gossypii* Glover. *Pesticide Biochemistry and Physiology* 62, 164–171.

Harrington, R. and Taylor, L.R. (1990) Migration for survival: fine scale population redistribution in an aphid, *Myzus persicae*. *Journal of Animal Ecology* 59, 1177–1193.

Hick, C.A., Field, L.M. and Devonshire, A.L. (1996) Changes in the methylation of amplified esterase DNA during loss and reselection of insecticide resistance in peach–potato aphids, *Myzus persicae*. *Insect Biochemistry and Molecular Biology* 26, 41–47.

Hohn, H., Hopli, H.U. and Graf, B. (1995) Mealy apple aphid: a problem pest. *Schweizerische Zeitschrift für Obst und Weinbau* 131, 204–206.

Hohn, H., Hopli, H.U. and Graf, B. (1996) Mealy apple aphid: attack development and control. *Schweizerische Zeitschrift für Obst und Weinbau* 132, 60–62.

Holtkamp, R.H., Edge, V.E., Dominiak, B.C. and Walters, P.J. (1992) Insecticide resistance in *Therioaphis trifolii* f. *maculata* (Hemiptera: Aphididae) in Australia. *Journal of Economic Entomology* 85, 1576–1582.

Hosoda, A., Hama, H., Suzuki, K. and Ando, Y. (1992) Insecticide resistance in cotton aphid, *Aphis gossypii* Glover (Homoptera: Aphididae) I. Aliesterase activity and organophosphorus-susceptibility of populations on eggplants and cucumbers. *Japanese Journal of Applied Entomology and Zoology* 36, 101–111.

Hrdy, I. (1975) Insecticide resistance in aphids. *Proceedings of the 8th Brighton Insecticides and Fungicides Conference, November 1975* 3, 737–749.

Hrdy, I., Kremheller, H.T., Kuldova, J., Luders, W. and Sula, J. (1986) Resistance to insecticides of the hop aphid, *Phorodon humuli*, in Bohemian, Bavarian and Baden-Wurttembergian hop growing areas. *Acta Entomologica Bohemoslovaca* 83, 131–139.

Ioannidis, P. (2000) Resistance of *Aphis fabae* and *Myzus persicae* to insecticides in sugarbeets. In: *Proceedings of the 63rd Congress of the International Institute for Beet Research, Interlaken, February 2000*. International Institute for Beet Research, Brussels, pp. 497–504.

Ishaaya, I. and Mendelson, Z. (1987) The susceptibility of the melon aphid, *Aphis gossypii*, to insecticides during the cotton growing season. *Hassadeh* 67, 1772–1773.

Kerns, D.L., Palumbo, J.C. and Byrne, D.N. (1998) Relative susceptibility of red and green color forms of green peach aphid to insecticides. *South Western Entomologist* 23, 17–24.

Laska, P. (1981) Low pirimicarb-susceptibility of buckthorn–potato aphid, *Aphis nasturtii* (Homoptera: Aphididae). *Acta Entomologica Bohemoslovaca* 78, 61–62.

Lewis, G.A. and Madge, D.S. (1984) Esterase activity and associated insecticide resistance in the damson–hop aphid, *Phorodon humuli* (Schrank) (Hemiptera: Aphididae). *Bulletin of Entomological Research* 74, 227–238.

Li, F. and Han, Z. (2004) Mutations in acetylcholinesterase associated with insecticide resistance in the cotton aphid, *Aphis gossypii* Glover. *Insect Biochemistry and Molecular Biology* 34, 397–405.

Lorriman, F. and Llewellyn, M. (1983) The growth and reproduction of hop aphid (*Phorodon humuli*) biotypes resistant and susceptible to insecticides. *Acta Entomologica Bohemoslovaca* 80, 87–95.

Loxdale, H.D., Brookes, C.P., Wynne, I.R. and Clark, S.J. (1998) Genetic variability within and between English populations of the damson–hop aphid, *Phorodon humuli* (Hemiptera: Aphididae), with special reference to esterases associated with insecticide resistance. *Bulletin of Entomological Research* 88, 513–526.

Martinez-Torres, D., Devonshire, A.L. and Williamson, M.S. (1997) Molecular studies of knockdown resistance to pyrethroids: cloning of domain II sodium channel gene sequences from insects. *Pesticide Science* 51, 265–270.

Martinez-Torres, D., Foster, S.P., Field, L.M., Devonshire, A.L. and Williamson, M.S. (1999) A sodium channel point mutation is associated with resistance to DDT and pyrethroid insecticides in the peach–potato aphid, *Myzus persicae* (Sulzer) (Hemiptera: Aphididae). *Insect Molecular Biology* 8, 1–8.

McKenzie, J.A. (1996) *Ecological and Evolutionary Aspects of Insecticide Resistance*. Academic Press, London, 185 pp.

Moores, G.D., Devine, G.J. and Devonshire, A.L. (1994a) Insecticide-insensitive acetylcholinesterase can enhance esterase-based resistance in *Myzus persicae* and *Myzus nicotianae*. *Pesticide Biochemistry and Physiology* 49, 114–120.

Moores, G.D., Devine, G.J. and Devonshire, A.L. (1994b) Insecticide resistance due to insensitive acetylcholinesterase in *Myzus persicae* and *Myzus nicotianae*. *Proceedings of the Brighton Crop Protection Conference, Pests and Diseases, November 1994* 1, 413–418.

Moores, G.D., Gao, X.W., Denholm, I. and Devonshire, A.L. (1996) Characterisation of insensitive acetyl-cholinesterase in insecticide-resistant cotton aphids, *Aphis gossypii* Glover (Homoptera: Aphididae). *Pesticide Biochemistry and Physiology* 56, 102–110.

Muggleton, J., Hockland, S., Thind, B.B., Lane, A. and Devonshire, A.L. (1996) Long-term stability in the frequency of insecticide resistance in the peach–potato aphid, *Myzus persicae*, in England. *Proceedings of the Brighton Crop Protection Conference, Pests and Diseases, November 1996* 2, 739–744.

Nabeshima, T., Kozaki, T., Takashi, T. and Kono, Y. (2003) An amino acid substitution on the second acetyl-cholinesterase in the pirimicarb-resistant strains of the peach–potato aphid, *Myzus persicae*. *Biochemical and Biophysical Research Communications* 307, 15–22.

Nauen, R. and Denholm, I. (2005) Resistance of insect pests to neonicotinoid insecticides: current status and future prospects. *Archives of Insect Biochemistry and Physiology* 58, 200–215.

Nauen, R., Strobel, J., Tietjen, K., Otso, Y., Erdelen, C. and Elbert, A. (1996) Aphicidal activity of imidacloprid against a tobacco feeding strain of *Myzus persicae* (Homoptera: Aphididae) from Japan closely related to *Myzus nicotianae* and highly resistant to carbamate and organophosphates. *Bulletin of Entomological Research* 86, 165–171.

Needham, P.H. and Sawicki, R.M. (1970) Diagnosis of resistance to insecticides in *Myzus persicae*. *Nature, London* 230, 125–126.

O'Brien, P.J., Abdel-Aal, Y.A., Ottea, J.A. and Graves, J.B. (1992) Relationship of insecticide resistance to carboxylesterases in *Aphis gossypii* (Homoptera: Aphididae) from midsouth cotton. *Journal of Economic Entomology* 85, 651–657.

Ono, M., Richman, J.S. and Siegfried, B.D. (1994a) Characterization of general esterases from susceptible and parathion-resistant strains of the greenbug (Homoptera, Aphididae). *Journal of Economic Entomology* 87, 1430–1436.

Ono, M., Richman, J.S. and Siegfried, B.D. (1994b) *In vitro* metabolism of parathion in susceptible and parathion-resistant strains of the greenbug, *Schizaphis graminum* (Rondani) (Homoptera: Aphididae). *Pesticide Biochemistry and Physiology* 49, 191–197.

Ono, M., Swanson, J.J., Field, L.M., Devonshire, A.L. and Siegfried, B.D. (1999) Amplification and methylation of an esterase gene associated with insecticide resistance in greenbugs, *Schizaphis graminum* (Rondani) (Homoptera: Aphididae). *Insect Biochemistry and Molecular Biology* 29, 1065–1073.

Peters, D.C., Wood, E.A., Jr. and Starks, K.J. (1975) Insecticide resistance in selections of the greenbug. *Journal of Economic Entomology* 68, 339–340.

Pickett, J.A., Wadhams, L.J., Woodcock, C.M. and Hardie, J. (1992) The chemical ecology of aphids. *Annual Review of Entomology* 37, 67–90.

Pimentel, D., Acquay, H., Biltonen, M., Rice, P., Silva, M., Nelson, J., Lipner, V., Giordano, S., Horowitz, A. and Damore, M. (1992) Environmental and economic costs of pesticide use. *Bioscience* 42, 750–760.

Rider, S.D. and Wilde, G.E. (1998) Variation in fecundity and sexual morph production among insecticide-resistant clones of the aphid *Schizaphis graminum* (Homoptera: Aphididae). *Journal of Economic Entomology* 91, 388–391.

Rider, S.D., Wilde, G.E. and Kambhampati, S. (1998) Genetics of esterase-mediated insecticide resistance in the aphid *Schizaphis graminum*. *Heredity* 81, 14–19.

Rufingier, C., Schoen, L., Martin, C. and Pasteur, N. (1997) Resistance of *Nasonovia ribisnigri* (Homoptera: Aphididae) to five insecticides. *Journal of Economic Entomology* 90, 1445–1449.

Rufingier, C., Pasteur, N., Lagnel, J., Martin, C. and Navajas, M. (1999) Mechanisms of insecticide resistance in the aphid *Nasonovia ribisnigri* (Mosley) (Homoptera: Aphididae) from France. *Insect Biochemistry and Molecular Biology* 29, 385–391.

Rui, C.H., Zhoa, Y.Q., Fan, X.I. and Wei, C. (1996) Resistance development of *Lipaphis erysimi* Kaltenbach to fenvalerate and cross-resistance. *Acta Phytophylacica Sinica* 23, 258–262.

Saito, T. (1989) Insecticide resistance of the cotton aphid, *Aphis gossypii* Glover (Homoptera: Aphididae). I. Susceptibility to several insecticides and esterase activity of field populations collected in Shizuoka Prefecture. *Japanese Journal of Applied Entomology and Zoology* 33, 204–210.

Saito, T. and Hama, H. (2000) Carboxylesterase isozymes responsible for organophosphate resistance in the cotton aphid, *Aphis gossypii* Glover (Homoptera: Aphididae). *Applied Entomology and Zoology* 35, 171–175.

Saito, T., Hama, H. and Suzuki, K. (1995) Insecticide resistance in clones of the cotton aphid, *Aphis gossypii* Glover (Homoptera: Aphididae), and synergistic effect of esterase and mixed-function oxidase inhibitors. *Japanese Journal of Applied Entomology and Zoology* 39, 151–158.

Schaub, L., Alame, M., Grandchamp, K. and Bloesch, B. (2001) Laboratory evaluation of the efficacy of neonicotinoids against rosy apple aphids. *Revue Suisse de Viticulture, Arboriculture et Horticulture* 33, 109–111.

Sekun, N.P., Kudel, K.A., Satsyuk, O.S., Mel'nikova, G.L., Zil'bermints, I.V. and Zhuravleva, L.M. (1990) A study on potential resistance of cereal aphids to insecticides. [In Russian.] *Zashchita Rastenii, Kiev* 37, 49–53.

Shufran, R.A., Wilde, G.E. and Sloderbeck, P.E. (1996) Description of three isozyme polymorphisms associated with insecticide resistance in greenbug (Homoptera: Aphididae) populations. *Journal of Economic Entomology* 89, 46–50.

Shufran, R.A., Wilde, G.E. and Sloderbeck, P.E. (1997a) Response of three greenbug (Homoptera: Aphididae) strains to five organophosphorous and two carbamate insecticides. *Journal of Economic Entomology* 90, 283–286.

Shufran, R.A., Wilde, G.E., Sloderbeck, P.E. and Morrison, W.P. (1997b) Occurrence of insecticide resistant greenbugs (Homoptera: Aphididae) in Kansas, Texas, Oklahoma, and Colorado and suggestions for management. *Journal of Economic Entomology* 90, 1106–1116.

Shufran, R.A., Wilde, G.E. and Sloderbeck, P.E. (1997c) Life history study of insecticide resistant and susceptible greenbug (Homoptera: Aphididae) strains. *Journal of Economic Entomology* 90, 1577–1583.

Siegfried, B.D. and Ono, M. (1995) Insecticide resistance mechanisms of the greenbug, *Schizaphis graminum* (Homoptera: Aphididae). *Resistant Pest Management* 7, 33–34.

Siegfried, B.D. and Zera, A.J. (1994) Partial purification and characterisation of a greenbug (Homoptera: Aphididae) esterase associated with resistance to parathion. *Pesticide Biochemistry and Physiology* 49, 132–137.

Siegfried, B.D., Swanson, J.J. and Devonshire, A.L. (1997) Immunological detection of greenbug (*Schizaphis graminum*) esterases associated with resistance to organophosphate insecticides. *Pesticide Biochemistry and Physiology* 57, 165–170.

Silver, A.R.J., van Emden, H.F. and Battersby, M.A. (1995) Biochemical mechanism of resistance to pirimicarb in two glasshouse clones of *Aphis gossypii*. *Pesticide Science* 43, 21–29.

Sloderbeck, P.E., Chowdhury, M.A., Depew, L.J. and Buschman, L.L. (1991) Greenbug (Homoptera, Aphididae) resistance to parathion and chlorpyrifos-methyl. *Journal of the Kansas Entomological Society* 64, 1–4.

Smirnova, A.A. and Ivanova, G.P. (1974) cited in Georgiou, G.P. and Lagunes-Tejeda (1991) *The Occurrence of Resistance to Pesticides in Arthropods: An Index of Cases Reported Through 1989*. FAO, Italy, 318pp.

Smith, I.M. (1999) Glasshouse quarantine pests for the EPPO region and measures recommended by EPPO and the EU to prevent their spread. *Bulletin OEPP* 29, 23–27.

Song, S.S., Oh, H.K. and Motoyama, N. (1995) Insecticide resistance mechanism in the spiraea aphid, *Aphis citricola* (van der Goot). *Korean Journal of Applied Entomology* 34, 89–94.

Stern, V.M. and Reynolds, H.T. (1958) Resistance of the spotted alfalfa aphid to certain organophoshorous insecticides in southern California. *Journal of Economic Entomology* 51, 312–316.

Stone, B.S., Shufran, R.A. and Wilde, G.E. (2000) Life history study of multiple clones of insecticide resistant and susceptible greenbug *Schizaphis graminum* (Homoptera : Aphididae). *Journal of Economic Entomology* 93, 971–974.

Stribley, M.F., Moores, G.D., Devonshire, A.L. and Sawicki, R.M. (1983) Application of the FAO-recommended method for detecting insecticide resistance in *Aphis fabae* Scopoli, *Sitobion avenae* (F.), *Metopolophium dirhodum* (Walker) and *Rhopalosiphum padi* (L.) (Hemiptera: Aphididae). *Bulletin of Entomological Research* 73, 107–115.

Sun, Y., Feng, G., Yuan, J., Zhu, P. and Gong, K. (1970) Biochemical mechanism of resistance of cotton aphids to organophosphorous insecticides. *Acta Entomologica Sinica* 30, 13–20.

Suzuki, K. and Hama, H. (1998) Carboxylesterase of the cotton aphid, *Aphis gossypii* Glover (Homoptera: Aphididae), isoelectric point variants in an organophosphorus insecticide-resistant clone. *Applied Entomology and Zoology* 33, 11–20.

Takada, H. and Murakami, Y. (1988) Esterase variation and insecticide resistance in Japanese *Aphis gossypii*. *Entomologia Experimentalis et Applicata* 48, 37–41.

Tang, Z.H. (1992) Insecticide resistance and countermeasures for cotton pests in China. *Resistant Pest Management* 4, 9–12.

Tang, Z.H., Gong, K.Y. and You, Z.P. (1988) Present status and countermeasures of insecticide resistance to agricultural pests in China. *Pesticide Science* 23, 189–198.

Udeaan, A.S. and Narang, D.D. (1988) A survey of mustard aphid, *Lipaphis erysimi* (Kalt.) populations for resistance to insecticides in Punjab. *Journal of Research Punjab Agricultural University* 25, 77–80.

Unruh, T., Night, A. and Bush, M.R. (1996) Green peach aphid (Homoptera: Aphididae) resistance to endosulfan in peach and nectarine orchards in Washington State. *Journal of Economic Entomology* 89, 1067–1073.

Vierbergen, G. (2001) *Thrips palmi*: pathways and possibilities for spread. *Bulletin OEPP* 31, 169–171.

Villatte, F., Auge, D., Touton, P., Delorme, R. and Fournier, D. (1999) Negative cross-insensitivity in insecticide-resistant cotton aphid *Aphis gossypii* Glover. *Pesticide Biochemistry and Physiology* 65, 55–61.

Vostrel, J. (1995) Effectiveness of insecticides against the hop aphid (*Phorodon humuli* Schrank) in laboratory conditions. *Rostlinna Vyroba* 41, 375–378.

Wachendorff, U. and Zoebelein, G. (1988) Diagnosis of insecticide resistance in *Phorodon humuli* (Homoptera: Aphididae). *Entomologia Generalis* 13, 145–155.

Walters, P.J. and Forrester, N. (1979) Resistant lucerne aphids at Tamworth. *Agricultural Gazette of New South Wales* 90, 5–7.

Wang, K.Y., Liu, T.X., Jiang, X.Y. and Yi, M.Q. (2001) Cross-resistance of *Aphis gossypii* to selected insecticides on cotton and cucumber. *Phytoparasitica* 29, 393–399.

Wei, C., Huang, S.N., Fan, X.L., Sun, X.P., Wang, W.L., Liu, Z.W. and Chen, G.Q. (1988) A study on resistance of grain aphid, *Sitobion avenae* Fab. to pesticides. *Acta Entomologica Sinica* 31, 148–156.

Wiesmann, R. (1955) Der heutige Stand des Insektizid-Resistenz Problem. *Mitteilungen Biologische Bundesanstalt für Land- und Forstwirtschaft Berlin-Dahlem* 83, 17–37.

Wolff, M.A., Abdel-Aal, Y.A.I., Goh, D.K.S., Lampert, E.P. and Roe, M. (1994) Organophosphate resistance in the tobacco aphid (Homoptera: Aphididae): purification and characterisation of a resistance-associated esterase. *Journal of Economic Entomology* 87, 1157–1164.

Wyatt, I.J. and White, P.F. (1977) Simple estimation of intrinsic increase rates for aphids and tetranychid mites. *Journal of Applied Ecology* 14, 757–766.

Zhu, K.Y. and Gao, J.R. (1999) Elevated esterases exhibiting arylesterase-like activity in an organophosphate-resistant clone of the greenbug, *Schizaphis graminum* (Homoptera: Aphididae). *Pesticide Biochemistry and Physiology* 67, 155–167.

Zhu, K.Y. and He, F.Q. (2000) Elevated esterases exhibiting arylesterase-like characteristics in an organophosphate-resistant clone of the greenbug, *Schizaphis graminum* (Homoptera: Aphididae). *Pesticide Biochemistry and Physiology* 67, 155–167.

Zhu, K., Gao, J. and Starkey, S.R. (2000) Organophosphate resistance mediated by alterations of acetylcholinesterase in a resistant clone of the greenbug, *Schizaphis graminum* (Homoptera: Aphididae). *Pesticide Biochemistry and Physiology* 68, 138–147.

11 Coping with Stress

Jeffrey S. Bale, Katherine L. Ponder and Jeremy Pritchard

School of Biosciences, University of Birmingham, Edgbaston, Birmingham, B15 2TT, UK

Introduction

Population growth of plants and animals is limited by the environmental stresses to which they are exposed. Aphids are particularly well adapted to coping with changes in their environment. They have highly flexible life cycles, which enable individual species to cope with a wide range of biotic and abiotic stresses through appropriate behavioural, physiological, and biochemical responses. Their high reproductive rates and telescoping of generations enable such responses to be transmitted rapidly through populations. As with other animals, aphid population growth is governed to a large extent by the quality of the food supply, and thus environmental stresses affect the insects both directly and indirectly through their effects on host plants. Many stresses will therefore only affect the aphid if the stress alters the plant. Stresses such as drought and temperature can have profound effects on the biochemical composition of the plant. Such changes may affect an aphid herbivore in three ways: by altering the suitability of the plant as diet; by changing the amount of resource the plant allocates to defence; or by modifying the form of any chemical cues used by the aphid to locate its feeding site in the phloem. The phloem tissue is dominated by companion cells, phloem parenchyma

and the sieve element (van Bel, 2003). Phloem refers to the whole tissue, while sieve element refers to the specific feeding site of the aphid.

In the context of this chapter, stress is defined as the effect of various environmental factors, which can lower the overall performance of an aphid in terms of population growth rate (Awmack and Leather, Chapter 6 this volume). The most common stresses experienced by aphids are poor nutritional quality and temperature extremes. Other limits to population growth such as predators and competition are dealt with elsewhere (Völkl *et al.*, Chapter 8 this volume and Kindlmann *et al.*, Chapter 12 this volume).

Plant Factors

The interaction between an aphid and its host plant has two main phases – phloem location and sieve element acceptance. Both can be affected by imposition of stress on the plant. Physical factors (surface waxes, leaf thickness, and toughness) or the presence of xenobiotics (such as glycocides) can affect the efficiency of phloem location. The proportion of resources allocated by the plant to these processes will be altered by stress and so, therefore, will aphid performance. Sieve element acceptance is modulated by defence responses within the sap; aphid saliva is one component of the mechanism by which aphids can overcome some of these defence responses. The outcome of the plant–aphid interaction will be affected by alterations in the allocation of resources to these defence processes when a plant is stressed, for example, by drought or altered nitrogen nutrition.

Nutritional stress

A major factor influencing the growth and survival of aphid populations is the nutritional quality of their host plants (Douglas and van Emden, Chapter 5 this volume). Plant composition affects aphid feeding in two ways: firstly, if components of the diet provide cues that aphids follow to locate their food source in the sieve element; and secondly, by the composition of the sieve element itself, which will influence both the volume of sap the aphid ingests and the nutritional value of this sap, to drive aphid growth and reproduction (Sandström and Pettersson (1994); Pettersson et al., Chapter 4 this volume). Currently, very little is known about the cues that aphids follow to locate the phloem or how they alter their feeding behaviour as sap composition is altered. Secondary plant compounds have received some attention (e.g. Tosh et al., 2003) but no individual cues have been identified; indeed, attempts to correlate specific metabolites with feeding behaviour have yielded contradictory results (Costa-Arbulu et al., 2001). Only sucrose and amino acid levels have received significant attention (see Douglas and van Emden, Chapter 5 this volume) but very little is known about the detailed metabolic composition of the sieve element or its regulation, and both of these dietary components are modified under drought. The recent acceleration in genomic and metabolic understanding of both plants (in particular the phloem) and aphids will increase our understanding of this area over the next few years. This area is addressed later in this chapter.

Although the proportion of nitrogen in sieve-element sap is lower than the proportion in leaf tissue as a whole (Table 11.1),

Table 11.1. Nitrogen content (% dry weight) of animals and different plant parts. Xylem and phloem sap values are expressed as nitrogen weight/volume (adapted from Mattson, 1980).

	Nitrogen content (%)
Animals	10
Seeds	0.5–5
Leaves (angiosperms)	1–5
Leaves (grasses)	0.5–3
Phloem sap	0.005–0.5
Xylem sap	0.0002–0.1

nitrogen in the sieve element occurs predominantly as amino acids and is therefore easily assimilated by the aphid. However, sieve-element sap also contains high concentrations of sugar, resulting in a high C:N ratio and high osmotic pressure. It is this high C:N ratio that is one of the problems with sieve element as a food source (see also Douglas and van Emden, Chapter 5 this volume).

The majority of information available on the influence of individual nutritional components, such as the various amino acids, on aphid performance has been gained via studies using artificial diets (e.g. Mittler, 1967; Srivastava and Auclair, 1975; Prosser and Douglas, 1992). These offer a convenient way to assess the impact of various dietary constituents on aphid performance and feeding behaviour, but results from such studies do not transfer readily to the more complex interactions between aphids and their host plants (discussed by Douglas and van Emden, Chapter 5 this volume).

Factors affecting phloem nitrogen

The amino acid composition of sieve-element sap varies around the plant and is subject to alteration by abiotic stresses such as drought and temperature. It is therefore a variable quality food source for aphids. Much of the variation is in response to fixed environmental changes such as season, allowing the aphid to predict the changes from reliable cues such as day length. However, aphids must also adapt to shorter-term, unpredictable changes in host quality, for example those caused by drought or disease. Many of these involve changes in nitrogen, but also in other phloem solutes such as sucrose.

Amino acids in the sieve-element sap of a variety of species show consistent changes with season (Weibull, 1987; Douglas, 1993) and plant age (Pate *et al.*, 1965; Weibull, 1987) as the positions of sources and sinks alter. Many plants, in particular annuals, accelerate senescence when stressed, with consequent effects on phloem amino acids, and therefore the diet available to aphids. Phloem amino acid concentrations were high in growing tissues, lower in mature and high in senescing leaves (Merritt, 1996), demonstrating temporal and spatial variation in diet.

Environmental factors

Plants acquire carbon *via* photosynthesis in the leaves, and mineral nutrients and water from uptake by the roots. These resources are apportioned to different plant parts, where they are allocated to biochemical fates associated with growth, development, maintenance, and defence (Jones and Coleman, 1991). The transport of photoassimilates around the plant is highly dynamic and can respond rapidly to the nutritional status of the plant (Pate and Atkins, 1983; Hayashi *et al.*, 1993). This enables plants to maintain a balance between rates of carbon fixation and export in source leaves, and carbon import and use in sinks (Geiger and Servaites, 1991). After imposition of stress, partitioning of carbon and nitrogen is altered in ways that help maintain their balance around some optimum value. For example, when exposed to a nitrogen or water deficit, plants usually allocate more resources to root growth (Bray, 1983; Hunt and Nicholls, 1986), which increases the surface area for absorption of water and nutrients. In this way, the effects of environmental stresses experienced by the plant can become buffered for the aphid, at least in the short term.

Chemical and physical properties of plants are influenced by their growing conditions, and this can affect the nitrogen composition of sieve-element sap. For example, plants grown with reduced nitrogen have lower concentrations of amino acids in the phloem compared to those grown with higher nitrogen concentrations (Weibull *et al.*, 1986; Ponder *et al.*, 2001). Conversely, sieve-element sap from drought-stressed plants has higher amino acid concentrations (Girousse and Bournoville, 1994; Hale *et al.*, 2003). Muskmelon (*Cucurbita pepo*) grown at reduced temperature has lower levels of sieve-element amino acids (Mitchell and Madore, 1992). However, under nitrogen and drought stress, aphid performance on barley is reduced (Fig. 11.1). In a similar comparison with barley and three species of native British grasses grown under drought stress, amino acid concentration increased in all plant species, but whereas aphid performance was reduced on three species, it was unaffected on *Arrhenatherum elatius* (Fig. 11.2, Hale *et al.*, 2003), suggesting that factors other than total sieve-element amino acid composition can affect aphid performance.

Substantial differences in the spectrum of amino acids in sieve-element sap are recorded between plant species or cultivars. Artificial diet studies have shown that amino acid *quality* is more important than *quantity*. It is currently unknown whether sieve-element sap from a stressed plant will have a different spectrum of amino acids to that from an unstressed plant of the same species. Where sieve-element amino acid composition or concentration is affected by growing conditions, it is the concentrations of non-essential rather than essential amino acids that are most affected (Weibull, 1987; Ponder *et al.*, 2000).

Non-nutritional plant factors

Secondary plant compounds are important components of plant defence systems against herbivores. These chemicals can affect aphids both when locating the phloem and in the subsequent acceptance and ingestion of sieve-element sap (Eschrich, 1970; Cole, 1994; Mayoral *et al.*, 1996; Merritt, 1996). Secondary compounds can be reduced when resource availability becomes restricted; for example, in spruce trees, experimentally induced increases in growth were accompanied by decreases in many secondary compounds

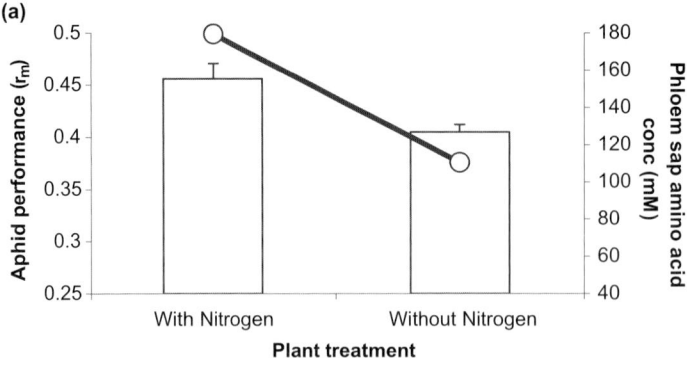

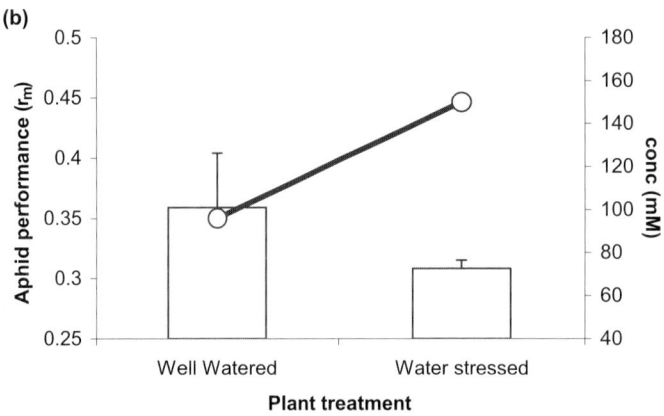

Fig. 11.1. Performance of *Rhopalosiphum padi*, measured by intrinsic rate of increase (r_m – black line) and the concentration of amino acids (open bars) in phloem sap of barley (*Hordeum vulgare*) on plants grown in different treatments. (a) Plants grown in hydroponic nutrient solutions containing 8 mM nitrogen (NH_4^+ and NO_3^-) or without nitrogen; (b) plants grown in a soil and vermiculite mix and either watered daily (well watered) or weekly (water stressed) (adapted from Ponder, 2000).

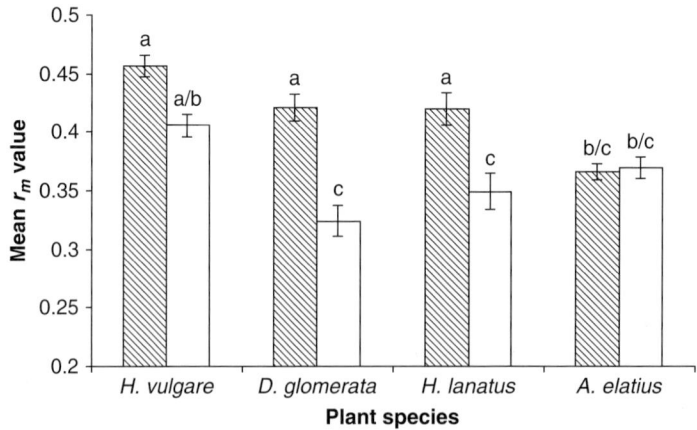

Fig. 11.2. Intrinsic rates of increase of *Rhopalosiphum padi* reared on four plant species (*Hordeum vulgare, Dactylis glomerata, Holcus lanatus, Arrhenatherum elatius*) grown under well-watered (shaded bars) and drought-stressed (open bars) conditions: means ± SE. Letters above columns indicate significant differences (Tukey-Kramer comparisons of means, $P < 0.05$). Sample sizes varied from 9 to 18 (modified from Hale et al., 2003).

(McKinnon *et al.*, 1998). Many studies have demonstrated alteration of plant secondary compounds following stress, such as the increase in leaf tannins in birch grown under elevated CO_2 (Lavola *et al.*, 2000). Variation in nitrogen nutrition, light levels, and elevated CO_2 can all affect levels of secondary compounds (Koricheva *et al.*, 1998). At the cellular level, secondary compounds in phloem can be affected by stress; under drought, rutin (a herbivore deterrent) decreased, while cyanide compounds increased in concentration (Calatayud *et al.*, 1994).

Hydroxamic acids (Hx) increased in maize leaves during drought (Richardson and Bacon, 1993). These compounds have been implicated in the resistance of some cereals to aphids by affecting their growth rates (Niemeyer and Perez, 1987) and disrupting normal probing behaviour. The feeding behaviour of five cereal aphids was monitored on wheat with varying levels of Hx. Only *Rhopalosiphum maidis* (corn leaf aphid) showed no disruption of feeding with increased levels of Hx, and it was suggested that this could be due to a feeding strategy that avoided contact with the compounds by decreasing the number of cellular punctures during stylet penetration (Givovich and Niemeyer, 1995). In barley, gramine occurs in epidermal and mesophyll tissues (Argandona *et al.*, 1987) and can affect the feeding behaviour of *Rhopalosiphum padi* (bird cherry–oat aphid) and *Schizaphis graminum* (greenbug) by increasing the time taken to reach the sieve elements. As a result, survival and reproductive output are reduced (Zuniga and Corcuera, 1986; Zuniga *et al.*, 1988). Gramine has been shown to accumulate following exposure to temperature stress (Hanson *et al.*, 1983).

Other secondary compounds have altered composition in stressed plants. Glucosinolates increase in Brassicas at elevated temperature (Pereira *et al.*, 2002) and following drought (Bouchereau *et al.*, 1996). Glucosinolates interfere with the aphid's probing and feeding behaviours by increasing the number of probes that fail to reach the sieve elements and the time required by aphids to commence sap ingestion (Dreyer *et al.*, 1985). The result is short or interrupted feeding bouts in the sieve element and a reduction in aphid development and reproduction (Cole, 1997a,b). However, the specialist aphid crucifer feeder, *Brevicoryne brassicae* (cabbage aphid), will feed from non-host plants to which the glucosinolate sinigrin has been applied (Wensler, 1962). Such apparently toxic chemicals are important in host recognition for some specialist aphid species, and they are subject to alteration by stress.

The ability of aphids to access and assimilate sieve-element sap, which could be related to non-nutritional components of the sap, is also crucial. Sieve-element sap contains a range of defence related proteins (Barnes *et al.*, 2004; Walz *et al.*, 2004). While little is currently known about their regulation by environmental factors, there is some evidence that some phloem defence compounds can be induced by drought (Walz *et al.*, 2004) and other stresses (Taylor *et al.*, 1996). Increased duration of watery saliva injections into sieve elements during aphid feeding have been correlated with overcoming phloem-specific defences (Caillaud *et al.*, 1995; Prado and Tjallingii, 1997; Ponder *et al.*, 2001), suggesting a role for these salivary secretions in counteracting poor sieve-element sap quality. Recently, sieve-element proteins of bean have been shown to change their conformation in the presence of elevated calcium and, as a consequence, have been implicated in the sealing of sieve elements since wounding is often accompanied by an influx of calcium ions (Knoblauch *et al.*, 2001).

While stress can affect the chemistry of the whole plant and phloem, it will also affect its physical parameters. The rise in osmotic pressure following drought increases sap viscosity, which could heighten resistance to sap ingestion. Sieve-element turgor pressure may also be reduced and this may make feeding easier or more difficult; however, this effect may not be significant as aphids are able to feed at pressures ranging from negative to fully turgid. A temperature change from 30 to 0°C will increase the viscosity of typical sieve-element sap so that flux would be reduced about 2.5 times for a similar pressure gradient. Again, this effect is unlikely to be important as aphid physiology probably has evolved to deal with these fluctuations and, in any case, the associated chemical changes probably have a larger effect.

How aphids cope with nutritional stress

The nutrient stress represented by altered nutrient composition can be dealt with by aphids in a number of ways, including compensatory feeding, altering the quality of the available diet, movement to new feeding sites, modifying reproductive output, and utilizing their bacterial endosymbionts.

Compensatory feeding

The relatively low concentrations of nitrogen in the sieve element require aphids to process large quantities of sap and to use the nitrogen content very efficiently. Aphids can respond to differing concentrations and profiles of amino acids in diets by adjusting the length and rate of ingestion and the number of probes they make (Mittler, 1958, 1967; Rahbé *et al.*, 1988; Montllor *et al.*, 1990; Prosser *et al.*, 1992). The mechanisms of regulation of food intake are not certain. *Acyrthosiphon pisum* (pea aphid) fed on various artificial diets appear to regulate sucrose intake by consumption (behaviourally), whilst amino acids are regulated post-ingestion (Abisgold *et al.*, 1994). Alteration in sieve-element amino acid composition in *Arabidopsis* modified the honeydew composition of *Myzus persicae* (peach–potato aphid), suggesting that metabolism was altered as a result of the changed diet (Zhu *et al.*, 2005). Reduced aphid performance on diets of different amino acid compositions was ascribed to a reduced aphid-feeding rate (Karley *et al.*, 2002).

Moving to a new feeding site

Drought has quantitative and qualitative effects on resource partitioning around plants, reducing leaf areas and slowing development, in particular that of leaves (Asch *et al.*, 2005). Growth itself imposes allocation stresses on the plant. On rapidly growing plants, aphids move regularly to new feeding sites (Hodgson, 1978). Inter-plant dispersal of aphids maximizes food intake and

increases the probability that nymphs are deposited on nutritionally favourable leaves (Jepson, 1983). This dispersal is determined largely by the aphids' response to abiotic conditions (Mann *et al.*, 1995) and to biotic factors such as plant growth stage, plant stress, aphid developmental stage, and population density (Jepson, 1983; Harrington and Taylor, 1990; Honek, 1991). Restlessness and dispersal by aphids have been correlated with cessation of plant growth (Hodgson, 1991) so that stress-induced cessation of growth reduces the opportunity to exploit new feeding sites, compounding any movement induced by lower levels of phagostimulants and/or nutrients such as amino acids (Harrewijn, 1978; Muller *et al.*, 2001).

Development of winged aphids has been correlated with increased tactile stimulation experienced by their mothers (Lees, 1967; Williams and Dixon, Chapter 3 this volume). On a poor quality host, increased movement of aphids as they search for a suitable feeding location results in increased tactile contact between individuals, and this has been suggested as a reason for increased production of winged forms on these plants. However, Chambers *et al.* (1985) provided evidence that the effects of crowding could be mediated through the host plant rather than by direct contact between aphids. The ratio of amino acids to sucrose and composition of mineral nutrients in the diet can exert a significant effect on wing determination (Raccah *et al.*, 1971). For example, omission of the amino acids methionine, histidine, and isoleucine resulted in an increase in the proportion of wingless aphids. On the other hand, lack of tyrosine and tryptophan derivatives in the diet has been shown to stimulate wing development (Harrewijn, 1976). Deteriorating plant quality, perhaps mediated by plant amino acid content, can result in a switch to alate production (Watt and Dixon, 1981), and early studies often reported that mature leaves (poor quality host) induce the development of more winged aphids than do seedlings (good quality host) (Johnson, 1966; Sutherland, 1969; Schaefers and Judge, 1971). The production of numerous winged individuals by aphids crowded on mature plants was rapidly halted by transfer to seedlings

(Sutherland, 1970). The fact that a rapid response was observed suggests that some factor encountered by aphids after transfer to the seedlings (or a behavioural change) prevented wing formation. This could be the presence or absence of a specific dietary or gustatory component acting on the aphid's neuroendocrine system (Kawada, 1987).

Altering reproductive output

When subjected to nutritional stresses, aphids show very flexible reproductive strategies that can anticipate predictable seasonal changes in habitat quality (Wellings *et al.*, 1980). For example, the number and rate of growth of embryos in each ovariole can be reduced on poor quality hosts (Wiktelius and Chiverton, 1985; Grüber and Dixon, 1988; Stadler, 1995), thereby re-allocating resources committed to many offspring, to produce fewer healthier progeny. *Megoura viciae* (vetch aphid) reared on poor quality hosts give birth to proportionally fewer offspring with a higher ovariole number than those reared on good quality plants (Walters *et al.*, 1988). Similarly, *Aphis fabae* (black bean aphid) reared throughout their larval development on poor quality hosts have a reduced fecundity (Leather *et al.*, 1983). Compensation for periods of poor nutrition occurs to varying extents in a number of other aphid species. *Sitobion avenae* (grain aphid) transferred from resistant to susceptible wheat are able to compensate for their poor nymphal growth, mainly through additional embryo development and an increase in the number of matured embryos in the first 10 days of their adult life. However, when transferred from susceptible to resistant wheat, most die within the first 10 days (Caillaud *et al.*, 1994). On a shorter timescale, van Emden (1977) found that *M. persicae* is unable to compensate for a period of poor nutrition lasting longer than 4 h. Starvation or poor nutrition for periods of less than this are compensated for, and relative growth rates are similar to those of aphids maintained on good hosts.

Enlisting symbionts

Bacterial endosymbionts contained in the aphid gut synthesize essential amino acids that are in short supply in the diet from surplus non-essential amino acids. The presence of endosymbionts is crucial to the survival of aphids on the unbalanced amino acid spectrum provided by sieve-element sap (see also Douglas and van Emden, Chapter 5 this volume). Symbionts reduce the impact on the aphid of a poor quality diet caused by stressed host plants, and can thus buffer the indirect plant-derived effect of environmental stress on the aphid. The aphid–symbiont–diet relationship is reviewed in detail by Douglas (2003) (see also Douglas and van Emden, Chapter 5 this volume).

Temperature Stresses

Temperature is the main abiotic factor affecting aphid bionomics (Harrington *et al.*, 1995). As with all other insects, there are three main effects of temperature on aphids, although these are closely interrelated in terms of their overall effect on performance. First, temperature affects rate-based processes such as development, reproduction, and movement. Second, acute or chronic exposure to low or high lethal temperatures causes mortality. Third, there is a range of sub-lethal (delayed) effects that can produce morphological abnormalities or modify the 'normal' response to temperatures within the so-called favourable zone. In recent years, there has been considerable interest in the effects of low temperatures on winter survival and the subsequent spring abundance and outbreaks of pest species, including the possible effects of climate warming (Bale, 1999).

Effects of temperature on development, reproduction and movement

For a relatively small taxonomic group with a predominantly temperate distribution, the lower developmental thresholds of different aphid species vary considerably, ranging from −2.2 and −3.6°C for *Metopolophium dirhodum* (rose–grain aphid) and *S. avenae*, respectively (Carter *et al.*, 1982; Zhou *et al.*, 1989) to 7.1°C for *B. brassicae* (Campbell *et al.*, 1974). However, given that temperatures around 0°C are stressful, and over time become lethal, estimates of sub-zero developmental thresholds should be treated with caution. Comparisons of the developmental threshold between species are also difficult for several reasons, including differences in the analytical methods used to estimate the threshold (Lamb, 1992; Hart *et al.*, 1997), different responses to constant and fluctuating temperature regimes (Michels and Behle, 1989), and variation attributable to regionally adapted clones or the use of different host plants. The accuracy of an estimated developmental threshold can be tested by maintaining aphids at and around the threshold and monitoring their ability to develop, but this is seldom done.

Aphid development and reproduction increase with temperature to a maximum between 20 and 25°C, depending on the species and location. The number of day-degrees above the developmental threshold required to complete one generation (from birth to first reproduction) varies from around 90 (*S. graminum*) to 250 (*Drepanosiphum platanoidis –* sycamore aphid) (see Harrington *et al.*, 1995 and references therein). In a winter field study with *S. avenae*, Knight and Bale (1987) observed both development and reproduction to be positively correlated with cumulative day-degrees above a threshold of 3°C.

Aphids undertake two forms of movement, walking and flight, both of which are influenced by temperature, among other factors. In recent studies on the walking speeds of *M. persicae* and *B. brassicae* (E. Hawkins and R. Harrington, unpublished results), the lower threshold for activity was around 2°C, with speed increasing up to 30°C and then declining. The importance of activity at lower temperatures was illustrated in a study of *M. persicae* on winter cabbage (Harrington and Cheng, 1984) in which aphids were observed to move from senescing leaves to younger leaves in sequence, needing to make such moves before the occupied leaf falls from the plant. Aphids that fail to transfer to a younger leaf can become trapped on the soil surface during periods of temperature below the activity threshold and may then die of starvation or drown (Harrington and Taylor, 1990).

Temperature thresholds for aphid flight vary between and within species, and with season and region. Flight thresholds for *R. padi* vary during the year, from 16–17°C in spring, to 13–14°C in summer and 9–10°C in autumn (Wiktelius, 1981); variation in these thresholds is attributed to the different functions of the various morphs examined, incorporating physiological as well as morphological polyphenism. Varying thresholds have also been recorded for different phases of flight activity in *A. fabae*, including 17°C for take-off, 15°C for sustained upward flight, 13°C for horizontal flight, and 6.5°C for wing beating (Johnson and Taylor, 1957; Cockbain, 1961). Upper flight thresholds are generally around 30°C, and can be over 40°C in *S. graminum*.

Cold stress

Components of cold stress

Low temperature is a potentially lethal stressor for all insects. Cold can kill insects in different ways, and the level of cold stress that can be tolerated varies greatly between species. In addition, cold stress can affect the aphid by altering plant composition, thus affecting sieve element localization and subsequent ingestion. Since sieve-element sap is a concentrated solution dominated by sucrose, low temperatures will significantly increase sieve-element sap viscosity, thus altering the rate of sap uptake and the associated energy demands. Despite the importance of diet on aphid performance, very little is known about how low temperatures affect phloem nutrient composition or the defence mechanisms in the sieve element.

From a biochemical and physiological perspective, insects are regarded generally as adopting one of two main overwintering strategies, freeze tolerance and freeze avoidance by supercooling. The roles of ice nucleating agents, antifreeze proteins and polyols in these two strategies are now well understood (at least at the cellular level), and increasing knowledge of their functional relationships has tended to strengthen the case for the importance of freeze tolerance and supercooling in insect overwintering.

On a worldwide scale, few insects are freeze tolerant, and few die only when they freeze (i.e. are freeze avoiding, Bale, 1987, 1996). Although the occurrence of mortality above the freezing temperature (or supercooling point, SCP) is not a recent discovery (see Salt, 1961), it is certainly the case that the importance of such 'pre-freeze mortality' as the dominant cause of low temperature death in the majority of insects has only been recognized much more recently, and studies on aphids (Knight *et al.*, 1986; Bale *et al.*, 1988) were pioneering in this respect. Bale (1996) suggested that all insects could be classified into one of five main groups (freeze tolerant, freeze avoiding, chill tolerant, chill susceptible, opportunistic survival), in which overwintering aphids were cited as the 'type' example for the chill susceptible class.

The cold tolerance of any species can be characterized by a combination of laboratory assessments. Once the SCP is known, it will be clear whether a species can survive the freezing event or is killed at or above the SCP. If 'pre-freeze' mortality is evident, the LTemp$_{50}$ (temperature in an exposure of fixed duration at which 50% of the sample is killed), or the LTime$_{50}$ (length of time at which 50% of a sample is killed in an exposure at a selected low or sub-zero temperature) can be calculated. These laboratory exposures can underestimate the duration of winter survival in natural environments, where insects experience variable conditions, including periods of 'favourable' temperatures, which can allow feeding and recovery from previous cold stress.

Lethal effects of low temperatures on aphids

Aphid eggs are much more cold hardy than the active stages. The freezing temperatures of eggs of temperate species such as *Rhopalosiphum insertum* (apple–grass aphid) and *R. padi* and polar species such as *Acyrthosiphon svalbardicum* and *Acyrthosiphon brevicorne* are similar, and also show an acclimation response, with a decrease in the mean SCP from around −35°C soon after oviposition to −40°C after a period of natural (field) or laboratory acclimation (James and Luff, 1982;

Strathdee *et al.*, 1995a). In constant exposures at −10 and −30°C for a period of 1 month, the survival of *R. padi* eggs decreased to 93 and 35%, respectively, whereas 80% of *A. svalbardicum* eggs survived 1 month at −30°C, indicating a greater level of cold tolerance in the Arctic species. In both species, egg cold hardiness is considerably greater than that required for survival in their respective winter environments (Strathdee *et al.*, 1995a; Bale, 1999). In effect, egg overwintering, at least in temperate climates, is largely independent of annual variations in winter minimum temperatures.

A range of studies on anholocyclic clones of *R. padi*, *S. avenae*, and *M. persicae* in both the field and laboratory has revealed a highly consistent pattern of cold tolerance among these species (see Bale, 1999 for a review). The SCPs of both nymphs and adults are consistently below −20°C, with little variation. However, substantial mortality occurs at much higher temperatures, even with exposures of only a few minutes. Typically, the first deaths occur at around −5°C and there are few survivors at −15 to −17.5°C, depending on the species; lethal effects would occur at higher temperatures in longer exposures. In an applied context, these data explain the relationship between winter temperatures and the timing of the spring migration and summer abundance of species such as *S. avenae* and *M. persicae* (Walters and Dewar, 1986; Bale *et al.*, 1988). Within the range of temperatures that characterize the mildest (−5°C) to the most severe (−15°C) of UK winters, aphid mortality increases from around 0 to approaching 100% (Bale *et al.*, 1988).

Causes of low temperature mortality

The freezing temperatures of aphid eggs are very low (−40°C), and low sub-zero temperatures (−10 to −30°C) can be tolerated for long periods of time (1 month) without substantial mortality, suggesting that instantaneous freezing would be rare, even in polar regions, where winter snow cover acts as a thermal buffer. For aphid eggs and overwintering mobile forms, temperatures above those that cause freezing are therefore the dominant cause of death, although the molecular, biochemical, or physiological processes invol-ved in this mortality are poorly understood.

Overwintering aphids feed on host plants that will freeze at regular intervals, leading to the idea that there is an 'ice front' that moves from the plant *via* the aphid stylets across a 'phloem bridge' to freeze and kill the aphid. However, this hypothesis is not true (Bale *et al.*, 1988). Aphids feed in the phloem and if the phloem freezes, the plant will be killed; in effect, the aphid utilizes a food resource that must remain unfrozen for the host to sur-vive. Furthermore, Butts *et al.* (1997) showed that survival of *R. padi* is higher when aphids are cooled in feeding contact with a host plant (barley) that becomes frozen, than when cooled to the same temperature (–5 or –10°C) in isolation. Contact with the host affects the post-exposure behaviour of the aphids; those that are cooled on the host plant readily settle on new leaves, but those that are cooled in isolation are subsequently 'hyperactive' and display no attraction to food sources. The pos-sibility that aphid pre-freeze mortality might be attributable to the effects of sub-zero tem-peratures on essential intestinal symbionts has also been investigated (Parish and Bale, 1991). There is evidence of structural changes in the symbionts immediately after aphid exposure to –10 and –17.5°C, and also for a 'repair mechanism' (normal appearance of symbionts) operating over the next 24 h.

Sub-lethal effects of low temperature exposure

A series of studies on *S. avenae* (Parish and Bale, 1993) and *R. padi* (Hutchinson and Bale, 1994) showed that exposure of first-instar nymphs and adult aphids to temperatures between –5 and –10°C for up to 6 h exerts a range of deleterious effects, increasing devel-opment time from birth to adult, whilst reducing daily fecundity and longevity. The scale of these effects increases at the lower temperature and in longer exposures. In both species, there is substantial post-natal mortal-ity in the progeny of aphids exposed to sub-zero temperatures as pre-reproductive adults. Together with similar observations on *A. pisum* (Harrison and Barlow, 1972) and *Macrosiphum albifrons* (lupin aphid) (Carter

and Nichols, 1989), these studies suggest that as winter temperatures become progressively lower, not only are more aphids killed, but the survivors suffer increasing cryo-injuries, affecting the key processes (rate of develop-ment, reproduction, longevity) that deter-mine the population dynamics.

Low temperatures can also suppress wing development in aphids, although this effect may not be detrimental. In an anholo-cylcic clone of *S. avenae*, over 75% of adults were routinely alate, even when aphids were reared under uncrowded conditions. Expo-sure of first-instar nymphs to –5°C for 1 or 6 h increased the proportion developing as wingless adults from <20% in the control to >40% in the exposed groups. Suppression of the alate condition was most evident in nymphs exposed within 6 h of birth (Parish and Bale, 1990).

Low temperature thus affects many aspects of aphid biology that are regarded as the most important factors contributing to the success of these insects as a group: rapid development, high fecundity, dispersal, and migration. In some respects, however, the apparently deleterious effects of cold on these processes can be viewed as a necessary trade-off between resource investment in devel-opment, reproduction, and survival during winter to enable exploitation of favourable conditions at other times.

Coping with cold

The mechanisms by which aphids cope with the cold are wide ranging, including adap-tation of life cycles, overwintering as an egg or mobile stages, differential cold tolerance between age groups and in the birth sequ-ence, production of specialist overwintering morphs, acclimation, reproduction in win-ter, and morphological variation.

Adaptation of life cycles

The life cycles of temperate aphids are under strong environmental regulation. Sexual morphs (ovipara and male) can be induced or averted by different photoperiodic regimes, and the number of generations per unit time is governed largely by the effects of temperature on development and reproduction. A further important feature of holocyclic species and clones is the interval timer mechanism (Bonnemaison, 1951; Lees, 1960, 1961). At the time of egg hatch in spring, the fundatrix and some successive generations experience night lengths that are decreasing, but are still above the critical threshold that induces the sexual morphs in autumn. The interval timer effectively inhibits these early generations from responding to this sexual morph-inducing cue, ensuring that asexual reproduction occurs in early spring when there is an abundance of nutritionally favourable host plants.

Studies on the Arctic aphid *A. svalbardicum* (Strathdee *et al.*, 1993a,b) revealed a highly adapted life cycle in which the number of generations per year is genetically fixed, and the absence of an interval timer enabling the fundatrix to give birth directly to both sexual morphs (Fig. 11.3). During the short Arctic summer, the aphid is routinely able to complete two generations of mobile stages (fundatrix and sexual morphs), enabling the oviparae and males to mate and lay the overwintering eggs before conditions deteriorate towards winter. The fundatrix also produces some viviparae that give rise exclusively to a further generation of sexual morphs, which, if conditions permit, mature and lay additional overwintering eggs. After egg hatch, the aphid therefore has a minimum of two (fundatrix and sexual morphs) and a maximum of three (fundatrix, vivipara, sexual morphs) generations per year. As a further adaptation to the restricted Arctic summer season, the fundatrix always produces the full complement of viviparae before any sexual

morphs, thus providing the greatest opportunity for the second generation of sexual morphs (progeny of the viviparae) to mature and lay the additional overwintering eggs. Application of a simulated climate-warming regime (Strathdee *et al.*, 1993c, 1995b) suggests that in summers when the aphid completes three full generations (compared with two), there is a 11-fold increase in the number of overwintering eggs. In such an extreme climate, the unusual ability of the fundatrix to give birth directly to the sexual morphs ensures a risk-free egg production by midsummer, guaranteeing the survival of the species to the next year.

In similar studies on the closely related *A. brevicorne* in sub-Arctic Sweden, the life cycle was found to be 'intermediate' between the high-Arctic *A. svalbardicum* and the life cycles of temperate holocyclic species (Strathdee and Bale, 1996). As with *A. svalbardicum*, the fundatrix can produce both sexual morphs, but the proportion of oviparae increases in successive generations and the last generation consists entirely of sexual morphs. Along a south to north latitudinal gradient, there is therefore a progressive reduction in environmental sensitivity, with an advantageously genetically inflexible life cycle in the most extreme high-Arctic environment.

Overwintering of holocyclic and anholocyclic clones

At the level of the individual clone, there are both advantages and disadvantages to holocycly and anholocycly with respect to overwintering. As examples, the eggs of holocyclic aphids are highly cold hardy and

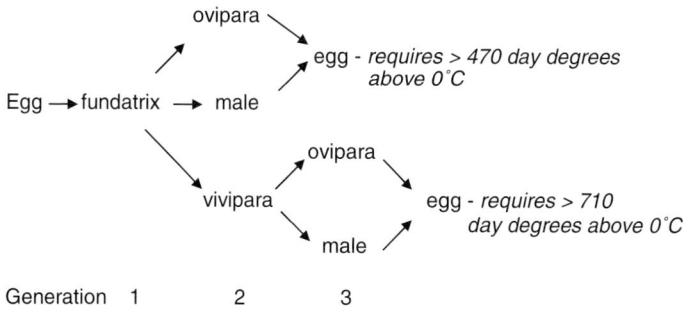

Fig. 11.3. Life cycle of the Arctic aphid *Acyrthosiphon svalbardicum*.

therefore largely immune from year-to-year variations in the severity of temperate winters. By contrast, overwintering viviparae are susceptible to moderately low temperatures (−5 to −10°C), hence the general abundance of these aphids in spring and summer is strongly determined by conditions in the preceding winter. The reciprocal outcomes are that viviparae that overwinter successfully are able to exploit the earliest and most nutritious growth of host plants in spring, at a time when the eggs of holocyclic clones may not have hatched. In terms of winter survival, a combination of holocycly and anholocycly therefore balances certainty with risk within a single species.

Differential cold hardiness between age groups and in the progeny sequence

Several studies have observed a greater level of 'inherent' cold tolerance in nymphal compared with adult aphids of anholocyclic clones; the LTemp$_{50}$ of first instar and adult *M. persicae* reared at 20°C were −8.1 and −6.9°C, respectively (Bale *et al.*, 1988; Clough *et al.*, 1990). The increased cold hardiness in nymphs also varies with their position in the birth sequence, decreasing from an LTemp$_{50}$ of −15.9°C in nymphs born on day 1 of the reproductive life of the parent to −8.3°C by day 4. The significance of the enhanced cold tolerance in first-born nymphs is highlighted by the lack of cold hardiness in other overwintering life cycle stages. Adults are the least cold tolerant stage and, as temperatures decrease in midwinter, reproduction will be reduced correspondingly. For these reasons, first-born nymphs may be the only offspring produced by aphids that mature to adult in winter, and are thus particularly important in ensuring population survival to the following spring.

Specialist overwintering morphs

Many insects produce a specialized overwintering morph in which cold tolerance is maximized by winter starvation, the synthesis of cryoprotectants, and entry into a diapause state. In holocyclic aphids, the egg is the cold-hardy diapausing morph. There are no equivalent stages in most anholocyclic clones but *Pemphigus bursarius* (lettuce root aphid) produces an apterous overwintering stage, the hiemalis (Judge, 1967, 1968), which is physiologically distinct from summer apterae. *P. bursarius* overwinters as eggs on poplar trees (particularly Lombardy poplar, *Populus nigra* var. *italica*) and migrates in summer to secondary hosts such as lettuce. A proportion of the population feeding on lettuce does not produce return migrants (sexuparae) to poplar in late summer and autumn, but remains in the soil and overwinters as asexual apterae, even though their host plants have died. This hiemalis morph can be found around decaying lettuce roots in winter and revived in the laboratory by maintaining populations at 10°C in a 12 h:12 h LD regime (Phillips *et al.*, 2000). Field-collected hiemalis were more cold tolerant than apterae from a laboratory colony reared under summer conditions with LTemp$_{50}$ values of −13.1 and 2.3°C and maximum survival times at 0°C of 18 days and 8 h, respectively. Hiemalis collected from the field in winter and revived in the laboratory had higher levels of triglycerides (lipid) than summer apterae: 12.8% fresh weight (39.9% dry weight) compared with 7.1% (32.5%). It has also been shown that hiemalis that overwinter successfully in the soil can produce alatae in spring that establish new apterous colonies on lettuce, enabling asexual populations to perpetuate and colonize new areas and host plants without the need to migrate back to poplar to complete the sexual cycle (Phillips *et al.*, 1999). Whilst the hiemalis is analagous to the foliar-feeding, overwintering apterae of anholocyclic clones, it differs by its ability to accumulate energy reserves and thus survive, if necessary, in a starved condition. Where populations of *P. bursarius* feed on perennial weeds such as *Taraxacum*, the aphids may be able to continue to feed through winter.

Acclimation

Most insects become more cold hardy from summer through autumn to winter through the synthesis of cryoprotectants induced by environmental cues, mainly decreasing temperature, but also photoperiod. In many cases, this acclimation process occurs within a particular stage of development. Eggs are usually the life cycle stage with the lowest freezing temperatures, often in the region of −40°C (Sømme, 1982). Aphid eggs contain glycerol (which is the commonest polyol) and mannitol (Sømme, 1969). There is a distinct seasonal pattern of supercooling in the eggs of *R. insertum* (James and Luff, 1982) and *R. padi* (Strathdee *et al.*, 1995a), with the SCP decreasing post-oviposition to the lowest value of −40°C in January. When field-collected eggs of *A. svalbardicum* were maintained for 1 month at −10°C, the mean freezing point was decreased from −36 to −38°C, with some eggs freezing at temperatures as low as −44°C (Strathdee *et al.*, 1995a).

Anholocyclic aphids also show acclimation responses at lower temperatures. When populations of first instar and adult *M. persicae* are reared at 20, 15, 10, and 5°C for a full generation prior to assessment, there is a progressive decrease in the LTemp$_{50}$ from −8.1 to −16.4°C for nymphs and −6.9 to −11.4°C for adults. Acclimation of nymphal aphids during their maturation to adult increases the cold hardiness of their progeny during embryonic development, but this enhanced cold tolerance is rapidly lost when adults are transferred back to higher temperatures (Clough *et al.*, 1990).

Rearing aphids in the laboratory at 10 or 5°C often produces increases in cold tolerance that are indicative of seasonal changes in cold hardiness that occur under field conditions. More recently, the phenomenon of rapid cold hardening has been discovered (Lee *et al.*, 1987), particularly in species in which the lethal temperature is above the freezing temperature, such as aphids. A rapid cold hardening ability has been found in both adults and nymphs of *S. avenae*, with survival at the 'discriminating temperature' (temperature at which 80–90% mortality occurs after direct transfer from the rearing temperature, in this case 3 h at −8°C) increasing from 18 to 85%, after an acclimation period of only 2 h at 0°C, or by cooling the aphids from 20 to 0°C at 0.05°C/min (Powell and Bale, 2004; Fig. 11.4). In a subsequent study (Powell and Bale, 2005) with a population acclimated for a generation at 10°C, the discriminating temperature was lowered to −11°C, and some aphids survived at −14°C. Exposure of aphids to a rapid cold-hardening regime also lowered the chill coma (activity) threshold, suggesting that aphids can respond to diurnal fluctuations in environmental temperatures, as recently described in natural populations of the fly *Drosophila melanogaster* (Kelty and Lee, 2001).

The additional cold tolerance acquired *via* rapid cold hardening is typically very limited, e.g. providing some survival in brief exposures at 3–4°C below the discriminating temperature, or increasing the period of

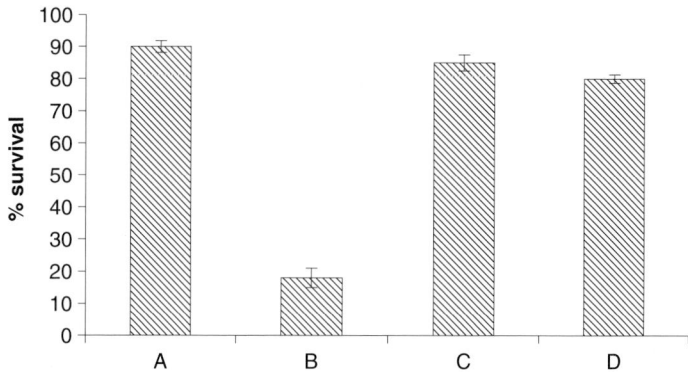

Fig. 11.4. Rapid cold hardening response in first-instar nymphs of *Sitobion avenae*: A, aphids maintained off host plant for 5 h; B, aphids transferred directly from 20 to −8°C for 3 h; C, aphids transferred from 20 to 0°C for 2.5 h and then −8°C for 3 h; D, aphids cooled from 10 to 0°C at a rate of 0.05°C/min and then exposed at −8°C for 3 h (from Powell and Bale, 2004).

survival at the discriminating temperature by a few hours. However, whilst these increases are small when compared with seasonal changes, in insects such as aphids, where mortality increases rapidly with each 1°C fall in temperature and generation times may be too short to allow seasonal acclimation, rapid cold hardening may be a particularly valuable response.

Reproduction in winter

In univoltine insects or those with only 2–3 generations per year, the life cycle stage that enters the winter is also the stage that has to survive to the end of winter; this is also the case with aphid eggs. With anholocyclic clones however, the aphids that colonize host plants in autumn will almost certainly die at some stage in winter, and overwintering success will then depend on their progeny, and probably on subsequent generations as well. It clearly is vital that nymphs born early in winter mature to adult and produce at least some progeny before they die. Overwintering is therefore another example of how aphids allocate different functions to different generations (and sometimes to a number of successive generations), whereas in most holometabolous insects, these 'tasks' are carried out by a particular stage of development within a generation.

Heat stress

There has been relatively little work on the effect of high temperatures on aphid performance. Aphid fecundity and longevity generally increase with increasing temperature up to a certain point, beyond which performance is rapidly reduced (Dean, 1974; Asín and Pons, 2001). The 'optimum' temperature varies between aphid species. Of the cereal aphids, in which most 'high temperature' work has been done, *R. padi* is particularly tolerant of high temperatures (de Barro, 1992; Asín and Pons, 2001). There are also differences between various geographical races of the same species. For example, populations of *R. padi* from Mediterranean regions can tolerate higher temperatures than populations

from Britain (Asín and Pons, 2001). The physiological mechanisms underlying differences in tolerance to high temperatures have not been investigated.

High temperatures can have indirect effects on aphid performance through effects on the host plant. Increased temperatures result in increased water loss and eventually lead to drought stress in the host plant. In most studies, the concentration of sieve-element sap (sugars, amino acids, or both) increases under drought stress (e.g. Tully and Hanson, 1979; Girousse *et al.*, 1996; Isaacs *et al.*, 1998; Ponder, 2000). However, only in some of these studies is there a concurrent reduction in aphid performance. It is likely that the degree of drought stress is critical in this respect, with the difficulties associated with probing and ingesting sieve-element sap from severely drought-stressed plants, possibly due to reduced sieve-element turgor (Wearing and van Emden, 1967; Wearing, 1972) being important (Ponder, 2000).

A Molecular Approach to Understanding Aphid Responses to Stress

It is evident that one of the most critical stresses experienced by aphids is that caused by variability in the quality of their diet. Our understanding of this area, and of aphid host plant relationships more generally, has been advanced significantly by novel molecular technologies.

In plants, we are now able to measure stress-induced changes in gene expression (transcriptomics), the protein complement (proteomics), and the consequent effect on metabolism (metabolomics). In an increasing number of cases, technical advances are allowing us to focus on specific tissues, or even individual cell types. In understanding the relationship between plants and their aphid herbivores, the crucial cell is the sieve element of the phloem from which the aphid derives the majority of its diet. Effects of stress can manifest themselves within the phloem as changes in gene transcripts. However, investigation into the regulation of the phloem is in its infancy, with relatively few studies currently published. Before the effect of a particular stress can be characterized, the basic regulation of unstressed tissue must be determined. Currently, gene transcripts found in the sieve element include those apparently involved in its function (e.g. transporters for sucrose, amino acids, and ions), structural and housekeeping roles (e.g. redox systems, profiling/actin). A few studies have examined the impact of aphid feeding on the plant transcriptome; sometimes, this is at the whole tissue level or can focus directly on the phloem. The former studies are informed by transcript annotation that indicates a cell-specific location for the gene expression, perhaps in cells overlying the vascular tissue, where a role in sieve element location is implicated, or directly in the phloem itself. Some transcripts have annotation that suggests a role in phloem-specific defence (lectins and protease inhibitors, for example), while other phloem transcripts could be the result of a non-adaptive damage response to aphid feeding. During initial probing of the tissue, physical damage is done by aphids to plant tissue, and structural cell wall gene transcripts are often seen to be upregulated in this pathway phase of aphid feeding.

Feeding by insect herbivores induces a range of transcriptional changes (Kessler and Baldwin, 2002). Whole tissue analysis reveals that defence-related genes such as those of the jasmonic and salicylic acid pathways are often induced (Moran and Thompson, 2001). Examination of the response of the whole *Arabidopsis* leaf transcriptome to aphid infestation indicates upregulation of a wide range of gene classes, including some common to chewing insects such as oxidative stress and other pathogenic responses (Moran *et al.*, 2002). However, the specialist nature of aphid feeding may induce a qualitatively different set of plant genes – those of the phloem. There is a growing body of information on genes expressed in the phloem. It is now possible to identify these genes using mRNA *in situ* analysis or immunolocalization (Xoconostle-Cazares *et al.*, 1999). An alternative method for localizing phloem genes used RT-PCR to demonstrate the presence of mRNA for a sucrose transporter, an aquaporin and a proton

ATPase in sieve-element sap (Doering-Saad *et al.*, 2002).

A further approach is to identify all genes expressed in phloem tissue, but such studies are limited by the relatively small proportion of phloem tissue in plants and difficulty in obtaining a pure sample. Gene expression has been studied in phloem-enriched samples from celery (Vilaine *et al.*, 2003) or by using laser capture microdissection (Asano *et al.*, 2002; Nakazono *et al.*, 2003). The data obtained on mRNA expression in sieve elements complement studies that have constructed protein libraries from the same cell types (Hayashi *et al.*, 2000; Hoffmann-Benning *et al.*, 2002; Barnes *et al.*, 2004; Walz *et al.*, 2004). Both approaches reveal that the sieve element–companion cell complex expresses a range of housekeeping genes, but also many involved in preventing oxidative damage, protein turnover, and the cytoskeleton.

Interestingly, many phloem expressed genes and proteins are potential candidates for the point of interaction with aphids. Several proteins consistent with a role in the metabolism of amino acids have been located to the phloem, including methionine synthase, glutamine synthase, and ornithine carbamoyltransferase, and were found in sieve-element sap (Hoffmann-Benning *et al.*, 2002). Transcripts for glutamine synthase were upregulated by aphid infestation in both tobacco leaves (Voelckel *et al.*, 2004) and celery phloem (Divol *et al.*, 2005). Gene transcripts or proteins implicated in defence have been localized to the phloem; a cysteine-proteinase inhibitor was found in a protein library from *Ricinus* (Barnes *et al.*, 2004), and a homologue was upregulated following aphid infestation in *Sorghum* (Zhu-Salzman *et al.*, 2004) and brown plant hopper infestation of rice (Yuan *et al.*, 2005). Messages for disease-resistance proteins were upregulated by aphid infestation in *Sorghum* leaves (Zhu-Salzman *et al.*, 2004), and a related transcript was found in the phloem of rice (Asano *et al.*, 2002). Proteins annotated as disease-resistant proteins have been found in phloem by other workers (Hoffmann-Benning *et al.*, 2002). Wound genes such as those involved in cell synthesis are often induced (Voelckel *et al.*, 2004;

Divol *et al.*, 2005), although the timing of the measurements is crucial to understanding plant interactions with aphids since the nature of the stress imposed will change during the infestation from one of mechanical damage during initial probing and phloem localization, to those of solute balance once sustained sap ingestion is established. In these later stages, since aphids represent an additional sink to the plant, it might be expected that genes regulating carbohydrate or solute metabolism and transport show altered expression. Sucrose synthase protein has been located to companion cells in maize (Nolte and Koch, 1993), and an aldolase and aldolase protein is present in cucumber sieve-element tubes (Chen *et al.*, 2004). The monosaccharide transporter STP4 was upregulated following aphid feeding in *Arabidopsis* leaves (Moran and Thompson, 2001).

Despite these tantalizing clues about the interaction between aphids and phloem, many of the genes identified by molecular genetics are of unknown function, underlining how much is unknown about the plant genome. Increasing use of functional genomics using loss of function mutants in *Arabidopsis* (Hunt, 2005; Zhu *et al.*, 2005) and emerging techniques such as RNAi will reveal the role of these genes in regulating the interaction between phloem and aphid and its modulation by abiotic stresses such as drought and temperature.

While we are at the beginning of the process of understanding the scope and regulation of the plant and phloem transcriptome in relation to plant interaction with the environment, both abiotic and biotic, we know almost nothing about corresponding changes in the aphid. Molecular studies of aphid biology focus largely on genetic fingerprinting approaches that examine evolutionary or ecological relationships between aphid species or populations (Loxdale and Lushai, Chapter 2 this volume). Occasionally, information is provided on genetic changes underlying adaptation or resistance of aphids to specific stresses such as insecticides (Loxdale *et al.*, 1998), but no general picture is yet emerging. What is required is a cataloguing of genes that are switched on and off when aphids are exposed to stresses such

as drought, poor diet, or low temperature. Apart from temperature, most of these stresses impact indirectly on the aphid through effects on the plant. Expressed sequence tags (ESTs) from specific regions (e.g. digestive system) will provide a focus on digestive and osmoregulatory responses. cDNA libraries are already being made from this material to understand insect digestive processes (Tagu *et al.*, 2004); however, at the present time, there is little or no published material on the changes in expression induced by stress of this emerging suite. The completion of the pea aphid genetic sequence (http://www.genome. gov/13014443) will provide new opportunities to correct the imbalance between knowledge of plant molecular responses to stress and that of aphids. Aphid genetic response to many stresses will be relatively easy to determine. The power of an EST/cDNA approach has been demonstrated by studies of the genetic changes that occur during transition of aphids from asexual to sexual forms (Tagu *et al.*, 2004). However, currently the only transcriptional analyses relevant to aphid responses to stress on non-ideal diets are those of the sequenced *Buchnera* symbiont – but these reveal few regulatory processes in the stripped down *Buchnera* genome (Moran *et al.*, 2002) – leaving the tantalizing suggestion that the regulation lies in the as yet unexplored aphid transcriptome.

Conclusions

Aphids exhibit highly flexible strategies to cope with adverse environmental conditions. They can compensate for nutritional stress by modifying their feeding behaviour, and respond to thermal stress by changing the allocation of resources between reproduction, morphology, and survival. Their high rates of reproduction allow such responses to spread rapidly through populations. The development of highly adaptive and genetically programmed life cycles enables aphids to survive in severe Arctic climates. The 'option' of overwintering as a cold hardy egg or in the mobile stage, together with the ability to cold harden both seasonally and rapidly, ensure survival through severe winters, with the opportunity to exploit more favourable conditions *via* anholocyclic clones. Further progress in understanding the relationship between aphids, their host plants, and the abiotic environment will emerge over the next few years as advances in molecular techniques are combined with an increasingly sophisticated physiological understanding of the aphid.

References

Abisgold, J.D., Simpson, S.J. and Douglas, A.E. (1994) Nutrient regulation in the pea aphid *Acyrthosiphon pisum*: application of a novel geometric framework to sugar and amino acid consumption. *Physiological Entomology* 19, 95–102.

Argandona, V.H., Zuniga, G.E. and Corcuera, L.J. (1987) Distribution of gramine and hydroxamic acids in barley and wheat leaves. *Phytochemistry* 26, 1917–1918.

Asano, T., Masumura, T., Kusano, H., Kikuchi, S., Kurita, A., Shimada, H. and Kadowaki, K. (2002) Construction of a specialized cDNA library from plant cells isolated by laser capture microdissection: toward comprehensive analysis of the genes expressed in the rice phloem. *Plant Journal* 32, 401–408.

Asch, F., Dingkuhn, M., Sow, A. and Audebert, A. (2005) Drought-induced changes in rooting patterns and assimilate partitioning between root and shoot in upland rice. *Field Crops Research* 93, 223–236.

Asín, L. and Pons, X. (2001) Effect of high temperature on the growth and reproduction of corn aphids (Homoptera: Aphididae) and implications for their population dynamics on the Northeastern Iberian Peninsula. *Environmental Entomology* 30, 1127–1134.

Bale, J.S. (1987) Insect cold hardiness: freezing and supercooling – an ecophysiological perspective. *Journal of Insect Physiology* 33, 899–908.

Bale, J.S. (1996) Insect cold hardiness: a matter of life and death. *European Journal of Entomology* 93, 369–382.

Bale, J.S. (1999) Impacts of climate warming on Arctic aphids: a comparative analysis. *Ecological Bulletins* 47, 38–47.

Bale, J.S., Harrington, R. and Clough, M.S. (1988) Low temperature mortality of the peach potato aphid *Myzus persicae*. *Ecological Entomology* 13, 121–129.

Barnes, A., Bale, J.S., Constantinidou, C., Ashton, P., Jones, A. and Pritchard, J. (2004) Determining protein identity from sieve element sap in *Ricinus communis* L. by quadrupole time of flight (Q-TOF) mass spectrometry. *Journal of Experimental Botany* 55, 1473–1481.

de Barro, P.J. (1992) The role of temperature, photoperiod, crowding and plant quality on the production of alate viviparous females of the bird cherry–oat aphid, *Rhopalosiphum padi*. *Entomologia Experimentalis et Applicata* 65, 205–214.

van Bel, A.J.E. (2003) The phloem, a miracle of ingenuity. *Plant Cell and Environment* 26, 125–149.

Bonnemaison, L. (1951) Contribution a l'étude des facteurs provoquant l'apparition des formes ailées et sexuées chez les Aphidinae. *Annales Epiphyties* 2, 1–380.

Bouchereau, A., ClossaisBesnard, N., Bensaoud, A., Leport, L. and Renard, M. (1996) Water stress effects on rapeseed quality. *European Journal of Agronomy* 5, 19–30.

Bray, C.M. (1983) *Nitrogen Metabolism in Plants.* Longman, New York, 214 pp.

Butts, R.A., Howling, G.G., Bone, W., Bale, J.S. and Harrington, R. (1997) Contact with the host plant enhances aphid survival at low temperatures. *Ecological Entomology* 22, 26–31.

Caillaud, C.M., Dedryver, C.A. and Simon, J.C. (1994) Development and reproductive potential of the cereal aphid *Sitobion avenae* on resistant wheat lines (*Triticum monococcum*). *Annals of Applied Biology* 125, 219–232.

Caillaud, C.M., Pierre, J.S., Chaubet, B. and Di Pietro, J.P. (1995) Analysis of wheat resistance to the cereal aphid *Sitobion avenae* using electrical penetration graphs and flow charts combined with correspondence analysis. *Entomologia Experimentalis et Applicata* 75, 9–18.

Calatayud, P.A., Tertuliano, M. and Leru, B. (1994) Seasonal-changes in secondary compounds in the phloem sap of cassava in relation to plant genotype and infestation by *Phenacoccus manihoti* (Homoptera, Pseudococcidae). *Bulletin of Entomological Research* 84, 453–459.

Campbell, A., Frazer, B.D., Gilbert, N., Gutierrez, A.P. and Mackauer, M. (1974) Temperature requirements of some aphids and their parasites. *Journal of Applied Ecology* 11, 431–438.

Carter, C.I. and Nichols, J.F.A. (1989) Winter survival of the lupin aphid *Macrosiphum albifrons* Essig. *Journal of Applied Entomology* 108, 213–216.

Carter, N., Dixon, A.F.G. and Rabbinge, R. (1982) *Cereal Aphid Populations: Biology, Simulation and Prediction.* PUDOC, Wageningen, 91 pp.

Chambers, R.J., Wellings, P.W. and Dixon, A.F.G. (1985) Sycamore aphid numbers and population density II. Some processes. *Journal of Animal Ecology* 54, 425–442.

Chen, Z.H., Walker, R.P., Tecsi, L.I., Lea, P.J. and Leegood, R.C. (2004) Phosphoenolpyruvate carboxykinase in cucumber plants is increased both by ammonium and by acidification, and is present in the phloem. *Planta* 219, 48–58.

Clough, M.S., Bale, J.S. and Harrington, R. (1990) Differential cold hardiness in adults and nymphs of the peach–potato aphid *Myzus persicae. Annals of Applied Biology* 116, 1–9.

Cockbain, A.J. (1961) Low temperature thresholds for flight in *Aphis fabae* Scop. *Entomologia Experimentalis et Applicata* 4, 211–219.

Cole, R. (1994) Locating a resistance mechanism to the cabbage aphid in two wild brassicas. *Entomologia Experimentalis et Applicata* 71, 23–31.

Cole, R. (1997a) Comparison of feeding behaviour of two brassica pests *Brevicoryne brassicae* and *Myzus persicae* on wild and cultivated brassica species. *Entomologia Experimentalis et Applicata* 85, 135–143.

Cole, R. (1997b) The relative importance of glucosinolates and amino acids to the development of two aphid pests *Brevicoryne brassicae* and *Myzus persicae* on wild and cultivated brassica species. *Entomologia Experimentalis et Applicata* 85, 121–133.

Costa-Arbulu, C., Gianoli, E., Gonzales, W.L. and, Niemeyer, H.M. (2001) Feeding by the aphid *Sipha flava* produces a reddish spot on leaves of *Sorghum halepense*: an induced defense? *Journal of Chemical Ecology* 27, 273–283.

Dean, G.J. (1974) Effect of temperature on the cereal aphids *Metopolophium dirhodum* (Wlk.), *Rhopalosiphum padi* (L.) and *Macrosiphum avenae* (F.) (Hem., Aphididae). *Bulletin of Entomological Research* 63, 401–409.

Divol, F., Vilaine, F., Thibivilliers, S., Amselem, J., Palauqui, J.C., Kusiak, C. and Dinant, S. (2005) Systemic response to aphid infestation by *Myzus persicae* in the phloem of *Apium graveolens. Plant Molecular Biology* 57, 517– 540.

Doering-Saad, C., Newbury, H.J., Bale, J.S. and Pritchard, J. (2002) Use of aphid stylectomy and RT-PCR for the detection of transporter mRNAs in sieve elements. *Journal of Experimental Botany* 53, 631–637.

Douglas, A.E. (1993) The nutritional quality of phloem sap utilised by natural aphid populations. *Ecological Entomology* 18, 31–38.

Douglas, A.E. (2003) The nutritional physiology of aphids. *Advances in Insect Physiology* 31, 73–140.

Dreyer, D.L., Jones, K.C. and Molyneux, R.J. (1985) Feeding deterrency of some pyrrolizidine indolizidine and quinolizidine alkaloids towards pea aphid (*Acyrthosiphon pisum*) and evidence for phloem transport of indolizidine alkaloid swainsonine. *Journal of Chemical Ecology* 11, 1045–1051.

van Emden, H.F. (1977) Failure of the aphid, *Myzus persicae*, to compensate for poor diet during early growth. *Physiological Entomology* 2, 53–58.

Eschrich, W. (1970) Biochemistry and fine structure of phloem in relation to transport. *Annual Review of Plant Physiology* 37, 193–214.

Geiger, D.R. and Servaites, J.C. (1991) Carbon allocation and response to stress. In: Mooney, H.A., Winner, W.E. and Pell, E.J. (eds) *Response of Plants to Multiple Stresses.* Academic Press, New York, pp. 104–127.

Girousse, C. and Bournoville, R. (1994) Role of phloem sap quality and exudation characteristics on performance of pea aphids grown on lucerne genotypes. *Entomologia Experimentalis et Applicata* 70, 227–235.

Girousse, C., Bournoville, R. and Bonnemain, J.L. (1996) Water deficit-induced changes in concentrations in proline and some other amino acids in the phloem sap of alfalfa. *Plant Physiology* 111, 109–113.

Givovich, A. and Niemeyer, H.M. (1995) Comparison of the effect of hydroxamic acids from wheat on five species of cereal aphid. *Entomologia Experimentalis et Applicata* 74, 115–119.

Grüber, K. and Dixon, A.F.G. (1988) The effect of nutrient stress on development and reproduction in an aphid. *Entomologia Experimentalis et Applicata* 47, 23–30.

Hale, B.K., Bale, J.S., Pritchard, J., Masters, G.J. and Brown, V.K. (2003) Effects of host plant drought stress on the performance of the bird cherry–oat aphid, *Rhopalosiphum padi* (L.): a mechanistic analysis. *Ecological Entomology* 28, 666–677.

Hanson, A.D., Ditz, K.M., Singletary, G.W. and Leland, T.J. (1983) Gramine accumulation in leaves of barley grown under high-temperature stress. *Plant Physiology* 71, 896–904.

Harrewijn, P. (1976) Host-plant factors regulating wing production in *Myzus persicae. Symposium Biologica Hungarica* 16, 79–83.

Harrewijn, P. (1978) The role of plant substances in polymorphism of the aphid *Myzus persicae*. *Entomologia Experimentalis et Applicata* 24, 198–214.

Harrington, R. and Cheng, X.-N. (1984) Winter mortality, development and reproduction in a field population of *Myzus persicae* (Sulz.) in England. *Bulletin of Entomological Research* 74, 633–640.

Harrington, R. and Taylor, L.R. (1990) Migration for survival: fine-scale population redistribution in an aphid *Myzus persicae*. *Journal of Animal Ecology* 59, 1177–1193.

Harrington, R., Bale, J.S. and Tatchell, G.M. (1995) Aphids in a changing climate. In: Harrington, R. and Stork, N.E. (eds) *Insects in a Changing Environment*. Academic Press, London, pp. 125–155.

Harrison, J.R. and Barlow, C.A. (1972) Population growth of the pea aphid *Acyrthosiphon pisum* (Homoptera: Aphididae) after exposure to extreme temperatures. *Annals of the Entomological Society of America* 65, 1011–1015.

Hart, A.J., Bale, J.S. and Fenlon, J. (1997) Developmental threshold, day degree requirements and voltinism of the aphid predator *Episyrphus balteatus* (Diptera: Syrphidae). *Annals of Applied Biology* 130, 427–437.

Hayashi, H., Nakamura, S., Ishiwatari, Y., Mori, S. and Chino, M. (1993) Changes in the amino acid composition and protein modification in phloem sap of rice. *Plant and Soil* 156, 171–174.

Hayashi, H., Fukuda, A., Suzui, N. and Fujimaki, S. (2000) Proteins in the sieve element–companion cell complexes: their detection, localization and possible functions. *Australian Journal of Plant Physiology* 27, 489–496.

Hodgson, C.J. (1978) The distribution and movement of apterous *Myzus persicae* on rapidly growing turnip plants. *Ecological Entomology* 3, 289–298.

Hodgson, C.J. (1991) Dispersal of apterous aphids (Homoptera: Aphididae) from their host plant and its significance. *Bulletin of Entomological Research* 81, 417–427.

Hoffmann-Benning, S., Gage, D.A., McIntosh, L., Kende, H. and Zeevaart, J.A.D. (2002) Comparison of peptides in the phloem sap of flowering and non-flowering Perilla and lupine plants using microbore HPLC followed by matrix-assisted laser desorption/ionization time-of-flight mass spectrometry. *Planta* 216, 140–147.

Honek, A. (1991) Environmental stress, plant quality and abundance of cereal aphids (Homoptera: Aphididae) on winter wheat. *Journal of Applied Entomology* 112, 71–75.

Hunt, E.J. (2005) Controlling aphid pests by manipulating their diet: physiological and genomic approaches. PhD thesis, University of Birmingham, Birmingham, UK.

Hunt, R. and Nicholls, A.D. (1986) Stress and coarse control of root–shoot partitioning in herbaceous plants. *Oikos* 47, 149–158.

Hutchinson, L.A. and Bale, J.S. (1994) Effects of sub-lethal cold stress on the aphid *Rhopalosiphum padi*. *Journal of Applied Ecology* 31, 102–108.

Isaacs, R., Byrne, D.N. and Hendrix, D.L. (1998) Feeding rates and carbohydrate metabolism by *Bemisia tabaci* (Homoptera: Aleyrodidae) on different quality phloem saps. *Physiological Entomology* 23, 241–248.

James, B.D. and Luff, M.L. (1982) Cold hardiness and the development of the eggs of *Rhopalosiphum insertum*. *Ecological Entomology* 7, 277–282.

Jepson, P.C. (1983) A controlled environment study of the effect of leaf physiological age on the movement of *Myzus persicae* on sugar-beet plants. *Annals of Applied Biology* 103, 173–183.

Johnson, B. (1966) Wing polymorphism in aphids III. The influence of the host plant. *Entomologia Experimentalis et Applicata* 9, 213–222.

Johnson, C.G. and Taylor, L.R. (1957) Periodism and energy summation with special reference to flight rhythms in aphids. *Journal of Experimental Biology* 34, 209–221.

Jones, C.G. and Coleman, J.S. (1991) Plant stress and insect herbivory: toward an integrated perspective. In: Mooney, H.A., Winner, W.E. and Pell, E.J. (eds) *Response of Plants to Multiple Stresses*. Academic Press, New York, pp. 249–280.

Judge, F.D. (1967) Overwintering in *Pemphigus bursarius* (L.). *Nature, London* 216, 1041–1042.

Judge, F.D. (1968) Polymorphism in a subterranean aphid, *Pemphigus bursarius*. 1. Factors affecting the development of sexuparae. *Annals of the Entomological Society of America* 61, 819–827.

Karley, A.J., Douglas, A.E. and Parker, W.E. (2002) Amino acid composition and nutritional quality of potato leaf phloem sap for aphids. *Journal of Experimental Biology* 205, 3009–3018.

Kawada, K. (1987) Polymorphism and morph determination. In: Minks, A.K. and Harrewijn, P. (eds) *Aphids. Their Biology, Natural Enemies and Control, Volume 2A*. Elsevier, Amsterdam, pp. 255–268.

Kelty, J.D. and Lee, R.E. (2001) Rapid cold-hardening of *Drosophila melanogaster* (Diptera: Drosophilidae) during ecologically-based thermoperiodic cycles. *Journal of Experimental Biology* 204, 1659–1666.

Kessler, A. and Baldwin, I.T. (2002) Plant responses to insect herbivory: the emerging molecular analysis. *Annual Review of Plant Biology* 53, 299–328.

Knight, J.D. and Bale, J.S. (1987) The effects of low temperature on the development, reproduction and survival of the grain aphid *Sitobion avenae*. *Proceedings of the 3rd European Congress of Entomology, Amsterdam, 1986*, 279–282.

Knight, J.D., Bale, J.S., Franks, F., Mathias, S.F. and Baust, J.G. (1986) Insect cold hardiness: supercooling points and pre-freeze mortality. *Cryo Letters* 7, 194–203.

Knoblauch, M., Peters, W.S., Ehlers, K. and van Bel, A.J.E. (2001) Reversible calcium-regulated stopcocks in legume sieve tubes. *Plant Cell* 13, 1221–1230.

Koricheva, J., Larsson, S., Haukioja, E. and Keinanen, M. (1998) Regulation of woody plant secondary metabolism by resource availability: hypothesis testing by means of meta-analysis. *Oikos* 83, 212–226.

Lamb, R.J. (1992) Development rates of *Acyrthosiphon pisum* (Homoptera: Aphididae) at low temperatures: implications for estimating rate parameters for insects. *Environmental Entomology* 21, 10–19.

Lavola, A., Julkunen-Tiitto, R., de la Rosa, T.M., Lehto, T. and Aphalo, P.J. (2000) Allocation of carbon to growth and secondary metabolites in birch seedlings under UV-B radiation and CO_2 exposure. *Physiologia Plantarum* 109, 260–267.

Leather, S.R., Ward, S.A. and Dixon, A.F.G. (1983) The effect of nutrient stress on some life history parameters of the black bean aphid, *Aphis fabae* Scop. *Oecologia* 57, 156–157.

Lee, R.E., Chen, C.-P. and Denlinger, D.L. (1987) A rapid cold-hardening response in insects. *Science, New York* 238, 1415–1417.

Lees, A.D. (1960) The role of photoperiod and temperature in the determination of the parthenogenetic and sexual forms in the aphid *Megoura viciae* Buckton 2. The operation of the interval timer in young clones. *Journal of Insect Physiology* 9, 153–164.

Lees, A.D. (1961) Clonal polymorphism in aphids. In: Kennedy, J.S. (ed.) *Insect Polymorphism. Proceedings of the Royal Entomological Society Symposium No. 1*. Royal Entomological Society of London, London, pp. 68–78.

Lees, A.D. (1967) The production of apterous and alate forms of the aphid *Megoura viciae* Buckton, with special reference to the role of crowding. *Journal of Insect Physiology* 3, 207–277.

Loxdale, H.D., Brookes, C.P., Wynne, I.R. and Clark, S.J. (1998) Genetic variability within and between English populations of the damson–hop aphid, *Phorodon humuli* (Hemiptera: Aphididae), with special reference to esterases associated with insecticide resistance. *Bulletin of Entomological Research* 88, 513–526.

Mann, J.A., Tatchell, G.M., Dupuch, M.J., Harrington, R., Clark, J.J. and McCartney, H.A. (1995) Movement of apterous *Sitobion avenae* (Homoptera: Aphididae) in response to leaf disturbances caused by wind and rain. *Annals of Applied Biology* 126, 417–427.

Mattson, W.J., Jr. (1980) Herbivory in relation to plant nitrogen content. *Annual Review of Ecology and Systematics* 11, 119–161.

Mayoral, A.M., Tjallingii, W.F. and Castañera, P. (1996) Probing behaviour of *Diuraphis noxia* on five cereal species with different hydroxamic levels. *Entomologia Experimentalis et Applicata* 78, 341–348.

McKinnon, M.L., Quiring, D.T. and Bauce, E. (1998) Influence of resource availability on growth and foliar chemistry within and among young white spruce trees. *Ecoscience* 5, 295–305.

Merritt, S.Z. (1996) Within-plant variation in concentrations of amino acids, sugar and sinigrin in phloem sap of black mustard, *Brassica nigra* (L.) Koch (Cruciferae). *Journal of Chemical Ecology* 22, 1133–1145.

Michels, G.J. and Behle, R.W. (1989) Influence of temperature on reproduction. Development, and intrinsic rate of interest of Russian wheat aphid, greenbug and bird-cherry oat aphid (Homoptera: Aphididae). *Journal of Economic Entomology* 82, 439–444.

Mitchell, D.E. and Madore, M.A. (1992) Patterns of assimilate production and translocation in muskmelon (*Cucumis melo* L.). 2. Low-temperature effects. *Plant Physiology* 99, 966–971.

Mittler, T.E. (1958) Studies on the feeding and nutrition of *Tuberolachnus salignus* (Gmelin) II. The nitrogen and sugar composition of ingested phloem sap and excreted honeydew. *Journal of Experimental Biology* 35, 74–84.

Mittler, T.E. (1967) Effect of amino acid and sugar concentrations on the food uptake of the aphid *Myzus persicae*. *Entomologia Experimentalis et Applicata* 10, 39–51.

Montllor, C.B., Campbell, B.C. and Mittler, T.E. (1990) Responses of *Schizaphis graminum* (Homoptera: Aphididae) to leaf excision in resistant and susceptible sorghum. *Annals of Applied Biology* 116, 189–198.

Moran, P.J. and Thompson, G.A. (2001) Molecular responses to aphid feeding in *Arabidopsis* in relation to plant defense pathways. *Plant Physiology* 125, 1074–1085.

Moran, P.J., Cheng, Y.F., Cassell, J.L. and Thompson, G.A. (2002) Gene expression profiling of *Arabidopsis thaliana* in compatible plant–aphid interactions. *Archives of Insect Biochemistry and Physiology* 51, 182–203.

Muller, C.B., Williams, I.S. and Hardie, J. (2001) The role of nutrition, crowding and interspecific interactions in the development of winged aphids. *Ecological Entomology* 26, 330–340.

Nakazono, M., Qiu, F., Borsuk, L.A. and Schnable, P.S. (2003) Laser-capture microdissection, a tool for the global analysis of gene expression in specific plant cell types: identification of genes expressed differentially in epidermal cells or vascular tissues of maize. *Plant Cell* 15, 583–696, 1049.

Niemeyer, H.M. and Perez, F.J. (1987) Hydroxamic acids from Graminae: their role in aphid resistance and their mode of action. In: Labiere, V., Fabres, G. and Lachaise, D. (eds) *Insects – Plants*. W. Junk, The Hague, pp. 49–52.

Nolte, K.D. and Koch, K.E. (1993) Companion-cell specific localization of sucrose synthase in zones of phloem loading and unloading. *Plant Physiology* 101, 899–905.

Parish, W.E.G. and Bale, J.S. (1990) Effects of short-term exposure to low temperature on wing development in the grain aphid *Sitobion avenae* (F.) (Hem., Aphididae). *Journal of Applied Entomology* 109, 175–181.

Parish, W.E.G. and Bale, J.S. (1991) Effects of low temperature on the intracellular symbionts of the grain aphid *Sitobion avenae* (F.) (Hemiptera: Aphididae). *Journal of Insect Physiology* 37, 339–345.

Parish, W.E.G. and Bale, J.S. (1993) Effects of brief exposures to low temperature on the development, longevity and fecundity of the grain aphid *Sitobion avenae* (Hem., Aphididae). *Annals of Applied Biology* 122, 9–21.

Pate, J.S. and Atkins, C.A. (1983) Xylem and phloem transport and the functional economy of carbon and nitrogen of a leaf legume. *Plant Physiology* 71, 835–840.

Pate, J.S., Walker, J. and Wallace, W. (1965) Nitrogen-containing compounds in the shoot system of *Pisum arvense* L. II. The significance of amino acids and amides released from the nodulated roots. *Annals of Botany* 29, 475–493.

Pereira, F.M.V., Rosa, E., Fahey, J.W., Stephenson, K.K., Carvalho, R. and Aires, A. (2002) Influence of temperature and ontogeny on the levels of glucosinolates in broccoli (*Brassica oleracea* var. italica) sprouts and their effect on the induction of mammalian phase 2 enzymes. *Journal of Agricultural and Food Chemistry* 50, 6239–6244.

Phillips, S.W., Bale, J.S. and Tatchell, G.M. (1999) Escaping an ecological dead-end; asexual overwintering and morph determination in the lettuce root aphid *Pemphigus bursarius* L. *Ecological Entomology* 24, 336–344.

Phillips, S., Bale, J.S. and Tatchell, G.M. (2000) Overwintering adaptations in the lettuce root aphid *Pemphigus bursarius* (L.). *Journal of Insect Physiology* 46, 353–363.

Ponder, K.L. (2000) Nitrogen and water stress in barley: influences on the performance and feeding behaviour of the aphid *Rhopalosiphum padi*. PhD thesis, University of Birmingham, Birmingham, UK.

Ponder, K.L., Pritchard, J., Harrington, R. and Bale, J.S. (2000) Difficulties in location and acceptance of phloem sap combined with reduced concentration of phloem amino acids explain lowered performance of the aphid *Rhopalosiphum padi* on nitrogen deficient barley (*Hordeum vulgare*) seedlings. *Entomologia Experimentalis et Applicata* 97, 203–210.

Ponder, K.L., Pritchard, J., Harrington, R. and Bale, J.S. (2001) Feeding behaviour of the aphid *Rhopalosiphum padi* (Hemiptera: Aphididae) on nitrogen and water stressed barley (*Hordeum vulgare*) seedlings. *Bulletin of Entomological Research* 91, 1–9.

Powell, S.J. and Bale, J.S. (2004) Cold shock injury and ecological costs of rapid cold hardening in the grain aphid *Sitobion avenae* (Hemiptera: Aphididae). *Journal of Insect Physiology* 50, 277–284.

Powell, S.J. and Bale, J.S. (2005) Low temperature acclimated populations of the grain aphid *Sitobion avenae* retain ability to rapidly cold harden with enhanced fitness. *Journal of Experimental Biology* 208, 2615–2620.

Prado, E. and Tjallingii, W.F. (1994) Aphid activities during sieve element punctures. *Entomologia Experimentalis et Applicata* 72, 157–165.

Prado, E. and Tjallingii, W.F. (1997) Effects of previous plant infestation on sieve element acceptance by two aphids. *Entomologia Experimentalis et Applicata* 82, 189–200.

Prosser, W.A. and Douglas, A.E. (1992) A test of the hypothesis that nitrogen is upgraded and recycled in an aphid (*Acyrthosiphon pisum*) symbiosis. *Journal of Insect Physiology* 38, 93–99.

Prosser, W.A., Simpson, S.J. and Douglas, A.E. (1992) How an aphid (*Acyrthosiphon pisum*) symbiosis responds to variation in dietary nitrogen. *Journal of Insect Physiology* 38, 301–307.

Raccah, B., Tahon, A.S. and Applebaum, S.W. (1971) Effect of nutritional factors in synthetic diets on increase of alate forms of *Myzus persicae*. *Journal of Insect Physiology* 17, 1385–1390.

Rahbé, Y., Febvay, G., Delobel, B. and Bournoville, R. (1988) *Acyrthosiphon pisum* performance in response to the sugar and amino acid composition of artificial diets, and its relation to lucerne varietal resistance. *Entomologia Experimentalis et Applicata* 48, 283–292.

Richardson, M.D. and Bacon, C.W. (1993) Cyclic hydroxamic acid accumulation in corn seedlings exposed to reduced water potentials before, during, and after germination. *Journal of Chemical Ecology* 19, 1613–1624.

Salt, R.W. (1961) Principles of insect cold-hardiness. *Annual Review of Entomology* 6, 55–74.

Sandström, J. and Pettersson, J. (1994) Amino acid composition of phloem sap and the relation to intraspecific variation in pea aphid (*Acyrthosiphon pisum*) performance. *Journal of Insect Physiology* 40, 947–955.

Schaefers, G.A. and Judge, F.D. (1971) Effects of temperature, photoperiod and host plant on alary polymorphism in the aphid *Chaetosiphon fragaefolii. Journal of Insect Physiology* 17, 365–379.

Sømme, L. (1969) Mannitol and glycerol in overwintering aphid eggs. *Norwegian Journal of Entomology* 16, 107–111.

Sømme, L. (1982) Supercooling and winter survival in terrestrial arthropods. *Comparative Biochemistry and Physiology* 73A, 519–543.

Srivastava, P.N. and Auclair, J.L. (1975) Role of single amino acids in phagostimulation, growth and survival of *Acyrthosiphon pisum. Journal of Insect Physiology* 21, 1865–1871.

Stadler, B. (1995) Adaptive allocation of resources and life history trade-offs in aphids relative to plant quality. *Oecologia* 102, 246–254.

Strathdee, A.T. and Bale, J.S. (1996) Life cycle and morph production in the Arctic aphid *Acyrthosiphon brevicorne. Polar Biology* 16, 293–300.

Strathdee, A.T., Bale, J.S., Hodkinson, I.D., Block, W., Webb, N.R. and Coulson, S.J. (1993a) Identification of three previously unknown morphs of *Acyrthosiphon svalbardicum* Heikinheimo (Hemiptera: Aphididae) on Spitsbergen. *Entomologica Scandinavica* 24, 43–47.

Strathdee, A.T., Bale, J.S., Block, W., Webb, N.R., Hodkinson, I.D. and Coulson, S.J. (1993b) Extreme adaptive life cycle in a high arctic aphid. *Ecological Entomology* 18, 254–258.

Strathdee, A.T., Bale, J.S., Block, W., Coulson, S.J., Hodkinson, I.D. and Webb, N.R. (1993c) Effects of temperature elevation on a field population of *Acyrthosiphon svalbardicum* (Hemiptera: Aphididae) on Spitsbergen. *Oecologia* 96, 457–465.

Strathdee, A.T., Howling, G.G. and Bale, J.S. (1995a) Cold hardiness of aphid eggs. *Journal of Insect Physiology* 41, 653–657.

Strathdee, A.T., Bale, J.S., Strathdee, F.C., Block, W., Coulson, S.J., Webb, N.R. and Hodkinson, I.D. (1995b) Climatic severity and the response to temperature elevation of Arctic aphids. *Global Change Biology* 1, 23–28.

Sutherland, O.R.W. (1969) The role of the host plant in the production of winged form by two strains of the pea aphid, *Acyrthosiphon pisum. Journal of Insect Physiology* 15, 2179–2201.

Sutherland, O.R.W. (1970) An intrinsic factor influencing alate production by two strains of the pea aphid *Acyrthosiphon pisum. Journal of Insect Physiology* 16, 1349–1354.

Tagu, D., Prunier-Leterme, N., Legeai, F., Gauthier, J.P., Duclert, A., Sabater-Munoz, B., Bonhomme, J. and Simon, J.C. (2004) Annotated expressed sequence tags for studies of the regulation of reproductive modes in aphids. *Insect Biochemistry and Molecular Biology* 34, 809–822.

Taylor, K.C., Albrigo, L.G. and Chase, C.D. (1996) Purification of a Zn-binding phloem protein with sequence identity to chitin-binding proteins. *Plant Physiology* 110, 657–664.

Tosh, C.R., Powell, G., Holmes, N.D. and Hardie, J. (2003) Reproductive response of generalist and specialist aphid morphs with the same genotype to plant secondary compounds and amino acids. *Journal of Insect Physiology* 49, 1173–1182.

Tully, R.E. and Hanson, A.D. (1979) Amino acids translocated from turgid and water-stressed barley leaves I. Phloem exudation studies. *Plant Physiology* 64, 460–466.

Vilaine, F., Palauqui, J.C., Amselem, J., Kusiak, C., Lemoine, R. and Dinant, S. (2003) Towards deciphering phloem: a transcriptome analysis of the phloem of *Apium graveolens. Plant Journal* 36, 67–81.

Voelckel, C., Weisser, W.W. and Baldwin, I.T. (2004) An analysis of plant–aphid interactions by different microarray hybridization strategies. *Molecular Ecology* 13, 3187–3195.

Walters, K.F.A. and Dewar, A.M. (1986) Overwintering strategy and the timing of the spring migration of the cereal aphids *Sitobion avenae* and *Sitobion fragariae. Journal of Applied Ecology* 23, 905–915.

Walters, K.F.A., Brough, C. and Dixon, A.F.G. (1988) Habitat quality and reproductive investment in aphids. *Ecological Entomology* 13, 337–345.

Walz, C., Giavalisco, P., Schad, M., Juenger, M., Klose, J. and Kehr, J. (2004) Proteomics of cucurbit phloem exudate reveals a network of defence proteins. *Phytochemistry* 65, 1795–1804.

Watt, A.D. and Dixon, A.F.G. (1981) The role of cereal growth stages and crowding in the induction of alatae in *Sitobion avenae* and its consequences for population growth. *Ecological Entomology* 6, 441–447.

Wearing, C.H. (1972) Selection of Brussels sprouts of different water status by apterous and alate *Myzus persicae* and *Brevicoryne brassicae* in relation to the age of the leaves. *Entomologia Experimentalis et Applicata* 15, 139–154.

Wearing, C.H. and van Emden, H.F. (1967) Studies on the relation of insect and host plant. I. Effects of water stress on infestation by *Aphis fabae* Scop., *Myzus persicae* (Sulz.) and *Brevicoryne brassicae* (L.). *Nature, London* 213, 1051–1052.

Weibull, J. (1987) Seasonal changes in the free amino acids of oat and barley phloem sap in relation to plant growth stage and growth of *Rhopalosiphum padi*. *Annals of Applied Biology* 111, 729–737.

Weibull, J., Brishammar, S. and Pettersson, J. (1986) Amino acid analysis of phloem sap from oats and barley: a combination of aphid stylet excision and high performance liquid chromatography. *Entomologia Experimentalis et Applicata* 42, 27–30.

Wellings, P.W., Leather, S.R. and Dixon, A.F.G. (1980) Seasonal variation in reproductive potential: a programmed feature of aphid life cycles. *Journal of Animal Ecology* 49, 975–985.

Wensler, R.J.D. (1962) Mode of host selection by an aphid. *Nature, London* 195, 830–831.

Wiktelius, S. (1981) Diurnal flight periodicities and temperature thresholds for flight for different migrant forms of *Rhopalosiphum padi* L. (Hom., Aphididae). *Zeitschrift für Angewandte Entomologie* 92, 449–457.

Wiktelius, S. and Chiverton, P.A. (1985) Ovariole number and fecundity for the two emerging generations of the bird cherry–oat aphid (*Rhopalosiphum padi*) in Sweden. *Ecological Entomology* 10, 349–355.

Xoconostle-Cazares, B., Yu, X., Ruiz-Medrano, R., Wang, H.L., Monzer, J., Yoo, B.C., McFarland, K.C., Franceschi, V.R. and Lucas, W.J. (1999) Plant paralog to viral movement protein that potentiates transport of mRNA into the phloem. *Science, New York* 283, 94–98.

Yuan, H.Y., Chen, X.P., Zhu, L.L. and He, G.C. (2005) Identification of genes responsive to brown planthopper *Nilaparvata lugens* Stal (Homoptera: Delphacidae) feeding in rice. *Planta* 221, 105–112.

Zhou, X.-L., Carter, N. and Mumford, J. (1989) A simulation model describing the population dynamics and damage potential of the rose grain aphid, *Metopolophium dirhodum* (Walker) (Hemiptera: Aphididae) in the UK. *Bulletin of Entomological Research* 79, 373–380.

Zhu, X.L., Shaw, P.N., Pritchard, J., Newbury, H.J., Hunt, E.J. and Barrett, D.A. (2005) Amino acid analysis by micellar electrokinetic chromatography with laser-induced fluorescence detection: application to nanolitre-volume biological samples from *Arabidopsis thaliana* and *Myzus persicae*. *Electrophoresis* 26, 911–919.

Zhu-Salzman, K., Salzman, R.A., Ahn, J.E. and Koiwa, H. (2004) Transcriptional regulation of sorghum defense determinants against a phloem-feeding aphid. *Plant Physiology* 134, 420–431.

Zuniga, G.E. and Corcuera, L.J. (1986) Effect of gramine in the resistance of barley seedings to the aphid *Rhopalosiphum padi*. *Entomologia Experimentalis et Applicata* 40, 259–262.

Zuniga, G.E., Varanda, E.M. and Corcuera, L.J. (1988) Effect of gramine on the feeding behaviour of the aphids *Schizaphis graminum* and *Rhopalosiphum padi*. *Entomologia Experimentalis et Applicata* 47, 161–165.

12 Population Dynamics

Pavel Kindlmann[1], Vojtěch Jarošík[2] and Anthony F.G. Dixon[3]

[1]Faculty of Biological Sciences, University of South Bohemia and Institute of Landscape Ecology, Czech Academy of Sciences, 37005 České Budějovice, Czech Republic; [2]Department of Ecology, Charles University, 12844 Prague 2, Czech Republic and Institute of Botany, Czech Academy of Sciences, 25243 Pruhonice, Czech Republic; [3]School of Biological Sciences, University of East Anglia, Norwich, NR4 7TJ, UK

Introduction

It is the abundance of many aphids that makes them such serious pests. Therefore, it is important for aphid pest management to have a good understanding of their population dynamics both in terms of theory and practice. Although this aspect of aphid biology is well studied, there has been a lack of long-term studies of the population dynamics of aphids living on herbaceous plants, including crops. This is because any single arable crop field supports only a small fraction of the shifting population in a region, and even a dramatic event there will have little or no impact on the regional population dynamics (Mackauer and Way, 1976). As a result, this chapter will draw on research on non-pest species where their study may provide insights relevant to crop pests.

Not surprisingly, the pest status of aphids and political concern over the prophylactic application of pesticides has attracted the attention of modellers since the 1960s (Hughes, 1963; Hughes and Gilbert, 1968; Gilbert and Hughes, 1971; Gosselke et al., 2001). Attempts were made to forecast the abundance of aphids and propose expert systems to help farmers optimize prophylactic measures and minimize their costs (Mann

et al., 1986; Gonzalez-Andujara, 1993; Ro and Long, 1999). These studies usually concluded that forecasting is a better strategy than either no control or prophylaxis, where yields are average and above (Watt, 1983; Watt et al., 1984). The advisory systems, however, did not receive general acceptance and disappointingly few forecasting systems are in use. Analysis of some of the existing models of aphid population dynamics reveals the reasons. For example, a model that describes the summer population dynamics of the Sitobion avenae (grain aphid) (Carter et al., 1982; Carter, 1985) was modified and extended to include the population dynamics of the aphidophagous predator Coccinella septempunctata (7-spot ladybird) (Skirvin et al., 1997a,b). It is claimed to give better predictions than the Carter et al. (1982) model, but there are few data against which it can be validated. The main weakness of the Skirvin et al. (1997a) model is that it gives the same prediction for identical initial conditions, which is contrary to what is observed in the field.

Early models of the population dynamics of Myzus persicae (peach–potato aphid) (Scopes, 1969; Tamaki and Weeks, 1972, 1973; Tamaki, 1973, 1984; DeLoach, 1974; Taylor, 1977; Whalon and Smilowitz, 1979;

Tamaki *et al.*, 1980, 1982; Mack and Smilowitz, 1981, 1982; Smilowitz, 1984; Ro and Long, 1998) were improved recently by Ro and Long (1999). However, even this model is not validated against data that were not used to derive the parameters, which devalues the claim that it gives a good prediction. In addition, it also makes the unwarranted assumption that the decline in aphid abundance is caused by predators.

A simulation model developed to investigate the interrelationship of factors influencing the population dynamics of *Rhopalosiphum padi* (bird cherry–oat aphid) in barley crops during autumn and winter (Morgan, 2000) accurately predicts outbreaks and peak aphid populations within 20% of that observed in all but one case. However, this model is not suitable for long-term predictions, as it requires the daily input of maximum and minimum temperatures, which invalidates its predictive value as those temperatures themselves cannot be predicted with sufficient accuracy. Another model for this species was developed by Wiktelius and Pettersson (1985) but not used for forecasting, and the need for further research stressed.

A whole family of models of *Aphis craccivora* (cowpea aphid) (Gutierrez *et al.*, 1974) and *Acyrthosiphum pisum* (pea aphid) population dynamics (Gutierrez and Baumgärtner, 1984a,b; Guttierez *et al.*, 1984), and that of their natural enemies (Gutierrez *et al.*, 1980, 1981) were developed by Gutierrez and his group, but even these were not used for long-term predictions. Similarly, a computer simulation model developed to investigate spatial and population dynamics of apterae of the *Diuraphis noxia* (Russian wheat aphid) on preferred (wheat) and non-preferred (oat) hosts by Knudsen and Schotzko (1991) is suitable only for short-term (14 and 21 days) predictions. A transition matrix model developed to simulate the population dynamics of *Aphis pomi* (green apple aphid) (Woolhouse and Harmsen, 1991) has also not been validated against an independent data set.

Recently, spatio-temporal or metapopulation models have been published (Weisser, 2000; Winder *et al.*, 2001). These are a promising development, but modellers employing this approach need to consider whether aphid migration, rather than predator-inflicted mortality, is the regulating factor. The question remains, whether aphid metapopulation dynamics are driven by predators or, as predicted by theory (Kindlmann and Dixon, 1996, 1999), the predators are responding to aphid abundance, which is self-regulated by migration.

In general, the failure of models to predict aphid population dynamics for practical purposes is due to the extremely wild oscillations in aphid numbers caused by intrinsic (size, fecundity, mortality, migration rate) and external (weather, especially temperature) factors. As a consequence, predictions are unlikely to be robust enough for reliable forecasting, mainly because they depend on the course of weather during the season, which cannot be predicted. In addition, most of the models tend to be very complex, which stems from the belief of their authors that complexity means better accuracy, which is not always the case (Stewart and Dixon, 1988). This is because the measuring errors associated with each of the large number of parameters yield highly variable predictions. Thus, there is a serious gap in our knowledge, which needs to be filled in order to confirm or refute the understanding arrived at mainly by studying aphids living on woody plants. For a further discussion of forecasting, see Harrington *et al.*, Chapter 19 this volume.

Biological Background

Aphid biology relevant to population dynamics

Most aphid species can reproduce both asexually and sexually, with several parthenogenetic generations between each period of sexual reproduction. This is known as cyclical parthenogenesis and, in temperate regions, sexual reproduction occurs in autumn and results in the production of overwintering eggs, which hatch the following spring and initiate another cycle. Many pest aphids, however, overwinter, not as an egg but as nymphs or adults, and others as both eggs and active stages (see Williams and Dixon, Chapter 3 this volume). For their size, the parthenogenetic individuals have very short developmental times and potentially prodigious rates of increase (de Réaumur, 1737; Huxley, 1858; Kindlmann and Dixon, 1989; Dixon, 1992). Thus, aphids show very complex and rapidly changing within-year dynamics, with each clone going through several generations during the vegetative season and being made up of many individuals, which can be widely scattered in space. The survival of the eggs and/or overwintering aphids determines the numbers of aphids present the following spring.

The study of the population dynamics of aphids living on herbaceous plants, including agricultural crops, is difficult because their host plants vary in abundance and distribution from year to year. Tree-living aphids, in addition to being very host-specific, live in a habitat that is both spatially and temporally relatively stable. Therefore, it is not surprising that most long-term population studies on aphids have been on such species (Dixon, 1963, 1966, 1969, 1970, 1971, 1975, 1979, 1990; Dixon and Barlow, 1979; Barlow and Dixon, 1980; Dixon and Mercer, 1983; Chambers *et al.*, 1985; Wellings *et al.*, 1985; Dixon *et al.*, 1993b, 1996). However, some of the theoretical results obtained from these studies are quite general and can be applied to other aphid species.

Within a year, aphid dynamics are very complicated and, in looking for the mechanism of regulation, this needs to be taken into consideration. An initial dramatic increase in population size in spring is typically followed by a steep decline in abundance during summer, and sometimes a further increase in autumn. During spring and summer, all the generations are parthenogenetic and short-lived (1–4 weeks). In autumn, sexual forms are also produced, which mate and give rise to the overwintering eggs from which fundatrices, the first parthenogenetic generation, hatch the following spring. The parthenogenetic generations overlap in time and environmental conditions change rapidly. Therefore, an individual

throughout its life, as well as individuals born at different but close instants in time, can experience quite different conditions, which results in aphids evolving different and varying reproductive strategies.

The within-year dynamics of aphids are determined largely by seasonal changes in host quality. Aphids do best when amino acids are actively translocated in the phloem. In spring, the leaves grow and import amino acids via the phloem; in summer, the leaves are mature and export mainly sugars; in autumn, the leaves senesce and export amino acids and other nutrients. Thus, on trees, the leaves are most suitable for aphids in spring and autumn. The differences in within-year population dynamics of aphids are due to differences in the effect these seasonal fluctuations in host plant quality have on the *per capita* rate of increase and intraspecific competition in each species. This annual cycle, of two short periods when the host plant is very favourable and a long intervening period when it is less favourable, is well documented for tree-dwelling aphids. This has greatly facilitated the modelling of their population dynamics. In general, the aphid-carrying capacity of annual crop plants tends to increase with the season until the plants mature, after which it tends to decrease very rapidly. Thus, the aphid-carrying capacity of trees tends to be high in spring and autumn and low in summer, whereas the carrying capacity of short-season crops in particular tends to be low early in the year, peaking mid-year, and then declining. This is an important point that will be returned to later in this chapter.

Much is known about the biology of the parthenogenetic generations of aphids, in particular the optimum behaviour for maximizing the instantaneous population growth rate, r_m, under various environmental conditions (Kindlmann and Dixon, 1989, 1992; Kindlmann *et al.*, 1992) and the optimal strategies for migration (Dixon *et al.*, 1993a). An individual-based model (Kindlmann and Dixon, 1996), which incorporated all that is known about the biology of tree-dwelling aphids, simulated most of the observed features of the population dynamics. It provided a theoretical background for the commonly observed phenomenon that the larger the numbers are at the beginning of a season, the larger and earlier the peak. Migration was shown to be the most important factor determining the summer decline in abundance, while changes in aphid size and food quality account for why the autumnal increase is less steep than in spring. Finally, the model suggests the possibility of a 'seesaw effect' (a negative correlation between spring and autumn peak numbers) in some cases, a phenomenon observed in census data (Dixon, 1970, 1971).

There is regularity in the population fluctuations from year to year. Very regular 2-year cycles, as indicated by suction trap catches, have proved very attractive to modellers, who have applied time series analysis to the data (e.g. Turchin, 1990; Turchin and Taylor, 1992). The conspicuous cyclicity observed in yearly totals of the number of some species of aphid on trees, however, is due mainly to the cyclicity in the peak numbers in spring, which are closely correlated with the yearly totals. It is driven by the inverse relationship between the size of the spring peak and the autumnal rate of increase, the 'seesaw effect' (Kindlmann and Dixon, 1992). This effect is present in some (Dixon, 1971), but not present or very weak in other (Dixon and Kindlmann, 1998), empirical data. In *Drepanosiphum platanoidis* (sycamore aphid), where the total numbers on the host tree are relatively constant from year to year, there is a within-year seesaw in the abundance of aphids in spring and autumn. As most of the aphids that migrate over long distances, rather than between trees, do so in autumn, the result is the 2-year cycles observed in the suction trap catches (Dixon and Kindlmann, 1998). Time series and correlation analyses reveal that the dynamics are often predictable in spring and late in autumn, but not during summer, as the size of the spring peak is not transferred into summer numbers of aphids (Kindlmann and Dixon, 1992).

It is argued that aphid population density is regulated by density-dependent processes acting within years, which is reflected in the year-to-year changes in overall abundance (Sequeira and Dixon, 1997). Some

results suggest a curvilinear density-dependence, with strong density-dependent regulation at low densities and weak at high densities (Jarošík and Dixon, 1999).

Biology of natural enemies relevant to aphid population dynamics

Aphid colonies are characterized by rapid increases and declines in abundance (Dixon, 1998) that are not synchronized in time, as they feed on different host plants with different phenologies (Galecka, 1966, 1977). On a large spatial scale, at any instant, populations of aphids exist as patches of prey, associated with patches of good host-plant quality (Kareiva, 1990). That is, aphid predators exploit patches of prey that vary greatly in quality both spatially and temporally.

The adult insect predator is winged and can move easily between patches, whereas its immature stages cannot. Thus, the best strategy for an adult is to distribute its offspring between patches in a way that maximizes the expected number and fecundity of offspring that survive to maturity. The developmental time of aphidophagous predators, like coccinellids, often spans several aphid generations. Thus, the ratio of generation time of these predators to that of their prey (generation time ratio, GTR) is large, and optimum foraging strategy therefore depends not only on the present state of an aphid colony, but also on its quality in the future. The optimum oviposition strategy is therefore likely to be determined by expectations of future bottlenecks in prey abundance.

In addition, predators like coccinellids and chrysopids are cannibalistic (Agarwala and Dixon, 1992, 1993). This behaviour is adaptive, as eating conspecific competitors will increase the fitness of a predatory larva. Therefore, eggs laid by predators late in the existence of a patch of prey are at a disadvantage, as they are highly likely to be eaten by larvae of predators that hatch from the first eggs to be laid. Avoiding laying eggs in patches already exploited by larvae is likely to reduce cannibalism and intraguild predation. Empirical data indicate that several different species of insect predator have evolved mechanisms that enable them to oviposit preferentially in patches of prey that are in an early stage of development and avoid those that are already being attacked by larvae (Hemptinne *et al.*, 1992, 1993, 2001). Laying eggs in the presence of conspecific larvae is strongly selected against in these predators, because to lay more eggs results in these eggs being eaten by older conspecific larvae. In addition, laying eggs late in the development of a patch of prey is maladaptive, as there is insufficient time for all the larvae to complete their development. The response to the presence of conspecific larvae reduces the number of eggs laid per patch.

Thus, oviposition commonly occurs only during a short 'egg window', early in the existence of each patch of prey (Hemptinne *et al.*, 1992). When predators are abundant and suitable, patches of prey are rare; however, many eggs may nevertheless be laid in a patch during the 'egg window'. In such circumstances, strong density-dependent cannibalism (Mills, 1982) greatly reduces the abundance of the predators relative to that of their prey. Therefore, these predators have little impact on aphid population dynamics (Dixon, 1992; Kindlmann and Dixon, 1993, 1999; Dixon *et al.*, 1995). They may, however, have short-term impact on local populations valuable to farmers. This is well illustrated by the fact that most of the IPM Case Histories (Chapters 21–30 this volume) recognize the value of natural enemies and the need to limit damage to them when selecting and applying insecticides. Powell and Pell (Chapter 18 this volume) discuss practical biological control of aphids entirely in terms of interventions to increase the enemy:pest ratio that occurs naturally. Without intervention, there is poor synchronization in time and numbers between natural enemies and their aphid prey on arable crops. In an international study of *M. persicae* populations on potato conducted by 16 workers over 2 years in 10 countries (Mackauer and Way, 1976), the majority of data sets recorded aphid population increases regardless of predator presence, and the latter only affected reductions at times when

the potential increase rate of the aphids was low.

Hymenopterous parasitoids can mature on one aphid and would appear to be potentially more likely to regulate aphid abundance. However, their effectiveness is often reduced by: (i) their longer developmental time relative to their host; (ii) the action of hyperparasitoids which, in many cases, are less specific than primary parasitoids; and (iii) their vulnerability to attack from aphid predators (Dixon and Russel, 1972; Hamilton, 1973, 1974; Holler *et al.*, 1993; Mackauer and Völkl, 1993). In addition, because of the risk of hyperparasitism, primary parasitoids are likely to cease ovipositing in a patch where many aphids are already parasitized, as high levels of primary parasitism make the patch attractive to hyperparasitoids. By continuing to oviposit in patches of aphids already attacked by conspecifics, these natural enemies may reduce their potential fitness (Ayal and Green, 1993; Kindlmann and Dixon, 1993).

In the initial phase of aphid population increase on the shorter-season arable crops such as spring-sown cereals, there are often slight dips or plateaux, followed by sudden acceleration. This is attributed to the activity of polyphagous predators (mainly carabid beetles, spiders, and earwigs), and referred to by Southwood and Comins (1976) as the 'natural enemy ravine'. They suggested that the outcome of a spring invasion of aphids is often determined by the balance between the number of invaders and the size of the autochthonous population of polyphagous predators. Carter and Dixon (1981) offered an alternative explanation: the lack of population growth in the initial phase of the population dynamics was attributed to the intermittent nature of aphid immigration, which is amplified by the pre-reproductive period of the offspring of the immigrant aphids. However, it is more likely that the ravine in population dynamics is a consequence of not being able to detect population increase at low population density using small sample sizes (Jarošík *et al.*, 2003). Small sample sizes were used in the studies cited by Southwood and Comins (1976) as evidence for a natural enemy ravine. In the study of

Smith and Hagen (1959), it was 200 lucerne stems. In that of van Emden (1965), it was 90 mustard plants. Wratten (1975) used 30 stems of wheat. The study of Carter and Dixon (1981), in which an alternative explanation for the ravine was proposed, was also based on small sample sizes, with the maximum sample size of 600 tillers of winter wheat. Honek and Jarošík (2000) and Honek *et al.* (2003) also found no evidence that polyphagous predators affect cereal aphid population dynamics in the field. In the habitat they studied, carabid beetles were the dominant guild of polyphagous predators. However, these carabids are mainly seed predators (Honek *et al.*, 2003) and their activity was only loosely correlated with aphid density (Honek and Jarošík, 2000). In addition, aphids have a low nutritional value and are not a preferred food of carabids (Bilde and Toft, 1999). However, in many crops other than cereals, there is a clear mid-season trough in aphid density between an early and a late peak similar to that which occurs on trees. Examples of this trough, attributed to unfavourable host plant condition with its depth influenced by natural enemy activity, can be found in several of the IPM Case Histories, e.g. brassicas (Fig. 21.1) and cotton (Fig. 23.3).

Theory of Aphid Population Dynamics

Features of aphid population dynamics that should be incorporated in models

If it is accepted that natural enemies do not regulate aphid populations, the modelling process is greatly simplified. The important features of any model are:

- Each year aphids show an initial dramatic increase in population size.
- This increase is typically followed by a steep decline in abundance.
- Sometimes there is a further increase in abundance.
- Migration is the most important factor determining the decline in abundance.
- Within-season aphid dynamics often show a 'seesaw effect' – a negative

correlation between initial and final peak numbers.

- The greater the initial aphid numbers, the larger and earlier the peak.
- Very regular, 2-year cycles are characteristic of aphid between-year population dynamics.
- Aphid population density is regulated by density-dependent processes acting within years, which can be potentially strong at low densities.
- Long-term aphid dynamics appear to be little affected by the activity of insect natural enemies.

We present several simple models based on estimates of relatively simple parameters, which may be useful for prediction of aphid dynamics complying with these rules.

Regression model

This model assumes one can divide the population dynamics into three periods: initial increase, subsequent decline, and late season increase. After converting population densities to logarithms, it is possible to assume a linear increase or decrease of density in time.

Let $x(t)$ be the natural logarithm of the population density at time t; let $x(0) = x_0$; let s_1, s_2, and s_3 be the rates of initial increase, subsequent decline, and late season increase, respectively; let x_1 be the natural logarithm of the population density when migration begins, which for simplicity coincides with the first peak in population density, and let x_2 be the population density after the decline in abundance (the trough). Then $x(t) = x_0 + s_1 t$ before the peak is reached, $x(t) = x_1 - s_2 t$ during the decline in abundance, $x(t) = x_2 + s_3 t$ late in the season. Because of the inverse relationships between the initial rates of increase, rates of decline, and late season rates of increase, and initial numbers, $s_1 = -k_1 x_0 + q_1$, $s_2 = k_2 x_1 - q_2$ and $s_3 = -k_3 x_2 + q_3$, where k_1, k_2, and k_3 are the slopes, and q_1, q_2, and q_3 the intercepts of the three rate relationships. Then it follows:

$$x_1 = x_0 + (-k_1 x_0 + q_1)T_1$$
$$= q_1 T_1 + (1 - k_1 T_1)x_0 \qquad (1)$$

$$x_2 = x_1 - (k_2 x_1 - q_2) T_2$$
$$= q_2 T_2 + (1 - k_2 T_2) x_1 \qquad (2)$$

$$x_3 = x_2 + (-k_3 x_2 + q_3) T_3$$
$$= q_3 T_3 + (1 - k_3 T_3) x_2 \qquad (3)$$

$$x_{0,\,n+1} = k_4 x_{3,n} \qquad (4)$$

where k_4 is the slope of the relationship between the numbers next spring and late this season, $x_{i,n}$ means x_i in year n (omitted if year is n and no confusion can arise) and T_1, T_2, and T_3 are the duration of the periods of initial increase, subsequent decline, and late season increase, respectively.

Assuming that for a species, k_i, q_i and T_i are the same each year, then from Equations (1)–(4) it follows:

$$x_{i,n+1} = K + k_4(1 - k_3 T_3)(1 - k_2 T_2)$$
$$\times (1 - k_1 T_1) x_{i,n}, \qquad (5)$$

where K is a constant and $i = 0, 1, 2, 3$.

If we make $Q = k_4(1 - k_3 T_3)(1 - k_2 T_2)(1 - k_1 T_1)$, then Equation (5) simplifies to:

$$x_{i,n+1} = K + Q \cdot x_{i,n}, \qquad (6)$$

which is a model for the between-year population dynamics. Equation (6) is a linear difference equation and therefore its predictions can be derived easily from the values of its parameters K and Q. The equilibrium of this system is $x = K/(1 - Q)$ and is positive (and therefore biologically realistic), if and only if either $K > 0$ and $Q < 1$, or if $K < 0$ and $Q > 1$. If the equilibrium of Equation (6) is not positive, then it predicts either an infinite increase in population size, or its eventual extinction. If the equilibrium of Equation (6) is positive, then it is stable, if and only if $-1 < Q < 1$. If $0 < Q < 1$, then Equation (6) tends not to oscillate, if $-1 < Q < 0$, then the system tends to oscillate.

Definition of Q allows many biologically interesting interpretations of these mathematical predictions, but it is beyond the scope of this text to list them all. For example, one interesting case is when the regulatory term in one of the three periods is large and the remaining two are small, so that one of the brackets defining Q is negative, but larger than -1, while the other two brackets are positive. Then Equation (6) predicts oscillations.

If the density-dependent terms are small in all three periods, so that all three brackets defining Q are positive, then Equation (6) is unlikely to predict oscillations, and there are lots of other scenarios.

Equations (1)–(3) accurately describe the within-year population trends observed in two species of aphids, *D. platanoidis* and *Myzocallis boerneri* (Turkey oak aphid) (Dixon and Kindlmann, 1998), and are therefore realistic. They are able to simulate most of the characteristic features of both within- and between-year aphid dynamics described in the preceding section. Their serious drawback is the large number of parameters that have to be estimated. In addition, the assumption that k_i, q_i, and T_i are the same each year is not always satisfied, especially the duration of the initial period, which is known to vary relative to the initial numbers of aphids: the greater the initial number of aphids, the larger and earlier the peak (Dixon and Kindlmann, 1998).

Regression model with stochasticity

The minimum number of aphids present during the trough does not depend on the peak numbers (Dixon and Kindlmann, 1998). If the peak is large, then there is intense competition for resources, and the aphids present at the beginning of autumn are small and have a low fecundity (Dixon, 1990), which affects the population rate of increase in autumn (Dixon, 1975). When the summer peak is small, the autumnal rate of increase can be either small or large (Dixon and Kindlmann, 1998). This variability may be a consequence of higher predation in some years, as predators attracted to large numbers of aphids in summer may have a marked effect on the few aphids that remain after the summer migration, or of meteorological factors like wind (Dixon, 1979). The autumnal rate of increase, however, is positively correlated with the size of the peak the following year (Dixon and Kindlmann, 1998). Summarizing, aphid dynamics are often not very deterministic. This is sometimes seen in the low values of the correlation coefficients of the relationships defining the individual phases in aphid dynamics (Dixon and Kindlmann, 1998). Thus, there are advantages in replacing Equations (2) and (3) with:

$$x_2 = x_1 - (k_2 x_1 - q_2) T_2 + r.\text{RND}_1$$
$$= q_2 T_2 + (1 - k_2 T_2) x_1 + r.\text{RND}_1 \quad (2a)$$

$$x_3 = x_2 + \text{RND}_2.(-k_3 x_2 + q_3) T_3$$
$$= \text{RND}_2 q_3 T_3 + (1 - \text{RND}_2 k_3 T_3) x_2 \quad (3a)$$

where RND_1 is a random number from $<-0.5; 0.5>$ and r is a constant, which together simulate the extremely low numbers, the stochasticity of migration, and the sampling error. RND_2 is a random number from $<0; 1>$, which takes into account that the population may be negatively affected in autumn.

This model has the advantage of taking into account the underlying stochasticity, but its predictions are stochastic. Interestingly, in the absence of the stochastic element, the cyclicity in the yearly totals of *M. boerneri* is more definite, but the amplitude of the fluctuations is smaller. That is, this model's prediction of the trend in yearly totals more closely reflects reality than does the model with no stochasticity (Kindlmann and Dixon, 1998).

Logistic model with variable 'carrying capacity'

A distinctive feature of aphid dynamics is that the decline in numbers is caused mainly by their own dynamics and not by other species. Thus, the commonly used logistic growth model cannot be used. The amount of soluble nitrogen in the leaves (an indicator of nutritive quality) is high in spring when the leaves are actively growing, falls to a low level in summer, and then increases in autumn prior to leaf fall (Dixon, 1963, 1971). Accepting that the concentration of soluble nitrogen in the leaves is a good indicator of changes in host quality, it is reasonable to assume that aphid-carrying capacity may show a similar trend. This would lead to a logistic model with carrying capacity varying in time, $x' = rx(1 - x/K(t))$. Figure 12.1 shows the predicted dynamics, when the

carrying capacity is assumed to vary between Kmax and Kmin following a cosine function: $K(t) = (K_{\max} - K_{\min}).((\cos(t\pi/d) + 1)/2) + K_{\min}$. The dynamics are similar to those observed – an initial increase followed by a steep decline and a further increase. The greater the initial numbers of aphids, the larger and earlier the peak. However, it can be shown that the population trajectories can never cross – in other words, if 2 years are compared in which the initial aphid numbers in one year are larger than those in the other, the same holds for autumn abundance. Thus, this model does not predict the seesaw effect.

Cumulative density model

In this model, it is assumed that the regulatory term that slows down the instantaneous rate of increase is cumulative density, rather than a term that is proportional to instantaneous density, as in logistic growth. This is based on the assumption that it is the sum of the numbers of individuals multiplied by their life span, which determines the slowing down of the instantaneous rate of increase. The logistic growth never yields a decline in population density with time – a phenomenon typical of aphid population dynamics. In the absence of an effect of natural enemies, an autoregulatory term that causes a decline in population density with time is needed. Cumulative density is a potential candidate, as it could influence food quality and hence slow down population rate of increase. Accepting this, aphid population dynamics can be described by the following set of differential equations:

$$\frac{dh}{dt} = ax, \qquad h(0) = 0, \qquad (7a)$$

$$\frac{dx}{dt} = (r - h)x, \qquad x(0) = x_0 \qquad (7b)$$

where $h(t)$ is cumulative density of aphids at time t, $x(t)$ is density of aphids at time t, a is a scaling constant relating aphid cumulative density to its own dynamics, and r is maximum potential growth rate of the aphids. Thus, while in the logistic model

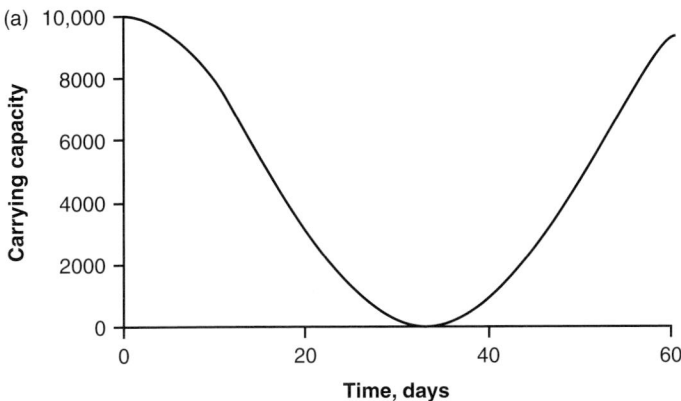

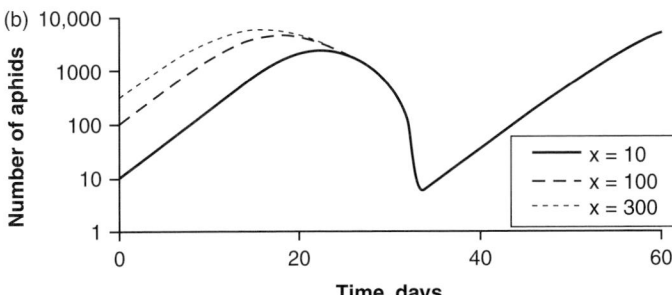

Fig. 12.1. (a) Carrying capacity and (b) aphid density as a function of time for various initial densities of the aphid, predicted by the logistic model using a variable carrying capacity: $r = 0.3$; $K_{max} = 10,000$, $K_{min} = 1$; $d = 33$. Initial aphid densities (x) in insets.

it is assumed that the growth rate linearly declines with population density, here it is assumed that the growth rate linearly declines with aphid cumulative density. This models the decline in population density with time, so typical of aphid populations.

This model predicts that with increasing initial aphid density, the peak density becomes larger, and is achieved earlier. This is illustrated in Fig. 12.2. This model also successfully simulates most of the features of aphid dynamics. One example of the fit of this model to empirical data is shown in Fig. 12.3.

Logistic model with variable carrying capacity and growth rate affected by cumulative density

Empirical data indicate that if the number of aphids present early in a season is large, then there is intense competition for resources and the aphids present at the beginning of autumn are small and have a low fecundity,

which affects the population rate of increase in autumn (Dixon, 1975, 1990). Simulation of this can be achieved by combination of the previous two models, which gives:

$$\frac{dh}{dt} = ax, \qquad\qquad h(0) = 0, \qquad (8a)$$

$$\frac{dx}{dt} = (r - h)x\left(1 - \frac{x}{K}\right), \qquad x(0) = x_0, \qquad (8b)$$

$$K(t) = (K_{max} - K_{min}).$$
$$\times ((\cos(t\pi d) + 1)/2) + K_{min} \qquad (8c)$$

This model, unlike the logistic model with a variable carrying capacity, yields a seesaw effect (Fig. 12.4). However, the basic practical problem of how to measure the time varying 'carrying capacity', K, remains.

Comparison of the different population models

The 'regression model' is descriptive, contains more parameters, and is thus more flexible, but is more data demanding than

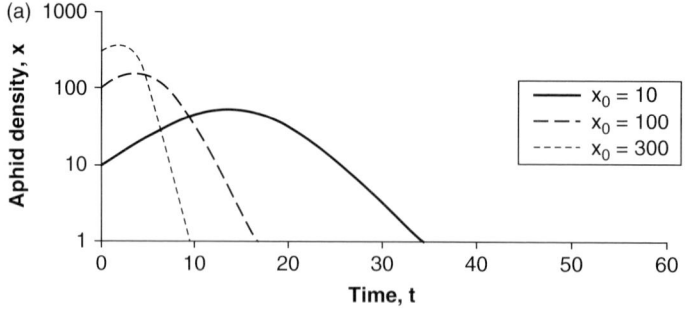

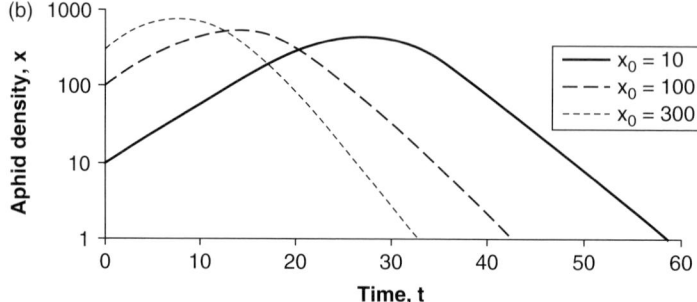

Fig. 12.2. Aphid density as a function of time for various initial densities of the aphid predictions of Equation (7): $r = 0.3$; (a) $a = 0.0005$, (b) $a = 0.00005$; initial aphid densities in insets.

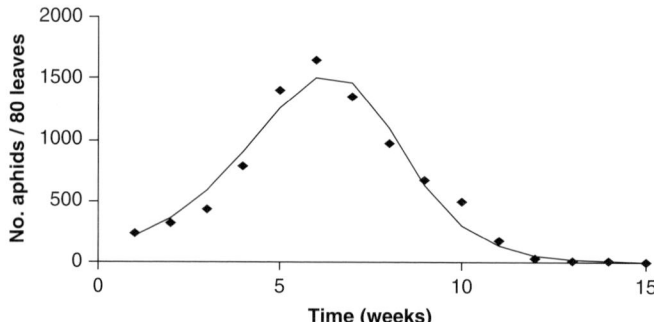

Fig. 12.3. Aphid density as a function of time predicted by Equation (7) (solid line), relative to the census data for the aphid on a Turkey oak tree in 1995; estimated parameters: $r = 0.71$; $a = 0.00015$, initial aphid density $x_0 = 215$.

the logistic growth model with variable carrying capacity. It divides a season into three periods: initial increase, subsequent decline, and final period of increase. Its parameters can be obtained easily from regressions of population densities on time, provided reliable data are available. Its serious drawback is that many of the parameters have to be estimated and the assumption that k_i, q_i, and T_i are the same each year is not always satisfied. Thus, this model is a good theoretical tool, but unlikely to be useful for predicting aphid dynamics.

The logistic growth model with variable carrying capacity is a logical consequence of the observation that food quality undergoes changes during the season. The problem with this model is that it requires an estimate of the trend in the carrying capacity during a season, which may be impossible to measure directly. An indirect measure, e.g. the concentration of soluble amino acids, is used instead. Its major drawback, however, is that it does not predict the seesaw effect.

The 'cumulative density model' is based on the intrinsic ecology of aphids, does not describe a specific scenario, and requires fewer data. However, it describes only the dynamics during the initial and

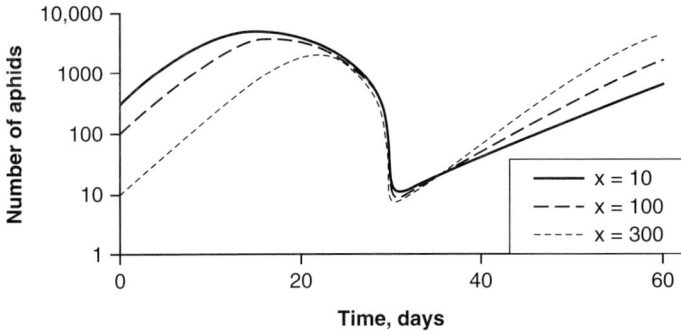

Fig. 12.4. Aphid density as a function of time for various initial densities of the aphid – prediction of the logistic model with variable carrying capacity: $r = 0.3$; $K_{max} = 10,000$, $K_{min} = 1$; $d = 33$. Initial aphid densities (x) in inset.

decline periods. The increase in abundance later in the season remains to be modelled.

The model with variable carrying capacity and growth rate affected by cumulative density combines the advantages of the two previous models and is the most flexible of the models. However, it also requires an estimate of the trend in the carrying capacity during a season. As stated above, it is likely to be V-shaped for tree-dwelling aphids and Λ-shaped for aphids infesting annual crops. Although a single peak is found with cereal aphids, it is rather atypical and due to the short time before plant decline that the aphid population increase starts. Thus, in case of multiple peaks occurring during one season, each peak has to be modelled separately.

Natural Enemies and Aphid Abundance

Although there is no evidence that natural enemies are capable of regulating the abundance of aphids, they may affect aphid abundance, and their activity may therefore be of economic benefit to the farmer. In some years with low cereal aphid numbers, they may even obviate the need to spray altogether (see Poehling *et al.*, Chapter 25 this volume). In killing aphids, natural enemies are therefore potentially capable of reducing the population rate of increase of aphids. This will depend, however, on the strength of the density-dependent processes acting on the aphid's rate of increase. If the natural enemy load associated with a pest aphid is high, then it might reduce its abundance. Thus, increasing the diversity and abundance of the natural enemies in crops might

reduce the abundance of pest aphids. The challenge is to show that it does and that it is a cost effective way of protecting crops.

Practical Problems

For forecasting pest aphid abundance and for making decisions in integrated control programmes, it may be necessary to have accurate estimates of aphid abundance and population growth rates.

Field estimates of abundance and population growth rate

The crucial parameters of the models, maximum potential growth rate of aphids and its parabolic decline, are difficult to assess for aphids living on herbaceous plants, including crops. The seasonal trends in aphid abundance on crop plants can be determined, however, providing there are no constraints on sample sizes. For example, this was done using census data from 268 winter wheat plots collected at 3- or 7-day intervals over a period of 10 years for aphids on wheat leaves and 6 years for those on the ears (Jarošík *et al.*, 2003). Aphids can become very abundant on wheat, which they colonize in spring. The initially very sparse populations of aphids grow and reach maximum densities, usually in the last days of June, and then decline sharply in abundance. An estimate of the maximum potential population growth rate of the aphids, i.e. the intrinsic rate of population increase, can be obtained by fitting linear regressions, in which population growth is linearized by log transformation of population size *N*, and by expressing time in day-degrees (*DD*) above the lower developmental threshold (Honek and Kocourek, 1990; Honek, 1996). The model corresponding to exponential population growth is then:

$$ln[N(DD)] = ln[N(0)] + rDD$$

where *ln[N(DD)]* is the response variable, *ln [N(0)]* is the intercept, *r* is the slope of the regression line and the population growth rate, and *DD* is an explanatory variable. The explanatory variable is the sampling date expressed in *DD*, which is from the beginning of immigration until the peak of aphid abundance. The population growth rate *r* is an estimate of the achieved intrinsic rate of increase (Jarošík *et al.*, 1996).

To test for the parabolic decline from exponential growth, the square of the explanatory variable can be calculated and subtracted from the regression:

$$ln[N(DD)] = ln[N(0)] + rDD - rDD^2$$

If this causes a significant reduction in deviance, there is evidence of parabolic decrease in population growth with increase in aphid density (e.g. Crawley, 1993).

The use of log aphid counts and normal distribution of errors in the statistical analyses is preferable to the use of generalized linear models with a Poisson or negative binomial distribution of errors (McCullagh and Nelder, 1989). The reason is that, in spite of the fact that aphid distributions are highly clumped, they usually differ significantly not only from a Poisson distribution, which describes a random distribution, but also from the negative binomial distribution, which is usually used to assess population densities where the distributions are clumped (Ekbom, 1985, 1987; Elliot and Kieckhefer, 1986; Krebs, 1989; Elliot *et al.*, 1990; Jarošík *et al.*, 2003).

Importance of sample size

In a study of cereal aphids (Jarošík *et al.*, 2003), no population growth was detected at very low densities on individual plots.

The probability of identifying exponential growth increased with aphid density and made it possible to determine the crucial density for the transition from undetectable population growth at low density to exponential growth as density increased. At high densities, the populations grew exponentially, and the growth rates did not show a parabolic decline as aphid density increased.

However, significant exponential growth was always detected in pooled data. This was strikingly different from the dynamics on individual plots, because significant exponential growth was revealed even at very low densities. The second important difference was that the growth rates decreased significantly at high densities, in accordance with the cumulative density model, illustrated in Figs. 12.2 and 12.3.

The detection of significant growth using pooled data, when no growth was detectable using data from individual plots, is attributable to the much larger sample size of the pooled data. The sample size on individual plots was 300 tillers or ears at low densities. The pooled sample sizes ranged from 2400 to 30,600 tillers (average 12,900) and from 1200 to 22,800 ears (average 10,500).

Sample size is important when assessing species abundance (e.g. Southwood and Henderson, 2000). Using the density for the transition to exponential growth, a sample size of 300 tillers or ears appeared clearly insufficient for the correct assessment of aphid abundance (Ekbom, 1985, 1987; Ward *et al.*, 1985a; Elliot and Kieckhefer, 1986; Elliot *et al.*, 1990). Thus, the random population fluctuation without apparent population growth, which occurred at low densities on individual plots, appears to be attributable to small sample size. It does not mean that these populations were not growing.

There is no simple solution to the problem of sampling low aphid densities. The required sample sizes are very large, and therefore time-consuming. The use of presence–absence counts, instead of counting the numbers of aphids, is not a solution, because the saving in time is associated with a decrease in accuracy, and hence a further increase in the required sample size (cf. Ward *et al.*, 1985a,b; Elliot *et al.*, 1990). The nigh impossibility of obtaining accurate estimates of aphid abundance when aphids are scarce further supports the notion that the natural enemy ravine may be an illusion. The studies cited in support of the ravine concept sampled aphid densities similar to those used in the study on cereal aphids. This is important, because at high densities it would be possible to detect the ravine using a smaller sample size. Therefore, there is no evidence from these studies that the ravine might be present when the densities are higher, as in the cereal aphid study, and could be detected by a smaller sample size.

Trap estimates of abundance

Another way of estimating the abundance of a pest aphid is by using traps (see Harrington *et al.*, Chapter 19 this volume). As in field sampling, it is important to ask the following question: 'What constitutes a series of trap catches that is suitable for statistical analysis?' In this respect, the selection procedures of several analyses are of interest. Redfearn and Pimm (1988) chose time series that 'contained at least 10 years in which at least one individual was collected', and Woiwod and Hanski (1992) and Hanski and Woiwod (1993) removed from their analyses all series for which the mean annual abundance was less than five individuals in order to remove time series with many zeros. However, it is not possible to state unequivocally that catches below ten differ from one another (Thacker, 1995). For this reason, the data selected by Redfearn and Pimm (1988) and Woiwod and Hanski (1992) could be considered too liberal. When assumptions of year-to-year differences in abundance are to be made, time series with many values below ten are suspect (Thacker, 1995). Variances of series containing many zeros are difficult to compare with certainty (Wolda and Marek, 1994). A selection procedure based solely on mean catches is also difficult to recommend for all series, as it does not consider the spread of the data values. Of two series with identical means, that with the higher variance is more suitable

for use, as conclusions about year-to-year changes in abundance can be more reliably drawn. Therefore, the selection process could be improved by including variance, unless a very conservative mean value cut-off is used. That is, if a catch is low, it is possible to say with certainty that the aphid population is low. However, the existence and magnitude of any population trends are rather more difficult to elicit. So, for example, if the catch sequence goes 4, 4, 4, the real population underlying that may have quadrupled or quartered over those three time periods. But if the sequence is 500, 500, 500, then the population being sampled is more abundant and very stable (Thacker, 1995).

The foregoing discussion assumes that the aerial population accurately reflects the abundance of an aphid on its host plant (cf. Howard and Dixon, 1990). This may not always be the case. For example, the yearly catches of *D. platanoidis* by a suction trap positioned 1 m above the ground and close to sycamore (*Acer pseudoplatanus*) trees accurately reflected the abundance of aphids on these trees and, like the total numbers on the trees, showed relatively little change from year to year (Dixon, 1990; Dixon and Kindlmann, 1998). In contrast, the yearly catches recorded by the Rothamsted Insect Survey (RIS) trap 106 km away fluctuated from year to year. The reason for this is that the size of the annual catch taken by the RIS trap is determined mainly by the aphids caught in June, October, and November (Fig. 9.13 in Dixon and Kindlmann, 1998). The sum of the catches in these 3 months taken by the suction trap close to the trees is well correlated with the yearly total catches of the RIS trap. That is, the RIS trap is, in this case, catching aphids mainly in the second half of a year. In addition, the ratio of the catches of the local and RIS traps each month did not remain constant throughout the year. Early in a year, the local trap caught many more aphids relative to the RIS trap than later in a year. That is, the flight behaviour of the aphid is changing during the course of a year and, as a consequence, the RIS trap, in this case, does not accurately reflect the trends in aphid abundance on the host tree (Dixon and Kindlmann, 1998).

In summary, providing it is known that the catches accurately reflect the abundance of a pest aphid on a crop plant, suction trap catches may reliably indicate year-to-year changes in abundance, and trends in abundance within years of abundant species. However, the catches may be a more reliable indicator of the timing of aphid migrations, which is important when interested mainly in virus transmission by aphids.

Consequences for Pest Management

If, at low aphid densities, population growth on cereals is undetectable, then the switch to exponential growth as density increases, assuming significant damage to this crop occurs mainly during the exponential phase of population increase, could be used to predict when, and if, the damage boundary (*sensu* Pedigo *et al.*, 1986) is likely to be exceeded. Distinguishing the early phase of population growth thus might have important consequences for aphid control (Jarosik *et al.*, 1996, 1997). However, pooled results showed that there is actually no transition from undetectable population growth to exponential increase in the case of cereal aphids (Jarošík *et al.*, 2003). The apparent transitions are just a consequence of low population density, when population increase is undetectable using small sample sizes. Small trap catches present similar problems.

In summary, for aphids on cereal crops, the required sample size at low densities for measuring exponential growth is more than 500 tillers and, in most cases, much larger sample sizes are required (cf. Ward *et al.*, 1985a; Elliot and Kieckhefer, 1986; Ekbom, 1987; Elliot *et al.*, 1990). Moreover, estimates of sample size are very unreliable at low densities (Ward *et al.*, 1985b), and the required sample size increases rapidly to infinity if the densities are less than, for example, an aphid per tiller. Such densities are typical of the onset of aphid population growth. This also applies to trap catches. Therefore, because the sample sizes needed for a reliable decision are extremely large, and therefore costly, it is likely that their use in forecasting will be limited.

Conclusions

In terms of theory, we have a good understanding of aphid population dynamics, and the models accurately predict the population trends observed in the field. This is particularly the case for aphids living on woody plants. On the practical side, progress has been disappointing. This is mainly because of the difficulty of obtaining accurate estimates of population size when aphids are scarce. In spite of this, there have been some successes in aphid pest management. Therefore, we should be optimistic and believe that a better understanding of the seasonal trends in aphid abundance in the field may result in better ways of determining when aphid control should be applied. In certain cases, it may even be possible to use them to forecast future trends.

References

Agarwala, B.K. and Dixon, A.F.G. (1992) Laboratory study of cannibalism and interspecific predation in ladybirds. *Ecological Entomology* 17, 303–309.

Agarwala, B.K. and Dixon, A.F.G. (1993) Kin recognition: egg and larval cannibalism in *Adalia bipunctata* (Coleoptera: Coccinellidae). *European Journal of Entomology* 90, 45–50.

Ayal, Y. and Green, R.F. (1993) Optimal egg distribution among host patches for parasitoids subject to attack by hyperparasitoids. *American Naturalist* 141, 120–138.

Barlow, N.D. and Dixon, A.F.G. (1980) *Simulation of Lime Aphid Population Dynamics*. Pudoc, Wageningen, 165 pp.

Bilde, T. and Toft, S. (1999) Prey consumption and fecundity of the carabid beetle *Calathus melanocephalus* on diets of three cereal aphids: high consumption rates of low quality prey. *Pedobiologia* 43, 422–429.

Carter, N. (1985) Simulation modelling of the population dynamics of cereal aphids. *Biosystems* 1, 111–119.

Carter, N. and Dixon, A.F.G. (1981) The 'natural enemy ravine' in cereal aphid population dynamics. *Journal of Animal Ecology* 50, 605–611.

Carter, N., Dixon, A.F.G. and Rabbinge, R. (1982) *Cereal Aphid Populations: Biology, Simulation and Predation.* Centre for Agricultural Publishing, Wageningen, 91 pp.

Chambers, R.J., Wellings, P.W. and Dixon, A.F.G. (1985) Sycamore aphid numbers and population density 11. Some processes. *Journal of Animal Ecology* 54, 425–442.

Crawley, M.J. (1993) *GLIM for Ecologists.* Blackwell, London, 379 pp.

DeLoach, C.J. (1974) Rate of increase of populations of cabbage, green peach, and turnip aphids at constant temperatures. *Annals of the Entomological Society of America* 67, 332–340.

Dixon, A.F.G. (1963) Reproductive activity of the sycamore aphid, *Drepanosiphum platanoides* (Schr.) (Hemiptera. Aphididae). *Journal of Animal Ecology* 32, 33–48.

Dixon, A.F.G. (1966) The effect of population density and nutritive status of the host on the summer reproductive activity of the sycamore aphid, *Drepanosiphum platanoides* (Schr.). *Journal of Animal Ecology* 35, 105–112.

Dixon, A.F.G. (1969) Population dynamics of the sycamore aphid *Drepanosiphum platanoides* (Schr.) (Hemiptera: Aphididae): migratory and trivial flight activity. *Journal of Animal Ecology* 38, 585–606.

Dixon, A.F.G. (1970) Stabilisation of aphid populations by an aphid induced plant factor. *Nature, London* 227, 1368–1369.

Dixon, A.F.G. (1971) The role of intra-specific mechanisms and predation in regulating the numbers of the lime aphid, *Eucallipterus tiliae* L. *Oecologia* 8, 179–193.

Dixon, A.F.G. (1975) Effect of population density and food quality on autumnal reproductive activity in the sycamore aphid, *Drepanosiphum platanoides* (Schr.). *Journal of Animal Ecology* 44, 297–304.

Dixon, A.F.G. (1979) Sycamore aphid numbers: the role of weather, host and aphid. In: Anderson, R.M., Turner, B.D. and Taylor, L.R. (eds) *Population Dynamics.* Blackwell, Oxford, pp. 105–121.

Dixon, A.F.G. (1990) Population dynamics and abundance of deciduous tree-dwelling aphids. In: Hunter, M., Kidd, N., Leather, S.R. and Watt, A.D. (eds) *Population Dynamics of Forest Insects.* Intercept, Andover, pp. 11–23.

Dixon, A.F.G. (1992) Constraints on the rate of parthenogenetic reproduction and pest status of aphids. *Invertebrate Reproduction and Development* 22, 159–163.

Dixon, A.F.G. (1998) *Aphid Ecology,* 2nd edn. Chapman and Hall, 300 pp.

Dixon, A.F.G. and Barlow, N.D. (1979) Population regulation in the lime aphid. *Zoological Journal of the Linnean Society* 67, 225–237.

Dixon, A.F.G. and Kindlmann, P. (1998) Population dynamics of aphids. In: Dempster, P.J. and McLean, I.F.G. (eds) *Insect Populations.* Kluwer, Dordrecht, pp. 207–230.

Dixon, A.F.G. and Mercer, D.R. (1983) Flight behaviour in the sycamore aphid: factors affecting take-off. *Entomologia Experimentalis et Applicata* 33, 43–49.

Dixon, A.F.G. and Russel, R.J. (1972) The effectiveness of *Anthocoris nemorum* and *A. confusus* (Hemiptera: Anthocoridae) as predators of the sycamore aphid, *Drepanosiphum platanoides* II. Searching behaviour and the incidence of predation in the field. *Entomologia Experimentalis et Applicata* 15, 35–50.

Dixon, A.F.G., Horth, S. and Kindlmann, P. (1993a) Migration in insects: costs and strategies. *Journal of Animal Ecology* 62, 182–190.

Dixon, A.F.G., Wellings, P.W., Carter, C. and Nichols, J.F.A. (1993b) The role of food quality and competition in shaping the seasonal cycle in the reproductive activity of the sycamore aphid. *Oecologia* 95, 89–92.

Dixon, A.F.G., Hemptinne, J.-L. and Kindlmann, P. (1995) The ladybird fantasy – prospects and limits to their use in the biological control of aphids. *Züchtungsforschung* 1, 395–397.

Dixon, A.F.G., Kindlmann, P. and Sequeira, R. (1996) Population regulation in aphids. In: Floyd, R.B., Sheppard, A.W. and de Barro, P.J. (eds) *Frontiers of Population Ecology.* CSIRO Publishing, Melbourne, pp. 103–114.

Ekbom, B.S. (1985) Spatial distribution of *Rhopalosiphum padi* (L.) (Homoptera: Aphididae) in spring cereals in Sweden and its importance for sampling. *Environmental Entomology* 14, 312–316.

Ekbom, B.S. (1987) Incidence counts for estimating densities of *Rhopalosiphum padi* (Homoptera: Aphididae). *Journal of Economic Entomology* 80, 933–935.

Elliot, N.C. and Kieckhefer, R.W. (1986) Cereal aphid populations in winter wheat: spatial distributions and sampling with fixed levels of precision. *Environmental Entomology* 15, 954–958.

Elliot, N.C., Kieckhefer, R.W. and Walgenbach, D.D. (1990) Binomial sequential sampling methods for cereal aphids in small grains. *Journal of Economic Entomology* 83, 1381–1387.

van Emden, H.F. (1965) The effect of uncultivated land on the distribution of cabbage aphid (*Brevicoryne brassicae*) on an adjacent crop. *Journal of Applied Ecology* 2, 171–196.

Galecka, B. (1966) The role of predators in the reduction of two species of potato aphids, *Aphis nasturtii* Kalt. and *A. frangulae* Kalt. *Ekologia Polska* 14, 245–274.

Galecka, B. (1977) Effect of aphid feeding on the water uptake by plants and on their biomass. *Ekologia Polska* 25, 531–537.

Gilbert, N. and Hughes, R.D. (1971) A model of an aphid population – three adventures. *Journal of Animal Ecology* 40, 525–534.

Gonzalez-Andujara, J.L., Garcia-de Cecab, J.L. and Fereresc, A. (1993) Cereal aphids expert system (CAES): identification and decision making. *Computers and Electronics in Agriculture* 8, 293–300.

Gosselke, U., Triltsch, H., Roßberg, D. and Freier, B. (2001) GETLAUS01 – the latest version of a model for simulating aphid population dynamics in dependence on antagonists in wheat. *Ecological Modelling* 145, 143–157.

Gutierrez, A.P. and Baumgärtner, J.U. (1984a) Multitrophic level models of predator–prey energetics. I. Age specific energetics models – pea aphid *Acyrthosiphum pisum* (Harris) (Homoptera: Aphididae) as an example. *Canadian Entomologist* 116, 924–932.

Gutierrez, A.P. and Baumgärtner, J.U. (1984b) Multitrophic level models of predator–prey energetics. II. A realistic model of plant–herbivore–parasitoid–predator interactions. *Canadian Entomologist* 116, 933–949.

Gutierrez, A.P., Havenstein, D.E., Nix, H.A. and Moore, P.A. (1974) Ecology of *Aphis craccivora* Koch and subterranean clover stunt virus in southeast Australia. 2. Model of cowpea aphid populations in temperate pastures. *Journal of Applied Ecology* 11, 1–20.

Gutierrez, A.P., Summers, C.G. and Baumgärtner, J. (1980) The phenology and distribution of aphids in California alfalfa as modified by ladybird beetle predation (Coleoptera, Coccinellidae). *Canadian Entomologist* 112, 489–495.

Gutierrez, A.P., Baumgärtner, J.U. and Hagen, K.S. (1981) A conceptual-model for growth, development, and reproduction in the ladybird beetle, *Hippodamia convergens* (Coleoptera, Coccinellidae). *Canadian Entomologist* 113, 21–33.

Gutierrez, A.P., Baumgärtner, J.U. and Summers, G.C. (1984) Multitrophic level models of predator–prey energetics. III. A case study of an alfalfa system. *Canadian Entomologist* 116, 950–963.

Hamilton, P.A. (1973) The biology of *Aphelinus flavus* (Hym. Aphelinidae), a parasite of the sycamore aphid *Drepanosiphum platanoidis* (Hemipt. Aphididae). *Entomophaga* 18, 449–462.

Hamilton, P.A. (1974) The biology of *Monoctonus pseudoplatani, Trioxys cirsii* and *Dyscritulus planiceps*, with notes on their effectiveness as parasites of the sycamore aphid, *Drepanosiphum platanoidis*. *Annales de la Société Entomologique de France* 10, 821–840.

Hanski, I. and Woiwod, I.P. (1993) Spatial synchrony in the dynamics of moth and aphid populations. *Journal of Animal Ecology* 62, 658–668.

Hemptinne, J.-L., Dixon, A.F.G. and Coffin, J. (1992) Attack strategy of ladybird beetles (Coccinellidae): factors shaping their numerical response. *Oecologia* 90, 238–245.

Hemptinne, J.L., Dixon, A.F.G., Doucet, J.L. and Petersen, J.E. (1993) Optimal foraging by hoverflies (Diptera: Syrphidae) and ladybirds (Coleoptera: Coccinellidae): mechanisms. *European Journal of Entomology* 90, 451–455.

Hemptinne, J.-L., Lognay, G., Doumbia, M. and Dixon, A.F.G. (2001) Chemical nature and persistence of the oviposition deterring pheromone in the tracks of the larvae of the two-spot ladybird, *Adalia bipunctata* (Coleoptera: Coccinellidae). *Chemoecology* 11, 43–47.

Holler, C., Borgemeister, C., Haardt, H. and Powell, W. (1993) The relationship between primary parasitoids and hyperparasitoids of cereal aphids: an analysis of field data. *Journal of Animal Ecology* 62, 12–21.

Honek, A. (1996) The relationship between thermal constants for insect development: a verification. *Acta Societatis Zoologicae Bohemoslovaca* 60, 115–152.

Honek, A. and Jarošík, V. (2000) The role of crop density, seed and aphid presence in diversification of field communities of Carabidae (Coleoptera). *European Journal of Entomology* 97, 517–525.

Honek, A. and Kocourek, F. (1990) Temperature and development time in insects: a general relationship between thermal constants. *Zoologische Jahrbücher. Abteilung für Systematik und Ökologie der Tiere* 117, 401–439.

Honek, A., Martinkova, J. and Jarošík, V. (2003) Carabids as seed predators. *European Journal of Entomology* 100, 531–544.

Howard, M.T. and Dixon, A.F.G. (1990) Forecasting of peak population density of the rose grain aphid *Metopolophium dirhodum* on wheat. *Annals of Applied Biology* 117, 9–19.

Hughes, R.D. (1963) Population dynamics of the cabbage aphid, *Brevicoryne brassicae* (L.). *Journal of Animal Ecology* 32, 393–424.

Hughes, R.D. and Gilbert, N. (1968) A model of an aphid population. *Journal of Animal Ecology* 37, 553–563.

Huxley, T.H. (1858) On the agamic reproduction and morphology of *Aphis* – Part 1. *Transactions of the Linnean Society* 22, 193–219.

Jarošík, V. and Dixon, A.F.G. (1999) Population dynamics of a tree-dwelling aphid: regulation and density independent processes. *Journal of Animal Ecology* 68, 726–732.

Jarošík, V., Honek, A., Lapchin, L. and Rabasse, J.-M. (1996) An assessment of time varying rate of increase of green peach aphid, *Myzus persicae*: its importance in IPM of commercial greenhouse peppers. *Plant Protection* 32, 269–276.

Jarošík, V., Kolias, M., Lapchin, L., Rochat, J. and Dixon, A.F.G. (1997) Seasonal trends in the rate of population increase of *Frankliniella occidentalis* (Thysanoptera, Thripidae) on cucumber. *Bulletin of Entomological Research* 87, 487–495.

Jarošík, V., Honek, A. and Dixon, A.F.G. (2003) Natural enemy ravine revisited: the importance of sample size for determining population growth. *Ecological Entomology* 28, 85–91.

Kareiva, P. (1990) Population dynamics in spatially complex environments: theory and data. *Philosophical Transactions of the Royal Society of London B* 330, 175–190.

Kindlmann, P. and Dixon, A.F.G. (1989) Developmental constraints in the evolution of reproductive strategies: telescoping of generations in parthenogenetic aphids. *Functional Ecology* 3, 531–537.

Kindlmann, P. and Dixon, A.F.G. (1992) Optimum body size: effects of food quality and temperature, when reproductive growth rate is restricted. *Journal of Evolutionary Biology* 5, 677–690.

Kindlmann, P. and Dixon, A.F.G. (1993) Optimal foraging in ladybird beetles (Coleoptera: Coccinellidae) and its consequences for their use in biological control. *European Journal of Entomology* 90, 443–450.

Kindlmann, P. and Dixon, A.F.G. (1996) Population dynamics of a tree-dwelling aphid: individuals to populations. *Ecological Modelling* 89, 23–30.

Kindlmann, P. and Dixon, A.F.G. (1998) Patterns in the population dynamics of the Turkey-oak aphid. In: Nieto Nafria, J.M. and Dixon, A.F.G. (eds) *Aphids in Natural and Managed Ecosystems.* Universidad de León, León, pp. 219–225.

Kindlmann, P. and Dixon, A.F.G. (1999) Generation Time Ratios – determinants of prey abundance in insect predator–prey interactions. *Biological Control* 16, 133–138.

Kindlmann, P., Dixon, A.F.G. and Gross, L.J. (1992) The relationship between individual and population growth rates in multicellular organisms. *Journal of Theoretical Biology* 157, 535–542.

Knudsen, G.R. and Schotzko, D.J. (1991) Simulation of Russian wheat aphid movement and population-dynamics on preferred and nonpreferred host plants. *Ecological Modelling* 57, 117–131.

Krebs, C.J. (1989) *Ecological Methodology.* Harper and Row, New York, 654 pp.

Mack, T.P. and Smilowitz, Z. (1981) The vertical distribution of green peach aphids and its effect on a model quantifying the relationship between green peach aphids and a predator. *American Potato Journal* 58, 345–353.

Mack, T.P. and Smilowitz, Z. (1982) Using temperature-mediated functional response models to predict the impact of *Coleomegilla maculata* (DeGeer) adults and 3rd-instar larvae on green peach aphids. *Environmental Entomology* 11, 46–52.

Mackauer, M. and Völkl, W. (1993) Regulation of aphid populations by aphidiid wasps: does parasitoid foraging behaviour or hyperparasitism limit impact? *Oecologia (Berlin)* 94, 339–350.

Mackauer, M. and Way, M.J. (1976) *Myzus persicae* Sulz., an aphid of world importance. In: Delucchi, V.L. (ed.) *Studies in Biological Control.* Cambridge University Press, Cambridge, pp. 51–119.

Mann, B.P., Wratten, S.D. and Watt, A.D. (1986) A computer-based advisory system for cereal aphid control. *Computers and Electronics in Agriculture* 1, 263–270.

McCullagh, P. and Nelder, J.A. (1989) *Generalized Linear Models.* Chapman and Hall, London, 511 pp.

Mills, N.J. (1982) Voracity, cannibalism and coccinellid predation. *Annals of Applied Biology* 101, 144–148.

Morgan, D. (2000) Population dynamics of the bird cherry–oat aphid, *Rhopalosiphum padi* (L.), during the autumn and winter: a modelling approach. *Agricultural and Forest Entomology* 2, 297–304.

Pedigo, L.P., Scott, H. and Higley, L.G. (1986) Economic injury levels in theory and practice. *Annual Review of Entomology* 31, 341–368.

de Réaumur, R.P. (1737) *Mémoires pour servir à l'histoire des insectes no. 111.* Imprimerie Royale, Paris, pp. 332–350.

Redfearn, A. and Pimm, S.L. (1988) Population variability and polyphagy in herbivorous insect communities. *Ecological Monographs* 58, 39–55.

Ro, T.H. and Long, G.E. (1998) Population dynamics pattern of green peach aphid (Homoptera: Aphididae) and its predator complex in a potato system. *Korean Journal of Biological Science* 2, 217–222.

Ro, T.H. and Long, G.F. (1999) GPA-Phenodynamics, a simulation model for the population dynamics and phenology of green peach aphid in potato: formulation, validation, and analysis. *Ecological Modelling* 119, 197–209.

Scopes, N.E.A. (1969) The potential of *Chrysopa carnea* as a biological control agent of *Myzus persicae* on glasshouse chrysanthemums. *Annals of Applied Biology* 64, 433–439.

Sequeira, R. and Dixon, A.F.G. (1997) Patterns and processes in the population dynamics of tree-dwelling aphids: the importance of time scales. *Ecology* 78, 2603–2610.

Skirvin, D.J., Perry, J.N. and Harrington, R. (1997a) A model describing the population dynamics of *Sitobion avenae* and *Coccinella septempunctata*. *Ecological Modelling* 96, 29–39.

Skirvin, D.J., Perry, J.N. and Harrington, R. (1997b) The effect of climate change on an aphid–coccinellid interaction. *Global Change Biology* 3, 1–11.

Smilowitz, Z. (1984) GPA-CAST: a computerized model for green peach aphid management on potatoes. In: Lashomb, J.H. and Casagrande, R. (eds) *Advances in Potato Pest Management.* Hutchinson Ross, Pennsylvania, pp. 193–203.

Smith, R.F. and Hagen, K.S. (1959) Impact of commercial insecticide treatments. *Hilgardia* 29, 131–154.

Southwood, T.R.E. and Comins, H.N. (1976) A synoptic population model. *Journal of Animal Ecology* 45, 949–965.

Southwood, T.R.E. and Henderson, P.A. (2000) *Ecological Methods*, 3rd edn. Blackwell, Oxford, 575 pp.

Stewart, L. and Dixon, A.F.G. (1988) Quantification of the effect of natural enemies on the abundance of aphids: precision and statistics. In: Niemczyk, E. and Dixon, A.F.G. (eds) *Ecology and Effectiveness of Aphidophaga.* Academic Publishing, The Hague, pp.187–197.

Tamaki, G. (1973) Spring populations of the green peach aphid on peach trees and the role of natural enemies in their control. *Environmental Entomology* 2, 186–191.

Tamaki, G. (1984) Biological control of potato pests. In: Lashomb, J.H. and Casagrande, R. (eds) *Advances in Potato Pest Management.* Hutchinson Ross, Pennsylvania, pp. 178–192.

Tamaki, G. and Weeks, R.E. (1972) Efficiency of three predators, *Geocoris bullatus*, *Nabis americoferous*, and *Coccinella transversoguttata*, used alone or in combination against three prey species, *Myzus persicae*, *Ceramica picta*, and *Mamestra configurata*, in a greenhouse study. *Environmental Entomology* 1, 258–263.

Tamaki, G. and Weeks, R.E. (1973) The impact of predators on populations of green peach aphids on field-grown sugar beets. *Environmental Entomology* 2, 345–349.

Tamaki, G., Weiss, M.A. and Long, G.E. (1980) Impact of high temperatures on the population dynamics of the green peach aphid in field cages. *Environmental Entomology* 9, 331–337.

Tamaki, G., Weiss, M.A. and Long, G.E. (1982) Effective growth units in population dynamics of the green peach aphid (Homoptera: Aphididae). *Environmental Entomology* 11, 1134–1136.

Taylor, L.R. (1977) Migration and the spatial dynamics of an aphid, *Myzus persicae*. *Journal of Animal Ecology* 46, 411–423.

Thacker, J.I. (1995) The role of density dependence and weather in determining aphid abundance. PhD thesis, University of East Anglia, Norwich, UK.

Turchin, P. (1990) Rarity of density dependence or population regulation with lags? *Nature, London* 344, 660–663.

Turchin, P. and Taylor, A.D. (1992) Complex dynamics in ecological time series. *Ecology* 73, 289–305.

Ward, S.A., Rabbinge, R. and Mantel, W.P. (1985a) The use of incidence counts for estimation of aphid populations. 1. Minimum sample size for required accuracy. *Netherlands Journal of Plant Pathology* 91, 93–99.

Ward, S.A., Rabbinge, R. and Mantel, W.P. (1985b) The use of incidence counts for estimation of aphid populations. 2. Confidence intervals from fixed sample sizes. *Netherlands Journal of Plant Pathology* 91, 100–104.

Watt, A.D. (1983) The influence of forecasting on cereal aphid control strategies. *Crop Protection* 2, 417–429.

Watt, A.D., Vickerman, G.P. and Wratten, S.D. (1984) The effect of the grain aphid, *Sitobion avenae* (F.), on winter wheat in England: an analysis of the economics of control practice and forecasting systems. *Crop Protection* 3, 209–222.

Weisser, W.W. (2000) Metapopulation dynamics in an aphid–parasitoid system. *Entomologia Experimentalis et Applicata* 97, 83–92.

Wellings, P.W., Chambers, R.J., Dixon, A.F.G. and Aikman, D.P. (1985) Sycamore aphid numbers and population density. 1. Some patterns. *Journal of Animal Ecology* 54, 411–424.

Whalon, M.E. and Smilowitz, Z. (1979) Temperature-dependent model for predicting field populations of green peach aphid, *Myzus persicae* (Sulzer) (Homoptera: Aphididae). *Canadian Entomologist* 111, 1025–1032.

Wiktelius, S. and Pettersson, J. (1985) Simulations of bird cherry–oat aphid population dynamics: a tool for developing strategies for breeding aphid-resistant plants. *Agriculture, Ecosystems and Environment* 14, 159–170.

Winder, L., Alexander, C.J., Holland, J.M., Woolley, C. and Perry, J.N. (2001) Modelling the dynamic spatio-temporal response of predators to transient prey patches in the field. *Ecology Letters* 4, 568–576.

Woiwod, I.P. and Hanski, I. (1992) Patterns of density dependence in aphids and moths. *Journal of Animal Ecology* 61, 619–630.

Wolda, H. and Marek, J. (1994) Measuring variation in abundance, the problem with zeros. *European Journal of Entomology* 91, 145–161.

Woolhouse, M.E.J. and Harmsen, R. (1991) Population dynamics of *Aphis pomi*: a transition matrix approach. *Ecological Modelling* 55, 103–111.

Wratten, S.D. (1975) The nature of the effects of aphids *Sitobion avenae* and *Metopolophium dirhodum* on the growth of wheat. *Annals of Applied Biology* 79, 27–34.

13 Feeding Injury

Sharron S. Quisenberry[1] and Xinzhi Ni[2]

[1]College of Agriculture and Life Sciences, Virginia Tech, Blacksburg, VA 24061, USA;
[2]USDA-ARS Crop Genetics and Breeding Research Unit, Coastal Plain
Experiment Station, Tifton, GA 31793, USA

Introduction

Aphids are a serious impediment to world food production, although some crops are injured more than others. All crops throughout the world are attacked by at least one species of aphid (Peters *et al.*, 1991). The importance of crop losses caused by insect herbivory in general has been described in these terms: 'The shift of energy from plants to insects rivals in scale mankind's own demands on the photosynthesizing world' (Southwood, 1997). The injury elicited by aphids and other sap-sucking arthropods can extract more energy per unit area than grazers and browsers, and they do so without consuming any of the plant structural tissues (Dixon, 1985). Aphids inflict injury *via* direct feeding and by vectoring plant pathogens. Aphid vectors and virus transmission are described elsewhere (Katis *et al.*, Chapter 14 this volume), while this chapter focuses on the feeding injury caused by Aphidoidea including aphids, adelgids, and phylloxerids.

Aphids are more than living plant sap syringes because they secrete and inject saliva into plants before ingesting plant sap, and have been described by van Emden (1972) as 'plant biochemists'. While phloem-feeding Aphidoidea usually cause limited symptoms of plant toxicosis, parenchyma-feeding Aphidoidea cause more profound species-specific effects such as the formation of true galls (Miles, 1990).

Some aphid species are known to be limiting biotic factors for crop yield when outbreaks occur, while other aphid species have caused profound ecological and sociological impacts. A well-known recent example is the introduction of *Diuraphis noxia* (Russian wheat aphid) into North America (Quisenberry and Peairs, 1998). Morrison and Peairs (1998) estimated that the economic losses caused by *D. noxia* to cereal crops in the USA were US$893 million between 1987 and 1993. One outstanding example that demonstrates the ecological impact of Aphidoidea is *Adelges (Dreyfusia) piceae* (balsam woolly adelgid) (Pimentel *et al.*, 1999). Over the past two decades, it has caused severe injury to the natural ecosystem of balsam fir (*Abies balsamea*) throughout the southern Appalachians, where it has destroyed 95% of the Fraser firs (*Abies fraseri*) (Alsop and Laughlin, 1991). As a result of adelgid-mediated Fraser fir death, Alsop and Laughlin (1991) reported the loss of two native bird species and the invasion of three other bird species in the region. Another example is the redoubtable *Viteus (=Daktulosphaira) vitifoliae* (grape phylloxera) (Granett *et al.*, 2001), which has been a formidable pest worldwide, not only in the history of entomology but also in viticulture and oenology.

The plague of *V. vitifoliae* on vines can be traced back to 1850 in Europe. Californian vineyards (at least 20,000 ha) are still succumbing to the notorious injury elicited by *V. vitifoliae*. The possible clearing, fumigation, and replanting of these vineyards will cost up to US$2000 million (Hill, 1997; Granett *et al.*, 2001).

Aphidoidea can adversely affect crop yield and quality in a number of ways other than by transmitting viruses, a form of indirect damage that is outside the scope of this chapter. The general topic of transmission of plant viruses by aphids is the topic of Katis *et al.*, Chapter 14 this volume. In addition to aphid injury to crop plants from feeding, excretion of body wax and honeydew can affect crop quality (Drees and Jackman, 1998). When abundant, aphids can excrete large amounts of honeydew that supports the growth of sooty mould, which is caused by filamentous ascomycetes (e.g. *Capnodium citri*) (Reynolds, 1999). The resulting discoloration of crop products significantly reduces their market value.

In this chapter, we focus on aphid feeding-elicited crop injury and damage. We propose a classification scheme for aphid-feeding injury in accordance with the symptoms that aphids elicit. Recent understandings of biochemical and physiological mechanisms of symptom formation as a result of aphid injury serve as the basis for the proposed classification scheme. Additionally, we discuss the ramifications of studying aphid injury in relation to improving crop resistance to aphid and other sap-feeding herbivores, as well as research frontiers in the understanding of etiology of aphid injury.

Injury Classification

In the entomological literature, the terms 'injury' and 'damage' had been used interchangeably until Peterson and Higley (2001) emphasized the importance of delineating the two, defining 'injury' as 'a stimulus producing an abnormal change in a physiological process' and 'damage' as 'a measurable reduction in plant growth, development, or reproduction (yield) as a result of injury'. We will therefore use throughout this chapter the two terms according to the definitions of Peterson and Higley (2001).

Plant response to aphid feeding can be asymptomatic (= free of obvious morphological aberrations) or symptomatic in nature. Aphid-elicited injury symptoms on plants range from development desistance to neoplasm. 'Desistance' is defined as the ceasing of plant growth (e.g. chlorosis and stunting), while neoplasm is defined as abnormal proliferation of plant tissues (e.g. leaf curling, galls from a variety of tissues). The nature of the injury depends on the site of aphid feeding and the sensitivity of the tissues there to that feeding. Desistance is elicited mainly by the phloem-feeding Aphidoidea guild, while neoplasm is elicited by the parenchyma-feeding guild. The contrasting terms of desistance and neoplasm elicited by Aphidoidea represent the result of a wide gamut of aphid–plant interactions. Thus, the ongoing study of aphid-elicited injury is an interdisciplinary, if not a pandisciplinary topic of all agricultural sciences.

Aphidoidea cause an array of plant injury symptoms ranging from apparent to unapparent as a result of their variable feeding behaviour (Miles, 1990), salivary factors (Miles, 1998, 1999), and host plants (Miles, 1989a,b). Thus, we propose a new scheme to categorize aphid injury according to symptoms (i.e. asymptomatic injury and symptomatic injury). Symptomatic injury can be divided further into desistance and deformation. Desistance includes plant stunting, chlorosis, and necrosis, while deformation contains misshapen fruits, leaf curling (or pseudogalling), and tissue galling (or neoplasm). The proposed classification of common crop injury is summarized in Table 13.1.

Asymptomatic injury

Although aphids usually cause obvious injury symptoms on crops, some aphid species do not produce such apparent symptoms, but nevertheless reduce crop growth. *Rhopalosiphum padi* (bird cherry–oat aphid),

(a)   (b)

Fig. 13.1. Asymptomatic injury. (a) *Rhopalosiphum padi* feeding on a wheat leaf (original colour photo courtesy of Kansas Department of Agriculture); (b) *Sitobion avenae* (English grain aphid) feeding on wheat heads (original colour photo from www.ipm.ucdavis.edu/PMG/M/I-HO-MAVE-IF.003.html, courtesy of Chet Fukushima, University of California).

plant tissue discoloration. The typical injury symptoms elicited by *D. noxia* include leaf chlorosis and rolling that lead to necrosis and death of seedlings under severe infestation (Burd *et al.*, 1998). The leaf chlorosis elicited by *D. noxia* also varies between host plants. Chlorotic streaks and reddish discoloration are common symptoms of *D. noxia* feeding on susceptible wheat and barley plants under field conditions (Fig. 13.2), although it is only chlorotic spots that are observed on resistant wheat and oat plants.

Therioaphis trifolii maculata (spotted alfalfa aphid) is an important pest of lucerne that leads to significant forage crop loss in Australia and the USA (Ehler, 1998). The typical injury symptom caused by *T. trifolii maculata* feeding is veinal chlorosis (also known as vein clearing) at the growing tips of lucerne leaves, irrespective of the aphids' feeding site (Madhusudhan and Miles, 1998). The involvement of viral infection in the development of the chlorotic symptom was excluded by research during the 1950s, and

a phytotoxin has been hypothesized to cause the specific chlorotic symptoms (Madhusudhan and Miles, 1998). However, whether or not the phytotoxin is a chemical has not been determined. Another chlorosis-eliciting aphid, but on trees, is *Elatobium abietium* (spruce aphid). Infestation by *E. abietium* is usually indicated by the yellowing and dropping of needles from spruce trees, *Picea* spp. Heavy infestation by *Pemphigus betae* (sugarbeet root aphid) also causes chlorosis and wilt of sugarbeet (*Beta vulgaris*) (Donahue *et al.*, 1998). The aphids often feed in circular or elliptical patches, in which the foliage is wilted or stunted and, in extreme cases, collapsed and dying (Hutchison and Campbell, 1994).

NECROSIS. The other common desistance elicited by Aphidoidea species is tissue necrosis, either on green leaf tissue or non-green stem tissue. *Schizaphis graminum* (greenbug) and *Sipha flava* (yellow sugarcane aphid) are two important necrosis-eliciting

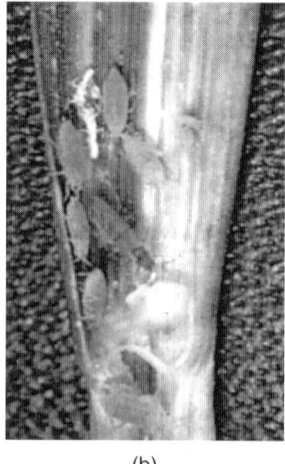

(a)   (b)

Fig. 13.2. Leaf chlorosis. (a) Veinal chlorosis by *Therioaphis trifolii maculata* (original colour photo from www.clay.agr.okstate.edu/alfalfa/images/insects/saaphotos.html, courtesy of John Caddel, Oklahoma State University); (b) chlorotic lesion and discoloration by *Diuraphis noxia* (original colour photo from Peairs, 2001, with permission).

pests on small grains and sorghum (*Sorghum bicolor*) worldwide. By feeding on the underside of sorghum leaves, *S. flava* causes purple-coloured leaves on seedling sorghum and yellow leaves on more mature plants (Breen and Teetes, 1990; Webster, 1990) (Fig. 13.3). Plants that are not killed by the aphids are severely stunted and maturity is delayed, which results in yield losses. *Schizaphis graminum* feeding on wheat leaves elicits necrotic spots surrounded by chlorotic halos (Ryan *et al.*, 1990; Flinn *et al.*, 2001). The chlorotic halos usually are several millimetres in diameter around aphid feeding sites. Kieckhefer and Kantack (1988) demonstrated that *S. graminum* caused 38% loss of kernel weight when compared with that of uninfested wheat seedlings, and annual losses caused by greenbug between 1951 and 1960 were estimated at over US$100 million (LeClerg *et al.*, 1965).

Tinocallis caryaefoliae (black pecan aphid) feeds on the underside of pecan (*Carya illinoinensis*) leaves. This feeding causes chlorotic and necrotic blotches that are rectangular in shape and end abruptly at a leaf vein. As a result, pecan leaves fall prematurely, and the loss of leaves on susceptible trees can significantly reduce tree vigour and growth (Wood and Reilly, 1998). *Macrosiphum euphorbiae* (potato aphid) also elicits black necrotic spots on the stems and leaf veins of potato (*Solanum tuberosum*) plants (Miles, 1990).

Deformation

Deformation of plant tissues (i.e. leaves, stems, roots, and fruits) is a common expression of symptomatic aphid injury.

MISSHAPEN FRUIT. Feeding by some aphids, for example, *Dysaphis plantaginea* (rosy apple aphid), causes deformation of fruits. Aphid feeding elicits localized inhibition of fruit growth that forms rough and deformed fruits (Fig. 13.4). *Dysaphis plantaginea* causes leaf, fruit, and systemic root injury, and can cause as much as 50% fruit damage (Varn and Pfeiffer, 1989). In wheat fields after the flowering stage, 'fish-hook'-like distortions with chlorotic lesions are ascribed to infestation by *D. noxia* (Peairs, 2001). The 'fish-hook' shape is caused by awns trapped in the tightly rolled flag leaves caused by *D. noxia* feeding preventing the head from emerging and straightening. The grain may be poorly formed or absent; sometimes, the entire head is destroyed.

PSEUDOGALLING. The 'pseudogall' symptom embraces all types of plant leaf tissue deformation (Fig. 13.5), except abnormal enlargement (neoplasm). The difference between pseudogalling and neoplasm (or tissue galling) (to be discussed in the next section) is that pseudogalling refers to all types of non-prolific tissue deformation, while tissue galling refers exclusively to the prolific growth of plant tissue. The most common pseudogalling symptoms of aphid injury include young shoot crinkling, leaf rolling, curling, and leaf folding (or 'balling').

The term 'pseudogall' was first used by Miles (1990) to describe leaf deformation of woody plants by aphids. Burd *et al.* (1993) used the term to describe the leaf rolling caused by *D. noxia* on wheat. *Diuraphis noxia* elicits one of two types of leaf deformation on wheat, depending on the maturity of the leaves. On seedlings, *D. noxia* prevents leaves from unfolding normally, so that these become convolutely rolled to form a pseudogall, and further growth is prevented (Burd *et al.*, 1993). On mature leaves, high densities of *D. noxia* can elicit leaf folding along the main vein. *Aphis gossypii* (cotton or melon aphid) usually feeds on the underside of leaves and causes a downward curl or distortion of leaves.

Although *Myzus persicae* (peach–potato aphid) is a vector for a number of plant viruses, direct feeding injury of *M. persicae* is also apparent on many of its host plants

Fig. 13.3. Leaf necrosis elicited by the *Schizaphis graminum* on a sorghum leaf (original colour photo from Mayo, 2000, with permission).

Fig. 13.4. Misshapen apple fruits caused by *Dysaphis plantaginea* feeding (original colour photo from www.ipm.ucdavis.edu/PMG/r4301511.html#DAMAGE, courtesy of Jack Kelly Clark, University of California).

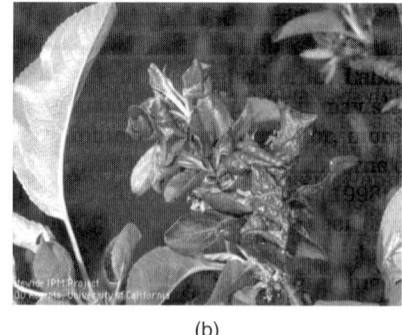

(a)   (b)

Fig. 13.5. Pseudogalling symptoms. (a) Cupped cotton leaves elicited by *Aphis gossypii*; (b) crinkling new apple leaves elicited by the *Dysaphis plantaginea* (original colour photos from www.ipm.ucdavis.edu/PMG/, courtesy of Jack Kelly Clark, University of California).

(estimated at over 875 plant species), including stone fruits and a number of vegetable crops. The direct feeding injury by *M. persicae* includes curling and distortion of new foliage at the growing point, and retarded growth of young peach shoots (Miles, 1990). *Acyrthosiphon kondoi* (blue alfalfa aphid) and *Aphis craccivora* (cowpea aphid) both also cause stunting and crinkling on terminal growth of legume crops as a result of their feeding (Drees and Jackman, 1998). *Lipaphis pseudobrassicae* (mustard aphid) and *B. brassicae* prefer young leaves and flowering parts of cruciferous plants (Boyd and Lentz, 1994). They feed on the underside of terminal leaves and cause the leaf to cup or curl and the plant to become stunted. *Brevicoryne brassicae* feeding on Brussels sprout plants causes leaf curling and stomata density reduction that reduces assimilates and plant weight (van Emden, 1990). Heavy infestation by *L. pseudobrassicae*, *M. persicae*, and *B. brassicae* on oilseed rape plants causes obviously stunted growth, reduced pod formation, and a stand of uneven maturity.

Macrosiphum euphorbiae is an important pest of tomato (*Lycopersicon esculentum*). The aphids prefer to colonize the tips of the plants and their feeding elicits crinkling of young leaves, which leads to significant yield and quality losses in fresh-market tomatoes (Walgenbach, 1997). In apple (*Malus domestica*) orchards, three aphid species are commonly

found. They are *D. plantaginea*, *Aphis pomi* (green apple aphid), and *Aphis spiraecola* (green citrus aphid). Although *D. plantaginea* feeding causes malformation of apple fruits and leaves, feeding by *A. pomi* and *A. spiraecola* usually causes severe curling of new foliage, as well as twisted growing terminals (Lawson and Weires, 1991; Blackman and Eastop, 1994).

In North America, feeding of the newly introduced pest on soybean *A. glycines* causes leaf distortions (e.g. crinkled or cupped leaves), with heavy infestations causing leaf chlorosis (Rice *et al.*, 2004). In addition, accumulation of aphid honeydew causes the growth of fungus, which reduces photosynthesis and stunted plants. Large populations of *A. glycines* can also lead to yield loss by abortion of flowers and young pods and reduction in the number and size of seeds per pod.

TISSUE GALL (NEOPLASM). 'Tissue galling' encompasses all forms of tissue proliferation or enlargement elicited by Aphidoidea feeding (Fig. 13.6). A number of insect species elicit neoplasm of plant tissues (Shorthouse and Rohfritsch, 1992; Doss *et al.*, 2000). Gall-inducing aphid species can be shoot- or root-feeding species on herbaceous or deciduous plants.

Leaf galls elicited by Aphidoidea vary in shape (Miles, 1989b; Hori, 1992). *Phylloxera* spp. elicit leaf and twig galls on

(a)   (b)

Fig. 13.6. Neoplasms (tissue galls). (a) Leaf galls elicited by *Viteus (=Daktulosphaira) vitifoliae* (original colour photo courtesy of R. Isaacs); (b) root galls elicited by the same species (original colour photo courtesy of K. Powell).

pecan trees. *Phylloxera notabilis* (pecan leaf phylloxera) is not particularly damaging, although infested leaves drop early. However, twig galls caused by *Phylloxera devastatrix* (pecan phylloxera) can lead to mid-season defoliation and twig breakage during windy weather or excessive weight (Drees and Jackman, 1998), resulting in reduced yield and deformed pecan trees. Feeding by *Phylloxera caryaecaulis* (hickory leaf stem gall aphid) on hickory (*Carya glabra*) twigs, leaf petioles, and at the bases of leaflet main veins often induces the formation of globular galls (50–250 mm in diameter) (Blackman and Eastop, 1994). The galls are usually yellowish-green, tinted with red. When the phylloxera mature, the galls split open and become leathery and black.

Like *P. caryaecaulis*, several species of Aphidoidea induce gall formation on more than one type of plant tissue. For example, overwintering *Pemphigus* spp. on poplar (*Populus* spp.) trees form leaf-petiole galls in the spring (Hill, 1997). Another closely related species, *Eriosoma lanigerum* (woolly apple aphid), colonizes the trunks, branches, and foliage of apple trees. Although gall-like proliferations can be found on arboreal parts of apple trees, most *E. lanigerum* injury is caused by its root feeding (Brown and Schmitt, 1994; Brown *et al.*, 1995). Two types (i.e. gallicoles and radicicoles) of grape phylloxera elicit galls on, respectively, the

leaves and roots of the grape vine (*Vitis vinifera*) (Granett *et al.*, 2001). The leaf galls formed by the grape phylloxera are wart-like, about 50 mm in diameter, on newly expanding leaves, shoots, and tendrils of grape vines. The galls extend below the leaf surface and fully enclose the feeding gallicoles. The galls open to the upper surface and allow the mobile first instars (called 'crawlers') to exit and move between roots and leaves. In addition, at the opening to the leaf galls, guard hairs maintain humidity and possibly reduce predation by restricting entry (Granett *et al.*, 2001).

Adelges (Sacchiphantes) abietis (spruce pineapple-gall adelgid) and *Adelges (Gilletteella) cooleyi* (Cooley spruce gall adelgid) induce shoot galls on spruce trees. These shoot galls are pineapple-like on the needle-bearing branchlets, and the adelgid offspring escape through lip-like slits above each needle in late summer (Armstrong, 1995). While *A. cooleyi* galls are at the end of new growth, *A. abietis* galls are at the base of new growth (Johnson and Lyon, 1988).

Among the galls formed by edaphic species, root galls formed by radicicoles of *V. vitifoliae* on grape roots (Granett *et al.*, 2001) and *E. lanigerum* on apple roots (Brown *et al.*, 1991) are most devastating. The root galls elicited by *V. vitifoliae* on grape are knot-like swellings on the rootlets, which are localized enlargements (called 'tuberosities') that result from the proliferation and expansion of phloem parenchyma cells (Granett *et al.*, 2001). Radicicoles elicit greater vine injury than gallicoles. While gallicoles only cause reduced cane growth and quality, radicicoles cause severe vine decline or death. *Eriosoma lanigerum* feeding elicits a proliferation of anomalous non-functional xylem that results in the reduction of water flow through the galled root tissue, and consequently reduction in the growth of the apple trees (Brown *et al.*, 1991).

Aetiological Agents in Aphidoidea Saliva

Aphid saliva has long been hypothesized to be the aetiological source of plant injury symptoms. Until recently, little progress had been made on identifying elicitors of injury symptoms in aphid saliva. Methods of collecting and analysing salivary enzymes, and the hypotheses on aetiological agents in aphid saliva, have been the subject of several studies (Ma *et al.*, 1990; Madhusudhan *et al.*, 1994; Cherqui and Tjallingii, 2000; Ni *et al.*, 2000) and reviews (Miles, 1990, 1998, 1999). Campbell and Dreyer (1985, 1990), using *S. graminum*, hypothesized that aphid-elicited necrosis was related to the activities of pectic enzyme in the saliva. The pectic enzyme hypothesis was further investigated and supported by Ma *et al.* (1990). Subsequently, using *T. t. maculata,* Jiang and Miles (1993) and Miles and Oertli (1993) concluded that the vein-clearing symptom on lucerne leaves was the result of aphid interruption of the redox balance in the plant. Ni *et al.* (2000) demonstrated that salivary enzyme profiles of symptom-inducing *D. noxia* and non-symptom-inducing *R. padi* differed in oxido-reductase activities. Only catalase was detected from *D. noxia* salivary gland tissue, while only peroxidase was detected from salivary gland tissue of *R. padi*. Additionally, Ni *et al.* (2001b) assayed the activity of chlorophyll degradation enzymes (i.e. chlorophyllase, oxidative bleaching, and Mg-dechelatase) from *D. noxia* and detected no activity for any of these

enzymes. Thus, it is likely that the aetiological agents of aphid injury, as far as they are salivary components, are aphid and plant species-specific.

The Biochemical and Physiological Bases for Symptoms of Injury

Most, if not all, of the literature postulates that aphid saliva contains the aetiological agent(s) that elicit plant injury symptoms, whereas the truth is that few studies have provided research evidence for such statements. We review recent advances that have been made in the understanding of the biochemical and physiological bases for injury to plants caused by aphids, and ultimately to unravel the conundrum of desistance and neoplasm.

Basis of asymptomatic injury

Asymptomatic aphid species arguably may be described as plant sap syringes because their draining of plant sap does not elicit apparent toxaemia symptoms (Miles, 1989a, 1990). Although such aphid species do not elicit apparent symptomatic injury, there is the hidden injury of plant biomass reduction.

Basis of symptomatic injury

In contrast to asymptomatic aphid species, injury symptom-eliciting species are anything but mere plant sap syringes. The injury symptoms, ranging from crinkling to neoplasm, are usually species- and plant tissue-specific.

Desistance

Most commonly observed desistant injury symptoms are stunting, chlorosis, and necrosis (Table 13.1). Three hypotheses have been proposed regarding aphid-elicited desistance in plants: the oligosaccharide hypothesis (Dreyer and Campbell, 1987; Ma *et al.*, 1990); the oxidation hypothesis (Miles and Oertli, 1993; Ni *et al.*, 2001a); and the

Mg-dechelatase hypothesis (Ni *et al.*, 2001b). Campbell (1986) and Dreyer and Campbell (1987) summarized the similarity between the penetration of aphid stylets during feeding and phytopathogen germination tubes into leaves. Certain polysaccharides are known as elicitors of pathogen injury symptoms, and Dreyer and Campbell (1987) found that oligosaccharides similar to these were detected in *S. graminum* honeydew. These findings indicated that aphid salivary pectic enzyme-elicited degradation of plant cell walls and the formation of the oligosaccharides were likely to be the underlying causes of *S. graminum*-elicited leaf necrosis.

Using *T. t. maculata*, Jiang and Miles (1993), Miles and Oertli (1993) and Jiang (1996) proposed that aphid feeding disturbed the plant redox balance. They described veinal chlorosis on young leaves as the result of oxidative interactions between aphids and lucerne plants. After demonstrating the differences in salivary enzyme profiles of *D. noxia* and *R. padi* (Ni *et al.*, 2000), Ni *et al.* (2001a) found that the magnitude of oxidative responses (measured by peroxidase, catalase, and polyphenol oxidase) of cereals (wheat, barley, and oat – *Avena sativa*) differed between the two aphid species. Although feeding by both species elicited an increase in total protein content in comparison with control plants, only *D. noxia* feeding elicited an increase in peroxidase activity on *D. noxia*-resistant 'Halt' wheat and susceptible 'Morex' barley, but caused no increase in activity on *D. noxia*-susceptible 'Arapahoe' wheat and resistant 'Border' oat. They concluded that the response of cereals to *D. noxia* feeding were cereal genotype-specific. Burd and Elliott (1996) and Haile *et al.* (1999) showed that *D. noxia* feeding on wheat had a negative effect on photosynthesis and chlorophyll fluorescence-transient kinetics. These findings support the hypothesis that *D. noxia*-elicited chlorosis is the result of a photo-oxidation process in susceptible cereal plants. When phloem sap was obtained by stylectomy of three cereal aphids, that of the two symptomatic aphid species (*D. noxia* and *S. graminum*) had a twofold higher concentration of amino acids and a higher proportion of essential

amino acids than that of the non-symptomatic aphid species (*R. padi*) (Sandström *et al.*, 2000). The authors concluded that the differences in phloem sap composition were likely to be nutritionally advantageous to the symptomatic aphid.

Although natural chlorophyll degradation is an annual event in temperate regions, with over one billion tonnes worldwide of chlorophyll biosynthesized and degraded annually, the mechanism of natural chlorophyll degradation is still not well understood (Janave, 1997; Matile *et al.*, 1999). The chlorophyll loss that accompanies aphid feeding is even more poorly understood. Ni *et al.* (2001b) demonstrated for the first time that aphid-elicited enzymatic chlorophyll degradation was a complex process that differed greatly from chlorophyll degradation during the natural senescence of plants. Using modified procedures for chlorophyllase, oxidative bleaching (or chlorophyll oxidase), and Mg-dechelatase activity assays, they found that *D. noxia* feeding elicited a significantly higher level only of Mg-dechelatase activity, while no change in either chlorophyllase or oxidative bleaching was detected (Ni *et al.*, 2001b). Mg-dechelatase is therefore likely to be involved in the development of *D. noxia*-elicited chlorotic lesions in wheat. In general, the biochemical and physiological bases for aphid-elicited desistance in plants are variable depending on aphid and plant species. There is not a universal hypothesis that explains aphid-elicited desistance that includes both stunting of growth and chlorosis.

Deformation

In comparison with desistance in the form of stunting and chlorosis, aphid-elicited plant deformation (i.e. crinkling and neoplasm) is more complicated. Burd *et al.* (1993) concluded that the leaf rolling elicited by *D. noxia* was the result of a reduction in leaf turgor. However, attempts to determine the role of plant hormones (ethylene, indole-3-acetic acid, gibberellin3, zeatin, kinetin, and abscisic acid) in *D. noxia* injury symptom formation failed (Miller *et al.*, 1994). In general, the involvement of plant hormones in crinkling and other distortions of plant tissues is still poorly understood.

Although plant gall formation is a fascinating natural process that reflects organism interactions, the study of plant gall formation (cecidology) represents a complex research area involving multiple disciplines (Mani, 1992; Doss *et al.*, 2000). A treatise by Shorthouse and Rohfritsch (1992) has focused exclusively on the biology of arthropod-induced plant galls. The formation of tissue galls demonstrates that plants can be reprogrammed with the right stimuli to produce localized prolific growth, as well as accumulate chemicals not normally associated with the infested organs. Shorthouse and Rohfritsch (1992) describe gall-inducing herbivores as 'the genetic engineers of the insect world'. Only a few studies in the recent literature of aphid-elicited plant galls have examined the resulting physiological and biochemical changes in plants. Hori *et al.* (1997) demonstrated significant decreases in phenoloxidase (i.e. peroxidase and polyphenol oxidase) activities in galls formed by *Tetraneura fusiformis* on the leaves of elm (*Ulmus davidiana* var. *japonica*). Although plant hormones (e.g. indole-3-acetic acid), amino acids, enzymes, and enzyme inhibitors were hypothesized as cecidogens as early as the 1940s (Miles, 1989a; Hori, 1992), the search for aetiological agents of aphid-elicited neoplasms is still an ongoing process. Brown *et al.* (1991) attempted to induce root galls on greenhouse-grown apple trees using 6-benzyl-aminopurine and indole-3-acetic acid. They found that growth anomalies on roots treated with 6-benzyl-aminopurine had typical xylem, but a proliferation of phloem tissue and limited internal or external deformation of roots. In contrast, root galls elicited by *E. lanigerum* result in the proliferation of non-functional xylem (Brown *et al.*, 1991).

Granett *et al.* (2001) proposed three mechanisms for understanding tissue galls elicited by *V. vitifoliae*: the loss of vigour caused by removal of photosynthates; premature mortality of roots caused by secondary pathogenesis; and physiological disruption of vascular systems. Using the preference of *V. vitifoliae* on vigorous shoots of canyon grape (*Vitis arizonica*), Kimberling *et al.* (1990) proposed the plant-vigour hypothesis. Using leaf size and the number of galls per leaf as indices of vigour, they concluded that herbivore species preferred vigorous to stressed plants. However, Rehill and Schultz (2001) found evidence that contradicted the plant-vigour hypothesis, using *Hormaphis hamamelidis,* another gall-forming aphid on witch hazel (*Hamamelis virginiana*). The number of *H. hamamelidis*-elicited leaf galls was not correlated with leaf size, and location on distal leaves conferred no increases in fecundity of the fundatrices.

Ecological Modulation of Aphid Injury

Previous studies on aphid injury have focused primarily on insect–plant interactions, while the impact of environmental factors (e.g. light and drought) on aphid-elicited plant injury has not been clearly delineated. Further, within the field of insect–plant interactions, many studies and treatises have documented the impact of host plants on insects in relation to selecting crops that are resistant to insect pests, whereas only a few studies have examined the responses of the susceptible crop plants to insect feeding. Therefore, ecological modulation of plant responses to aphid-feeding injury symptoms on susceptible genotypes is a relatively new area of research that could significantly benefit our understanding of aphid toxicosis on plants. The unravelling of the mechanisms of ecological factor-modulated plant responses to piercing/sucking insect feeding would advance our knowledge of the dynamic nature of plant–aphid interactions. It would ultimately benefit breeding of aphid-resistant cultivars by blocking the known mechanisms of plant responses to injury from aphids (see 'tolerance' in van Emden, Chapter 17 this volume).

Abiotic factors

Photoperiod

Although a wealth of entomological literature has focussed on either aphid–host plant interactions or on the effect of photoperiod and temperature on aphid biology, the influence of environmental factors on aphid-elicited plant response and injury symptom formation has not been examined until recently. Using five wheat genotypes ('Arapahoe', 'Halt', PI 137739, PI 225245, and PI 262660), Ni and Quisenberry (1997) showed that the probing behaviour of *D. noxia* exhibited varied circadian rhythms on resistant and susceptible wheat genotypes. The probing duration per aphid probe was significantly different among the five genotypes and between diurnal and nocturnal periods. Aphids had the shortest probes on PI 137739, a line antibiotic to *D. noxia*, among the five genotypes, both diurnally and nocturnally. Irrespective of genotype, total aphid probing duration recorded in an 8 h monitoring period was significantly longer nocturnally than diurnally. This suggests that *D. noxia* injury symptoms on wheat plants are probably the result of light- or photoperiod-modulated plant–aphid–environment interactions.

Macedo *et al.* (2003) have also demonstrated that photoperiod is an important factor in the formation of chlorosis on *D. noxia*-susceptible wheat. They demonstrated that chlorophyll *a* fluorescence parameters, measured by non-variable (F_0), variable (F_v), and maximum fluorescence (F_m), and photochemical efficiency of photosystem II (F_v/F_m), were significantly higher in plants exposed to light than in plants kept in darkness. This showed that the development of *D. noxia*-elicited leaf rolling and chlorosis on susceptible wheat seedlings was a light-activated process.

Temperature

The impact of temperature on aphid development and aphid biology has been studied by Butts (1992), Butts and Schaalje (1997), and Butts *et al.* (1997). However, the impact of temperature on aphid-elicited plant injury has not been defined precisely because distinguishing between the impact of temperature on aphids and plants is difficult. For example, there is anecdotal evidence that *D. noxia* infestation elicits much less leaf

chlorosis at 10°C than at 21°C under the same light regime, 16:8 (L:D) (X. Ni and S.S. Quisenberry, unpublished results). However, it is difficult to determine whether the lower chlorosis ratings at 10°C were caused by less aphid feeding, lower plant photosynthetic rate, or a synergistic function of the aphid–plant interaction.

Drought

Drought affects the formation of aphid injury symptoms. Higher numbers of galls, but smaller ones, of *A. abietis* were recorded on drought-stressed Norway spruce trees (*Picea abies*) (Björkman, 2000) than on unstressed trees. The enhancement of aphid injury by drought stress has also been demonstrated on cereal crops (Riedell, 1989; Archer *et al.*, 1995; de Farias *et al.*, 1995; Oswald and Brewer, 1997; Johnson *et al.*, 1998). *Diuraphis noxia* feeding elicited symptoms of drought stress in barley, even with ample soil moisture (Riedell, 1989). Several studies have confirmed that drought stress enhances *D. noxia* injury on susceptible wheat (Johnson *et al.*, 1998) and barley plants (Oswald and Brewer, 1997). In addition, Johnson *et al.* (1998) reported that aphid infestation at the two-leaf stage caused significantly greater leaf rolling and cholorosis rating than the four-leaf stage.

Soil nutrients

Another important abiotic factor in agricultural ecosystems is soil nutrients. In particular, nitrogen fertilizers enhance aphid populations (Latimer and Oetting, 1999); however, we have found no reports on the correlation between fertilizer levels and aphid injury ratings on plants.

By monitoring both drought stress and fertilization, Archer *et al.* (1995) demonstrated higher densities of *D. noxia* on drought-stressed plants, although nitrogen fertilization did not affect aphid population development under field conditions. However, Ponder *et al.* (2001) demonstrated that *R. padi* feeding behaviour on barley seedlings was affected by nitrogen fertilizer level, but not by drought stress. It took

longer for aphids to reach the sieve elements of nitrogen-deficient than of normal
seedlings, while drought did not have any
effect on aphid feeding behaviour. In addition, the type of fertilizer also affects aphid
populations. For example, when maize plants
were grown in plots treated with organic and
synthetic fertilizers, fewer *Rhopalosiphum
maidis* (corn leaf aphid) were detected
when organic fertilizer was used (Morales
et al., 2001). These differences were attributed to concentrations of foliar nitrogen in
maize leaves, with higher nitrogen found in
leaves of plants grown on plots treated with
synthetic fertilizers. Aphid populations did
not cause a significant yield loss because
feeding injury was compensated by fertilizer treatments. The application of fertilizer
to an 11-year-old stand of Sitka spruce (*Picea
sitchesis*) hastened the recovery of diameter
growth after defoliation by *E. abietium*
(Thomas and Miller, 1994). Thus, the impact
of fertilizer on aphid population increase
and, in turn, plant injury is likely to be
aphid species-specific.

Elevated CO_2

The impact of elevated CO_2 on insect–plant
interactions has been reported in more than
40 studies (Hughes and Bazzaz, 2001).
Chewing insects fed on foliage grown at elevated CO_2 levels had much slower growth
rates, took longer to mature, and suffered
greater mortality than at normal CO_2 levels,
although the insects had a higher consumption rate in elevated CO_2 than in the experimental control. In contrast, increased CO_2
concentrations may not have an adverse
effect on phloem-feeding insects, including
aphids (Salt *et al.*, 1996; Diaz *et al.*, 1998;
Hughes and Bazzaz, 2001). Short generation
times and the capacity for quick population
increases give aphids the potential for rapid
responses to global climate change and
could result in the acceleration of some aphid
species to pest status (Harrington *et al.*,
1995). When the effect on broad bean growth
of the initial inoculation of a single *A. pisum*
and *Aulacorthum solani* (glasshouse and
potato aphid) was compared under ambient
and elevated CO_2 levels, *A. pisum* and its

progeny caused greater shoot and root
weight losses (7 and 10%, respectively) under
elevated CO_2 than ambient levels, while
flower numbers decreased by 73% (Awmack
and Harrington, 2000). In comparison, *A.
solani* caused greater shoot and root weight
losses (20 and 18%, respectively) and a
60% decline in flower numbers under similar conditions (Awmack and Harrington,
2000). Percy *et al.* (2002) showed that the
injury to American aspen (*Populus tremuloides*) trees from outbreaks of a sap-feeding
insect, the aphid *Chaitophorus stevensis*,
increased under elevated CO_2 and O_3 levels,
while damage by a chewing insect, *Malacosoma disstria* (forest tent caterpillar), was
potentially reduced. This difference in
damage was attributed to induced changes
in the physical and chemical defensive
characteristics of the aspen trees.

Biotic factors

Induced resistance

Karban and Baldwin (1997) and Agrawal
et al. (1999) reviewed the mechanisms of
induced responses to herbivory and pathogenesis on plants. The impact of biotic factors on aphid–plant interactions should be
considered when assessing induced plant
resistance to aphid feeding. The signal
transduction pathways that plants use to
activate induced responses to herbivores
differ from those that elicit responses to
pathogens, and Felton *et al.* (1999) demonstrated that pathogenesis and herbivory had
an inverse relationship in eliciting systemic
resistance in plants. While insect herbivory
elicited an increase in one signalling compound, jasmonic acid, in plants, pathogenesis
elicited an increase in another signalling
compound, salicylic acid. However, in a
comparison of whiteflies and aphids with
chewing insects, Walling (2000) concluded
that phloem-feeding herbivores were perceived by the plant as pathogens and activated
salicylic acid-dependent and jasmonic acid/
ethylene-dependent signalling pathways
(see also Pickett and Glinwood, Chapter 9
this volume). In contrast, chewing insects

(e.g. caterpillars and beetles) and cell-content feeders (e.g. mites and thrips) elicited extensive tissue injury and activated wound-signalling pathways. Herbivore feeding is, however, not equivalent to mechanical wounding, because salivary factors are often elicitors for the development of plant injury symptoms (Walling, 2000). Thus, understanding induced resistance mechanisms in crops would help us to reduce aphid injury to crop plants.

Symbiosis

Like plant pathogens, microbial symbionts in plants also affect aphid–plant interactions. Quisenberry and Joost (1990), Barbosa et al. (1991), Joost and Quisenberry (1993), and Clement et al. (1994) have all reviewed the impact of the fungal endophytes Neotyphodium spp. (formerly Acremonium spp.) in grasses on both chewing and piercing/sucking herbivores. Fungal endophytes provide grasses with enhanced protection from herbivory of aphids, drought, and pathogens (Clement et al., 1994; Wilkinson et al., 2000). Recently, several studies have demonstrated the induction of resistance to insect-transmitted diseases by plant growth-promoting rhizobacteria in cucumber (Cucumis sativus) and tomato (Zehnder et al., 1999; Murphy et al., 2000). Zehnder et al. (2001) demonstrated that plant growth-promoting rhizobacteria elicited induced systemic resistance to Cucumber mosaic virus infection under both field and greenhouse conditions, although the level of the induced resistance was variable. However, the modulation of plant growth-promoting rhizobacteria on aphid injury in crop plants is not well understood.

Multiple pest resistance

Aphid-elicited plant injury symptoms are usually correlated with the susceptibility of the host plant. Thus, understanding aphid and host-plant resistance (i.e. single or multiple pest resistance) interactions is critical in mitigating aphid-feeding injury. Rossi et al. (1998) described a Meloidogyne incognita (root-knot nematode) resistant Mi-gene

in tomato that also conferred resistance to M. euphorbiae. This demonstrated for the first time that a single plant gene could confer resistance to biotic stresses from different phyla. However, a contrasting phenomenon is observed on cereal crops. The resistance to two cereal aphids (S. graminum and D. noxia) in wheat involves different wheat chromosomes (Castro et al., 2001). In addition, while D. noxia resistance in general is related mainly to chromosome 7D (Castro et al., 2001; Liu et al., 2001), antixenotic resistance to S. graminum is related mainly to chromosome 2B (Castro et al., 2001). Although feeding by Heterodera avenae (cereal cyst nematode) elicits an increase in esterase, peroxidase, and superoxide dismutase activities in a resistant wheat/Aegilops ventricosa (barbed goat grass) introgression line (H-93-8) compared with the susceptible 'Anza' wheat (Andres et al., 2001), the correlation between nematode and aphid resistance in wheat has not been assessed. Therefore, the molecular mechanism of resistance to injury by aphids and its application in multiple pest resistance in crop plants is one of the intriguing research areas for pest management.

Ramifications of Aphid Injury Research for Crop Protection

Basic research into the injury resulting from feeding by aphids ultimately will benefit crop production. Entomologists have worked on mechanisms of plant resistance to insect feeding injury and insect pathogenesis caused by Bacillus thuringiensis (Bt), while plant pathologists have worked on the formation of the crown gall inflicted by Agrobacterium tumefaciens (crown gall). Technological advances over the past 50 years have enabled scientists to use this knowledge to introduce Bt genes into a number of crop plants using A. tumefaciens plasmid vectors. This has led to the production of transgenic crop plants with insect resistance (Patlak, 1998). The current impact of these advances on humanity has considerably surpassed the imagination of the researchers who initiated the original basic research programmes to understand the injury symptoms of plants elicited by insect feeding.

Similarly, understanding physiological and molecular mechanisms, in particular the elicitors of aphid injury symptoms, irrespective of whether these take the form of leaf desistance or neoplasm, will provide us with essential information needed to develop more efficient procedures for screening the global plant genetic resources for resistance to insects, including aphids (Clement and Quisenberry, 1999). Such improved procedures would significantly shorten the programme for developing aphid- and other insect-resistant crop cultivars. Because aphids are a group of serious insect pests on cereal, forage, fruit, and vegetable crops, results obtained from enhanced basic research relevant to plant injury should significantly benefit integrated pest management programmes, and through this, the sustainability of global agricultural ecosystems.

Conclusions on Research Frontiers in the Understanding of Injury from Aphid Feeding

Although the injury symptoms elicited by aphids have been recognized for a long time, the precise measurement of aphid-elicited plant injury symptoms is still lagging behind comparable work in other disciplines like plant pathology and plant physiology. We can benefit greatly by learning from these other disciplines. For example, the use of imaging techniques developed for recording stress-induced changes in plants (Chaerle and van der Straeten, 2001) to quantify and compare aphid injury symptoms could provide us with details about temporal and spatial patterns of chlorosis formation. The technique is likely also to facilitate the correlation of physiological and biochemical signal transduction pathways with aphid injury elicitation.

Aphids and other phloem-feeding insects are known to adversely affect the transport of plant nutrients and growth regulatory

molecules, such as hormones in the plant's vascular system. Stylectomy of aphids has been used for collecting and studying plant phloem physiology. In addition to the current thinking on phloem transport of photosynthates of plants, there is mounting evidence indicating that protein and RNA molecules use this transport pathway to regulate developmental and physiological processes (Crawford and Zambryski, 1999). Madhusudhan and Miles (1998) compared the mobility of different salivary components in phloem and found that the contrasting responses of lucerne plants to *T. t. maculata* and *A. pisum* were elicited by differences in the mobility of their salivary components. Therefore, studies on aphid–plant interactions, in particular analysis for injury-eliciting factors in aphid saliva, need to include both small and large phloem-mobile molecules. The results from phloem mobility studies using aphid–plant systems will benefit the discipline of plant physiology, as well as that of crop protection.

New advances in other areas of plant protection, such as multiple pest resistance, have stimulated a fast-growing research area. For instance, the nematode-resistant *Mi* gene in tomato has been shown to confer aphid resistance in tomato plants (Rossi *et al.*, 1998). Nombela *et al.* (2001) and Jiang *et al.* (2001) demonstrated that *Mi*-gene transgenic tomato plants were also resistant to *Bemisia tabaci* (sweet potato whitefly). This finding indicates that multiple pest resistance is likely to be a common phenomenon within a feeding guild of herbivores. Thus, the mechanism and relationship between aphid injury symptom development and other biotic or abiotic factor resistances (e.g. cold tolerance, pathogen, and herbivory resistance) in crop plants is an intriguing and fascinating research area with a potential for reducing aphid-elicited injury on a variety of crops.

A limited number of studies have demonstrated the effect of asymptomatic aphids on plant growth and biomass partitioning (van Emden, 1990; van Emden and Hadley, 1994; Riedell and Kieckhefer, 1995). The general allometric model provided by Enquist and Niklas (2002) showed that allocation patterns of biomass partitioning in seed-producing plants were sensitive to local environmental conditions. van Emden (1990) reported significant dry matter loss in Brussels sprouts at low *B. brassicae* infestation that coincided with a reduction in dry matter loss of roots and stomata density. Using the Enquist and Niklas (2002) study as a baseline, the impact of herbivory by asymptomatic aphids on crop biomass partitioning should provide us with new insights on assessing feeding injury. To delineate plant injury, it is important that the impact of aphid feeding on the physiological and biochemical mechanisms of plant biomass accumulation and its partitioning among leaves, stems, fruits, and roots be clearly understood.

Ecological modulation of aphid-elicited plant injury is also not well understood. To date, most literature on biochemical and physiological mechanisms of aphid injury symptom formation has focused exclusively on aphid–plant interactions, while the role of ecological factors (e.g. photoperiod, temperature, soil moisture) has received little attention. Further, although we have considerable information on how aphids respond to different plants, we do not understand variation in plant responses to the feeding by different aphid species in varied agricultural ecosystems. This type of research could improve our understanding of aphid-elicited desistance and contribute significantly to aphid management.

It is well known that death of a specific set of cells is an essential part of the paradox of growth and development in many eukaryotic organisms (Dangl *et al.*, 2000). Senescence and hypersensitive responses are two examples that demonstrate the diverse range of programmed cell death in plants (Dangl *et al.*, 2000). Although apoptosis is a well-known term in the study of animal cell death, it is ironically a Greek word used to describe the dropping of the petals of flowers and leaves from plants. In general, the mechanisms of apoptotic cell death in plants have not yet been determined (Dangl *et al.*, 2000). Aphid-elicited plant desistance is commonly categorized as chlorosis, hypersensitive responses, and necrosis, while plant-tissue deformation (or neoplasm) elicited by aphids is often described as tissue galls or abnormal growth. These symptoms are likely to be part of cell death that is triggered by aphid feeding. Therefore, comparative studies of desistance and neoplasm in plant and apoptosis in animal systems could provide us with new information that ultimately unravels plant injury symptoms elicited by aphid feeding, as well as being of potential benefit to cancer research for human and animal health.

References

Agrawal, A.A., Tuzun, S. and Bent, E. (1999) *Induced Plant Defenses Against Pathogens and Herbivores: Biochemistry, Ecology, and Agriculture.* The American Phytopathological Society, St Paul, 390 pp.

Alsop, F.J. and Laughlin, T.F. (1991) Changes in spruce-fir avifauna of Mt. Guyout, Tennessee 1967–1985. *Journal of the Tennessee Academy of Science* 66, 207–209.

Andres, M.F., Melillo, M.T., Delibes, A., Romero, M.D. and Bleve-Zacheo, T. (2001) Changes in wheat root enzymes correlated with resistance to cereal cyst nematodes. *New Phytologist* 52, 343–354.

Archer, T.L., Bynum, E.D., Jr., Onken, A.B. and Wendt, C.W. (1995) Influence of water and nitrogen fertilizer on biology of the Russian wheat aphid (Homoptera: Aphididae) on wheat. *Crop Protection* 14, 165–169.

Armstrong, W.P. (1995) 'To be or not to be a gall.' *Pacific Horticulture* 56, 39–45.

Awmack, C.S. and Harrington, R. (2000) Elevated CO_2 affects the interactions between pests and host plant flowering. *Agricultural and Forest Meteorology* 2, 57–61.

Barbosa, P., Krischik, V.A. and Jones, C.G. (1991*) Microbial Mediation of Plant–Herbivore Interactions.* Wiley, New York, 530 pp.

Bing, J.W., Novak, M.G., Obrycki, J.J. and Guthrie, W.D. (1991) Stylet penetration and feeding sites of *Rhopalosiphum maidis* (Homoptera: Aphididae) on two growth stages of maize. *Annals of the Entomological Society of America* 84, 549–554.

Björkman, C. (2000) Interactive effects of host resistance and drought stress on the performance of a gall-making aphid living on Norway spruce. *Oecologia* 123, 223–231.

Blackman, R.L. and Eastop, V.F. (1994) *Aphids on the World's Trees: An Identification and Information Guide.* CAB International, Wallingford, 987 pp.

Boyd, M.L. and Lentz, G.L. (1994) Seasonal incidence of aphids and the aphid parasitoid *Diaeretiella rapae* (M'Intosh) (Hymenoptera: Aphidiidae) on rapeseed in Tennessee. *Environmental Entomology* 23, 349–353.

Breen, J.P. and Teetes, G.L. (1990) Economic injury levels for yellow sugarcane aphid (Homoptera: Aphididae) on seedling sorghum. *Journal of Economic Entomology* 83, 1008–1014.

Brown, M.W. and Schmitt, J.J. (1994) Population dynamics of woolly apple aphid (Homoptera: Aphididae) in West Virginia apple orchards. *Environmental Entomology* 23, 1182–1188.

Brown, M.W., Glenn, D.M. and Wisniewski, M.E. (1991) Functional and anatomical disruption of apple roots by the woolly apple aphid (Homoptera: Aphididae). *Journal of Economic Entomology* 84, 1823–1826.

Brown, M.W., Schmitt, J.J., Ranger, S. and Hogmire, H.W. (1995) Yield reduction in apple by edaphic woolly apple aphid (Homoptera: Aphididae) populations. *Journal of Economic Entomology* 88, 127–133.

Budak, S., Quisenberry, S.S. and Ni, X. (1999) Comparison of *Diuraphis noxia* resistance between wheat isolines and corresponding plant introduction lines. *Entomologia Experimentalis et Applicata* 92, 157–164.

Burd, J.D. and Elliott, N.C. (1996) Changes in chlorophyll *a* fluorescence induction kinetics in cereals infested with Russian wheat aphid (Homoptera: Aphididae). *Journal of Economic Entomology* 89, 1332–1337.

Burd, J.D., Burton, R.L. and Webster, J.A. (1993) Evaluation of Russian wheat aphid (Homoptera: Aphididae) damage on resistant and susceptible hosts with comparisons of damage ratings to quantitative plant measurements. *Journal of Economic Entomology* 86, 974–980.

Burd, J.D., Butts, R.A., Elliott, N.C. and Shufran, K.A. (1998) Seasonal development, overwintering biology, and host plant interactions of Russian wheat aphid (Homoptera: Aphididae) in North America. In: Quisenberry, S.S. and Peairs, F.B. (eds) *A Response Model for an Introduced Pest – The Russian Wheat Aphid.* Thomas Say Publications in Entomology, Entomological Society of America, Lanham, pp. 65–99.

Butts, R.A. (1992) Cold hardiness and its relationship to overwintering of the Russian wheat aphid (Homoptera: Aphididae) in South Alberta. *Journal of Economic Entomology* 85, 1140–1145.

Butts, R.A. and Schaalje, G.B. (1997) Impact of subzero temperatures on survival, longevity, and natality of adult Russian wheat aphid (Homoptera: Aphididae). *Environmental Entomology* 26, 661–667.

Butts, R.A., Howling, G.G., Bone, W., Bale, J.S. and Harrington, R. (1997) Contact with the host plant enhances aphid survival at low temperatures. *Ecological Entomology* 22, 26–31.

Campbell, B.C. (1986) Host-plant oligosaccharins in the honeydew of *Schizaphis graminum* (Rondani) (Insecta: Aphididae). *Experientia* 42, 451–452.

Campbell, B.C. and Dreyer, D.L. (1985) Host-plant resistance of sorghum: differential hydrolysis of sorghum pectic substances by polysaccharases of greenbug biotypes, *Schizaphis graminum* (Homoptera: Aphididae). *Archives of Insect Biochemistry and Physiology* 2, 203–215.

Campbell, B.C. and Dreyer, D.L. (1990) The role of plant matrix polysaccharides in aphid–plant interactions. In: Campbell, R.K. and Eikenbary, R.D. (eds) *Aphid–Plant Genotype Interactions.* Elsevier, Amsterdam, pp. 149–170.

Castro, A.M., Ramos, S., Vasicek, A., Worland, A., Gimenez, D., Clua, A.A. and Suarez, E. (2001) Identification of wheat chromosomes involved with different types of resistance against greenbug (*Schizaphis graminum* Rond.) and the Russian wheat aphid (*Diuraphis noxia* Mordvilko). *Euphytica* 118, 321–330.

Chaerle, L. and van der Straeten, D. (2001) Seeing is believing: imaging techniques to monitor plant health. *Biochimica et Biophysica Acta* 1519, 153–166.

Cherqui, A. and Tjallingii, W.F. (2000) Salivary proteins of aphids, a pilot study on identification, separation and immunolocalisation. *Journal of Insect Physiology* 46, 1177–1186.

Clement, S.L. and Quisenberry, S.S. (1999) *Global Plant Genetic Resources for Insect Resistant Crops.* CRC Press, Boca Raton, 295 pp.

Clement, S.L., Kaiser, W.J. and Eichenseer, H. (1994) *Acremonium* endophytes in germplasms of major grasses and their utilization for insect resistance. In: Bacon, C.W. and White, J.F., Jr. (eds) *Biotechnology of Endophytic Fungi of Grasses.* CRC Press, Boca Raton, pp. 185–199.

Crawford, K.M. and Zambryski, P.C. (1999) Phloem transport: are you chaperoned? *Current Biology* 9, 281–285.

Dangl, J.L., Dietrich, R.A. and Thomas, H. (2000) Senescence and programmed cell death. In: Buchanan, B.B., Gruissem, W. and Jones, R.L. (eds) *Biochemistry and Molecular Biology of Plants.* American Society of Plant Physiologists, Rockville, pp. 1044–1100.

de Farias, A.M.I., Hopper, K.R. and Leclant, F. (1995) Damage symptoms and abundance of *Diuraphis noxia* (Homoptera: Aphididae) for four wheat cultivars at three irrigation levels. *Journal of Economic Entomology* 88, 169–174.

Diaz, S., Fraser, L.H., Grime, J.P. and Falczuk, V. (1998) The impact of elevated CO_2 on plant–herbivore interactions: experimental evidence of moderating effects at the community level. *Oecologia* 117, 177–186.

Dixon, A.F.G. (1985) *Aphid Ecology*. Blackie, Glasgow, 157 pp.

Donahue, J.D., Brewer, M.J., Peairs, F.B. and Hein, G.L. (1998) *High Plains Integrated Pest Management Guide for Colorado–Western Nebraska–Wyoming, No. 564A*. Colorado State University, Fort Collins, 406 pp.

Doss, R.P., Oliver, J.E., Proebsting, W.M., Potter, S.W., Kuy, S., Clement, S.L., Williamson, R.T., Carney, J.R. and DeVilbiss, E.D. (2000) Bruchins: insect-derived plant regulators that stimulate neoplasm formation. *Proceedings of National Academy of Sciences of the United States of America* 97, 6218–6223.

Drees, B.M. and Jackman, J.A. (1998) *A Field Guide to Common Texas Insects*. Gulf Publishing Company, Houston, 359 pp.

Dreyer, D.L. and Campbell, B.C. (1987) Chemical basis of host-plant resistance to aphids. *Plant, Cell and Environment* 10, 353–361.

Ehler, L.E. (1998) Invasion biology and biological control. *Biological Control* 13, 127–133.

van Emden, H.F. (1972) Aphids as phytochemists. In: Harborne, J.B. (ed.) *Phytochemical Ecology*. Academic Press, London, pp. 25–43.

van Emden, H.F. (1990) The effect of *Brevicoryne brassicae* on leaf area, dry matter distribution and amino acids of Brussels sprout. *Annals of Applied Biology* 116, 199–204.

van Emden, H.F. and Hadley, P. (1994) The application of the concepts of resource capture to the effect of pest incidence on crops. In: Monteith, J.L., Scott, R.K. and Unsworth, M.H. (eds) *Resource Capture by Crops*. Nottingham University Press, Nottingham, pp. 149–165.

Enquist, B.J. and Niklas, K.J. (2002) Global allocation rules for patterns of biomass partitioning in seed plants. *Science, New York* 295, 1517–1520.

Felton, G.W., Korth, K.L., Bi, J.L., Wesley, S.V., Huhman, D.V., Mathews, M.C., Murphy, J.B., Lamb, C. and Dixon, R.A. (1999) Inverse relationship between systemic resistance of plants to microorganisms and to insect herbivory. *Current Biology* 9, 317–320.

Flinn, M., Smith, C.M., Reese, J.C. and Gill, B. (2001) Categories of resistance to greenbug (Homoptera: Aphididae) biotype I in *Aegilops tauschii* germplasm. *Journal of Economic Entomology* 94, 558–563.

Granett, J., Walker, M.A., Kocsis, L. and Omer, A.D. (2001) Biology and management of grape phylloxera. *Annual Review of Entomology* 46, 387–412.

Haile, F.J., Higley, L.G., Ni, X. and Quisenberry, S.S. (1999) Physiological and growth tolerance in wheat to Russian wheat aphid (Homoptera: Aphididae) injury. *Environmental Entomology* 28, 787–794.

Halbert, S.E. and Brown, L.G. (1996) *Toxoptera citricida (Kirkaldy), Brown Citrus Aphid – Identification, Biology and Management Strategies*. Florida Department of Agriculture and Consumer Services, Entomology Circular No. 374, 6 pp.

Harrington, R., Bale, J.S. and Tatchell, G.M. (1995) Aphids in a changing climate. In: Harrington, R. and Stork, N.E. (eds) *Insects in a Changing Environment*. Academic Press, London, pp. 126–157.

Hill, D.S. (1997) *The Economic Importance of Insects*. Chapman and Hall, London, 395 pp.

Hori, K. (1992) Insect secretions and their effect on plant growth, with special reference to Hemipterans. In: Shorthouse, J.D. and Rohfritsch, O. (eds) *Biology of Insect-Induced Galls*. Oxford University Press, New York, pp. 157–170.

Hori, K., Wada, A. and Shibuta, T. (1997) Changes in phenoloxidase activities of the galls on leaves of *Ulmus davidana* formed by *Tetraneura fusiformis* (Homoptera: Eriosomatidae). *Journal of Applied Entomology and Zoology* 32, 365–371.

Hughes, L. and Bazzaz, F.A. (2001) Effects of elevated CO_2 on five plant–aphid interactions. *Entomologia Experimentalis et Applicata* 99, 87–96.

Hutchison, W.D. and Campbell, C.D. (1994) Economic impact of sugarbeet root aphid (Homoptera: Aphididae) on sugarbeet yield and quality in southern Minnesota. *Journal of Economic Entomology* 87, 465–475.

Janave, M.T. (1997) Enzymatic degradation of chlorophyll in cavendish bananas: *in vitro* evidence for two independent degradative pathways. *Plant Physiology and Biochemistry* 35, 837–846.

Jiang, Y. (1996) Oxidative interactions between the spotted alfalfa aphid (*Therioaphis trifolii maculata*) (Homoptera: Aphididae) and the host plant *Medicago sativa*. *Bulletin of Entomological Research* 86, 533–540.

Jiang, Y. and Miles, P.W. (1993) Responses of a compatible lucerne variety to attack by spotted alfalfa aphid: changes in the redox balance in affected tissues. *Entomologia Experimentalis et Applicata* 67, 263–274.

Jiang, Y.X., Nombela, G. and MuZiz, M. (2001) Analysis by DC-EPG of the resistance to _Bemisia tabaci_ on an _Mi_-tomato line. _Entomologia Experimentalis et Applicata_ 99, 295–302.

Johnson, G.D., Ni, X., McLendon, M.E., Jacobsen, J.S. and Wraith, J.W. (1998) Impact of Russian wheat aphid (Homoptera: Aphididae) on drought-stressed spring wheat. In: Quisenberry, S.S. and Peairs, F.B. (eds) _A Response Model for an Introduced Pest – The Russian Wheat Aphid._ Thomas Say Publications in Entomology, Entomological Society of America, Lanham, pp. 109–121.

Johnson, W.T. and Lyon, H.H. (1988) _Insects that Feed on Trees and Shrubs_, 2nd edn. Comstock Publishing Associates, Ithaca, 556 pp.

Joost, R.E. and Quisenberry, S.S. (1993) _Acremonium/Grass Interactions._ Elsevier, Amsterdam, 324 pp.

Karban, R. and Baldwin, I.T. (1997) _Induced Responses to Herbivory._ University of Chicago Press, Chicago, 319 pp.

Kieckhefer, R.W. and Kantack, B.H. (1988) Yield losses in winter grains caused by cereal aphids (Homoptera: Aphididae) in South Dakota. _Journal of Economic Entomology_ 81, 317–321.

Kimberling, D.N., Scott, E.R. and Price, P.W. (1990) Testing a new hypothesis: plant vigor and phylloxera distribution on wild grape in Arizona. _Oecologia_ 84, 1–8.

Kindler, S.D. and Hammon, R.W. (1996) Comparison of host suitability of western wheat aphid with the Russian wheat aphid. _Journal of Economic Entomology_ 89, 1621–1630.

Latimer, J.G. and Oetting, R.D. (1999) Conditioning treatments affect insect and mite populations on bedding plants in the greenhouse. _HortScience_ 34, 235–238.

Lawson, D.S. and Weires, R.W. (1991) Management of European red mite (Acari: Tetranychidae) and several aphid species on apple with petroleum oils and an insecticidal soap. _Journal of Economic Entomology_ 84, 1550–1557.

LeClerg, E.L., Cook, H.T., Van Hauweling, C.D., Anderson, R.J., Vance, A.M., Dorwood, K. and Thomas, H.R. (1965) Injurious crop insects. In: _Losses in Agriculture._ US Department of Agriculture, Agricultural Research Service, Agricultural Handbook, No. 291, pp. 43–44.

Liu, X.M., Smith, C.M., Gill, B.S. and Tolmay, V. (2001) Microsatellite markers linked to six Russian wheat aphid resistance genes in wheat. _Theoretical and Applied Genetics_ 102, 504–510.

Losey, J.E. and Eubanks, M.D. (2000) Implications of pea aphid host-plant specialization for the potential colonization of vegetables following post-harvest emigration from forage crops. _Environmental Entomology_ 29, 1283–1288.

Ma, R., Reese, J.C., Black IV, W.C. and Bramel-Cox, P. (1990) Detection of pectinesterase and polygalacturonase from salivary secretions of living greenbugs, _Schizaphis graminum_ (Homoptera: Aphididae). _Journal of Insect Physiology_ 36, 507–512.

Macedo, T., Higley, L., Ni, X. and Quisenberry, S.S. (2003) Light activation of Russian wheat aphid-elicited physiological responses in susceptible wheat. _Journal of Economic Entomology_ 96, 194–201.

Madhusudhan, V.V. and Miles, P.W. (1998) Mobility of salivary components as a possible reason for differences in the responses of alfalfa to the spotted alfalfa aphid and pea aphid. _Entomologia Experimentalis et Applicata_ 86, 25–39.

Madhusudhan, V.V., Taylor, G.S. and Miles, P.W. (1994) The detection of salivary enzymes of phytophagous Hemiptera: a compilation of methods. _Annals of Applied Biology_ 124, 405–412.

Mani, M.S. (1992) Introduction to cecidology. In: Shorthouse, J.D. and Rohfritsch, O. (eds) _Biology of Insect-Induced Galls._ Oxford University Press, New York, pp. 3–7.

Matile, P., Hörtensteiner, S. and Thomas, H. (1999) Chlorophyll degradation. _Annual Review of Plant Physiology and Plant Molecular Biology_ 50, 67–95.

Miles, P.W. (1989a) The responses of plants to the feeding of Aphidoidea: principles. In: Minks, A.K. and Harrewijn, P. (eds) _Aphids. Their Biology, Natural Enemies and Control, Volume 2C._ Elsevier, Amsterdam, pp. 1–21.

Miles, P.W. (1989b) Specific responses and damage caused by Aphidoidea. In: Minks, A.K. and Harrewijn, P. (eds) _Aphids. Their Biology, Natural Enemies and Control, Volume 2C._ Elsevier, Amsterdam, pp. 23–47.

Miles, P.W. (1990) Aphid salivary secretions and their involvement in plant toxicoses. In: Campbell, R.K. and Eikenbary, R.D. (eds) _Aphid–Plant Genotype Interactions._ Elsevier, Amsterdam, pp. 131–147.

Miles, P. (1998) Aphid salivary functions: the physiology of deception. In: Nieto Nafria, J.M. and Dixon, A.F.G. (eds) _Aphids in Natural and Managed Ecosystems._ Universidad de León, León, pp. 255–263.

Miles, P.W. (1999) Aphid saliva. _Biological Reviews_ 74, 41–85.

Miles, P.W. and Oertli, J.J. (1993) The significance of antioxidants in the aphid–plant interaction: the redox hypothesis. _Entomologia Experimentalis et Applicata_ 67, 275–283.

Miller, H., Neese, P.A., Ketring, D.L. and Dillwith, J.W. (1994) Involvement of ethylene in aphid infestation of barley. *Journal of Plant Growth Regulation* 13, 167–171.

Morales, H., Perfecto, I. and Ferguson, B. (2001) Traditional fertilization and its effect on corn insect populations in the Guatemalan highlands. *Agricultural Ecosystems and Environment* 84, 145–155.

Morrison, W.P. and Peairs, F.B. (1998) Response model concept and economic impact. In: Quisenberry, S.S. and Peairs, F.B. (eds) *A Response Model for an Introduced Pest – The Russian Wheat Aphid*. Thomas Say Publications in Entomology, Entomological Society of America, Lanham, pp. 1–11.

Murphy, J.F., Zehnder, G.W., Schuster, D.J., Sikora, E.J., Polston, J.E. and Kloepper, J.W. (2000) Plant growth-promoting rhizobacterial mediated protection in tomato against tomato mottle virus. *Plant Disease* 84, 779–784.

Ni, X. and Quisenberry, S.S. (1997) Distribution of Russian wheat aphid salivary sheaths on resistant and susceptible wheat leaves. *Journal of Economic Entomology* 90, 848–853.

Ni, X., Quisenberry, S.S., Pornkulwat, S., Figarola, J.L., Skoda, S.R. and Foster, J.E. (2000) Hydrolase and oxido-reductase activities in *Diuraphis noxia* and *Rhopalosiphum padi* (Hemiptera: Aphididae). *Annals of the Entomological Society of America* 93, 595–601.

Ni, X., Quisenberry, S.S., Heng-Moss, T., Markwell, J., Sarath, G., Klucas, R. and Baxendale, F. (2001a) Oxidative responses of resistant and susceptible cereal leaves to symptomatic and nonsymptomatic cereal aphid (Hemiptera: Aphididae) feeding. *Journal of Economic Entomology* 94, 743–751.

Ni, X., Quisenberry, S.S., Markwell, J., Heng-Moss, T., Higley, L., Baxendale, F., Sarath, G. and Klucas, R. (2001b) *In vitro* enzymatic chlorophyll catabolism in wheat elicited by cereal aphid feeding. *Entomologia Experimentalis et Applicata* 101, 159–166.

Nombela, G., Beitia, F. and MuZiz, M. (2001) A differential interaction study of *Bemisia tabaci* Q-biotype on commercial tomato varieties with or without the *Mi* resistance gene, and comparative host responses with the B-biotype. *Entomologia Experimentalis et Applicata* 98, 339–344.

Oswald II, C.J. and Brewer, M.J. (1997) Aphid–barley interactions mediated by water stress and barley resistance to Russian wheat aphid (Homoptera: Aphididae). *Environmental Entomology* 26, 591–602.

Patlak, M. (1998) Designer seeds. In: *Beyond Discovery: The Path from Research to Human Benefit*. National Academy of Sciences of the Unites States of America, Washington, D.C., 8 pp.

Peairs, F.B. (2001) *Aphids in Small Grains*. Colorado State University Cooperative Extension, Publication No. 5.568, 7 pp.

Percy, K.E., Awmack, C.S., Lindroth, R.L., Kubiske, M.E., Kopper, B.J., Isebrands, J.G., Pregitzer, K.S., Hendrey, G.R., Dickson, R.E., Zak, D.R., Oksanen, E., Sober J., Harrington, R. and Karnosky, D.F. (2002) Altered performance of forest pests under atmospheres enriched by CO_2 and O_3. *Nature, London* 420, 403–407.

Peters, D.C., Webster, J.A. and Chlouber, C.S. (1991) *Aphid–Plant Interactions: Populations to Molecules*. Oklahoma Agricultural Experimental Station, Stillwater, 335 pp.

Peterson, R.K.D. and Higley, L.G. (2001) Illuminating the black box: the relationship between injury and yield. In: Peterson, R.K.D. and Higley, L.G. (eds) *Biotic Stress and Yield Loss*. CRC Press, Boca Raton, pp. 1–12.

Pimentel, D., Lach, L., Zuniga, R. and Morrison, D. (1999) *Environmental and Economic Costs Associated with Non-Indigenous Species in the United States*. College of Agriculture and Life Sciences, Cornell University, Ithaca, 22 pp.

Ponder, K.L., Pritchard, J., Harrington, R. and Bale, J.S. (2001) Feeding behaviour of the aphid *Rhopalosiphum padi* (Hemiptera: Aphididae) on nitrogen and water-stressed barley (*Hordeum vulgare*) seedlings. *Bulletin of Entomological Research* 91, 125–130.

Quisenberry, S.S. and Joost, R.E. (1990) *The Proceedings of the International Symposium on Acremonium/ Grass Interactions*. Louisiana Agricultural Experiment Station, Baton Rouge, LA, 289 pp.

Quisenberry, S.S. and Peairs, F.B. (1998) *A Response Model for an Introduced Pest – The Russian Wheat Aphid*. Thomas Say Publications in Entomology, Entomological Society of America, Lanham, 442 pp.

Rehill, B.J. and Schultz, J.C. (2001) *Hormaphis hamamelidis* and gall size: a test of the plant vigor hypothesis. *Oikos* 95, 94–104.

Reynolds, D.E. (1999) *Capnodium citri*: the sooty mold fungi comprising the taxon concept. *Mycopathologia* 148, 141–147.

Rice, M.E., O'Neal, M. and Pederson, P. (2004) *Soybean Aphid in Iowa – 2004*. Iowa State University Extension, Publication No. SP237, 8 pp.

Riedell, W.E. (1989) Effects of Russian wheat aphid infestation on barley plant response to drought stress. *Physiologia Plantarum* 77, 587–592.

Riedell, W.E. and Kieckhefer, R.W. (1995) Feeding damage effects of three aphid species on wheat root growth. *Journal of Plant Nutrition* 18, 1881–1891.

Riedell, W.E., Kieckhefer, R.W., Haley, S.D., Langham, M.A.C. and Evenson, P.D. (1999) Winter wheat responses to bird cherry–oat aphids and barley yellow dwarf virus infection. *Crop Science* 39, 158–163.

Rossi, M., Goggin, F.L., Milligan, S.B., Kaloshian, I., Ullman, D.E. and Williamson, V.M. (1998) The nematode resistance gene *Mi* of tomato confers resistance against the potato aphid. *Proceedings of the National Academy of Sciences of the United States of America* 95, 9750–9754.

Ryan, J.D., Morgham, A.T., Richardson, P.E., Johnson, R.C., Mort, A.J. and Eikenbary, R.D. (1990) Greenbugs and wheat: a model system for the study of phytotoxic Homoptera. In: Campbell, R.K. and Eikenbary, R.D. (eds) *Aphid–Plant Genotype Interactions*. Elsevier, Amsterdam, pp. 171–186.

Salt, D.T., Brooks, G.L. and Whittaker, J.B. (1996) Interspecific herbivore interactions in a high CO_2 environment: root and shoot aphids feeding on *Cardamine*. *Oikos* 77, 326–339.

Sandström, J., Telang, A. and Moran, N.A. (2000) Nutritional enhancement of host plants by aphids – a comparison of three aphid species on grasses. *Journal of Insect Physiology* 46, 33–40.

Shorthouse, J.D. and Rohfritsch, O. (1992) *Biology of Insect-Induced Galls*. Oxford University Press, New York, 285 pp.

Southwood, R. (1997) Foreword. In: Schoonhoven, L.M., Jermy, T. and van Loon, J.J.A. (eds) *Insect–Plant Biology: From Physiology to Evolution*. Chapman and Hall, London, p. xi.

Thomas, R.C. and Miller, H.G. (1994) The interaction of green spruce aphid and fertilizer applications on the growth of Sitka spruce. *Forestry* 67, 329–341.

Varn, M. and Pfeiffer, D.G. (1989) Effect of rosy apple aphid and spirea aphid (Homoptera: Aphididae) on dry matter accumulation and carbohydrate concentration in young apple trees. *Journal of Economic Entomology* 82, 565–569.

Voss, T.S., Kieckhefer, R.W., Fuller, B.W., McLeod, M.J. and Beck, D.A. (1997) Yield losses in maturing spring wheat caused by cereal aphids (Homoptera: Aphididae) under laboratory conditions. *Journal of Economic Entomology* 90, 1346–1350.

Walgenbach, J.F. (1997) Effect of potato aphid (Homoptera: Aphididae) on yield, quality, and economics of staked-tomato production. *Journal of Economic Entomology* 90, 996–1004.

Walling, L.L. (2000) The myriad plant responses to herbivores. *Journal of Plant Growth Regulation* 19, 195–216.

Webster, J.A. (1990) Yellow sugarcane aphid (Homoptera: Aphididae): detection and mechanisms of resistance among Ethiopian sorghum lines. *Journal of Economic Entomology* 83, 1053–1057.

Wilkinson, H.H., Siegel, M.R., Blankenship, J.D., Mallory, A.C., Bush, L.P. and Schardl, C.L. (2000) Contribution of fungal loline alkaloids to protection from aphids in a grass-endophyte mutualism. *Molecular Plant–Microbe Interaction* 13, 1027–1033.

Wilson, H.K. and Quisenberry, S.S. (1986) Impact of feeding by alfalfa weevil larvae (Coleoptera: Curculionidae) and pea aphid (Homoptera: Aphididae) on yield and quality of first and second cuttings of alfalfa. *Journal of Economic Entomology* 79, 785–789.

Wood, B.W. and Reilly, C.C. (1998) Susceptibility of pecan to black pecan aphids. *HortScience* 33, 798–801.

Zehnder, G.W., Yao, C., Murphy, J.F., Sikora, E.J., Kloepper, J.W., Schuster, D.J. and Polston, J.E. (1999) Microbe-induced resistance against pathogens and herbivores: evidence of effectiveness in agriculture. In: Agrawal, A.A., Tuzun, S. and Bent, E. (eds) *Induced Plant Defenses Against Pathogens and Herbivores: Biochemistry, Ecology, and Agriculture*. The American Phytopathological Society, St Paul, pp. 335–355.

Zehnder, G.W., Murphy, J.F., Sikora, E.J. and Kloepper, J.W. (2001) Application to rhizobacteria for induced resistance. *European Journal of Plant Pathology* 107, 39–50.

14   Transmission of Plant Viruses

Nikos I. Katis[1], John A. Tsitsipis[2], Mark Stevens[3] and Glen Powell[4]

[1]Aristotle University of Thessaloniki, Department of Agriculture, Laboratory of Plant Pathology, 541 24, Thessaloniki, Greece; [2]University of Thessaly, Department of Agriculture, Crop Production and Rural Environment, Laboratory of Entomology and Agricultural Zoology, 384 46, Nea Ionia Magnissias, Greece; [3]Broom's Barn Research Station, Higham, Bury St Edmunds, Suffolk, IP28 6NP, UK; [4]Division of Biology, Imperial College London, Wye campus, Ashford, Kent, TN25 5AH, UK

Introduction

As obligate parasites, plant viruses need to move from infected to healthy plants in order to survive. This is achieved either by mechanical means or, in the case of most plant viruses, by exploiting biological vectors such as arthropods, nematodes, and fungi. Of the 700 or more plant viruses (van Regenmortel et al., 2000), about 70% are known, or suspected, to be transmitted by arthropod, nematode, or fungal vectors (Nault, 1997). Sap-feeding insects in the Auchenorrhyncha and Sternorryhncha are particularly important vectors, transmitting more than 380 viruses (Nault, 1997). The aphids (Aphididae) are by far the most important family among these vectors, transmitting many more viruses than whiteflies (Aleyrodidae), leafhoppers (Cicadellidae), or planthoppers (Delphacidae) (Fig. 14.1). More than 4700 aphid species have been described (Remaudière and Remaudière, 1997) and of these, over 190 have been reported to transmit plant viruses (Nault, 1997), with many species able to transmit more than one virus (Eastop, 1983; Nault, 1997; Hull, 2002). The majority of reported aphid virus vectors belong to the genera Myzus, Aphis, Acyrthosiphon, and Macrosiphum in the subfamily Aphidinae (Kennedy et al., 1962).

Only a small proportion of aphid species (approximately 300 – Kennedy et al., 1962; Eastop, 1983; Nault, 1997) have been tested as virus vectors and therefore the potential number of aphid vectors is probably very much larger. For example, in a study of the transmission of Zucchini yellow mosaic virus (ZYMV) by previously unrecorded vectors, 16 out of 19 species tested were found to transmit the virus (Katis et al., 2006).

Aphid-transmitted viruses belong to 19 of the 70 recognized virus genera and comprise approximately 275 virus species (i.e. about 50% of insect-borne plant viruses – Nault, 1997). Many of these viruses cause diseases of major economic importance in crops (Hull, 2002) such as cereals (Plumb, 2002), potatoes (de Bokx and van der Want, 1987; Salazar, 1996; Brunt and Loebenstein, 2001), and sugar beet (Stevens et al., 2004). In this chapter, we review virus transmission by aphids, with particular emphasis on the attributes that make aphids such welladapted vectors, the different modes of transmission, determinants of transmissibility, factors affecting virus acquisition and inoculation, and control methods. Table 14.1 shows the names and standard acronyms for viruses used as examples throughout the

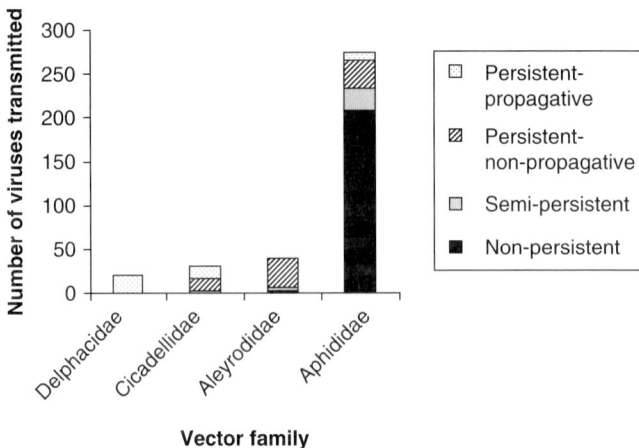

Fig. 14.1. Number of viruses transmitted by the four major homopteran vector families, divided into the four transmission categories (data from Nault, 1997).

chapter, as well as their genome type and classification to genus and family.

Virus–Vector Interactions

Virus transmission by an aphid (or any other vector) involves the transfer of virions (virus particles, complete with the RNA or DNA core and protein coat) from infected to healthy plants. The transmission cycle comprises up to four phases:

1. *Acquisition* – the process by which the aphid takes up virions from an infected plant;
2. *Retention* – the carriage of virions at specific sites in or on the vector;
3. *Latency* – an inability to inoculate immediately following acquisition (the aphid is able to transmit the virus after the 'latent period' (LP) has passed); and
4. *Inoculation* – the release of retained virions into the tissues of a susceptible plant in such a way that they are able to establish a new infection.

Based on the period of retention by the vector, virus transmission by aphids has been divided into three major categories (Table 14.2):

1. *Non-persistent transmission* – where acquisition and inoculation require only very brief stylet penetration (< 1 min). There is

no LP and the entire transmission cycle may therefore be completed within a few minutes. Viruses transmitted in this way have also been referred to as 'stylet-borne' and aphids rapidly lose the ability to inoculate them following acquisition (e.g. *Potato virus Y* (PVY)).
2. *Semi-persistent transmission* – where efficient acquisition and inoculation requires longer periods of plant access than for non-persistent viruses, often at least 15 min. There is no LP and the aphids retain the ability to inoculate for longer periods, and may continue to transmit for up to 2 days following acquisition (e.g. *Citrus tristeza virus* (CTV)).
3. *Persistent transmission* – where virus acquisition and inoculation require relatively long periods of plant access. A LP occurs between acquisition and inoculation, but once it has passed, the aphid can remain infective for life (e.g. *Cereal yellow dwarf virus*-RPV (CYDV-RPV)).

These categories are properties of the relationship between viruses and their vectors, but will often be used herein to describe the viruses. A given virus is always in the same category, regardless of its vector. A fundamental distinction regarding the mode of transmission is whether virions are circulative or non-circulative within the aphid vector. All viruses transmitted non-persistently and semi-persistently are non-circulative,

Table 14.1. Abbreviations of plant viruses included in the text and their classification to genus/family.

Acronym	Virus name	Genus/Family
AMV	*Alfalfa mosaic virus*	*Alfamovirus*/Bromoviridae
AYV	*Anthriscus yellows virus*	*Waikavirus*/Sequiviridae
BBTV	*Banana bunchy top virus*	*Babuvirus*/Nanoviridae
BCMV	*Bean common mosaic virus*	*Potyvirus*/Potyviridae
BMYV	*Beet mild yellowing virus*	*Polerovirus*/Luteoviridae
BtMV	*Beet mosaic virus*	*Potyvirus*/Potyviridae
BWYV	*Beet western yellows virus*	*Polerovirus*/Luteoviridae
BYDV-MAV	*Barley yellow dwarf virus*-MAV	*Luteovirus*/Luteoviridae
BYDV-PAV	*Barley yellow dwarf virus*-PAV	*Luteovirus*/Luteoviridae
BYMV	*Bean yellow mosaic virus*	*Potyvirus*/Potyviridae
BYV	*Beet yellows virus*	*Closterovirus*/Closteroviridae
CABYV	*Cucurbit aphid-borne yellows virus*	*Polerovirus*/Luteoviridae
CaMV	*Cauliflower mosaic virus*	*Caulimovirus*/Caulimoviridae
CtVY	*Carrot virus Y*	*Potyvirus*/Potyviridae
CeMV	*Celery mosaic virus*	*Potyvirus*/Potyviridae
CMV	*Cucumber mosaic virus*	*Cucumovirus*/Bromoviridae
CTV	*Citrus tristeza virus*	*Closterovirus*/Closteroviridae
CYDV-RPV	*Cereal yellow dwarf virus*-RPV	*Polerovirus*/Luteoviridae
FBNYV	*Faba bean necrotic yellows virus*	*Nanovirus*/Nanoviridae
LMV	*Lettuce mosaic virus*	*Potyvirus*/Potyviridae
LYSV	*Leek yellow stripe virus*	*Potyvirus*/Potyviridae
MDMV	*Maize dwarf mosaic virus*	*Potyvirus*/Potyviridae
OYDV	*Onion yellow dwarf virus*	*Potyvirus*/Potyviridae
PEMV-1	*Pea enation mosaic virus-1*	*Enamovirus*/Luteoviridae
PEMV-2	*Pea enation mosaic virus-2*	*Umbravirus*/Luteoviridae
PLRV	*Potato leaf roll virus*	*Polerovirus*/Luteoviridae
PPV	*Plum pox virus*	*Potyvirus*/Potyviridae
PSbMV	*Pea seed-borne mosaic virus*	*Potyvirus*/Potyviridae
PVY	*Potato virus Y*	*Potyvirus*/Potyviridae
PYFV	*Parsnip yellow fleck virus*	*Sequivirus*/Sequiviridae
SbDV	*Soybean dwarf virus*	*Luteovirus*/Luteoviridae
SMV	*Soybean mosaic virus*	*Potyvirus*/Potyviridae
SYVV	*Sowthistle yellow vein virus*	*Nucleorhabdovirus*/Rhabdoviridae
TBV	*Tulip breaking virus*	*Potyvirus*/Potyviridae
TEV	*Tobacco etch virus*	*Potyvirus*/Potyviridae
TuMV	*Turnip mosaic virus*	*Potyvirus*/Potyviridae
WMV	*Watermelon mosaic virus*	*Potyvirus*/Potyviridae
ZYMV	*Zucchini yellow mosaic virus*	*Potyvirus*/Potyviridae

residing on or in the aphid stylets and foregut, whereas those transmitted persistently are known as circulative, as they pass from the gut into the haemocoel and then to the accessory salivary glands (ASGs) before they can be inoculated. Persistent viruses may be propagative, i.e. able to multiply in the aphid, but are mostly non-propagative (Table 14.2).

Table 14.2. Transmission and other characteristics of non-persistent, semi-persistent, and persistent aphid-borne plant viruses.

Transmission characteristic	Transmission mode			
	NP	SP	PC	PP
Optimal acquisition time	Seconds (>5,) minutes	Minutes to hours	Hours to days	Hours to days
Retention half life	Minutes	Hours	Days to weeks	Weeks to months (often for lifetime)
Transtadial passage	No	No	Yes	Yes
Virus in vector haemolymph	No	No	Yes	Yes
Latent period	No	No	Hours to days	Weeks ('incubation')
Virus multiplication in the aphid's body	No	No	No	Yes
Transovarial transmission	No	No	No	Rarely
Aphid specificity	Low	Medium	High	High
Effect of pre-acquisition fasting	Enhances transmission	No effect	No effect	No effect
Mechanical transmission under lab conditions	All	Some with difficulty (Caulimo-, Sequi-, clostero- with difficulty, Waika: no)	Only few (PEMV-2), Umbravirus* (uncapsidated RNA is stable)	A number of rhabdoviruses infecting dicots**
Seed transmission	Some (species belonging to Cucumo-, Alfamo-, Poty-)	No	No	No
Common symptomatology	Mo, M, VN, LD, FB	Y, LR	Y, LR	Y, LR
Tissue acquisition	(Parenchyma) epidermis	Phloem	Phloem	Phloem
Inoculation	Parenchyma	Parenchyma/ phloem	Phloem	Phloem

FB, flower breaking; LD, leaf deformation; LR, leaf rolling; M, mottling; Mo, mosaic; NP, non-persistent; PC, persistent-circulative (non-propagative); PP, persistent-propagative; SP, semi-persistent; VN, vein necrosis; Y, leaf yellowing; *, PEMV-2 is transmitted mechanically when plants are also infected with PEMV-1 *Luteovirus*. Co-infection of plants with PEMV-2 facilitates mechanical transmission of some luteoviruses such as PLRV and BMYV; **, some rhabdoviruses infecting monocots are only transmitted mechanically by vascular puncture. Modified from Hull (2002), Nault (1997).

1

2

Plate 1. *Acyrthosiphon pisum* – pea aphid (courtesy of B. Chaubet and INRA).

Plate 2. *Aphis craccivora* – cowpea aphid (courtesy of B. Chaubet and INRA).

3

4

Plate 3. *Aphis fabae* – black bean aphid (courtesy of Syngenta).

Plate 4. *Aphis gossypii* – cotton or melon aphid (left, courtesy of Bayer CropScience ImageBank: right, courtesy of B. Chaubet and INRA).

5

6

Plate 5. *Aphis spiraecola* – green citrus aphid (courtesy of B. Chaubet and INRA).

Plate 6. *Diuraphis noxia* – Russian wheat aphid (courtesy of B. Chaubet and INRA).

7

8

Plate 7. *Lipaphis pseudobrassicae* – mustard aphid (courtesy of B. Chaubet and INRA).

Plate 8. *Macrosiphum euphorbiae* – potato aphid (left, courtesy of B. Chaubet and INRA: right, courtesy of Bayer CropScience ImageBank).

9

10

Plate 9. *Myzus persicae* – peach–potato aphid (courtesy of U. Wyss; inset courtesy of A. M. Dewar).

Plate 10. *Rhopalosiphum maidis* – corn leaf aphid (courtesy of S. Barbagallo).

11

12

Plate 11. *Rhopalosiphum padi* – bird cherry–oat aphid (courtesy of Rothamsted Research; inset courtesy of U.Wyss).

Plate 12. *Schizaphis graminum* – greenbug (courtesy of B. Chaubet and INRA).

13

14

Plate 13. *Sitobion avenae* – grain aphid (left, courtesy of A. M. Dewar; centre, courtesy of S. D. Wratten; right, courtesy of U. Wyss).

Plate 14. *Therioaphis trifolii maculata* – spotted alfalfa aphid (courtesy of B. Chaubet and INRA).

15

16

Plate 15. Coccinellidae. Eggs of *Adalia bipunctata* (courtesy of the late S. P. Hopkin and Ardea, London).

Plate 16. Coccinellidae. Larva of *Coccinella 7-punctata* (courtesy of A. Stewart-Jones).

17

18

Plate 17. Coccinellidae. Pupa of *Coccinella 7-punctata* (courtesy of the late S. P. Hopkin).

Plate 18. Coccinellidae. Adults of *Coccinella 7-punctata* (courtesy of the late S. P. Hopkin and Ardea, London).

19

20

Plate 19. Syrphidae. Syrphid eggs, species not known (courtesy of B. Freier and the Biologische Bundesanstal für Land- und Forstwirtschaft, Kleinmachnow).

Plate 20. Syrphidae. Larva of *Episyrphus balteatus* (courtesy of the late S. P. Hopkin and Ardea, London).

21

22

Plate 21. Syrphidae. Puparium of *Episyrphus balteatus* (courtesy of U. Wyss).

Plate 22. Syrphidae. Adult of *Episyrphus balteatus* (courtesy of the late S. P. Hopkin and Ardea, London).

23

24

Plate 23. Chrysopidae. Eggs of *Chrysoperla carnea* (courtesy of U. Wyss).

Plate 24. Chrysopidae. Larva of *Chrysoperla carnea* (courtesy of U. Wyss).

25

26

Plate 25. Chrysopidae. Adult of *Chrysoperla carnea* (courtesy of Rothamsted Research).

Plate 26. Cecidomyiidae. Larva of *Aphidoletes aphidimyza* (courtesy of J. Bennison and ADAS).

27

28

Plate 27. Cecidomyiidae. Adult of *Aphidoletes aphidimyza* (courtesy of U. Wyss).

Plate 28. Aphidiidae. Adult of *Aphidius rhopalosiphi* (courtesy of G. R. Gowling).

29

30

Plate 29. Aphidiidae. Mummy of *Metopolophium dirhodum* parasitized by *Aphidius rhopalosiphi* (courtesy of G. R. Gowling).

Plate 30. Anthocoridae. Adult *Anthocoris nemorum* (courtesy of U. Wyss).

31

Plate 31. Entomopathogenic fungi. *Acyrthosiphon pisum* infected by *Pandora neoaphidis* (courtesy of Rothamsted Research).

Aphids as Effective Virus Vectors

Certain characteristics of aphids predispose them to being effective virus vectors. Host selection and feeding behaviour are particularly important factors in virus epidemiology, and the host range and life cycle characteristics of aphid species are also key in determining the rate of spread of viruses. In this chapter, we consider the host-selection processes specifically related to virus transmission. Several reviews have previously covered feeding behaviour (Montllor, 1991; Pettersson *et al.*, Chapter 4 this volume), host-plant selection (Dixon, 1998; Powell *et al.*, 2006; Pettersson *et al.*, Chapter 4 this volume; Pickett and Glinwood, Chapter 9 this volume), and virus transmission by aphids (Sylvester, 1989; Harris, 1990; Nault, 1997; Gray and Banerjee, 1999; Pirone and Perry, 2002; Gray and Gildow, 2003; Ng and Perry, 2004).

Host selection related to virus acquisition and inoculation

A variety of sensory stimuli may influence aphid host-selection behaviour before plant contact and stylet insertion. Migrant aphids locate plant material primarily by responding to colour (Moericke, 1955; Kennedy *et al.*, 1959a,b; Kring, 1972). Olfactory stimuli may also play a role at this stage (Pettersson, 1970, 1973; Chapman *et al.*, 1987; Visser and Taanman, 1987; Pickett *et al.*, 1992; Park and Hardie, 2004; Pettersson *et al.*, Chapter 4 this volume), but only a small fraction of the dispersing populations succeed in finding a suitable host plant (Ward *et al.*, 1998). Host recognition continues after alighting on plants, where physical (plant surface colour, texture, and topology) and chemical (olfactory and gustatory) stimuli are evaluated by various sensilla located on the head, antennae, tarsi, and labium (Pettersson, 1971; Wensler, 1974; Tjallingii, 1978; Eisenbach and Mittler, 1980; Yan and Visser, 1982; Anderson and Bromley, 1987; Backus, 1988; Nottingham *et al.*, 1991; Pickett *et al.*, 1992; Park and Hardie, 2004; Pettersson *et al.*, Chapter 4 this volume). However, aphids attempt to make brief (< 1 min)

stylet insertions ('probes') as a behavioural reflex that follows tarsal contact with any solid surface (Powell *et al.*, 1999), even when repellent or deterrent cues are present (Griffiths *et al.*, 1982; Phelan and Miller, 1982). The aphids' propensity for landing on green surfaces and initiating probing behaviour helps explain their extraordinary capacity for transmitting plant viruses.

The aphid stylet bundle comprises two pairs of long stylets that taper to a sharp point at the tip. The two inner, maxillary stylets are tightly interlocked by ridges and grooves, forming the food canal (FC) and salivary canal (SC) (Fig. 14.2a). Although the FC and SC are distinct canals for almost the entire length of the aphid stylet bundle, they merge to form a common duct (CD) at the maxillary stylet tips (terminal 2–4 µm region – Forbes, 1969). The outer pair of 'mandibular' stylets plays an important role in physical penetration of plant cell walls, but it is the maxillary stylets that enter plant cells (Tjallingii and Hogen Esch, 1993). The maxillary FC, SC, and CD are therefore the channels through which aphids acquire and inoculate virions.

Probing behaviour is a particularly important feature of host-plant selection by aphids, because it provides a means of assessing internal plant chemistry (Powell and Hardie, 2000; Powell *et al.*, 2006). The mandibular stylets have been found to be innervated, each containing a pair of dendrites (Fig. 14.2a; Forbes, 1966; Parrish, 1967), but these probably function as proprioreceptors and allow the insect to monitor stylet movement and position (Wensler, 1974, 1977). The stylets therefore apparently lack chemoreceptors, and aphids need to ingest plant sap through the maxillary FC to the pharyngeal area of the foregut to allow chemosensory assessment using a gustatory organ (Fig. 14.2 b; Wensler and Filshie, 1969; Ponsen, 1972; McLean and Kinsey, 1984).

Stylet penetration activities can be monitored by attaching a fine gold wire to the aphid, making the insect and plant part of an electrical circuit (McLean and Kinsey, 1965; Tjallingii, 1988; Pettersson *et al.*, Chapter 4 this volume) to produce an electrical penetration graph (EPG) signal. When this

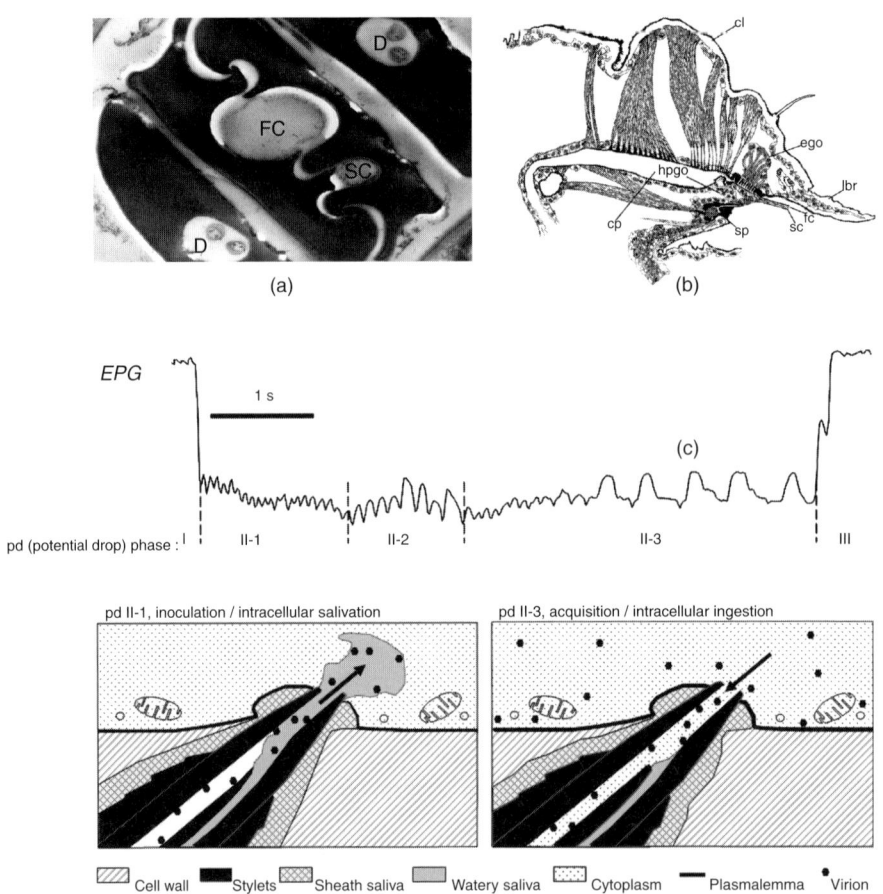

Fig. 14.2. Diagrammatic representations of aphid mouthparts. (a) Transmission electron micrograph showing transverse section of the stylet bundle of *Myzus persicae*. The inner pair of stylets (maxillae) forms the larger food canal (FC; approximately 0.7 µm diameter) and smaller salivary canal (SC; approximately 0.3 µm diameter) (cf. Forbes, 1969). D, ducts in the mandibular stylets contain dendrites. (b) Longitudinal section through the head of an aphid (modified from Ponsen, 1987; cl, clypeus; cp, cibarial pump; ego, epipharyngeal gustatory organ; fc, food canal; hpgo, hypopharyngeal gustatory organ; lbr, labrum; sc, salivary canal; sp, salivary pump. (c) Longitudinal sections of the stylet tips introduced into an epidermal or mesophyll plant cell (bottom) and associated electrical penetration graph (EPG) waveforms (top). The EPG signal shows the three successive intracellular sub-phases: II-1, II-2, and II-3. The first sub-phase (II-1) is associated with inoculation of non-persistent viruses, possibly involving release of virions from the common duct during salivation (bottom left). The third sub-phase (II-3) is linked with acquisition, and therefore represents ingestion of cytosol (bottom right) (modified from Martin *et al.*, 1997).

technique is based on a direct current (DC) circuit, it enables researchers to discriminate stylet penetration of the apoplast (the extracellular environment) from puncture of the symplast (the intracellular compartment). The fluid-filled FC and SC enable the maxillary stylets to function as microelectrodes, recording the transmembrane potential of plant cells. Maxillary puncture of a cell membrane is therefore recorded as a distinct drop to a lower (intracellular) voltage level of the EPG (Tjallingii, 1988).

Since plant viruses have no independent means of movement into or out of the symplast, their acquisition and inoculation by aphids occurs during intracellular punctures by the maxillary stylets. Aphids usually initiate stylet penetrations in the anticlinal grooves between epidermal cells (Nault and Gyrisco, 1966; Yoshii, 1966), and much of the stylet pathway remains intercellular (apoplastic) (Tjallingii and Hogen Esch, 1993). However, DC-EPG reveals that aphids regularly puncture the symplast during their host selection, phloem location, and feeding behaviours. Even initial epidermal probes, lasting less than 30 s, usually include a brief (5–10 s) maxillary puncture of one cell (Powell, 1991; Martin *et al.*, 1997; Powell, 2005). Such brief cell punctures always include ingestion of a cytosolic sample (Powell *et al.*, 1995; Martin *et al.*, 1997), presumably allowing chemosensory evaluation *via* the gustatory organ in the foregut (Wensler and Filshie, 1969).

Virions of non-persistent viruses are acquired and inoculated during these brief epidermal cell punctures (Powell, 1991; Powell *et al.*, 1992). Aphids typically make several brief probes following contact with a new plant and, regardless of whether the plant is a host or non-host species, this behaviour is often followed by plant rejection (departure). Alate aphids will then often make a short flight before landing on and probing another plant. Flight and landing/probing continue to alternate as antagonistic reflexes (Kennedy, 1965), leading to brief probes on a series of plants. If one of the plants in the series is already infected with a non-persistent virus, the result may be very rapid spread to one or more plants visited later. In newly inoculated plants, virions delivered into cells multiply and spread to neighbouring cells through plasmodesmata and eventually become systemic *via* the phloem, infecting the whole plant (Carrington *et al.*, 1996) and providing a new source of virus inoculum and further spread.

The virus acquisition and inoculation processes occur as a direct consequence of various stylet activities that enable aphids to select appropriate host plants, suppress plant defensive processes, and extract nutrients. Depending on the mode of transmission, virions may adhere to the cuticular lining of the stylets and the foregut, or be ingested and pass through the gut into the circulatory system and salivary glands and then be inoculated *via* salivation.

Transmission Modes of Plant Viruses by Aphids

Non-persistent transmission

Approximately 75% of viruses with aphid vectors are transmitted in a non-persistent manner (Fig. 14.1; Nault, 1997). Non-persistently transmitted viruses cause a number of important diseases in cultivated plants (Raccah, 1986; Nault, 1997; Hull, 2002) and belong to six genera (Murphy *et al.*, 1995; Ng and Perry, 2004) (Table 14.3). These viruses are usually distributed in many tissues of their host plants, including the epidermis, and they cause stunting, leaf distortion, vein clearing, vein necrosis, leaf mosaic and mottling, and in some cases, flower breaking (e.g. *Tulip breaking virus* (TBV) and *Turnip mosaic virus* (TuMV) in certain hosts). In some cases, as with *Alfalfa mosaic virus* (AMV) and some virus species of the genus *Carlavirus*, they do not cause any symptoms (latent infections). All non-persistent viruses are readily transmitted mechanically under laboratory conditions, and some are seed-borne (e.g. *Lettuce mosaic virus* (LMV) and *Cucumber mosaic virus* (CMV)). For transmission and other characteristics see Table 14.2.

Transmission characteristics

Non-persistent transmission is virtually exclusive to aphids (Fig. 14.1), and acquisition and inoculation occur optimally during very brief stylet penetrations (< 1 min). During these probes, aphid stylets penetrate the epidermal layer, where virus concentration can be high. As the stylet insertion period increases beyond approximately 30 s–1 min (representing penetration beyond the epidermal cell layer), the subsequent transmission rate decreases rapidly. Pre-acquisition fasting of aphids for 15 min to 4 h can result in increased transmission efficiency (Hull, 2002). Initially, this 'pre-acquisition starvation' (PAS) effect was attributed to behavioural changes as starved aphids have the propensity to make short probes, favouring acquisition and inoculation of non-persistent viruses (Bradley, 1952). However, more recent studies have indicated that non-behavioural

Table 14.3.   Aphid-transmitted viruses: families, genera, particle morphology, and type of genome.

Mode of transmission	Genus/family	Particle morphology	Genome*
Non-persistent	*Alfamovirus* /Bromoviridae	Quasi-isometric to bacilliform (18 nm diameter, 30–56 nm long)	Tp/+ssRNA
	Carlavirus /Flexiviridae	Slightly flexuous filamentous (470–580 nm long, 13 nm diam.)	Mp/+ssRNA
	Cucumovirus /Bromoviridae	Isometric to bacilliform (26–35 nm)	Tp/+ssRNA
	Fabavirus /Comoviridae	Isometric (22–32 nm)	Bp/+ssRNA
	Macluravirus /Potyviridae	Short flexuous rods (650–675 nm long, 11–15 nm diam.)	Mp/+ssRNA
	Potyvirus /Potyviridae	Flexuous rods (680–900 nm long, 12–15 nm diam.)	Mp/+ssRNA
Semi-persistent	*Caulimovirus* /Caulimoviridae	Isometric (35–50 nm)	circular dsDNA
	Closterovirus /Closteroviridae	Flexuous filamentous (1250–2000 nm long, 12 nm diam.)	Mp/+ssRNA
	Sadwavirus /Not assigned to a family	Isometric (26–30 nm)	Bp/+ssRNA
	Sequivirus /Sequiviridae	Isometric (31 nm)	Mp/+ssRNA
	Waikavirus /Sequiviridae	Isometric (30 nm)	Mp/+ssRNA
Persistent-non-propagative	*Enamovirus* /Luteoviridae	Isometric (25–30 nm)	Mp/+ssRNA
	Luteovirus /Luteoviridae	Isometric (25–30 nm)	Mp/+ssRNA
	Polerovirus /Luteoviridae	Isometric (24 nm)	Bp/+ssRNA
	Umbravirus /Not assigned to a family	Do not form conventional virions except in coat protein of a helper luteovirus	Mp/+ssRNA
	*Babuvirus*** /Nanoviridae	Isometric (18–20 nm)	Mc/ssDNA
	Nanovirus /Nanoviridae	Isometric (17–26 nm)	Mc/ssDNA
	*Sobemovirus**** /Not assigned to a family	Isometric (about 30 mm)	Mp/+ssRNA
Persistent-propagative	*Cytorhabdovirus* / Rhabdoviridae	Bullet-shaped or bacilliform, membrane enveloped (42–130 nm diameter, 100–360 long)	Mp/–ssRNA
	Nucleorhabdovirus / Rhabdoviridae	Bullet-shaped or bacilliform, membrane enveloped (43–100 nm diameter, 95–500 long)	Mp/–ssRNA

*Viruses with monopartite (Mp) or undivided genomes have only one molecule of nucleic acid whereas viruses with divided or segmented genomes (multipartite) may have two (bipartite) (Bp) or three (tripartite) (Tp) molecules.
Multi component (Mc) genome: with more RNA segments (6–11 in the case of nanoviruses).
dsDNA: double stranded DNA; ssDNA, single stranded DNA; + ssRNA, single stranded RNA of positive polarity acting as messenger RNA; – ssRNA, single stranded RNA of negative polarity.
**BBTV is the sole member of the genus.
***Most members of the genus are beetle-transmitted; one member is transmitted by mirids.

factor(s) also contribute to the PAS effect (Powell, 1993; Powell *et al.*, 1995; Wang and Pirone, 1996). Aphids lose their infectivity after moulting since the stylets and cuticular lining of the foregut are shed with any retained virions. The aphids remain viruliferous for only a few minutes to hours, and lose the ability to inoculate more rapidly when given access to a plant (irrespective of whether it is a host for either the aphid or virus) than when fasted (Watson and Roberts, 1940; Bradley, 1959). Although the rate of loss tends to be exponential in fasted aphids, it is positively correlated with temperature (Kassanis, 1941). In a few cases, aphids can retain their vectoring ability for a longer period if they are not feeding on plants, for at least 18 h in the case of *Maize dwarf mosaic virus* (MDMV) (Berger *et al.*, 1987).

Mechanisms of transmission

ACQUISITION. The retention of virions in the aphid foregut, particularly the maxillary FC (see below), indicates that non-persistent viruses are acquired *via* ingestion (Pirone and Perry, 2002) during maxillary puncture of an epidermal cell. The DC-EPG technique has revealed that three successive intracellular activities occur after the maxillary stylet tips puncture the plasma membrane, designated sub-phases II-1, II-2, and II-3 (Powell *et al.*, 1995; Martin *et al.*, 1997; Powell, 2005; Fig. 14.2c). The third intracellular activity, occurring immediately before stylet withdrawal from the cell, is associated with efficient acquisition of both PVY and CMV (Martin *et al.*, 1997), and therefore represents active ingestion of cytosol by the aphid (Powell and Hardie, 2000; Fig. 14.2c).

RETENTION. Localization studies have focused particularly on potyviruses, where a virally encoded helper component (HC; or 'helper factor' or 'aphid transmission factor' – reviewed by Pirone and Blanc, 1996) accessory protein is required for transmission. Present evidence suggests that the CD and FC of the maxillary stylets and the cibarium (Berger and Pirone, 1986; Ammar *et al.*, 1987; Childress and Harris, 1989; Ammar and Nault, 1991; Ammar *et al.*, 1994; Wang *et al.*, 1996) are the sites where virions of potyviruses are attached to the cuticular surfaces of the stylets. However, virions are particularly associated with the distal third of the maxillary stylets (Wang *et al.*, 1996). HC acts as a bridge to facilitate binding between a putative aphid cuticular receptor (ACR) and the viral capsid-protein (CP) subunit (Raccah *et al.*, 2001). With cucumoviruses, such as

CMV, there is no viral HC and only an ACR–CP bridge exists (Perry, 2001). Therefore, among non-persistently transmitted viruses, two strategies have developed, either helper-dependent or helper-independent transmission. Attached virions provide the inoculum for viral infection of healthy plants.

INOCULATION. In common with the acquisition process, inoculation of virions requires maxillary puncture of the plasma membrane (Powell, 1991). Cell puncture interruption experiments have been performed to investigate which of the three intracellular sub-phases is associated with efficient inoculation by viruliferous aphids. This approach showed that only the first intracellular activity (II-1) is necessary for delivery of PVY and CMV virions into the cell (Martin *et al.*, 1997). Virions are therefore released very rapidly (within 2 s) after puncture of the plasma membrane. However, the inoculation mechanism has been the subject of controversy and two hypotheses have been advocated, presenting alternative scenarios.

The 'ingestion–egestion' hypothesis suggests that virions carried in the maxillary FC and more proximal areas of the foregut are expelled into the inoculated cell during egestion (regurgitation) (Watson and Plumb, 1972; Garrett, 1973; Harris and Bath, 1973; Harris, 1977; Harris and Harris, 2001). Microscopic observations of aphid stylets during penetration of an *in vitro* artificial feeding system suggested that the aphid cibarial pump may function bidirectionally: ink particles were reported to move both into and out of the FC (Harris and Bath, 1973). However, while aphids may be capable of egestion under such artificial conditions, it is not clear whether or when egestion occurs during penetration of plant tissues. There is no evidence that the behavioural event associated with non-persistent virus inoculation (sub-phase II-1) represents egestion. It has been argued that experiments demonstrating transfer of radioisotopes from labelled plants to unlabelled plants provide evidence for transfer of plant sap *via* ingestion and subsequent egestion (Garrett, 1973; Harris, 1977). However, efficient acquisition and transfer of the label required longer probes by the aphids (3–5 min; Garrett, 1973) than those that typically characterize the non-persistent acquisition and inoculation processes (< 1 min).

Martin *et al.* (1997) have put forward an alternative ('ingestion–salivation') hypothesis whereby virions are inoculated *via* salivation. They pointed out that the anatomy of the aphid stylets, with the FC and SC merging near the maxillary tips (forming the CD), provides a point of convergence where retained virions are exposed directly to the flow of saliva. Aphids are able to secrete two types of saliva (Miles, 1999). Gelling saliva is produced during penetration of the apoplast, and forms a sheath around the stylet bundle. This sheath material is excluded from the cell when the maxillary stylets puncture the plasma membrane (Tjallingii and Hogen Esch, 1993), and intracellular salivation involves the production of non-gelling, 'watery' saliva (Cherqui and Tjallingii, 2000). Initiation of the II-1 EPG waveform immediately following cell entry may therefore represent a sudden switch from the production of gelling saliva to watery saliva. Virions retained at the common duct may be released at this point if factors present in watery saliva have the ability to reverse binding between virions and their cuticular retention sites (either direct or HC-mediated).

Support for the ingestion–salivation hypothesis comes from recent EPG experiments investigating inoculation of the persistent (circulative) *Pea enation mosaic virus* (PEMV) (Powell, 2005). Unlike many other persistent, non-propagative viruses, PEMV is not phloem-limited and can be inoculated to superficial plant tissues. Inoculation of this virus can therefore be used as a marker for early intracellular salivation by aphids. During brief epidermal probes, inoculation of PEMV was linked to the occurrence and duration of sub-phase II-1, but not to the other two subsequent intracellular sub-phases. When considered together with the results of the previous study associating II-1 with inoculation of non-persistent viruses (Martin *et al.*, 1997), these results suggest that the non-persistent inoculation process involves the injection of watery saliva into the cell cytoplasm, with the implication that the CD represents the likely functional retention site. The CD occupies an extremely small (2–4 μm) region of the stylets, i.e. < 1% of the stylet bundle (Forbes, 1969), and virions retained here will be a very small proportion of those acquired. However, the retention of virions at more proximal sites does not provide evidence that they function in transmission (Pirone and Perry, 2002).

Semi-persistent transmission

Viruses transmitted by aphids in a semi-persistent manner occur in five genera (Nault, 1997) (Table 14.3). *Cauliflower mosaic virus* (CaMV), the type member of the genus *Caulimovirus*, was considered initially to be an atypical non-persistent virus (Hamlyn, 1955; van Hoof, 1958). Later, it was thought to be transmitted bimodally, i.e. both in a non- and a semi-persistent manner (Chalfant and Chapman, 1962; Bouchery *et al.*, 1990), but it is now considered to be transmitted semi-persistently (Markham *et al.*, 1987).

Transmission characteristics

Similar to the non-persistent viruses, some semi-persistent viruses may be acquired within minutes, but the efficiency to inoculate virions subsequently to a new host increases with the length of acquisition access (Palacios *et al.*, 2002). Efficient virus acquisition appears to be a function of feeding rather than probing. In common with non-persistent viruses, semi-persistent viruses do not require a LP in the vector, are not present in vector haemolymph, and cannot be transmitted after injection into the vector haemocoel. Infectivity may be retained for a few days but it is lost after moulting (Fereres and Collar, 2001). However, unlike non-persistent viruses, pre-acquisition fasting does not affect transmission efficiency.

Mechanisms of transmission

ACQUISITION. Using the DC-EPG technique, Palacios *et al.* (2002) showed that the principal vectors of CaMV (*Myzus persicae* – peach–potato aphid and *Brevicoryne brassicae* – cabbage aphid) are able to acquire virions during brief punctures of epidermal cells. Such rapid acquisition probably occurs *via* the same process as acquisition of non-persistent viruses, with virions ingested into the FC during intracellular sub-phase II-3. However, the efficiency of acquisition (measured by subsequent transmission to test plants) increased significantly if aphids reached a phloem sieve element and ingested phloem sap for at least 15 min on CaMV-infected plants (Palacios *et al.*, 2002). Other semi-persistent viruses may be limited to the phloem (particularly closteroviruses; Hull, 2002), and their acquisition therefore may have a more rigid requirement for phloem sap ingestion (Limburg *et al.*, 1997).

RETENTION. The transmission characteristics of semi-persistent viruses suggest the functional retention of a much larger number of virions than for non-persistent viruses. Harris and Harris (2001) pointed out that the surface area in the proximal part of the maxillary FC and the pre- and post-cibarial area is extensive, providing for the binding of a large number of virions. Transmission electron microscopy has indicated that semi-persistent viruses are retained in the aphid foregut (Murant *et al.*, 1976), whereas Lopez-Abella *et al.* (1988) suggested that they may use two binding sites, one in the aphid stylets and another in the foregut. The exact location of cuticular receptors for semi-persistent viruses in aphids remains unclear and, as with non-persistent viruses, retention of virions at particular sites does not prove that they are functional in the transmission process.

INOCULATION. It is widely accepted that semi-persistent virus transmission occurs *via* an ingestion–egestion mechanism (Fereres and Collar, 2001; Harris and Harris, 2001). However, further research is required to determine when aphids egest during plant penetration. Moreno *et al.* (2005b) recently investigated inoculation of CaMV using the DC-EPG technique. Both *M. persicae* and *B. brassicae* were able to inoculate the virus very rapidly,

during the first (epidermal) cell puncture. Both vector species inoculated CaMV more efficiently when allowed several short cell punctures during a longer (5 min) period of plant access, and inoculation efficiency increased further when aphids made sustained (>15 min) contact with a phloem sieve element. However, the occurrence of inoculation during initial brief cell punctures is a finding that has important implications for the CaMV inoculation process and retention site. Since sub-phases II-2 or II-3 have not been linked with the ejection of stylet contents into the cell (Martin *et al.*, 1997; Powell, 2005), the likely mechanism of such rapid CaMV inoculation is *via* intracellular salivation (during sub-phase II-1 – Moreno *et al.*, 2005b). The maxillary CD may therefore function as a retention site for semi-persistent viruses, as seems to be the case for non-persistent ones. However, a definitive association between II-1 and CaMV inoculation will require interrupting cell punctures during the various sub-phases (Martin *et al.*, 1997), an approach that has not yet been reported with CaMV. The study by Moreno *et al.* (2005b) suggests that aphids are able to inoculate CaMV during early brief cell punctures and later sustained sieve element punctures, but it is not clear whether egestion occurs during such behaviours. There is currently no experimental evidence that aphids egest during penetration of superficial cells (Powell, 2005) or the phloem (Prado and Tjallingii, 1994).

Virus–aphid molecular interactions during transmission of non- and semi-persistent (non-circulative) viruses

As all non- and semi-persistent viruses have a simple structure with their nucleic acid (RNA or DNA) encapsidated in simple virus particles (Hull, 2002), it is the CP that is available for interaction with cuticular surfaces of the aphid stylets and foregut. Two forms of interaction have been identified between the CP and aphid retention sites: direct interaction (called the capsid protein strategy), and indirect interaction (the helper component(s) strategy) in which one (e.g. potyviruses) or two (e.g. caulimoviruses) non-structural virus-coded protein(s) are involved.

CAPSID PROTEIN (CP) STRATEGY. The transmission of cucumoviruses, alfamoviruses, and carlaviruses does not require a helper component, and aphids can therefore acquire virions from purified virus preparations through an artificial membrane feeding system and then transmit them to host plants (Pirone and Megahed, 1966; Pirone, 1977; Weber and Hampton, 1980). In the case of CMV and other cucumoviruses, transmissibility depends solely on the CP, as was found in *in vitro* dissociation and re-assembly experiments with highly-aphid-transmissible and poorly-aphid-transmissible virus isolates (Gera *et al.*, 1979; Chen and Francki, 1990). Although CMV transmission by aphids does not require a helper component, it is considered that the transmission mechanism(s) involved are very similar to those of potyviruses and other viruses, transmission of which is mediated by HC.

HELPER STRATEGY (NON-PERSISTENT). Aphid transmission of potyviruses involves the so-called helper strategy (Fig. 14.3). In addition to virions, the acquisition by aphids of a virus-coded helper protein is required for successful potyvirus transmission. For this reason, aphids cannot transmit purified potyviruses (Pirone

and Megahed, 1966) unless they have previous or simultaneous access to HC (Kassanis and Govier, 1971a,b; Govier and Kassanis, 1974a,b).

The viral HC-Pro (to denote the dual functional role of HC in both aphid transmission and proteolysis) has been identified in many potyvirus–host-plant systems and seems to be a general phenomenon (Raccah *et al.*, 2001). Potyviral HCs can assist transmission of virions belonging to the same (homologous) or different (heterologous) virus species, although transmission efficiency is often higher with homologous HC (Pirone, 1981; Sako and Ogata, 1981; Sako *et al.*, 1984; Lecoq *et al.*, 1991). Potyviral HC appears to act as a link between virions and aphid mouthparts (Govier and Kassanis, 1974a,b). The 'bridge' hypothesis (Pirone and Blanc, 1996) has been supported experimentally by introducing mutations within the highly conserved proline–threonine–lysine (PTK) motif of HC-Pro (Peng *et al.*, 1998). This hypothesis presumes that HC-Pro contains two domains: one specifically binding to the aphid stylets and the second binding to the CP of the virion. Experimental data have identified the two domains in HC-Pro: the N-terminal domain lysine–isoleucine–threonine–cysteine motif, which might be involved in the specific binding to the aphid's stylets (Blanc *et al.*, 1998), and the

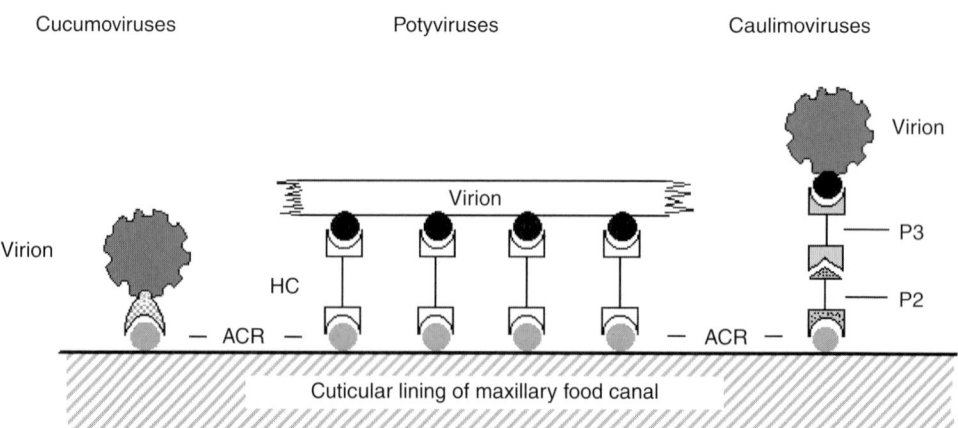

Fig. 14.3. Diagrammatic representation of the attachment of different categories of non-circulative viruses to the cuticular lining of maxillary food canal/common duct of aphids. ACR, aphid cuticular receptors; HC, helper component protein; P2, P3, proteins.

C-terminal domain proline–lysine– thronine motif, which has been shown to be involved either directly or indirectly in HC binding to the amino acid triplet aspartic acid–alanine–glycine, located very close to the N-terminus of CP (Peng *et al.*, 1998). Recent experimental evidence supports the bridging role of HC (Blanc *et al.*, 1998; Wang *et al.*, 1998). The inability of aphids to acquire different potyviruses from purified preparations has been attributed to their failure to retain the virus in the stylets (Revers *et al.*, 1999), and retention in the maxillary FC and foregut occurs only when active HC is present (Ammar *et al.*, 1994; Wang *et al.*, 1996).

HC-Pro seems to be a multifunctional protein (Maia *et al.*, 1996) and, apart from mediating aphid transmission (Govier *et al.*, 1977), it has also been found to be involved in different processes of the viral cycle such as cell-to-cell (Rojas *et al.*, 1997) and long-distance potyvirus movement in the plant (Kasschau and Carrington, 2001; Saenz *et al.*, 2002), pathogenicity (Revers *et al.*, 1999; Redondo *et al.*, 2001), suppression of gene silencing (Llave *et al.*, 2000), and also synergism between co-infecting viruses (Vance *et al.*, 1995; Pruss *et al.*, 1997).

HELPER STRATEGY (SEMI-PERSISTENT). Studies with caulimoviruses suggest that these also depend on a helper strategy, but instead of a single HC protein, there are two proteins mediating interactions between the ACR and CP (Blanc *et al.*, 2001). Transmission of CaMV involves the virus-encoded proteins P2 and P3, together with the CP (P4) (Espinoza *et al.*, 1992; Leh *et al.*, 1999) (Fig. 14.3). Aphid transmission of CaMV requires a bridge between P2 and P3 (Drucker *et al.*, 2002) to link virions to the aphid's cuticular surfaces. The P2–P3-virion complex is formed within the aphid alimentary canal and not in infected plant tissues, as was believed originally (Drucker *et al.*, 2002). Aphids may acquire P2 first during stylet penetration of infected mesophyll cells, and subsequently ingest P3 and virions from other mesophyll or phloem cells (Drucker *et al.*, 2002). The C-terminal domain of P3 is responsible for the interaction with the virion (Leh *et al.*, 2001), whereas the N-terminus of P3 interacts with the C-terminus of P2 (Leh *et al.*, 1999). An additional interaction is required to ensure the formation of the stable P2–P3 complex. There are three possible roles for P2 and P3 in the aphid transmission of CaMV (Blanc *et al.*, 2001): (i) P2 is indeed the only helper, and P3 would be more correctly referred to as a capsid component; (ii) both P2 and P3 equally can be considered helpers; (iii) neither P2 nor P3, but rather the complex P2–P3, is the helper. More research is needed to clarify the functional role of both P2 and P3 in aphid transmission of CaMV, and possibly other caulimoviruses.

Helper components are also implicated in the semi-persistent transmission of sequiviruses, and are attributed to a helper virus belonging to the genus *Waikavirus* (Harrison and Murant, 1984). Aphid transmission of *Parsnip yellow fleck virus* (PYFV) is dependent upon acquisition of a helper produced by plants infected with *Anthriscus yellows virus* (AYV) (Elnagar and Murant, 1976). Helper components may also be implicated in the transmission of closteroviruses (Murant *et al.*, 1998). The evidence for helper component dependency is based on the observation that, as in the case of potyviruses and caulimoviruses, aphids cannot acquire and transmit them from purified preparations. However, such factors have not been identified and further work is needed to investigate this.

Vector specificity in non- and semi-persistent (non-circulative) viruses

'Vector specificity' refers to the degree of interdependence between a specific virus and a particular aphid vector species. As would be expected, the vector specificity of non-persistent viruses, because of their relatively transient relationship with the vector, is lower than that of the other aphid-borne viruses (Pirone and Harris, 1977). Aphids acquire and inoculate non-persistent viruses during brief probes, and many different aphid species (including some that are unable to feed on virus host plants) may be vectors (e.g. Radcliffe and Ragsdale, 2002; Jones *et al.*, 2006).

A given virus may be vectored with varying efficiency according to the aphid species, clone, or biotype. Aphid species or clones may or may not transmit CMV and, if they do, have varying vector efficiencies (Bhargava, 1951; Badami, 1958; Conti et al., 1979). Clones of *Acyrthosiphon pisum* (pea aphid) differ considerably in their ability to transmit *Bean yellow mosaic virus* (BYMV), and this is not related to the geographic origin or the colour of the clone (Jurík et al., 1980). Also, *M. persicae* transmits efficiently both *Beet mosaic virus* (BtMV) and PVY, whereas *Rhopalosiphum padi* (bird cherry–oat aphid) transmits both viruses inefficiently (Katis and Gibson, 1984, 1985).

Virus strains may differ in the efficiency with which they are transmitted by a particular aphid species. For example, strains of CMV and CTV differ in the efficiency with which they are transmitted by aphids (Simons, 1957a; Raccah et al., 1976; Gera et al., 1979). The CP determines transmission specificity at least in the case of CMV (e.g. Gera et al., 1979), and a conserved CP surface domain of CMV that was essential for efficient aphid-vector transmission was identified (Liu et al., 2002).

For non-circulative viruses with a helper strategy, it is likely that helper proteins play a very important role in vector specificity. It was shown that in the case of potyvirus transmission, specificity was conferred by HC and transmissibility was highly correlated with retention of virus in the stylets (Wang et al., 1998). In the P2 helper component of CaMV, changing a single amino acid residue can alter the spectrum of aphid-vector species (Moreno et al., 2005a).

The question remains as to what makes an aphid such as *M. persicae* an efficient vector of many potyviruses, whereas, for example, *Myzus ascalonicus* (shallot aphid) cannot transmit, or transmits very inefficiently, potyviruses such as TuMV, *Tobacco etch virus* (TEV), PVY, LMV, and BtMV (Doncaster and Kassanis, 1946; Wang et al., 1998).

Factors affecting transmission of non- and semi-persistent viruses

Apart from interactions between virus and aphid genotypes, abiotic factors such as temperature and humidity, and the presence of satellite RNA (satRNA), can greatly affect virus transmission by aphids (Robert and Lemaire, 1999). Temperature affects aphid behaviour during acquisition and/or inoculation (Robert et al., 2000) and also affects host plants, both as virus sources and new virus hosts, and has direct effects on virus survival (Syller, 1987). The transmission rate of ZYMV by *M. persicae* is greater at 21°C than at 8°C, and *Aphis gossypii* (cotton aphid or melon aphid) failed to transmit it at lower temperatures when relative humidity was less than 80% (Fereres et al., 1992). High humidity was correlated positively with the transmission efficiency of PVY by different aphid species (Singh et al., 1988).

Persistent transmission

Persistent viruses are usually, but not exclusively, confined to the phloem tissue (sieve elements and companion cells) of their host plants and cause symptoms such as stunting, leaf discoloration, and leaf rolling (Table 14.2). Such viruses either replicate (propagative viruses) or not (non-propagative viruses) within the aphid vector.

Non-propagative circulative transmission

LUTEOVIRIDAE. Non-propagative, aphid-transmitted viruses belong mainly to the family Luteoviridae, which comprises four genera: *Luteovirus*, *Polerovirus*, *Enamovirus*, and *Umbravirus*. Viruses belonging to the family Nanoviridae (*Babuvirus*, *Nanovirus*) and the unassigned genus *Sobemovirus* are also transmitted in this manner (Ng and Perry, 2004). The Luteoviridae comprises 8 fully recognized and assigned viruses, along with 12 unclassified and 20 tentative members that have many characteristics of the type species of the four genera (Smith and Barker, 1999). PEMV represents an obligate symbiosis between an enamovirus (PEMV-1, a member of the family Luteoviridae) and an umbravirus (PEMV-2) (de Zoeten and Skaf, 2001). PEMV-1 confers transmission by aphids in a circulative (non-propagative) manner and, as in the case of luteoviruses, acquired PEMV virions accumulate in the

ASGs and are inoculated *via* salivation (de Zoeten and Skaf, 2001; Reavy and Mayo, 2002; Gray and Gildow, 2003). PEMV-2 is not known to play any role in determining interactions with aphid vectors, but influences the site of successful virus inoculation within plants by conferring cell-to-cell movement and escape from phloem limitation.

Transmission characteristics – As luteoviruses are phloem limited, aphids need to ingest phloem sap in order to acquire them (Prado and Tjallingii, 1994). The minimum recorded virus acquisition access period (AAP) therefore reflects the time taken for aphid stylets to reach the phloem, and has been determined as 15 min for *Barley yellow dwarf virus* (BYDV) and 1 h for *Potato leaf roll virus* (PLRV) (Gray *et al.*, 1991; Taliansky *et al.*, 2003). However, transmission efficiency of both viruses increases with a rise in the AAP to 2 or more days, as in the case of PLRV. The LP is at least 24 h and may extend to as much as 4 days. LP length depends on several factors such as the aphid species, virus concentration in the host plant tissue, and environmental factors such as temperature (van der Broek and Gill, 1980; Power and Gray, 1995). An inoculation access period (IAP) of about 10–30 min is required for most virus species, but efficiency is higher when IAP is increased (Power *et al.*, 1991). EPG experiments with *R. padi* indicate that some aphids are able to inoculate BYDV-PAV without showing sustained contact with a phloem sieve element. All such aphids showed numerous short cell punctures (presumably including brief puncture of sieve elements or companion cells where the virus could be inoculated) and virions were therefore probably delivered during the very brief initial intracellular salivation activity (subphase II-1). Transmission of BYDV-PAV occurred at a higher frequency when aphids showed sustained puncture of a sieve element, and efficient inoculation was linked with the sieve-element salivation EPG waveform E1 (Prado and Tjallingii, 1994). Aphids retain the capability to transmit luteoviruses for several days, and in some cases throughout their life span. The interactions between luteoviruses and their aphid vectors have been reviewed by Gray and Gildow (2003).

Uptake route in the vector – For successful transmission, virions have to pass at least two barriers: the gut wall and the salivary gland membranes. Virions first enter the aphid's body *via* the ingestion of phloem sap, then are transported across the gut wall to the haemocoel, and finally accumulate in the ASGs. The particles are delivered into the SC, from where they are injected into a plant during subsequent stylet penetration (Gildow, 1999). Endocytosis and exocytosis at the two epithelial barriers (alimentary tract and ASG) involve specific interactions between virus CPs and receptors present in these two locations within the aphid (Gildow, 1987, 1993; Stevens *et al.*, 2005).

Luteovirus–aphid vector molecular interactions and vector specificity – Viruses transmitted by aphids in a persistent manner are usually vectored by one, or only a few, aphid species that colonize the virus host plants. The molecular interactions determining transmission and vector specificity have been researched in some detail, particularly with the *Luteovirus* species BYDV and CYDV (Gray and Banerjee, 1999). As aphids can acquire and transmit purified luteoviruses, it is likely that no additional proteins, other than the CP, are involved. Studies with different luteoviruses have suggested that both the CP and the minor CP (CPm) structural proteins play key roles in the aphid transmission process (Mayo and Ziegler-Graff, 1996), as point mutations in or near conserved domains of *Beet western yellows virus* (BWYV) CPm affect aphid transmissibility (Brault *et al.*, 2000).

There is good evidence that symbionin, a 60 kDa protein produced by the endosymbiotic bacterium *Buchnera* sp. and found in *M. persicae* haemolymph, interacts with PLRV and other *Polerovirus* species to assist virus retention in the haemolymph (van den Heuvel, 1999). Symbionin possibly mediates the longevity of virus particles by binding and protecting them from proteolysis (Hogenhout *et al.*, 1998). However, the binding of symbionin to virus particles or to P5 (the coat protein readthrough structural proteins) has only been demonstrated *in vitro*, and the exact role of symbionin in virus transmission has yet to be elucidated (Reavy and Mayo, 2002). As symbionin homologues from vector and non-vector aphid species bind, *in vitro*, to six different purified luteoviruses (van den Heuvel, 1999) and the binding capacity is not correlated with transmission ability or efficiency, it has been suggested that if symbionin plays a role in virus transmission, it does not play a major role in determining vector specificity (van den Heuvel *et al.*, 1997).

Barriers within the aphid act as 'filters' to the passage of luteoviruses and thereby contribute to vector specificity. The key barriers are the gut and the ASG walls. After ingestion, the virions in the gut lumen cross the gut wall, *via* receptor-mediated endocytosis, at two possible regions: the posterior midgut and the hindgut. Previous studies have shown that BYDV-PAV passes through the hindgut of *Sitobion avenae* (grain aphid) and *R. padi*, as does BYDV-MAV in *S. avenae* (Gildow, 1999) and *Soybean dwarf virus* (SbDV) in *Aulacorthum solani* (glasshouse and potato aphid) and *M. persicae* (Gildow *et al.*, 1994). In contrast, PLRV (Garret *et al.*, 1993) and BWYV (Reinbold *et al.*, 2001) penetrate through the posterior midgut of *M. persicae*. In some aphid–virus combinations, such as *Cucurbit aphid-borne yellows virus* (CABYV) in *M. persicae* and *A. gossypii*, both areas of the gut allow virus transport (Reinbold *et al.*, 2003). Also, specificity can occur within the same aphid species. For example, in *M. persicae*, SbDV passes through the hindgut whilst PLRV passes through the posterior midgut. Having crossed the gut wall, viruses are transported *via* the haemolymph to the ASG. Specificity at the ASG involves two barriers: the extracellular basal lamina (basement membrane) and the basal plasmalemma (cell membrane). Strains of SbDV have been found to have different properties with regard to ASG membrane penetration. Virions of the SbDV-D strain can penetrate the basal lamina in *A. solani* but not in *M. persicae*. The SbDV-Va20 strain can penetrate the basal lamina in both species, but the basal plasmalemma membrane can be penetrated in *M. persicae* only (Gildow *et al.*, 2000). In the ASG, the basal plasmalemma forms the physical barrier, in contrast to the gut wall, where the apical plasmalemma membrane forms the barrier. The mechanism of the passage through the cellular membrane presumably involves a receptor-mediated endocytosis. In the basement membrane, however, attachment of the capsid protein of the virus is postulated to cause loosening of crosslinking bonds of the collagen and laminin matrix, allowing the passage of virions larger than the membrane pore (Peiffer *et al.*, 1997; Gray and Gildow, 2003). Little is known about the molecular mechanisms of receptor-mediated endocytosis. It is believed that parts of the capsid protein are involved. Evidence has been obtained supporting the hypothesis of involvement of the N-terminal half of the readthrough domain encoded by open reading frame 5 of SbDV (Terauchi *et al.*, 2003). A review of virus-vector specificity has been given by Gildow (1999). The nature of the viral determinants governing tissue specificity (posterior midgut, hindgut, or both) in the aphid body during virus acquisition has not been identified, and sequence comparisons of viral structural proteins do not show any obvious sequence motifs that could be correlated with tissue specificity (Reinbold *et al.*, 2003). However, the minor CP of BWYV plays an important role in the transport of virions through midgut cells and is also necessary for the maintenance of virions in the haemolymph and their passage through ASG cells (Reinbold *et al.*, 2001).

Factors affecting luteovirus transmission – As with non- and semi-persistently transmitted viruses, transmission of persistent viruses such as luteoviruses may also be affected by the aphid species, clone or biotype and/or virus strain. This has been reviewed by Gray (1999). It has been shown that transmission efficiency of PLRV and BYDV is affected not only by the aphid species but also by the clone, morph, and instar (Björling and Ossiannilsson, 1958; Hinz, 1966; Robert and Maury, 1970; Robert, 1971; Upreti and Nagaich, 1971; Tamada *et al.*, 1984; Zhou and Rochow, 1984; Bourdin *et al.*, 1998; Gray and Gildow, 2003). A twofold difference in transmission efficiencies of BYDV-PAV was found between the least and most effective clones of *R. padi*, and an 8-fold difference was found between

S. avenae clones tested (Guo *et al.*, 1997). Similarly, Bourdin *et al.* (1998) showed that some PLRV isolates are poorly transmitted by clones of *M. persicae* and *M. persicae nicotianae* (tobacco aphid), and this could not be related to intrinsic properties of either the virus particle, as was suggested initially (Tamada *et al.*, 1984), or the aphid clone; the transmission process and its specificity depends on interactions between them. Variation in the percentage of *S. avenae* individuals that transmitted BYDV-PAV ranged from 3.7 to 92.5% and this was affected by the aphid clone, the plant species on which clones were collected, and the reproductive mode of the clones (Dedryver *et al.*, 2005). Populations of *A. solani* originating from Japan were more efficient vectors of SbDV than those originating from the USA and New Zealand (Damgsteeght and Hewings, 1986).

Abiotic factors such as temperature and humidity may also affect the ability of aphids to transmit luteoviruses. Temperature affects aphid behaviour during acquisition and/or inoculation (see Robert *et al.*, 2000) and also host plants, both as virus sources and new virus hosts (Syller, 1987). The effect of temperature on BYDV transmission is related to the aphid species involved, and seven clones of *R. padi* have been shown to differ in their vectoring ability at 5 or 10°C but not at 15°C (Smyrnioudis *et al.*, 2001b), whereas no interclonal variation in the transmission efficiency by *S. avenae* was observed at any temperature. Humidity was positively correlated with the transmission efficiency of PLRV by different aphid species (Singh *et al.*, 1988).

Umbraviruses do not encode for a CP and therefore require a helper *Luteovirus* CP for aphid transmission. Seven *Umbravirus* species have been described so far, each associating with a different *Luteovirus* (Robinson and Murant, 1999). The system has the following characteristics: (i) both the *Umbravirus* and its *Luteovirus* helper are non-propagative; (ii) the *Umbravirus* is mechanically transmitted *via* sap, but the helper *Luteovirus* is not; (iii) aphids can only transmit the *Umbravirus* if they acquire it from plants dually infected with both the *Umbravirus* and the helper *Luteovirus*; (iv) the *Umbravirus* is transmitted by the aphid only when its RNA is packaged in a protein shell comprising the helper virus protein (Harrison and Murant, 1984); and (v) the *Umbravirus* may be helped by several different luteoviruses.

NANOVIRUSES. The minimum AAP is 15–30 min for *Faba bean necrotic yellows virus* (FBNYV) and 4 h for *Banana bunchy top virus* (BBTV) (Hu *et al.*, 1996), with a 5–15 min IAP required for both viruses. Aphids retain the viruses throughout the aphid life span. Purified FBNYV is not transmitted by its vectors, which indicates that possibly a helper component is involved. It has been shown that this regulates virus transport across the haemocoel–salivary gland interface (Franz *et al.*, 1999).

Propagative viruses

Rhabdoviruses have a unique ability to propagate in the body of aphid vectors (Jackson *et al.*, 1999). Indirect evidence for replication in aphids is the requirement for long LPs and virus retention throughout vector life. Transovarial passage also occurs (see below) and is consistent with propagation in the insect. More direct evidence for virus replication has been obtained by continued transmission after repeated serial dilution passages from aphid to aphid (Sylvester *et al.*, 1974; Sylvester and Richardson, 1981). It is assumed that the virus particles produced in aphid cells are released into the haemolymph, find their way to the salivary glands, and finally are injected into the plant along with saliva. Bacilliform particles of *Sowthistle yellow vein virus* (SYVV) have been observed in the nucleus and cytoplasm of cells in the brain, suboesophageal ganglion, salivary glands, ovaries, fat bodies, mycetome, and muscles of aphid vectors (Sylvester and Richardson, 1970), and the virus is assembled in the nucleus.

The aphid gut appears to be a potential barrier to transmission because bypassing it by injection of the virus into the haemolymph increases transmission efficiency and often allows transmission by non-vector aphid species (Sylvester and Richardson, 1992).

Conversely, *Macrosiphum euphorbiae* (potato aphid) remained an inefficient vector of SYVV when the virus was acquired *via* experimental injection into the haemolymph, indicating that additional factors (other than a gut barrier) constrain transmission in this case (Behncken, 1973).

As with non-propagative viruses, acquisition and inoculation efficiency of propagative viruses increases when the plant access periods (AAP and IAP) are increased. Usually, they have much longer mean LPs (latent periods of 1 week) than non-propagative viruses (1 or 2 days), and LP may be affected by temperature (Hull, 2002). Plant rhabdoviruses are not normally transmitted from parent vectors to their offspring (Redinbaugh and Hogenhout, 2005): only about 1% of the offspring of viruliferous *Hyperomyzus lactucae* (black currant–sowthistle aphid) were able to transmit SYVV (Sylvester and Richardson, 1971; Sylvester and McClain, 1978). Plant rhabdoviruses usually do not cause disease in their insect vectors (Redinbaugh and Hogenhout, 2005), but serial transmission of some SYVV isolates from aphid to aphid (by haemolymph injection) was associated with increased aphid mortality (Sylvester, 1973). However, as infected aphids lived beyond the period of maximum reproduction, the intrinsic rate of population growth was little affected (Sylvester, 1973).

Epidemiology

Patterns of spread of aphid-borne viruses depend on interactions between the virus, its host plant(s), the aphid vector(s), the environmental conditions, and the cultural practices undertaken by the growers. Epidemiological aspects have been reviewed many times, including by Harrison (1981), Maramorosch and Harris (1981), Raccah (1986), Nault (1997), Burgess *et al.* (1999), Dewar and Smith (1999), Lecoq (1999), Robert (1999), Robert and Lemaire (1999), and Ragsdale *et al.* (2001). Evaluation and knowledge of the impact of various factors, discussed below, on the epidemiology of a specific virus allows the design of the most effective control measures against that virus

(Madden *et al.*, 1987; Jones, 2001, 2004) (for reviews see Chapter 9 in Smith and Barker, 1999).

Biological factors

Virus

HOST RANGE AND SURVIVAL IN NATURE (VIRUS SOURCES). Viruses differ in their host range and this may affect their spread. For example, more than 1000 host species, both cultivated and non-cultivated, are susceptible to CMV, whereas *Leek yellow stripe virus* (LYSV) infects a very limited number of species. The presence of virus sources is usually a prerequisite for the appearance and spread of viral disease in a specific area, although aphids are able to carry viruses, even non-persistent ones (see MDMV example above), over long distances and provide new infection sources. Infection of a crop by a specific virus may originate from:

1. *Infected seed* – seed transmission plays an important role in the epidemiology of non-persistently transmitted virus species that belong to the *Potyvirus, Cucumovirus,* and *Alfamovirus* genera (Brunt *et al.*, 1996).
2. *Perpetuating sources (tubers, groundkeepers, and volunteer plants)* – infected propagules play an important role in epidemiology of viruses with a narrow host range such as *Onion yellow dwarf virus* (OYDV), LYSV in different *Allium* species (Dovas *et al.*, 2001), PLRV and PVY in potato (see Robert *et al.*, 2000), and fruit tree viruses (Garsney and Lee, 1987; Thresh, 1988). Groundkeepers left behind after the harvest and fodder beet or mangold clamps kept too long are all important reservoirs of *Beet mild yellowing virus* (BMYV) (Smith, 1986). Volunteer potato plants are potential sources for all potato viruses (Wright and Bishop, 1981; Thomas, 1983; Robert *et al.*, 2000) and for *Carrot virus Y* (CtVY) (Jones *et al.*, 2005).
3. *Other susceptible crops in the vicinity* – nearby sugarbeet crops, for example, are important sources of aphid-borne beet viruses (Peters, 1988; Dewar and Smith, 1999; Dusi, 1999), and potato crops act as sources of all

aphid-borne potato viruses (Radcliffe *et al.*, Chapter 26 this volume) and of PVY for other solanaceous crops such as tobacco, pepper, and tomato (Chatzivassiliou *et al.*, 2004). In many countries, vegetable crops are grown successively in the field and in these cases the older crops can act as major virus reservoirs. Overlapping or successive crops play important roles in the epidemiology of viruses that are not seed-borne and have a narrow host range such as OYDV, where overlapping of onion and shallot crops results in serious epidemics (Chamberlain and Bayliss, 1948; van Dijk, 1993).

4. *Weeds of arable crops* – weeds are implicated in the ecology and epidemiology of most aphid-borne viruses (Duffus, 1971; Adlerz, 1981; Radcliffe *et al.*, Chapter 26 this volume) as they may be hosts of both the virus and the aphid vectors, and in many cases are the most important overwintering or summer hosts. Their importance is related to the mode of virus transmission by aphids. They play a crucial role in the epidemiology of non-persistent viruses such as CMV (Quiot *et al.*, 1979 a,b), a virus with a wide host range, but are less important for PVY (Orsenigo and Zitter, 1971; Chatzivassiliou *et al.*, 2004), which has a relatively narrow host range. They also act as serious sources of persistently transmitted viruses such as the BYDV/ CYDV complex (D'Arcy, 1995) and the beet poleroviruses (Stevens *et al.*, 1994).

VIRUS MUTABILITY AND STRAIN SELECTION. New virus strains appear readily in nature and this is the result of mutation and selective adaptation, heterologous encapsidation, pseudorecombination or genetic recombination due to direct interaction between strains of a virus or between different viruses in the host plant (Hull, 2002). Mutation has been the primary source of genetic variation in plant viruses, particularly RNA viruses (Garcia-Arenal *et al.*, 2001). Viruses differ in their rate of mutation, with CMV showing high genetic diversity (Palukaitis and Garcia-Arenal, 2003), whereas others such as PLRV are more stable genetically (Hull, 2002). The new virus strains that appear in nature may differ in their host range, pathogenicity, and transmissibility by their aphid

vectors, characteristics that may affect their epidemiology.

The effect of environmental conditions on the diversity of plant viruses has not been studied extensively. However, temperature seems to be an important factor in strain selection, and CMV thermosensitive strains dominate in spring tomato and pepper crops in southern France, whereas thermoresistant strains are prevalent in the summer (Quiot *et al.*, 1979b).

DISPERSAL. Field spread of aphid-borne viruses takes place in two phases. Primary infections are brought into the field by either winged aphids and/or infected seed. Secondary infections within the field may then result from alatae moving locally and/ or apterae walking from plant to plant, especially when they are in contact. Frequently, spread of non- and semi-persistent viruses is by alatae on their host-selection flights, whereas only colonizing aphid species contribute to the spread of persistent viruses. The rate and pattern of spread of an aphid-borne virus within a specific field depends on many factors, including: the source and the amount of inoculum; its relationship (whether non-persistent, semi-persistent, or persistent) with the aphid vectors; the time at which the aphids appear in relation to the host's age; and weather conditions, which affect both aphid biology and host physiology.

In early studies, attempts were made to relate aphid populations to virus spread by counting aphids on host plants (Doncaster and Gregory, 1948) but more recently, different types of traps have been used to assess the number of flying aphids (see Irwin and Ruesink, 1986; Harrington *et al.*, Chapter 19 this volume). However, in order to understand virus epidemiology, it is necessary to know the number of infective aphids that will visit the crop, and this can be done either by using sensitive detection techniques (Harrington *et al.*, Chapter 19 this volume) or by placing trapped aphids onto indicator plants (Harrington *et al.*, 1986). The 'infection pressure' to which a crop is exposed can be assessed by exposing indicator plants within the field for successive periods (van Hoof, 1977).

Virus spread over long distances can be the result of either human activities moving infected propagative material (potato, fruit trees) and/or the movement of aphid vectors, especially in the case of persistent viruses (Thresh, 1983) such as BYDV, which moves into the northern United States and Canada during the spring migration of aphids in jet streams from southern States (Irwin and Thresh, 1988).

Cultural practices

Cultural practices such as planting date, crop rotation, field size and other characteristics, plant population density, and plant size may have pronounced effects on dispersal of aphid-borne viruses (Thresh, 1982a). Planting date in relation to the build-up of aphid populations affects spread of aphid-borne viruses such as BYDV (Wyatt et al., 1988; Foster et al., 2004) and potato viruses (Ragsdale et al., 2001). Crop rotation may also affect the incidence of viruses that overwinter in weeds and/or volunteer plants, as in the case of potato viruses (Doncaster and Gregory, 1948). The effect of field size and other physical characteristics of the field on viruses and their vectors is little studied and probably complex, although persistent viruses have been reported as more common in larger fields, and non-persistent viruses in smaller ones (Watson, 1967). This may be due to the greater perimeter/area ratio in small fields and the exposure of plants around the margin to greater risk from vectors introducing virus from adjacent sources (Vanderplank, 1948). However, Foster et al. (2004), who studied BYDV incidence in 623 untreated autumn-sown cereal crops in the UK in relation to a wide range of field characteristics, reported aphid and virus incidence to be greatest in fields of 2–9 ha and lower in both smaller and larger fields. They also found a virus strain-dependent effect of field aspect, and more virus in crops near the coast and in areas with little arable land, the latter probably because it is associated with an increased area of grassland, which can act as a reservoir of the virus and aphids at times when crops are absent or unsuitable. Crop spacing and plant size may also affect the landing behaviour and response of migrating aphids, and therefore the incidence of the viruses they carry. Heathcote (1974) noted that incidence of viruses causing yellowing symptoms in sugarbeet was higher in low-density crops. Similarly, large cauliflower plants in the seed beds become infected more readily with insect-borne viruses than do small ones, because they are visited more often by aphid vectors (Broadbent, 1957).

Weather

Air temperature, rainfall, and wind are key weather components in aphid-borne pathosystems. Air temperature is particularly influential, as it has marked effects on the rate of development and activity of aphid vectors and the phenology of the virus hosts (either cultivated or arable weeds and wild hosts). For example, cool conditions do not favour the flights of aphids, and very high temperatures may reduce aphid populations and the incidence of the viruses they are vectoring. Temperature also affects the transmission efficiency of aphid-borne viruses (see above). Heavy rainfall may reduce the aphid population, with consequences for virus spread (Wallin and Loonan, 1971). Wind also affects virus incidence as winged aphids do not fly when wind speed is too high and flight direction is affected by the prevailing wind. MDMV epidemics in the USA were favoured by low-level jet winds and resulted in the rapid transport of the virus from southern areas north to Minnesota (Zeyen et al., 1987).

Disease Forecasting

Various virus forecasting systems and warning schemes have been developed. For further details see Harrington et al., Chapter 19 this volume; Knight and Thackray, Chapter 31 this volume.

Disease Management Strategies

Direct control of viral diseases is not possible. Control is based on preventative measures and virus-vector relations are of major importance in determining the best strategies. As already mentioned, cropping practices play an important role in virus spread, and their manipulation is often used to achieve some degree of control (Zitter and Simons, 1980). However, control of viruses, particularly those transmitted in a non-persistent manner, has proved to be difficult to accomplish by chemical control of aphids, as insecticides usually do not act quickly enough to stop transmission (Rains and Christensen, 1983; Jayasena and Randles, 1985; Collar et al., 1995). In general, preventative measures aim at reducing virus sources (virus-free propagative material, weed control, etc.) and also at reducing and/or avoiding aphid-vector populations. Conventional breeding has also been effective in successfully combating a number of aphid-borne virus diseases of potato, tobacco, etc. (Valkonen, 1994). In addition, introducing genes for virus resistance into existing crop varieties may produce transgenic crop plants resistant to viruses. Genes derived from the viruses themselves in a concept referred to as 'pathogen-derived resistance' have been used for this purpose (Beachy, 1993, 1997; Fitchen and Beachy, 1993; Barker, and Waterhouse, 1998; Kaniewski and Lawson, 1998; Berger and German, 2001; Hull, 2002). Below, we describe conventional preventative measures for the control of aphid-borne viruses. Effective control should be based on the integration of different control measures as, in most virus–host systems, the adoption of only one method is usually insufficient (Jones, 2001).

Use of genotypes resistant to the virus and/or to the aphid vectors

The use of resistant cultivars has been an attractive, effective, and environmentally friendly way of combating viral diseases (Hammond, 1998), particularly those caused by non-persistent viruses (Walker et al., 1982), as these are more difficult to control by other means. Resistance to CMV and Papaya ringspot virus (PRSV) has been incorporated into commercial melon cultivars, and resistance to potato viruses has been incorporated into potato (Berger and German, 2001; Brown and Corsini, 2001; Solomon-Blackburn and Barker, 2001). Genotypes resistant to seed transmission (Khetarpal and Maury, 1987; Wang et al., 1993) have been selected and used for the control of Soybean mosaic virus (SMV) and LMV in soybean and lettuce, respectively.

Breeding for resistance to the aphid vector(s) may be effective (van Emden, Chapter 17 this volume), particularly if there is only one or a few vector species involved in transmission of several viruses of a given crop. If the aphid vectors are also serious crop pests in their own right, then resistance is of double benefit. Host resistance to aphids has been studied extensively and exploited in different plant–aphid systems such as cowpea, cucurbits, maize, Solanum berthaultii, tobacco, peach, and tomato (Kennedy, 1976; Gibson and Plumb, 1977; Jones, 1987).

Although resistance to one or a few aphid species may not prevent transmission of non-persistent viruses, since non-colonizing aphids also transmit them, this can be an important component in an integrated control programme. For example, muskmelon (Cucurbita pepo) genotypes resistant to A. gossypii, an efficient vector of PRSV, Watermelon mosaic virus (WMV), and ZYMV, show resistance to these viruses when A. gossypii is their main vector (Lecoq et al., 1980; Risser et al., 1981). However, the plants are not protected against these viruses if other aphid vectors are also involved. Some peach selections resistant to M. persicae and Myzus varians (cigar-rolling peach aphid) are also resistant to Plum pox virus (PPV) transmission by these species (Massonie and Maison, 1980; Maison and Massonie, 1982; Massonie et al., 1982), and this resistance can be useful under field conditions if these aphid species are the only species involved in PPV epidemiology, a rather unusual situation. However, in some crops such as lupins, resistance to aphids also confers field resistance to the viruses they transmit (Wilcoxson and Peterson, 1960; Jones, 1979, 1987; Gray et al., 1986).

Elimination of virus sources (see also
Radcliffe *et al.,* Chapter 26 this volume)

*Use of virus-free seed and vegetative
propagative material through seed certification
programmes*

Seed transmission provides early within-
crop foci of infection from which subsequent
secondary spread by aphids can occur early
in the growing season when the crop plants
are more susceptible to virus infection. A
number of non-persistently transmitted viruses such as LMV, BYMV, and *Bean common
mosaic virus* (BCMV), are transmitted by
seed. The use of virus-free lettuce seed minimized both LMV dissemination and virus-
induced yield losses in California and the
UK (Grogan *et al.,* 1952; Tomlinson, 1962;
Grogan, 1980). Seed certification programmes have also been the fundamental management tools for the control of tuber-borne
potato viruses (Hall, 1993; Slack, 1993; Slack
and Singh, 1998).

Weed control

Weeds serve as both virus reservoirs and
hosts for their insect vectors (Duffus, 1971;
Bos, 1981; Thresh, 1982b). Weed removal and/
or destruction can contribute to the prevention or elimination of crop infection by the
virus(es) they harbour, if they are important
sources of primary infection. However,
weed control may be expensive and will
only stop short-range spread, such as in that
of the non-persistent viruses CMV (Rist and
Lorbeer, 1989), and *Celery mosaic virus*
(CeMV) and PVY in celery and pepper crops,
respectively (Orsenigo and Zitter, 1971).

Roguing of diseased crop plants

Eradication of infected plants is a long-
established method of controlling the spread
of many virus diseases (Thresh, 1988). The
approach has been applied more widely,
and with more success, in woody perennial
crops, but has also been effective in some
annual herbaceous crops. In general, roguing
should begin as soon as symptoms appear
and in potato is easier to accomplish before

the aphids arrive and the canopy closes
(Salazar, 1996; Woodford and Gordon, 1990).

Eradication of volunteer plants

Volunteer plants play an important role in
the epidemiology of PVY and PLRV in
potato crops (Robert, 1999), and also of PVY
in tobacco and possibly other solanaceous
crops in Greece (Chatzivassiliou *et al.,*
2004). Therefore, eradication of volunteer
plants can reduce greatly the spread of these
viruses.

Isolation from virus sources

Isolation of crops in time or space from
virus sources might give a useful reduction
in virus incidence. Production of virus-free
seed potatoes is carried out frequently in
areas isolated from ware potatoes (Ragsdale
et al., 2001). A crop-free period may help
reduce the incidence of viruses with limited host range such as OYDV in onions
(Chamberlain and Bayliss, 1948) and CtVY
in carrots (Latham and Jones, 2004).

Prevention or reduction of virus spread

Plant spacing

The spread of plant viruses may also be
affected by plant density and this has been
documented, for example, by Hull (1964)
and A'Brook (1964, 1968) for groundnuts and
rosette virus disease, respectively. A'Brook
demonstrated that higher plant populations
in groundnut crops decreased the incidence
of rosette disease as well as aphid populations. However, high plant density was costly
and also decreased crop yield. It is recommended that the planting rate should
achieve complete ground cover as soon as
possible without reducing yield due to
competition. Similarly, incidence of BMYV
and *Beet yellows virus* (BYV) in sugarbeet
was reduced when plants were closely
spaced (Johnstone *et al.,* 1982), whereas low
plant density resulted in considerably
higher incidence of BYMV in narrow-
leaved lupin (*Lupinus angustifolius*) (Jones,
1993, 1994).

Control or avoidance of vectors

CHEMICAL CONTROL OF APHID VECTORS (see also Dewar, Chapter 15 this volume). Insecticides reducing aphid populations may also prevent or reduce the spread of viruses they transmit (Loebenstein *et al.*, 1964; Broadbent, 1969). However, vector control strategies are often associated with adverse effects on beneficials, aphid resistance, environmental pollution, recolonization of the crop, and induction of vector movement, causing further virus spread (Irwin and Thresh, 1990). Field application of pyrethroids, which have a knock-down effect, can reduce PVY spread in potato crops (Gibson *et al.*, 1982; Gibson, 1983; Gibson and Cayley, 1984) and spread of a range of potyviruses in flower bulbs, provided that aphid populations are not resistant to the insecticides. However, in other cases, insecticides do not act quickly enough to stop transmission of non-persistent viruses, and insecticides may actually increase vector activity and the incidence of potyviruses (Ferro *et al.*, 1980). On the other hand, spread of persistent viruses can be managed effectively by aphid control (Plumb and Johnstone, 1995; Perring *et al.*, 1999; Mowry, 2005). Imidacloprid seed treatment reduced incidence of yellowing viruses of sugarbeet by 72%, but yield benefits were recorded only when virus incidence in untreated plots was above 10% (Dewar *et al.*, 1996).

USE OF OIL SPRAYS. It has been known for more than 40 years that mineral oils can reduce substantially PVY transmission (Bradley *et al.*, 1962). Field application has been used successfully for the control of many non-persistent viruses, especially in high-value ornamental bulb crops (lilies, hyacinths) and PVY in seed potato crops (Simons, 1982; Simons and Zitter, 1980; Kerlan *et al.*, 1987; Radcliffe *et al.*, Chapter 26 this volume). Although mineral oils generally are considered effective in inhibiting transmission of non-persistent viruses, they failed to control some non-persistent viruses such as MDMV in maize (Szatmari-Goodman and Nault, 1983), TuMV, BCMV, and BYMV (Walkey and Dance, 1979).

Oil applications do not reduce transmission of persistent and semi-persistent viruses such as PEMV (Vanderveken, 1977), PLRV (Hein, 1971), and BYV (Walkey and Dance, 1979). Generally, their use for virus control is rather limited due to phytotoxicity, inadequate efficacy, volatility or viscosity, oil removal by rain and/or irrigation, and non-covering of new vegetation (de Wijs, 1980). Combining insecticides with mineral oils has improved control of non-persistently transmitted viruses such as MDMV, PVY, and BtMV (Ferro *et al.*, 1980; Gibson and Rice, 1986).

BIOLOGICAL CONTROL OF APHID VECTORS. Although biological control has been used as an alternative strategy for aphid control, the effect on virus spread has not been well documented. Parasitoids are thought to have little effect on virus transmission and spread (e.g. Carter and Harrington, 1991). Parasitization of *S. avenae* by *Aphidius ervi* did not obstruct BYDV transmission and in some experiments, parasitized aphids were more efficient vectors (Christiansen-Weniger *et al.*, 1998). Parasitism may also cause increased aphid movement (Höller, 1991) and perhaps exacerbate virus spread in the field (Weber *et al.*, 1996). Fewer plants were infected with BYDV by *R. padi* when the parasitoid *Aphidius rhopalosiphi* was present, whereas more plants were infected in the presence of the predator *Coccinella septempuncata* (7-spot ladybird) (Smyrnioudis *et al.*, 2001a). In South America, the introduction and establishment of several coccinellid predators and aphelinid and aphidiid parasitoids decreased BYDV transmission in the field (Zuniga, 1990). Further work is needed to clarify the role of beneficials in the epidemiology of aphid-borne viruses in different virus-host systems.

SOIL COVERAGE BY USING COLOURED OR REFLECTIVE MULCHES (see also Wratten *et al.*, Chapter 16 this volume). Aphids, like most insects, respond preferentially to certain wavelengths of light. Alatae are attracted initially to short wavelength light and therefore show a dispersive phase of flight behaviour (Irwin *et al.*, Chapter 7 this volume). Their responses subsequently change and they are attracted to yellow and green wavelengths, searching for a suitable host (Moericke, 1950; Kennedy, 1960; Johnson, 1969, Irwin *et al.*, Chapter 7 this volume). This aphid behaviour led Smith *et al.* (1964) to investigate the possibility of using different reflective mulches in order to deter winged aphids from landing on virus-susceptible crops. Coloured or reflective mulches consisting of aluminium foil, or aluminium-coated paper, polythene, or straw have been used with some success to reduce the incidence of non-persistent viruses such as WMV in cucurbits (McLean *et al.*, 1982), *Pea seed-borne mosaic virus* (PSbMV) in broad bean (Tachibana, 1981), and BYMV in narrow-leaved lupin crops (Jones, 1991). Straw mulch has also been used to reduce aphid infestation and the incidence of non-persistent viruses in several crops (Heimbach *et al.*, 2002; Jones, 2004). Although mulching is more effective in controlling non-persistent viruses than others, it also delayed the spread of CABYV, a persistent virus (Lecoq, 1999).

The use of coloured or reflective mulches has limitations. Their effectiveness is reduced as the crop plants grow and cover the mulches. Placement and removal increase the cost of the crop, and mulches are ineffective in cloudy weather. Therefore, the practice is only economically justifiable in high-value crops or germplasm collections (Loebenstein and Raccah, 1980; Zitter and Simons, 1980; Jones, 1991), and even in these cases, only as part of an integrated control programme (Lecoq and Pitrat, 1983; Jones, 1991).

BARRIER CROPS AND INTERCROPPING.   Use of barriers of immune plants surrounding virus-susceptible crops is another cultural practice to reduce/eliminate the spread of non-persistent viruses (Broadbent, 1952; Jenkinson, 1955; Simons, 1957b; Fereres, 2000; Jones, 2005). Different host plants such as sunflower, barley, wheat, maize, okra (*Abelmoschus esculentus*), sorghum, cotton, oats, marigold (*Calendula officinalis*), potato, and kale have been used as barrier crops (Fereres, 2000). In Florida, a barrier of sunflower plants reduced the incidence of PVY in pepper

crops (Simons, 1957b). Similarly, a wheat barrier protected muskmelon crops from non-persistent viruses in California (Toba *et al.*, 1977). The effectiveness of this strategy depends on various factors such as the height of the host used as a barrier crop at the time of maximum risk of infection and the extent of competition between the barrier and the protected crop (Fereres, 2000). Barrier crops may act as a sink for non-persistent viruses by cleaning the mouthparts of the infective aphids that land on the barrier crop (Fereres, 2000) and hence reducing the aphids' ability to transmit the virus to the adjacent, protected crop (Toba *et al.*, 1977; Difonzo *et al.*, 1996).

In some cases, instead of having a barrier crop as a border, the immune crop is integrated into the susceptible crop, thus providing camouflage and impeding both aphid movement and virus spread, and possibly acting as a source of natural enemies of the aphid vectors (Thresh, 1982a).

MANIPULATION OF PLANTING DATES.   As spread of aphid-borne viruses is greatly affected by aphid population size, the avoidance of high populations, especially at the beginning of the growing season when the crop is more susceptible to virus infection, may decrease virus incidence. Early sowing greatly reduced the incidence of groundnut rosette virus disease (Naidu *et al.*, 1998), whereas infection of autumn-sown cereal crops by BYDV can be avoided by delaying sowing (Plumb and Johnstone, 1995; Knight and Thackray, Chapter 31 this volume).

CULTIVATION IN ISOLATED AREAS UNFAVOURABLE TO APHIDS.   This practice has been used for the production of virus-free potato seed in areas with relatively cool and windy conditions during the growing season and where the relatively low aphid populations appear rather late in the growing season (Todd, 1961; de Bokx and van der Want, 1987). Such climatic conditions do not favour the spread of aphid-borne potato viruses. In such areas, aphid populations are monitored during the cultivation period and, in years when an early aphid invasion is recorded, the entire potato crop is sprayed with a foliar desiccant in order to prevent tuber infection (de Bokx and van der Want, 1987).

Conclusions

Aphids have an extraordinary ability to transmit plant viruses. The virus-vector interactions vary widely from superficial binding with aphid cuticular surfaces associated with a rapid transmission cycle (non-persistent/non-circulative), to more long-term processes involving selective transport across gut and salivary gland membranes (persistent/circulative), and in some cases, multiplication within the aphid (circulative/ propagative). Since persistent viruses accumulate in accessory salivary glands, the link between their inoculation and salivation is very well established, but salivation seems also to provide a means of intracellular delivery of non-circulative viruses. The immediate injection of watery saliva into punctured cells probably allows aphids to subvert early defensive plant signalling processes. If aphids are able to use salivary components to manipulate plant physiology and suppress cellular defences, these changes may also benefit inoculated viruses by facilitating successful establishment of infection. Interactions between aphid salivary components, aphid-transmitted viruses, and plant cell biology will be an important area for future research.

Acknowledgements

We thank Alberto Fereres for permission to use Fig. 14.2c.

References

A'Brook, J. (1964) The effect of planting date and spacing on the incidence of groundnut rosette disease and of the vector *Aphis craccivora* Koch, at Mokawa, Northern Nigeria. *Annals of Applied Biology* 54, 199–208.

A'Brook, J. (1968) The effect of plant spacing on the numbers of aphids trapped over the groundnut crop. *Annals of Applied Biology* 61, 289–294.

Adlerz, W.C. (1981) Weed hosts of aphid-borne viruses of vegetable crops in Florida. In: Thresh, J.M. (ed.) *Pests, Pathogens and Vegetation*. Pitman, London, pp. 467–478.

Ammar, E.D. and Nault, R.L. (1991) Maize chlorotic dwarf virus-like particles associated with the foregut in vector and non-vector leafhopper species. *Phytopathology* 81, 444–448.

Ammar, E.D., Gingery, R.E. and Nault, R.L. (1987) Interactions between maize mosaic and maize stripe viruses in their insect vector, *Peregrinus maidis*, and in maize. *Phytopathology* 77, 1051–1056.

Ammar, E.D., Jarfors, U. and Pirone, T.P. (1994) Association of potyvirus helper component protein and the cuticle lining the maxillary food canal and foregut of an aphid vector. *Phytopathology* 84, 1054–1060.

Anderson, M. and Bromley, A.K. (1987) Sensory system. In: Minks, A.K. and Harrewijn, P. (eds) *Aphids. Their Biology, Natural Enemies and Control, Volume 2A*. Elsevier, Amsterdam, pp. 153–162.

Backus, E.A. (1988) Sensory systems and behaviours which mediate hemipteran plant-feeding, a taxonomic overview. *Journal Insect Physiology* 34, 151–157.

Badami, R.S. (1958) Changes in the transmissibility by aphids of a strain of cucumber mosaic virus. *Annals of Applied Biology* 46, 554–562.

Barker, H. and Waterhouse, P.M. (1998) The development of resistance to luteoviruses mediated by host genes and pathogen-derived transgenes. In: Smith, H.G. and Barker, H. (eds) *The Luteoviridae*. CAB International, Wallingford, pp. 169–210.

Beachy, R.N. (1993) Transgenic resistance to plant viruses. *Seminars in Virology* 4, 327–416.

Beachy, R.N. (1997) Mechanisms and applications of pathogen-derived resistance in transgenic plants. *Current Topics in Biotechnology* 8, 215–220.

Behncken, G.M. (1973) Evidence of multiplication of sowthistle yellow vein virus in an inefficient aphid vector, *Macrosiphum euphorbiae*. *Virology* 53, 405–412.

Berger, P. and German, T. (2001). Biotechnology and resistance to potato viruses. In: Loebenstein, G., Berger, P.H., Brunt, A.A. and Lawson, R.H. (eds) *Virus and Virus-like Diseases of Potatoes and Production of Seed-Potatoes*. Kluwer, Dordrecht, pp. 341–363.

Berger, P.H. and Pirone, T.P. (1986) The effect of helper component on the uptake and localization of potyviruses in *Myzus persicae. Virology* 153, 256–261.

Berger, P.H., Zeyen, R.J. and Groth, J.V. (1987) Aphid retention of maize dwarf mosaic virus (potyvirus): epidemiological implications. *Annals of Applied Biology* 111, 337–344.

Bhargava, K.S. (1951) Some properties of four strains of cucumber mosaic virus. *Annals of Applied Biology* 38, 377–388.

Björling, K. and Ossiannilsson, F. (1958) Investigations on individual variations in the virus-transmitting ability of different aphid species. *Socker* 14, 1–13.

Blanc, S., Ammar, E.D., Garcia-Lampasona, S., Dolja, V.V., Llave, C., Baker, J. and Pirone, T.P. (1998) Mutations in the potyvirus helper component protein: effects on interactions with virions and aphid stylets. *Journal of General Virology* 79, 3119–3122.

Blanc, S., Hebrard, E., Drucker, M. and Froisard, R. (2001) Molecular basis of vector transmission: cauli-moviruses. In: Harris, K.F., Smith, O.P. and Duffus, J.E. (eds) *Virus–Insect–Plant Interactions.* Academic Press, New York, pp. 143–166.

de Bokx, J.A. and van der Want, J.P.H. (eds) (1987) *Viruses of Potatoes and Seed-Potato Production.* Pudoc, Wageningen, 259 pp.

Bos, L. (1981) Wild plants in the ecology of virus diseases. In: Maramorosch, K. and Harris, K.F. (eds) *Plant Diseases and Vectors: Ecology and Epidemiology.* Academic Press, New York, pp. 1–33.

Bouchery, Y., Givord, L. and Monestiez, P. (1990) Comparison of short- and long-feed transmission of the cauliflower mosaic virus Cabb-S strain and S delta II hybrid by two species of aphid: *Myzus persicae* (Sulzer) and *Brevicoryne brassicae* (L.). *Research in Virology* 141, 677–683.

Bourdin, D., Rouzé, J., Tanguy, S. and Robert, Y. (1998) Variation among clones of *Myzus persicae* and *Myzus nicotianae* in the transmission of a poorly and a highly aphid-transmissible isolate of potato leafroll luteovirus (PLRV). *Plant Pathology* 47, 794–800.

Bradley, R.H.E. (1952) Studies on the aphid transmission of a strain of henbane mosaic virus. *Annals of Applied Biology* 39, 78–97.

Bradley, R.H.E. (1959) Loss of virus from the stylets of aphids. *Virology* 8, 308–318.

Bradley, R.H.E., Wade, C.V. and Wood, F.A. (1962) Aphid transmission of potato virus Y inhibited by oils. *Virology* 18, 327–328.

Brault, V., Mutterer, J., Scheidecker, D., Simonis, M.T., Herrbach, E., Richards, K. and Ziegler-Graff, V. (2000) Effects of point mutations in the readthrough domain of the beet western yellows virus minor capsid protein on virus accumulation in plants and on transmission by aphids. *Journal of Virology* 74, 1140–1148.

Broadbent, L. (1952) Barrier crops may help to reduce cauliflower mosaic. *Grower* 38, 1140.

Broadbent, L. (1957) *Investigation of Virus Diseases of Brassica Crops. Agricultural Research Council Report Series, No 14.* Cambridge University Press, Cambridge, 94 pp.

Broadbent, L. (1969) Disease control through vector control. In: Maramorosch, K. (ed.) *Virus, Vectors and Vegetation.* Interscience, New York, pp. 593–630.

van der Broek, L.J. and Gill, C.C. (1980) The median latent periods for three isolates of barley yellow dwarf virus in aphid vectors. *Phytopathology* 70, 644–646.

Brown, C.R. and Corsini, D. (2001) Genetics and breeding of virus resistance: traditional methods. In: Loebenstein, G., Berger, P.H., Brunt, A.A. and Lawson, R.H. (eds) *Virus and Virus-like Diseases of Potatoes and Production of Seed-Potatoes.* Kluwer, Dordrecht, pp. 323–340.

Brunt, A.A. and Loebenstein, G. (2001) The main viruses infecting potato crops. In: Loebenstein, G., Berger, P.H., Brunt, A.A. and Lawson, R.H. (eds) *Virus and Virus-like Diseases of Potatoes and Production of Seed-Potatoes.* Kluwer, Dordrecht, pp. 65–134.

Brunt, A.A., Crabtree, K., Dallwitz, M.J., Gibbs, A.J. and Watson, L. (1996) *Viruses of Plants. Descriptions and Lists from the VIDE Database.* CAB International, Wallingford, 1484 pp.

Burgess, A.J., Harrington, R. and Plumb, R.T. (1999) Barley and cereal dwarf virus epidemiology and control strategies. In: Smith, H.G. and Barker, H. (eds) *The Luteoviridae.* CAB International, Wallingford, pp. 248–279.

Carrington, J.C., Kasschau, K.D., Mahajan, S.K. and Schaad, M.C. (1996) Cell-to-cell and long distance transport of viruses in plants. *Plant Cell* 8, 1669–1681.

Carter, N. and Harrington, R. (1991) Factors influencing aphid population dynamics and behavior and the consequences for virus spread. In: Harris, K.F. (ed.) *Advances in Disease Vector Research, Volume 7.* Springer, New York, pp. 19–43.

Chalfant, R.B. and Chapman, R.K. (1962) Transmission of cabbage viruses A and B by the cabbage aphid and the green peach aphid. *Journal of Economic Entomology* 55, 584–590.

Chamberlain, E.E. and Bayliss, G.T.S. (1948) Onion yellow dwarf. Successful cultivation. *New Zealand Journal of Science and Technology* A29, 300–301.

Chapman, R.F., Bernays, E.A. and Simpson, S.J. (1987) Attraction and repulsion of the aphid, *Cavariella aegopodii*, by plant odors. *Journal of Chemical Ecology* 7, 881–888.

Chatzivassiliou, E.K., Efthimiou, K., Drossos, E., Papadopoulou, A., Poimenidis, G. and Katis, N.I. (2004) A survey of tobacco viruses in tobacco crops and native flora in Greece. *European Journal of Plant Pathology* 110, 1011–1023.

Chen, B. and Francki, R.I.B. (1990) Cucumovirus transmission by the aphid *Myzus persicae* is determined solely by the viral coat protein. *Journal of General Virology* 71, 939–944.

Cherqui, A. and Tjallingii, W.F. (2000) Salivary proteins of aphids, a pilot study on identification, separation and immunolocalisation. *Journal of Insect Physiology* 46, 1177–1186.

Childress, S.A. and Harris, K.F. (1989) Localization of virus-like particles in the foreguts of viruliferous *Graminella nigrifrons* leafhoppers carrying the semipersistent maize chlorotic dwarf virus. *Journal of General Virology* 70, 247–251.

Christiansen-Weniger, P., Powell, G. and Hardie, J. (1998) Plant virus and parasitoid interactions in a shared insect vector/host. *Entomologia Experimentalis et Applicata* 86, 205–213.

Collar, J.L., Avilla, C., Duque, M. and Fereres, A. (1995) Assessment of potato virus Y (PVY) spread in bell peppers treated with different insecticides. In: *Abstracts of the 6th International Plant Virus Epidemiology Symposium: Epidemiological Aspects of Plant Virus Control, Jerusalem, April 1995*. Ortra, Tel Aviv, p. 3.

Conti, M., Caciagli, P. and Casetta, A. (1979) Infection sources and aphid vectors in relation to the spread of cucumber mosaic virus in pepper crops. *Phytopathologia Mediterranea* 18, 123–128.

Damgsteeght, V.D. and Hewings, A.D. (1986) Comparative transmission of soybean dwarf virus by three geographically diverse populations of *Aulacorthum* (= *Acyrthosiphum*) *solani*. *Annals of Applied Biology* 109, 453–463.

D'Arcy, C.J. (1995) Symptomatology and host range of barley yellow dwarf. In: D'Arcy, C.J. and Burnett, P.A. (eds) *Barley Yellow Dwarf: 40 Years of Progress*. APS, St Paul, pp. 9–28.

Dedryver, C.-A., Riault, G., Tanguy, S., Le Gallic, J.F., Trottet, M. and Jacquot, E. (2005) Intra-specific transmission variation and inheritance of BYDV-PAV transmission in the aphid *Sitobion avenae*. *European Journal of Plant Pathology* 111, 341–354.

Dewar, A.M. and Smith, H.G. (1999) Forty years of forecasting virus yellows incidence in sugar beet. In: Smith, H.G. and Barker, H. (eds) *The Luteoviridae*. CAB International, Wallingford, pp. 229–243.

Dewar, A.M., Haylock, L.A. and Ecclestone, P.M.J. (1996) Strategies for controlling virus yellows in sugar beet. *British Sugar Beet Review* 64, 44–48.

Difonzo, C.D., Ragsdale, D.W., Radcliffe, E.B., Gudmestad, N.C. and Secor, G.A. (1996) Crop borders reduce potato virus Y in seed potato. *Annals of Applied Biology* 129, 289–302.

van Dijk, P. (1993) Survey and characterization of potyviruses and their strains of *Allium* species. *Netherlands Journal of Plant Pathology* 99 (supplement 2), 1–48.

Dixon, A.F.G. (1998) *Aphid Ecology*, 2nd edn. Chapman and Hall, London, 300 pp.

Doncaster, J.P. and Gregory, P.H. (1948) *The Spread of Virus Diseases in the Potato Crop. Agricultural Research Council Report Series, No. 7*, 189 pp.

Doncaster, J.P. and Kassanis, B. (1946) The shallot aphis, *Myzus ascalonicus* Doncaster, and its behaviour as a vector of plant viruses. *Annals of Applied Biology* 33, 66–69.

Dovas, C.I., Hatziloukas, E., Salomon, R., Barg, E., Shiboteth, Y. and Katis, N.I. (2001) Incidence of viruses infecting *Allium* spp. in Greece. *Journal of Phytopathology* 107, 677–684.

Drucker, M., Froissart, R., Hébrard, E., Uzest, M., Ravallac, M., Esperandieu, P., Mani, J.-L., Pugniere, M., Roquet, F., Fereres, A. and Blanc, S. (2002) Intracellular distribution of viral gene products regulates a complex mechanism of cauliflower mosaic virus acquisition by its vector. *Proceedings of the National Academy of Sciences of the United States of America* 99, 2422–2427.

Duffus, J.E. (1971) Role of weeds in the incidence of virus diseases. *Annual Review of Phytopathology* 9, 319–340.

Dusi, A.N. (1999) Beet mosaic virus: epidemiology and damage. PhD thesis, The Agricultural University, Wageningen, The Netherlands.

Eastop, V. (1983) The biology of the principal aphid virus vectors. In: Plumb, R.T. and Thresh, J.M. (eds) *Plant Virus Epidemiology*. Blackwell, Oxford, pp. 115–132.

Eisenbach, J. and Mittler, T.E. (1980) An aphid circadian rhythm: factors affecting the release of sex pheromone by oviparae of the green-bug, *Schizaphis graminum*. *Journal of Insect Physiology* 26, 511–515.

Elnagar, S. and Murant, A.F. (1976) The role of the helper virus, anthriscus yellows, in the transmission of parsnip yellow fleck virus by the aphid *Cavariella aegopodii*. *Annals of Applied Biology* 84, 169–181.

Espinoza, A.M., Usmany, M., Pirone, T.P., Harvey, M., Woolston, C.J., Medina, V., Vlak, J.M. and Hull, R. (1992) Expression of cauliflower mosaic virus ORF II in a baculovirus system. *Intervirology* 34, 1–12.

Fereres, A. (2000) Barrier crops as a cultural control measure of non-persistently transmitted aphid-borne viruses. *Virus Research* 71, 221–231.

Fereres, A. and Collar, J.L. (2001) Analysis of noncirculative transmission by electrical penetration graphs. In: Harris, K.F., Smith, O.P. and Duffus, J.E. (eds). *Virus–Insect–Plant Interactions*. Academic Press, New York, pp. 87–109.

Fereres, A., Blua, M.J. and Perring, T.M. (1992) Retention and transmission characteristics of zucchini yellow mosaic virus by *Aphis gossypii* and *Myzus persicae* (Homoptera: Aphididae). *Journal of Economic Entomology* 73, 730–735.

Ferro, D.N., Mackenzie, J.D. and Margolies, D.C. (1980) Effect of mineral oil and a systemic insecticide on field spread of aphid-borne maize dwarf mosaic virus in sweet corn. *Journal of Economic Entomology* 73, 730–733.

Fitchen, J.H. and Beachy, R.N. (1993) Genetically engineered protection against viruses in transgenic plants. *Annual Review of Microbiology* 47, 739–763.

Forbes, A.R. (1966) Electron microscope evidence for nerves in the mandibular stylets of the green peach aphid. *Nature, London* 212, 726.

Forbes, A.R. (1969) The morphology and fine structure of the gut of the green peach aphid, *Myzus persicae* (Sulzer) (Homoptera: Aphididae). *Memoirs of the Entomological Society of Canada* 36, 1–74.

Foster, G.N., Blake, S., Tones, S.J., Barker, I. and Harrington, R. (2004) Occurrence of barley yellow dwarf virus in autumn-sown cereal crops in the United Kingdom in relation to field characteristics. *Pest Management Science* 60, 113–125.

Franz, A.W.E., van der Wilk, F., Verbeek, M., Dullemans, A.M. and van den Heuvel, J.F.J.M. (1999) Faba bean necrotic yellows virus (genus *Nanovirus*) requires a helper factor for its aphid transmission. *Virology* 262, 210–219.

Garcia-Arenal, F., Fraile, A. and Malpica, J. (2001) Variability and genetic structure of plant virus populations. *Annual Review of Phytopathology* 39, 156–186.

Garret, A., Kerlan, C. and Thomas, D. (1993) The intestine is a site of passage for potato leafroll virus from the gut lumen to the haemocoel in the aphid vector, *Myzus persicae* Sulz. *Archives of Virology* 131, 377–392.

Garrett, R.G. (1973) Non-persistent aphid-borne viruses. In: Gibbs, A.J. (ed.) *Viruses and Invertebrates*. North-Holland, Amsterdam, pp. 476–492.

Garsney, S.M. and Lee, R.F. (1987) Tristeza. In: Whiteside, J.O., Garsney, S.M. and Timmer, T.W. (eds.) *Compendium of Citrus Diseases*. APS, St Paul, pp. 48–50.

Gera, A., Loebenstein, G. and Raccah, B. (1979) Protein coats of two strains of cucumber mosaic virus affect transmission by *Aphis gossypii*. *Phytopathology* 69, 396–399.

Gibson, R.W. (1983) The ability of different pyrethroids to control spread of potato viruses by aphids. *Proceedings of the 10th International Congress of Plant Protection, Brighton, November1983* 3, 1192.

Gibson, R.W. and Cayley, G.R. (1984) Improved control of potato virus Y by mineral oil plus the pyrethroid cypermethrin applied electrostatically. *Crop Protection* 3, 469–478.

Gibson, R.W. and Plumb, R.T. (1977) Breeding plants for resistance to aphid infestation. In: Harris, K.F. and Maramorosch, K. (eds) *Aphids as Virus Vectors*. Academic Press, New York, pp. 473–500.

Gibson, R.W. and Rice, A.D. (1986) The combined use of mineral oils and pyrethroids to control plant viruses transmitted non- and semi-persistently by *Myzus persicae*. *Annals of Applied Biology* 109, 465–472.

Gibson, R.W., Rice, A.D. and Sawicki, R.M. (1982) Effects of the pyrethroid deltamethrin on the acquisition and inoculation of viruses by *Myzus persicae*. *Annals of Applied Biology* 100, 49–54.

Gildow, F.E. (1987) Virus-membrane interactions involved in circulative transmission of luteoviruses by aphids. *Current Tropical Vector Research* 4, 93–120.

Gildow, F.E. (1993) Evidence for receptor-mediated endocytosis regulating luteovirus acquisition by aphids. *Phytopathology* 83, 270–277.

Gildow, F.E. (1999) Luteovirus transmission and mechanisms regulating vector specificity. In: Smith, H.G. and Barker, H. (eds) *The Luteoviridae*. CAB International, Wallingford, pp. 88–113.

Gildow, F.E., Damsteegt, V.D., Smith, O.P. and Gray, S.M. (1994) Cellular mechanisms regulating circulative transmission and aphid vector specificity of soybean dwarf luteoviruses. *Phytopathology* 84, 1155–1156.

Gildow, F.E., Damsteegt, V.D., Stone, A.L., Smith, O.P. and Gray, S.M. (2000) Virus-vector cell interactions regulating transmission specificity of soybean dwarf luteorviruses. *Journal of Phytopathology* 148, 333–342.

Govier, D.A. and Kassanis, B. (1974a) Evidence that a component other than the virus particle is needed for aphid transmission of potato virus Y. *Virology* 57, 285–286.

Govier, D.A. and Kassanis, B. (1974b) A virus-induced component of plant sap needed when aphids acquire potato virus Y from purified preparations. *Virology* 61, 420–426.

Govier, D.A., Kassanis, B. and Pirone, T.P. (1977) Partial purification and characterization of the potato virus Y helper component. *Virology* 78, 306–314.

Gray, J.M., Moyer, J.W., Kennedy, G.G. and Campell, C.L. (1986) Virus suppression and aphid resistance effects on spatial and temporal spread of watermelon mosaic virus-2. *Phytopathology* 76, 1254–1259.

Gray, S.M. (1999) Intraspecific variability of luteovirus transmission within aphid vector populations. In: Smith, H.G. and Barker, H. (eds). *The Luteoviridae*. CAB International, Wallingford, pp. 119–123.

Gray, S.M. and Banerjee, N. (1999) Mechanisms of arthropod transmission of plant and animal viruses. *Microbiology and Molecular Biology Reviews* 63, 128–148.

Gray, S.M. and Gildow, F.E. (2003) Luteovirus–aphid interactions. *Annual Review of Phytopathology* 41, 539–566.

Gray, S.M., Power, A.G., Smith, D.M., Seaman, A.J. and Altman, N.S. (1991) Aphid transmission of barley yellow dwarf virus: acquisition access periods and virus concentration requirements. *Phytopathology* 81, 539–545.

Griffiths, D.C., Pickett, J.A. and Woodcock, C. (1982) Behaviour of alatae of *Myzus persicae* (Sulzer) (Hemiptera: Aphididae) on chemically-treated surfaces after tethered flight. *Bulletin of Entomological Research* 72, 687–693.

Grogan, R.G. (1980) Control of lettuce mosaic with virus-free seed. *Plant Disease* 64, 446–449.

Grogan, R.G., Welch, J.E. and Bardin, R. (1952) Common lettuce mosaic and its control by the use of mosaic-free seed. *Phytopathology* 42, 573–578.

Guo, J.Q., Lapierre, H. and Moreau, J.P. (1997) Clonal variations and virus regulation by aphids in transmission of a French PAV-type isolate of barley yellow dwarf virus. *Plant Disease* 81, 570–575.

Hall, T.D. (1993) Seed potato certification in the UK. In: Ebbels, D. (ed.) *Plant Health and the Single Market. BCPC Monograph No. 54*. BCPC, Farnham, pp. 77–82.

Hamlyn, B.M.G. (1955) Aphid transmission of cauliflower mosaic virus on turnips. *Plant Pathology* 4, 13–16.

Hammond, J. (1998) Resistance to plant viruses: an overview. Breeding for resistance to plant viruses. In: Hadidi, A., Khetapal, R.K. and Koganewaza, H. (eds) *Plant Virus Disease Control*. APS, St Paul, pp. 163–171.

Harrington, R., Katis, N. and Gibson, R.W. (1986) Field assessment of the relative importance of different aphid species in the transmission of potato virus Y. *Potato Research* 29, 67–76.

Harris, K.F. (1977) An ingestion–egestion hypothesis of non-circulative virus transmission by aphids. In: Harris, K.F. and Maramorosch, K. (eds) *Aphids as Virus Vectors*. Academic Press, New York, pp. 165–220.

Harris, K.F. (1990) Aphid transmission of plant viruses. In: Mandahar, C.L. (ed.) *Plant Viruses, Volume 2. Pathology*. CRC, Boca Raton, pp. 177–204.

Harris, K.F. and Bath, J.E. (1973) Regurgitation by *Myzus persicae* during membrane feeding: its likely function in transmission of nonpersistent plant viruses. *Annals of the Entomological Society of America* 66, 793–796.

Harris, K.F. and Harris, L.J. (2001) Ingestion–egestion theory of cuticular-borne virus transmission. In: Harris, K.F., Smith, O.P. and Duffus, J.E. (eds) *Virus–Insect–Plant Interactions*. Academic Press, New York, pp. 111–132.

Harrison, B.D. (1981) Plant virus ecology: ingredients, interactions and environmental influences. *Annals of Applied Biology* 99, 195–209.

Harrison, B.D. and Murant, A.F. (1984) Involvement of plant virus-coded proteins in transmission of plant viruses by vectors. In: Mayo, M.A. and Herrap, K.R. (eds) *Vectors in Virus Biology*. Academic Press, London, pp. 1–36.

Heathcote, G.D. (1974) The effect of plant spacing, nitrogen fertilizer and irrigation on the appearance of symptoms and spread of virus yellows in sugar beet crops. *Journal of Agricultural Science* 82, 53–60.

Heimbach, U., Eggers, C. and Thieme, T. (2002) Weniger Blattläuse durch Mulchen. *Gesunde Pflanzen* 54, 119–125.

Hein, A. (1971) Zur Wirkung von Öl auf die Virusübertragung durch Blattläuse. *Phytopathologische Zeitschrift* 71, 42–48.

van den Heuvel, J.F.J.M. (1999) Fate of a luteovirus in the haemolymph of an aphid. In: Smith, H.G. and Barker, H. (eds) *The Luteoviridae*. CAB International, Wallingford, pp. 112–119.

van den Heuvel, J.F.J.M., Bruyere, A., Hogenhout, A., Ziegler Graff, V., Braunt, V., Verbeek, M., van der Wilk, F. and Richards, K. (1997) The N-terminal region of the luteovirus readthrough domain determines virus binding to *Buchnera* GroEL and is essential for virus persistence in the aphid. *Journal of Virology* 71, 7258–7265.

Hinz, B. (1966) Beiträge zur Analyse der Vektoreignung eineger wirtschaftlich wichtiger Blattlausarten und –rassen. I. Versuche zur Ermittlung der Vektoreigenchaften für das Blattrovirus der Kartoffel bei Rassen von *Myzus persicae* (Sulz.). *Phytopathologische Zeitschrift* 56, 54–77.

Hogenhout, S.A., van der Wilk, F., Verbeek, M., Goldbach, R.W. and van der Heuvel, J.F.J.M. (1998) Potato leafroll virus binds to the equatorial domain of the aphid endosymbiotic Groel homolog. *Journal of Virology* 72, 358–365.

Höller, C. (1991) Movement away from the feeding site in parasitized aphids: host suicide or an attempt by the parasitoid to escape hyperparasitism? In: Polgár, L., Chambers, R.J., Dixon, A.F.G. and Hodek, I. (eds) *Behaviour and Impact of Aphidophaga*. Academic Publishing, The Hague, pp. 45–49.

van Hoof, H.A. (1958) An investigation of the biological transmission of a nonpersistent virus. PhD thesis, The Agricultural University, Wageningen, The Netherlands.

van Hoof, H.A. (1977) Determination of the infection pressure of potato virus YN. *Netherlands Journal of Plant Pathology* 83, 123–127.

Hu, J.S., Wang, M., Sether, D., Xie, W. and Leonhardt, K.W. (1996) Use of polymerase chain reaction (PCR) to study transmission of banana bunchy top virus by the banana aphid (*Pentolonia nigronervosa*). *Annals of Applied Biology* 128, 55–64.

Hull, R. (1964) Spread of groundnut rosette virus by *Aphis craccivora* (Koch). *Nature, London* 202, 213–214.

Hull, R. (2002) *Matthews' Plant Virology*, 4th edn. Academic Press, London, 1001 pp.

Irwin, M.E. and Ruesink, W.G. (1986) Vector intensity: a product of propensity. In: McLean, G.D., Garrett, R.G. and Ruesink, W.G. (eds) *Plant Virus Epidemics: Monitoring, Modelling and Predicting Outbreaks*. Academic Press, Sydney, pp. 13–33.

Irwin, M.E. and Thresh, J.M. (1988) Long-range dispersal of cereal aphids as virus vectors in North America. *Philosophical Transactions of the Royal Society of London B* 321, 421–446.

Irwin, M.E. and Thresh, J.M. (1990) Epidemiology of barley yellow dwarf: a study in ecological complexity. *Annual Review of Phytopathology* 28, 393–424.

Jackson, A.O., Goodin, M., Moreno, I., Johnson, J. and Lawrence, D.M. (1999) Plant rhabdoviruses. In: Granoff, A. and Webster, R.G. (eds) *Encyclopedia of Virology*. Academic Press, San Diego, pp. 1531–1541.

Jayasena, K.W. and Randles, J.W. (1985) The effect of insecticides and a plant barrier row on aphid populations and the spread of bean yellow mosaic potyvirus and subterranean clover red leaf luteovirus in *Vicia faba* in South Australia. *Annals of Applied Biology* 107, 355–364.

Jenkinson, J.G. (1955) The incidence and control of cauliflower mosaic in broccoli in south-west England. *Annals of Applied Biology* 43, 409–422.

Johnson, C.G. (1969) *Migration and Dispersal of Insects by Flight*. Methuen, London, 763 pp.

Johnstone, G.R., Koen, T.B. and Colney, H.L. (1982) Incidence of yellows in sugar beet as affected by variation in plant density and arrangement. *Bulletin of Entomological Research* 72, 289–294.

Jones, A.T. (1979) Further studies on the effect of resistance to *Amphorophora idaei* in raspberry (*Rubus idaeus*) on the spread of aphid-borne virus. *Annals of Applied Biology* 92, 119–123.

Jones, A.T. (1987) Control of virus infection in crop plants through vector resistance. A review of achievements, prospects and problems. *Annals of Applied Biology* 111, 745–772.

Jones, R.A.C. (1991) Reflective mulch decreases the spread of two non-persistently aphid transmitted viruses to narrow-leafed *Lupinus angustifolius*. *Annals of Applied Biology* 118, 79–85.

Jones, R.A.C. (1993) Effects of cereal borders, admixture with cereals and plant density on the spread of bean yellow mosaic potyvirus into narrow-leafed lupins (*Lupinus angustifolius*). *Annals of Applied Biology* 122, 501–518.

Jones, R.A.C. (1994) Effect of mulching with cereal straw and row spacing on spread of bean yellow mosaic potyvirus into narrow-leafed lupins (*Lupinus angustifolius*). *Annals of Applied Biology* 124, 45–58.

Jones, R.A.C. (2001) Developing integrated disease management strategies against non-persistently aphid-borne viruses: a model programme. *Integrated Pest Management Reviews* 6, 15–46.

Jones, R.A.C. (2004) Using epidemiological information to develop effective integrated virus disease management strategies. *Virus Research* 100, 5–30.

Jones, R.A.C. (2005) Patterns of spread of two non-persistently aphid-borne viruses in lupin stands under four different infection scenarios. *Annals of Applied Biology* 146, 337–350.

Jones, R.A.C., Smith, L.J., Gajda, B.E. and Latham, L.J. (2005) Patterns of spread of carrot virus Y in carrot plantings and validation of control measures. *Annals of Applied Biology* 147, 57–67.

Jones, R.A.C., Smith, L.J., Smith, T.N. and Latham, L.J. (2006) Relative abilities of different aphid species to act as vectors of carrot virus Y. *Australasian Plant Pathology* 35, 23–27.

Jurík, M., Mucha, V. and Valenta, V. (1980) Intraspecies variability in transmission efficiency of stylet-borne viruses by the pea aphid (*Acyrthosiphon pisum*). *Acta Virologica* 24, 351–357.

Kaniewski, W. and Lawson, C. (1998) Coat protein and replicase mediated resistance to plant viruses. In: Hadidi, A., Khetarpal, R.K. and Koganezawa, H. (eds) *Plant Virus Disease Control*. APS, St Paul, pp. 65–78.

Kassanis, B. (1941) Transmission of tobacco etch viruses by aphids. *Annals of Applied Biology* 28, 238–243.

Kassanis, B. and Govier, D.A. (1971a) New evidence on the mechanism of transmission of potato virus C and potato aucuba mosaic viruses. *Journal of General Virology* 10, 99–101.

Kassanis, B. and Govier, D.A. (1971b) The role of the helper virus in aphid transmission of potato aucuba mosaic virus and potato virus C. *Journal of General Virology* 13, 221–228.

Kasschau, K.D. and Carrington, J.C. (2001) Long-distance movement and replication maintenance functions correlate with silencing suppression activity of potyviral HC-Pro. *Virology* 285, 71–81.

Katis, N. and Gibson, R.W. (1984) The transmission of beet mosaic virus by cereal aphids. *Plant Pathology* 33, 425–427.

Katis, N. and Gibson, R.W. (1985) Transmission of potato virus Y by cereal aphids. *Potato Research* 28, 65–70.

Katis, N.I., Tsitsipis, J.A., Lykouressis, D.P., Papapanayotou, A., Gargalianou, I., Kokinis, G., Perdikis, D.Ch., Manoussopoulos, J. and Margaritopoulos, J.T. (2006) Zucchini yellow mosaic virus (ZYMV) affected by colonizing and non-colonizing aphids. New aphid species vectors of the virus. *Journal of Phytopathology* 154, 293–302.

Kennedy, G.G. (1976) Host plant resistance and the spread of plant viruses. *Environmental Entomology* 5, 827–832.

Kennedy, J.S. (1960) The behavioural fitness of aphids as field vectors of viruses. *Report of the 7th Commonwealth Entomological Conference, London, July 1960*, pp. 165–168.

Kennedy, J.S. (1965) Co-ordination of successive activities in the aphid. Reciprocal effects of settling on flight. *Journal of Experimental Biology* 43, 489–509.

Kennedy, J.S., Booth, C.O. and Kershaw, W.J.S. (1959a) Host finding by aphids in the field. I. Gynoparae of *Myzus persicae* (Sulzer). *Annals of Applied Biology* 47, 410–423.

Kennedy, J.S., Booth, C.O. and Kershaw, W.J.S. (1959b) Host finding by aphids in the field. II. *Aphis fabae* Scop. (gynoparae) and *Brevicoryne brassicae* L.; with a re-appraisal of the role of host-finding behaviour in virus spread. *Annals of Applied Biology* 47, 424–444.

Kennedy, J.S., Day, M.F. and Eastop, V.F. (1962) *A Conspectus of Aphids as Vectors of Plant Viruses*. Commonwealth Institute of Entomology, London, 114 pp.

Kerlan, C., Robert, Y., Perennec, P. and Guillery, E. (1987) Survey of the level of infection by PVY-0 and control methods developed in France for potato seed production. *Potato Research* 30, 651–667.

Khetarpal, R.K. and Maury, Y. (1987) Pea seed-borne mosaic virus – a review. *Agronomie* 7, 215–224.

Kring, J.B. (1972) Flight behaviour of aphids. *Annual Review of Entomology* 17, 461–492.

Latham, L.J. and Jones, R.A.C. (2004) Carrot virus Y: symptoms, losses, incidence, epidemiology and control. *Virus Research* 100, 89–99.

Lecoq, H. (1999) Epidemiology of Cucurbit aphid-borne yellows virus. In: Smith, H.G. and Barker, H. (eds) *The Luteoviridae*. CAB International, Wallingford, pp. 243–248.

Lecoq, H. and Pitrat, M. (1983) Field experiments on the integrated control of aphid-borne viruses in muskmelons. In: Plumb, R.T. and Thresh, J.M. (eds) *Plant Virus Epidemiology*. Blackwell, Oxford, pp. 169–176.

Lecoq, H., Labonne, G. and Pitrat, M. (1980) Specificity of resistance to virus transmission by aphids in *Cucumis melo*. *Annales de Phytopathologie* 12, 139–144.

Lecoq, H., Bourdin, D., Raccah, B., Hiebert, E. and Purcifull, D.E. (1991) Characterization of a zucchini yellow mosaic virus isolate with a deficient helper component. *Phytopathology* 81, 1087–1091.

Leh, V., Jacquot, E., Geldreich, A., Hermann, T., Leclerc, D., Cerruti, M., Yot, P., Keller, M. and Blanc, S. (1999) Aphid transmission of cauliflower mosaic virus requires the viral PIII protein. *EMBO Journal* 18, 7077–7085.

Leh, V., Jacquot, E., Geldreich, A., Kaas, M., Blanc, S., Keller, M. and Yot, P. (2001) Interaction between the open reading frame III product and the coat protein is required for transmission of cauliflower mosaic virus by aphids. *Journal of Virology* 75, 100–106.

Limburg, D.D., Mauk, P.A. and Godfrey, L.D. (1997) Characteristics of beet yellows closterovirus transmission to sugar beets by *Aphis fabae*. *Phytopathology* 87, 766–771.

Liu, S., He, X., Park, G., Josefsson, C. and Perry, K.L. (2002) A conserved protein surface domain of cucumber mosaic virus is essential for efficient aphid vector transmission. *Journal of Virology* 76, 9756–9762.

Llave, C., Kasschau, K.D. and Carrington, J.C. (2000) Virus-encoded suppressor of posttranscriptional gene silencing targets a maintenance step in the silencing pathway. *Proceedings of the National Academy of Sciences of the United States of America* 97, 13401–13406.

Loebenstein, G. and Raccah, B. (1980) Management of non-persistently transmitted aphid-borne viruses. *Phytoparasitica* 8, 21–35.

Loebenstein, G., Alper, M. and Deutsch, M. (1964) Preventing aphid-spread cucumber mosaic virus with oils. *Phytopathology* 54, 960–962.

Lopez-Abella, D., Bradley, R.H.E. and Harris, K.F. (1988) Correlation between stylet paths made during superficial probing and the ability to transmit nonpersistent viruses. *Advances in Disease Vector Research* 5, 251–285.

Madden, L.V., Pirone, T.P. and Raccah, B. (1987) Temporal analysis of two viruses increasing in the same tobacco fields. *Phytopathology* 77, 974–980.

Maia, I.G., Haenni, A.-L. and Bernardi, F. (1996) Potyviral HC-Pro: a multifunctional protein. *Journal of General Virology* 77, 1335–1341.

Maison, P. and Massonie, G. (1982) Premières observations sur la spécificité de la résistance du pecher à la transmission aphidienne du virus de la sharka. *Agronomie* 2, 681–683.

Maramorosch, K. and Harris, K.F. (eds) (1981) *Plant Diseases and Vectors: Ecology and Epidemiology*. Academic Press, New York, 360 pp.

Markham, P.G., Pinner, M.S., Raccah, B. and Hull, R. (1987) The acquisition of a caulimovirus by different aphid species: comparison with a potyvirus. *Annals of Applied Biology* 111, 571–587.

Martin, B., Collar, J.L., Tjallingii, W.F. and Fereres, A. (1997) Intracellular ingestion and salivation by aphids may cause the acquisition and inoculation of nonpersistently transmitted plant viruses. *Journal of General Virology* 78, 2701–2705.

Massonie, G. and Maison, M. (1980) Peach resistant to aphid vectors of plum pox virus. *Acta Phytopathologica Academiae Scientiarum Hungaricae* 15, 89–95.

Massonie, G., Maison, P., Monet, R. and Grasselly, C. (1982) Résistance au puceron vert du pecher, *Myzus persicae* Sulzer (Homoptera: Aphididae) chez *Prunus persica* (L.) Batsch et d'autres espèces de *Prunus*. *Agronomie* 2, 63–70.

Mayo, M.A. and Ziegler-Graff, V. (1996) Recent developments in luteovirus molecular biology. *Advances in Virus Research* 46, 413–460.

McLean, G. and Kinsey, M.G. (1965) Identification of electronically recorded curve patterns associated with aphid salivation and ingestion. *Nature, London* 205, 1130–1131.

McLean, G. and Kinsey, M.G. (1984) The precibarial valve and its role in the feeding behavior of the pea aphid, *Acyrthosiphon pisum*. *Bulletin of the Entomological Society of America* 30, 26–31.

McLean, G., Burt, J.R., Thomas, D.W. and Sproul, A.N. (1982) The use of reflective mulch to reduce the incidence of watermelon mosaic virus in Western Australia. *Crop Protection* 1, 491–496.

Miles, P.W. (1999) Aphid saliva. *Biological Reviews* 74, 41–85.

Moericke, V. (1950) Über das Farbensehen der Pfirsichblattlaus (*Myzodes persicae* Sulz.) *Zeitschrift für Tierpsychologie* 7, 265–274.

Moericke, V. (1955) Über die Lebensgewohnheiten der geflügelten Blattläuse (Aphidina) unter besonderer Berücksichtigung des Verhaltens beim Landen. *Zeitschrift für Angewandte Entomologie* 37, 29–91.

Montllor, C.B. (1991) The influence of plant chemistry on aphid feeding behavior. In: Bernays, E. (ed.) *Insect–Plant Interactions, Volume 3*. CRC, Boca Raton, pp. 125–173.

Moreno, A., Hebrard, E., Uzest, M., Blanc, S. and Fereres, A. (2005a) A single amino acid position in the helper component of cauliflower mosaic virus can change the spectrum of transmitting aphid species. *Journal of Virology* 79, 13587–13593.

Moreno, A., Palacios, I., Blanc, S. and Fereres, A. (2005b) Intracellular salivation is the mechanism involved in the inoculation of cauliflower mosaic virus by its major vectors *Brevicoryne brassicae* and *Myzus persicae*. *Annals of the Entomological Society of America* 98, 763–769.

Mowry, T.M. (2005) Insecticidal reduction of potato leafroll virus by *Myzus persicae. Annals of Applied Biology* 146, 81–88.

Murant, A.F., Roberts, I.M. and Elnagar, S. (1976) Association of virus-like particles with the foregut of the aphid *Cavariella aegopodii* transmitting the semipersistent viruses anthriscus yellows and parsnip yellow fleck. *Journal of General Virology* 31, 47–57.

Murant, T., Raccah, B. and Pirone, T.P. (1998) Transmission by vectors. In: Milne, R.G. (ed.) *The Plant Viruses, Volume 4. The Filamentous Plant Viruses and Bipartite RNA Genomes.* Plenum, New York, pp. 237–273.

Murphy, F.A., Fauquet, C.M., Bishop, D.H.L., Ghabrial, S.A., Jarvis, A.W., Martelli, G.P., Mayo, M.A. and Summers, M.D. (1995) *Virus Taxonomy. Classification and Nomenclature of Viruses. Sixth Report of the International Committee on Taxonomy of Viruses.* Springer, Vienna, 586 pp.

Naidu, R.A., Bottenberg, H., Subrahmany, P., Kimmins, F.M., Robinson, D.J. and Thresh, J.M. (1998) Epidemiology of groundnut rosette virus disease: current status and future research needs. *Annals of Applied Biology* 132, 525–548.

Nault, L.R. (1997) Arthropod transmission of plant viruses: a new synthesis. *Annals of the Entomological Society of America* 90, 521–541.

Nault, L.R. and Gyrisco, G.G. (1966) Relation of the feeding process of the pea aphid to the inoculation of pea enation mosaic virus. *Annals of the Entomological Society of America* 61, 180–185.

Ng, J.C.K. and Perry, K.L. (2004) Transmission of plant viruses by aphid vectors. *Molecular Plant Pathology* 5, 505–511.

Nottingham, S.F., Hardie, J., Dawson, G.W., Hick, A.J., Pickett, J.A., Wadhams, L.J. and Woodcock, C.M. (1991) Behavioural and electrophysiological responses of aphids to host and non-host plant volatiles. *Journal of Chemical Ecology* 17, 1231–1242.

Orsenigo, J.R. and Zitter, T.A. (1971) Vegetable virus problems in South Florida as related to weed science. *Proceedings of the Florida Station of the Horticultural Society* 69, 38–42.

Palacios, I., Drucker, M., Blanc, S., Leite, S. and Moreno, A. (2002) Cauliflower mosaic virus is preferentially acquired from the phloem by its aphid vectors. *Journal of General Virology* 83, 3163–3171.

Palukaitis, P. and Garcia-Arenal, F. (2003) Cucumoviruses. *Advances in Virus Research* 62, 241–323.

Park, K.C. and Hardie, J. (2004) Electrophysiological characterization of olfactory sensilla in the black bean aphid, *Aphis fabae. Journal of Insect Physiology* 50, 647–655.

Parrish, W.B. (1967) The origin, morphology and innervation of aphid stylets (Homoptera). *Annals of the Entomological Society of America* 60, 273–276.

Peiffer, M.L., Gildow, F.E. and Gray, S.M. (1997) Two distinct mechanisms regulate luteovirus transmission efficiency and specificity at the aphid salivary gland. *Journal of General Virology* 78, 495–503.

Peng, Y., Kadoury, D., Gal-On, A., Huet, H., Wang, Y. and Raccah, B. (1998) Mutations in the HC-Pro gene of zucchini yellow mosaic potyvirus: effects on aphid transmission and binding to purified virions. *Journal of General Virology* 79, 897–904.

Perring, T.M., Gruenhagen, N.M. and Farrar, C.A. (1999) Management of plant viral diseases through chemical control of insect vectors. *Annual Review of Entomology* 44, 457–481.

Perry, K.L. (2001) Cucumoviruses. In: Harris, K.F., Smith, O.P. and Duffus, J.E. (eds) *Virus–Insect–Plant Interactions.* Academic Press, New York, pp. 167–180.

Peters, D. (1988) A conspectus of plant species as host for viruses causing beet yellows diseases. In: *Virus Yellows Monograph.* Institut International de Recherches Betteravières, Brussels, pp. 87–117.

Pettersson, J. (1970) Studies on *Rhopalosiphum padi* (L.). I. Laboratory studies on olfactometric responses to the winter host *Prunus padus* L. *Lantbrukshögskolans Annaler* 36, 381–389.

Pettersson, J. (1971) An aphid sex attractant. II. Histological, ethological and comparative studies. *Entomologica Scandinavica* 2, 81–93.

Pettersson, J. (1973) Olfactory reactions of *Brevicoryne brassicae* (L.) (Homoptera, Aphididae). *Swedish Journal of Agricultural Research* 3, 95–103.

Phelan, P.L. and Miller, J.R. (1982) Post-landing behaviour of alate *Myzus persicae* as altered by (*E*)-β-farnesene and three carboxylic acids. *Entomologia Experimentalis et Applicata* 32, 46–53.

Pickett, J.A., Wadhams, L.J., Woodcock, C.M. and Hardie, J. (1992) The chemical ecology of aphids. *Annual Review of Entomology* 37, 67–90.

Pirone, T.P. (1977) Accessory factors in nonpersistent virus transmission. In: Harris, K.F. and Maramorosch, K. (eds) *Aphids as Virus Vectors.* Academic Press, New York, pp. 221–235.

Pirone, T.P. (1981) Efficiency and selectivity of the helper-component-mediated aphid transmission of some potyviruses. *Phytopathology* 71, 922–924.

Pirone, T.P. and Blanc, S. (1996) Helper-dependent vector transmission of plant viruses. *Annual Review of Phytopathology* 34, 227–247.

Pirone, T.P. and Harris, K.F. (1977) Nonpersistent transmission of plant viruses by aphids. *Annual Review of Phytopathology* 15, 5–73.

Pirone, T.P. and Megahed, E.-S. (1966) Aphid transmissibility of some purified viruses and viral RNAs. *Virology* 30, 631–637.

Pirone, T.P. and Perry, K.L. (2002) Aphids: non-persistent transmission. *Advances in Botanical Research* 36, 1–20.

Plumb, R.T. (2002) Viruses of Poaceae: a case history in plant pathology. *Plant Pathology* 51, 673–683.

Plumb, R.T. and Johnstone, G.R. (1995) Cultural, chemical, and biological methods for the control of barley yellow dwarf. In: D'Arcy, C.J. and Burnett, P.A. (eds) *Barley Yellow Dwarf: 40 Years of Progress*. APS, St Paul, pp. 307–319.

Ponsen, M.B. (1972) The site of potato leafroll virus multiplication in its vector, *Myzus persicae*. An anatomical study. *Mededelingen Landbouwhogeschool Wageningen* 72- 16, 1–147.

Ponsen, M.B. (1987) Anatomy and physiology. In: Minks, A.K. and Harrewijn, P. (eds) *Aphids. Their Biology, Natural Enemies and Control, Volume 2A*. Elsevier, Amsterdam, pp. 79–97.

Powell, G. (1991) Cell membrane punctures during epidermal penetrations by aphids: consequences for the transmission of two potyviruses. *Annals of Applied Biology* 119, 313–321.

Powell, G. (1993) The effect of pre-acquisition starvation on aphid transmission of potyviruses during observed and electrically recorded stylet penetrations. *Entomologia Experimentalis et Applicata* 66, 255–260.

Powell, G. (2005) Intracellular salivation is the aphid activity associated with inoculation of non-persistently transmitted viruses. *Journal of General Virology* 86, 469–472.

Powell, G. and Hardie, J. (2000) Host-selection behaviour by genetically identical aphids with different plant preferences. *Physiological Entomology* 25, 54–62.

Powell, G., Harrington, R. and Spiller, N.J. (1992) Stylet activities and potato virus Y vector efficiencies by the aphids *Brachycaudus helichrysi* and *Drepanosiphum platanoidis*. *Entomologia Experimentalis et Applicata* 72, 157–165.

Powell, G., Pirone, T. and Hardie, J. (1995) Aphid stylet activities during potyvirus acquisition from plants and an *in vitro* system that correlate with subsequent transmission. *European Journal of Plant Pathology* 101, 411–420.

Powell, G., Maniar, S.P., Pickett, J.A. and Hardie, J. (1999) Aphid responses to non-host epicuticular lipids. *Entomologia Experimentalis et Applicata* 91, 115–123.

Powell, G., Tosh, C.R. and Hardie, J. (2006) Host plant selection by aphids: behavioural, evolutionary and applied perspectives. *Annual Review of Entomology* 51, 309–330.

Power, A.G. and Gray, S.M. (1995) Aphid transmission of barley yellow dwarf viruses: interactions between viruses, vectors, and host plants. In: Burnett, P.A. (ed.) *Barley Yellow Dwarf: 40 Years of Progress*. APS, St Paul, pp. 259–291.

Power, A.G., Seaman, A.J. and Gray, S.M. (1991) Aphid transmission of barley yellow dwarf virus: inoculation access periods and epidemiological implications. *Phytopathology* 81, 545–548.

Prado, E. and Tjallingii, W.F. (1994) Aphid activities during sieve element punctures. *Entomologia Experimentalis et Applicata* 72, 157–165.

Pruss, G., Ge, X., Shi, X.M., Carrington, J.C. and Bowman Vance, V. (1997) Plant viral synergism: the potyviral genome encodes a broad-range pathogenicity enhancer that transactivates replication of heterologous viruses. *Plant Cell* 9, 859–868.

Quiot, J.B., Marchoux, G., Douine, L. and Vigouroux, A. (1979a) Écologie et épidémiologie du virus de la mosaïque du concombre dans la sud-est de la France 5. Rôle des espèces spontanées dans la conservation du virus. *Annales de Phytopathologie* 11, 325–348.

Quiot, J.B., Devegne, J.C., Marchoux, G., Cardin, L. and Douine, L. (1979b) Écologie et épidémiologie du virus de la mosaïque du concombre dans la sud-est de la France 6. Conservation de deux types de populations virales dans les plantes sauvage. *Annales de Phytopathologie* 11, 349–357.

Raccah, B. (1986) Non-persistent viruses. Epidemiology and control. *Advances in Virus Research* 31, 387–429.

Raccah, B., Loebenstein, G. and Bar-Joseph, M. (1976) Transmission of citrus tristeza virus by the melon aphid. *Phytopathology* 66, 1102–1104.

Raccah, B., Huet, H. and Blanc, S. (2001) Potyviruses. In: Harris, K.F., Smith, O.P. and Duffus, J.E. (eds) *Virus–Insect–Plant–Interactions*. Academic Press, New York, pp. 181–206.

Radcliffe, E.B. and Ragsdale, D.W. (2002) Aphid-transmitted potato viruses: the importance of understanding vector biology. *American Journal of Potato Research* 79, 353–386.

Ragsdale, D.W., Radcliffe, E.B. and Dofonzo, C.D. (2001) Epidemiology and field control of PVY and PLRV. In: Loebenstein, G., Berger, P.H., Brunt, A.A. and Lawson, R.H. (eds) *Virus and Virus-like Diseases of Potatoes and Production of Seed-Potatoes.* Kluwer, Dordrecht, pp. 237–270.

Rains, B.D. and Christensen, C.M. (1983) Effects of soil-applied carbofuran on transmission of maize chlorotic dwarf virus and maize dwarf mosaic virus to susceptible field corn hybrid. *Annals of Applied Biology* 74, 290–293.

Reavy, B. and Mayo, M.A. (2002) Persistent transmission of luteoviruses by aphids. *Advances in Botanical Research* 36, 21–46.

Redinbaugh, M.G. and Hogenhout, S.A. (2005) Plant rhabdoviruses. In: Fu, Z.F. (ed.) *The World of Rhabdoviruses.* Springer, Berlin, pp. 143–163.

Redondo, E., Krause-Sakate, R., Yang, S.J., Lot, H., Le Gall, O. and Candresse, T. (2001) Lettuce mosaic virus pathogenicity determinants in susceptible and tolerant lettuce varieties map to different regions of the viral genome. *Molecular Plant–Microbe Interactions* 14, 804–810.

van Regenmortel, M.H.V., Fauquet, C.M., Bishop, D.H.L., Carstens, E., Estes, M.K., Lemon, S., Manillof, J., Mayo, M.A., McGeoth, D.J., Pringle, C.R. and Wicker, R. (eds) (2000) *Virus Taxonomy. Classification and Nomenclature of Viruses. Seventh Report of the International Committee on Taxonomy of Viruses.* Academic Press, San Diego, 1162 pp.

Reinbold, C., Gildow, F.E., Herrbach, E., Ziegler-Graff, V., Goncalves, M.C. and van den Heuvel, J.F.J.M. (2001) Studies on the role of the minor capsid protein in transport of beet western yellows virus through *Myzus persicae. Journal of General Virology* 82, 1995–2007.

Reinbold, C., Herrbach, E. and Brault, V. (2003) Posterior midgut and hindgut are both sites of acquisition of cucurbit aphid-borne yellows virus in *Myzus persicae* and *Aphis gossypii. Journal of General Virology* 84, 3473–3484.

Remaudière, G. and Remaudière, M. (1997) *Catalogue des Aphididae du Monde, Homoptera Aphidoidea.* INRA, Paris, 473 pp.

Revers, F., LeGall, O., Candresse, T. and Maule, A. (1999) New advances in understanding the molecular biology of plant/potyvirus interactions. *Molecular Plant–Microbe Interactions* 12, 367–376.

Risser, G., Pitrat, M., Lecoq, H. and Rode, J.C. (1981) Sensibilité variétale du melon (*Cucumis melo* L.) au virus du rabougrissement jaune du melon (MYSV) et à sa transmission par *Aphis gossypii. Agronomie* 1, 835–838.

Rist, D.L. and Lorbeer, J.W. (1989) Occurrence and overwintering of cucumber mosaic virus and broad bean wilt virus in weeds growing near commercial lettuce fields in New York. *Phytopathology* 79, 65–69.

Robert, Y. (1971) Epidémiologie de l'enroulement de la pomme de terre: capacité vectrice de stades et de formes des pucerons *Aulacorthum solani* Kltb, *Macrosiphum euphorbiae* Thomas et *Myzus persicae* (Sulz.). *Potato Research* 14, 130–139.

Robert, Y. (1999) Epidemiology and control strategies: epidemiology of potato leafroll disease. In: Smith, H.G. and Barker, H. (eds) *The Luteoviridae.* CAB International, Wallingford, pp. 221–228.

Robert, Y. and Lemaire, O. (1999) Epidemiology and control strategies. In: Smith, H.G. and Barker, H. (eds) *The Luteoviridae.* CAB International, Wallingford, pp. 211–279.

Robert, Y. and Maury, Y. (1970) Capacités vectrices comparées de plusiers souches de *Myzus persicae* (Sulz.), *Aulacorthum solani* Kltb et *Macrosiphum euphorbiae* Thomas dans l'étude de la transmission de l'enroulement de la pomme de terre. *Potato Research* 13, 199–209.

Robert, Y., Woodford, J.A.T. and Ducray-Bourdin, D.G. (2000) Some epidemiological approaches to the control of aphid-borne virus diseases in seed potato crops in northern Europe. *Virus Research* 71, 33–47.

Robinson, D.J. and Murant, A.F. (1999) Umbraviruses. In: Granoff, A. and Webster, R.G. (eds) *Encyclopedia of Virology*, 2nd edn. Academic Press, San Diego, pp.1855–1859.

Rojas, M.R., Zerbini, F.M., Allison, R.F., Gilbertson, R.L. and Lucas, W.J. (1997) Capsid protein and helper component-proteinase function as potyvirus cell-to-cell movement proteins. *Virology* 237, 283–295.

Saenz, P., Salvador, B., Simon-Mateo, C., Kasschau, K.D., Carrington, J. and Garcia, J.A. (2002) Host-specific involvement of the HC protein in the long-distance movement of potyviruses. *Journal of Virology* 76, 1922–1931.

Sako, N. and Ogata, K. (1981) Different helper factors associated with aphid transmission of some poty-viruses. *Virology* 112, 762–765.

Sako, N., Yoshioka, K. and Eguchi, K. (1984) Mediation of helper component in aphid transmission of some potyviruses. *Annals of the Phytopathological Society of Japan* 50, 515–521.

Salazar, L.F. (1996) *Potato Viruses and their Control*. International Potato Center, Lima, 214 pp.

Simons, J.N. (1957a) Three strains of cucumber mosaic virus affecting bell pepper in the everglades area of South Florida. *Phytopathology* 47, 145–150.

Simons, J.N. (1957b) Effects on insecticides and physical barriers on field spread of pepper vein-banding mosaic virus. *Phytopathology* 47, 139–145.

Simons, J.N. (1982) Use of oil sprays and reflective surfaces for control of insect-transmitted plant viruses. In: Harris, K.F. and Maramorosch, K. (eds) *Pathogens, Vectors, and Plant Diseases: Approaches to Control*. Academic Press, New York, pp. 71–93.

Simons, J.N. and Zitter, T.A. (1980) Use of oils to control aphid-borne viruses. *Plant Disease* 64, 542–546.

Singh, M.N., Khurana, S.M.P., Nagaich, B.B. and Agrawal, H.O. (1988) Environmental factors affecting aphid transmission of potato virus Y and potato leafroll virus. *Potato Research* 31, 501–509.

Slack, S.A. (1993) Seed certification and seed improvement programs. In: Rowe, R.C. (ed.) *Potato Health Management*. American Phytopathological Society, St Paul, pp. 61–65.

Slack, S.A. and Singh, R.P. (1998) Control of viruses affecting potatoes through seed potato certification programs. In: Hadidi, A., Khetarpal, R.K. and Koganezawa, H. (eds) *Plant Virus Disease Control*. American Phytopathological Society, St Paul, 684 pp.

Smith, F.F., Johnson, G.V., Kahn, R.P. and Bing, A. (1964) Repellancy of reflective aluminium to transient aphid virus vectors. *Phytopathology* 54, 748.

Smith, H.G. (1986) Fodder beet clamps as a source of virus yellows. *British Sugar Beet Review* 54, 38–39.

Smith, H.G. and Barker, H. (eds) (1999) *The Luteoviridae*. CAB International, Wallingford, 297 pp.

Smyrnioudis, I.N., Harrington, R., Clark, S.J. and Katis, N. (2001a) The effect of natural enemies on the spread of barley yellow dwarf virus (BYDV) by *Rhopalosiphum padi*. *Bulletin of Entomological Research* 91, 301–306.

Smyrnioudis, I.N., Harrington, R., Hall, M., Katis, N. and Clark, S.J. (2001b) The effect of temperature on variation in transmission of a BYDV PAV-like isolate by clones of *Rhopalosiphum padi* and *Sitobion avenae*. *European Journal of Plant Pathology* 107, 167–173.

Solomon-Blackburn, R.M. and Barker, H. (2001) Breeding virus resistant potatoes (*Solanum tuberosum*): a review of traditional and molecular approaches. *Heredity* 86, 17–35.

Stevens, M., Smith, H.G. and Hallsworth, P.B. (1994) The host range of beet yellowing viruses among common arable weed species. *Plant Pathology* 43, 579–588.

Stevens, M., Hallsworth, P.B. and Smith, H.G. (2004) The effects of beet mild yellowing virus and beet chlorosis virus on the yield of field-grown sugar beet, 1997, 1999 and 2000. *Annals of Applied Biology* 144, 113–119.

Stevens, M., Freeman, B., Liu, H.-Y., Herrbach, E. and Lemaire, O. (2005). Beet poleroviruses: close friends or distant relatives? *Molecular Plant Pathology* 6, 1–9.

Syller, J. (1987) The influence of temperature on transmission of potato leafroll virus by *Myzus persicae* (Sulz.). *Potato Research* 30, 47–58.

Sylvester, E.S. (1973) Reduction of excretion, reproduction and survival in *Hyperomyzus lactucae* fed on plants infected with isolates of sowthistle yellow vein virus. *Virology* 56, 632–635.

Sylvester, E.S. (1989) Viruses transmitted by aphids. In: Minks, A.K. and Harrewijn, P. (eds) *Aphids. Their Biology, Natural Enemies and Control, Volume 2 C*. Elsevier, Amsterdam, pp. 65–87.

Sylvester, E.S. and McClain, E. (1978) Rate of transovarial passage of sowthistle yellow vein virus in selected subclones of the aphid *Hyperomyzus lactucae* (Hemiptera: Homoptera Aphididae). *Journal of Economic Entomology* 71, 17–20.

Sylvester, E.S. and Richardson, J. (1970) Infection of *Hyperomyzus lactucae* by sowthistle yellow vein virus. *Virology* 42, 1023–1042.

Sylvester, E.S. and Richardson, J. (1971) Decreased survival of *Hyperomyzus lactucae* inoculated with serially passed sowthistle yellow vein virus. *Virology* 46, 310–317.

Sylvester, E.S. and Richardson, J. (1981) Inoculation of the aphids *Hyperomyzus lactucae* and *Chaetosiphon jacobi* with isolates of sowthistle yellow vein virus and strawberry crinkle virus. *Phytopathology* 71, 598–602.

Sylvester, E.S. and Richardson, J. (1992) Aphid-borne rhabdoviruses – relationships with their vectors. *Advances in Disease Vector Research* 9, 313–341.

Sylvester, E.S., Richardson, J. and Frazier, N.W. (1974) Serial passage of strawberry crinkle virus in the aphid *Chaetosiphon jacobi*. *Virology* 59, 301–306.

Szatmari-Goodman, G. and Nault, L.R. (1983) Test of oil-sprays for suppression of aphid-borne maize dwarf mosaic in Ohio sweet corn. *Journal of Economic Entomology* 76, 144–149.

Tachibana, Y. (1981) Control of aphid-borne viruses in faba bean by mulching with silver polyethylene film. *FABIS Newsletter* 3, 56.

Taliansky, M., Mayo, M.A. and Barker, H. (2003) Potato leafroll virus: a classic pathogen shows some new tricks. *Molecular Plant Pathology* 4, 81–89.

Tamada, T., Harrison, B.D. and Roberts, I.M. (1984) Variation among British isolates of potato leafroll virus. *Annals of Applied Biology* 104, 107–116.

Terauchi, H., Honda, K.-I., Yamagishi, N., Kanematsu, S. and Hidaka, S. (2003) The N-terminal region of the readthrough domain is closely related to aphid vector specificity of soybean dwarf virus. *Phytopathology* 93, 1560–1564.

Thomas, P.E. (1983) Sources and dissemination of potato viruses in the Columbia Basin of the Northwestern United States. *Plant Disease* 67, 744–747.

Thresh, J.M. (1982a) Cropping practices and virus spread. *Annual of Review of Phytopathology* 20, 193–218.

Thresh, J.M. (1982b) The role of weeds and wild plants in the epidemiology of plant virus diseases. In: Thresh, J.M. (ed.) *Pests, Pathogens and Vegetation*. Pitman, London, pp. 53–70.

Thresh, J.M. (1983) The long-range dispersal of plant viruses by arthropod vectors. *Philosophical Transactions of the Royal Society of London B* 302, 497–528.

Thresh, J.M. (1988) Eradication as a virus disease control measure. In: Clifford, B.C. and Lester, E. (eds) *Control of Plant Diseases: Costs and Benefits*. Blackwell, Oxford, pp. 155–194.

Tjallingii, W.F. (1978) Mechanoreceptors of the aphid labium. *Entomologia Experimentalis et Applicata* 24, 531–537.

Tjallingii, W.F. (1988) Electrical recording of stylet penetration activities. In: Minks, A.K. and Harrewijn, P. (eds) *Aphids. Their Biology, Natural Enemies and Control, Volume 2 B*. Elsevier, Amsterdam, pp. 95–108.

Tjallingii, W.F. and Hogen Esch, Th. (1993) Fine structure of aphid stylet routes in plant tissue in correlation with EPG signals. *Physiological Entomology* 18, 317–328.

Toba, H.H., Kishaba, A.N., Bohn, G.W. and Hield, H. (1977) Protecting muskmelons against aphid-borne viruses. *Phytopathology* 67, 1418–1423.

Todd, J.M. (1961) The incidence and control of aphid-borne potato virus diseases in Scotland. *European Potato Journal* 4, 316–329.

Tomlinson, J.A. (1962) Control of lettuce mosaic by the use of healthy seed. *Plant Pathology* 11, 61–64.

Upreti, G.C. and Nagaich, B.B. (1971) Variations in the ability of *Myzus persicae* (Sulz.) to transmit potyviruses. I. Leaf roll. *Phytopathologische Zeitschrift* 71, 163–168.

Valkonen, J.R.T. (1994) Natural genes and mechanisms for resistance to viruses in cultivated and wild potato species (*Solanum* spp.). *Plant Breeding* 112, 1–16.

Vance, V.B., Berger, P.H., Carrington, J.C., Hunt, A.G. and Shi, X.M. (1995) 5′ proximal potyviral sequences mediate potato virus X/potyviral synergistic disease in transgenic tobacco. *Virology* 206, 583–590.

Vanderplank, J.E. (1948) The relation between the size of the field and the spread of plant-disease into them. Part I. Crowd diseases. *Empire Journal of Experimental Agriculture* 16, 134–142.

Vanderveken, J. (1977) Oils and other inhibitors of nonpersistent virus transmission. In: Harris, K.F. and Maramorosch, K. (eds) *Aphids as Virus Vectors*. Academic Press, New York, pp. 435–454.

Visser, J.H. and Taanman, J.W. (1987) Odour-conditioned anemotaxis of apterous aphids (*Cryptomyzus korschelti*) in response to host plants. *Physiological Entomology* 12, 473–479.

Walker, D.G.A., Tomlinson, J.A., Innes, N.L. and Pink, D.A.C. (1982) Breeding for virus resistance in British vegetable crops. *Acta Horticulturae* 127, 125–135.

Walkey, D.G.A. and Dance, M.C. (1979) The effect of oil sprays on aphid transmission of turnip mosaic, beet yellows, bean common mosaic and bean yellow mosaic virus. *Plant Disease Reporter* 63, 877–881.

Wallin, J.R. and Loonan, D.V. (1971) Low-level jet winds, aphid vectors, local weather and barley yellow dwarf virus outbreaks. *Phytopathology* 61, 1068–1070.

Wang, D., Woods, R.D., Cockbain, A.J., Maule, A.J. and Biddle, A.J. (1993) The susceptibility of pea cultivars to pea seed-borne mosaic virus infection and virus seed transmission in the UK. *Plant Pathology* 42, 42–47.

Wang, R.Y. and Pirone, T.P. (1996) Potyvirus transmission is not increased by pre-acquisition fasting of aphids reared on artificial diet. *Journal of General Virology* 77, 3145–3148.

Wang, R.Y., Ammar, E.D., Thornbury, D.W., Lopez-Moya, J.J. and Pirone, T.P. (1996) Loss of potyvirus transmissibility and helper component activity correlates with nonretention of virions in aphid stylets. *Journal of General Virology* 77, 861–867.

Wang, R.Y., Powell, G., Hardie, J. and Pirone, T.P. (1998) Role of the helper component in vector-specific transmission of potyviruses. *Journal of General Virology* 79, 1519–1524.

Ward, S.A., Leather, S.R., Pickup, J. and Harrington, R. (1998) Mortality during dispersal and the cost of host-specificity in parasites: how many aphids find hosts? *Journal of Animal Ecology* 67, 763–773.

Watson, M.A. (1967) Epidemiology of aphid-transmitted plant-virus diseases. *Outlook on Agriculture* 5, 155–166.

Watson, M.A. and Plumb, R.T. (1972) Transmission of plant pathogenic viruses by aphids. *Annual Review of Entomology* 17, 425–452.

Watson, M.A. and Roberts, F.M. (1940) Evidence against the hypothesis that certain plant viruses are transmitted mechanically by aphids. *Annals of Applied Biology* 27, 227–233.

Weber, C.A., Godfrey, L.D. and Mauk, P.A. (1996) Effects of parasitism by *Lysiphlebus testaceipes* (Hymenoptera: Aphidiidae) on transmission of beet yellows closterovirus by bean aphid (Homoptera: Aphididae). *Journal of Economic Entomology* 89, 1431–1437.

Weber, K.A. and Hampton, R.O. (1980) Transmission of two purified carlaviruses by the pea aphid. *Phytopathology* 70, 631–633.

Wensler, R.J.D. (1974) Sensory innervation monitoring movement and position in the mandibular stylets of the aphid *Brevicoryne brassicae*. *Journal of Morphology* 143, 349–363.

Wensler, R.J.D. (1977) The fine structure of distal receptors on the aphid *Brevicoryne brassicae* L. (Homoptera) – implications for current theories of sensory transduction. *Cell Tissue Research* 181, 409–422.

Wensler, R.J.D. and Filshie, B. (1969) Gustatory sense organs in the food canal of aphids. *Journal of Morphology* 129, 473–492.

de Wijs, J.J. (1980) The characteristics of mineral oil in relation to their inhibitory activity on the aphid transmission of potato virus Y. *Netherlands Journal of Plant Pathology* 86, 291–300.

Wilcoxson, R.D. and Peterson, A.G. (1960) Resistance of dollard red clover to the pea aphid, *Macrosiphum pisi*. *Journal of Economic Entomology* 53, 863–865.

Woodford, J.A.T. and Gordon, S.C. (1990) New approaches for restricting spread of potato leafroll virus by different methods of eradicating infected plants from potato crops. *Annals of Applied Biology* 116, 477–487.

Wright, G.C. and Bishop, G.W. (1981) Volunteer potatoes as a source of potato leafroll virus and potato virus X. *American Potato Journal* 58, 603–609.

Wyatt, S.D., Seybert, L.J. and Mink, G. (1988) Status of the barley yellow dwarf problem of winter wheat in eastern Washington. *Plant Disease* 72, 110–113.

Yan, F.S. and Visser, J.H. (1982) Electroantenogram responses of the cereal aphid *Sitobion avenae* to plant volatile compounds. *Proceedings of the 5th International Symposium of Insect–Plant Relationships, Wageningen, March 1982*. Pudoc, Wageningen, pp. 387–388.

Yoshii, H. (1966) Transmission of turnip mosaic virus by *Myzus persicae* (Sulz.): mode of stylet insertion and infection site of the virus. *Annals of the Phytopathological Society of Japan* 31, 46–51.

Zeyen, R.J., Stromberg, E.L. and Kuehnast, E.L. (1987) Long-range aphid transport hypothesis for maize dwarf mosaic virus: history and distribution in Minnesota, USA. *Annals of Applied Biology* 111, 325–336.

Zhou, G. and Rochow, W.F. (1984) Differences among five stages of *Schizaphis graminum* in transmission of a barley yellow dwarf luteovirus. *Phytopathology* 74, 1450–1453.

Zitter, T.A. and Simons, J.N. (1980) Management of viruses by alteration of vector efficiency and by cultural practices. *Annual Review of Phytopathology* 18, 289–310.

de Zoeten, G.A. and Skaf, J.S. (2001) Pea enation mosaic and the vagaries of a plant virus. *Advances in Virus Research* 57, 323–350.

Zuniga, E. (1990) Biological control of cereal aphids in the southern cone of South America. In: Burnett, P.A. (ed.) *World Perspectives on Barley Yellow Dwarf*. CIMMYT, Mexico, D.F., pp. 362–367.

15 Chemical Control

Alan M. Dewar

*Entomology Research Group, Broom's Barn Research Station, Higham,
Bury St Edmunds, Suffolk, IP28 6NP, UK Correspondence address: Dewar Crop
Protection Ltd, Drumlanrig, Great Saxham, Bury St Edmunds, Suffolk, IP29 5JR, UK*

Introduction

Aphids are one of the main target groups for development of new insecticides by the ever-shrinking number of pesticide manufacturers. At the time of publication of the last comprehensive book on aphids (Minks and Harrewijn, 1989), the insecticides dominating the aphid control market were mostly organophosphates (OPs) and carbamates, with pyrethroids coming in a poor third (Schepers, 1989; Jeschke *et al.*, 2002). Although aphicides in the former two groups were systemic and relatively persistent, most were highly toxic, not just to the target pests but also to many beneficial insects, and many have since been withdrawn under pressure from the environmental lobby (Pesticides Safety Directorate (PSD), http://www.pesticides.gov.uk/fg_ec.asp?id=648). In many cases, pyrethroid insecticides have replaced OPs, but their lack of systemic activity, and broad-spectrum effects on many non-target insects, makes them even less suitable candidates than OPs as aphicides, even though their rapid action can sometimes prevent primary infection with some viruses. The 'Golden Age' of these insecticides was reviewed recently by Casida and Quistad (1998), and those compounds with novel modes of action are described by Ishaaya and Horowitz (1998).

In the past 10 years, the demand for safer insecticides has stimulated the development of some novel groups. Some have properties that are ideal for aphid control, including the neonicotinoids, pymetrozine and triazamate (no longer available in Europe). This chapter reviews the most recent research conducted on these novel products from an aphid control point of view, and compares their potential with that of the older established products. By contrast, the IPM Case Studies (Chapters 20–30 this volume) describe current chemical control practice and use of economic thresholds in a variety of often region-specific crop scenarios.

Choice of Aphicide

Table 15.1 lists the range of insecticides approved for use against aphids in at least one crop in the UK in 2005 (Whitehead, 2007). This list differs from that for other countries around the world, most notably Africa and the Far East, but it does illustrate the choice available these days, even if products are not approved for all aphid-affected crops. Carbamates, OPs, and pyrethroids are the mainstays of aphid control in many crops, and this is reflected (especially with the last group) in the choice available. However, the neonicotinoids are now approved for use in an increasing number of crops in many countries around the world, but use of pymetrozine

©CAB International 2007. *Aphids as Crop Pests*
(eds H. van Emden and R. Harrington)

391

Table 15.1. List of approved aphicides in the UK as notified in the 2007 UK Pesticide Guide (Whitehead, 2007) with later additions.

Insecticide	Class	Properties	Aphicidal activity
Aldicarb	Carbamate	Systemic	****
Carbosulfan	Carbamate	Systemic	**
Oxamyl	Carbamate	Systemic	**
Pirimicarb	Carbamate	Fumigant	****
Chlorpyriphos	OP	Systemic	**
Dimethoate	OP	Systemic	***
Malathion	OP	Contact	*
Pirimiphos-methyl	OP	Fumigant	***
Alpha-cypermethrin	Pyrethroid	Contact	***
Bifenthrin	Pyrethroid	Contact	**
Cyfluthrin	Pyrethroid	Contact	**
Cypermethrin	Pyrethroid	Contact	**
Deltamethrin	Pyrethroid	Contact	***
Esfenvalerate	Pyrethroid	Contact	***
Lambda-cyhalothrin	Pyrethroid	Contact	***
Pyrethrin	Natural pyrethroid	Contact	**
Tau fluvalinate	Pyrethroid	Contact	**
Zeta cypermethrin	Pyrethroid	Contact	**
Lambda-cyhalothrin + pirimicarb	Pyrethroid + carbamate	Contact – fumigant	****
Nicotine	Alkaloid	Contact	**
Acetamiprid	Neonicotinoid	Systemic	***
Imidacloprid	Neonicotinoid	Systemic	****
Imidacloprid + beta-cyfluthrin	Neonicotinoid + pyrethroid	Systemic + contact	****
Clothianidin + beta-cyfluthrin	Neonicotinoid + pyrethroid	Systemic + contact	****
Thiacloprid	Neonicotinoid	Systemic	****
Thiamethoxam	Neonicotinoid	Systemic	****
Flonicamid	Pyridine carboxamide	Systemic	? ***
Pymetrozine	Azomethine	Systemic	***
Others			
Tebufenpyrad	Pyrazole METI	Contact	**
Fatty acids		Contact	**
Rotenone		Contact	**

and triazamate is still restricted to just a few.

Neonicotinoids were first discovered in the early 1970s, but they were not developed for use in agriculture until 1991, when imidacloprid (Elbert *et al.*, 1990; Altmann and Elbert, 1992; Shiokawa *et al.*, 1994) was introduced to the market as the first of the second-generation neonicotinoids (Jeschke *et al.*, 2002). Imidacloprid had the required photostability, insecticidal activity, and residual persistence to be marketed for a wide range of uses and, in the past decade, has become the largest selling insecticide worldwide for crop protection and animal health. It is a broad-spectrum insecticide, but its excellent systemic action makes it ideal for controlling aphids. Other insecticides developed within this group include acetamiprid (Takahashi *et al.*, 1992), clothianidin (Ohkawara *et al.*, 2002; Jeschke *et al.*, 2003), dinotefuran (Wakita *et al.*, 2005), nitenpyram (Kashiwada, 1996), thiacloprid (Elbert *et al.*, 2000; Jeschke *et al.*, 2001), and thiamethoxam (Senn *et al.*, 1998; Hofer *et al.*, 2001; Maienfisch *et al.*, 2001). All have aphicidal properties, but some are more active than others.

Pymetrozine was first reported in 1992 as CGA 215'944 (Flückiger *et al.*, 1992a). It is a pyrimidine azomethine (Fig. 15.1) and thus has a novel mode of action; it is highly active against aphids and whiteflies in vegetables, ornamentals, cotton, field crops, deciduous fruits, and citrus, including where these pests have developed resistance to other insecticides.

Triazamate was first reported as RH 7988 in 1988 as a highly selective aphicide, with activity against many aphid species on a wide variety of crops (Murray *et al.*, 1988). Unfortunately, following a recent review, it was withdrawn from Europe in 2005.

Modes of Action

To understand the differences in physical properties of the various novel insecticides now being used for aphid control, it is useful to recapitulate on the modes of action of the older ones.

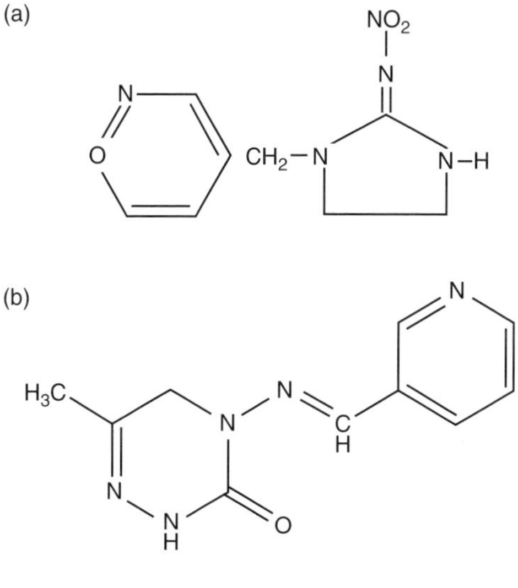

(a)

(b)

Fig. 15.1. Chemical structure of (a) imidacloprid; (b) pymetrozine.

Organophosphates and carbamates

Organophosphate and carbamate insecticides are both acetylcholinesterase (AchE) inhibitors, i.e. they interfere with the transmission of nerve impulses across the synaptic gap between two nerve cells by preventing the breakdown of the predominant neurotransmitter, acetylcholine (Tomizawa and Casida, 2003). This results in tetanic paralysis that destroys the ability of insects and other animals (including man) to respond to external stimuli. Both groups of insecticides tend to have high mammalian toxicity (Plapp, 1991).

Pyrethroids

Pyrethroids, typified by deltamethrin, act on the voltage-gated sodium channel, which governs the movement of sodium ions into the nerve cell and enables impulses to travel along the nerve axon. These insecticides have low mammalian toxicity but high toxicity to fish. Pyrethroids frequently have repellent activity against colonizing aphids, which makes them potentially valuable in controlling non-persistent viruses (Foster *et al.*, 1994; Bedford *et al.*, 1998). However, to have maximum effect they need to be applied before the aphids arrive, and it is very difficult to forecast the arrival of aphids with the required precision (Harrington *et al.*, Chapter 19 this volume).

Neonicotinoids

Neonicotinoids, typified by imidacloprid, act on the post-synaptic nicotinic acetylcholine-receptors (nAChR) in the central and peripheral nervous systems, resulting in excitation and paralysis, followed by death (Ishaaya and Horowitz, 1998; Nauen *et al.*, 2001, 2003; Tomizawa and Casida, 2003). These chemicals are xylem-mobile and hence suitable for seed treatment and soil application. They have low mammalian toxicity due to fundamental differences between the nAChRs of insects and those of mammals

(Tomizawa and Casida, 2003), but some have high toxicity to birds (Anonymous, 1993).

Imidacloprid and acetamiprid, applied topically to the upper surface of leaves, showed translaminar and acropetal (movement towards leaf margins) activity against *Myzus persicae* (peach–potato aphid) and *Aphis gossypii* (cotton or melon aphid) in a laboratory study on feeding behaviour respectively on cabbage and cotton (Bucholz and Nauen, 2002). It also suppressed honeydew production by *M. persicae* feeding on cabbage leaves treated with low sublethal concentrations, suggesting that feeding was interrupted even though the aphids were not killed (Nauen, 1995). In studies on wheat seedlings with radioactive (^{14}C) imidacloprid, high toxicity to aphids persisted 195 days after sowing; a concentration gradient from the oldest to the youngest leaves implicated apoplastic translocation (Stein-Dönecke *et al.*, 1994). Imidacloprid strongly reduced the sucking activity of *Rhopalosiphum padi* (bird cherry–oat aphid) as measured by electrical penetration graphs (EPG) in laboratory studies on maize (Jaschewski, 1997; Abraham and Epperlein, 1999); transmission of *Barley yellow dwarf virus* (BYDV) was also reduced by 65% in the latter study. However, imidacloprid did not affect probing behaviour or transmission of the nonpersistent *Potato virus Y* (PVY) (Collar *et al.*, 1997). *Macrosiphum euphorbiae* (potato aphid), feeding on terminal potato leaflets placed in solutions of imidacloprid, showed a significant reduction in speed of movement, distance moved, and then propensity to fly compared to aphids feeding on untreated plants; however, no repellence was detected in colonization behaviour (Boiteau *et al.*, 1997). Radioactive imidacloprid, when applied topically to sugarbeet pelleted seeds, metabolized to several metabolites, of which the olefinic fraction was the most important (Westwood *et al.*, 1998) as it has greater aphicidal activity than the parent compound (Nauen *et al.*, 1998). This perhaps explains the long persistence of this insecticide. In the former study, most of the parent compound was metabolized within 97 days. Little was then present in the roots, but 23% was still in the soil, and most (44.5 % of the

metabolites plus 4.5% of the parent compound) in the leaves.

Thiamethoxam has notably lower binding affinities with insect nAChRs than the other nicotinoids, but, as it is likely to be a nicotinoid precursor for clothianidin, to which it can be rapidly metabolized (Nauen *et al.*, 2003), this does not appear to be a disadvantage. Thiamethoxam is transported in the xylem to cotyledons and leaves acropetally when applied as a seed treatment to maize, oilseed rape and other plants (Senn *et al.*, 1998; Fischer and Widmer, 2001). The ability of *M. persicae* to penetrate the phloem of potato leaves was substantially reduced by this insecticide in EPG studies (Harrewijn *et al.*, 1998). Its presence on the leaf surface did not affect probing behaviour, thus discounting a repellent effect, but very often stylets were withdrawn, and feeding not resumed, after penetration of the mesophyll cells. After drench applications, a large number of electric potential drops (pds) were observed on the EPGs on treated plants. This showed that increased xylem penetration, and thus exposure to higher concentrations of the insecticide, had occurred; once feeding was terminated, there was no recovery and death by starvation followed.

Clothianidin shares a common mode of action with the other neonicotinoids. It has very low vapour pressure and volatility, and does not dissociate under acidic to slightly basic conditions (Jeschke *et al.*, 2003). It has specific activity against insect nervous systems, and very low toxicity to vertebrates (Ohkawara *et al.*, 2002). Clothianidin demonstrated translaminar activity against *A. gossypii* on cucumbers, excellent translocation from the roots to the leaves, and also between adjacent leaves. Such translocation makes it suitable as a seed and soil treatment (Ohkawara *et al.*, 2002). It is more evenly distributed over the whole leaf lamina than the other neonicotinoids, which tend to accumulate in the leaf margins (Jeschke *et al.*, 2002).

Thiacloprid is an acute contact and stomach poison with systemic properties. It has a short half-life with a good safety margin for bees (Elbert *et al.*, 2000, 2002), making it ideal for spray application. The mode of action is similar to that of the other neonicotinoids.

Pymetrozine (Fig. 15.1)

Pymetrozine has a very specific mode of action: it affects the nerves controlling the salivary pump of sucking pests and causes irreversible cessation of feeding within a few hours of application, followed eventually by starvation and death (Schwinger *et al.*, 1994). Pymetrozine inhibits feeding in aphids immediately after application without producing visible neurotoxic effects, and acts *via* a novel mechanism that is linked to the signalling pathway of serotonin (Kaufmann *et al.*, 2004). Pymetrozine has systemic and translaminar activity, ideal for use in soil and foliar application; however, it can take a little time for the aphids to die (Flückiger *et al.*, 1992a,b; Kayser *et al.*, 1994; Harrewijn and Kayser, 1997). Low concentrations on wheat of CGA215944 (pymetrozine) and imidacloprid had antifeedant effects on *Diuraphis noxia* (Russian wheat aphid); aphids on treated plants spent significantly longer in non-probing activities, but nevertheless penetrated the leaf more frequently, because each penetration was followed by a shorter period of phloem ingestion (Burd *et al.*, 1996).

Diafenthiuron

Diafenthiuron is a new type of thiourea derivative that affects respiration in insects (Ishaaya *et al.*, 2001) by inhibiting ATPase through its reactive intermediate metabolite and photoproduct, the corresponding carbodiimide (Casida and Quistad, 1998).

Triazamate

Triazamate is a carbamoyl triazole, with similar insecticidal properties to the carbamates. It is a fast-acting cholinesterase inhibitor that has both contact and systemic activity, ideal for controlling aphids, and can also be downwardly translocated to the roots, in contrast to most carbamates (Murray *et al.*, 1988). Unfortunately, the mechanism conferring resistance to pirimicarb – modified acetyl cholinesterase (MACE) (Moores *et al.*, 1994a,b) – also affects the efficacy of triazamate (Foster *et al.*, 2002). In the absence of this resistance mechanism, triazamate gives excellent control of aphids, even those with high esterase levels that confer resistance to OPs, pyrethroids, and other carbamates (Dewar *et al.*, 1994).

Adjuvants and synergists

The efficacy of some insecticides can be enhanced by the use of adjuvants in the mix. For example, ethyl fatty acid ester adjuvant (EOP) enhanced the efficacy of methomyl and endosulfan against aphids on maize, and of imidacloprid against aphids on cotton (Killick and Schulteis, 1998). This was attributed to a slowing of the recrystallization of the active ingredients in spray droplets, thus maintaining for longer a liquid state for increased contact. It was also suggested that the addition of EOP softened the cuticle of the insects, allowing increased uptake of the active ingredients.

The synergist MB-599 (proposed common name 'verbutin') was reported to enhance the activity of some aphicides, including pirimicarb, triazamate, and imidacloprid, 2–4-fold against *A. gossypii*, *Acyrthosiphum pisum* (pea aphid), and *R. padi* (Szekely *et al.*, 1996).

Application

Most aphicides in the 1980s were applied as sprays (Fig. 15.2), often in response to spray thresholds (e.g. Hull, 1968; George and Gair, 1979; Oakley and Walters, 1994), but sometimes according to prescribed programmes, e.g. in potatoes for control of viruliferous aphids (Foster *et al.*, 1994). Some insecticides, particularly formulations of carbamates, are applied as granules to the soil (e.g. aldicarb for control of *M. persicae* in sugarbeet and potatoes). This is often

Fig. 15.2.   Spray application to sugarbeet (original colour photo courtesy of Broom's Barn Research Station).

done prophylactically at sowing (Fig. 15.3), before the need for insecticides can be assessed.

However, forecasting schemes are sometimes available to guide such usage (Harrington *et al.*, 1989; Werker *et al.*, 1998; Qi *et al.*, 2004). It is difficult to judge how much use farmers have made of these, as usage of insecticides has tended to remain constant from year to year, irrespective of the forecast level of disease (Dewar and Smith, 1999). The advent of the xylem-mobile neonicotinoids has allowed these to be considered as seed treatments (Fig. 15.4) to replace the older chemicals, albeit on a prophylactic basis.

Crops that are protected with seed treatments include temperate cereals such as wheat and barley (Schmeer *et al.*, 1990; Hofer *et al.*, 2001), sugarbeet (Dewar and Read, 1990; Proft *et al.*, 1999; Dewar *et al.*, 2002), oilseed rape (Birch and Nicholson, 2001), brassicas (Kumar and Dikshit, 2001), lettuce (Ester and Brantjes, 1999; Parker *et al.*, 2002), cotton (Kumar and Santharam, 1999; Zang *et al.*, 1999; Leite *et al.*, 2001), maize (Proft *et al.*, 1999; Hofer *et al.*, 2001), and sorghum (Krauter *et al.*, 2001). The insecticides may be applied in several ways

(Brandl, 2001): directly to the seed either as a dressing or a film coat (as with oilseed rape and cereals); partially encrusted (as with sugarbeet in Turkey); or to the outside of pelleted seed (as with sugarbeet in most of Western Europe). When applied in this last manner, usually as a film coat (Halmer, 1994), the quantities of active ingredient applied to the land are much reduced. For example, the recommended rate of application of imidacloprid to sugarbeet in the UK is 90 g a.i. per unit of seeds (1 unit = 100,000 seeds), which equates to just under 100 g a.i./ha, much less than the rate of aldicarb granules (760 g a.i./ha for aphid control) or sprays (e.g. pirimicarb at 140 g a.i./ha). In addition, with seed treatments the insecticide is placed only round the seed, whereas granules are applied along the whole furrow, and sprays over the whole soil and plant surface, thus potentially endangering more non-target organisms. On the other hand, with seed treatments, the insecticides have to be applied when the seed is being prepared, often months before aphid attack, and this leads to prophylaxis.

With cereal seed treatments, the same principles apply, but with less clear benefits to the environment. Imidacloprid is

Fig. 15.3. Application of imidacloprid granules through a curved spike (a) that injects the treatment below the seed during sowing (b) (original colour photos courtesy of Bayer PlantScience).

Fig. 15.4. A production plant for insecticide treated seeds (original copyright colour photo courtesy of Germain's Technology Group).

applied to wheat and barley at 0.7–1.5 g a.i./ kg seed. When seed is sown at 50–150 kg/ha, this is 17–52 g a.i./ha (Ahmed *et al.*, 2001; Miles *et al.*, 2001). The alternative control for BYDV is well-timed pyrethroid sprays (Miles *et al.*, 2001), applied at very low rates (c. 7.5 g a.i./ha). However, pyrethroids are very active compounds that are applied overall, thus potentially affecting any non-target organisms present at the time.

Soil drench or granule treatments are an alternative for ensuring that the neonicotinoid active ingredients are transported to the most appropriate plant organs – the stems and leaves. Imidacloprid is applied as a soil drench to hops to control *Phorodon humuli* (damson–hop aphid) (Martin *et al.*, 1992), and is included in compost for controlling vine weevils and, by default, also aphids in ornamentals in glasshouses (Pasian *et al.*, 1997). Soil drenches of imidacloprid to transplanted Tabasco pepper (*Capsicum frutescens*) gave complete control of *M. persicae* 96 h after application and persisted for 5 weeks, covering the most susceptible stage (Diaz and McLeod, 2005).

Application can also be *via* irrigation systems (Fig. 15.5), e.g. for aphid control in vegetables in Spain (Hernandez *et al.*, 1999). However, by far the most commonly used method is foliar spraying, which can be timed for maximum efficacy, in relation to need rather than as a prophylactic.

Thresholds for Control (see also the IPM section of this volume)

Thresholds for control of aphids are useful to encourage rational use of pesticides and discourage prophylaxis. However, the use of thresholds requires a reasonably comprehensive monitoring scheme.

Wheat and barley

In these cereals, there has been a long-standing threshold of five aphids per ear prior to flowering (GS 61) for control of *Sitobion avenae* (grain aphid) (George and Gair, 1979). This threshold still applies, but

Fig. 15.5. Drip irrigation of imidacloprid to outdoor vegetables planting (original colour photos courtesy of Bayer PlantScience).

has been converted to the proportion of tillers infested (75%) (Oakley and Walters, 1994), and is more likely to be adopted by growers because this is quicker to assess than counting aphids on individual tillers. A similar system called EPIPRE was set up in the Netherlands in the late 1970s (reviewed by Ankersmit, 1989; see also Poehling *et al.*, Chapter 25 this volume) and is still operational (Bouma, 2003).

Significant increases in the use of pyrethroid sprays to control the vectors of BYDV in the autumn (usually *R. padi* and/or *S. avenae* in the UK) occurred after an epidemic of the disease in England following the mild winter of 1988/89 (Oakley and Young, 2000). Prior to this, control was often instigated in response to an 'Infectivity Index' (Plumb and Lennon, 1982), which took account of the number of vector aphids migrating (in the Rothamsted Insect Survey suction traps) and the proportion of them carrying virus. The latter was measured by collecting live aphids in the autumn and testing them for virus transmission on indicator test plants (oats), and subsequently by ELISA (see also Harrington *et al.*, Chapter 19 this volume). Unfortunately, in the mild winter of 1988/89, significant spread of virus occurred following a low forecast (Harrington *et al.*, 1994), and the ensuing loss of confidence in the forecasting system resulted in a much greater prophylactic approach to aphid control in winter cereals, encouraged by the low cost of pyrethroids (Oakley and Young, 2000). More recently, the Infectivity Index has been linked to winter weather records in an attempt to model overwintering spread of the virus (Foster *et al.*, 2004).

Pea

Trials in combining peas conducted between 1985 and 1989 suggested that economic yield responses to aphicides could be obtained when control was applied, if 20% or more plants were infested with *A. pisum* and populations were increasing, at any time between flowering and when the pods were fully formed on the fourth truss growth stage

(Lane and Walters, 1991). This was quite complicated, and required crop inspection at least twice to determine whether or not aphid numbers were increasing. Thus, it is not surprising that in 30% of sites examined, this threshold was not used (Walters *et al.*, 1994). At these sites, aphid populations crashed dramatically even in the absence of insecticides. It was suggested that data on the timing of both the colonization of crops and the start of rapid increase of infestation are needed to forecast the need for and timing of insecticide applications.

In vining peas, the growing season is much shorter and, prior to 1994, no specific threshold had been determined. Control was routinely applied irrespective of growth stage or infestation level (Biddle *et al.*, 1994). Subsequent trials showed that a single spray of pirimicarb at either the visible bud stage or at first flower, when the proportion of plants infested was over 50%, could give economic yield responses; later sprays at first pod did not produce significant responses.

Field bean (*Vicia faba*)

Current guidelines for the control of *Aphis fabae* (black bean aphid) in field beans in the UK, based on the work of Way *et al.* (1977), suggest that control would be economically justified if more than 5% of plants were infested. However, more recent varieties may be more tolerant of aphid colonization, as single or multiple applications of pirimicarb at a range of growth stages and infestations did not produce significant yield increases (Parker and Biddle, 1998).

Brassicas

Routine spraying of brassica vegetable crops to control *Brevicoryne brassicae* (cabbage aphid) is widespread in the UK because there are no established thresholds (Blood-Smyth *et al.*, 1992). Supervised treatments (i.e. applied according to need rather than in a programme), in this case a tank mix of pirimicarb and deltamethrin, were applied when more than 10% of plants were infested. Although the number of treatments was reduced from an average of 6.1 on the routinely sprayed plots to 1.6 on the supervised plots, the percentage of marketable produce also declined at four of the seven supervised sites. Most of the damage was caused by various caterpillars, which were controlled by the routine sprays.

By contrast, action thresholds have been developed for control of *B. brassicae* on spring and winter oilseed rape (Ellis *et al.*, 1999). These authors suggest that when more than 10% of the racemes in a field are infested, control may be worthwhile. However, it is stressed that cabbage aphid is a sporadic pest and rarely reaches such levels in crops.

Cotton

Thresholds of 100 aphids per plant have been implemented for carbofuran against mid-season epidemics of cotton aphids in Oklahoma (Karner *et al.*, 1997).

Sugarbeet

The threshold for control of the vectors of virus yellows symptoms in sugarbeet in the UK was established as one green wingless aphid per four plants when a spray-warning scheme was introduced in 1958 (Hull, 1968). The threshold has remained at that level since, although it has been modified to take account of the reduced susceptibility of plants to both aphids and virus infection with plant maturity (Dewar and Smith, 1999). After the 12–14-leaf stage, the threshold reduces to one aphid per plant. After the 16-leaf stage, usually achieved by early July, no further control measures are necessary because the plants become unpalatable to green aphids (Kift *et al.*, 1997). At this stage, *A. fabae* usually becomes more prominent, but does not transmit the most abundant virus, *Beet mild yellowing virus*, and is often controlled by the large number of predators and parasitoids at that time of year. Control measures are therefore not usually recommended.

Top fruit

Few control thresholds exist for aphids with the multitude of other pests that can occur on fruit trees. However, the threshold for *Dysaphis plantaginea* (rosy apple aphid) on apples in Nova Scotia in Canada is 1.5 colonies/m of tree height (Hardman, 1992), though the paper does not make it clear what this sampling unit actually is.

Efficacy

The reasons for controlling aphids are two-fold – to prevent direct feeding damage and/or to prevent or reduce transmission of plant pathogenic viruses. Virus diseases can be spread by relatively few aphids, and control measures to prevent this often need to be applied early. If viruses are not of concern, control to prevent feeding damage is often best applied when a threshold for damage has been exceeded, rather than when aphids are first detected.

Poaceae

Wheat and barley

Aphid control in cereals has two main targets – aphids carrying strains of BYDV, and those causing direct feeding damage. The most important aphids carrying BYDV are *R. padi* and *S. avenae*. Control of the former is the focus of attention in many countries. Direct feeding damage is caused mostly by *S. avenae*, *Rhopalosiphum maidis* (corn leaf aphid), *Metopolophium dirhodum* (rose–grain aphid), *Schizaphis graminum* (greenbug), and *D. noxia*.

CONTROL OF BYDV (SEE ALSO KNIGHT AND THACKRAY, CHAPTER 31 THIS VOLUME). BYDV is most devastating when crops are infected as seedlings. Treatments applied when plants are very young therefore are likely to give the greatest economic return. However, sprays applied to growing crops are often not very persistent, and further sprays may be necessary to give sufficient protection when crops are most vulnerable, especially in spring-sown crops but less so in autumn-sown crops. Treatments in the autumn just need to be active until the autumn migrations of aphids cease, usually in November (in the northern hemisphere).

The advent of seed treatments with enough persistence to cover the susceptible period has offered excellent alternative control strategies, particularly in autumn-sown wheat and barley crops in Europe. Although primary colonization by *S. avenae* or *R. padi* is not prevented, secondary spread of the viruses they may be carrying is reduced significantly by seed treatments such as imidacloprid (Huth, 1992; Knaust and Poehling, 1992, 1995; Gourmet *et al.*, 1996; Hofer *et al.*, 2001; Royer *et al.*, 2005), thiamethoxam (Senn *et al.*, 1998; Hofer *et al.*, 2001), or clothianidin (Meredith *et al.*, 2002), often with consequent increases in yield. It is only the relative cost of such treatments compared to pyrethroid sprays that has limited their uptake by growers.

Even where aphid activity is not halted by the weather, these seed treatments can be effective, replacing at least some sprays in the early growing season. For example, in Kenya the incidence of BYDV was reduced significantly in plots of barley sown with imidacloprid-treated seed and subsequently sprayed with cypermethrin; the main aphids controlled were *M. dirhodum* and *R. padi*, but *R. maidis*, *S. avenae*, *S. graminum*, and *Hysteroneura setariae* were also present (Wangai *et al.*, 2000). Imidacloprid seed dressing, followed by foliar sprays of the pyrethroid alpha-cypermethrin, reduced the incidence of BYDV and increased yields of autumn-sown wheat in Western Australia, largely by controlling *R. padi* (McKirdy and Jones, 1996, 1997), and imidacloprid seed dressings to wheat and oats gave significant control of *R. maidis* and *R. padi* in trials in New York State, USA, reducing the incidence of some strains of BYDV in some years (Gray *et al.*, 1996).

CONTROL OF FEEDING DAMAGE. Prevention of feeding damage in the early growth stages is particularly important in warmer climates, where rapid increases in aphid populations are possible in a very short time. Seed treatments again can offer more persistent protection than short-lived sprays. For example, imidacloprid applied as a seed treatment gave excellent control of both *R. maidis* and *S. graminum* for 6–8 weeks after sowing barley in the Sudan (Ahmed *et al.*, 2001), for up to 100 days on barley in India (Singh and Venkateswarlu, 2000), and for 27–85 days after planting wheat and barley in the USA (Pike *et al.*, 1993), the latter against *D. noxia*. Indeed, imidacloprid gave best control of *D. noxia* in many countries in Africa, where this aphid is especially prominent (Westhuizen *et al.*, 1994; Tolmay *et al.*, 1997; Macharia *et al.*, 1999), and protection often covered the whole season. The seed treatments were regarded as environmentally safer than broad-spectrum foliar sprays of the OPs monocrotophos and dimethoate (Liu *et al.*, 1999).

Maize

Maize is colonized mostly by *M. dirhodum* at an early growth stage, or by *R. padi* or *R. maidis* when plants are more mature. Again, the neonicotinoid seed treatments, imidacloprid, thiamethoxam, or clothianidin, gave significant improvements in yield by reducing feeding damage and/or BYDV infection in Belgium (Haesaert *et al.*, 1998; Proft *et al.*, 1999), Germany (Epperlein *et al.*, 1997), France (Naibo, 1994), and the USA (Ohkawara *et al.*, 2002; Andersch and Schwarz, 2003). However, the long persistence of imidacloprid when applied to maize and sunflowers has been blamed by French and Italian beekeepers for the sharp decline in honey production, especially in the south of the country, and the insecticide has been banned in these crops (Schmuck, 1999; Schnier *et al.*, 2003), despite research having shown no effect on bee mortality or behaviour (Maus *et al.*, 2003; Stadler *et al.*, 2003; Faucon *et al.*, 2005).

Sorghum

The same early control by imidacloprid seed treatments of infestations of *S. graminum* on sorghum, and in some cases reductions in virus infection with sugarcane mosaic virus, was demonstrated in several experiments in the USA (Harvey *et al.*, 1996; Sloderbeck *et al.*, 1996; Wilde *et al.*, 1999).

Soft fruit

The main pest aphids in soft fruits are *Chaetosiphon fragaefolii* (strawberry aphid) in strawberries and *Amphorophora rubi* (Rubus aphid) in European red raspberries (*Rubus idaeus* subsp. *idaeus*) and blackberries (*Rubus fructicosus* agg.). Very little research has been done with the newer insecticides against these aphids. Most control relies on encouraging natural enemies or developing resistant varieties (Easterbrook *et al.*, 1997; Gordon *et al.*, 1997; Meesters *et al.*, 1998; Birch *et al.*, 2003), although neem seed oil was reported to give some control of *C. fragaefolii* on strawberries (Lowery *et al.*, 1993). More recently, sprays of thiacloprid have given very good control of *Hyperomyzus lactucae* (blackcurrant–

sowthistle aphid) and *Aphis schneideri* on blackcurrants in Poland, comparable to that given by acetamiprid (Labanowska, 2004).

Top fruit

Apple and pear

The most important aphids colonizing apples are *D. plantaginea* and *Aphis pomi* (green apple aphid). These two species can occur together, but the former causes greater damage. *Eriosoma lanigerum* (woolly apple aphid) can cause problems in some countries.

Only foliar applications of insecticides (Fig. 15.6), and sometimes topical applications to tree trunks, are available for aphid control in fruit trees. The relative efficacy and persistence of products is determined by the timing of application in relation to petal fall and development of the fruit. Timing is limited by the need to protect pollinators from potential adverse effects.

Fig. 15.6.   High-volume pesticide application to top fruit in Australia (original colour photo courtesy of H.F. van Emden).

In most studies, imidacloprid sprays gave better control of the target aphids than the OPs and pirimicarb that preceded them (Barbieri and Cavallini, 1997; Pollini and Bariselli, 1997; Civolani *et al.*, 2000; Helsen, 2001), especially when the target aphids were resistant to the older insecticides (Pasqualini *et al.*, 1996, 1998, 2001; Delorme *et al.*, 1997; de Fanti, 1997; Pasqualini and Vergnani, 1997; Bylemans, 2000). Other neonicotinoids, such as acetamiprid and thiacloprid, have also been investigated, giving good control of *A. pomi* or *D. plantaginea* in Spain, France, and Germany (Elbert *et al.*, 2000), Poland (Badowska-Czubik *et al.*, 1999), and the Ukraine (Yanovskii and Larcheva, 2000).

The systemic properties of triazamate have been exploited in many studies on fruit trees. For example, triazamate gave good control of *D. plantaginea* on apples and *Dysaphis pyri* (pear–bedstraw aphid) on pears in studies in France (Capou and Renou, 1999), Switzerland, (Schaub *et al.*, 1999), and Israel (Yehuda and Izhar, 1992). However, some populations of aphids are resistant to this carbamoyl product (Delorme *et al.*, 1997), but it is no longer available in Europe following a recent review by the EU.

Imidacloprid as a drench (1 ml a.i. per tree) to the roots of 'Granny Smith' apple trees in South Africa gave good control of subterranean populations of *E. lanigerum*, and also resulted in reduced aerial populations (Pringle, 1998). This method of application has least effect of non-target organisms such as pollinators, but is probably not as fast acting as sprays when pest populations develop quickly.

Citrus fruits

Toxoptera citricidus (tropical citrus aphid) was well controlled by foliar applications of imidacloprid to oranges (*Citrus sinensis*) in China (Zhong and Gan, 2000), but applications of this insecticide or acetamiprid to the trunks of citrus trees gave best control of this aphid in Brazil (Yamamoto *et al.*, 2000) and Spain (Mansanet *et al.*, 1999).

Other aphid species, such as those attacking Japanese apricots (*Prunus mume*), or *Aphis spiraecola* (green citrus aphid), *A. gossypii*, and *Toxoptera aurantii* (black citrus aphid) on clementines (*Citrus reticulata*) in the Pignataro Maggiore area of Italy, were also well controlled by imidacloprid sprays, which generally performed better than pirimicarb (de Biase and Russo, 1996; Xue, 2000). Pymetrozine gave as good control of *M. persicae* on peaches as pirimicarb, but was much more persistent, giving over 90% control for 21 days compared to only 7 days for pirimicarb (Flückiger *et al.*, 1992a).

Plum (Prunus domestica)

The three most important aphid species attacking plums are *Brachycaudus helichrysi* (leaf-curling plum aphid), *Hyalopterus pruni* (mealy plum aphid), and *P. humuli*.

Pirimicarb gave effective control of spring populations of *B. helichrysi* on plums and showed no damaging effects on natural enemies; as a result, the surviving predators kept mid- to late season populations of *P. humuli* below economic thresholds (Hartfield and Campbell, 1996).

Solanaceae

Potato

The most important aphid pests of potatoes are *M. persicae*, *M. euphorbiae*, *Aphis nasturtii* (buckthorn–potato aphid), *Aulacorthum solani* (glasshouse and potato aphid), and *Rhopalosiphoninus latysiphon* (potato root aphid). In some countries, other aphids such as *A. gossypii* and *A. fabae* can also cause problems in this crop (Rongai *et al.*, 1998), although this is relatively rare. Most use of insecticides to control aphids on potatoes is to reduce infection by viruses such as *Potato virus Y* (PVY) or *Potato leaf roll virus* (PLRV); direct feeding damage is relatively rare. Control of PVY is difficult with systemic products because the vector aphids (of which there are many species – see Radcliffe *et al.*, Chapter 26 this volume) can transmit the virus very quickly (Collar *et al.*, 1997) before this type of insecticide can exert an effect; rapid-action insecticides such as pyrethroids, or oils that interfere with probing behaviour, are more effective in preventing PVY spread (Foster *et al.*, 1994; Powell and Hardie, 1994).

Reduction in PLRV incidence is more achievable with the newer neonicotinoids as only feeding aphids, which then become exposed to toxic doses, persistently transmit this virus. Transmission of PLRV was reduced substantially on plants treated with imidacloprid, painted on to potato skins in the laboratory (Woodford, 1992) and, in other laboratory tests, imidacloprid, thiamethoxam, and pymetrozine gave greater reductions in PLRV transmission than older insecticides such as esfenvalerate, methamidophos, or oxamyl (Mowry, 2005); pymetrozine also substantially reduced acquisition of virus from infected plants. In field trials, granular or spray applications gave control of the aphid vectors of PLRV better than or equivalent to aldicarb granules (Meredith and Heatherington, 1992; Woodford and Mann, 1992; Turska and Pawinska, 1995; Boiteau *et al.*, 1997; Boiteau and Singh, 1999).

Resistance to insecticides is a constant threat in the potato crop, particularly with *M. persicae*, so treatments that have different modes of action are likely to be most effective. So far, the neonicotinoid insecticides have proven to be remarkably resilient to the development of resistance in aphids, even when susceptibility varies up to 20-fold (Nauen and Denholm, 2005). Lambda-cyhalothrin and imidacloprid sprays gave very effective control of *A. gossypii* and *M. persicae* that were insensitive to pirimicarb in laboratory studies on seed potatoes (Rongai *et al.*, 1998), but pymetrozine gave over 90% control for 14 days of mixed populations of *M. persicae* and *M. euphorbiae* on potatoes in Italy. By comparison, the persistence of pirimicarb was less than 8 days (Flückiger *et al.*, 1992a). An oil-dispersion formulation of thiacloprid improved control of *M. persicae* and *M. euphorbiae*, including clones that were resistant to other classes of insecticides (Bardsley and Lankford, 2005).

Tobacco

Control of virus diseases such as *Tobacco mosaic virus* is the main objective of aphid control in tobacco since aphids can open the plant to infection when they break the cuticular hairs as they move about (Harris and Bradley, 1973). Although TMV is not transmitted by aphids when probing, wounding allows the virus to enter. However, as with potatoes, many biotypes of both target species, *M. persicae* and *M. persicae nicotianae*, are resistant to OPs and carbamates. Consequently, the neonicotinoids provide alternative control.

In India, an imidacloprid spray gave more persistent control (16 days) of *M. p. nicotianae* on tobacco than acetamiprid (14 days), pymetrozine (12 days), and the OPs, acephate and oxydemeton-methyl (only 8 days) (Sreedhar *et al.*, 1999). Plant hole treatments of imidacloprid or acephate also gave excellent control of this species (Patil *et al.*, 1999), sometimes season long (Ramaprasad *et al.*, 1998). In contrast, oxydemeton methyl sprays gave only 8 days effective control. In Italy, the most effective aphicides controlling *M. persicae* on tobacco were foliar applications of pirimicarb plus endosulfan and imidacloprid; granular formulations of benfuracarb and liquid formulations of azadirachtin gave inadequate control (Sannino, 1997).

Tomato

Triazamate, used experimentally, gave good control of *M. euphorbiae* on glasshouse tomatoes in North Carolina (Walgenbach, 1997). In another study, pymetrozine gave poor control of *M. persicae* on beet but good control on tomatoes, due to the different distribution of this insecticide in chenopodiaceous plants compared to solanaceous plants (Wyss and Bolsinger, 1997).

Aubergine (Solanum melongena)

The neonicotinoids have proved to be more reliable than the older chemicals for controlling aphid pests of aubergines, as with other Solanaceae. Imidacloprid reduced aphid numbers on aubergines in India, increasing seedling height and total leaf chlorophyll (Jarande and Dethe, 1994), while clothianidin sprays gave control of both *A. gossypii* and *M. persicae* for more than 3 weeks without any phytotoxicity (Ohkawara *et al.*, 2002). Even longer persistence (56 days) was given by granules applied to the planting hole. As an alternative, pymetrozine gave as good control of *M. persicae* as pirimicarb for up to 20 days on aubergines in Spain (Flückiger *et al.*, 1992a).

Chenopodiaceae

Sugarbeet

The most important aphids in the various beet crops (*Beta vulgaris*) are *A. fabae*, *M. persicae*, *M. euphorbiae*, and last but not least in some countries (e.g. the USA) *Pemphigus fuscicornis*, a root aphid. Aphid control in beet aims primarily to reduce the spread of virus yellows caused by *Beet yellows virus* (BYV), BMYV (Smith and Hallsworth, 1990) and *Beet chlorosis* virus (Stevens *et al.*, 1994). Aphids feeding, rather than just probing, spread both of these viruses; therefore, insecticides with systemic activity are likely to give the best control. However, not all these aphids (e.g. *A. fabae*) can transmit BMYV (San Román *et al.*, 1995), so it is important to determine the most important vectors before devising control measures.

Imidacloprid applied as a seed treatment gives excellent control of aphids for up to 10 weeks after sowing, resulting in significant reductions in virus yellows infection and in consequent increases in sugar yield (Dewar and Read, 1990; Schmeer *et al.*, 1990; Merkens and Groeneweg, 1991; Muchembled, 1991; Schaufele, 1991; Courbon, 1992; Dewar, 1992; Dewar *et al.*, 1992; Heatherington and Meredith, 1992; Tossens *et al.*, 1992; Wauters, 1993; Bosch and Schaufele, 1994; Kuthe, 1995; Ayala *et al.*, 1996; Wauters and Dewar, 1996; Heltbech *et al.*, 1999; Proft *et al.*, 1999; Barcic *et al.*, 2000; Cioni *et al.*, 2001). In trials comparing control strategies, seed treatment with imidacloprid reduced virus infection to just 7% compared to 12% with two spray applications of pirimicarb or triazamate, and 25% with aldicarb granules; infection in untreated plots was 70% (Dewar *et al.*, 1996). The persistence of the seed treatment was the main factor in such good control. More recent trials showed that clothianidin and thiamethoxam gave as good or better control of aphids and virus yellows than imidacloprid, reflecting their reported greater activity (Dewar *et al.*, 2001, 2002; Meredith *et al.*, 2002; Meredith and Morris, 2003).

In other trials with spray treatments only, triazamate gave good control of aphids compared to pirimicarb alone or mixed with deltamethrin (Dewar *et al.*, 1996; Dewar and Haylock, 1997). The persistence of triazamate (at least 6 days) was much greater than pirimicarb against both *M. persicae* and *A. fabae* (Westwood *et al.*, 1997). However, when *M. persicae* with the MACE resistance mechanism were present, only imidacloprid as a seed treatment gave effective control (Dewar *et al.*, 1998) (see Foster *et al.*, Chapter 10 this volume).

CONTROL OF FEEDING DAMAGE IN BEET. Although *A. fabae* transmits only one of the three yellowing viruses – the relatively uncommon BYV (San Román *et al.*, 1995) – populations can become quite high in beet crops in some years, causing substantial direct damage. Often, the crops can compensate once the aphid populations have crashed due to predation, parasitization, and disease in late July/early August (in the northern hemisphere), so control measures are often unnecessary. However, some yield losses from this species have been recorded in Sweden and in glasshouse trials in The Netherlands (Hurej and van der Werf, 1993a,b), illustrating the potential for damage. The neonicotinoid seed treatments and carbamate granules do not give good control of *A. fabae* as their effect has usually worn off when the main migrations to beet occur in early July (Dewar *et al.*, 2002). Control then, if necessary, must rely on sprays, of which pirimicarb and triazamate give the best control while also preserving some natural enemies. In Spain, chlorpyriphos-methyl plus cypermethrin, oxydemeton-methyl, imidacloprid, and thiometon were the most efficient at controlling very high populations of *A. fabae* (Ayala *et al.*, 1996).

Control of root aphids is particularly difficult with conventional insecticides. However, triazamate is downwardly translocated and so resulted in good control of *P. fuscicornis* on sugarbeet roots in Greece (Ioannidis, 1996).

Leguminosae

Aphis fabae and *A. pisum* are the most important aphid pests of beans and peas. Unfortunately, because these legumes are regarded as minor crops in a world context, there is the problem in many vegetable crops that the aphicides approved for use tend to be the older OPs, carbamates and pyrethroids. The newer neonicotinoids have not yet been introduced on any significant scale. Nevertheless, experiments have been conducted on some legume crops to assess the efficacy of potential products.

Imidacloprid applied as a seed dressing protected broad bean (*Vicia faba*) plants throughout the whole season against *A. fabae* in Poland (Narkiewicz-Jodko and Rogowska, 1999). In contrast, *Macrosiphum albifrons* (lupin aphid), *Acyrthosiphon kondoi* (blue-green alfalfa aphid), and *Aphis craccivora* (cowpea aphid) on narrow-leaved lupin (*Lupinus angustifolius*) in Western Australia were controlled most effectively by alpha-cypermethrin, while *M. persicae* was best controlled by imidacloprid sprays (Thackray *et al.*, 2000). The greatest reduction of the non-persistent *Cucumber mosaic virus* in lupins was given by the contact activity of alpha-cypermethrin; the systemic insecticides tested, including imidacloprid, methamidophos, and triazamate, were ineffective.

In India, a foliar spray of imidacloprid gave better control of *A. craccivora* and thrips on groundnuts (*Arachis hypogaea*) than the OP insecticides, dimethoate and demeton-S-methyl (Babu and Santharam, 2001).

In Korea, *A. solani* and *Aphis glycines* (soybean aphid) were well controlled by imidacloprid, applied as a granule, for 52 days after sowing of soybean (*Glycine max*) crops, with consequent reductions in the incidence of *Soybean mosaic virus*. Benfuracarb and acephate also gave good control (Kim *et al.*, 2000).

Brassicas

The main aphid species that damage vegetable brassicas are *M. persicae*, *B. brassicae*, and *Lipaphis pseudobrassicae* (mustard aphid). The newer insecticides, as with the legumes, have not yet been introduced on a large scale for aphid control, but some studies on their control potential have been carried out using different formulations and application methods.

For example, thiacloprid sprayed on cabbage foliage at 50 mg a.i./l gave complete control of *M. persicae* for 18 days in glasshouse trials (Elbert *et al.*, 2000). Two applications of some doses of triazamate, lambda-cyhalothrin, imidacloprid, and thiamethoxam suppressed populations of *L. pseudobrassicae* on cabbage crops for the entire season in Texas (Liu *et al.*, 2001). In India, one spray of imidacloprid was sufficient to give 99% control, although root dips of both imidacloprid and diafenthiuron were ineffective (Zhu *et al.*, 1996; Sreekanth and Babu, 2001). Acetamiprid gave the highest mortality of *L. pseudobrassicae* in trials on mustard in India (Chinnabbai *et al.*, 1999), and imidacloprid and pirimicarb sprays gave at least 90% control of *M. persicae* and *L. pseudobrassicae* on cabbage for up to 17 days in Chinese trials (Zhu *et al.*, 1996). Pymetrozine applied at half field rate (0.0106 g a.i./ha) gave 99.8% control of *B. brassicae* on broccoli in laboratory studies, prompting a recommendation to reduce the rate even further to allow integrated management, including that given by natural enemies (Acheampong and Stark, 2004).

With other application methods, granules of imidacloprid controlled *M. persicae* and *B. brassicae* on Brussels sprout plants for 12 weeks after sowing (Schoonejans *et al.*, 1991), while drenches to cauliflower root plugs gave as good control of *M. persicae* and *B. brassicae* as in-furrow applications in Californian trials (Natwick *et al.*, 1996). Granules of 2% acetamiprid in the planting hole gave excellent control of *M. persicae* on cabbage (Takahashi *et al.*, 1992), and imidacloprid, applied as a seed treatment in a formulated mix with beta-cyfluthrin to control flea beetles in oilseed rape, also controlled virus-carrying aphids (probably *M. persicae*), reduced the incidence of *Beet western yellows virus*, and increased yield by up to 16% (Birch and Nicholson, 2001).

Lettuce (*Lactuca sativa*)

Aphids are the most serious pest of outdoor lettuce in Europe (Ester *et al.*, 1993; Ellis *et al.*, 1996; Martin *et al.*, 1996; Monnet and Ricateau, 1997) and North America (Alleyne and Morrison, 1977; Forbes and Mackenzie, 1982; Toscano *et al.*, 1990). The presence of any aphid on this high-value crop is not tolerated in today's supermarket culture, though small numbers have no effect on yield. The public's aversion to insect presence has resulted in very rigorous control measures (Parker *et al.*, 2002). In addition, lettuce is subject to several yield-reducing aphid-transmitted virus diseases such as *Lettuce mosaic virus* (see Tatchell, Chapter 24 this volume).

The aphid species that colonize lettuce include *Pemphigus bursarius* (lettuce root aphid), *Nasonovia ribisnigri* (currant–lettuce aphid), *M. euphorbiae*, and the ubiquitous *M. persicae* (Reinink and Dieleman, 1993; Collier *et al.*, 1999). The latter three species can be controlled by foliar sprays of aphicides such as pirimicarb, demeton-s-methyl, pymetrozine, and triazamate (Parker and Blood-Smyth, 1996), although *N. ribisnigri* is particularly difficult to control as it resides in the heartleaves. *Pemphigus bursarius* is also difficult to control as it lives on the roots, and none of the approved sprays is downwardly translocated.

More recently, the neonicotinoids (especially imidacloprid or thiamethoxam) have been used. Applied as seed treatments, they have given aphid control for more than 7 weeks after sowing (Martin *et al.*, 1996; Ester and Brantjes, 1998, 1999). Imidacloprid, applied in the seed furrow 7.6 cm below the seed, provided control of aphids on lettuce in Arizona for up to 100 days after sowing, resulting in 90% marketable heads compared to just 20% from untreated plots (Palumbo and Kerns, 1994). Foliar sprays of acetamiprid, imidacloprid, pymetrozine, and pirimicarb gave good control 5 days after application against *N. ribisnigri* in Italy, but only a mixture of imidacloprid and cyfluthrin gave longer persistence (up to 19 days) (Gengotti, 2001).

Imidacloprid has given effective control of light infestations of *P. bursarius* (Parker and Blood-Smyth, 1996). However, trials have shown downwardly translocated triazamate sprays gave better control (Parker *et al.*, 2002). In the absence of either seed treatments or triazamate sprays, as many as 16 pirimicarb sprays were necessary to reduce infestation of plants by both foliar- and root-infesting aphids to commercially acceptable levels in these trials. Use of triazamate reduced the number of applications by at least half. Unfortunately, this highly efficacious insecticide has been withdrawn from the market in Europe.

Other vegetables

Registration of novel insecticides in many countries these days is so expensive that many small-area vegetable crops are not target crops for major manufacturers. Eventual use of new insecticides on these crops is often undertaken under an 'off-label' approval, the cost of which is borne by the users (the growers or trade organizations) rather than by the manufacturers. Nevertheless, the novel insecticides often give superior control of target pests than the older alternatives and, due to ever more stringent safety requirements, are often much safer for the consumer. In this economic climate, studies on the efficacy of novel insecticides are relatively rare; a few are mentioned here.

Cavariella aegopodii (willow–carrot aphid) was better and more rapidly controlled by lambda-cyhalothrin than imidacloprid in laboratory studies in The Netherlands (Vercruysse *et al.*, 2000). However, imidacloprid and a formulated mixture of lambda-cyhalothrin with pirimicarb gave adequate control of all aphid vectors of carrot motley dwarf (a joint infection by *Carrot mottle virus* and *Carrot red leaf virus*) in parsley (*Petroselinum crispum*) in field trials. The root aphid *Pemphigus phenax* was well controlled by imidacloprid, either as a seed treatment or foliar spray to carrots in Poland (Szwejda, 1997, 1999). Imidacloprid as a spray or *via* a drip irrigation system gave 'outstanding' control of aphids in vegetable crops in Spain (Hernandez *et al.*, 1999). The latter application method had no effect on bees and other beneficial insects present in the crops. Imidacloprid, applied to glasshouse sweet and sharp pepper plants (*Capsicum* spp.) as a spray or *via* a hydroponic system, eradicated *M. persicae* for at least 8 weeks, depending on dose rate (Vantornhout *et al.*, 1999). Thiacloprid gave best control of aphids on seedlings of chilli (*Capsicum annuum*) up to 14 days after each spray (Walunj and Pawar, 2004). Another neonicotinoid, acetamiprid (NI-25), showed residual activity against *A. gossypii* on glasshouse cucumbers for 2 weeks, compared to only 7 days for acephate (Takahashi *et al.*, 1992).

Malvaceae

Cotton

The main pest aphid on cotton is *A. gossypii.* Though this is not usually the main target of insecticide applications in cotton, the advent of genetically modified cotton varieties with built-in resistance to caterpillar pests ('Bt varieties') may increase its importance in the future. The exploitation of the novel insecticides in this crop has been investigated around the world.

In many countries, imidacloprid, thiamethoxam, and diafenthiuron have shown good efficacy against cotton aphids as either a seed treatment (Almand, 1996; Careme *et al.*, 1996; Lucas *et al.*, 1999; dos Santos and dos Santos; 1999; Scarpellini and Nakamura, 1999a,b; Zang *et al.*, 1999) or sprays (Almand, 1996; Furr and Harris, 1996; Hopkins and Donaldson, 1996; Layton *et al.*, 1996; Wright *et al.*, 1997; d'Albuquerque

et al., 1999; Calafiori *et al.*, 1999; Kharboutli and Allen, 2000; Koenig *et al.*, 2000), especially when mixed with a pyrethroid (Hopkins *et al.*, 2000). Some discrepancies in the reported efficacy of imidacloprid against *A. gossypii* were attributed to its feeding behaviour; it can avoid the internal glands of the cotton plants when it penetrates the leaf in its search for phloem. This means that it sometimes avoids lethal doses of insecticide, and is not controlled quickly enough (Nauen and Elbert, 1994).

Of the other neonicotinoids, thiacloprid gave much better control than methamidophos of *A. gossypii* on cotton in USA trials (Elbert *et al.*, 2000), while acetamiprid applied alone or mixed with endosulfan gave efficient control of *A. gossypii* for up to 8 days compared to only 5 days for endosulfan plus thiamethoxam (Franco, 1999).

Pymetrozine has also been tested for its potential against cotton aphids, especially in the USA (Koenig *et al.*, 2000). Pymetrozine sprays followed by cyhalothrin effectively reduced aphid numbers in unirrigated or early irrigated cotton in Texas, but was less effective in late irrigated plots (Slosser *et al.*, 2001). Over 80% of *A. craccivora* were controlled by pymetrozine in cotton trials in Egypt (Flückiger *et al.*, 1992a).

Some evidence of phenotypic variability in aphid susceptibility to insecticides was reported in California. Here, *A. gossypii* reared on cotton under early season temperature and photoperiod conditions was less susceptible to bifenthrin, chlorpyriphos, and triazamate than those reared under late season conditions (Godfrey and Fuson, 2001). The same species reared on melon plants was more susceptible to bifenthrin and chlorpyrifos than genetically identical clones reared on cotton. However, in leaf dip bioassays, diafenthiuron gave slower control of *A. gossypii* than conventional OPs and pyrethroids, but there was no evidence of cross-resistance in strains that were resistant to the latter (Denholm *et al.*, 1995).

Okra (Abelmoschus esculentus)

Imidacloprid seed treatment of okra increased plant height, leaf area, and yield by giving effective control of aphids and jassids (Sreelatha Divakar, 1997). It was also the most effective spray treatment against *A. gossypii* in India when applied twice at 0.006% at a 15-day interval (Sunitha *et al.*, 2005).

Others

Various methods of applying the novel insecticides have shown good efficacy in some very minor crops, and especially against *P. humuli* on its secondary host, hop. Imidacloprid applied to hops through a drip system gave control of *P. humuli* for up to 2 years compared to only 1 year with the OP disulfoton (Wright and Cone, 1999). In another study in Poland, brushing imidacloprid onto plants at a low dose was as effective as spraying against *P. humuli* at higher doses (Solarska and Jastrzebski, 1998). Imidacloprid sprays in the Czech Republic gave 100% control of *P. humuli* compared to only 81% with cyfluthrin, due to the development of resistance to the latter (Vostrel, 1998). Surface granule applications to the plant crowns in April successfully controlled *P. humuli* in England; also, a spray to the base of the bines in May was as effective as the standard mephosfolan (Martin *et al.*, 1992).

Good control of *Adelges (Adelges) laricis* (larch adelgid) and *Adelges (Sacchiphantes) abietis* (spruce pineapple-gall adelgid) on trees was achieved in laboratory tests in Poland using sprays of acetamiprid, imidacloprid, and other older insecticides (Labanowski and Soika, 1999). Imidacloprid gave as good control as autumn treatments of insecticidal soap or horticultural oil of *A. (Aphrastasia) funitecta* (hemlock woolly adelgid) on *Tsuga canadensis* and *Tsuga caroliniana* in eastern USA; the effects lasted for at least 1 year (Rhea, 1996). In later experiments, microinjections of imidacloprid into the active sapwood gave good control (Doccola *et al.*, 2003). Injection (Fig. 15.7), implantation, or stem painting (Fig. 15.8) of amenity and forest trees with imidacloprid and methamidophos gave

worthwhile control of some insect pests, including *Aphis grossulariae* (gooseberry aphid) (Scholz and Wulf, 1998). Of a range of insecticides tested, foliar applications of omethoate, dimethoate plus chlorpyriphos, vamidothion, imidacloprid, and soil applications of disulfoton gave best control of *Myzocallis coryli* (hazel aphid) on hazelnut (*Gevuina avellana*) in Chile (Aquilera *et al.*, 1996). In preliminary tests on pecan trees (*Carya illinoinensis*) in Cyprus, imidacloprid showed promising results against *Monellia caryella* (black-margined yellow pecan aphid) (Krambias, 1994). Best control of *Phyllaphis fagi* (woolly beech aphid) was given by imidacloprid sprays in German trials comparing 48 insecticides in a tree nursery (Losing, 1993).

Monocrotophos, imidacloprid, and dimethoate continued to give complete control of *Pentalonia nigronervosa* (banana aphid) for 40 days in New South Wales, Australia (Treverrow, 1996).

Tinocallis kahawaluokalani (crape myrtle aphid) was prevented from causing serious damage to crape myrtle (*Lagerstroemia indica* – a landscape and nursery plant) by acephate, imidacloprid, and thiamethoxam in studies in the USA (Pettis *et al.*, 2005).

Conclusions

The battle to control aphids will continue as long as these pests compete with mankind for food resources. The new generation of neonicotinoids, and other insecticides with novel modes of action such as pymetrozine, will provide more effective control as their potential is examined in a wider spectrum of crops in the future. Their application methods will achieve less exposure to potentially hazardous chemicals of non-target organisms and operators than was usual with older insecticides from the last century. In future, it is likely that genetic modification of plants to resist pests, including aphids, will provide yet more options for control. Certainly, it is now possible to devise useful integrated pest management

Fig. 15.8. Injecting imidacloprid into banana stems in the Philippines (original colour photos courtesy of Bayer PlantScience).

Fig. 15.7. Application of imidacloprid directly to the trunks of citrus in South Africa (original colour photos courtesy of Bayer PlantScience).

protocols for aphid problems. However, such protocols may often increase prophylactic use of insecticides, because their application as seed treatments or at sowing comes before an assessment of risk can be made.

References

Abraham, K. and Epperlein, K. (1999) Laboruntersuchungen mittels Electrical Penetration Graph zum Einfluss von Imidacloprid als Saatgutbehandlungsmittel am Mais auf das Saugverhalten der Traubenkirschenlaus (*Rhopalosiphum padi* L.) und die Virusübertragung von BYDV. *Gesunde Pflanzen* 51, 90–94.

Acheampong, S. and Stark, J.D. (2004) Can reduced rates of pymetrozine and natural enemies control the cabbage aphid, *Brevicoryne brassicae* (Homoptera: Aphididae), on broccoli? *International Journal of Pest Management* 50, 275–279.

Ahmed, N.E., Kanan, H.O., Inanaga, S., Ma, Y.Q. and Sugimoto, Y. (2001) Impact of pesticide seed treatments on aphid control and yield of wheat in the Sudan. *Crop Protection* 20, 929–934.

d'Albuquerque, F.A., Ros, A.B., Mendes, S.C. and Weber, L.F. (1999) Control of cotton aphid *Aphis gossypii* (Glover, 1877) (Hemiptera: Aphididae) using different insecticides as sprays. *Anais 20 Congresso Brasileiro de Algodao: O Algodao no Seculo 2⁰, Perspectivar para o Seculo 21, Ribeirao Preto, San Paulo, September 1999*, 233–235.

Alleyne, E.H. and Morrison, F.O. (1977) The lettuce root aphid, *Pemphigus bursarius* (L.) (Homoptera: Aphidoidea) in Quebec, Canada. *Annals of the Entomological Society of Quebec* 22, 171–180.

Almand, L.K. (1996) The importance of Provado for earliness management in cotton. In: *Proceedings Beltwide Cotton Conferences, Nashville, January 1996, Volume 2*, pp. 948–954.

Altmann, R. and Elbert, A. (1992) Imidacloprid – ein Neues Insektizid für die Saatgutbehandlung in Zuckerrüben, Getreide und Mais. *Mitteilungen der Deutschen Gesellschaft für Allgemeine und Angewandte Entomologie* 8, 212–221.

Andersch, W. and Schwarz, M. (2003) Clothianidin seed treatment (Poncho) – the new technology for control of corn rootworms and secondary pests in US corn production. *Pflanzenschutz-Nachrichten Bayer* 56, 147–172.

Ankersmit, G.W. (1989) Integrated control of cereal aphids. In: Minks, A.K. and Harrewijn, P. (eds) *Aphids. Their Biology, Natural Enemies and Control, Volume 2 C.* Elsevier, Amsterdam, pp. 273–278.

Anonymous (1993) *Evaluation on Imidacloprid. Disclosure Document for Imidacloprid in the Products 'Gaucho' and 'Zelmone'.* Advisory Committee on Pesticides Issue No. 118. Stationery Office, London, 233 pp.

Aquilera, P.A., Pacheco, V.C. and Guerrero, C.J. (1996) Eficacia de diez insecticidas aplicados al follaje y uno al suelo, en avellano europeo, para el control de *Myzocallis coryli* (Goeze) (Hemiptera: Homoptera: Aphidoidea). *Agricultura Técnica, Santiago* 56, 183–186.

Ayala, J., Pérez de San Román, C., Ortiz, A. and Juanche, J. (1996) Control químico de *Myzus persicae* (Sulz.) y *Aphis fabae* (Scop.) (Homoptera: Aphididae) en remolacha azucarera mediante aplicación de aficidas en siembra y foliares. *Boletín de Sanidad Vegetal, Plagas* 22, 731–740.

Babu, K.R. and Santharam, G. (2001) Bioefficacy of imidacloprid against thrips and aphids on groundnut, *Arachis hypogaea* L. *Madras Agricultural Journal* (2000) 871, 605–608.

Badowska-Czubik, T., Pala, E. and Olszak, R.W. (1999) Ocena skuteczności wybranych insektycydów w zwalczaniu mszyc (Homoptera: Aphidoidea) na jabłoni. *Zeszyty Naukowe Instytutu Sadownictwa i Kwiaciarstwa w Skierniewicach* 6, 53–64.

Barbieri, R. and Cavallini, G. (1997) L'Imidacloprid nella difesa del melo da afidi. *L'Informatore Agrario* 53 (supplement), 18.

Barcic, J.I., Dobrincic, R., Sarec, V. and Kristek, A. (2000) Studies on insecticide seed dressing of sugarbeet. *Poljoprivredna Znanstvena Smotra, Agriculturae Conspectus Scientificus* 65, 89–97.

Bardsley, E. and Lankford, B. (2005) The development of a new aphicide for use in potatoes. In: Champion, G., Dale, M.F.B., Jaggard, K., Parker, W.E., Pickup, J. and Stevens, M. (eds) *Production and Protection of Sugar Beet and Potatoes. Aspects of Applied Biology, No. 76*, pp. 175–180.

Bedford, I.D., Kelly, A., Banks, G.K., Fuog, D. and Markham, P.G. (1998) The effect of pymetrozine, a feeding inhibitor of Homoptera, in preventing transmission of cauliflower mosaic caulimovirus by the aphid species *Myzus persicae* (Sulzer). *Annals of Applied Biology* 132, 453–462.

de Biase, L.M. and Russo, L.F. (1996) Prove di lotta contro gli afidi degli agrumi in agro di Pignataro Maggiore (Caserta). *Informatore Fitopatologico* 46 (11), 60–61.

Biddle, A.J., Blood-Smyth, J.A. and Talbot, G. (1994) Determination of pea aphid thresholds in vining peas. *Proceedings of the Brighton Crop Protection Conference, Pests and Diseases, November 1994* 2, 713–716.

Birch, A.N.E., Jones, A.T., Fenton, B., Malloch, G., Geoghegan, I., Gordon, S.C., Hillier, J. and Begg, G. (2003) Resistance-breaking raspberry aphid biotypes: a challenge for plant breeding. *Bulletin OILB/SROP* 26 (2), 51–54.

Birch, P.A. and Nicholson, T. (2001) A new insecticidal seed treatment for oilseed rape. In: *Seed Treatment: Challenges and Opportunities. Proceedings of the 2001 British Crop Council Protection Symposium No. 76.* British Crop Protection Council, Farnham, pp. 27–32.

Blood-Smyth, J.A., Emmett, B.J. and Mead, A. (1992) Supervised control of foliar pests in brassica crops. *Proceedings of the Brighton Crop Protection Conference, Pests and Diseases, 1992* 3, 1015–1070.

Boiteau, G. and Singh, R.P. (1999) Field assessment of imidacloprid to reduce the spread of PVY and PLRV in potato. *American Journal of Potato Research* 76, 31–36.

Boiteau, G., Osborn, W.P.L. and Drew, M.E. (1997) Residual activity of imidacloprid controlling Colorado potato beetle (Coleoptera: Chrysomelidae) and three species of potato colonizing aphids (Homoptera: Aphidae). *Journal of Economic Entomology* 90, 309–319.

Bosch, U. and Schaufele, W.R. (1994) Zur Bekämpfung von Schädlingen der Zuckerrübe durch Saatgutbehandlung mit Imidacloprid. *Gesunde Pflanzen* 46, 44–55.

Bouma, E. (2003) Decision support systems used in the Netherlands for reduction of input of active substances in agriculture. *Bulletin OEPP* 33, 461–466.

Brandl, F. (2001) Seed treatment technologies: evolving to achieve crop genetic potential. In: *Seed Treatment: Challenges and Opportunities. Proceedings of the 2001 British Crop Council Protection Symposium No. 76.* British Crop Protection Council, Farnham, pp. 3–18.

Buchholz, A. and Nauen, R. (2002) Translocation and translaminar bioavailability of two neonicotinoid insecticides after foliar application to cabbage and cotton. *Pest Management Science* 58, 10–16.

Burd, J.D., Elliott, N.C. and Reed, D.K. (1996) Effects of the aphicides 'Gaucho' and CGA-215'944 on feeding behaviour and tritrophic interactions of Russian wheat aphids. *Southwestern Entomologist* 21, 145–152.

Bylemans, D. (2000) Recent experiences and opinions on rosy apple aphid control in IPM managed orchards. *Acta Horticulturae* 525, 291–297.

Calafiori, M.H., Barbieri, A.A. and de Salvo, S. (1999) Eficiência de inseticidas no controle de tripes, *Thrips tabaci* (Linderman, 1876) e pulgão, *Aphis gossypii* Glover 1888, em algodoeiro, *Gossypium hirsutum* L. *Anais 2⁰ Congresso Brasileiro de Algodão: O Algodão no Século 20, Perspectivas para o Século 21, Ribeirão Preto, São Paulo, September 1999*, 208–211.

Capou, J. and Renou, C. (1999) Le triazamate: une nouvelle matière active dans la lutte contre le puceron cendré du pommier (*Dysaphis plantaginea*) et le puceron mauve du poirier (*Dysaphis piri*). *Annales de la 5e ANPP Conférence Internationale sur les Ravageurs en Agriculture, Montpellier, December 1999* 2, 261–268.

Careme, C., Mergeai, G., Ydraiou, F. and Schiffers, B.C. (1996) Comparaison des effets de différents types de traitement phytosanitaire des semences du cotonnier au Burundi et en Grèce. *Tropicultura* 14, 45–53.

Casida, J.E. and Quistad, G.B. (1998) Golden age of insecticide research: past, present, or future. *Annual Review of Entomology* 43, 1–16.

Chinnabbai, C.H., Devi, C.H.R. and Venkataiah, M. (1999) Bio-efficacy of some new insecticides against the mustard aphid, *Lipaphis erysimi* (Kalt.) (Aphididae, Homoptera). *Pest Management and Economic Zoology* 7, 47–50.

Cioni, F., Tugnoli, V., Giunchedi, L. and Pollini, A. (2001) Attività dei geodisinfestanti della barbabietola da zucchero. *L'Informatore Agrario* 57 (4), 57–59.

Civolani, S., Vergnani, S., Ardizzoni, M., Cavazza, C. and Pasqualini, E. (2000) Valutazione di alcuni aficidi impiegati nella difesa del melo. *L'Informatore Agrario* 56 (13), 62–64.

Collar, J.L., Avilla, C., Duque, M. and Fereres, A. (1997) Behavioral response and virus vector ability of *Myzus persicae* (Homoptera: Aphididae) probing on pepper plants treated with aphicides. *Journal of Economic Entomology* 90, 1628–1634.

Collier, R.H., Tatchell, G.M., Ellis, P.R. and Parker, W.E. (1999) Strategies for the control of aphid pests of lettuce. *Bulletin OILB/SROP* 22 (5), 25–35.

Courbon, R. (1992) Potentialités de l'imidaclopride. *Phytoma* 441, 25–27.

Delorme, R., Auge, D., Touton, P. and Villatte, F. (1997) Résistance de *Dysaphis plantaginea* à divers produits insecticides en France. *Annales de la 4e ANPP Conférence Internationale sur les Ravageurs en Agriculture, Montpellier, January 1997* 1, 45–52.

Denholm, I., Rollett, A.J., Cahill, M.R. and Ernst, G.H. (1995) Response of cotton aphids and whiteflies to diafenthiuron and pymetrozine in laboratory bioassays. In: *Proceedings Beltwide Cotton Conferences, San Antonio, January 1995, Volume 2.* National Cotton Council, Memphis, pp. 991–994.

Dewar, A.M. (1992) The effects of imidacloprid on aphids and virus yellows in sugar beet. *Pflanzenschutz-Nachrichten Bayer* 45, 423–442.

Dewar, A.M. and Haylock, L. (1997) New insecticide sprays for sugar beet. *British Sugar Beet Review* 65, 37–39.

Dewar, A.M. and Read, L.A. (1990) Evaluation of an insecticidal seed treatment, imidacloprid, for controlling aphids on sugar beet. *Proceedings of the Brighton Crop Protection Conference, Pests and Diseases, November 1990* 2, 721–726.

Dewar, A.M. and Smith, H.G. (1999) Forty years of forecasting virus yellows incidence in sugar beet. In: Smith, H.G. and Barker, H. (eds) *The Luteoviridae.* CAB International, Wallingford, pp. 228–243.

Dewar, A.M., Read, L.A., Hallsworth, P.B. and Smith, H.G. (1992) Effect of imidacloprid on transmission of viruses by aphids in sugar beet. *Proceedings of the Brighton Crop Protection Conference, Pests and Diseases, November 1992* 2, 563–568.

Dewar, A.M., Haylock, L.A., Chapman, J., Devine, G.J., Harling, Z. and Devonshire, A.L. (1994) Effect of triazamate on resistant *Myzus persicae* on sugarbeet under field cages. *Proceedings of the Brighton Crop Protection Conference, Pests and Diseases, November 1994* 1, 407–412.

Dewar, A.M., Haylock, L.A. and Ecclestone, P.M.J. (1996) Strategies for controlling aphids and virus yellows in sugarbeet. *Proceedings of the Brighton Crop Protection Conference, Pests and Diseases, November 1996* 1, 185–190.

Dewar, A.M., Haylock, L.A., Campbell, J., Harling, Z., Foster, S.P. and Devonshire, A.L. (1998) Control in sugarbeet of *Myzus persicae* with different insecticide-resistance mechanisms. In: Dale, M.F.B., Dewar, A.M., Fisher, S.J., Haydock, P.P.J., Jaggard, K.W., May, M.J., Smith, H.G., Storey, R.M.J. and Wiltshire, J.J.J. (eds) *Protection and Production of Sugar Beet and Potatoes. Aspects of Applied Biology, No. 52*, pp. 407–414.

Dewar, A.M., Haylock, L.A., Bean, K.M., Garner, B.H. and Sands, R.J.N. (2001) Novel seed treatments to control aphids and virus yellows in sugar beet. In: *Seed Treatment: Challenges and Opportunities. Proceedings of the 2001 British Crop Council Protection Symposium No. 76.* British Crop Protection Council, Farnham, pp. 33–40.

Dewar, A.M., Haylock, L., Garner, B.H., Baker, P. and Sands, R.J.N. (2002) The effect of clothianidin on aphids and virus yellows in sugar beet. *Proceedings of the British Crop Protection Council Conference, Pests and Diseases, Brighton, November 2002* 2, 647–652.

Diaz, F.J. and McLeod, P. (2005) Movement, toxicity, and persistence of imidacloprid in seedling Tabasco pepper infested with *Myzus persicae* (Hemiptera: Aphididae). *Journal of Economic Entomology* 98, 2095–2099.

Doccola, J.J., Wild, P.M., Ramasamy, I., Castillo, P. and Taylor, C. (2003) Efficacy of Arborjet VIPER microinjections in the management of hemlock woolly adelgid. *Journal of Arboriculture* 29, 327–330.

Easterbrook, M.A., Crook, A.M.E., Cross, J.V. and Simpson, D.W. (1997) Progress towards integrated pest management on strawberry in the United Kingdom. *Acta Horticulturae* 439, 899–904.

Elbert, A., Overbeck, H., Iwaya, K. and Tsuboi, S. (1990) Imidacloprid, a novel systemic nitromethylene analogue insecticide for crop protection. *Proceedings of the Brighton Crop Protection Conference, Pests and Diseases, November 1990* 1, 21–28.

Elbert, A., Erdelen, C., Kuhnhold, J., Nauen, R., Schmidt, H.W. and Hattori, Y. (2000) Thiacloprid, a novel neonicotinoid insecticide for foliar application. *Proceedings of the British Crop Protection Council Conference, Pests and Diseases, Brighton, November 2000* 1, 21–26.

Elbert, A., Ebbinghaus, D., Maeyer, L., de Nauen, R., Comparini, S., Pitta, L. and Brinkmann, R. (2002) CalypsoReg., a new foliar insecticide for berry fruit. *Acta Horticulturae* 585, 337–341.

Ellis, P.R., Tatchell, G.M., Collier, R.H. and Parker, W.E. (1996) Assessment of several components that could be used in an integrated programme for controlling aphids on field crops of lettuce. In: *Integrated Control of Field Vegetable Pests. IOBC/WRPS* 19, 91–97.

Ellis, S.A., Oakley, J.N., Parker, W.E. and Raw, K. (1999) The development of an action threshold for cabbage aphid (*Brevicoryne brassicae*) in oilseed rape in the UK. *Annals of Applied Biology* 134, 153–162.

Epperlein, K., Schmidt, H.W. and Schwalbe, R. (1997) Untersuchungen zum Einfluss von Gaucho(R) als Saatgutbeize am Mais auf das Virusauftreten, Schadinsekten und die epigäische Bodenfauna. *Archives of Phytopathology and Plant Protection* 31, 185–200.

Ester, A. and Brantjes, N.B.M. (1998) Pelleting the seed of iceberg lettuce (*Lactuca sativa* L.) and butterhead lettuce (*Lactuca sativa* L. var. *capitata* L.) with imidacloprid to control aphids. *Mededelingen Faculteit Landbouwkundige en Toegepaste Biologische Wetenschappen Universiteit Gent* 63, 563–570.

Ester, A. and Brantjes, N.B.M. (1999) Controlling aphids in iceberg lettuce by pelleting the seeds with insecticides. *Mededelingen Faculteit Landbouwkundige en Toegepaste Biologische Wetenschappen Universiteit Gent* 64, 3–10.

Ester, A., Gut, J., van Oosten, A.M. and Pijnenburg, H.C.H. (1993) Controlling aphids in iceburg lettuce by alarm pheromone in combination with an insecticide. *Journal of Applied Entomology* 115, 432–440.

de Fanti, L. (1997) Esperienze di lotta contro l'afide grigio del melo. *Informatore Agrario Supplemento* 53 (11), 12–13.

Faucon, J.P., Aurieres, C., Drajnudel, P., Mathieu, L., Ribiere, M., Martel, A.C., Zeggane, S., Chauzat, M.P. and Aubert, M.F.A. (2005) Experimental study on the toxicity of imidacloprid given in syrup to honey bee (*Apis mellifera*) colonies. *Pest Management Science* 61, 111–125.

Fischer, W. and Widmer, H. (2001) Chemodynamic behaviour of the new insecticide thiamethoxam as seed treatment. In: *Seed Treatment: Challenges and Opportunities. Proceedings of the 2001 British Crop Council Protection Symposium No. 76.* British Crop Protection Council, Farnham, pp. 203–208.

Flückiger, C.R., Kristinsson, H., Senn, R., Rindlisbacher, A., Bulholzer, H. and Voss, G. (1992a) CGA 215'944 – a novel agent to control aphids and whiteflies. *Proceedings of the Brighton Crop Protection Conference, Pests and Diseases, November 1992* 1, 43–50.

Flückiger, C.R., Senn, R. and Bulholzer, H. (1992b) CGA 215'944 – opportunities for use in vegetables. *Proceedings of the Brighton Crop Protection Conference, Pests and Diseases, November 1992* 3, 1187–1192.

Forbes, A.R. and Mackenzie, J.R. (1982) The lettuce aphid, *Nasonovia ribisnigri* (Homoptera:Aphididae) damaging lettuce crops in British Columbia. *Journal of the Entomological Society of British Columbia* 79, 28–31.

Foster, G.N., Pallett, D. and Woodford, J.A.T. (1994) Suppression of spread of potato leaf roll virus and potato virus Y by aphicide sprays. *Proceedings of the Brighton Crop Protection Conference, Pests and Diseases, November 1994* 1, 223–228.

Foster, G., Blake, S., Barker, I., Harrington, R., Taylor, M., Walters, K., Northing, P. and Morgan, D. (2004) Decision support for BYDV control in the United Kingdom: can a regional forecast be made field specific? In: Simon, J.C., Dedryver, C.A., Rispe, C. and Hullé, M. (eds) *Aphids in a New Millennium. Proceedings of the 6th International Aphid Symposium, Rennes, France, September 2001.* Institut National de la Recherche Agronomique, Paris, pp. 287–291.

Foster, S.P., Harrington, R., Dewar, A.M., Denholm, I. and Devonshire, A. (2002) Temporal and spatial dynamics of insecticide resistance in *Myzus persicae* (Hemiptera: Aphididae). *Pest Management Science* 58, 895–907.

Franco, G.V. (1999) Controle químico de *Aphis gossypii* Glover, 1876 (Homoptera: Aphididae) em cultivar suscetível a viroses. *Anais 2⁰ Congresso Brasileiro de Algodão: O Algodão no Século 20, Perspectivas para o Século 21, Ribeirão Preto, São Paulo, September 1999*, 195–197.

Furr, R.E. and Harris, F.A. (1996) Cotton aphid insecticide efficacy trials in the Mississippi delta in 1995. In: *Proceedings Beltwide Cotton Conferences, San Antonio, January 1995, Volume 2.* National Cotton Council, Memphis, pp. 891–892.

Gengotti, S. (2001) Valutazione dell'efficacia di alcuni aficidi sistemici é della resistenza di una varieta di lattuga nella lotta agli afidi in Emilia-Romagna. *Informatore Fitopatologico* 51, 67–72.

George, K.S. and Gair, R. (1979) Crop loss assessment on winter wheat attacked by the grain aphid, *Sitobion avenae. Plant Pathology* 28, 143–149.

Godfrey, L.D. and Fuson, K.J. (2001) Environmental and host plant effects on insecticide susceptibility of the cotton aphid (Homoptera: Aphididae). *Journal of Cotton Science* 5, 22–29.

Gordon, S.C., Woodford, J.A.T. and Birch, A.N.E. (1997) Arthropod pests of *Rubus* in Europe: pest status, current and future control strategies. *Journal of Horticultural Science* 72, 831–862.

Gourmet, C., Kolb, F.L., Smyth, C.A. and Pedersen, W.L. (1996) Use of imidacloprid as a seed-treatment insecticide to control barley yellow dwarf virus (BYDV) in oat and wheat. *Plant Disease* 80, 136–141.

Gray, S.M., Bergstrom, G.C., Vaughan, R., Smith, D.M. and Kalb, D.W. (1996) Insecticidal control of cereal aphids and its impact on the epidemiology of the barley yellow dwarf luteoviruses. *Crop Protection* 15, 687–697.

Haesaert, G., Latre, J., Derycke, V. and Goen, K. (1998) Effect of imidacloprid as seed treatment on yield and yield characteristics of silage maize. *Mededelingen Faculteit Landbouwkundige en Toegepaste Biologische Wetenschappen Universteit Gent* 63, 555–561.

Halmer, P. (1994) The development of quality seed treatments in commercial practice – objectives and achievements. In: *Seed Treatment: Progress and Prospects. Proceedings of the 1994 British Crop Council Protection Symposium No. 57.* British Crop Protection Council, Farnham, pp. 363–374.

Hardman, J.M. (1992) Apple pest management in North America. *Proceedings of the Brighton Crop Protection Conference, Pests and Diseases, November 1992* 2, 507–516.

Harrewijn, P. and Kayser, H. (1997) Pymetrozine, a fast-acting and selective inhibitor of aphid feeding. *In-situ* studies with electronic monitoring of feeding behaviour. *Pesticide Science* 49, 130–140.

Harrewijn, P., de Kogel, W.J. and Piron, P.G.M. (1998) CGA 293'343 effects on *Myzus persicae*: electrical penetration graph studies and effect on non-persistent virus transmission. *Proceedings of the Brighton Crop Protection Conference, Pests and Diseases, November 1998* 3, 813–818.

Harrington, R., Dewar, A.M. and George, B. (1989) Forecasting the incidence of virus yellows in sugarbeet in England. *Annals of Applied Biology* 114, 459–469.

Harrington, R., Mann, J.A., Plumb, R.T., Smith, A.J., Taylor, M.S., Foster, G.N., Holmes, S.J., Masterman, A.J., Tones, S.J., Knight, J.D., Oakley, J.N., Barker, I. and Walters, K.F.A. (1994) Monitoring and forecasting BYDV – the way forward? In: Brain, P., Hockland, S.M., Lancashire, P.D. and Sim, L.C. (eds) *Sampling to Make Decisions. Aspects of Applied Biology, No. 37*, pp. 197–206.

Harris, K.F. and Bradley, R.H. (1973) Importance of leaf hairs in the transmission of tobacco mosaic virus by aphids. *Virology* 52, 295–300.

Hartfield, C.M. and Campbell, C.A.M. (1996) The use of the selective insecticide pirimicarb for integrated pest management of plum aphids in UK orchards. *Proceedings of the Brighton Crop Protection Conference, Pests and Diseases, 1996* 3, 879–884.

Harvey, T.L., Seifers, D.L. and Kofoid, K.D. (1996) Effect of sorghum hybrid and imidacloprid seed treatment on infestations by corn leaf aphid and greenbug (Homoptera: Aphididae) and the spread of sugarcane mosaic virus strain MDMV-B. *Journal of Agricultural Entomology* 13, 9–15.

Heatherington, P.J. and Meredith, R.H. (1992) United Kingdom field trials with Gaucho for pest and virus control in sugarbeet, 1989–1991. *Pflanzenschutz-Nachrichten Bayer* 45, 491–526.

Helsen, H. (2001) Roze appelluis ook in herfst te bestrijden. *Fruitteelt, Den Haag* 91 (28), 12–13.

Heltbech, K., Husby, J. and Højer, P. (1999) Gaucho(R) WS-70 new high-efficacy insecticidal seed-treatment for beets. *DJF Rapport, Markburg* 10, 237–247.

Hernandez, D., Mansanet, V. and Puiggros Jove, J.M. (1999) Use of Confidor(R) 200 SL in vegetable cultivation in Spain. *Pflanzenschutz-Nachrichten Bayer* 52, 374–385.

Hofer, D., Brandl, F., Druebbisch, B., Doppmann, F. and Zang, L. (2001) Thiamethoxam (CGA 293'343) – a novel insecticide for seed delivered insect control. In: *Seed Treatment: Challenges and Opportunities. Proceedings of the 2001 British Crop Council Protection Symposium No. 76*. British Crop Protection Council, Farnham, pp. 41–46.

Hopkins, J.A. and Donaldson, F.S. (1996) Early-season insect control with Provado in the Mississippi Delta. In: *Proceedings Beltwide Cotton Conferences, Nashville, January 1996, Volume 2*. National Cotton Council, Memphis, pp. 945–948.

Hopkins, J.A., Donaldson, F.S., Bell, C.R. and Sweeden, M.B. (2000) Performance of Leverage 2.7 SE in the Mississippi delta. In: *Proceedings Beltwide Cotton Conferences, San Antonio, January 2000, Volume 2*. National Cotton Council, Memphis, pp. 1094–1097.

Hull, R. (1968) The effect of infection with beet yellows virus on the growth of sugar beet. *Journal of the American Society of Sugar Beet Technologists* 15, 192–199.

Hurej, M. and van der Werf, W. (1993a) The influence of black bean aphid, *Aphis fabae* Scop., and its honeydew on the photosynthesis of sugar beet. *Annals of Applied Biology* 122, 189–200.

Hurej, M. and van der Werf, W. (1993b) The influence of black bean aphid, *Aphis fabae* Scop., and its honeydew on leaf growth and dry matter production of sugar beet. *Annals of Applied Biology* 122, 201–214.

Huth, W. (1992) Einfluss verschiedener Insektizide auf die Übertragung von BYDV-PAV durch *Rhopalosiphum padi*. *Nachrichtenblatt des Deutschen Pflanzenschutzdienstes* 44, 243–247.

Ioannidis, P.M. (1996) L'effét du puceron des racines *Pemphigus fuscicornis* Koch sur la betterave sucrière. *Annales du 59e Congrès Institut International de Recherches Betteravières, Bruxelles, February 1996*. Institut International de Recherches Betteravières, Brussels, pp. 269–276.

Ishaaya, I. and Horowitz, A.R. (1998) Insecticides with novel modes of action: an overview. In: Ishaaya, I. and Degheele, D. (eds) *Insecticides with Novel Modes of Action. Mechanisms and Application*. Springer, Berlin, pp. 1–24.

Ishaaya, I., Kontsedalov, S., Mazirov, D. and Horowitz, A.R. (2001) Biorational agents – mechanism and importance in IPM and IRM programs for controlling agricultural pests. *Mededelingen Faculteit Landbouwkundige en Toegepaste Biologische Wetenschappen Universiteit Gent* 66, 363–374.

Jarande, N.T. and Dethe, M.D. (1994) Effective control of brinjal sucking pests by imidacloprid. *Plant Protection Bulletin, Faridabad* 46, 43–44.

Jaschewski, K. (1997) Zum Einfluss von Gaucho(R) (imidacloprid) als Saatgutbeize am Mais auf die Saugaktivität der Traubenkirschenlaus (*Rhopalosiphum padi* L.). *Mitteilungen der Deutschen Gesellschaft für Allgemeine und Angewandte Entomologie* 11, 315–318.

Jeschke, P., Moriya, K., Lantzsch, R., Seifert, H., Lindner, W., Jelich, K., Gohrt, A., Beck, M.E. and Etzel, W. (2001) Thiacloprid (Bay YRC 2894) a new member of the chloronicotinyl insecticide (CNI) family. *Pflanzenschutz-Nachrichten Bayer* 54, 147–160.

Jeschke, P., Schindler, M. and Beck, M.E. (2002) Neonicotinoid insecticides – retrospective consideration and prospects. *Proceedings of the British Crop Protection Council Conference, Pests and Diseases, Brighton, November 2002* 1, 137–144.

Jeschke, P., Uneme, H., Benet-Buchholz, J., Stölting, J., Sirges, W., Beck, M.E. and Etzel, W. (2003) Clothianidin (TI-435) – the third member of the chloronicotinyl insecticide (CNI™) family. *Pflanzenschutz-Nachrichten Bayer* 56, 5–25.

Karner, M.A., Goodson, J.R. and Payton, M. (1997) Efficacy of various insecticide treatments to control cotton aphids and prevent economic loss in Oklahoma in 1995. In: *Proceedings Beltwide Cotton Conferences, New Orleans, January 1997, Volume 2*. National Cotton Council, Memphis, pp. 1054–1057.

Kashiwada, Y. (1996) Bestguard(R) (nitenpyram, TI-304) – a new systemic insecticide. *Agrochemicals Japan* 68, 18–19.

Kaufmann, L., Schürmann, F., Yiallouros, M., Harrewijn, P. and Kayser, H. (2004) The serotogenic system is involved in feeding inhibition by pymetrozine. Comparative studies on a locust (*Locusta migratoria*) and an aphid (*Myzus persicae*). *Comparative Biochemistry and Physiology C* 38, 469–483.

Kayser, H., Kaufmann, L., Schürmann, F. and Harrewijn, P. (1994) Pymetrozine (CGA 215'944): a novel compound for aphid and whitefly control. An overview of its mode of action. *Proceedings of the Brighton Crop Protection Conference, Pests and Diseases, November 1994* 2, 737–742.

Kharboutli, M.S. and Allen, C.T. (2000) Comparison of insecticides for cotton aphid control. In: *Special Report Arkansas Agricultural Experiment Station*. Arkansas Agricultural Experiment Station, University of Arkansas, Fayetteville, pp. 128–131.

Kift, N.B., Dewar, A.M. and Dixon, A.F.G. (1997) The effect of plant age and infection with virus yellows on the survival of *Myzus persicae* on sugar beet. *Annals of Applied Biology* 129, 371–378.

Killick, R.W. and Schulteis, D.T. (1998) The toxicity response from insecticides with an ethyl fatty ester-based adjuvant. *Proceedings of the Brighton Crop Protection Conference, Pests and Diseases, November 1998* 1, 115–120.

Kim, Y., Roh, J., Kim, M., Im, D. and Hur, I. (2000) Seasonal occurrence of aphids and selection of insecticides for controlling aphids transmitting soybean mosaic virus [in Korean]. *Korean Journal of Crop Science* 45, 353–355.

Knaust, H.J. and Poehling, H.M. (1992) Effect of imidacloprid on cereal aphids and their efficiency as vectors of BYD virus. *Pflanzenschutz-Nachrichten Bayer* 45, 381–408.

Knaust, H.J. and Poehling, H.M. (1995) Studies on the action of Gaucho (imidacloprid) on cereal aphids in spring and summer. *Mitteilungen der Deutschen Gesellschaft für Allgemeine und Angewandte Entomologie* 10, 461–466.

Koenig, J.P., Lawson, D.S., Ngo, N., Minton, B., Ishida, C., Lovelace, K. and Moore, S. (2000) 1999 field trial results with pymetrozine (Fulfill(R)) and thiamethoxam (CentricTM/ActaraTM) for control of cotton aphid (*Aphis gossypii*). In: *Proceedings Beltwide Cotton Conferences, San Antonio, January 2000, Volume 2*. National Cotton Council, Memphis, pp. 1335–1337.

Krambias, A. (1994) Improvements for the control of pecan aphid (*Monellia caryella*) in Cyprus. *Acta Horticulturae* 365, 151–153.

Krauter, P.C., Sansone, C.G. and Heinz, K.M. (2001) Assessment of Gaucho(R) seed treatment effects on beneficial insect abundance in sorghum. *Southwestern Entomologist* 26, 143–146.

Kumar, K. and Santharam, G. (1999) Effect of imidacloprid against aphids and leafhoppers on cotton. *Annals of Plant Protection Sciences* 7, 248–250.

Kumar, R. and Dikshit, A.K. (2001) Assessment of imidacloprid in brassica environment. *Journal of Environmental Science and Health. Part B, Pesticides, Food Contaminants & Agricultural Wastes* 36, 619–629.

Kuthe, K. (1995) Beobachtungen über die Auswirkung verschiedener Insecktizide und Zuckerrübensaatgut auf Schädlinge und Ertrag von 1989 bis 1994 in MittelHessen. *Gesunde Pflanzen* 47, 139–150.

Labanowska, B.H. (2004) Pest control in blackcurrant IFP in Poland using the new neonicotinoid – thiacloprid as Calypso 480 SC. *Bulletin IOBC/WPRS* 27 (4), 101–106.

Labanowski, G. and Soika, G. (1999) Badania laboratoryjne nad zwalczaniem ochojników i czerwców. *Progress in Plant Protection* 39, 165–171.

Lane, A. and Walters, K.F.A. (1991) Effect of pea aphid (*Acyrthosiphon pisum*) on the yield of combining peas. In: Froud-Williams, R.J., Gladders, P., Heath, M.C., Jenkyn, J.F., Knott, C.M., Lane, A. and Pink, D. (eds) *Production and Protection of Legumes. Aspects of Applied Biology, No. 27*, pp. 363–368.

Layton, M.B., Smith, H.R. and Andrews, G. (1996) Cotton aphid infestations in Mississippi: efficacy of selected insecticides and impact on yield. In: *Proceedings Beltwide Cotton Conferences, Nashville, January 1996, Volume 2*. National Cotton Council, Memphis, pp. 892–893.

Leite, O. de C., Brandl, F., Hofer, D., Aramaki, P., Gehmann, K. and Weissenberg, J. (2001) Seed treatment – an emerging technology in agriculture in Latin America demonstrated by the development of thiamethoxam. In: *Seed Treatment: Challenges and Opportunities. Proceedings of the 2001 British Crop Council Protection Symposium No. 76*. British Crop Protection Council, Farnham, pp. 209–214.

Liu, A., Li, S. and Hao, S. (1999) The effectiveness of imidacloprid and pirimicarb against wheat aphids and their effects on their predators. *Journal of Henan Agricultural Sciences* 4, 25–26.

Liu T., Sparks, A.N. and Yue, B. (2001) Toxicity and efficacy of triazamate against turnip aphid (Homoptera: Aphididae) on cabbage. *Journal of Entomological Science* 36, 244–250.

Losing, H. (1993) Bekämpfung der Buchenblatt-Baumlaus. *Allgemeine Forst Zeitschrift* 48, 356–357.

Lowery, D.T., Isman, M.B. and Brard, N.L. (1993) Laboratory and field evaluation of neem for the control of aphids (Homoptera: Aphididae). *Journal of Economic Entomology* 86, 864–870.

Lucas, M.B., da Silveira, C.A., de Rezende, A.C. and de Lucas, R.V. (1999) Estudo de eficiência agronômica do inseticida imidacloprid no controle das pragas iniciais na cultura do algodão. *Anais 2⁰ Congresso Brasileiro de Algodão: O Algodão no Século 20, Perspectivas para o Século 21, Ribeirão Preto, São Paulo, September 1999*, 149–151.

Macharia, M., Muthangya, P.M. and Wanjama, J.K. (1999) Response to seed-dressing aphicides in commercial varieties for preventing Russian wheat aphid (*Diuraphis noxia*) damage in Kenya. *Proceedings of the 10th Regional Wheat Workshop for Eastern, Central and Southern Africa, University of Stellenbosch, September 1998*. CIMMYT, Addis Ababa, pp. 418–425.

Maienfisch, P., Angst, M., Brandl, F., Fischer, W., Hofer, D., Kayser, H., Kobel, W., Rindlisbacher, A., Senn, R., Steinemann, A. and Widmer, H. (2001) Chemistry and biology of thiamethoxam: a second generation neonicotinoid. *Pest Management Science* 57, 906–913.

Mansanet, V., Sanz, J.V., Izquierdo, J.I. and Puiggros Jove, J.M. (1999) Imidacloprid: a new strategy for controlling the citrus leaf miner (*Phyllocnistis citrella*) in Spain. *Pflanzenschutz-Nachrichten Bayer* 52, 350–363.

Martin, C., Schoen, L., Rufingier, C. and Pasteur, N. (1996) A contribution to the integrated pest management of the aphid *Nasonovia ribisnigri* in salad crops. *Bulletin IOBC/WPRS* 19 (11), 98–101.

Martin, T.J., Birch, P.A. and Bluett, D.J. (1992) Damson–hop aphid control in UK trials with imidacloprid, a nitroguanidine insecticide. *Proceedings of the Brighton Crop Protection Conference, Pests and Diseases, November 1992* 3, 1211–1216.

Maus, C., Cure, G. and Schmuck, R. (2003) Safety of imidacloprid seed dressings to honey bees: a comprehensive overview and compilation of the current state of knowledge. *Bulletin of Insetology* 56, 51–57.

McKirdy, S.J. and Jones, R.A.C. (1996) Use of imidacloprid and newer generation synthetic pyrethroids to control the spread of barley yellow dwarf luteovirus in cereals. *Plant Disease* 80, 895–901.

McKirdy, S.J. and Jones, R.A.C. (1997) Effect of sowing time on barley yellow dwarf virus infection in wheat: virus incidence and grain yield losses. *Australian Journal of Agricultural Research* 48, 199–206.

Meesters, P., Sterk, G. and Latet, G. (1998) Aspects of integrated production of raspberries and strawberries in Belgium. *Bulletin OILB/SROP* 21 (10), 45–50.

Meredith, R.H. and Heatherington, P.J. (1992) Aphid control in potatoes from imidacloprid, a new systemic insecticide for application to seed tubers or in furrow at planting. *Proceedings of the Brighton Crop Protection Conference, Pests and Diseases, November 1992* 2, 551–556.

Meredith, R.H. and Morris, D.B. (2003) Clothianidin on sugarbeet: field trial results from Northern Europe. *Pflanzenschutz-Nachrichten Bayer* 56, 111–126.

Meredith, R.H., Heatherington, P.J. and Morris, D.B. (2002) Clothianidin – a new chloronicotinyl seed treatment for use on sugar beet and cereals: field trial experiences from Northern Europe. *Proceedings of the British Crop Protection Council Conference, Pests and Diseases, Brighton, November 2002* 2, 691–696.

Merkens, W.S.W. and Groeneweg, H. (1991) Gaucho, een nieuw insekticide ter bestrijding van onder meer bietekevers en bladluizen door behandeling van suikerbietenzaad. *Gewasbescherming* 22, 20.

Miles, E.J., Bluett, D.J. and Mann, D.H. (2001) The influence of seed rate on the efficacy of imidacloprid seed treatment against BYDV in winter cereals. In: *Seed Treatment: Challenges and Opportunities. Proceedings of the 2001 British Crop Council Protection Symposium No. 76*. British Crop Protection Council, Farnham, pp. 47–52.

Minks, A.K. and Harrewijn, P. (1989) *Aphids. Their Biology, Natural Enemies and Control, Volume 2 C.* Elsevier, Amsterdam, 312 pp.

Monnet, Y. and Ricateau, J.F. (1997) La lutte aphicide raisonnée en cultures de laitues de plein champ: bilan de trois années de pratique. *Annales de la 4e ANPP Conférence Internationale sur les Ravageurs en Agriculture, Montpellier, January 1997* 2, 497–504.

Moores, G.D., Devine, G.J. and Devonshire, A.L. (1994a) Insecticide-insensitive acetylcholinesterase can enhance esterase-based resistance in *Myzus persicae* and *Myzus nicotianae*. *Pesticide Biochemistry and Physiology* 49, 114–120.

Moores, G.D., Devine, G.J. and Devonshire, A.L. (1994b) Insecticide resistance due to insensitive acetylcholinesterase in *Myzus persicae* and *Myzus nicotianae*. *Proceedings of the Brighton Crop Protection Conference, Pests and Diseases, November 1994* 1, 413–418.

Mowry, T.M. (2005) Insecticidal reduction of potato leaf roll virus transmission by *Myzus persicae*. *Annals of Applied Biology* 146, 81–88.

Muchembled, C. (1991) Development of insecticidal treatments in beet. *Pflanzenschutz-Nachrichten Bayer* 44, 175–182.

Murray, A., Siddi, G. and Vietto, M. (1988) RH-7988: a new selective systemic aphicide. *Proceedings of the Brighton Crop Protection Conference, Pests and Diseases, November 1988* 1, 73–80.

Naibo, B. (1994) Polyvalence d'un traitement de semences à l'imidaclopride en culture de maïs. *Phytoma* 465, 27–31.

Narkiewicz-Jodko, J. and Rogowska, M. (1999) Using imidacloprid as part of an integrated system for controlling the black bean aphid *Aphis fabae* Scop. *Bulletin OILB/SROP* 22, 219–222.

Natwick, E.T., Palumbo, J.C. and Engle, C.E. (1996) Effects of imidacloprid on colonization of aphids and silverleaf whitefly and growth, yield and phytotoxicity in cauliflower. *Southwestern Entomologist* 21, 283–292.

Nauen, R. (1995) Behaviour modifying effects of low systemic concentrations of imidacloprid on *Myzus persicae* with special reference to an antifeeding response. *Pesticide Science* 44, 145–153.

Nauen, R. and Denholm, I. (2005) Resistance of insect pests to neonicotinoid insecticides: current status and future prospects. *Archives of Insect Biochemistry and Physiology* 58, 200–215.

Nauen, R. and Elbert, A. (1994) Effect of imidacloprid on aphids after seed treatment of cotton in laboratory and greenhouse experiments. *Pflanzenschutz-Nachrichten Bayer* 47, 177–210.

Nauen, R., Tietjen, K., Wagner, K. and Elbert, A. (1998) Efficacy of plant metabolites of imidacloprid against *Myzus persicae* and *Aphis gossypii* (Homoptera: Aphididae). *Pesticide Science* 52, 53–57.

Nauen, R., Ebbinghaus-Kintscher, U. and Schmuck, R. (2001) Toxicity and nicotinic acetylcholine receptor interaction of imidacloprid and its metabolites in *Apis mellifera* (Hymenoptera: Apidae). *Pest Management Science* 57, 577–586.

Nauen, R., Ebbinghaus-Kintscher, U., Salgado, V.L. and Kaussman, M. (2003) Thiamethoxam is a neonicotinoid precursor converted to clothianidin in insects and plants. *Pesticide Biochemistry and Physiology* 76, 55–69.

Oakley, J.N. and Walters, K.F.A. (1994) A field evaluation of different criteria for determining the need to treat winter wheat against the grain aphid *Sitobion avenae* and the rose–grain aphid *Metopolophium dirhodum*. *Annals of Applied Biology* 124, 195–211.

Oakley, J.N. and Young, J.E.B. (2000) Economics of pest control in cereals in the UK. *Proceedings of the British Crop Protection Council Conference, Pests and Diseases, Brighton, November 2000* 2, 663–670.

Ohkawara, Y., Akayama, A., Matsuda, K. and Andersch, W. (2002) Clothianidin: a novel broad-spectrum neonicotinoid insecticide. *Proceedings of the British Crop Protection Council Conference, Pests and Diseases, Brighton, November 2002* 1, 51–58.

Palumbo, J.C. and Kerns, D.L. (1994) Effects of imidacloprid as a soil treatment on colonization of green peach aphid and marketability of lettuce. *Southwestern Entomologist* 19, 339–346.

Parker, W.E. and Biddle, A.J. (1998) Assessing the damage caused by black bean aphid (*Aphis fabae*) on spring beans. *Proceedings of the Brighton Crop Protection Conference, Pests and Diseases, November 1998* 3, 1077–1082.

Parker, W.E. and Blood-Smyth, J.A. (1996) Insecticidal control of foliar and root aphids on outdoor lettuce. *Proceedings of the Brighton Crop Protection Conference, Pests and Diseases, November 1996* 3, 861–866.

Parker, W.E., Collier, C.R., Ellis, P.R., Mead, A., Chandler, D., Blood-Smyth, J.A. and Tatchell, G.M. (2002) Matching control options to a pest complex: the integrated pest management of aphids in sequentially-planted crops of outdoor lettuce. *Crop Protection* 21, 235–248.

Pasian, C.C., Lindquist, R.K. and Struve, D.K. (1997) A new method of applying imidacloprid to potted plants for controlling aphids and whiteflies. *HortTechnology* 7, 265–269.

Pasqualini, E. and Vergnani, S. (1997) Recenti esperienze nel controllo dell'afide grigio del melo. *Informatore Agrario* 53 (supplement), 14–17.

Pasqualini, E., Antropoli, A. and Zecchini, G. (1996) Prospettive di lotta contro l'afide grigio del melo (*Dysaphis plantaginea* Pass., Rhynchota Aphididae). *Informatore Fitopatologico* 46, 50–55.

Pasqualini, E., Civolani, S., Vergnani, S., and Natale, D. (1998) Neem and chloronicotinyl compounds in integrated pest management on apple orchards in Emilia-Romagna. *Informatore Fitopatologico* 48, 52–57.

Pasqualini, E., Civolani, S., Vergnani, S. and Ardizzoni, M. (2001) Efficiency of some insecticides used for apple protection. *L'Informatore Agrario* 57, 97–99.

Patil, C.S., Lingappa, S. and Bhat, B.N. (1999) Bioefficacy of insecticides and botanicals against tobacco aphid, *Myzus nicotianae* Blackman. *Tobacco Research* 25, 23–32.

Pettis, G.V., Braman, S.K., Guillebeau, L.P. and Sparks, B. (2005) Evaluation of insecticides for suppression of Japanese beetle, *Popillia japonica* Newman, and crapemyrtle aphid, *Tinocallis kahawaluokalani* Kirkaldy. *Journal of Environmental Horticulture* 23, 145–148.

Pike, K.S., Reed, G.L., Graf, G.T. and Allison, D. (1993) Compatability of imidacloprid with fungicides as a seed-treatment control of Russian wheat aphid (Homoptera: Aphididae) and effect on germination, growth, and yield of wheat and barley. *Journal of Economic Entomology* 86, 586–593.

Plapp, F.W. (1991) The nature, modes of action, and toxicity of insecticides. In: Pimentel, D. (ed.) *CRC Handbook of Pest Management in Agriculture, Volume 2*, 2nd edn. CRC, Boston, pp. 447–459.

Plumb, R.T. and Lennon, E. (1982) Aphid infectivity and the infectivity index. In: *Report of Rothamsted Experimental Station, 1981, Part 1*, pp. 195–197.

Pollini, A. and Bariselli, M. (1997) Attivita di nuovi aficidi utilizzati in frutticoltura. *Informatore Agrario Supplemento* 53, 3–8.

Powell, G. and Hardie, J. (1994) Effects of mineral oil applications on aphid behaviour and transmission of potato virus Y. *Proceedings of the Brighton Crop Protection Conference, Pests and Diseases, November 1994* 1, 229–234.

Pringle, K.L. (1998) The use of imidacloprid as a soil treatment for the control of *Eriosoma lanigerum* (Hausmann) (Hemiptera: Aphididae). *Journal of the Southern African Society for Horticultural Sciences* 8, 55–56.

Proft, M., de Ryckel, B., de Ducat, N., Pigeon, O. and Bernes, A. (1999) Seed treatment with thiamethoxam for the protection of sugarbeet, maize and cereals against insect pests. *Mededelingen Faculteit Landbouwkundige en Toegepaste Biologische Wetenschappen Universiteit Gent* 64, 327–341.

Qi, A., Dewar, A.M. and Harrington, R. (2004) Decision making in controlling virus yellows of sugar beet in the UK. *Pest Management Science* 60, 727–732.

Ramaprasad, G., Sreedhar, U., Sitaramaiah, S., Rao, S.N. and Satyanarayana, S.V.V. (1998) Efficacy of imidacloprid, a new insecticide for controlling *Myzus nicotianae* on flue cured Virginia tobacco (*Nicotiana tabacum*). *Indian Journal of Agricultural Sciences* 68, 165–167.

Reinink, K. and Dieleman, F.L. (1993) Survey of aphid species on lettuce. *Bulletin IOBC/WPRS* 16 (5), 56–68.

Rhea, J.R. (1996) Preliminary results for the chemical control of hemlock woolly adelgid in ornamental and natural settings. In: *Proceedings of the 1st Hemlock Woolly Adelgid Review, Charlottesville, October 1995*. USDA Forest Service, Morgantown, West Virginia, pp. 113–125.

Rongai, D., Cerato, C., Martelli, R. and Ghedini, R. (1998) Aspects of insecticide resistance and reproductive biology of *Aphis gossypii* Glover on seed potatoes. *Potato Research* 41, 29–37.

Royer, T.A., Giles, K.L., Nyamanzi, T., Hunger, R.M., Krenzer, E.G., Elliott, N.C., Kindler, S.D. and Payton, M. (2005) Economic evaluation of the effects of planting date and application rate of imidacloprid for management of cereal aphids and barley yellow dwarf virus in winter wheat. *Journal of Economic Entomology* 98, 95–102.

Sannino, L. (1997) La difesa del tabacco dagli afidi con insetticidi fogliari è sistemici radicali. *L'Informatore Agrario* 53, 101–103.

San Román, C.P., Ortiz, A. and Ayala, J. (1995) Efecto del imidacloprid en el control de pulgones y en la transmisión de los virus de la amarillez de la remolacha. *Boletín de Sanidad Vegetal, Plagas* 21, 507–515.

dos Santos, W.J. and dos Santos, K.B. (1999) Controle de pulgões, *Aphis gossypii*, e tripes, *Frankliniella schultzei*, em algodoeiro. *Anais 2º Congresso Brasileiro de Algodão: O Algodão no Século 20, Perspectivas para o Século 21, Ribeirão Preto, SãoPaulo, September 1999*, 175–177.

Scarpellini, J.R. and Nakamura, G. (1999a) Controle do tripes *Frankliniella schultzei* Tribon, 1920 (Thysanoptera: Thripidae) e do pulgão *Aphis gossypii* Glover, 1877 (Homoptera. Aphididae) com thiamethoxam e diafenthiuron na cultura do algodoeiro *Gossypium hirsutum* L. *Anais 2⁰ Congresso Brasileiro de Algodão: O Algodão no Século 20, Perspectivas para o Século 21, Ribeirão Preto, São Paulo, September 1999*, 338–340.

Scarpellini, J.R. and Nakamura, G. (1999b) Thiamethoxam em tratamento de sementes de algodão no controle de *Aphis gossypii* Glover, 1877 (Homoptera: Aphididae). *Anais 2⁰ Congresso Brasileiro de Algodão: O Algodão no Século 20, Perspectivas para o Século 21, Ribeirão Preto, São Paulo, September 1999*, 345–348.

Schaub, L., Granchamp, K. and Bloesch, B. (1999) Sensibilité du puceron cendré au pirimicarbe et au triazamate. *Revue Suisse de Viticulture Arboriculture Horticulture* 31, 73–76.

Schaufele, W.R. (1991) Vektorenbekämpfung in Zuckerrüben. *Gesunde Pflanzen* 43, 152–155.

Schepers, A. (1989) Control of aphids. In: Minks, A.K. and Harrewijn, P. (eds) *Aphids. Their Biology, Natural Enemies and Control, Volume 2 C*. Elsevier, Amsterdam, pp. 89–122.

Schmeer, H.E., Bluett, D.J., Meredith, R. and Heatherington, P.J. (1990) Field evaluation of imidacloprid as an insecticidal seed treatment in sugar beet and cereals with particular reference to virus vector control. *Proceedings of the Brighton Crop Protection Conference, Pests and Diseases, November 1990* 1, 29–36.

Schmuck, R. (1999) No causal relationship between GauchoReg. seed dressing in sunflowers and harm to bees in France. *Pflanzenschutz-Nachrichten Bayer* 52, 257–299.

Schnier, H.F., Wenig, G., Laubert, F., Simon, V. and Schmuck, R. (2003) Honey bee safety of imidacloprid corn seed treatment. *Bulletin of Insectology* 56, 73–75.

Scholz, D. and Wulf, A. (1998) Ansätze zur selektiven Bekämpfung von Baumschädlingen im offentlichen Grün und im Forst mittels Stammapplikation systemischer Pflanzenschutzmittel. *Gesunde Pflanzen* 50, 1–6.

Schoonejans, T., Maeyer, L., de Tossens, H., D'Hollander, R., Sysmans, J., Baets, D. and Vincinaux, C. (1991) Étude de l'insecticide systémique imidacloprid en betteraves, céréales, cultures maraîchères et ornementales en Belgique. *Mededelingen van de Faculteit Landbouwwetenschappen, Rijksuniversiteit Gent* 56, 1161–1179.

Schwinger, M., Harrewijn, P. and Kayser, H. (1994) Effect of pymetrozine (CGA 15'944), a novel aphicide, on feeding behaviour of aphids. *Proceedings of 8th IUPAC International Congress of Pesticide Chemistry, Washington DC, July 1994* 1, 230.

Senn, R., Hofer, D., Hoppe, T., Angst, M., Wyss, P., Brandl, F., Maienfisch, P., Zang, L. and White, S. (1998) CGA 293'343: a novel broad-spectrum insecticide supporting sustainable agriculture worldwide. *Proceedings of the Brighton Crop Protection Conference, Pests and Diseases, November 1998* 1, 27–36.

Shiokawa, K., Tsuboi, S., Iwaya, K. and Moriya, K. (1994) Development of chloronicotinyl insecticide, imidacloprid [in Japanese]. *Journal of Pesticide Science* 19, 329–332.

Singh, V.S. and Venkateswarlu, N.C. (2000) Evaluation of certain insecticides and neem against cereal aphids on barley. *Shashpa* 7, 67–75.

Sloderbeck, P.E., Witt, M.D. and Buschman, L.L. (1996) Effects of imidacloprid seed treatment on greenbug (Homoptera: Aphididae) infestations on three sorghum hybrids. *Southwestern Entomologist* 21, 181–187.

Slosser, J.E., Parajulee, M.N., Idol, G.B. and Rummel, D.R. (2001) Cotton aphid response to irrigation and crop chemicals. *Southwestern Entomologist* 26, 1–14.

Smith, H.G. and Hallsworth, P.B. (1990) The effects of yellowing viruses on yield of sugar beet in field trials, 1985 and 1987. *Annals of Applied Biology* 116, 503–511.

Solarska, E. and Jastrzebski, A. (1998) A new method of applying Confidor 200SL against the hop aphid, *Phorodon humuli*. In: *Aphids in Natural and Managed Ecosystems. Proceedings of the 5th International Symposium on Aphids, León, Spain, September 1997*. Universidad de León Secretariado de Publicaciones, León, Spain, pp. 619–622.

Sreedhar, U., Prasad, G.R., Sitaramajah, S., Swathi, P. and Rao, S.N. (1999). Persistent toxicity of new insecticides to *Myzus nicotianae* Blackman on flue cured Virginia tobacco. *Tobacco Research* 25, 33–36.

Sreekanth, M. and Babu, T.R. (2001) Evaluation of certain new insecticides against the aphid *Lipaphis erysimi* (Kalt.) on cabbage. *International Pest Control* 43, 242–244.

Sreelatha Divakar, B.J. (1997) Impact of imidacloprid seed treatment on insect pest incidence in okra. *Indian Journal of Plant Protection* 25, 52–55.

Stadler, T., Martinez Gines, D. and Buteler, M. (2003) Long-term toxicity assessment of imidacloprid to evaluate side effects on honey bees exposed to treated sunflower in Argentina. *Bulletin of Insectology* 56, 77–81.

Stein-Dönecke, U., Fuhr, F. and Wieneke, J. (1994) Dressing zone formation, uptake, translocation and action of [14C]imidacloprid for winter wheat after seed treatment and under the influence of various soil moisture levels. In: *Seed Treatment: Progress and Prospects. Proceedings of the 1994 British Crop Council Protection Symposium No. 57*. British Crop Protection Council, Farnham, pp.135–140.

Stevens, M., Smith, H.G. and Hallsworth, P.B. (1994) Identification of a second distinct strain of BMYV using monoclonal antibodies and transmission sudies. *Annals of Applied Biology* 125, 515–520.

Sunitha, P., Rao, G.R. and Rao, P.A. (2005) Bio-efficacy of certain eco-friendly insecticides against sucking pests on okra *Abelmoschus esculentus* (Linn.). *Journal of Applied Zoological Researches* 16, 186–187.

Szekely, I., Pap, L. and Bertok, B. (1996) MB-599, a new synergist in pest control. *Proceedings of the Brighton Crop Protection Conference, Pests and Diseases, November 1996* 2, 473–480.

Szwejda, J. (1997) Nowa metoda ochrony marchwi przed bawenica topolowo-marchwiana (Pemphigus phenax B. & B.). *Progress in Plant Protection* 37 (2), 19–21.

Szwejda, J.H. (1999) The occurrence and control of *Pemphigus phenax*, an aphid that infests the roots of carrots. *Bulletin OILB/SROP* 22, 229–233.

Takahashi, J., Mitsui, J., Takakusa, N., Matsuda, M., Yoneda, H., Suzuki, J., Ishimitsu, K. and Kishimoto, T. (1992) NI-25, a new type of systemic and broad spectrum insecticide. *Proceedings of the Brighton Crop Protection Conference, Pests and Diseases, November 1992* 1, 89–96.

Thackray, D.J., Jones, R.A.C., Bwye, A.M. and Coutts, B.A. (2000) Further studies on the effects of insecticides on aphid vector numbers and spread of cucumber mosaic virus in narrow-leafed lupins (*Lupinus angustifolius*). *Crop Protection* 19, 121–139.

Tolmay, V.L., van Lill, D. and Smith, M.F. (1997) The influence of demeton-S-methyl/parathion and imidacloprid on the yield and quality of Russian wheat aphid resistant and susceptible wheat cultivars. *South African Journal of Plant and Soil* 14, 107–111.

Tomizawa, M. and Casida, J.E. (2003) Selective toxicity of neonicotinoids attributable to specificity of insect and mammalian nicotinic receptors. *Annual Review of Entomology* 48, 339–364.

Toscano, N.C., Kido, K. and Davis, R.M. (1990) Lettuce pest management guidelines. *UCPMG Publication No. 15*. IPM Education and Publications, University of California, Davis, 6 pp.

Tossens, H., Schoonejans, T., Sysmans, J., D'hollander, R., Vermeulen, R. and Vincinaux, C. (1992) Étude de l'insecticide systémique imidacloprid en traitement des semences de betteraves sucrières en Belgique en 1991. *Mededelingen van de Faculteit Landbouewwetenschappen, Rijksuniversiteit Gent* 57, 759–773.

Treverrow, N. (1996) Controlling banana aphids by insecticide injection. *BGF Bulletin* 60, 22–23.

Turska, E. and Pawinska, M. (1995) Zastosowanie imidaklopridu w ochronie plantacji ziemniaka. *Materialy Sesji Instytutu Ochrony Roslin* 35, 296–302.

Vantornhout, I., Veire, M. and van de Tirry, L. (1999) Toxicity of imidacloprid to *Myzus persicae* and the predatory bugs *Orius laevigatus* and *Macrolophus caliginosus*. *Mededelingen Faculteit Landbouwkundige en Toegepaste Biologische Wetenschappen Universiteit Gent* 64, 49–57.

Vercruysse, P., Meert, F., Tirry, L. and Hofte, M. (2000) Evaluation of insecticides for control of *Cavariella aegopodii* and carrot motley dwarf disease in parsley. *Mededelingen Faculteit Landbouwkundige en Toegepaste Biologische Wetenschappen Universiteit Gent* 65, 9–18.

Vostřel, J. (1998) Ověřování biologické účinnosti vybraných insekticidu a akaricidu na rezistentní populace mšice a svilušky. *Chmelařství* 71, 32–34.

Wakita, T., Yasui, N., Yamada, E. and Kishi, D. (2005) Development of a novel insecticide, dinotefuran. *Journal of Pesticide Science* 30, 122–123.

Walgenbach, J.F. (1997) Effect of potato aphid (Homoptera: Aphididae) on yield, quality, and economics of staked-tomato production. *Journal of Economic Entomology* 90, 996–1004.

Walters, K.F.A., Lane, A., Oakley, J.N. and Heath, M.C. (1994) Control of pea aphid on combining peas and improved management strategies. *Proceedings of the Brighton Crop Protection Conference, Pests and Diseases, November 1994* 1, 211–216.

Walunj, A.R. and Pawar, S.A. (2004) An evaluation of thiacloprid against pests of chilli. *Tests of Agrochemicals and Cultivars* 25, 6–7.

Wangai, A.W., Plumb, R.T. and van Emden, H.F. (2000) Effects of sowing date and insecticides on cereal aphid populations and barley yellow dwarf virus on barley in Kenya. *Journal of Phytopathology* 148, 33–37.

Wauters, A. (1993) Rapport de l'imidacloprid comme traitement de semences en culture betteravière en Belgique. *Mededelingen van de Faculteit Landbouwwetenschappen Universiteit Gent* 58, 641–651.

Wauters, A. and Dewar, A.M. (1996) The effect of insecticide seed treatments on pests of sugar beet in Europe: results of the IIRB co-operative trials with pesticides added to pelleted seed in 1991, 1992 and 1993. *Parasitica* 51, 143–173.

Way, M.J., Cammell, M.E., Alford, D.V., Gould, H.J., Graham, C.W., Lane, A., Light, W.I. St G., Rayner, J.M., Heathcote, G.D., Fletcher, K.E. and Seal, K. (1977) Use of forecasting in chemical control of black bean aphid, *Aphis fabae* Scop., on spring-sown field beans, *Vicia faba* L. *Plant Pathology* 26, 1–7.

Werker, A.R., Dewar, A.M. and Harrington, R. (1998) Analysis of virus yellows incidence in sugar beet in relation to migrations of the vector, *Myzus persicae. Journal of Applied Ecology* 35, 811–818.

Westhuizen, M.C., van der Jager, J., de Mason, M.H. and Deall, M.W. (1994) Cost benefit analysis of Russian wheat aphid (*Diuraphis noxia*) control in the Republic of South Africa using a seed treatment. *Pflanzenschutz-Nachrichten Bayer* 47, 81–90.

Westwood, F., Dewar, A.M., Bean, K.M. and Haylock, L.A. (1997) Persistence of the selective aphicide triazamate in sugar beet in relation to its efficacy against aphids. *Annales du 60e Congres Institut International de Recherches Betteravieres, Cambridge, July 1997*. Institut International de Recherches Betteravieres, Brussels, pp. 585–588.

Westwood, F., Bean, K.M., Dewar, A.M., Bromilow, R.H. and Chamberlain, K. (1998) Movement and persistence of (14 C)imidacloprid in sugar-beet plants following application to pelleted sugar-beet seed. *Pesticide Science* 52, 97–103.

Whitehead, R. (2007) *The UK Pesticide Guide 2007*. CAB International, Wallingford, Oxon, 685 pp.

Wilde, G., Roozeboom, K., Claassen, M., Sloderbeck, P., Witt, M., Janssen, K., Harvey, T., Kofoid, K., Brooks, L. and Shufran, R. (1999) Does the systemic insecticide imidacloprid (Gaucho) have a direct effect on yield of grain sorghum. *Journal of Production Agriculture* 12, 382–389.

Woodford, J.A.T. (1992) Effects of systemic applications of imidacloprid on the feeding behaviour and survival of *Myzus persicae* on potatoes and on transmission of potato leafroll virus. *Pflanzenschutz-Nachrichten Bayer* 45, 527–546.

Woodford, J.A.T. and Mann, J.A. (1992) Systemic effects of imidacloprid on aphid feeding behaviour and virus transmission on potatoes. *Proceedings of the Brighton Crop Protection Conference, Pests and Diseases, November 1992* 2, 557–562.

Wright, L.C. and Cone, W.W. (1999) Carryover of imidacloprid and disulfoton in subsurface drip-irrigated hop. *Journal of Agricultural and Urban Entomology* 16, 59–64.

Wright, S.D., Godfrey, L.D., Jimenez, M.R., Weinholds, P. and Wood, J.P. (1997) Aphid control screening studies in San Joaquin Valley. In: *Proceedings Beltwide Cotton Conferences, New Orleans, January 1997, Volume 2*. National Cotton Council, Memphis, pp.1053–1054.

Wyss, P. and Bolsinger, M. (1997) Plant-mediated effects on pymetrozine efficacy against aphids. *Pesticide Science* 50, 203–210.

Xue, D. (2000) Effect of several insecticides for control of Japanese apricot aphids. *South China Fruits* 29 (1), 43.

Yamamoto, P.T., Roberto, S.R. and Pria, W.D. (2000) Systemic insecticides applied on citrus tree trunk to control *Oncometopia facialis, Phyllocnistis citrella* and *Toxoptera citricida. Scientia Agricola* 57, 415–420.

Yanovskii, Y.P. and Larcheva, E.I. (2000) New insecticides for apple pest control [in Russian]. *Zashchita i Karantin Rastenii* 11, 25.

Yehuda, S.B. and Izhar, Y. (1992) Preliminary experiments with new aphicides in apple orchards [in Hebrew]. *Alon Hanotea* 46, 281–284.

Zang, L., Ngo, N. and Minto, B. (1999) AdageTM (thiamethoxam) seed treatment for cotton. In: *Proceedings Beltwide Cotton Conferences, Orlando, January 1999, Volume 2*. National Cotton Council, Memphis, pp. 1104–1106.

Zhong, J.-Q. and Gan, F.-L. (2000) Experiment to control citrus aphid by spraying Admire insecticide [in Chinese]. *South China Fruits* 29 (5), 24.

Zhu, G., Xu, B.Y., Qian, H. and Ju, Z. (1996) Efficacy of imidacloprid for control of *Myzus persicae* and *Lipaphis erysimi* on cabbage [in Chinese]. *Plant Protection* 22, 39–40.

16 Cultural Control

Steve D. Wratten[1], Geoff M. Gurr[2], Jason M. Tylianakis[3] and
Katherine A. Robinson[1]

[1]National Centre for Advanced Bio-Protection Technologies, Lincoln University,
Canterbury 7647, New Zealand; [2]School of Rural Management, Charles Sturt
University, PO Box 883, Orange, New South Wales 2800, Australia; [3]School of
Biological Sciences, University of Canterbury, Private Bag 4800,
Christchurch 8020, New Zealand

Introduction

In this chapter, a variety of techniques is described under the heading of cultural control. The unifying theme is the reduction of aphid damage through management of the physical or biological environment of the crop, either at establishment or during its growth. However, several mechanisms are involved. These range from preventing aphid access by physically protecting crops, to enhancing aphid mortality by providing neighbouring habitats that provide out-of-season refuges for aphid predators. Some methods are widely known, pre-dating intensive agrochemical use (Gurr *et al.*, 2000). Others are the product of more recent research activity (Gurr *et al.*, 2004). Some are used in conjunction with insecticides, and others are considered as more benign alternatives (Harrewijn and Minks, 1989). The mechanisms may be well understood, or they may involve multitrophic-level interactions between species that are very difficult to predict (Tscharntke and Hawkins, 2002; Brewer and Elliott, 2004).

The following section provides an overview of methods of cultural control and the mechanisms through which they effect aphid control. Subsequent sections review

approaches that have attracted recent interest among researchers, drawing on studies from a range of crop types and climates. The discussion focuses on the ecological mechanisms by which the control techniques operate rather than on the full range of socioeconomic factors, which must be evaluated before the methods can be commercialized.

Overview

Figure 16.1 lists the techniques of cultural control of aphids, based on the way in which the technique is interpreted in this chapter. In addition to the environmental factors included in the figure, genetic features of the crop have an important influence on susceptibility to aphid damage. Manipulating genetic variation, including choice of crop species for a particular climate and variation between cultivars in aphid resistance, might be considered as cultural methods of aphid control. However, host-plant resistance is discussed by van Emden (Chapter 17 this volume). Techniques that involve enhancement of natural enemies through release programmes are discussed by Powell and Pell (Chapter 18 this volume).

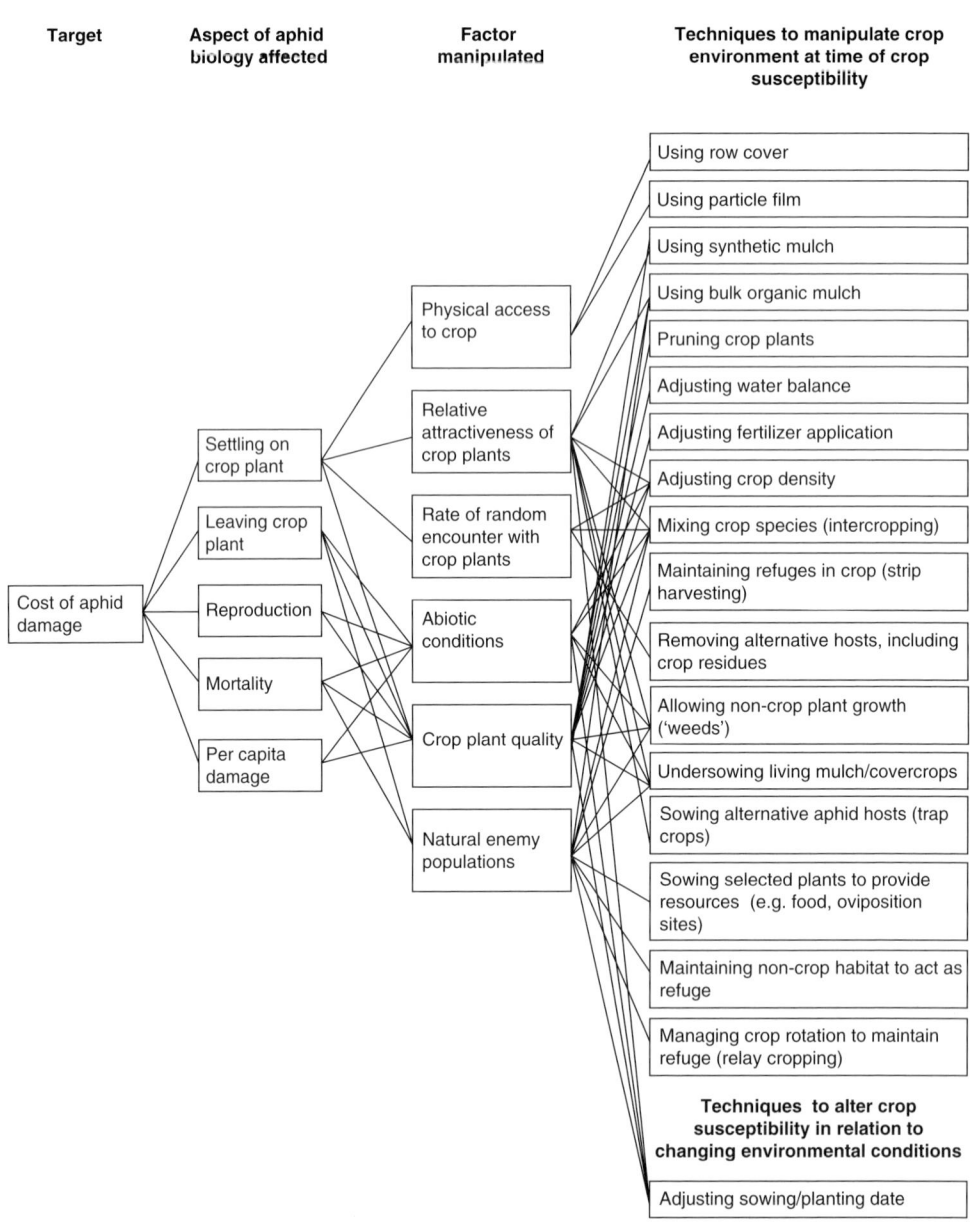

Fig. 16.1. Cultural methods for managing aphid populations and the mechanisms by which they affect levels of aphid damage. These methods may be applied within the cropped area or in the surrounding landscape. The influences represented by linking lines can be positive or negative and may be deliberate or inadvertent effects of employing the technique.

In Fig. 16.1, a diversity of approaches is apparent, and this arises in two ways. First, a reduction in aphid damage can be achieved *via* effects on one or more of the processes determining aphid population size, or on the damage imposed by each aphid. Second, these aspects of aphid biology can be affected directly by cultural practice, or the effects can be achieved through manipulation of abiotic conditions, host-plant quality, or natural enemy populations.

More important are the various ways in which any particular technique operates. If it has multiple effects, each of these may increase or decrease aphid damage and the net effect will depend on the state of other components of the system. For example, almost all techniques have the potential to affect aphids indirectly by altering host-plant quality. Furthermore, the techniques will affect yield in ways other than *via* aphid control, and these must also be considered.

Although there are interacting factors to be considered, manipulations of the physical environment are relatively amenable to experimental research. The use of synthetic mulches can affect many determinants of plant growth, including soil temperature, water balance, and competitive weed growth (e.g. Farias-Larios *et al.*, 1994; Orozco-Santos *et al.*, 1995; Brault *et al.*, 2002). These factors can affect both susceptibility of the crop to damage by a given aphid population and the degree to which the crop supports aphid population growth. However, research has shown that, in many situations, the overriding effect is that on the ability of the pests to find the host plants (Kennedy *et al.*, 1961; Kennedy and Booth, 1963). The reflectivity of the mulch alters the aphid's visual perception of the crop and, consequently, the rate at which alatae find host plants (Lu and Feng, 1989; Chiang *et al.*, 1992; Kring and Schuster, 1992; Pinese *et al.*, 1994; Csizinszky *et al.*, 1995; Orozco-Santos *et al.*, 1995; Brown *et al.*, 1996; Farias-Larios and Orozco-Santos, 1997a,b).

Several cultural methods of control involve manipulating the biotic environment by increasing plant diversity. These are discussed under 'Intercropping', 'Trap Crops', 'Provision of Resources for Natural Enemies' and 'Provision of Refuges for Natural Enemies'. The complexity of interacting mechanisms involved in approaches using increased plant diversity means that theoretical community ecology has often been used to inform experimental research.

An inverse relationship between herbivore populations and plant diversity is found in many ecological communities. As discussed in the relevant sections below, the various mechanisms by which increasing plant diversity can reduce herbivore populations may operate directly through the plants themselves, or *via* their effect on predators. In an agroecological context, Root (1973) summarized these perspectives as two hypotheses. First, the 'enemies' hypothesis places emphasis on natural enemies of the herbivore and 'top-down' control (Root, 1973). According to this hypothesis, the natural enemies receive some benefit from non-crop plants that increases their impact on herbivore populations in diversified systems. In contrast, the 'resource concentration' hypothesis emphasizes 'bottom-up' influences on community dynamics. It states that herbivores are more likely to aggregate in monocultures because the concentration of resources in pure stands facilitates the finding and/or recognition of the host (Kennedy *et al.*, 1961; Kennedy and Booth, 1963). Numerous studies have addressed this subject and have provided evidence for both hypotheses (van Emden and Williams, 1974; van Emden, 1990; Andow, 1991; Wratten and Powell, 1991; Gurr *et al.*, 1998; Wratten *et al.*, 1998; Landis *et al.*, 2000). Together with work addressing the ecology of key species in particular crop systems, this reinforces the need for a multiplicity of factors to be considered in the design and use of biocultural techniques in pest management (Dent, 2000).

In discussing the efficacy of methods of aphid control, it should be recognized that there is not a simple relationship between aphid populations and 'satisfactory control'. The degree to which populations must be suppressed to keep damage at acceptable levels depends on the nature of the damage (see Quisenberry and Ni, Chapter 13 this volume). For example, low population densities may cause significant yield losses through transmission of viruses that may cause severe plant diseases. Where damage arises solely through the removal of nutrients through feeding, in contrast, even high aphid densities may cause only modest yield loss. In addition, the reduction in yield that a producer will accept will vary with the value of the crop.

Mulches

Mulches (ground coverings) are a widely recognized cultural technique for improving yield of horticultural crops by conserving water (Farias-Larios et al., 1994), inhibiting weed growth (Brault et al., 2002), improving biological and chemical composition of the soil (Lal et al., 1980), and increasing soil temperature (Orozco-Santos et al., 1995; Farias-Larios and Orozco-Santos, 1997a,b). Altering microclimate and host-plant quality can also indirectly affect aphid damage, but recent attention has focused on the deterrence of alate aphids through reflectance of short-wavelength radiation by the mulch surface. The effect is the result of diminishing the conspicuousness of crop plants against the background of bare soil, leading to disruption of host-plant location by settling alatae (Kennedy et al., 1961), for which visual stimulation is known to be an important factor (Prokopy and Owens, 1983). Undersown plants may have a similar affect on crop perception (Dempster, 1969).

Many studies have investigated the optical properties of different mulches alongside their effect on aphid colonization and incidence of aphid-borne viruses (Lu and Feng, 1989; Chiang et al., 1992; Kring and Schuster, 1992; Pinese et al., 1994; Csizinszky et al., 1995; Orozco-Santos et al., 1995; Brown et al., 1996; Farias-Larios and Orozco-Santos, 1997a,b; Stapleton and Summers, 2002; Greer and Dole, 2003; Jenni et al., 2003; Saucke and Doring, 2004; Summers et al., 2004). These studies have compared plastics of various colours, metallic surfaces, spray-on latexes, and organic materials such as straw. For example, transparent and aluminium-painted plastic mulches have been found to reduce population densities of aphids better than black or blue plastic mulches (Lu and Feng, 1989; Lamont et al., 1990; Kring and Schuster, 1992; Csizinszky et al., 1995; Brown et al., 1996; Farias-Larios and Orozco-Santos, 1997a,b). Kring and Schuster (1992) found that aluminium-painted plastic mulch reflected approximately four times as much UV light (< 390 nm) than black or white mulches.

This increased reflectance led to lower densities of thrips and aphids on pepper and tomato plants grown in Florida. More evidence for the importance of reflectance comes from the observation that the aphid-repelling effects of white mulches diminish as they become duller or yellow with time (Summers et al., 1995).

Polyethylene mulches appear to have their greatest impact on aphid colonization early in the season (Cartwright et al., 1990; Liotta and di Trapani, 1994; Csizinszky et al., 1995; Brown et al., 1996). A study in Alabama showed that early season aphid densities in summer squash (Cucurbita pepo) plants were low and Zucchini yellow mosaic virus and Cucumber mosaic virus were absent in treatments with aluminium-painted plastic mulch. However, as the season progressed, aphid numbers began to increase, leading to an increase in the incidence of virus diseases (Brown et al., 1996). This delay in the onset of mosaic virus disease was attributed to the short-wavelength light reflected from the mulch decreasing in intensity as the plants grew over the mulch.

Water-miscible silver spray mulches also show promise for management of aphids. Not only do they confer reflectivity to the soil surface on which they are sprayed, thereby repelling aphids, but they have advantages over polyethylene in that they are biodegradable and can be incorporated into the soil at the end of the season, requiring no removal or landfill (Summers et al., 1995). Biodegradable plastics have been developed, but may degrade prematurely (Greer and Dole, 2003).

By altering the microclimate and habitat at ground level, mulches also have the potential to affect populations of epigeal predators. The use of organic materials such as straw and compost is particularly likely to have this effect, and may also affect host quality by providing nutrients. In some crops, the observed impacts on aphid populations have been attributed to these mechanisms rather than to the optical properties of the materials. For example, in an experiment comparing the arthropod fauna of plots with and without compost in an apple orchard, compost was associated with lower numbers of the nymphs of *Eriosoma lanigerum* (woolly apple aphid) on some sampling dates (Brown and Tworkoski, 2004). Aphid populations on sampled tree roots, however, were not significantly different. In a study where aphids were added to cereal plots (so the results were not due to the effect of mulch on immigration rates), Schmidt *et al.* (2004) found straw mulch led to lower aphid populations later in the season, and mulching was associated with higher spider populations. However, against expectations, the effect of mulch on aphids was stronger when ground predators were excluded.

Row Covers

Lightweight row covers of spun-bonded polyester or polyethylene, designed to rest ('float') on the surface of crop plants, provide protection by excluding insects. Floating row covers can reduce aphid density and virus incidence in potatoes (*Solanum tuberosum*) (Harrewijn *et al.*, 1991), cantaloupe (*Cucumis melo*) (Espinoza and McLeod, 1994), squash/muskmelon (*Cucurbita pepo*) (Webb and Linda, 1992; Farias-Larios *et al.*, 1999; Walters, 2003), and bell pepper (*Capsicum annuum*) (Avilla *et al.*, 1997). They act by repelling alate aphids and preventing alighting aphids from inserting their stylets (Harrewijn *et al.*, 1991; Webb and Linda, 1992). Tunnels have been used to similar effect (Jenni *et al.*, 2003).

Covers made from needle-punched wool have also been examined for crop protection. Although such covers carry the advantage of being biodegradable, they offer significantly poorer protection against aphids on cabbages than do synthetic fibres (Evans *et al.*, 1997).

Plastic mulches have been used in conjunction with floating row covers (e.g. Webb and Linda, 1992; Farias-Larios *et al.*, 1999). However, high intensity light reflected from the mulch may accelerate degradation of the covers, without providing any aphid management benefits beyond those of the covers alone (Webb and Linda, 1992).

Particle Films

Spraying crops with inert reflective materials can reduce aphid populations. By altering the visual cues presented to alatae, as occurs with reflective mulches, the sprays reduce settling rates. Early trials found that spraying peppers and potatoes with various whitewash formulations reduced rates of aphid-transmitted virus infection on some sampling dates (Marco, 1986, 1993). In the potato study, aerial aphid populations were monitored and although catches of one aphid species, *Aphis gossypii* (cotton or melon aphid), were higher in treated plots, the principal species (*Sinomegoura citricola*) was less abundant (Marco, 1986). However, spraying reduced yields of peppers and potatoes, possibly because the shading effect of whitewash outweighed the benefits of reduced viral infection.

Recent improvements in technology have allowed the development of more sophisticated particle films (Fig. 16.2) based on kaolin $(Al_4Si_4O_{10}(OH)_8)$ (Glenn and Puterka, 2005). These are multifunctional, reducing disease and solar injury as well as insect damage. As described above, the high reflectivity acts as a visual repellent. In addition, those insects that do colonize the sprayed crop are adversely affected by kaolin particles covering the plant surface or adhering to their exoskeleton. Particles on their tarsi impair insects' mobility and grip, foraging behaviour can be disrupted by excessive time spent grooming, and particles interfere with feeding and oviposition. These effects combine to increase mortality and reduce the rate of reproduction (Glenn and Puterka, 2005).

Laboratory studies have shown that *Tinocallis caryaefoliae* (black pecan aphid) prefers unsprayed to sprayed pecan (*Carya illinoiensis*) foliage, and longevity and nymph production are lower on treated seedlings (Cottrell *et al.*, 2002). Mortality of *Aphis spiraecola* (green citrus aphid) was increased by particle film in a test on apple (*Malus domestica*) leaves (Glenn *et al.*, 1999). This effect was partly attributed to aphids falling from the sprayed leaves.

Fig. 16.2. Kaolin applied as two coats to swedes by knapsack sprayer (original colour photo courtesy of A. Evans).

In apple and pear orchards, field trials of kaolin applications have successfully controlled a range of arthropod pests (Glenn *et al.*, 1999; Glenn and Puterka, 2005). However, their efficacy against aphids has been relatively low. Knight *et al.* (2001) sampled apple trees on six dates over two years and found that the aphid (*Aphis* spp.) populations on trees that had received kaolin applications were not lower than on unsprayed trees. This could be explained by the concentration of aphids at the growing tips where, despite repeated spraying, the leaves were sufficiently particle-free for the treatment to have no significant effect. In the short-term laboratory studies of Glenn *et al.* (1999) and Cottrell *et al.* (2002), this effect would not have been apparent. On two dates, the aphid infestations were actually higher on sprayed trees. Populations of spiders and other natural enemies were often reduced by spraying, suggesting that aphids may have benefited from a negative effect of the particle film on their predators (Knight *et al.*, 2001).

Sowing and Planting Date

In seasonal climates, there is natural variation in the rate of growth, reproduction, and dispersal of aphids and of their natural enemies. Similarly, in rotational cropping systems where there is a cycle of crops in the landscape, the availability of source populations for a particular field will vary. By altering the time that a crop is established, the grower can control the time of crop susceptibility in relation to these fluctuations. Sowing date also affects the size of plants at the time of aphid colonization, with potential effects on between-plant movement of aphids and perception of the crop by incoming alatae. The degree to which aphid damage can be managed in this way is clearly limited by the direct effect of the weather on plant growth, but it has proved a successful strategy in some situations.

For example, work with oilseed rape (*Brassica napus*) in Pakistan examined the effect of six sowing dates ranging from 1 October to 20 November (Shafique *et al.*,

1999). Aphid (species unstated) density ranged from 94 per plant in the early-sown treatment to 392 in the latest sown treatment. Yields declined markedly with later sowing dates, ranging from 496 to 21 g of seed per 2.7 m² plot. A similar effect was evident in Indian work (Saha and Baral, 1999). Late-sown rape was attacked more heavily by *Lipaphis pseudobrassicae* (mustard aphid) and was lower yielding than were treatments sown up to a month earlier (in early November). Another study of this system used regression techniques to determine that each 10 days of delay in sowing resulted in an increase of 36 aphids per 5 cm of stem length and a yield loss of 93 kg/ha (Bhadauria *et al.*, 1992). A broadly similar trend was evident in Bangladesh for the effect of *L. pseudobrassicae* on Indian mustard (*Brassica juncea*) and rape (Rahman *et al.*, 1989).

In tobacco production in Georgia, USA, planting date again influenced infestation by *Myzus persicae nicotianae* (tobacco aphid) (McPherson *et al.*, 1993). Plants sown in late or mid-April rather than in late March had higher aphid population peaks in two of the three years of study. In contrast, delayed sowing was associated with low aphid densities in barley grown in northern England (McGrath and Bale, 1990). However, the potential value arising from late sowing of a cereal may not be economically viable because of the yield penalty resulting from the shorter growing season (Selman, 1980). In a comparison of wheat production systems in the UK, Holland (1997) concluded that the gross margin for later, spring-sown crops was £170/ha lower than for autumn-sown crops, despite substantially lower densities of *Sitobion avenae* (grain aphid) and *Sitodiplosis mosellana* (orange wheat blossom midge).

High densities of *Aphis craccivora* (cowpea aphid) on cowpea (*Vigna unguiculata*) in Uganda tended to be associated with late planting (6 weeks after the onset of rains) (Karungi *et al.*, 2000). However, in the case of *A. gossypii* on cotton in Texas, USA, planting date had no effect on abundance (Parajulee *et al.*, 1999).

Incidence of aphid-borne virus infections can vary independently of aphid populations and may also show seasonal fluctuations. Sowing date effects may lead to differences in the incidence of plant diseases such as *Barley yellow dwarf virus* (BYDV) (Snidaro and Delogu, 1990). In temperate cereals, a high level of aphid activity during autumn may cause early-sown crops to suffer particularly heavy attack. From these crops, there is then more transmission to young plants in the latest planted fields in the subsequent spring (Hunger *et al.*, 1992). For this reason, delayed sowing of winter wheat has been practised in Australia and has led to a decline in BYDV infestations (Johnstone *et al.*, 1990). A study of *Diuraphis noxia* (Russian wheat aphid) in western USA, where it had been a pest since 1986 (Stoetzel, 1987), concluded that a delay in planting of just 12 days from an early September sowing date was sufficient to manage the viral diseases vectored by *D. noxia* and to increase grain yield (Hammon *et al.*, 1996). In a multivariate analysis of field characteristics and agronomic practice, Foster *et al.* (2004) found sowing date to be a key risk factor for BYDV in UK cereals.

Incidence of aphid-borne *Potato virus Y* (PVY) in potatoes increases when the stage of greatest crop susceptibility coincides with the main period of aphid flight. Where soil conditions prevent earlier sowing, crops can be established before aphid activity by pre-sprouting. In late varieties however, potatoes that have not been pre-sprouted will emerge after peak aphid flight, and pre-sprouting may then increase virus infection rates (Saucke and Doring, 2004).

Plant Density

The density of crop plants at the time of peak aphid susceptibility affects the appearance of the crop to alatae, the dispersal of aphids within the crop, and the prevailing microclimatic conditions. The effects of inter-plant competition on host-plant quality may also influence aphids. These mechanisms may have conflicting effects on aphid damage.

High densities of *A. craccivora* on cow-pea in Uganda tended to be associated with low planting density (Karungi *et al.*, 2000). The same pattern has been observed with *A. gossypii* in cotton crops in Texas, USA (Parajulee *et al.*, 1999). Where cotton was planted in rows with two empty rows in between, the mean number of aphids per leaf was 4.71, significantly higher than where cotton was planted in every row (1.82 aphids per leaf), with intermediate densities where cotton was planted in alternate rows. This effect is consistent with the greater visibility of crop plants against the background of bare soil in unplanted rows, leading to easier location by settling alatae, as discussed above. A similar effect was seen in Australian studies of narrow-leaved lupins (*Lupinus angustifolius*); wide row spacings resulted in higher incidence of *Bean yellow mosaic virus* (BYMV) (Jones, 1993). It was considered that delayed canopy closure resulted in prolonged exposure of bare earth between rows and led to a higher population of vector aphids. More recent studies of this system suggest scope for avoiding high aphid immigration in crops at low densities by retaining the stubble of the previous crop (Berlandier and Bwye, 1998; Bwye *et al.*, 1999).

In contrast, wider tree spacing in young plantations of Sakhalin fir (*Abies sachalinensis*) resulted in lower infestation levels of the aphid *Cinara todocola* in Japan (Furuta and Aloo, 1994). This is to be expected since in this system tree-to-tree dispersal of aphids within plantations is largely by apterous females, not alatae.

Crop Plant Pruning

Aphids often feed preferentially on certain parts of a plant (e.g. Wratten, 1974), and some parts may be more susceptible to damage. In some situations, pruning the crop plants may provide an appropriate means of reducing the effect of aphids.

In developing countries, relatively low labour costs can allow techniques that would be uneconomical in developed countries.

Work in central Africa investigated the effectiveness of hand removal of the terminal shoots of cotton plants at the end of the growing season (Deguine *et al.*, 2000). This was important because *A. gossypii* otherwise feed upon the terminal shoot and secrete honeydew that falls onto the open bolls and causes 'sticky cotton'. Removal of terminal shoots using a pruning knife in large-scale field experiments on seven sites in Cameroon had no effect at sites with low aphid densities, but did reduce aphid densities and the proportion of leaves infested at the most heavily infested site. This method was considered inexpensive and well suited to local conditions.

Under other conditions, however, host-plant pruning can exacerbate aphid infestation because it encourages the growth of lush new shoots favoured by the herbivores. The aphid *Hyadaphis tataricae* feeds on the succulent new growth of honeysuckle (*Lonicera* spp.). Fieldwork in Nebraska, USA, indicated that plants that were pruned and watered had greater aphid densities than plants that were either just watered or just pruned or untreated (Coffelt and Jones, 1989).

Irrigation and Fertilizer Management

Rates of aphid growth and reproduction are influenced strongly by the quality of sap they obtain from their host plants. This varies with the stage of plant development, soil fertility, and water status. While management of water balance and nutrient availability is determined primarily by grower requirements for optimal crop quality and yield, the impact on aphid damage also should be considered. Other agronomic practices that improve plant quality can have comparable indirect effects.

It is well known that high soil nitrogen concentrations can lead to host plants being more heavily attacked by aphids. *Metopolophium dirhodum* (rose–grain aphid), for example, was favoured by nitrogen applications to winter wheat (Zhou and Carter, 1991). Increased aphid performance has been associated with high amino acid concentrations in plant phloem sap (van Emden, 1972; Wratten, 1974; Jansson and Smilowitz, 1986; Weibull, 1987), but the relationship between sap composition and pest populations is not always clear (Bi *et al.*, 2003). The effect of water balance can also be complicated, with factors such as the periodicity of water stress determining whether aphids benefit (from increased sap nitrogen concentrations) or suffer (from reduced turgor pressure) (Archer *et al.*, 1995; Huberty and Denno, 2004).

Aphis gossypii developed into one of the most serious pests of Californian cotton during the 1990s. This period coincided with the use of a plant growth regulator (mepiquat chloride) as a means of limiting early-season vegetative growth. Before this, a combination of managed nitrogen and water deficits was used, making conditions less favourable for aphids (Godfrey *et al.*, 2000). Small plot field experiments showed that aphid densities were three times higher under high applications of nitrogen fertilizer (218 kg/ha) than where a low nitrogen input (55 kg/ha) was used (Cisneros and Godfrey, 1998). A similar association was evident in work by Slosser *et al.* (1997). Larger-scale work monitoring cotton under a range of nitrogen regimes revealed a consistent trend for greater aphid densities with high levels of nitrogen fertilizer (Godfrey *et al.*, 2000). Furthermore, cage tests using laboratory-reared aphids placed into the field plots showed that the first cohort of aphids had shorter generation times and greater fecundity under a high nitrogen (218 kg/ha) treatment. The effect of high nitrogen availability was reduced in a treatment in which high nitrogen was 'balanced' with potassium (109 kg/ha K_2O). In contrast, no differences were observed with *L. pseudobrassicae* between potassium applications rates of 30, 45, 60, and 75 kg/ha to oilseed rape in India (Saha and Baral, 1999).

High soil nitrogen concentrations may not always enhance aphid densities. In the Lebanon, *Rhopalosiphum maidis* (corn leaf aphid) densities on maize at the silk stage of growth were lower under the most fertilized treatment. This treatment was ammonium nitrate (100 kg N/ha) at planting and again at

flowering. A second treatment consisted of a single application of 100 kg N/ha at planting and a third received no nitrogen fertilizer (Atiyeh *et al.*, 1996). It was postulated that plants grown without nitrogen supplementation remobilized nitrogen from leaves at the time of silking, an effect that would have been delayed in the treatment that received fertilizer at the time of flowering. Proteolysis and hydrolysis of structural proteins would have increased concentrations of amino acids in sap as they were translocated to the maize ear. Such an effect would not, however, explain the low aphid densities in the treatment that received 100 kg N/ha at planting, in which amino acid concentrations would be expected to be higher than in the zero nitrogen treatment.

Organic and synthetic fertilizers may differ in their effects on aphid damage. These differences can arise from their nutrient status or the additional effects that mulching fertilizers such as composts can have on the soil habitat, as described above. Yardim and Edwards (2003) recorded aphid populations on tomatoes treated with synthetic fertilizer or composted cow manure applied at an 'equivalent' rate. The populations were higher in the organic treatment in the first year but lower in the second.

Intercropping, Living Mulches and Cover Crops

Intercropping is the growing of multiple crops in a field, spatially integrated in such a way as to alter the environment of the plants of each crop relative to a monoculture. When aphid-susceptible crops are mixed with non-hosts, a key aspect of this environment is the combination of cues available to aphids arriving at, and moving within, the field. Undersowing a non-crop species or allowing weed cover to develop has a comparable effect to intercropping on plant diversity. These practices may also have benefits such as fixing nitrogen, suppressing undesirable weeds, retaining soil moisture, or reducing erosion.

It has been suggested that non-host plants may interfere with the host-finding ability of specialist herbivores such as aphids (Root, 1973). This hypothesis is in accordance with the frequent observation that reduced aphid densities in high diversity crops are the result of slower colonization of host plants (Smith, 1969; Horn, 1981; Costello and Altieri, 1994; Lehmhus *et al.*, 1996; Vidal, 1997; but see Helenius, 1997) rather than differences in rates of reproduction or survival (Costello and Altieri, 1995; Lehmhus *et al.*, 1996). When acting this way, the effect of increased vegetation diversity on aphid colonization rates would depend on the host-finding mechanism employed by a particular species of aphid. If an aphid species locates its host using chemosensory information, volatiles produced by non-crop plants may mask the chemical cues required by aphids for host location (Tahvanainen and Root, 1972). However, chemosensory information may only be sufficient to arrest a dispersing aphid in the general vicinity of a host plant, rather than provide explicit directional information on its location (Finch and Collier, 2000). At this point, visual information may be of primary importance for host location.

If an aphid species relies on visual information for host location, increased crop diversity may provide a confusing landscape mosaic, making it difficult for winged aphids to identify host plants.

In an intercropped system, aphid attack on the 'primary' crop may also be affected by the density of the 'secondary' crop (Fig. 16.3). For example, when maize was sown at differing densities within beans, the positive relationship between density and population of *Aphis fabae* (black bean aphid) was partly attributed to variation in the amount of bare soil and the impact of this on aphid landing responses (Ogenga-Latigo *et al.*, 1992). A more important factor reducing the incidence of aphids on the intercropped beans, however, was disruption of the aphid's location of the host plants. Trap data indicated that few aphids penetrated the maize canopy to reach and settle on the beans. Host-location by alatae involves landing on potential host plants and probing (van Emden, 1973; Irwin *et al.*, Chapter 7 this volume). This is followed by settling, or rejection, and take-off. *Aphis fabae* may have tended to land on the taller maize plants in the intercropped bean system and dispersed from the area after probing had detected the

Fig. 16.3.   An intercrop of maize and beans (*Phaseolus vulgaris*) in northwest Peru.

non-host status of maize. A final contributing factor may have been adverse microclimatic conditions, particularly low light levels, inhibiting colonization of alatae on the beans (Ogenga-Latigo *et al.*, 1992).

Many phytophagous insects avoid landing on brown surfaces such as soil (Kostal and Finch, 1994), but if non-crop plants are present, rather than bare soil, winged aphids arrested by olfactory stimuli may still have difficulty locating their host. This confusion will be particularly important if an insect requires contact stimuli from more than one 'appropriate' landing (i.e. host plant) before colonization (Finch and Collier, 2000). Consistent with this hypothesis is the observation that aphids prefer to colonize low-density crops (discussed above; Heathcote, 1970) as well as monocultures. In addition, the presence of cover crops between rows of broccoli reduces light reflectance at certain spectral wavebands (Costello and Altieri, 1994). As aphid colonization rates (and numbers of winged aphids trapped in yellow water pan traps) are positively correlated with reflected light intensity in the blue-green, yellow, and orange wavebands (Costello and Altieri, 1994), the increased reflectance offered by clean cropped or widely spaced crop plants appears to be more attractive to incoming alate aphids (Kennedy *et al.*, 1961; Costello and Altieri, 1994). If the mechanism of visual interference prevails, any species of non-crop plant, or even increased densities of crop plants (of the same or mixed species), potentially could be used to reduce aphid colonization rates.

Horn (1981) examined the effects of a weedy background on colonization of collards (*Brassica oleracea* var. *viridis*) by *Myzus persicae* (peach–potato aphid) and its predators. In both study years (1978 and 1979), *M. persicae* preferentially invaded (and developed higher densities on) weed-free collards. However, this difference in aphid density was considerably more pronounced in 1979, when weedless collards averaged nearly twice the height of weedy collards than in 1978, when there was no apparent difference in collard size between treatments. Preference for larger plants (perhaps

due to easier location or improved nutrition) may therefore have been partially responsible for increased aphid abundance on weedless collards in 1979. Showler and Greenberg (2003) found higher peak populations of aphids in weed-free than in weedy cotton plots. This was associated with lower populations of some (but not all) predatory orders. However, the implications of weeds for crop damage were not clear because some herbivores were positively affected by weeds, and the effect of weeds varied through the season.

Increasing crop diversity may have an indirect effect on aphid abundance by altering crop plant quality. Quality of the crop plant may be improved (e.g. by intercropping with a legume) or impaired (if the non-crop plant competes with the crop plant for resources) by increased diversity, leading to variation in aphid abundance. In autumn 1990, Costello and Altieri (1994) found an increased rate of broccoli infestation by *Brevicoryne brassicae* (cabbage aphid) in clean cultivated plots (55%) compared with plots where strawberry clover (*Trifolium fragiferum*) was used as a living mulch (7.5%). However, this increase in infestation was apparent only in treatments with low soil nitrogen (i.e. where compost, rather than synthetic fertilizer was applied), where nitrogen fixed by the strawberry clover was more likely to affect plant quality. In summer 1991, this increase in infestation rates was evident in both fertilizer treatments, although it was still more severe in low nitrogen (compost) treatments. Although post-colonization aphid population growth was unaffected by nitrogen levels, *B. brassicae* did appear to be more attracted to nitrogen-stressed flower buds (Costello and Altieri, 1994). Likewise, Lehmhus *et al.* (1996) found that although undersowing white cabbage with clover significantly reduced colonization by *B. brassicae* and *M. persicae*, the natural increase in aphid numbers over time was unaffected, or even enhanced, by crop diversification. Plant quality may therefore affect host location/choice more than population growth of aphids.

Although plant quality potentially is important, it is unlikely to be a sole cause of differences in aphid colonization between mono and polycultures as Horn (1981) and Costello and Altieri (1994) both still found differences, though less pronounced, in aphid infestation in experiments where plant quality did not appear to be a major factor. Moreover, when broccoli plants were grown in containers to reduce competition for soil nutrients and water, undersowing with rye grass significantly reduced densities of *B. brassicae*, despite there being no difference in leaf water, nitrogen content, or specific dry weight between treatments (Vidal, 1997).

It is logical that many of the mechanisms discussed above in the context of their effect on aphids should also have an analogous impact on predatory arthropods. For example, enemies that search for prey using visual, tactile, or host-plant chemical cues are likely to be less effective at locating, and hence reducing, aphid populations in diverse systems (Sheehan, 1986). A complex set of tritrophic interactions operates whereby a predator may not only be affected directly by variation in diversity, but also indirectly by the effects of diversity on its prey (Smith, 1969). Understanding these relationships is also necessary if the effects of these cultural methods of control are to be predicted.

Trap Crops

If non-crop plants that are attractive to aphids are planted near to crop plants, they can act as a sink or barrier known as a trap crop. Trap crops may be used either to prevent pests from reaching the target crop, or to localize them in an area that allows easier chemical or other control such as plant destruction (Hokkanen, 1991). Although trap crops may be a different species from the target crop, this is not a prerequisite. If the pest has a limited host range, use of an area of the main crop (e.g. borders) as a trap may be preferable. For aphids, earlier sowing of the barrier will give it a height differential particularly effective at filtering out some of the immigrating aphids and

concentrating the deposition of others in the edge of the main crop (Lewis, 1965).

Even if the trap crop does not act as a permanent sink for pests, it may slow colonization of the target crop or, more importantly, act as a sink for aphid-borne diseases (Jones, 1993; Thieme *et al.*, 1998; Fereres, 2000). Trap crops have been shown to reduce the spread of non-persistent viruses PVY and BYMV either by acting as barriers to viruliferous aphid movement, or because aphids lost infectivity by probing the non-crop plant (Jones, 1993; Thieme *et al.*, 1998).

Trap crops can also act as a sink for aphidophages. Thus, unless they also operate as a source, overall they may have a negative effect on biological control. For example, when a population of the parasitoid *Aphidius ervi* was maintained on a population of *Acyrthosiphon pisum* (pea aphid) outside the target cereal crop, the parasitoids did not readily transfer later to the cereal aphid pest *S. avenae* (Cameron *et al.*, 1984). After four to five generations on *S. avenae*, *A. ervi* began to parasitize this species more readily, and this adaptation to the new host was associated with the loss of several esterase bands revealed during electrophoretic examination of parasitoid proteins (Cameron *et al.*, 1984). The time taken for this species to adapt to new hosts may make ineffective the use of alternative hosts to accumulate reservoirs of parasitoids.

Provision of Resources for Natural Enemies

If predators or parasitoids require nutritional resources other than those provided by aphids or the target crop, these can often be provided by sowing flowering plants among or alongside the crop (Figs 16.4, 16.5, 16.6) (Wratten, 1992; Hickman and Wratten, 1996; Baggen and Gurr, 1998; Nentwig, 1998; Stephens *et al.*, 1998; Wratten *et al.*, 1998). Sugars from floral and extrafloral nectaries can have beneficial effects on parasitoids (Heimpel and Jervis, 2005), and pollen can provide an important

Fig. 16.4. Flowering buckwheat (*Fagopyrum esculentum*) strips beneath apples to increase biological control of a range of aphid species by parasitoid wasps, New Zealand.

Fig. 16.5. Flowering phacelia (*Phacelia tanacetifolia*) in maize to increase biological control of *Rhopalosiphum padi* (bird cherry–oat aphid) by providing nectar for parasitoid wasps *Aphidius* spp. (Hymenoptera: Aphidiidae), New Zealand.

protein source for female predators (e.g. hover flies) approaching sexual maturity (Andow and Risch, 1985; Jervis *et al.*, 1993; Hickman *et al.*, 1995; Irvin *et al.*, 2000). If flowers are to be used to enhance natural enemy populations, the floral architecture must be compatible with the feeding morphology and behaviour of the target biological control agent (Gilbert, 1985; Patt *et al.*, 1997a,b).

Fig. 16.6. Buckwheat (*Fagopyrum esculentum*) strips sown in a vineyard to improve biological control of pests, New Zealand.

Enemies may also need alternative hosts or prey to maintain high population densities before the establishment of pest populations (van Emden, 1990; Corbett and Rosenheim, 1996; Landis *et al.*, 2000; Langer and Hance, 2004). Where an aphid is not a pest in a particular system, the aphid and its host plant can be introduced as a 'banker plant system' to a glasshouse where a different crop is to be grown later. Thus, *Schizaphis graminum* (greenbug) on wheat and its associated poly- or oligophagous parasitoids (*Aphidius colemani* and *Lysiphlebus testaceipes*) provide a banker system for glasshouse beans (see also Powell and Pell, Chapter 18 this volume). This 'alternative host and parasitoid in first' method provides a reservoir of natural enemies in case the target pest (*M. persicae*) is accidentally introduced (Starý, 1993). In a more open environment, *Aphis helianthi* on sunflowers has been used as an alternative host for *L. testaceipes* when the pest *S. graminum* is absent from the sorghum crop (Powell, 1986).

Aggregation of natural enemies around resource-providing plants has frequently been recorded (van Emden, 1963; Root, 1973; Hickman and Wratten, 1996; Hooks *et al.*, 1998; Berndt *et al.*, 2002). This can lead to

higher, early-season densities of predators and parasitoids in mixed crops (Smith, 1969; Horn, 1981; Bugg *et al.*, 1991; Costello and Altieri, 1995; Theunissen *et al.*, 1995; Hickman and Wratten, 1996; Lehmhus *et al.*, 1996, 1999; Montandon and Slosser, 1996; Goller *et al.*, 1997; Vidal, 1997; but see Moreby and Sotherton, 1997; also Hooks *et al.*, 1998) that may (Horn, 1981) or may not (White *et al.*, 1995) disperse readily to monocultures later in the season. It should be noted, however, that an increase in predator abundance does not always translate to reduced aphid density (Bugg *et al.*, 1991; Theunissen *et al.*, 1995; Lehmhus *et al.*, 1996; Goller *et al.*, 1997), particularly if predator numbers are high only in response to high aphid population densities, rather than to floral or other resources in mixed crops. For example, although border planting with phacelia (*Phacelia tanacetifolia*) enhanced biological control of brassica pests by hover flies, the flies did not disperse far from the pollen source (White *et al.*, 1995). Increased diversity can enhance (Letourneau, 1987), reduce (Helenius, 1993; Costello and Altieri, 1995), or have no effect on (Letourneau, 1990) rates of parasitism, making it difficult to predict a general effect of increasing diversity on aphid natural enemies.

As an alternative to managing non-crop plants, supplementary foods can be provided as foliar applications. Formulations containing sugars, proteins, and yeasts have been developed to attract or retain natural enemies in fields while pest populations are still low (Wade *et al.*, 2004). This technique has been found to cause short-term, localized increases of aphidophagous ladybird, lacewing, and hover fly populations in lucerne (*Medicago sativa*) (Evans and Swallow, 1993; Evans and Richards 1997). It has not, however, been developed fully as a method for aphid control.

Provision of Refuges for Natural Enemies

Monocultural crops or their associated field margins rarely provide a consistently favourable habitat for natural enemies (Figs 16.7, 16.8). At one extreme, natural enemy populations in annual crops are likely to suffer high mortality or enforced dispersal at the time of harvesting, although this can be managed. However, many perennial crops will also vary seasonally. Independently of changes in the crop itself, the requirements of natural enemies may change over time. The agricultural landscape can be managed to provide alternative habitats that act as refuges for natural enemies at times when crops or the within-field microclimate are unfavourable.

Maintenance of grass strips around or within fields have been shown to increase populations of predatory beetles, spiders, and parasitoids in adjacent crops, and some studies have shown reduced aphid populations in association with such 'beetle banks' (van Emden, 1990; Thomas *et al.*, 1992; Kidd and Jervis, 1996; Gurr *et al.*, 1998; Landis *et al.*, 2000; Collins *et al.*, 2002; Langer and Hance, 2004; MacLeod *et al.*, 2004; Levie *et al.*, 2005). Additional techniques may be used to encourage dispersal of natural enemies. For example, it may be profitable in the following spring to set out sources of aphid sex pheromone to attract emerging parasitoids that overwintered as mummies in grass strips. The parasitoids are attracted to the pheromone but at this time it has no effect on aphids (Powell, 2000; see also Pickett and Glinwood, Chapter 9 this volume; also Powell and Pell, Chapter 18 this volume). Additionally, predator populations may increase between years, reaching levels

Fig. 16.7. An extreme example of a field boundary with no margin vegetation. The line of traps is for sampling hover fly and parasitoid populations.

Fig. 16.8. Monocultural cotton in New South Wales, Australia.

that would otherwise have been unlikely in a wholly disturbed cropping system (Thomas *et al.*, 1992).

Where different stages of a crop are present simultaneously, appropriate spatial arrangements of these can allow dispersal of natural enemies from harvested areas to susceptible areas. Strip harvesting employs this technique and can enhance predator populations in, for example, lucerne (Smith and van den Bosch, 1967; Hossain *et al.*, 2002). Strategically located crops that have a complementary growth cycle can also serve as refuges, a system known as relay cropping. In an experimental study with cotton grown adjacent to lucerne, aphid numbers were lowered in the cotton by cutting the lucerne, and the numbers of spider and lacewing predators showed a corresponding increase (Lin *et al.*, 2003).

As with natural enemy populations enhanced by the provision of food resources or alternative hosts, the effect on aphid populations of increased populations in refuges may be limited either by low enemy dispersal or a weak correlation between aphid mortality and enemy population. Nevertheless, landscape scale patterns of habitat have been shown to be important for several aphid predators and parasitoids (Elliott *et al.*, 2002; Thies *et al.*, 2003, 2005).

Conclusions

The potential for achieving control of aphid damage by manipulating the physical and biological environment of the crop is enormous. However, the potential for causing unintended effects on crop yields is similarly large. Research has investigated the mechanisms by which new and traditional cultural methods influence aphid populations. There is still much to be learned about the complex relationships between the many components of agroecosystems and, as this knowledge grows, so should our ability to predict which techniques are appropriate in particular pest management scenarios.

Many cultural methods of control have intrinsic benefits over alternative means of reducing aphid damage. In particular, they do not impose the externalized costs often associated with insecticide use; that is, environmental damage and health risks (Rajendran, 2002). As Tilman *et al.* (2002)

have said: 'Further increases in agricultural output are essential for global political and social stability and equity. Doubling food production again, and sustaining food production at this level, are major challenges. Doing so in ways that do not compromise environmental integrity and public health is a greater challenge still.' Cultural pest management has an important role in meeting this challenge.

References

Andow, D.A. (1991) Vegetational diversity and arthropod population response. *Annual Review of Entomology* 36, 561–586.

Andow, D.A. and Risch, S.J. (1985) Predation in diversified agroecosystems: relations between a coccinellid predator *Coleomegilla maculata* and its food. *Journal of Applied Ecology* 22, 357–372.

Archer, T.L., Bynum, E.D., Onken, A.B. and Wendt, C.W. (1995) Influence of water and nitrogen fertilizer on biology of the Russian wheat aphid (Homoptera: Aphididae) on wheat. *Crop Protection* 14, 165–169.

Atiyeh, R., Aslam, M. and Baalbaki, R. (1996) Nitrogen fertilizer and planting date effects on insect pest populations of sweet corn. *Pakistan Journal of Zoology* 28, 163–167.

Avilla, C., Collar, J.L., Duque, M., Perez, P. and Fereres, A. (1997) Impact of floating rowcovers on bell pepper yield and virus incidence. *Hortscience* 32, 882–883.

Baggen, L.R. and Gurr, G.M. (1998) The influence of food on *Copidosoma koehleri*, and the use of flowering plants as a habitat management tool to enhance biological control of potato moth, *Phthorimaea operculella*. *Biological Control* 11, 9–17.

Berlandier, F.A. and Bwye, A.M. (1998) Cultural practices to control aphid landings in narrow-leaved lupin crops in Western Australia. In: Zaluki, M.P., Drew, R.A.I. and White, G.C. (eds) *Proceedings of the Sixth Australian Applied Entomological Research Conference.* University of Queensland Press, Brisbane, pp. 289–293.

Berndt, L.A., Wratten, S.D. and Hassan, P.G. (2002) Effects of buckwheat flowers on leafroller (Lepidoptera: Tortricidae) parasitoids in a New Zealand vineyard. *Agricultural and Forest Entomology* 4, 39–45.

Bhadauria, N.S., Bahadur, J., Dhamdhere, S.V. and Jakhmola, S.S. (1992) Effect of different sowing dates of mustard crop on infestation by the mustard aphid, *Lipaphis erysimi* (Kalt.). *Journal of Insect Science* 5, 37–39.

Bi, J.L., Toscano, N.C. and Madore, M.A. (2003) Effect of urea fertilizer application on soluble protein and free amino acid content of cotton petioles in relation to silverleaf whitefly (*Bemisia argentifolii*) populations. *Journal of Chemical Ecology* 29, 747–761.

Brault, D., Stewart, K.A. and Jenni, S. (2002) Optical properties of paper and polyethylene mulches used for weed control in lettuce. *Hortscience* 37, 87–91.

Brewer, M.J. and Elliott, N.C. (2004) Biological control of cereal aphids in North America and mediating effects of host plant and habitat manipulations. *Annual Review of Entomology* 49, 219–242.

Brown, J.E., Yates, R.P., Stevens, C., Khan, V.A. and Witt, J.B. (1996) Reflective mulches increase yields, reduce aphids and delay infection of mosaic viruses in summer squash. *Journal of Vegetable Crop Production* 2, 55–60.

Brown, M.W. and Tworkoski, T. (2004) Pest management benefits of compost mulch in apple orchards. *Agriculture, Ecosystems and Environment* 103, 465–472.

Bugg, R.L., Dutcher, J.D. and McNeill, P.J. (1991) Cool-season cover crops in the pecan orchard understorey: effects on Coccinellidae (Coleoptera) and pecan aphids (Homoptera: Aphididae). *Biological Control* 1, 8–15.

Bwye, A.M., Jones, R.A.C. and Proudlove, W. (1999) Effects of different cultural practices on spread of cucumber mosaic virus in narrow-leafed lupins (*Lupinus angustifolius*). *Australian Journal of Agricultural Research* 50, 985–996.

Cameron, P.J., Powell, W. and Loxdale, H.D. (1984) Reservoirs for *Aphidius ervi* Haliday (Hymenoptera: Aphidiidae), a polyphagous parasitoid of cereal aphids (Hemiptera: Aphididae). *Bulletin of Entomological Research* 74, 647–656.

Cartwright, B., Roberts, B.W., Hartz, T.K. and Edelson, J.V. (1990) Effects of mulches on the population increase of *Myzus persicae* (Sulzer) on bell peppers. *Southwestern Entomologist* 15, 475–479.

Chiang, J.K., Chou, D.S. and Huang, M.C. (1992) Field trials on the control of aphid-borne viruses by PE mulching in the fall tobacco planting [in Chinese]. *Bulletin of the Tobacco Research Institute, Taiwan Tobacco and Wine Monopoly Bureau* 36, 71–77.

Cisneros, J.J. and Godfrey, L.D. (1998) Agronomic and environmental factors influencing control of cotton aphids with insecticides. In: *Proceedings Beltwide Cotton Conferences, San Diego, January, 1998, Volume 2*. National Cotton Council, Memphis, pp. 1242–1246.

Coffelt, M.A. and Jones, J.A. (1989) Bionomics of *Hyadaphis tataricae* (Homoptera: Aphididae). *Environmental Entomology* 18, 46–50.

Collins, K.L., Boatman, N.D., Wilcox, A., Holland, J.M. and Chaney, K. (2002) Influence of beetle banks on cereal aphid predation in winter wheat. *Agriculture, Ecosystems and Environment* 93, 337–350.

Corbett, A. and Rosenheim, J.A. (1996) Impact of a natural enemy overwintering refuge and its interaction with the surrounding landscape. *Ecological Entomology* 21, 155–164.

Costello, M.J. and Altieri, M.A. (1994) Living mulches suppress aphids in broccoli. *California Agriculture* 48, 24–28.

Costello, M.J. and Altieri, M.A. (1995) Abundance, growth rate and parasitism of *Brevicoryne brassicae* and *Myzus persicae* (Homoptera: Aphididae) on broccoli grown in living mulches. *Agriculture, Ecosystems and Environment* 52, 187–196.

Cottrell, T.D., Wood, B.W. and Reilly, C.C. (2002) Particle film affects black pecan aphid (Homoptera: Aphididae) on pecan. *Journal of Economic Entomology* 95, 782–788.

Csizinszky, A.A., Schuster, D.J. and Dring, J.B. (1995) Color mulches influence yield and insect pest populations in tomatoes. *Journal of the American Society for Horticultural Science* 120, 778–784.

Deguine, J.P., Gozé, E. and Leclant, F. (2000) The consequences of late outbreaks of the aphid *Aphis gossypii* in cotton growing in central Africa: towards a possible method for the prevention of cotton stickiness. *International Journal of Pest Management* 46, 85–89.

Dempster, J.P. (1969) Some effects of weed control on the numbers of the small cabbage white (*Pieris rapae* L.) on Brussels sprouts. *Journal of Applied Ecology* 6, 339–345.

Dent, D. (2000) *Insect Pest Management*. CAB International, Wallingford, Oxon, 432 pp.

Elliott, N.C., Kieckhefer, R.W., Michers, G.J. and Giles, K.Z. (2002) Predator abundance in alfalfa fields in relation to aphids, within field vegetation and landscape matrix. *Environmental Entomology* 31, 253–260.

van Emden, H.F. (1963) Observations on the effects of flowers on the activity of parasitic Hymenoptera. *Entomologists Monthly Magazine* 98, 265–270.

van Emden, H.F. (1972) Aphids as phytochemists. In: Harborne, J.B. (ed.) *Phytochemical Ecology*. Academic Press, London, pp. 25–43.

van Emden, H.F. (1973) Aphid host plant relationships. Some recent studies. In: *Perspectives in Aphid Biology*. The Entomological Society of New Zealand, Auckland, pp. 54–64.

van Emden, H.F. (1990) Plant diversity and natural enemy efficiency in agroecosystems. In: Mackauer, M., Ehler, L.E. and Roland, J. (eds) *Critical Issues in Biological Control*. Intercept, Andover, pp. 63–80.

van Emden, H.F. and Williams, G.F. (1974) Insect stability and diversity in agro-ecosystems. *Annual Review of Entomology* 19, 455–475.

Espinoza, H.R. and McLeod, P.J. (1994) Use of row cover in cantaloupe (*Cucumis melo* L.) to delay infection of aphid-transmitted viruses in Honduras. *Turrialba* 44, 179–183.

Evans, A., Wratten, S.D., Frampton, C., Causer, S. and Hamilton, M. (1997) Row covers: effects of wool and other materials on pest numbers, microclimate and crop quality. *Journal of Economic Entomology* 90, 1661–1664.

Evans, E.W. and Richards, D.R. (1997) Managing the dispersal of ladybird beetles (Col: Coccinellidae): use of artificial honeydew to manipulate spatial distributions. *Entomophaga* 42, 93–102.

Evans, E.W. and Swallow, J.G. (1993) Numerical responses of natural enemies to artificial honeydew in Utah alfalfa. *Environmental Entomology* 22, 1392–1401.

Farias-Larios, J. and Orozco-Santos, M. (1997a) Color polyethylene mulches increase fruit quality and yield in watermelon and reduce insect pest populations in dry tropics. *Gartenbauwissenschaft* 62, 255–260.

Farias-Larios, J. and Orozco-Santos, M. (1997b) Effect of polyethylene mulch colour on aphid populations, soil temperature, fruit quality, and yield of watermelon under tropical conditions. *New Zealand Journal of Crop and Horticultural Science* 25, 369–374.

Farias-Larios, J., Orozco-Santos, M., Guzman, S. and Aguilar, S. (1994) Soil temperature and moisture under different plastic mulches and their relation to growth and cucumber yield in a tropical region. *Gartenbauwissenschaft* 59, 249–252.

Farias-Larios, J., Orozco-Santos, M. and Perez, J. (1999) Effect of plastic mulch, floating row cover and microtunnels on insect populations and yield of muskmelon. *Plasticulture* 118, 6–13.

Fereres, A. (2000) Barrier crops as a cultural control measure of non-persistently transmitted aphid-borne viruses. *Virus Research* 71, 221–231.

Finch, S. and Collier, R.H. (2000) Host-plant selection by insects – a theory based on 'appropriate/inappropriate landings' by pest insects of cruciferous plants. *Entomologia Experimentalis et Applicata* 96, 91–102.

Foster, G.N., Blake, S., Tones, S.J., Barker, I. and Harrington, R. (2004) Occurrence of barley yellow dwarf virus in autumn-sown cereal crops in the United Kingdom in relation to field characteristics. *Pest Management Science* 60, 113–125.

Furuta, K. and Aloo, I.K. (1994) Between-tree distance and spread of the Sakhalin fir aphid (*Cinara todocola* Inouye) (Hom., Aphididae) within a plantation. *Journal of Applied Entomology* 117, 64–71.

Gilbert, F.S. (1985) Ecomorphological relationships in hoverflies (Diptera: Syrphidae). *Proceedings of the Royal Society of London B* 224, 91–95.

Glenn, D.M. and Puterka, G.J. (2005) Particle films: a new technology for agriculture. In: Janick, J. (ed.) *Horticultural Reviews, Volume 31.* John Wiley and Sons, Hoboken, New Jersey, pp. 1–44.

Glenn, D.M., Puterka, G.J., Vanderzwet, T., Byers, R.E. and Feldhake, C. (1999) Hydrophobic particle films: a new paradigm for suppression of arthropod pests and plant diseases. *Journal of Economic Entomology* 92, 759–771.

Godfrey, L.D., Cisneros, J.J., Keillor, K.E. and Hutmacher, R.B. (2000) Influence of cotton nitrogen fertility on cotton aphid, *Aphis gossypii*, population dynamics in California. In: *Proceedings Beltwide Cotton Conferences, San Antonio, January 2000, Volume 2.* National Cotton Council, Memphis, pp. 1162–1165.

Goller, E., Nunnenmacher, L. and Goldbach, H.E. (1997) Faba beans as a cover crop in organically grown hops: influence on aphids and aphid antagonists. *Biological Agriculture and Horticulture* 15, 279–284.

Greer, L. and Dole, J.M. (2003) Aluminum foil, aluminum-painted, plastic, and degradable mulches increase yields and decrease insect-vectored viral diseases of vegetables. *HortTechnology* 13, 276–284.

Gurr, G.M., van Emden, H.F. and Wratten, S.D. (1998) Habitat manipulation and natural enemy efficiency: implications for the control of pests. In: Barbosa, P. (ed.) *Conservation Biological Control.* Academic Press, San Diego, pp. 155–183.

Gurr, G.M., Barlow, N.D., Memmott, J., Wratten, S.D. and Greathead, D.J. (2000) A history of methodological, theoretical and empirical approaches to biological control. In: Gurr, G. and Wratten, S. (eds) *Biological Control: Measures of Success.* Kluwer, Dordrecht, pp. 3–37.

Gurr, G.M., Wratten, S.D. and Altieri, M.A. (2004) (eds) *Ecological Engineering for Pest Management: Advances in Habitat Manipulation for Arthropods.* CSIRO, Collingwood, Australia and CAB International, Wallingford, 232 pp.

Hammon, R.W., Pearson, C.H. and Peairs, F.B. (1996) Winter wheat planting date effect on Russian wheat aphid (Homoptera: Aphididae) and a plant virus complex. *Journal of the Kansas Entomological Society* 69, 302–309.

Harrewijn, P. and Minks, A.K. (1989) Integrated aphid management: general aspects. In: Minks, A.K. and Harrewijn, P. (eds) *Aphids. Their Biology, Natural Enemies and Control, Volume 2 C.* Elsevier, Amsterdam, pp. 267–272.

Harrewijn, P., den Ouden, H. and Piron, P.G.M. (1991) Polymer webs to prevent virus transmission by aphids in seed potatoes. *Entomologia Experimentalis et Applicata* 58, 101–107.

Heathcote, G.D. (1970) Effect of plant spacing and time of sowing of sugar beet on aphid infestation and spread of virus yellows. *Plant Pathology* 19, 32–39.

Heimpel, G.E. and Jervis, M.A. (2005) Does floral nectar improve biological control by parasitoids? In: Wäckers, F.L., van Rijn, P.C.J. and Bruin, J. (eds) *Plant-Provided Food and Herbivore–Carnivore Interactions.* Cambridge University Press, Cambridge, pp. 267–304.

Helenius, J. (1993) Incidence of specialist natural enemies of *Rhopalosiphum padi* (L.) (Hom., Aphididae) on oats in monocrops and mixed intercrops with faba bean. *Journal of Applied Entomology* 109, 136–143.

Helenius, J. (1997) Spatial scales in ecological pest management (EPM): importance of regional crop rotations. *Biological Agriculture and Horticulture* 15, 163–170.

Hickman, J.M. and Wratten, S.D. (1996) Use of *Phacelia tanacetifolia* strips to enhance biological control of aphids by hoverfly larvae in cereal fields. *Journal of Economic Entomology* 89, 832–840.

Hickman, J.M., Lövei, G.L. and Wratten, S.D. (1995) Pollen feeding by adults of the hoverfly *Melanostoma fasciatum* (Diptera: Syrphidae). *New Zealand Journal of Zoology* 22, 387–392.

Hokkanen, H.M.T. (1991) Trap cropping in pest management. *Annual Review of Entomology* 36, 119–138.

Holland, J.M. (1997) Impact of integrated farming husbandry practices on cereal pests and yield. In: *Optimising Cereal Inputs: Its Scientific Basis. Part 2. Crop Protection and Systems. Aspects of Applied Biology, No. 50,* pp. 305–311.

Hooks, C.R.R., Valenzuela, H.R. and Defrank, J. (1998) Incidence of pests and arthropod natural enemies in zucchini grown with living mulches. *Agriculture, Ecosystems and Environment* 69, 217–231.

Horn, D.J. (1981) Effect of weedy backgrounds on colonization of collards by green peach aphid, *Myzus persicae*, and its major predators. *Environmental Entomology* 10, 285–289.

Hossain, Z., Gurr, G.M., Wratten, S.D. and Raman, A. (2002) Habitat manipulation in lucerne (*Medicago sativa* L.): arthropod population dynamics in harvested and 'refuge' crop strips. *Journal of Applied Ecology* 39, 445–454.

Huberty, A.F. and Denno, R.F. (2004) Plant water stress and its consequences for herbivorous insects: a new synthesis. *Ecology* 85, 1383–1398.

Hunger, R.M., Sherwood, J.L., Evans, C.K. and Montana, J.R. (1992) Effects of planting date and inoculation date on severity of wheat streak mosaic in hard red winter wheat cultivars. *Plant Disease* 76, 1056–1060.

Irvin, N.A., Wratten, S.D. and Frampton, C.M. (2000) Understorey management for the enhancement of the leafroller parasitoid *Dolichogenidea tasmanica* (Cameron) in orchards at Canterbury, New Zealand. In: Austin, A.D. and Dowton, M. (eds) *Hymenoptera: Evolution, Biodiversity and Biological Control.* CSIRO, Collingwood, Australia, pp. 396–403.

Jansson, R.K. and Smilowitz, Z. (1986) Influence of nitrogen on population parameters of potato insects: abundance, population growth, and within-plant distribution of the green peach aphid, *Myzus persicae* (Homoptera: Aphididae). *Environmental Entomology* 15, 49–55.

Jenni, S., Dubuc, J.F. and Stewart, K.A. (2003) Plastic mulches and row covers for early and midseason crisphead lettuce produced on organic soils. *Canadian Journal of Plant Science* 83, 921–929.

Jervis, M.A., Kidd, N.A.C., Fitton, M.G., Huddleston, T. and Dawah, H.A. (1993) Flower-visiting by hymenopteran parasitoids. *Journal of Natural History* 27, 67–105.

Johnstone, G.R., Sward, R.J., Farrell, J.A., Greber, R.S., Guy, P.L., McEwan, J.M. and Waterhouse, P.M. (1990) Epidemiology and control of barley yellow dwarf viruses in Australia and New Zealand. In: Burnett, P.A. (ed.) *World Perspectives on Barley Yellow Dwarf* (Proceedings of an International Workshop, Udine, Italy, July 1987). CIMMYT, Texcoco, Mexico, pp. 228–239.

Jones, R.A.C. (1993) Effects of cereal borders, admixture with cereals and plant density on the spread of bean yellow mosaic potyvirus into narrow-leafed lupins (*Lupinus angustifolius*). *Annals of Applied Biology* 122, 501–518.

Karungi, J., Adipala, E., Ogenga-Latigo, M.W., Kyamanywa, S. and Oyobo, N. (2000) Pest management in cowpea. Part 1. Influence of planting time and plant density on cowpea field pests infestation in eastern Uganda. *Crop Protection* 19, 231–236.

Kennedy, J. and Booth, C.O. (1963) Co-ordination of successive activities in an aphid. The effect of flight on the settling responses. *Journal of Experimental Biology* 40, 351–369.

Kennedy, J.S., Booth, C.O. and Kershaw, W.J.S. (1961) Host finding by aphids in the field. III. Visual attraction. *Annals of Applied Biology* 49, 1–21.

Kidd, N.A.C. and Jervis, M.A. (1996) Population dynamics. In: Jervis, M.A. and Kidd, N.A.C. (eds) *Insect Natural Enemies.* Chapman and Hall, London, pp. 293–374.

Knight, A.L., Christianson, B.A., Unruh, T.R., Puterka, G. and Glenn, D.M. (2001) Impacts of seasonal kaolin particle films on apple pest management. *Canadian Entomologist* 133, 413–428.

Kostal, V. and Finch, S. (1994) Influence of background on host plant selection and subsequent oviposition by the cabbage root fly (*Delia radicum*). *Entomologia Experimentalis et Applicata* 70, 153–163.

Kring, J.B. and Schuster, D.J. (1992) Management of insects on pepper and tomato with UV-reflective mulches. *Florida Entomologist* 75, 119–129.

Lal, R., de Vleeschauver, D. and Malafa-Nganje, R. (1980) Changes in properties of a newly cleared tropical alfisol as affected by mulching. *Soil Science Society of America Journal* 44, 827–833.

Lamont, W.J., Sorensen, K.A. and Averre, C.W. (1990) Painting aluminium strips of black plastic mulch reduces mosaic symptoms on summer squash. *Hortscience* 25, 1305.

Landis, D.A., Wratten, S.D. and Gurr, G.M. (2000) Habitat management to conserve natural enemies of arthropod pests in agriculture. *Annual Review of Entomology* 45, 175–201.

Langer, A. and Hance, T. (2004) Enhancing parasitism of wheat aphids through apparent competition: a tool for biological control. *Agriculture, Ecosystems and Environment* 102, 205–212.

Lehmhus, J., Vidal, S. and Hommes, M. (1996) Population dynamics of herbivorous and beneficial insects found in plots of white cabbage undersown with clover. *Bulletin OILB/SROP* 19, 115–121.

Lehmhus, J., Hommes, M. and Vidal, S. (1999) The impact of different intercropping systems on herbivorous pest insects in plots of white cabbage. *Bulletin OILB/SROP* 22, 163–169.

Letourneau, D.K. (1987) The enemies hypothesis: tritrophic interaction and vegetational diversity in tropical agroecosystems. *Ecology* 68, 1616–1622.

Letourneau, D.K. (1990) Abundance patterns of leafhopper enemies in pure and mixed stands. *Environmental Entomology* 19, 505–509.

Levie, A., Legrand, M.A., Dogot, P., Pels, C., Baret, P.V. and Hance, T. (2005) Mass releases of *Aphidius rhopalosiphi* (Hymenoptera: Aphidiinae), and strip management to control of wheat aphids. *Agriculture, Ecosystems and Environment* 105, 17–21.

Lewis, T. (1965) The effects of shelter on the distribution of insect pests. *Scientific Horticulture* 17, 74–84.

Lin, R., Liang, H., Zhang, R., Tian, C. and Ma, Y. (2003) Impact of alfalfa/cotton intercropping and management on some aphid predators in China. *Journal of Applied Entomology* 127, 33–36.

Liotta, G. and di Trapani, L. (1994) The influence of mulching and mineral fertilization of winter-melon aphid infestations [in Italian]. *Phytophaga* 4, 69–92.

Lu, H.L. and Feng, J.M. (1989) Effects of plastic film mulch for dispelling *Myzus persicae* and controlling CMV disease in tobacco [in Chinese]. *Zhejiang Agricultural Science* 1, 44–46.

MacLeod, A., Wratten, S.D., Sotherton, N.W. and Thomas, M.B. (2004) 'Beetle banks' as refuges for beneficial arthropods in farmland: long-term changes in predator communities and habitat. *Agricultural and Forest Entomology* 6, 147–154.

Marco, S. (1986) Incidence of aphid-transmitted virus infections reduced by whitewash sprays on plants. *Phytopathology* 76, 1344–1348.

Marco, S. (1993) Incidence of nonpersistently transmitted viruses in pepper sprayed with whitewash, oil, and insecticide, alone or combined. *Plant Disease* 77, 1119–1122.

McGrath, P.F. and Bale, J.S. (1990) The effects of sowing date and choice of insecticide on cereal aphids and barley yellow dwarf virus epidemiology in northern England. *Annals of Applied Biology* 117, 31–43.

McPherson, R.M., Bondari, K., Stephenson, M.G., Severson, R.F. and Jackson, D.M. (1993) Influence of planting date on the seasonal abundance of tobacco budworms (Lepidoptera: Noctuidae) and tobacco aphids (Homoptera: Aphididae) on Georgia flue-cured tobacco. *Journal of Entomological Science* 28, 156–167.

Montandon, R. and Slosser, J.E. (1996) Relay intercropping: effect on predators in cotton. In: *Proceedings Beltwide Cotton Conferences, Nashville, January 1996, Volume 2*. National Cotton Council, Memphis, pp. 786–787.

Moreby, S.J. and Sotherton, N.W. (1997) A comparison of some important chick-food insect groups found in organic and conventionally-grown winter wheat fields in southern England. *Biological Agriculture and Horticulture* 15, 51–60.

Nentwig, W. (1998) Weedy plant species and their beneficial arthropods: potential for manipulation in field crops. In: Pickett, C.H. and Bugg, R.L. (eds) *Enhancing Biological Control: Habitat Management to Promote Natural Enemies of Agricultural Pests*. University of California Press, Berkeley, pp. 49–71.

Ogenga-Latigo, M.W., Ampofo, J.K.O. and Baliddawa, C.W. (1992) Influence of maize row spacing on infestation and damage of intercropped beans by the bean aphid (*Aphis fabae* Scop.). I. Incidence of aphids. *Field Crops Research* 30, 111–121.

Orozco-Santos, M., Perez-Zamora, O. and Lopez-Arriaga, O. (1995) Effect of transparent mulch on insect populations, virus diseases, soil temperature, and yield of cantaloup in a tropical region. *New Zealand Journal of Crop and Horticultural Science* 23, 199–204.

Parajulee, M.N., Slosser, J.E. and Bordovsky, D.G. (1999) Cultural practices affecting the abundance of cotton aphids and beet armyworms in dryland cotton. In: *Proceedings Beltwide Cotton Conferences, Orlando, Florida, USA, January 1999, Volume 2*. National Cotton Council, Memphis, pp. 1014–1016.

Patt, J.M., Hamilton, G.C. and Lashomb, J.H. (1997a) Foraging success of parasitoid wasps on flowers: interplay of insect morphology, floral architecture and searching behavior. *Entomologia Experimentalis et Applicata* 83, 21–30.

Patt, J.M., Hamilton, G.C. and Lashomb, J.H. (1997b) Impact of strip-insectary intercropping with flowers on conservation biological control of the Colorado potato beetle. *Advances in Horticultural Science* 11, 175–181.

Pinese, B., Lisle, A.T., Ramsey, M.D., Halfpapp, K.H. and de Faveri, S. (1994) Control of aphid-borne papaya ringspot potyvirus in zucchini marrow (*Cucurbita pepo*) with reflective mulches and mineral oil-insecticide sprays. *International Journal of Pest Management* 40, 81–87.

Powell, W. (1986) Enhancing parasitoid activity in crops. In: Waage, J. and Greathead, D. (eds) *Insect Parasitoids*. Academic Press, London, pp. 201–206.

Powell, W. (2000) The use of field margins in the manipulation of parasitoids for aphid control in arable crops. *Proceedings of the British Crop Protection Council Conference, Pests and Diseases, Brighton, November 2000* 2, 579–584.

Prokopy, R.J. and Owens, E.D. (1983) Visual detection of plants by herbivorous insects. *Annual Review of Entomology* 28, 337–362.

Rahman, M.M., Uddin, M.M. and Khan, M.R. (1989) Influence of seeding date on the occurrence of aphids (*Lipaphis pseudobrassicae* Davis) and seed yield of mustard and rape seed. *Bangladesh Journal of Scientific and Industrial Research* 24, 116–122.

Rajendran, B. (2002) Cultural controls. In: Pimentel, D. (ed.) *Encyclopedia of Pest Management.* Dekker, New York, pp. 174–178.

Root, R.B. (1973) Organization of a plant–arthropod association in simple and diverse habitats: the fauna of collards (*Brassica oleracea*). *Ecological Monographs* 43, 94–125.

Saha, C.S. and Baral, K. (1999) Effect of dates of sowing and potash levels on incidence of mustard aphid *Lipaphis erysimi* (Kaltenbach). *Environment and Ecology* 17, 211–213.

Saucke, H. and Doring, T.F. (2004) Potato virus Y reduction by straw mulch in organic potatoes. *Annals of Applied Biology* 144, 347–355.

Schmidt, M.H., Thewes, U., Thies, C. and Tscharntke, T. (2004) Aphid suppression by natural enemies in mulched cereals. *Entomologia Experimentalis et Applicata* 113, 87–93.

Selman, M. (1980) Ways of increasing the yield of winter barley. *Arable Farming* 7, 82–85.

Shafique, M., Anwar, M., Ashraf, M. and Bux, M. (1999) The impact of sowing time on aphid management and yield of canola varieties. *Pakistan Journal of Zoology* 31, 361–363.

Sheehan, W. (1986) Response by specialist and generalist natural enemies to agroecosystem diversification: a selective review. *Environmental Entomology* 15, 456–461.

Showler, A.T. and Greenberg, S.M. (2003) Effects of weeds on selected arthropod herbivore and natural enemy populations, and on cotton growth and yield. *Environmental Entomology* 32, 39–50.

Slosser, J.E., Montandon, R., Pinchak W.E. and Rummel, D.R. (1997) Cotton aphid response to nitrogen fertility in dryland cotton. *Southwestern Entomologist* 22, 1–10.

Smith, J.G. (1969) Some effects of crop background on populations of aphids and their natural enemies on Brussels sprouts. *Annals of Applied Biology* 63, 326–333.

Smith, R.F. and van den Bosch, R. (1967) Integrated control. In: Kilgore, W.W. and Doutt, R.L. (eds) *Pest Control: Biological, Physical and Selected Chemical Methods.* Academic Press, New York, pp. 295–340.

Snidaro, M. and Delogu, G. (1990) Agronomic techniques for preventing barley yellow dwarf damage in winter cereals. In: Burnett, P.A. (ed.) *World Perspectives on Barley Yellow Dwarf.* CIMMYT, Texcoco, Mexico, pp. 457–463.

Stapleton, J.J. and Summers, C.G. (2002) Reflective mulches for management of aphids and aphid-borne virus diseases in late-season cantaloupe (*Cucumis melo* L. var. *cantalupensis*). *Crop Protection* 21, 891–898.

Starý, P. (1993) Alternative host and parasitoid in first method in aphid pest management in glasshouses. *Journal of Applied Entomology* 116, 187–191.

Stephens, M.J., France, C.M., Wratten, S.D. and Frampton, C. (1998) Enhancing biological control of leafrollers (Lepidoptera: Tortricidae) by sowing buckwheat (*Fagopyrum esculentum*) in an orchard. *Biocontrol Science and Technology* 8, 547–558.

Stoetzel, M.B. (1987) Information on and identification of *Diuraphis noxia* (Homoptera: Aphididae) and other aphid species colonizing leaves of wheat and barley in the United States. *Journal of Economic Entomology* 80, 696–704.

Summers, C.G., Stapleton, J.J., Newton, A.S., Duncan, R.A. and Hart, D. (1995) Comparison of sprayable and film mulches in delaying the onset of aphid-transmitted virus diseases in zucchini squash. *Plant Disease* 79, 1126–1131.

Summers, C.G., Mitchell, J.P. and Stapleton, J.J. (2004) Management of aphid-borne viruses and *Bemisia argentifolii* (Homoptera: Aleyrodidae) in zucchini squash by using UV reflective plastic and wheat straw mulches. *Environmental Entomology* 33, 1447–1457.

Tahvanainen, J.O. and Root, R.B. (1972) The influence of vegetational diversity on the population ecology of a specialized herbivore, *Phyllotreta cruciferae* (Coleoptera: Chrysomelidae). *Oecologia* 10, 321–346.

Theunissen, J., Booij, C.J.H. and Lotz, L.A.P. (1995) Effects of intercropping white cabbage with clovers on pest infestation and yield. *Entomologia Experimentalis et Applicata* 74, 7–16.

Thieme, T., Heimbach, U., Thieme, R. and Weidemann, H.L. (1998) Introduction of a method for preventing transmission of potato virus Y (PVY) in Northern Germany. In: Dale, M.F.B., Dewar, A.M., Fisher, S.J.,

Haydock, P.P.J., Jaggard, K.W., May, M.J., Smith, H.G., Storey, R.M.J. and Wiltshire, J.J.J. (eds) *Protection and Production of Sugar Beet and Potatoes. Aspects of Applied Biology, No. 52*, pp. 25–29.

Thies, C., Steffan-Dewenter, I. and Tscharntke, T. (2003) Effects of landscape context on herbivory and parasitism at different spatial scales. *Oikos* 101, 18–25.

Thies, C., Roschewitz, I. and Tscharntke, T. (2005) The landscape context of cereal aphid–parasitoid interactions. *Proceedings of the Royal Society of London B* 272, 203–210.

Thomas, M.B., Wratten, S.D. and Sotherton, N.W. (1992) Creation of 'island' habitats in farmland to manipulate populations of beneficial arthropods: predator densities and species composition. *Journal of Applied Ecology* 29, 524–531.

Tilman, D., Cassman, G., Matson, P.A., Naylor, R. and Polasky, S. (2002) Agricultural sustainability and intensive production practices. *Nature, London* 418, 671–677.

Tscharntke, T. and Hawkins B.A. (eds) (2002) *Multitrophic Level Interactions.* Cambridge University Press, Cambridge, 282 pp.

Vidal, S. (1997) Factors influencing the population dynamics of *Brevicoryne brassicae* in undersown Brussels sprouts. *Biological Agriculture and Horticulture* 15, 285–295.

Wade, M.R., Zalucki, M.P. and Wratten, S.D. (2004) Use of artificial food supplements in conservation biological control. *Proceedings of the 4th California Conference on Biological Control, Berkeley, July 2004.* Center for Biological Control, University of California, Berkeley, pp. 145–149.

Walters, S.A. (2003) Suppression of watermelon mosaic virus in summer squash with plastic mulches and rowcovers. *HortTechnology* 13, 352–357.

Webb, S.E. and Linda, S.B. (1992) Evaluation of spunbonded polyethylene row covers as a method of excluding insects and viruses affecting fall-grown squash in Florida. *Journal of Economic Entomology* 85, 2344–2352.

Weibull, J. (1987) Seasonal changes in the free amino acids of oat and barley phloem sap in relation to plant growth stage and growth of *Rhopalosiphum padi. Annals of Applied Biology* 111, 719–737.

White, A.J., Wratten, S.D., Berry, N.A. and Weigmann, U. (1995) Habitat manipulation to enhance biological control of *Brassica* pests by hover flies (Diptera: Syrphidae). *Journal of Economic Entomology* 88, 1171–1176.

Wratten, S.D. (1974) Aggregation in the birch aphid, *Euceraphis punctipennis* (Zett.) in relation to food quality. *Journal of Animal Ecology* 43, 191–198.

Wratten, S.D. (1992) Farmers weed out the cereal killers. *New Scientist* 1835, 31–35.

Wratten, S.D. and Powell, W. (1991) Cereal aphids and their natural enemies. In Firbank, L.G., Carter, N., Darbyshire, J.F. and Potts, G.R. (eds) *The Ecology of Temperate Cereal Fields. Proceedings of the British Ecological Society Symposium No. 32.* Blackwells, Oxford, pp. 233–257.

Wratten, S.D., van Emden, H.F. and Thomas, M.B. (1998) Within-field and border refugia for the enhancement of natural enemies. In: Pickett, C.H. and Bugg, R.L. (eds) *Enhancing Biological Control: Habitat Management to Promote Natural Enemies of Agricultural Pests.* University of California Press, Berkeley, pp. 375–404.

Yardim, E.N. and Edwards, C.A. (2003) Effects of organic and synthetic fertilizer sources on pest and predatory insects associated with tomatoes. *Phytoparasitica* 31, 324–329.

Zhou, X. and Carter, N. (1991) The effects of nitrogen and fungicide on cereal aphid population development and the consequences for the aphid-yield relationship in winter wheat. *Annals of Applied Biology* 119, 433–441.

17 Host-plant Resistance

Helmut F. van Emden

*School of Biological Sciences, University of Reading, Whiteknights, Reading, Berks,
RG6 6AJ, UK*

Introduction

In this section, the word 'variety' will be used as an umbrella word for the many types of variation involved. Thus, the word should be read in that sense; varieties, cultivars, accessions, breeders' lines may all be included. The practical use of host-plant resistance (HPR) for aphid control in different crops is explored in the pest management section of this volume (Chapters 21–30).

It might appear that the role of their symbionts in nitrogen metabolism (Douglas and van Emden, Chapter 5 this volume) could enable aphids to compensate for nutritional differences between crop varieties, and that the phloem-feeding habit means aphids would avoid many surface characters and allelochemicals in plants. However, the literature probably has more examples of HPR to aphids than to any other group of crop pests, and examples relevant to aphids can be found for nearly all of the mechanisms of HPR known for insect pests in general. Moreover, HPR to aphids can be highly effective and dramatic; for example, the resistance of 'Avoncrisp' lettuce to *Pemphigus bursarius* (lettuce root aphid) (Fig. 17.1).

Plant resistance to aphids tends to increase with age ('age-related resistance') (van Emden and Bashford, 1971). The rate of this increase can differ between crop varieties, and it is not uncommon for

susceptibility to fall with age at a faster rate the more susceptible the young plants are (Dodd and van Emden, 1979), so that resistance rankings between varieties can change with age.

Resistance to one aphid species may not give resistance to another. HPR to *Diuraphis noxia* (Russian wheat aphid) in wheat and barley seems especially specific, and has no effect on the several other species of cereal aphid (Robinson, 1992; Schotzko and Bosque-Perez, 2000; Webster and Porter, 2000a; Messina and Bloxham, 2004).

Before the advent of genetic engineering for resistance to aphids (see later), wild relatives within the natural crossing barriers were often the source of resistance genes then transferred by traditional breeding. Table 17.1 lists some of the wild relatives of crops that have often been involved in resistance breeding for aphids, usually entailing the transfer of deterrents and toxins (normally at much lower concentrations than in the wild plants) that had been lost in the process of domesticating the wild species (van Emden, 1997).

Types of Host-plant Resistance to Aphids

The classification of Painter (1951), modified by Kogan and Ortman (1978), still has great value.

Fig. 17.1. Junction of experimental lettuce plots testing for host plant resistance to *Pemphigus bursarius*. The impact of variety as an aphid control measure is seen clearly in the contrast between the susceptible variety 'Mildura' in the foreground and the resistant 'Avoncrisp' behind (courtesy of J.A. Dunn).

Table 17.1. Some examples of wild relatives as sources of genes in traditional breeding for host plant resistance to aphids.

'Wild' relative	Crop	Resistance character	Sample reference(s)
Avena macrostachya and *Avena barbata*	Oats		Weibull, 1987
Lycopersicon hirsutum f. *glabratum* and *Solanum pennellii*	Tomato	Glandular trichomes	Kok-Yokomi, 1978; Simmons *et al.*, 2005
Lycopersicon peruvianum	Tomato	Glandular hairs and high tomatine	Kok-Yokomi, 1978
Solanum etuberosum	Potato		Novy *et al.*, 2002
Brassica fruticolosa and *Brassica spinescens*	Cabbage	Gluconapin; lectin	Cole, 1994a,b
Triticum monococcum	Wheat		Sotherton and van Emden, 1982; Deol *et al.*, 1995; Di Pietro *et al.*, 1998; Migui and Lamb, 2004
Triticum monococcum subsp. *aegilopoides*	Wheat		Di Pietro *et al.*, 1998
Triticum urartu			
Solanum berthaultii	Potato	Glandular trichomes	Gibson, 1976

Note: The same source of resistance may influence several aphid species attacking the same crop; hence, names of aphids affected are not included in the table. However, aphid names are usually given in the reference titles.

Antixenosis (close to Painter's 'non-preference')

This is resistance to colonization by aphids, and antixenosis shows in the high proportion of immigrating alatae which do not remain on the plant but take off again within a few hours. At the immigration stage, aphids seem to exhibit a high degree of non-preference, even on susceptible plants. Müller (1958) showed that the fourfold greater population of *Aphis fabae* (black bean aphid) that developed on the susceptible bean (*Vicia faba*) variety Schlandstedt compared with the resistant Rastatt was based on respective re-take-off rates of 95% and an only slightly higher 99%. Antixenosis is usually assessed by liberating alatae over different varieties, both in choice and no-choice situations

(Adams and van Emden, 1972), although van Emden *et al.* (1991) devised a test (involving systemic insecticide with a correction for any detection thereof by the aphids) that identifies which aphids have ingested from the plants.

Tjallingii's (1988) electrical penetration graph (EPG) technique (see also Pettersson *et al.*, Chapter 4 this volume) has added a new dimension to the study of antixenosis, as it enables the continuous monitoring of aphid penetration and probing activity (Fig. 17.2) (e.g. Mentink *et al.*, 1984).

Antibiosis

This negatively affects the multiplication of aphids that have colonized the plant by, for

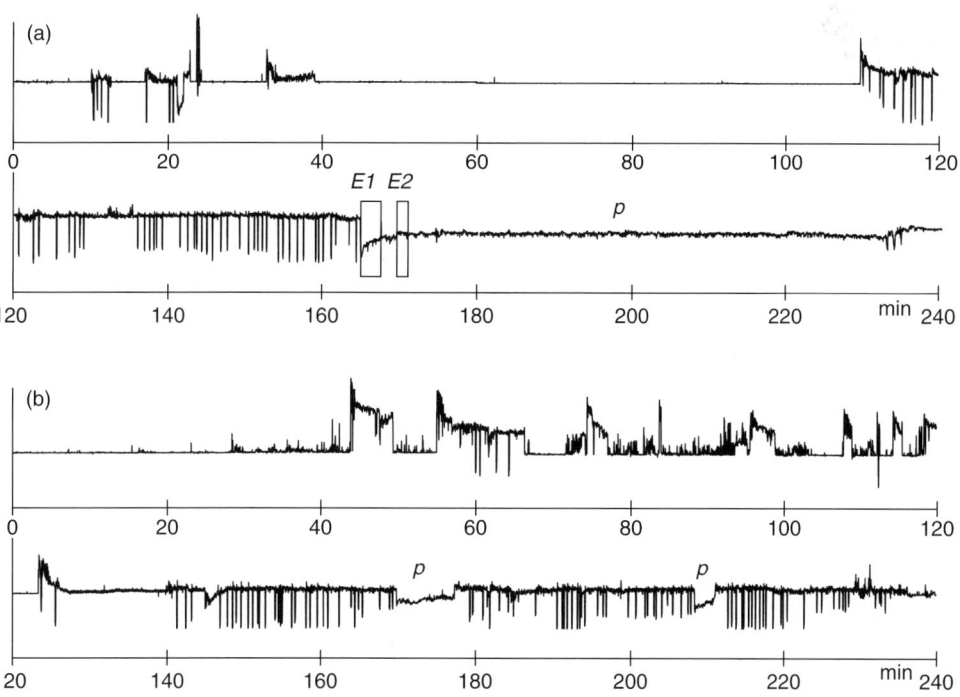

Fig. 17.2. Electrical penetration graph (EPG) recordings for *Nasonovia ribisnigri* (currant–lettuce aphid) on susceptible (a) and nearly isogenic resistant (b) lettuce varieties. On susceptible lettuce, beginning at about 170 min, one long continuous phloem phase (*p*) is shown. This starts with sieve element (watery) salivation (waveform *E1*) and continues for the rest of the 4 h recording (only the first 2 h are shown) with sieve element ingestion (waveform *E2*). In contrast, the graph from the resistant lettuce shows only two short phloem-feeding periods (*p*), mainly of E1 waveform with some very brief switches to E2. The average start of the first phloem phase does not differ between the graphs from the two varieties; a phloem factor prevents sieve element ingestion (courtesy of W.F. Tjallingii).

example, reducing survival, growth and fecundity, and extending development time. These characteristics are measured easily on individuals confined on leaves in small clip cages (Adams and van Emden, 1972), although care needs to be taken to ensure any apparent antibiosis is no more than a first-generation expression of transfer to a novel plant. Survival, development time, and fecundity can be combined into a single statistic, the intrinsic rate of increase (r_m) (Birch, 1948; Awmack and Leather, Chapter 6 this volume). This can be converted into population doubling time. The growth effects of HPR can be measured on individuals in a short time (e.g. 3 or 4 days) by weighing nymphal aphids at either end of a time interval and calculating the mean relative growth rate (MRGR) as:

MRGR ($\mu g/\mu g/day$)

$$= \frac{\log_e \text{ final weight } (\mu g) - \log_e \text{ initial weight } (\mu g)}{\text{number of days between weighings}}$$

The statistic MRGR is reasonably independent of initial weight or the number of days over which it is measured (van Emden, 1969; Awmack and Leather, Chapter 6 this volume).

Tolerance

Tolerance of a variety shows as a better yield when suffering the same aphid burden, both in numbers and duration. Tolerance is difficult to quantify since any antibiosis defeats the 'equal burden' requirement, and intolerance to infestation will result in plant damage, which will cause the aphid population to crash and defeat the 'equal duration' requirement. Havlickova (1997) has proposed a covariance model in which the main component is the slope of the relationship between the weight of infested and control seedlings. However, tolerance may perhaps be better quantified by the slopes of regressions, in which the yield or plant biomass is regressed on several levels of initial aphid infestation.

What is the preferred type of resistance?

Antixenosis is rarely effective on its own in a no-choice situation. Antibiosis puts pressure on the population for selection of genotypes not affected by the antibiosis and so may break down (see 'biotypes' later), a danger not applicable to tolerance. However, farmers growing tolerant varieties would not need to control the aphids on them and so would breed populations to infest their neighbours' crops. Thus, there is general agreement (van Emden, 1997) that the preferred resistance is antibioisis coupled with some antixenosis.

Webster and Porter (2000b) devised a 'resistance index' to combine the levels of antixenosis, antibiosis, and tolerance of a variety into a single number, after normalizing the data for the three criteria.

Mechanisms of Host-plant Resistance to Aphids

This chapter reviews HPR to aphids by the various anatomical, physiological, and biochemical methods that have been correlated with HPR phenomena and Fig. 17.7 shows these mechanisms in relation to Painter's (1951) classification of resistance, together with an indication of the order in which the aphid would encounter them. However, it must be readily admitted that correlations are not the same as 'cause and effect'. Ruggle and Gutierrez (1995) have pointed out that 40 years of research has failed to reveal the mechanism of HPR of lucerne varieties to *Therioaphis trifolii* (yellow clover aphid), and Kazemi and van Emden (1992) found that some chemical correlations with HPR in wheats to *Rhopalosiphum padi* (bird cherry–oat aphid) proposed in the literature broke down when the range of wheats tested covered a wider spread of origin. Similarly, Weibull (1994) found that correlations of chemistry with resistance to *R. padi* in parental barley (*Hordeum vulgare*) lines broke down when the segregating offspring lines were tested. A study of the literature over many years suggests that, if only a few varieties are being studied, it is not difficult to correlate resistance ranking with the chemical group representing one's own particular interest.

Mechanisms of antixenosis

Colour

Colour can influence the preference of immigrating aphids. Alatae of *Brevicoryne brassicae* (cabbage aphid) do not settle well on red cabbage varieties, even though apterae caged on such varieties have a better growth rate than those on normal green varieties (Radcliffe and Chapman, 1965, 1966).

Palatability

That so many common and even Latin names of aphids associate the insects with particular host plants (e.g. lupin aphid, *B. brassicae*, etc.) shows that aphids often accept or reject a plant on the basis of the secondary chemistry before reaching the phloem. Thus, toxins (which would otherwise reflect antibiosis) become deterrents and contribute to antixenosis. Polyphagous aphids such as *Myzus persicae* (peach–potato aphid) will tolerate levels of secondary compounds such as glucosinolates in brassicas to become pests of such plants, but elevating levels still tolerable to the specialist *B. brassicae* will lead to antibiosis (van Emden, 1978; Cole, 1997b). In screening wild and cultivated brassicas for resistance to these two aphids, Cole (1997a) found that most of the variation in HPR was explained by four glucosinolates (2-OH-3-butenyl, 2-propenylglucosinolate, 3-methoxyindolyl, and 4-pentylglucosinolate). Hydroxamic acids (Hx) in cereals have often been associated with antibiosis to aphids, but the compound is also a deterrent, leading to antixenosis; for example, to *Metopolophium dirhodum* (rose–grain aphid), *Rhopalosiphum maidis* (corn leaf aphid), *Sitobion avenae* (grain aphid), and *Schizaphis graminum* (greenbug). Hx is antibiotic to all these species in artificial diet, as it is to *R. maidis*, which, however, is not deterred in the field. This is because it avoids Hx in high Hx plants by reducing puncturing of the cells by the stylets on their way to the phloem (Givovich and Niemeyer, 1995). In lucerne, other Hx compounds (the 3-amyl-4-hydroxycoumarins such as coumestrol) are deterrent to *Acyrthosiphon pisum* (pea aphid), and deterrence increases with the size of the 3-amyl group (Dreyer *et al.*, 1987).

In tobacco, antixenosis to *M. persicae* can be caused by sugar ester levels and alpha and beta monols on the leaf surface (Johnson *et al.*, 2002), though not all the resistant lines among the 62 tested had these chemical characteristics.

Waxiness

Waxiness of the leaf surface (Fig. 17.3) has often been linked with antixenosis to aphids. However, it is often the less waxy (glossy) varieties that are resistant, and the resistance then appears to be based in higher levels of deterrent chemicals, albeit in a reduced wax layer. Glossy wheats, antixenotic to *S. avenae*, contain dihydroketones as the deterrent compounds (Lowe *et al.*, 1985). With *A. pisum* studied on seven pea (*Pisum sativum*) varieties, the glossy ones again were resistant (White and Eigenbrode, 2000), and the epicuticle of some lucerne varieties resistant to *Therioaphis trifolii maculata* (spotted alfalfa aphid) had wax ester levels 50% higher than those in susceptible lucerne (Bergman *et al.*, 1991). Glossy brassica varieties are also antixenotic to *B. brassicae* (Thompson, 1963; Ellis *et al.*, 1996), with reductions in aphid populations as high as 95% having been reported (Stoner, 1992). Leaf surface wax components have also been implicated in the resistance of European red raspberries to *Amphorophora idaei* conferred by gene A_{10} (Robertson *et al.*, 1991; Shepherd *et al.*, 1999a,b).

Ni *et al.* (1998) tested whether, in relation to *D. noxia*, the resistance ranking of wheat (most susceptible), oat (intermediate), and barley (most resistant) had a basis in surface wax by removing this with ethyl ether, but the resistance rankings remained as before.

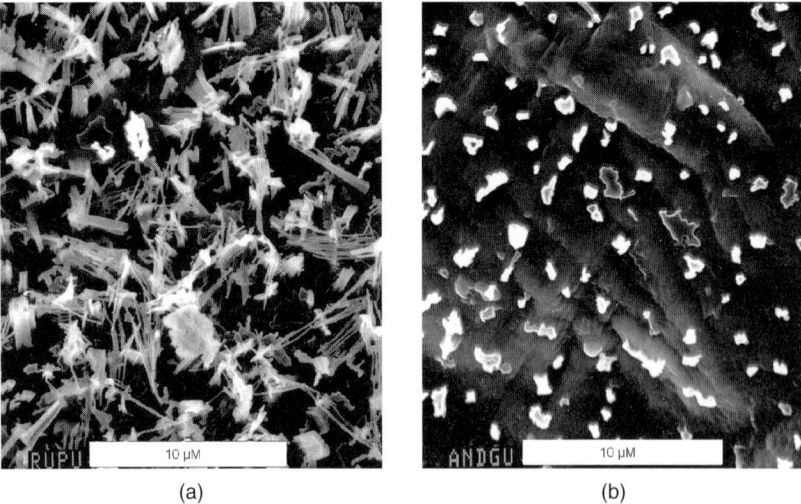

Fig. 17.3. Electronmicrographs of wax quantity and its distribution on normal (a) and glossy (b) cabbage leaf surfaces (courtesy of S.G. Eigenbrode).

Mechanical

Hardness of the plant surface, a common resistance mechanism to chewing insects, appears not to be a source of antibiosis to aphids. For aphids, a more common cause of mechanical antixenotic resistance is difficulties in reaching the phloem. The role of pectin in the cellular middle lamella in hindering aphid access to the phloem has been reviewed by Dreyer and Campbell (1987). In sorghum resistant to *S. graminum*, the resistance is due to increased methylation of the middle lamellar pectin (Dreyer and Campbell, 1984). A variety of melon resistant to *Aphis gossypii* (cotton or melon aphid) deposits callose in attacked leaf veins (Shinoda, 1993).

A major qualitative chemical feature of two wild brassicas resistant to *B. brassicae* was the presence of gluconapin rather than the more usual glucobrassicin, but since the former had no effect on the aphid in artificial diet, Cole (1994a) proposed that the resistance mechanism was probably stylet blocking.

Local necrosis

Local necrosis, i.e. the death of cells with the production of deterrent polyphenols wherever the aphid tries to insert its stylets, is a valuable resistance mechanism since it also gives protection against other piercing organisms such as fungal hyphae and nematodes (van Emden, 1987). The polyphenols are produced by the aphid damage bringing substrate and enzyme into contact. This is a 'hypersensitive response', perhaps an unusual concept of 'resistance'. It is the mechanism of resistance of apples to *Eriosoma lanigerum* (woolly apple aphid) (Wartenberg, 1953) and *Dysaphis plantaginea* (rosy apple aphid) (Alston and Briggs, 1970). Another example is the hypersensitive cell death response of some resistant barleys to attack by *D. noxia* (Belefantmiller *et al.*, 1994), though some aphid species may have reducing compounds in their saliva that counter the hypersensitive response of the plant (Miles, 1999).

Trichomes (non-glandular)

Trichomes (non-glandular) are barriers causing antixenosis to many small insects, including aphids. For example, high trichome density on wheat leaves deters *Sipha flava* (yellow sugarcane aphid), but any effect on *S. graminum* is questionable (Webster *et al.*, 1994). High trichome density also deters *M. persicae* on crosses of tomato with wild potato (Simmons *et al.*, 2005). Pubescent broccoli is as resistant to *B. brassicae* as glossy varieties (see earlier) (Stoner, 1992), and the pubescent lime *Tilia platyphyllos* has fewer *Eucallipterus tiliae* (lime aphid) than the less pubescent *Tilia cordata* and *Tilia europaea* (European lime) (Zuparko and Dahlsten, 1994). However, with *A. gossypii* on cotton, it is the glabrous varieties that have fewer aphids (Weathersbee and Hardee, 1994; Weathersbee *et al.*, 1994, 1995).

Mechanisms of antibiosis

Glandular trichomes

Glandular trichomes are of two types: long slender ones that secrete drops of fluid from the tip, and short ones bearing a spherical glandular head (Fig. 17.4). The head is divided into compartments by septa separating substrate and enzyme which, when brought together when the head is ruptured by contact with an aphid, combine to form polyphenols (see 'local necrosis' above). These polyphenols are not only distasteful to the aphid, but also harden and disable the mouthparts and tarsi. Great interest has focused on the presence of such trichomes on wild potato (especially *Solanum berthaultii*) and tomato crossed with *Solanum pennellii* (Simmons *et al.*, 2005), and on

breeding this character into cultivars for resistance to aphids (e.g. Gibson, 1976). However, no such resistant varieties were ever commercialized. Any *Macrosiphum euphorbiae* (potato aphid) on the stems multiply faster than on varieties without glandular trichomes (Ashouri *et al.*, 2001). The wild tomatoes *Lycopersicon hirsutum* f. *glabratum* and *Lycopersicon peruvianum* have a dense pubescence with both types of trichomes (Kok-Yokomi, 1978), and Dreyer and Campbell (1987) observed that the resistance attributable to the glandular trichomes decreased as the density of non-glandular trichomes increased.

Toxins

These are often the same chemicals that cause antixenosis if the aphid does not avoid ingesting them. Ansari *et al.* (1989) discovered some varieties of cowpea (*Vigna unguiculata*) in which the leaves produced a powerful graft-transmissible but unidentified toxin that killed *Aphis craccivora* (cowpea aphid) very quickly, but did not seem to be detected by the aphid.

Hydroxamic acids (Hx) are a deterrent to aphids (see earlier), but also decrease feeding in the diet of, for example, *S. avenae* (Cambier *et al.*, 2001). Also in diet,

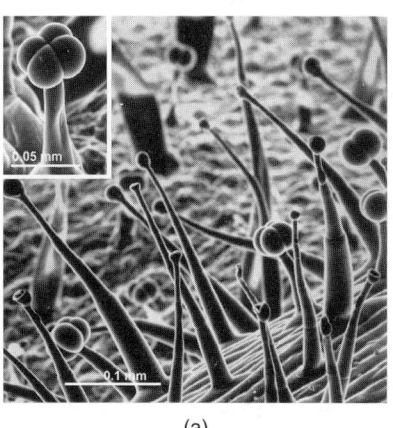

(a)

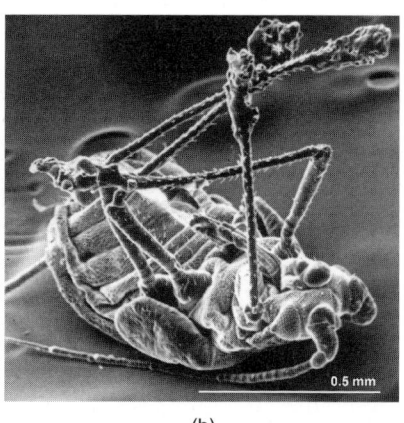

(b)

Fig. 17.4. Glandular trichomes. (a) Scanning electron micrograph of the leaf surface of *Solanum berthaultii* showing both trichomes with the four-celled head (magnified on insert) and those surmounted by a glandular vesicle. (b) Scanning electron micrograph of an aphid that was on *Solanum polyadenium*, showing the material from the glandular trichomes hardened on the feet (courtesy of R.W. Gibson).

2,4-dihydroxy-7-methoxy-1,4-benzoxazin-3-one (DIMBOA) decreases the survival of *M. dirhodum* (Cambier *et al.*, 2001), as do low concentrations of its decomposition product 6-methoxybenzoxazolin-2-one (MBOA) for *S. avenae* (Hansen, 2006). In whole plant studies, a wheat cultivar with high Hx was found to be antibiotic to *S. avenae* (Fuentes-Contreras and Niemeyer, 1998), and the growth rate of the aphid on three wheat and two oat cultivars was negatively correlated with Hx levels (Fuentes-Contreras *et al.*, 1996). Also, 26 Hungarian cultivars showed a negative correlation between Hx levels and populations of *R. padi* (Gianoli *et al.*, 1996).

Legumes are also well known to contain toxins, as in the example of cowpeas above. Another example concerns lupins; the alkaloids in narrow-leaved lupins (*Lupinus angustifolius*) depress the fecundity of *Macrosiphum albifrons* (lupin aphid) (Berlandier, 1996). *Myzus persicae* is less affected by these alkaloids (gramine, sparteine, lupanine, lupinine, 13-hydroxylupanine, and angustifoline) than *A. craccivora*, both in plants and in artificial diet (Risdill-Smith *et al.*, 2004).

Cotton varieties have been bred with different levels of the polyphenol gossypol, and *A. gossypii* showed shorter longevity and lower fecundity on a high gossypol cultivar than on two with lower levels (Du Li *et al.*, 2004).

It has been realized relatively recently that where a toxin gives resistance to aphids in grasses, the toxin is often actually produced by fungal endophytes, most in or related to the genus *Neotyphodium* (Fig. 17.5). In work with *R. padi*, the toxins were identified as loline alkaloids (Bultman and Bell, 2003), though the aphid can survive on the senescing leaves that contain less loline (Eichenseer *et al.*, 1991). On tall fescue (*Festuca arundinacea*), the toxin gives 100% mortality of *D. noxia* in only 4 days (Clement *et al.*, 1996), and endophyte-produced toxins have been identified also as the basis of HPR to this aphid in some wild barleys (Clement *et al.*, 1997), as well as in turf grasses (Kindler *et al.*, 1991). With *S. graminum* on perennial rye grass (*Lolium perenne*), it has been deduced from plant–endophyte associations that the endophyte producing the toxin peramine may be *Neotyphodium lolii* (Breen, 1992). However, the potential for using endophytes for HPR is bedevilled by the variability of the association, with efficacy dependent on the exact genotypes of both insect and endophyte. Thus, *D. noxia* and *S. graminum* are reduced by endophyte-produced toxins both in tall fescue and in perennial rye grass, *R. padi* only on tall fescue, and *R. maidis* only on perennial rye grass. *Metopolophium dirhodum*, *S. avenae*, and *Sitobion fragariae* (blackberry–cereal aphid) are not affected by endophytes in either grass species (Clement

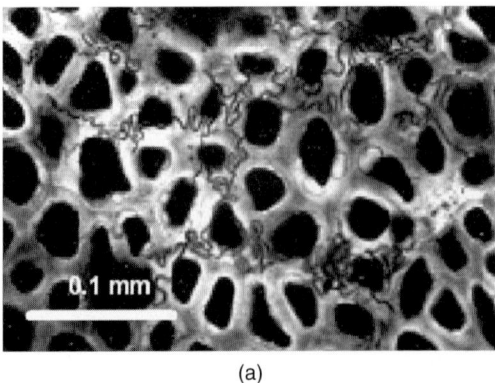
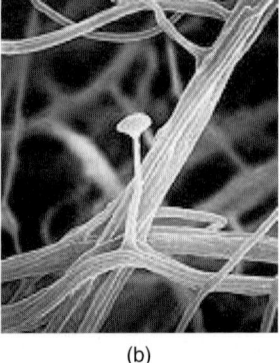

(a)   (b)

Fig. 17.5. The endophytic fungus *Neotyphodium*. (a) Section through a tall fescue (*Festuca arundinacea*) seed showing mycelium between the aleurone cells. (b) Electronmicrograph of hyphae, conidiophore and conidium (4 μm long) in culture. Host is wild barley (*Hordeum* sp.) (courtesy of S.L. Clement).

et al., 1994). Although endophytic fungi are also found frequently in dicotyledenous plants (Saikkonen *et al.*, 1998), any role they may have in HPR to aphids (or indeed any other insect) has yet to be established.

Of toxins currently being investigated for transgenically transferable aphid resistance, lectins seem the main focus. *Bacillus thuringiensis* (Bt) toxins used for GM Lepidoptera resistance are not regarded as aphicides, but a GM potato cultivar with Bt against *Leptinotarsa decemlineata* (Colorado beetle) also depressed the growth and fecundity of *M. euphorbiae* (Ashouri *et al.*, 2001). Usually, lectins give only partial resistance to aphids. When 30 were tested in artificial diet against *A. pisum* (Rahbe *et al.*, 1995), most caused only low toxicity. However, those from jack-pine (*Pinus banksiana*), grain amaranth (*Amaranthus caudatus*), lentil (*Lens culinaris*), and snowdrop (*Galanthus nivalis*) induced worthwhile mortality. Concanavilin A (the jack-pine lectin) was then tested on five other aphid species, but with very variable results. A chitin-binding lectin in two wild brassicas shows some insecticidal activity against *B. brassicae* (Cole, 1994b).

There have also been studies with proteinase inhibitors. Oryzacystatin 1, effective against leaf-chewing insects in transgenic oilseed rape, inhibited growth of *A. gossypii*, *A. pisum*, and *M. persicae* (Rahbe *et al.*, 2003).

More recently, compounds (jasmonic and salicylic acids) that form part of the defensive and signalling system of plants when damaged have been shown to affect aphids negatively (Pickett and Glinwood, Chapter 9 this volume). Work with tomato suggests that such compounds explain the resistance to *M. euphorbiae* of tomatoes carrying the Mi-1.2 gene (Cooper *et al.*, 2004).

Nutritional factors

Given the limiting levels of nitrogen dietary requirements for aphids in the phloem (Douglas and van Emden, Chapter 5 this volume), one might expect nutritional factors to be a valuable basis for HPR. Auclair *et al.* (1957) studied several pea varieties and found a negative correlation between their resistance to *A. pisum* and the soluble and total amino nitrogen content of the leaves. As mentioned earlier, Kazemi and van Emden (1992) repeated the analyses for Hx and total phenolics that other workers had been able to correlate with HPR to aphids, each in a few and different wheats. Kazemi and van Emden analysed a wider provenance range of wheats (European, Iranian, and the 'ancient' diploid wheat Einkorn (*Triticum monococcum*)). There was little correlation of HPR to *R. padi* across this range of wheats for any of the previously proposed correlates, but a stepwise multiple regression involving the leaf concentrations of three amino acids (alanine, histidine, and threonine) accounted for over 95% of the variation in the fecundity of the aphid. Weibull (1994) analysed EDTA (ethylenediaminetetraacetic acid) leaf exudates as a better reflection of phloem content than whole leaf analyses, and found a negative instead of a positive correlation between several amino acids and the performance of *R. padi* on wild barley (*Hordeum spontaneum*). However, since the segregating offspring of crosses showed no such relationship, no cause and effect was attached to the correlation.

Although nutritional factors are a mechanism of HPR to aphids, such resistance is particularly likely to fail under different environmental conditions such as different soils and fertilizer regimes.

Extrinsic factors

Factors extrinsic to the plant may be responsible for greater mortality on the apparently 'resistant' than on the 'susceptible' variety. In the absence of such factors, true HPR may be much less, or even absent. Thus, Gowling and van Emden (1994) showed that the number of *M. dirhodum* falling from wheat plants was doubled when parasitoids were added, and doubled again if the wheat variety was partially aphid resistant. The partial variety ('Rapier') was only 25% resistant compared with the susceptible comparison (the variety 'Armada'). However, the increased falling of aphids in the presence of searching parasitoids made the variety appear 86% resistant.

A very good example of extrinsic resistance is the work of Kareiva and Sahakian (1990) with 'leafless' pea varieties bred for low leaf area to be resistant to powdery mildew (*Erysiphe polygoni*) and, instead of leaves, a profusion of photosynthesizing tendrils (Fig. 17.6). They found that the apparent resistance to *A. pisum* in the field was considerably greater than the true intrinsic resistance of only some 8%. The reason lay in the ability of the ladybirds *Coccinella septempunctata* and *Hippodamia convergens* to grip the tendrils, whereas a high proportion fell off the shiny leaves of normal peas. The reduction in falling was from 47–26% for *C. septempunctata* and from 32–9% for *H. convergens*.

Mechanisms of tolerance

Rather little is known about the mechanisms in varieties tolerant to aphids. Tolerance is often coupled with antibiosis, and the latter may explain the better yield sufficiently for tolerance not being identified. So, the examples of tolerance reported in the literature tend to be dramatic – large aphid populations are observed; yet crop growth and yield are less affected than expected. For example, Hesler (2005) found that three triticale (× *Triticosecale* spp.) accessions showed no difference in shoot length whether infested or uninfested with *R. padi*.

Compensation

Compensation is one tolerance mechanism, and some lines of wheat show much enhanced growth when attacked by *S. graminum* or *D. noxia* (Castro *et al.*, 2001). Such compensatory growth as a response to aphid attack may, however, lead to a delay in harvest.

Symptom expression

Another mechanism for tolerance relates to the fact that symptoms arising as a result of the plant reaction to aphid saliva can vary with the particular genotypes of both aphid and plant involved. Thus, some genotypes of wheat are specifically tolerant to the E biotype of *S. graminum* (Morgham *et al.*, 1994). An interesting mechanism for tolerance in wheat varieties to *S. graminum* was proposed by Maxwell and Painter (1962), who compared the two cultivars 'Dickson' (tolerant) and 'Pawnee' (intolerant). These two cultivars differed as to the timing of the cessation of further expansion of the flag leaf, on the photosynthesis of which much of the grain yield depends. Since the aphids tend to feed in an aggregation on the underside of the leaf, growth there is inhibited, while the upper leaf surface continues growing. The leaf therefore curls strongly, reducing the effective leaf area exposed to solar radiation. In 'Dickson', leaf growth stops

(a)   (b)

Fig. 17.6.   (a) Normal and (b) 'leafless' cultivars of peas ((b) courtesy of C. Coyne).

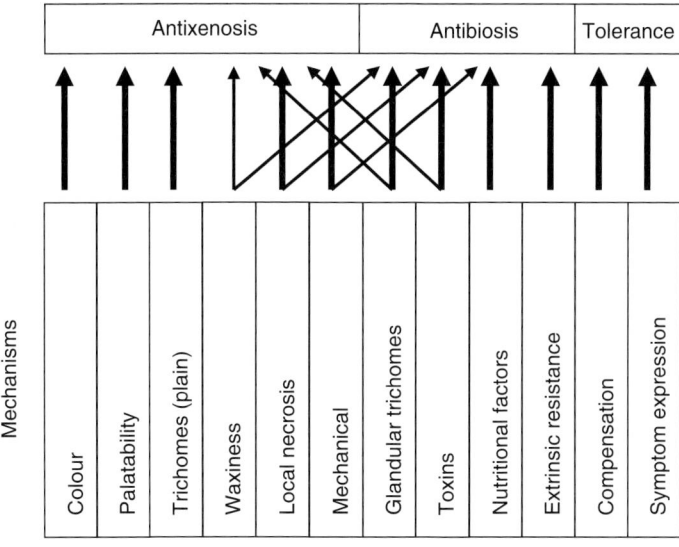

Fig. 17.7. Mechanisms of host plant resistance to aphids, from left to right approximately in the sequence in which they will affect colonizing aphids. Thickness of arrows indicates the likely outcome of the interaction between mechanism and aphid in terms of plant antixenosis, antibiosis, or tolerance.

earlier, and feeding by the aphid aggregate therefore causes no curling of the leaf.

Further Considerations

Yield drag or other fitness costs

The majority of HPR mechanisms involve the production by the plant of more of a chemical compound or additional plant tissue. Entomologists involved in plant breeders' trials regularly claim that crop lines bred for high yield are always pest-susceptible, yet fitness costs of pest-resistance have rarely been reported. Thus, Assad *et al.* (2004) found no negative correlations between the height, grains per spike or total biomass of 14 wheat genotypes and their antixenosis or tolerance to *D. noxia*. Reasons as to why yield drag may not show have been discussed by van Emden (1987) (see also Quisenberry and Ni, Chapter 13 this volume), but such arguments do not explain why genetic modification to express Bt toxin in otherwise isogenetic lines of maize has proceeded with no detectable yield drag. Data of Gershenzon (1994) suggest that different secondary compounds can account for between 0.01 and 30% of the glucose in leaf tissue, but such static concentrations are negligible in comparison with the glucose production by photosynthesis over a period of time. Thus, if there were any fitness costs for varieties resistant to aphids, these would reveal themselves only under conditions of extreme plant stress.

Negative effects on natural enemies

Mortality or sterility of natural enemies can arise from feeding on aphids on plant species containing toxins. Elder (*Sambucus nigra*) with sambunigrin, oleander (*Nerium oleander*) and milkweed (*Asclepias curassivica*), both with cardenolides, are examples of such plants (van Emden, 1987). DIMBOA, a hydroxamic acid studied extensively in relation to HPR to cereal aphids (see earlier), is toxic to the ladybird predator *Eriopis connexa*. Paradoxically, maize varieties with high DIMBOA result in less ladybird mortality than varieties with intermediate levels, since the predator can detect the compound in the aphids and can therefore avoid ingesting a toxic dose (Martos *et al.*, 1992).

Parasitoids (*Aphidius rhopalosiphi*) reared on the smaller individuals of *M. dirhodum*, which develop on the partially resistant wheat variety 'Rapier', showed progressively smaller size and reduced egg load on emergence. After ten generations on the resistant variety, size and egg load had decreased by 15 and over 50%, respectively. However, parasitoids returned to an aphid-susceptible variety regained their fitness in one generation (Fig. 17.8, Salim Jan and H.F. van Emden, unpublished results). Reductions in size and fecundity of *Aphidius nigripes* emerging from *M. euphorbiae* reared on Bt potato have also been attributed to the

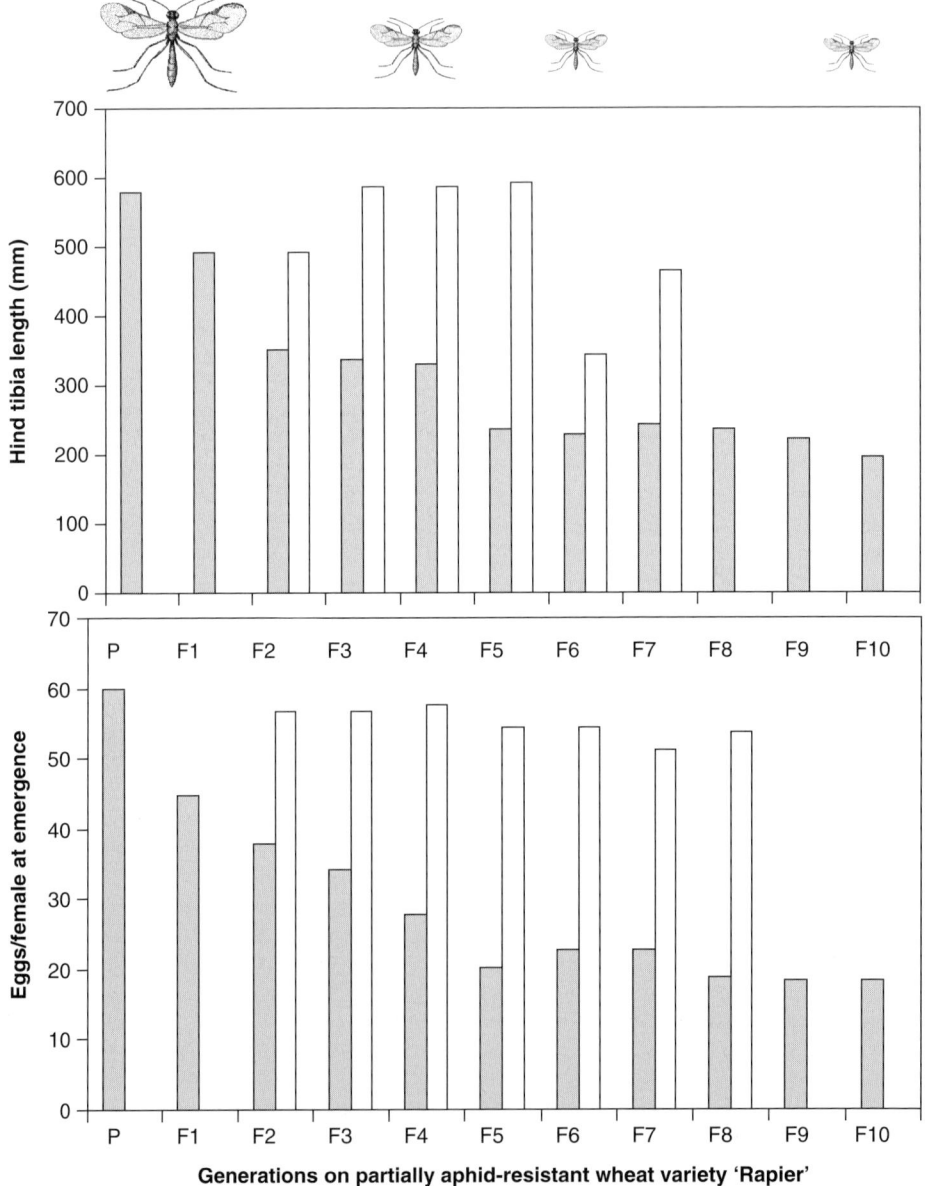

Fig. 17.8. Decline in size (hind tibia length) and number of eggs at emergence in *Aphidius rhopalosiphi* from *Metopolophium dirhodum* reared continuously on the partially aphid-resistant wheat variety 'Rapier' (grey columns). P, parent generation on the aphid-susceptible variety 'Armada'; white columns, change if parasitoids in the previous generation were transferred to parasitize aphids on 'Armada'. Images of parasitoid illustrate the reduction in size (Salim Jan and H.F. van Emden, unpublished results).

poorer quality of the host aphids (Ashouri *et al.*, 2001; Ashouri, 2004).

Problem trading

HPR involves modifying some anatomical, morphological, physiological, or chemical attribute of the plant. There is therefore always the possibility that this will make the plant more susceptible to another damaging organism.

Early examples were the hairy South African cottons, resistant to leafhoppers but especially susceptible to *A. gossypii* (Dunnam and Clark, 1938). When lucerne varieties that were resistant to lucerne wilt (*Verticillium albo-atrum*) were released, they proved especially susceptible to *T. trifolii* (van den Bosch and Messenger, 1973).

Examples with different insects are that glossy brassica varieties resistant to *B. brassicae* have proved to suffer from increased *Phyllotreta cruciferae* (flea beetle) damage (Stoner, 1992; Eigenbrode *et al.*, 2000), and that glossy peas, resistant to *A. pisum*, are more heavily attacked by *Sitona lineatus* (pea and bean weevil) (White and Eigenbrode, 2000).

Surprisingly, problem trading has also been reported in tall fescue plants resistant to *R. padi* because of alkaloids produced by the fungal endophyte *Neotyphodium coenophialum*. The noctuid *Spodoptera frugiperda* (armyworm) showed enhanced performance on the endophyte-infected plants (Bultman and Bell, 2003).

In breeding F2 hybrid willows as biomass fuels, resistance to a *Chaitophorus* sp. aphid and the eriophyid mite *Aculus tetanothrix* were found to be inversely related (Czesak *et al.*, 2004).

Biotypes

Species of aphid often show genotypic variation expressed through different behavioural traits, including sensitivity or tolerance to varieties bred for HPR. Tolerant variants are often referred to as 'resistance-breaking biotypes', but Blackman and Eastop (Chapter 1 this volume) express doubts about the use of the word because such genotypes may be just a single clone subject to recombination by sexual reproduction. The tolerance trait they display may occur therefore in many genetic backgrounds. However, the word 'biotype' is so well embedded in the literature of HPR that probably it would be misleading to the reader to use an alternative terminology here. An example of a resistance-breaking biotype was found when the near immunity of some cowpea varieties producing a toxin strongly aphicidal to *A. craccivora* collected at the International Institute for Tropical Agriculture (IITA) in Nigeria was broken by a later collection of aphids at IITA, though feeding more on pods than the first aphids collected (Ansari, 1984). The literature suggests that the 'biotype problem' with pests is particularly common with aphids, and 10% of references between 1972 and 2003 concerning HPR and aphids are about this aspect. Table 17.2 gives more detail of the reports of actual or potential resistance-breaking biotypes of different aphid species. It is almost certainly not complete, and it is impossible to know how far the same biotypes were involved in different reports for the same species, although for *S. graminum* the characteristic biotypes are distinguished by letters of the alphabet and can be identified by their responses to a clearly defined set of plant varieties. Another point that needs to be made in regard to Table 17.2 is that many of the reports relate to the identification of resistance-breaking biotypes in the glasshouse or laboratory – only relatively few of the listed examples have been accompanied by a problem in the field. The resistance-breaking biotypes of *S. graminum* certainly have proved a major headache in the field; for example, biotype E breaks the resistance to other biotypes incorporated in some sorghums (Eisenbach and Mittler, 1987). No single resistance gene in either sorghum or wheat is effective against all biotypes of *S. graminum*. By contrast, much of the biotype work on *D. noxia* had been in anticipation of possible problems. A cautionary example is *Viteus* (= *Daktulosphaira*) *vitifoliae* (grape phylloxera).

Table 17.2. Aphid biotypes in relation to host-plant resistance.

Aphid	Crop	No. of biotypes	Reference
Acyrthosiphon pisum	Lucerne	4	Cartier *et al.*, 1965; Frazer, 1972; Auclair, 1978; Badenhausser *et al.*, 2005
	Pea	3	Markkula and Roukka, 1971
		5	Frazer, 1972
Amphorophora agathonica	Raspberry	2	Daubeny and Anderson, 1993
Amphorophora rubi	Raspberry	4	Keep and Knight, 1967
Aphis craccivora	Cowpea	2	Ansari, 1984
Brevicoryne brassicae	Sprouts	2	Lammerink, 1968
	Sprouts	7	Dunn and Kempton, 1972
Viteus (= *Daktulosphaira*) *vitifoliae*	Grapes (USA)	2	Stevenson, 1970
	Grapes (Europe)	2	Anonymous, 1994
Diuraphis noxia	Wheat	7	Puterka *et al.*, 1992
	Wheat	2	Basky, 2003
Dysaphis devecta	Apple	3	Briggs and Alston, 1969
Eriosoma lanigerum	Apple	?	Knight *et al.*, 1962
	Apple	3	Sen Gupta, 1969; Briggs and Alston, 1969
Macrosiphum euphorbiae	Tomato	2	Goggin *et al.*, 2001
Rhopalosiphum maidis	Sorghum and maize	5	Painter and Pathak, 1962; Wilde and Feese, 1973
Schizaphis graminum	Sorghum	2	Campbell and Dreyer, 1985; Eisenbach and Mittler, 1987
	Sorghum	3	Bowling *et al.*, 1998
	Sorghum	5	Kindler *et al.*, 2001
	Sorghum	4	Katsar *et al.*, 2002
	Sorghum	8	Zhu-Salzman *et al.*, 2003
	Wheat	8	Wood *et al.*, 1969; Puterka *et al.*, 1988
	Wheat	4	Bowling *et al.*, 1998; Anstead *et al.*, 2003
	Wheat	5	Porter *et al.*, 2000
Sitobion avenae	Wheat	3	Lowe, 1981
Therioaphis trifolii maculata	Lucerne	9	Manglitz *et al.*, 1966; Nielson and Don, 1974

Since the late 19th century, grapes in most parts of the world have been grafted on to rootstocks of phylloxera-resistant old communion wine varieties brought from Europe to North America by the early settlers. This resistance lasted for a long time, but was finally broken by a biotype appearing in Germany as recently as 1994 (Anonymous, 1994). Thus, in spite of the potential for 'biotypic breakdown' of resistance, HPR to

aphids can be quite durable. Ten years after the introduction of *D. noxia* into the USA, there was still only one biotype (Shufran *et al.*, 1997), although a resistance-breaking biotype now appears to occur in Colorado (J.P. Michaud, 2004, personal communication). Di Pietro and Caillaud (1998) reared *S. avenae* on resistant wheat (including the highly resistant 'Einkorn') for 2 years and could find no selection for breaking the

HPR, and Tonet and daSilva (1995) claimed there was no clear evidence that growing resistant wheat creates new biotypes. Indeed, Anstead *et al.* (2003) located several biotypes of *S. graminum* on wild grasses, including one new biotype, and proposed that new biotypes are formed on wild grasses rather than on cereal crops.

The occurrence of biotypes is variable both between species and within species (in relation to geographical distribution). In 1992, Puterka *et al.* recorded seven biotypes of *D. noxia* worldwide (Puterka *et al.*, 1992). In 1993, the populations in the USA, Mexico, Chile, France, and South Africa were still all of the same single biotype (Robinson *et al.*, 1993). However, the biotype in Hungary differed from that in South Africa (Basky, 2003). A detailed study of *Elatobium abietinum* (spruce aphid) compared 40 populations each from the UK and New Zealand (Nicol *et al.*, 1998). There were 28 genotypes (but of course not necessarily 28 biotypes) in the UK collections, but only one in New Zealand.

There is little information as to the mechanisms whereby aphid biotypes break HPR. It is, however, known that resistance to *S. graminum* in sorghums, based on increased methylation of the middle lamellar pectin, is defeated in resistance-breaking biotypes of the aphid by their saliva possessing enhanced pectin methylesterase activity (Dreyer and Campbell, 1984). There is also the intriguing hypothesis that it is the symbionts that enable biotypes of *T. t. maculata* to break the resistance of some lucerne varieties (Ruggle and Gutierrez, 1995).

Coping with the biotype problem involves a continual plant breeding effort to have varieties with new resistance genes available if control begins to fail, as with *S. graminum* (Puterka and Peters, 1990). The breakdown of plant resistance is far more serious with fungal diseases, and plant pathologists have developed strategies for deploying resistance genes so as to delay such breakdown. These strategies are equally applicable to HPR against aphids. One such strategy is combining more than one resistance gene in a variety ('pyramiding'). Porter *et al.* (2000) used a model for *S. graminum* to test whether pyramiding or sequential release of genes would maintain HPR for longer. The model predicted that pyramiding was the longer-lasting strategy, but Porter *et al.* pointed out that experimental evidence suggested the opposite was true. Bush *et al.* (1991) studied the value of variety mixtures against greenbug biotypes C and E. They used three varieties of wheat, one susceptible to the aphid (S) and one resistant to each of the biotypes C and E (CR and ER). The mixtures 3R:1S in separate experiments for each biotype, or 1CR:1ER:1S with both biotypes, managed the biotypes successfully, and the two-variety mix worked as well as the one with three varieties.

Spread of virus (see also Katis *et al.*, Chapter 14 this volume)

Some workers have argued that, because aphids on resistant varieties are more restless, the spread of virus in such varieties is likely to be accelerated, particularly of non-persistent virus diseases. Atiri *et al.* (1984) reported such greater spread of *Cowpea mosaic virus* in small cage experiments with *A. craccivora* on cowpeas. Other work in Africa appears to confirm this. In field experiments on chemical control of the same aphid on cowpeas with pyrethroids, Roberts *et al.* (1993) reported that secondary spread of virus was greater on an aphid-resistant than on a susceptible variety. It is also possible that the phenomenon may occur with aphids in watermelon (Webb and Linda, 1993) and with *M. persicae* in sugarbeet. *Myzus persicae* shows high mortality on sugarbeet, especially as the plants get older, but survival is improved on plants infected with either *Beet yellows virus* or *Beet mild yellowing virus* (Kift *et al.*, 1996).

However, such reports are in the minority, perhaps partly because much of the spread of non-persistent viruses is by immigrants not associated with the crop. The crop is totally resistant to these immigrants, regardless of its resistance status to the aphid species for which it is a host plant. Even in respect of the latter species, the overwhelming evidence is that HPR to aphids leads to less virus spread (van Emden, 1987). Thus, for example, Kishaba *et al.* (1992) compared *Watermelon mosaic virus 2* transmission by *A. gossypii* in one aphid-susceptible and three resistant lines that were all equally good virus sources, and yet transmission by the aphid on the three resistant lines was reduced by 31, 71, and 74%. The number of aphids needed to give 59% infection was only 20 per plant on the susceptible line, but varied from 60–400 on the three aphid-resistant lines. Chen *et al.* (1997) also found that, for the water melon/ *A. gossypii* combination, levels of non-persistent virus were correlated with resistance to the aphid.

Interactions with other control measures

The aphid literature provides many examples of HPR having a positive effect on the impact of natural enemies. HPR also affects the susceptibility of aphids to insecticides. Such interactions are not discussed here, but will be found in a later chapter (van Emden, Chapter 20 this volume).

HPR as a single-component control measure, and have therefore screened for virtual immunity to aphids. Such immunity is likely to result from high levels of toxic allelochemicals, often the expression of a single gene. This has certainly been the approach with the industrial production of insect-resistant transgenic crops, although as yet no transgenic event effective against aphids has been commercialized.

It will be evident from this account of HPR that such monogenetic and high expression of allelochemicals is likely to accentuate the potential disadvantages of this control method such as yield drag, damaging side-effects on natural enemies, and biotypic breakdown. The increased tolerance to insecticides that such resistance may cause is discussed by van Emden, Chapter 20 this volume.

It is more broadly-based, partial HPR to aphids that has real potential for the future, especially given the worldwide move towards the greater integration of several control methods, which is the feature of IPM (van Emden, Chapter 20 this volume).

Conclusions

Host-plant resistance has far greater potential for reducing populations of aphids than has as yet been exploited. This is partly because plant breeders have sought to use

Acknowledgements

I am very grateful to Dr Stephen Clement and Professor Christine Foyer for their advice on endophytes and yield drag, respectively.

References

Adams, J.B. and van Emden, H.F. (1972) The biological properties of aphids and their host plant relationships. In: van Emden, H.F. (ed.) *Aphid Technology*. Academic Press, London, pp. 47–104.

Alston, F.H. and Briggs, J.B. (1970) Inheritance of hypersensitivity to rosy apple aphid *Dysaphis plantaginea* in apples. *Canadian Journal of Genetics and Cytology* 12, 257–258.

Anonymous (1994) Die Rückkehr der Reblaus. *Profil, October* 1994, 11.

Ansari, A.K. (1984) Biology of *Aphis craccivora* (Koch) and varietal resistance of cowpeas. PhD thesis, The University of Reading, Reading, UK.

Ansari, A.K., van Emden, H.F. and Singh, S.R (1989) Graft-transmissibility of resistance to cowpea aphid, *Aphis craccivora* Koch (Hemiptera: Aphididae), in six highly antibiotic cowpea varieties. *Bulletin of Entomological Research* 79, 393–399.

Anstead, J.A., Burd, J.D. and Shufran, K.A. (2003) Over-summering and biotypic diversity of *Schizaphis graminum* (Homoptera: Aphididae) populations on noncultivated grass hosts. *Environmental Entomology* 32, 662–667.

Ashouri, A. (2004) Transgenic-Bt potato plant resistance to the Colorado potato beetle affect the aphid parasitoid *Aphidius nigripes*. *Mededelingen Faculteit Landbouwkundige en Toegepaste Biologische Wetenschappen Universiteit Gent* 69, 185–189.

Ashouri, A., Michaud, D. and Cloutier, C. (2001) Unexpected effects of different potato resistance factors to the Colorado potato beetle (Coleoptera: Chrysomelidae) on the potato aphid (Homoptera: Aphididae). *Environmental Entomology* 30, 524–532.

Assad, M.T., Mehrabi, A.M., Pakniyat, H. and Nematollahy, M.R. (2004) The effect of resistance components on reducing yield and its related characters in wheat as infected by *Diuraphis noxia* (Hemiptera: Aphididae). In: *Cereal Research Communcations, No. 32*. Cereal Research Non-Profit Company, Zged, Hungary, pp. 69–73.

Atiri, G.I., Ekpo, E.J.A. and Thottappilly, G. (1984) The effect of aphid-resistance in cowpea on infestation and development of *Aphis craccivora* and the transmission of cowpea aphid-borne mosaic virus. *Annals of Applied Biology* 104, 339–346.

Auclair, J.L. (1978) Biotypes of the pea aphid *Acyrthosiphon pisum* in relation to host plants and chemically-defined diets. *Entomologia Experimentalis et Applicata* 24, 212–216.

Auclair, J.L., Maltais, J.B. and Cartier, J.J. (1957) Factors in resistance of peas to the pea aphid, *Acyrthosiphon pisum* (Harr.) (Homoptera: Aphididae). II. Amino acids. *Canadian Entomologist* 89, 457–464.

Badenhausser, I., Bournoville, R., Carré, S. and Fortini, D. (2005) Identification de sources de résistance au puceron du pois chez des *Medicago* annuelles. *Fourrages* 184, 567–578.

Basky, Z. (2003) Biotypic and pest status differences between Hungarian and South African populations of Russian wheat aphid, *Diuraphis noxia* (Kurdjumov) (Homoptera: Aphididae). *Pest Management Science* 59, 1152–1158.

Belefantmiller, H., Porter, D.R., Pierce, M.L. and Mort, A.J. (1994) An early indicator of resistance in barley to Russian wheat aphid. *Plant Physiology* 105, 1289–1294.

Bergman, D.K., Dillwith, J.W., Zarrabi, A.A., Caddel, J.L. and Berberet, R.C. (1991) Epicuticular lipids of alfalfa relative to its susceptibility to spotted alfalfa aphids (Homoptera, Aphididae). *Environmental Entomology* 20, 781–785.

Berlandier, F.A. (1996) Alkaloid level in narrow-leafed lupin, *Lupinus angustifolius*, influences green peach aphid reproductive performance. *Entomologia Experimentalis et Applicata* 79, 19–24.

Birch, L.C. (1948) The intrinsic rate of natural increase of an insect population. *Journal of Animal Ecology* 17, 15–26.

van den Bosch, R. and Messenger, P.S. (1973) *Biological Control*. Intext, New York, 137 pp.

Bowling, R., Wilde, G. and Margolies, D. (1998) Relative fitness of greenbug (Homoptera: Aphididae) biotypes E and I on sorghum, wheat, rye, and barley. *Journal of Economic Entomology* 91, 1219–1223.

Breen, J.P. (1992) Temperature and seasonal effects on expression of *Acremonium* endophyte enhanced resistance to *Schizaphis graminum* (Homoptera, Aphididae). *Environmental Entomology* 21, 68–74.

Briggs, J.B. and Alston, F.H. (1969) Sources of pest resistance in apple cultivars. *Report of East Malling Research Station, 1966*, 170–171.

Bultman, T.L. and Bell, G.D. (2003) Interaction between fungal endophytes and environmental stressors influences plant resistance to insects. *Oikos* 103, 182–190.

Bush, L., Slosser, J.E., Worrall, W.D. and Horner, N.V. (1991) Potential of wheat cultivar mixtures for greenbug (Homoptera, Aphididae) management. *Journal of Economic Entomology* 84, 1619–1624.

Cambier, V., Hance, T. and De Hoffmann, E. (2001) Effects of 1,4-benzoxazin-3-one derivatives from maize on survival and fecundity of *Metopolophium dirhodum* (Walker) on artificial diet. *Journal of Chemical Ecology* 27, 359–370.

Campbell, B.C. and Dreyer, D.L. (1985) Host plant resistance of sorghum: differential hydrolysis of sorghum pectic substances by polysaccharases of greenbug biotypes (*Schizaphis graminum*, Homoptera: Aphididae). *Archives of Insect Biochemistry and Physiology* 2, 203–215.

Cartier, J.J., Isaak, A., Painter, R.H. and Sorensen, E.L. (1965) Biotypes of pea aphid *Acyrthosiphon pisum* (Harris) in relation to alfalfa clones. *Canadian Entomologist* 97, 754–760.

Castro, A.M., Ramos, S., Vasicek, A., Worland, A., Gimenez, D., Clua, A.A. and Suarez, E. (2001) Identification of wheat chromosomes involved with different types of resistance against greenbug (*Schizaphis graminum*, Rond.) and the Russian wheat aphid (*Diuraphis noxia*, Mordvilko). *Euphytica* 118, 321–330.

Chen, J.Q., Rahbe, Y., Delobel, B., Sauvion, N., Guillaud, J. and Febvay, G. (1997) Melon resistance to the aphid *Aphid gossypii*: behavioural analysis and chemical correlations with nitrogenous compounds. *Entomologia Experimentalis et Applicata* 85, 33–44.

Clement, S.L., Kaiser, W.J. and Eichenseer, H. (1994) *Acremonium* endophytes in germplasms of major grasses and their utilization for insect resistance. In: Bacon, C.W. and White, J.M., Jr. (eds) *Biotechnology of Endophytic Fungi of Grasses*. CRC, Boca Raton, pp. 185–199.

Clement, S.L., Lester, D.G., Wilson, A.D., Johnson, R.C. and Bouton, J.H. (1996) Expression of Russian wheat aphid (Homoptera: Aphididae) resistance in genotypes of tall fescue harbouring different isolates of *Acremonium* endophyte. *Journal of Economic Entomology* 89, 766–770.

Clement, S.L., Wilson, A.D., Lester, D.G. and David, C.M. (1997) Fungal endophytes of wild barley and their effects on *Diuraphis noxia* population development. *Entomologia Experimentalis et Applicata* 82, 275–281.

Cole, R.A. (1994a) Locating a resistance mechanism to the cabbage aphid in two wild brassicas. *Entomologia Experimentalis et Applicata* 71, 23–31.

Cole, R.A. (1994b) Isolation of a chitin-binding lectin, with insecticidal activity in chemically defined synthetic diets, from two wild brassica species with resistance to cabbage aphid *Brevicoryne brassicae*. *Entomologia Experimentalis et Applicata* 72, 181–187.

Cole, R.A. (1997a) The relative importance of glucosinolates and amino acids to the development of two aphid pests *Brevicoryne brassicae* and *Myzus persicae* on wild and cultivated *Brassica* species. *Entomologia Experimentalis et Applicata* 85, 121–133.

Cole, R.A. (1997b) Comparison of feeding behaviour of two *Brassica* pests *Brevicoryne brassicae* and *Myzus persicae* on wild and cultivated *Brassica* species. *Entomologia Experimentalis et Applicata* 85, 135–141.

Cooper, W.C., Jia, L. and Goggin, F.L. (2004) Acquired and R-gene-mediated resistance against the potato aphid in tomato. *Journal of Chemical Ecology* 30, 2527–2542.

Czesak, M.E., Knee, M.J., Gale, R.G., Bodach, S.D. and Fritz, R.S. (2004) Genetic architecture of resistance to aphids and mites in a willow hybrid system. *Heredity* 93, 619–626.

Daubeny, H.A. and Anderson, A.K. (1993) The British Columbia red raspberry breeding programme – achievements and prospects. *Acta Horticulturae* 352, 285–293.

Deol, G.S., Wilde, G.E. and Gill, B.S. (1995) Host plant resistance in some wild wheats to the Russian wheat aphid, *Diuraphis noxia* (Mordvilko) (Homoptera: Aphididae). *Plant Breeding* 114, 545–546.

Di Pietro, J.P. and Caillaud, C.M. (1998) Response to intracolonal selection for adaptation to resistant wheat in the English grain aphid (Homoptera: Aphididae). *Environmental Entomology* 27, 80–85.

Di Pietro, J.P., Caillaud, C.M., Chaubet, P., Pierre, J.S. and Trottet, M. (1998) Variation in resistance to the grain aphid, *Sitobion avenae* (Sternorrhynca: Aphididae), among diploid wheat genotypes. *Plant Breeding* 117, 407–412.

Dodd, G.D. and van Emden, H.F. (1979) Shifts in host plant resistance to the cabbage aphid (*Brevicoryne brassicae*) exhibited by Brussels sprout plants. *Annals of Applied Biology* 91, 251–262.

Dreyer, D.L. and Campbell, B.C. (1984) Association of the degree of methylation of intercellular pectin with plant resistance to aphids and with induction of aphid biotypes. *Experientia* 40, 224–226.

Dreyer, D.L. and Campbell, B.C. (1987) Chemical basis of host-plant resistance to aphids. *Plant, Cell and Environment* 10, 353–361.

Dreyer, D.L., Jones, K.C., Jurd, L. and Campbell, B.C. (1987) Feeding deterrency of some 4-hydroxycoumarins and related compounds: relationship to host-plant resistance of alfalfa towards pea aphid (*Acyrthosiphon pisum*). *Journal of Chemical Ecology* 13, 925–930.

Du Li, Ge Feng, Zhu SanRong and Parajulee, M.N. (2004) Effect of cotton cultivar on development and reproduction of *Aphis gossypii* (Homoptera: Aphididae) and its predator *Propylaea japonica* (Coleoptera: Coccinellidae). *Journal of Economic Entomology* 97, 1278–1283.

Dunn, J.A. and Kempton, D.P.H. (1972) Resistance to attack by *Brevicoryne brassicae* (L.) on Brussels sprouts. *Annals of Applied Biology* 72, 1–11.

Dunnam, E.W. and Clark, J.C. (1938) The cotton aphid in relation to the pilosity of cotton leaves. *Journal of Economic Entomology* 31, 663–666.

Eichenseer, H., Dahlman, D.L. and Bush, L.P. (1991) Influence of endophyte infection, plant-age and harvest interval on *Rhopalosiphum padi* survival and its relation to quantity of N-formyl and N-acetyl loline in tall fescue. *Entomologia Experimentalis et Applicata* 60, 29–38.

Eigenbrode, S.D., Kabalo, N.N. and Rutledge, C.E. (2000) Potential of reduced-waxbloom oilseed *Brassica* for insect pest resistance. *Journal of Agricultural and Urban Entomology* 17, 53–63.

Eisenbach, J. and Mittler, T.E. (1987) Extra-nuclear inheritance in a sexually produced aphid: the ability to overcome host plant resistance by biotype hybrids of the greenbug, *Schizaphis graminum*. *Experientia* 43, 332–334.

Ellis, P.R., Singh, R., Pink, D.A.C., Lynn, J.R. and Shaw, P.L. (1996) Resistance to *Brevicoryne brassicae* in horticultural brassicas. *Euphytica* 88, 85–96.

van Emden, H.F. (1969) Plant resistance to *Myzus persicae* induced by a plant regulator and measured by aphid relative growth rate. *Entomologia Experimentalis et Applicata* 12, 125–131.

van Emden, H.F. (1978) Insects and secondary substances – an alternative viewpoint with special reference to aphids. In: Harborne, J.B. (ed.) *Phytochemical Aspects of Plant and Animal Co-evolution.* Academic Press, London, pp. 309–323.

van Emden, H.F. (1987) Cultural methods: the plant. In: Burn, A.J., Coaker, T.H. and Jepson, P.C. (eds) *Integrated Pest Management.* Academic Press, London, pp. 27–68.

van Emden, H.F. (1997) Host-plant resistance to insect pests. In: Pimentel, D. (ed.) *Techniques for Reducing Pesticide Use.* Wiley, Chichester, pp. 130–152.

van Emden, H.F. and Bashford, M.A. (1971) The performance of *Brevicoryne brassicae* and *Myzus persicae* in relation to plant age and leaf amino acids. *Entomologia Experimentalis et Applicata* 14, 349–360.

van Emden, H.F., Vidyasagar, P. and Kazemi, M.H. (1991) Use of systemic insecticide to measure antixenosis to aphids in plant choice experiments. *Entomologia Experimentalis et Applicata* 58, 69–74.

Frazer, B.D. (1972) Population dynamics and recognition of biotypes in the pea aphid (Homoptera: Aphididae). *Canadian Entomologist* 104, 1729–1733.

Fuentes-Contreras, E. and Niemeyer, H.M. (1998) DIMBOA glucoside, a wheat chemical defense, affects host acceptance and suitability of *Sitobion avenae* to the cereal aphid parasitoid *Aphidius rhopalosiphi. Journal of Chemical Ecology* 24, 371–381.

Fuentes-Contreras, J.E., Powell, W., Wadhams, L.J., Pickett, J.A. and Niemeyer, H.M. (1996) Influence of wheat and oat cultivars on the development of the cereal aphid parasitoid *Aphidius rhopalosiphi* and the generalist aphid parasitoid *Ephedrus plagiator. Annals of Applied Biology* 129, 181–187.

Gershenzon, J. (1994) The cost of plant chemical defense against herbivory: a biochemical perspective. In: Bernays, E.A. (ed.) *Insect–Plant Interactions, Volume 5.* CRC, Boca Raton, pp. 173–205.

Gianoli, E., Papp, M. and Niemeyer, H.M. (1996) Costs and benefits of hydroxamic acids-related resistance on winter wheat against the bird cherry–oat aphid, *Rhopalosiphum padi* L. *Annals of Applied Biology* 129, 83–90.

Gibson, R.W. (1976) Glandular hairs are a possible means of limiting aphid damage to the potato crop. *Annals of Applied Biology* 82, 143–146.

Givovich, A. and Niemeyer, H.M. (1995) Comparison of the effect of hydroxamic acids from wheat on five species of cereal aphids. *Entomologia Experimentalis et Applicata* 74, 115–119.

Goggin, F.I., Williamson, V.M. and Ullman, D.E. (2001) Variability in the response of *Macrosiphum euphorbiae* and *Myzus persicae* (Hemiptera: Aphididae) to the tomato resistance gene Mi. *Environmental Entomology* 30, 101–106.

Gowling, G.R. and van Emden, H.F. (1994) Falling aphids enhance impact of biological control by parasitoids on partially aphid-resistant plant varieties. *Annals of Applied Biology* 125, 233–242.

Hansen, L.M. (2006) Effect of 6-methoxybenzoxazolin-2-one (MBOA) on the reproductive rate of the grain aphid (*Sitobion avenae* F.). *Journal of Agriculture and Food Chemistry* 54, 1031–1035.

Havlickova, H. (1997) Differences in level of tolerance to cereal aphids in five winter wheat cultivars. *Rostlinna Vyroba* 43, 593–596.

Hesler, L.S. (2005) Resistance to *Rhopalosiphum padi* (Homoptera: Aphididae) in three triticale accessions. *Journal of Economic Entomology* 98, 603–610.

Johnson, A.W., Sisson, V.A., Snook, M.F., Fortnum, B.A. and Jackson, D.M. (2002) Aphid resistance and leaf surface chemistry of sugar ester producing tobaccos. *Journal of Entomological Research* 37, 154–165.

Kareiva, P. and Sahakian, R. (1990) Tritrophic effects of a simple architectural mutation in pea plants. *Nature, London* 345, 433–434.

Katsar, C.S., Paterson, A.H., Teetes, G.L. and Peterson, G.C. (2002) Molecular analysis of sorghum resistance to the greenbug (Homoptera: Aphididae). *Journal of Economic Entomology* 95, 448–457.

Kazemi, M.H. and van Emden, H.F. (1992) Partial antibiosis to *Rhopalosiphum padi* in wheat and some phytochemical correlations. *Annals of Applied Biology* 121, 1–9.

Keep, E. and Knight, R.L. (1967) A new gene from *Rubus occidentalis* L. for resistance to strains 1, 2 and 3 of the Rubus aphid, *Amphorophora rubi* Kalt. *Euphytica* 16, 209–214.

Kift, N.B., Dewar, A.M., Werker, A.R. and Dixon, A.F.G. (1996) The effect of plant age and infection with virus yellows on the survival of *Myzus persicae* on sugar beet. *Annals of Applied Biology* 129, 371–378.

Kindler, S.D., Breen, J.P. and Springer, T.L. (1991) Reproduction and damage by Russian wheat aphid (Homopotera: Aphididae) as influenced by fungal endophytes and cool-season turfgrasses. *Journal of Economic Entomology* 84, 685–692.

Kindler, S.D., Harvey, T.L., Wilde, G.E., Shufran, R.A., Brooks, H.L. and Solderbeck, P.E. (2001) Occurrence of greenbug biotype K in the field. *Journal of Agricultural and Urban Entomology* 18, 23–34.

Kishaba, A.N., Castle, S.J., Coudriet, D.L., McCreight, J.D. and Bohn, G.W. (1992) Virus transmission by *Aphis gossypii* Glover to aphid-resistant and susceptible muskmelons. *Journal of the American Society for Horticultural Science* 117, 248–254.

Knight, R.L., Briggs, J.B., Massee, A.M. and Tydeman, H.M. (1962) The inheritance of resistance to woolly aphid, *Eriosoma lanigerum* (Hsmnn.), in the apple. *Journal of Horticultural Science* 37, 207–218.

Kogan, M. and Ortman, E.F. (1978) Antixenosis – a new term proposed to define Painter's 'non-preference' modality of resistance. *Bulletin of the Entomological Society of America* 24, 175–176.

Kok-Yokomi, M.L. (1978) Mechanisms of host plant resistance in tomato to the potato aphid *Macrosiphum euphorbiae* (Thomas). *Dissertation Abstracts International B* 39, 1121.

Lammerink, J. (1968) A new biotype of cabbage aphid (*Brevicoryne brassicae* (L.)) on aphid resistant rape (*Brassica napus* L.). *New Zealand Journal of Agricultural Research* 11, 341–344.

Lowe, H.J.B. (1981) Resistance and susceptibility to colour forms of the aphid *Sitobion avenae* in spring and winter wheats (*Triticum aestivum*). *Annals of Applied Biology* 99, 87–98.

Lowe, H.J.B., Murphy, G.J.P. and Parker, M.I. (1985) Non-glaucousness, a probable aphid-resistance character of wheat. *Annals of Applied Biology* 106, 555–560.

Manglitz, G.R., Calkins, C.O., Walstrom, R.J., Hintz, S.D., Kindler, S.D. and Peters, L.L. (1966) Holocylcic strains of the spotted alfalfa aphid in Nebraska and adjacent States. *Journal of Economic Entomology* 59, 636–639.

Markkula, M. and Roukka, K. (1971) Resistance of plants to the pea aphid *Acyrthosiphon pisum* Harris (Hom. Aphididae). III. Fecundity on different pea varieties. *Annales Agriculturae Fenniae* 10, 33–37.

Martos, A., Givovich, A. and Niemeyer, H.M. (1992) Effect of DIMBOA, and aphid resistance factor in wheat, on the aphid predator *Eriopsis connexa* Germar (Coleoptera, Coccinellidae). *Journal of Chemical Ecology* 18, 469–479.

Maxwell, F.G. and Painter, R.H. (1962) Auxin content of extracts of certain tolerant and susceptible host plants of *Toxoptera graminum*, *Macrosiphum pisi*, and *Therioaphis maculata*, and relation to host plant resistance. *Journal of Economic Entomology* 55, 46–56.

Mentink, P.J.M., Kimmins, F.M., Harrewijn, P., Dieleman, F.L., Tjallingi, W.F., van Rheenen, B. and Ennink, A.H. (1984) Electrical penetration graphs combined with stylet cutting in the study of host plant resistance to aphids. *Entomologia Experimentalis et Applicata* 35, 210–213.

Messina, F.J. and Bloxham, A.J. (2004) Plant resistance to the Russian wheat aphid: effects on a nontarget aphid and the role of induction. *Canadian Entomologist* 136, 129–137.

Migui, S.M. and Lamb, R.J. (2004) Seedling and adult plant resistance to *Sitobion avenae* (Hemiptera: Aphididae) in *Triticum monoccum* (Poaceae), an ancestor of wheat. *Bulletin of Entomological Research* 94, 35–46.

Miles, P.W. (1999) Aphid saliva. *Biological Reviews of the Cambridge Philosophical Society* 74, 41–85.

Morgham, A.T., Richardson, P.E., Campbell, R.K., Burd, J.D., Eikenbary, R.D. and Sumner, L.C. (1994) Ultrastructural responses of resistant and susceptible wheat to infestation by greenbug biotype E (Homoptera: Aphididae). *Annals of the Entomological Society of America* 87, 908–917.

Müller, H.J. (1958) The behaviour of *Aphis fabae* in selecting its host plants, especially different varieties of *Vicia faba*. *Entomologia Experimentalis et Applicata* 1, 66–72.

Ni, X.Z., Quisenberry, S.S., Siegfried, B.D. and Lee, K.W. (1998) Influence of cereal leaf epicuticular wax on *Diuraphis noxia* probing behaviour and nymphoposition. *Entomologia Experimentalis et Applicata* 89, 111–118.

Nicol, D., Armstrong, K.F., Wratten, S.D., Walsh, P.J., Straw, N.A., Cameron, C.M., Lahmann, C. and Frampton, C.M. (1998) Genetic diversity of an introduced pest, the green spruce aphid *Elatobium abietinum* (Hemiptera: Aphididae) in New Zealand and the United Kingdom. *Bulletin of Entomological Research* 88, 537–543.

Nielson, M.W. and Don, H. (1974) A new virulent biotype of the spotted alfalfa aphid in Arizona. *Journal of Economic Entomology* 67, 64–66.

Novy, R.G., Nasruddin, A., Ragsdale, D.W. and Radcliffe, E.B. (2002) Genetic resistance to potato leafroll virus, potato virus Y, and green peach aphid in progeny of *Solanum etuberosum*. *American Journal of Potato Research* 79, 9–18.

Painter, R.H. (1951) *Insect Resistance in Crop Plants*. Macmillan, New York, 520 pp.

Painter, R.H. and Pathak, M.D. (1962) The distinguishing features and significance of the four biotypes of the corn leaf aphid, *Rhopalosiphum maidis* (Fitch). *Proceedings of the 11th International Congress of Entomology, Vienna, 1960* 2, 110–115.

Porter, D.R., Burd, J.D., Shufran, K.A. and Webster, J.A. (2000) Effect of pyramiding greenbug (Homoptera: Aphididae) resistance genes in wheat. *Journal of Economic Entomology* 93, 1315–1318.

Puterka, G.J. and Peters, D.C. (1990) Sexual reproduction and inheritance of virulence in the greenbug, *Schizaphis graminum* (Rondani). In: Campbell, R.K. and Eikenbary, R.D. (eds) *Aphid–Plant Genotype Interactions*. Elsevier, Amsterdam, pp. 289–318.

Puterka, G.J., Peters, D.C., Kerns, D.L., Slosser, J.E., Bush, L., Worrall, D.W. and McNew, R.W. (1988) Designation of two new greenbug (Homoptera: Aphididae) biotypes G and H. *Journal of Economic Entomology* 81, 1754–1759.

Puterka, G.J., Burd, J.D. and Burton, R.L. (1992) Biotypic variation in a worldwide collection of Russian wheat aphid (Homoptera: Aphididae). *Journal of Economic Entomology* 85, 1497–1506.

Radcliffe, E.B. and Chapman, R.K. (1965) The relative resistance to insect attack of three cabbage varieties at different stages of plant maturity. *Annals of the Entomological Society of America* 58, 897–902.

Radcliffe, E.B. and Chapman, R.K. (1966) Varietal resistance to insect attack in various cruciferous crops. *Journal of Economic Entomology* 59, 120–125.

Rahbe, Y., Sauvion, N., Febvay, G., Peumans, W.J. and Gatehouse, A.M.R. (1995) Toxicity of lectins and processing of ingested proteins in the pea aphid *Acyrthosiphon pisum*. *Entomologia Experimentalis et Applicata* 76, 143–155.

Rahbe, Y., Deraison, C., Bonade-Bottino, M., Girard, C., Nardon, C. and Jouanin, L. (2003) Effects on the cysteine protease inhibitor oryzacystatin (OC-1) on different aphids and reduced performance of *Myzus persicae* on OC-1 expressing transgenic oilseed rape. *Plant Science* 164, 441–450.

Risdill-Smith, J., Edwards, O., Wang ShaoFang, Ghisalberti, E. and Reidy-Crofts, J. (2004) Aphid response to plant defensive compounds in plants. In: *Aphids in a New Millennium. Proceedings of the 6th International Symposium on Aphids, Rennes, September 2001*. INRA, Paris, pp. 491–497.

Roberts, J.M.F., Hodgson, C.J., Jackai, L.E.N., Thottappilly, G. and Singh, S.R. (1993) Interaction between two synthetic pyrethroids and the spread of two nonpersistent viruses in cowpea. *Annals of Applied Biology* 122, 57–67.

Robertson, G.W., Griffiths, D.W., Birch, A.N.E., Jones, A.T., McNicol, J.W. and Hall, J.E. (1991) Further evidence that resistance in raspberry to the virus vector aphid, *Amphorophora idaei*, is related to the chemical composition of the leaf surface. *Annals of Applied Biology* 119, 443–449.

Robinson, J. (1992) Modes of resistance in barley seedlings to six aphid (Homoptera, Aphididae) species. *Journal of Economic Entomology* 85, 2510–2515.

Robinson, J., Fischer, M. and Hoisington, D. (1993) Molecular characterization of *Diuraphis* spp. (Homoptera, Aphididae) using random amplified polymorphic DNA. *Southwestern Entomologist* 18, 121–127.

Ruggle, P. and Gutierrez, A.P. (1995) Use of life-tables to assess host-plant resistance in alfalfa to *Therioaphis trifolii* f. *maculata* (Homoptera: Aphididae) – hypothesis for maintenance of resistance. *Environmental Entomology* 24, 313–325.

Saikkonen, K., Faeth, S.H., Helander, M.L. and Sullivan T.J. (1998) Fungal endophytes: a continuum of interactions with host plants. *Annual Review of Ecology and Systematics* 29, 319–343.

Schotzko, D.J. and Bosque-Perez, N.A. (2000) Seasonal dynamics of cereal aphids on Russian wheat aphid (Homoptera: Aphididae) susceptible and resistant wheats. *Journal of Economic Entomology* 93, 975–981.

Sen Gupta, G.C. (1969) The recognition of biotypes of the woolly aphid, *Eriosoma lanigerum* (Hausmann), in South Australia by their differential ability to colonise varieties of apple rootstock, and an investigation of some possible factors in the susceptibility of varieties to these insects. PhD thesis, University of Adelaide, Adelaide, Australia.

Shepherd, T., Robertson, G.W., Griffiths, D.W. and Birch, A.N.E. (1999a) Epicuticular wax composition in relation to aphid infestation and resistance in red raspberry (*Rubus idaeus* L.). *Phytochemistry* 52, 1239–1254.

Shepherd, T., Robertson, G.W., Griffiths, D.W. and Birch, A.N.E. (1999b) Epicuticular wax ester and triacylglycerol composition in relation to aphid infestation and resistance in red raspberry (*Rubus idaeus* L.). *Phytochemistry* 52, 1255–1267.

Shinoda, T. (1993) Callose reaction induced in melon leaves by feeding of melon aphid, *Aphis gossypii* Glover as possible aphid resistant factor. *Japanese Journal of Applied Entomology and Zoology* 73, 145–152.

Shufran, K.A., Burd, J.D. and Webster, J.A. (1997) Biotypic status of Russian wheat aphid (Homoptera: Aphididae) populations in the United States. *Journal of Economic Entomology* 90, 1684–1689.

Simmons, A.T., McGrath, D. and Gurr, G.M. (2005) Trichome characteristics of F1 *Lycopersicon esculentum* × *L. pennellii* hybrids and effects on *Myzus persicae*. *Euphytica* 144, 313–320.

Sotherton, N.W. and van Emden, H.F. (1982) Laboratory assessments of the resistance to the aphids, *Sitobion avenae* and *Metopolophium dirhodum* in three *Triticum* species and two modern wheat cultivars. *Annals of Applied Biology* 101, 99–107.

Stevenson, A.B. (1970) Strains of the grape phylloxera in Ontario with different effects on the foliage of certain grape cultivars. *Journal of Economic Entomology* 63, 135–138.

Stoner, K.A. (1992) Density of imported cabbageworms (Lepidoptera, Pieridae), cabbage aphids (Homoptera, Aphididae), and flea beetles (Coleoptera, Chrysomelidae) on glossy and trichome-bearing lines of *Brassica oleracea*. *Journal of Economic Entomology* 85, 1023–1030.

Thompson, K.F. (1963) Resistance to the cabbage aphid (*Brevicoryne brassicae*) in *Brassica* plants. *Nature, London* 198, 209–210.

Tjallingii, W.F. (1988) Electrical recording of stylet penetration activities. In: Minks, A.K. and Harrewijn, P. (eds) *Aphids. Their Biology, Natural Enemies and Control, Volume 2B*. Elsevier, Amsterdam, pp. 95–108.

Tonet, G.L. and daSilva, R.F.P. (1995) Resistance of wheat genotypes for C biotype of *Schizaphis graminum* (Rondani, 1852) (Homoptera, Aphididae). *Pesquisa Agropecuaria Brasiliera* 30, 1283–1287.

Wartenberg, H. (1953) Über pflanzenphysiologische Ursachen des Massenwechsels der Apfelblattlaus (*Eriosoma lanigerum*) auf *Malus pumila*. *Mitteilungen der Biologischen Zentralanstalt Berlin-Dahlem* 75, 53–56.

Weathersbee, A.A. and Hardee, D.D. (1994) Abundance of cotton aphids (Homoptera, Aphididae) and associated biological control agents on six cotton cultivars. *Journal of Economic Entomology* 87, 258–265.

Weathersbee, A.A., Hardee, D.D. and Meredith, W.R. (1994) Effects of cotton genotype on seasonal abundance of cotton aphid (Homoptera, Aphididae). *Journal of Agricultural Entomology* 11, 29–37.

Weathersbee, A.A., Hardee, D.D. and Meredith, W.R. (1995) Differences in yield response to cotton aphids (Homoptera, Aphididae) between smooth-leaf and hairy-leaf isogenic cotton lines. *Journal of Economic Entomology* 88, 749–754.

Webb, S.E. and Linda, S.B. (1993) Effect of oil and insecticide on epidemics of potyviruses in watermelon in Florida. *Plant Disease* 77, 869–874.

Webster, J.A., Inayatullah, C., Hamissou, M. and Mirkes, K.A. (1994) Leaf pubescence effects in wheat on yellow sugarcane aphids and greenbugs (Homoptera, Aphididae). *Journal of Economic Entomology* 87, 231–240.

Webster, J.A. and Porter, D.R. (2000a) Reaction of four aphid species on a Russian wheat aphid resistant wheat. *Southwestern Entomologist* 25, 83–90.

Webster, J.A. and Porter, D.R. (2000b) Plant resistance components of two greenbug (Homoptera: Aphididae) resistant wheats. *Journal of Economic Entomology* 93, 1000–1004.

Weibull, J. (1987) Work on plant resistance to *Rhopalosiphum padi* (L.) in oats and barley – present status. *Bulletin OILB/SROP* 10, 160–161.

Weibull, J. (1994) Glutamic acid content of phloem sap is not a good predictor of plant resistance to *Rhopalosiphum padi*. *Phytochemistry* 35, 601–602.

White, C. and Eigenbrode, S.D. (2000) Effects of surface wax variation in *Pisum sativum* on herbivorous and entomophagous insects in the field. *Environmental Entomology* 29, 773–780.

Wilde, G. and Feese, H. (1973) A new corn leaf aphid biotype and its effect on some cereal and small grains. *Journal of Economic Entomology* 66, 570–571.

Wood, E.A., Jr., Chada, H.L. and Saxena, P.N. (1969) Reactions of small grains and grain sorghum to three greenbug biotypes. *Report of the Oklahoma Agriculture Experiment Station, No. 618*, 7 pp.

Zhu-Salzman, K., Li, H.W., Klein, P.E., Gorena, R.L. and Salzman, R.A. (2003) Using high-throughput amplified fragment length polymorphism to distinguish sorghum greenbug (Homoptera: Aphididae) biotypes. *Agricultural and Forest Entomology* 5, 311–315.

Zuparko, R.L. and Dahlsten, D.L. (1994) Host plant resistance and biological control for linden aphids. *Journal of Arboriculture* 20, 278–281.

18 Biological Control

Wilf Powell and Judith K. Pell

*Plant and Invertebrate Ecology Division, Rothamsted Research,
Harpenden, Herts, AL5 2JQ, UK*

Introduction

Biological pest control strategies fall into three main categories (Van Driesche and Bellows, 1996): (i) 'classical', involving introductions of natural enemies into geographic areas where they did not previously occur and usually directed against introduced pests; (ii) 'augmentation', involving mass-rearing and release of natural enemies that already exist in the system but which do not occur naturally in sufficient numbers, or at the optimum time, to control the target pest; and (iii) 'conservation' biological control, which includes the enhancement of naturally occurring, wild populations of natural enemies by means of habitat management or manipulation of their behaviour. In this chapter, we focus on attempts to control aphids using the first two of these strategies, while Wratten *et al.* (Chapter 16 this volume) describe habitat management for conservation biological control. Under augmentation, we include both inoculative releases, which aim to establish a self-perpetuating population of the natural enemy, at least over the period of crop risk, and inundative releases, which are not expected to establish and therefore not to produce further generations to any significant extent. The biological control of aphids

to reduce damage on crops is a different topic from the role of natural enemies in the year-to-year population dynamics of aphids. We accept the view of Kindlmann *et al.* (Chapter 12 this volume) that natural enemies may not be an important component in the long-term regulation of aphid populations, but there is no doubt they can be the deciding factor in preventing pest levels of aphids developing on a particular cropped area within a growing season.

Biological control of aphids was reviewed in the late 1980s (Carver, 1989), but interest in this topic has increased since that time and commercial use of natural enemies to control aphids is now widespread in the protected crop industry. Progress has been slower in outdoor crops, which present a greater challenge, but research trials assessing augmentation strategies have been conducted in a wide variety of crops. Economic constraints associated with mass-rearing and release methods have frequently hindered the uptake of augmentation techniques for field crops. However, continuing technological advances in both mass-rearing and field release are addressing these constraints, and a growing number of natural enemy species is becoming commercially available (Van Driesche and Bellows, 1996; Nordlund *et al.*, 2001; Copping, 2004) (Table 18.1).

Table 18.1. Commercial availability of biological control agents for aphid control (extracted from Copping, 2004).

Type	Species	No. of suppliers	Countries with suppliers
Braconid parasitoid	*Aphidius colemani*	13	UK, USA, Belgium, Canada, Germany, The Netherlands, Thailand
Braconid parasitoid	*Aphidius ervi*	7	UK, USA, Belgium, Germany, The Netherlands
Braconid parasitoid	*Aphidius matricariae*	4	USA, Canada
Braconid parasitoid	*Lysiphlebus testaceipes*	2	USA, Italy
Aphelinid parasitoid	*Aphelinus abdominalis*	9	UK, USA, Belgium, Germany, The Netherlands
Aphelinid parasitoid	*Aphelinus mali*	1	USA
Predatory midge	*Aphidoletes aphidimyza*	18	UK, USA, Belgium, Canada, Germany, The Netherlands
Ladybird	*Adalia bipunctata*	2	UK, Belgium
Ladybird	*Coleomegilla maculata*	3	USA, The Netherlands
Ladybird	*Harmonia axyridis*	5	UK, USA, Belgium, Canada, Finland
Ladybird	*Hippodamia convergens*	14	UK, USA, Belgium, Canada, The Netherlands
Lacewing	*Chrysoperla carnea*	11	UK, USA, Belgium, Germany
Lacewing	*Chrysoperla rufilabris*	1	USA
Hoverfly	*Episyrphus balteatus*	2	UK, Belgium
Predatory bug	*Anthocoris nemoralis*	3	UK, Belgium, Italy
Predatory bug	*Deraeocoris brevis*	2	UK, Canada
Predatory bug	*Geocoris punctipes*	3	USA
Predatory bug	*Orius species*	18	UK, USA, Canada, Belgium, Germany, Italy, Japan, The Netherlands, Poland
Fungal pathogen	*Beauveria bassiana*	5	USA, Colombia, Czech Republic, France, India, Italy, Switzerland, Russia
Fungal pathogen	*Paecilomyces fumosoroseus*	1	USA
Fungal pathogen	*Lecanicillium* spp.	9	USA, Belgium, India, Mexico, The Netherlands

Parasitoids

Aphid parasitoids (Hymenoptera: Braconidae and Aphelinidae) have been used in biological control and integrated pest management (IPM) programmes much more often than other aphid natural enemies (Fig. 18.1). They have the advantage of preying exclusively on aphids, although many will attack a wide range of species. Several are produced commercially in large numbers, particularly for use in glasshouses, where some important aphid species have developed resistance to chemical insecticides. In addition, several parasitoid species have been moved around the world and used in classical introductions to combat major aphid pests of outdoor crops (Table 18.2).

Cereal aphids

Cereal crops are attacked by a variety of aphid species, most of which have spread throughout the major cereal-growing regions of the world. The most notorious recent example is the spread of *Diuraphis noxia* (Russian wheat aphid) to new regions, especially North America and South Africa, where it has become a serious pest (Prinsloo, 2000; Brewer *et al.*, 2001). Because cereal aphids are introduced pests in many regions, biological control efforts have focused on the introduction and establishment of exotic parasitoids, usually collected from the geographical regions where the aphids are believed to have originated (Table 18.2). However, there is also an increasing emphasis on conservation biological control (see Wratten *et al.*, Chapter 16 this volume). Interest has also developed recently in the use of semiochemicals, such as aphid sex pheromones, to manipulate parasitoids and enhance their impact in integrated crop management systems (Powell *et al.*, 1998, 2003; Powell, 2000; Powell and Pickett, 2003).

In 1969–1970, *Aphelinus varipes* from France and *Aphelinus asychis* from Iran were introduced into North America against *Schizaphis graminum* (greenbug), although the latter species apparently did not establish (Jackson *et al.*, 1970, 1971; Archer *et al.*, 1974). These early introductions had little impact in comparison with that of the existing native parasitoid *Lysiphlebus testaceipes* (Archer *et al.*, 1974; Carver, 1989), which is now considered to be one of the most important

Fig. 18.1. The braconid parasitoid *Diaeretiella rapae* ovipositing in a young *Myzus persicae* nymph (original colour photo courtesy of Rothamsted Research).

Table 18.2. Introductions of parasitoids for biological control of aphids.

Species	Target	Crop	Origin	Introduced	Year	Established
Aphelinus mali	*Eriosoma lanigerum*	Apple	USA	51 countries	1920 onwards	Y (42 countries)
Trioxys complanatus *Praon exsoletum* *Aphelinus asychis*	*Therioaphis trifolii maculata*	Lucerne (alfalfa)	Middle East, Europe	USA	1955–1957	Y Y Y
Aphidius smithi *Aphidius ervi*	*Acyrthosiphon pisum*	Lucerne; peas	India, Europe	USA, Canada	1958–1963	Y Y
Trioxys pallidus	*Chromaphis juglandicola*	Walnut	France, Iran	USA	1959, 1968	Y
Aphidius matricariae	range of species		France	Brazil, Chile	1961	Y
Aphidius salicis	*Cavariella aegopodii*	Carrot	USA	Australia	1962	Y
Aphidius asychis *Aphelinus varipes*	*Schizaphis graminum*	Sorghum	Iran France	USA	1970	N Y
Trioxys curvicaudus *Trioxys tenuicaudus*	*Eucallipterus tiliae* *Tinocallis platani*	Urban trees	Europe	USA	1970s	Y Y
Aphidius smithi	*Acyrthosiphon pisum*	Legumes	USA	Chile	1971	Y
Aphidius ervi	*Acyrthosiphon pisum* *Acyrthosiphon kondoi*	Legumes	Europe	Argentina	1972, 1978	Y
Lysiphlebus testaceipes	*Toxoptera aurantii*	Citrus	Cuba	France	1973–1974	Y
Lysiphlebus testaceipes	*Schizaphis graminum*	Cereals	USA	Chile	1974	Y
Monoctonus nervosus	*Sitobion avenae*	Cereals	USA	Chile	1976	N
Aphidius ervi *Aphidius rhopalosiphi* *Aphidius uzbekistanicus* *Praon gallicum* *Praon volucre*	*Sitobion avenae* *Metopolophium dirhodum*	Cereals	France, Iran	Chile	1976–1982	Y Y Y Y Y
Trioxys complanatus *Praon exsoletum* *Aphelinus asychis*	*Therioaphis trifolii maculata*	Lucerne (alfalfa)	USA, Iran, Cyprus Pakistan, France	Australia	1977–1979	Y Y Y
Aphidius smithi *Aphidius eadyi* *Aphidius pisivorus*	*Acyrthosiphon pisum*	Lucerne; peas	USA, Canada	Australia New Zealand	1977–1979	Tasmania only Y ?

Parasitoid	Host aphid	Host plant	Origin	Release location	Year	Established
Aphidius ervi	Acyrthosiphon kondoi	Lucerne	Europe	Australia, New Zealand	1978–1982	Y
Ephedrus plagiator						N
Praon barbatum						N
Aphidius sonchi	Hyperomyzus lactucae	Lettuce	Mediterranean, Japan	Australia	1981	Y
Praon volucre			Mediterranean			Y-Tasmania
Pauesia cedrobii	Cinara (Cedrobium) laportei	Cedar trees	Morocco	France	1981	Y
Lysiphlebus testaceipes	Aphis craccivora	Beans	USA	Australia	1982	Y
Lysiphlebus fabarum			Mediterranean			N
Pauesia bicolor	Cinara cronartii	Pine trees	USA	S. Africa, Kenya, Malawi	1983	Y
Trioxys pallidus	Myzocallis coryli	Hazel (filbert)	Europe	USA	1984	Y
Lysiphlebus testaceipes	Schizaphis graminum	Cereals	USA	Argentina	1984	Y
Aphidius rhopalosiphi	Metopolophium dirhodum	Cereals	England, France, Chile, New Zealand	New Zealand, Australia	1985–1986	Y, N
Binodoxys indicus	Aphis craccivora	Lupin	India	Australia	1986	Y
Diaeretiella rapae	Diuraphis noxia	Cereals	Czech Republic	USA	1989	Y
Aphidius colemani	Diuraphis noxia	Cereals	Chile	Czech Republic	1996–1997	Y
Aphelinus albipodus	Diuraphis noxia	Cereals	Eurasia, Morocco	USA	1988–1997	Y
Aphelinus asychis						Y
Aphelinus varipes						Y
Aphidius colemani						Y
Aphidius matricariae						?
Aphidius picipes						?
Diaeretiella rapae						Y
Ephedrus plagiator						?
Aphidius colemani	Pentalonia nigronervosa	Banana	Australia	Tonga	1990–1991	Y
Aphelinus varipes	Diuraphis noxia	Cereals	Ukraine, Germany	Australia	1990	N
Aphelinus sp. nr. varipes	Diuraphis noxia	Cereals	South Africa	South Africa	1991–1994	Y
Aphidius picipes	Sitobion avenae	Cereals	Ukraine	Chile	1991–1992	?
Ephedrus cerasicola	Metopolophium dirhodum					?
Ephedrus nacheri	Diuraphis noxia					?
Ephedrus plagiator	Sitobion fragariae					?
	Metopolophium festucae					

biological control agents of *S. graminum* in the USA (Fernandes *et al.*, 1998). Subsequently, *L. testaceipes* was introduced into Chile and Argentina against *S. graminum* on sorghum (Botto *et al.*, 1991; Starý, 1993a).

In Chile, an extensive parasitoid introduction programme in the late 1970s led to the establishment of several species (Table 18.2), resulting in significant aphid control in cereal ecosystems (Starý, 1993a; Starý *et al.*, 1993). The small size of crop fields within an ecosystem that incorporates high plant and habitat diversity is thought to have contributed to this success in Chile (Starý *et al.*, 1993). This diversity allows polyphagous parasitoids to utilize other aphids as alternative hosts at times when cereal aphids are scarce. The presence of *Acyrthosiphon pisum* (pea aphid) and *Acyrthosiphon kondoi* (blue alfalfa aphid) on a variety of legume crops within the mixed cropping system is believed to contribute to the success of *Aphidius ervi* in cereals in Chile, where it is often more abundant than the cereal aphid specialists *Aphidius rhopalosiphi* and *Aphidius uzbekistanicus* (Starý *et al.*, 1993).

In the 1980s, *Metopolophium dirhodum* (rose–grain aphid) and *Sitobion avenae* (grain aphid) invaded Australia and New Zealand, leading to the introduction into the latter of *A. rhopalosiphi* from Europe (Farrell and Stufkens, 1990). Successful establishment resulted in a 10–20-fold reduction in *M. dirhodum* populations on barley crops between 1984 and 1989, with parasitism reaching 100% on some spring-sown crops (Farrell and Stufkens, 1988, 1990).

In North America, the Russian wheat aphid (RWA), *D. noxia*, was first recorded in 1986 (Stoetzel, 1987; Sipes, 1997), since when it has spread across the cereal-growing areas of the western United States and Canada. In 1989, the 'National RWA Integrated Pest Management Programme' was initiated and adopted a strategy involving the search for and importation of aphid natural enemies, principally from the putative geographical origin of RWA in Eurasia (Tanagoshi *et al.*, 1995). By 1997, more than 11.8 million individuals of 11 parasitoid species, comprising over 80 geographical strains from 25 countries had been released, and many of these became

established (Michels and Whitaker-Deerberg, 1993; Elliott *et al.*, 1995a; Tanagoshi *et al.*, 1995; Prokrym *et al.*, 1998; Burd *et al.*, 2001) (Table 18.2). At most release sites, all natural enemy species currently available were released, including simultaneous releases of all available geographic strains of imported species (Prokrym *et al.*, 1998). However, full economic and biological evaluation of these releases was not practicable because it was not possible to identify positively many of the released or recovered parasitoids.

While these attempts at 'classical' biological control have had some impact on RWA in the USA, the aphid still remains a significant pest and one of the reasons for the failure of parasitoids to control RWA fully could be the very large field sizes and monoculture-cropping pattern prevalent in many cereal production areas. Burd *et al.* (2001) suggested that cropping system diversification could be the key to improving biological control of cereal aphids in the southern Great Plains region, providing refuges for enhancing parasitoid survival during the summer months. Lack of summer crops with available aphid hosts may be a constraint on the efficiency of aphelinids which, unlike many of the aphidiine braconids, do not undergo a summer diapause. This view is supported by the contrasting situation in Chile, where biological control programmes against cereal aphids, including RWA, have been more successful (see above). *Aphidius ervi*, which also utilizes aphids on diverse legume crops in the cereal ecosystem, has become abundant on RWA in Chile, although it rarely attacks this aphid in Eurasia (Starý *et al.*, 1993).

It has long been known that small differences in biological attributes amongst different strains of imported natural enemy species can significantly affect their success, especially when releases occur over a wide geographic area (Bernal and Gonzalez, 1997). High thermal thresholds for development result in the late appearance of *Aphelinus* species in the field, often preventing them from hitting RWA populations at an early stage of crop colonization, which is essential for successful aphid control (Bernal and

Gonzalez, 1993, 1996; Bernal *et al.*, 1997, 2001; Lee and Elliott, 1998a,b). However, if they are part of a diverse parasitoid fauna, late season parasitoids can still help to reduce the overwintering pest populations, which colonize crops the following year (Bernal and Gonzalez, 1996; Lee and Elliott, 1998b). Some aphidiine species, such as *Diaeretiella rapae* and *Aphidius colemani*, seem to be much more adapted to climatic conditions in some important regions of RWA infestation (Elliott *et al.*, 1995b). The importance of 'climate-matching' the collection regions with intended release locations, linked with laboratory studies of temperature effects on parasitoid biology, has now been emphasized strongly in classical biological control programmes (Bernal *et al.*, 2001). Nevertheless, comparisons of developmental thresholds and day-degree requirements for development between a parasitoid and its host do not necessarily provide an accurate prediction of temporal synchrony (Lee and Elliott, 1998a).

RWA has also extended its range within Europe, and a Chilean strain of *A. colemani* was introduced into the Czech Republic to help with control, even though native parasitoids were showing a high degree of adaptation to RWA (Starý, 1999). Ironically, *D. rapae* had been collected previously from a non-pest aphid, *Hayhurstia atriplicis*, in the Czech Republic and tested for its acceptance of both RWA and greenbug before being shipped to the USA as part of the RWA control programme (Starý and Gonzalez, 1991). In South Africa, an aphelinid has been introduced as part of an IPM programme for RWA on wheat (Prinsloo, 1998). Originally introduced as *Aphelinus varipes*, it was later identified as *Aphelinus hordei* (Prinsloo and Neser, 1994), but more recently has been referred to as *Aphelinus* sp. nr. *varipes* (Prinsloo, 2000). During initial attempts to establish this parasitoid, the light conditions prevalent during rearing induced a diapause soon after release, but more recent attempts have been more successful (Prinsloo, 1998).

In Australia, initial attempts to introduce and establish this same *Aphelinus* parasitoid on *Rhopalosiphum padi* (bird cherry–oat aphid), in anticipation of the arrival of *D. noxia*, appear to have failed as it was not recovered during subsequent surveys in the release areas. Possible reasons for the failure include climatic incompatibility of the released strains and their distinct preference for ovipositing in RWA as opposed to other cereal aphids (Hughes *et al.*, 1994; Prinsloo, 2000).

Legume aphids

Legume crops, particularly lucerne (alfalfa), are also attacked by several aphid species that have spread throughout regions where these crops are grown, frequently prompting the importation and release of parasitoids (Table 18.2).

A search for natural enemies of *Therioaphis trifolii maculata* (spotted alfalfa aphid) in Europe, the Middle East, Africa, and Asia led to the importation and release into the USA of *Praon exsoletum*, *Trioxys complanatus*, and *Aphelinus asychis* (originally recorded as *Aphelinus semiflavus*) (van den Bosch, 1957; Barnes, 1960; van den Bosch *et al.*, 1964). This resulted in a considerable decline in aphid abundance, and hence reduced crop damage, attributed to the successful establishment of all three imported parasitoids, which varied in their climatic requirements and tolerances. The ability to enter a facultative aestival diapause allowed *T. complanatus* to dominate in the hotter, more arid areas, whereas the non-diapausing *A. asychis* preferred more humid regions with less severe temperature extremes (van den Bosch *et al.*, 1964). It is thought that the spotted alfalfa aphid would have escaped control in some regions, such as California, if only one or two of the three parasitoids had been introduced (van den Bosch *et al.*, 1964).

This success stimulated the importation of the same three parasitoid species into Australia, after *T. t. maculata* invaded that country and subsequently developed resistance to a broad range of insecticides (Hughes *et al.*, 1987; Holtkamp *et al.*, 1992). *Trioxys complanatus* was the only parasitoid to become widely established throughout the lucerne-growing regions and was considered to be the main factor responsible for the subsequent decline of the aphid (Walters and Dominiak, 1984; Hughes *et al.*, 1987), although native arthropod predators and the introduction of more aphid-resistant lucerne cultivars also played a role (Carver, 1989). Hughes *et al.* (1987) considered that the success and persistence of the parasitoid was aided by the inherent spatial and temporal patchiness of the irrigated lucerne environment, reducing the impact of local extinctions of both aphid and parasitoid.

In 1989, an aphid that appeared to be indistinguishable from the spotted alfalfa aphid, but which subsequently proved to be genetically distinct, began to cause considerable damage to clover pastures in parts of Australia (Milne, 1997). This aphid was rarely parasitized by *T. complanatus*, which strongly preferred to forage on lucerne rather than clover (Milne, 1997). As *T. complanatus* was imported into Australia from California, to where it had originally been introduced from Iran, it is probable that the population consists of a narrow genetic strain strongly adapted to the spotted alfalfa aphid.

Aphidius smithi and *A. ervi* were imported from India and Europe, respectively, and released against *A. pisum* in pea- and lucerne-growing regions of North America between 1958 and 1961 (Hagen and Schlinger, 1960; Mackauer and Kambhampati, 1986). Although the two parasitoid species appeared to occupy different climatic zones, *A. smithi* declined during the 1970s and *A. ervi* became dominant. This raised questions about the reliability of laboratory screening before release, as this had suggested that *A. smithi* was the more promising species for legume aphid control (Mackauer and Kambhampati, 1986). In California, *A. smithi* could not survive long periods when aphids were absent from the hotter, drier interior regions during summer as it did not enter diapause, but it established well in coastal areas (Hagen and Schlinger, 1960).

A range of parasitoids has been imported into New Zealand and Australia to combat *A. pisum* and *A. kondoi* (Cameron *et al.*, 1981; Carver, 1989). *Aphidius smithi* failed to establish and attempts to introduce *Ephedrus plagiator* and *Praon barbatum* into Australia

also failed (Carver, 1989). However, *Aphidius eadyi* and *A. ervi* did establish and have had a significant impact on *A. pisum* and *A. kondoi*, respectively, in both countries. *Aphidius eadyi* appears to be specific to *A. pisum* and in New Zealand is well synchronized with its host, becoming active early in spring when aphid numbers are still low, which contributes to its early success (Cameron *et al.*, 1981). *Aphidius ervi* took longer to establish, but the introduction of a number of different geographic strains over several years eventually paid off (Cameron and Walker, 1989). Interestingly, *A. ervi* has since taken over from *A. eadyi* as the dominant species attacking *Acyrthosiphon* spp. in New Zealand, reminiscent of the way it replaced *A. smithi* in North America. Unlike *A. eadyi*, *A. ervi* can utilize a number of other aphid species in the ecosystem, and it has been suggested that it is the more efficient of the two at low aphid densities, partly explaining why it became more dominant when aphid population densities on legume crops declined (Cameron and Walker, 1989). *Aphidius ervi* has also been successfully introduced into Chile and Argentina, and in both countries it contributes to the control of aphids on both legumes and cereals (Carver, 1989; Starý *et al.*, 1993).

Aphids on orchard crops

Trioxys pallidus was introduced into California from France in 1959 in an attempt to control *Chromaphis juglandicola* (walnut aphid) on walnuts (Brown *et al.*, 1992; Edwards and Hoy, 1995a). The parasitoid became established only in the more humid climatic areas, especially along the coastal plain. An Iranian ecotype, introduced in 1969, rapidly became established throughout the drier Central Valley of California, but did not appear to interbreed with the previously introduced French ecotype (van den Bosch *et al.*, 1970; Edwards and Hoy, 1995a). Thus, two ecotypes occupied distinct climatic niches, emphasizing the importance of matching climatic conditions when sourcing and introducing biological control agents.

The established *T. pallidus* failed to control another known host, *Myzocallis coryli* (hazel aphid), and so a further ecotype, originating from *M. coryli* in Europe, was introduced in 1984 and proved successful (Messing and Aliniazee, 1988, 1989). Tolerance of high densities of *M. coryli* by filbert nut (*Corylus maxima*) trees and the retention of fallen leaves in the orchard to provide an overwintering habitat for *T. pallidus* mummies allowed synchrony between host and parasitoid early in the following season, and probably contributed to this success (Messing and Aliniazee, 1989; Aliniazee, 1998). However, biological control of *C. juglandicola* has subsequently been hindered by the use of chemical insecticides against lepidopteran pests in walnut orchards, frequently leading to secondary outbreaks of the aphid (Hoy and Cave, 1988; Edwards and Hoy, 1995a). This prompted the laboratory selection of parasitoids for resistance to the insecticide azinphos-methyl (Hoy and Cave, 1988; Edwards and Hoy, 1995a). Early field releases of these were promising as they established successfully in walnut orchards, survived field rate applications of azinphos-methyl, and spread to nearby orchards (Hoy *et al.*, 1989, 1990). They also proved to be cross-resistant to other insecticides used in IPM in walnut orchards (Hoy and Cave, 1989a,b; Edwards and Hoy, 1995a). However, molecular studies showed that the released, pesticide-resistant parasitoids did not completely replace the wild-type populations (Edwards and Hoy, 1995a,b).

Aphelinus mali, a native of North America, has been introduced into many countries in efforts to control *Eriosoma lanigerum* (woolly apple aphid) (Howard, 1929; Carver, 1989). In some regions, the parasitoid introductions were very successful, and it is considered to be one of Western Australia's most outstanding biological control successes, though insecticide use against other apple pests has sometimes hindered its impact (Sproul, 1981; Asante and Danthanarayana, 1992). Again, climate seems to be a critical factor and introductions have been much less successful in cooler climatic regions (Mueller *et al.*, 1992; Cross *et al.*, 1999b), possibly because the parasitoid's threshold temperature for development in comparison to that of its host prevents synchronization between host and parasitoid early in the season (Asante and Danthanarayana, 1992; Mueller *et al.*, 1992; Blommers, 1994; Mols, 1996, 1997).

Banana aphids

Pentalonia nigronervosa (banana aphid) is an important virus vector (Hu *et al.*, 1996) and attempts at biological control have been initiated on Tonga (Starý and Stechmann, 1990). *Lysiphlebus testaceipes* and *A. colemani* were introduced from Europe and Australia (Table 18.2), but only *A. colemani* became established, principally on *Aphis gossypii* (cotton or melon aphid) on taro (*Colocasia esculenta*) crops (Stechmann and Völkl, 1988; Carver *et al.*, 1993; Wellings *et al.*, 1994; Stechmann *et al.*, 1996). The two parasitoids have different search patterns on banana plants, *A. colemani* searching concealed locations that are avoided by *L. testaceipes*, so that many aphids escape the attentions only of the latter (Stadler and Völkl, 1991; Starý and Stechmann, 1991). Studies of aphid behaviour on banana plants revealed a significant degree of niche alternation, with aphids moving between leaves in response to growth, maturation and senescence as the host plant grew, leading to population redistribution and therefore greater exposure to parasitoids (Starý and Stechmann, 1991). However, the biggest constraint to biological control of banana aphids on Tonga appears to be ant attendance of aphid colonies, protecting the aphids from parasitoid attack (Wellings *et al.*, 1994; Stechmann *et al.*, 1996) (see Völkl *et al.*, Chapter 8 this volume).

Aphids on forest and urban trees

Cinara cronartii (black pine aphid), a native of southeastern USA, became a pest in South Africa in the 1970s (van Rensberg, 1979; Marsh, 1991). In 1983, *Pauesia cinaravora* was introduced into South Africa from the USA and became established throughout

the pine-growing areas in the summer rainfall regions of the country, where it had a
major impact on *C. cronartii* populations
(Kfir *et al.*, 1985; Kfir and Kirsten, 1991).

Several other aphids have invaded
conifer plantations in Africa and these are
considered to be suitable targets for classical biological control (Mills, 1990; Murphy
et al., 1994). *Pauesia juniperorum* was introduced into Malawi from Europe and has
had a significant impact on *Cinara* sp. nov.
(initially thought to be *Cinara cupressi*), a
recent pest of cypress and juniper trees in
East Africa (Murphy *et al.*, 1994; Chilima,
1995; Kairo and Murphy, 1999). *Cinara
(Cedrobium) laportei* (cedar aphid) invaded
southern France from northern Africa in the
1960s (Remaudière and Starý, 1993). Its
parasitoid *Pauesia cedrobii* was collected
from Morocco and released in southeastern
France in 1981 (Fabre and Rabasse, 1987),
where it became established and spread as
far north as Paris within 12 years (Remaudière
and Starý, 1993).

In California, *Eucallipterus tiliae* (lime
aphid), originally a Palaearctic species, causes
a nuisance problem on urban lime (linden)
trees. In 1970, four European parasitoids were
released in California, although only *Trioxys
curvicaudus* became established, contributing to local control of the aphid, aided by
the development of aphid resistance in host
trees (Zuparko, 1983; Zuparko and Dahlsten,
1995, 1996). Interestingly, it did not spread
further than 40 km from the release sites
over a 20-year period, possibly due to the
patchy and temporally sporadic distribution
of its host, combined with local variations in
climatic conditions (Zuparko and Dahlsten,
1995).

Glasshouse aphids

Aulacorthum solani (glasshouse and potato
aphid), *A. gossypii*, *Macrosiphum euphorbiae*
(potato aphid), and *Myzus persicae* (peach–
potato aphid) are widespread pests in glasshouses (van Lenteren *et al.*, 1997). Inoculative
or inundative release of aphid parasitoids is
now commonly used on protected crops,
particularly in Europe, and several species

are commercially available (Table 18.1, see
also Tatchell, Chapter 24 this volume).
Commercial rearing of *Aphidius matricariae* was developed after early trials (Wyatt,
1970; Tremblay, 1974; Rabasse *et al.*, 1983;
Shijko, 1989), but more recently, *A. colemani* has become the main species used in
glasshouses. The two species differ in their
ability to control different aphid hosts;
M. euphorbiae, for example, is not a suitable
host for *A. matricariae*, but is readily attacked by *A. colemani* (Blümel and Hausdorf,
1996).

Timing is critical, since the parasitoids
need to impact on the aphid population very
soon after crop infestation in order to prevent the aphids reaching damaging levels.
Open rearing 'systems' (banker plants) have
been developed to ensure an appropriate
aphid/parasitoid ratio at an early stage of
crop infestation. This strategy uses alternative host aphids on non-crop plants to build
up parasitoid populations in the glasshouse
prior to invasion of the crop by the target
pest. This approach was first developed for
establishing populations of the predatory
cecidomyiid midge *Aphidoletes aphidimyza*
to control *M. persicae* populations on sweet
peppers, using either *Megoura viciae* (vetch
aphid) on broad bean plants (Hansen, 1983)
or cereal aphids on wheat seedlings (Kuo-
Sell, 1989). The use of banker plants was then
adapted for establishing early populations
of *A. matricariae* and *A. colemani*, particularly to combat *A. gossypii* on glasshouse
cucumber crops (Bennison, 1992; van Steenis,
1992; Bennison and Corless, 1993) (Fig. 18.2).
This strategy is based on the 'multilateral
control' approach advocated by Starý (1967,
1972, 1993b). Banker plants are now widely
used and the strategy also appears to reduce
problems with hyperparasitoids later in the
season (Kuehne, 1998).

Aphis gossypii has become a major
problem on protected crops (Fig. 18.3), exa-
cerbated by its resistance to a range of insecticides (van Steenis, 1992, 1995). Early
attempts at biological control of *A. gossypii*
on cucumber often failed because releases
of parasitoids and predators were made too
late to prevent the rapid increase of the
aphid (van Steenis, 1992). The banker plant

Fig. 18.2. Barley seedlings infested with the cereal aphid *Rhopalosiphum padi* being used in a banker plant system to build up populations of parasitoids for control of aphid pests colonizing glasshouse cucumber plants (original colour photo courtesy of S.Corless).

system, using *R. padi* on trays of cereal seedlings to establish early populations of *A. colemani* and *A. aphidimyza* (Bennison, 1992; Bennison and Corless, 1993), greatly increased success rates and by 1998, 30% of Dutch cucumber growers were using the system (Mulder *et al.*, 1999). Van Steenis and El-Khawass (1995) advocate the use of cucumber cultivars that show partial resistance to *A. gossypii* to facilitate biocontrol by *A. colemani*, suggesting that plant breeders should routinely screen new cultivars for aphid resistance. *Lysiphlebus testaceipes* and *Ephedrus cerasicola* have also been proposed for use in glasshouses against *A. gossypii*, using *M. persicae* on swedes in a banker plant system for *E. cerasicola* (van Steenis, 1994; Hågvar and Hofsvang, 1995).

Banker plants also work well against aphids on sweet peppers, especially early in the year before secondary parasitoids build up and before shading by the crop affects growth of the banker plants (Bennison, 1992; van Schelt, 1999). To combat the latter problem, wheat plants can be placed in buckets hung between the tops of the pepper plants (van Schelt, 1999). The efficiency of *A.*

colemani in controlling *A. gossypii* can vary between crops, its foraging behaviour on crops such as melons being influenced by plant growth habit and architecture (Burgio *et al.*, 1997). However, satisfactory control on glasshouse melons was obtained when *A. colemani* was introduced twice on banker plants (Schoen and Martin, 1997). The banker plant approach, using *M. euphorbiae* on potato shoots, was also successful in trial releases of *Aphelinus abdominalis* to control aphids on glasshouse roses (Blümel and Hausdorf, 1996).

Examples of trials involving augmentative release of parasitoids to control aphids in glasshouses are listed in Table 18.3.

Aphidoletes aphidimyza (Predatory Gall Midge)

The potential of *A. aphidimyza* (Diptera; Cecidomyiidae) for aphid control was recognized in the 1970s, but by 1985 it was still only being used regularly in the former USSR and parts of Scandinavia (Meadow *et al.*, 1985; Havelka and Zemek, 1988, 1999).

Fig. 18.3. *Aphis gossypii* damage to glasshouse melons (original colour photo courtesy of W. Powell).

However, early successes in these countries (Morse and Croft, 1987) led to much wider use in recent years, especially in glasshouses. Examples of trials involving augmentative release of *A. aphidimyza* against glasshouse aphids are listed in Table 18.4.

Glasshouse aphids

Mayr (1973) proposed the mass-rearing and augmentative release of *A. aphidimyza*, and early trials against *M. persicae* on glasshouse brassicas showed promise (El Titi, 1974). By the early 1980s, it was being used against *M. persicae* by glasshouse vegetable growers in Finland, where commercial rearing of the midge began in 1978, and against *A. gossypii* on cucumbers in the former USSR (Nijveldt, 1988; Tiittanen and Markkula, 1989). However, in comparative trials in Norway, the parasitoid *E. cerasicola* controlled *M. persicae*

on paprika plants as well as, or even better than, *A. aphidimyza*, especially later in the year when the midge began to enter diapause (Hofsvang and Hågvar, 1982). Also, *A. aphidimyza* failed to control aphids on lettuces in German glasshouses, even when pupae were applied three times, probably due to local climatic conditions (Quentin *et al.*, 1995). Winter glasshouse trials in the USA, in which the midge appeared to reproduce faster than *M. persicae*, were successful when a release rate of one midge per ten aphids was used on sweet peppers (Gilkeson and Hill, 1987). Bondarenko (1989) recommended a release rate of 1 : 5 in glasshouses and 1 : 9 in plastic tunnel houses, with subsequent additional releases to maintain these ratios if necessary. The initiation of diapause at the prepupal larva stage was recognized as a constraint in the use of *A. aphidimyza* in temperate-region glasshouses, but it is possible to select for a low

Table 18.3. Examples of trials involving augmentative release of parasitoids to control aphids on glasshouse crops.

Species	Target	Crop	Banker plants	Country	Effect	References
Aphidius matricariae	*Myzus persicae*	Sweet pepper		France	Significant impact on aphids	Rabasse *et al.*, 1983
Aphidius matricariae	*Myzus persicae*	Sweet pepper		Former USSR	Good control	Shijko, 1989
Aphidius matricariae	*Aphis gossypii*	Cucumber	*Rhopalosiphum padi* on cereals	UK	Good control using banker plants combined with predatory midge	Bennison, 1992
Aphidius colemani	*Aphis gossypii*	Cucumber	*Rhopalosiphum padi* on cereals	UK	Good control using banker plants combined with predatory midge	Bennison and Corless, 1993
Aphidius colemani	*Myzus persicae*	Beans	*Schizaphis graminum* on cereals	Czech Republic	Good control using banker plants	Starý, 1993b
Aphidius colemani	*Aphis gossypii*	Cucumber	*Rhopalosiphum padi* on cereals	The Netherlands	Good control using banker plants combined with predatory midge	Mulder *et al.*, 1999
Aphidius colemani	*Aphis gossypii*	Cucumber and melon		Italy	Efficient control of aphids on cucumber but not on melon	Burgio *et al.*, 1997
Aphidius colemani	*Aphis gossypii*	Melon	*Rhopalosiphum padi* on cereals	France	Satisfactory control when released twice on banker plants	Schoen and Martin, 1997
Aphidius colemani	*Myzus persicae nicotianae*	Sweet pepper	*Rhopalosiphum padi* on cereals	The Netherlands	Good control using banker plants	van Schelt, 1999

(*Continued*)

Table 18.3. Continued.

Species	Target	Crop	Banker plants	Country	Effect	References
Aphidius colemani	*Aulacorthum solani*	Sweet pepper	*Rhopalosiphum padi* on cereals	The Netherlands	Good control using banker plants	van Schelt, 1999
Lysiphlebus testaceipes	*Myzus persicae*	Beans	*Schizaphis graminum* on cereals	Czech Republic	Some effect using banker plants	Starý, 1993b
Ephedrus cerasicola	*Aphis gossypii*	Cucumber	*Myzus persicae* on swedes	Norway	Significant effect using banker plants but *Myzus persicae* a potential risk	Hågvar and Hofsvang, 1995
Aphelinus abdominalis	*Macrosiphum euphorbiae*	Cut roses	*Macrosiphum euphorbiae* on potato	Austria	Good control using banker plants combined with selective aphicide	Blümel and Hausdorf, 1996
Aphelinus abdominalis	*Myzus persicae*	Cut roses	*Macrosiphum euphorbiae* on potato	Austria	Good control using banker plants combined with selective aphicide	Blümel and Hausdorf, 1996
Aphelinus asychis	*Macrosiphum euphorbiae*	Cut roses		USA	Significant impact on aphids combined with a ladybird beetle	Snyder et al., 2004a

Table 18.4. Examples of trials involving augmentative release of the predatory midge *Aphidoletes aphidimyza* to control aphids on glasshouse crops.

Target	Crop	Banker plants	Country	Effect	References
Myzus persicae	Brassicas		Germany	Good control within 7 weeks	El Titi, 1974
Myzus persicae	Paprika		Norway	Good control early in season Midge entered diapause later in season	Hofsvang and Hågvar, 1982
Myzus persicae	Sweet pepper		USA	Good control with release rate of 1:10 aphids	Gilkeson and Hill, 1987
Myzus persicae	Sweet pepper	*Megoura viciae*, broad bean	Denmark	Good control using banker plants Supplementary releases later in season sometimes needed	Hansen, 1983
Myzus persicae	Sweet pepper	Cereal aphids, cereals	Germany	Good control using banker plants	Kuo-Sell, 1989
Myzus persicae	Sweet pepper		Germany	Good control in 2 years out of 4 combined with lacewing egg releases	Hommes, 1992
Aulacorthum solani	Lettuce		Germany	Failed to control aphids	Quentin *et al.*, 1995
Aphis gossypii	Cucumber	*Rhopalosiphum padi*, cereals	UK	Good control using banker plants combined with parasitoid releases	Bennison, 1992; Bennison and Corless, 1993

incidence of diapause using material with appropriate geographic origins (Gilkeson and Hill, 1986).

Banker plants have proved beneficial for establishing early populations of *A. aphidimyza* and for allowing midges emerging from diapause in overwintering cocoons to build up on alternative hosts before invasion by the target pest (Hansen, 1983: Kuo-Sell, 1989; Bennison, 1992). *Megoura viciae* on broad beans and cereal aphids on cereal seedlings have both been used as banker plants for *A. aphidimyza* in glasshouses (Hansen, 1983; Kuo-Sell, 1989; Bennison, 1992). Cereal aphids have been suggested as a convenient host for the mass-rearing of *A. aphidimyza* (Bondarenko, 1989; Kuo-Sell, 1989) although aphid-infested sweet pepper plants are also frequently used (Van Driesche and Bellows, 1996).

Aphidoletes aphidimyza, released with lacewing eggs, successfully controlled aphids on sweet peppers in two years out of four, failures occurring when introductions were made too late or when too few were released (Hommes, 1992). Good control of *A. gossypii* on cucumbers was obtained by the combined release of the midge *A. aphidimyza*, the lacewing *Chrysoperla carnea* (green lacewing), and the parasitoid *A. matricariae*, and by *A. aphidimyza* and the parasitoid *A. colemani*, using a banker plant system (Bennison, 1992; Bennison and Corless, 1993). In the Czech Republic, a combination of *A. aphidimyza* and *Hippodamia convergens* (convergent ladybird) successfully controlled glasshouse aphids (Navratilova, 1999). However, Harizanova and Ekbom (1997) studied the interaction between *A. aphidimyza* and the parasitoid *A. colemani* and concluded that releasing two biological control agents together may not always be significantly better than one.

Aphidoletes aphidimyza is now widely available commercially (Table 18.1) and can be very effective, although failures sometimes occur (van Schelt and Mulder, 2000). The presence of spider webs in the glasshouse increases the rate of successful mating because the midges use horizontal webs as a mating platform (van Schelt and Mulder, 2000). These authors successfully released

the midge from bottles containing cocoons in vermiculite placed at the end of crop rows. The emerging midges were able to move through at least 15 cm depth of vermiculite and rapidly located aphid colonies up to 35 m away.

Aphids on field and orchard crops

Successes have been reported for the release of *A. aphidimyza* on to field-grown peppers and outdoor roses to control *M. persicae* and *Macrosiphum rosae* (rose aphid), respectively (Meadow *et al.*, 1985; Nijveldt, 1988). However, outdoor releases of the midge have been attempted most often in apple orchards (Bouchard *et al.*, 1988; Grasswitz and Burts, 1995; Wyss *et al.*, 1999b). In northeastern USA, natural populations of *A. aphidimyza* are not always synchronized with prey populations early in spring because its eggs need slightly higher temperatures than those of the aphids before they hatch (Havelka, 1980; Morse and Croft, 1987). This seems to be less of a problem in Canada (Stewart and Walde, 1997). Bouchard *et al.* (1988) concluded that the midge was an effective predator of apple aphids in Quebec and that natural populations in orchards should be encouraged, but that mass releases could not be economically justified. In field cage assessments of *A. aphidimyza* releases to control *Dysaphis plantaginea* (rosy apple aphid) in Europe, the midge was adversely affected early in the season by low temperatures and night frosts (Wyss *et al.*, 1999b). Augmentative releases of *A. aphidimyza*, together with lacewings, in apple orchards in the USA had little impact on aphid populations, possibly because significant natural populations of the midge already existed at the release sites (Grasswitz and Burts, 1995). Nevertheless, *A. aphidimyza* can be a valuable component of IPM programmes since it shows some resistance to insecticides used in apple orchards (Warner and Croft, 1982; Croft, 1990).

Considerable intraspecific variability in a number of biological traits occurs amongst different geographical populations of *A. aphidimyza*, highlighting the importance of considering such variability when exploiting biological control agents (Havelka and Zemek, 1988, 1999). Midge larvae were significantly smaller on aphid-resistant cabbage cultivars than on a fully susceptible cultivar, but this did not reduce the aphid populations because, although the smaller larvae consumed aphids at a slower rate, aphid populations increased more slowly on the partially resistant cultivars (Verkerk *et al.*, 1998).

Coccinellidae (Ladybird Beetles)

Many ladybirds prey extensively on aphids and there have been frequent attempts to use these beetles in biological control and IPM programmes (Obrycki and Kring, 1998). However, success has been poor compared with that achieved against scale insects (Dixon *et al.*, 1997; Iperti, 1999; Dixon, 2000). Several reasons for the poor performance of aphidophagous compared with coccidophagous coccinellids have been proposed, including the much slower development rate of aphidophagous ladybirds in relation to that of their prey, their oviposition strategies, which optimize use of an often ephemeral

food resource, and their greater mobility and dispersal abilities as adults (Hemptinne *et al.*, 1995; Dixon *et al.*, 1997; Iperti, 1999; Dixon, 2000). Nevertheless, they are still regarded as important components of the natural enemy complex and their presence is actively encouraged in conservation biological control strategies.

Since 1900, attempts have been made to introduce 179 coccinellid species into North America but only 18 have become established, sometimes taking many years (Angelet *et al.*, 1979; Gordon, 1987; Wheeler, 1993; LaMana and Miller, 1996; Obrycki and Kring, 1998). In addition, several species have become established after accidental introduction, including *Coccinella septempunctata* (7-spot ladybird) (Fig. 18.4), *Harmonia axyridis* (harlequin ladybird), and *Propylea quatuordecimpunctata* (14-spot ladybird). More recently, a South American species, *Eriopis connexa*, has been introduced into the USA for biological control of RWA (Miller, 1995). Introduced species have been implicated in the decline of some native species in the USA and elsewhere (Staines *et al.*, 1990; Wheeler and Hoebeke, 1995; Elliott *et al.*, 1996; Obrycki and Kring, 1998; Calunga-Garcia and Gage, 1999; Dixon,

Fig. 18.4.   Adult 7-spot ladybird *Coccinella septempunctata* eating a pea aphid *Acyrthosiphon pisum* (original colour photo courtesy of Rothamsted Research).

2000; Snyder *et al.*, 2004b). The biology of ladybirds and their potential use in biological control is reviewed in more detail by Majerus (1994), Obrycki and Kring (1998), and Dixon (2000). Examples of trials involving augmentative release of ladybirds to control aphids on glasshouse and field crops are listed in Table 18.5.

Glasshouse aphids

Few attempts have been made to control aphids in glasshouses using ladybirds, but trials were undertaken in the 1970s in Finland using *C. septempunctata* and *Adalia bipunctata* (2-spot ladybird) against *M.persicae* on sweet peppers and chrysanthemums and *M. rosae* on cut roses (Hämäläinen, 1980). Neither species managed to establish self-perpetuating populations and repeated releases of larvae were necessary, making their use uneconomic. More recently, El Habi *et al.* (1999, 2000) released *C. septempunctata* and *Hippodamia variegata* (variegated ladybird) against *A. gossypii* on glasshouse cucumbers in Morocco. Both adult and larval releases could provide satisfactory control, but efficiency was dependent on the timing of releases in relation to aphid population levels and on a high predator:aphid ratio. Adults of *Coccinella undecimpunctata undecimpunctata* were released into a glasshouse, at rates of up to 200 per 140 plants, to control *Aphis fabae* (black bean aphid) on soybeans, with some success (Zaki *et al.*, 1999). High densities of *H. convergens*, together with the parasitoid *A. colemani*, were effective against *A. gossypii* on strawberries, although control of *M. euphorbiae* was attributed to another parasitoid, *A. ervi* (Sterk and Meesters, 1997). Trial results have also suggested that *H. axyridis* can be released to complement control of *M. euphorbiae* on glasshouse-grown roses (Snyder *et al.*, 2004a).

Aphids on field crops and amenity plants

A few aphidophagous ladybirds overwinter in mountainous areas in huge adult aggregations (Hagen, 1966; Iperti and Buscarlet,

1986; Dixon, 2000). For many years, *H. convergens* has been collected from aggregations in the USA and sold to crop growers (Fenton and Dahms, 1951; Dreistadt and Flint, 1996), but released adults usually disperse rapidly (Davidson, 1924; DeBach and Hagen, 1964; Kieckhefer and Olson, 1974). Dreistadt and Flint (1996) preconditioned aggregation-collected *H. convergens* by allowing them to fly in screen cages and providing them with water before release on to aphid-infested, potted plants. Although they dispersed more slowly than unconditioned ladybirds and had an impact on the aphids, most had left after 3 days and the preconditioning process was deemed uneconomic. However, inundative releases of aggregation-collected *H. convergens* may be useful in nurseries and glasshouses (Flint *et al.*, 1995) and have also been used successfully on chilli peppers grown in netting cages (Votava and Bosland, 1996). Nevertheless, Obrycki and Kring (1998) believe that there is no biological basis for the release of *H. convergens* collected from overwintering aggregations, and that the practice could deplete local populations, spread ladybird parasitoids and diseases, and increase vectoring of plant pathogens.

A chemical mutagen followed by selective breeding has been used to produce a flightless form of *H. axyridis* (Tourniaire *et al.*, 1999, 2000). This ladybird was introduced into Europe from China in 1982 and is mass-reared on eggs of the pyralid moth *Ephestia kuehniella* (Tourniaire *et al.*, 1999), but due to the dispersal tendencies of the adults, only *H. axyridis* larvae are used for augmentative releases (Ferran *et al.*, 1996; Trouvé *et al.*, 1997). The flightless mutant form of *H. axyridis*, together with the production method, has now been patented within Europe (Tourniaire *et al.*, 1999). Also, some success has been achieved in Japan with the release of egg masses and flightless adults of *H. axyridis* against *A. gossypii* on glasshouse cucumbers (Kuroda and Miura, 2003). However, the mutation reduces adult reproductive fitness and even though significantly more flightless adults than normal adults remained on glasshouse cucumbers infested with *A. gossypii* following

Table 18.5. Examples of trials involving augmentative release of ladybirds to control aphids.

Species	Target	Crop	Country	Effect	References
Glasshouse:					
Coccinella septempunctata *Adalia bipunctata*	*Myzus persicae* *Macrosiphum rosae*	Sweet pepper Chrysanthemum Cut roses	Finland	Control with repeated releases of larvae Uneconomic	Hämäläinen, 1980
Coccinella septempunctata *Hippodamia variegata*	*Aphis gossypii*	Cucumber	Morocco	Satisfactory control but dependent on timing of release and initial aphid density	El Habi *et al.*, 1999, 2000
Coccinella undecimpunctata undecimpunctata	*Aphis fabae*	Soybean	Egypt	Successful control at release rate of 200 adults per 140 plants	Zaki *et al.*, 1999
Hippodamia convergens	*Aphis gossypii* *Macrosiphum euphorbiae*	Strawberry	Belgium	Effective against *A. gossypii* when high densities released together with parasitoids	Sterk and Meesters, 1997
Harmonia axyridis	*Macrosiphum euphorbiae*	Cut roses	USA	Significant impact on aphids Combined with aphelinid parasitoid	Snyder *et al.*, 2004a
Harmonia axyridis	*Aphis gossypii*	Cucumber	Japan	Some effect with release of egg masses and flightless adults	Kuroda and Miura, 2003
Field:					
Hippodamia convergens	*Schizaphis graminum*	Sorghum	USA	Ineffective due to dispersal after release	Starks *et al.*, 1975
Hippodamia convergens	*Capitophorus elaeagni*	*Elaeagnus* hedge	USA	77% reduction in aphid densities after 15 days	Raupp *et al.*, 1994
Hippodamia convergens	*Monellia caryella* *Monelliopsis pecanis*	Pecan orchard	USA	Good control Released with lacewings	LaRock and Ellington, 1996

(Continued)

Table 18.5. Continued.

Species	Target	Crop	Country	Effect	References
Hippodamia convergens	*Diuraphis noxia*	Wheat	USA	Little effect and considered uneconomic	Randolph *et al.*, 2002
Harmonia axyridis	*Macrosiphum rosae*	Amenity roses	France	Combined with lacewings Good control but timing of release critical	Ferran *et al.*, 1996
Harmonia axyridis	*Phorodon humuli*	Hops	France	Early release of 50 larvae per plant reduced aphid populations Resultant adults quickly dispersed	Trouvé, 1995 Trouvé *et al.*, 1997
Coccinella septempunctata Propylea japonica	*Aphis gossypii*	Cotton	China	Good control with repeated releases Overwintering adults collected to mass-rear larvae for release	Zhang, 1992
Coccinella septempunctata brucki	*Acyrthosiphon kondoi*	Lucerne (alfalfa)	Japan	Good control Overwintering adults collected to mass-rear larvae for release	Takahashi, 1997
Coccinella septempunctata Cheilomenes sexmaculatus	*Lipaphis pseudobrassicae*	Mustard	India	Successful control when larvae released at a rate of 1000 per ha	Shenhmar and Brar, 1995
Coccinella undecimpunctata undecimpunctata	*Aphis gossypii*	Okra	Egypt	Successful control when larvae released at rates of 1:50 aphids	Zaki *et al.*, 1999
Adalia bipunctata	*Dysaphis plantaginea*	Apples	Switzerland	Releases in spring limited aphid build up Releases in autumn reduced overwintering eggs	Wyss *et al.*, 1999a Kehrli and Wyss, 2001
Synonycha grandis Coleophora biplagiata	*Ceratovacuna lanigera*	Sugarcane	China	Good control by release of adults collected over winter	Deng *et al.*, 1987

release, they produced significantly fewer offspring despite ovipositing for a longer period (Ferran *et al.*, 1998). Therefore, they may have some potential for inundative release but are unlikely to be effective in inoculative release strategies.

Early summer releases of 50 H. *axyridis* larvae per plant to control *Phorodon humuli* (damson–hop aphid) gave encouraging results, but the emerging adult population rapidly dispersed (Trouvé, 1995; Trouvé *et al.*, 1997). A natural population of *A. bipunctata* was also present and contributed to the aphid reduction. *Harmonia axyridis* larvae were also released against aphids on amenity roses in Paris, performing as well as conventional insecticide treatments (Ferran *et al.*, 1996). However, if they were released when aphid numbers were too few (< 30 per bush), the ladybirds dispersed quickly. Experimental releases of adult *H. convergens* on to an ornamental *Elaeagnus* hedge reduced the abundance of *Capitophorus elaeagni* by 77% after 15 days (Raupp *et al.*, 1994).

In China, *C. septempunctata* (Fig. 18.4) and *Propylea japonica* were mass-reared from field-collected, overwintered adults and released as larvae (repeated several times) to control *A. gossypii* on cotton (Zhang, 1992). In northern China, *C. septempunctata* is collected by hand from wheat crops and translocated to cotton, a strategy that has led to a reduction in pesticide use on cotton, thereby conserving natural enemies of other pests such as bollworms (Zhang, 1992). A similar approach has also been used to control *Ceratovacuna lanigera* (sugarcane woolly aphid) by augmentative releases of *Synonycha grandis* and *Coelophora biplagiata* (Deng *et al.*, 1987). In Japan, overwintering adults of *C. septempunctata brucki* were collected from tussocks of Japanese pampas grass (*Miscanthus sinensis*), bred in the laboratory on *A. pisum*, and released as larvae into lucerne crops to control *A. kondoi* (Takahashi, 1997). The optimum time for release was early March when the maximum temperature was around 10°C. Pampas grass was planted around lucerne fields to provide overwintering and aestivation sites for the ladybirds.

In Egypt, releases of *C. u. undecimpunctata* larvae successfully controlled *A. gossypii*

in a field of okra (*Abelmoschus esculentus*) (Zaki *et al.*, 1999), whilst in India, larvae of *C. septempunctata* and *Cheilomenes sexmaculatus* released on to a mustard field effectively controlled *Lipaphis pseudobrassicae* (mustard aphid) (Shenhmar and Brar, 1995).

Aphids on orchard crops

Adalia bipunctata is commonly associated with trees and can be an important natural enemy of aphid pests in orchards. In field cage experiments in Switzerland, *A. bipunctata* was more consistent and efficient than the gall midge *A. aphidimyza* and the hover fly *Episyrphus balteatus* when released against *D. plantaginea* on apples, partly because it was less affected by cold temperatures early in the year (Wyss *et al.*, 1999b). When tested together, *A. bipunctata* and *E. balteatus* appeared to have an additive rather than a synergistic effect and reduced aphid densities to 5% of those in the control cages. April releases of *A. bipunctata* larvae were effective against the spring population of *D. plantaginea* in apple orchards (Wyss *et al.*, 1999a), while autumn releases significantly reduced the number of fundatrices in the following spring, although the necessity for a high release rate made this approach expensive (Kehrli and Wyss, 2001).

Adult *H. convergens*, together with lacewings, have been released against aphids in pecan orchards in New Mexico, with some success (LaRock and Ellington, 1996), while *H. axyridis* and *Cycloneda sanguinea* have been evaluated for release against citrus aphids (Michaud, 2000).

Lacewings

Lacewings (Fig. 18.5) have received considerable attention as potential biological control agents for a number of pests, including aphids and lepidopteran caterpillars (Senior and McEwen, 2001). Improvements in mass-rearing techniques have led to their availability from a growing number of commercial suppliers, principally selling *Chrysoperla*

Fig. 18.5. Adult green lacewing *Chrysoperla carnea* (original colour photo courtesy of Rothamsted Research).

species (Nordlund *et al.*, 2001; Copping, 2004) (Table 18.1). Augmentative releases of lacewing eggs or young larvae have been used in a range of experimental or commercial aphid control programmes in orchards, field crops, and glasshouses. Lacewings are regarded as particularly valuable for use in IPM schemes because of their compatibility with a number of chemical insecticides (Bartlett, 1964; Grafton-Cardwell and Hoy, 1985; Senior and McEwen, 2001), including insecticidal soaps (Heinz and Parrella, 1990; Breene *et al.*, 1992; Ehler and Kinsey, 1995). The biology of lacewings and their potential as biological control agents is discussed in more detail by McEwen *et al.* (2001). Examples of trials involving augmentative release of lacewings to control aphids on glasshouse and field crops are listed in Table 18.6.

Glasshouse aphids

In early trials involving *M. persicae* on potted chrysanthemums, releases of *C. carnea* larvae significantly reduced aphid numbers (Scopes, 1969). Releases of *C. carnea* larvae every 2–5 weeks at ratios of 1 : 20 aphids or less were required to maintain significant reductions of *M. persicae* on aubergine (*Solanum melongena*), but a ratio of 1 : 5 almost completely eliminated them (Hassan, 1978). Similar experimental releases against the same aphid on young sugarbeet plants completely eliminated the pest for 3–4 weeks at ratios up to 1 : 40 aphids and for 5–6 weeks at ratios up to 1 : 10 (Hassan *et al.*, 1985). The efficiency of *Chrysoperla* larvae varies according to the crop and appears to be influenced by leaf structure, which affects foraging efficiency, and the rate of aphid increase (Hassan *et al.*, 1985). In trials in the former USSR in the 1970s, on both glasshouse and field crops, releases of second-instar larvae were sometimes effective in glasshouse trials but needed repeating at regular intervals because populations did not become established and reproduce (Beglyarov and Smetnik, 1977). The release methods were also labour intensive, often making their use uneconomic at this time.

Releases of *C. carnea* eggs were tested in Scandanavian glasshouses against several aphid pests on green peppers (*Capsicum* spp.) and parsley (*Petroselinum crispum*),

Table 18.6. Examples of trials involving augmentative release of lacewings to control aphids.

Species	Target	Crop	Country	Effect	References
Glasshouse:					
Chrysoperla carnea	*Myzus persicae*	Chrysanthemum	UK	Significant reduction in aphids with 1–2 larvae released per plant	Scopes, 1969
Chrysoperla carnea	*Myzus persicae*	Aubergine	Germany	Good control with release of 1 larva per 5 aphids Repeated releases needed with 1 larva per 20 aphids	Hassan, 1978
Chrysoperla carnea	Several species	Sweet pepper, parsley	Finland	Effective on pepper with release of 1 egg per 3 aphids and on parsley with 1 egg per 27 aphids Uneconomic at time of trials	Tulisalo and Tuovinen, 1975; Tulisalo *et al.*, 1977
Chrysoperla carnea	Several species	Lettuce	Germany	Good control when eggs placed on young plants before transplanting, then 25-30 eggs per m² weekly	Quentin *et al.*, 1995
Chrysoperla carnea	*Macrosiphum euphorbiae Chaetosiphon fragaefolii*	Strawberry	Italy	Good control Release of larvae better than release of eggs	Tommasini and Mosti, 2001
Field:					
Chrysoperla carnea	*Myzus persicae Macrosiphum euphorbiae*	Potato	USA	Some effects on *M. persicae* when eggs and larvae released	Shands *et al.*, 1972a,b,c
Chrysoperla carnea	*Acyrthosiphon pisum Brevicoryne brassicae Myzus persicae Aphis craccivora*	Pea, brassicas, tomato, aubergine, sweet pepper	Former USSR	Releases of larvae ineffective on peas and brassicas Effective control with release of 1 larva per 5 aphids on tomato, aubergine, and sweet pepper	Beglyarov and Smetnik, 1977
Chrysoperla carnea	*Aphis fabae*	Sugarbeet	Germany	Good control for 2 weeks with release of 1 egg per 5 aphids	Sengonca *et al.*, 1995 Sengonca and Löchte, 1997

(Continued)

Table 18.6. Continued.

Species	Target	Crop	Country	Effect	References
Chrysoperla carnea Chrysoperla rufilabris	Aphis fabae	Sugarbeet	USA	Effect difficult to assess because aphids declined naturally on control plots after treatments applied	Ehler et al., 1997
Chrysoperla carnea	Aphis gossypii	Okra	Egypt	Good control with two releases of 1 larva per 5 aphids	Zaki et al., 1999
Chrysoperla carnea	Aphis gossypii	Cotton	Egypt	Good control with repeated releases of eggs and larvae	El Arnaouty and Sewify, 1998
Chrysoperla carnea	Aphis pomi	Apple orchard	Canada	Significant reduction in aphids with egg releases	Hagley, 1989
Chrysoperla carnea	Illinoia liriodendri	Amenity tulip trees	USA	Release of eggs ineffective due to low egg viability and predation by ants	Dreistadt et al., 1986
Chrysoperla lucasina	Aphis gossypii	Melon	France	Good control with release of 1 larva per 20 aphids	Malet et al., 1994
Chrysoperla lucasina	Aphis fabae	Artichoke	France	Satisfactory control with repeated releases of larvae	Maisonneuve, 2001
Chrysoperla rufilabris	Diuraphis noxia	Wheat	USA	Little effect and considered uneconomic Combined with ladybirds	Randolph et al., 2002
Chrysoperla rufilabris	Monellia caryella Monelliopsis pecanis	Pecan orchard	USA	Good control when eggs and larvae released with ladybirds	LaRock and Ellington, 1996
Chrysoperla rufilabris	Mindarus kinseyi	White fir seedlings	USA	Good control with release of 2 larvae per seedling Combined with insecticidal soap treatments	Ehler and Kinsey, 1995

with much higher ratios being needed for effective control on the former than on the latter crop (Tulisalo and Tuovinen, 1975; Tulisalo *et al.*, 1977). The release of young larvae was considered to be a better option because of low hatch rates and delayed hatching of eggs, but it was concluded that augmentative releases of lacewings would not be economic until mass-rearing methods were improved. Also, the presence of ants in the glasshouses hindered the effectiveness of the lacewings in these trials. More recently, application of *C. carnea* eggs to glasshouse lettuces has given better aphid control than releases of either the parasitoid *A. matricariae* or the predatory midge *A. aphidimyza* (Quentin *et al.*, 1995).

Aphids on field crops and amenity plants

In early field experiments, releases of eggs and newly hatched larvae of both *C. carnea* and the ladybird *C. septempunctata* had a significant impact on *M. persicae* on potatoes but were less effective against *M. euphorbiae*, although interplot movement of released predators sometimes masked treatment effects (Shands *et al.*, 1972a,b,c). Releases of *Chrysoperla* larvae against aphids on outdoor crops in the former USSR also gave mixed results, being more effective on sweet peppers, tomatoes, and eggplants than on peas and brassicas (Beglyarov and Smetnik, 1977).

Second-instar larvae of *Chrysoperla lucasina*, released at a ratio of 1 : 20 aphids on to melons that had been inoculated with *A. gossypii*, kept the aphids below the economic damage threshold and comparable to populations on plots treated with a recommended chemical insecticide (Malet *et al.*, 1994). Releases of *C. carnea* eggs into sugarbeet crops in Germany controlled *A. fabae* for 2 weeks (Sengonca *et al.*, 1995; Sengonca and Löchte, 1997), but similar trials in California, using *C. carnea* and *Chrysoperla rufilabris* eggs, were less successful, although aphid populations declined naturally on all plots soon after treatments were applied, hindering adequate impact assessment (Ehler *et al.*, 1997). Similarly, augmentative releases of *C. rufilabris* and ladybirds against RWA

had very limited success and were considered uneconomic (Randolph *et al.*, 2002).

Improved methods of mass-rearing and field application of lacewings have stimulated an increasing number of trials (Sengonca and Löchte, 1997; Nordlund *et al.*, 2001), including releases of *C. carnea* larvae against *M. euphorbiae* and *Chaetosiphon fragaefolii* (strawberry aphid) on outdoor strawberries in Italy (Tommasini and Mosti, 2001), *C. lucasina* larvae against *Capitophorus horni* and *A. fabae* on globe artichokes (*Cynara scolymus*) in France (Maisonneuve, 2001), *C. carnea* larvae against *A. gossypii* on okra in Egypt (Zaki *et al.*, 1999), *C. carnea* eggs against *Illinoia liriodendri* on amenity tulip trees (*Liriodendron tulipifera*) in California (Dreistadt *et al.*, 1986), and *C. rufilabris* eggs against *A. gossypii* on cotton (Knutson and Tedders, 2002). These last authors concluded that high larval mortality, adult dispersal, and high cost were major constraints to augmentative releases of biocontrol agents in cotton.

Adult lacewings are not suitable for field releases and are also problematical in glasshouses because of their dispersal behaviour (Nordlund *et al.*, 2001; Tommasini and Mosti, 2001). However, Sengonca and Henze (1992) used lacewing hibernation boxes to collect overwintering adults in arable fields in Germany, subsequently re-releasing them into wheat fields to coincide with the appearance of cereal aphids. This resulted in lacewing eggs being recorded a month earlier than in control fields, but subsequent effects on aphid populations were small. In China, overwintering adults of *Chrysoperla sinica* were collected from fields and used to produce larvae for inoculative release the following spring against a range of pests in mixed cropping systems, especially cotton intercropped with wheat (Wang and Nordlund, 1994). The wheat served as a reservoir habitat early in the year, from which the lacewings moved on to other adjacent crops as the wheat ripened. In Egypt, *C. carnea* larvae, bred from adults collected from clover fields, were released on to okra at rates of 1 : 5, 1 : 10, and 1 : 20 aphids and reduced *A. gossypii* populations by 99%, 89%, and 85%, respectively, after 12 days (Zaki *et al.*, 1999).

Also in Egypt, releases of either eggs or larvae of *C. carnea*, over several consecutive weeks, both reduced *A. gossypii* populations by > 95% in experimental field plots of cotton, matching the level of control in insecticide treatments (El Arnaouty and Sewify, 1998).

Problems have been encountered with augmentative releases of lacewings, including natural predation of eggs and young larvae (Dreistadt *et al.*, 1986; Daane and Yokota, 1997), poor egg hatching rates and post-hatching cannibalism (Tulisalo and Tuovinen, 1975), and competition with other, naturally occurring predators (Maisonneuve, 2001). In California, 98% of *C. carnea* eggs released on to tulip trees were removed from some trees by the ant *Linepithema humile* (Dreistadt *et al.*, 1986). There have also been concerns expressed about the effects of prolonged commercial culturing on lacewing performance (Ehler *et al.*, 1997).

Aphids on orchard crops and tree seedlings

Trial releases of *C. carnea* eggs on to apple trees in Canada significantly reduced *Aphis pomi* (green apple aphid) densities, but the method of releasing the eggs from cardboard containers placed manually in the trees was not very cost effective (Hagley, 1989). Although larvae of *Chrysoperla plorabunda* ate *Toxoptera citricidus* (tropical citrus aphid) in the laboratory, very few were observed feeding on this aphid after inundative releases in subsequent field trials (Michaud, 2001). In early trials in pecan orchards in the USA, eggs or newly hatched larvae of *C. rufilabris* were released from aircraft, but in later trials this was replaced by manual placement on to selected trees (LaRock and Ellington, 1996). These releases formed part of an IPM programme that also included releases of adult ladybirds (*H. convergens*), cultural management strategies, and limited insecticide inputs. *Monelliopsis pecanis* (yellow pecan aphid) and *Monellia caryella* (black-margined pecan aphid) were kept below economic damage levels, but control of *Tinocallis caryaefoliae* (black pecan aphid) often required a single insecticide application per year.

Effective control of *Mindarus kinseyi* on white fir (*Abies concolor*) seedlings in California was attained using releases of two *C. rufilabris* larvae per seedling in conjunction with insecticidal soap treatments, although treatments had to be repeated later in the season (Ehler and Kinsey, 1995).

Syrphidae (Hover Flies)

Some hover fly larvae are voracious and effective aphid predators. However, attempts to enhance their impact on pests have concentrated on the manipulation of habitats around field crops to provide floral food resources for adult hover flies prior to oviposition as part of conservation biological control strategies, a topic covered by Wratten *et al.*, Chapter 16 this volume. There have been few attempts to use hover flies in classical or augmentation biological control programmes, but the potential of *E. balteatus* (Fig. 18.6) for augmentative releases against *D. plataginea* was assessed using laboratory and field cage experiments in Switzerland (Wyss *et al.*, 1999b). Releases of larvae, together with ladybird larvae, on to apple seedlings reduced aphid densities to 5% of those on untreated control seedlings. Larvae of *E. balteatus*, released in vegetable fields in China at a ratio of 1 : 180 aphids reduced aphid populations by over 90% in 3 days (Yang *et al.*, 2002).

Entomopathogenic Fungi

Entomopathogenic fungi are the most common pathogens attacking aphids and are therefore potential biological control agents (Inglis *et al.*, 2001; Pell *et al.*, 2001; Evans, 2003; Shah and Pell, 2003; Völkl *et al.*, Chapter 8 this volume). Most species, including those attacking aphids, are from the fungal divisions Ascomycota (order Hypocreales, e.g. *Lecanicillium longisporum*, *Beauveria bassiana*, *Metarhizium anisopliae*, and *Paecilomyces fumosoroseus*) and Zygomycota (order Entomophthorales, e.g. *Pandora neoaphidis*, *Zoophthora radicans*, and *Neozygites fresenii*).

Fig. 18.6. Adult hover fly *Episyrphus balteatus* (original colour photo courtesy of A.M. Dewar).

Most entomopathogenic fungi penetrate the cuticle of their host directly, without the requirement for ingestion, and have evolved to exploit the resources provided by their insect hosts, ultimately killing them. However, their ecology, physiology, and life cycles are highly variable and this influences the biocontrol strategies in which they may be used (Inglis *et al.*, 2001; Pell *et al.*, 2001; Völkl *et al.*, Chapter 8 this volume). The asexual (anamorphic) forms of the Ascomycetes (compared with the sexual or teleomorphic forms) generally are readily mass-produced *in vitro* and the small infective spores (conidia or blastospores) can be formulated and applied as sprays (e.g. Feng *et al.*, 1994; Bateman, 1997; Burges, 1998; Jenkins *et al.*, 1998; Chapple *et al.*, 2000; Wraight *et al.*, 2001; Jackson *et al.*, 2003). For this reason, they have been developed commercially as mycoinsecticides for use in inundative augmentation programmes against aphids (Copping, 2004). In contrast, entomophthoralean species are difficult, or impossible, to mass-produce *in vitro* (McCabe and Soper, 1985; Li *et al.*, 1993; Shah *et al.*, 1998; Freimoser *et al.*, 2001). Therefore, these have been used more successfully in small-scale classical introductions, inoculative augmentation (often using *in vivo* produced material) and, more recently, conservation biocontrol approaches (Pell *et al.*, 2001; Shah and Pell, 2003; Ekesi *et al.*, 2005).

Glasshouse aphids

Ascomycetes

Lecanicillium longisporum (previously *Verticillium lecanii*) (Gams and Zare, 2001), as the product Vertalec® (produced by Koppert, the Netherlands, since 1990), was the first fungus to be developed as an inundative mycoinsecticide against aphids in glasshouses and is widely available in Europe (Fig. 18.7). The isolate forming the active ingredient has good efficacy against many aphid species, particularly highly mobile species such as *M. persicae* (e.g. Hall and Burges, 1979; Hall, 1981; Harper and Huang, 1986; Quinlan, 1988; Milner, 1997; Burges, 2000). The commercial product is produced by liquid fermentation as blastospores, formulated with a nutrient source in a wettable powder containing 10^9 blastospores per g (Milner, 1997). Nutrient formulations allow satellite colonies to grow on leaf surfaces, thereby increasing coverage significantly (Hall and Papierok, 1982).

Vertalec is used exclusively in glasshouses where optimal conditions for infection, particularly humidity, can be maintained. For example, in the past, chrysanthemums were timed to flower in succession by controlling day length using polythene blackout covers that, provided they were not water vapour porous, maintained high humidity

Fig. 18.7. Aphid infected by the entomopathogenic fungus *Lecanicillium longisporum*, which is available as the commercial mycoinsecticide Vertalec® for use against aphid pests on glasshouse crops. The fungus can be seen emerging from the body of the infected aphid (photo courtesy of D. Chandler).

and therefore ensured fungal infection. A single application prior to covering provided good control of *M. persicae* (Hall and Burges, 1979). In the UK, aphids nearly always infest chrysanthemums at some time and so prophylactic sprays were encouraged 2 weeks prior to planting (Quinlan, 1988; Curtis, 1998). Frequent, low dose applications have also provided improved control of *M. persicae* and *A. gossypii* (Helyer and Wardlow, 1987), as has the use of electrostatic spraying (Sopp *et al.*, 1989, 1990) and combining the fungus with synthetic aphid alarm pheromone ((*E*)-ß-farnesene (Hockland *et al.*, 1986). On green peppers, application in conjunction with sublethal doses of imidacloprid also improved efficacy against *M. persicae* (Roditakis *et al.*, 2000).

However, even in the glasshouse, temperatures and humidities can fall outside the activity range of the Vertalec isolate (Hall, 1980, 1982; Hall and Papierok, 1982; Milner and Lutton, 1986; Malais and Ravensberg, 1992; Curtis, 1998; Yeo *et al.*, 2003). Even in chrysanthemums, the blackout method has been changed, reducing humidity sufficiently to make Vertalec ineffective (D. Chandler, personal communication) and field use remains impossible (Milner, 1997; Yeo, 2000). Recent advances by Koppert in formulation technology have resulted in an adjuvant ('Addit') to enhance activity of their product Mycotal® (*Lecanicillium muscarium*) at low humidities and extend its usefulness to the field. Unfortunately, the adjuvant is not

compatible with the Vertalec isolate, limiting its use to crops with canopies that naturally favour fungal infection. These problems have been overcome by formulation for several species of fungi and this may yet be possible for Vertalec (Pfrommer and Mendgen, 1992; Burges, 1998; Curtis *et al.*, 2003) or for other isolates of *Lecanicillium* spp. that are also known to attack aphids (Feng *et al.*, 1990; Etzel and Petitt, 1992).

Commercial products based on *B. bassiana* are also available for the control of aphids and other insect pests in glasshouses. They include Mycotrol® and Botanigard® from Emerald BioAgriculture Corporation (recently, the active ingredient, isolate GHA, was sold to Laverlam SA of Colombia, and so its current status is unclear) and Naturalis® from Troy Biosciences. These products are generally mass-produced by solid-state fermentation processes, followed by extraction and formulation of conidia (Bradley *et al.*, 1992; Copping, 2004). All formulations contain between 2.3×10^7 and 2×10^{10} conidia/ml (in emulsifiable suspensions or suspension concentrates) or 4×10^8 conidia/g (in wettable powders) and are mixed with water before spraying using conventional equipment (Copping, 2004).

Zygomycetes

Inoculative augmentation has been attempted experimentally to establish *N. fresenii* and *P. neoaphidis* for aphid control in glasshouses in Europe. Under optimal conditions, Dedryver (1979) achieved some success releasing *N. fresenii* in living infected aphids amongst bean plants infested with *A. fabae*. Similarly, Latgé *et al.* (1983) and Silvie *et al.* (1990) applied *in vitro* mycelium of *P. neoaphidis* as an aqueous spray and, although the fungus was able to infect some aphids (*A. solani* and *M. euphorbiae*), population suppression was not achieved. More recently, glasshouse trials of *P. neoaphidis* formulated as dried mycelium or in alginate beads have shown some promise for control of *M. euphorbiae* (Shah *et al.*, 2000).

Aphids on field crops

Ascomycetes

Isolates of the fungi *Lecanicillium* spp., *B. bassiana*, *M. anisopliae*, and *P. fumosoroseus* all have virulence against a number of aphid species found on field crops (e.g. Feng and Johnson, 1990; Feng *et al.*, 1990; Miranpuri and Khachatourians, 1993; Butt *et al.*, 1994; Vandenberg, 1996). However, for aphid control, only products based on the GHA isolate of *B. bassiana*, e.g. Mycotrol® and MycotrolO®, are currently available and these are primarily only registered for use in the United States (Shah and Goettel, 1999; Shah and Pell, 2003). Although products based on this isolate are registered against numerous pests on many crops, including cereals, legumes, cotton, vegetables, orchards, and turf, the majority of published literature on aphid control on field crops is experimental and has focused on control of *D. noxia*.

Vandenberg (1996) demonstrated that both the GHA isolate of *B. bassiana* and a number of isolates of *P. fumosoroseus* were virulent against *D. noxia*. In small plot field evaluations in the USA, isolates were applied by backpack sprayer to infested aphid-susceptible and aphid-resistant wheat plants and, in most experiments, reductions in aphid density were observed within 14 days. Even though the crop was irrigated, variable abiotic conditions in the field were implicated as the cause of the slow speed of kill compared to previous laboratory assays (Vandenberg *et al.*, 2001). Aphid control by fungi was poorer on the aphid-resistant plants, perhaps because the architecture of the plant resulted in a microclimatre less favourable to infection. Following initial field tests, large-scale application of *B. bassiana* through the irrigation system demonstrated the practical feasibility of this delivery system. Problematic commercial production remains a contributory factor to the limited availability of commercial products based on *P. fumosoroseus*, but this is being addressed by ongoing research (e.g. Vandenberg *et al.*, 1998a; Sandoval-Coronado *et al.*, 2001; Jackson *et al.*, 2003). *Beauveria bassiana* and *P. fumosoroseus* are yet to fulfil their potential for aphid control in the open field, particularly in low-value, broad acre crops such as cereals. However, they represent a potentially useful tool for integrated control of aphids and other pests (e.g. Feng *et al.*, 1994; Shelton *et al.*, 1998; Vandenberg *et al.*, 1998b; Furlong and Groden, 2001; Evans, 2003).

Zygomycetes

In the 1970s, *T. t. maculata* was a significant introduced pest of pasture legumes in Australia. Entomopathogenic fungi were recorded at extremely low levels, but *Z. radicans*, an important fungal natural enemy of *T. t. maculata* in other countries, was absent (Hall and Dunn, 1957; Kenneth and Olmert, 1975; Milner *et al.*, 1980). An Israeli isolate from a region with a similar climate was introduced either as *in vitro* cultures or laboratory-infected living and dead aphids. Infection levels were initially very low, but within 5 weeks, epizootics had established (Milner *et al.*, 1982). Even in the absence of rain, humidity remained high at night, which may have contributed to initial establishment, but the production of resting spores by the fungus ensured year-to-year persistence. The fungus subsequently spread by itself and the aphid was effectively controlled (Milner *et al.*, 1982).

Inoculative augmentation in cereals and legumes in Europe has been attempted experimentally on several occasions to encourage the early impact of *P. neoaphidis* and *N. fresenii* where they occur indigenously. Latteur and Godefroid (1983) introduced *in vitro* produced mycelium of *P. neoaphidis* into aphid populations on cereals in France, but did not achieve population suppression. Under field conditions in the UK, Wilding (1981) released *P. neoaphidis* and *N. fresenii* as infected aphid cadavers into *A. fabae* populations and was able to cause an early-season crash in aphid populations with both fungal species in 2 out of 4 years, with yield improvements recorded in 1 year. Both the successful years had below average temperatures and above average rainfall. In a further trial, irrigation increased the proportion of infected aphids, confirming that the fungus can be limited by dry conditions (Wilding *et al.*, 1986). Similar effects were seen in cereals (Wilding *et al.*, 1990), with *P. neoaphidis* causing increased infection in treated plots but acting too slowly and unpredictably. More recently, Poprawski and Wraight (1998) placed *in vitro*-produced mycelium of *P. neoaphidis* into rolled wheat leaves with large colonies of *D. noxia*. Aphids became infected, but the spread of the fungus to uninoculated tillers was slow. Conservation approaches, including habitat manipulation, are currently becoming more important for improving the impact of naturally occuring entomophthoralean fungi in field crops without the requirement for augmentation (Pell *et al.*, 2001; Ekesi *et al.*, 2005).

Cotton aphids

Epizootics of *N. fresenii* in cotton aphids have been documented commonly in the mid-south of the USA and can be predicted by diagnosis of aphid samples (Hollingsworth *et al.*, 1995; Steinkraus *et al.*, 1995). When fungal prevalence is 15% or higher, aphid populations are likely to decline within a week and if it is 50% or higher, then declines occur within a few days, making insecticide applications unnecessary (Hollingsworth *et al.*, 1995). An extension-based advisory service has been established at the University of Arkansas to determine the prevalence of the fungus in cotton aphids collected by farmers, providing information *via* a dedicated website (www.uark.edu/misc/aphid/). This reduces farmer costs, preserves beneficial insects, and reduces environmental contamination by pesticides (Steinkraus *et al.*, 1996, 1998; Steinkraus and Boys, 1997).

In contrast, cotton aphids in the San Joaquin Valley of California lacked fungal pathogens, leaving potential for classical biological control introductions of *N. fresenii* from Arkansas. Infection was established on inoculated plants and reached levels higher than those used to predict initiation of epizootics in southeastern USA. However, population suppressing epizootics did not develop (Steinkraus *et al.*, 2002), suggesting that environmental conditions in the San Joaquin Valley were too extreme for the isolates from Arkansas. Only isolates from ecologically similar environments are likely to achieve long-term suppression of cotton aphid populations.

Aphids in orchards

Cross *et al.* (1999a) concluded that entomopathogenic fungi have potential for control of pests in orchards, particularly Lepidoptera, but that this potential has yet to be realized. In the case of aphids, Mycotrol® is registered for use in orchards, and laboratory and field efficacy of the GHA isolate has been demonstrated against *T. citricidus* in the USA (Poprawski *et al.*, 1999). Dorschner *et al.* (1991) also demonstrated some efficacy of a different, aphid-derived isolate of *B. bassiana* against hop aphids on potted plants but not in commercial fields. Two studies have identified the potential to use autodissemination or assisted autodissemination to distribute *L. longisporum* to aphids on hops and apples (Hartfield *et al.*, 2001; Bird *et al.*, 2004). In these strategies, the target aphids (Hartfield *et al.*, 2001), or non-target vectors such as ants (Bird *et al.*, 2004), are attracted to inoculation devices containing the fungal pathogen. They become contaminated with infective conidia and, on exiting from the inoculation device, spread the disease to the remaining aphid population.

Aphids on vegetables

Ascomycetes

Mycotrol® is registered for use against pests on field vegetables in the USA, and recent laboratory and field trials with the GHA isolate suggest some potential for foliar aphid control on brassicas in the UK (D. Chandler, personal communication). A virulent isolate of *M. anisopliae* has been identified by laboratory bioassay (Chandler, 1997) and evaluated as part of IPM strategies to control *Pemphigus bursarius* (lettuce root aphid) on sequentially planted lettuce (Parker *et al.*, 2002). Although field studies were inconclusive, it was suggested that *M. anisopliae* could be a useful tool in IPM if registration was financially and practically feasible (Parker *et al.*, 2002).

Zygomycetes

To our knowledge, the only work considering entomophthoralean fungi was a conservation approach demonstrating that irrigation in spinach in the USA significantly increased the impact of *P. neoaphidis* against *M. persicae* (McLeod and Steinkraus, 1999). As with arable field crops, conservation by habitat manipulation also has potential to improve the impact of indigenous entomophthoralean fungi in vegetable crops (Pell *et al.*, 2001).

Conclusions

There is an increasing interest in developing biological methods of aphid control, driven not only by the desire to reduce reliance on chemical pesticides for environmental and health reasons but also by the continuing expansion of insecticide resistance problems among aphids and the withdrawal of registration approval for an increasing number of insecticide active ingredients. So far, most success has been achieved with the use of hymenopteran parasitoids, but significant successes have also been achieved with the predatory midge *A. aphidimyza*, especially in the protected crop industry. Entomopathogenic fungi similarly have great potential within IPM strategies, although there is no single criterion that guarantees their successful uptake, and difficulties to be overcome not only are practical but economic, social, and political (Shah and Pell, 2003). Currently, the major practical limitation for the uptake of ascomycete fungi as mycoinsecticides against aphids is ensuring their efficacy under variable abiotic conditions in the open field. These issues have already been overcome using formulation technology in other systems, notably for the control of grasshoppers using *M. anisopliae* var. *acridum* (Burges, 1998), where oil formulations of conidia applied by ultra-low volume equipment ensure infection of the target insect even at extremely low ambient humidities (Bateman, 1992). Another significant limitation is the prohibitive costs of registration, particularly when the target pest is on a low-value crop. This requires legislative advances, based on scientific input, at the national and international level. The ecological impact of ascomycete species with broad host ranges must be better understood to be able to inform the registration authorities, and research is currently ongoing to achieve this (e.g. Strasser *et al.*, 2000; Castrillo *et al.*, 2003, 2004; van Lenteren *et al.*, 2003; Wang *et al.*, 2004). For entomophthoralean fungi, the challenges for mass-production are also still technically prohibitive.

For aphid control on field crops, emphasis has moved towards the development of conservation biological control strategies (Wratten *et al.*, Chapter 16 this volume) as part of integrated crop management schemes that seek to exploit naturally occurring communities of aphid natural enemies through habitat and behaviour manipulation. Extensive studies of aphid and natural enemy ecology and population dynamics in agricultural ecosystems, such as cereals in Northern Europe, have highlighted the importance of encouraging a diverse community of antagonists in order to promote natural aphid control mechanisms (Wratten and Powell, 1991; Sunderland et al., 1997). Nevertheless, some examples of significant reductions in aphid populations on field crops through inundative releases of predators have been recorded, particularly in countries in Asia and North Africa where labour intensive methods are less of an economic constraint.

Two strong messages emerge from the large and varied amount of work that has been carried out on aphid biological control. Firstly, because of the rapid exponential growth rate of aphid populations, the timing of natural enemy impact is critical. This needs to occur early in the population growth curve if it is to be successful, and this 'window of opportunity' is often narrow. Achieving a high predator or parasitoid: aphid ratio, or high level of fungal infection, before the aphid population growth rate

becomes too great is essential. This 'window of opportunity' could be extended by combining biological control with other strategies such as the breeding of crop varieties with partial resistance to the pest (van Emden, Chapter 17 this volume) and ecological manipulation of agricultural ecosystems to conserve and enhance natural enemy populations. The second message, the need to match the climatic adaptation of the biological control agent with climatic conditions in the target region, is particularly relevant to classical introduction programmes and to augmentative releases of commercially bred antagonists. Considerable progress is being made in our understanding of aphid/natural enemy interactions, leading to improved strategies for biological control, but practical and economic constraints still exist and need to be overcome, particularly in relation to mass-rearing and augmentative releases on outdoor crops.

References

Aliniazee, M.T. (1998) Ecology and management of hazelnut pests. *Annual Review of Entomology* 43, 395–419.

Angelet, G.W., Tropp, J.M. and Eggert, A.N. (1979) *Coccinella septempunctata* in the United States: recolonizations and notes on its ecology. *Environmental Entomology* 8, 896–901.

Archer, T.L., Cate, R.H., Eikenbary, R.D. and Starks, K.J. (1974) Parasitoids collected from greenbugs and corn leaf aphids in Oklahoma in 1972. *Annals of the Entomological Society of America* 67, 11–14.

Asante, S.K. and Danthanarayana, W. (1992) Development of *Aphelinus mali*, an endoparasitoid of woolly apple aphid *Eriosoma lanigerum* at different temperatures. *Entomologia Experimentalis et Applicata* 65, 31–37.

Barnes, O.L. (1960) Establishment of imported parasites of the spotted alfalfa aphid in Arizona. *Journal of Economic Entomology* 53, 1094–1096.

Bartlett, B.R. (1964) Toxicity of some pesticides to eggs, larvae and adults of the green lacewing *Chrysopa carnea*. *Journal of Economic Entomology* 57, 366–369.

Bateman, R.P. (1992) Controlled application of mycopesticides to locusts. In: Lomer, C.J. and Prior, C. (eds) *Biological Control of Locusts and Grasshoppers*. CAB International, Wallingford, pp. 249–254.

Bateman, R. (1997) The development of a mycoinsecticide for the control of locusts and grasshoppers. *Outlook on Agriculture* 26, 13–18.

Beglyarov, G.A. and Smetnik, A.I. (1977) Seasonal colonization of entomophages in the USSR. In: Ridgeway, R.L. and Vinson, S.B. (eds) *Biological Control by Augmentation of Natural Enemies*. Plenum, New York, pp. 283–328.

Bennison, J.A. (1992) Biological control of aphids on cucumbers: use of open rearing systems or 'banker plants' to aid establishment of *Aphidius matricariae* and *Aphidoletes aphidimyza*. *Mededelingen van de Faculteit Landbouwwetenschappen Universiteit Gent* 57, 457–466.

Bennison, J.A. and Corless, S.P. (1993) Biological control of aphids on cucumbers: further development of open rearing units or 'banker plants' to aid establishment of aphid natural enemies. *Bulletin IOBC/WPRS* 16 (2), 5–8.

Bernal, J.S. and Gonzalez, D. (1993) Temperature requirements of four parasites of the Russian wheat aphid *Diuraphis noxia*. *Entomologia Experimentalis et Applicata* 69, 173–182.

Bernal, J.S. and Gonzalez, D. (1996) Thermal requirements of *Aphelinus albipodus* (Hayat and Fatima) (Hym., Aphelinidae) on *Diuraphis noxia* (Mordwilko) (Hom., Aphididae) hosts. *Journal of Applied Entomology* 120, 631–637.

Bernal, J.S. and Gonzalez, D. (1997) Reproduction of *Diaeretiella rapae* on Russian wheat aphid hosts at different temperatures. *Entomologia Experimentalis et Applicata* 82, 159–166.

Bernal, J.S., Waggoner, M. and Gonzalez, D. (1997) Reproduction of *Aphelinus albipodus* (Hymenoptera: Aphelinidae) on Russian wheat aphid (Hemiptera: Aphididae) hosts. *European Journal of Entomology* 94, 83–96.

Bernal, J.S., Gonzalez, D. and David-DiMarino, E. (2001) Overwintering potential in California of two Russian wheat aphid parasitoids (Hymenoptera: Aphelinidae and Aphidiidae) imported from central Asia. *Pan-Pacific Entomologist* 77, 28–36.

Bird, A.E., Hesketh, H., Cross, J.V. and Copland, M. (2004) The common black ant, *Lasius niger* (Hymenoptera: Formicidae), as a vector of the entomopathogen *Lecanicillium longisporum* to rosy apple aphid, *Dysaphis plantaginea* (Homoptera: Aphididae). *Biocontrol Science and Technology* 14, 757–767.

Blommers, L.H.M. (1994) Integrated pest management in European apple orchards. *Annual Review of Entomology* 39, 213–241.

Blümel, S. and Hausdorf, H. (1996) Greenhouse trials for the control of aphids on cut-roses with the chalcid *Aphelinus abdominalis* Dalm. (Aphelinidae, Hymen.). *Anzeiger für Schädlingskunde, Pflanzenschutz, Umweltschutz* 69, 64–69.

Bondarenko, N.V. (1989) Observations on rearing and use of the predatory gall midge *Aphidoletes aphidimyza* (Diptera, Cecidomyiidae) against aphids in greenhouses. *Acta Entomologica Fennica* 53, 7–10.

van den Bosch, R. (1957) The spotted alfalfa aphid and its parasites in the Mediterranean region, Middle East, and East Africa. *Journal of Economic Entomology* 50, 352–356.

van den Bosch, R., Schlinger, E.I., Dietrick, E.J., Hall, J.C. and Puttler, B. (1964) Studies on succession, distribution, and phenology of imported parasites of *Therioaphis trifolii* (Monell) in southern California. *Ecology* 45, 602–621.

van den Bosch, R., Frazer, R.D., Davis, C.S., Messenger, P.S. and Hom, R. (1970) *Trioxys pallidus*. An effective new walnut aphid parasite from Iran. *California Agriculture* 24, 8–10.

Botto, E.N., Monetti, N.C. and de Saluso, A.R. (1991) Introduction, colonization and establishment of *Lysiphlebus testaceipes* (Hymenoptera: Aphidiidae), in Argentina. *Entomophaga* 36, 323–324.

Bouchard, D., Hill, S.B. and Pilon, J.G. (1988) Control of green apple aphid populations in an orchard achieved by releasing adults of *Aphidoletes aphidimyza* Rondani (Diptera, Cecidomyiidae). In: Niemczyk, E. and Dixon, A.F.G. (eds) *Ecology and Effectiveness of Aphidophaga*. SPB Academic Publishing, The Hague, pp. 257–260.

Bradley, C.A., Black, W.E., Kearns, R. and Wood, P. (1992) Role of production technology in mycoinsecticide development. In: Leatham, G.F. (ed.) *Frontiers in Industrial Mycology* Chapman and Hall, New York, pp. 160–173.

Breene, R.G., Meagher, R.L., Nordlund, D.A. and Yin-Tung Wang (1992) Biological control of *Bemisia tabaci* (Homoptera:Aleyrodidae) in a greenhouse using *Chrysopa rufilabris* (Neuroptera: Chrysopidae). *Biological Control* 2, 9–14.

Brewer, M.J., Nelson, D.J., Ahern, R.G., Donahue, J.D. and Prokrym, D.R. (2001) Recovery and range of expansion of parasitoids (Hymenoptera: Aphelinidae and Braconidae) released for biological control of *Diuraphis noxia* (Homoptera: Aphididae) in Wyoming. *Environmental Entomology* 30, 578–588.

Brown, E.J., Cave, F.E. and Hoy, M.A. (1992) Mode of inheritance of azinphosmethyl resistance in a laboratory-selected strain of *Trioxys pallidus*. *Entomologia Experimentalis et Applicata* 63, 229–236.

Burd, J.D., Shufran, K.A., Elliott, N.C., French, B.W. and Prokrym, D.A. (2001) Recovery of imported hymenopterous parasitoids released to control Russian wheat aphids (Homoptera: Aphididae) in Colorado. *Southwestern Entomologist* 26, 23–31.

Burges, H.D. (1998) *Formulation of Microbial Biopesticides: Beneficial Microorganisms, Nematodes and Seed Treatments*. Kluwer, Dordrecht, 496 pp.

Burges, H.D. (2000) Techniques for testing microbials for control of arthropod pests in greenhouses. In: Lacey, L.A. and Kaya, H.K. (eds) *Field Manual of Techniques in Invertebrate Pathology: Application and Evaluation of Pathogens for Control of Insects and Other Invertebrate Pests*. Kluwer, Dordrecht, pp. 505–526.

Burgio, G., Ferrari, R. and Nicoli, G. (1997) Biological and integrated control of *Aphis gossypii* Glover (Hom., Aphididae) in protected cucumber and melon. *Bolletino dell' Istituto di Entomologia 'Guido Grandi' della Universita degli Studi di Bologna* 51, 171–178.

Butt, T.M., Ibrahim, L., Ball, B.V. and Clark, S.J. (1994) Pathogenicity of the entomogenous fungi *Metarhizium anisopliae* and *Beauveria bassiana* against crucifer pests and the honey bee. *Biocontrol Science and Technology* 4, 207–214.

Calunga-Garcia, M. and Gage, S.H. (1999) Arrival, establishment, and habitat use of the multicolored asian lady beetle (Coleoptera: Coccinellidae) in a Michigan landscape. *Environmental Entomology* 27, 1574–1580.

Cameron, P.J. and Walker, G.P. (1989) Release and establishment of *Aphidius* spp. (Hymenoptera: Aphidiidae), parasitoids of pea aphid and blue green aphid in New Zealand. *New Zealand Journal of Agricultural Research* 32, 281–290.

Cameron, P.J., Walker, G.P. and Allan, D.J. (1981) Establishment and dispersal of the introduced parasite *Aphidius eadyi* (Hymenoptera: Aphidiidae) in the North Island of New Zealand, and its initial effect on pea aphid. *New Zealand Journal of Zoology* 8, 105–112.

Carver, M. (1989) Biological control of aphids. In: Minks, A.K. and Harrewijn, P. (eds) *Aphids. Their Biology, Natural Enemies and Control, Volume 2C.* Elsevier, Amsterdam, pp. 141–165.

Carver, M., Hart, P.J. and Wellings, P.W. (1993) Aphids (Hemiptera: Aphididae) and associated biota from the Kingdom of Tonga, with respect to biological control. *Pan-Pacific Entomologist* 69, 250–260.

Castrillo, L.A., Vandenberg, J.D. and Wraight, S.P. (2003) Strain-specific detection of introduced *Beauveria bassiana* in agricultural fields by use of sequence-characterized amplified region markers. *Journal of Invertebrate Pathology* 82, 75–83.

Castrillo, L.A., Griggs, M.H. and Vandenberg, J.D. (2004) Vegetative compatibility groups in indigenous and mass-released strains of the entomopathogenic fungus *Beauveria bassiana*: likelihood of recombination in the field. *Journal of Invertebrate Pathology* 86, 26–37.

Chandler, D. (1997) Selection of an isolate of the insect pathogenic fungus *Metarhizium anisopliae* virulent to the lettuce root aphid *Pemphigus bursarius*. *Biocontrol Science and Technology* 7, 95–104.

Chapple, A.C., Downer, R.A. and Bateman, R.P. (2000) Theory and practice of microbial insecticide application. In: Lacey, L.A. and Kaya, H.K. (eds) *Field Manual of Techniques in Invertebrate Pathology: Application and Evaluation of Pathogens for Control of Insects and Other Invertebrate Pests.* Kluwer, Dordrecht, pp. 5–37.

Chilima, C.Z. (1995) Cypress aphid control: first African release of *Pauesia juniperorum*. *Forestry Research Institute of Malawi Newsletter, No. 74*, p. 2.

Copping, L.G. (ed.) (2004) *The Manual of Biocontrol Agents.* British Crop Protection Council, Alton, 702 pp.

Croft, B.A. (1990) *Arthropod Biological Control Agents and Pesticides.* Wiley, New York, 723 pp.

Cross, J.V., Solomon, M.G., Chandler, D., Jarrett, P., Richardson, P.N., Winstanley, D., Bathon, H., Huber, J., Keller, B., Langenbruch, G.A. and Zimmerman, G. (1999a) Biocontrol of pests of apples and pears in northern and central Europe: 1. Microbial agents and nematodes. *Biocontrol Science and Technology* 9, 125–149.

Cross, J.V., Solomon, M.G., Babandreier, D., Blommers, L., Easterbrook, M.A., Jay, C.N., Jenser, G., Jolly, R.L., Kuhlmann, U., Lilley, R., Olivella, E., Toepfer, S. and Vidal, S. (1999b) Biocontrol of pests of apples and pears in northern and central Europe: 2. Parasitoids. *Biocontrol Science and Technology* 9, 277–314.

Curtis, J.E. (1998) Enhancing the effectiveness of *Verticillium lecanii* against the green peach aphid *Myzus persicae*. PhD thesis, La Trobe University, Bundoora, Australia.

Curtis, J.E., Price, T.V. and Ridland, P.M. (2003) Initial development of a spray formulation which promotes germination and growth of the fungal pathogen *Verticillium lecanii* (Zimmerman) Viegas (Deuteromycotina: Hyphomycetes) on capsicum leaves (*Capsicum annuum grossum* Sendt. var. California Wonder) and infection of *Myzus persicae* Sulzer (Homoptera: Aphididae). *Biocontrol Science and Technology* 13, 35–46.

Daane, K.M. and Yokota, G.Y. (1997) Release strategies affect survival and distribution of green lacewings (Neuroptera: Chrysopidae) in augmentation programs. *Environmental Entomology* 26, 455–464.

Davidson, W.M. (1924) Observations and experiments on the dispersion of the convergent lady-beetle (*Hippodamia convergens* Guérin-Méneville) in California. *Transactions of the American Entomological Society* 50, 163–175.

DeBach, P. and Hagen, K.S. (1964) Manipulation of entomophagous species. In: DeBach, P. (ed.) *Biological Control of Insect Pests and Weeds.* Reinhold, New York, pp. 429–458.

Dedryver, C.A. (1979) Déclanchement en serre d'une épizootie à *Entomophthora fresenii* Nowak. sur *Aphis fabae* Scop. par introduction d'inoculum et régulation de l'humidité relative. *Entomophaga* 24, 443–453.

Deng, G., Yang, H. and Jin, M. (1987) Augmentation of coccinellid beetles for controlling sugarcane woolly aphid. *Chinese Journal of Biological Control* 3, 166–168.

Dixon, A.F.G. (2000) *Insect Predator–Prey Dynamics: Ladybird Beetles and Biological Control.* Cambridge University Press, Cambridge, 257 pp.

Dixon, A.F.G., Hemptinne, J.L. and Kindlmann, P. (1997) Effectiveness of ladybirds as biological control agents: patterns and processes. *Entomophaga* 42, 71–83.

Dorschner, K.W., Feng, M.G. and Baird, C.R. (1991) Virulence of an aphid-derived isolate of *Beauveria bassiana* (Fungi: Hyphomycetes) to the hop aphid, *Phorodon humuli* (Homoptera: Aphididae). *Environmental Entomology* 20, 690–693.

Dreistadt, S.H. and Flint, M.L. (1996) Melon aphid (Homoptera: Aphididae) control by inundative convergent lady beetle (Coleoptera: Coccinellidae) release on chrysanthemum. *Environmental Entomology* 25, 688–697.

Dreistadt, S.H., Hagen, K.S. and Dahlsten, D.L. (1986) Predation by *Iridomyrmex humulis* (Hym.: Formicidae) on eggs of *Chrysoperla carnea* (Neu.: Chrysopidae) released for inundative control of *Illinoia liriodendri* (Hom.: Aphididae) infesting *Liriodendron tulipifera*. *Entomophaga* 31, 397–400.

van Driesche, R.G. and Bellows, T.S. (1996) *Biological Control.* Chapman and Hall, New York, 539 pp.

Edwards, O.R. and Hoy, M.A. (1995a) Monitoring laboratory and field biotypes of the walnut aphid parasite, *Trioxys pallidus*, in population cages using RAPD-PCR. *Biocontrol Science and Technology* 5, 313–327.

Edwards, O.R. and Hoy, M.A. (1995b) Random amplified polymorphic DNA markers to monitor laboratory-selected, pesticide-resistant *Trioxys pallidus* (Hymenoptera, Aphidiidae) after release into three California walnut orchards. *Environmental Entomology* 24, 487–496.

Ehler, L.E. and Kinsey, M.G. (1995) Ecology and management of *Mindarus kinseyi* Voegtlin (Aphidoidea: Mindaridae) on white-fir seedlings at a California nursery. *Hilgardia* 62, 1–62.

Ehler, L.E., Long, R.F., Kinsey, M.G. and Kelley, S.K. (1997) Potential for augmentative biological control of black bean aphid in California sugarbeet. *Entomophaga* 42, 241–256.

Ekesi, S., Shah, P.A., Clark, S.J. and Pell, J.K. (2005) Conservation biological control with the fungal pathogen *Pandora neoaphidis*: implications of aphid species, host plant and predator foraging. *Agricultural and Forest Entomology* 7, 21–30.

El Arnaouty, S.A. and Sewify, G.H. (1998) A pilot experiment for using eggs and larvae of *Chrysoperla carnea* (Stephens) against *Aphid gossypii* (Glover) on cotton in Egypt. *Acta Zoologica Fennica* 209, 103–106.

El Habi, M., El Jadd, L., Sekkat, A. and Boumezzough, A. (1999) Lutte contre *Aphis gossypii* Glover (Homoptera: Aphididae) sur concombre sous serre par *Coccinella septempunctata* Linnaeus (Coleoptera: Coccinellidae). *Insect Science and its Application* 19, 57–63.

El Habi, M., Sekkat, A., El Jadd, L. and Boumezzough, A. (2000) Biology of *Hippodamia variegata* Goeze (Col., Coccinellidae) and its suitability against *Aphis gossypii* Glov. (Hom., Aphididae) on cucumber under greenhouse conditions. *Journal of Applied Entomology* 124, 365–374.

Elliott, N.C., Burd, J.D., Armstrong, J.S., Walker, C.B., Reed, D.K. and Peairs, F.B. (1995a) Release and recovery of imported parasitoids of the Russian wheat aphid in eastern Colorado. *Southwestern Entomologist* 20, 125–129.

Elliott, N.C., Burd, J.D., Kindler, S.D. and Lee, J.H. (1995b) Temperature effects on development of three cereal aphid parasitoids (Hymenoptera: Aphidiidae). *Great Lakes Entomologist* 28, 199–204.

Elliott, N., Kieckhefer, R. and Kauffman, W. (1996) Effects of an invading coccinellid on native coccinellids in an agricultural landscape. *Oecologia* 105, 537–544.

El Titi, A. (1974) Auswirkung von der räuberischen Gallmücke *Aphidoletes aphidimyza* (Rond.) (Itonididae: Diptera) auf Blattlauspopulationen unter Glas. *Zeitschrift für Angewandte Entomologie* 76, 406–417.

Etzel, R.W. and Petitt, F.L. (1992) Association of *Verticillium lecanii* with population reduction of red rice root aphid (*Rhopalosiphum rufiabdominalis*) on aeroponically grown squash. *Florida Entomologist* 75, 605–606.

Evans, H.C. (2003) Use of clavicipitalean fungi for the biological control of arthropod pests. In: White, J.F., Bacon, C.W., Hywel-Jones, N.L. and Spatafora, J.W. (eds) *Clavicipitalean Fungi: Evolutionary Biology, Chemistry, Biocontrol and Cultural Impacts.* Dekker, New York, pp. 517–548.

Fabre, J.P. and Rabasse, J.M. (1987) Introduction dans le sud-est de la France d'un parasite: *Pauesia cedrobii* (Hym.: Aphidiidae) du puceron: *Cedrobium laportei* (Hom.: Lachnidae) du cèdre de l'atlas: *Cedrus atlantica*. *Entomophaga* 32, 127–141.

Farrell, J.A. and Stufkens, M.W. (1988) Abundance of the rose–grain aphid, *Metopolophium dirhodum*, on barley in Canterbury, New Zealand. *New Zealand Journal of Zoology* 15, 499–505.

Farrell, J.A. and Stufkens, M.W. (1990) The impact of *Aphidius rhopalosiphi* (Hymenoptera: Aphidiidae) on populations of the rose–grain aphid (*Metopolophium dirhodum*) (Hemiptera: Aphididae) on cereals in Canterbury, New Zealand. *Bulletin of Entomological Research* 80, 377–383.

Feng, M.G. and Johnson, J.B. (1990) Relative virulence of six isolates of *Beauveria bassiana* on *Diuraphis noxia* (Homoptera: Aphididae). *Environmental Entomology* 19, 785–790.

Feng, M.G., Johnson, J.B. and Kish, L.P. (1990) Virulence of *Verticillium lecanii* and an aphid-derived isolate of *Beauveria bassiana* (Fungi: Hyphomycetes) for six species of cereal infesting aphids (Homoptera: Aphididae). *Environmental Entomology* 19, 815–820.

Feng, M.G., Poprawski, T.J. and Khachatourians, G.G. (1994) Production, formulation and application of the entomopathogenic fungus *Beauveria bassiana* for insect control: current status. *Biocontrol Science and Technology* 4, 3–24.

Fenton, F.A. and Dahms, R.G. (1951) Attempts at controlling the greenbug by the importation and release of lady beetles in Oklahoma. *Proceedings of the Oklahoma Academy of Science* 32, 49–51.

Fernandes, O.A., Wright, R.J. and Mayo, Z.B. (1998) Parasitism of greenbugs (Homoptera: Aphididae) by *Lysiphlebus testaceipes* (Hymenoptera: Braconidae) in grain sorghum: implications for augmentative biological control. *Journal of Economic Entomology* 91, 1315–1319.

Ferran, A., Niknam, H., Kabiri, F., Picart, J.L., DeHerce, C., Brun, J., Iperti, G. and Lapchin, L. (1996) The use of *Harmonia axyridis* larvae (Coleoptera: Coccinellidae) against *Macrosiphum rosae* (Hemiptera: Sternorrhyncha: Aphididae) on rose bushes. *European Journal of Entomology* 93, 59–67.

Ferran, A., Giuge, L., Tourniaire, R., Gambier, J. and Fournier, D. (1998) An artificial non-flying mutation to improve the efficiency of the ladybird *Harmonia axyridis* in biological control of aphids. *BioControl* 43, 53–64.

Flint, M.L., Dreistadt, S.H., Rentner, J. and Parrella, M.P. (1995) Lady beetle release controls aphids on potted plants. *California Agriculture* 49, 5–8.

Freimoser, F.M., Jensen, A.B., Tuor, U., Aebi, M. and Eilenberg, J. (2001) Isolation and *in vitro* cultivation of the aphid pathogenic fungus *Entomophthora planchoniana*. *Canadian Journal of Microbiology* 47, 1082–1087.

Furlong, M.J. and Groden, E. (2001) Evaluation of synergistic interactions between the Colorado potato beetle (Coleoptera: Chrysomelidae) pathogen *Beauveria bassiana* and the insecticides, imidacloprid and cyromazine. *Journal of Economic Entomology* 92, 344–356.

Gams, W. and Zare, R. (2001) A revision of *Verticillium* sect. *Prostrata*. III. Generic classification. *Nova Hedwigia* 72, 329–337.

Gilkeson, L.A. and Hill, S.B. (1986) Genetic selection for and evaluation of nondiapause lines of predatory midge, *Aphidoletes aphidimyza* (Rondani) (Diptera: Cecidomyiidae). *Canadian Entomologist* 118, 869–879.

Gilkeson, L.A. and Hill, S.B. (1987) Release rates for control of green peach aphid (Homoptera, Aphididae) by the predatory midge *Aphidoletes aphidimyza* (Diptera, Cecidomyiidae) under winter greenhouse conditions. *Journal of Economic Entomology* 80, 147–150.

Gordon, R.D. (1987) The first North American records of *Hippodamia variegata* (Goeze) (Coleoptera: Coccinellidae). *Journal of the New York Entomological Society* 95, 307–309.

Grafton-Cardwell, E.E. and Hoy, M.A. (1985) Short-term effects of permethrin and fenvalerate on oviposition by *Chrysoperla carnea* (Neuroptera: Chrysopidae). *Journal of Economic Entomology* 78, 955–959.

Grasswitz, T.R. and Burts, E.C. (1995) Effect of native natural enemies and augmentative releases of *Chrysoperla rufilabris* Burmeister and *Aphidoletes aphidimyza* (Rondani) on the population dynamics of the green apple aphid *Aphis pomi* DeGeer. *International Journal of Pest Management* 41, 176–183.

Hagen, K.S. (1966) Suspected migratory flight behaviour of *Hippodamia convergens*. In: Hodek, I. (ed.) *Ecology of Aphidophagous Insects*. Academia, Prague, pp. 135–136.

Hagen, K.S. and Schlinger, E.I. (1960) Imported Indian parasite of pea aphid established in California. *California Agriculture* 14, 5–6.

Hagley, E.A.C. (1989) Release of *Chrysoperla carnea* Stephens (Neuroptera: Chrysopidae) for control of the green apple aphid, *Aphis pomi* De Geer (Homoptera: Aphididae). *Canadian Entomologist* 119, 205–206.

Hågvar, E.B. and Hofsvang, T. (1995) Colonization behavior and parasitization by *Ephedrus cerasicola* (Hym., Aphidiidae) in choice studies with two species of plants and aphids. *Journal of Applied Entomology* 118, 23–30.

Hall, I.M. and Dunn, P.H. (1957) Entomogenous fungi on the spotted alfalfa aphid. *Hilgardia* 27, 159–181.

Hall, R.A. (1980) Effect of relative humidity on survival of washed and unwashed conidiospores of *Verticillium lecanii*. *Acta Oecologia Applicata* 1, 265–274.

Hall, R.A. (1981) The fungus *Verticillium lecanii* as a microbial insecticide against aphids and scales. In: Burges, H.D. (ed.) *Microbial Control of Pests and Plant Diseases 1970–1980*. Academic Press, London, pp. 483–498.

Hall, R.A. (1982) Control of the glasshouse whitefly, *Trialeurodes vaporariorum*, and the cotton aphid, *Aphis gossypii*, by two isolates of the fungus, *Verticillium lecanii*. *Annals of Applied Biology* 101, 1–11.

Hall, R.A. and Burges, H.D. (1979) Control of aphids in glasshouses with the fungus *Verticillium lecanii*. *Annals of Applied Biology* 93, 235–246.

Hall, R.A. and Papierok, B. (1982) Fungi as biological control agents of arthropods of agricultural and medical importance. *Parasitology* 84, 205–240.

Hämäläinen, M. (1980) Evaluation of two native coccinellids for aphid control in glasshouses. *Bulletin IOBC/WPRS* 3 (3), 59–64.

Hansen, L.S. (1983) Introduction of *Aphidoletes aphidimyza* (Rond.) (Diptera: Cecidomyiidae) from an open rearing unit for the control of aphids in glasshouses. *Bulletin IOBC/WPRS* 6 (3), 146–150.

Harizanova, V. and Ekbom, B. (1997) An evaluation of the parasitoid *Aphidius colemani* Viereck (Hymenoptera: Braconidae) and the predator *Aphidoletes aphidimyza* Rondani (Diptera: Cecidomyiidae) for biological control of *Aphis gossypii* Glover (Homoptera: Aphididae) on cucumber. *Journal of Entomological Science* 32, 17–24.

Harper, A.M. and Huang, H.C. (1986) Evaluation of the entomogenous fungus *Verticillium lecanii* (Moniliales: Moniliaceae) as a control agent for insects. *Environmental Entomology* 15, 281–284.

Hartfield, C.M., Campbell, C.A.M., Hardie, J., Pickett, J.A. and Wadhams, L.J. (2001) Pheromone traps for the dissemination of an entomopathogen by the damson–hop aphid, *Phorodon humuli*. *Biocontrol Science and Technology* 11, 401–410.

Hassan, S.A. (1978) Releases of *Chrysopa carnea* Steph. to control *Myzus persicae* (Sulzer) on eggplant in small greenhouse plots. *Zeitschrift für Pflanzenkrankheiten und Pflanzenschutz* 85, 118–123.

Hassan, S.A., Klingauf, F. and Shahin, F. (1985) Role of *Chrysopa carnea* as an aphid predator on sugar beet and the effect of pesticides. *Zeitschrift für Angewandte Entomologie* 100, 163–174.

Havelka, J. (1980) Effect of temperature on the development rate of preimaginal stages of *Aphidoletes aphidimyza* (Diptera, Cecidomyiidae). *Entomologia Experimentalis et Applicata* 27, 83–90.

Havelka, J. and Zemek, R. (1988) Intraspecific variability of aphidophagous gall midge *Aphidoletes aphidimyza* (Rondani) (Diptera, Cecidomyiidae) and its importance for biological control of aphids. 1. Ecological and morphological characteristics of populations. *Journal of Applied Entomology* 105, 280–288.

Havelka, J. and Zemek, R. (1999) Life table parameters and oviposition dynamics of various populations of the predacious gall midge *Aphidoletes aphidimyza*. *Entomologia Experimentalis et Applicata* 91, 481–484.

Heinz, K.M. and Parrella, M.P. (1990) Biological control of insect pests on greenhouse marigolds. *Environmental Entomology* 19, 825–835.

Helyer, N. and Wardlow, L.R. (1987) Aphid control on chrysanthemums using frequent, low dose applications of *Verticillium lecanii*. *Bulletin IOBC/WPRS* 10 (2), 62–65.

Hemptinne, J.L., Doumbia, M. and Gaspar, C. (1995) The reproductive strategy of predators is a major constraint to the implementation of biological control in the field. *Mededelingen Faculteit Landbouwkundige en Toegepaste Biologische Wetenschappen Universiteit Gent* 60, 735–741.

Hockland, S.H., Dawson, G.W., Griffiths, D.C., Marples, B., Pickett, J.A. and Woodcock, C.M. (1986) The use of aphid alarm pheromone ((*E*)-ß-farnesene) to increase effectiveness of the entomophilic fungus *Verticillium lecanii* in controlling aphids on chrysanthemums. *Proceedings of the 4th International Colloquium of Invertebrate Pathology, Veldhoven, August 1986*, 252.

Hofsvang, T. and Hågvar, E.B. (1982) Comparison between the parasitoid *Ephedrus cerasicola* Starý and the predator *Aphidoletes aphidimyza* (Rondani) in the control of *Myzus persicae* (Sulzer). *Zeitschrift für Angewandte Entomologie* 94, 412–419.

Hollingsworth, R.G., Steinkraus, D.C. and McNew, R.W. (1995) Sampling to predict fungal epizootics on cotton aphids (Homoptera: Aphididae). *Environmental Entomology* 24, 1414–1421.

Holtkamp, R.H., Edge, V.E., Dominiak, B.C. and Walters, P.J. (1992) Insecticide resistance in *Therioaphis trifolii* f. *maculata* (Hemiptera, Aphididae) in Australia. *Journal of Economic Entomology* 85, 1576–1582.

Hommes, M. (1992) Biological control of aphids on *Capsicum*. *Bulletin OEPP, No. 22*, 421–427.

Howard, L.O. (1929) *Aphelinus mali* and its travels. *Annals of the Entomological Society of America* 22, 341–368.

Hoy, M.A. and Cave, F.E. (1988) Guthion-resistant strain of walnut aphid parasite. *California Agriculture* 42, 4–5.

Hoy, M.A. and Cave, F.E. (1989a) Parasite tolerates other pesticides. *California Agriculture* 43, 24–26.

Hoy, M.A. and Cave, F.E. (1989b) Toxicity of pesticides used on walnuts to a wild and azinphosmethyl-resistant strain of *Trioxys pallidus* (Hymenoptera, Aphidiidae). *Journal of Economic Entomology* 82, 1585–1592.

Hoy, M.A., Cave, F.E., Beede, R.H., Grant, J., Krueger, W.H., Olson, W.H., Spollen, K.M., Barnett, W.W. and Hendricks, L.C. (1989) Guthion-resistant walnut aphid parasite release dispersal and recovery in orchards. *California Agriculture* 43, 21–23.

Hoy, M.A., Cave, F.E., Beede, R.H., Grant, J., Krueger, W.H., Olson, W.H., Spollen, K.M., Barnett, W.W. and Hendricks, L.C. (1990) Release, dispersal and recovery of a laboratory-selected strain of the walnut aphid parasite *Trioxys pallidus* (Hymenoptera, Aphidiidae) resistant to azinphosmethyl. *Journal of Economic Entomology* 83, 89–96.

Hu, J.S., Wang, M., Sether, D., Xie, W. and Leonhardt, K.W. (1996) Use of polymerase chain reaction (PCR) to study transmission of banana bunchy top virus by the banana aphid (*Pentalonia nigronervosa*). *Annals of Applied Biology* 128, 55–64.

Hughes, R.D., Woolcock, L.T., Roberts, J.A. and Hughes, M.A. (1987) Biological control of the spotted alfalfa aphid *Therioaphis trifolii* f. *maculata* on lucerne crops in Australia by the introduced parasitic hymenopteran *Trioxys complanatus*. *Journal of Applied Ecology* 24, 515–538.

Hughes, R.D., Hughes, M.A., Aeschlimann, J.-P., Woolcock, L.T. and Carver, M. (1994) An attempt to anticipate biological control of *Diuraphis noxia* (Hom., Aphididae). *Entomophaga* 39, 211–223.

Inglis, G.D., Goettel, M.S., Butt, T.M. and Strasser, H. (2001) Use of hyphomycetous fungi for managing insect pests. In: Butt, T.M., Jackson, C. and Magan, N. (eds) *Fungi as Biocontrol Agents: Progress, Problems and Potential*. CAB International, Wallingford, pp. 23–69.

Iperti, G. (1999) Biodiversity of predaceous Coccinellidae in relation to bioindication and economic importance. *Agriculture, Ecosystems and Environment* 74, 323–342.

Iperti, G. and Buscarlet, L.A. (1986) Seasonal migration of the ladybird *Semiadalia undecimnotata*. In: Hodek, I. (ed.) *Ecology of Aphidophaga*. Academia, Prague, pp. 199–204.

Jackson, H.B., Coles, L.W., Wood, E.A., Jr. and Eikenbary, R.D. (1970) Parasites reared from the greenbug and corn leaf aphid in Oklahoma in 1968 and 1969. *Journal of Economic Entomology* 63, 733–736.

Jackson, H.B., Rogers, C.E. and Eikenbary, R.D. (1971) Colonisation and release of *Aphelinus asychis*, an imported parasite of the greenbug. *Journal of Economic Entomology* 64, 1435–1438.

Jackson, M.A., Cliquet, S. and Iten, L.B. (2003) Media and fermentation processes for the rapid production of high concentrations of stable blastospores of the bioinsecticidal fungus *Paecilomyces fumosoroseus*. *Biocontrol Science and Technology* 13, 23–33.

Jenkins, N.E., Heviefo, G., Langewald, J., Cherry, A.J. and Lomer, C.J. (1998) Development of mass production technology for aerial conidia for use as mycoinsecticides. *Biocontrol News and Information* 19, 21–31.

Kairo, M.T.K. and Murphy, S.T. (1999) Temperature and plant nutrient effects on the development, survival and reproduction of *Cinara* sp. nov., an invasive pest of cypress trees in Africa. *Entomologia Experimentalis et Applicata* 92, 147–156.

Kehrli, P. and Wyss, E. (2001) Effects of augmentative releases of the coccinellid *Adalia bipunctata*, and of insecticide treatments in autumn on the spring population of aphids of the genus *Dysaphis* in apple orchards. *Entomologia Experimentalis et Applicata* 99, 245–252.

Kenneth, R., and Olmert, I. (1975) Entomopathogenic fungi and their insect hosts in Israel: additions. *Israel Journal of Entomology* 10, 105–112.

Kfir, R. and Kirsten, F. (1991) Seasonal abundance of *Cinara cronartii* (Homoptera, Aphididae) and the effect of an introduced parasite, *Pauesia* sp. (Hymenoptera, Aphidiidae). *Journal of Economic Entomology* 84, 76–82.

Kfir, R., Kirsten, F. and van Rensberg, N.J. (1985) *Pauesia* sp. (Hymenoptera, Aphidiidae) – a parasite introduced into South Africa for biological control of the black pine aphid, *Cinara cronartii* (Homoptera, Aphididae). *Environmental Entomology* 14, 597–601.

Kieckhefer, R.W. and Olson, G.A. (1974) Dispersal of marked adult coccinellids from crops in South Dakota. *Journal of Economic Entomology* 67, 52–54.

Knutson, A.E. and Tedders, L. (2002) Augmentation of green lacewing, *Chrysoperla rufilabris*, in cotton in Texas. *Southwestern Entomologist* 27, 231–239.

Kuehne, S. (1998) Open rearing of generalist predators: a strategy for improvement of biological pest control in greenhouses. *Phytoparasitica* 26, 277–281.

Kuo-Sell, H.L. (1989) Getreideblattläuse als Grundlage zur biologischen Bekämpfung der Pfirsichblattlaus, *Myzus persicae* (Sulz.), mit *Aphidoletes aphidimyza* (Rond.) (Dipt., Cecidomyiidae) in Gewächshäusern. *Journal of Applied Entomology* 107, 58–64.

Kuroda, T. and Miura, K. (2003) Comparison of the effectiveness of two methods for releasing *Harmonia axyridis* (Pallas) (Coleoptera: Coccinellidae) against *Aphis gossypii* Glover (Homoptera: Aphididae) on cucumbers in a greenhouse. *Applied Entomology and Zoology* 38, 271–274.

LaMana, M.L. and Miller, J.C. (1996) Field observations on *Harmonia axyridis* Pallas (Coleoptera: Coccinellidae) in Oregon. *Biological Control* 6, 232–237.

LaRock, D.R. and Ellington, J.J. (1996) An integrated pest management approach, emphasizing biological control, for pecan aphids. *Southwestern Entomologist* 21, 153–166.

Latgé, J.P., Silvie, P., Papierok, B., Remaudière, G., Dedryver, C.A. and Rabasse, J.M. (1983) Advantages and disadvantages of *Conidiobolus obscurus* and *Erynia neoaphidis* in the biological control of aphids. In: Cavalloro, R. (ed.) *Aphid Antagonists*. Balkema, Rotterdam, pp. 20–32.

Latteur, G. and Godefroid, J. (1983) Trial of field treatments against cereal aphids with mycelium of *Erynia neoaphidis* (Entomophthorales) produced *in vitro*. In: Cavalloro, R. (ed.) *Aphid Antagonists*. Balkema, Rotterdam, pp. 2–10.

Lee, J.H. and Elliott, N.C. (1998a) Temperature effects on development in *Aphelinus albipodus* (Hymenoptera: Aphelinidae) from two geographic regions. *Great Lakes Entomologist* 31, 173–179.

Lee, J.H. and Elliott, N.C. (1998b) Comparison of developmental responses to temperature in *Aphelinus asychis* (Walker) from two different geographic regions. *Southwestern Entomologist* 23, 77–82.

van Lenteren, J.C., Drost, Y.C., van Roermund, H.J.W. and Posthuma-Doodeman, C.J.A.M. (1997) Aphelinid parasitoids as sustainable biological control agents in glasshouses. *Journal of Applied Entomology* 121, 473–485.

van Lenteren, J.C., Babendreier, D., Bigler, F., Burgio, G., Hokkanen, H.M.T., Kuske, S., LoomAns, A.J.M., Menzler-Hokkanen, I., Van Rijn, P.C.J., Thomas, M.B., Tommasini, M.G. and Zeng, Q.-Q. (2003) Environmental risk assessment of exotic natural enemies used in inundative biological control. *Biocontrol* 48, 3–38.

Li, Z., Butt, T.M., Beckett, A. and Wilding, N. (1993) The structure of dry mycelia of the entomophthoralean fungi *Zoophthora radicans* and *Erynia neoaphidis* following different preparation treatments. *Mycological Research* 97, 1315–1323.

Mackauer, M. and Kambhampati, S. (1986) Structural changes in the parasite guild attacking the pea aphid in North America. In: Hodek, I. (ed.) *Ecology of Aphidophaga*. Academia, Prague, pp. 347–356.

Maisonneuve, J.C. (2001) Biological control with *Chrysoperla lucasina* against *Aphis fabae* on artichoke in Brittany (France). In: McEwen, P.K., New, T.R. and Whittington, A.E. (eds) *Lacewings in the Crop Environment*. Cambridge University Press, Cambridge, pp. 513–517.

Majerus, M.E.N. (1994) *Ladybirds*. Harper Collins, London, 367 pp.

Malais, M. and Ravensberg, W.I. (1992) *Knowing and Recognising. The Biology of Glasshouse Pests and their Natural Enemies*. Koppert, Berkel en Rodenrijs, 109 pp.

Malet, J.C., Noyer, C., Maisonneuve, J.C. and Canard, M. (1994) *Chrysoperla lucasina* (Lacroix) (Neur., Chrysopidae), a potential predator of the Mediterranean *Chrysoperla* Steinmann complex: first experiment to control *Aphis gossypii* Glover (Hom., Aphididae) on melon in France. *Journal of Applied Entomology* 118, 429–436.

Marsh, P.M. (1991) A new species of *Pauesia* (Hymenoptera: Braconidae, Aphidiinae) from Georgia and introduced into South Africa against the black pine aphid (Homoptera: Aphididae). *Journal of Entomological Science* 26, 81–84.

Mayr, L. (1973) Möglichkeiten und Grenzen des Einsatzes von *Aphidoletes aphidimyza* (Rond.) (Diptera, Cecidomyiidae) gegen Blattläuse im Gewächshaus. *Zeitschrift für Angewandte Entomologie* 73, 255–260.

McCabe, D. and Soper, R.S. (1985) *Preparation of an Entomopathogenic Fungal Insect Control Agent*. US Patent No. 4530834.

McEwen, P., New, T.R. and Whittington, A.E. (2001) *Lacewings in the Crop Environment*. Cambridge University Press, Cambridge, 546 pp.

McLeod, P.J. and Steinkraus, D.C. (1999) Influence of irrigation and fungicide sprays on prevalence of *Erynia neoaphidis* (Entomophthorales: Entomophthoraceae) infections of green peach aphid (Homoptera: Aphididae) on spinach. *Journal of Agricultural and Urban Entomology* 16, 279–284.

Meadow, R.H., Kelly, W.C. and Shelton, A.M. (1985) Evaluation of *Aphidoletes aphidimyza* (Diptera, Cecidomyiidae) for control of *Myzus persicae* (Homoptera, Aphididae) in greenhouse and field experiments in the USA. *Entomophaga* 30, 385–392.

Messing, R.H. and Aliniazee, M.T. (1988) Hybridization and host suitability of two biotypes of *Trioxys pallidus* (Hymenoptera Aphidiidae). *Annals of the Entomological Society of America* 81, 6–9.

Messing, R.H. and Aliniazee, M.T. (1989) Introduction and establishment of *Trioxys pallidus* (Hymenoptera, Aphidiidae) in Oregon, USA for control of filbert aphid *Myzocallis coryli* (Homoptera, Aphididae). *Entomophaga* 34, 153–164.

Michaud, J.P. (2000) Development and reproduction of ladybeetles (Coleoptera: Coccinellidae) on the citrus aphids *Aphis spiraecola* Patch and *Toxoptera citricida* (Kirkaldy) (Homoptera: Aphididae). *Biological Control* 18, 287–297.

Michaud, J.P. (2001) Evaluation of green lacewings, *Chrysoperla plorabunda* (Fitch) (Neurop., Chrysopidae), for augmentative release against *Toxoptera citricida* (Hom., Aphididae) in citrus. *Journal of Applied Entomology* 125, 383–388.

Michels, G.J. and Whitaker-Deerberg, R.L. (1993) Recovery of *Aphelinus asychis*, an imported parasitoid of Russian wheat aphid, in the Texas panhandle. *Southwestern Entomologist* 18, 11–17.

Miller, J.C. (1995) A comparison of techniques for laboratory propagation of a South American ladybeetle, *Eriopis connexa* (Coleptera: Coccinellidae). *Biological Control* 5, 462–465.

Mills, N.J. (1990) Biological control of forest aphid pests in Africa. *Bulletin of Entomological Research* 80, 31–36.

Milne, W.M. (1997) Studies on the host-finding ability of the aphid parasitoid, *Trioxys complanatus* (Hym.: Braconidae), in lucerne and clover. *Entomophaga* 42, 173–183.

Milner, R.J. (1997) Prospects for biopesticides for aphid control. *Entomophaga* 42, 227–239.

Milner, R.J. and Lutton, G.G. (1986) Dependence of *Verticillium lecanii* (Fungi: Hyphomycetes) on high humidities for infection and sporulation using *Myzus persicae* (Homoptera: Aphididae) as host. *Environmental Entomology* 15, 380–382.

Milner, R.J., Teakle, R.E., Lutton, G.G. and Dare, F.M. (1980) Pathogens of the blue green aphid *Acyrthosiphon kondoi* Shinji and other aphids in Australia. *Australian Journal of Botany* 28, 601–619.

Milner, R.J., Soper, R.S. and Lutton, G.G. (1982) Field release of an Israeli strain of the fungus *Zoophthora radicans* (Brefeld) Batko for biological control of *Therioaphis trifolii* (Monell) f. *maculata*. *Journal of the Australian Entomological Society* 21, 113–118.

Miranpuri, G.S. and Khachatourians, G.G. (1993) Application of entomopathogenic fungus *Beauveria bassiana* against green peach aphid *Myzus persicae* (Sulzer) infesting canola. *Insect Science* 6, 287–289.

Mols, P.J.M. (1996) Do natural enemies control woolly apple aphid? *Bulletin IOBC/WPRS* 19 (4), 203–207.

Mols, P.J.M. (1997) Simulation of population dynamics of woolly apple aphid and its natural enemies. In: Powell, W. (ed.) *Arthropod Natural Enemies in Arable Land III. The Individual, the Population and the Community*. Acta Jutlandica 72, 113–126.

Morse, J.G. and Croft, B.A. (1987) Biological control of *Aphis pomi* (Hom.: Aphididae) of *Aphidoletes aphidimyza* (Dip.: Cecidomyiidae): a predator–prey model. *Entomophaga* 32, 339–356.

Mueller, T.F., Blommers, L.H.M. and Mols, P.J.M. (1992) Woolly apple aphid (*Eriosoma lanigerum* Hausm., Hom., Aphididae) parasitism by *Aphelinus mali* Hal. (Hym., Aphelinidae) in relation to host stage and host colony size, shape and location. *Journal of Applied Entomology* 114, 143–154.

Mulder, S., Hoogerbrugge, H., Altena, K. and Bolckmans, K. (1999) Biological pest control in cucumbers in the Netherlands. *Bulletin IOBC/WPRS* 22 (1), 177–180.

Murphy, S.T., Chilima, C.Z., Cross, A.E., Abraham, Y.J., Kairo, M.T.K., Allard, G.B. and Day, R.K. (1994) Exotic conifer aphids in Africa: ecology and biological control. In: Leather, S.R., Watt, A.D., Mills, N.J. and Walters, K.F.A. (eds) *Individuals, Populations and Patterns in Ecology*. Intercept, Andover, pp. 233–242.

Navratilova, M. (1999) Results of the efficacy evaluation of biological control agents in glasshouses in the Czech Republic. *Bulletin OEPP No. 29*, 69–72.

Nijveldt, W. (1988) Cecidomyiidae. In: Minks, A.K. and Harrewijn, P. (eds) *Aphids. Their Biology, Natural Enemies and Control, Volume 2B*. Elsevier, Amsterdam, pp. 271–277.

Nordlund, D.A., Cohen, A.C. and Smith, R.A. (2001) Mass-rearing, release techniques, and augmentation. In: McEwen, P.K., New, T.R. and Whittington, A.E. (eds) *Lacewings in the Crop Environment*. Cambridge University Press, Cambridge, pp. 303–319.

Obrycki, J.J. and Kring, T.J. (1998) Predaceous Coccinellidae in biological control. *Annual Review of Entomology* 43, 295–321.

Parker, W.E., Collier, R.H., Ellis, P.R., Mead, A., Chandler, D., Blood Smythe, J.A. and Tatchell, G.M. (2002) Matching control options to a pest complex: the integrated pest management of aphids in sequentially-planted crops of outdoor lettuce. *Crop Protection* 21, 235–248.

Pell, J.K., Eilenberg, J., Hajek, A.E. and Steinkraus, D.C. (2001) Biology, ecology and pest management potential of Entomophthorales. In: Butt, T.M., Jackson, C. and Magan, N. (eds) *Fungi as Biocontrol Agents: Progress, Problems and Potential*. CAB International, Wallingford, pp. 71–153.

Pfrommer, W. and Mendgen, K. (1992) Control of the cabbage aphid (*Brevicoryne brassicae*) with the entomopathogenic fungus *Verticillium lecanii* in the laboratory and field. *Zeitschrift für Pflanzenkrankheiten und Pflanzenschutz* 99, 209–217.

Poprawski, T.J. and Wraight, S.P. (1998) Fungal pathogens of Russian wheat aphid (Homoptera: Aphididae). In: Quisenberry, S.S. and Peairs, F.B. (eds) *Response Model for an Introduced Pest – the Russian Wheat Aphid*. Thomas Say Publications in Entomology, Entomological Society of America, Lanham, pp. 209–233.

Poprawski, T.J., Parker, P.E. and Tsai, J.H. (1999) Laboratory and field evaluation of hyphomycete insect pathogenic fungi for control of brown citrus aphid (Homoptera: Aphididae). *Environmental Entomology* 28, 315–321.

Powell, W. (2000) The use of field margins in the manipulation of parasitoids for aphid control in arable crops. *Proceedings of the British Crop Protection Conference, Pests and Diseases, Brighton, November 2000* 2, 579–584.

Powell, W. and Pickett, J.A. (2003) Manipulation of parasitoids for aphid pest management: progress and prospects. *Pest Management Science* 59, 149–155.

Powell, W., Pennacchio, F., Poppy, G.M. and Tremblay, E. (1998) Strategies involved in the location of hosts by the parasitoid *Aphidius ervi* Haliday (Hymenoptera: Braconidae: Aphidiinae). *Biological Control* 11, 104–112.

Powell, W., Walters, K., A'Hara, S., Ashby, J., Stevenson, H. and Northing, P. (2003) Using field margin diversification in agri-environment schemes to enhance aphid natural enemies. In: Rossing, W.A.H., Poehling, H.M. and Burgio, G. (eds) *Landscape Management for Functional Biodiversity. Bulletin IOBC/WPRS* 26 (4), 123–128.

Prinsloo, G.J. (1998) *Aphelinus hordei* (Kurdjumov) (Hymenoptera: Aphelinidae), a parasitoid released for the control of Russian wheat aphid, *Diuraphis noxia* (Kurdjumov) (Homoptera: Aphididae), in South Africa. *African Entomology* 6, 147–156.

Prinsloo, G.J. (2000) Host and host instar preference of *Aphelinus* sp. nr. *varipes* (Hymenoptera: Aphelinidae), a parasitoid of cereal aphids (Homoptera: Aphididae) in South Africa. *African Entomology* 8, 57–61.

Prinsloo, G.J. and Neser, O.C. (1994) The southern African species of *Aphelinus* Dalman (Hymenoptera: Aphelinidae), parasitoids of aphids (Homoptera: Aphidoidea). *Journal of African Zoology* 108, 143–162.

Prokrym, D.R., Pike, K.S. and Nelson, D.J. (1998) Biological control of *Diuraphis noxia* (Homoptera: Aphididae): implementation and evaluation of natural enemies. In: Quisenberry, S.S. and Peairs, F.B. (eds) *Response Model for an Introduced Pest: the Russian Wheat Aphid*. Thomas Say Publications in Entomology, Entomological Society of America, Lanham, pp. 183–208.

Quentin, U., Hommes, M. and Basedow, T. (1995) Studies on the biological control of aphids (Hom., Aphididae) on lettuce in greenhouses. *Journal of Applied Entomology* 119, 227–232.

Quinlan, R.J. (1988) Use of fungi to control insects in glasshouses. In: Burge, M.N. (ed.) *Fungi in Biological Control Systems*. Manchester University Press, Manchester, pp. 19–36.

Rabasse, J.M., Lafont, J.P., Delpuech, I. and Silvie, P. (1983) Progress in aphid control in protected crops. *Bulletin IOBC/WPRS* 6 (3), 151–162.

Randolph, T.L., Kroening, M.K., Rudolph, J.B., Peairs, F.B. and Jepson, R.F. (2002) Augmentative releases of commercial biological control agents for Russian wheat aphid management in winter wheat. *Southwestern Entomologist* 27, 37–44.

Raupp, M.J., Hardin, M.R., Braxton, S.M. and Bull, B.B. (1994) Augmentative releases for aphid control on landscape plants. *Journal of Arboriculture* 20, 241–249.

Remaudière, G. and Starý, P. (1993) Spontaneous settlement in the Paris area of the Hymenoptera Aphidiidae *Pauesia cedrobii*, parasite of the Cedrus aphid *Cedrobium laportei*. *Revue Française d'Entomologie (Nouvelle Serie)* 15, 157–158.

van Rensberg, N.J. (1979) *Cinara cronartii* on the roots of pine trees (Homoptera: Aphididae). *Journal of the Entomological Society of South Africa* 42, 151–152.

Roditakis, E., Couzin, I.D., Balrow, K., Franks, N.R. and Charnley, A.K. (2000) Improving the secondary pick up of insect pathogen conidia by manipulating host behaviour. *Annals of Applied Biology* 137, 329–335.

Sandoval-Coronado, C.F., Luna-Olvera, H.A., Arevalo-Nino, K., Jackson, M.A., Poprawski, T.J. and Galan-Wong, L.J. (2001) Drying and formulation of blastospores of *Paecilomyces fumosoroseus* (Hyphomycetes) produced in two different liquid media. *World Journal of Microbiology and Biotechnology* 17, 423–428.

van Schelt, J. (1999) Biological control of sweet pepper pests in the Netherlands. *Bulletin IOBC/WPRS* 22 (1), 217–220.

van Schelt, J. and Mulder, S. (2000) Improved methods of testing and release of *Aphidoletes aphidimyza* (Diptera: Cecidomyiidae) for aphid control in glasshouses. *European Journal of Entomology* 97, 511–515.

Schoen, L. and Martin, C. (1997) Controle du puceron *Aphis gossypii* Glover en culture de melon sous abri par le système des plantes relais avec *Aphidius colemani* en Rouissillon (sud France). *Annales de la 4e ANPP Conférence Internationale sur les Ravageurs en Agriculture, Montpellier, 1997*, 759–765.

Scopes, N.E.A. (1969) The potential of *Chrysopa carnea* as a biological control agent of *Myzus persicae* on glasshouse chrysanthemums. *Annals of Applied Biology* 64, 433–439.

Sengonca, C. and Henze, M. (1992) Conservation and enhancement of *Chrysoperla carnea* (Stephens) (Neuroptera, Chrysopidae) in the field by providing hibernation shelters. *Journal of Applied Entomology* 114, 497–501.

Sengonca, C. and Löchte, C. (1997) Development of a spray and atomizer technique for applying eggs of *Chrysoperla carnea* (Stephens) in the field for biological control of aphids. *Zeitschrift für Pflanzenkrankheiten und Pflanzenschutz* 104, 214–221.

Sengonca, C., Griesbach, M. and Löchte, C. (1995) Geeignete Räuber-Beute-Verhältnisse für den Einsatz von *Chrysoperla carnea* (Stephens) – Eiern zur Bekämpfung von Blattläusen an Zuckerrüben unter Labor- and Freilandbedingungen. *Zeitschrift für Pflanzenkrankheiten und Pflanzenschutz* 102, 113–120.

Senior, L.J. and McEwen, P.K. (2001) The use of lacewings in biological control. In: McEwen, P.K., New, T.R. and Whittington, A.E. (eds) *Lacewings in the Crop Environment*. Cambridge University Press, Cambridge, pp. 296–302.

Shah, P.A. and Goettel, M.S. (1999) *Directory of Microbial Control Products and Services*, 2nd edn. Society for Invertebrate Pathology, Gainesville, 31 pp.

Shah, P.A. and Pell, J.K. (2003) Entomopathogenic fungi as biological control agents. *Applied Microbiology and Biotechnology* 61, 413–423.

Shah, P.A., Aebi, M. and Tuor, U. (1998) Method to immobilise the aphid-pathogenic fungus *Erynia neoaphidis* in an alginate matrix for biocontrol. *Applied and Environmental Microbiology* 64, 4260–4263.

Shah, P.A., Aebi, M. and Tuor, U. (2000) Infection of *Macrosiphum euphorbiae* with mycelial preparations of *Erynia neoaphidis* in a greenhouse trial. *Mycological Research* 104, 645–652.

Shands, W.A., Simpson, G.W. and Brunson, M.H. (1972a) Insect predators for controlling aphids on potatoes. 1. In small plots. *Journal of Economic Entomology* 65, 511–514.

Shands, W.A., Gordon, C.C. and Simpson, G.W. (1972b) Insect predators for controlling aphids on potatoes. 6. Development of a spray technique for applying eggs in the field. *Journal of Economic Entomology* 65, 1099–1103.

Shands, W.A., Simpson, G.W. and Storch, R.H. (1972c) Insect predators for controlling aphids on potatoes. 3. In small plots separated by aluminium flashing strip – coated with a chemical. *Journal of Economic Entomology* 65, 799–805.

Shelton, A.M., Vandenberg, J.D., Ramos, M. and Wilsey, W.T. (1998) Efficacy and persistence of *Beauveria bassiana* and other fungi for control of the diamondback moth (Lepidoptera: Plutellidae) on cabbage seedlings. *Journal of Entomological Science* 33, 142–151.

Shenhmar, M. and Brar, K.S. (1995) Biological control of mustard aphid, *Lipaphis erysimi* (Kaltenbach) in the Punjab. *Journal of Biological Control* 9, 9–12.

Shijko, E.S. (1989) Rearing and application of the peach aphid parasite, *Aphidius matricariae* (Hymenoptera, Aphidiidae). *Acta Entomologica Fennica* 53, 53–56.

Silvie, P., Dedryver, C.A. and Tanguy, S. (1990) Application expérimentale de mycelium d'*Erynia neoaphidis* (Zygomycetes: Entomophthorales) dans des populations de pucerons sur laitues en serre maraîchère étude du suivi de l'inoculum par caractérisation enzymatique. *Entomophaga* 35, 375–384.

Sipes, D. (1997) Control of the Russian wheat aphid. *Journal of Natural Resources, Life Science and Education* 26, 78–79.

Snyder, W.E., Ballard, S.N., Yang, S., Clevenger, G.M., Miller, T.D., Ahn, J.J., Hatten, T.D. and Berryman, A.A. (2004a) Complementary biocontrol of aphids by the ladybird beetle *Harmonia axyridis* and the parasitoid *Aphelinus asychis* on greenhouse roses. *Biological Control* 30, 229–235.

Snyder, W.E., Clevenger, G.M. and Eigenbrode, S.D. (2004b) Intraguild predation and successful invasion by introduced ladybird beetles. *Oecologia* 140, 559–565.

Sopp, P.I., Gillespie, A.T. and Palmer, A. (1989) Application of *Verticillium lecanii* for the control of *Aphis gossypii* by a low-volume electrostatic rotary atomiser and a high-volume hydraulic sprayer. *Entomophaga* 34, 417–428.

Sopp, P.I., Gillespie, A.T. and Palmer, A. (1990) Comparison of ultra-low volume electrostatic and high-volume hydraulic application of *Verticillium lecanii* for aphid control on chrysanthemums. *Crop Protection* 9, 177–184.

Sproul, A.N. (1981) Biological success against woolly aphis. *Western Australia Journal of Agriculture* 22, 75.

Stadler, B. and Völkl, W. (1991) Foraging patterns of two aphid parasitoids, *Lysiphlebus testaceipes* and *Aphidius colemani* on banana. *Entomologia Experimentalis et Applicata* 58, 221–229.

Staines, C.L., Rothschild, M.J. and Trumbule, R.B. (1990) A survey of the Coccinellidae (Coleoptera) associated with nursery stock in Maryland. *Proceedings of the Entomological Society of Washington* 92, 310–313.

Starks, K.J., Wood, E.A., Jr., Burton, R.L. and Somsen, H.W. (1975) *Behavior of convergent lady beetles in relation to greenbug control in sorghum. Observations and preliminary tests. USDA-ARS ARS-S-53.* USDA-ARS, Stillwater, Oklahoma, 10 pp.

Starý, P. (1967) Multilateral aphid control concept. *Annales de la Société Entomologique de France* 3, 221–225.

Starý, P. (1972) Host range of parasites and ecosystem relations, a new viewpoint in multilateral control concept (Hom., Aphididae; Hym., Aphidiidae). *Annales de la Société Entomologique de France* 8, 351–358.

Starý, P. (1993a) The fate of released parasitoids (Hymenoptera, Braconidae, Aphidinae) for biological control of aphids in Chile. *Bulletin of Entomological Research* 83, 633–639.

Starý, P. (1993b) Alternative host and parasitoid in first method in aphid pest management in glasshouses. *Journal of Applied Entomology* 116, 187–191.

Starý, P. (1999) Parasitoids and biocontrol of Russian wheat aphid, *Diuraphis noxia* (Kurdj.) expanding in central Europe. *Journal of Applied Entomology* 123, 273–279.

Starý, P. and Gonzalez, D. (1991) The Chenopodium aphid, *Hayhurstia atriplicis* (L.) (Hom., Aphididae), a parasitoid reservoir and a source of biocontrol agents in pest management. *Journal of Applied Entomology* 111, 243–248.

Starý, P. and Stechmann, D.H. (1990) *Ephedrus cerasicola* Starý (Hym, Aphidiidae), a new biocontrol agent of the banana aphid, *Pentalonia nigronervosa* Coq. (Hom, Aphididae). *Journal of Applied Entomology* 109, 457–462.

Starý, P. and Stechmann, D.H. (1991) Niche distribution and dispersal of the banana aphid, *Pentalonia nigronervosa* in relation to parasitization by biocontrol agents (Hom., Aphididae; Hym., Aphidiidae). *Acta Entomologica Bohemoslovaca* 88, 187–195.

Starý, P., Gerding, M., Norambuena, H. and Remaudière, G. (1993) Environmental research on aphid parasitoid biocontrol agents in Chile (Hym., Aphidiidae, Hom., Aphidoidea). *Journal of Applied Entomology* 115, 192–306.

Stechmann, D.H. and Völkl, W. (1988) Introduction of *Lysiphlebus testaceipes* (Cresson) (Hym.: Aphidiidae) into the Kingdom of Tonga, Oceania. In: Niemczyk, E. and Dixon, A.F.G. (eds) *Ecology and Effectiveness of Aphidophaga.* SPB Academic Publishing, The Hague, pp. 271–273.

Stechmann, D.H., Völkl, W. and Starý, P. (1996) Ant-attendance as a critical factor in the biological control of the banana aphid *Pentalonia nigronervosa* Coq. (Hom, Aphididae) in Oceania. *Journal of Applied Entomology* 120, 119–123.

van Steenis, M.J. (1992) Biological control of the cotton aphid, *Aphis gossypii* Glover (Hom., Aphididae) – preintroduction evaluation of natural enemies. *Journal of Applied Entomology* 114, 362–380.

van Steenis, M.J. (1994) Intrinsic rate of increase of *Lysiphlebus testaceipes* Cresson (Hym., Braconidae), a parasitoid of *Aphis gossypii* Glov. (Hom., Aphididae), at different temperatures. *Journal of Applied Entomology* 118, 399–406.

van Steenis, M.J. (1995) Evaluation of four aphidiine parasitoids for biological control of *Aphis gossypii. Entomologia Experimentalis et Applicata* 75, 151–157.

van Steenis, M.J. and El-Khawass, K.A.M.H. (1995) Life history of *Aphis gossypii* on cucumber – influence of temperature, host-plant and parasitism. *Entomologia Experimentalis et Applicata* 76, 121–131.

Steinkraus, D.C. and Boys, G.O. (1997) Update on prediction of epizootics with extension-based sampling service. *Proceedings of the Beltwide Cotton Conferences, New Orleans, January 1997, Volume 2.* National Cotton Council of America, Memphis, pp. 1047–1048.

Steinkraus, D.C., Hollingsworth, R.G. and Slaymaker, P.H. (1995) Prevalence of *Neozygites fresenii* (Entomophthorales: Neozygitaceae) on cotton aphids (Homoptera: Aphididae) in Arkansas cotton. *Environmental Entomology* 24, 465–475.

Steinkraus, D.C., Boys, G.O., Hollingsworth, R.G., Bacheler, J.S., Durant, J.A., Freeman, B.L., Gaylor, M.J., Harris, F.A., Knutson, A., Lentz, G.L., Leonard, B.R., Luttrell, R., Parker, D., Powell, J.D., Ruberson, J.R. and Sorenson, C. (1996) Multistate sampling for *Neozygites fresenii* in cotton. *Proceedings of the Beltwide Cotton Research Conferences, Nashville, January 1996, Volume 2.* National Cotton Council of America, Memphis, pp. 735–738.

Steinkraus, D.C., Boys, G.O., Bagwell, R.D., Johnson, D.R., Lorenz, G.M., Meyers, H., Layton, M.B. and O'Leary, P.F. (1998) Expansion of extension-based aphid fungus sampling service to Louisiana and Mississippi. *Proceedings of the Beltwide Cotton Conferences, San Diego, January 1998, Volume 2.* National Cotton Council of America, Memphis, pp. 1239–1242.

Steinkraus, D.C., Boys, G.O. and Rosenheim, J.A. (2002) Classical biological control of *Aphis gossypii* (Homoptera: Aphididae) with *Neozygites fresenii* (Entomophthorales: Neozygitaceae) in California cotton. *Biological Control* 25, 297–304.

Sterk, G. and Meesters, P. (1997) IPM on strawberries in glasshouses and plastic tunnels in Belgium, new possibilities. *Acta Horticulturae* 439, 905–911.

Stewart, H.C. and Walde, S.J. (1997) The dynamics of *Aphis pomi* De Geer (Homoptera: Aphididae) and its predator, *Aphidoletes aphidimyza* (Rondani) (Diptera: Cecidomyiidae), on apple in Nova Scotia. *Canadian Entomologist* 129, 627–636.

Stoetzel, M.B. (1987) Information on and identification of *Diuraphis noxia* (Homoptera: Aphididae) and other aphid species colonizing leaves of wheat and barley in the United States. *Journal of Economic Entomology* 80, 696–704.

Strasser, H., Vey, A. and Butt, T.M. (2000) Are there any risks in using entomopathogenic fungi for pest control, with particular reference to the bioactive metabolites of *Metarhizium*, *Tolypocladium* and *Beauveria* species? *Biocontrol Science and Technology* 10, 717–735.

Sunderland, K.D., Axelsen, J.A., Dromph, K., Freier, B., Hemptinne, J.-L., Holst, N.H., Mols, P.J.M., Petersen, M.K., Powell, W., Ruggle, P., Triltsch, H. and Winder, L. (1997) Pest control by a community of natural enemies. *Acta Jutlandica* 72, 271–326.

Takahashi, K. (1997) Use of *Coccinella septempunctata brucki* Mulsant as a biological agent for controlling alfalfa aphids. *Japan Agricultural Research Quarterly* 31, 101–108.

Tanagoshi, L.K., Pike, K.S., Miller, R.H., Miller, T.D. and Allison, D. (1995) Search for, and release of, parasitoids for the biological control of Russian wheat aphid in Washington State (USA). *Agriculture, Ecosystems and Environment* 52, 25–30.

Tiittanen, K. and Markkula, M. (1989) Biological control of pests on Finnish greenhouse vegetables. *Acta Entomologica Fennica* 53, 57–60.

Tommasini, M.G. and Mosti, M. (2001) Control of aphids by *Chrysoperla carnea* on strawberry in Italy. In: McEwen, P.K., New, T.R. and Whittington, A.E. (eds) *Lacewings in the Crop Environment*. Cambridge University Press, Cambridge, pp. 481–486.

Tourniaire, R., Ferran, A., Gambier, J., Guige, L. and Bouffault, F. (1999) Locomotor behaviour of flightless *Harmonia axyridis* Pallas (Col., Coccinellidae). *Journal of Insect Behavior* 12, 545–558.

Tourniaire, R., Ferran, A., Guige, L., Piotte, C. and Gambier, J. (2000) A natural flightless mutation in the ladybird, *Harmonia axyridis*. *Entomologia Experimentalis et Applicata* 96, 33–38.

Tremblay, E. (1974) Possibilities for utilization of *Aphidius matricariae* against *Myzus persicae* in small glasshouses. *Zeitschrift für Pflanzenkrankheiten und Pflanzenschutz* 81, 612–619.

Trouvé, C. (1995) Tests on biological pest control against the hop aphid *Phorodon humuli* (Schrank) (Homoptera: Aphididae) in northern France: results from 1993 to 1994. *Mededelingen Faculteit Landbouwkundige en Toegepaste Biologische Wetenschappen Universiteit Gent* 60, 781–792.

Trouvé, C., Ledee, S., Ferran, A. and Brun, J. (1997) Biological control of the damson–hop aphid, *Phorodon humuli* (Hom.: Aphididae), using the ladybeetle *Harmonia axyridis* (Col.: Coccinellidae). *Entomophaga* 42, 57–62.

Tulisalo, U. and Tuovinen, T. (1975) The green lacewing, *Chrysopa carnea* Steph. (Neuroptera, Chrysopidae), used to control the green peach aphid, *Myzus persicae* Sulz., and the potato aphid, *Macrosiphum euphorbiae* Thomas (Homoptera, Aphididae), on greenhouse green peppers. *Annales Entomologici Fennici* 41, 94–102.

Tulisalo, U., Tuovinen, T. and Kurppa, S. (1977) Biological control of aphids with *Chrysopa carnea* on parsley and green pepper in the greenhouse. *Annales Entomologici Fennici* 43, 97–100.

Vandenberg, J.D. (1996) Standardised bioassay and screening of *Beauveria bassiana* and *Paecilomyces fumosoroseus* against the Russian wheat aphid (Homoptera: Aphididae). *Journal of Economic Entomology* 89, 1418–1423.

Vandenberg, J.D., Jackson, M.A. and Lacey, L.A. (1998a) Relative efficacy of blastospores and aerial conidia of *Paecilomyces fumosoroseus* against the Russian wheat aphid. *Journal of Invertebrate Pathology* 72, 181–183.

Vandenberg, J.D., Shelton, A.M., Wilsey, W.T. and Ramos, M. (1998b) Assessment of *Beauveria bassiana* for control of diamondback moth (Lepidoptera: Plutellidae) on crucifers. *Journal of Economic Entomology* 91, 624–630.

Vandenberg, J.D., Sandovol, L.E., Jaronski, S.T., Jackson, M.A., Souza, E.J. and Halbert, S.E. (2001) Efficacy of fungi for control of Russian wheat aphid (Homoptera: Aphididae) in irrigated wheat. *Southwestern Entomologist* 26, 73–85.

Verkerk, R.H.J., Neugebauer, K.R., Ellis, P.R. and Wright, D.J. (1998) Aphids on cabbage: tritrophic and selective insecticide interactions. *Bulletin of Entomological Research* 88, 343–349.

Votava, E.J. and Bosland, P.W. (1996) Use of ladybugs to control aphids in *Capsicum* field isolation cages. *HortScience* 31, 1237.

Walters, P.J. and Dominiak, B.C. (1984) The establishment in New South Wales of *Trioxys complanatus* (Hymenoptera, Aphidiidae) an imported parasite of *Therioaphis trifilii* f. *maculata* spotted alfalfa aphid. *General and Applied Entomology* 16, 65–67.

Wang, C., Fan, M., Li, Z. and Butt, T.M. (2004) Molecular monitoring and evaluation of the application of the insect-pathogenic fungus *Beauveria bassiana* in southeast China. *Journal of Applied Microbiology* 96, 861–870.

Wang, R. and Nordlund, D.A. (1994) Use of *Chrysoperla* spp. (Neuroptera: Chrysopidae) in augmentative release programmes for control of arthropod pests. *Biocontrol News and Information* 15, 52 N–57 N.

Warner, L.A. and Croft, B.A. (1982) Toxicities of azinphosmethyl and selected orchard pesticides to an aphid predator, *Aphidoletes aphidimyza*. *Journal of Economic Entomology* 75, 410–415.

Wellings, P.W., Hart, P.J., Kami, V. and Morneau, D.C. (1994) The introduction and establishment of *Aphidius colemani* Viereck (Hym., Aphidiinae) in Tonga. *Journal of Applied Entomology* 118, 419–428.

Wheeler, A.G. (1993) Establishment of *Hippodamia variegata* and new records of *Propylea quatuordec-impunctata* (Coleoptera: Coccinellidae) in the eastern United States. *Entomology News* 105, 228–243.

Wheeler, A.G. and Hoebeke, E.R. (1995) *Coccinella novemnotata* in northeastern North America: historical occurrence and current status (Coleoptera: Coccinellidae). *Proceedings of the Entomological Society of Washington* 97, 701–716.

Wilding, N. (1981) The effect of introducing aphid-pathogenic Entomophthoraceae into field populations of *Aphis fabae*. *Annals of Applied Biology* 99, 11–23.

Wilding, N., Mardell, S.K. and Brobyn, P.J. (1986) Introducing *Erynia neoaphidis* into a field population of *Aphis fabae*: form of the inoculum and effect of irrigation. *Annals of Applied Biology* 108, 373–385.

Wilding, N., Mardell, S.K., Brobyn, P.J., Wratten, S.D. and Lomas, J. (1990) The effect of introducing the aphid-pathogenic fungus *Erynia neoaphidis* into populations of cereal aphids. *Annals of Applied Biology* 117, 683–691.

Wraight, S.P., Jackson, M.A. and Kock, S.L. de (2001) Production, stabilization and formulation of fungal biocontrol agents. In: Butt, T.M., Jackson, C. and Magan, N. (eds) *Fungi as Biocontrol Agents: Progress, Problems and Potential*. CAB International, Wallingford, pp. 253–287.

Wratten, S.D. and Powell, W. (1991) Cereal aphids and their natural enemies. In: Firbank, L.G., Carter, N., Darbyshire, J.F. and Potts, G.R. (eds) *The Ecology of Temperate Cereal Fields. Proceedings of the British Ecological Society Symposium No. 32*. Blackwell, Oxford, pp. 233–257.

Wyatt, I.J. (1970) The distribution of *Myzus persicae* (Sulz.) on year round chrysanthemums. II Winter season: the effect of parasitism by *Aphidius matricariae* Hal. *Annals of Applied Biology* 65, 31–41.

Wyss, E., Villiger, M. and Muller-Scharer, H. (1999a) The potential of three native insect predators to control the rosy apple aphid, *Dysaphis plantaginea*. *Biocontrol* 44, 171–182.

Wyss, E., Villiger, M., Hemptinne, J.L. and Muller-Scharer, H. (1999b) Effects of augmentative releases of eggs and larvae of the ladybird beetle, *Adalia bipunctata*, on the abundance of the rosy apple aphid, *Dysaphis plantaginea*, in organic apple orchards. *Entomologia Experimentalis et Applicata* 90, 167–173.

Yang, Y., Wang, H., Wang, Q., Cao, L., Liang, D. and Zhang, Z. (2002) Control of aphids in vegetable fields with syrphid flies. *Chinese Journal of Biological Control* 18, 124–127.

Yeo, H. (2000) Mycoinsecticides for aphid management: a biorational approach. PhD thesis, University of Nottingham, Nottingham, UK.

Yeo, H., Pell, J.K., Alderson, P.G., Clark, S.J. and Pye, B.J. (2003) Laboratory evaluation of temperature effects on the germination and growth of entomopathogenic fungi and on their pathogenicity to two aphid species. *Pest Management Science* 59, 156–165.

Zaki, F.N., El-Shaarawy, M.F. and Farag, N.A. (1999) Release of two predators and two parasitoids to control aphids and whiteflies. *Journal of Pest Science* 72, 19–20.

Zhang, Z.Q. (1992) The natural enemies of *Aphis gossypii* Glover (Hom, Aphididae) in China. *Journal of Applied Entomology* 114, 251–262.

Zuparko, R. (1983) Biological control of *Eucallipterus tiliae* (Hom.: Aphididae) in San Jose, Calif., through establishment of *Trioxys curvicaudus* (Hym.: Aphidiidae). *Entomophaga* 28, 325–330.

Zuparko, R. and Dahlsten, D.L. (1995) Parasitoid complex of *Eucallipterus tiliae* (Homoptera, Drepanosiphidae) in northern California. *Environmental Entomology* 24, 730–737.

Zuparko, R. and Dahlsten, D.L. (1996) New potential for classical biological control of *Eucallipterus tiliae* (Homoptera, Drepanosiphidae). *Biological Control* 6, 407–408.

19 Monitoring and Forecasting

Richard Harrington[1], Maurice Hullé[2] and Manuel Plantegenest[2]

[1]Plant and Invertebrate Ecology Division, Rothamsted Research, Harpenden, Herts, AL5 2JQ, UK; [2]INRA, UMR BiO3P, 35653 Le Rheu, France

Introduction

This chapter covers the monitoring and forecasting of aphids: why, what, where, when, and how? Attention to the first four questions is a prerequisite for proper consideration of the fifth.

Why Monitor and Forecast?

Within the context of aphids as crop pests, monitoring and forecasting are aimed ultimately at optimizing the nature, location, and timing of control measures, although a series of strategic programmes will be necessary ahead of those that directly support the decision making of individual growers. Optimization of control measures involves economic and environmental considerations to varying degrees, depending on the nature of the problem and of the individual trying to solve it. For example, prophylactic use of chemicals may be more desirable for the control of aphids causing virus spread in seed potato crops, where there may be very low tolerance levels for the virus and hence for the vectors, than for control of aphids causing direct feeding damage in a ware potato crop, where tolerance levels will be higher. A prophylactic approach may suit a wealthy, risk-averse grower with little concern for the environment or longer-term problems such as the build-up of insecticide resistance, whereas advice to reduce inputs may be more readily adopted by those at the other end of the wealth and environmental concern spectra. The degree to which monitoring and forecasting influence the grower community depends ultimately on the proven robustness of the method, and on political agendas that impact upon the cost of control.

More recently, there has been interest in longer-term forecasting in order to assess the likely impacts of global environmental change. Here, instead of trying to predict values of population variables a few days to months ahead on the basis of known, measured explanatory variables, the need is to predict mean population measures decades ahead on the basis of environmental change scenarios provided by experts in that field.

What Should be Monitored and Forecast?

Monitoring aphid abundance can lead directly to control decisions provided that economic threshold levels in the crop and their relationship with the monitoring data are known. These thresholds and relationships vary according to a range of factors associated with, for example, location, weather, soil conditions, cultivar, aphid species, and aphid genotype, and are therefore not known for a majority of situations. This limits the value of some monitoring programmes. However, in many cases, especially in seed crops where aphids are important as virus vectors, any aphid activity may be unacceptable during the susceptible stages of crop growth. Thus, as soon as aphids appear, control is necessary and hence monitoring simply for presence can provide an immediate return. Forecasting the time of the start of aerial activity can prevent monitoring in the crop being undertaken too early or too late.

As well as distribution, abundance, and phenology of aphids, it may be useful to monitor or predict certain aphid attributes such as whether they are carrying viruses or are resistant to certain insecticides. It may also be useful to know the morph of the aphids as, for example, winged gynoparae of heteroecious species feed on different host plants to winged virginoparae of the same species. As more becomes known about differences in the pest potential of different aphid clones, it will become more useful to understand the population structure at a sub-specific level and techniques are developing rapidly to achieve this. For example, Dedryver *et al.* (2005) found different clones of *Sitobion avenae* (grain aphid) to transmit a particular isolate of *Barley yellow dwarf virus* (BYDV) with efficiencies varying from 3.7–92.5%.

Where Should Monitoring and Forecasting be Carried Out?

Monitoring and forecasting may be carried out at various spatial scales depending on objectives and available resources. Crop monitoring provides data that are of more local value than aerial monitoring but, because of the generally aggregated distribution patterns of aphids in crops, multiple samples are needed in single fields to assess the variance. When aerial monitoring, the

higher above ground the sample is taken, the less will be the aggregation. As a result, the area represented will be greater, especially where there is little topographical variation, but samples will not take account of local variation in crops (Taylor, 1979).

When Should Monitoring and Forecasting be Undertaken?

For most purposes relating to aphids as crop pests, aerial monitoring is of more value when aphids are entering crops than when leaving them. However, numbers leaving may give an indication of the potential risk to subsequent crops, especially when several crops are planted sequentially or the time gap between one crop becoming unsuitable and another emerging is short.

It is necessary to tailor crop-specific monitoring programmes to crop phenology. For example, new infections of BYDV do not damage small grain cereal crops once these reach growth stage 31 (Doodson and Saunders, 1970). As potato crops mature, the chance of newly acquired virus reaching the tubers is reduced and vectors become less important (Beemster, 1972; Sigvald, 1986).

In general, the greater the frequency of sampling, the better. Inevitably, costs dictate limitations.

How Can Aphids be Monitored and Forecast?

Monitoring

Aphid monitoring remains, generally, a 'low tech' activity. Table 19.1 lists a range of methods commonly used for sampling aphids, together with some of the major considerations when assessing their suitability for the intended purpose of the sampling programme. Heathcote (1972) gives more detail on sampling, extraction, and counting techniques for evaluating populations on plants, and Taylor and Palmer (1972) provide the same for aerial sampling. Taylor and Palmer (1972) give an account of

Table 19.1. Advantages and disadvantages of sampling methodologies.

Methodology	Advantages (+) and disadvantages (–)
Crop sampling	
In situ counts on plants	Measure absolute abundance per plant or plant part
	+ Part of plant colonized can be recorded
	+ Can monitor progress on individual plants
	+ Suitable in a sequential sampling programme
	+ Provides immediate results
	– Time-consuming in the field
	– Difficult and sometimes uncomfortable
	– Results will be weather dependent
	– Results will depend on observer efficiency
	– Cannot adequately record non-colonizers
	– Problem with inaccessible aphids (between leaves, on roots. . .)
Destructive counts on plants	Measure absolute abundance per plant part
	+ Part of plant colonized can be recorded if bagged separately
	+ Quick in the field
	+ Exhaustive
	– Time-consuming in the laboratory, although counting and identification can be postponed to a convenient time provided that plants are stored at 3°C or lower to prevent further aphid development
Vacuuming from plants	Measures absolute abundance per unit area
	+ Quick in the field
	– Time-consuming in the laboratory, although counting and identification can be postponed to a convenient time
	– Tends to bias towards older/larger and winged aphids
	– Results will be weather dependent
	– Reliable quantification is difficult
Sweeping	Measures numbers per sweep
	+ Quick in the field
	+ Suitable for presence/absence sampling
	– Time-consuming in the laboratory, although counting/identification can be postponed to a convenient time
	– Results will be weather dependent
	– Appropriate quantification is virtually impossible, although it may be possible to compare different samples made under similar conditions in similar crop species
	– No use for tall plants
	– Only suitable to assess presence/absence
Beating	+ Quick in field and laboratory
	– Results will be weather dependent
	– Quantification is not possible
	– Only suitable to assess presence/absence

(Continued)

Table 19.1. Continued.

Methodology	Advantages (+) and disadvantages (−)

Aerial sampling

Impaction traps. Sample size depends on wind speed and shape of trap surface. Colour of the surface will affect the sample size in a species-specific manner.

Coloured water traps	Measure relative abundance and can be used to compare catches of a given species over space or time, assuming similar behaviour of different clones, but not to compare across species
	+ Many individuals caught if yellow used
	− Species bias depending on colour preferences
	− Aphids cannot be collected alive
	− Reflections may affect alighting behaviour
	− May dry out in hot weather unless emptied regularly (overflowing in wet weather can be prevented by using mesh drains on the sides)
Clear/plant-coloured water traps	Measure absolute numbers alighting per unit area
	+ Sample should be representative of individuals landing in crop
	+ Non-colonizing species can be recorded
	− Very few individuals caught compared to yellow traps
	− Reflections may affect alighting behaviour
	− Aphids cannot be collected alive
Sticky traps	Measure relative abundance
	+ Can be vertical, horizontal, laminar, or cylindrical
	− Dealing with samples is difficult
	− Samples may be in poor condition
	− Aphids cannot be collected alive
Sex pheromones	Measure relative local abundance of males
	+ Can be used in conjunction with many of the above methods
	+ Can select species
	− Synthetic pheromone is not available for many species

Filter traps. These traps collect air and filter aphids from it.

Nets (pivoted or non-pivoted)	Measure relative abundance of different species flying at a given height
	+ Aphids can easily be trapped alive and transferred immediately to plants to test for virus infectivity
	− The volume of air sampled depends on wind run, but can be controlled if vehicle mounted
	− Need constant attendance
Suction traps	Measure absolute abundance per unit volume of air
	+ Constant, known volumes of air can be sampled
	+ Efficiency is almost independent of wind speed
	+ Aphids can be collected alive or dead
	+ Samples can be automatically divided according to time
	− Require electricity
	− Sampling is usually at a single height
	Low level traps sample aphids that are not randomly distributed Aphids are randomly distributed above about 10 m

factors relating to the physical properties of traps and to aphid behaviour that affect interpretation of data from various aerial sampling systems. This account is still essential reading prior to expenditure on resources for aerial sampling.

Aphids must still be identified by eye. However, automated methods of identifying flying insects are progressing rapidly. It is possible, by using radar, to identify certain large insects to order and, occasionally, lower taxa (Smith *et al.*, 2000; Chapman *et al.*, 2002), but the equipment required to make advances with respect to smaller insects such as aphids is, currently, prohibitively expensive. Promising advances are being made using optical sensors that measure wing beat waveforms (Moore and Miller, 2002). These waveforms can be used as species-specific signatures for automated identification. In the laboratory, automated species identification involving, for example, immunological techniques or pattern analysis is, currently, impracticable when dealing with the large number of individuals sampled in most monitoring programmes. Perhaps the most significant recent advance is the ability to monitor the genetic structure of populations, and to screen individuals in this respect with ever-increasing speed (see Loxdale and Lushai, Chapter 2 this volume). Such molecular methods are particularly useful with respect to strategic research and are not yet used in more immediately applied programmes, except in relation to monitoring insecticide resistance status. Advances have also been made as a result of the phenomenal increase in computing power, which has facilitated great improvements in database management, data analysis, developing forecasting models, and information dissemination. Here we summarize, only briefly, techniques that have changed little since the reviews mentioned above and concentrate more on areas where subsequent progress has been evident.

Monitoring in crops

Direct assessment of aphid densities in crops must be made by researchers attempting to determine economic control thresholds and by agronomists and growers wishing to apply them. The use of such thresholds is fraught with difficulties and they provide, at best, rather crude guidelines. This is because they are inevitably based on work carried out under a limited set of conditions but applied under a much wider range of conditions. The aphid density at which control becomes justified is affected by biotic and abiotic factors, and by complex interactions between them. For example, susceptibility to damage will vary with crop growth stage, planting density, variety, and environmental stress. Although quantifying these effects is not inherently difficult, it is very laborious and only a narrow range of each variable is usually considered in published thresholds. In general, younger plants are more susceptible than older plants to yield loss from aphids, including that caused by viruses. However, damage to product quality or appearance may occur at specific growth stages, depending on the product. For example, the milling quality of grain can be greatly affected by *S. avenae* feeding on the ears of cereal crops. Crop variety and density will affect susceptibility to damage and the degree of compensatory growth that occurs. Damage will also vary with aphid species. Assessors who have had little training can identify species that look very different, but often different species look very similar. For example, agronomists examining sugarbeet crops in the UK for potential virus vectors are asked to look for 'green aphids', which may comprise *Myzus persicae* (peach–potato aphid), *Macrosiphum euphorbiae* (potato aphid), or *Aulacorthum solani* (glasshouse and potato aphid). It is also likely to be the case that there will be differences in the damage caused by different aphid genotypes, which cannot be identified during routine field counts. For example, only certain genotypes of *M. euphorbiae* contain the toxin in the saliva that causes false top roll in potatoes, and different genotypes of *S. avenae* vary in their ability to transmit BYDV (Dedryver *et al.*, 2005). Weather conditions subsequent to sampling cannot easily be predicted, but may have a profound effect on the need for control and the ability to carry it out.

Strategies for making control decisions almost inevitably involve *in situ* sampling. They must take into account the spatial distribution of aphids. This varies with time, often quite rapidly, because of reproduction, movement of apterae, dispersal of alatae, and mortality caused by, for example, natural enemies. Knowledge of the form of the frequency distribution of aphid counts, or of the form of the relationship between the mean and its variance, enables sequential sampling programmes to be devised whereby sampling continues only until a preset probability that the threshold for control will, or will not, be exceeded is reached, or until a fixed level of precision in determining the mean count is reached. A classical statistical text on this topic is that of Binns *et al.* (2000). Robert *et al.* (1988) provide a good summary of the theory and practice of sequential sampling relating specifically to aphids. Determination of a summary statistic to describe spatial pattern here requires random sampling of aphids, and the statistic does not require the spatial co-ordinates of the data points. Perry *et al.* (1999) devised a methodology for assessing the spatial distribution of organisms, which utilized two-dimensionally spatially referenced count data. Termed SADIE (Spatial Analysis of Distance Indices), the methodology takes direct account of behavioural mechanisms leading to distributions. It has been used to investigate the spatial distribution of *S. avenae* in winter wheat (Winder *et al.*, 1998). Such methods that describe spatial pattern help to inform choice of sampling scheme and thus provide considerable improvements in accuracy (Alexander *et al.*, 2005). They have the potential for use in precision spraying programmes, where only areas at particular risk are targeted.

Aerial monitoring

The methods listed in Table 19.1 for sampling flying aphids do not, on their own, provide any information on the distances flown by aphids. Indeed, it is still remarkably difficult to estimate this because most species of interest are very widespread and do not provide clues that help to distinguish between local and distant arrivals in trap samples (Loxdale *et al.*, 1993). Where discrete centres of aphid population are known, the distribution of migratory ranges of individuals can be estimated by trapping at different distances from the sources (Taylor *et al.*, 1979). Some evidence for long-distance migration has been obtained by placing traps a long way from the nearest possible host plants (e.g. Wiktelius, 1984). Molecular techniques may enable estimation of the rate of spread of clones and, in some cases, may hint at the origin of individuals. Loxdale and Lushai (Chapter 2 this volume) and Irwin *et al.* (Chapter 7 this volume) discuss these matters further.

MONITORING NETWORKS. Several trap networks have been permanently or occasionally operated in order to monitor and forecast aphid flight dynamics. The most extensive involves aerial monitoring using 12.2 m suction traps (design described by Macaulay *et al.*, 1988, see Fig. 19.1d) in Europe (Fig. 19.2). There are also long-running networks of 8 m suction traps in some states of the USA (Halbert *et al.*, 1990, 1998), and a few traps in other locations such as New Zealand (Teulon *et al.*, 2004). The first trap began operating in 1965 in Europe and in 1983 in the USA. During the 1980s, the European Union (EU) sponsored meetings under the title 'EURAPHID', reports from which show the variety of uses to which the European trap data have been put (Taylor, 1981; Bernard, 1982; Cavalloro, 1987, 1989). However, the data were stored independently and in different formats by different individuals. In 2000, an EU Thematic Network ('EXAMINE') was set up in order to bring the data together in a single, integrated database to facilitate Europe-wide analyses (Harrington *et al.*, 2004). In 2005, 72 traps were operated in 19 countries (Fig. 19.2). Fifteen sites have a data run of at least 30 years and a further 22 sites of at least 20 years. In total, by the end of 2005, more than 1800 site-years of data were available. Most traps are emptied daily during the aphid season. For some, all aphids are identified to species, species group, or genus. At other sites, only a few species are identified.

Fig. 19.1. Examples of methods for trapping flying aphids: (a) yellow water pan traps (courtesy of Broom's Barn Research Station); (b) sticky wire trap; (c) 'mist net'; (d) suction trap (b–d courtesy of Rothamsted Research).

The area represented by a trap varies according to topography, but Taylor (1974) found that there was an extremely strong correlation between the composition of samples in 12.2 m traps 80 km apart, and still a significant correlation in traps nearly 400 km apart. Cocu *et al.* (2005a) found abundance of *M. persicae* in northwest Europe to be correlated between sites over distances up to 700 km. In agronomically and topographically homogeneous regions, the correlation between traps is linearly related to the distance between them (Hullé and Gamon, 1989). Halbert *et al.* (1998) showed that collections from the 8 m traps in Idaho reflected aphid activity within a 30 km radius of the traps.

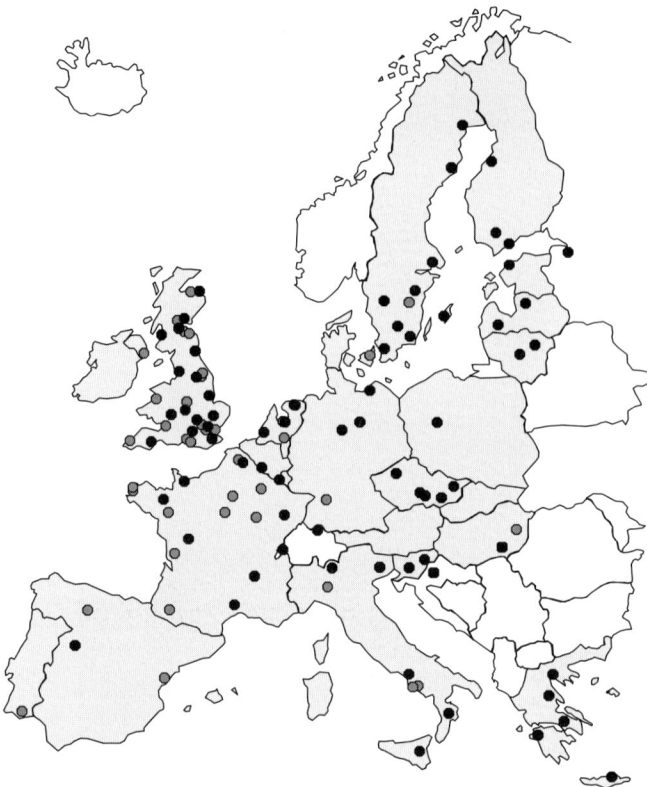

● Trap still operated
◐ Trap no longer operated
Shaded countries are EU members

Fig. 19.2. Sites of 12.2 m suction traps 2005.

Monitoring aphid attributes

MORPH. Aerial sampling, of course, deals only with winged aphids, whereas crop sampling can be used to gather information on winged and wingless aphids. Wingless aphids may be fundatrices, other virginoparae, males, oviparae, or sexuparae. The first four forms are generally morphologically easily distinguishable. Sexuparae are usually morphologically similar to virginoparae but, if trapped alive, may be distinguished on the basis of their morphologically distinctive offspring. Winged aphids may be virginoparae, gynoparae, or males. Males generally are easy to distinguish from the two female forms by their heavily pigmented genitalia. They are also often morphologically distinct from conspecific females in other ways. For example, they tend to have thinner abdomens and more rhinaria on their antennae. However, gynoparae and virginoparae are usually very difficult to distinguish morphologically. This is an important practical issue when, for example, monitoring *Rhopalosiphum padi* (bird cherry–oat aphid) in order to assess the risk of BYDV. Whilst all three winged forms are capable of introducing the virus to an autumn-sown cereal crop, only the virginoparae will colonize the crop and produce offspring that may spread the virus. BYDV risk assessment programmes that use information on the aerial abundance of vectors thus require a quick, cheap, and reliable method for distinguishing virginoparae from gynoparae in regions where both forms are known to be present. Morphometric analysis (Simon *et al.*, 1991; Taylor *et al.*, 1994) and microscopic observation of embryos (Hardie and Tatchell, 1989) are extremely time-consuming and impracticable as a routine when many aphids must be tested. Trapping aphids alive, offering them a choice of the primary or secondary host plant, and noting on which they reproduce (Tatchell and Parker, 1990), is also time-consuming and causes a delay in provision of data on aphid morph. Lowles (1995) found that, if embryos of *R. padi* are teased out into alcohol, those of gynoparae are yellow-green and those of virginoparae are red-brown. This test is cheap, rapid, and reliable but requires living, or very recently killed, aphids and is not suitable for specimens stored in alcohol, or for other species so far tested, including the closely related *Rhopalosiphum insertum* (apple–grass aphid).

INSECTICIDE RESISTANCE. Routine monitoring of the insecticide resistance status of aphids can provide valuable information for control programmes involving insecticides (Foster *et al.*, 1998, 2002, Chapter 10 this volume; Foster, 2000). The esterases responsible for conferring the most prevalent resistance mechanism in *M. persicae* can be detected by immunoassay (Devonshire *et al.*, 1986) using aphids trapped alive or trapped and stored for up to 2 weeks in a purpose-designed, glycerol-based solution (Tatchell *et al.*, 1988). Aphids from this solution can also be tested for target-site ('MACE') (see Foster *et al.*, Chapter 10 this volume) resistance using the method of Moores *et al.* (1994). Point mutations in the sodium channel that confer knockdown resistance to pyrethroids and DDT (Martinez-Torres *et al.*, 1999) can be detected from live or alcohol-preserved specimens by using real-time fluorescence PCR.

VIRUSES IN APHIDS. Persistent viruses can readily be detected routinely in individual aphids using double antibody sandwich enzyme-linked immunosorbent assay (DAS-ELISA). This is much quicker and more sensitive than the previously used infectivity and electron microscopy assays, and was first achieved by Tamada and Harrison (1981) to detect *Potato leaf roll virus* (PLRV)

in *M. persicae* and *M. euphorbiae*. Paliwal (1982) used serologically-specific electron microscopy to detect BYDV in *S. avenae*. Torrance (1987) was the first to use enzyme amplification to increase the sensitivity of ELISA in detecting BYDV in individual aphids. Smith *et al.* (1991) used the same method to distinguish between the closely related *Beet mild yellowing virus* and *Beet western yellows virus* in *M. persicae* and *M. euphorbiae*. Detection of persistent viruses can be improved by the use of molecular techniques such as standard- or real-time reverse transcription-polymerase chain reaction (RT-PCR) (Canning *et al.*, 1996; Singh *et al.*, 1996; Stevens *et al.*, 1997; Fabre *et al.*, 2003b).

Non-persistent viruses, because of the very small titre of virus in the aphids, present more of a challenge than persistent viruses when it comes to detection in individual aphids, even though the first example of the use of ELISA to detect virus in aphids was for the non-persistent *Cucumber mosaic virus* (CMV) in *Aphis gossypii* (cotton or melon aphid) (Gera *et al.*, 1978). However, this and subsequent attempts to use ELISA for non-persistent viruses have met with far more limited success than for persistent viruses. Molecular techniques appear more promising. López-Moya *et al.* (1992) detected the non-persistent DNA virus, *Cauliflower mosaic virus*, in *M. persicae* using PCR. Singh *et al.* (1996) used duplex RT-PCR and a southern blot analysis to detect the RNA virus, *Potato virus Y* (PVY), in individual aphids. There was no difference in the efficiency of virus detection in aphids whether they were tested immediately or had been stored for up to 45 days in ethanol at room temperature. However, virus was not detected in aphids that had been stored in water with 0.002% detergent for 1–3 days, a problem that does not seem to occur when trying to detect persistent viruses. Olmos *et al.* (1997) described a print-capture (PC)-hemi-nested-PCR technique, involving two PCR rounds, which successfully detected a range of isolates of *Plum pox virus* in *A. gossypii*. Varveri (2000) used a combined immunological and molecular technique (PC-RT-PCR-ELISA) to detect PVY in individual aphids. Detection of the virus by this method closely matched the transmission efficiency of a parallel batch of aphids. This is the only method described so far for non-persistent viruses that may be quick and reliable enough for routinely testing large numbers of aphids.

The presence of a virus in an aphid does not necessarily mean that the aphid is capable of transmitting that virus (Plumb, 1989; Barker and Torrance, 1990; Smith *et al.*, 1991). The only way to get round this problem, essential in epidemiological studies, is to revert to time-consuming bioassays, or to ensure that work has been done to relate numbers that are viruliferous to numbers that are infective so that an appropriate conversion can be made.

Databases and data handling

The monitoring networks provide a large amount of data. In the case of EXAMINE, more than 25 million aphids have been recorded and the database comprises more that 3000 megabytes. The EXAMINE consortium is keen that these data are used as widely as possible, which requires that the data are processed to ensure easy handling and use. This became possible from the 1990s with the Client–Server architecture, which brings together the power of the database and the universality of the World-wide Web (WWW). The EXAMINE database is now interfaced to an Internet Information Server (www.rothamsted.bbsrc.ac.uk/examine/), and the clients, i.e. the EXAMINE data users, can access the database *via* the WWW on acceptance of certain conditions. Unrestricted access to the database can cause problems because the data are difficult to interpret by people not involved in the data collection, and errors can result.

Recent progress in data handling and dissemination through the WWW has enlarged use of the information by the farming community. For instance, weekly bulletins from the Scottish Agricultural Science Agency provide information on the abundance of known vectors of PVY and PLRV, as well as appropriate commentaries and

comparisons with previous years (www. sasa.gov.uk/seed_potatoes/aphids/ bulletins/index.cfm). Agronomists in Scotland use this information to provide advice to growers. More complicated systems, based on statistical or simulation models, constitute decision support systems (see Knight and Thackray, Chapter 31 this volume).

Forecasting

Two categories of forecasting tool have been developed. First, large data sets have been used to develop phenomenological models based on a purely statistical approach. Second, process-based simulation models have been produced by making use of accumulated scientific knowledge on the biology of the target species. Examples of forecasting models are shown in Table 19.2 and some of them are described briefly below. Kindlmann *et al.* (Chapter 12 this volume) provide a critique of process-based models against a background of population dynamics theory.

Phenomenological models

Phenomenological models are easy to build from a technical point of view, but require a huge data set. The most common approach is based on regression analysis, relating aspects of aphid phenology or abundance to environmental variables, often climatic. This implies that the predicted variable is a simple scalar variable and that the relationship between the predicted variable and the predictors is linear. Non-linearity between predicted variables and predictors can be explored by the use of artificial neural networks (for example, Lankin *et al.*, 2001; Cocu *et al.*, 2005c). In order to have confidence in such models for forecasting, it is important that the statistical relationships between predicted and explanatory variables can be explained, at least in qualitative terms. If the model fails, it may be difficult to pinpoint why and hence revise it. For example, Pierre and Dedryver (1985) have shown that the regression model they developed from data collected in Brittany,

Table 19.2. Modelling approaches for aphid forecasting.

Methodology	Example
Phenomenological models	
Linear regressions	Prediction of peak density
	Pierre and Dedryver, 1984, 1985; Entwistle and Dixon, 1986, 1987; Latteur and Oger, 1987; Dedryver *et al.*, 1987; Parker, 1997; Thacker *et al.*, 1997
	Prediction of migration phenology
	Turl, 1980; Way *et al.*, 1981; Wiktelius, 1982; Walters and Dewar, 1986; Harrington *et al.*, 1990; Worner *et al.*, 1995; Rispe *et al.*, 1998
Linear mixed model	Predicting impacts of environmental change on phenology
	Harrington *et al.*, 2006
Artificial neural networks	Predicting impacts of environmental change on abundance
	Cocu *et al.*, 2005c
Multivariate descriptive analysis	Study of patterns of flight phenology
	Hullé *et al.*, 1994; Masterman *et al.*, 1996
Time series analysis	Search for density-dependent effects
	Woiwod and Hanski, 1992
Process-based models	
Ordinary differential equations	Using a global rate of increase to predict aphid population dynamics
	Derron and Forrer, 1989; Bommarco and Ekbom, 1995; Onstad *et al.*, 2005
	Detailed models incorporating the main biological traits to predict aphid population dynamics
	Carter and Rabbinge, 1980; Carter, 1985, 1987; Rossberg *et al.*, 1986; Zhou *et al.*, 1989; Zhou and Carter, 1990; Mann and Wratten, 1991; Morgan, 2000
	Forecasting of aphid-transmitted virus diseases
	Sigvald, 1986; Morgan, 1990; Kendall *et al.*, 1992; Nemecek *et al.*, 1995; Werker *et al.*, 1998; Leclercq-Le Quillec *et al.*, 2000; Thackray *et al.*, 2004
Partial derivative equations	Detailed models incorporating aphid age-structure to predict field population dynamics
	Plantegenest *et al.*, 1994, 1997, 2001
Combination of approaches	
Statistical + process-based model	Forecasting of aphid-transmitted virus diseases
	Northing *et al.*, 2004
Bayesian approach	Forecasting of aphid-transmitted virus diseases
	Fabre *et al.*, 2003a, 2006

France, was not able to predict accurately aphid densities in the Parisian basin, probably because of differences in the biology of the aphids and their natural enemies in the two areas. Worner *et al.* (1995) found similar disparities when using predictive sample re-use and day-degree methodologies to forecast the time of 50% migration of *Phorodon humuli* (damson–hop aphid) at two sites in the UK. Thacker *et al.* (1997) have also shown that, although linear regressions can describe the existing data well, they are not always good at predicting future patterns. However, as variables change, provided that the monitoring system is still operational, forecast equations can be updated to take account of the change.

To be of potential value in statistical forecasting models, data must have been collected using a standardized methodology for a long period of time (for example, A'Brook (1983) considers 13 years insufficient). Twenty years is, perhaps, about the minimum period necessary to reduce adequately the probability that explanatory variables – in a year in which aphid populations are to be forecast – will be sufficiently outside the range used to construct the forecast as to render it unreliable. Unfortunately, there are very few aphid data sets that meet both these criteria of standardization and duration, the suction trap networks being notable exceptions. Some short-term data have shown strong correlations, but these have not been tested over a long time period. For example, the timing of peak aphid populations in potato crops has been related to temperature in January and February (Parker, 1997). Peak density of *S. avenae* on wheat has been related to aphid density and rate of increase (Entwistle and Dixon, 1986, 1987), to aphid density, percentage mummified and number of syrphids at flowering (Latteur and Oger, 1987), and to aphid density at ear emergence, temperature in February and number of rainy days in May (Pierre and Dedryver, 1984, 1985; Dedryver *et al.*, 1987). Thacker *et al.* (1997) used the same methodology to predict the summer population density of *Aphis fabae* (black bean aphid), *M. persicae*, and *Acyrthosiphon pisum* (pea aphid).

One of the earliest successful uses of long-term aphid data in forecasting was for *A. fabae* on spring-sown field beans in southern England (Way *et al.*, 1981). Numbers of *A. fabae* in suction trap samples in spring gave the most accurate estimate of the size and timing of crop infestation, but autumn trap samples, numbers of eggs, and numbers of fundatrigeniae provided useful, though less accurate, earlier warnings. Egg counts have been used to forecast outbreaks of *R. padi* in areas where it is largely dependent on the egg stage for successful overwintering (Leather, 1983; Kurppa, 1989).

The phenology of primary host-plant flowering, and temperature, has been shown to be very strongly correlated with migration phenology of *P. humuli* in England (Worner *et al.*, 1995). However, the model used temperature right up to the migration event, so prediction of migration phenology was not early enough to be useful in planning control programmes. Slightly less accurate predictions using temperature alone were used by Thomas *et al.* (1983) for the start and end of *P. humuli* migrations into hops far enough ahead to be used to recommend when control should begin and end, without the need for laborious crop inspection.

Strong relationships have been found between winter or early spring temperature and the timing and size of spring migrations of aphids, as recorded by suction traps, particularly for species that are largely anholocyclic in the region of interest (Turl, 1980; Wiktelius, 1982; Walters and Dewar, 1986; Harrington *et al.*, 1990; Fig. 19.3). Such relationships can be particularly useful on a regional basis in warning of the likely timing and abundance of aphids with respect to crop colonization. Strong relationships have been found, for example, between numbers of *R. padi* in suction traps and numbers alighting on cereal crops in autumn (Fig. 19.4).

Pierre *et al.* (1986) developed a general statistical method for searching for critical climatic periods that are related to the timing of seasonal events in the annual cycle of aphids. Rispe *et al.* (1998) applied this method to the study of the autumn flight of *R. padi*. Software called 'Criticor' has been developed by these authors and can be used on the web (http://www.rennes.inra.fr/bio3p/criticor_accueil.htm). The Windowpane software (Calvero *et al.*, 1994) serves the same purpose.

Other statistical approaches have been used less frequently to produce phenomenological models, such as multivariate descriptive analysis for the study of global patterns of flight phenology (Hullé *et al.*, 1994; Masterman *et al.*, 1996), or time series analysis in search of density-dependent effects (Woiwod and Hanski, 1992). A linear mixed model using Residual Maximum Likelihood (REML) has been used by Harrington

(a) *Myzus persicae* at Rothamsted 1965–2005

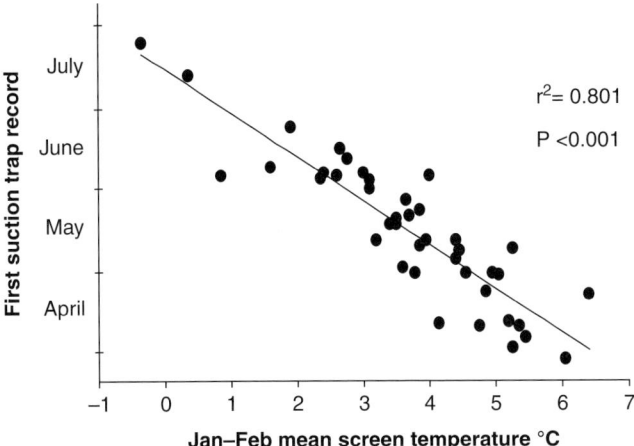

$r^2 = 0.801$

$P < 0.001$

(b) *Myzus persicae* at Rothamsted 1965–2005

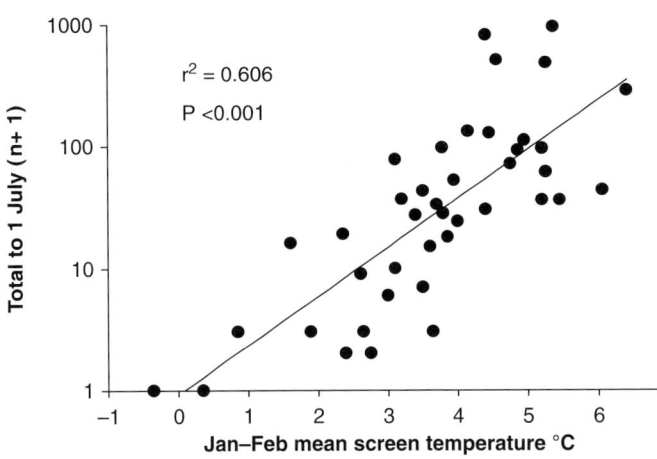

$r^2 = 0.606$

$P < 0.001$

Fig. 19.3. Relationship between mean January and February temperature at Rothamsted, UK, and (a) time of first record and (b) abundance to 1 July of *Myzus persicae* in the Rothamsted suction trap.

et al. (2007) to predict the impact of environmental changes on the flight phenology of several aphid species throughout Europe.

Process-based models

Mechanistic models generally attempt to simulate the entire aphid dynamics. They allow short-term forecasting, but can also be used to evaluate the *a priori* risk level through the study of particular scenarios (e.g. climatic scenarios), or to improve aphid management strategies (decision tools). They have the advantage that they can potentially be built to account quantitatively for impacts of specific, individual variables and interactions between them. However, in practice, because of the huge array of complex interactions within and between the biotic and abiotic components of ecosystems, it may be impossible to carry out the necessary experimentation over a sufficient range for all the variables to parameterize the model adequately. Furthermore, it is not easy to predict the most influential primary and lower order interactions and hence to simplify the model. As variables (e.g. crop variety) change, models may need to be parameterized afresh, and resources are often not available to achieve this.

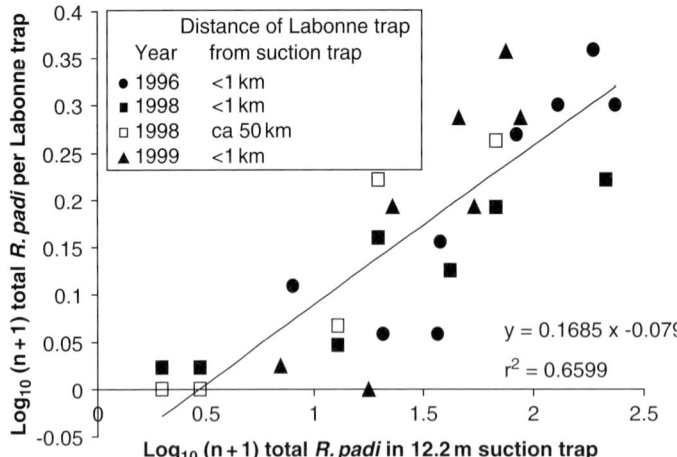

Fig. 19.4. Correlation between abundance of *Rhopalosiphum padi* in a 12.2 m suction trap and on sticky wire (Labonne) traps. Each point represents aphid totals for 7 days. The Labonne trap figures are means of 18 traps, except for 1998 when they are means of 12 traps at <1 km from the suction trap and of 6 traps at *c.* 50 km from the suction trap.

Cereal aphids have been the subject of the greatest effort with respect to producing forecasts. Carter (1994) reviewed the history and various approaches for modelling these species, some of which are re-iterated below.

Rossing (1991b) carried out a comparative study of the accuracy of the prediction of the yield losses caused to wheat by *S. avenae* obtained by the models of Vereijken (1979), Rabbinge and Mantel (1981), Entwistle and Dixon (1987), and Rossing (1991a). He concluded that models that take into account the duration and intensity of aphid infestation provide much more reliable predictions than those that simply characterize population dynamics by peak density.

Various modelling strategies have been used, and they can be classified on the basis of their relative complexity. The simplest approach is to use accurate degree-day summation to detect the timing of an event. Bommarco and Ekbom (1995), for instance, applied such a method to the determination of the date of occurrence of peak density of *A. pisum* on peas. More complex models can be designed in order to simulate entire population dynamics. They are based either on the use of ordinary differential equations (ODE), or on the use of difference equations. The simplest form of such equations considers only a global rate of increase of the population that can be put under the dependence of various environmental variables.

Such a model has been applied, among others, to the simulation of the population dynamics of *S. avenae* by Derron and Forrer (1989), of *A. pisum* by Bommarco and Ekbom (1995), and of *Aphis glycines* (soybean aphid) on soybean by Onstad *et al.* (2005). The next step in complexity is to build detailed models incorporating explicitly the main biological traits (fecundity, mortality, development, morphogenesis, etc.) of the target species. Several attempts have been devoted to the simulation of the population dynamics of various species of cereal aphid. ODE models have been designed for forecasting the population dynamics of *S. avenae* in the UK and The Netherlands (Carter and Rabbinge, 1980; Carter, 1985, 1987; Mann and Wratten, 1991) and in Germany (Rossberg *et al.*, 1986), the latter being integrated with a wheat growth model so that interactions between the plants and the aphids can be explored. Holz (1991) used this last model to build a simplified system for pest management. Similarly, models have been designed for *Metopolophium dirhodum* (rose–grain aphid) in spring and summer (Zhou *et al.*, 1989; Zhou and Carter, 1990) and *R. padi* in autumn (Morgan, 2000).

In many cases, the variable of agronomic interest is the incidence of an aphid-transmitted virus, adding a further complication to an already difficult modelling problem. Simulation models have been designed for forecasting, for example, the incidence in sugarbeet of viruses causing yellows symptoms transmitted mainly by *M. persicae* (Werker *et al.*, 1998; Qi *et al.*, 2004; see 'Uptake and outcomes' below), BYDV transmitted by *R. padi* (Morgan, 1990; Kendall *et al.*, 1992; Leclercq-Le Quillec *et al.*, 2000), PVY transmitted by a range of vectors (Sigvald, 1986), and CMV transmitted to lupin crops by numerous vectors (Thackray *et al.*, 2004). A system developed by Nemecek *et al.* (1995) coupled a crop growth model to an epidemic model to forecast virus infection and yield in seed potato crops. Although these models have not always been successful or stood the test of time, they demonstrate the principles involved and provide a base on which to build.

Models based on partial derivative equations (PDE) have also been used to describe population dynamics not only in terms of densities, but also in terms of age structure. This approach was developed for simulation of the population dynamics of *S. avenae* on wheat in spring and summer (Plantegenest *et al.*, 1994, 1997, 2001). As stressed by Morgan (2000) and Plantegenest *et al.* (2001), the factors that are often most difficult to account for in predictive simulation are those influencing aphid mortality, especially that caused by the activity of natural enemies. A model by Ro and Long (1999) successfully simulated the phenology of *M. persicae* in potato on the basis of the interactions between the aphid, its predators, and temperature.

The main weakness of classical process-based models for decision-making purposes is that they do not consider the wide uncertainty associated with ecological data and processes. Bayesian probability theory provides an accurate general framework to take into account explicitly various sources of uncertainty. The recent development of efficient algorithms and software has resulted in a rapid increase of the use of Bayesian approaches in ecology. For example, they have been applied to the assessment of BYDV risk in cereal fields based on the population dynamics of its main vector in Western Europe, *R. padi* (Fabre *et al.*, 2003a, 2006).

Uptake and outcomes

Numerous industry bulletins report results from monitoring and forecasting programmes, but there are few published assessments of their impact on farming practice. Examples of thresholds used for aphids in different crops are given in the IPM Case Studies in Chapters 21–30 this volume. In the case of UK sugarbeet, the whole crop is grown under contract to British Sugar plc, which employs agronomists to advise its growers. The agronomists are provided with pre-season forecasts of the incidence of aphids and aphid-transmitted viruses causing sugarbeet yellows (Qi *et al.*, 2004). The forecasts are issued in early March (e.g. Stevens and Dewar, 2004) and are based on the relationships between virus incidence and winter temperature, the timing and size of the spring aphid migration as recorded by suction traps, crop emergence date, and the use of insecticides, including seed treatments (Qi *et al.*, 2004). The widespread use of imidacloprid-treated seed (74% in 2005 in the UK) has reduced the crop area to which the forecasts are relevant and, even though it is now possible to provide forecasts by the middle of February, this cannot influence the amount of seed that is treated as, for practical reasons, the treatment process is completed by the previous November. Use of imidacloprid-treated seed is only economically justifiable if virus incidence is expected to be at least 12% (Dewar *et al.*, 2001). Running the model of Qi *et al.* (2004) assuming no seed treatment shows that, with an average temperature of 4°C from January to mid-February, virus yellows incidence would remain well below the 12% threshold, even if only 25% of fields were treated. The forecast is supplemented by season-long information on the incidence of vector aphids in suction traps and yellow pan water traps (Smith *et al.*, 1997), their insecticide resistance status, and the viruses they are carrying. This information helps growers who have not used treated seed to decide on the value of a spray programme. As a result of the information provided and resulting action by growers, incidence of virus was reduced by over 50%

between 1984 and 1999, giving a net annual benefit of about 1.7% (worth approximately £5.5 million) (Dewar and Smith, 1999).

Other examples of the use of suction trap data in monitoring and forecasting for crop protection include tracking the range expansion of *Diuraphis noxia* (Russian wheat aphid) and alerting growers to potentially damaging infestations (Halbert *et al.*, 1998; Lukásová *et al.*, 1999). Numbers in suction traps are correlated very strongly with numbers in crops. Moreover, *D. noxia* are found in suction traps before economically significant populations develop in spring-sown cereal crops. Peak population density of *M. dirhodum* has been forecast using the strong relationships between numbers in suction traps and in fields up to 80 km away, and between winter and spring temperature and the time that the crop becomes unsuitable for the aphid (Howard and Dixon, 1990).

A system is being developed to support decisions on the need to control the aphid vectors of BYDV in autumn-sown cereal crops in the UK. This attempts to make forecasts field-specific and combines a regional statistical model with a process-based stochastic simulation model (Northing *et al.*, 2004). Suction traps are used to monitor the arrival of vector aphid species. On the basis of: (i) relationships between numbers in the traps and numbers in the field; (ii) estimates of the proportion of aphids carrying the virus; and (iii) the proportion of those aphids that will inoculate the virus, numbers of vectors in suction traps are converted to numbers of foci of infection per unit area of crop. A simulation model takes these foci and spreads them out using algorithms describing the relationship between weather (mainly temperature) and aphid development, reproduction, survival, and movement. The effects of temperature on virus acquisition, inoculation, and latent period in the aphid and plant are also included. This regional estimate of virus spread is made field-specific on the basis of an extensive survey of BYDV incidence in unsprayed crops and relationships found between incidence and a range of field characteristics (Foster *et al.*, 2004).

Forecasting the impacts of environmental changes

Statistical and processed-based modelling approaches have been used to assess the potential impacts of environmental changes on aphid dynamics. Statistical approaches range from simple linear regressions of mean temperature at single sites (Harrington *et al.*, 1990) to pan-European models considering many meteorological and land-use variables (Cocu *et al.*, 2005a,b,c; Harrington *et al.*, 2006). These models all suggest that most aphids will begin flying earlier in the season than at present and be more abundant early in the season, especially those that are largely anholocyclic, although the rate that phenology advances with time varies with species, location, model, and change scenario. Total annual abundance of alatae is not expected to change greatly, partly because large numbers early in the season tend to result in increased numbers of natural enemies to keep later populations lower than they otherwise would be.

Manipulative experiments examining the potential impact of elevated CO_2 levels on aphid abundance have suggested a range of responses. Bezemer *et al.* (1999) found the complete range of responses (increase, decrease, no change) in *M. persicae* and *Brevicoryne brassicae* (cabbage aphid) depending on aphid and host plant combination and duration of the experiments. They review the results of other work that, collectively, shows a great range of responses.

Newman (2005) produced a mechanistic model that predicted a decline in the abundance of *R. padi* in the face of increasing CO_2 emissions and attempted to explain the various responses of aphid species by considering their requirements for nitrogen and their density-dependent response in alata production.

The results described above illustrate that no one methodology is a panacea. The multivariate statistical approaches account implicitly for variability in all relevant conditions experienced over the period of data collection, but involve extrapolation to new combinations of values for the variables. The manipulative experiments and the mechanistic model take account of only a very limited range of environmental and biological variation. All provide important pieces in a jigsaw that will take some time to complete.

Conclusions

Much effort, building on fundamental and strategic studies, has gone into monitoring and forecasting programmes dealing with various aphid and crop species around the world. For many reasons, few of these programmes have, as yet, had a sustained impact. This is sometimes because technology transfer has not been planned properly or implemented, or because the scientists involved in development have moved on to other problems, and resources to update systems are not available. This is especially true in the case of process-based approaches, which are often used in specific short-term projects aimed at one problem. As Kindlmann *et al.* (Chapter 12 this volume) point out, particular problems arise because of the lack of standardized, long-term population data

from crops and because of the limitations imposed by the need to predict weather. In the case of statistical approaches using data sets of adequate length, some stability is possible as a result of the wide range of strategic and applied applications to which the data can be put. This does not make such approaches inherently superior, and the desirability of complementary approaches has already been discussed. However, the reality is that for any forecasting system to be adopted widely, it must be proven to have economic benefits over space and time. In relation to climate change issues, the importance of a long-term outlook is further emphasized.

Acknowledgements

We thank Jon Pickup and Joe Perry for useful information, and Ian Barker and Lesley Torrance for help in compiling references. Rothamsted Research receives grant-aided support from the Biotechnology and Biological Sciences Research Council of the UK.

References

A'Brook, J. (1983) Forecasting the incidence of aphids using weather data. *EPPO Bulletin* 13, 229–233.

Alexander, C.J., Holland, J.M., Winder, L., Wooley, C. and Perry, J.N. (2005) Performance of sampling strategies in the presence of known spatial patterns. *Annals of Applied Biology* 146, 361–370.

Barker, I. and Torrance, L. (1990) The relationship between barley yellow dwarf virus content in aphids and their ability to transmit. In: Burnett, P. (ed.) *World Perspectives on Barley Yellow Dwarf.* CIMMYT, Mexico, pp. 166–168.

Beemster, A.B.R. (1972) Virus translocation in potato plants and mature plant resistance. In: de Bokx, J.A. (ed.) *Potatoes and Seed-Potato Production.* PUDOC, Wageningen, pp. 144–151.

Bernard, J. (ed.) (1982) *Utilisation du piège a succion en vue de prévoir les invasions aphidiennes.* CEC, Luxembourg, 91 pp.

Bezemer, T.M., Knight, K.J., Newington, J.E. and Jones, T.H. (1999) How general are aphid responses to elevated atmospheric CO_2? *Annals of the Entomological Society of America* 92, 724–730.

Binns, M.R., Nyrop, J.P. and van der Werf, W. (2000) *Sampling and Monitoring in Crop Protection: The Theoretical Basis of Designing Practical Decision Guides.* CAB International, Wallingford, 296 pp.

Bommarco, R. and Ekbom, B. (1995) Phenology and prediction of pea aphid infestations on peas. *International Journal of Pest Management* 41, 109–113.

Calvero, S.B., Coakley, S.M., McDaniel, L.R. and Teng, P.S. (1994) A weather factor searching program for plant pathological studies: Windowpane Version W1B00003. *IRRI Discussion Paper Series, No. 5.* IRRI, Manila, 43 pp.

Canning, E.S.G., Penrose, M.J., Barker, I. and Coates, D. (1996) Improved detection of barley yellow dwarf virus in single aphids using RT-PCR. *Journal of Virological Methods* 56, 191–197.

Carter, N. (1985) Simulation modelling of the population dynamics of cereal aphids. *Biosystems* 18, 111–119.

Carter, N. (1987) Monitoring, forecasting and control of cereal aphids: some recent developments in England. *Proceedings of the 1st International Conference on Pests in Agriculture, Paris, December 1987*, 105–111.

Carter, N. (1994) Cereal aphid modelling through the ages. In: Leather, S.R., Watt, A.D., Mills, N.J. and Walters, K.F.A. (eds) *Individuals, Populations and Patterns in Ecology*. Intercept, Andover, pp. 129–138.

Carter, N. and Rabbinge, R. (1980) Simulation models of the population development of *Sitobion avenae*. *Bulletin OILB/SROP* 3, 93–98.

Cavalloro, R. (ed.) (1987) *Aphid Migration and Forecasting 'Euraphid' Systems in European Community Countries*. CEC, Luxembourg, 264 pp.

Cavalloro, R. (ed.) (1989) *'Euraphid' Network: Trapping and Aphid Prognosis*. CEC, Luxembourg, 331 pp.

Chapman, J.W., Smith, A.D., Woiwod, I.P., Reynolds, D.R. and Riley, J.R. (2002) Development of vertical-looking radar technology for monitoring insect migration. *Computers and Electronics in Agriculture* 35, 95–110.

Cocu, N., Harrington, R., Hullé, M. and Rounsevell, M.D.A. (2005a) Spatial autocorrelation as a tool for identifying the geographical patterns of aphid annual abundance. *Agricultural and Forest Entomology* 7, 31–43.

Cocu, N., Harrington, R. and Rounsevell, M.D.A. (2005b) Geographic location, climate and land use influences on the distribution and abundance of the aphid *Myzus persicae* in Europe. *Journal of Biogeography* 32, 615–632.

Cocu, N., Harrington, R., Rounsevell, M.D.A., Worner, S.P., Hullé, M. and the EXAMINE project participants (2005c) Geographical location, climate and land use influences on the phenology and numbers of the aphid, *Myzus persicae*, in Europe. *Journal of Biogeography* 32, 615–632.

Dedryver, C.-A., Fougeroux, A., De La Messeliere, C., Pierre, J.S. and Taupin, P. (1987) Typologie des courbes de fluctuation des populations de pucerons des céréales sur le blé d'hiver dans le Bassin Parisien et élaboration d'un modèle de prévision des pullulations de *Sitobion avenae*. *Annales de l'Association Nationale de Protection des Plantes* 6, 93–104.

Dedryver, C.-A., Riault, G., Tanguy, S., Le Gallic, J.F., Trottet, M. and Jacquot, E. (2005) Intra-specific variation and inheritance of BYDV-PAV transmission in the aphid *Sitobion avenae*. *European Journal of Plant Pathology* 111, 341–354.

Derron, J.O. and Forrer, H.R. (1989) Opportunité des traitements contre les pucerons des céréales: un programme d'aide à la décision sur Videotex. *Revue Suisse d'Agriculture* 21, 133–136.

Devonshire, A.L., Moores, G.D. and ffrench-Constant, R.H. (1986) Detection of insecticide resistance by immunological estimation of carboxylesterase activity in *Myzus persicae* (Sulzer) and cross reaction of the antiserum with *Phorodon humuli* (Schrank) (Hemiptera: Aphididae). *Bulletin of Entomological Research* 76, 97–106.

Dewar, A.M. and Smith, H.G. (1999) Forty years of forecasting virus yellows incidence in sugar beet. In: Smith, H.G. and Barker, H. (eds) *The Luteoviridae*. CAB International, Wallingford, pp. 229–243.

Dewar, A.M., Qi, A., Werker, A.R. and Harrington, R. (2001) Virus yellows forecasting in sugar beet and the impact of Gaucho. *British Sugar Beet Review* 69 (1), 36–39.

Doodson, J.K. and Saunders, P.J.W. (1970) Some effects of barley yellow dwarf virus on spring and winter cereals in field trials. *Annals of Applied Biology* 66, 361–374.

Entwistle, J.C. and Dixon, A.F.G. (1986) Short-term forecasting of peak population density of the grain aphid (*Sitobion avenae*) in summer. *Annals of Applied Biology* 109, 215–222.

Entwistle, J.C. and Dixon, A.F.G. (1987) Short-term forecasting of wheat yield loss caused by the grain aphid (*Sitobion avenae*) in summer. *Annals of Applied Biology* 111, 489–508.

Fabre, F., Dedryver, C.A., Leterrier, J.L. and Plantegenest, M. (2003a) Aphid abundance on cereals in autumn predicts yield losses caused by Barley yellow dwarf virus. *Phytopathology* 93, 1217–1222.

Fabre, F., Kervarrec, C., Mieuzet, L., Riault, G., Vialatte, A. and Jacquot, E. (2003b) Improvement of *Barley yellow dwarf virus*-PAV detection in single aphids using a fluorescent real time RT-PCR. *Journal of Virological Methods* 110, 51–60.

Fabre, F., Pierre, J.S., Dedryver, C.A. and Plantegenest, M. (2006) Barley yellow dwarf disease risk assessment based on Bayesian modelling of aphid population dynamics. *Ecological Modelling* 193, 457–466.

Foster, G.N., Blake, S., Tones, S.J., Barker, I. and Harrington, R. (2004) Occurrence of barley yellow dwarf virus in autumn-sown cereal crops in the United Kingdom in relation to field characteristics. *Pest Management Science* 60, 113–125.

Foster, S.P. (2000) Knock-down resistance (kdr) to pyrethroids in peach–potato aphids (*Myzus persicae*) in the UK: a cloud with a silver lining? *Proceedings of the British Crop Protection Conference, Pests and Diseases, Brighton, November 2000* 1, 465–472.

Foster, S.P., Denholm, I., Harling, Z.K., Moores, G.D. and Devonshire, A.L. (1998) Intensification of insecticide resistance in UK field populations of the peach–potato aphid, *Myzus persicae* (Hemiptera: Aphididae) in 1966. *Bulletin of Entomological Research* 88, 127–130.

Foster, S.P., Harrington, R., Dewar, A.M., Denholm, I. and Devonshire, A.L. (2002) Temporal and spatial dynamics of insecticide resistance in *Myzus persicae* (Hemiptera: Aphididae). *Pest Management Science* 58, 895–907.

Gera, A., Loebenstein, G. and Raccah, B. (1978) Detection of cucumber mosaic virus in viruliferous aphids by enzyme-linked immunosorbent assay. *Virology* 86, 542–545.

Halbert, S., Connelly, J. and Sandvol, L. (1990) Suction trapping of aphids in western North America. *Acta Phytopathologica et Entomologica Hungarica* 25, 411–422.

Halbert, S.E., Elberson, L.R., Feng, Ming-guang, Poprawski, T.J., Wraight, S., Johnson, J.B. and Quisenberry, S.S. (1998) Suction trap data: implications for crop protection forecasting. In: *Russian Wheat Aphid*. Thomas Say Publications in Entomology, Entomological Society of America, Lanham, pp. 412–428.

Hardie, J. and Tatchell, G.M. (1989) A method for separating summer and autumn migrants of host-alternating aphids. *Entomologia Experimentalis et Applicata* 52, 451–458.

Harrington, R., Tatchell, G.M. and Bale, J.S. (1990) Weather, life cycle strategy and spring populations of aphids. *Acta Phytopathologica et Entomologica Hungarica* 25, 423–432.

Harrington, R., Verrier, P., Denholm, C., Hullé, M., Maurice, D., Bell, N., Knight, J., Rounsevell, M., Cocu, N., Barbagallo, S., Basky, Z., Coceano, P.-G., Derron, J., Katis, N., Lukášová, H., Marrkula, I., Mohar, J., Pickup, J., Rolot, J.-L., Ruszkowska, M., Schliephake, E., Seco-Fernandez, M.-V., Sigvald, R., Tsitsipis, J. and Ulber, B. (2004) 'EXAMINE' (EXploitation of Aphid Monitoring in Europe): an EU Thematic Network for the study of global change impacts on aphids. In: Simon, J.C., Dedryver, C.A., Rispe, C. and Hullé, M. (eds) *Aphids in a New Millennium*. INRA, Paris, pp. 45–49.

Harrington, R., Clark, S.J., Welham, S.J., Verrier, P.J., Denholm, C.H., Hullé, M., Maurice, D., Rounsevell, M.D.A., Cocu, N. and European Union EXAMINE Consortium (2007) Environmental change and the phenology of European aphids. *Global Change Biology* (in press).

Heathcote, G.D. (1972) Evaluating aphid populations on plants. In: van Emden, H.F. (ed.) *Aphid Technology*. Academic Press, London, pp. 105–145.

Holz, F. (1991) A model-based catalogue of case studies as a tool for decision making in the control of the cereal aphid *Macrosiphum (Sitobion) avenae* in winter wheat. *Bulletin OILB/SROP* 14, 35–41.

Howard, M.T. and Dixon, A.F.G. (1990) Forecasting of peak population density of the rose–grain aphid *Metopolophium dirhodum* on wheat. *Annals of Applied Biology* 117, 9–19.

Hullé, M. and Gamon, A. (1989) Relations entre les captures des pièges à succion du réseau Agraphid. In: Cavalloro, R. (ed.) *'Euraphid' Network: Trapping and Aphid Prognosis*. CEC, Luxembourg, pp. 165–177.

Hullé, M., Coquio, S. and Laperche, V. (1994) Patterns in flight phenology of a migrant cereal aphid species. *Journal of Applied Ecology* 31, 49–58.

Kendall, D.A., Brain, P. and Chinn, N.E. (1992) A simulation model of the epidemiology of barley yellow dwarf virus in winter sown cereals and its application to forecasting. *Journal of Applied Ecology* 29, 414–426.

Kurppa, S. (1989) Predicting outbreaks of *Rhopalosiphum padi* in Finland. *Annales Agriculturae Fenniae* 28, 333–347.

Lankin, G., Worner, S.P., Samarasinghe, S. and Teulon, D.A.J. (2001) Can artificial neural network systems be used for forecasting aphid flight patterns? *New Zealand Plant Protection* 54, 188–192.

Latteur, G. and Oger, R. (1987) Principes de base du système d'avertissement relatif à la lutte contre les pucerons des froments d'hiver en Belgique. *Annales de l'Association Nationale de Protection des Plantes* 6, 149–167.

Leather, S.R. (1983) Forecasting aphid outbreaks using winter egg counts: an assessment of its feasibility and an example of its application in Finland. *Zeitschrift für Angewandte Entomologie* 96, 282–287.

Leclercq-Le Quillec, F., Plantegenest, M., Riault, G. and Dedryver, C.A. (2000) Analyzing and modeling temporal disease progress of barley yellow dwarf virus serotypes in barley fields. *Phytopathology* 90, 860–866.

López-Moya, J.J., Cubero, J., López-Abella, D. and Díaz-Ruíz, J.R. (1992) Detection of cauliflower mosaic virus (CaMV) in single aphids by the polymerase chain reaction (PCR). *Journal of Virological Methods* 37, 129–138.

Lowles, A. (1995) A quick method for distinguishing between the two autumn winged female morphs of the aphid *Rhopalosiphum padi*. *Entomologia Experimentalis et Applicata* 74, 95–99.

Loxdale, H.D., Hardie, J., Halbert, S., Foottit, R., Kidd, N.A. and Carter, C.I. (1993) The relative importance of short- and long-range movement of flying aphids. *Biological Reviews* 68, 291–311.

Lukásová, H., Basky, Z. and Starý, P. (1999) Flight patterns of Russian wheat aphid, *Diuraphis noxia* (Kurdj.) *Journal of Pest Science* 72, 41–44.

Macaulay, E.D.M., Tatchell, G.M. and Taylor, L.R. (1988) The Rothamsted Insect Survey '12-metre' suction trap. *Bulletin of Entomological Research* 78, 121–129.

Mann, B.P. and Wratten, S.D. (1991) A computer-based advisory system for cereal aphids: field testing the model. *Annals of Applied Biology* 118, 503–512.

Martinez-Torres, D., Foster, S.P., Field, L.M., Devonshire, A.L. and Williamson, M.S. (1999) A sodium channel point mutation is associated with resistance to DDT and pyrethroid insecticides in the peach–potato aphid, *Myzus persicae* (Sulzer) (Hemiptera: Aphididae). *Insect Molecular Biology* 8, 339–346.

Masterman, A.J., Foster, G.N., Holmes, S.J. and Harrington, R. (1996) The use of the Lamb daily weather types and the indices of progressiveness, southerliness and cyclonicity to investigate the autumn migration of *Rhopalosiphum padi. Journal of Applied Ecology* 33, 23–30.

Moore, A. and Miller, R.H. (2002) Automated identification of optically sensed aphid (Homoptera: Aphididae) wingbeat waveforms. *Annals of the Entomological Society of America* 95, 1–8.

Moores, G.D., Devine, G.J. and Devonshire, A.L. (1994) Insecticide-insensitive acetylcholinesterase can enhance esterase-based resistance in *Myzus persicae* and *Myzus nicotianae. Pesticide Biochemistry and Physiology* 49, 114–120.

Morgan, D. (1990) Simulation model of barley yellow dwarf virus epidemiology. In: Burnett, P.A. (ed.) *World Perspectives on Barley Yellow Dwarf.* CIMMYT, Mexico, pp. 300–304.

Morgan, D. (2000) Population dynamics of the bird cherry–oat aphid, *Rhopalosiphum padi* (L.) during the autumn and winter: a modelling approach. *Agricultural and Forest Entomology* 2, 297–304.

Nemecek, T., Derron, J.O., Fischlin, A. and Roth, O. (1995) Use of a crop growth model coupled to an epidemic model to forecast yield and virus infection in seed potatoes. In: Haverkort, A.J. and Mackeron, D.K.L. (eds) *Potato Ecology and Modelling of Crops Under Conditions Limiting Growth.* Kluwer, Dordrecht, pp. 281–289.

Newman, J.A. (2005) Climate change and the fate of cereal aphids in southern Britain. *Global Change Biology* 11, 940–944.

Northing, P., Walters, K., Barker, I., Foster, G., Harrington, R., Taylor, M., Tones, S. and Morgan, D. (2004) Use of the internet for provision of user specific support for decisions on the control of aphid-borne viruses. In: Simon, J.C., Dedryver, C.A., Rispe, C. and Hullé, M. (eds) *Aphids in a New Millennium.* INRA, Paris, pp. 331–336.

Olmos, A., Cambra, M., Dasi, M.A., Candresse, T., Esteban, O., Gorris, M.T. and Asensio, M. (1997) Simultaneous detection and typing of plum pox potyvirus (PPV) isolates by heminested-PCR and PCR-ELISA. *Journal of Virological Methods* 68, 127–137.

Onstad, D.W., Fang, S. and Voegtlin, D.J. (2005) Forecasting seasonal population growth of *Aphis glycines* (Hemiptera: Aphididae) in soybean in Illinois. *Journal of Economic Entomology* 98, 1157–1162.

Paliwal, Y.C. (1982) Detection of barley yellow dwarf virus in aphids by serologically specific electron microscopy. *Canadian Journal of Botany* 60, 179–185.

Parker, W.E. (1997) Forecasting the timing and size of field populations of aphids on potato in England and Wales. *Annales de l'Association Nationale de Protection des Plantes* 3, 1087–1094.

Perry, J.N., Winder, L., Holland, J.M. and Alston, R.D. (1999) Red-blue plots for detecting clusters in count data. *Ecology Letters* 2, 106–113.

Pierre, J.-S. and Dedryver, C.A. (1984) Un modèle de régression multiple appliqué à la prévision des pullulations d'un puceron des céréales, *Sitobion avenae* F., sur blé d'hiver. *Acta Oecologica* 5, 153–172.

Pierre, J.-S. and Dedryver, C.A. (1985) Un modèle de prévision des pullulations du puceron *Sitobion avenae* sur blé d'hiver. *Phytoma – Défense des cultures.* Rutalia Publications, Boulogne, pp. 13–17.

Pierre, J.-S., Guillôme, M. and Querrien, M.-T. (1986) Une méthode statistique et graphique de recherche des périodes de l'année, où les populations animales sont particulièrement sensibles à une composante donnée du climat (périodes critiques). Application au cas des pucerons des céréales. *Acta Oecologica* 7, 365–380.

Plantegenest, M., Mattioda, H., Pierre, J.-S. and de Tourdonnet, S. (1994) An optimization of the effect of deltamethrin against the grain aphid *Sitobion avenae* F. *Bulletin OILB SROP* 17, 41–47.

Plantegenest, M., Pierre, J.-S. and van Waetermeulen, X. (1997) Développement opérationnel d'un modèle de prévision des pullulations et d'optimisation de traitements insecticides contre le puceron des épis, *Sitobion avenae. Annales de la 4e ANPP Conférence Internationale sur les Ravageurs en Agriculture, Montpellier, 1997* 3, 1095–1103.

Plantegenest, M., Pierre, J.S., Dedryver, C.A. and Kindlmann, P. (2001) Assessment of relative impact of different natural enemies species on population dynamics of the grain aphid, *Sitobion avenae* (Homoptera: Aphididae) in the field. *Ecological Entomology* 26, 404–410.

Plumb, R.T. (1989) Detecting plant viruses in their vectors. In: Harris, K.F. (ed.) *Advances in Disease Vector Research, Volume 6*. Springer, New York, pp. 191–208.

Qi, A., Dewar, A. and Harrington, R. (2004) Decision making in controlling virus yellows of sugar beet in the UK. *Pest Management Science* 60, 727–732.

Rabbinge, R. and Mantel, W.P. (1981) Monitoring for cereal aphids in winter wheat. *Netherlands Journal of Plant Pathology* 87, 25–29.

Rispe, C., Hullé, M., Gauthier, J.-P., Pierre, J.-S. and Harrington, R. (1998) Effect of climate on the proportion of males in the autumn flight of the aphid *Rhopalosiphum padi* L. (Hom., Aphididae). *Journal of Applied Entomology* 122, 129–136.

Ro, T.H. and Long, G.E. (1999) GPA-Phenodynamics, a simulation model for the population dynamics and phenology of green peach aphid in potato: formulation, validation and analysis. *Ecological Modelling* 119, 197–209.

Robert, Y., Dedryver, C.A. and Pierre, J.S. (1988) Sampling techniques. In: Minks, A.K. and Harrewijn, P. (eds) *Aphids. Their Biology, Natural Enemies and Control, Volume B*. Elsevier, Amsterdam, pp. 1–20.

Rossberg, D., Holz, F., Freier, B. and Wenzel, V. (1986) PESTSIM-MAC. A model for simulation of *Macrosiphum avenae* Fabr. Populations. *Tagungsbericht, Akademie der Landwirtschaftlichen Wissenschaften, DDR, Berlin* 242, 87–100.

Rossing, W.A.H. (1991a) Simulation of damage in winter wheat caused by the grain aphid *Sitobion avenae*. 2. Construction and evaluation of a simulation model. *Netherlands Journal of Plant Pathology* 97, 25–54.

Rossing, W.A.H. (1991b) Simulation of damage in winter wheat caused by the grain aphid *Sitobion avenae*. 3. Calculation of damage at various attainable yield levels. *Netherlands Journal of Plant Pathology* 97, 87–103.

Sigvald, R. (1986) Forecasting the incidence of potato virus Y. In: McLean, G.D., Garrett, R.G. and Ruesink, W.G. (eds) *Plant Virus Epidemics*. Academic Press, Sydney, pp. 419–441.

Simon, J.C., Blackman, R.L. and Le Gallic, J.F. (1991) Local variability in the life cycle of the bird cherry–oat aphid, *Rhopalosiphum padi* (Homoptera: Aphididae) in western France. *Bulletin of Entomological Research* 81, 315–322.

Singh, R.P., Kurz, J. and Boiteau, G. (1996) Detection of stylet-borne and circulative potato viruses in aphids by duplex reverse transcription polymerase chain reaction. *Journal of Virological Methods* 59, 189–196.

Smith, A.D., Reynolds, D.R. and Riley, J.R. (2000) The use of vertical-looking radar to continuously monitor the insect fauna flying at altitude over southern England. *Bulletin of Entomological Research* 90, 265–277.

Smith, H.G., Stevens, M. and Hallsworth, P.B. (1991) The use of monoclonal antibodies to detect beet mild yellowing virus and beet western yellows virus in aphids. *Annals of Applied Biology* 119, 295–302.

Smith, H., Hallsworth, P. and Stevens, M. (1997) Aphid infectivity and virus yellows forecasting. *British Sugar Beet Review* 65 (1), 20–22.

Stevens, M. and Dewar, A.M. (2004) A review of pest problems in 2003: was there a sting in the tale? *British Sugar Beet Review* 72 (1), 12–17.

Stevens, M., Hull, R. and Smith, H.G. (1997) Comparison of ELISA and RT-PCR for the detection of beet yellows closterovirus in plants and aphids. *Journal of Virological Methods* 68, 9–16.

Tamada, T. and Harrison, B.D. (1981) Quantitative studies on the uptake and retention of potato leafroll virus by aphids in laboratory and field conditions. *Annals of Applied Biology* 98, 261–276.

Tatchell, G.M. and Parker, S.J. (1990) Host plant selection by migrant *Rhopalosiphum padi* in autumn and the occurrence of an intermediate morph. *Entomologia Experimentalis et Applicata* 54, 237–244.

Tatchell, G.M., Thorn, M., Loxdale, H.D. and Devonshire, A.L. (1988) Monitoring for insecticide resistance in migrant populations of *Myzus persicae*. *Proceedings of the Brighton Crop Protection Conference, Pests and Diseases, November 1988* 1, 439–444.

Taylor, L.R. (1974) Monitoring change in the distribution and abundance of insects. *Rothamsted Experimental Station Report for 1973, Part 2*, 202–239.

Taylor, L.R. (1979) The Rothamsted Insect Survey – an approach to the theory and practice of synoptic pest forecasting in agriculture. In: Rabb, R.L. and Kennedy, G.G. (eds) *Movement of Highly Mobile Insects: Concepts and Methodology in Research*. North Carolina State University, Raleigh, pp. 148–185.

Taylor, L.R. (ed.) (1981) *Aphid Forecasting and Pathogens and A Handbook for Aphid Identification*. Rothamsted Experimental Station, Harpenden, 219 pp.

Taylor, L.R. and Palmer, J.M.P. (1972) *Aerial sampling*. In: van Emden, H.F. (ed.) *Aphid Technology*. Academic Press, London, pp. 189–234.

Taylor, L.R., Woiwod, I.P. and Taylor, R.A.J. (1979) The migratory ambit of the hop aphid and its significance in aphid population dynamics. *Journal of Animal Ecology* 48, 955–972.

Taylor, M.S., Tatchell, G.M. and Clark, S.J. (1994) Morph identification in natural populations of alate female bird cherry aphids, *Rhopalosiphum padi* L., by multivariate methods. *Annals of Applied Biology* 125, 1–11.

Teulon, D.A.J., Stufkens, M.A.W. and Fletcher, J.D. (2004) Crop infestation by aphids is related to flight activity detected with 7.5 metre high suction traps. *New Zealand Plant Protection* 57, 227–232.

Thacker, J.I., Thieme, T. and Dixon, A.F.G. (1997) Forecasting of periodic fluctuations in annual abundance of the bean aphid: the role of density dependence and weather. *Journal of Applied Entomology* 121, 137–145.

Thackray, D.J., Diggle, A.J., Berlandier, F.A. and Jones, R.A.C. (2004) Forecasting aphid outbreaks and epidemics of Cucumber mosaic virus in lupin crops in a Mediterranean-type environment. *Virus Research* 100, 57–82.

Thomas, G.G., Goldwin, G.K. and Tatchell, G.M. (1983) Associations between weather factors and the spring migration of the damson–hop aphid, *Phorodon humuli. Annals of Applied Biology* 102, 7–17.

Torrance, L. (1987) Use of enzyme amplification in an ELISA to increase sensitivity of detection of barley yellow dwarf virus in oats and in individual vector aphids. *Journal of Virological Methods* 15, 131–138.

Turl, L.A.D. (1980) An approach to forecasting the incidence of potato and cereal aphids in Scotland. *EPPO Bulletin* 10, 135–141.

Varveri, C. (2000) Potato Y potyvirus detection by immunological and molecular techniques in plants and aphids. *Phytoparasitica* 28, 141–148.

Vereijken, P.H. (1979) Feeding and multiplication of three cereal aphid species and their effect on yield of winter wheat. PhD thesis, Wageningen Agricultural University, The Netherlands.

Walters, K.F.A. and Dewar, A.M. (1986) Overwintering strategy and the timing of the spring migration of the cereal aphids *Sitobion avenae* and *Sitobion fragariae. Journal of Applied Ecology* 23, 905–915.

Way, M.J., Cammell, M.E., Taylor, L.R. and Woiwod, I.P. (1981) The use of egg counts and suction trap samples to forecast the infestation of spring-sown field beans, *Vicia faba*, by the black bean aphid, *Aphis fabae. Annals of Applied Biology* 98, 21–34.

Werker, A.R., Dewar, A.M. and Harrington, R. (1998) Modelling the incidence of virus yellows in sugar beet in the UK in relation to numbers of migrating *Myzus persicae. Journal of Applied Ecology* 35, 811–818.

Wiktelius, S. (1982) Flight phenology of cereal aphids and possibilities of using suction trap catches as an aid in forecasting outbreaks. *Swedish Journal of Agricultural Research* 12, 9–16.

Wiktelius, S. (1984) Long-range migration of aphids into Sweden. *International Journal of Biometeorology* 28, 185–200.

Winder, L., Holland, J.M. and Perry, J.N. (1998) The within-field spatial and temporal distribution of the grain aphid (*Sitobion avenae*) in winter wheat. *Proceedings of the Brighton Crop Protection Conference, Pests and Diseases, November 1998* 3, 1089–1094.

Woiwod, I.P. and Hanski, I. (1992) Patterns of density dependence in moths and aphids. *Journal of Animal Ecology* 61, 619–629.

Worner, S.P., Tatchell, G.M. and Woiwod, I.P. (1995) Predicting spring migration of the damson–hop aphid *Phorodon humuli* (Homoptera: Aphididae) from historical records of host-plant flowering phenology and weather. *Journal of Applied Ecology* 32, 17–28.

Zhou, X. and Carter, N. (1990) A simulation study of aphid damage and control strategies in cereals. *Proceedings of the Brighton Crop Protection Conference, Pests and Diseases, November 1990* 2, 697–702.

Zhou, X., Carter, N. and Mumford, J. (1989) A simulation model describing the population dynamics and damage potential of the rose–grain aphid, *Metopolophium dirhodum* (Walker) (Hemiptera: Aphididae), in the UK. *Bulletin of Entomological Research* 79, 373–380.

20 Integrated Pest Management and Introduction to IPM Case Studies

Helmut F. van Emden

School of Biological Sciences, University of Reading, Whiteknights, Reading, Berks, RG6 6AJ, UK

Introduction

Most people would agree that the foundation of modern IPM is to be found in the 'Integrated Control' (IC) concept of Stern *et al.* (1959). This means it all began with aphids. The concept was formulated on the integration of chemical and biological control of *Therioaphis trifolii maculata* (spotted alfalfa aphid) on lucerne (alfalfa), *Medicago sativa.* By the late 1950s, *T. t. maculata* was destroying the lucerne crop in California, following the arrival of the aphid from Europe in 1954 and rapid subsequent appearance of resistance to organophosphate insecticides (OPs), compounds that also killed the indigenous natural enemies. To solve the OP-resistant *T. t. maculata* problem, the Californian workers integrated a reduced dose of an OP insecticide with the biological control that the low dose now allowed to survive. Integration at that time referred to integration of control methodologies, as opposed to integration of control of a range of pests; therefore, integrated control of an aphid pest was a meaningful concept. The term 'Pest Management' (PM) as a successor to IC dates from a conference at Raleigh, North Carolina, USA, in 1970 (Beirne, 1970). PM embraced the use of both single and multiple control measures. 'Integrated Pest Management' (IPM) emerged later in the 1970s (Apple and Smith, 1976) and is described as follows: 'The concept of pest management has now been broadened to include all classes of pests (pathogens, insects, nematodes, and weeds) and in this context is commonly referred to as IPM'. Thus, the 'I' of IPM originally included an integration of *crop protection disciplines* (i.e. entomology, plant pathology, nematology, weed science, etc.), and Apple and Smith (1976) would have regarded IPM of aphids as a contradiction in terms. Yet, since 1976, definitions have loosened, and today IPM seems indistinguishable from PM, or even from IC.

A good way of looking at IPM stems from its reciprocal, pest mismanagement. The drivers behind IPM were indeed the mismanagement of pest populations in the 1940s and 1950s through:

- overdosing with pesticides and the resulting appearance of tolerant pest populations;
- loss of biological control through use of broad-spectrum pesticides and loss of habitat diversity in agroecosystems;
- introduction of genetically uniform, high-yielding but pest-susceptible crop cultivars in large monocultures;
- abandonment of labour-intensive cultural controls.

These sources of mismanagement translate easily into the main components of IPM systems for aphids:

- decisions on chemical control are guided by economic thresholds, and selective materials are chosen where they are available (Dewar, Chapter 15 this volume; Harrington *et al.*, Chapter 19 this volume);
- biological control (Völkl *et al.*, Chapter 8 this volume; Powell and Pell, Chapter 18 this volume) is conserved by selective pesticides and promoted by habitat modification, including planned biodiversity in farm management (Wratten *et al.*, Chapter 16 this volume); agents may be re-colonized where they have disappeared or new agents, especially from overseas, introduced;
- the use of partially aphid-resistant crop varieties (van Emden, Chapter 17 this volume);
- introduction or re-introduction of cultural controls, especially to improve conditions for natural enemies of aphids (Wratten *et al.*, Chapter 16 this volume).

Apart from these four reciprocals of pest mismanagement, techniques for using semiochemicals modifying the behaviour of aphids and their natural enemies (Pickett and Glinwood, Chapter 9 this volume) represent a more recent contribution of considerable potential for inclusion in the IPM armoury.

IPM as the Use of Multiple Control Measures

The golden rules of IPM

Using more than one control method is really implicit in the concept of IPM. From this there follow two golden rules for IPM (van Emden, 2002):

- *If a single method gives adequate control on its own, then there is the danger of a tolerant pest strain increasing in gene frequency and no opportunity to use a second method in addition. The method therefore needs to be made less*

efficient (reduced dose of pesticide, partial host-plant resistance rather than immunity) for there to be value in introducing another control method to supplement it.
- *Methods are increasingly worth combining to the extent that the control then achieved exceeds the additive effects of the two methods in isolation.*

A variable amount is known about the different interactions that can occur between the main IPM components of chemical control, biological control, host-plant resistance, and cultural control of aphids. The information is summarized below.

Interaction between chemical and biological control

The prevalent expression of this interaction in the literature is that most insecticides are toxic to natural enemies of aphids. This is often misinterpreted as meaning that damage to biological control of aphids is inevitable. However, assuming there are no harmful sublethal effects of the pesticide on the surviving natural enemies, biological control is only damaged if, as a result of applying the pesticide, the ratio of aphids to natural enemies increases. If it decreases (i.e. the application is even marginally selective in favour of the natural enemies), there is then the potential for improved biological control, even though some, or even many, of the natural enemies are killed (van Emden and Service, 2004). The following sources of such selectivity are known:

- Use of a selective active ingredient (a famous example being the carbamate pirimicarb, to which only the acetylcholinesterase in the nervous system of aphids and Diptera is sensitive, Silver *et al.*, 1995). It is worth noting that natural enemies of aphids are not necessarily more susceptible to insecticides than their aphid prey. Croft and Brown's (1975) literature review identified that, for 36 aphid–coccinellid combinations, the coccinellid was more tolerant to insecticide than the aphid in 31 cases, the extreme being a 43-fold difference. Acheampong and Stark (2004) found that pymetrozine was not only non-toxic to the parasitoid *Diaeretiella rapae* at 0.212 g a.i./ha, but that the r_m of the parasitoid increased by 11%. However, adding the adjuvant Sylgard 309 caused a 39% reduction in fecundity compared to the control.

- Selectivity in space. Even broad-spectrum aphicides can be applied selectively in space if they are applied as band sprays or soil treatments, or if systemic compounds applied to the foliage are rapidly withdrawn into the plant.
- Selectivity in time. Early sprays may reduce aphid populations before natural enemies appear. Thus, Hull and Sterner (1983) found that a single early application of pesticide gave good control of *Dysaphis plantaginea* (rosy apple aphid) on apples without disrupting later predation by important natural enemies. The lowered aphid numbers may, however, result in fewer natural enemies colonizing the crop. Morse (1989) therefore suggested allowing *Aphis craccivora* (cowpea aphid) to attract coccinellids but, once the beetles had laid their eggs, reducing aphid numbers with an ephemeral insecticide while the coccinellid embryos/larvae were still protected by the egg shell.

It has similarly been suggested that parasitoid larvae within mummies often survive the application of insecticide provided only low residues are still on the mummy cuticle when the adult parasitoids emerge.

- Reduced pesticide dose. This was the approach to selectivity of a broad-spectrum OP insecticide reported by Stern *et al.* (1959) and mentioned at the start of this chapter. It can be expected that per cent kill of carnivores (natural enemies) will reduce faster than that of herbivores (aphids) as pesticide dose reduces (Plapp, 1981). This implies a steeper slope of the probit mortality regression on toxin concentration for carnivores than herbivores (Fig. 20.1, comparison of solid and dashed line), i.e. individuals in a population of carnivores show less variation in tolerance to insecticides. Plapp's (1981) reasoning was that herbivores require a diverse armoury of enzymes for detoxifying foreign toxins (i.e. plant defensive compounds) to an extent carnivores do not.

Interaction between chemical control and host-plant resistance (HPR)

Aphids are usually (but not invariably) smaller on resistant plants. Since toxicity of an insecticide is a function of body weight,

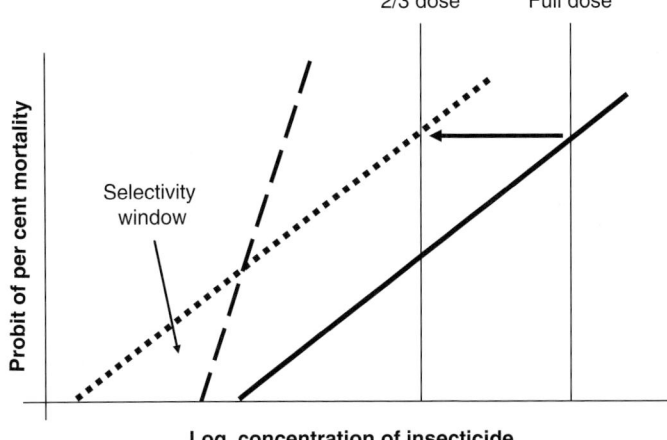

Fig. 20.1. The interaction of partial host-plant resistance, biological control, and insecticides. Solid line, mortality response of aphids treated with normal dose on susceptible variety; dotted line, mortality response of aphids treated with a dose reduction (arrow) of one-third on a resistant variety. The mortality response of a natural enemy (dashes) is assumed to be unaffected by plant resistance.

one would expect aphids on resistant plants to show enhanced susceptibility to toxins. The first report of such enhanced susceptibility concerned *Myzus persicae* (peach–potato aphid), *Aphis gossypii* (cotton or melon aphid), and *Aulacorthum solani* (glasshouse and potato aphid) on chrysanthemum (*Dendranthema × grandiflorum*) (Selander *et al.*, 1972). The LD$_{50}$ (the dose adjusted for aphid weight required to kill 50% of the aphids) of malathion, dimethoate, and lindane on the resistant variety 'Princess Anne' was between 50 and 66% lower than on the susceptible variety 'Tuneful'. Nicol *et al.* (1993) compared the tolerance to deltamethrin of *Sitobion avenae* (grain aphid) on two wheat varieties, one of which ('Altar') possessed resistance to aphids based on high levels of DIMBOA (see van Emden, Chapter 17 this volume). On 'Altar', deltamethrin showed three times the relative toxicity shown on the susceptible wheat ('Dollarbird'). The enhanced susceptibility to insecticides on resistant plants cannot be accounted for by reduced size alone. Selander *et al.* (1972) presented their data corrected for differences in aphid weight between the varieties and, when this correction was similarly applied in the DIMBOA example, the LD$_{50}$ was still reduced by over 90%. With *M. persicae* and Brussels sprout, Mohamad and van Emden (1989) calculated that the 45% increase in mortality from malathion on the only slightly aphid-resistant variety 'Early Half Tall' (compared with 'Winter Harvest') was still as large as 42% after correcting for differences in aphid weight on the two varieties. Similarly, with *Metopolophium dirhodum* (rose–grain aphid) on the susceptible wheat variety 'Maris Kinsman' (*Triticum aestivum*) and the partially aphid-resistant 'Emmer' (*Triticum turgidum*), Attah and van Emden (1993) found that the increase in mortality of over 50% on the resistant variety was only reduced by about 5% following a correction for weight. Some stress of HPR on the aphids, perhaps poorer nutrition and lower fat levels in the body, appears far more important than body weight differences.

In general, it would appear that the minimum reduction in aphicide concentration likely to be effective on aphids on even a variety only slightly aphid-resistant is about 30%. However, it is important to add that aphids on resistant plants may show the converse phenomenon, i.e. greater tolerance to insecticides. Ahmad and Shakoori (2001) found a higher mortality from demeton-S-methyl of *Brevicoryne brassicae* (cabbage aphid) on the aphid-susceptible Ethiopian mustard (*Brassica carinata*) than on four aphid-resistant accessions of Indian mustard (*Brassica juncea*). With caterpillars (Lepidoptera), it has been shown that such results arise from the induction by secondary plant compounds of insecticide-detoxifying enzymes in the insect (Kennedy, 1984; Yu and Hsu, 1985).

Interaction between biological control and host-plant resistance

HPR may synergize beneficially with biological control, leave its impact unaffected, or be damaging, as discussed by van Emden, Chapter 17 this volume. In this chapter, however, the emphasis will be on the contribution positive synergy may make to IPM of aphids, together with an assessment of how frequently such synergism may be expected. There are several data sets (Fig. 20.2) where measurements have been made of the effect on an aphid population of plant resistance without biological control (*a*, see legend to Fig. 20.2), biological control without plant resistance (*b*) (i.e. on the aphid-susceptible variety), and the combination of plant resistance and biological control (*c*). From these data sets, it is possible to compare the actual outcome (*c*) with the expected outcome assuming no synergy (starting population on susceptible comparison × (1–*a*) × (1–*b*), i.e. converting *a* and *b* to survival quotients) (van Emden, 2003). In five of the seven data sets, there is very strong positive synergism – the population reduction is between twice and 20 times that expected, with no synergism between the two restraints. The phenomena for such positive synergism can be divided into numerical and functional responses of the natural enemies.

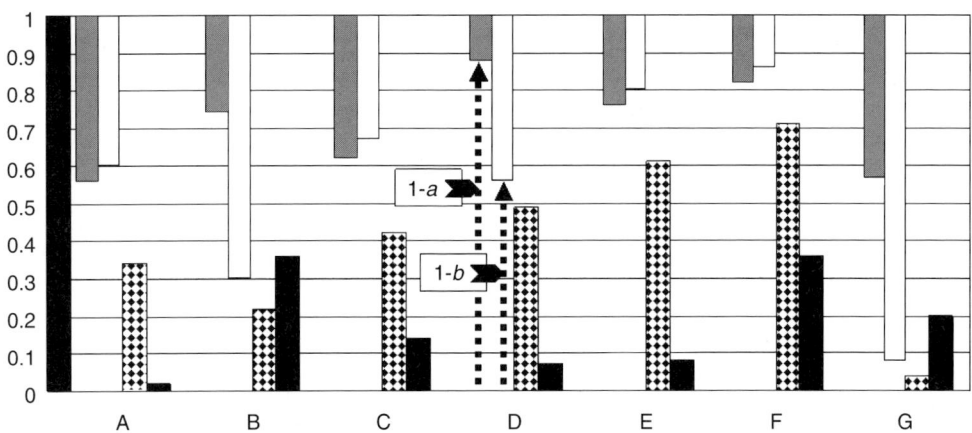

Fig. 20.2. Interaction of partial host-plant resistance to aphids with biological control. Aphid populations are expressed as a proportion of that on susceptible varieties without biological control (column at extreme left reaching unity). Histograms in each block from left to right: reduction (shown as a descending column) by plant resistance alone (grey, *a* – see text), reduction from biological control on susceptible variety (white, *b*), population (shown as an ascending column) predicted on resistant variety with biological control (chequered), observed on resistant variety with biological control (black, *c*). The chequered column is obtained by multiplying the two quotients (1-*a*) and (1-*b*) indicated by arrows using data set D as an example. A, *Schizaphis graminum* parasitized by *Lysiphlebus testaceipes* on barley (Starks *et al.*, 1972); B, *Sitobion avenae* parasitized by *Aphelinus abdominalis* on wheat (Lykouressis, 1982); C and D, *Metopolophium dirhodum* parasitized by *Aphidius rhopalosiphi* on wheat (Gowling, 1989); E, *Brevicoryne brassicae* and natural predation in brassicas (Gowling, 1989); F, *Brevicoryne brassicae* and natural predation in brassicas (Dodd, 1973); G, *Brevicoryne brassicae* and parasitization by *Diaeretiella rapae* on Brussels sprout (van Emden, 1978).

Numerical responses

- Slower reproduction of aphids on resistant varieties increases the potential of natural enemies to contain the aphid population (van Emden and Wearing, 1965).
- Aphids on resistant varieties usually show increased development times (e.g. Sotherton and Lee, 1988). This increases their chance of being predated before they reproduce.
- Parasitoids may show constancy to variety (Wickremasinghe and van Emden, 1992) and so will continue searching on resistant varieties even though aphid numbers are reduced.

Functional responses

- Natural enemies can often detect the locations of aphid colonies on the plant by plant-emitted chemical cues (Storeck *et al.*, 2000), so searching time may not be increased by lower pest densities.
- Predators will eat smaller aphids (as typical on resistant varieties) in greater numbers before becoming satiated (Fig. 20.3). Hassell *et al.* (1977) have shown that a positive density-dependent voracity of *Coccinella septempunctata* (7-spot ladybird) extended to higher densities of prey if the latter (different instars of *B. brassicae*) were smaller (Fig. 20.3).
- Smaller aphids on resistant varieties are less able to escape natural enemies by rapid locomotion or effective kicking. Dixon (1985) showed that *Microlophium carnosum* (nettle aphid) were able to survive encounters with larvae of *Adalia decempunctata* (10-spot ladybird), and interpreted Mackauer's (1973) results as a similar phenomenon.
- The activity of natural enemies searching in aphid colonies disturbs aphids and

causes them to fall from the plant; this is considerably more pronounced on resistant varieties (Gowling and van Emden, 1994; Fig. 20.4). Note in Fig. 20.4 that total per cent parasitization on the resistant wheat 'Rapier' was higher than on 'Armada', partly because more aphids that fell were parasitized and mummified

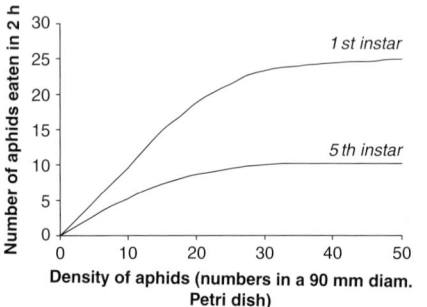

Fig. 20.3. Predation by *Coccinella septempunctata* on first- and fifth-instar *Brevicoryne brassicae* at different aphid densities (data from two different graphs at two different scales several pages apart in Hassell *et al.*, 1977).

on the soil, and that, when the data are expressed in the form of Fig. 20.2, the outcome in terms of increased impact of biological control on 'Rapier' is almost the same as the overall comparison of the two varieties (Fig. 20.2, D). Hatting *et al.* (2004) suggested that the greater restlessness of aphids on resistant varieties, and therefore their exposure to fungal spores, explained the improved control of *Diuraphis noxia* (Russian wheat aphid) on aphid-resistant wheat by the fungus *Beauveria bassiana*.

• Plant structure may interact with biological control. Resistant varieties may have less deformation in the form of leaf rolling. Natural enemies then find their aphid prey more easily (Reed *et al.*, 1992). Lower amounts of leaf surface wax give the coccinellids a better grip (Eigenbrode *et al.*, 1998).

• Natural enemies may spend less time cleaning off wax particles on aphid-resistant varieties with low surface wax (Eigenbrode *et al.*, 1998). Parasitoids will also divert searching time to cleaning

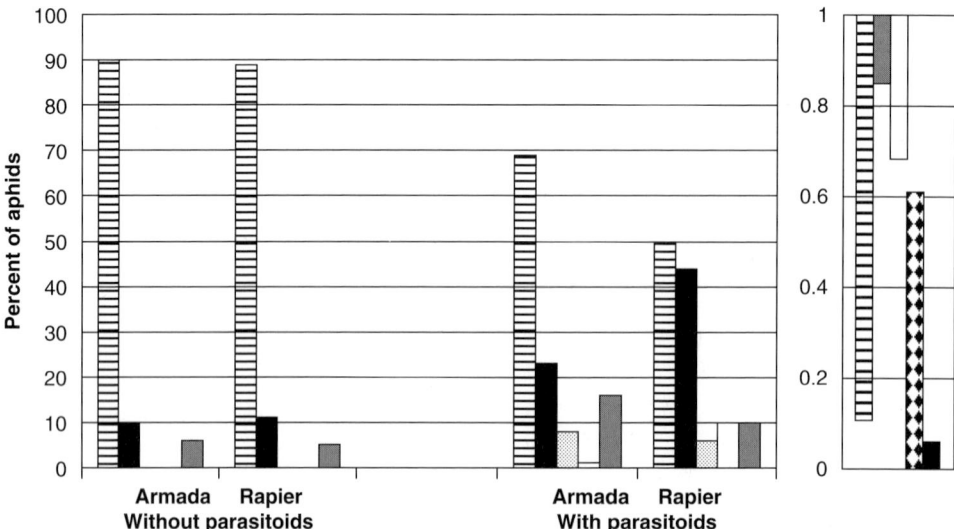

Fig. 20.4. Percentage fate of *Metopolophium dirhodum* on 'Armada' (susceptible) and 'Rapier' (partially aphid-resistant) wheat after 6 days with and without activity of the parasitoid *Aphidius rhopalosiphi*. Columns from left to right in each data set: remaining alive on plant (barred); fallen from plant (black); mummified on plant (dotted); mummified on the soil (white); regained plant after falling (grey) (data of Gowling, 1989). The histogram to the right summarizes these data in the form of, and for comparison with, Fig. 20.2, D.

activity if there is copious aphid honeydew, as is more characteristic of aphid-susceptible than of aphid-resistant varieties (Wickremasinghe, 1989).

The literature

The literature for insect pests in general contains at least 41 examples (many of which relate to aphids) of the interaction of HPR with biological control. Of these, 21 show positive synergism, 8 show simple additivity, and 12 show a negative interaction (though 4 of the latter refer only to effects on development time and/or size of the natural enemy). Not surprisingly, four of the examples of negative interaction (including the example from van Emden, 1978 in Fig. 20.2, G) are known to relate to toxic secondary plant compounds, and it is likely the deleterious effects of such compounds on natural enemies may also apply to other examples. However, the ladybird *Propylea japonica*, when fed *A. gossypii* reared on an aphid-resistant high gossypol cotton variety, had a shorter development time and greater adult weight (Du Li *et al.*, 2004). Another example of positive synergism relates to the size and fecundity of the aphid parasitoid *Aphidius nigripes* being increased when reared on *Macrosiphum euphorbiae* (potato aphid) on a transgenic potato expressing the protease inhibitor rice cystatin (Ashouri *et al.*, 2001). Kalule and Wright (2005) showed that the performance of the generalist *Aphidius colemani* reared on *B. brassicae* or *Myzus persicae* (peach–potato aphid) was never poorer and often improved on cabbage varieties partially resistant to the aphids.

Three-way interaction between chemical control, host-plant resistance and biological control

Taking together two phenomena already mentioned – that insecticide dose can often be reduced on aphid-resistant varieties and that dosage reductions are likely to increase the selectivity of a pesticide application to the benefit of natural enemies – a three-way interaction seems to be indicated. This interaction

is shown in Fig. 20.1. As yet, the only experimental test to confirm this interaction in relation to aphids stems from laboratory work on cereal aphids, parasitoids, and coccinellids (Tilahun and van Emden, 1997; Fig. 20.5), when both *Aphidius rhopalosiphi* and *C. septempunctata* actually showed greater tolerance to malathion when reared on *M. dirhodum* on the partially aphid-resistant wheat 'Rapier' than on the aphid-susceptible 'Maris Huntsman'.

Interaction between cultural control and biological control

Although there is considerable interest in using cultural measures directly to promote biological control of aphids (Powell and Pell, Chapter 18 this volume; Wratten *et al.*, Chapter 16 this volume), the interaction between cultural measures primarily for controlling aphids and biological control seems relatively unexplored. However, such interactions almost certainly exist and therefore should not be ignored in designing IPM programmes. An attempt to investigate such interactions was made by Ul-Haq (1997) in glasshouse experiments on the effects of fertilizer applications, water stress, and wheat/pea (*Pisum sativum*) 'intercrops' on aphids and the size and fecundity of parasitoids. He found that 'cultural treatments' which decreased the size of aphids, also decreased the size and fecundity of the parasitoids (as reported with partial plant resistance to aphids by van Emden, 1995).

In the absence of experimental evidence for interaction between cultural control of aphids and biological control, Table 20.1 lists the principal approaches to cultural control of aphids and aphid-transmitted viruses with speculation on how biological control may be affected.

The IPM Case Studies

This introduction to IPM of aphids is followed by ten case studies, where scientists working on the control of aphids in particular

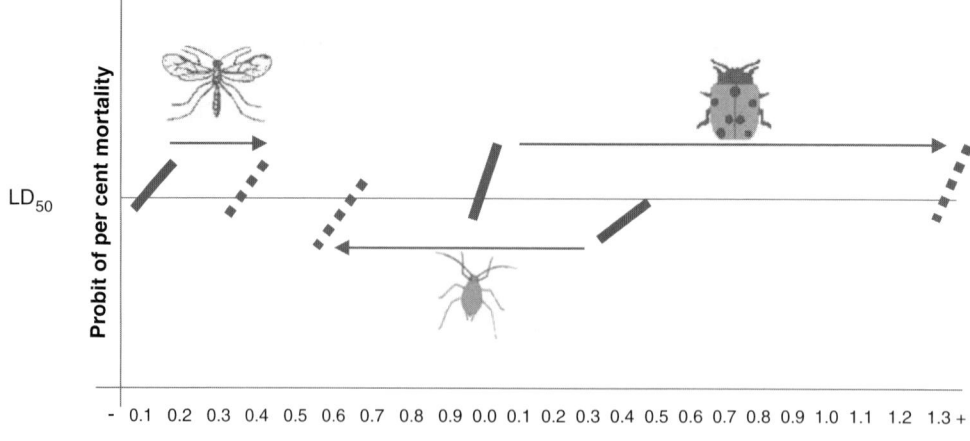

Fig. 20.5.   Effect of a partial aphid-resistant wheat ('Rapier') on susceptibility to malathion of *Metopolophium dirhodum*, the parasitoid *Aphidius rhopalosiphi*, and the coccinellid *Coccinella septempunctata*. Solid lines, LD_{50} and slope of probit regression on the aphid-susceptible wheat 'Maris Huntsman'; dotted line, LD_{50} and slope of probit regression on 'Rapier'; arrows, direction and degree of change in susceptibility to malathion.

crop scenarios were asked to report on the state of IPM in their specialized area of interest. The case studies have been chosen to represent a wide diversity, with the result that the various case studies could not be presented to a formula. At one end, cotton is a single crop with one important aphid species; at the other, the case study on temperate fruit trees and stone fruits considers over ten crops with over 25 pest aphid species. With crops such as seed potatoes and several types of cucurbit, where virus transmission is the key aphid problem, IPM cannot be a serious proposition, whereas in cotton, brassicas, cereals, and fruits, economic thresholds are reasonably high. Prophylactic spraying can therefore be reduced and compounds chosen to give maximum selectivity in favour of natural enemies, which often give an inadequate, but nonetheless valuable, restraint. Other contrasts include that between the importance of host-plant resistance in berry fruits and the virtual lack of it in tree fruit crops, where plant breeding is a very long-term process.

In order to make it easier to compare and contrast IPM in the case studies, each finishes with an 'executive summary', recapitulating the main points in a consistent order.

Each case study is written by a scientist working in a particular region of the world, mostly in developed agriculture. Thus, the case studies are both crop-specific and, to a large extent, location-specific, though often reference has been made to contrasts with other regions. This specificity to location is seen clearly in the active ingredients of the insecticides mentioned; products mentioned may well be banned or have been withdrawn in other regions of the world.

Conclusions

The case studies show in a striking way that implementation of practical IPM, however desirable and accepted in terms of 'lip-service', is still at a very early stage in respect to aphids on most crops. It is very much 'stage one' IPM, with some economic thresholds worked out and choice of pesticide usually seeking to avoid damage to indigenous biological control as far as possible. This is really the 'Integrated Control' of Stern *et al.* (1959). Indeed, apart from provision of nectar plants and the release of

Table 20.1. Cultural control of aphids and potential effects on natural enemies.

Control measure	Potential effects on natural enemies	Supporting evidence (if any)
Limited N fertilization	As for partial plant resistance; greater impact, but parasitoids smaller and less fecund	Ul-Haq, 1997
Avoidance of intermittent drought stress	As for partial plant resistance; greater impact, but parasitoids smaller and less fecund	Ul-Haq, 1997
Reducing late leaf area by techniques such as termination of irrigation and early harvest	Greater impact at the stage of reducing the aphid population at the end of the crop season	
High plant density	Effects of plant ground cover? Larger numbers of anthocorids, syrphids, and epigeal predators	Smith, 1969, 1976; Powell *et al.*, 1981
Earlier sowing	As plant resistance (increases with plant age); greater impact, but parasitoids smaller and less fecund, and poorer temporal synchronization with natural enemies	
Delayed sowing (especially for reduction of virus problems)	Better synchronization between natural enemies and aphids	van Emden, 1966
Hand removal of terminal shoots	Probably some partial resistance, therefore greater impact, but parasitoids smaller and less fecund	
Intercropping	Effects of plant ground cover? Larger numbers of anthocorids, syrphids, and epigeal predators	
Trap crops	May form sink for natural enemies and delay their movement to the commercial crop	Wratten *et al.*, Chapter 16 this volume
Removal of weed sources of virus	Unlikely to have a major effect?	
Crop isolation	Specific predators and parasitoids may be lacking or scarce in the new areas	
Reflective mulches	May affect colonizing natural enemies less than aphids, and thus increase natural enemy:aphid ratio	
Crop covers	Likely to exclude natural enemies as well as aphids, but any reaching the crop will remain confined over it	

natural enemies in some glasshouse crops, the scenarios presented involve virtually no other manipulation or purposeful introduction of biological control agents. 'Stage two' IPM, where host- plant resistance and/or cultural control are key additions, is poorly represented (except perhaps in cotton for host-plant resistance and in sorghum for cultural control by increasing plant density). Although aphid-resistant varieties have been bred for many of the crops, the pressure to exploit their potential for reducing aphid problems appears, in most cases, to be insufficient to overcome the growers' preference for the agronomic properties of other varieties. Aphid-resistance obtained by genetically modifying grower-acceptable varieties are not yet available for aphid control and, in any case, probably would not synergize beneficially with chemical or biological control (van Emden, 2003). 'Stage three' IPM, where it is the consequence of synergism between methods that gives the IPM output, is still a significant way in the future, as is introducing the considerable potential of manipulating the behaviour of aphids and their natural enemies with semiochemicals.

References

Acheampong, S. and Stark, J.D. (2004) Effects of the agricultural adjuvant Sylgard 309 and the insecticide pymetrozine on demographic parameters of the aphid parasitoid, *Diaeretiella rapae. Biological Control* 31, 133–137.

Ahmad, M. and Shakoori, A.R. (2001) Integration of chemical control and host plant resistance in winter planted rapeseed and mustard against aphid *Brevicoryne brassicae* L. *Proceedings of the Pakistan Congress of Zoology, 2001* 21, 53–60.

Apple, J.L. and Smith, R.F. (1976) *Integrated Pest Management.* Plenum, New York, 200 pp.

Ashouri, A., Michaud, D. and Cloutier, C. (2001) Recombinant and classically selected factors of potato plant resistance to the Colorado potato beetle, *Leptinotarsa decemlineata*, variously affect the potato aphid parasitoid *Aphidius nigripes. BioControl* 46, 401–418.

Attah, P.K. and van Emden, H.F. (1993) The susceptibility to malathion of *Metopolophium dirhodum* on two wheat varieties at two growth stages and the effect of plant growth regulators on this susceptibility. *Insect Science and its Application* 14, 101–106.

Beirne, B.P. (1970) The practical feasibility of pest management systems. In: Rabb, R.L. and Guthrie, F.E. (eds) *Concepts of Pest Management.* North Carolina State University, Raleigh, pp. 158–169.

Croft, B.A. and Brown, A.W.A. (1975) Responses of arthropod natural enemies to insecticides. *Annual Review of Entomology* 20, 285–335.

Dixon, A.F.G. (1985) *Aphid Ecology.* Blackie, Glasgow, 157 pp.

Dodd, G.D. (1973) Integrated control of the cabbage aphid (*Brevicoryne brassicae* (L.)). PhD thesis, The University of Reading, Reading, UK.

Du Li, Ge Feng, Zhu SanRong and Parajulee, M.N. (2004) Effect of cotton cultivar on development and reproduction of *Aphis gossypii* (Homoptera: Aphididae) and its predator *Propylaea japonica* (Coleoptera: Coccinellidae). *Journal of Economic Entomology* 97, 1278–1283.

Eigenbrode, S.D., White, C., Rhode, M. and Simon, C.J. (1998) Behavior and effectiveness of adult *Hippodamia convergens* (Coleoptera: Coccinellidae) as a predator of *Acyrthosiphon pisum* (Homoptera: Aphididae) on a wax mutant of *Pisum sativum. Environmental Entomology* 27, 902–909.

van Emden, H.F. (1966) The effectiveness of aphidophagous insects in reducing aphid populations. In: Hodek, I. (ed.) *Ecology of Aphidophagous Insects.* Academia, Prague, pp. 227–235.

van Emden, H.F. (1978) Insects and secondary substances – an alternative viewpoint with special reference to aphids. In: Harborne, J.B. (ed.) *Phytochemical Aspects of Plant and Animal Co-evolution.* Academic Press, London, pp. 309–323.

van Emden, H.F. (1995) Host plant–aphidophaga interactions. *Agriculture, Ecosystems and Environment* 52, 3–11.

van Emden, H.F. (2002) Integrated pest management. In: Pimentel, D. (ed.) *Encyclopedia of Pest Management.* Dekker, New York, pp. 413–415.

van Emden, H.F. (2003) GM crops: a potential for pest mismanagement. *Acta Agriculturae Scandinavica, Section B, Plant Soil Science* 53 (supplement 1), 26–33.

van Emden, H.F. and Service, M.W. (2004) *Pest and Vector Control.* Cambridge University Press, Cambridge, 349 pp.

van Emden, H.F. and Wearing, C.H. (1965) The role of the aphid host plant in delaying economic damage levels in crops. *Annals of Applied Biology* 56, 323–324.

Gowling, G.R. (1989) Field and glasshouse studies of aphids on the interaction of partial plant resistance and biological control. PhD thesis, The University of Reading, Reading, UK.

Gowling, G.R. and van Emden, H.F. (1994) Falling aphids enhance impact of biological control by parasitoids on partially aphid-resistant plant varieties. *Annals of Applied Biology* 125, 233–242.

Hassell, M.P., Lawton, J.H. and Beddington, J.R. (1977) Sigmoid functional responses by invertebrate predators and parasitoids. *Journal of Animal Ecology* 46, 249–262.

Hatting, J.L., Wraight, S.P. and Miller, R.M. (2004) Efficacy of *Beauveria bassiana* (Hyphomycetes) for control of Russian wheat aphid (Homoptera: Aphididae) on resistant wheat under field conditions. *Biocontrol Science and Technology* 14, 459–473.

Hull, L.A. and Sterner, V.R. (1983) Effectiveness of insecticide applications timed to correspond with the development of rosy apple aphid on apple. *Journal of Economic Entomology* 76, 594–598.

Kalule, T. and Wright, D.J. (2005) Effect of cultivars with varying levels of resistance to aphids on development time, sex ratio, size and longevity of the parasitoid *Aphidius colemani*. *BioControl* 50, 235–246.

Kennedy, G.G. (1984) 2-tridecanone, tomatoes and *Heliothis zea*: potential incompatibility of plant antibiosis with insecticidal control. *Entomologia Experimentalis et Applicata* 35, 305–311.

Lykouressis, D. (1982) Studies under controlled conditions on the effects of parasites on the population dynamics of *Sitobion avenae* (F.). PhD thesis, The University of Reading, Reading, UK.

Mohamad, B.M. and van Emden, H.F. (1989) Host plant modification to insecticide susceptibility in *Myzus persicae* (Sulz.). *Insect Science and its Application* 10, 699–703.

Mackauer, M. (1973) Host selection and host suitability in *Aphidius smithi* (Homoptera: Aphidiidae). In: Lowe, A.D. (ed.) *Perspectives in Aphid Biology*. New Zealand Entomological Society, Auckland, pp. 20–29.

Morse, S. (1989) The integration of partial plant resistance with biological control by an indigenous natural enemy complex in affecting populations of cowpea aphid (*Aphis craccivora* Koch). PhD thesis, The University of Reading, Reading, UK.

Nicol, D., Wratten, S.D., Eaton, N. and Copaja, S.V. (1993) Effects of DIMBOA levels in wheat on the susceptibility of the grain aphid (*Sitobion avenae*) to deltamethrin. *Annals of Applied Biology* 122, 427–433.

Plapp, F.W., Jr. (1981) Ways and means of avoiding or ameliorating resistance to insecticides. *Proceedings of the 9th International Congress of Plant Protection, Washington D.C., August 1979* 2, 244–249.

Powell, W., Dean, G.J., Dewar, A. and Wilding, N. (1981) Towards integrated control of cereal aphids. *Proceedings of the British Crop Protection Conference, Pests and Diseases, Brighton, November 1989* 1, 201–206.

Reed, D.K., Kindler, S.D. and Springer, T.L. (1992) Interactions of Russian wheat aphid, a hymenopterous parasitoid and resistant and susceptible slender wheatgrasses. *Entomologia Experimentalis et Applicata* 64, 239–246.

Selander, J.M., Markkula, M. and Tiittanen, K. (1972) Resistance of the aphids *Myzus persicae* (Sulz.), *Aulacorthum solani* (Kalt.) and *Aphis gossypii* Glov. to insecticides and the influence of the host plant on this resistance. *Annales Agriculturae Fenniae* 11, 141–145.

Silver, A.R.J., van Emden, H.F. and Battersby, M. (1995) A biochemical mechanism of resistance to pirimicarb in two glasshouse clones of *Aphis gossypii*. *Pesticide Science* 43, 21–29.

Smith, J.G. (1969) Some effects of crop background on populations of aphids and their natural enemies on Brussels sprouts. *Annals of Applied Biology* 63, 326–329.

Smith, J.G. (1976) Influence of crop background on natural enemies of aphids on Brussels sprouts. *Annals of Applied Biology* 83, 15–29.

Sotherton, N.W. and Lee, G. (1988) Field assessments of resistance to the aphids *Sitobion avenae* and *Metopolophium dirhodum* in old and modern spring-sown wheats. *Annals of Applied Biology* 112, 239–248.

Starks, K.J., Muniappan, R. and Eikenbary, R.D. (1972) Interaction between plant resistance and parasitism against the greenbug on barley and sorghum. *Annals of the Entomological Society of America* 65, 650–655.

Stern, V.M., Smith, R.F., van den Bosch, R. and Hagen, K.S. (1959) The integrated control concept. *Hilgardia* 29, 81–101.

Storeck, A., Poppy, G.M., van Emden, H.F. and Powell, W. (2000) The role of plant chemical cues in determining host preference in the generalist aphid parasitoid *Aphidius colemani*. *Entomologia Experimentalis et Applicata* 97, 41–46.

Tilahun, D.A. and van Emden, H.F. (1997) The susceptibility of rose–grain aphid (Homoptera: Aphididae) and its parasitoid (Hymenoptera: Aphidiidae) and predator (Coleoptera: Coccinellidae) to malathion on aphid susceptible and resistant wheat cultivars. *Annales de la 4ᵉ ANPP Conférence Internationale sur les Ravageurs en Agriculture, Montpellier, January 1997*, 1137–1148.

Ul-Haq, E. (1997) Interaction of cultural control with biological control of rose–grain aphid, *Metopolophium dirhodum* (Walker) (Aphididae: Hemiptera) on wheat: the potential of glasshouse simulations. PhD thesis, The University of Reading, Reading, UK.

Wickremasinghe, M.G.V. (1989) Behaviour of *Aphidius rhopalosiphi* (Hymenoptera: Aphidiidae) in relation to potential host community location. PhD thesis, The University of Reading, Reading, UK.

Wickremasinghe, M.G.V. and van Emden, H.F. (1992) Reactions of female parasitoids, particularly *Aphidius rhopalosiphi*, to volatile chemical cues from the host plants of their aphid prey. *Physiological Entomology* 17, 291–304.

Yu, S.J. and Hsu, E.L. (1985) Induction of hydrolases by allelochemicals and host plants in fall armyworm (Lepidoptera: Noctuidae) larvae. *Environmental Entomology* 14, 512–515.

21   IPM Case Studies: Brassicas

Rosemary H. Collier and Stan Finch

Warwick HRI, The University of Warwick, Wellesbourne, Warwick, CV35 9EF, UK

Introduction

All members of the family Brassicaceae, which includes the genus *Brassica*, contain specific secondary plant compounds known as glucosinolates. Although plants containing these chemicals are toxic to most aphids, some aphids have overcome the toxins and feed exclusively on such plants (Finch, 1980). The aphid species that have adapted in this way can colonize many of the species of the 220 genera of Cruciferae found throughout the world. Hence, many cruciferous wild plants and garden flowers are sources of pest aphids of brassica crops. Until about 25 years ago, cultivars of *Brassica oleracea* (e.g. cabbage, cauliflower, Brussels sprouts) and *Brassica napus* var. *napobrassicae* (swede) were the commonest crucifers grown on a field scale in the UK. However, following expansion in the area of land drilled with oilseed rape during the mid-1970s, the oil-bearing cultivars of *B. napus* and *Brassica campestris* have become the commonest (Finch and Thompson, 1992). Not only are such plants grown at a much higher density (80–120 plants/m) than the *B. oleracea* cultivars (2–20 per m), but they are also grown on an area of about 500,000 ha, approximately ten times the area used for growing cruciferous vegetable crops.

The main aphids found on brassica crops (e.g. cabbage and mustard) worldwide are the foliage-feeding *Brevicoryne brassicae* (cabbage aphid), *Lipaphis pseudobrassicae*, *Myzus ascalonicus* (shallot aphid), *Myzus persicae* (peach–potato aphid), *Macrosiphum euphorbiae* (potato aphid), *Aphis maidiradicis* (corn root aphid), and the root-feeding *Pemphigus populitransversus* and *Smynthurodes betae* (Blackman and Eastop, 2000). Of these, *B. brassicae*, *M. persicae*, and *L. pseudobrassicae* are the species of most economic importance. This chapter will focus on IPM strategies used for aphid control in the UK, drawing comparisons with strategies employed in other countries where information is available.

Biology of Pest Aphids in the UK

In the UK, *B. brassicae* (Fig. 21.1) often causes severe crop damage. This aphid infests the leaves and shoots of many brassica crops and remains on herbaceous cruciferous plants throughout its life cycle. In most northern areas of its distribution, *B. brassicae* overwinters as an egg on the stems of crop plants that are in the field during the winter period (e.g. autumn-drilled oilseed rape and overwintering vegetable brassica crops). The eggs hatch in February/March and the

Fig. 21.1. Cabbage aphid – *Brevicoryne brassicae.*

resulting aphids colonize nearby host plants. Although *B. brassicae* still overwinters in the egg stage in the UK (N. Kift, personal communication), overwintering eggs have been hard to find during the past 20 years (Chua, 1977; P.R. Ellis, personal communication). It appears that the overwintering population of *B. brassicae* is now made up largely of anholocyclic nymphs and adults (Harrington *et al.*, 1990; R. Collier, unpublished results). During late spring and early summer, alate *B. brassicae* leave their overwintering sites and, after dispersing (Finch and Thompson, 1992), some find new host plants. Once the aphids colonize horticultural brassica crops, aphid numbers increase rapidly. This increase is usually followed by a mid-season 'crash', in which the aphid population is often nearly annihilated (Fig. 21.2). It is not known whether environmental conditions *per se*, host plant maturation, natural biological control agents, or a combination of the three, contribute most to this mid-season crash. Following the crash, aphid numbers increase again in early autumn and then decline at the onset of winter. The early peak in *B. brassicae* numbers usually occurs between mid-July and mid-August, and the late peak between mid-September and mid-December. There are considerable variations in the levels and patterns (Dixon, 1998) of aphid infestations from year to year. For example, although *M. persicae* numbers were low in most brassica crops grown during the 10 years from 1992 to 2001, large infestations were found during the autumns of 1996 and 2001. Unfortunately, it is often difficult to destroy large infestations of aphids in the autumn, even with insecticides that are highly effective earlier in the year (because low temperatures do not favour the activity of certain insecticides). Therefore, late infestations cause major problems for growers, as retailers and consumers have an extremely low tolerance to the presence of aphids or aphid damage in the marketed produce. In addition to direct damage and crop contamination, *B. brassicae* and *M. persicae* can cause further problems by transmitting *Turnip mosaic virus* and *Cauliflower mosaic virus* (Kennedy *et al.*, 1962). *Lipaphis pseudobrassicae* is not a major pest in UK brassica crops.

Integrated Crop Management

In the UK, most *Brassica* crops are grown according to the standards described in

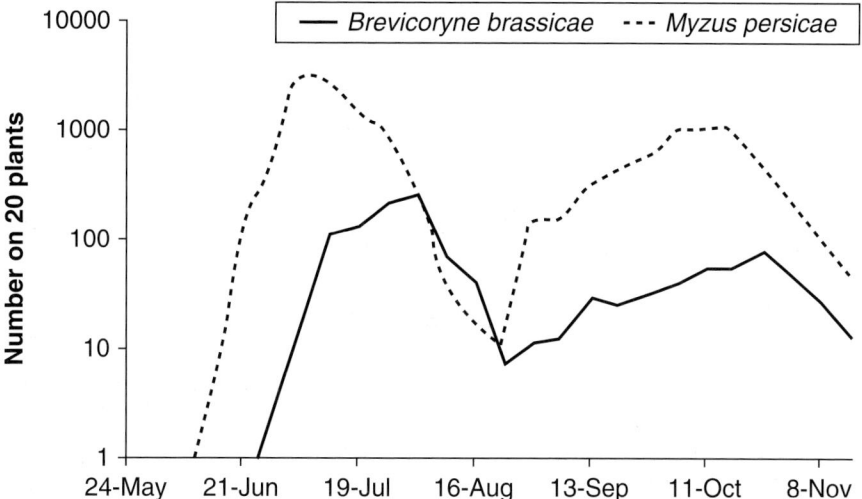

Fig. 21.2. Numbers of *Brevicoryne brassicae* and *Myzus persicae* found on insecticide-free Brussels sprout plants at Kirton, Lincolnshire, UK in 2001 (after Collier *et al.*, 2003). The mid-season crash occurred during August.

the Assured Produce Scheme (www. assuredproduce.co.uk/), a scheme supported by both the major supermarkets and the major vegetable processors. The aim of the scheme is to promote the production of 'safe and environmentally responsible' crops of vegetables. The crops within the scheme should be grown in accordance with the principles of integrated crop management, in which the use of pesticides and fertilizers must be kept to a minimum. Each year, the participating growers are provided with crop-specific protocols that describe 'best' practice. In addition, each grower is required to complete a self-assessment questionnaire and to make his/her premises available for inspection by an independent auditor. The assured produce protocols are not prescriptive; they are simply to advise growers of the best practice and to provide useful sources of information. A similar set of standards and procedures is being used in continental Europe and is based on a EU initiative called EurepGAP (www.eurepgap.org). The Euro Retailer Produce Working Group (EUREP) represents leading European food retailers and uses GAP (Good Agricultural Practice) as a framework for verification. The initiative is being extended to include crop production worldwide.

Chemical Control

In the UK, aphids in commercial *Brassica* crops are controlled exclusively with insecticide. The insecticides can be applied as seed treatments, as granules at sowing or planting, or as foliar sprays. In the recent past, UK growers have relied heavily on organophosphorus (OP), carbamate (particularly the selective aphicide, pirimicarb) and pyrethroid insecticides. However, the UK Department of the Environment, Food and Rural Affairs/Pesticide Safety Directorate (DEFRA/PSD) has been reviewing all compounds with anticholinesterase activity and asking manufacturers to provide additional information on the biological activity of such compounds. As a consequence, insecticide manufacturers have withdrawn some OP and carbamate compounds, of which the current sales would not justify the expense of collecting the new data. In addition, the European Commission (EC) is reviewing all pesticides used for plant protection purposes to ensure that the safety of all pesticides used throughout Europe is evaluated to modern standards. The basis of the EC review programme is set out in Article 8.2 of Council Directive 91/414/EEC. This may also lead to the withdrawal of certain pesticides.

Recent research has shown that *B. brassicae* and *M. persicae* can be controlled using the newer active ingredients imidacloprid and pymetrozine (R. Collier, unpublished results), which are favoured because they are relatively selective. Imidacloprid and pymetrozine are approved in the UK for use as seed treatments and foliar sprays, respectively.

To date, no population of *B. brassicae* has developed resistance to any insecticide in the UK, or elsewhere. In contrast, populations of *M. persicae* have become resistant to various carbamate, organophosphorus, and pyrethroid insecticides (Foster *et al.*, 2000, Chapter 10 this volume). Some *M. persicae* populations have lower susceptibility to imidacloprid, which in most cases has been correlated with an increased tolerance to nicotine (Devine *et al.*, 1996). However, there is no firm evidence that neonicotinoids such as imidacloprid are failing to control *M. persicae* effectively under field conditions (Nauen and Denholm, 2005). With the plasticity exhibited by *M. persicae* populations, it is important to adopt strategies that will prevent the build-up of resistance. Consequently, an Insecticide Resistance Action Group has been formed and has published resistance management guidelines for *M. persicae*. Although no formal system exists for testing aphids for their resistance to insecticides, *M. persicae* populations are tested regularly by insecticide manufacturers and by some research organizations.

Monitoring and Forecasting

Brevicoryne brassicae and *M. persicae* are both caught in the network of suction traps run by the Rothamsted Insect Survey (see Harrington *et al.*, Chapter 19 this volume). Until recently, the current information from this network was available only to a restricted number of researchers and advisors. However, the Horticultural Development Council (HDC) – a statutory body responsible for collecting an 'industry levy' used to fund research/development of direct benefit to UK growers – enabled the information collected during 2001 to be made available also to growers by e-mail and through the Internet. This information included the numbers of *B. brassicae* and *M. persicae* caught each week in the suction traps, together with some interpretation of what the numbers indicate. Since 2004, this information has been available as a 'bulletin' on a website funded by the HDC.

Analyses of long-term data sets from the Rothamsted Insect Survey have shown that, for several aphid species, there is a negative correlation between winter temperature and the date the first alate of a given species is trapped (Harrington *et al.*, 1990). The relationship with winter temperature is particularly pronounced for *M. persicae*, a species that is largely anholocyclic in the UK (Harrington *et al.*, 1990). There is also often a strong positive correlation between winter temperature and aphid abundance, especially for *M. persicae*, up until early July. This information is used to alert growers to the likely need for, and timing of, control of *M. persicae* in sugarbeet (*Beta vulgaris*) crops. Similar forecasts, combined with current information on aerial aphid abundance, should help growers of vegetable brassica crops to decide when it is worthwhile to look for aphids. A similar forecast for *B. brassicae* is currently being developed (R. Harrington, personal communication).

Day-degree models for *B. brassicae* and *M. persicae* are also available from the University of California Statewide Integrated Pest Management Project (www.ipm.ucdavis.edu). Although such models are based on published data and are supported by a comprehensive list of references, they have not been validated in the UK.

In the UK, crop consultants and representatives of pesticide companies provide local information on the development of aphid infestations by walking and inspecting on a regular basis ('scouting') the brassica crops under their jurisdiction. Between 1996 and 2001, scientists from Horticulture Research International (now Warwick HRI, The University of Warwick) supplied growers in South Lincolnshire (the area of eastern England in which most vegetable brassica crops are grown) with information on aphid infestations. Such information was collected by taking weekly records of the numbers of aphids found on insecticide-free brassica plots. Although the information collected was faxed weekly to subscribers, the service ceased after 2001 because of insufficient grower funding. In Germany, members of the extension service warn growers, by fax or telephone, as soon as large numbers of aphids start to migrate into brassica crops (M. Hommes, personal communication). In the USA, control advice and information about crop scouting are available on a number of websites managed by the extension services (e.g. University of California (UC IPM Online)).

Sampling and Decision Making

In the UK, vegetable brassica growers spend considerable time inspecting their crops for aphids. The information collected is used to make decisions about applying sprays to their crops and is also documented carefully as part of the Assured Produce Scheme. Unfortunately, no standard method has been developed for estimating aphid numbers in commercial crops. Therefore, the spray thresholds used by growers vary from region to region, as they are based on the method used there for estimating aphid numbers.

Systems of supervised control, based on spray thresholds for pest aphids of vegetable brassica crops, have been available for some time in Northern Europe (Ellis *et al.*, 1988; Hommes *et al.*, 1988; Theunissen, 1988; Freuler *et al.*, 1991; Blood Smyth *et al.*, 1992, 1994; Paterson *et al.*, 1994; Hildenhagen and Hommes, 1997). However, no figures have been published to indicate how many commercial growers use spray thresholds. In Germany, although the extension service circulates spray thresholds (Hildenhagen and Hommes, 1997) to most growers, they are used by only 10–25% of growers (M. Hommes, personal communication). Similarly, in Switzerland, spray thresholds are used by less than 5% of growers (R. Baur, S. Fischer, personal communications), and in Sweden (B. Jonsson, personal communication) and

France (F. Villeneuve, personal communication) they are not used at all. Therefore, despite considerable inputs into developing sampling and decision-making systems, uptake by growers has been poor, possibly because the original systems were too inflexible and/or too time-consuming. Research has also been carried out to develop an action threshold for *B. brassicae* in oilseed rape in the UK (Ellis *et al.*, 1999).

However, as applying sprays on a prophylactic basis is no longer acceptable, some form of supervised control will have to be adopted. Therefore, some of the methods developed in The Netherlands (Theunissen, 1988) are now being revisited in the UK (Perry *et al.*, 1998; Collier, 1999; Collier and Mead, 1999; Collier *et al.*, 2003) in the hope that they can be made user-friendlier. Obviously, the reliability of the spray thresholds on which any new system is based is of paramount importance. Hence, the major thrust of the current research is to ensure that the sampling methods used to obtain the underlying data are as robust as possible.

In the USA, the major pests of brassica crops are the caterpillars of Lepidoptera, as aphids are regarded only as sporadic pests. Nevertheless, information on crop walking ('scouting') and treatment thresholds for aphids is available. For example, the University of California Management Guidelines includes a presence/absence sequential sampling programme for deciding how best to control *B. brassicae* infestations in Brussels sprout crops. In the midwest USA, treatment for *B. brassicae* infestations is suggested if 1–2% of the plants are infested. If *M. persicae* is the infesting species, then populations of 100 aphids per plant can be tolerated (Eastman *et al.*, 2005).

In some brassica crops, aphids are allowed to build up to relatively high numbers before a spray is applied. In Germany, for example, when brassica plants are infested with less than 100 aphids/plant, sprays are applied only when 20% of the plants have become infested. However, when more than 100 aphids are found on any plant, the spray threshold is lowered to 10% of infested plants (M. Hommes, personal communication). Similarly, in the USA, insecticides are applied to the vegetative stage of broccoli and cauli-flower crops only when infestation exceeds 100 aphids/plant. In contrast, once such crops start to form heads, sprays are applied as soon as 5 aphids/plant are found (Eastman *et al.*, 1995). This, again, is because com-pletely removing any aphids close to harvest is difficult.

Research has been undertaken in Asia to establish treatment thresholds for the pest aphids of brassica crops (e.g. Singh and Malik, 1998; Chen YongNian *et al.*, 2000 – both for *L. pseudobrassicae*). However, how widely such thresholds are used in practice cannot be determined.

Biological Control

Aphid infestations on brassica crops are reduced by specific predators and para-sitoids (Hafez, 1961; Dunn and Kempton, 1971; Raworth *et al.*, 1984; Hart *et al.*, 1997), by a range of polyphagous predators (Sunderland *et al.*, 1987), and by entomo-pathogenic fungi (Dunn and Kempton, 1971; Milner, 1997; McLeod *et al.*, 1998) (see also Völkl *et al.*, Chapter 8 this volume). Most pest management guidelines recognize the importance of natural enemies and propose that natural enemies should be considered when making treatment decisions. However, nobody has indicated how the application of insecticide sprays against aphids should be modified to take account of the control contribution from natural enemies.

In addition, there is no concerted effort to conserve natural enemies by (i) ensuring that only selective insecticides are sprayed, or (ii) altering the cropping system to pro-vide predators and parasitoids with addi-tional sources of food and shelter. The latter is normally achieved by growing flowering plants alongside crop plants. For example, van Emden (1965) showed that growing flow-ers alongside a Brussels sprout crop inc-reased the hover fly (Syrphidae) population on the crop plants adjacent to the flowers, giving a 15–20% reduction in aphid num-bers. Similarly, in New Zealand, fewer aphids (*B. brassicae* and *M. persicae*) were recorded in plots of cabbage surrounded by

phacelia (*Phacelia tanacetifolia*), a plant grown specifically to provide hover flies with a plentiful supply of pollen (White *et al.*, 1995). In both cases, however, aphid num-bers were not reduced sufficiently to give commercially acceptable levels of control. To date, no detailed research has been done in the UK on the biological control of aphids in field brassica crops. Although some poten-tial arthropod biological agents have been considered elsewhere in the world (Raworth, 1984; Makhmoor and Verma, 1989; Nirmala Devi *et al.*, 1996), control of aphids using arthropods has not been demonstrated in any field brassica crop. However, biopesticides based on entomopathogenic fungi are now approved for use on field brassica crops in certain countries, e.g. the USA (for example, 'BotaniGard' containing *Beauveria bassiana*).

Host-plant Resistance

Certain species and cultivars of *Brassica* are more resistant to aphid infestation than oth-ers (Singh and Ellis, 1993; Ellis *et al.*, 1998). Unfortunately, the current levels of resis-tance to aphids of such cultivars are not suf-ficient to ensure that the 'partially resistant' crops remain aphid free. Therefore, insecti-cides still have to be applied. It has been suggested that the benefits from partially resistant cultivars, natural enemies, and insecticides could be combined to improve overall aphid control (Verkerk *et al.*, 1998). However, such an approach has not been tested in the field. There is no evidence that growers select cultivars because of their resistance to pest aphids.

Cultural Control

Crop covers (Fig. 21.3)

Fine-mesh covers can be used to keep aphids off most crop plants (Bedford *et al.*, 1994; Evans *et al.*, 1997). In the UK, and in Switzerland (R. Baur, personal communica-tion), crop covers are considered too expensive for large-scale use. However, a significant proportion of the UK swede crop

Fig. 21.3. Crop covers being used to exclude pest insects in a field trial at Warwick HRI, Wellesbourne, UK.

used for human consumption is covered in fine-mesh netting to exclude *Delia radicum* (cabbage root fly), for which, since the withdrawal of chlorfenvinphos, there is no effective approved insecticide. Crop covers are also used routinely in Germany on cauliflower and Chinese cabbage crops, as the plants command a high price at market (M. Hommes, personal communication). In contrast to crop covers, reflective mulches (supposed to deter aphids from landing on brassica crops) have not been effective anywhere in Europe.

(Fig. 21.4) reduced *B. brassicae* infestations by 78–95% when compared to cabbage grown in bare soil. The clover simply disrupted host-plant colonization by the aphids (Finch and Edmonds, 1994; Finch and Collier, 2000). If, as Finch and Collier (2000) state, it is just the number of green objects surrounding a host plant that reduces colonization by aphids, then it should not be too difficult to identify the type of background plants needed in future to reduce aphid numbers in any given brassica crop.

Increasing crop diversity

Many researchers have shown that the numbers of aphids found in brassica crops are reduced considerably when the crop background is allowed to become weedy (Smith, 1969; Dempster and Coaker, 1974), when the crop is intercropped with another plant species (Altieri *et al.*, 1985; Andow *et al.*, 1986), or when the crop is undersown with clover (Finch and Kienegger, 1997). In field experiments, Finch and Kienegger (1997) showed that undersowing cabbage plants with clover

Executive Summary

Insecticides are still really the sole tool used by the vast majority of UK brassica growers, although considerable progress has been made in establishing spray thresholds for the two main pest aphids, *B. brassicae* and *M. persicae*. At present, however, prophylactic spraying is still common; only a tiny percentage of growers use spray thresholds. Systems for monitoring and forecasting aphid populations on behalf of the grower are well developed and sophisticated, and growers

Fig. 21.4. Cabbage plants undersown with clover to reduce colonization by pest insects.

receive considerable help from extension services about when spraying is likely to be necessary. This reduces substantially the numbers of sprays that would otherwise be applied prophylactically by the majority of growers.

This dependence on pesticides could change if, as seems likely, organophosphate, carbamate, and some pyrethroid insecticides are withdrawn, and should resistance to the newer insecticides, such as imidacloprid, then occur.

Except in some supermarket protocols, insecticides are rarely selected on the basis of being less harmful to natural enemies. No predatory arthropods are released to reduce aphid numbers in brassica crops. Biopesticides based on entomopathogenic fungi are now available in the USA and could be used in future IPM in the UK.

Although plant cultivars are available that are partially resistant to aphids, the resistance levels are still not high enough to induce producers to grow such cultivars, even though supermarket protocols recommend them.

Of the methods of cultural control available, some growers use crop covers, but the remaining methods are 'still on the research bench'.

At present, semiochemicals are not applied to disrupt pest aphids or to manipulate natural enemies in brassica crops.

Apart from using aphid forecasts and crop scouting to reduce the number of prophylactic sprays, the only other aspects of IPM being considered currently are those in which the supermarkets indicate in their protocols that, whenever possible, they prefer partially-resistant cultivars and selective pesticides to be used.

References

Altieri, M.A., Wilson, R.C. and Schmidt, L.L. (1985) The effects of living mulches and weed cover on the dynamics of foliage- and soil-arthropod communities in three crop systems. *Crop Protection* 4, 201–213.

Andow, D.A., Nicholson, A.G., Wien, H.C. and Willson, H.R. (1986) Insect populations on cabbage grown with living mulches. *Environmental Entomology* 15, 293–299.

Bedford, I.D., Markham, P.G. and Strauss, P.A. (1994) A study of the effectiveness of crop covering within IPM, using an Amoco non-woven fleece as a barrier to aphids, whiteflies and their associated plant viruses. *Proceedings of the Brighton Crop Protection Conference, Pests and Diseases, November 1994* 3, 1163–1168.

Blackman, R.L. and Eastop, V.F. (2000) *Aphids on the World's Crops. An Identification and Information Guide.* Wiley, Chichester, 466 pp.

Blood Smyth, J.A., Davies, J., Emmett, B.J., Lole, M., Paterson, C. and Powell, V. (1992) Supervised control of aphid and caterpillar pests in Brassica crops. *Bulletin IOBC/WPRS* 15 (4), 9–15.

Blood Smyth, J.A., Emmett, B.J., Mead, A., Davies, J.S., Paterson, C.D. and Runham, S. (1994) Supervised control of foliar pests of Brussels sprouts and calabrese crops. *Bulletin IOBC/WPRS* 17 (8), 41–50.

Chen YongNian, Chen Chan and Ma Jun (2000) Study on the action threshold for aphids in optimised pest insect management of cabbage. *Plant Protection* 26, 10–13.

Chua, T.H. (1977) Population studies of *Brevicoryne brassicae* (L.) its parasites and hyperparasites in England. *Researches on Population Ecology* 19, 125–139.

Collier, R.H. (1999) Integrated control of aphid pests of lettuce and brassica crops in the UK. *Mededelingen Faculteit Landbouwkundige en Toegepaste Biologische Wetenschappen Universiteit Gent* 64 (3a), 3–9.

Collier, R.H. and Mead, A. (1999) Simulating sampling strategies for aphid and caterpillar pests of brassica crops. In: *Integrated Control in Field Vegetable Crops. Bulletin IOBC/WPRS* 22 (5), 1–7.

Collier, R.H., Mead, A., Parker, W.E. and Ellis, S.A. (2003) A risk management system for controlling the foliar pests of *Brassica* crops. *Proceedings of the British Crop Protection Council International Congress on Crop Science and Technology, Glasgow, November 2003* 1, 335–340.

Dempster, J.P. and Coaker, T.H. (1974) Diversification of crop ecosystems as a means of controlling pests. In: Price Jones, D. and Solomon, M.E. (eds) *Biology in Pest and Disease Control.* Wiley, New York, pp. 106–114.

Devine, G.J., Harling, Z.K., Scarr, A.W. and Devonshire, A.L. (1996) Lethal and sublethal effects of imidacloprid on nicotine-tolerant *Myzus nicotianae* and *Myzus persicae*. *Pesticide Science* 48, 57–62.

Dixon, A.F.G. (1998) *Aphid Ecology.* Chapman and Hall, London, 300 pp.

Dunn, J.A. and Kempton, D.P.H. (1971) Seasonal changes in aphid populations on Brussels sprouts. *Annals of Applied Biology* 68, 233–244.

Eastman, C., Mahr, S., Wyman, J., Radcliffe, E., Hoy, C. and Oloumi-Sadeghi, H. (1995) Cabbage, broccoli and cauliflower. In: Foster, R. and Flood, B. (eds) *Vegetable Insect Management. With Emphasis on the Midwest.* Meister Publishing Company, Willoughby, Ohio, pp. 99–112.

Eastman, C., Barrido, R., Oloumi-Sadeghi, H., Hoy, C., Wyman, J., Palumbo, J., Leibee, G. and Flood, B.R. (2005) Cabbage, broccoli and cauliflower. In: Foster, R. and Flood, B.R. (eds) *Vegetable Insect Management.* Meister Media Worldwide, Willoughby, Ohio, pp. 156–173.

Ellis, P.R., Hardman, J.A., Hommes, M., Dunne, R., Fischer, S., Freuler, J., Kahrer, A. and Terretaz, C. (1988) An evaluation of supervised systems for applying insecticide treatments to control aphid and foliage caterpillar pests of cabbage. *Proceedings of the Brighton Crop Protection Conference, Pests and Diseases, November 1988* 1, 269–274.

Ellis, P.R., Pink, D.A.C., Phelps, K., Jukes, P.L., Breeds, S.E. and Pinnegar, A.E. (1998) Evaluation of a core collection of *Brassica oleracea* accessions for resistance to *Brevicoryne brassicae*, the cabbage aphid. *Euphytica* 103, 149–160.

Ellis, S.A., Oakley, J.N., Parker, W.E. and Raw, K. (1999) The development of an action threshold for cabbage aphid (*Brevicoryne brassicae*) in oilseed rape in the UK. *Annals of Applied Biology* 134, 153–162.

van Emden, H.F. (1965) The effect of uncultivated land on the distribution of the cabbage aphid (*Brevicoryne brassicae*) on an adjacent crop. *Journal of Applied Ecology* 2, 171–196.

Evans, A., Wratten, S., Frampton, C., Causer, S. and Hamilton, M. (1997) Row covers: effects of wool and other materials on pest numbers, microclimate, and crop quality. *Journal of Economic Entomology* 90, 1661–1664.

Finch, S. (1980) Chemical attraction of plant-feeding insects to plants. In: Coaker, T.H. (ed.) *Applied Biology, Volume 5.* Academic Press, London, pp. 67–143.

Finch, S. and Collier, R.H. (2000) Host-plant selection by insects – a theory based on 'appropriate/inappropriate landings' by pest insects of cruciferous plants. *Entomologia Experimentalis et Applicata* 96, 91–102.

Finch, S. and Edmonds, G.H. (1994) Undersowing cabbage crops with clover – the effects on pest insects, ground beetles and crop yield. *Bulletin IOBC/WPRS* 17 (8), 159–167.

Finch, S. and Kienegger, M. (1997) A behavioural study to help clarify how undersowing with clover affects host-plant selection by pest insects of brassica crops. *Entomologia Experimentalis et Applicata* 84, 165–172.

Finch, S. and Thompson, A.R. (1992) Pests of cruciferous crops. In: McKinlay, R.G. (ed.) *Vegetable Crop Pests*. Macmillan, Basingstoke, pp. 87–138.

Foster, S.P., Denholm, I. and Devonshire, A.L. (2000) The ups and downs of insecticide resistance in peach–potato aphids (*Myzus persicae*) in the UK. *Crop Protection* 19, 873–879.

Freuler, J., Fischer, S., Hurni, B. and Stadler, E. (1991) Kontrollmethoden und Anwendung von Schaden-schwellen fur die Schädlinge im Freilandgemüsebau. *Landwirtschaft Schweiz* 4, 341–360.

Hafez, M. (1961) Seasonal fluctuations of population density of the cabbage aphid, *B. brassicae* (L.), in the Netherlands and the role of its parasite, *Aphidius (Diaeretiella) rapae* (Curtis). *Tijdschrift voor Planten-ziekten* 67, 445–548.

Harrington, R., Tatchell, G.M. and Bale, J.S. (1990) Weather, life cycle strategy and spring populations of aphids. *Acta Phytopathologica et Entomologica Hungarica* 25, 423–432.

Hart, A.J., Bale, J.S. and Fenlon, J.S. (1997) Developmental threshold, day-degree requirements and voltinism of the aphid predator *Episyrphus balteatus* (Diptera: Syrphidae). *Annals of Applied Biology* 130, 427–437.

Hildenhagen, R. and Hommes, M. (1997) Mehlige Kohlblattlaus nach Schwellenwerten bekämpfen. *Gemüse* 33, 316–320.

Hommes, M., Dunne, R., Ellis, P.R., Fischer, S., Freuler, J., Kahrer, A. and Terretaz, C. (1988) Testing damage thresholds for caterpillars and aphids on cabbage in five European countries – report on a collaborative project done in 1985 and 1986. *Bulletin IOBC/WPRS* 11 (1), 118–126.

Kennedy, J.S., Day, M.F. and Eastop, V.F. (1962) *A Conspectus of Aphids as Vectors of Plant Viruses*. Commonwealth Institute of Entomology, London, 114 pp.

Makhmoor, H.D. and Verma, A.K. (1989) The intrinsic rate of natural increase of *Metasyrphus confrater* (Wiedemann) (Diptera: Syrphidae), a predator of the cabbage aphid (Homoptera: Aphididae). *Proceedings of the Indian National Science Academy B* 55, 79–84.

McLeod, P.J., Steinkraus, D.C., Correll, J.C. and Morelock, T.E. (1998) Prevalence of *Erynia neoaphidis* (Entomophthorales: Entomophthoraceae) infections of green peach aphid (Homoptera: Aphididae) on spinach in the Arkansas River Valley. *Environmental Entomology* 27, 796–800.

Milner, R.J. (1997) Prospects for biopesticides for aphid control. *Entomophaga* 42, 227–239.

Nauen, R. and Denholm, I. (2005) Resistance of insect pests to neonicotinoid insecticides: current status and future prospects. *Archives of Insect Biochemistry and Physiology* 58, 200–215.

Nirmala Devi, Desh Raj and Verma, S.C. (1996) Biology and feeding potential of *Coccinella septempunctata* Linn. (Coccinellidae: Coleoptera) on cabbage aphid *Brevicoryne brassicae* Linn. *Journal of Entomological Research* 20, 23–25.

Paterson, C., Mead, A., Blood Smyth, J.A., Davies, J. and Runham, S. (1994) Using a sequential sampling method for supervised control of pests in Brussels sprouts and calabrese. *Aspects of Applied Biology* 37, 73–82.

Perry, J.N., Parker, W.E., Alderson, L., Korie, S., Blood-Smyth, J.A., McKinlay, R. and Ellis, S.A. (1998) Simulation of counts of aphids over two hectares of Brussels sprout plants. *Computers and Electronics in Agriculture* 21, 33–51.

Raworth, D.A. (1984) Population dynamics of the cabbage aphid, *Brevicoryne brassicae* (Homoptera: Aphididae), at Vancouver, British Columbia IV. Predation by *Aphidoletes aphidimyza* (Diptera: Cecidomyiidae). *Canadian Entomologist* 116, 889–893.

Raworth, D.A., Frazer, B.D., Gilbert, N. and Wellington, W.G. (1984) Population dynamics of the cabbage aphid, *Brevicoryne brassicae* (Homoptera: Aphididae), at Vancouver, British Columbia I. Sampling methods and population trends. *Canadian Entomologist* 116, 861–870.

Singh, R. and Ellis, P.R. (1993) Sources, mechanisms and bases of resistance in Cruciferae to the cabbage aphid, *Brevicoryne brassicae*. In: *Breeding for Resistance to Insects and Mites*. *Bulletin IOBC/WPRS* 16 (5), 21–35.

Singh, S.V. and Malik, Y.P. (1998) Population dynamics and economic thresholds of *Lipaphis erysimi* on mustard. *Indian Journal of Entomology* 60, 43–49.

Smith, J.G. (1969) Some effects of crop background on populations of aphids and their natural enemies on Brussels sprouts. *Annals of Applied Biology* 63, 326–330.

Sunderland, K.D., Crook, N.E., Stacey, D.L. and Fuller, B.J. (1987) A study of feeding by polyphagous predators on cereal aphids using ELISA and gut dissection. *Journal of Applied Ecology* 24, 907–933.

Theunissen, J. (1988) Sequential sampling of insect pests in Brussels sprouts. *Bulletin IOBC/WPRS* 11 (1), 111–115.

Verkerk, R.H.J., Neugebauer, K.R., Ellis, P.R. and Wright, D.J. (1998) Aphids on cabbage: tritrophic and selective insecticide interactions. *Bulletin of Entomological Research* 88, 343–349.

White, A.J., Wratten, S.J., Berry, N.A. and Weigmann, U. (1995) Habitat manipulation to enhance biological control of *Brassica* pests by hover flies (Diptera: Syrphidae). *Journal of Economic Entomology* 88, 1171–1176.

22 IPM Case Studies: Berry Crops

Rufus Isaacs[1] and J.A. Trefor Woodford[2]

[1]*Department of Entomology, Michigan State University, East Lansing, MI 48824, USA;*
[2]*Scottish Crop Research Institute, Invergowrie, Dundee, DD2 5DA, UK*

Introduction

The perennial crops referred to as berries, soft fruits, or small fruits are grown throughout the world, with primary regions of production in Western and Eastern Europe, the Americas, and Australasia. Economically significant berry crops include grape (*Vitis* spp.), strawberry (*Fragaria* × *ananassa*), blueberry (*Vaccinium* spp.), American cranberry (*Vaccinium macrocarpon*), and species and hybrids in the genus *Rubus*. These include European red raspberry (*Rubus idaeus* subsp. *idaeus*), North American red raspberry (*Rubus idaeus* subsp. *strigosus*), black raspberry (*Rubus occidentalis*), blackberry (*Rubus fruticosus* agg.), cut-leaved blackberry (*Rubus laciniatus*), California blackberry (*Rubus ursinus*), and loganberry and boysenberry (the *Rubus* × *loganbaccus* hybrids between European red raspberry and blackberry). They are all high-value crops, often with a long establishment period incurring high capital costs. Increasingly, berry crops are grown under complete or partial cover to improve yields and quality, and to manipulate the timing of harvest. This change will have implications for many aspects of their pest management, and is expected to provide an environment more suitable for releasing aphid natural enemies (Cross *et al.*, 2001).

Aphids infesting berry crops (Blackman and Eastop, 2000) are rarely abundant enough to cause feeding damage. More often, their primary economic impact is as virus vectors. Infection with these diseases can cause debilitating losses of productivity, and infected plants must be removed, leading to increased replant costs. Aphids may also contaminate harvested fruit and thus reduce fruit quality. The demand for high quality fruit has continued to drive the use of insecticides for aphid control, but pesticide options are limited in minor crops such as berry crops. Food safety concerns, increasing pre-harvest intervals for pesticides, and pest resistance to pesticides have all contributed to the development of integrated management strategies that minimize the use of toxicants and maintain fruit quality. This chapter provides a general overview of successful strategies for management of aphids in berry crops, with focus on aphid IPM programmes in raspberry and blueberry.

Aphid IPM in Raspberry

Raspberry (*R. idaeus*) is a high-value crop, grown commercially in Europe and temperate regions of the Americas and Australasia for the fresh and processing markets. World production in 1999 was estimated to be

352,000 t, mainly from Europe and North America (FAO, 2002). In the North American literature, this crop is referred to as red raspberry to distinguish it from the black raspberry (*R. occidentalis*), a species indigenous to North America. Blackman and Eastop (2000) list 11 aphid species occurring on European red and black raspberries, but only four cause economic damage (*Amphorophora idaei* (Fig. 22.1), *Amphorophora agathonica*, *Aphis idaei*, and *Aphis rubicola*). Aphids can cause contamination of machine-harvested raspberries (Kieffer *et al.*, 1983; Gordon *et al.*, 1997a), but are more important as vectors of a number of viruses (Jones, 1986; Converse, 1987) (Table 22.1). Virus diseases reduce yield and fruit quality and produce a range of symptoms depending on the virus, virus complex, and cultivar. European and North American red raspberry cultivars are susceptible to aphid-borne viruses transmitted by *Amphorophora* spp. and *Aphis* spp. (Table 22.1). However, many are tolerant and can be infected without showing clear symptoms. Mosaic diseases resulting from viruses transmitted by *Amphorophora* species are widespread, but cause symptomless decline in most cultivars (Converse, 1987; Jones and

McGavin, 1998). However, sensitive cultivars, e.g. 'Glen Clova', introduced to Britain in 1970, show severe symptoms and may die within 2–4 years of infection (Jones, 1986).

The costs of establishing a raspberry plantation are high, but fruit can be harvested for many years if the crop remains free from disease (Gordon *et al.*, 1997b). Control depends on planting certified virus-free stocks and preventing them from becoming infected (Jones, 1986). Insecticides have proved ineffective in preventing the introduction of the most important viruses, vectored by *Amphorophora* species (Taylor and Chambers, 1969; Freeman and Stace-Smith, 1970). Because these viruses are transmitted in a semi-persistent manner (Jones, 1986), they can be spread before immigrant viruliferous alatae are killed. Although insecticides may limit virus spread within plantations, aphid control is not usually the primary reason for insecticide use in raspberry crops. In addition, insecticides cannot be applied at harvest time, when aphid populations usually peak, because of the risk of chemical residues on the fruit.

In contrast to methods employed in blueberries (see below), the removal of

Fig. 22.1.   Apterous virginopara and nymph of the large raspberry aphid, *Amphorophora idaei*, feeding on a raspberry leaf (from colour photo copyright Scottish Crops Research Institute, reproduced with permission).

Table 22. 1. Vectors of the major aphid-borne viruses of berry crops. Information from Converse (1987) and Blackman and Eastop (2000).

Natural vector	Crop	Virus	Disease
Amphorophora idaei	European red raspberry	*Black raspberry necrosis virus (BRNV)*	Symptomless decline
		Raspberry leaf mottle virus (RLMV)	Raspberry leaf spot mosaic
		Raspberry leaf spot virus (RLSV)	Raspberry leaf spot mosaic
			Yellow mosaic
			Raspberry vein banding mosaic
Amphorophora agathonica	North American red raspberry	BRNV	Symptomless decline
		BRNV + *Rubus yellow net virus (RYNV)*	Yellow mosaic
		RLMV ?, RLSV ?	Raspberry leaf spot mosaic
Amphorophora agathonica	Black raspberry	BRNV	Tip necrosis
Aphis idaei	European red raspberry	*Raspberry vein chlorosis virus (RVCV)*	Vein chlorosis
Aphis rubicola	North American red raspberry	*Raspberry leaf curl virus (RLCV)*	Leaf curl
	Black raspberry	RLCV	Leaf curl
	Blackberry	RLCV	Leaf curl
Aphis gossypii	Strawberry	*Strawberry mottle virus (SMoV)*	Strawberry mottle
Chaetosiphon fragaefolii	Strawberry	SMoV	Strawberry mottle
and other *Chaetosiphon* spp. including *C. fragaefolii*		*Strawberry crinkle virus (SCV)*	Strawberry crinkle
		Strawberry latent C virus (SLcV)	Strawberry latent C disease
		Strawberry mild yellow edge associated virus (SMYEaV)	Strawberry mild yellow-edge
		Strawberry vein banding virus (SVBV)	Strawberry vein banding
Illinoia pepperi	Blueberry	*Blueberry shoestring virus (BSSV)*	Blueberry shoestring
Ericaphis (= Fimbriaphis) fimbriata	Blueberry	*Blueberry scorch virus (BlSV)*	Blueberry scorch

infected plants (roguing) is of little value in preventing the spread of aphid-transmitted raspberry viruses because they induce no symptoms in most modern cultivars. Fortunately, strong and heritable sources of resistance to *Amphorophora* species have been used by plant breeders to diminish the importance of viruses transmitted by these aphids. All modern commercial raspberry cultivars contain resistance to *A. idaei* and *A. agathonica*, provided either by major or minor genes (Keep and Knight, 1967; Daubeny, 1972; Jones, 1986). The existence of heritable resistance to colonization by *A. agathonica* was shown over 60 years ago in the USA (Huber and Schwartze, 1938). This resistance was derived from European raspberry cultivars (mainly cv. 'Lloyd George') and resulted from a single dominant gene, Ag_1 (Daubeny, 1966). An additional benefit is that *A. agathonica* acquires virus less readily from raspberry genotypes resistant to that vector than from aphid-susceptible genotypes (Stace-Smith, 1960). This resistance has been effective in controlling the spread of raspberry mosaic (a disease caused by mixed infection by up to four viruses) in the Pacific Northwest of North America for many decades (Daubeny, 1980).

Similarities in the gross morphology and biology of *Amphorophora* species in Europe and North America delayed progress in the identification of further sources of resistance. In both parts of the world, these large, green aphids are holocyclic and monoecious on *Rubus*, producing several parthenogenetic generations over the summer (Dicker, 1940; Kennedy and Schaefers, 1974). Their taxonomic identity was confused for many years (Kennedy *et al.*, 1962) and, prior to the early 1970s, the species *A. agathonica* was referred to as a North American race of *Amphorophora rubi* (Daubeny, 1972). However, whereas the British cultivar 'Lloyd George' was resistant to *A. rubi* in North America and remained free from mosaic, it was highly susceptible to *A. rubi* and virus infection in Europe (Hill, 1956, 1957; Rautapää, 1967; Jones, 1976). Blackman *et al.* (1977) showed that the species on European raspberry is *A. idaei* (2n = 18) and that it can be separated by morphological characters and cytology

from *A. rubi* (2n = 20), a species confined largely to *R. fruticosus* agg. and other host plants in the blackberry group. They also confirmed that *A. agathonica*, which originates from native red raspberry *R. idaeus* subsp. *strigosus*, is a distinct species with a chromosome complement of 2n = 14. Knight *et al.* (1959) identified a single dominant gene, A_1, from cv. 'Baumforth A' as a source of strong resistance to *A. idaei*. The ability to select resistant progeny plants accurately and rapidly has led to a succession of aphid-resistant cultivars becoming available to European growers since the early 1970s. Twelve dominant resistance genes have been identified from these cultivars (Keep, 1989), although some of these genes are ineffective against virulent biotypes of *A. idaei* (Briggs, 1959, 1965). The 'universal' resistance gene, A_{10}, from *R. occidentalis*, cv. 'Cumberland', which prevents colonization by most biotypes, has been incorporated into most new cultivars (Keep, 1989).

Resistance to *Amphorophora* provides an outstanding example of the value and durability of vector resistance for controlling aphid-borne virus spread in the field (Jones, 1987). Virus infection in the aphid-susceptible cultivar 'Malling Jewel' reached 100% in four growing seasons, compared with < 20% in aphid-resistant genotypes (Fig. 22.2). Those with the highest resistance to *A. idaei* were infected least rapidly; almost half the number of plants containing A_1 or A_{10} resistance genes were still uninfected after seven growing seasons (Jones, 1976, 1979). Genotypes with minor gene resistance (Jennings, 1963) also became infected less rapidly than aphid-susceptible genotypes. Both forms of resistance to *A. idaei* have proved effective in limiting the spread of viruses transmitted by this vector in commercial crops (0.5–1 ha), with virus incidence rarely exceeding 5% after 10 years cultivation (Jones, 1987).

Molecular analysis, using rDNA IGS markers, showed that UK populations of *A. idaei* contained a wide range of genotypes (Birch *et al.*, 1994; Fenton *et al.*, 1994). The widespread cultivation of cultivars containing gene A_1 or A_{10} has imposed strong selection pressure on *A. idaei* (Jones, 1988). In a recent survey, 77% of *A. idaei* from plantations in

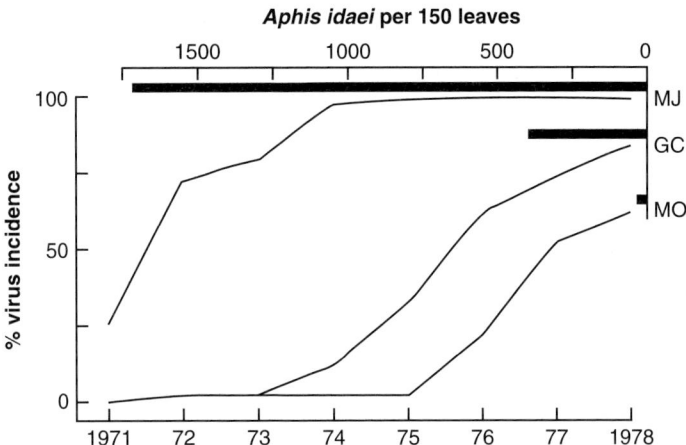

Fig. 22.2. Effect of resistance to *Amphorophora idaei* in raspberry on incidence of viruses transmitted by this vector in Scotland during 1971–1978. Histograms indicate relative numbers of *A. idaei* found on 150 leaves during aphid population peaks, averaged across all years. Cultivars and their resistance to *A. idaei* are: MJ, 'Malling Jewel' (susceptible); GC, 'Glen Clova' (minor gene resistance); MO, 'Malling Orion' (A_1 major gene resistance) (data from Jones, 1976, 1979; courtesy of the Association of Applied Biologists).

England and Scotland were found to be virulent and capable of colonizing cultivars containing gene A_1. A new virulent biotype was found on two cultivars containing A_1 (Birch *et al.*, 1994; Jones *et al.*, 2000), and an A_{10}-breaking biotype of *A. idaei* has been detected recently in England (Birch *et al.*, 1997, 2002). In North America, resistance to *A. agathonica* has been durable for many years. This may, in part, result from the absence of intense selection pressure because cultivars immune to *A. agathonica* have not been grown extensively in North America (Converse *et al.*, 1971). However, a resistance-breaking biotype was found recently in raspberry breeding plots in British Columbia (Daubeny and Anderson, 1993).

The development of cultivars resistant to *A. idaei* has decreased the use of aphicides in raspberry plantations in the UK over the past 40 years (Jones, 1988; Gordon *et al.*, 1997b). However, the development of economic thresholds, a fundamental concept in the introduction of IPM (Kennedy, 2000), is inappropriate for aphids as virus vectors, since these must be controlled even at low population densities to prevent virus infection. Nevertheless, the use of aphid-resistant cultivars fits well into the overall management of the crop (Gordon *et al.*, 1993) and frees growers

from the need to make aphid control decisions. It is uncertain, however, whether aphid-resistant raspberry cultivars will continue to contribute substantially to integrated crop management strategies in the future. Apart from the problem of new resistance-breaking biotypes of *A. idaei*, two cultivars, 'Glen Ample', which contains gene A_1, and the North American cultivar 'Tulameen', which is susceptible to *A. idaei*, now account for most raspberry production in the UK because their fruit quality best meets supermarket standards (J. Allen, 2002, personal communication). Increasingly, raspberries (and other soft fruit) are being grown under protective covers to meet year-round consumer demand, and this will bring a new set of challenges for aphid management. Research to integrate plant resistance and natural enemies in raspberry plantations will be needed in the future to meet the requirements of consumers for high quality fruit with a minimum of pesticide residues (Gordon *et al.*, 1990).

Aphid IPM in Blueberry

Aphids recorded from blueberry include *Aphis fabae, Aphis gossypii, Aphis vaccinii*,

Ericaphis (= Fimbriaphis) scammelli, E. (= Fimbriaphis) fimbriata, Illinoia azaleae, Illinoia borealis, and the blueberry aphid, *Illinoia pepperi* (Fig. 22.3) (Blackman and Eastop, 2000; Polavarapu and Peng, 2000a,b). *Ericaphis fimbriata* was identified recently as a species of economic importance on highbush blueberries (*Vaccinium corymbosum*) in the eastern and western USA (Raworth, 2004) because it is a vector of *Blueberry scorch virus.* Immediate responses to this new virus have focused on aphid ecology, chemical control, and characterizing virus transmission dynamics (Bristow *et al.*, 2000; Raworth, 2004; S. Polavarapu, personal communication).

In the past 20 years, aphid management programmes in the North American highbush blueberry industry have emphasized control of *I. pepperi,* due to its distribution throughout most of the major production regions and its role as a virus vector. *Blueberry shoestring virus* (BSSV) is a circulative, persistently transmitted viral disease of highbush blueberry (Morimoto *et al.*, 1985; Terhune *et al.*, 1991) that has been recorded from most regions of highbush blueberry production in North America (Ramsdell, 1987). Infection leads to poor plant vigour (Fig. 22.4), and losses to the Michigan blueberry crop from this virus have been estimated at US$3 million per year (Ramsdell, 1979, 1987). BSSV is also distributed across most regions of highbush blueberry production in North America (Ramsdell, 1987) and has been detected in wild highbush blueberry and lowbush blueberries (*Vaccinium angustifolium*) (Lockhart and Hall, 1962; Hancock *et al.*, 1986, 1993). Symptoms include elongated reddish streaks on current year and one-year-old stems, and strap-like or crescent-shaped leaves (Stretch and Hilborn, 1970; Ramsdell, 1987). Petals may be reddened and fruit on infected bushes may become reddish-purple instead of the typical blue. Urban *et al.* (1989) show that the virus spreads slowly within plants, and field observations are consistent with a latent period before visible symptoms of between 2 and 4 years. Through a combination of variety selection, regular monitoring, cultural controls, and accurate timing of effective insecticides, blueberry growers can minimize the level of BSSV and protect their plantings.

Variation in resistance of 16 highbush blueberry cultivars to *I. pepperi* revealed four varieties with significantly lower aphid population growth than the others (Hancock *et al.*, 1982). Some aphids were still present on these varieties, however, suggesting that resistant cultivars were likely to provide only partial prevention of virus transmission. A more promising approach is the finding by Schulte *et al.* (1985) that some varieties are resistant to infection by BSSV, either because the virus cannot replicate, or because the plant actively defends itself against development of the virus. The varieties 'Bluecrop' and 'Atlantic' were found to have particularly

Fig. 22.3. Apterous virginopara of the blueberry aphid, *Illinoia pepperi,* feeding on a blueberry leaf (photo Rufus Isaacs).

Fig. 22.4. During bloom, highbush blueberry bushes show poor leaf and shoot growth indicative of *Blueberry shoestring virus* infection. The plant on the right is infected and highly symptomatic, with thin, strap-like leaves and stunted shoots (photo Rufus Isaacs).

effective field resistance and are recommended for planting where virus pressure is great.

Morimoto and Ramsdell (1985) used trap plants to show that infestation by apterous blueberry aphids started at around 700 day-degrees (base 3.4°C) after 1 March. Immigration was almost exclusively by apterae, and coloured water traps revealed that the primary source of infected aphids was from within infected fields. Movement of the vector within plantings was predominantly by local dispersal from plant to plant and, because aphids overwintered as eggs on blueberry bud scales (Elsner and Kriegel, 1989), removal of plants showing symptoms was recommended as an effective foundation of BSSV eradication programmes.

Soon after the relationship between *A. vaccinii* and BSSV had been verified, Whalon and Elsner (1982) evaluated the impact of insecticides on the vector and its biological control agents. Pirimicarb and acephate provided the greatest control of *I. pepperi*, while methomyl, malathion, and diazinon were less effective and were most damaging to biological control agents. Despite these findings, methomyl and diazinon have been the primary labelled insecticide options for blueberry aphid control (Gut *et al.*, 2002). Applications of carbamate and organophosphate insecticides for control of primary lepidopteran and coleopteran pests have helped to reduce the importance of aphids in commercial blueberry production in some regions, but these insecticides are at risk of restriction in response to the Food Quality Protection Act of 1986. Greater reliance on pyrethroids may be expected to lead to outbreaks of secondary pests such as aphids, but the recent registration of the neonicotinoid insecticides imidacloprid and thiamethoxam for use in US blueberries provides effective tools for aphid control with low impact on predators. For example, foliar and soil applications of these insecticides both provided complete control of *I. pepperi* and *E. fimbriata* (Polavarapu and Peng, 2000a,b). While their impact on virus transmission remains to be determined, prevention of population development and subsequent spread is likely to make this insecticide class preferred in future aphid IPM programmes.

Current IPM recommendations for management of *I. pepperi* and BSSV include monitoring bushes annually for viral infection and aphid infestation. If virus symptoms are absent, minimal use of aphicides is desirable to maintain the activity of natural enemies, since these can cause significant aphid mortality in blueberry (Whalon and Elsner, 1982; Elsner and Kriegel, 1989). Bushes showing symptoms should be removed and disposed of away from the field. If BSSV and aphids were detected in the previous year, insecticides should be applied with thorough coverage beginning at 700 day-degrees (base 3.4°C) after 1 March and re-applied, if needed, to maintain a very low aphid population (Morimoto and Ramsdell, 1985). Aphid control prior to harvest is particularly important in fields with a history of BSSV, to prevent mechanical harvesters spreading the vectors. In Michigan, where 70% of the 6800 ha of blueberry are machine harvested, rubidium-labelled aphids were spread at least 64 bushes along the row by the mechanical harvester (M.E. Whalon, personal communication). Washing harvesters before moving to the next field is a simple strategy to help reduce the spread of BSSV within and between blueberry farms.

Aphid IPM in Strawberry

Aphids are pests in most regions of strawberry production (Maas, 1998), in part due to transmission of viruses (Table 22.1). Viral diseases remain a threat to nurseries and perennial plantings, but have become somewhat less of a problem as annual strawberry production has increased (Hancock, 1999). In addition to the aphids of the *Chaetosiphon fragaefolii* group, *A. gossypii* has also become a direct pest in California, where aphids cause fruit contamination from cast cuticles and honeydew on which sooty moulds develop. Selection of insecticides that minimize the impact on predators in strawberry fields (Easterbrook, 1997) has been emphasized in the development of IPM programmes for

strawberry (Zalom *et al.*, 2001). Recommendations for production in southern California call for treatment with insecticidal soap if more than 30% of the oldest trifoliates on 100 plants/ha are infested with aphids. It is expected that broad spectrum and selective aphicides will continue to be components of aphid and virus management (e.g. Raworth and Clements, 1990), unless host-plant resistance can be bred into marketable strawberry varieties (Easterbrook *et al.*, 1997).

Aphid IPM in Grape

Host-plant resistance has been used for over 120 years as the foundation of crop protection against *Viteus (= Daktulosphaira) vitifoliae* (grape phylloxera) (Grannett *et al.*, 2001). However, a biotype capable of breaking this long-lasting resistance has appeared in Europe (van Emden, Chapter 17 this volume). *Viteus vitifoliae* is by far the most economically important aphid pest of berry crops worldwide, with the potential to devastate *Vitis vinifera* vineyards by causing galls on roots. To protect vines from this pest, viticulturalists plant desired cultivars grafted on to resistant rootstocks derived from *Vitis rupestris*, *Vitis riparia*, or *Vitis berlandieri*. Grapevines also support a complex of seven other aphid species (Blackman and Eastop, 2000), including *Aphis illinoisensis* (grape vine aphid), but populations of these foliar-feeding species are rarely large enough to require control.

Acknowledgements

We thank Zsofia Szendrei and Michelle Carlson for assistance with the preparation of this manuscript, which was improved by comments on an earlier draft by Eric Hanson, Stuart Gordon, and Nick Birch. Thanks to Stuart Gordon for supplying raspberry aphid images. Funding was provided by the Michigan Agricultural Experiment Station to R.I., and by the Scottish Executive Environment and Rural Affairs Department to J.A.T.W.

Executive Summary

In most berry crops, aphid transmission of viruses is the primary problem and economic thresholds are often inappropriate. The exceptional quality standards demanded by consumers of fruit increasingly constrain berry growers' adoption of IPM approaches.

Use of chemicals is the main control tactic in blueberry and strawberry crops, primarily to control non-aphid pests. Selective aphicides, such as the neonicotinoids and insect growth regulators, are promising for future control programmes against aphid pests, but these products are expensive options and their impact on virus transmission remains unclear. Their gradual registration for use in berry crops may enable increased activity of biocontrol agents and reduced incidence of secondary pest outbreaks.

There are relatively few releases of biocontrol agents in commercial production but where other pest problems allow it, pesticides are chosen for their selectivity in favour of natural enemies. The increasing production of berries in protected environments is expected to provide greater potential for inundative releases of natural enemies for aphid control.

Host-plant resistance has been the mainstay of control in several berry crops, and blueberry growers avoid planting cultivars susceptible to BSSV in infected regions. The apparently durable resistances to aphids in raspberries and to phylloxera in grapes have both been broken recently, and intense chemical control will be required in both crops unless resistance breeding can incorporate new genes. Even so, it is clear that for many berry growers (as exemplified by the recent attitude of raspberry growers), a new variety with attractive yield and quality traits will be adopted in spite of susceptibility to aphids.

The main cultural control practice is removal and destruction of plants showing symptoms of virus disease.

There is no use of semiochemicals for aphid management.

Truly integrated management programmes that enable the synergism possible between different control methods are seen only in the use of selective insecticides and host-plant resistance. The further step of integrating the management of aphids with that of other insects, together with diseases and weeds, into unified programmes remains a future goal for most berry crops. For these high-value crops, adoption of some IPM practices may be considered too risky in the context of the pressures to deliver near-perfect fruit. The minor crop status of berry crops in most regions and societal pressure to reduce food residues and worker exposure means that pesticide use is being increasingly restricted. In response, berry IPM programmes must aim to find effective sustainable alternatives with reduced chemical inputs that maintain fruit quality.

References

Birch, A.N.E., Fenton, B., Jones, A.T., Malloch, G., Phillips, M.S., Harrower, B., Woodford, J.A.T. and Catley, M.A. (1994) Ribosomal spacer length variability in the large raspberry aphid *Amphorophora idaei*. *Insect Molecular Biology* 3, 239–245.

Birch, A.N.E., Geoghegan, I.E., Majerus, M.E.N., Hackett, C. and Allen, J. (1997) Interactions between plant resistance genes, pest aphid populations and beneficial aphid predators. *Annual Report of the Scottish Crop Research Institute 1996*, 68–72.

Birch, A.N.E., Jones, A.T., Fenton, B., Malloch, G., Geoghegan, I., Gordon, S.C., Hillier, J. and Begg, G. (2002) Resistance-breaking raspberry aphid biotypes: constraints to sustainable control through plant breeding. *Acta Horticulturae* 585, 315–317.

Blackman, R.L. and Eastop, V.F. (2000) *Aphids on the World's Crops: An Identification and Information Guide*. Wiley, Chichester, 466 pp.

Blackman, R.L., Eastop, V.F. and Hills, M. (1977) Morphological and cytological separation of *Amphorophora* feeding on European raspberry and blackberry (*Rubus* spp.). *Bulletin of Entomological Research* 67, 285–296.

Briggs, J.B. (1959) Three new strains of *Amphorophora rubi* (Kalt.) on cultivated raspberries in England. *Bulletin of Entomological Research* 50, 81–87.

Briggs, J.B. (1965) The distribution, abundance and genetic relationships of four strains of the rubus aphid (*Amphorophora rubi* (Kalt.)) in relation to raspberry breeding. *Journal of Horticultural Science* 40, 109–117.

Bristow, P.R., Martin, R.R. and Windom, G.E. (2000) Transmission, field spread, cultivar response and impact on yield in highbush blueberry infected with blueberry scorch virus. *Phytopathology* 90, 474–479.

Converse, R.H. (1987) *Virus Diseases of Small Fruits*. United States Department of Agriculture, Agriculture Handbook, No. 631, 277 pp.

Converse, R.H., Daubeny, H.A., Stace-Smith, R., Russell, L.M., Koch, E.J. and Wiggans, S.C. (1971) Search for biological races in *Amphorophora agathonica* Hottes on red raspberries. *Canadian Journal of Plant Science* 51, 81–85.

Cross, J.V., Easterbrook, M.A., Crook, A.M., Crook, D., Fitzgerald, J.D., Innocenzi, P.J., Jay, C.N. and Solomon, M.G. (2001) Natural enemies and biocontrol of pests of strawberry in northern and central Europe. *Biocontrol Science and Technology* 11, 165–216.

Daubeny, H.A. (1966) Inheritance of immunity in the red raspberry to the North American strain of the aphid *Amphorophora rubi* Kltb. *Proceedings of the American Society of Horticultural Science* 88, 344–351.

Daubeny, H.A. (1972) Screening red raspberry cultivars and selections for immunity to *Amphorophora agathonica* Hottes. *HortScience* 7, 265–266.

Daubeny, H.A. (1980) Red raspberry cultivar development in British Columbia with special reference to pest response and germplasm exploitation. *Acta Horticulturae* 112, 59–67.

Daubeny, H.A. and Anderson, A.K. (1993) The British Columbia red raspberry breeding programme – achievements and prospects. *Acta Horticulturae* 352, 285–293.

Dicker, G.H.L. (1940) The biology of the *Rubus* aphides. *Journal of Pomology* 18, 1–33.

Easterbrook, M.A. (1997) A field assessment of the effects of insecticides on the beneficial fauna of strawberry. *Crop Protection* 16, 147–152.

Easterbrook, M.A., Crook, A.M.E., Cross, J.V. and Simpson, D.W. (1997) Progress toward integrated pest management on strawberry in the United Kingdom. *Acta Horticulturae* 439, 899–904.

Elsner, E.A. and Kriegel, R.D. (1989) Distribution and seasonal phenology of blueberry aphids (*Illinoia pepperi* (MacG.)) in Michigan. *Acta Horticulturae* 241, 330–332.

FAO (2002) *Production Yearbook. FAO Statistics Series, No. 176*. FAO, Rome, 336 pp.

Fenton, B., Birch, A.N.E., Malloch, G., Woodford, J.A.T. and Gonzalez, C. (1994) Molecular analysis of ribosomal DNA from the aphid *Amphorophora idaei* and an associated fungal organism. *Insect Molecular Biology* 3, 183–189.

Freeman, J.A. and Stace-Smith, R. (1970) Effects of raspberry mosaic viruses on yield and growth of red raspberries. *Canadian Journal of Plant Science* 50, 521–527.

Gordon, S.C., Woodford, J.A.T. and Barrie, I.A. (1990) Monitoring pests of red raspberry in the United Kingdom and the possible implementation of an integrated pest management system. In: Bostanian, N.J., Wilson, L.T. and Dennehy, T.J. (eds) *Monitoring and Integrated Management of Arthropod Pests of Small Fruit Crops*. Intercept, Andover, pp. 1–26.

Gordon, S.C., Brennan, R.M., Lawson, H.M., Birch, A.N.E., McNicol, R.J. and Woodford, J.A.T. (1993) Integrated crop management in *Rubus* and *Ribes* crops in Europe – the present and prospects for the future. *Acta Horticulturae* 352, 539–545.

Gordon, S.C., Cormack, M.R. and Hackett, C.A. (1997a) Arthropod contamination of red raspberry (*Rubus ideaus* L.) harvested by machine in Scotland. *Journal of Horticultural Science* 72, 677–685.

Gordon, S.C., Woodford, J.A.T. and Birch, A.N.E. (1997b) Arthropod pests of *Rubus* in Europe: pest status, current and future control strategies. *Journal of Horticultural Science* 72, 831–862.

Grannett, J., Walker, M.A., Kocsis, L. and Omer, A.D. (2001) Biology and management of grape phylloxera. *Annual Review of Entomology* 46, 387–412.

Gut, L.J., Isaacs, R., Wise, J.C., Jones, A.L., Schilder, A.M.C., Zandstra, B. and Hanson, E. (2002) *2002 Fruit Spraying Calendar*. Michigan State University Extension Bulletin E-154, 133 pp.

Hancock, J.F. (1999) *Strawberries*. CAB International, Wallingford, 237 pp.

Hancock, J.F., Schulte, N.L., Siefker, J.H., Pritts, M.P. and Roueche, J.M. (1982) Screening highbush blueberry cultivars for resistance to the aphid *Illinoia pepperi*. *Hortscience* 17, 362–363.

Hancock, J.F., Morimoto, K.M., Pritts, M.P. and Ramsdell, D.C. (1986) Blueberry shoestring virus in natural populations of highbush and lowbush blueberry. *HortScience* 21, 1059–1060.

Hancock, J.F., Callow, P.W., Krebs, S.L., Ramsdell, D.C., Ballington, J.R., Lareau, M.J., Luby, J.J., Pavlis, G.P., Pritts, M.P. and Smagula, J.M. (1993) Blueberry shoestring virus in eastern North American populations of native *Vaccinium. HortScience* 28, 175–176.

Hill, A.R. (1956) Observations on the North American form of *Amphorophora rubi* Kalt. (Homoptera, Aphididae). *Canadian Entomologist* 88, 89–91.

Hill, A.R. (1957) Observations on the reproductive behaviour of *Amphorophora rubi* (Kalt.), with special reference to the phenomenon of insect resistance in raspberries. *Bulletin of Entomological Research* 48, 467–476.

Huber, G.A. and Schwartze, C.D. (1938) Resistance in the red raspberry to the mosaic vector *Amphorophora rubi* Kalt. *Journal of Agricultural Research* 57, 623–633.

Jennings, D.L. (1963) Preliminary studies on breeding raspberries for resistance to mosaic disease. *Horticultural Research* 2, 82–96.

Jones, A.T. (1976) The effectiveness of resistance to *Amphorophora rubi* in raspberry (*Rubus idaeus*) on the spread of aphid-borne viruses. *Annals of Applied Biology* 82, 503–510.

Jones, A.T. (1979) Further studies on the effect of resistance to *Amphorophora idaei* in raspberry (*Rubus idaeus*) on the spread of aphid-borne viruses. *Annals of Applied Biology* 92, 119–123.

Jones, A.T. (1986) Advances in the study, detection and control of viruses and virus diseases of *Rubus*, with particular reference to the United Kingdom. *Crop Research* 26, 127–171.

Jones, A.T. (1987) Control of virus infection in crop plants through vector resistance: a review of achievements, prospects and problems. *Annals of Applied Biology* 111, 745–772.

Jones, A.T. (1988) The influence of cultivating new raspberry varieties on the incidence of viruses in raspberry crops in the UK. *Environmental Aspects of Applied Biology. Aspects of Applied Biology No. 17*, 179–186.

Jones, A.T. and McGavin, W.J. (1998) Infectivity and sensitivity of UK raspberry, blackberry and hybrid berry cultivars to *Rubus* viruses. *Annals of Applied Biology* 132, 229–251.

Jones, A.T., McGavin, W.J. and Birch, A.N.E. (2000) Effectiveness of resistance genes to the large raspberry aphid, *Amphorophora idaei* (Börner), in different raspberry (*Rubus idaeus* L.) genotypes and under different environmental conditions. *Annals of Applied Biology* 136, 107–113.

Keep, E. (1989) Breeding red raspberry for resistance to diseases and pests. *Plant Breeding Reviews* 6, 245–321.

Keep, E. and Knight, R.L. (1967) A new gene from *Rubus occidentalis* L. for resistance to strains 1, 2 and 3 of the *Rubus* aphid, *Amphorophora rubi* Kalt. *Euphytica* 16, 209–214.

Kennedy, G.G. (2000) Perspectives on progress in IPM. In: Kennedy, G.G. and Sutton, T.B. (eds) *Emerging Technologies for Integrated Pest Management: Concepts, Research and Implementation*. American Phytopathological Society, St Paul, pp. 2–11.

Kennedy, G.G. and Schaefers, G.A. (1974) The distribution and seasonal history of *Amphorophora agathonica* Hottes on 'Latham' red raspberry. *Annals of the Entomological Society of America* 67, 356–358.

Kennedy, J.S., Day, M.F. and Eastop, V.F. (1962) *A Conspectus of Aphids as Vectors of Plant Viruses*. Commonwealth Institute of Entomology, London, 114 pp.

Kieffer, J.N., Shanks, C.H. and Turner, W.J. (1983) Populations and control of insects and spiders contaminating mechanically harvested red raspberries in Washington and Oregon. *Journal of Economic Entomology* 76, 649–653.

Knight, R.L., Keep, E. and Briggs, J.B. (1959) Genetics of resistance to *Amphorophora rubi* (Kalt.) in the raspberry. I. The gene A1 from Baumforth A. *Journal of Genetics* 56, 261–280.

Lockhart, C.N. and Hall, I.V. (1962) Note on an indication of shoestring virus in the lowbush blueberry, *Vaccinium angustifolium* Ait. *Canadian Journal of Botany* 40, 1561–1562.

Maas, J.L. (1998) *Compendium of Strawberry Diseases*. American Phytopathological Society, St Paul, 98 pp.

Morimoto, K.M. and Ramsdell, D.C. (1985) Aphid vector population dynamics and movement relative to field transmission of blueberry shoestring virus. *Phytopathology* 75, 1217–1222.

Morimoto, K.M., Ramsdell, D.C., Gillet, J.M. and Chaney, W.G. (1985) Acquisition and transmission of blueberry shoestring virus by its aphid vector *Illinoia pepperi. Phytopathology* 75, 709–712.

Polavarapu, S. and Peng, H. (2000a) Efficacy of soil applied insecticides against blueberry aphids on blueberries, 1998. *Arthropod Management Tests* 25, 1.

Polavarapu, S. and Peng, H. (2000b) Evaluation of foliar applications of insecticides against blueberry aphids on blueberries, 1998. *Arthropod Management Tests*, 25, 1.

Ramsdell, D.C. (1979) Physical and chemical properties of blueberry shoestring virus. *Phytopathology* 69, 1087–1091.

Ramsdell, D.C. (1987) Blueberry shoestring. In: Converse, R.H. (ed.) *Virus Diseases of Small Fruits*. United States Department of Agriculture, Agriculture Handbook No. 631, pp. 103–105.

Rautapää, J. (1967) Studies on the host plant relationships of *Aphis idaei* v.d. Goot and *Amphorophora rubi* (Kalt.) (Hom. Aphididae). *Annales Agriculturae Fenniae* 6, 174–190.

Raworth, D.A. (2004) Ecology and management of *Ericaphis fimbriata* (Hemiptera: Aphididae) in relation to the potential for spread of blueberry scorch virus. *Canadian Entomologist* 136, 711–718.

Raworth, D.A. and Clements, S.J. (1990) Lowering incidence of a virus complex dominated by strawberry mottle virus by reducing numbers of the aphid vector with oxydemeton-methyl. *Plant Disease* 74, 365–367.

Schulte, N.L., Hancock, J.F. and Ramsdell, D.C. (1985) Development of a screen for resistance to blueberry shoestring virus. *Journal of the American Society for Horticultural Science* 110, 343–346.

Stace-Smith, R. (1960) Studies on *Rubus* virus diseases in British Columbia. VI. Varietal susceptibility to aphid infestation in relation to virus acquisition. *Canadian Journal of Botany* 38, 283–285.

Stretch, A.W. and Hillborn, M.T. (1970) Blueberry shoestring. In: Frazier, N.W., Fulton, J.P., Thresh, J.M., Converse, R.H., Varney, E.H. and Hewitt, W.B. (eds) *Virus Diseases of Small Fruits and Grapevines*. University of California, Berkeley, pp. 186–188.

Taylor, C.E. and Chambers, J. (1969) Effects of insecticide treatments on aphid populations and on spread of latent viruses in raspberry cane nurseries. *Horticultural Research* 9, 37–43.

Terhune, B.T., Ramsdell, D.C., Klomparens, K.L. and Hancock, J.F. (1991) Quantification of blueberry shoestring virus RNA and antigen in its aphid vector, *Illinoia pepperi*, during acquisition, retention, and transmission. *Phytopathology* 81, 1096–1102.

Urban, L.A., Ramsdell, D.C., Klomparens, K.L., Lynch, T. and Hancock, J.F. (1989) Detection of blueberry shoestring virus in xylem and phloem tissues of highbush blueberry. *Phytopathology* 79, 488–493.

Whalon, M.E. and Elsner, E.A. (1982) Impact of insecticides on *Illinoia pepperi* and its predators. *Journal of Economic Entomology* 75, 356–358.

Zalom, F.G., Phillips, P.A., Toscano, N.C. and Udayagiri, S. (2001) *Strawberry Aphids*. University of California Department of Agriculture and Natural Resources Publication No. 3339, 4 pp.

23 IPM Case Studies: Cotton

Jean-Philippe Deguine[1], Maurice Vaissayre[1]
and the late François Leclant[2]

[1]Centre de Coopération Internationale en Recherche Agronomique pour le
Développement, Avenue Agropolis, 34398 Montpellier Cedex 5, France; [2]Ecole
Nationale Supérieure Agronomique, 34060 Montpellier Cedex 5, France

Introduction

Cotton is an industrial crop of prime importance in numerous producer countries and regions such as the USA, Australia, and Central Asia and in consumer countries including China, India, Pakistan, and Brazil (Munro, 1994; ICAC, 2000). The crop has both production and fibre quality requirements.

Aphids hold a special position in the cotton pest complex because of their biological characteristics and the diversity of their relations with the cotton plant. Although several species may be found on cotton, *Aphis gossypii* (cotton or melon aphid) is by far the most important species economically (Leclant and Deguine, 1994).

Numerous reviews have been published on the relations between *A. gossypii* and cotton (Leclant and Deguine, 1994; Deguine and Leclant, 1997).

Biological and Ecological Characteristics of *Aphis gossypii*

Aphis gossypii is highly polyphagous. Hundreds of host plants have been recorded worldwide (Leonard *et al.*, 1971; Roy and Behura, 1983; Deguine *et al.*, 1999).

The cotton aphid displays great morphological variability (Fig. 23.1) and various colours associated with a complex biology (Nevo and Coll, 2001; Liu *et al.*, 2002). There are several types of life cycle according to geographical zone (Leclant and Deguine, 1994). Parthenogenesis is the main mode of reproduction in the tropics. Generations take only 5 days under optimal temperature conditions (26–28°C), and there can be over 50 generations per year.

Infestation

In the colonization of cotton fields by alatae, along with volatile cues released by the cotton plant (Powell and Hardie, 2001), the contrast between plants and soil and the particular colour of seedlings appear important (Favret and Voegtlin, 2001). Host selection by different strains occurs after probing by alatae. Specialized clones of apterae (Brévault *et al.*, 2005) then colonize new plants (Fig. 23.2) until feeding conditions deteriorate through overpopulation and alatae are produced.

Populations of *A. gossypii* in cotton crops are reduced by rain, by deteriorating host plants, and by the abundant fauna of predators and parasitoids to be found in cotton in the absence of heavy insecticide use. Epizootics of entomopathogenic fungi reduce populations very effectively (Jones *et al.*, 2003).

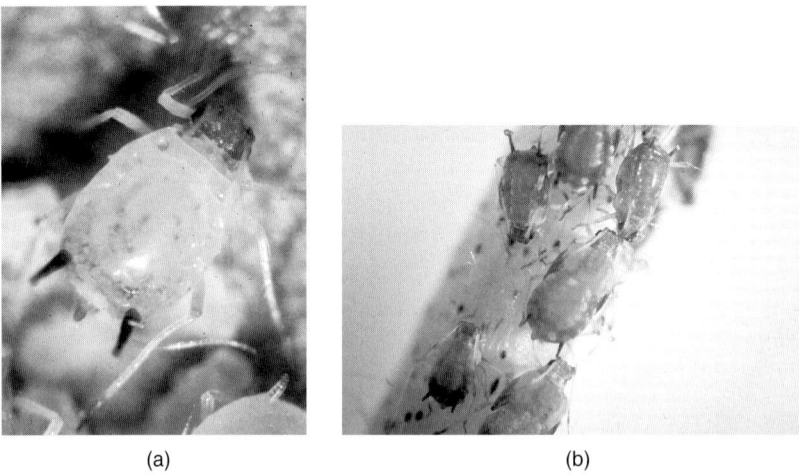

(a)   (b)

Fig. 23.1.   *Aphis gossypii.* (a) Apterous viviparous female; (b) nymph (original colour photos: J.-P. Deguine).

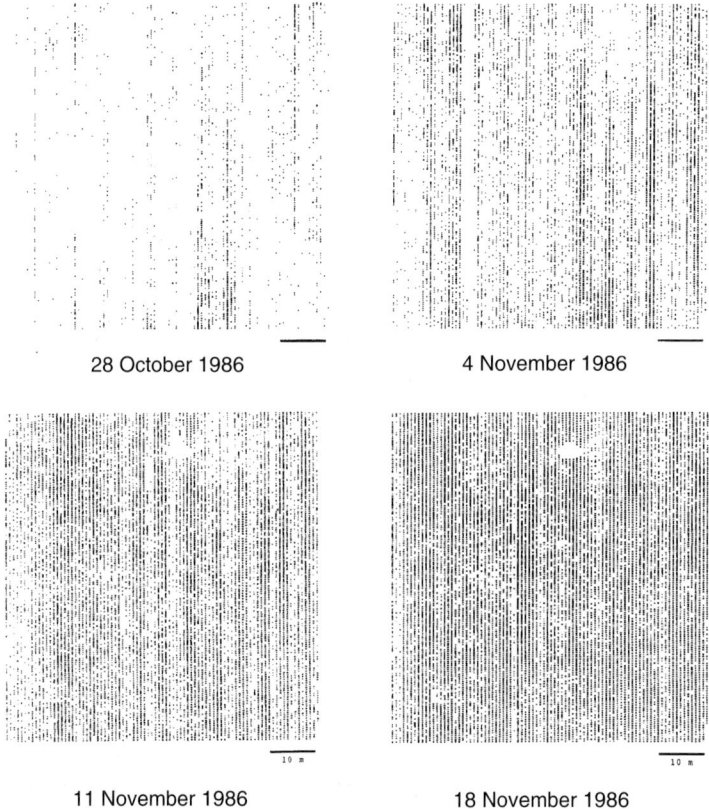

28 October 1986   4 November 1986

11 November 1986   18 November 1986

Fig. 23.2.   Four maps of *Aphis gossypii* infestation on different dates in 1986 in a half-hectare farmer's field in Bitanda (Chad). 13,886 cotton plants per observation. Symbol size is proportional to the number of infested leaves among the five terminal leaves (from Deguine, 1995).

Damage and its Economic Importance

The importance of *A. gossypii* in the cotton pest complex has grown steadily over the past two decades in most of the cotton-growing areas in the world, e.g. in China (Luo and Gan, 1986), the Middle East (Broza, 1986), the USA (King *et al.*, 1987; Grafton-Cardwell, 1991), and Africa (Leclant and Deguine, 1994).

Like other aphids, *A. gossypii* causes three types of damage to cotton: the spread of virus diseases, damage by sap removal, and honeydew fouling of the cotton fibre (Fig. 23.3). *Aphis gossypii* has been identified as the vector of 'blue disease' of cotton in Africa (Cauquil and Vaissayre, 1971), a suspected viral disease probably very close to *Cotton leaf roll dwarf virus* in South-east Asia and 'mosaico da nevuras' in Brazil (Cauquil and Follin, 1983). It was previously also known to be the vector of another disease ('vermelhao'), believed to be caused by *Cotton anthocyanosis virus*, in Brazil (Costa, 1956) and India (Mali, 1978).

Removal of sap at the beginning of the season can result in a substantial delay of the harvest and in yield loss (Andrews and Kitten, 1989; Harris *et al.*, 1992; Deguine *et al.*, 1994), both being accentuated when drought periods occur before flowering. After the start of flowering, losses from sap removal become negligible, partly because the plant has developed sufficient foliage and also, and above all, because natural factors cause the population to crash.

However, quality is seriously affected by late attacks due to the production of honeydew. At the end of the cotton cycle, a build-up of the aphid population occurs after the last rains and a proportion of the honeydew falls on to the opening bolls and soils the fibre. This contamination is of importance in spinning (Héquet *et al.*, 2000) and 'sticky' cotton suffers substantial price depreciation.

(a)

(b)

(c)

Fig. 23.3. Symptoms of damage by *Aphis gossypii*. (a) Virus; (b) general stunting as a result of sap removal and salivation; (c) sticky bolls as a result of honeydew (original colour photos: J.-P. Deguine).

Control and its Evolution

The limits of chemical control of *A. gossypii*

Total reliance on chemical control soon showed that, in many places, there was a limit to its effectiveness and profitability, either because *A. gossypii* developed resistance to the insecticides (Grafton-Cardwell *et al.*, 1992; Gubran *et al.*, 1992; Furk and Hines, 1993) or because the timing, application technique, or insecticide were unsuitable for aphid control (Matthews, 2000).

Evolution towards integrated control

Control techniques in most countries are now based on the concept of integrated control, making the most of the rich natural enemy fauna that can develop in cotton fields and working to economic thresholds (OILB, 1977). All these techniques require methods for *in situ* scouting (Fig. 23.4a) and trapping alatae (Fig. 23.4b). In various regions, and particularly in the USA, the evolution of control measures has progressed to IPM – the judicious combination of several procedures – and this is used widely (Hardee and O'Brien, 1990; Leser *et al.*, 1992; Graves *et al.*, 1993; Hardee, 1993; Norman *et al.*, 1993). IPM is also used widely in Asia (Ingram *et al.*, 1989) and Africa (Kiss, 1991; Zethner, 1995; Morse and Buhler, 1997). Various aspects of

IPM in cotton have been reviewed by Bottrell and Adkisson (1977) and more recently by Frisbie *et al.* (1989), Green and Lyon (1989), Matthews and Tunstall (1994), Mengech *et al.* (1995), and Fitt (2000).

IPM of *Aphis gossypii* in the USA

Although the cotton aphid was already known as a cotton pest in the mid-19th century (Slosser *et al.*, 1989), it only recently gained its status as one of the major cotton pests in the USA (Hardee and O'Brien, 1990; Godfrey and Leser, 1999), and especially in the western and southwestern parts of the cotton belt. 'Blind' chemical control, destroying the beneficial fauna and leading to resistance to insecticides (King *et al.*, 1988), is often considered to have been responsible for this emergence of *A. gossypii* as a pest, like *Bemisia tabaci* (sweet potato whitefly) or *Frankliniella occidentalis* (western flower thrips) (Luttrell, 1994).

Until the 1990s, the IPM programmes developed for cotton took little or no account of the management of aphid populations (Frisbie *et al.*, 1989; Pendergrass, 1989). Since then, chemical control applied specifically for the control of aphids has been found to be expensive and unreliable, in particular because ULV (ultra low volume) and VLV (very low volume) sprays do not provide satisfactory coverage of the aphid populations

(a)

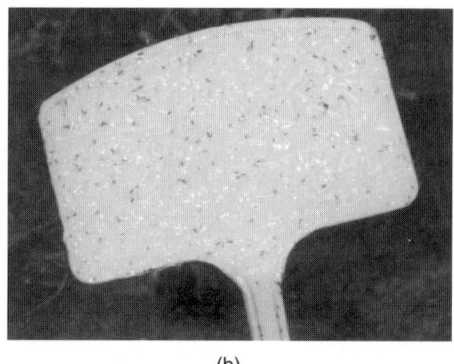

(b)

Fig. 23.4.   Monitoring *Aphis gossypii* populations. (a) *In situ* scouting for aphids; (b) trapping system (sticky yellow plate) for alatae (original colour photos: J.-P. Deguine).

located on the undersides of leaves (Hardee, 1993; Deguine *et al.*, 2004).

From 1991 onwards, *A. gossypii* has been considered to be a major pest responsible for yield losses exceeding 2%, i.e. nearly US$20 million nationally (Head, 1992), together with fibre soiling by honeydew. A Cotton Aphid Task Force was set up and recommended control principles based on cultural techniques, rational fertilization, genotypes with a degree of tolerance, and the conservation of biological control agents (Hardee *et al.*, 1994).

The cultural practices recommended call for sowing concentrated within a period of a few days in the same field, or on the same cultivation block, and for a uniform stand (Parajulee *et al.*, 1999). Higher densities of aphids at the beginning of the season have been associated with no-till or reduced tillage (Jones *et al.*, 2001), whereas late irrigation enhances outbreaks and honeydew production at the end of the cycle (Slosser and Parajulee, 2001).

Aphidophagous species are abundant, but little is known of their role (Kerns and Gaylor, 1993). However, there is emphasis on augmenting predator populations (Knutson and Tedders, 2002) and conserving the entomophagous fauna by judicious choice of spray dates and using soft or selective insecticides (King *et al.*, 1987; Abney *et al.*, 2001). The beneficial role of certain types of relay cropping (Parajulee and Slosser, 1999), and even food sprays (Slosser *et al.*, 2000), to enhance the action of natural enemies is being researched.

The economic threshold proposed in the USA is ten aphids per leaf, after the examination of 50 developed leaves (Leser, 1994). More complex approaches involve different leaf stages and aphid population classes, and most authors believe that thresholds should be lowered when there is a risk of plant stress – generally water stress.

Pesticides that enhance aphids directly or indirectly must be avoided (Rummel *et al.*, 1995). Chemical control must under no circumstances hinder the action of the main aphid mortality factor observed in the USA, *Neozygites freseni* (Entomophthorales) (Steinkraus *et al.*, 1991), which alone regularly causes 75% mortality of aphid populations

(Steinkraus and Rosenheim, 1995; Abney *et al.*, 2001; Steinkraus *et al.*, 2002).

Cotton resistance to aphids in the USA is based on several mechanisms. Weathersbee *et al.* (1995) showed that hairiness confers resistance, whereas El-Zik and Thaxton (1989) reported the advantage of the red foliage character. Resistance of the species *Gossypium arboreum* to aphids has also been demonstrated (Reed *et al.*, 1999).

Careful fertilization is important because late nitrogenous fertilization increases the leaf area used by aphid populations and prolongs the cotton plant cycle, with the risk of abundant honeydew (Slosser *et al.*, 1997; Godfrey *et al.*, 2000; Cisneros and Godfrey, 2001).

Most IPM practices recommended today by universities in a number of USA states (Texas A & M, University of California at Davis) are relevant to aphid control. All include:

- early, uniform planting;
- stands or skip-row planting patterns;
- avoidance of practices promoting lush succulent growth such as excessive irrigation and high nitrogen levels; protection of seedlings from early stress by using seed or in-furrow treatments against thrips and diseases, by avoiding foliar insecticides early in the season, by postponing treatment if at all possible (especially with pyrethroids) and, while spraying against aphids, using aphicides at full rate and maximizing insecticide coverage.

The introduction and spectacular success among USA farmers of Bt cotton that has been genetically modified for resistance to the *Heliothis/Helicoverpa* bollworm complex will reinforce research on pests, that include the cotton aphid, which escape the effects of *Bacillus thuringiensis* toxins (Harris *et al.*, 1996).

IPM of *Aphis gossypii* in Brazil and elsewhere in South America

In Brazil, cotton aphids first became a problem in the northeast (the Nordeste), where both *Gossypium hirsutum* and *Gossypium barbadense* are grown. The latter is grown as a semi-perennial crop and the importance of intercropping for reducing aphid infestation was soon observed in *A. gossypii* population dynamics (Viera *et al.*, 1983; Araujo and de Sales, 1985). In this region, harvest losses can be 20–40% of production potential (Vendramim and Nakano, 1981). In contrast, aphid populations in the more humid southern part of Brazil are limited by epizootics caused by Entomophthorales fungi (Latteur, 1982); serious damage by removal of sap does not occur and the economic effects are limited. In the zones into which cotton growing has most recently extended (Parana, Mato Grosso), the cotton aphid is seen primarily as the vector of 'doença azul', a virus disease that can cause spectacular damage (Papa *et al.*, 2001; Correa *et al.*, 2005). Insecticide spraying thresholds are set very low and farmers do not hesitate to use seed treatment with a systemic insecticide, followed by numerous foliar sprays (Franco, 1999). Here, the introduction of disease-tolerant varieties (DeltaOpal, CD 401) is essential for achieving an economic balance for the crop (Freire *et al.*, 1999; Ortiz *et al.*, 1999) and also to protect beneficial fauna and the environment.

A preliminary approach to IPM for cotton in Brazil was proposed by Hambleton (1937), but it was years before it was implemented, using the recommendations of the research sector (Bleicher *et al.*, 1981). From the early 1980s, aphids began to be considered an important component of the pest complex, responsible for significant crop losses (Leclant and Deguine, 1994). A new pest, *Anthonomus grandis* (boll weevil), appeared at the same time and caused significant changes, not only to the perception of pests but also to treatment practices (Ramalho, 1994).

The practices in Brazil today first of all involve the adoption of a cultivar resistant to diseases, especially those spread by *A. gossypii*, and with rapid fruiting and a short cycle; also, the development of diversified agroecosystems rich in natural enemies, especially in the Nordeste (Ramalho *et al.*, 1989). Direct sowing mulch-based systems appear to reduce aphid incidence at the beginning of the season (Morita *et al.*, 1999). The use of economic thresholds for the main pests (such

as bollworms) nevertheless enables natural enemies to impact on the aphid population.

IPM of *Aphis gossypii* in South-east Asia and China

In South-east Asian countries, and in particular in Vietnam and Thailand, *A. gossypii* is only a major pest when it spreads the *Cotton leaf roll dwarf virus*. This does not apply to China, however, where it has major pest status because of the direct damage caused, especially in southern production regions. Early attacks slow plant growth and cause a delay in harvesting that is incompatible with the agricultural calendar in regions where autumn temperatures are a limiting factor. Chinese farmers responded with massive insecticide applications and were soon confronted by resistance of the pest (JianGo *et al.*, 1987; Tang, 1992), as evinced by the spectacular infestations observed in Xinjiang during the 2001 season. This crisis created interest in the natural regulation factors of aphid populations (Luo and Gan, 1986; Zhao and Zhou, 1988; Mu *et al.*, 1993), and led to the modelling of the effects (Zhang *et al.*, 1987) and demonstration of the negative effect of early insecticide treatments (Zhang and Chen, 1991).

In China, those in favour of IPM proposed cultural practices, with particular stress on intercropping (Dong *et al.*, 1992), in addition to the necessary conservation of natural enemies (Zhou and Wang, 1989). This is understandable in farming systems with small fields and a variety of crops.

The recent growing of Bt cotton has reduced bollworm pressure, and hence reduced the number of pesticide sprays. This has resulted in the emergence of *A. gossypii* as a key insect pest (Wu and Guo, 2003), leading to an economic effect on the intercropping systems used in central and eastern China (Cui and Xia, 2000).

IPM of *Aphis gossypii* in Africa and Madagascar

For many years, *A. gossypii* was not considered a major pest of cotton, except in central

Africa (the Sudan, Central African Republic) and Madagascar, where intervention was justified. In the Sudan, Abdelrahman *et al.* (1998) showed that natural enemies were capable of containing aphid populations until an advanced stage of flowering of the crop, and Hillocks (1995) showed that outbreaks of the aphid had probably been induced by pesticide used to control other pests.

In West Africa, aphid problems were recognized from the 1980s onwards, first for their honeydew fouling and then as a cause of economic damage at the beginning of the season (Gahukar, 1991). Several hypotheses have been put forward to account for the new importance of *A. gossypii* in the region, but it should first be remembered that this is not a geographically isolated phenomenon. Three hypotheses are (i) climate changes, and especially drought pockets that have appeared during the season, that clearly enhance infestation, (ii) the introduction of higher-yielding cultivars, which are not as hardy as earlier ones, has proved favourable for pests, and (iii) the systematic use of pyrethroids in spraying programmes has contributed to imbalances in the fauna resulting from the destruction of numerous beneficials.

Both extension services and farmers believed that including aphicides in foliar spraying programmes would control the aphid problem. This was not the case for several reasons, first and foremost because of the ineffectiveness of the spray technique (Deguine, 1992) and also perhaps because of the emergence of resistance to certain aphicides such as dimethoate (Deguine, 1996; Villate *et al.*, 1999).

Detailed studies have been made of the infestation of cotton fields in West Africa (Duviard and Mercadier, 1973; Deguine, 1995). Natural enemies have been shown to play a significant role, but they cannot contain the aphid explosions in favourable climatic conditions. These studies also showed that, as elsewhere in the world, the main factor in the control of aphids is the entomopathogenic fungus *N. fresenii* (Silvie and Papierok, 1991).

Likewise, there seems to be little information on the plant characters giving resistance to the aphid in Africa. Only the okra leaf character seems to hinder infestation, but the reduction in aphid populations resulting from such limiting of the leaf area does not always lead to a lessening of damage symptoms such as sticky fibre (Deguine, 1995). *Aphis gossypii* is less affected than *B. tabaci* (Deguine *et al.*, 2004). Furthermore, it appears that there may be a certain microclimatic trade-off (unfavourable for aphids but also for entomopathogenic fungi).

Efforts to control honeydew in West Africa have employed two features of the 'insect–leaf–open boll' trilogy (Fig. 23.5): shortening exposure to fouling by encouraging

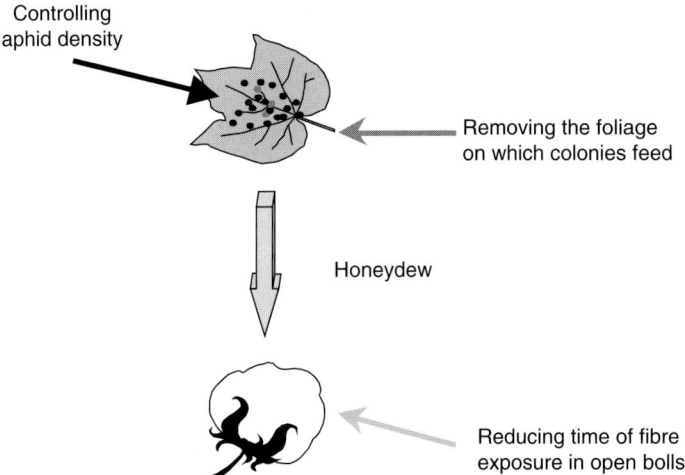

Controlling aphid density

Removing the foliage on which colonies feed

Honeydew

Reducing time of fibre exposure in open bolls

Fig. 23.5. Pathways for the reduction of the sticky cotton problem.

early harvesting, coupled with going as far as instigating a cotton seed quality bonus, and systematic destruction of the sites of residual aphid populations. The latter can be performed with a defoliant or by systematic plant topping in zones known to be favourable for the continuation of vegetation at the end of the cycle (Deguine *et al.*, 2000).

The chemical solution to the cotton aphid problem is therefore being seriously questioned in West Africa today. A true integrated approach is recommended for aphid population management, with a number of cultural practices likely to limit the incidence of end-of-cycle infestations, including early sowing, rational fertilization, topping, early picking (Deguine, 1995), and spraying to economic thresholds (Fig. 23.6).

In order to implement these IPM techniques, scouting, sampling, and trapping populations of *A. gossypii* are required. In Cameroon, an attractant panel technique for trapping winged forms of *A. gossyppi* has been developed (Deguine and Leclant, 1996). A trapping unit consists of four small sticky yellow plates installed in the middle of the borders of a 5 m square in a 100-m² area that is weeded regularly (Fig. 23.4b). The system is effective, reliable, inexpensive, and simple to establish in the field.

Alatae activity can be monitored at the beginning of the season, making it possible to incorporate appropriate control measures in IPM programmes.

Executive Summary

Cotton in the USA is the textbook example of problems arising from the overuse of pesticides, and therefore has now become perhaps the best example of a well-developed IPM system. Most other parts of the world are still well behind in such developments for cotton, especially in relation to aphid rather than caterpillar control, though increasingly, economic threshold-based selection of insecticides less damaging to indigenous natural enemies and cultural controls are being introduced.

The introduction of Bt cotton against caterpillars has made the choice of any insecticides used governed more by aphid problems than previously and, as well as seeking 'softer' chemicals, selectivity often can be obtained by replacing foliar insecticides with soil and in-furrow treatments, though these of course have the disadvantage of being prophylactic.

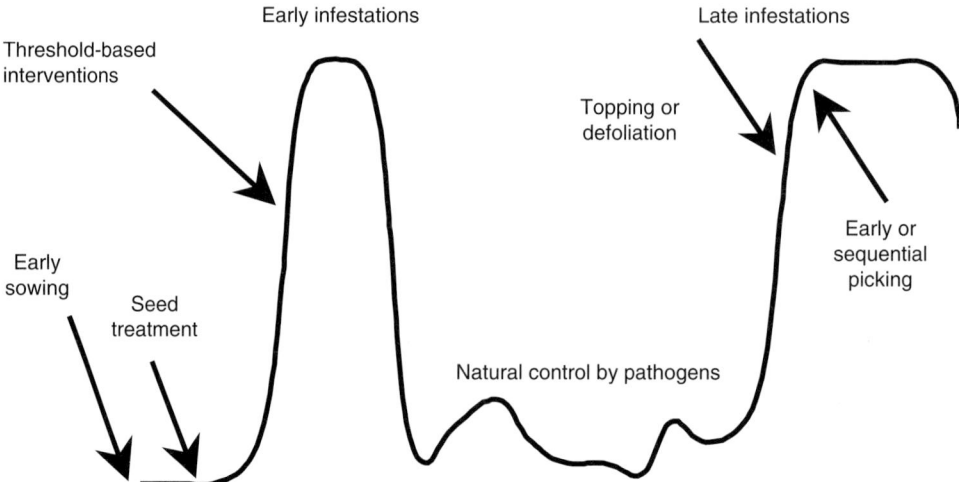

Fig. 23.6. The pattern of infestation of the crop and basic IPM techniques for *Aphis gossypii* population management in West Africa.

Economic thresholds for aphids are used extensively in the USA, and here complications such as a change of threshold during the season have been introduced. In South America, *A. gossypii* mainly causes damage as a virus vector, and thresholds have been set very low compared with thresholds elsewhere when control has relied largely on virus-tolerant varieties. In West Africa, a simple aphid monitoring system based on sticky traps has been proposed.

No releases of biocontrol agents against aphids are made, and interest in biological control lies in conserving the rich natural enemy fauna characteristic of cotton fields in the absence of heavy use of broad-spectrum insecticides.

Several sources of host-plant resistance to aphids are available in the USA, particularly varieties with dense pubescence. Little use of plant resistance for aphid control seems to be made elsewhere.

Cultural control of aphids concentrates on reducing late leaf area in the crop by techniques such as avoiding late fertilizer applications and ensuring an early harvest by termination of irrigation (introduced primarily for bollworm control).

There is no purposeful use of semiochemicals to manipulate the behaviour of either the aphids or their natural enemies.

Interactions between control methods are only purposeful in relation to conserving biological control by modifying insecticide use.

References

Abdelrahman, A.A., Al-Saffar, A., Munir, B. and Stam, P.A. (1998) Cotton integrated pest management in central Sudan. *Arab Journal of Plant Protection* 16, 52–54.

Abney, M.R., Ruberson, J.R., Herzog, G.A., Kring, T.J. and Steinkraus, D.C. (2001) Impact of natural enemies on the cotton aphid: implications for control. *Proceedings of the Beltwide Cotton Conferences, Anaheim, January 2001, Volume 2.* National Cotton Council, Memphis, pp. 1029–1031.

Andrews, G.L. and Kitten, W.F. (1989) How cotton yields are affected by aphid populations which occur during boll set. *Proceedings Cotton Production and Research Conferences, Nashville, January 1989, Volume 1.* National Cotton Council, Memphis, pp. 291–293.

Araujo, P.A.C.B. and de Sales, F.J.M. (1985) Influencia do clima e da fenologia do algodoeiro na dinamica populacional do pulgao. *Fitosanidade* 6, 57–72.

Bleicher, E., Silva, A.L., Santos, W.J., Gravena, S., Nakano, O. and Ferreira, L. (1981) *Manual de manejo integrado de pragas do algodoeiro.* Boletim Técnico No. 2, EMBRAPA/CNPA, Campina Grande, 11 pp.

Bottrell, D.G. and Adkisson, P.L. (1977) Cotton insect pest management. *Annual Review of Entomology* 22, 451–481.

Brévault, T., Carletto, J., Linderme, D. and Vanlerberghe-Masutti, F. (2005) Pucerons du coton: enquête d'identité. *Résumés de la 7e AFPP Conférence Internationale sur les Ravageurs en Agriculture, Montpellier, October 2005,* 198–199.

Broza, M. (1986) An aphid outbreak in cotton fields in Israel. *Phytoparasitica* 14, 81–85.

Cauquil, J. and Follin, J.C. (1983) Presumed virus and mycoplasma-like organism diseases in sub-Saharan Africa and in the rest of the world. *Coton et Fibres Tropicales* 38, 309–317.

Cauquil, J. and Vaissayre, M. (1971) La 'maladie bleue' du cotonnier en Afrique: transmission de cotonnier à cotonnier par *Aphis gossypii* Glover. *Coton et Fibres Tropicales* 26, 463–466.

Cisneros, J.J. and Godfrey, L.D. (2001) Midseason pest status of the cotton aphid (Homoptera: Aphididae) in California cotton: is nitrogen a key factor? *Environmental Entomology* 30, 501–510.

Correa, R.L., Silva, T.F., Simões-Araùjo, J.L., Barroso, P.A.V., Vidal, M.S. and Vaslin, M.F.S. (2005) Molecular characterization of a virus from the family *Luteoviridae* associated with cotton blue disease. *Archives of Virology* 67, 1357–1367.

Costa, A.S. (1956) Anthocyanosis, a virus disease of cotton in Brazil. *Phytopathologische Zeitschrift* 28, 167–186.

Cui, J.J. and Xia, J.Y. (2000) Effects of Bt transgenic cotton variety on the dynamics of pest populations and their enemies. *Acta Phytophylactica Sinica* 27, 141–145.

Deguine, J.-P. (1992) Considérations pour une lutte intégrée vis-à-vis du puceron *Aphis gossypii* Glover en culture cotonnière en Afrique centrale. *Revue Scientifique du Tchad* 2, 74–82.

Deguine, J.-P. (1995) Bioécologie et épidémiologie du puceron *Aphis gossypii* Glover, 1877 (Hemiptera, Aphididae) sur cotonnier en Afrique centrale. Vers une évolution de la protection phytosanitaire. PhD thesis, Ecole Nationale Supérieure Agronomique, Montpellier, France, 138 pp.

Deguine, J.-P. (1996) The evolution of insecticide resistance in *Aphis gossypii* Glover in Cameroon. *Resistant Pest Management Newsletter* 8 (1), 13–14.

Deguine, J.-P. and Leclant, F. (1996) Description et mode d'emploi d'un dispositif de piégeage des formes ailées du puceron du cotonnier *Aphis gossypii* Glover (Hemiptera: Aphididae) au Cameroun. *Annales de la Société Entomologique de France* 32, 427–443.

Deguine, J.-P. and Leclant, F. (1997) *Aphis gossypii* Glover 1877 (Hemiptera, Aphididae). *Coton et Fibres Tropicales, Série: Les déprédateurs du cotonnier en Afrique tropicale et dans le reste du monde, No. 11.* CIRAD, Montpellier, 112 pp.

Deguine, J.-P., Gozé, E. and Leclant, F. (1994) Incidence of early outbreaks of the aphid *Aphis gossypii* Glover in cotton growing in Cameroon. *International Journal of Pest Management* 40, 132–140.

Deguine, J.-P., Martin, J. and Leclant, F. (1999) Extreme polyphagy of *Aphis gossypii* Glover (Hemiptera: Aphididae) during the dry season in Northern Cameroon. *Insect Science and its Application* 19, 23–36.

Deguine, J.-P., Gozé, E. and Leclant, F. (2000) The consequences of late outbreaks of the aphid *Aphis gossypii* Glover in cotton growing in central Africa – towards a possible method for the prevention of cotton stickiness. *International Journal of Pest Management* 46, 85–89.

Deguine, J.-P., Vaissayre, M. and Ferron, P. (2004) Aphid and whitefly management in cotton growing: review and challenges for the future. *Proceedings of the 3rd World Cotton Research Conference, Cape Town, March 2003.* ICAC, Washington, pp. 1177–1194.

Dong, Z.Q., Feng, C.T. and Il, G.L. (1992) Study on the control of intercropping cotton with watermelon on cotton aphids. *China Cottons* 1, 36–37.

Duviard, D. and Mercadier, G. (1973) Les invasions saisonnières de pucerons en culture cotonnière: origine et mécanismes. *Coton et Fibres Tropicales* 28, 483–491.

El-Zik, K.M. and Thaxton, P.M. (1989) Genetic improvement for resistance to pests and stresses in cotton. In: Frisbie, R.E., El-Zik, K.M. and Wilson, T.L. (eds) *Integrated Pest Management Systems and Cotton Production.* Wiley, New York, pp. 191–224.

Favret, C. and Voegtlin, D.J. (2001) Migratory aphid (Hemiptera: Aphididae) habitat selection in agricultural and adjacent natural habitats. *Environmental Entomology* 30, 371–379.

Fitt, G.P. (2000) A future for IPM in cotton: the challenge of integrating new tools to minimise pesticide dependence. *Proceedings of the 2nd World Cotton Research Conference, Athens, September, 1998.* ICAC, Washington, pp. 75–84.

Franco, G.V. (1999) Contrôle quimico de *Aphis gossypii* Glover em cultivar sucetivel a viroses. *Anais 2do Congresso Brasileiro de Algodao, Ribeirão Preto, September 1999.* Embrapa, Campina Grande, Brazil, pp. 195–197.

Freire, E.C., Farias, F.J.C., Aguiar, P.H., Siquieri, F. and Dos Reis, C.R. (1999) Reduçao nos custos de produçao do algodoao obtidos com uso de cultivares resistentes a viroses no cerrado. *Anais 2do Congresso Brasileiro de Algodao, Ribeirão Preto, September 1999.* Embrapa, Campina Grande, Brazil, pp. 4–6.

Frisbie, R.E., El-Zik, K.M. and Wilson, T.L. (eds) (1989) *Integrated Pest Management Systems and Cotton Production.* Wiley, New York, 437 pp.

Furk, C. and Hines, C.M. (1993) Aspects of insecticide resistance in the melon and cotton aphid *Aphis gossypii* (Homoptera, Aphididae). *Annals of Applied Biology* 123, 9–17.

Gahukar, R.T. (1991) Control of cotton insect and mite pests in subtropical Africa: current status and future needs. *Insect Science and its Application* 12, 313–338.

Godfrey, L.D. and Leser, J.F. (1999) Cotton aphid management: status and needs. *Proceedings of the Beltwide Cotton Conferences, Orlando, January 1999, Volume 2.* National Cotton Council, Memphis, pp. 37–40.

Godfrey, L.D., Cisneros, J.J., Keillor, K.E. and Hutmacher, R.B. (2000) Influence of cotton nitrogen fertility on cotton aphid population dynamics in California. *Proceedings of the Beltwide Cotton Conferences, San Antonio, January 2000, Volume 2.* National Cotton Council, Memphis, pp. 1162–1165.

Grafton-Cardwell, E.E. (1991) Geographical and temporal variation responses of insecticides in various stages of *Aphis gossypii* (Homoptera: Aphididae) infesting cotton in California. *Journal of Economic Entomology* 84, 741–749.

Grafton-Cardwell, E.E., Leight, T.F., Bentley, W.J. and Goodell, P.B. (1992) Cotton aphids have become resistant to commonly used pesticides. *California Agriculture* 46, 4–7.

Graves, J.B., Ottea, J.A., Leonard, B.R., Burris, E. and Macinski, S. (1993) Insecticide resistance management: an integral part of IPM in cotton. *Louisiana Agriculture* 36, 3–5.

Green, M.B. and Lyon, D.J. de B. (1989) *Pest Management in Cotton*. Ellis Horwood, Chichester, 259 pp.

Gubran, E.E., Delorme, R., Augé, D. and Moreau, J.P. (1992) Insecticide resistance in cotton aphid *Aphis gossypii* Glov. in the Sudan Gezira. *Pesticide Science* 35, 101–107.

Hambleton, E.J. (1937) Em defesa da cultura algodoeira: a broca do algodoeiro e a lagarta rosada. *Revista da Sociedad Rural Brasileira* 17, 40–43.

Hardee, D.D. (1993) Resistance in aphids and whiteflies: principles and keys to management. *Proceedings of the Beltwide Cotton Conferences, New Orleans, January 1993, Volume 2*. National Cotton Council, Memphis, pp. 20–23.

Hardee, D.D. and O'Brien, P.J. (1990) Cotton aphids: current status and future trends in management. *Proceedings of the Beltwide Cotton Conferences, Las Vegas, January 1990*. National Cotton Council, Memphis, pp. 169–171.

Hardee, D.D., Weathersbee, A.A. and Smith, M.T. (1994) Biological control of the cotton aphid. *Proceedings of the Beltwide Cotton Conferences, San Diego, January 1994, Volume 2*. National Cotton Council, Memphis, pp. 132–133.

Harris, F.A., Andrews, G.L., Caillavet, D.F. and Furr, R.E. (1992) Cotton aphid effect on yield, quality and economics of cotton. *Proceedings of the Beltwide Cotton Conferences, Nashville, January 1992, Volume 2*. National Cotton Council, Memphis, pp. 652–656.

Harris, F.A., Furr, R.E., Jr. and Calhoun, D.S. (1996) Cotton insect management in transgenic Bt cotton in the Mississippi Delta, 1992–1995. *Proceedings of the Beltwide Cotton Conferences, Nashville, January 1996, Volume 2*. National Cotton Council, Memphis, pp. 854–858.

Head, R.B. (1992) Cotton insect losses. *Proceedings of the Beltwide Cotton Conferences, Nashville, January 1992, Volume 2*. National Cotton Council, Memphis, pp. 621–625.

Héquet, E., Ethridge, D. and Wyatt, B. (2000) Update on cotton stickiness measurement. *Proceedings of the 2nd World Cotton Research Conference, Athens, September 1998, Volume 2*. ICAC, Washington, pp. 976–980.

Hillocks, R.J. (1995) Integrated management of insect pests, diseases and weeds of cotton in Africa. *Integrated Pest Management Reviews* 1, 31–47.

ICAC (2000) *Outlook for Cotton Supply in 2000/2001*. International Cotton Advisory Committee, Washington, 23 pp.

Ingram, W.R., Sutherland, J.A., Haque, H. and Lyon, D.J. de B. (1989) Pesticide management in cotton in Pakistan. In: Green, M.B. and Lyon, D.J. de B. (eds) *Pest Management in Cotton*. Ellis Horwood, Chichester, pp. 16–26.

JianGo, Y., FuJie, T. and ZiPing, Y. (1987) Preliminary study on the insecticide resistance of cotton aphids in north China. *Journal of the Nanjing Agricultural University* 12, 13–21.

Jones, R.H., Leonard, B.R., Boquet, D.J. and Emfinger, K. (2001) Influence of agronomic practices on cotton aphid, *Aphis gossypii* Glover, densities in Louisiana cotton. *Proceedings of the Beltwide Cotton Conferences, Anaheim, January 2001, Volume 2*. National Cotton Council, Memphis, pp. 1097–1099.

Jones, R.H., Leonard, B.R. and Bagwell, R.D. (2003) Fungus helps control Louisiana cotton aphids. *Louisiana Agriculture* 46, 24–25.

Kerns, D.L. and Gaylor, M.J. (1993) Induction of cotton aphid outbreaks by insecticides in cotton. *Crop Protection* 12, 387–393.

King, E.G., Philips, J.R. and Head, R.B. (1987) 40th annual conference report on cotton insect research and control. *Proceedings of the Beltwide Cotton Conferences, Dallas, January 1987*. National Cotton Council, Memphis, pp. 170–192.

King, E.G., Phillips, J.R. and Head, R.B. (1988) 41st annual conference report on cotton insect research and control. *Proceedings of the Beltwide Cotton Conferences, New Orleans, January 1988*. National Cotton Council, Memphis, pp. 188–202.

Kiss, A. (1991) The cotton integrated pest management program in Sudan. In: Kiss, A. and Meerman, F. (eds) *Integrated Pest Management and African Agriculture*. World Bank, Washington, pp. 67–75.

Knutson, A.E. and Tedders, L. (2002) Augmentation of green lacewing, *Chrysopa rufilabris*, in cotton in Texas. *Southwestern Entomologist* 27, 231–239.

Latteur, G. (1982) Quelques Entomophthoraceae parasites de pucerons trouvées au Brésil, dans l'état du Rio Grande do Sul. *Parasitica* 38, 85–89.

Leclant, F. and Deguine, J.-P. (1994) Cotton aphids. In: Matthews, G.A. and Tunstall, J.P. (eds) *Insect Pests of Cotton*. CAB International, Wallingford, pp. 285–323.

Leonard, M.D., Walker, H.G. and Enari, L. (1971) Host plants of *Aphis gossypii* at Los Angeles State and county Arboretum, Arcadia, California. *Proceedings of the Entomological Society of Washington* 74, 9–16.

Leser, J.F. (1994) Management of cotton aphids: Texas style. *Proceedings of the Beltwide Cotton Conferences, San Diego, January 1994, Volume 2*. National Cotton Council, Memphis, pp. 137–141.

Leser, J.F., Allen, C.T. and Fuchs, T.W. (1992) Cotton aphid infestation in West Texas: a growing management problem. *Proceedings of the Beltwide Cotton Conferences, Nashville, January 1992, Volume 2*. National Cotton Council, Memphis, pp. 823–827.

Liu, X.D., Zhang, X.X. and Zhai, B.P. (2002) On the variation law of body color and reproductive characteristics of cotton aphid, *Aphis gossypii* Glover (Homoptera: Aphididae), possessing different body colors. *Acta Phytophylactica Sinica* 29, 153–157.

Luo, Z.H. and Gan, G.P. (1986) Population dynamics of cotton aphids on cotton during square-boll stage and the relation between population, age, structure and parasitization. *Acta Entomologica Sinica* 29, 156–161.

Luttrell, R.G. (1994) Cotton pest management: 2. US perspective. *Annual Review of Entomology* 39, 527–542.

Mali, V.R. (1978) Anthocyanosis virus disease of cotton a new record for India. *Current Science* 47, 235–237.

Matthews, G.A. (2000) Application techniques to meet the requirements of IPM, IRM and transgenics. *Proceedings of the 2nd World Cotton Research Conference, Athens, September 1998. Volume 2*. ICAC, Washington, pp. 976–980.

Matthews, G.A. and Tunstall, J.P. (1994) *Insect Pests of Cotton*. CAB International, Wallingford, 593 pp.

Mengech, A.N., Saxena, K.N. and Gopalan, H.N.B. (1995) *Integrated Pest Management in the Tropics. Current Status and Future Prospects*. Wiley, Chichester, 171 pp.

Morita, M.S., Brunetta, E., Takizawa, E.K. and Degrande, P.E. (1999) Pragas do algodoeiro na regiao de Novo Sao Joachim. *Anais 2do Congresso Brasileiro de Algodao*. Embrapa, Campina Grande, pp. 272–275.

Morse, S. and Buhler, W. (1997) *Integrated Pest Management: Ideals and Realities in Developing Countries*. Lynne Rienner, Boulder, 171 pp.

Mu, J.Y., Li, Z.H., Mu, S.M. and Fan, G.H. (1993) Study on control criteria of *Aphis gossypii* Glover in a field of directly-sown spring cotton. *Journal of Shandong Agricultural University* 24, 183–187.

Munro, J.M. (1994) Cotton and its production. In: Matthews, G.A. and Tunstall, J.P. (eds) *Insect Pests of Cotton*. CAB International, Wallingford, pp. 3–26.

Nevo, E. and Coll, M. (2001) Effect of nitrogen fertilization on *Aphis gossypii* (Homoptera: Aphididae): variation in size, color and reproduction. *Journal of Economic Entomology* 94, 27–32.

Norman, J.W., Jr., Riley, D.G., Sparks, A.N. and Leser, J.F. (1993) Texas suggestions for managing sweetpotato whitefly and aphids in cotton. *Proceedings of the Beltwide Cotton Conferences, New Orleans, January 1993, Volume 2*. National Cotton Council, Memphis, pp. 36–37.

OILB (1977) Vers la Production Intégrée par la Lutte Intégrée. *Bulletin OILB/SROP* 4, 163 pp.

Ortiz, A.C.S., Genezine, F.A., Trenhago, G. da S. and Degrande, P.E. (1999) Pragas do algodoeiro na regiao de Sapezal. *Anais 2do Congresso Brasileiro de Algodao*. Embrapa, Campina Grande, pp. 155–158.

Papa, G., Tomquelski, G.V. and Silva, R.B. (2001) Strategies for management of the cotton aphid, *Aphis gossypii* (Homoptera: Aphididae), under 'cerrado' conditions in Brazil. *Proceedings of the Beltwide Cotton Conferences, Anaheim, January 2001, Volume 2*. National Cotton Council, Memphis, pp. 1127–1129.

Parajulee, M.N. and Slosser, J.E. (1999) Evaluation of potential relay strip crops for predator enhancement in Texas cotton. *International Journal of Pest Management* 45, 275–286.

Parajulee, M.N., Slosser, J.E. and Bordovsky, D.G. (1999) Cultural practices affecting the abundance of cotton aphids and beet armyworm in dryland cotton. *Proceedings of the Beltwide Cotton Conferences, Orlando, January 1999, Volume 2*. National Cotton Council, Memphis, pp. 1014–1016.

Pendergrass, J. (1989) An overview of pest management in cotton in the USA. In: Green, M.B. and de Lyon, D.J. de B. (eds) *Pest Management in Cotton*. Ellis Horwood, Chichester, pp. 11–15.

Powell, G. and Hardie, J. (2001) The chemical ecology of aphid–host interaction: how do return migrants find the primary host-plant? *Applied Entomology and Zoology* 36, 259–267.

Ramalho, F.S. (1994) Cotton pest management: 4. A Brazilian perspective. *Annual Review of Entomology* 39, 563–578.

Ramalho, F.S., Jesus, F.M.M. and Bleicher, E. (1989) Manejo integrado de pragas e viabilidade do algodoeiro herbaceo no Nordeste. *Seminario sobre Contrôle de Insetos*. Sociedade Entomologica do Brasil, Campinas, pp. 112–123.

Reed, B., Gannaway, J., Rummel, D.R. and Thorvilson, H.G. (1999) Screening for resistance in cotton genotypes to *Aphis gossypii* Glover. *Proceedings of the Beltwide Cotton Conferences, Orlando, January 1999, Volume 2*. National Cotton Council, Memphis, pp. 1002–1007.

Roy, D.K. and Behura, B.K. (1983) Notes of host-plants, feeding behaviour, infestation and ant attendance of cotton aphids, *Aphis gossypii* Glov. *Journal of the Bombay Natural History Society* 80, 654–656.

Rummel, D.R., Arnold, M.D., Slosser, J.E., Neece, K.C. and Pinchak, W.E. (1995) Cultural factors influencing the abundance of *Aphis gossypii* Glover in Texas high plains cotton. *Southwestern Entomologist* 20, 395–406.

Silvie, P. and Papierok, B. (1991) Les ennemis naturels d'insectes du cotonnier au Tchad: premières données sur les champignons de l'ordre des entomophthorales. *Coton et Fibres Tropicales* 46, 293–303.

Slosser, J.E. and Parajulee, M.N. (2001) Factors contributing to late-season cotton aphid infestations. *Proceedings of the Beltwide Cotton Conferences, Anaheim, January 2001, Volume 2*. National Cotton Council, Memphis, pp. 995–957.

Slosser, J.E., Pinchak, W.E. and Rummel, D.R. (1989) A review of known and potential factors affecting the population dynamics of the cotton aphid. *Southwestern Entomologist* 14, 302–313.

Slosser, J.E., Montandon, W.E., Pinchak, W.E. and Rummel, D.R. (1997) Cotton aphid response to nitrogen fertility in dryland cotton. *Southwestern Entomologist* 22, 1–10.

Slosser, J.E., Parajulee, M.N. and Bordovsky, D.G. (2000) Evaluation of food sprays and relay strip crops for enhancing biological control of bollworms and cotton aphids in cotton. *International Journal of Pest Management* 46, 267–275.

Steinkraus, D.C. and Rosenheim, J. (1995) Biological factors influencing the epizootiology of cotton aphid fungus. *Proceedings of the Beltwide Cotton Conferences, San Antonio, January 1995, Volume 2* National Cotton Council, Memphis, pp. 887–889.

Steinkraus, D.C., Kring, T.J. and Tugwell, N.P. (1991) *Neozygites fresenii* in *Aphis gossypii* on cotton. *Southwestern Entomologist* 16, 118–122.

Steinkraus, D.C., Boys, G.O. and Rosenheim, J. (2002) Classical biological control of *Aphis gossypii* (Homoptera: Aphididae) with *Neozygites fresenii* (Entomophthorales: Neozygitaceae) in California cotton. *Biological Control* 25, 297–304.

Tang, Z.H. (1992) Insecticide resistance and countermeasures for cotton pests in China. *Resistant Pest Management Newsletter* 4 (2), 9–12.

Vendramim, J.D. and Nakano, O. (1981) Distribuiçao de *Aphis gossypii* Glover em plantas de algodoao. *Poliagro* 3, 1–7.

Viera, F.V., Santos, J.H.R. and Oliveira, F.J. (1983) Influence of intercropping on the incidence of the aphid *Aphis gossypii* Glover on semi-perennial cotton. *Fitosanidade* 5, 26–30.

Villate, F., Augé, D., Touton, P., Delorme, R. and Fournier, D. (1999) Negative cross-insensivity in insecticide-resistant cotton aphid. *Pesticide Biochemistry and Physiology* 65, 55–61.

Weathersbee, A.A., Hardee, D.D. and Meredith, W.R., Jr. (1995) Differences in yield response to cotton aphids (Homoptera: Aphididae) between smooth-leaf and hairy-leaf isogenic cotton lines. *Journal of Economic Entomology* 88, 749–754.

Wu, K.M. and Guo, Y.Y. (2003) Influences of *Bacillus thuringiensis* cotton planting on population dynamics of the cotton aphid, *Aphis gossypii* Glover, in Northern China. *Environmental Entomology* 32, 312–318.

Zethner, O. (1995) Practice of integrated pest management in tropical and sub-tropical Africa: an overview of two decades (1970–1990). In: Mengech, A.N., Saxena, K.N. and Gopalan, H.N.B. (eds) *Integrated Pest Management in the Tropics: Current Status and Future Prospects*. Wiley, Chichester, pp. 1–67.

Zhang, L.X., Niu, X.T. and Wu, M. (1987) A dynamic simulation model of the biological control system of cotton aphid. In: *Proceedings of the 1st International Conference on Agricultural Systems Engineering*, Changchun, 1987. Jilin University Press, Changchun, China, pp. 456–464.

Zhang, Z.Q. and Chen, P.R. (1991) Spring populations of *Aphis gossypii* in cotton fields: to spray or not to spray? *Agriculture, Ecosystems and Environment* 35, 349–351.

Zhao, D.X. and Zhou, L. (1988) A preliminary study on the population of *Aphis gossypii* at the seedling stage of cotton. *Insect Knowledge* 25, 79–80.

Zhou, Y. and Wang, H. (1989) Evaluation of the dominant predator of *Aphis gossypii* in the seedling stage of cotton. *Journal of Anhui Agricultural College* 16, 39–44.

24 IPM Case Studies: Leafy Salad Crops

G. Mark Tatchell

Department of Biological Sciences, University of Warwick, Gibbet Hill Road, Coventry, CV4 7AL, UK

Introduction

What are leafy salad crops?

Leafy salads across the world show great diversity, dependent on local cultures and plant biodiversity, and include both species that are cultivated intensively and those that are gathered from the wild. For the purposes of this review, discussion will be confined to the commercial production of the Asteraceae, which include many varieties of lettuce, *Lactuca sativa* (Fig. 24.1). However, the principles developed here can be applied to any vegetable with a short growing season and leaves that are harvested and eaten raw. These crops may be grown outdoors or under different degrees of protection, ranging from fleece or polythene covers to glasshouses. The different levels of environmental protection are used both to extend cropping seasons and to protect crops from invasion by aphids, as well as by moths and leaf miners (Ester *et al.*, 1994). The greatest areas of these crops are grown outdoors. Nearly all species and varieties are characterized by a short growing period, sometimes only 3 or 4 weeks. The species of aphid encountered on crops will differ between plants from different families and regions of the world.

The IPM challenge

Leafy salad crops are very different to many food crops in that they are eaten fresh and raw, with little or no processing. The rapidly expanding market for 'bagged salads', often containing a number of different plant species that are labelled 'Washed and ready to eat', illustrates the extreme of such products. The consequence of this development is that the total yield is of little relevance; rather, it is the proportion of harvested units (usually heads) that are free from insects that is key to market success. The challenge to the grower of leafy salads is to deliver to retailers and the food service industry a continuous supply of product that is free from contamination from insects, pest or vagrant, and other foreign bodies, and with minimum or no insecticide residues. Such contamination can be eliminated only by attention to detail.

Year-round supply is achieved by growing crops in different regions of the world and freighting them to the areas of consumption. Growers that produce on a large scale may transfer their own production activities to different climatic regions in different seasons to achieve continuous supply of product to their own processing plants and markets; they sometimes also transfer the processing activity. For example, farmers growing

Fig. 24.1. Lettuce grown in beds maximizes crop uniformity and enables access by machinery (original colour photo: G.M. Tatchell).

lettuce crops in southern England in summer will move their production to southern Spain in winter. Similarly, production in the USA focuses on the Central Coast areas of Salinas and Santa Maria in California from May to September, but shifts to the Imperial and Yuma Valleys on the California and Arizona border from November to March. In different locations, growing conditions and the pest spectrum will differ. The delivery of crops free from pests under such circumstances requires high skill levels.

Leafy salads are grown on a small land area and, as such, constitute a small market for the sale of agrochemicals. The consequence is that the armoury of agrochemicals available for use is more limited than that available to the producers of broad acre crops, so alternatives to agrochemicals need to be developed and deployed if the quality of produce demanded by retailers and consumers is to be delivered routinely.

Pemphigus bursarius (lettuce root aphid) may colonize the roots of lettuce, while the specialist *Nasonovia ribisnigri* (currant–lettuce aphid) (Fig. 24.2) and the generalists *Myzus persicae* (peach–potato aphid) and *Macrosiphum euphorbiae* (potato aphid) may colonize the foliage (Reinink and Dieleman, 1991). It is the infestation of the heart of the lettuce by *N. ribisnigri* that is of greatest economic concern (Forbes and Mackenzie, 1982; Mackenzie and Vernon, 1988). In North America, similar spectrums of aphid species are pests (Alleyne and Morrison, 1977; Forbes and Mackenzie, 1982), though *N. ribisnigri*, along with *Aulocorthum solani* (glasshouse and potato aphid), was only introduced to California in 1998 (Palumbo, 1999). In South America (Brazil), *M. persicae* has been identified as the most important species (dos Santos *et al.*, 1992). *N. ribisnigri* has recently been introduced to New Zealand and Australia (Stufkens and Teulon, 2003).

The Aphids

Twenty-one species of aphid have been recorded from lettuce (Blackman and Eastop, 2000). However, of these only a few are the focus for control on leafy salads. In Europe,

Components for Aphid IPM

Chemicals and resistance to insecticides

The intolerance of consumers in many regions of the world to any contamination of fresh,

Fig. 24.2. The currant–lettuce aphid (*Nasonovia ribisnigri*) colonizes the hearts of lettuce and is therefore very difficult to control (original colour photo courtesy Warwick HRI).

unprocessed food by insect pests has led growers to minimize the risk of downgrading or rejection of their crops by the frequent use of prophylactic applications of insecticides. The consequence of this unsustainable approach, both on lettuce and other crops, has been the development of resistance to insecticides in at least three species of aphid that infest lettuce. *Myzus persicae* is exposed to insecticides on a wide range of crops and has shown resistance to different classes of insecticides for many years (reviewed by Field *et al.*, 1997; Foster *et al.*, 2000, Chapter 10 this volume). The frequency of different insecticide-resistant genotypes varies between years (Foster *et al.*, 2002b). More recently, broad-spectrum resistance to the organophosphate acephate and the cyclodiene endosulfan has been identified in *N. ribisnigri* in southern France and Spain (Rufingier *et al.*, 1997, 1999). Resistance to pirimicarb, and to a lesser extent pyrethroids and organophosphates, in *N. ribisnigri* has also been identified in the UK (Barber *et al.*, 1999, 2002). Interestingly, insecticide-resistant *N. ribisnigri* were less aggregated on plants than susceptible aphids, so that a greater proportion of plants were infested, with the potential for a greater rejection rate of plants at harvest (Kift *et al.*, 2004). More recently, limited resistance to pirimicarb and lambda-cyhalothrin has been identified in the laboratory in *M. euphorbiae* (Foster *et al.*, 2002a). The complexity of controlling three species of aphid with different mechanisms of resistance to insecticides on a single crop is bound to drive the approach to aphid management away from agrochemicals to more biologically based approaches. This direction is further driven by the impact of insecticide-based control strategies for *M. persicae* on the wide range of crops that are also hosts for this aphid and over which the manager of leafy salad crops has no control.

Reducing insecticides

More careful, and therefore selective, placement of insecticides in space and time has been achieved by the use of insecticide seed treatments. Imidacloprid has been used successfully on lettuce in many regions, including the UK (Parker and Blood Smyth, 1996; Parker *et al.*, 2002), and this is thought to have contributed to a reduction in the quantities of insecticide applied to lettuce (Garthwaite *et al.*, 2001). In addition, insecticide applications to lettuce foliage can be timed better through the use of day-degree forecasts that predict immigration to crops (Collier *et al.*, 1994, 1999), though in practice these may be of greater use in strategic planning than for the tactical control of aphids in a single crop.

Biological control

The very short growing season of leafy salad crops does not allow for the development of classical approaches to biological control, though those natural enemies that occur outdoors undoubtedly do achieve some reduction of aphid numbers, particularly where insecticide usage is limited (Nunnenmacher and Goldbach, 1996). There are no reported attempts to introduce biological control agents into field-grown lettuce for aphid control.

The biological control of aphids has, by contrast, been very successful on peppers and cucumbers grown in glasshouses where aphids are particularly troublesome. Effective control of *Aphis gossypii* (melon aphid) on cucumbers has been achieved by either the weekly release of small numbers of the parasitoid *Aphidius colemani* or by rearing the parasitoid on *Rhopalosiphum padi* (bird cherry–oat aphid) living on cereal plants grown in pots in the glasshouse, from which the parasitoids disperse to attack aphids on cucumbers (Jacobson and Croft, 1998; Webb, Chapter 28 this volume). *Myzus persicae* is the primary aphid pest on peppers and can be controlled by the repeated release of *A.*

colemani (Foster *et al.*, 2003) or by the larvae of *Chrysoperla carnea* (Kift *et al.*, 2005). It should be noted that the fruit of peppers and cucumbers are not infested directly and that the plants have a long growing season, which provides sufficient time for equilibrium to be established between predator and prey. Efforts have been made to transfer these approaches to control aphids on glasshouse-grown lettuce crops, but as yet with limited success. In Germany, the parasitoid *Aphidius matricariae* was of little use in aphid control in lettuce because its main host, *M. persicae*, was less abundant than *N. ribisnigri, M. euphorbiae,* and *A. solani*, which were seldom, if ever, parasitized by this species. However, repeated applications of *C. carnea* eggs did give good control (Quentin *et al.*, 1995). Recently, the parasitoid *Aphidius hieraciorum* has been identified from *N. ribisnigri* in the UK and is being explored as a potential biological control agent (P. Croft, personal communication). Yet the use of arthropod biological control agents has to be undertaken with care, as the customer will not discriminate between an aphid pest and its natural enemies (including aphid mummies) and the presence of either can result in rejection at point of sale.

Opportunities do, however, arise for the inundative release of natural enemies as biological insecticides. To date, fungal pathogens have shown the greatest potential for aphid control, particularly for the smaller areas of crops grown under protection. A formulation of the fungus *Lecanicillium lecanii sensu lato*, which is available commercially for the control of a number of different pests of glasshouse crops, has been shown to reduce the numbers of *M. persicae*, *M. euphorbiae*, and *N. ribisnigri* on lettuce foliage in laboratory and glasshouse experiments, though the relative efficacy against each aphid species differed between experiments and the level of control achieved did not satisfy market requirements alone (Fournier and Brodeur, 2000). This example amply illustrates the challenges of using biological approaches where only complete elimination of aphids and cadavers on the marketable parts of the plants is acceptable. This contrasts with infestations of *P. bursarius* on the roots of lettuce, where effective control does not mean elimination. Adequate control of *P. bursarius* can be achieved by raising lettuce plants in modules in which the potting mix contains conidia of the fungus *Metarhizium anisopliae* (Chandler, 1997).

Host-plant resistance

Resistance to both diseases and aphids in lettuce has been the target of many plant-breeding programmes and resistant varieties hold much promise for the future. As yet, only a few cultivars with resistance to aphids are available commercially. A single dominant gene conferring resistance to *N. ribisnigri* has been identified in accessions of the wild species *Lactuca virosa* and introgressed into *L. sativa* by backcrossing (Eenink *et al.*, 1982a,b; Eenink and Dieleman, 1983; van Helden *et al.*, 1993). This resistance has been deployed successfully in integrated control programmes (Aarts *et al., 1999; Parker et al.*, 2002). More than one source of resistance has also been identified to *P. bursarius* in the old lettuce varieties 'Avoncrisp' and 'Lakeland' (Dunn, 1974; Ellis *et al.*, 1994) and in *L. virosa* (Ellis *et al.*, 2002).

Modifying aphid behaviour

A surprising number of unconnected studies have explored aspects of aphid behaviour to improve control on lettuce. For example, Ester *et al.* (1993) reduced the amounts of carbamate and organophosphate insecticides to one tenth of that recommended and achieved successful control when the alarm pheromone (*E*)-β-farnesene was included with the insecticides.

Volatile plant chemicals known to be involved in the process of plant colonization by aphids include chemicals from an aphid's primary host plant that seem to interrupt colonization by the aphid of its secondary host (Pettersson *et al.*, 1994). *cis*-Jasmone has been isolated from *Ribes nigrum*, the primary host of *N. ribisnigri*, and shown to be electrophysiologically active to *N. ribisnigri*. *cis*-Jasmone is also a well-known plant volatile produced in response to insect herbivory.

In olfactometer studies, *cis*-jasmone repels *N. ribisnigri* in the presence of lettuce (Birkett *et al.*, 2000) and, when included in a mixture of other electrophysiologically active compounds from *R. nigrum*, reduced the numbers of *N. ribisnigri* alatae colonizing lettuce in field-cage experiments (G.M. Tatchell and L.J. Wadhams, unpublished results). The production and deployment of such plant volatile chemicals could provide components for a novel push–pull strategy, though much work is still required.

IPM in Practice

Assurance and accreditation schemes

Accreditation and assurance schemes of crop production are being developed in Europe. Euro Retailer Produce Working Group Good Agricultural Practice (EurepGAP) Fruit and Vegetables has been established, initially by retail organizations, to set the standards of production by their suppliers. In the UK, the protocols of the Assured Produce Scheme (APS) have been revised to incorporate the EurepGAP standard. The APS is an industry-wide initiative that addresses the key issues of production, including aphid control, of fruit, salads, and vegetables through the promotion of integrated crop management. It is a prerequisite that producers in the UK be accredited to these schemes if they are to supply the multiple retailers and major players in the food services industry. However, these protocols provide very little detail of how aphids should be controlled beyond the safe use of approved insecticides. This perhaps reflects the tensions between growers, who do not wish to be constrained in the approaches available to them to deliver insect-free produce to their markets, and the retailers, who need to demonstrate to customers that the food purchased in their stores is safe.

Practical control

Outlined above is a fairly traditional list of individual approaches to aphid control developed for the aphid pests of leafy salads, but that could be combined in any number of ways to achieve effective season-long control. Yet it is the dichotomy between season-long control and the often very short season of an individual salad crop that now needs to be resolved. In practice, a series of crops is planted sequentially throughout the growing season, but each crop will be present only for a few weeks of the whole annual cycle of climatic seasons, and will be exposed to a different risk of aphid infestation (driven by changes in the seasonal abundance of aphids). IPM is therefore considered on two different temporal scales, i.e. that of a single crop and that of the season. The season effect will differ between different climatic regions and will require identification for each region.

In the UK, low temperature and light in the winter months confine outdoor lettuce production to the period from February–October. The earliest crops may be grown under fleece, primarily to protect them from frosts, but also to prevent aphid infestation once aphids begin to fly in spring.

The monitoring of aphid numbers over a number of years on plots of lettuce planted sequentially throughout the growing season, and in water traps, has identified two periods of peak abundance. These two periods of high

risk are characterized by different aphid species. In midsummer (late June and July), the three species that infest the foliage, *N. ribisnigri*, *M. persicae*, and *M. euphorbiae*, as well as the root-infesting *P. bursarius*, are all abundant. In contrast, in early autumn (September), only *N. ribisnigri* is abundant. These two periods of high risk are separated by a period of low risk, irrespective of the growth stage of the plants (Fig. 24.3) (Parker *et al.*, 2002). A similar pattern was identified in Spain (Nebreda *et al.*, 2005) and in Switzerland where it stimulated the application of control measures when 10% of plants were infested during the high-risk period compared with 40% during the low-risk period (Fischer and Terrettaz, 1999).

Growers are aware of the seasonality of risk, though perhaps not as precisely as research monitoring programmes have identified, and seasonal control strategies are modified accordingly. The effort devoted to crop walking to assess aphid populations on crops is modified through the season.

The IPM strategies adopted within the UK are focused around the optimum use of effective insecticides. The introduction of an imidacloprid seed treatment in the mid-1990s has resulted in its almost ubiquitous use. However, this does not protect the crop

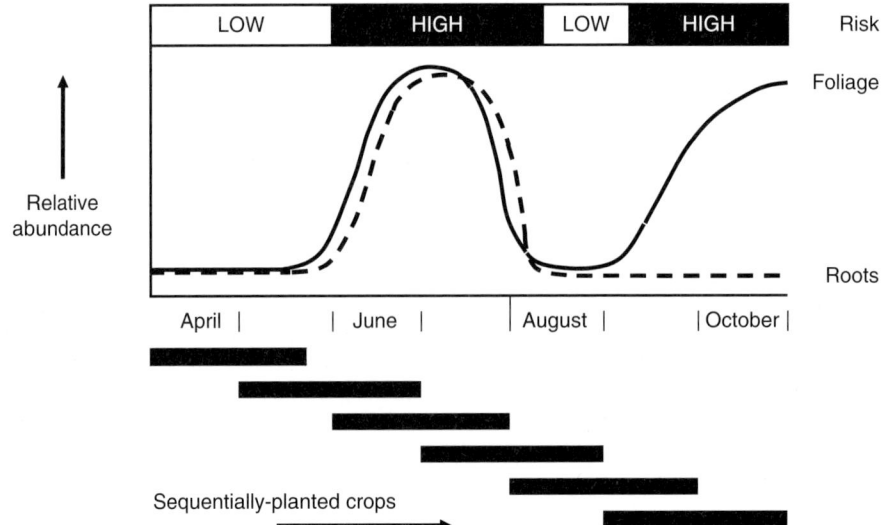

Fig. 24.3. The risk of infestation by aphids of leafy salad crops alters with planting date (adapted from Parker *et al.*, 2002). The relative risk to foliage and the roots informs strategic IPM decisions.

for its duration and further aphid control is required as harvest approaches. In the UK, the number of active ingredients that are available for use on leafy salads is limited due to the small commercial market for use on specialist crops and recent reviews of registrations. The tactical selection of any one chemical for use is related to previous applications and concerns over control failures associated with insecticide resistance.

Varieties resistant to aphids are in practical use. However, the heads produced on such varieties are said not always to meet the quality requirements of retailers and processors. Such quality considerations may override the preferred IPM approach. Resistant varieties are, however, deployed effectively in multi-cropping production systems where the growing of more than one crop per hectare limits the use of imidacloprid-treated seed to only one crop. In such circumstances, seed treatments are applied only to crops grown during the high-risk period in summer, when more than one aphid species is present. Varieties resistant to *N. ribisnigri* are deployed in late summer and autumn, when it may be the only species present.

The increasing number of plant species being grown as leafy salads for inclusion in bags of prepared salads has increased the IPM challenge. Some of these species are grown for very short periods prior to harvest, and the approved harvest intervals of insecticides may restrict their use. Many growers are adapting the IPM methods they have developed on lettuce to the novel crops as best they can.

Executive Summary

The growers of leafy salads that supply major retailers and food processors have taken significant steps towards the use of effective IPM compared to the prophylactic control strategies adopted a decade or more ago. However, despite research demonstrating that a range of components can be deployed within effective IPM strategies (Parker *et al.*, 2002), practice remains chemically based with limited use of resistant varieties. At present, growers cannot risk selling crops that have any insects on them (whether pests, natural

enemies, or cadavers) for fear of crop rejection and the consequent financial loss. Thus, insecticides remain the principal control measure, though resistance to insecticides is a current and future problem. However, seed treatments of imidacloprid reduce the need for foliar applications of pesticide, applications which damage natural biological control in field-grown crops. Applications of foliar insecticides are applied only in response to the results of detailed crop monitoring.

The use of biological alternatives is unlikely to replace insecticides until their efficacies are matched, or the availability of insecticides is limited still further. At present, no releases of insect natural enemies, as now routine in other long-season protected crops, are made on lettuce, though the fungus *L. lecanii* is used commercially.

Under such circumstances, the use of resistant varieties is likely to increase considerably, particularly as more varieties deliver good agronomic characteristics favoured by growers.

Cultural control is limited to the use of crop covers to protect crops from insect infestation, particularly early in the season, or those crops harvested as 'baby leaf'.

Research has suggested that semiochemicals have a practical potential for manipulating aphid behaviour, perhaps in the form of a 'push–pull' strategy involving sacrificial or less valuable plants, but much further research is required before commercial implementation.

Since the annual periodicity of aphid life cycles is superimposed on the short growth period of individual leafy salad crops, any individual crop is 'exposed' to a different component of the aphid life cycle. The grower managing the production of these crops uses an IPM approach across the annual periodicity of aphid life cycles, applying a different range of control options to the short growing period of each crop, though the range used on a single crop may be limited. Much progress has been made to develop and implement IPM in leafy salad crops, but the focus is currently limited to the more effective deployment of insecticides and resistant varieties based on a better-informed knowledge of aphid biology.

References

Aarts, R., Schut, J.W., Driessen, R. and Reinink, K. (1999) Integrated control of aphids on lettuce varieties resistant to *Nasonovia ribisnigri*. *Mededelingen Faculteit Landbouwkundige en Toegepaste Biologische Wetenschappen Gent* 64 (3a), 11–15.

Alleyne, E.H. and Morrison, F.O. (1977) The lettuce root aphid, *Pemphigus bursarius* (L.) (Homoptera: Aphidoidea) in Quebec, Canada. *Annals of the Entomological Society of Quebec* 22, 171–180.

Barber, M.D., Moores, G.D., Tatchell, G.M., Vice, W.E. and Denholm, I. (1999) Insecticide resistance in the currant–lettuce aphid, *Nasonovia ribisnigri* (Hemiptera: Aphididae) in the UK. *Bulletin of Entomological Research* 89, 17–23.

Barber, M.D., Moores, G.D., Denholm, I., Kift, N.B. and Tatchell, G.M. (2002) Resistance to insecticides in the currant–lettuce aphid, *Nasonovia ribisnigri*: laboratory and field evidence. *Proceedings of the British Crop Protection Conference, Pests and Diseases, Brighton, November 2002* 2, 817–822.

Birkett, M.A., Campbell, C.A.M., Chamberlain, K., Guerrieri, E., Hick, A.J., Martin, J.L., Matthes, M., Napier, J.A., Petterson, J., Pickett, J.A., Poppy, G.M., Pow, E.M., Pye, B.J., Smart, L.E., Wadhams, G.H., Wadhams, L.J. and Woodcock, C.M. (2000) New role for *cis*-jasmone as an insect semiochemical and in plant defense. *Proceedings of the National Academy of Science* 97, 9329–9334.

Blackman, R.L. and Eastop, V.F. (2000) *Aphids on the World's Crops: An Identification and Information Guide.* Wiley, Chichester, 466 pp.

Chandler, D. (1997) Selection of an isolate of the insect pathogenic fungus *Metarhizium anisopliae* virulent to the lettuce root aphid, *Pemphigus bursarius*. *Biocontrol Science and Technology* 7, 95–104.

Collier, R.H., Davies, J., Roberts, M., Leatherland, M., Runham, S. and Blood Smyth, J. (1994) Monitoring and forecasting the times of attacks of the lettuce root aphid, *Pemphigus bursarius* L. *Bulletin IOBC/WPRS* 17 (8), 31–40.

Collier, R.H., Tatchell, G.M., Ellis, P.R. and Parker, W.E. (1999) Strategies for the control of aphid pests of lettuce. *Bulletin IOBC/WPRS* 22 (5), 25–35.

Dunn, J.A. (1974) Study on inheritance of resistance to root aphid, *Pemphigus bursarius,* in lettuce. *Annals of Applied Biology* 76, 9–18.

Eenink, A.H. and Dieleman, F.L. (1983) Inheritance of resistance to the leaf aphid *Nasonovia ribisnigri* in wild lettuce species. *Euphytica* 32, 691–695.

Eenink, A.H., Groenwold, R. and Dieleman, F.L. (1982a) Resistance of lettuce (*Lactuca*) to the leaf aphid *Nasonovia ribisnigri*. 1. Transfer of resistance from *L. sativa* by interspecific crosses and selection of resistant breeding lines. *Euphytica* 31, 291–300.

Eenink, A.H., Dieleman, F.L. and Groenwold, R. (1982b) Resistance of lettuce (*Lactuca*) to the leaf aphid *Nasonovia ribisnigri*. 2. Inheritance of the resistance. *Euphytica* 31, 301–304.

Ellis, P.R., Pink, D.A.C. and Ramsey, A.D. (1994) Inheritance of resistance to lettuce root aphid in the lettuce cultivars Avoncrisp and Lakeland. *Annals of Applied Biology* 124, 141–151.

Ellis, P.R., McClement, S.J., Saw, P.L., Phelps, K., Vice, W.E., Kift, N.B. and Pink, D.A.C. (2002) Identification of sources of resistance in lettuce to the lettuce root aphid, *Pemphigus bursarius*. *Euphytica* 125, 305–315.

Ester, A., Gut, J., van Oosten, A.M. and Pijnenburg, H.C.H. (1993) Controlling aphids in iceberg lettuce by alarm pheromone in combination with an insecticide. *Journal of Applied Entomology* 115, 432–440.

Ester, A., van de Zande, J.C. and Frost, A.J.P. (1994) Crop covering to prevent pest damage to field vegetables, and the feasibility of pesticide application through polythene nets. *Proceedings of the Brighton Crop Protection Conference, Pests and Diseases, November 1994* 2, 761–766.

Field, L.M., Anderson, A.P., Denholm, I., Foster, S.P., Harling, Z.K., Javad, N., Martinez-Torres, D., Moores, G.D., Williamson, M.S. and Devonshire, A.L. (1997) Use of biochemical and DNA diagnostics for characterising multiple mechanisms of insecticide resistance in the peach–potato aphid, *Myzus persicae* (Sulzer). *Pesticide Science* 51, 283–289.

Fischer, S. and Terrattaz, C. (1999) Pucerons sur laitue et seuils d'intervention. *Revue Suisse Viticulture, Arboriculture et Horticulture* 31, 135–138.

Forbes, A.R. and Mackenzie, J.R. (1982) The lettuce aphid, *Nasonovia ribisnigri* (Homoptera: Aphididae), damaging lettuce crops in British Columbia. *Journal of the Entomological Society of British Columbia* 79, 28–31.

Foster, S.P., Denholm, I. and Devonshire, A.L. (2000) The ups and downs of insecticide resistance in peach–potato aphids (*Myzus persicae*) in the UK. *Crop Protection* 19, 873–879.

Foster, S.P., Hackett, B., Mason, N., Moores, G.D., Cox, D.M., Campbell, J. and Denholm, I. (2002a) Resistance to carbamate, organophosphate and pyrethroid insecticides in the potato aphid (*Macrosiphum euphorbiae*). *Proceedings of the Brighton Crop Protection Conference, Pests and Diseases, November 1994* 2, 811–816.

Foster, S.P., Harrington, R., Dewar, A.M., Denholm, I. and Devonshire, A.L. (2002b) Temporal and spatial dynamics of insecticide resistance in *Myzus persicae* (Hemiptera: Aphididae). *Pest Management Science* 58, 895–907.

Foster, S.P., Kift, N.B., Baverstock, J., Sime, S., Reynolds, K., Jones, J.E., Thompson, R. and Tatchell, G.M. (2003) Association of MACE-based insecticide resistance in *Myzus persicae* with reproduction rate, response to alarm pheromone and vulnerability to attack by *Aphidius colemani*. *Pest Management Science* 59, 1169–1178.

Fournier, V. and Brodeur, J. (2000) Dose-response susceptibility of pest aphids (Homoptera: Aphididae) and their control on hydroponically grown lettuce with the entomopathogenic fungus *Verticillium lecanii*, azadirachtin, and insecticide soap. *Environmental Entomology* 29, 568–578.

Garthwaite, D.G., Thomas, M.R. and Dean, S. (2001) *Outdoor Vegetable Crops in Great Britain 1999. Pesticide Usage Survey, Report No.163*. Department for Environment, Food and Rural Affairs and Scottish Executive Environment and Rural Affairs Department, London, 61 pp.

van Helden, M., Tjallingii, W.F. and Dieleman, F.L. (1993) The resistance of lettuce (*Lactuca sativa* L.) to *Nasonovia ribisnigri*: bionomics of *N. ribisnigri* on near isogenic lettuce lines. *Entomologia Experimentalis et Applicata* 66, 53–58.

Jacobson, R.J. and Croft, P. (1998) Strategies for the control of *Aphis gossypii* Glover (Hom.: Aphididae) with *Aphidius colemani* Viereick (Hym.: Braconidae) in protected cucumbers. *Biocontrol Science and Technology* 8, 377–387.

Kift, N.B., Mead, A., Reynolds, K., Sime, S., Barber, M.D., Denholm, I. and Tatchell, G.M. (2004) The impact of insecticide resistance in the currant–lettuce aphid, *Nasonovia ribisnigri*, on pest management in lettuce. *Agricultural and Forest Entomology* 6, 295–309.

Kift, N.B., Sime, S., Reynolds, K.A., Jones, J.E. and Tatchell, G.M. (2005) *Chrysoperla carnea* (Neuroptera: Chrysopidae) controls *Myzus persicae* (Homoptera: Aphididae) despite behavioural differences between aphid clones that are associated with MACE-based insecticide resistance. *Biocontrol Science and Technology* 15, 97–103.

Mackenzie, J.R. and Vernon, R.S. (1988) Sampling for the distribution of the lettuce aphid, *Nasonovia ribisnigri* (Homoptera: Aphididae), in fields and within heads. *Journal of the Entomological Society of British Columbia* 85, 10–14.

Nebreda, M., Michelena, J.M. and Fereres, A. (2005) Seasonal abundance of aphid species on lettuce crops in Central Spain and identification of their main parasitoids. *Zeitschrift für Pflanzenkrankheiten und Pflanzenschutz* 112, 405–415.

Nunnenmacher, L. and Goldbach, H.E. (1996) Aphids on lettuce: the effects of excluding aphid predators. In: *Integrated Control of Field Vegetables. Bulletin IOBC/WPRS* 19 (11), 38–47.

Palumbo, J.C. (1999) Preliminary examination of the population dynamics and control of the lettuce aphid in Romaine. In: Byrne, D.N. and Baciewicz, P. (eds) *Vegetable Report. University of Arizona College of Agriculture Series P-117, AZ1143*. University of Arizona College of Agriculture, Yuma, pp. 130–135.

Parker, W.E. and Blood Smyth, J.A. (1996) Insecticidal control of foliar and root feeding aphids on outdoor lettuce. *Proceedings of the Brighton Crop Protection Conference, Pests and Diseases, November 1994* 3, 861–866.

Parker, W.E., Collier, R.H., Ellis, P.R., Mead, A., Chandler, D., Blood Smyth, J.A. and Tatchell, G.M. (2002) Matching control options to a pest complex: the integrated management of aphids in sequentially planted crops of outdoor lettuce. *Crop Protection* 21, 235–248.

Pettersson, J., Pickett, J.A., Pye, B.J., Quiroz, A., Smart, L.E., Wadhams, L.J. and Woodcock, C.M. (1994) Winter host component reduces colonization by bird-cherry–oat aphid, *Rhopalosiphum padi* (L.) (Homoptera, Aphididae), and other aphids in cereal fields. *Journal of Chemical Ecology* 20, 2565–2574.

Quentin, U., Hommes, M. and Basedow, Th. (1995) Untersuchungen zur biologischen Bekämpfung von Blattläusen (Hom., Aphididae) an Kopfsalat im Unterglasanbau. *Journal of Applied Entomology* 119, 227–232.

Reinink, K. and Dieleman, F.L. (1991) Survey of aphid species on lettuce. *Bulletin IOBC/WPRS* 16 (5), 56–68.

Rufingier, C., Schoen, L., Martin, C. and Pasteur, N. (1997) Resistance of *Nasonovia ribisnigri* (Homoptera: Aphididae) to five insecticides. *Journal of Economic Entomology* 90, 1445–1449.

Rufingier, C., Pastuer, N., Lagnel, J., Martin, C. and Navajas, M. (1999) Mechanisms of insecticide resistance in the aphid *Nasonovia ribisnigri* (Mosley) from France. *Insect Biochemistry and Molecular Biology* 29, 385–391.

dos Santos, B.B., Casmo, P.C. and Polack, S.W. (1992) Insetos associados à cultura da alface em Campo Largo, Paraná, Brasil. *Revieu de Agricultura, Piracicaba* 67, 83–88.

Stufkens, M.A.W. and Teulon, D.A.J. (2003) Distribution and host range of the lettuce aphid in New Zealand and its effect on the lettuce industry. *New Zealand Plant Protection* 56, 27–32.

25 IPM Case Studies: Grain

Hans-Michael Poehling[1], Bernd Freier[2] and A. Michael Klüken[1]

[1]Institute of Plant Protection and Plant Diseases, University of Hannover, 30419 Hannover, Germany; [2]Institute for Integrated Plant Protection, BBA, 14532 Kleinmachnow, Germany

Introduction

Aphids are major insect pests in many cereal-growing regions of the world, feeding on phloem sap and infecting the plants with harmful viruses. Among the numerous aphid species occurring in cereals, the following six are considered to be major pests (Blackman and Eastop, 2000):

- *Sitobion avenae* (grain aphid), attacking wheat, oats, and rye in Europe, North and South America, Africa, and Asia (Fig. 25.1; in Australia and New Zealand the related *Sitobion miscanthi* and *Sitobion* sp. nr. *fragariae* are pests, but not important ones).
- *Rhopalosiphum padi* (bird cherry–oat aphid), a cosmopolitan pest of wheat, oats, and barley, especially in temperate regions.
- *Rhopalosiphum maidis* (corn leaf aphid), infesting maize and various other cereals in warmer regions of the world.
- *Metopolophium dirhodum* (rose–grain aphid), a widely distributed pest of many cereal crops, except for the tropics (Fig. 25.2).
- *Schizaphis graminum* (greenbug), a cosmopolitan pest attacking many cereal crops, including sorghum; found only in southern regions in Europe.

- *Diuraphis noxia* (Russian wheat aphid), infesting mainly barley and wheat in the Middle East, parts of Eastern Europe, North America, South America, and some parts of southern and eastern Africa.

In Central Europe, the focus of this chapter, mainly *S. avenae*, *R. padi*, and *M. dirhodum* attack cereal crops, particularly winter wheat. The first important outbreaks, associated with the increase in the area of winter wheat, were recorded in 1966, 1968, and 1969 (Rautapää, 1966; Kolbe, 1969; Latteur, 1970; Ressel, 1970). Since then, cereal aphids have increasingly become major pests in Central and Western Europe (Carter *et al.*, 1980; Kröber and Carl, 1991). Among the factors that have contributed to the increasing pest status of cereal aphids are more susceptible varieties, increasing N-fertilization, and a decreasing impact of natural enemies due to the side effects of pesticides and habitat changes.

Consequently, since the 1970s, cereal aphids have become a major focus of agricultural research and extension services to estimate damage potential, to re-establish natural control by predators and/or parasitoids, and to make pesticide use safer for the environment. The cereal aphid system is now being used as a model for the development

Fig. 25.1. *Sitobion avenae* infesting a wheat ear, with black moulds associated with honeydew fouling. Natural enemies (parasitoid and coccinellid larva) are also present (from original colour photo, courtesy of Syngenta).

of integrated plant protection strategies in arable farming.

Population Dynamics

Cereal aphid populations are generally holocyclic, with anholocycly common only in regions with mild winters. Holocyclic clones of *S. avenae* produce eggs on Poaceae, *R. padi* on bird cherry (*Prunus padus*), and *M. dirhodum* on *Rosa* spp. Holocyclic clones appear on cereal crops in late April. High population levels are rarely reached before the end of May, corresponding to the start of heading in winter wheat. All three species usually occur simultaneously and can form mixed colonies. *Sitobion avenae*, however, preferentially feeds on the ears, whereas *R. padi* and *M. dirhodum* prefer the leaves. Population build-up is correlated closely

with the maturation of host plants, and reproduction is greatly reduced after the soft dough stage. Therefore, in winter wheat, the critical period for aphid infestations in spring and early summer is generally not longer than 6 weeks (Wratten *et al.*, 1979; Freier and Wetzel, 1984; Rossing, 1991). Population density frequently remains low until flowering; thereafter, there is often an exponential increase, though this can vary widely between sites and years. Infestation levels of > 50 aphids per tiller are relatively rare. Population growth is typically a single peak curve with a short period of high density, followed rapidly by a crash caused by plant deterioration and natural enemies. After this crash, cereal aphids re-appear in autumn on newly sown winter barley and winter wheat. Generally, early sowing and, consequently, early emergence of the plants can lead to high infestation levels in autumn. During this

Fig. 25.2. Nymphs of *Metopolophium dirhodum* (from original colour photo, courtesy of U. Wyss).

period, *R. padi* in particular transmits *Barley yellow dwarf virus* (BYDV). Populations of virginoparous (i.e. anholocyclic) *S. avenae* can overwinter in cereal fields if weather conditions remain mild.

Damage

Direct damage

Sitobion avenae, *R. padi*, and *M. dirhodum* feed on phloem sap, reducing the transport of phloem to storage sites, particularly the grains. Yield losses occur mainly because of loss of phloem sap and excretion of honeydew, which occludes the stomata and leads to secondary fungal infections (sooty mould) reducing photosynthetic efficacy of the plants (Rossing, 1991). Aphid infestations not only reduce yield, but also spoil the baking quality of the grains by inducing changes in protein composition (Oakley *et al.*, 1993).

The most widely used measurements for evaluating the impact of aphids on grain yield (in t/ha or g/tiller) are peak density (i.e. individuals per tiller) and/or aphid index (i.e. aphid days per tiller). The amount of damage depends on:

- *Aphid species and site of attack.* With mainly leaf attack, as in *R. padi* or *M. dirhodum*, the yield loss per aphid is only about half as much as if ears are attacked by *S. avenae* (Niehoff and Stäblein, 1998).
- *Sensitive plant stage.* During flowering, the yield loss per aphid or aphid day is about twice as high as during green ripeness or later (Wratten *et al.*, 1979; Entwistle and Dixon, 1987; Möwes *et al.*, 1997).
- *Level of infestation.* Yield loss with few individuals per tiller is generally undetectable, mainly because of plant tolerance. It has been argued that yield might even be stimulated (e.g. Jahn and Mehrbach, 1984). Though yield loss rises with increasing pest density, yield loss per aphid (which can range between 0–0.4 mg grain weight per aphid day) decreases.

Indirect damage

Cereal aphids, in particular *R. padi* and *S. avenae*, are vectors of different strains of BYDV, i.e. BYDV-MAV, BYDV-PAV, and of *Cereal yellow dwarf virus*-RPV. Barley, oats, and wheat are highly susceptible to infection. Because of its later sowing date, winter wheat is usually less colonized by aphids than winter barley. Frequent outbreaks of BYDV occur in Europe, particularly in regions with mild Atlantic weather.

In late summer, usually only a small fraction of the winged aphid population inhabiting wild grasses, maize, and volunteer cereals is viruliferous (Plumb, 1990). First infections of winter cereals occur during the autumn migration, and early disease incidence is related to the proportion of winged aphids, mostly *R. padi*, carrying the virus (Fabre *et al.*, 2005). BYDV-epidemiology depends largely on weather conditions following this immigration. During warm autumn, winter, or spring periods, reproduction and movement (especially secondary spread of aphids, i.e. within-field) favour virus spread (Huth and Lauenstein, 1991; Knaust and Poehling, 1996). BYDV infections result in fewer ears, less grain per tiller, and lower thousand grain weights. Unfortunately, so many factors affect the epidemiology of the pathogen that no functional relationship between aphid infestation and virus-related damage has yet been found, preventing the development of reliable aphid thresholds for BYDV infection.

Control Possibilities

Chemical control

If aphid densities exceed action thresholds (see below), an insecticide treatment is recommended. In European countries, several efficient insecticides are registered for control of cereal aphids, both as BYDV vectors and as direct pests. The decision to treat is often based on simple economic considerations regarding both cost of application and amount of work involved. Hence, combined applications with fungicides before flowering are popular. Of major concern are the possible side effects of drift on non-target organisms (both in the crop and on adjacent non-cereal areas) since, with cereals, large areas are treated simultaneously.

Wick and Freier (2000) showed that broad-spectrum insecticides like λ-cyhalothrin negatively affected several non-target organisms immediately after application. However, such effects were often ephemeral, with rapid recovery and recolonization (e.g. Kühne *et al.*, 2002; Langhof *et al.*, 2003). Good agricultural practice (e.g. spraying at low wind speed,

using drift-reducing nozzles) can protect non-crop areas, even from highly toxic chemicals. Some European countries place restrictions on using pesticides in and near field margins to protect arthropod communities from spray drift (Campbell, 1998).

Inherently more selective compounds like pirimicarb can be used; alternatively, some potent insecticides for cereal aphid control can be applied successfully at considerably reduced dose rates, especially if infestation is late and levels only marginally exceed the action threshold, as shown by Poehling (1989) and Niehoff (1996) for λ-cyhalothrin and pirimicarb. Low dosage strategies are a very important element of IPM in cereals because of the considerably reduced side-effects on predators and parasitoids.

Forecasting systems

Longer-term forecasts of aphid outbreaks have been attempted for wheat, based on suction trap and weather data, or aphid field counts in autumn and early spring (Carter *et al.*, 1982, see also Harrington *et al.*, Chapter 19 this volume). However, the resulting predictions have not proved sufficiently accurate for practical application to individual fields.

Monitoring and decision-making
systems (see also Harrington *et al.*, Chapter 19 this volume)

Lately, much research has focused on developing monitoring, including short-term forecasts. Data on population dynamics of cereal aphids and their natural enemies, as well as infestation–damage relationships, have been used to develop a monitoring and decision-making system for Germany that involves counting aphids during the early build-up of the infestation (Fig. 25.3). Monitoring should start at ear emergence. Most often, the end of flowering is a suitable date for decision making in Germany, based on action thresholds, but it could be earlier, e.g. in the UK (Oakley and Walters, 1994). Several field experiments with insecticide treatments

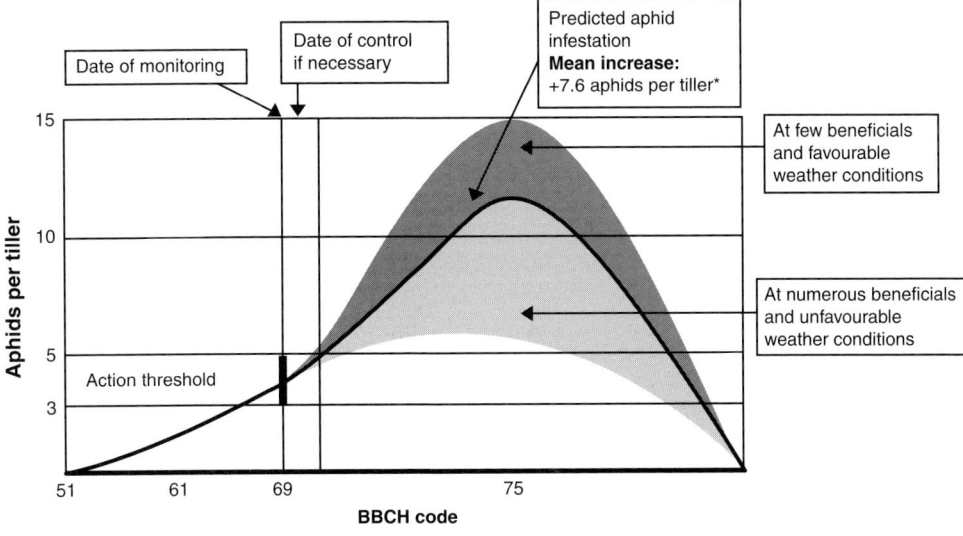

Action threshold

*Germany, Magdeburger Boerde, 1993–2000

Fig. 25.3. Principle of monitoring and decision making in integrated control of cereal aphids in winter wheat.

(Basedow *et al.*, 1994) and statistical analyses of aphid density trends and infestation–damage relationships (Holz *et al.*, 1994) came to the identical action threshold; namely, three to five aphids either per ear and flag leaf, similar to between 60 and 70% ears infested and quite similar to the UK threshold of 66% (Oakley and Walters, 1994). This threshold assumes that aphid densities are likely to reach the economic threshold level in the following 1–3 weeks. Basic knowledge on the biology and population dynamics of cereal aphids, together with a consideration of the prevalence of natural enemies, can be combined into more flexible thresholds (Rappaport and Freier, 2001) or cereal aphid expert systems (Gonzalez Andujar *et al.*, 1993). However, the higher work load makes these seldom accepted by farmers.

Autumn action thresholds for aphids as BYDV vectors have also been recommended (but see earlier concerning limitations). Depending on the regional advisory service, thresholds between 10 and 15% aphid-infested cereal plants have been recommended, e.g. in France (Bayon *et al.*, 1982). Recently, a stochastic population model was developed to predict the risk of *R. padi*

vectoring BYDV (Fabre *et al.*, 2006). However, these thresholds are based on experience rather than on experimentation. Decision support systems for advising on the control of aphid vectors of BYDV are in use in Australia and New Zealand (Knight and Thackray, Chapter 31 this volume).

Biological control

Table 25.1 lists the most important natural enemies of cereal aphids that occur in the cereal ecosystem. In recent years, an impressive body of knowledge on the biology and ecology of natural enemies of cereal aphids has been accumulated, yet the only resulting IPM activities relate to careful choice/application of pesticide (see earlier) and habitat modifications (see later).

Entomopathogenic fungi (Entomophthorales) are the most important microorganisms attacking cereal aphids. Their efficacy depends mainly on weather conditions during late spring and early summer: under favourable conditions, they can cause epizootics, and hence a fast crash of aphid populations (Dedryver, 1982; Ardisson *et al.*, 1997; Powell

Table 25.1.   Efficient natural antagonists of cereal aphids in Europe (only important species listed).

Taxon	References
Entomopathogenic fungi (Entomophthorales): *Conidiobolus obscurus, Entomophthora planchoniana, Pandora neoaphidis*	Dedryver, 1982; Freier *et al.*, 1986
Parasitoids (Aphididae): *Aphidius rhopalosiphi, Aphidius uzbekistanicus, Aphidius picipes, Aphidius ervi, Praon volucre*	Hågvar and Hofsvang, 1991; Kröber and Carl, 1991; Höller *et al.*, 1993
Aphid-specific predators:	
Coccinellids (Coccinellidae): *Coccinella septempunctata, Propylea quatuordecimpunctata, Adalia bipunctata*	Triltsch et al., 1996; Hemptinne and Naisse, 1988; Kröber and Carl, 1991
Syrphids (Syrphidae): *Episyrphus balteatus, Eupeodes (Metasyrphus) corollae, Sphaerophoria scripta*	Chambers and Adams, 1986; Stechmann, 1990; Storck-Weyhermüller, 1988
Chrysopids (Chrysopidae): *Chrysoperla carnea*	Kröber and Carl, 1991
Polyphagous predators:	
Spiders (Linyphiidae, Lycosidae and others): *Oedothorax apicatus, Erigone atra, Pachygnatha degeeri*	Sunderland, 1987; Volkmar *et al.*, 1994
Staphylinids (Staphylinidae): *Tachyporus hypnorum, Tachyporus chrysomelinus, Tachyporus obtusus*	Dennis *et al.*, 1991; Bothe and Heimbach, 1995
Carabids (Carabidae): *Platynus dorsalis, Bembidion lampros, Trechus quadristriatus, Harpalus rufipes, Pterostichus melanarius, Poecilus cupreus*	Sunderland and Vickerman, 1980; Ekbom *et al.*, 1992; Volkmar *et al.*, 1994

and Pell, Chapter 18 this volume). The role of parasitoids is contradictory in the literature and often overestimated. Their advantage is often a good synchronization with their aphid hosts in time and space due to their close relation to the hosts, even during winter. However, only high parasitization rates during an early stage of the aphid infestation have a substantial impact. In addition, hyperparasitoids usually build up rapidly and limit parasitoid efficacy. In most years, rates of parasitism of cereal aphids during critical periods (e.g. the start of exponential population growth) are below 10%, much too low to influence aphid population dynamics (Borgemeister, 1992; Höller *et al.*, 1993). Mass releases, habitat management, and increased diversity of the landscape can enhance locally the effectiveness of parasitoids (Levie *et al.*, 2005; Roschewitz *et al.*, 2005).

Polyphagous, and particularly aphid-specific, predators are more often the main factors in the natural mortality of cereal aphids than parasitoids. The predator community (in terms of species and guilds) in the cereal ecosystem is highly diverse, and its impact depends on the temporal and spatial synchronization, community composition, prey preferences, and predatory potential at given temperature conditions. Some species of polyphagous predators are spatially associated with cereal aphids through predation (Winder *et al.*, 2005). Under optimal conditions, polyphagous predators can reduce late aphid infestations in cereals by up to 31% (Ekbom *et al.*, 1992; Winder *et al.*, 1994; Holland and Thomas, 1997; Holland, 1998; Östman *et al.*, 2003). Carabids and most spiders show continuous, but relatively low, aphid consumption rates compared to aphid-specific predators (Sunderland *et al.*, 1987; Triltsch, 2001; Nienstedt and Poehling, 2004; Harwood *et al.*, 2005). Similarly, Bilde and Toft (1997, 2001) and Toft (2005) showed (with seven potential generalist aphid predators) low or non-preference of epigeal predators for aphids and low food quality of cereal aphids. Thus, through early predation when alternative prey is scarce (Madsen *et al.*, 2004; Harwood *et al.*, 2005), polyphagous predators can reduce the initial density of cereal aphids. Later, however, their voracity fails to keep up with increasing aphid densities, or they switch to more convenient prey (Sunderland and Vickerman, 1980).

The high voracity of aphid-specific predators (Freier and Triltsch, 1996; Tenhumberg, 1997) and their good synchronization in time and space with cereal aphids can greatly reduce the rate of population increase (Chambers and Adams, 1986; Poehling, 1988; Latteur and Oger, 1991; Poehling *et al.*, 1991; Wetzel, 1995; Elliott *et al.*, 2000). Syrphids and coccinellids particularly have a strong potential to regulate cereal aphid populations, showing both numerical and functional responses to their prey (Poehling and Borgemeister, 1989; Hemptinne *et al.*, 1992; Tenhumberg and Poehling, 1995; Freier *et al.*, 2001).

Despite a huge body of data, a proper assessment of the effects of single predator species on cereal aphid populations remains very difficult (Sunderland *et al.*, 1997). A simple addition of different predators with their varying aphid consumption rates is not possible. Freier *et al.* (1998) therefore devised the 'predator unit', whereby different predators can be assessed for their potential for aphid consumption. For instance, a female *Coccinella septempunctata* receives the value 1.0, a male 0.88, and a green lacewing *Chrysoperla carnea* larva only 0.14. Multiplying these values with densities of each predator quantifies the impact of a predator community on their prey.

Simulation models have been developed not only to describe the population dynamics of cereal aphids, but also to assess the impact of natural enemies (e.g. Carter *et al.*, 1982; Skirvin *et al.*, 1997; Gosselke *et al.*, 2001). Running such models with and without the presence of natural enemies provides a better understanding of regulation capacities (e.g. the model GETLAUS01 – Gosselke *et al.*, 2001).

Host-plant resistance

Differences in susceptibility/resistance and sensitivity/tolerance to cereal aphids occur between cereal cultivars (Sengonca *et al.*, 1994; Havlickova, 1997; Hesler and Tharp, 2005). Some plant chemicals like hydroxamic and phenolic acids are involved in pre- or post-infectional resistance (Thackray *et al.*, 1990; Havlickova *et al.*, 1996; Niraz *et al.*, 1996; Eleftherianos *et al.*, 2006; see also van Emden, Chapter 17 this volume). In addition, there are possibilities of induced resistance and/or tolerance through use of microbial products or chemicals derived from microbes (Galler *et al.*, 1998). So far, these approaches have not yielded truly reliable and efficient tools for aphid control; control is usually only partial.

Cultural control

Options for cultural control of cereal aphids are limited. Reducing the proportion of winter wheat in the rotation is possibly an option. However, for economic reasons, and because of a lack of scientific evidence of the effects of different proportions of winter wheat in the rotation and/or in a region on aphid infestation levels, this option seems unrealistic.

The autumn sowing date of winter barley and winter wheat strongly affects the establishment of aphid populations as vectors of BYDV. In Central Europe, wheat is sown between the middle of September and November. If the winter cereals are sown later than 20 September, aphid colonization and virus spread are significantly reduced and insecticide applications are normally not required. Thus, late sowing is a very useful cultural control for BYDV.

Various studies indicate a strong positive relationship between nitrogen fertilization and aphid development, especially of *M. dirhodum*. High amounts of nitrogen influence positively the intrinsic rate of increase in cereal aphids (Hanisch, 1980; Honek, 1991; Hasken and Poehling, 1995; Gash *et al.*, 1996; Duffield *et al.*, 1997). Hence, limiting nitrogen fertilization to moderate levels can lower aphid densities.

Habitat management practices

There are many such approaches to improving or stabilizing the naturally available regulation potential of aphid antagonists. Under the headlines of 'habitat manipulation', 'ecological engineering', and 'functional biodiversity', several scientific papers and guidelines have been published (Boller *et al.*, 2004; Gurr *et al.*, 2004). In addition, the International Organization for Biological Control (IOBC) recently established a new working group on this topic (Rossing *et al.*, 2003, 2006). At the field level, habitat management techniques can increase predator and parasitoid densities in the cereal ecosystem. For instance, reduced weed control favours some carabid species and other ground-dwelling predators (Andersen and Eltun, 2000). Holland *et al.* (1999, 2004) observed that the spatial distribution of different carabids and lycosid spiders is associated with the degree of weed cover. Yet the actual impact of tillage, particularly ploughing or reduced tillage, on beneficials

is lower than previously assumed (Baguette and Hance, 1997; Andersen, 1999).

Organic or integrated systems with low or moderate input can help to maintain high levels of beneficials in contrast to intensive farming involving frequent pesticide applications (e.g. Hasken and Poehling, 1994; Moreby *et al.*, 1994; Holland and Thomas, 1997; Holland *et al.*, 1998).

At the farm and landscape level, high diversity (e.g. with expanded and connected field margins and hedgerows) can promote the abundance and diversity of predators and parasitoids (Rossing *et al.*, 2006; Wratten *et al.*, Chapter 16 this volume). Field margins with high floral diversity are attractive habitats with food resources and hibernation sites for polyphagous epigeal arthropods, syrphids, coccinellids, and other predators (Dennis *et al.*, 1994; Nicoli *et al.*, 1995; Thomas and Marshall, 1999; Ländis *et al.*, 2000; Boller *et al.*, 2004). Such boundary effects can be magnified if crop edges are kept free of herbicide (Fig. 25.4), leading to reduced aphid infestation near the edges. Sowing wild flowers around the edges of wheat fields (Harwood *et al.*, 1992) and drilling phacelia (*Phacelia tanacetifolia*) in margins can promote adult syrphids (Hickman and Wratten, 1996). Moreover, sown herbaceous strips (strip management) and so-called 'beetle banks' (grassy banks, Fig. 25.5) can favour

Fig. 25.4.   Floral diversity at the edge of a wheat field, created by shutting off the outside spray boom when spraying broad-leaf herbicides at the edges of the crop (from original colour photo, courtesy of N.W. Sotherton).

Fig. 25.5. A beetle bank in a cereal field (from original colour photo, courtesy of N.W. Sotherton).

beneficials (Nentwig, 1992; Lys *et al.*, 1994; Collins *et al.*, 1996; Frank, 1997; Lemke and Poehling, 2002), but often the impact on aphid population growth during the summer is weak and spatially limited. Studies on higher scales such as the landscape level demonstrate the importance of regional diversity for conservation biocontrol of cereal aphids (Östman *et al.*, 2001; Roschewitz *et al.*, 2005; Tscharntke *et al.*, 2005).

Conclusions

For 30 years, aphids have been the most important pests of cereal crops in Central Europe. Research has provided very detailed knowledge on their biology and their interactions with the host plant and natural enemies. Since damaging infestation levels occur only during a relatively short time in summer, chemical control is highly economic. Frequent applications are not required and, at least in Central Europe, insecticide resistance has so far never been an issue with cereal aphids. However, cereals cover large areas in Central Europe, and so even a single insecticide application per year can be highly detrimental to non-target organisms.

Thus, environmental concerns have been the main driving force for the development of IPM in cereals.

Executive Summary

Chemical control of cereal aphids, especially using a selective material like pirimicarb or applying at reduced rates, is a viable option and recommended economic thresholds for cereal aphids have been available for some time. However, farmers find counting aphids difficult and time-consuming, as well as complicated if estimating natural enemy abundance is part of the calculation. Central forecasting is therefore a highly desirable future development.

There is no release of biological control agents, but conserving indigenous natural enemies by using softer insecticides and habitat manipulations, such as providing nectar plants, and modified headland spraying can greatly reduce, or even eliminate, the need for insecticide application in some years. Thus, there is wide interest in maintaining or recreating sufficient refuge areas for important natural enemies of aphids in cereal ecosystems. Financial incentives for purposeful biodiversity, such as the set-aside and country stewardship schemes in Europe, have made the provision of such refuges acceptable to farmers.

Thus, the main sustainable control options are based on integrated farming systems, with reduced inputs of pesticides, fertilizers, and perhaps even the use of less susceptible varieties, but also with more consideration given to the important effects of broader landscape diversity.

References

Andersen, A. (1999) Plant protection in spring cereal production with reduced tillage. II. Pests and beneficial insects. *Crop Protection* 18, 651–657.

Andersen, A. and Eltun, R. (2000) Long-term developments in the carabid and staphylinid (Col., Carabidae and Staphylinidae) fauna during conversion from conventional to biological farming. *Journal of Applied Entomology* 124, 51–56.

Ardisson, C.N., Pierre, J.S., Plantegenest, M. and Dedryver, C.A. (1997) Parameter estimation for a descriptive epizootiological model of the infection of cereal aphid population by a fungal pathogen (Entomophthorale). *Entomophaga* 42, 575–591.

Baguette, M. and Hance, T. (1997) Carabid beetles and agricultural practices: influence of soil ploughing. *Biological Agriculture and Horticulture* 15, 185–190.

Basedow, T., Poehling, H.M. and Lauenstein, G. (1994) Untersuchungen zur Anpassung der Bekämpfungsschwelle der Getreideblattläuse (Hom., Aphididae) (Saugschäden an Weizen im Sommer) an die veränderten ökonomischen Rahmenbedingungen im Ackerbau. *Journal of Plant Diseases and Protection* 101, 337–349.

Bayon, F., Ayrault, J.P. and Pichon, P. (1982) Epidemiologie de la jaunisse nanisante de l'orge (B.Y.D.V.) en Poitou-Charentes. *Mededelingen van de Faculteit Landbouwwetenschappen, Rijksuniversiteit Gent* 47, 1039–1052.

Blackman, R.L. and Eastop, V.F. (2000) *Aphids on the World's Crops: An Identification and Information Guide.* Wiley, Chichester, 466 pp.

Bilde, T. and Toft, S. (1997) Limited predation capacity by generalist arthropod predators on the cereal aphid, *Rhopalosiphum padi. Biological Agriculture and Horticulture* 15, 143–150.

Bilde, T. and Toft, S. (2001) The value of three cereal aphid species as food for a generalist predator. *Physiological Entomology* 26, 58–68.

Boller, E.F., Häni, F. and Poehling, H.M. (2004) *Ecological Infrastructures: Ideabook on Functional Biodiversity at the Farm Level. IOBC WPRS Commission on Integrated Production Guidelines and Endorsement.* Swiss Centre for Agricultural Extension and Rural Development (LBL), Eschikon, Switzerland, 212 pp.

Borgemeister, C. (1992) Primär- und Hyperparasitoide von Getreideblattläusen: Interaktionen und Beeinflussungen durch Insektizide. *Agrarökologie* 3, 1–191.

Bothe, S, and Heimbach, U. (1995) Untersuchungen zur Erfassung und Bedeutung von Kurzflügelkäfern (Coleoptera, Staphylinidae) unter Berücksichtigung der Blattlauspopulation in Winterweizen. *Archives of Phytopathology and Plant Protection* 29, 429–436.

Campbell, P.J. (1998) Labelling and risk management strategies for pesticides and terrestrial non-target arthropods: a UK proposal. In: Haskell, P.T. and McEwen, P. (eds) *Ecotoxicology, Pesticides and Beneficial Organisms.* Kluwer, Dordrecht, pp. 232–240.

Carter, N., McLean, J.F.G., Watt, A.D. and Dixon, A.F.G. (1980) Cereal aphids: a case study and review. *Applied Biology* 5, 271–348.

Carter, N., Dixon, A.F.G. and Rabbinge, R. (1982) *Cereal Aphid Populations: Biology, Simulation and Prediction.* Pudoc, Wageningen, 97 pp.

Chambers, R.J. and Adams, T.H.L. (1986) Quantification of the impact of hoverflies (Diptera: Syrphidae) on cereal aphids in winter wheat: an analysis of field populations. *Journal of Applied Ecology* 23, 895–904.

Collins, K.L., Wilcox, A., Chaney, K. and Boatman, N.D. (1996) Relationships between polyphagous predator density and overwintering habitat within arable field margins and beetle banks. *Proceedings of the Brighton Crop Protection Conference, Pests and Diseases, November 1996* 2, 635–640.

Dedryver, C.A. (1982) Biologie des pucerons des cereales dans l'ouest de la France. II. Répartition spatio-temporelle et action limitative de trois espèces d'Entomophthoraceae. *Entomophaga* 26, 381–393.

Dennis, P., Wratten, S.D. and Sotherton, N.W. (1991) Mycophagy as a factor limiting predation of aphids (Hemiptera: Aphididae) by staphylinid beetles (Coleoptera: Staphylinidae) in cereals. *Bulletin of Entomological Research* 81, 25–31.

Dennis, P., Thomas, M.B. and Sotherton, N.W. (1994) Structural features of field boundaries which influence the overwintering densities of beneficial arthropod predators. *Journal of Applied Ecology* 31, 361–370.

Duffield, S.J., Bryson, R.J., Young, J.E.B., Sylvester-Bradley, R. and Scott, R.K. (1997) The influence of nitrogen fertiliser on the population development of the cereal aphids *Sitobion avenae* (F.) and *Metopolophium dirhodum* (Wlk.) on field grown winter wheat. *Annals of Applied Biology* 130, 13–26.

Ekbom, B.S., Wiktelius, S. and Chiverton, P.A. (1992) Can polyphagous predators control the bird cherry–oat aphid (*Rhopalosiphum padi*) in spring cereals? A simulation study. *Entomologia Experimentalis et Applicata* 65, 215–223.

Eleftherianos, I., Vamvatsikos, P., Ward, D. and Gravanis, F. (2006) Changes in the levels of plant total phenols and free amino acids induced by two cereal aphids and effects on aphid fecundity. *Journal of Applied Entomology* 130, 15–19.

Elliott, N.C., Kieckhefer, R.W. and Beck, D.A. (2000) Adult coccinellid activity and predation on aphids in spring cereals. *Biological Control* 17, 218–226.

Entwistle, J.C. and Dixon, A.F.G. (1987) Short-term forecasting of wheat yield loss caused by the grain aphid (*Sitobion avenae*) in summer. *Annals of Applied Biology* 111, 489–508.

Fabre, F., Plantegenest, M., Mieuzet, L., Dedryver, C.A., Leterrier, J.L. and Jacquot, E. (2005) Effects of climate and land use on the occurrence of viruliferous aphids and the epidemiology of barley yellow dwarf disease. *Agriculture, Ecosystems and Environment* 106, 49–55.

Fabre, F., Pierre, J.S., Dedryver, C.A. and Plantegenest, M. (2006) Barley yellow dwarf disease risk assessment based on Bayesian modelling of aphid population dynamics. *Ecological Modelling* 193, 457–466.

Frank, T. (1997) Species diversity of ground beetles (Carabidae) in sown weed strips adjacent fields. *Biological Agriculture and Horticulture* 15, 297–307.

Freier, B. and Triltsch, H. (1996) Climate chamber experiments and computer simulations on the influence of increasing temperature on wheat–aphid–predator interactions. In: Edis, D., Hull, M.R., Cobb, A.H. and Sanders-Mill, G.E. (eds) *Implications of 'Global Environmental Change' for Crops in Europe. Aspects of Applied Biology, No. 45*, pp. 293–298.

Freier, B. and Wetzel, T. (1984) Abundanzdynamik von Schadinsekten im Winterweizen. *Zeitschrift für Angewandte Entomologie* 98, 483–494.

Freier, B., Trothe, G. and Wetzel, T. (1986) Fungal infection of cereal aphids and its detection. *Nachrichtenblatt für den Pflanzenschutz in der DDR* 40, 62–64.

Freier, B., Möwes, M. and Triltsch, H. (1998) Beneficial thresholds for *Coccinella 7-punctata* L. (Col., Coccinellidae) as a predator of cereal aphids in winter wheat – results of population investigations and computer simulations. *Journal of Applied Entomology* 122, 213–217.

Freier, B., Triltsch, H. and Gosselke, U. (2001) Potential and limitations of long term field data to identify numerical and functional responses of predators to aphid density in wheat. *Bulletin IOBC/WPRS* 24 (6), 65–71.

Galler, M., Wittmann, J. and Poehling, H.-M. (1998) Induced tolerance and induced resistance against biotrophic pathogens and cereal aphids in wheat. *Bulletin IOBC/WPRS* 21 (8), 193–199.

Gash, A.F., Carter, N. and Bale, J.S. (1996) The influence of nitrogen fertiliser applications on the cereal aphids *Metopolophium dirhodum* and *Sitobion avenae*. *Proceedings of the Brighton Crop Protection Conference, Pests and Diseases, November 1996* 1, 209–214.

Gonzalez Andujar, J.L., Garcia de Ceca, J.L. and Fereres, A. (1993) Cereal aphids expert system (CAES): identification and decision making. *Computers and Electronics in Agriculture* 8, 293–300.

Gosselke, U., Triltsch, H., Rossberg, D. and Freier, B. (2001) GETLAUS01 – the latest version of a model for simulating aphid population dynamics in dependence on antagonists in wheat. *Ecological Modelling* 145, 143–157.

Gurr, G.M., Wratten, S.D. and Altieri, M.A. (2004) *Ecological Engineering for Pest Management.* CAB International, Wallingford, 232 pp.

Hågvar, E.B. and Hofsvang, T. (1991) Aphid parasitoids (Hymenoptera: Aphidiidae): biology, host selection and use in biological control. *Biocontrol News and Information* 12, 13–42.

Hanisch, H.-C. (1980) Untersuchungen zum Einfluss unterschiedlich hoher Stickstoffdüngung zu Weizen auf die Populationsentwicklung von Getreideblattläusen. *Journal of Plant Diseases and Protection* 87, 546–556.

Harwood, R.W.J., Wratten, S.D. and Nowakowski, M. (1992) The effect of managed field margins on hoverfly (Diptera: Syrphidae) distribution and within-field abundance. *Proceedings of the Brighton Crop Protection Conference, Pests and Diseases, November 1992*, 1033–1037.

Harwood, J.D., Sunderland, K.D. and Symondson, W.O.C. (2005) Monoclonal antibodies reveal the potential of the tetragnathid spider *Pachygnatha degeeri* (Araneae: Tetragnathidae) as an aphid predator. *Bulletin of Entomological Research* 95, 161–167.

Hasken, K.H. and Poehling, H.-M. (1994) Some effects of low input agriculture on cereal aphids and aphid specific predators in winter wheat. *Bulletin IOBC/WPRS* 17 (4), 137–147.

Hasken, K.H. and Poehling, H.-M. (1995) Effects of different intensities of fertilisers and pesticides on aphids and aphid predators in winter wheat. *Agriculture, Ecosystems and Environment* 52, 45–50.

Havlickova, H. (1997) Character and extent of damage to winter wheat cultivars caused by cereal aphids. *Rostlinná Výroba* 43, 113–116.

Havlickova, H., Cvirova, M. and Eder, J. (1996) Phenolic acids in wheat cultivars in relation to plant suitability for and response to cereal aphids. *Journal of Plant Diseases and Protection* 103, 535–542.

Hemptinne, J.-L. and Naisse, J. (1988) Life cycle strategy of *Adalia bipunctata* (L.) (Col., Coccinellidae) in a temperate country. In: Niemczyk, E. and Dixon, A.F.G. *Ecology and Effectiveness of Aphidophaga*. SPB Academic Publishing, The Hague, pp. 71–77.

Hemptinne, J.-L., Dixon, A.F.G. and Coffin, J. (1992) Attack strategy of ladybird beetles (Coccinellidae): factors shaping their numerical response. *Oecologia* 90, 238–245.

Hesler, L.S. and Tharp, C.I. (2005) Antibiosis and antixenosis to *Rhopalosiphum padi* among triticale accessions. *Euphytica* 143, 153–160.

Hickman, J.M. and Wratten, S.D. (1996) Use of *Phacelia tanacetifolia* strips to enhance biological control of aphids by hoverfly larvae in cereal fields. *Journal of Economic Entomology* 89, 832–840.

Holland, J.M. (1998) The effectiveness of exclusion barriers for polyphagous predatory arthropods in wheat. *Bulletin of Entomological Research* 88, 305–310.

Holland, J.M. and Thomas, S.R. (1997) Assessing the role of beneficial invertebrates in conventional and integrated farming systems during an outbreak of *Sitobion avenae*. *Biological Agriculture and Horticulture* 15, 73–82.

Holland, J.M., Cook, S.K., Drysdale, A.D., Hewitt, M.V., Spink, J. and Turley, D.B. (1998) The impact on non-target arthropods of integrated compared to conventional farming: results from the LINK integrated farming systems project. *Proceedings of the Brighton Crop Protection Conference, Pests and Diseases, November 1998* 2, 625–630.

Holland, J.M., Perry, J.N. and Winder, L. (1999) The within-field spatial and temporal distribution of arthropods in winter wheat. *Bulletin of Entomological Research* 89, 499–513.

Holland, J.M., Winder, L., Woolley, C., Alexander, C.J. and Perry, J.N. (2004) The spatial dynamics of predatory arthropods and their aphid prey in winter wheat. *Bulletin of Entomological Research* 94, 419–443.

Höller, C., Borgemeister, C., Haardt, H. and Powell, W. (1993) The relationship between primary parasitoids and hyperparasitoids of cereal aphids: an analysis of field data. *Journal of Animal Ecology* 62, 12–21.

Holz, F., Wetzel, T. and Freier, B. (1994) 3 bis 5 Blattläuse pro Ähre im Winterweizen – eine neue Bekämpfungsschwelle? *Gesunde Pflanzen* 46, 8–12.

Honěk, A. (1991) Nitrogen fertilization and abundance of the cereal aphids *Metopolophium dirhodum* and *Sitobion avenae* (Homoptera, Aphididae). *Journal of Plant Diseases and Protection* 98, 655–660.

Huth, W. and Lauenstein, G. (1991) Zum Problem der Gelbverzwergung in Getreidebeständen. *Gesunde Pflanzen* 43, 139–148.

Jahn, B. and Mehrbach, W. (1984) Einfluss der Getreidelaus (*Macrosiphum avenae*[Fabr.]) auf die Ertragsbildung von Winterweizen. *Tagungsberichte der Akademie der Landwirtschaftswissenschaften der DDR* 224, 437–441.

Knaust, H.-J. and Poehling, H.-M. (1996) Studies on the movement and dispersal of apterous *Sitobion avenae* in winter barley and a new simulation model on secondary spread. *Bulletin IOBC/WPRS* 19 (3), 117–130.

Kolbe, W. (1969) Untersuchungen über das Auftreten verschiedener Blattlausarten als Ursache von Ertrags- und Qualitätsminderungen im Getreidebau. *Bayer Pflanzenschutznachrichten* 22, 177–211.

Kröber, T. and Carl, K. (1991) Cereal aphids and their natural enemies in Europe – a literature review. *Biocontrol News and Information* 12, 357–371.

Kühne, S., Freier, B., Kaul, P., Jüttersonke, B., Schenke, D., Forster, R., Baier, B. and Moll, E. (2002) Auswirkung der Abdrift von Insektiziden in einem Saumbiotop. *Agrarökologie* 42, 1–121.

Ländis, D., Wratten, S.D. and Gurr, G.M. (2000) Habitat management for natural enemies. *Annual Review of Entomology* 45, 175–201.

Langhof, M., Gathmann, A., Poehling, H.M. and Meyhöfer, R. (2003) Impact of insecticide drift on aphids and their parasitoids: residual toxicity, persistence and recolonisation. *Agriculture, Ecosystems and Environment* 94, 265–274.

Latteur, G. (1970) Les pucerons des cereales. *Revue de l'Agriculture* 11, 1633–1646.

Latteur, G. and Oger, R. (1991) Winter wheat aphids in Belgium: prognosis and dynamics of their populations. *Bulletin IOBC/WPRS* 14 (4), 13–34.

Lemke, A. and Poehling, H.-M. (2002) Sown weed strip in cereal fields: overwintering site and 'Source' habitat for *Oedothorax apicatus* (Blackwall) and *Erigone atra* (Blackwall) (Araneae: Erigonida). *Agriculture, Ecosystems and Environment* 90, 67–80.

Levie, A., Legrand, M.A., Dogot, P., Pels, C., Baret, P.V. and Hance, T. (2005) Mass releases of *Aphidius rhopalosiphi* (Hymenoptera: Aphidiinae), and strip management to control of wheat aphids. *Agriculture, Ecosystems and Environment* 105, 17–21.

Lys, J.-A., Zimmermann, M. and Nentwig, W. (1994) Increase in activity density and species number of carabid beetles in cereals as a result of strip-management. *Entomologia Experimentalis et Applicata* 73, 1–9.

Madsen, M., Terkildsen, S. and Toft, W. (2004) Microcosm studies on control of aphids by generalist arthropod predators: effects of alternative prey. *Biocontrol*, 49, 483–504.

Meier, U. (1997) *Growth Stages of Plants*. Blackwell, Berlin, 622 pp.

Moreby, S.J., Aebischer, N.J. and Southway, S.E. (1994) A comparison of the flora and arthropod fauna of organically and conventionally grown winter wheat in southern England. *Annals of Applied Biology* 125, 13–27.

Möwes, M., Freier, B. and Heimann, J. (1997) Variation in yield loss per aphid-day due to *Sitobion avenae*-infestation in high yielding winter wheat. *Journal of Plant Diseases and Protection* 104, 569–575.

Nentwig, W. (1992) Die nützlingsfördernde Wirkung von Unkräutern in angesäten Unkrautstreifen. *Journal of Plant Diseases and Protection*, Special Issue No. 13, 33–40.

Nicoli, G., Limonta, L., Cavazzuti, C. and Pozzati, M. (1995) Il ruolo delle siepi nell'ecologia del campo coltivato. I. Prime indagini sui Coccinellidi predatori di afidi. *Informatore Fitopatologico* 45, 58–64.

Niehoff, B. (1996) Untersuchungen zum Einfluss gestaffelter Aufwandmengen der Insektizide Pirimor und Karate auf die Populationsdynamik von Getreideblattläusen in Winterweizen unter besonderer Berücksichtigung von Nebenwirkungen auf ausgewählte Nutzarthropoden. PhD thesis, University of Göttingen, Germany.

Niehoff, B. and Stäblein, J. (1998) Vergleichende Untersuchungen zum Schadpotential der Getreideblattlausarten *Metopolophium dirhodum* (Wlk.) und *Sitobion avenae* (F.) in Winterweizen. *Journal of Applied Entomology* 122, 223–229.

Nienstedt, K.M. and Poehling, H.M. (2004) Invertebrate predation of ^{15}N-marked prey in semi-field wheat enclosures. *Entomologia Experimentalis et Applicata* 112, 191–200.

Niraz, S., Leszczynski, B., Urbanska, A., Matok, H. and Ciepiela, A. (1996) Biochemical mechanism of aphid resistance in cereals – 20 years of research. *Plant Breeding and Seed Science* 40, 87–91.

Oakley, J.N. and Walters, K.F.A. (1994) A field evaluation of different criteria for determining the need to treat winter wheat against the grain aphid *Sitobion avenae* and the rose grain aphid *Metopolophium dirhodum*. *Annals of Applied Biology* 124, 195–211.

Oakley, J.N., Ellis, S.A., Walters, K.F.A. and Watling, M. (1993) The effects of cereal aphid feeding on wheat quality. In: Starling, W. and Richards, M.C. (eds) *Quality of Commercial Samples of Organically-Grown Wheat. Aspects of Applied Biology, No. 36*, pp. 383–390.

Östman, Ö., Ekbom, B. and Bengtsson, J. (2001) Landscape heterogeneity and farming practice influence biological control. *Basic and Applied Ecology* 2, 365–371.

Östman, Ö., Ekbom, B. and Bengtsson, J. (2003) Yield increase attributable to aphid predation by ground-living polyphagous natural enemies in spring barley in Sweden. *Ecological Economics* 45, 149–158.

Plumb, R.T. (1990) The epidemiology of barley yellow dwarf in Europe. In: Burnett, P.A. (ed.) *World Perspectives on Barley Yellow Dwarf*. CIMMYT, Mexico City, pp. 215–227.

Poehling, H.-M. (1988) Zum Auftreten von Syrphiden- und Coccinellidenlarven in Winterweizen von 1984–1987 in Relation zur Abundanz von Getreideblattläusen. *Mitteilungen der Deutschen Gesellschaft für Allgemeine und Angewandte Entomologie* 6, 248–254.

Poehling, H.-M. (1989) Selective application strategies for insecticides in agricultural crops. In: Jepson, P. (ed.) *Pesticides and Non-Target Invertebrates*. Intercept, Wimborne, pp. 151–175.

Poehling, H.-M. and Borgemeister, C. (1989) Abundance of coccinellids and syrphids in relation to cereal aphid density in winter wheat fields in northern Germany. *Bulletin IOBC/WPRS* 12 (1), 99–107.

Poehling, H.-M., Tenhumberg, B. and Groeger, U. (1991) Different pattern of cereal aphid population dynamics in northern (Hannover-Göttingen) and southern areas of West Germany. *Bulletin IOBC/WPRS* 14 (4), 1–12.

Rappaport, V. and Freier, B. (2001) Erprobung eines flexiblen Schwellenswertkonzepts für Getreideblattläuse an Winterwiezen unter Berücksichtigung der naturlichen Gegenspieler. *Nachrichtenblatt des Deutschen Pflanzenschutzdienstes* 53, 113–119.

Rautapää, J. (1966) The effect of the English grain aphid *Macrosiphum avenae* (F.) (Hom., Aphididae) on the yield and quality of wheat. *Annales Agriculturae Fenniae* 5, 334–341.

Ressel, F. (1970) Blattlausbekämpfung im Getreidebau 1969 im Bezirk Halle. *Nachrichtenblatt des Deutschen Pflanzenschutzdienstes* 24, 72–75.

Roschewitz, I., Hucker, M., Tscharntke, T. and Thies, C. (2005) The influence of landscape context and farming practices on parasitism of cereal aphids. *Agriculture, Ecosystems and Environment* 108, 218–227.

Rossing, W.A.H. (1991) Simulation of damage in winter wheat caused by the grain aphid *Sitobion avenae*. 3. Calculation of damage at various attainable yield levels. *Netherlands Journal of Plant Pathology* 97, 87–103.

Rossing, A.H., Poehling, H.M. and Burgio, G. (2003) Landscape management for functional biodiversity. *Bulletin IOBC/WPRS* 26 (4), 220 pp.

Rossing, A.H., Eggenschwiler, L. and Poehling, H.M. (2006) Landscape management for functional biodiversity. *Bulletin IOBC/WPRS* 29 (6), 168 pp.

Sengonca, C., Josch, H. and Kleinhenz, B. (1994) Einfluss verschiedener Wintergerste- und Winterweizensorten auf die Besiedlung und Populationsentwicklung von Getreideblattläusen. *Gesunde Pflanzen* 46, 3–7.

Skirvin, D.J., Perry, J.N. and Harrington, R. (1997) A model describing the population dynamics of *Sitobion avenae* and *Coccinella septempunctata*. *Ecological Modelling* 96, 29–40.

Stechmann, D.-H. (1990) Getreideblattläuse und aphidophage Insekten – Zur tierökologischen Funktion von Hecken in der Kulturlandschaft. Habilitation thesis, University of Bayreuth, Germany.

Storck-Weyhermüller, S. (1988) Einfluss natürlicher Feinde auf die Populationsdynamik der Getreideblattläuse im Winterweizen Mittelhessens (Homoptera: Aphididae). *Entomologia Generalis* 13, 189–206.

Sunderland, K.D. (1987) Spiders and cereal aphids in Europe. *Bulletin IOBC/WPRS* 10 (1), 82–102.

Sunderland, K.D. and Vickerman, G.P. (1980) Aphid feeding by some polyphagous predators in relation to aphid density in cereal fields. *Journal of Applied Ecology* 17, 389–396.

Sunderland, K.D., Crook, N.E., Stacey, D.L. and Fuller, B.J. (1987) A study of feeding by polyphagous predators on cereal aphids using ELISA and gut dissection. *Journal of Applied Ecology* 24, 907–933.

Sunderland, K.D., Axelsen, J.A., Dromph, K., Freier, B., Hemptinne, J.-L., Holst, N.H., Mols, P.J.M., Petersen, M.K., Powell, W., Ruggle, P., Triltsch, H. and Winder, L. (1997) Pest control by community of natural enemies. *Acta Jutlandica* 72, 271–326.

Tenhumberg, B. (1997) Estimating predatory efficiency of *Episyrphus balteatus* (Diptera: Syrphidae) in cereal fields. *Environmental Entomology* 24, 687–691.

Tenhumberg, B. and Poehling, H.-M. (1995) Syrphids as natural enemies of cereal aphids in Germany: aspects of their biology and efficacy in different years and regions. *Agriculture, Ecosystems and Environment* 52, 39–43.

Thackray, D.J., Wratten, S.D., Edwards, P.J. and Niemeyer, H.M. (1990) Resistance to the aphids *Sitobion avenae* and *Rhopalosiphum padi* in Gramineae in relation to hydroxamic acid levels. *Annals of Applied Biology* 116, 573–582.

Thomas, C.F.G. and Marshall, E.J.P. (1999) Arthropod abundance and diversity in differently vegetated margins of arable fields. *Agriculture, Ecosystems and Environment* 72, 131–144.

Toft, S. (2005) The quality of aphids as food for generalist predators: implications for natural control of aphids. *European Journal of Entomology* 102, 371–383.

Triltsch, H. (2001) A study of aphid predation by *Coccinella septempunctata* L. (Coleoptera: Coccinellidae) using gut dissection. *Bulletin IOBC/WPRS* 24 (6), 147–151.

Triltsch, H., Freier, B. and Möwes, M. (1996) Marienkäfer (Coleoptera, Coccinellidae) als Nützlinge in agrarischen Ökosystemem. *Mitteilungen der Biologischen Bundesanstalt für Land- und Forstwirtschaft* 323, 1–96.

Tscharntke, T., Rand, T.A. and Bianchi, F.J.J.A. (2005) The landscape context of tritrophic interactions: insect spillover across the crop–noncrop interface. *Annales Zoologici Fennici* 42, 421–432.

Volkmar, C., Bothe, S., Kreuter, T., Lübke-Al Hussein, M., Richter, L., Heimbach, U. and Wetzel, T. (1994) Epigäische Raubarthropoden in Winterweizenbeständen Mitteldeutschlands und ihre Beziehung zu Blattläusen. *Mitteilungen der Biologischen Bundesanstalt für Land- und Forstwirtschaft* 299, 1–134.

Wetzel, T. (1995) Getreideblattläuse im Pflanzenschutz und im Agroökosystem (Übersichtsbeitrag). *Archives of Phytopathology and Plant Protection* 29, 437–469.

Wick, M. and Freier, B. (2000) Long-term effects of an insecticide application on non-target arthropods in winter wheat – a field study over two seasons. *Journal of Pest Science* 73, 61–69.

Winder, L., Hirst, D.J., Carter, N., Wratten, S.D. and Sopp, P.I. (1994) Estimating predation of the grain aphid *Sitobion avenae* by polyphagous predators. *Journal of Applied Ecology* 31, 1–12.

Winder, L., Alexander, C.J., Holland, J.M., Symondson, W.O.C., Perry, J.N. and Woolley, C. (2005) Predatory activity and spatial pattern: the response of generalist carabids to their aphid prey. *Journal of Animal Ecology* 74, 443–454.

Wratten, S.D., Lee, G. and Stevens, D.J. (1979) Duration of cereal aphid populations and the effects on wheat yield and quality. *Proceedings of the Brighton Crop Protection Conference, Pests and Diseases, November 1979* 1, 1–8.

26 IPM Case Studies: Seed Potato

Edward B. Radcliffe[1], David W. Ragsdale[1] and Robert A. Surányi[2]

[1]Department of Entomology, University of Minnesota, St Paul, MN 55108, USA;
[2]McLaughlin Gormley King Company, 8810 Tenth Ave. N, Minneapolis,
MN 55427, USA

Introduction

Potato, *Solanum tuberosum*, is clonally propagated; therefore, 'seed piece' tubers are only as healthy as their mother plant. Access to high quality, disease-free seed potatoes has been described as 'the single most important integrated pest management practice available to potato growers' (Gutbrod and Mosley, 2001) and essential for successful commercial potato production (Horváth, 1990). Among the most important tuber-borne diseases of potato are aphid-transmitted viruses (Salazar, 1996).

Virus can be introduced to a seed potato field by immigrating aphids or spread from point sources within. Thus, virus management should be implemented on both a regional and individual farm basis. Tactics at the regional level tend to focus on seed potato certification schemes that provide the regulatory framework for enforcing standards and are the central component of virus management programmes (Slack, 1993). Farm-level management will vary in its implementation based on the manager's understanding of the epidemiological processes involved.

Certification

Potato certification programmes originated in Holland and Germany in the early 1900s and in Canada and the USA around 1914 (Slack, 1993). Most seed potato certification programmes use a limited generation production system (Allen *et al.*, 1992; Franc, 2001). Seed lots derived from tissue culture are required to be virus free. Tolerances usually are relaxed incrementally with successive generations (Hiddema, 1972; Gutbrod and Mosley, 2001). If virus levels exceed tolerances, seed lots are downgraded to a more advanced generation, or rejected outright if infection exceeds tolerance for certified seed. Typically, virus tolerances for seed lots to be increased another year range from 0–1%, and for ware fields from 1–5% tubers infected (Woodford, 1988).

Certification programmes sometimes are so successful in eliminating inoculum that seed potato growers become careless about vector management. Whenever seed lot rejections for virus rise, vector control, especially with insecticides, becomes a priority concern. High levels of virus inoculum can negate any benefit of aphid control (Sigvald, 1989). Often, it can take several years to reduce the presence of virus inoculum in the seed production system sufficiently to end an epidemic (Harrison, 1971).

Worldwide, the primary reason seed potato lots are rejected or downgraded in certification programmes is that the incidence of *Potato virus Y* (PVY) (Fig. 26.1) or *Potato leaf*

©CAB International 2007. *Aphids as Crop Pests*
(eds H. van Emden and R. Harrington)

Fig. 26.1. Potato infected with *Potato virus Y.*

roll virus (PLRV) (Fig. 26.2) exceeds established tolerances. These viruses differ in that PLRV is circulative in the insect and transmitted in a persistent manner (Katis *et al.*, Chapter 14 this volume), whereas all other known aphid transmitted potato viruses are stylet-borne and transmitted non-persistently.

A latent period of 8–24 h exists between PLRV acquisition by the vector and the onset of its ability to transmit (Tanaka and Shiota, 1970). Once a vector acquires PLRV, it usually remains infective for life and capable of transmission, even following a moult. By contrast, acquisition and inoculation of PVY occurs within seconds and, once acquired, aphids are immediately capable of transmitting to a susceptible host. Aphids lose their ability to transmit PVY after one to several feeding probes and must feed again on an infected plant to re-acquire the virus (Bradley and Rideout, 1953).

Most viruses occur as variants that, if they differ sufficiently from type, are designated as strains. PVY strains include PVY^O (ordinary or common), PVY^N (tobacco veinal necrosis), PVY^C (stipple streak), and PVY^{NTN} (potato tuber necrotic ringspot). PVY^O is worldwide in its distribution and

can produce severe mosaic, leaf drop, and stem necrosis in susceptible cultivars. In potato, symptoms caused by PVY^N are usually mild compared to PVY^O, and some cultivars are essentially asymptomatic, making detection difficult except by serological assay or RT–PCR (reverse transcriptase–polymerase chain reaction). Presence of asymptomatic cultivars tends to compromise the effectiveness of current seed certification virus screening procedures because these rely primarily on visual recognition of infected plants. Among PVY strains, there are no obvious differences in vector associations, but PVY^N or PVY$^{O:N}$ recombinants are increasingly common worldwide and predominant in much of Europe (Weidemann, 1988) and North America (Piche *et al.*, 2004).

Vector Species

Myzus persicae (peach–potato aphid) is the most efficient vector of PLRV and worldwide is generally the most abundant PLRV vector (Hille Ris Lambers, 1972; Raman and Radcliffe, 1992; Robert and Bourdin, 2001).

Fig. 26.2. Potato infected with *Potato leaf roll virus.*

Most *Macrosiphum euphorbiae* (potato aphid) populations transmit PLRV poorly, if at all (Robert and Maury, 1970; Tamada *et al.*, 1984), but this species has been implicated in early season spread of PLRV in New Brunswick, USA (Singh and Boiteau, 1986) and sometimes is an important PLRV vector in Scotland (Woodford *et al.*, 1995). Other potential PLRV vectors include *Aphis nasturtii* (buckthorn–potato aphid) (Loughnane, 1943), the *Aphis gossypii* group (cotton or melon aphid) and *Aphis frangulae* (Foster and Woodford, 1997), *Aulacorthum solani* (glasshouse and potato aphid) (Robert and Rouzé-Jouan, 1971), *Myzus ascalonicus* (shallot aphid) (Hille Ris Lambers, 1972), and *Aulacorthum circumflexum* (mottled arum aphid) (Heinze, 1960).

More than 50 species of aphid are known to transmit PVY (Harrington *et al.*, 1986; Sigvald, 1987, 1989; Heimbach *et al.*, 1998; Ragsdale *et al.*, 2001). Most of these are considered transients in the sense that they never colonize potato. *Myzus persicae* is probably the most efficient vector of PVY (MacGillivray, 1981), but the greater abundance of some less efficient vector species,

or their propensity to develop alatae, can make them more important in PVY epidemiology (Piron, 1986; Sigvald, 1987; Boiteau *et al.*, 1988; Weidemann, 1988; Harrington and Gibson, 1989). Common PVY vectors include *Acyrthosiphon pisum* (pea aphid), *Aphis fabae* (black bean aphid), *Aphis glycines* (soybean aphid) (Davis *et al.*, 2005), *A. gossypii* (Raccah *et al.*, 1985), *Aphis nasturtii* (Harrington and Gibson, 1989), *Brachycaudus helichrysi* (leaf-curling plum aphid) (Harrington *et al.*, 1986), and *M. euphorbiae* (Singh and Boiteau, 1986), as well as several species of cereal aphid, most notably *Rhopalosiphum padi* (bird cherry–oat aphid) (van Hoof, 1977, 1980; Harrington *et al.*, 1986; Piron, 1986; Harrington and Gibson, 1989).

Most seed production schemes begin with virus-free tubers derived from tissue culture (see earlier). In the field, winged aphids then introduce virus from outside (Broadbent, 1950; Broadbent and Tinsley, 1951; Boiteau, 1997). When seed potatoes are grown in close proximity to later generation seed potato or ware-production fields, this outside source of inoculum is most often potato. Volunteer potatoes and perennial weeds also can be important sources of inoculum (Thomas, 1983; Hanafi *et al.*, 1989). Within-field spread of PLRV is often by apterae walking from plant to plant (Ribbands, 1963; Hanafi *et al.*, 1989; Flanders *et al.*, 1991; Hodgson, 1991), and localized centres of infection can appear. Winged *M. persicae* tend to be less efficient at transmitting PLRV than are apterae; typically, nymphs are more efficient than adults (Robert, 1971). Spread of PVY appears to be almost exclusively by alatae (Ragsdale *et al.*, 1994).

Economic Thresholds

Thresholds based on *M. persicae* apterae are widely accepted by growers producing potatoes for fresh market or processing, but this concept may not be applicable to seed potatoes because of the biological complexities involved and the stringent phytosanitary standards that must be achieved. Static thresholds in the range of 20–100 aphids per 100 leaves have been proposed for ware production (Davies, 1934; Byrne and Bishop, 1979; Cancelado and Radcliffe, 1979; Shields *et al.*, 1984). Differences in susceptibility to PLRV among cultivars suggest that higher thresholds may be appropriate for some (DiFonzo *et al.*, 1995). Proposed action thresholds to minimize within-field spread of PLRV in cultivars susceptible to 'net necrosis' (a tuber defect associated with PLRV infection) and for use in seed potato production have ranged from 1–10 *M. persicae* apterae per 100 leaves (Cancelado and Radcliffe, 1979; Hanafi *et al.*, 1989; Flanders *et al.*, 1991; Mowry, 2001).

Mathematical Modelling

Aphid-vectored pathosystems are immensely complex because many of the variables required to parameterize mathematical models are not measured easily. A model for forecasting incidence of PLRV in Scottish seed potatoes used virus incidence in the seed tubers planted and captures of *M. persicae* the previous year in suction traps (Pickup and Brewer, 1994). Current season spread of

PLRV in Minnesota is correlated strongly with the proportion of seed lots rejected for recertification because of PLRV the previous winter and cumulative duration of spring wind events (low level jets) originating from south central USA (Zhu *et al.*, 2006). Because more potential vector species are involved in PVY transmission, modelling that pathosystem is inherently more challenging than for PLRV. Models to predict PVY spread in Sweden were driven by two variables: aphid capture and relative transmission efficiencies of the key vector species (Sigvald, 1986, 1987, 1992). In the UK and The Netherlands, a simulation model, EPOVIR, uses the incidence of inoculum in seed lots to forecast current season PVY infection levels (Nemecek *et al.*, 1995, 1996). Models have rarely been used for managing aphid-transmitted potato viruses, but they provide insights into which variables need to be controlled to limit virus spread.

IPM Tactics

Preventing virus spread in potato requires integration of multiple tactics based on knowledge of the complex biology of insect-vectored pathosystems. Approaches to management of virus spread in potato can be categorized as preventive or therapeutic (Ragsdale *et al.*, 2001). Prevention focuses on reducing inoculum, a primary objective of all seed potato certification programmes, with therapeutic action focused on vector reduction. Because PLRV and PVY differ in their transmission characteristics, different approaches are required to manage these diseases. Use of aphicides to control vectors is often effective in limiting the spread of PLRV, but is seldom so for PVY (Ragsdale *et al.*, 1994; Radcliffe and Ragsdale, 2002). PVY management, therefore, tends to rely on cultural tactics.

Monitoring aphid flight

The association of virus spread with aphid flight activity is well documented (Hille Ris Lambers, 1972; van Harten, 1983). In many seed-producing areas, aphid-trapping networks monitor aphid flight activity with management advice issued when risk thresholds are reached. This is a longstanding practice in the UK (Woiwod *et al.*, 1984), The Netherlands (Hille Ris Lambers, 1972), Germany (Müller, 1987), Sweden (Sigvald, 1992), Canada (Boiteau and Parry, 1985), the USA (Halbert *et al.*, 1990), and elsewhere. However, it is difficult to evaluate the data rapidly enough to provide real-time advice to seed potato growers.

Chemical control

Although much of the insecticide used on potato is targeted against other pests, more than one-third of all applications in the USA are specifically for aphid control (Guenthner *et al.*, 1999; NASS, 2000) (Fig. 26.3). Insecticides are the only practical means of suppressing colonizing aphids on the crop, but are of inconsistent benefit in controlling virus spread. Among reported successes in controlling virus spread with insecticides (all crops and insect vectors), 94 of 119 cases

involved persistent and semi-persistent viruses (Perring *et al.*, 1999), whereas most failures, 32 of 48 cases, involved non-persistent viruses. Viruliferous alatae are not killed quickly enough to prevent PVY transmission (Shanks and Chapman, 1965). In contrast, spread of PLRV from within-field sources can be interrupted because of the extended post-acquisition latent period before an aphid can transmit (Leonard and Holbrook, 1978; Flanders *et al.*, 1991; DiFonzo *et al.*, 1995).

Systemic insecticides applied at planting or plant emergence can reduce within-field spread of PLRV significantly (Woodford *et al.*, 1983, 1988; Flanders *et al.*, 1991; Woodford, 1992; DiFonzo *et al.*, 1995; Boiteau and Singh, 1999). Such timing gives the greatest benefit in locations where migrant aphids are rarely viruliferous. However, in the Pacific Northwest, PLRV infection rates can approach 100% if *M. persicae* is not controlled with insecticides (Thomas *et al.*, 1997a).

Insecticide resistance often severely limits a grower's choice of aphicides (Radcliffe *et al.*, 1991). With *M. persicae*, this is a worldwide problem (Sawicki *et al.*, 1978, 1983), and resistance has developed to all

Fig. 26.3. Aeroplane applying pesticide to a Minnesota potato field.

major insecticide classes except (as of 2006) neonicotinoids (Devonshire and Moores, 1982; Devonshire et al., 1998; Dewar et al., 1998; Foster et al., Chapter 10 this volume). Bradley et al. (1962, 1966) demonstrated that non-toxic mineral oils applied to plants substantially reduced PVY transmission. It is unclear why field control is generally inferior to that obtained in laboratory studies (Shands, 1977; Boiteau and Singh, 1982; Bell, 1989), but reasons probably include weathering of oil deposits (Boiteau and Wood, 1982), new plant growth between applications, and incomplete coverage.

Biological control

The contribution of parasitoids, predators, and fungal entomopathogens to suppression of aphid populations is little recognized by seed potato growers. It might seem unlikely that biological control agents could be effective given the intensive use of pesticides in seed potato production. However, the tremendous outbreaks that can be induced by insecticides when the aphids have developed resistance are indirect evidence of the importance of natural enemies (ffrench-Constant et al., 1988; Harrington et al., 1989; Lagnaoui and Radcliffe, 1998).

Host-plant resistance

Current potato cultivars differ too little in aphid susceptibility for host-plant resistance to be useful. However, many wild potato species are highly resistant to aphids (Flanders et al., 1992; van Emden, Chapter 17 this volume). Yet only limited use has been made of wild potato species in developing insect-resistant cultivars (Flanders et al., 1999).

Various Agrobacterium-mediated transformations have produced potato lines expressing genes that confer pathogen-derived resistance to viruses. Transgenic lines have been developed that are highly resistant, but not immune, to infection by PLRV, PVY, and PVX (Brown et al., 1995). While aphids can still acquire virus from low titre plants, efficiency of transmission is greatly reduced (Thomas et al., 1997b). Transgenic cultivars were released in the USA that expressed the Leptinotarsa decemlineata (Colorado beetle) specific toxin Bacillus thuringiensis var. tenebrionis (Bt) combined with PLRV replicase (Thomas et al., 2000), and other cultivars expressed Bt and PVY coat protein (Berger and German, 2001). This technology was far more effective than any presently used tactic, but these cultivars have been withdrawn because of concerns over a public backlash against genetically modified food.

Cultural control

Crop isolation

Ideally, seed potato increase should occur in localities well isolated from potential sources of disease inoculum. Thus, many USA states and Canadian provinces have established seed farms in geographically isolated areas or designated growing regions where potato production is limited to seed (Slack, 1993). However, in most countries, and generally in North America, there is little isolation of late-generation seed and commercial potato production. The question then is not what degree of isolation would be optimal, but what is the minimum separation – from sources of virus inoculum or crops producing large vector populations – that is required to reduce risk of virus spread to acceptable levels? Spread of PLRV in southeast Scotland was found to be largely from inoculum sources within the crop (Cadman and Chambers, 1960; Howell, 1974; Woodford et al., 1983). Based on vector flight behaviour as evidenced by captures in suction traps in eastern Idaho, Halbert et al. (1990) suggested 400 m to 5 km could provide effective isolation from known PVY sources, but that 32 km might be required for isolation from PLRV sources because the vector remains infective for life. In England, minimum separation of 800 m from potential sources of PVY is recommended (Harrington et al., 1986), though just 40 m reduced spread of PVY in Denmark

(Hiddema, 1972). Seed potato growers generally have limited flexibility in locating their fields, and thus other cultural control methods and vector management assume greater importance. Isolation can also be achieved by early planting and haulm destruction ('vine-kill'), which have proved effective in maintaining the health of elite seed stock in The Netherlands since 1810 (Hille Ris Lambers, 1972).

Roguing

Physical removal of symptomatic plants from seed potato fields has long been an important virus management tactic (Thresh, 1988), and is most practical when virus incidence is low and the field is small enough that every plant can be inspected several times during the growing season. Roguing is easiest to accomplish before the canopy closes (Woodford and Gordon, 1990) and should begin as soon as secondary (tuber-borne) infection symptoms can be seen, typically when plants are 15–20 cm tall. The goal of rouging must be to remove all infected plants before winged aphids arrive. If the seed field is heavily infected (>~1% virus), roguing is often ineffective because some infected plants will be missed and

will remain in the field as sources of inoculum.

Barriers

Polymer webs can provide a high degree of protection against aphid-transmitted viruses (Hemphill *et al.*, 1988; Harrewijn *et al.*, 1991; Avilla *et al.*, 1997). However, the cost and inconvenience of row covers limit their application to seed potato fields of very high value and small size (e.g. the first field increase following propagation in the greenhouse or laboratory). Barrier crops (Fig. 26.4) are more widely adaptable than mulches or floating row covers; they are easier to install and keep in place, and do not lose effectiveness due to weathering or when the canopy closes. Barrier crops should have a fallow outside border with no gap between the barrier crop and the potato field, since winged aphids tend to alight at the interface of fallow ground and green crop. If immigrating alatae carrying PVY feed first on the border crop, they will probably lose their virus inoculum before moving into the potatoes (DiFonzo *et al.*, 1996). Barrier crops need be only a few metres wide to be effective. In Lower Saxony, oat borders just 1 m wide reduced the number of winged

Fig. 26.4. Experiment in which soybean 'crop borders' were used to protect a virus-free potato seed lot from the spread of *Potato virus Y* from nearby sources of inoculum (the potato rows are outside the fallow strips).

aphids (especially *R. padi*) caught in potato fields and were more effective in reducing PVY spread than intensive use of insecticides (Thieme *et al.*, 1998).

Interventions on non-potato hosts

The most vulnerable period in the life of a holocyclic aphid is that passed on the primary host, since this is where spring frosts occur and the number of potential invaders leaving for secondary hosts is determined. In the Columbia Basin, defoliants have been applied to peach (*Prunus persica*) before *M. persicae* oviparae mature (Tamaki and Weeks, 1968; Tamaki and Powell, 1972), tree-bands provided to shelter predators (Tamaki and Halfhill, 1968), peach pruned to remove overwintering eggs (Tamaki and Powell, 1968), and weeds that serve as early season aphid hosts removed in orchards and irrigation ditches (Wallis and Turner, 1969; Tamaki and Olsen, 1979; Tamaki *et al.*, 1980).

In temperate climates, most of the known weed hosts of PVY or PLRV are annuals and, since transmission *via* true seed does not occur (Salazar, 1996), these species are not virus sources. In the Souss Valley of Morocco, however, jimsonweed (*Datura stramonium*) supports *M. persicae* before the winter potato crop emerges. This weed and volunteer potatoes were implicated as being principal sources of both viruliferous aphids and PLRV inoculum (Hanafi *et al.*, 1995). Volunteer potatoes, which emerge as weeds in rotation crops in the Columbia Basin, are important PLRV reservoirs (Thomas, 1983), as are potatoes sprouting in cull piles in Canada (Frazer, 1987). In regions where winter frost extends below the root zone of potatoes, volunteer potatoes are rare and appear to be inconsequential sources of potato viruses (DiFonzo *et al.*, 1997). Because ware producers often plant tubers with 1–5% virus (375 to 1800 infected plants/ha), potato generally overshadows perennial weeds as a source of potato viruses.

Cropping practices

Changes in cropping patterns may change the effective isolation of a seed production area. In Scotland, with increased production of winter oilseed rape (*Brassica napus*), virginoparous *M. persicae* successfully overwintered on this crop and resulted in earlier colonization of potato (Woodford, 1988). In the Columbia Basin, three distinct flights of *M. persicae* occurred (Thomas *et al.*, 1997a). Spring migrants from peach appeared to be PLRV-free, but summer migrants, presumably from volunteer potatoes, winter oilseed rape, and weed hosts, were PLRV-infected. Fall migrants arrived so late they did not affect potato production.

Executive Summary

Seed potato growers rely on sophisticated and complex, multi-tactic pest management programmes to limit the spread of potato viruses.

Seed potato certification programmes remain, universally, the primary defence against introduction of virus into the production system.

Worldwide, the most important vector of PLRV is *M. persicae*. The virus is spread by both apterae and alatae. Growers rely heavily on aphicides to limit current-season PLRV spread, whereas PVY spread is by alatae of many aphid species and cannot be prevented by the use of insecticides. Near zero tolerance to virus makes economic thresholds irrelevant in seed potato production, except in situations where PLRV spread is exclusively by apterae. Seed potato insect pest management is often disparagingly equated with 'potato pesticide management'.

Biological control and host-plant resistance, key components of IPM programmes for many other crops, have yet to find practical application in seed potato IPM.

Cultural control and crop oils have proven most useful in limiting PVY spread.

There is no commercial use of semiochemicals for aphid control.

In few, if any, seed potato virus/vector management programmes are the various pest management tactics knowingly deployed in a complementary manner. Rather, tactics are simply pyramided in the hope of achieving maximum benefit.

The regional nature of potato virus management problems presents a compelling argument for industry-wide cooperation between seed potato growers, seed certification agencies, growers of potatoes for fresh market and processing, the processing industry, and all parties involved in potato pest management decisions to effect reduction of virus inoculum and vector abundance.

References

Allen, E.J., O'Brien, P.J. and Firman, D. (1992) Seed tuber production and management. In: Harris, P. (ed.) *The Potato Crop*, 2nd edn. Chapman and Hall, London, pp. 247–291.

Avilla, C., Collar, J.L., Duque, M., Perez, P. and Fereres, A. (1997) Impact of floating rowcovers on bell pepper yield and virus incidence. *Horticultural Science* 32, 882–883.

Bell, A.C. (1989) Use of oil and pyrethroid sprays to inhibit the spread of potato virus Y[N] in the field. *Crop Protection* 8, 37–39.

Berger, P.H. and German, T.L. (2001) Biotechnology and resistance to potato viruses. In: Loebenstein, G., Berger, P.H., Brunt, A.A. and Lawson, R.H. (eds) *Virus and Virus-Like Diseases of Potatoes and Production of Seed-Potatoes*. Kluwer, Dordrecht, pp. 341–363.

Boiteau, G. (1997) Comparative propensity for dispersal of apterous and alate morphs of three potato-colonizing aphid species. *Canadian Journal of Zoology* 75, 1396–1403.

Boiteau, G. and Parry, R.H. (1985) Monitoring of inflights of green peach aphids, *Myzus persicae* (Sulzer), in New Brunswick potato fields by yellow pans from 1974 to 1983: results and degree-day simulation. *American Potato Journal* 62, 489–496.

Boiteau, G. and Singh, R.P. (1982) Evaluation of mineral oil sprays for reduction of virus Y spread in potatoes. *American Potato Journal* 59, 253–262.

Boiteau, G. and Singh, R.P. (1999) Field assessment of imidacloprid to reduce the spread of PVY[O] and PLRV in potato. *American Journal of Potato Research* 76, 31–36.

Boiteau, G. and Wood, F.A. (1982) Persistence of mineral oil spray deposits on potato leaves. *American Potato Journal* 59, 55–63.

Boiteau, G., Singh, R.P., Parry, R.H. and Pelletier, Y. (1988) The spread of PVY[O] in New Brunswick potato fields: timing and vectors. *American Potato Journal* 65, 639–649.

Bradley, R.H.E. and Rideout, D.W. (1953) Comparative transmission of potato virus Y by four aphid species that infest potato. *Canadian Journal of Zoology* 31, 333–341.

Bradley, R.H.E., Wade, C.V. and Wood, F.A. (1962) Aphid transmission of potato virus Y inhibited by oils. *Virology* 18, 327–329.

Bradley, R.H.E., Moore, C.A. and Pond, D.D. (1966) Spread of potato virus Y curtailed by oil. *Nature, London* 209, 1370–1371.

Broadbent, L. (1950) The correlation of aphid numbers with the spread of leaf roll and rugose mosaic in potato crops. *Annals of Applied Biology* 37, 58–65.

Broadbent, L. and Tinsley, T.W. (1951) Experiments on the colonization of potato plants by apterous and by alate aphids in relation to the spread of virus diseases. *Annals of Applied Biology* 38, 411–424.

Brown, C.R., Smith, O.P., Damsteegt, V.D., Yang, C.-P., Fox, L. and Thomas, P.E. (1995) Suppression of PLRV titer in transgenic Russet Burbank and Ranger Russet. *American Potato Journal* 72, 589–597.

Byrne, D.N. and Bishop, G.W. (1979) Relationship of green peach aphid numbers to spread of potato leaf roll virus in southern Idaho. *Journal of Economic Entomology* 72, 809–811.

Cadman, C.H. and Chambers, J. (1960) Factors affecting the spread of aphid-borne viruses in potato in Eastern Scotland. III. Effects of planting date, roguing and age of crop on the spread of potato leaf-roll and Y viruses. *Annals of Applied Biology* 48, 729–738.

Cancelado, R.E. and Radcliffe, E.B. (1979) Action thresholds for green peach aphid on potatoes in Minnesota. *Journal of Economic Entomology* 72, 606–609.

Davies, W.M. (1934) Studies of aphides infesting the potato crop. II. Aphis survey: its bearing upon the selection of districts for seed potato production. *Annals of Applied Biology* 21, 283–299.

Davis, J.A., Radcliffe, E.B. and Ragsdale, D.W. (2005) Soybean aphid (*Aphid glycines*), a new vector of *Potato virus Y* in potato. *American Journal of Potato Research* 82, 197–205.

Devonshire, A.L. and Moores, G.D. (1982) A carboxylesterase with broad substrate specificity causes organophosphorus, carbamate, and pyrethroid resistance in peach–potato aphids (_Myzus persicae_). _Pesticide Biochemistry and Physiology_ 18, 235–246.

Devonshire, A.L., Field, L.M., Foster, S.P., Moores, G.D., Williamson, M.S. and Blackman, R.L. (1998) The evolution of insecticide resistance in the peach–potato aphid, _Myzus persicae. Philosophical Transactions of the Royal Society of London B_ 1376, 1677–1684.

Dewar, A., Haylock, L., Foster, S., Devonshire, A. and Harrington, R. (1998) Three into one will go – the evolution of resistance in the peach–potato aphid, _Myzus persicae. British Sugar Beet Review_ 66, 14–19.

DiFonzo, C.D., Ragsdale, D.W. and Radcliffe, E.B. (1995) Potato leafroll virus spread in differentially resistant potato cultivars under varying aphid densities. _American Potato Journal_ 72, 119–132.

DiFonzo, C.D., Ragsdale, D.W., Radcliffe, E.B., Gudmestad, N.C. and Secor, G.A. (1996) Crop borders reduce potato virus Y incidence in seed potato. _Annals of Applied Biology_ 129, 289–302.

DiFonzo, C.D., Ragsdale, D.W., Radcliffe, E.B., Gudmestad, N.C. and Secor, G.A. (1997) Seasonal abundance of aphid vectors of potato virus Y in the Red River Valley of Minnesota and North Dakota. _Journal of Economic Entomology_ 90, 824–831.

ffrench-Constant, R.H., Clark, S.J. and Devonshire, A.L. (1988) Effect of decline of insecticide residues on selection for insecticide resistance in _Myzus persicae_ (Sulzer) (Hemiptera: Aphididae). _Bulletin of Entomological Research_ 78, 19–29.

Flanders, K.L., Radcliffe, E.B. and Ragsdale, D.W. (1991) Potato leafroll virus spread in relation to densities of green peach aphid (Homoptera: Aphididae): implications for management thresholds for Minnesota seed potatoes. _Journal of Economic Entomology_ 84, 1028–1036.

Flanders, K.L., Hawkes, J.G., Radcliffe, E.B. and Lauer, F.I. (1992) Insect resistance in potatoes: sources, evolutionary relationships, morphological and chemical defenses, and ecogeographic associations. _Euphytica_ 61, 83–111.

Flanders, K.L., Arnone, S. and Radcliffe, E.B. (1999) The potato: genetic resources and insect resistance. In: Clement, S.L. and Quisenberry, S.S. (eds) _Global Plant Genetic Resources for Insect-Resistant Crops._ CRC, Boca Raton, pp. 207–239.

Foster, G.N. and Woodford, J.A.T. (1997) Melon and cotton aphid (_Aphis gossypii_ Glover: (Homoptera: Aphididae)) as a potential new virus vector of potatoes in Scotland. _The Entomologist_ 116, 175–178.

Franc, G.D. (2001) Seed certification as a virus management tool. In: Loebenstein, G., Berger, P.H., Brunt, A.A. and Lawson, R.H. (eds) _Virus and Virus-Like Diseases of Potatoes and Production of Seed-Potatoes._ Kluwer, Dordrecht, pp. 407–420.

Frazer, B.D. (1987) Aphid and virus management in potatoes in British Columbia. In: Boiteau, G., Singh, R.P. and Parry, R.H. (eds) _Potato Pest Management in Canada: Proceedings of a Symposium on Improving Potato Production._ Canada/New Brunswick Agreement on Food and Agriculture Development, Fredericton, pp. 23–29.

Guenthner, J.F., Wiese, M.V., Pavlista, A.D., Sieczka, J.B. and Wyman, J.A. (1999) Assessment of pesticide use in the U.S. potato industry. _American Journal of Potato Research_ 76, 25–29.

Gutbrod, O.A. and Mosley, A.R. (2001) Common seed potato certification schemes. In: Loebenstein, G., Berger, P.H., Brunt, A.A. and Lawson, R.H. (eds) _Virus and Virus-Like Diseases of Potatoes and Production of Seed-Potatoes._ Kluwer, Dordrecht, pp. 421–438.

Halbert, S.E., Connelly, J. and Sandvol, L.E. (1990) Suction trapping of aphids in western North America (emphasis on Idaho). _Acta Phytopathologica et Entomologica Hungarica_ 25, 411–422.

Hanafi, A., Radcliffe, E.B. and Ragsdale, D.W. (1989) Spread and control of potato leafroll virus in Minnesota. _Journal of Economic Entomology_ 82, 1201–1206.

Hanafi, A., Radcliffe, E.B. and Ragsdale, D.W. (1995) Spread and control of potato leafroll virus in the Souss Valley of Morocco. _Crop Protection_ 14, 145–153.

Harrewijn, P., Ouden, H.D. and Piron, P.G.M. (1991) Polymer webs to prevent virus transmission by aphids in seed potatoes. _Entomologia Experimentalis et Applicata_ 58, 101–107.

Harrington, R. and Gibson, R.W. (1989) Transmission of potato virus Y by aphids trapped in potato crops in southern England. _Potato Research_ 32, 167–174.

Harrington, R., Katis, N. and Gibson, R.W. (1986) Field assessment of the relative importance of different aphid species in the transmission of potato virus Y. _Potato Research_ 29, 67–76.

Harrington, R., Bartlet, E., Riley, D.K., ffrench-Constant, R.H. and Clark, S.J. (1989) Resurgence of insecticide-resistant _Myzus persicae_ on potatoes treated repeatedly with cypermethrin and mineral oil. _Crop Protection_ 8, 340–348.

Harrison, B.D. (1971) Potato viruses in Britain. In: Western, J.H. (ed.) _Diseases of Crop Plants._ Macmillan, London, pp. 123–159.

van Harten, A. (1983) The relation between aphid flights and the spread of potato virus Y^N (PVY^N) in the Netherlands. *Potato Research* 26, 1–15.

Heimbach, U., Thieme, T., Weidemann, H.-L. and Thieme, R. (1998) Transmission of potato virus Y by aphid species which do not colonise potatoes. In: Nieto Nafría, J.M. and Dixon, A.F.G. (eds) *Aphids in Natural and Managed Ecosystems*. Universidad de León, León, Spain, pp. 555–559.

Heinze, K. (1960) Versuche zur Übertragung nichtpersistenter und persistenter Viren durch Blattläuse. *Nachrichtenblatt des Deutschen Pflanzenschutzdienstes* 12, 119–121.

Hemphill, D.D., Jr., Reed, G.L., Wilson, R.C., Gutbrod, O. and Allen, T.C. (1988) Prevention of potato virus Y transmission in potato seed stock with direct covers. *Plasticulture* 79, 31–36.

Hiddema, J. (1972) Inspection and quality grading of seed potatoes. In: de Bokx, J.A. (ed.) *Viruses of Potatoes and Seed-Potato Production*. Centre for Agricultural Publishing and Documentation, Wageningen, pp. 206–215.

Hille Ris Lambers, D. (1972) Aphids: their life cycles and their role as virus vectors. In: de Bokx, J.A. (ed.) *Viruses of Potatoes and Seed-Potato Production*. Centre for Agricultural Publishing and Documentation, Wageningen, pp. 36–56.

Hodgson, C. (1991) Dispersal of apterous aphids (Homoptera: Aphididae) from their host plant and its significance. *Bulletin of Entomological Research* 81, 417–427.

van Hoof, H.A. (1977) Determination of the infection pressure of potato virus Y^N. *Netherlands Journal of Plant Pathology* 83, 123–127.

van Hoof, H.A. (1980) Aphid vectors of potato virus Y^N. *Netherlands Journal of Plant Pathology* 86, 159–162.

Horváth, J. (1990) A magyarországi burgonyatermesztés virologiai problémái: eredmények és kudarcok fél évszázados története, a jövo kilátásai. In: Kovács, G. (ed.) *Szántóföldi Növénytermesztés VII. Országos Növényvédelmi és Agrokémiai Tanácskozása*. Növény és Talajvédelmi Szolgálat, Budapest, pp. 83–108.

Howell, P.J. (1974) Field studies of potato leaf roll virus spread in southeastern Scotland, 1962–1969, in relation to aphid populations and other factors. *Annals of Applied Biology* 76, 187–197.

Lagnaoui, A. and Radcliffe, E.B. (1998) Potato fungicides interfere with entomopathogenic fungi impacting population dynamics of green peach aphid. *American Journal of Potato Research* 75, 19–25.

Leonard, S.H. and Holbrook, F.R. (1978) Minimum acquisition and transmission times for potato leaf roll virus by the green peach aphid. *Annals of the Entomological Society of America* 71, 493–495.

Loughnane, J.B. (1943) *Aphis rhamni* Boyer; its occurrence in Ireland and its efficiency as a vector of potato viruses. *Journal of the Department of Agriculture of Eire (Ireland)* 40, 291–298.

MacGillivray, M.E. (1981) Aphids. In: Hooker, W.J. (ed.) *Compendium of Potato Diseases*. American Phytopathological Society, St Paul, pp. 101–103.

Mowry, T.M. (2001) Green peach aphid (Homoptera: Aphididae) action thresholds for controlling the spread of potato leafroll virus in Idaho. *Journal of Economic Entomology* 94, 1332–1339.

Müller, H.L. (1987) Yellow water traps – a support for monitoring aphid flights in seed potatoes in Baden-Württemberg. In: Cavalloro, R. (ed.) *Aphid Migration and Forecasting 'Euraphid' Systems in European Community Countries*. Office for Official Publications of the European Communities, Luxembourg, pp. 169–171.

NASS (U.S. National Agricultural Statistics Service) (2000). *Agricultural Chemical Usage, 1999 Field Crops Summary* (http://usda.mannlib.cornell.edu/reports/nassr/other/pcu-bb/).

Nemecek, T., Derron, J.O., Fischlin, A. and Roth, O. (1995) Use of a crop-growth model coupled to an epidemic model to forecast yield and virus infection in seed potatoes. In: Haverkort, A.J. and MacKerron, D.K.L. (eds) *Potato Ecology and Modelling of Crops Under Conditions Limiting Growth*. Kluwer, Dordrecht, pp. 281–291.

Nemecek, T., Derron, J.O., Roth, O. and Fischlin, A. (1996) Adaptation of a crop-growth model and its extension by a tuber size function for use in a seed potato forecasting system. *Agricultural Systems* 52, 419–437.

Perring, T.M., Gruenhagen, N.M. and Farrar, C.A. (1999) Management of plant viral diseases through chemical control of insect vectors. *Annual Review of Entomology* 44, 457–481.

Piche, L.M., Singh, R.P., Nie, X. and Gudmestad, N.C. (2004) Diversity among potato virus Y isolates obtained from potatoes grown in the United States. *Phytopathology* 94, 1368–1375.

Pickup, J. and Brewer, A.M. (1994) The use of aphid suction-trap data in forecasting the incidence of potato leafroll virus in Scottish seed potatoes. *Proceedings of the British Crop Protection Conference, Pests and Diseases, Brighton, November 1994* 1, 351–358.

Piron, P.G.M. (1986) New aphid vectors of potato virus Y^N. *Netherlands Journal of Plant Pathology* 92, 223–229.

Raccah, B., Gal-on, A. and Eastop, V.F. (1985) The role of flying aphid vectors in the transmission of cucumber mosaic virus and potato virus Y to peppers in Israel. *Annals of Applied Biology* 106, 451–460.

Radcliffe, E.B. and Ragsdale, D.W. (2002) Invited review. Aphid transmitted potato viruses: the importance of understanding vector biology. *American Journal of Potato Research* 79, 353–386.

Radcliffe, E.B., Flanders, K.L., Ragsdale, D.W. and Noetzel, D.M. (1991) Pest management systems for potato insects. In: Pimentel, D. (ed.) *CRC Handbook of Pest Management in Agriculture, Volume 3*, 2nd edn. CRC, Boca Raton, pp. 587–621.

Ragsdale, D.W., Radcliffe, E.B., DiFonzo, C.D. and Connelly, M.S. (1994) Action thresholds for an aphid vector of potato leafroll virus. In: Zehnder, G.W., Powelson, M.L., Jansson, R.K. and Raman, K.V. (eds) *Advances in Potato Pest Biology and Management*. American Phytopathological Society, St Paul, pp. 99–110.

Ragsdale, D.W., Radcliffe, E.B. and DiFonzo, C.D. (2001) Epidemiology and field control of PVY and PLRV. In: Loebenstein, G., Berger, P.H., Brunt, A.A. and Lawson, R.H. (eds) *Virus and Virus-Like Diseases of Potatoes and Production of Seed-Potatoes*. Kluwer, Dordrecht, pp. 237–270.

Raman, K.V. and Radcliffe, E.B. (1992) Pest aspects of potato production, Part 2. Insect pests. In: Harris, P.M. (ed.) *The Potato Crop, The Scientific Basis for Improvement*, 2nd edn. Chapman and Hall, London, pp. 476–506.

Ribbands, C.R. (1963) The spread of apterae of *Myzus persicae* (Sulz.) and of yellows viruses within a sugar-beet crop. *Bulletin of Entomological Research* 54, 267–283.

Robert, Y. (1971) Épidémiologie de l'enroulement de la pomme de terre: capacité vectrice de stades et de formes des pucerons *Aulacorthum solani* Kltb. *Macrosiphum euphorbiae* Thomas et *Myzus persciae* Sulz. *Potato Research* 14, 130–139.

Robert, Y. and Bourdin, D. (2001) Aphid transmission of potato viruses. In: Loebenstein, G., Berger, P.H., Brunt, A.A. and Lawson, R.H. (eds) *Virus and Virus-Like Diseases of Potatoes and Production of Seed-Potatoes*. Kluwer, Dordrecht, pp. 195–225.

Robert, Y. and Maury, Y. (1970) Capacités vectrices comparées de plusieurs souches de *Myzus persicae* Sulz., *Aulacorthum solani* Kltb. et *Macrosiphum euphorbiae* Thomas dans l'étude de la transmission de l'enroulement de la pomme de terre. *Potato Research* 13, 199–209.

Robert, Y. and Rouzé-Jouan, J. (1971) Premières observations sur le rôle de la température au moment de la transmission de l'enroulement par *Aulacorthum solani* Kltb., *Macrosiphum euphorbiae* Thomas et *Myzus persicae* Sulz. *Potato Research* 14, 154–157.

Salazar, L.F. (1996) *Potato Viruses and their Control*. International Potato Center (CIP), Lima, 214 pp.

Sawicki, R.M., Devonshire, A.L., Rice, A.D., Moores, G.D., Petzing, S.M. and Cameron, A. (1978) The detection and distribution of organophosphorous and carbamate insecticide-resistant *Myzus persicae* (Sulz.) in Britain in 1976. *Pesticide Science* 9, 189–201.

Sawicki, R.M., Rice, A.D. and Gibson, R.W. (1983) Insecticide-resistance in the peach–potato aphid *Myzus persicae* and the prevention of virus spread. *Aspects of Applied Biology, No. 2*, pp. 29–33.

Shands, W.A. (1977) Control of aphid-borne potato virus Y in potatoes with oil emulsions. *American Potato Journal* 54, 179–187.

Shanks, C.H., Jr. and Chapman, R.K. (1965) The effects of insecticides on the behavior of the green peach aphid and its transmission of potato virus Y. *Journal of Economic Entomology* 58, 79–83.

Shields, E.J., Hygnstrom, J.R., Curwen, D., Stevenson, W.R., Wyman, J.A. and Binning, L.K. (1984) Pest management for potatoes in Wisconsin – a pilot program. *American Potato Journal* 61, 508–516.

Sigvald, R. (1986) Forecasting the incidence of potato virus Y[O]. In: McLean, G.D., Garrett, R.G. and Ruesink, W.G. (eds) *Plant Virus Epidemics: Monitoring, Modelling and Predicting Outbreaks*. Academic Press, Sydney, pp. 419–441.

Sigvald, R. (1987) Aphid migration and the importance of some aphid species as vectors of potato virus Y[O] (PVY[O]) in Sweden. *Potato Research* 30, 267–283.

Sigvald, R. (1989) Relationship between aphid occurrence and spread of potato virus Y[O] (PVY[O]) in field experiments in southern Sweden. *Journal of Applied Entomology* 108, 35–43.

Sigvald, R. (1992) Progress in aphid forecasting systems. *Netherlands Journal of Plant Pathology* 98, 55–62.

Singh, R.P. and Boiteau, G. (1986) Re-evaluation of the potato aphid, *Macrosiphum euphorbiae* (Thomas), as vector of potato virus Y. *American Potato Journal* 63, 335–340.

Slack, S.A. (1993) Seed certification and seed improvement programs. In: Rowe, R.C. (ed.) *Potato Health Management*. American Phytopathological Society Press, St Paul, pp. 61–65.

Tamada, T., Harrison, B.D. and Roberts, I.M. (1984) Variation among British isolates of potato leafroll virus. *Annals of Applied Biology* 104, 107–116.

Tamaki, G. and Halfhill, J.E. (1968) Bands on peach trees as shelters for predators of the green peach aphid. *Journal of Economic Entomology* 61, 707–711.

Tamaki, G. and Olsen, D. (1979) Evaluation of orchard weed hosts of green peach aphid and the production of winged migrants. *Environmental Entomology* 8, 314–317.

Tamaki, G. and Powell, D.M. (1968) Egg distribution as a factor in suppressing the green peach aphid by pruning. *Journal of Economic Entomology* 61, 1437–1439.

Tamaki, G. and Powell, D.M. (1972) An insecticide and a defoliant evaluated for use in a program of integrated control designed to suppress the green peach aphid on peach trees. *Journal of Economic Entomology* 65, 271–275.

Tamaki, G. and Weeks, R.E. (1968) Use of chemical defoliants on peach trees in integrated program to suppress populations of green peach aphids. *Journal of Economic Entomology* 61, 431–435.

Tamaki, G., Fox, L. and Chauvin, R.L. (1980) Green peach aphid: orchard weeds are host to fundatrix. *Environmental Entomology* 9, 62–66.

Tanaka, S. and Shiota, H. (1970) Latent period of potato leaf roll virus in the green peach aphid (*Myzus persicae* Sulzer). *Annals of the Phytopathological Society of Japan* 36, 106–111.

Thieme, T., Heimbach, U., Thieme, R. and Weidemann, H.-L. (1998) Introduction of a method for preventing transmission of potato virus Y (PVY) in Northern Germany. *Aspects of Applied Biology* 52, 25–29.

Thomas, P.E. (1983) Sources and dissemination of potato viruses in the Columbia Basin of the Northwestern United States. *Plant Disease* 67, 744–747.

Thomas, P.E., Kaniewski, W.K. and Lawson, E.C. (1997a) Reduced field spread of potato leafroll virus in potatoes transformed with the potato leafroll virus coat protein gene. *Plant Disease* 81, 1447–1453.

Thomas, P.E., Pike, K.S. and Reed, G.L. (1997b) Role of green peach aphid flights in the epidemiology of potato leaf roll disease in the Columbia Basin. *Plant Disease* 81, 1311–1316.

Thomas, P.E., Lawson, E.C., Zalewski, J.C., Reed, G.L. and Kaniewski, W.K. (2000) Extreme resistance to *Potato leafroll virus* in potato cv. Russet Burbank mediated by the viral replicase gene. *Virus Research* 71, 49–62.

Thresh, J.M. (1988) Eradication as a virus control measure. In: Clifford, B.C. and Lester, E. (eds) *Control of Plant Diseases: Costs and Benefits*. Blackwell, Oxford, pp. 155–194.

Wallis, R.L. and Turner, J.E. (1969) Burning weeds in drainage ditches to suppress populations of green peach aphids and the incidence of beet western yellows disease in sugarbeets. *Journal of Economic Entomology* 62, 307–309.

Weidemann, H.-L. (1988) Importance and control of potato virus Y^N (PVY^N) in seed potato production. *Potato Research* 31, 85–94.

Woiwod, I.P., Tatchell, G.M. and Barrett, A.M. (1984) A system for the rapid collection, analysis and dissemination of aphid-monitoring data from suction traps. *Crop Protection* 3, 273–288.

Woodford, J.A.T. (1988) The impact of cropping policy on methods to control potato leafroll virus. *Aspects of Applied Biology* 17, 163–171.

Woodford, J.A.T. (1992) Effects of systemic applications of imidacloprid on the feeding behaviour and survival of *Myzus persicae* on potatoes and on transmission of potato leafroll virus. *Pflanzenschutz-Nachrichten Bayer* 45, 527–546.

Woodford, J.A.T. and Gordon, S.C. (1990) New approaches for restricting spread of potato leafroll virus by different methods of eradicating infected plants from potato crops. *Annals of Applied Biology* 116, 477–487.

Woodford, J.A.T., Harrison, B.D., Aveyard, C.S. and Gordon, S.C. (1983) Insecticidal control of aphids and the spread of potato leafroll virus in potato crops in eastern Scotland. *Annals of Applied Biology* 103, 117–130.

Woodford, J.A.T., Gordon, S.C. and Foster, G.N. (1988) Side-band application of systemic granular pesticides for the control of aphids and potato leafroll virus. *Crop Protection* 7, 96–105.

Woodford, J.A.T., Jolly, C.A. and Aveyard, C.S. (1995) Biological factors influencing the transmission of potato leafroll virus by different aphid species. *Potato Research* 38, 133–141.

Zhu, M., Radcliffe, E.B., Ragsdale, D.W., MacRae, I.V. and Seeley, M.W. (2006) Low-level jet streams associated with spring aphid migration and current season spread of potato viruses in the U.S. northern Great Plains. *Agricultural and Forest Meteorology* 138, 192–202.

27 IPM Case Studies: Sorghum

Gerald J. Michels, Jr.[1] and John D. Burd[2]

[1]Texas Agricultural Experiment Station, Bushland, TX 79012, USA; [2]Plant Science and Water Conservation Laboratory, USDA-ARS, Stillwater, OK 74075, USA

Introduction

Sorghum, *Sorghum bicolour* (Fig. 27.1), also known as great millet and Guinea corn, originated in Africa and is cultivated throughout the tropical, subtropical, and warm temperate areas of the world. Sorghum is grown for animal feed and forage, human consumption, and for fibre (FAO, 1979; Dibb, 1983). Worldwide, 41.5 million ha of sorghum were harvested in 2001 (FAO, 2001).

There are four important worldwide aphid pests of sorghum: *Schizaphis graminum* (greenbug), *Rhopalosiphum maidis* (corn leaf aphid), *Melanaphis sorghi* (sugarcane aphid), and *Sipha flava* (yellow sugarcane aphid) (Young and Teetes, 1977). *S. graminum* is the key cosmopolitan aphid pest of sorghum listed by Young and Teetes (1977). The other three species are considered occasional pests.

There is an extensive literature base on IPM components for the greenbug on sorghum. However, few studies have compared integrated and single-method management systems, including an economic analysis. Due to the startling lack of IPM case studies addressing other sorghum-feeding aphids, this review will concentrate on greenbug IPM practices, especially since greenbug has been studied for many years and has a cosmopolitan distribution. It is hoped that this review will

stimulate research comparing single and multiple-component control strategies for aphid pests of sorghum.

A Short History of *Schizaphis graminum* on Sorghum

Schizaphis graminum (Figs 27.2 and 27.3) has been a key pest of sorghum for a relatively short period of time. Greenbugs were known to utilize sorghum as a host as early as 1863 in Italy (Hunter, 1909) and were noted on sorghum in Africa by Matthee (1962), and again in Europe by Barbulescu (1964). The greenbug was first described as a significant pest of sorghum in the USA in 1968 (USDA, 1968) when serious damage to sorghum was reported in the southwestern and Great Plains regions. This outbreak resulted in millions of hectares being treated with insecticides to control the pest. Harvey and Hackerott (1969) reported that over 400,000 ha of sorghum in Kansas were infested with greenbugs and that the pest destroyed over 100,000 ha of the crop. Harvey and Hackerott (1969) designated this sorghum-damaging greenbug as biotype C, based on its ability to damage 'Piper' Sudan grass (*Sorghum* × *drummondii*), which was resistant to biotype B greenbugs.

Fig. 27.1. Healthy, irrigated grain sorghum at the Texas Agricultural Experiment Station North Plains Research Field, Etter, Texas (photo by G.J. Michels, Jr.).

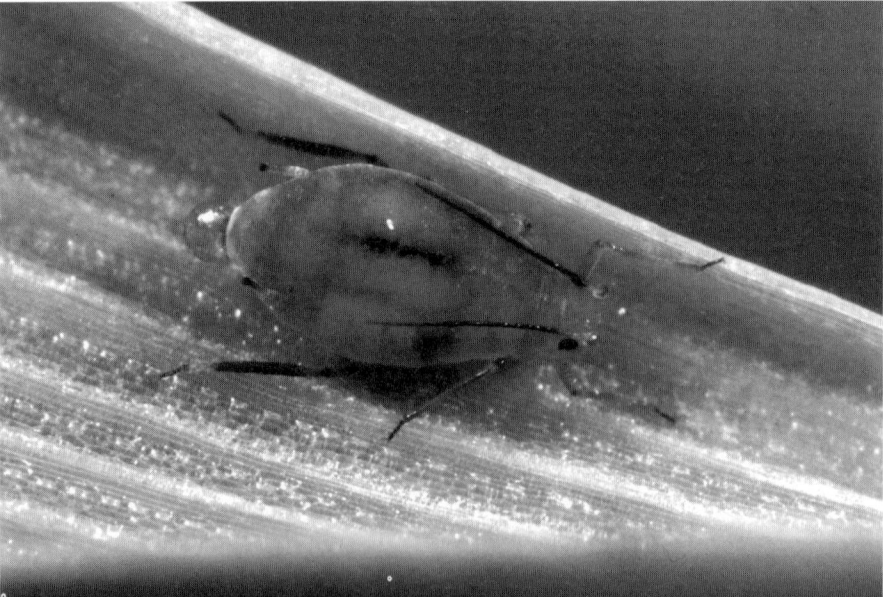

Fig. 27.2. The greenbug, *Schizaphis graminum* (original colour photo courtesy of B. Chaubet and INRA).

By 1981, greenbugs were estimated to be responsible for 2.5% of the 9.0% annual loss in grain sorghum due to insects, and it ranked as the second most damaging insect pest of sorghum in the USA (USDA, 1981). As late as 1992, 31% of sorghum in the USA was treated with insecticides to control greenbugs (Webster *et al.*, 1995).

Fig. 27.3. Greenbug damage to grain sorghum at the Texas Agricultural Experiment Station North Plains Research Field, Etter, Texas (photo by G.J. Michels, Jr.).

Current Greenbug Management Practices

In 1981, the US Department of Agriculture (USDA, 1981) stated: 'The controls for reducing greenbug damage are as complete as those for any insect.' At the time of the report, IPM components, especially resistant hybrids, were available for use by producers.

Chemical control

Research examining the efficacy of foliar and soil-applied insecticides to control greenbug began in 1968 and articles appeared in the early 1970s (DePew, 1971, 1974; Cate *et al.*, 1973a,b). Foliar organophosphates, carbamates, and even one organochlorine (endrin) gave excellent control (DePew, 1971; Cate *et al.*,

1973a). Soil-applied organophosphates also provided control, while soil-applied carbamates varied in their efficacy (DePew, 1974).

Within a few years, however, greenbug resistance to organophosphates, specifically disulfoton, was reported (Peters *et al.*, 1975; Teetes *et al.*, 1975). Additional organophosphates and carbamates were developed throughout the 1970s and 1980s, most notably chlorpyrifos, which became the chemical of choice in the 1980s. However, greenbug resistance to chlorpyrifos has been documented (Niemczyk and Moser, 1982; Sloderbeck *et al.*, 1991). Research conducted by Shufran *et al.* (1996, 1997a,b) and Rider *et al.* (1998) showed that greenbug genotypes differed in their levels of insecticide resistance. Shufran *et al.* (1997b) and Stone *et al.* (2000) demonstrated that life history parameters of these genotypes differed, with the most prevalent insecticide-resistant one having none of the reproductive disadvantages often associated with insecticide-resistant arthropods. Archer *et al.* (1999) concluded that combinations of chlorpyrifos and malathion applied to a 3 : 1 mixture of insecticide-resistant and -susceptible greenbug populations resulted in a significant net return when compared to untreated sorghum, or even when the greenbugs received a pretreatment application of chlorpyrifos, which supposedly increased the percentage of resistant aphids in the population, 4 days prior to the chlorpyrifos–malathion mixture treatment.

Buschman and DePew (1990) showed that sorghum sprayed with chlorpyrifos and parathion for greenbug control had significantly higher densities of *Oligonychus pratensis* (Banks grass mite) than untreated fields, but the cause of the mite outbreaks was not determined.

Other insecticides have been developed in the past 10 years, such as the soil-applied systemic chloro-nicotinyl, imidacloprid, and the newly developed nicotinamide, N-cyanomethyl-4-trifluromethyl nicotinamide (FMC, 2002). These insecticides, especially when formulated as seed treatments, have gained wide acceptance, and in some areas of the Great Plains have replaced chlorpyrifos as the insecticide of choice (C.D. Patrick, personal communication).

Biological control

Using predators and parasites to control greenbug began long before the aphid became a key sorghum pest. Fenton and Dahms (1951) attempted inundative releases of *Hippodamia convergens* (convergent ladybird) in wheat, but concluded that these were ineffective. Attempts have been made to import and establish predaceous coccinellids and parasitic Hymenoptera to control greenbugs in sorghum and wheat (Jackson *et al.*, 1971; Cartwright *et al.*, 1977; Gilstrap *et al.*, 1984; Michels and Bateman, 1986). However, little success has been achieved in the past 30 years.

Conservation biological control, 'the use of tactics and approaches that involve the manipulation of the environment of natural enemies so as to enhance their survival, and/or physiological and behavioural performance' (Barbosa, 1998), is the current practice of choice. In the southern Great Plains area of the USA, Kring *et al.* (1985) and Rice and Wilde (1988) demonstrated conclusively that indigenous coccinellids (*H. convergens*, *Hippodamia sinuata*, *Coleomegilla maculata lengi*, and *Scymnus* spp.) were key to suppressing greenbugs on sorghum early in the season. The impact of the most abundant parasitoid, *Lysiphlebus testaceipes*, in these experiments was sporadic, typically occurring late in the season. In Nebraska, Fernandes *et al.* (1998) concluded that *L. testaceipes* could control greenbugs effectively in an inundative biocontrol programme at a release rate of 24,000–36,000 wasps/ha, and suggested that planting alternating strips of greenbug-resistant sorghum hybrids with greenbug-susceptible hybrids as banker plants for *L. testaceipes* may be an economically feasible way to produce the desired numbers of wasps.

There is strong evidence that, at least in the US Great Plains region, greenbug biocontrol by predaceous coccinellids is enhanced when *R. maidis* is present early in the growing season. Kring and Gilstrap (1986) noted that corn leaf aphids helped maintain *Hippodamia* spp. in sorghum, and Michels and Behle (1992) found in a 3-year experiment that greenbugs did not reach economic thresholds in 2 years when *R. maidis* were present early in the season, but severely damaged sorghum in the 1 year that *R. maidis* populations failed to develop. Field data (G.J. Michels, unpublished results) showed that in 31 irrigated and dryland sorghum fields sampled from 1988–2000, peak greenbug density never reached damaging levels when corn leaf aphid densities reached 100 or more per plant prior to the sorghum reaching boot stage (approximately 15 July) of a given year (Fig. 27.4). Peak coccinellid density was significantly correlated to corn leaf aphid density rather than greenbug density. Thus, the predators' impact on mid-season greenbug infestations may be

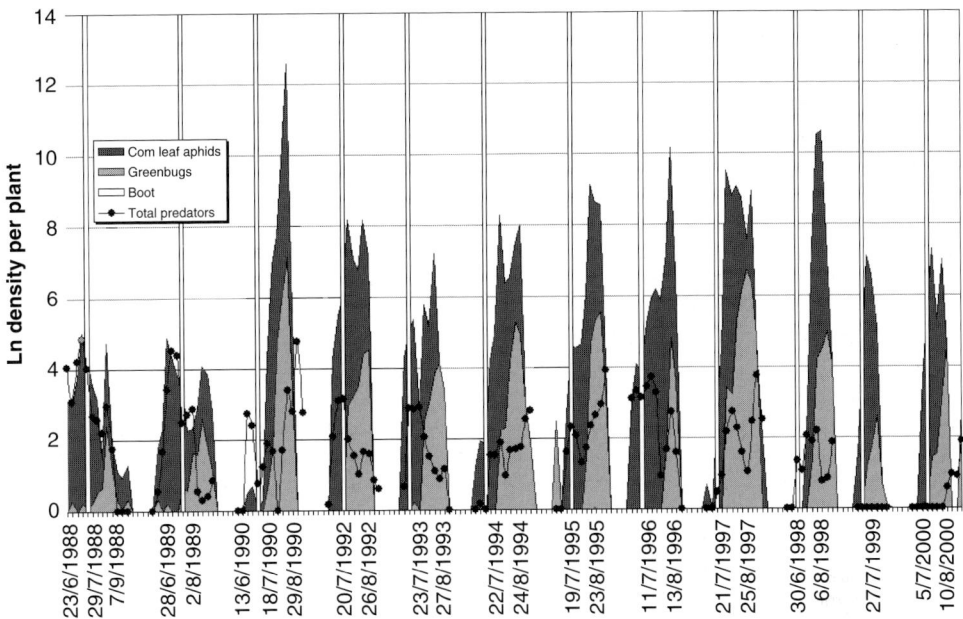

Fig. 27.4. Seasonal relationship of corn leaf aphids *(Rhopalosiphum maidis)*, greenbugs *(Schizaphis graminum)*, and predaceous coccinellids in irrigated grain sorghum from 1988–1990 and 1992–2000, Texas Agricultural Experiment Station, Bushland, Texas.

predetermined by early-season corn leaf aphid populations. As a bonus, corn leaf aphids seldom cause economic damage to sorghum, even when present in large numbers (G.J. Michels, unpublished results). Wilde and Ohiagu (1976) concluded that chemical control of corn leaf aphid in sorghum did not increase yield and, in light of its potential impact on biological control, was usually unwarranted.

Host-plant resistance

Development of greenbug-resistant sorghum hybrids began shortly after the advent of biotype C (Hackerott *et al.*, 1969; Wood, 1971; Teetes and Johnson, 1974). Commercial sorghum hybrids became available in 1976 (Morgan *et al.*, 1980). However, greenbug resistance in sorghum, as in wheat, has been ephemeral due to the development of host races or biotypes (Puterka and Peters, 1995). The ability of greenbugs to 'overcome' previously resistant sorghum hybrids has resulted in the designation of greenbug

biotypes I (Harvey *et al.*, 1991; Bowling *et al.*, 1994) and K (Harvey *et al.*, 1997), although previous biotypes, such as biotype E, had also overcome biotype C-resistant sorghum hybrids. Porter *et al.* (1997) and Anstead *et al.* (2003) provide reviews of the greenbug biotype concept in wheat and sorghum. Both papers concluded that biotypes were not engendered by resistant hybrids, but rather that resistant hybrids selected for greenbug genotypes that already existed.

Regardless of the ability of greenbugs to overcome previously resistant sorghum genotypes, research continues to develop new sources of resistance (Wilde and Tuinstra, 2000). Antibiosis, antixenosis, and tolerance mechanisms have all been identified in sorghum hybrids either singly or in combination (see Bowling and Wilde, 1996 for a review). Harvey *et al.* (1997) suggested that breeding efforts should concentrate on multigenic greenbug resistance since there was evidence that such hybrids could maintain their resistance to an array of greenbug genotypes. Harvey *et al.* (1994) and Thindwa and Teetes (1994) noted that temperature played a role in resistance expression. Higher temperatures resulted in delayed development, reduced fecundity, and shorter overall lifespan of greenbugs on resistant sorghum hybrids than on susceptible hybrids or resistant hybrids at lower temperatures.

Unfortunately, Rice and Wilde (1989) noted a negative interaction between resistant sorghum hybrids and predaceous coccinellids preying on greenbugs. They observed that *H. convergens* larval-pupal survival was reduced and eclosion to pupation time was increased when the beetles fed on aphids reared on antibiotic plants. They concluded that the widely accepted concept that host-plant resistance and biological control are compatible was probably too broad a generalization, and that understanding the effects of resistant sorghum hybrids at the third trophic level was essential.

Cultural control

Like most crop-feeding insects, greenbugs respond to plant condition and agronomic practices. Schweissing and Wilde (1979) and Archer *et al.* (1982) found that increasing N fertilizer improved sorghum for the greenbug. Greenbug densities were higher on plants that received N than on those that did not (Schweissing and Wilde, 1979). Archer *et al.* (1982) noted that greenbugs responded positively to increasing N (0, 45, 90, 135, 180, and 225 kg N/ha), and that plant damage

remained essentially the same, even though more plant biomass was produced. The general conclusions were that applying a higher than normal rate of N to sorghum in an attempt to raise crop tolerance to infestation was not a positive greenbug management tool.

Harvey and Thompson (1988) reported that greenbug densities were consistently lower on sorghum grown at high plant density than on plants grown at low density. Burton *et al.* (1987) demonstrated that increasing plant residues on the soil surface by reduced tillage resulted in decreased greenbug density and decreased plant damage. A dense plant canopy that obscured furrows also reduced greenbug infestation.

Kindler and Staples (1981) concluded that the greenbug economic injury level was lower in water-stressed than in well-watered sorghum. Michels and Undersander (1986) demonstrated that water stress negatively influenced greenbug reproduction in sorghum when water potential on stressed plants fell below –0.3 MPa. In a 3-year study comparing plant populations and irrigation regimes, Michels *et al.* (2002) found that peak greenbug densities were highest in well-watered fields with low plant populations, and significantly lower in well-watered fields with high plant populations. The results also indicated that low plant populations coupled with heavy irrigation created more of a greenbug problem than higher plant populations and moderate irrigation amounts (Fig. 27.5).

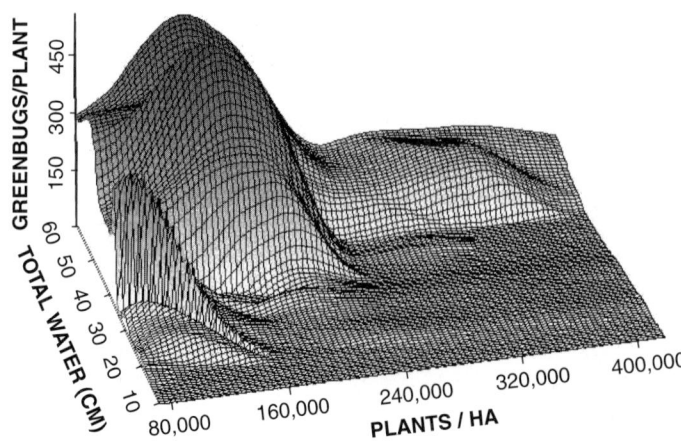

Fig. 27.5. Graphic representation of greenbug densities associated with varying plant population and total water applied through the season in grain sorghum at Bushland, Texas, 1998–2000.

Putting the pieces together

It is evident from the above discussion that the components for greenbug IPM are available, if somewhat scattered. It is also apparent that the majority of the research into greenbug IPM components has centred in the Great Plains states of North America. A lack of research from other countries may be due to the fact that more important pests have superseded the greenbug (Abate *et al.*, 2000).

In the USA, several researchers have published papers that have incorporated at least two IPM components. Starks *et al.* (1972) conducted a greenhouse experiment that addressed the interaction between resistant sorghum and parasitism by *L. testaceipes*. They concluded that host-plant resistance and parasitism were complementary factors in reducing greenbug numbers. These results were supported by Dogramaci (1998), who concluded that *L. testaceipes* parasitism and the use of currently available resistant hybrids were compatible and complimentary control strategies for biotype I greenbugs.

In field experiments comparing greenbug densities on resistant and susceptible sorghum hybrids grown at differing plant densities over 2 years, Harvey and Thompson (1988) found that growing resistant hybrids in thick stands reduced greenbug densities significantly over any other combination of hybrid and stand. Susceptible hybrids grown in thick stands yielded similarly to resistant hybrids in thin stands; therefore, the authors concluded that any practice that reduced plant stand would intensify greenbug problems.

Burton *et al.* (1990) extrapolated their previous work (Burton *et al.*, 1987) to compare tillage practices and the inclusion of resistant sorghum hybrids. In conventional tillage plots, greenbug density was significantly lower (50%) on the resistant hybrid. When the resistant hybrid was grown in no-tillage plots, greenbug density was significantly lower than on the resistant hybrid grown in conventional tillage plots.

A survey and economic analysis of the results by Dharmaratne *et al.* (1986) encompassed 5 years of sorghum production by farmers in the Texas Blacklands focusing on the use of resistant hybrids and insecticides to control greenbugs. Texas Blacklands farmers preferred greenbug-resistant sorghums and reported that, without insecticides, there was a net return of approximately US$165/ha using susceptible hybrids without insecticides as against US$200/ha using resistant hybrids. When an average of 1.3 insecticide applications were made, the net returns for susceptible and resistant hybrids were almost the same, US$199 and US$203/ha, respectively. The results demonstrated that the only significant difference in net return occurred between the use of greenbug-susceptible hybrids and all other management strategies. Therefore, farmers would be better off by planting resistant sorghum hybrids and foregoing insecticide treatments. The study also noted that decreasing insecticide use could forestall insecticide resistance, thus preventing increased insecticide use in the future, which would erode net profits. Beneficial effects to the environment through decreased insecticide use were noted, but no monetary value was associated with the effects.

The 'Current Model' and the Future

The best current IPM strategy to control greenbugs on sorghum would be to plant dense stands of resistant hybrids. The plants should receive only the N fertilizer normally used for sorghum production, and water stress should be minimized. *Rhopalosiphum maidis*, predators, and parasitoids should be monitored to determine if sufficient naturally occurring biological control will be present midway through the growing season. Chemical control should not be prophylactic and should not be employed for corn leaf aphid control. If chemical control does become necessary (determined by local-area economic injury levels), a single application of chlorpyrifos plus malathion could be made to prevent yield loss.

Greenbug IPM in sorghum will change over time. Resistant hybrids seem to be the base strategy. However, the history of greenbug biotypes almost guarantees that presently unknown greenbug genotypes will be selected for by the currently available resistant hybrids, and breeding new hybrids will continue. New hybrids will undoubtedly have differing agronomic characteristics that may change planting densities and fertilization rates. We would not expect changes in hybrids to influence the benefits accrued through no-till operations. New hybrids may influence biological control, and monitoring the impact new hybrids may have on the third trophic level is important. In our opinion, it would be of great value if this monitoring were incorporated routinely in developing new hybrids rather than an aspect to be studied only after hybrids are released.

Chemical control will change dramatically. Organophosphate insecticides will disappear over time and replacements utilizing new chemistries will appear. Given continual environmental concerns, it is likely that these new insecticides will be more specific, less toxic to non-target organisms, and have little residual activity. Therefore, when chemical control is selected, timing will be very important.

Conclusions

Although there are limited case studies addressing greenbug IPM for sorghum, there is a rich literature describing numerous strategies that form a productive whole. There is a need for new research in the form of case studies where single and multiple tactic IPM strategies are compared. When such research is attempted, it is crucial to include economic evaluations and farmer input in the planning stage. With the explosive growth of the Internet, and the resultant availability of information, transfer of research results is easier than in the past, and should be utilized. Research scientists must engage extension and consultant personnel to convey information to producers, and actively seek feedback regarding success, failure, and acceptance of recommended strategies.

Executive Summary

In the Great Plains region of the USA, there is currently a fairly complete IPM programme for *S. graminum* on sorghum for farmers who wish to use it, though adaptation by the aphid to pesticides and resistant varieties is bound to lead to future modifications.

Economic thresholds have been established so that prophylactic use of pesticides can be avoided, though the organophosphates – for which such an approach is eminently possible – are likely to be phased out. Farmers are already using the newer neonicotinoids, and here the simplicity of in-furrow and seed treatment is likely to increase the use of prophylaxis.

Indigenous natural enemies can be supported by providing reserve prey in the form of *R. maidis*, which itself is not a pest problem.

Host-plant resistance is the foundation of the IPM programme for greenbug on sorghum, especially when grown at high plant density and irrigation. Cultural control therefore focuses on planting density and water management.

There is no use of semiochemicals for control of aphids.

These approaches exploit interactions between them, particularly between plant resistance and cultural measures. Unfortunately, it appears that aphid-resistant sorghums may have negative effects on coccinellids by reducing their survival and slowing their development.

References

Abate, T., van Huis, A. and Ampofo, J.K.O. (2000) Pest management strategies in traditional agriculture: an African perspective. *Annual Review of Entomology* 45, 631–659.

Anstead, J., Burd, J.D. and Shufran, K.A. (2003) Over-summering and biotypic diversity of *Schizaphis graminum* (Homoptera: Aphididae) populations on non-cultivated grass hosts. *Environmental Entomology* 32, 662–667.

Archer, T.L., Onken, A.B., Matheson, R.L. and Bynum, E.D., Jr. (1982) Nitrogen fertilizer influence on greenbug (Homoptera: Aphididae) dynamics and damage to sorghum. *Journal of Economic Entomology* 75, 695–698.

Archer, T.L., Segarra, E. and Bynum, E.D., Jr. (1999) Greenbug resistance management on sorghum with insecticide mixtures: a biological and economic analysis. *Journal of Economic Entomology* 92, 794–803.

Barbosa, P. (1998) *Conservation Biological Control.* Academic Press, San Diego, 396 pp.

Barbulescu, A. (1964) Research on bionomics and control of aphids injurious to sorghum. *Problems in Agriculture* 16, 29–37.

Bowling, R. and Wilde, G. (1996) Mechanisms of resistance in three sorghum cultivars resistant to greenbug (Homoptera: Aphididae) biotype I. *Journal of Economic Entomology* 89, 558–561.

Bowling, R., Wilde, G., Harvey, T., Sloderbeck, P., Bell, K.O., Morrison, W.P. and Brooks, H.L. (1994) Occurrence of greenbug (Homoptera: Aphididae) biotypes E and I in Kansas, Texas, Nebraska, Colorado, and Oklahoma. *Journal of Economic Entomology* 87, 1696–1700.

Burton, R.L., Jones, O.R., Burd, J.D., Wicks, G.A. and Krenzer, E.G., Jr. (1987) Damage by greenbug (Homoptera: Aphididae) to grain sorghum as affected by tillage, surface residues, and canopy. *Journal of Economic Entomology* 80, 792–798.

Burton, R.L., Burd, J.D., Jones, O.R. and Wicks, G.A. (1990) Crop production strategies for managing greenbug (Homoptera: Aphididae) in sorghum. *Journal of Economic Entomology* 83, 2476–2479.

Buschman, L.L. and DePew, L.J. (1990) Outbreaks of Banks grass mite (Acari: Tetranychidae) in grain sorghum following insecticide applications. *Journal of Economic Entomology* 83, 1570–1574.

Cartwright, B.O., Eikenbary, R.D., Johnson, J.W., Farris, T.N. and Morrison, R.D. (1977) Field release and dispersal of *Menochilus sexmaculatus*, and imported predator of the greenbug, *Schizaphis graminum.* *Environmental Entomology* 6, 699–704.

Cate, J.R., Bottrell, D.G. and Teetes, G.L. (1973a) Management of the greenbug on grain sorghum. 1. Testing foliar treatments of insecticides against greenbugs and corn leaf aphids. *Journal of Economic Entomology* 66, 945–951.

Cate, J.R., Bottrell, D.G. and Teetes, G.L. (1973b) Management of the greenbug on grain sorghum. 2. Testing seed and soil treatments for greenbug and corn leaf aphid control. *Journal of Economic Entomology* 66, 953–959.

DePew, L.J. (1971) Evaluation of foliar and soil treatments for greenbug control on sorghum. *Journal of Economic Entomology* 64, 169–172.

DePew, L.J. (1974) Controlling greenbugs in grain sorghum with foliar and soil insecticides. *Journal of Economic Entomology* 67, 553–555.

Dharmaratne, G.S., Lacewell, R.D., Stoll, J.R. and Teetes, G.L. (1986) Economic impact of greenbug resistant grain sorghum varieties: Texas Blacklands. *Texas Agricultural Experiment Station Miscellaneous Publication* MP-1585, 43 pp.

Dibb, D.W. (1983) Agronomic systems to feed the next generation. *Crops and Soils Magazine* November 1983, 5–6.

Dogramaci, M. (1998) Interaction of *Lysiphlebus testaceipes* (Cresson) and greenbug (*Schizaphis graminum* (Rondani)) on resistant and susceptible sorghums. MSc thesis, University of Nebraska, Lincoln, USA.

FAO (1979) *Production Yearbook, Volume 33.* FAO, Rome, 309 pp.

FAO (2001) *FAOSTAT Database Collections* (www.apps.fao.org/).

Fenton, F.A. and Dahms, R.G. (1951) Attempts at controlling the greenbug by the importation and release of lady beetles in Oklahoma. *Proceedings of the Oklahoma Academy of Science* 32, 1–2.

Fernandes, O.A., Wright, R.J. and Mayo, Z.B. (1998) Parasitism of greenbug (Homoptera: Aphididae) by *Lysiphlebus testaceipes* (Hymenoptera: Braconidae) in grain sorghum: implications for augmentative biological control. *Journal of Economic Entomology* 91, 1315–1319.

FMC (2002) Flonicamid. *FMC Technical Bulletin.* FMC, Philadelphia, 6 pp.

Gilstrap, F.E., Brooks, G.W. and Kring, T.J. (1984) Status of greenbug biological control in Texas sorghum. *Texas Agricultural Experiment Station Miscellaneous Publication* MP-1585, 6 pp.

Hackerott, H.L., Harvey, T.L. and Ross, W.M. (1969) Greenbug resistance in sorghums. *Crop Science* 9, 656–658.

Harvey, T.L. and Hackerott, H.L. (1969) Recognition of a greenbug biotype injurious to sorghum. *Journal of Economic Entomology* 62, 776–779.

Harvey, T.L. and Thompson, C.A. (1988) Effects of sorghum density and resistance on infestations of greenbug, *Schizaphis graminum* (Homoptera: Aphididae). *Journal of the Kansas Entomological Society* 61, 68–71.

Harvey, T.L., Kofoid, K.D., Martin, T.J. and Sloderbeck, P.E. (1991) A new greenbug virulent to E biotype resistant sorghum. *Crop Science* 31, 1689–1691.

Harvey, T.L., Wilde, G.E., Kofoid, K.D. and Bramel-Cox, P.J. (1994) Temperature effects on resistance to greenbug (Homoptera: Aphididae) biotype I in sorghum. *Journal of Economic Entomology* 87, 500–503.

Harvey, T.L., Wilde, G.E. and Kofoid, K.D. (1997) Designation of a new greenbug, biotype K, injurious to resistant sorghum. *Crop Science* 37, 989–991.

Hunter, S.J. (1909) The greenbug and its enemies. *Bulletin of the University of Kansas, No. 9*, 213 pp.

Jackson, H.B., Rogers, C.E. and Eikenbary, R.D. (1971) Colonization and release of *Aphelinus asychis*, an imported parasite of the greenbug. *Journal of Economic Entomology* 64, 1435–1438.

Kindler, S.D. and Staples, R. (1981) *Schizaphis graminum*: effect on grain sorghum exposed to severe drought stress. *Environmental Entomology* 10, 247–248.

Kring, T.J. and Gilstrap, F.E. (1986) Beneficial role of corn leaf aphid, *Rhopalosiphum maidis* (Fitch) (Homoptera: Aphididae), in maintaining *Hippodamia* spp. (Coleoptera: Coccinellidae) in grain sorghum. *Journal of Crop Protection* 5, 125–128.

Kring, T.J., Gilstrap, F.E. and Michels, G.J., Jr. (1985) Role of indigenous coccinellids in regulating greenbugs (Homoptera: Aphididae) on Texas grain sorghum. *Journal of Economic Entomology* 78, 269–273.

Matthee, J.J. (1962) Guard against aphids on kafficorn. *Farming in South Africa* 37, 27–29.

Michels, G.J., Jr. and Bateman, A.C. (1986) Larval biology of two imported predators of the greenbug, *Hippodamia variegata* Goetz and *Adalia flavomaculata* DeGeer under constant temperatures. *Southwestern Entomologist* 11, 23–30.

Michels, G.J., Jr. and Behle, R.W. (1992) Evaluation of sampling methods for lady beetles (Coleoptera: Coccinellidae) in grain sorghum. *Journal of Economic Entomology* 85, 2251–2257.

Michels, G.J., Jr. and Undersander, D.J. (1986) Temporal and spatial distribution of the greenbug (Homoptera: Aphididae) on sorghum in relation to water stress. *Journal of Economic Entomology* 79, 1221–1225.

Michels, G.J., Jr., Rush, C.M., Piccinni, G., Fritts, D.A. and Jones, D. (2002) Effect of irrigation regimes and plant population on greenbug (Homoptera: Aphididae) density in grain sorghum. *Southwestern Entomologist* 27, 135–147.

Morgan, J., Wilde, G. and Johnson, D. (1980) Greenbug resistance in commercial sorghum hybrids in the seedling stage. *Journal of Economic Entomology* 73, 510–514.

Niemczyk, H.D. and Moser, J.R. (1982) Greenbug occurrence and control on turfgrasses in Ohio. In: Niemczyk, H.D. and Joyner, B.G. (eds) *Advances in Turfgrass Entomology*. Hammen Graphics, Piqua, Ohio, pp. 105–111.

Peters, D.C., Wood, E.A., Jr. and Starks, K.J. (1975) Insecticide resistance in selections of the greenbug. *Journal of Economic Entomology* 68, 339–340.

Porter, D.R., Burd, J.D., Shufran, K.A., Webster, J.A. and Teetes, G.L. (1997) Greenbug (Homoptera: Aphididae) biotypes: selected by resistant cultivars or preadapted opportunists? *Journal of Economic Entomology* 90, 1055–1065.

Puterka, G.J. and Peters, D.C. (1995) Genetics of greenbug (Homoptera: Aphididae) virulence to resistance in sorghum. *Journal of Economic Entomology* 88, 421–429.

Rice, M.E. and Wilde, G.E. (1988) Experimental evaluation of predators and parasitoids in suppressing greenbugs (Homoptera: Aphididae) in sorghum and wheat. *Environmental Entomology* 17, 836–841.

Rice, M.E. and Wilde, G.E. (1989) Antibiosis effect of sorghum on the convergent lady beetle (Coleoptera: Coccinellidae), a third-trophic level predator of the greenbug (Homoptera: Aphididae). *Journal of Economic Entomology* 82, 570–573.

Rider, S.D., Dobesh, S.M. and Wilde, G.E. (1998) Insecticide resistance in a novel laboratory-produced greenbug (Homoptera: Aphididae) clone. *Journal of Economic Entomology* 91, 30–33.

Schweissing, F.C. and Wilde, G. (1979) Temperature and plant nutrient effects on resistance of seedling sorghum to the greenbug. *Journal of Economic Entomology* 72, 20–23.

Shufran, R.A., Wilde, G.E. and Sloderbeck, P.E. (1996) Descriptions of three isozyme polymorphisms associated with insecticide resistance in greenbug (Homoptera: Aphididae) populations. *Journal of Economic Entomology* 89, 46–50.

Shufran, R.A., Wilde, G.E. and Sloderbeck, P.E. (1997a) Response of three greenbug (Homoptera: Aphididae) strains to five organophosphorus and two carbamate insecticides. *Journal of Economic Entomology* 90, 283–286.

Shufran, R.A., Wilde, G.E. and Sloderbeck, P.E. (1997b) Life history of insecticide resistant and susceptible greenbug (Homoptera: Aphididae) strains. *Journal of Economic Entomology* 90, 1577–1583.

Sloderbeck, P.E., Chowdhury, M.A., DePew, L.J. and Buschman, L.L. (1991) Greenbug (Homoptera: Aphididae) resistance to parathion and chlorpyrifos-methyl. *Journal of the Kansas Entomological Society* 64, 1–4.

Starks, K.J., Muniappan, R. and Eikenbary, R.D. (1972) Interaction between plant resistance and parasitism against the greenbug on barley and sorghum. *Annals of the Entomological Society of America* 65, 650–655.

Stone, B.S., Shufran, R.A. and Wilde, G.E. (2000) Life history study of multiple clones of insecticide resistant and susceptible greenbug *Schizaphis graminum* (Homoptera: Aphididae). *Journal of Economic Entomology* 93, 971–974.

Teetes, G.L. and Johnson, J.W. (1974) Assessment of damage by the greenbug in grain sorghum hybrids of different maturities. *Journal of Economic Entomology* 67, 514–516.

Teetes, G.L., Schaefer, C.A., Gipson, J.R., McIntyre, R.C. and Latham, E.E. (1975) Greenbug resistance to organophosphorus insecticides on the Texas High Plains. *Journal of Economic Entomology* 68, 214–216.

Thindwa, H.P. and Teetes, G.L. (1994) Effect of temperature and photoperiod on sorghum resistance to biotype C and E greenbug (Homoptera: Aphididae). *Journal of Economic Entomology* 87, 1366–1372.

USDA (1968) Greenbug (*Schizaphis graminum*). *Plant Pest Control Division Cooperative Insect Report, No. 18,* 781 pp.

USDA (1981) Components for management of field corn, and grain sorghum insects and mites in the United States. *USDA-ARS ARM-S-18.* USDA-ARS, New Orleans, 18 pp.

Webster, J.A., Amosson, S., Brooks, L., Hein, G., Johnson, G., Legg, D., Massey, B., Morrison, P., Peairs, F. and Weiss, M. (1995) Economic impact of the greenbug in the western United States: 1992–1993. *Great Plains Council Publication, No. 155,* 16 pp.

Wilde, G.E. and Ohiagu, C. (1976) Relation of corn leaf aphid to sorghum yield. *Journal of Economic Entomology* 69, 195–197.

Wilde, G.E. and Tuinstra, M.R. (2000) Greenbug (Homoptera: Aphididae) resistance in sorghum: characterization of KS 97. *Journal of Agricultural and Urban Entomology* 17, 15–19.

Wood, E.A. (1971) Designation and reaction of three biotypes of the greenbug cultures on resistant and susceptible species of sorghum. *Journal of Economic Entomology* 64, 183–185.

Young, W.R. and Teetes, G.L. (1977) Sorghum entomology. *Annual Review of Entomology* 22, 193–218.

28 IPM Case Studies: Cucurbits

Susan E. Webb

Entomology and Nematology Department, University of Florida, Gainesville, FL 32611, USA

Introduction

Cucurbits (family *Cucurbitaceae*) were among the earliest cultivated plants, dating back to 10,000 BC or earlier (Robinson and Decker-Walters, 1997). Today, watermelon (*Citrullus lanatus*), cucumber (*Cucumis sativus*), melon (*Cucumis melo*, including muskmelon, cantaloupe, and honeydew melon), squash and pumpkin (*Cucurbita pepo*, *Cucurbita moschata*, and *Cucurbita maxima*) are the most commonly grown cucurbits on a worldwide basis (FAO, 2001) (Fig. 28.1).

Aphids attack cucurbits wherever they are grown. *Aphis gossypii* (melon aphid) is by far the most important of the direct pests, but *Myzus persicae* (peach–potato aphid) and *Aphis craccivora* (cowpea aphid) also attack cucurbits. Leaves of watermelon, melons, and cucumber become crumpled and distorted when melon aphid feeds on them (Goff and Tissot, 1932) and heavy infestations can result in yield losses (Webb, 1996). Sooty mould develops on leaves and fruit coated with honeydew excreted by aphids. For cucurbit crops throughout the world (Dahal *et al.*, 1997; Luis-Arteaga *et al.*, 1998; Abou-Jawdah *et al.*, 2000; Sevik and Arli-Sokmen, 2003), aphids are most damaging in their role as virus vectors.

There are relatively few published IPM programmes for cucurbits. Multiple tactics are generally recommended for control of aphids and the viruses they transmit. This chapter will focus primarily on aphid pest management in the USA, but will also refer to strategies used in other countries, particularly for management of aphids on cucumbers under protected cultivation.

Aphid-Vectored Viruses Affecting Cucurbits

Lecoq *et al.* (1998) list ten major aphid-transmitted viruses that affect cucurbits. All but two, the non-persistently transmitted *Cucumber mosaic virus* (CMV) (genus *Cucumovirus*) and *Cucurbit aphid-borne yellows virus* (CABYV), a persistently transmitted luteovirus, are potyviruses. Of the potyviruses, the watermelon strain of *Papaya ringspot virus* (PRSV-W), *Zucchini yellow mosaic virus* (ZYMV) (Fig. 28.2), and *Watermelon mosaic virus* (WMV) or *Watermelon mosaic virus 2* (WMV-2) are the most common on a worldwide basis. PRSV-W is found in subtropical and tropical regions and WMV is more often found in temperate and Mediterranean climates (as is CMV). Other potyviruses cause economic damage on a more regional basis, such as *Moroccan watermelon mosaic virus* in Africa and the Mediterranean basin, and *Zucchini yellow*

Fig. 28.1.   Field of young watermelon (*Citrullus lanatus*) plants in north central Florida, USA.

Fig. 28.2.   *Zucchini yellow mosaic virus* symptoms in cantaloupe (*Cucumis melo*).

fleck virus in the Mediterranean basin. More aphid-transmitted potyviruses of unknown economic importance continue to be found and described wherever cucurbits are grown; for example, *Watermelon leaf mottle virus* (Purcifull *et al.*, 1998) in Florida. All are non-persistently transmitted.

Many species of aphids that do not reproduce on cucurbits have been found capable of transmitting the above viruses. The important vector species will vary from region to region (Adlerz, 1987; Castle *et al.*, 1992; Basky *et al.*, 2001). Understanding which aphids are important in virus spread in a particular region may help in management decisions, such as deciding when to apply oils or whether to remove aphid host plants around the crop.

Management Options for Aphids and Aphid-Vectored Viruses in Cucurbits

Chemical control

Aphis gossypii has become resistant to organophosphate, carbamate, organochlorine (O'Brien *et al.*, 1992), and pyrethroid insecticides (Sun *et al.*, 1994) in various parts of the world, particularly in cotton, but also in field-grown (Hollingsworth *et al.*, 1994) and glasshouse cucurbits. In field-grown cucurbits in Hawaii, resistance levels can vary from one field to another, depending on the history of insecticide use (against all insect pests) in a particular field (Hollingsworth *et al.*, 1994). Decreasing insecticide use helps reduce resistance, and furthermore the use of pirimicarb can actually stimulate reproduction in populations of melon aphid that are resistant to this carbamate (Rongai and Cerato, 1996).

Where resistance to insecticides is not a problem, melon aphid has been relatively easy to manage as a direct pest of cucurbits. Newer insecticides, such as the neonicotinoids (imidacloprid, thiamethoxam, and dinotefuran are registered for use on cucurbits in the USA) and pymetrozine give long-lasting control, resulting in increased yields (e.g. Webb, 1996) (Fig. 28.3). So far, no stable resistance to imidacloprid has been found in melon aphid (Nauen and Elbert, 2003) and no information exists on resistance to pymetrozine. In the USA, neonicotinoids are used mainly to control *Bemisia tabaci* (sweet potato whitefly), with aphid control a secondary benefit. No experimental evidence exists for reduction of secondary spread of non-persistent viruses in cucurbits from the use of these insecticides. In general, little effect on the spread of these viruses has been found for any of the older insecticides (Loebenstein and Raccah, 1980) and some evidence exists that residues increase virus spread in cucurbits, presumably by causing increased movement of aphids (Ullman *et al.*, 1991; Webb and Linda, 1993). Bees are essential for pollination of cucurbits (except for parthenocarpic varieties of cucumber grown under protected cultivation) and their preservation must be considered in any pest management programme that makes use of insecticides.

Oils

In Florida, Georgia, and South Carolina, where much of the spread of aphid-transmitted viruses in cucurbits is within the field (secondary spread), the use of a highly refined mineral oil (JMS Stylet Oil, Vero Beach, Florida) is recommended to reduce spread of virus in squash (Kucharek and Purcifull, 1997; Smith, 2001; Boyhan, 2004), although it does not control aphids developing on the plants (Webb and Linda, 1993). Mineral oil has been shown to interfere with retention of virus in the aphid's stylets (Wang and Pirone, 1996). Finding additional chemicals that interfere with transmission would be of great benefit.

Sampling and thresholds

There is little information available about the actual use of specific sampling schemes and thresholds for aphids on field-grown cucurbits. The New York State IPM Program publishes scouting procedures for cucurbits (Zitter *et al.*, 2000). Sampling patterns such as a V, W, or X are suggested, with a choice of five sites, varied from week to week, along

(a)   (b)

Fig. 28.3.   (a) Cantaloupe, 'Athena', treated with pymetrozine; (b) untreated cantaloupe showing severe crumpling of leaves due to a heavy infestation of melon aphid. 1996 field trial at University of Florida, IFAS, Research and Education Center, Leesburg.

the path chosen through the field. Early in the season, five whole plants are examined at each site. As vine crops grow and it becomes impossible to distinguish separate plants, ten leaves and five fruits are examined in areas 3.1 m × 3.1 m (= 10 ft × 10 ft). For aphids, the number of leaves with five or more aphids is recorded and if more than 20% of the leaves have aphids, an insecticide treatment may be needed. If many aphids are parasitized or infected with a fungal pathogen, treatment may not be necessary.

Lam and Foster (2002) recommend flagging areas of the field where natural enemies are present in aphid colonies and returning in several days to see if the population of aphids is increasing or decreasing before making a decision to use an insecticide.

In Hawaii (Johnson *et al.*, 1989), density treatment levels (DTLs) were guessed for watermelon, based on type of injury, potential for aphid population increase, and the grower's judgement regarding the amount of damage that could be tolerated. The DTL for *A. gossypii* was set at 20 aphids per leaf in the absence of viruses, and zero aphids per leaf in their presence. The lower threshold was not practical, however, and later work in Hawaii showed that it was rarely effective for reducing virus spread (Ullman *et al.*, 1991).

Under protected cultivation, growers use yellow sticky traps to monitor for melon aphid. Action must be taken as soon as aphids are found because of their reproductive capacity. Within a week, populations can increase as much as 10- or 12-fold on glasshouse-grown cucumber (Wyatt and Brown, 1977).

Biological control

Under field conditions, naturally occurring predaceous coccinellid beetles, hover fly larvae, and braconid wasp parasitoids reduce the number of aphids on cucurbit crops (summarized by Capinera, 2001 for the USA). Most recommendations point out the need to conserve these natural enemies. With more specific insecticides available for managing other pests (USA), this is currently feasible.

The introduction of insect predators and parasitoids to control aphids is limited to crops grown under protected cultivation and is most common in The Netherlands and British Columbia (Ramakers and Rabasse, 1995; Shipp, 2004). The braconid *Aphidius colemani* is used to control melon aphid (Shipp, 2004; Tatchell, Chapter 24 this volume). It can be introduced by weekly releases of parasitized aphid mummies or reared in a banker plant system, using *Rhopalosiphum padi* (bird cherry–oat aphid) on cereal plants. The predatory midge *Aphidoletes aphidimyza* is also recommended for aphid control in cucumber (Shipp, 2004).

Host-plant resistance

Resistant and tolerant varieties can provide excellent control of aphid-vectored viruses (e.g. Clough and Hamm, 1995; Tricoli *et al.*, 1995; Webb and Tyson, 1997; Walters, 2004). Zitter (2002) has listed the commercially available cucumber, zucchini, and yellow summer squash varieties that have resistance or tolerance to one or more viruses, including genetically modified varieties that contain the coat protein genes of one or more viruses (Fuchs and Gonsalves, 1995; Tricoli *et al.*, 1995). A new cantaloupe variety, 'Hannah's Choice', developed in the USA by M. Jahn, a plant breeder at Cornell University, has resistance to WMV, PRSV-W, and ZYMV (Gordon, 2004). Recently, Harris Moran released the first pumpkin (*Cucurbita pepo*) variety, 'Magician F1', with tolerance to ZYMV (Anonymous, 2004). In Australia, a 'Jarrahdale' type pumpkin (*C. maxima*) has been released that is highly resistant to ZYMV, PRSV-W, and WMV (Herrington *et al.*, 2004). At the present time, there are no commercially available virus-resistant or virus-tolerant varieties of watermelon in the USA.

Resistance to *A. gossypii* and its transmission of viruses has been identified in muskmelon germplasm from India (Kishaba *et al.*, 1971) and Korea (Pitrat and Lecoq, 1980). However, examples of the practical use of this resistance are lacking. In Bangladesh, local genotypes of ash gourd (*Benincasa hispida*), also known as wax gourd or winter melon, are relatively resistant to *A. gossypii* (Khan *et al.*, 2000). The density of trichomes on leaves was negatively correlated with the number of aphids per leaf.

Cross-protection

Cross-protection, the inoculation of plants with a mild strain of a virus to protect against a severe strain (Fletcher, 1978; Fulton, 1986), is another strategy that has been used with some success to control ZYMV in cucurbits (Lecoq *et al.*, 1991; Wang *et al.*, 1991; Perring *et al.*, 1995). In Israel, since 1996, over 2500 ha of cucurbits have been protected from infection with ZYMV by inoculation with a mild strain (Yarden *et al.*, 2000). Rezende and Pacheco (1998) reported control of PRSV-W in zucchini by cross-protection in Brazil, and developed practical methods that growers could use for inoculating plants. The disadvantages of cross-protection include potential mutation of the mild strain of the virus to a severe form and possible synergistic effects with other viruses (Fulton, 1986).

Cultural control

The role of weeds in the control of aphid-vectored viruses

Two cucurbitaceous weeds have been shown to be sources of PRSV-W in south Florida. Creeping cucumber (*Melothria pendula*) is often heavily infected with PRSV-W. A perennial, it grows back quickly after relatively brief periods of freezing weather and is present as a source of virus for aphids as soon as crops are planted (Adlerz, 1969, 1972a, 1974). A second wild cucurbit, balsam apple (*Momordica charantia*), also serves as a source of PRSV-W in south Florida (Adlerz, 1972b). The efforts of growers to remove either *M. pendula* or *M. charantia*, however, have not been effective at reducing incidence of the virus on the cultivated crop (Adlerz, 1981).

In contrast to the importance of weed virus sources in Florida, it is the crops in a desert region in Israel that are the major source of aphids and the viruses they transmit (Ucko *et al.*, 1998). Regulations in the Arava region stipulate the removal of vegetable crops and all weeds growing in vegetable crop fields at the end of the growing season. All perennial crop-growing areas must also be kept free of weeds during a one-month crop-free or phytosanitary period. Instituting this one-month phytosanitary period during the latter half of June and the beginning of July eliminates virus epidemics in melons and other vegetables in the Arava region.

Mulches

Work by Smith *et al.* (1964) and Kring (1964) first demonstrated the repellent effects of reflective aluminium. Reflective mulches have been improved in the intervening years and are more widely available to growers. New studies (Stapleton and Summers, 2002) continue to show the benefits of reflective mulches for reducing early season virus spread, and their use is being recommended, except in situations where the use of reflective mulch would delay warming of the soil in spring.

Floating row covers

Floating row covers act as a physical barrier to aphids, both those that colonize cucurbits and those that land temporarily and transmit virus. Lightweight covers can be left on squash and cantaloupe plants until flowering, when they must be removed for pollination (Natwick *et al.*, 1988; Perring *et al.*, 1989; Webb and Linda, 1992), and have the added advantage of excluding all insects, not just viruliferous aphids (Fig. 28.4). Heavier row covers, used for frost protection, allow earlier planting in temperate climates. Earlier planting and delaying the removal of covers may help the crop escape infection

with aphid-transmitted viruses. However, cost and disposal problems may limit the use of covers in commercial plantings to situations in which cold protection makes possible earlier harvest and higher prices for the crop.

Living mulches and intercropping

Aphids may be less likely to land on a crop grown with a living mulch because the mulch eliminates the contrast with bare ground that aphids respond to when locating host plants (Kennedy *et al.*, 1959, 1961). Living mulches (yellow mustard – *Sinapis alba*, or buckwheat – *Fagopyrum esculentum*) have been tried in Hawaii with courgette (a *Cucurbita pepo*) (Hooks *et al.*, 1998), with the result that aphids, both apterae and alatae, were greatly reduced and infection with PRSV-W delayed in mulched plots.

Increased plant diversity may increase the abundance of insect predators and parasitoids (Andow, 1991). Marcovitch (1935) found that the number of melon aphids was much reduced on watermelon when it was grown in a mixed planting with turnips. Turnips were planted a month before the watermelon and became heavily infested with *Lipaphis pseudobrassicae* (mustard aphid). Natural enemies

Fig. 28.4. Floating row covers over courgette (*Cucurbita pepo*), field trial at University of Florida, IFAS, Research and Education Center, Leesburg.

became very numerous on the turnips and presumably kept melon aphids from becoming abundant enough to damage the watermelon. *Aphis gossypii* severely affected a solid planting of watermelon, only 110 m away.

Although research continues (Frank and Liburd, 2005), the use of living mulches and intercropping has been difficult to implement in commercial cucurbit production.

Sanitation and removal of infected plants

Additional cultural recommendations for managing aphid-vectored viruses include destroying plants showing symptoms in the early stages of an epidemic (Blancard *et al.*, 1994), ploughing-in the crop immediately after harvest, removing volunteer cucurbits in the vicinity of production fields, and avoiding sequential plantings when virus is a problem (Kucharek and Purcifull, 1997).

For glasshouse crops, recommended practices include screening vents and keeping nearby areas free of weeds (Gilkeson, 1994).

IPM Programmes for Cucurbits that Include Aphid Management

In the USA, recommendations to growers for managing aphid-borne viruses include multiple tactics, either as a list of possible measures or a specific combination. Smith (2001) recommended that South Carolina squash growers used a combination of virus-resistant varieties (depending on what their buyers would accept), reflective mulch for early season control, and application of highly refined mineral oils to interfere with transmission. In Arizona and southern California (Palumbo and Kerns, 1996), both *M. persicae* and *A. gossypii* attack cucurbits. Recommendations include eliminating alternate host plants such as mustards for *M. persicae*, using row covers over the seedbed until first bloom, and using reflective mulches. Monitoring flight activity and aphid species composition with yellow sticky traps is recommended. It is suggested that melons planted in January or February, before peak aphid flights occur in California, be treated at planting with a soil-applied systemic insecticide such as imidacloprid or thiamethoxam. The same tactics are suggested for *A. gossypii*. Conservation of naturally occurring predators and parasitoids may keep this aphid under control if foliar insecticides are avoided.

The University of California cucurbit IPM guidelines for insect pests (Coviello *et al.*, 2005) provide no threshold for aphids, but note that naturally occurring predators and parasitoids can limit populations. Reflective mulches, row covers, and weed control are suggested. Growers are encouraged to plant sweet alyssum (*Lobularia maritima*) around the field to provide a nectar source for parasitoids. Delaying planting until the spring aphid flight is over is another option. Growers are cautioned not to apply excess nitrogen fertilizer.

Shipp (2004) outlines an IPM programme for cucumber grown in protected culture. Components of this programme include monitoring with yellow sticky traps, releases or banker plant production of *A. colemani*, physical exclusion of pests, sanitation, and selective use of pesticides when pests increase too quickly for biological control to work. In The Netherlands, practices have been developed that allow growers to use an IPM label for their crop (Ramakers and Rabasse, 1995). Specialized advisors are seen as essential for the success of biologically based IPM programmes in protected cultivation (Shipp, 2004).

Executive Summary

Especially where virus transmission is the major problem of aphid presence, chemical control with newer products such as imidacloprid is normal. Thresholds for aphids are available in many situations, though the threshold varies from place to place. Monitoring with sticky traps is common practice in protected cultivation.

Biological control is the main form of aphid control on cucurbits in glasshouses where *A. gossypii* has frequently developed resistance to insecticides. There is no release of natural enemies on field crops, though many growers aim to conserve local natural enemies from damage from foliar sprays as far as possible.

Resistant varieties are available, though usually as tolerance to virus rather than resistance to aphids. In the USA, zucchini varieties with virus resistance transferred by traditional plant breeding are gaining in acceptance. However, in other parts of the world, aphid- or virus-resistant varieties are little used, as susceptible varieties have more desirable agronomic properties.

Cultural control involves the use of reflective mulches and crop covers, with some growers promoting biological control by providing nectar plants. Removal of weed sources of cucurbit viruses is also recommended. There are many other cultural measures available for inclusion in IPM programmes, such as delaying planting and limiting nitrogen applications.

There is no use of semiochemicals for control of aphids.

Several IPM programmes are available to growers. They involve a combination of control methods, though such combinations do not seem to be designed intentionally to exploit any synergism between them.

References

Abou-Jawdah, Y., Sobh, H., El-Zammar, S., Fayyad, A. and Lecoq, H. (2000) Incidence and management of virus diseases of cucurbits in Lebanon. *Crop Protection* 19, 217–224.

Adlerz, W.C. (1969) Distribution of watermelon mosaic viruses 1 and 2 in Florida. *Proceedings of the Florida State Horticultural Society* 81, 160–165.

Adlerz, W.C. (1972a) *Melothria pendula* plants infected with watermelon mosaic virus 1 as a source of inoculum for cucurbits in Collier County, Florida. *Journal of Economic Entomology* 65, 1303–1306.

Adlerz, W.C. (1972b) *Momordica charantia* as a source of watermelon mosaic virus 1 for cucurbit crops in Palm Beach County, Florida. *Plant Disease Reporter* 56, 563–564.

Adlerz, W.C. (1974) Wind effects on spread of watermelon mosaic virus 1 from local virus sources to watermelon. *Journal of Economic Entomology* 67, 361–364.

Adlerz, W.C. (1981) Weed hosts of aphid-borne viruses of vegetable crops in Florida. In: Thresh, J.M. (ed.) *Pests, Pathogens and Vegetation*. Pitman, London, pp. 467–478.

Adlerz, W.C. (1987) Cucurbit potyvirus transmission by alate aphids (Homoptera: Aphididae) trapped alive. *Journal of Economic Entomology* 80, 87–92.

Andow, D.A. (1991) Vegetational diversity and arthropod population response. *Annual Review of Entomology* 36, 561–586.

Anonymous (2004) Magician F1 (HMX) 0683. *Harris Seeds – Product Detail*. Harris Seeds, Rochester, New York (www.growers.harrisseeds.com/cart/detail.asp?product_id=1678).

Basky, Z., Perring, T.M. and Tóbiás, I. (2001) Spread of zucchini yellow mosaic potyvirus in squash in Hungary. *Journal of Applied Entomology* 125, 271–275.

Blancard, D., Lecoq, H. and Pitrat, M. (1994) *A Colour Atlas of Cucurbit Diseases: Observation, Identification, and Control*. Wiley, New York, 299 pp.

Boyhan, G. (2004) Dealing with disease. *American Vegetable Grower* 52 (6), 16–17.

Capinera, J.L. (2001) *Handbook of Vegetable Pests*. Academic Press, San Diego, 729 pp.

Castle, S.J., Perring, T.M., Farrar, C.A. and Kishaba, A.N. (1992) Field and laboratory transmission of watermelon mosaic virus 2 and zucchini yellow mosaic virus by various aphid species. *Phytopathology* 82, 235–240.

Clough, G.H. and Hamm, P.B. (1995) Coat protein transgenic resistance to watermelon mosaic and zucchini yellows mosaic virus in squash and cantaloupe. *Plant Disease* 79, 1107–1109.

Coviello, R.L., Natwick, E.T., Godfrey, L.D., Fouche, C.B., Summers, C.G. and Stapleton, J.J. (2005) Insects and mites. In: Ohlendorf, B. and Flint, M.L. (eds) *UC IPM Pest Management Guidelines: Cucurbits, Publication No. 3445*. University of California, Davis, pp. 4–42.

Dahal, G., Lecoq, H. and Albrechtsen, S.E. (1997) Occurrence of papaya ringspot potyvirus and cucurbit viruses in Nepal. *Annals of Applied Biology* 130, 491–502.

FAO (2001) *Production Yearbook, Volume 53 (1999)*. FAO, Rome, 328 pp.

Fletcher, J.T. (1978) The use of avirulent virus strains to protect plants against the effects of virulent strains. *Annals of Applied Biology* 89, 110–114.

Frank, D.L. and Liburd, O.E. (2005) Effects of living and synthetic mulch on the population dynamics of whiteflies and aphids, their associated natural enemies, and insect-transmitted plant diseases in zucchini. *Environmental Entomology* 34, 857–865.

Fuchs, M. and Gonsalves, D. (1995) Resistance of transgenic hybrid squash ZW-20 expressing the coat protein genes of zucchini yellow mosaic virus and watermelon mosaic virus 2 to mixed infections by both potyviruses. *Bio/Technology* 13, 1466–1473.

Fulton, R.W. (1986) Practices and precautions in the use of cross protection for plant virus disease control. *Annual Review of Phytopathology* 24, 67–81.

Gilkeson, L.A. (1994) Greenhouse cucumber: melon (cotton) aphid. In: Howard, R.J., Garland, J.A. and Seaman, W.L. (eds) *Diseases and Pests of Vegetable Crops in Canada*. Canadian Phytopathology Society and Entomological Society of Canada, Ottawa, pp. 321–322.

Goff, C.C. and Tissot, A.N. (1932) The melon aphid. *Bulletin of the University of Florida Agricultural Experiment Station, Gainesville, Florida, No. 252*, 23 pp.

Gordon, R.O. (2004) Breeding for quality. *American Vegetable Grower* 52 (6), 18–19.

Herrington, M., Wright, R., Walker, I., Prytz, S. and Persley, D. (2004) *New Virus Resistant Pumpkin Variety: 'Dulong QHI'*. Department of Primary Industries, Queensland (www.dpi.qld.gov.au/horticulture/5281.html).

Hollingsworth, R.G., Tabashnik, B.E., Ullman, D.E., Johnson, M.W. and Messing, R. (1994) Resistance of *Aphis gossypii* (Homoptera: Aphididae) to insecticides in Hawaii: spatial patterns and relation to insecticide use. *Journal of Economic Entomology* 87, 293–300.

Hooks, C.R.R., Valenzuela, H.R. and DeFrank, J. (1998) Incidence of pests and arthropod natural enemies in zucchini grown with living mulches. *Agriculture, Ecosystems and Environment* 69, 217–231.

Johnson, M.W., Mau, R.F.L., Martinez, A.P. and Fukuda, S. (1989) Foliar pests of watermelon in Hawaii. *Tropical Pest Management* 35, 90–96.

Kennedy, J.S., Booth, C.O. and Kershaw, W.J.S. (1959) Host finding by aphids in the field. II. *Aphis fabae* Scop. (gynoparae) and *Brevicoryne brassicae* L.; with a reappraisal of the role of host-finding behaviour in virus spread. *Annals of Applied Biology* 47, 424–444.

Kennedy, J.S., Booth, C.O. and Kershaw, W.J.S. (1961) Host finding by aphids in the field. III. Visual attraction. *Annals of Applied Biology* 49, 1–21.

Khan, M.M.H., Kundu, R. and Alam, M.Z. (2000) Impact of trichome density on the infestation of *Aphis gossypii* Glover and incidence of virus disease in ashgourd [*Benincasa hispida* (Thunb.)]. *International Journal of Pest Management* 46, 201–204.

Kishaba, A.N., Bohn, G.W. and Toba, H.H. (1971) Resistance to *Aphis gossypii* in muskmelon. *Journal of Economic Entomology* 64, 935–937.

Kring, J.B. (1964) New ways to repel aphids. *Frontiers of Plant Science* 17, 6–7.

Kucharek, T. and Purcifull, D. (1997) Aphid-transmitted viruses of cucurbits in Florida. *Florida Cooperative Extension Service Circular, No. 1184*. University of Florida, Gainesville, 11 pp.

Lam, R. and Foster, R. (2002) Vegetables: cucurbit insect management. *Department of Entomology, Purdue University Cooperative Extension Service Publication E-30-W*. West Lafayette, Indiana, 6 pp.

Lecoq, H., Lemaire, J.M. and Wipf-Scheibel, C. (1991) Control of zucchini yellow mosaic virus in squash by cross protection. *Plant Disease* 75, 208–211.

Lecoq, H., Wisler, G. and Pitrat, M. (1998) Cucurbit viruses: the classic and the emerging. In: McCreight, J.D. (ed.) *Cucurbitaceae '98: Evaluation and Enhancement of Cucurbit Germplasm*. ASHS Press, Alexandria, Virginia, pp. 126–142.

Loebenstein, G. and Raccah, B. (1980) Control of non-persistently transmitted aphid-borne viruses. *Phytoparasitica* 8, 221–235.

Luis-Arteaga, M., Alvarez, J.M., Alonso-Prados, J.L., Bernal, J.J., García-Arenal, F., Laviña, A., Batlle, A. and Moriones, E. (1998) Occurrence, distribution, and relative incidence of mosaic viruses infecting field-grown melon in Spain. *Plant Disease* 82, 979–982.

Marcovitch, S. (1935) Control of the melon louse by intercropping. *University of Tennessee Agricultural Experiment Station Circular, No. 55*, 4 pp.

Natwick, E., Durazo 3rd, A. and Laemmlen, F. (1988) Direct row covers for insects and virus diseases protection in desert agriculture. *Plasticulture* 78, 35–46.

Nauen, R. and Elbert, A. (2003) European monitoring of resistance to insecticides in *Myzus persicae* and *Aphis gossypii* (Hemiptera: Aphididae) with special reference to imidacloprid. *Bulletin of Entomological Research* 93, 47–54.

O'Brien, P.J., Abdel-Aal, Y.A., Ottea, J.A. and Graves, J.B. (1992) Relationship of insecticide resistance to carboxylesterases in *Aphis gossypii* (Homoptera: Aphididae) from midsouth cotton. *Journal of Economic Entomology* 85, 651–657.

Palumbo, J.C. and Kerns, D.L. (1996) Melon IPM: Southwestern USA. In: Radcliffe, E.B. and Hutchison, W.D. (eds) *Radcliffe's IPM World Textbook* (online). University of Minnesota, St Paul (www.ipmworld.umn.edu).

Perring, T.M., Royalty, R.N. and Farrar, C.A. (1989) Floating row covers for the exclusion of virus vectors and the effect on disease incidence and yield of cantaloupe. *Journal of Economic Entomology* 83, 1709–1715.

Perring, T.M., Farrar, C.A., Blua, M.J., Wang, H.L. and Gonsalves, D. (1995) Cross protection of cantaloupe with a mild strain of zucchini yellow mosaic virus: effectiveness and application. *Crop Protection* 14, 601–606.

Pitrat, M. and Lecoq, H. (1980) Inheritance of resistance to cucumber mosaic virus transmission by *Aphis gossypii* in *Cucumis melo*. *Phytopathology* 70, 958–961.

Purcifull, D.E., Hiebert, E., Petersen, M.A., Simone, G.W., Kucharek, T.A., Gooch, M.D., Crawford, W.E., Beckham, K.A. and De Sa, P.B. (1998) Partial characterization of a distinct potyvirus isolated from watermelon in Florida. *Plant Disease* 82, 1386–1390.

Ramakers, P.M.J. and Rabasse, J.-M. (1995) Integrated pest management in protected cultivation. In: Reuveni, R. (ed.) *Novel Approaches to Integrated Pest Management*. CRC, Boca Raton, pp. 199–229.

Rezende, J.A.M. and Pacheco, D.A. (1998) Control of papaya ringspot virus–type W in zucchini squash by cross-protection in Brazil. *Plant Disease* 82, 171–175.

Robinson, R.W. and Decker-Walters, D.S. (1997) *Cucurbits*. CAB International, Wallingford, Oxon, UK, 226 pp.

Rongai, D. and Cerato, C. (1996) Insecticide-stimulated reproduction of cotton aphid, *Aphis gossypii* Glover, resistant to pirimicarb. *Resistant Pest Management Newsletter No. 8* (online). Istituto Sperimentale per le Colture Industriali, Bologna (www.msstate.edu/Entomology/v8n2/art03.html).

Sevik, M.A. and Arli-Sokmen, M. (2003) Viruses infecting cucurbits in Samsun, Turkey. *Plant Disease* 87, 341–344.

Shipp, J.L. (2004) IPM program for cucumber. In: Heinz, K.M., Van Driesche, R.G. and Parrella, M.P. (eds) *Biocontrol in Protected Culture*. Ball Publishing, Batavia, Illinois, pp. 419–437.

Smith, F.F., Johnson, G.V., Kahn, R.P. and Bing, G. (1964) Repellency of reflective aluminium to transient aphid virus-vectors. *Phytopathology* 54, 748.

Smith, P. (2001) Producing summer squash during the virus season. *Vegetable Production Fact Sheet of the Clemson University Cooperative Extension Service*. Clemson, South Carolina, 2 pp.

Stapleton, J.J. and Summers, C.G. (2002) Reflective mulches for management of aphids and aphid-borne virus diseases in late-season cantaloupe (*Cucumis melo* L. var. *cantalupensis*). *Crop Protection* 21, 891–898.

Sun, Y.Q., Feng, G.L., Yuan, J.G. and Gong, K.Y. (1994) Insecticide resistance of cotton aphid in North China. *Entomologica Sinica* 1, 242–250.

Tricoli, D.M., Carney, K.J., Russell, P.F., McMaster, J.R., Groff, D.W., Hadden, K.C., Himmel, P.T., Hubbard, J.T., Boeshore, M.L. and Quemada, H.D. (1995) Field evaluation of transgenic squash containing single or multiple coat protein gene constructs for resistance to cucumber mosaic virus, watermelon mosaic virus 2, and zucchini yellow mosaic virus. *Bio/Technology* 13, 1458–1465.

Ucko, O., Cohen, S. and Ben-Joseph, R. (1998) Prevention of virus epidemics by a crop-free period in the Arava region of Israel. *Phytoparasitica* 26, 313–321.

Ullman, D.E., Cho, J.J. and Ebesu, R.H. (1991) Strategies for limiting the spread of aphid-transmitted viruses in zucchini. In: Johnson, M.W., Ullman, D.E. and Vargo, A. (eds) *Proceeding of a Conference on Agricultural Development in the American Pacific – Crop Protection, Honolulu, May 1989, University of Hawaii Research Extension Series No. 134*, pp. 3–5.

Walters, S.A. (2004) Influence of watermelon mosaic virus on slicing cucumber farm gate revenues. *HortTechnology* 14, 144–148.

Wang, H.L., Gonsalves, D., Provvidenti, R. and Lecoq, H.L. (1991) Effectiveness of cross protection by a mild strain of zucchini yellow mosaic virus in cucumber, melon, and squash. *Plant Disease* 75, 203–207.

Wang, R.Y. and Pirone, T.P. (1996) Mineral oil interferes with retention of tobacco etch virus potyvirus in the stylets of *Myzus persicae*. *Phytopathology* 86, 820–823.

Webb, S.E. (1996) Management of melon aphid on muskmelon and watermelon with insecticides specific for Homoptera. *Proceedings of the Florida State Horticultural Society* 109, 202–205.

Webb, S.E. and Linda, S.B. (1992) Evaluation of spunbonded polyethylene row covers as a method of excluding insects and viruses affecting fall-grown squash in Florida. *Journal of Economic Entomology* 85, 2344–2352.

Webb, S.E. and Linda, S.B. (1993) Effect of oil and insecticide on epidemics of potyviruses in watermelon in Florida. *Plant Disease* 77, 869–874.

Webb, S.E. and Tyson, R.V. (1997) Evaluation of virus-resistant squash varieties. *Proceedings of the Florida State Horticultural Society* 110, 299–302.

Wyatt, I.J. and Brown, S.J. (1977) The influence of light intensity, day length and temperature on increase rates of four glasshouse aphids. *Journal of Applied Ecology* 14, 391–399.

Yarden, G., Hemo, R., Livne, H., Maoz, E., Lev, E., Lecoq, H. and Raccah, B. (2000) Cross-protection of Cucurbitaceae from zucchini yellow mosaic potyvirus. In Katzir, N. and Paris, H.S. (eds) *Cucurbitaceae 2000. Proceedings of the 7th EUCARPIA Meeting on Cucurbit Breeding and Genetics, Ma'ale Ha Hamisha, Israel, March, 2000. Acta Horticulturae* 510, 349–356.

Zitter, T.A. (2002) Diseases of cucurbits and their control with genetic resistance. *Vegetable MD Online*, Cornell University, Ithaca (http://vegetablemdonline.ppath.cornell.edu/NewsArticles/Cuc_Genetic.htm).

Zitter, T.A., Hoffman, M.P., McGrath, M.T., Petzoldt, C.H., Seaman, A.J. and Pedersen, L.H. (2000) 2000 – Cucurbit IPM scouting procedures. *IPM Bulletin, No. 113*. New York State Integrated Pest Management Program, Cornell University, Ithaca, 18 pp., also (www.nysipm.cornell.edu/scouting/scoutproc/cuke00.pdf).

29   IPM Case Studies: Deciduous Fruit Trees

Sebastiano Barbagallo[1], Giuseppe Cocuzza[1], Piero Cravedi[2]
and Shinkichi Komazaki[3]

[1]*Dipartimento di Scienze e Tecnologie Fitosanitarie, University of Catania, Italy;*
[2]*Istituto di Entomologia e Patologia Vegetale, University 'Cattolica Sacro Cuore',*
Piacenza, Italy; [3]*Grape and Persimmon Research Station, National*
Institute of Fruit Tree Science, NARO, Hiroshima, Japan

Introduction

Fruit tree aphids are here defined as those pest species that affect crop plants belonging to the Rosaceae subfamilies Pyroideae, mainly apple and pear trees, and Prunoideae, commonly known as stone-fruit trees. Comprehensive accounts of these aphids, including identification keys, and much updated literature records have recently been given by Blackman and Eastop (2000). Barbagallo *et al.* (1997) have provided some epidemiological notes and colour illustrations of the most damaging species.

Apple and Pear Aphids (Fig. 29.1)

Apple (*Malus domestica*) and pear (*Pyrus communis*) are the most economically important crops in the Pyroideae worldwide. Other cultivated plants belonging to the same group include: Japanese pear or 'Nashi' (*Pyrus pyrifolia*), quince (*Cydonia oblonga*), loquat (*Eryobotria japonica*), medlar (*Mespilus germanica*), service tree (*Sorbus domestica*), and azarole (*Crataegus azarolus*). There are about 60 aphid species recorded worldwide as pests of these plants; the most economically damaging are some species infesting both apple and pear, as listed in Table 29.1.

Damage and virus transmission

Apple and pear aphids mostly cause direct damage through leaf deformation and shoot distortion, coupled with the effects of inoculated saliva and sap draining. Honeydew excretion causes further damage.

Eriosoma lanigerum (woolly apple aphid) is responsible for cankers on apple trunks and branches; infestations of roots and branches also cause formation of nodules and tumours, a reduction of plant tissue weight, and disturbance of the nutrient balance (Weber and Brown, 1988; Brown *et al.*, 1995).

Aphis pomi (green apple aphid) and *Aphis spiraecola* (green citrus aphid) attack the leaves throughout the season, while *Rhopalosiphum insertum* (apple–grass aphid) infests only the leaves in early spring.

Severe injury to both shoots and fruits can result from *Dysaphis plantaginea* (rosy apple aphid) on apple and from *Dysaphis pyri* (pear–bedstraw aphid) on pear; similar damage can be caused on pear by *Melanaphis pyraria* (brown pear aphid).

Aphanostigma piri (pear phylloxera) can directly damage fruits, causing necrosis around the calyx. *Schizaphis piricola* (and a few other similar species, such as *Schizaphis pyri* and *Schizaphis rotundiventris*) causes deformations of leaves and defoliation,

Table 29.1. Aphid species damaging to deciduous fruit trees (Pyroideae).

Aphid	Life cycle	Host	Geographical distribution	Damage	Notes
Aphis pomi – Fig. 29.1a	Holocyclic, monoecious	Most Pyroideae	Northern hemisphere, particularly west Palaearctic	Young shoots in spring	Quite injurious pest
Aphis spiraecola	Holocyclic, heteroecious or anholocyclic	Polyphagous	Worldwide, mostly tropics and subtropics	Young shoots in spring	Minor pest, injurious in East Asia
Aphis gossypii	Mostly anholocyclic	Polyphagous	Worldwide, mostly tropics and subtropics	Young shoots in spring	Minor pest
Melanaphis pyraria	Holocyclic, heteroecious	Pear (Poaceae as secondary host)	Southern Europe, Middle East	Young shoots in spring	Occasionally injurious
Anuraphis farfarae	Holocyclic, heteroecious	Pear (*Tussilago* [root collar] as secondary host)	Europe and Central Asia	Pseudogalls on leaves	Negligible injury
Rhopalosiphum insertum	Holocyclic, heteroecious	Apple (Poaceae [roots] as secondary host)	Holarctic and Australia	Leaves in spring	Minor pest
Schizaphis piricola	Holocyclic, heteroecious	Pear (secondary host unknown)	East Asia	Pseudogalls on leaves	Minor pest
Dysaphis plantaginea – Fig. 29.1b	Holocyclic, heteroecious	Apple (*Plantago* spp. [roots] as secondary host)	Europe, Middle East, Central Asia, introduced to parts of Africa and America	Leaf pseudogalls in spring, also flowers and young fruits	Most injurious species in Western Europe
Dysaphis devecta group – Fig. 29.1c	Holocyclic, monoecious (*devecta*) or heteroecious (other species)	Apple and other plants (heteroecious species)	Europe	Red pseudogalls on leaves	Usually minor pest

Species	Life cycle	Host	Distribution	Damage	Pest status
Dysaphis pyri	Holocyclic, heteroecious	Pear (*Galium* spp. as secondary host)	Palaearctic	Leaf pseudogalls; less injurious than *D. plantaginea*	Occasionally quite injurious pest
Dysaphis reaumuri	Holocyclic, heteroecious	Pear (*Galium* spp. as secondary host)	Southern Europe and Central Asia	Leaf distortions	Occasionally on cultivated pear, while common on wild ones
Eriosoma lanigerum	Anholocyclic	Apple, rarely other Pyroideae	North America, introduced into Europe and now worldwide	Cankers on stems, branches and roots (collar parts)	Very injurious pest
Eriosoma lanuginosum	Holocyclic, heteroecious	European elm (galls) and pear (roots)	Europe and the Palaearctic	Poorly studied	Minor pest
Eriosoma pyricola	Holocyclic, heteroecious	Elms (leaf gall) and pear (roots)	Southern Europe, Americas, Australasia	Poorly studied	Minor pest
Nippolachnus piri	Holocyclic, heteroecious	Pear and loquat	East Asia	Colonies on leaves	Minor pest
Pyrolachnus pyri	Holocyclic, monoecious or anholocyclic	Pear and other Pyroideae	Middle East and East Asia (China, Korea), including India	Colonies on branches	Minor pest
Aphanostigma piri – Fig. 29.1d	Holocyclic, monoecious or anholocyclic	Pear (mostly fruits and twigs)	Southwest Europe, Middle East	2- to 3-year-old twigs, inner parts of calyx in summer	Localized as injurious pest
Aphanostigma iaksuiense	Holocyclic, monoecious or anholocyclic	Japanese pear (mostly fruits and twigs)	East Asia	2- to 3-year-old twigs, inner parts of calyx in summer	Localized as injurious pest

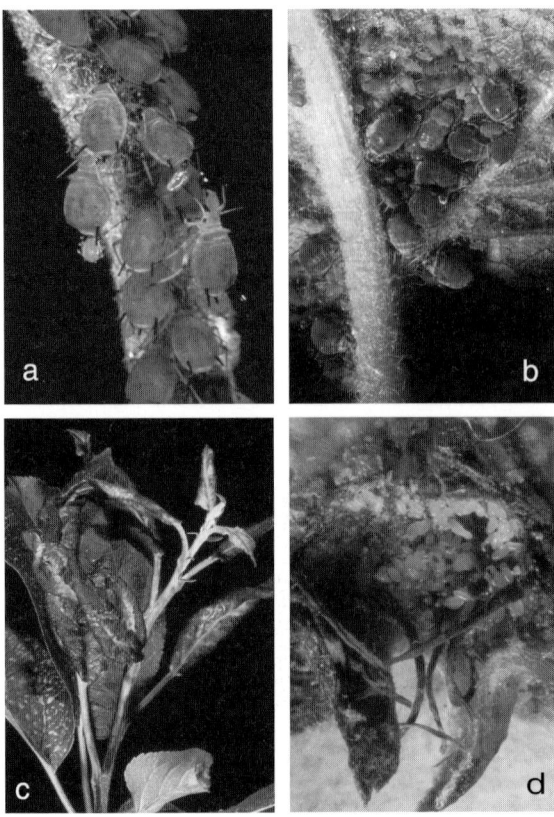

Fig. 29.1. Apple and pear aphids (colour original from Barbagallo *et al.*, 1997, courtesy of Bayer). (a) Colony of *Aphis pomi* on an apple shoot; (b) colony of *Dysaphis plantaginea* on an apple leaf; (c) pseudogalls of *Dysaphis anthrisci*, a member of the group causing red leaf pseudogalls on apple; (d) a section through a pear fruit, showing a colony of *Aphanostigma piri* inside the calyx area.

whereas dense populations of *Aphis gossypii* (cotton or melon aphid) lead to sooty mould development.

No virus transmission seems to be linked to aphids on apples, pears, or other Pyroideae.

Monitoring and economic thresholds

Visual counts are recommended for evaluating aphid population density on apple and pear at the appropriate times (Pasqualini *et al.*, 1981). Milaire *et al.* (1974) have suggested a simple methodology for different phenological periods, as well as economic thresholds for the main apple-infesting aphids in Western Europe.

Pre-flowering is the most critical phase because early attacks of *D. plantaginea* during flowering can cause severe losses. The presence of *A. pomi* (or *A. spiraecola*) either on apple or pear can be tolerated at higher levels because its reproductive rate is lower and the aphid causes no direct injury to fruits; in summer, damage by this aphid is negligible.

As a consequence, 1–2% of shoots infested is considered as a tolerance threshold for *D. plantaginea*, while attack by *A. pomi* can be tolerated up to a threshold of 8–10% of shoots infested during the most critical period in spring. The same thresholds can be used on pear against *D. pyri* and *A. pomi* or *M. pyraria*, respectively. Although there are no data for pear phylloxera, we believe that 1–2% of fruits infested can be considered as a reasonable threshold.

Against *E. lanigerum* in Massachusetts (USA), standard inspection involves examination of five pruning cuts on one tree per 1.5 ha, with a provisional threshold for treatment of 50% infected pruning cuts (CAB International, 2001). In Western Europe, however, the suggested threshold is 8–12 colonies per 100 shoots (Milaire *et al.*, 1974).

Parasitization by *Aphelinus mali* (see later), which can keep *E. lanigerum* at a low level, can be promoted and preserved from destruction by avoiding or the judicious use of insecticides (see below) (Castellari, 1967).

Chemical and supervised control

Chemical control is applied frequently against *D. plantaginea* on apple, though it is usually not necessary against other aphids such as *A. pomi* and *R. insertum*. Fluvalinate and thiamethoxam are currently used against *D. plantaginea* pre-flowering; the more selective imidacloprid and pirimicarb are mostly used post-flowering.

Eriosoma lanigerum* is normally kept under biological control, but chemical measures are sometimes necessary. Some effective systemic insecticides are chlorpyrifos, phosphamidon, pirimicarb, and demeton-methyl. Post-flowering is the best time for such treatments because apple trees are at their maximum vegetative growth and natural antagonists have not yet become very active. Even though chlorpyrifos is highly effective, it is very toxic to *A. mali*; pirimicarb is more selective, but has low persistence.

Therefore, in apple orchards under IPM programmes in New Zealand, mineral oil or oil and buprofezin are applied (Shaw *et al.*, 1996).

On pear, insecticide application is occasionally necessary against *D. pyri* and other shoot-infesting species, while locally the pear phylloxera, as well as *A. gossypii* and *A. spiraecola* on Japanese pear, can require spraying.

The main goal in applying aphicides on both apple and pear orchards should be to avoid their side effects on entomophagous and non-target pest species as much as possible, so that the former are preserved and undue population increase of the latter is prevented.

Biological control

Although several species of predators (mainly coccinellids, syrphids, and chrysopids) are often found in or near aphid colonies, their late appearance means that chemical control usually cannot be avoided. Neither are parasitoids very effective, though *A. mali* usually keeps *E. lanigerum* under biological control. The efficiency of *A. mali* can be promoted easily whenever necessary by bringing apple shoots containing mummified aphid colonies from other orchards into the infested orchard, and the use of broad-spectrum pesticides on that crop can then be avoided (Pasqualini *et al.*, 1981; Mueller *et al.*, 1992).

Stone-fruit Tree Aphids (Figs 29.2 and 29.3)

Cultivated Prunoideae are affected by about 26 aphid species (see Blackman and Eastop, 2000), only some of which are of economic importance either for the direct or indirect damage they cause. Peach (*Prunus persica*) and nectarine (*P. persica* var. *nucipersica*) host not less than a dozen more or less harmful species, among which the well-known *Myzus persicae* (peach–potato aphid) is undoubtedly the most frequent and injurious (Mackauer and Way, 1976). As well as spring colonies derived from overwintering eggs, spring colonies on the primary host can also arise from alatae arriving from anholocyclic populations overwintering on secondary host plants in warm climates or sheltered sites. On plum (*Prunus domestica*), the most injurious aphid pest is *Brachycaudus helichrysi* (leaf-curling plum aphid). Cherries (*Prunus avium* and *Prunus cerasus*) have just one serious aphid pest, *Myzus cerasi* (cherry blackfly). The above-mentioned and other aphids that are economically important on stone fruits, including apricot (*Prunus armeniaca*), Japanese apricot (*Prunus mume*), and almond (*Prunus dulcis*), are listed in Table 29.2.

Damage and virus transmission

Myzus persicae* causes leaf twisting on infested buds; on nectarines, it also affects fruits, causing typical pitting and discoloration. Symptoms are similar to those produced by *Brachycaudus schwartzi* (brown peach aphid) and sometimes these two

Table 29.2. Aphid species damaging to stone-fruit trees (Prunoideae).

Aphid	Life cycle	Host	Geographical distribution	Damage	Notes
Myzus persicae – Fig. 29.2c	Holocyclic, heteroecious on peach, or anholocyclic	Peach, also almond, plum, apricot, Japanese apricot and others (many different herbaceous and woody hosts)	Worldwide	Young shoots in spring	Most injurious species
Myzus amygdalinus	? holocyclic	Almond	Central Asia	Young leaves	Minor pest
Myzus cerasi – Fig. 29.3b	Heteroecious	Cherry (*Galium* and *Veronica* as secondary hosts)	Worldwide	Serious leaf curling	Quite injurious pest
Myzus mumecola	Probably heteroecious	Plum, apricot and Japanese apricot	India and East Asia	Leaf curling	Occasionally injurious
Myzus siegesbeckiae	Holocyclic, heteroecious	Cherry, Japanese apricot, peach (Lamiaceae, i.e. *Plectantrus* and *Salvia*, as facultatively secondary hosts)	East Asia	Leaf curling	Occasionally injurious
Myzus varians – Fig. 29.3a	Holocyclic, heteroecious	Peach (*Clematis* as secondary host)	Native in East Asia, now widespread in Western Europe	Leaf curling	Quite injurious pest
Brachycaudus persicae – Fig. 29.2a	Anholocyclic or holocyclic, heteroecious	Peach, apricot, Japanese apricot, almond (Scrophulariaceae as secondary hosts)	West Palaearctic, Australasia and Americas	Dense colonies on twigs, buds, flowers, and fruits	Overwinters on roots and collar area as virginoparae; a quite injurious pest
Brachycaudus schwartzi – Fig. 29.2b	Monoecious and usually holocyclic	Peach	West Palaearctic and parts of America	Deforms leaves	Sometimes overwinters as virginoparae; occasionally injurious

Species	Life cycle	Hosts	Distribution	Damage	Injuriousness
Brachycaudus amygdalinus	Holocyclic, heteroecious	Almond (*Polygonum aviculare* as secondary host)	Mediterranean and Middle East	Serious leaf curling	Occasionally injurious
Brachycaudus helichrysi	Holocyclic, heteroecious or anholocyclic where conditions allow	Plum, almond (many Asteraceae and other plants as secondary hosts)	Worldwide	Dense colonies on undersides of leaves, causing twisting	A serious pest of plum
Brachycaudus prunicola	Holocyclic, monoecious	Plum, also sloe (*Prunus spinosa*) and damson (*Prunus insititia*)	Europe	Deforms leaves	Occasionally injurious
Hyalopterus amygdali – Fig. 29.2d	Holocyclic, heteroecious	Peach, almond (*Phragmites* and *Arundo* as secondary hosts)	West Palaearctic	Dense colonies on leaf undersides, but no deformation	Most injurious species in Western Europe with *H. pruni*
Hyalopterus pruni	Holocyclic, heteroecious	Many *Prunus* (*Phragmites* and *Arundo* as secondary hosts)	Worldwide	Dense colonies on leaf undersides, but no deformation	Most injurious species in Western Europe with *H. amygdali*
Rhopalosiphum rufiabdominale	Holocyclic, heteroecious	Peach, plum, Japanese apricot and cherry (Poaceae and other monocotyledons as secondary hosts)	Worldwide	Dense colonies on young shoots	Occasionally injurious in East Asia
Phorodon humuli	Holocyclic, heteroecious	Plum (hop as secondary host)	West-Palaearctic parts of America and New Zealand	Deformation of young shoots	Only occasionally damaging to plums
Pterochloroides persicae – Fig. 29.3c,d	Usually anholocyclic	Peach, plum, apricot, Japanese apricot, occasionally cherry and almond	Indian sub-continent to western Mediterranean	Heavy colonies on trunk and branches	Quite injurious

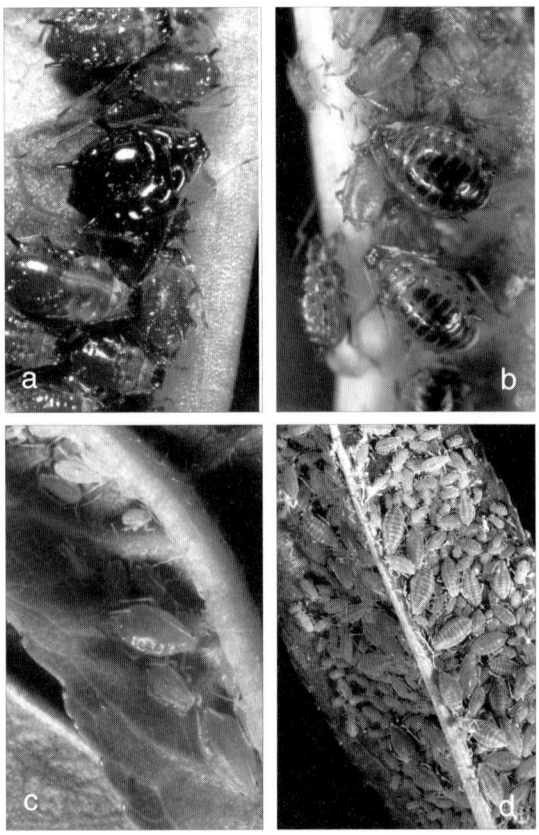

Fig. 29.2. Stone-fruit tree aphids. (a) Colony (one aptera and nymphs) of *Brachycaudus persicae* on peach; (b) colony of *Brachycaudus schwartzi* on peach (colour original from Barbagallo *et al.*, 1997, courtesy of Bayer); (c) colony of *Myzus persicae* on a peach leaf; (d) dense colony of *Hyalopterus amygdali* on a peach leaf.

species form mixed colonies. *Myzus varians* (cigar-rolling peach aphid) is economically less important, but causes some leaf rolling.

Hyalopterus amygdali (mealy peach aphid) and *Hyalopterus pruni* (mealy plum aphid) produce copious honeydew. *Brachycaudus persicae* (black peach aphid) colonizes shoots and young fruits; during winter, it lives on the roots.

Brachycaudus helichrysi can cause severe deformations on leaves and the shedding of flowers and young fruits. *M. cerasi* causes severe curling and distortion of leaves.

Myzus persicae is considered as the most important virus vector on many plants. On stone fruits it is an efficient vector of *Plum pox virus*, causing 'Sharka'. Other vectors of the same virus disease are *Phorodon humuli* (damson–hop aphid), *B. helichrysi*, *H. pruni*, *M. varians*, and some species that occasionally infest Prunoideae; these are *Brachycaudus cardui*, *A. spiraecola*, and *Aphis craccivora* (cowpea aphid) (Leclant, 1973; OEPP, 1974; Conti *et al.*, 1985; Davino *et al.*, 1991; Avinent *et al.*, 1994; Wallis *et al.*, 2005).

Monitoring and economic thresholds

Control on peaches and nectarines must begin before the overwintering eggs hatch, therefore before flowering.

The pre-flowering threshold for *M. persicae* (as well as for *M. varians* and *B. schwartzi*) is estimated as 3% of buds infested. After flowering, the threshold stays at 3% for nectarines, but is raised on peach to 10% of shoots infested (Malavolta *et al.*, 1995). In summer, the effectiveness of natural enemies increases; at the same time, winged morphs develop and leave the tree. Monitoring during this period should evaluate both these population-reducing factors in order to avoid unnecessary insecticide treatments. It is

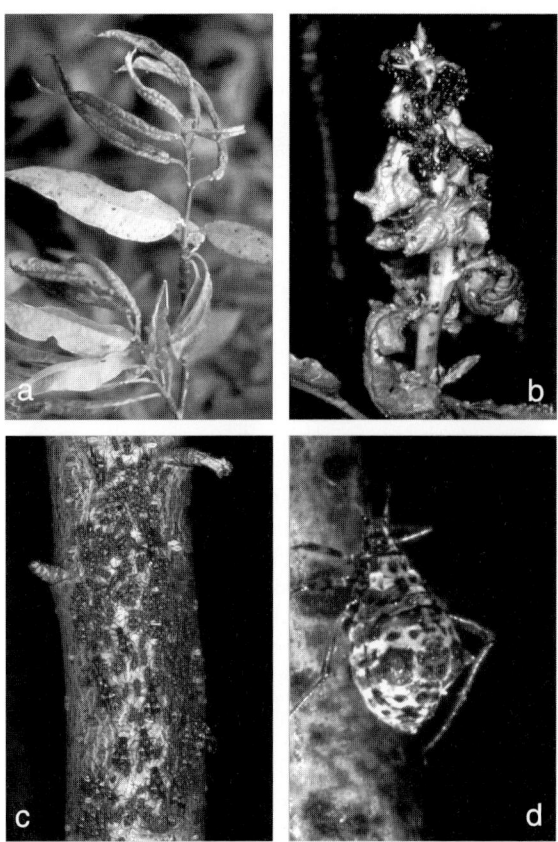

Fig. 29.3. Stone-fruit tree aphids (colour original from Barbagallo *et al.*,1997, courtesy of Bayer). (a) Typical damage of *Myzus varians* on peach leaves; (b) a dense colony of *Myzus cerasi* on a sweet cherry shoot; (c) colony of *Pterochloroides persicae* on an almond twig; (d) apterous viviparous female of *Pterochloroides persicae*.

important to monitor *B. persicae* at the end of winter when temperatures start to rise because this species migrates from the roots to the shoots well before flowering. Late attacks are less harmful and easier to control because the sensitivity of *B. persicae* to insecticides is then greater than earlier in the year. By contrast, with *H. amygdali* on peach and *H. pruni* on plum, later attacks are more harmful than earlier ones and spraying with spot applications following visual monitoring is advised (Malavolta *et al.*, 1996).

The threshold used in Italy on plum trees for *B. helichrysi* and for *P. humuli* is 10% of shoots infested. In California, against the same aphids, a threshold of three colonies per tree is suggested from April to May (UC IPM, 2005). Against *M. cerasi* on cherries, a threshold of 3% shoots infested is recommended, and aphicide should be applied shortly after petal fall if the threshold is reached or exceeded (Malavolta *et al.*, 1995).

Chemical and supervised control

Growers frequently apply oil or chlorpyrifos sprays to dormant buds against overwintering eggs of *M. persicae*, but such materials are not selective and therefore they are not recommended.

During spring and summer, several pesticides are currently applied against aphids. Pirimicarb is a selective aphicide, but resistance to it has now often been found. Many other insecticides such as diazinon, endosulfan, and formetanate have the disadvantage of being broad spectrum. Neem oil is used in 'organic' farming, and it has given good results against *M. persicae*. Synthetic pyrethroids are effective, but they are very toxic to beneficials; they also increase mite populations. New active ingredients like imidacloprid, thiamethoxam, and pymetrozine have shown high efficiency and selectivity.

Myzus persicae is a particularly frequent problem, both as regards damage and resistance to insecticides. The first mechanism of resistance discovered in the 1980s for organophosphates, carbamates, and pyrethroids was an increase in the production of carboxylesterase E4/FE4 in the aphids. At the beginning of the 1990s, a modification of the acetylcholinesterase that made the insect insensitive to dimethylcarbamates like pirimicarb and triazamate was discovered, and an amino acid substitution on the acetylcholinesterase was detected in resistant strains (Nabeshima *et al.*, 2003). Recently, other resistance mechanisms (kdr = knockdown resistance and rdl = resistance to dieldrin) have been identified: they are the result of mutations in the target sites for pyrethroids (sodium channel) and for cyclodienes (GABA receptor) (Field *et al.*, 1997; Foster *et al.*, 1998; Chapter 10 this volume).

Biological control

Several parasitoids, predators, and pathogens are known to affect stone fruit aphids, particularly *M. persicae.*

Unfortunately, natural enemies tend to appear too late in the season when large aphid colonies have already developed, whereas damage by aphids is highest well before then in early spring. As the impact of natural enemies is usually low until late spring and summer, they are unable to prevent damage.

Acknowledgements

Many thanks are due to the editor (Bayer S.p.A., Milan) of the book *Aphids of the Principal Fruit-Bearing Crops* (Barbagallo *et al.*, 1997) for permission to reproduce Fig. 29.1, Fig. 29.2b and Fig. 29.3b from colour plates.

Executive Summary

Deciduous fruit trees are attacked by a very large number of aphid species, and economic thresholds and the appropriate timings of insecticides have been established for the majority of the most economically important species. Chemical control may often prove necessary, and choice of active ingredient from a large selection of possible compounds is influenced by the insecticide-resistance status of the aphid population, with the aim of causing least damage to non-targets (particularly natural enemies), and by the difference between countries as to which effective compounds are still permitted (particularly among the organophosphates and carbamates). Winter oils and synthetic pyrethroids should be avoided.

Biological control of aphids is seen mainly as a useful general restraint of inadequate strength, but to be preserved as far as possible. However, biological control of *E. lanigerum* by the parasitoid *A. mali* is often effective and can be manipulated by bringing prunings with aphid mummies into the orchard.

There is almost no use of host-plant resistance in such long-lived crops with long breeding programmes, though some varieties of apple in use are reported as resistant to *E. lanigerum* and sources of resistance are known for some other aphids (see van Emden, Chapter 17 this volume).

No cultural controls are practised; neither is there any use of semiochemicals for monitoring aphids or controlling their behaviour or that of their natural enemies.

There are no purposeful synergistic interactions between control methods, other than a general aim when considering the use of insecticides to avoid damage to the beneficial fauna.

References

Avinent, L., Hermoso de Mendoza, A. and Llacer, G. (1994) Transmission of plum pox potyvirus in Spain. *Bulletin OEPP* 24, 669–674.
Barbagallo, S., Cravedi, P., Pasqualini, E. and Patti, I. (1997) *Aphids of the Principal Fruit-Bearing Crops*. Bayer, Milan, 123 pp.

Blackman, R.L. and Eastop, V.F. (2000) *Aphids on the World's Crops. An Identification and Information Guide*, 2nd edn. Wiley, Chichester, 466 pp.

Brown, M.W., Schmitt, J.J., Ranger, S. and Hogmire, H.W. (1995) Yield reduction in apple by edaphic woolly apple aphid (Homoptera: Aphididae) populations. *Journal of Economic Entomology* 88, 127–133.

CAB International (2001) *Eriosoma lanigerum*. In: *Crop Protection Compendium, Global Module,* 3rd edn. [CD-ROM]. CAB International, Wallingford.

Castellari, P.L. (1967) Ricerche sulla etologia e sull'ecologia dell'*Eriosoma lanigerum* Haus. e del suo parassita *Aphelinus mali* Hald. in Emilia, con particolare riguardo agli effetti secondari della lotta chimica. *Bollettino Istituto Entomologia Università di Bologna* 28, 177–231.

Conti, M., Luisoni, E. and Giunchedi, L. (1985) La Sharka delle Drupacee. *L'Italia Agricola* 122 (April–June), 183–193.

Davino, M., Patti, I., Areddia, R., Tirrò, A. and D'Urso, F. (1991) Diffusione del virus della vaiolatura del susino in Sicilia. *Tecnica Agricola* 43 (1/2), 3–14.

Field, L.M., Anderson, A.P., Denholm, I., Foster, S.P., Harling, Z.K., Javed, N., Martinez-Torres, D., Moores, G.D., Williamson, M.S. and Devonshire, A.L. (1997) Use of biochemical and DNA diagnostic for characterising multiple mechanisms of insecticide resistance in the peach–potato aphid, *Myzus persicae* (Sulzer). *Pesticide Science* 51, 283–289.

Foster, S.P., Denholm, I., Harling, Z.K., Moores, G.D. and Devonshire, A.L. (1998) Intensification of insecticide resistance in UK field populations of the peach–potato aphid, *Myzus persicae* (Hemiptera: Aphididae) in 1996. *Bulletin of Entomological Research* 88, 127–130.

Leclant, F. (1973) Aspect ecologique de la transmission de la Sharka (plum pox) dans le sud-est de la France. Mise en évidence de nouvelles espèces d'aphides vectrices. *Annales de Phytopathologie* 5, 431–439.

Mackauer, M. and Way, M.Y. (1976) *Myzus persicae* Sulz., an aphid of world importance. In: Delucchi, V.F. (ed.) *Studies in Biological Control*. Cambridge University Press, Cambridge, pp. 51–119.

Malavolta, C., Ponti, I., Pollini, A., Galassi, T., Cravedi, P., Molinari, F., Brunelli, A., Pasini, F., Missere, D., Scudellari, D. and Pissi, M. (1995) The application of integrated production on stone fruits in Emilia-Romagna (Italy). *IOBC/WPRS Bulletin* 18 (2), 55–59.

Malavolta, C., Ponti, I., Pollini, A., Cravedi, P., Molinari, F., Mazzoni, E., Brunelli, A. and Scudellari, D. (1996) The application of integrated production on plum and apricot in Emilia-Romagna (Italy). *Proceedings of the IOBC/ISHS International Conference on Integrated Fruit Production, Cedzyna, Poland, August 1995. IOBC/WPRS Bulletin* 19 (4), 108–112.

Milaire, H.G., Baggiolini, M., Gruys, P. and Steiner, H. (eds) (1974) *Les organismes auxiliaires en verger de pommiers*. Groupe de travail pour la lutte intégrée en arboriculture, brochure no. 3, OILB/SROP, Wageningen, 242 pp.

Mueller, T.F., Blommers, L.H.M. and Mols, P.J.M. (1992) Woolly apple aphid (*Eriosoma lanigerum* Hausm., Hom., Aphididae) parasitism by *Aphelinus mali* Hal. (Hym., Aphelinidae) in relation to host stage and host colony size, shape and location. *Journal of Applied Entomology* 114, 143–154.

Nabeshima, T., Kozaki, T., Tomita, T. and Kono, T. (2003) An amino acid substitution on the second acetyl-cholinesterase in the pirimicarb-resistant strains of the peach–potato aphid, *Myzus persicae*. *Biochemical and Biophysical Research Communications* 307, 15–22.

OEPP (1974) Progrès réalisés dans la connaissance de la sharka. *Bulletin OEPP* 4 (1), 125.

Pasqualini, E., Briolini, G., Memmi, M. and Monari, S. (1981) Prove di lotta guidata contro gli Afidi del melo. *Bollettino Istituto Entomologia Università di Bologna* 36, 160–171.

Shaw, P.W., Walker, J.T.S. and O'Callaghan, M. (1996) Biological control of woolly apple aphid by *Aphelinus mali* in an integrated fruit production programme in Nelson. *Proceedings of the 49th New Zealand Plant Protection Conference, Nelson, New Zealand, August 1996*. New Zealand Plant Protection Society, Rotorua, pp. 59–63.

UC IPM (2005) *Pest Management Guidelines: Prune*. UC ANR publication No. 3464, Insects and Mites (www.ipm.ucdavis.edu).

Wallis, C.M., Fleischer, S.J., Luster, D. and Gildow, F.E. (2005) Aphid (Hemiptera: Aphididae) species composition and potential aphid vectors of plum pox virus in Pennsylvania peach orchards. *Journal of Economic Entomology* 98, 1441–1450.

Weber, D.C. and Brown, M.W. (1988) Impact of woolly apple aphid (Homoptera: Aphididae) on the growth of potted apple trees. *Journal of Economic Entomology* 81, 1170–1177.

30 IPM Case Studies: Tropical and Subtropical Fruit Trees

Sebastiano Barbagallo[1], Giuseppe Cocuzza[1], Piero Cravedi[2] and Shinkichi Komazaki[3]

[1]Dipartimento di Scienze e Tecnologie Fitosanitarie, University of Catania, Catania, Italy; [2]Istituto di Entomologia e Patologia Vegetale, University 'Cattolica Sacro Cuore', Piacenza, Italy; [3]Grape and Persimmon Research Station, National Institute of Fruit Tree Science, NARO, Hiroshima, Japan

Introduction

This chapter deals briefly with some applied aspects of the epidemiology, damage, and IPM control measures against pest aphid species affecting citrus trees and other widely grown tropical fruit tree crops. The more important aphid species are illustrated (Figs 30.1, 30.2 and 30.3).

Citrus Aphids

About 20 aphid species are recorded on citrus crops worldwide (Barbagallo and Patti, 1986; Blackman and Eastop, 2000), but not more than six or seven species occur widely and only four or five strongly affect citrus trees, either by direct or indirect damage (Table 30.1). For most of them, illustrations and/or identification keys are provided in comprehensive papers (Stroyan, 1961; Barbagallo, 1966; Meliá, 1982; Barbagallo et al., 1997; Blackman and Eastop, 2000). The most widespread and harmful species are Aphis spiraecola (green citrus aphid), Aphis gossypii (cotton or melon aphid), Toxoptera aurantii (black citrus aphid), and Toxoptera citricidus (tropical citrus aphid).

Aphis spiraecola has a cosmopolitan distribution (CIE, 2001) and appears to be mostly invasive and widespread around tropical and subtropical areas, while its presence in temperate climates is sometimes limited to sheltered or relatively warm habitats. In Japan, the aphid can also lay overwintering eggs on citrus (Komazaki, 1983, 1991, 1998), and its populations in the warm temperate areas (Mediterranean and comparable latitudes) show two main peaks of infestation, during spring and the early autumn months, synchronized with flushing of the shoots. Lemon is not usually infested.

Aphis gossypii (CIE, 1968) is another species which very commonly infests citrus in most areas where this is grown. While the aphid is usually anholocyclic, holocyclic populations are detected in a few of its distribution areas, including Japan, where winter eggs can be laid on citrus (Komazaki et al., 1979), as well as on several other unrelated plant species (Inaizumi, 1981). It is a highly polyphagous and genetically complex species. The most affected citrus are cultivars of orange (Citrus sinensis), mandarin orange and clementine (both Citrus reticulata), and others, while lemon is very little infested.

Table 30.1. Aphid species damaging to citrus.[*]

Aphid	Life cycle	Host	Geographical distribution	Damage	Notes
Aphis spiraecola – Fig. 30.1a,b	Anholocyclic on citrus in warmer habitats; often holocyclic and heteroecious elsewhere	Polyphagous (citrus, Asteraceae, Rosaceae, etc.)	Cosmopolitan	Strong deformation of young shoots of orange, mandarin orange, clementine, grapefruit; negligible on lemon	Very injurious to citrus; vector of *Citrus tristeza virus* (CTV)
Aphis gossypii – Fig. 30.1c,d	Usually anholocyclic; rarely holocyclic and heteroecious in some areas	Highly polyphagous (citrus, Cucurbitaceae, Malvaceae, etc.)	Cosmopolitan	Infests young shoots (no deformation) of orange, mandarin orange, clementine, and grapefruit	Commonly injurious on citrus; vector of CTV
Toxoptera aurantii – Fig. 30.2a,b	Anholocyclic everywhere	Polyphagous (citrus as main host)	Cosmopolitan	Infests young shoots of all citrus species	Commonly injurious on citrus; vector of CTV; dominant where *T. citricidus* is not distributed
Toxoptera citricidus – Fig. 30.2c	Usually anholocyclic, rarely holocyclic, monoecious in Japan	Mostly citrus and other plants	Southern hemisphere, but up to 40°N in East Asia; also recently spread to Caribbean and Florida	Infests young shoots of all citrus species	Commonly injurious on citrus; highly efficient vector of CTV
Sinomegoura citricola	Anholocyclic	Polyphagous	Widespread from South-east Asia to Australia	Infests young shoots	Minor pest

Aulacorthum magnoliae	Mainly anholocyclic	Rather polyphagous	East Asia	Infests young shoots	Minor pest
Toxoptera odinae - Fig. 30.2d	Anholocyclic	Rather polyphagous	East Asia, sub-Saharan Africa	Infests young shoots	Minor pest
Myzus persicae	Holocyclic, heteroecious or anholocyclic	Highly polyphagous; occasionally on citrus	Cosmopolitan	Sometimes only single shoots	Occasional problem; unusual outbreak on mandarin orange in Spain
Macrosiphum euphorbiae	Holocyclic, heteroecious or anholocyclic	Polyphagous; occasionally on citrus	Cosmopolitan	Sporadically on tender shoots	Minor pest
Aphis fabae	Holocyclic, heteroecious or anholocyclic	Polyphagous; occasionally on citrus	Cosmopolitan	Sporadically on tender shoots	Minor pest
Aphis craccivora	Mostly anholocyclic	Polyphagous; occasionally on citrus	Cosmopolitan	Sporadically on tender shoots	Minor pest

*The literature also contains records of a few other species (e.g. *Brachycaudus cardui*, *Brachycaudus helichrysi*, and *Pterochloroides persicae*), but it is unlikely these were actually living on citrus.

Fig. 30.1. Tropical and subtropical fruit tree aphids (colour original from Barbagallo *et al.*, 1997, courtesy of Bayer). (a) Colony of *Aphis spiraecola* (apterae and nymphs) on a clementine leaf; (b) strong deformations following a severe attack of *A. spiraecola* on the shoot apex of clementine; (c) colony of *Aphis gossypii* on an orange leaf; (d) colonies of *A. gossypii* on orange shoot: despite the severe attack leaves usually remain undeformed.

Toxoptera aurantii (CIE, 1961) affects all citrus species and cultivars. While polyphagous, mostly around the intertropical areas, citrus trees represent the main hosts of economic interest around the Mediterranean. It is considered to be strictly anholocyclic. Therefore, it overwinters as parthenogenetic morphs only, usually in sheltered environments (e.g. greenhouses) in those temperate zones that have a rather cold winter.

Toxoptera citricidus is widespread all over the citrus-growing areas of the southern hemisphere, though in East Asia it is found as far north as 40° of latitude (CIE, 1998). Recently, it has spread quickly around the Caribbean and Florida (Halbert and Brown, 1996) and it has also been detected very recently in Southern Europe in several northern localities of Portugal and Spain (Ilharco *et al.*, 2005). Nevertheless, the aphid has, as yet, not been recorded around the inner Mediterranean basin, where it would pose a serious threat to citrus, since it is a highly

efficient vector of *Citrus tristeza virus* (CTV). For a literature review on this aphid, see Michaud (1998).

All other citrus aphids listed in Table 30.1 are usually of secondary economic importance.

Damage and virus transmission

Direct damage varies with aphid species and the density of their populations, as well as with the species and age of the infested citrus groves. In addition to the sap and nutrient drain by aphid colonies, there is variable injury due to the stunting of young shoots and the rolling and twisting of young leaves. The most harmful such species is *A. spiraecola*, high-density colonies of which can lead to serious detrimental effects, especially on the growth of young orange and mandarin orange trees. Even at high population density, *A. gossypii* causes only minor deformations, while both *Toxoptera*

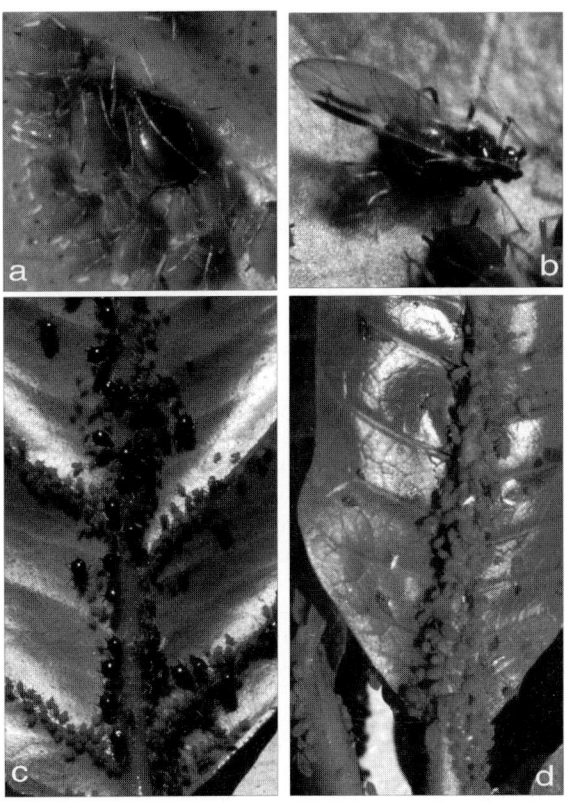

Fig. 30.2 Tropical and subtropical fruit tree aphids. (a) Small colony (aptera with nymphs) of *Toxoptera aurantii* on lemon (colour original from Barbagallo *et al.*, 1997, courtesy of Bayer); (b) alate viviparous female of *T. aurantii* (colour original from Barbagallo *et al.*, 1997, courtesy of Bayer); (c) colony of *Toxoptera citricidus* on a grapefruit (*Citrus × paradisi*) leaf (from Alcântara Santos and Barbagallo, 1989; original colour photo by S. Barbagallo); (d) colony of *Toxoptera odinae* on a tender cashew leaf (from Barbagallo and Alcântara Santos, 1989; original colour photo by S. Barbagallo).

species cause intermediate levels of damage. There are other well-known damaging effects of aphid outbreaks: i.e. honeydew excretion, followed by the development of sooty moulds, and problems with ants where aphids are myrmecophilous. However, indirect damage is mostly due to virus transmission. CTV, including both the 'stem pitting' and 'seedling yellows' strains, is a destructive disease and it has been known for many years that *T. citricidus* is the most efficient vector (see above). A direct correlation between the spread of this aphid and the spread of CTV has been demonstrated across all citrus-growing areas. Other aphids infesting citrus (mainly *A. gossypii* and, to a lesser extent, *A. spiraecola* and *T. aurantii*) have also tested positively as vectors of tristeza, indicating that transmission efficiency is correlated with different bioecological combinations of interacting factors, such as virus strains, host plant, and climatic conditions (Bar-Joseph *et al.*, 1977; Roistacher *et al.*, 1984; Davino *et al.*, 1990;

Komazaki, 1994; Halbert *et al.*, 2004; Yokomi and DeBorde, 2005). *Aphis gossypii*, *Myzus persicae* (peach–potato aphid), and *T. citricidus* may also be involved in the transmission of *Citrus vein enation–woody gall virus* (Wallace, 1975).

Monitoring and economic thresholds

Aphid populations on citrus can be evaluated directly in the field by inspection. A general sampling methodology has been suggested (Cavalloro and Prota, 1983), consisting of a weekly check of a convenient number of plants (usually between 5 and 10% of an orchard) during the period of exponential aphid population increase (i.e. mostly during the spring months and occasionally in early autumn in temperate climates, such as the Mediterranean). Both aphid-infested and uninfested shoots are counted inside a frame of about 0.25 m² held up against the canopy on four sides of each of the randomly selected plants. The threshold is expressed as an overall percentage of shoots that are infested.

In relation to this sampling methodology, the suggested economic thresholds (at least for Mediterranean areas) for the application of chemical control measures against citrus aphids are: 25% of shoots infested by *T. aurantii* and *A. gossypii*; 5 and 10% of shoots infested by *A. spiraecola* for clementine/mandarin oranges and orange trees, respectively.

Chemical and supervised control

IPM methods can be suggested for controlling citrus aphid populations, and several factors should be considered before deciding to apply an insecticide. In addition to the threshold level of per cent shoots infested, it is useful to look at the density (or size) of aphid colonies, the aphid species involved (since different species cause different levels of injury), the species and age of the host plants (younger orchards need a higher degree of protection than older ones) and, not least, the density of biocontrol agents in the aphid colonies (Barbagallo and Patti, 1986). It is important to bear in mind that some cultural practices – such as irrigation, pruning, and particularly nitrogen fertilization of the soil, and even more so as foliage sprays (as currently used by farmers during spring) – increase the flush of new growth, which promotes aphid populations both by accelerating their reproduction and increasing the time tender shoots are available to them.

Once the decision has been taken to apply a chemical spray against aphids, the selection of an appropriate pesticide is of paramount importance in citrus orchards in order to minimize negative side effects on non-target arthropods (particularly natural enemies). The use of selective and preferably also short-persistence aphicides is therefore recommended. The most frequently used compounds are pirimicarb, ethiofencarb, imidacloprid, and a few others that have a similar range of biological activity. Other insecticides, belonging to the organophosphate or carbamate groups, are frequently used by farmers against some aphid species, either because the aphids are thought to be resistant to other common aphicides (to pirimicarb, for example), or because of the comparatively lower cost of these compounds. However, these broad-spectrum pesticides should not be applied for aphid control by citrus growers using IPM programmes. Similarly, pyrethroid compounds should be avoided in citrus orchards because of their general negative side effects in increasing mite populations and in killing beneficial arthropods.

Insecticide resistance has been shown in several aphid species on citrus. *Aphis gossypii* is undoubtedly the aphid that seems most adept at showing resistance to different pesticides, including selective aphicides such as pirimicarb. Such resistance in *A. gossypii* has been detected on several crops and in different areas worldwide. Komazaki and Osakabe (1998) have detected resistance in strains collected from different crops (including citrus) in Japan. Recently, a change in the acetylcholinesterase gene has been identified in pirimicarb-resistant strains (Toda *et al.*, 2004; Andrews *et al.*, 2004). In Korea, Song *et al.* (1995) found resistance to pirimicarb in *A. spiraecola*. In any case, this latter species – independently of any resistance – is quite difficult to control effectively because the heavy deformation of infested citrus shoots protects the colonies of this aphid from being reached directly by sprays, unless systemic compounds are used.

Biological control

Predators and parasitoids can reduce citrus aphid populations significantly under different ecological conditions. Among predators, several species of coccinellids, syrphids, and chrysopids are the most efficient groups (Barbagallo and Patti, 1983; Tremblay and Barbagallo, 1983). Some species of parasitoids in the Aphidiinae – such as *Lysiphlebus testaceipes*, *L. fabarum*, and *L. confusus* – can reach levels of parasitism as high as 90–100% on *T. aurantii* and *A. gossypii* in the Mediterranean countries (Tremblay *et al.*, 1980; Starý *et al.*, 1988; Meliá, 1993).

Unfortunately, these parasitoids seem unable to parasitize *A. spiraecola*, against which the aphelinid wasp *Aphelinus spiraecolae* has recently proved promising in Florida (Tang and Yokomi, 1996). Natural biological control of citrus aphids is therefore usually more effective against *A. gossypii* and *T. aurantii* than against *A. spiraecola*. The efficiency of biological control agents of *T. citricidus* is variable (Michaud, 1998). Some natural enemies of this species are known from some areas (De Huiza and Ortiz, 1981; Nickel and Klingauf, 1985; Takanashi, 1990; Komazaki, 1994), whereas none is known from other large regions, such as sub-Saharan Africa (Remaudière *et al.*, 1985). In Japan, *Lysiphlebia japonica* can parasitize up to 70% of *T. citricidus* late in the season. Predators and fungal diseases (Rondón *et al.*, 1983) are apparently commonly found infecting this aphid, but are not very effective.

On the whole, natural enemies only have an impact on citrus aphid populations late in the season; as a consequence, farmers often cannot avoid using chemical control.

Tropical Fruit Tree Aphids

Aphid pests can affect several cultivated tropical fruit trees. This section focuses on the main species infesting the following crops, all selected as being fairly extensively grown in tropical countries (Calabrese, 1993): avocado (*Persea americana*), mango (*Mangifera indica*), guava (*Psidium guajava*), custard apple (*Annona cherimola*) and other congeneric species (i.e. soursop – *Annona muricata*, sugar apple – *Annona squamosa*, atemoya – a cross between *A. squamosa* and *A. cherimola*, bullock's heart – *Annona reticulata*), banana (*Musa paradisiaca*), cashew nut (*Anacardium occidentale*), and pecan (*Carya illinoiensis*), though the latter is also cultivated outside the tropics.

Blackman and Eastop (2000) updated and keyed all aphids recorded on these plants, while Martin (1983) gave a general illustrated identification key for common aphid pests in the tropics. A variable number of 4–9 aphid species is recorded for each of the named cultivated plants (Table 30.2); most are polyphagous. The commonest are well-known representatives of the genus *Aphis* (*A. gossypii*, *A. spiraecola*, *A. craccivora* – cowpea aphid) and *Toxoptera* (*T. aurantii*, *T. citricidus*, *T. odinae*): most of these have already been mentioned as citrus pests. These aphids, as well as a few other polyphagous species (i.e. *M. persicae*, *Macrosiphum euphorbiae* – the potato aphid, and occasionally *Sinomegoura citricola*) are commonly reported on different tropical fruit crops (i.e. avocado, cashew nut, litchi – *Litchi chinensis*, macadamia nut – *Macadamia integrifolia*, mango, passion fruit – *Passiflora edulis*). These aphids are permanently anholocyclic in the tropics, and their seasonal outbreaks are linked mainly to the effective growth periods of their host plants.

In addition, there are several oligophagous or, more rarely, monophagous aphids injurious to some of the above-named tropical crops. Among these species should be mentioned *Pentalonia nigronervosa* (banana aphid) and a few *Rhopalosiphum* infesting banana, some *Greenidea* species recorded on different host plants (e.g. *Annona* species, guava, mango, litchi), and mainly a group of species linked to pecan, as shown in Table 30.2.

Damage and virus transmission

Direct damage varies within the different crops and is correlated with aphid population density. Thus, *P. nigronervosa* can form quite high infestations on banana, as can other aphids (*Monellia caryella* and *Monelliopsis pecanis* – the yellow pecan aphids, see Table 30.2) on pecan. *Aphis gossypii* is relatively common on several other fruit crops. Most aphid species affecting tropical crops develop colonies on young shoots and leaves, but occasionally also on flowers and young fruits. *Pentalonia nigronervosa* infesting banana develops myrmecophilous colonies on the inner leaf sheaths and folded apical leaves, including flower bracts. *Longistigma caryae* is a bark-feeder, developing

Table 30.2. Aphid species damaging to tropical fruit trees.

Aphid	Life cycle	Host (main tropical fruits)	Geographical distribution	Damage	Notes
Aphis spiraecola – Fig. 30.1a,b	Anholocyclic in the intertropical zone	Avocado, custard apple, cashew, papaya, guava, mango, etc.	Widespread in the intertropical zone	Infests young shoots and leaves	Commonly recorded as a pest
Aphis gossypii – Fig. 30.1c,d	Anholocyclic in the intertropical zone	Avocado, custard apple, cashew, papaya, guava, mango, etc.	Widespread in the intertropical zone	Infests young shoots and leaves	Most common species affecting guava and other crops
Aphis craccivora	Anholocyclic in the intertropical zone	Papaya, guava, litchi, avocado, etc.	Widespread in the intertropical zone	Infests young shoots and leaves	Occasionally injurious
Toxoptera aurantii – Fig. 30.2a,b	Anholocyclic in the intertropical zone	Cashew, mango, guava, litchi, avocado, etc.	Widespread in the intertropical zone	Infests young shoots and leaves	Commonly recorded
Toxoptera citricidus – Fig. 30.2c	Anholocyclic in the intertropical zone	Cashew, mango, guava, litchi, avocado, etc.	Widespread in the intertropical zone	Infests young shoots and leaves	Commonly recorded
Toxoptera odinae – Fig. 30.2d	Anholocyclic in the intertropical zone	Mango, cashew, etc.	East Asia, sub-Saharan Africa	Infests young shoots and leaves	Occasionally injurious
Myzus persicae	Anholocyclic in the intertropical zone	Avocado, guava, papaya, passion fruit, etc.	Cosmopolitan	Infests young shoots and leaves	Commonly recorded
Sinomegoura citricola	Anholocyclic in the intertropical zone	Mango and avocado	South-east Asia, Australia	Infests young shoots and leaves	Minor pest
Macrosiphum euphorbiae	Anholocyclic in the intertropical zone	*Annona* spp., papaya, etc.	Cosmopolitan	Infests young shoots and leaves	Occasionally recorded
Pentalonia nigronervosa	Anholocyclic in the intertropical zone	Banana and other Musaceae	Throughout the intertropical zone; occasionally in Europe and North America (in greenhouses)	Infests leaf sheaths and folded leaves	Efficient vector of *Banana bunchy top virus*
Rhopalosiphum rufiabdominale	Anholocyclic or holocyclic, heteroecious	Occasionally on banana	Nearly worldwide	Occasionally infests roots	Minor pest - there are a few other species of the same

					genus infesting banana and vectoring *Banana mosaic virus*
Greenidea anonae	Anholocyclic	*Annona* spp. and others	Widespread from India to Japan and Indonesia	Infests young leaves	Minor pest
Greenidea mangiferae	Anholocyclic	Mango	Recorded from Taiwan	Infests young shoots	Minor pest
Greenidea formosana	Anholocyclic	Guava and several other Myrtaceae	Widespread from India to East Asia	Infests young shoots	Occasionally injurious
Greenidea ficicola - Fig. 30.3c,d	Anholocyclic	Guava, litchi and other Myrtaceae and Moraceae	Mediterranean, South-east Asia, Africa, Australia, North America	Infests young shoots	Minor pest
Taiwanaphis kalipadi	Anholocyclic	Sugar apple	India (West Bengal)	Infests young shoots	Minor pest
Monelliopsis pecanis - Fig 30.3a,b	Holocyclic, monoecious	Pecan	North and Central America, Mediterranean, South Africa	Infests leaves	Commonly injurious
Monellia caryella	Holocyclic, monoecious	Pecan	North America; Israel and Spain (introduced)	Infests leaves	Commonly injurious
Tinocallis caryaefoliae	Holocyclic, monoecious	Pecan	North America	Infests leaves	Commonly injurious
Phylloxera spp. (group of five species)*	Holocyclic (mostly monoecious, except one species)	Pecan	North America (one species introduced in Southeastern Europe: Georgia)	Galls on leaflets, or sometimes on twigs and nuts	Commonly injurious

*For details see Blackman and Eastop, 2000.

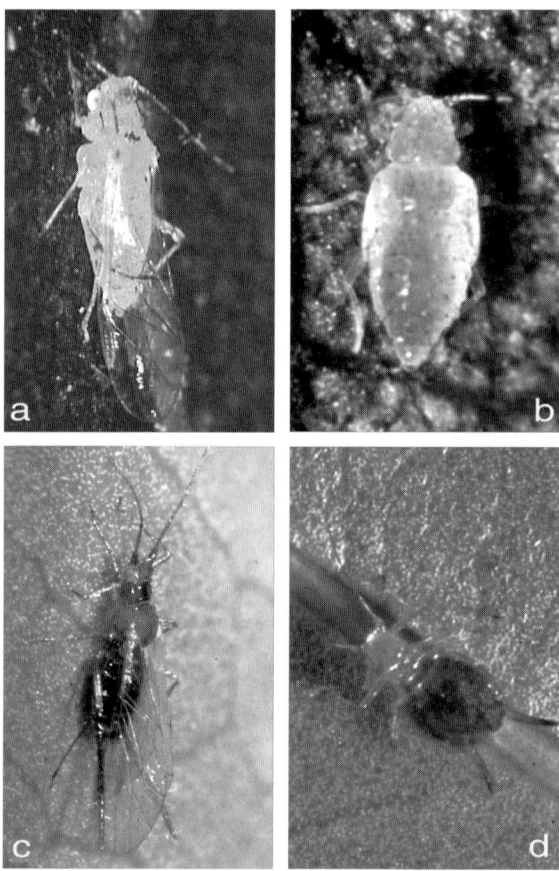

Fig. 30.3. Tropical and subtropical fruit tree aphids. (a) alate viviparous female of *Monelliopsis pecanis* (from Barbagallo and Suma, 1999; original photo by S. Barbagallo); (b) 4th-instar nymph of *M. pecanis* (from Barbagallo and Suma, 1999; original photo by S. Barbagallo); (c) alate viviparous female of *Greenidea ficicola* (from Barbagallo *et al.*, 2005; courtesy of *Informatore Fitopatologico*); (d) apterous viviparous female of *G. ficicola* (from Barbagallo *et al.*, 2005; courtesy of *Informatore Fitopatologico*).

large colonies on branches of pecan, while *Phylloxera* species develop leaf-galls on the same host plant. Sap drain and saliva injection lead to loss of productivity and general debilitation of affected plants. Development of sooty mould and ant attendance on colonies of myrmecophilous aphids are additional sources of trouble and inconvenience for growers. Nevertheless, virus transmission by aphids represents the most important source of losses for several tropical crops. Non-persistent viruses can even be transmitted by the probes of vagrant alatae on receptive hosts, i.e. by aphid species which usually do not develop colonies on those plants. The widespread *Banana mosaic virus* (a variant strain of *Cucumber mosaic virus*) is transmitted non-persistently by several aphid species including *A. gossypii*, *M. persicae*, and *Rhopalosiphum* spp.

(Hughes and Eastop, 1991). *Pentalonia nigronervosa* infesting banana is responsible for semi-persistent transmission (Hu *et al.*, 1996) of the more severe *Banana bunchy top virus* in several regions (Africa, East Asia, India, Pacific Regions, Australia). *Papaya ringspot virus* is another destructive disease transmitted by polyphagous aphids (mainly *A. gossypii* and *M. persicae*, but also *A. spiraecola* and *M. euphorbiae*); it is widespread in Eastern Asia, the Caribbean, Hawaii, and South America (Gonsalves, 1998; Valdes-Estrada *et al.*, 2005). The same aphids, particularly the first two species, are also vectors of various virus diseases of passion fruits (*Passion fruit woodiness virus*, *Passion fruit mottle virus*, *Sri Lanka passion fruit mottle virus*) in Asia and Australasia (Chang and Lin, 1992; Parry *et al.*, 2004).

Monitoring and economic thresholds

Although aphid population monitoring can be carried out easily by direct inspection of the plants, very little is known about damage thresholds for aphids for most of the tropical crops referred to here. In areas of high humidity in southeast USA (Alabama, Georgia), a variable threshold for pecan of 10–20 aphids in the 'yellow aphid' complex (*M. pecanis* and *M. caryella*) per leaf in June, and an average of three *Tinocallis caryaefoliae* (black pecan aphid) per leaf later in the season are suggested (Harris, 1983). Against the 'yellow aphid' complex in California, thresholds of an average of more than 20 aphids/leaf (from June to mid-August), or more than 10 aphids/leaf (after mid-August to autumn), have been recommended (Sibbett and van Steenwyk, 2000).

Supervised chemical control

Chemical control, if it is necessary, should follow the methodology of supervised control as suggested for citrus orchards (i.e. based on economic thresholds and therefore on the economic benefits of treatment). However, for most situations such thresholds have not yet been determined. As for citrus orchards (see earlier), it is necessary to pay attention to the correct choice of pesticides, with emphasis on some selectivity in order to reduce undesirable side effects, particularly on entomophagous and other non-target organisms in the agroecosystem. This means that the more broad-spectrum insecticides, still applied (because they are usually cheaper) against aphids on these crops (especially pecan and banana), should be replaced by specific aphicides whenever possible.

Biological control

In tropical climates, biological control plays an important role in limiting the population increase of aphids. Several species of predators and parasitoids have been recorded,

but their efficiency is not well known (see, for example, Remaudière *et al.*, 1985, for a general overview on the efficiency of Aphidiinae in tropical Africa; also Verghese and Tandon, 1987).

In general, some parasitoid wasps, like *L. testaceipes*, play an important role in the natural biological control on different aphid pests (i.e. *A. gossypii* infesting several tropical fruit tree crops, *P. nigronervosa* on banana) in several countries. On pecan, against the 'yellow aphids' complex, the two parasitoids *Aphelinus perpallidus* and *Trioxys pallidus*, in combination with predators, make a useful contribution to limiting aphid population growth in the USA and Israel (Watterson and Stone, 1982; Monsour, 1988).

Acknowledgements

Many thanks are due to the editor (Bayer S.p.A., Milan) of the book *Aphids of the Principal Fruit-Bearing Crops* (Barbagallo *et al.*, 1997) for permission to reproduce Fig. 30.1 (a–d) and Fig. 30.2 (a and b) from colour plates.

Executive Summary

The extensive research effort that has been devoted to citrus means that economic thresholds have been proposed for most of the injurious aphid species on this crop; in contrast, thresholds on tropical fruit trees have still to be worked out.

For both citrus and tropical fruits, choice of a selective and ephemeral insecticide is crucial to preserve natural enemies that, although they may appear too late to prevent the first increase of aphid numbers in the spring, often provide good control later in the season. Currently, any biological control relies on the indigenous entomophagous populations; there is no release of natural enemies in the Mediterranean region.

No use is made of host-plant resistance.

Cultural control is involved in IPM in the sense that farmer practices such as irrigation and nitrogen fertilization can predispose crops for rapid aphid increase and therefore the need to monitor with a view to timing sprays in citrus.

Semiochemicals do not form part of aphid IPM in citrus or tropical fruit tree crops.

Interactions between control measures only involve care in the use of insecticides to avoid damage to biological control.

References

Alcântara Santos, L. and Barbagallo, S. (1989) Afidios dos citrinos em Moçambique. *Phytophaga* 3 (1985–1989), 199–217.

Andrews, M.C., Callaghan, A., Field, L.M., Williamson, M.S. and Moores, G.D. (2004) Identification of mutations conferring insecticide-insensitive AChE in the cotton–melon aphid, *Aphis gossypii* Glover. *Insect Molecular Biology* 13, 555–561.

Barbagallo, S. (1966) L'afidofauna degli agrumi in Sicilia. *Entomologica* 2, 201–260.

Barbagallo, S. and Alcântara Santos, L. (1989) *Toxoptera odinae* (v.d.G.) (Hom.: Aphididae), uma nova praga do cajueiro (*Anacardium occidentale* L.) em Moçambique. *Phytophaga* 3 (1985–1989), 163–171.

Barbagallo, S. and Patti, I. (1983) Citrus aphids and their entomophagous [*sic*] in Italy. In: Cavalloro, R. (ed.) *Aphid Antagonists*. Balkema, Rotterdam, pp. 116–119.

Barbagallo, S. and Patti, I. (1986) The citrus aphids: behaviour, damages and integrated control. In: Cavalloro, R. and Di Martino, E. (eds) *Integrated Pest Control in Citrus-Groves*. Balkema, Rotterdam, pp. 67–75.

Barbagallo, S. and Suma, P. (1999) Recenti infestazioni in Sicilia dell'afide giallo del Pecan, *Monelliopsis pecanis* Bissell. *Bollettino Zoologia Agraria e Bachicoltura, Series 2* 31, 241–249.

Barbagallo, S., Cravedi, P., Pasqualini, E. and Patti, I. (1997) *Aphids of the Principal Fruit-Bearing Crops*. Bayer, Milan, 123 pp.

Barbagallo, S., Bella, S. and Cocuzza, G. (2005) Rinvenimento dell'afide orientale *Greenidea ficicola* su *Ficus* ornamentali in Italia meridionale. *Informatore Fitopatologico* 2, 25–29.

Bar-Joseph, M., Raccah, B. and Loebenstein, G. (1977) Evaluation of the main variables that affect citrus tristeza transmission by aphids. *Proceedings of International Society of Citriculture* 3, 958–961.

Blackman, R.L. and Eastop, V.F. (2000) *Aphids on the World's Crops. An Identification and Information Guide*, 2nd edn. Wiley, Chichester, 466 pp.

Calabrese, F. (1993) *Frutticoltura Tropicale e Subtropicale, Volumes 1 and 2*. Edagricole, Bologna, 247 and 471 pp.

Cavalloro, R. and Prota, R. (1983) *Integrated Control in Citrus Orchards: Sampling Methodology and Threshold for Intervention Against the Principal Phytophagous Pests*. CEC, Luxembourg, 63 pp.

Chang, C.A. and Lin, Y.D. (1992) Ecology of passionfruit virus transmission and the evaluation of the effectiveness of using virus-free seedlings to control passionfruit virus diseases in Taiwan. *Plant Pathology Bulletin* 1, 140–146.

CIE (Commonwealth Institute of Entomology) (1961) *Distribution Maps of Pests, No. 131*. CAB International, Wallingford, Oxon, 1 p.

CIE (1968) *Distribution Maps of Pests, No. 18*. CAB International, Wallingford, Oxon, 1 p.

CIE (1998) *Distribution Maps of Pests, No. 132*. CAB International, Wallingford, Oxon, 1 p.

CIE (2001) *Distribution Maps of Pests, No. 256*. CAB International, Wallingford, Oxon, 1 p.

Davino, M., Areddia, R., Polizzi, G. and Patti, I. (1990) Aphid transmissibility of some isolates of citrus tristeza virus (CTV) under restricted environment in Sicily. In: Cavalloro, R. (ed.) *'Euraphid' Network: Trapping and Aphid Prognosis*. CEC, Luxembourg, pp. 237–243.

De Huiza, J.R. and Ortiz, P.M.S. (1981) Algunos Aphidiinae (Hymenopt.: Braconidae) parasitoides de afidos (Homopt.: Aphididae) en el Perù. *Revista Peruana de Entomologia* 23, 129–132.

Gonsalves, D. (1998) Control of papaya ringspot virus in papaya: a case study. *Annual Review of Phytopathology* 36, 415–437.

Halbert, S.E. and Brown, L.G. (1996) *Toxoptera citricida* (Kirkaldy), brown citrus aphid. Identification, biology and management strategies. *Entomology Circular (Gainesville), No. 374*. Florida Department of Agriculture and Consumer Services, Gainesville, 6 pp.

Halbert, S.E., Genc, H., Cevik, B., Brown, L.G., Rosales, I.M., Manjunath, K.L., Pomerinke, M., Davison, D.A., Lee, R.F. and Niblett, C.L. (2004) Distribution and characterization of citrus tristeza virus in South Florida following establishment of *Toxoptera citricida*. *Plant Disease* 88, 935–941.

Harris, M.K. (1983) Integrated pest management of pecans. *Annual Review of Entomology* 28, 291–318.

Hu, J.S., Wang, M., Sether, D., Xie, W. and Leonhardt, K.W. (1996) Use of polymerase chain-reaction (PCR) to study transmission of banana bunch-top virus by the banana aphid (*Pentalonia nigronervosa*). *Annals of Applied Biology* 128, 55–64.

Hughes, M.A. and Eastop, V.F. (1991) Aphids associated with damage to banana plants and a corrected identity for a banana mosaic virus vector (Hemiptera: Aphididae). *Journal of the Australian Entomological Society* 30, 278.

Ilharco, F.A., Sousa-Silva, C.R. and Aluneda, A.A. (2005) First report on *Toxoptera citricidus* (Kirkaldy) in Spain and continental Portugal (Homoptera, Aphidoidea). *Agronomia Lusitana* 51 (2003–2005), 19–21.

Inaizumi, M. (1981) Life cycle of *Aphis gossypii* Glover (Homoptera, Aphididae) with special reference to biotype differentiation on various host plants. *Kontyu* 49, 219–240.

Komazaki, S. (1983) Overwintering of the spirea aphid, *Aphis citricola* van der Goot (Homoptera: Aphididae) on citrus and spirea plants. *Applied Entomology and Zoology* 18, 301–307.

Komazaki, S. (1991) Studies on the biology of the spiraea aphid *Aphis spiraecola* Patch with special reference to biotypic differences. *Bulletin of the Fruit Tree Research Station 1991* (Supplement 2), 60 pp.

Komazaki, S. (1994) Ecology of citrus aphids and their importance to virus transmission. *Japan Agricultural Research Quarterly* 28, 177–184.

Komazaki, S. (1998) Difference of egg diapause in two host races of the spirea aphid. *Entomologia Experimentalis et Applicata* 89, 201–205.

Komazaki, S. and Osakabe, Mh. (1998) Variation of Japanese *Aphis gossypii* clones in the life-cycle, host suitability and insecticide susceptibility, and estimation of their genetic variation by DNA analysis. In: Nieto Nafría, J.M. and Dixon, A.F.G. (eds) *Aphids in Natural and Managed Ecosystems*. Universidad de León, Spain, pp. 83–89.

Komazaki, S., Sakagami, Y. and Korenaga, R. (1979) Overwintering of aphids on citrus trees. *Japanese Journal of Applied Entomology and Zoology* 23, 246–250.

Martin, J.H. (1983) The identification of common aphid pests of tropical agriculture. *Tropical Pest Management* 29, 395–411.

Meliá, A. (1982) Prospeccion de pulgones (Homoptera, Aphidoidea) sobre citricos en España. *Boletín del Servicio de Plagas* 8, 159–168.

Meliá, A. (1993) Evolución poblacional de *Toxoptera aurantii* (Boyer de Fonscolombe) (Homoptera: Aphididae) en los últimos quince años y su relación a la aparición de *Lysiphlebus testaceipes* (Cresson) (Hymenoptera: Aphididae). *Boletín de Sanidad Vegetal-Plagas* 19, 609–617.

Michaud, J.P. (1998) A review of the literature on *Toxoptera citricida* (Kirkaldy) (Homoptera: Aphididae). *Florida Entomologist* 81, 39–61.

Monsour, F. (1988) Parasites of *Monellia caryella* (Hom.: Aphididae): phenology and effect on the aphid population in pecan orchards in Israel. *Entomophaga* 33, 371–375.

Nickel, O. and Klingauf, F. (1985) Biologie und Massenwechsel der tropischen Citrus-Blattlaus *Toxoptera citricidus* in Beziehung zu Nützlingsaktivität und Klima in Misiones Argentinien (Homoptera: Aphididae). *Entomologia Generalis* 10, 231–240.

Parry, J.N., Davis, R.I. and Thomas, J.E. (2004) Passiflora virus Y, a novel virus infecting *Passiflora* spp. in Australia and the Indonesian province of Papua. *Australasian Plant Pathology* 33, 423–427.

Remaudière, G., Autrique, A., Eastop, V.F., Starý, P., Aymonin, G., Kafurera, J. and Dedonder, R. (1985) *Contribution à l'Écologie des Aphides Africains*. FAO, Rome, 214 pp.

Roistacher, C.N., Bar-Joseph, M. and Gumpf, D.J. (1984) Transmission of tristeza and seedling yellows tristeza virus by small populations of *Aphis gossypii*. *Plant Disease* 68, 494–496.

Rondón, A., Arnal, E. and Godoy, F. (1983) Comportamiento del *Verticillium lecanii* (Zimm.) Viegas, patogeno del afido *Toxoptera citricidus* (Kirk.) en fincas citricolas de Venezuela. *Agronomia Tropical* 30, 201–212.

Sibbett, G.S. and van Steenwyk, R.A. (2000) *Pest Management Guidelines – Pecan. Yellow Aphid Complex*. University of California, ANR Publication No. 3456, 2 pp.

Song, S.S., Oh, H.K. and Motoyama, N. (1995) Insecticide resistance mechanism in the spiraea aphid, *Aphis citricola* (van der Goot). *Korean Journal of Applied Entomology* 34, 89–94.

Starý, P., Lyon, J.P. and Leclant, F. (1988). Biocontrol of aphids by the introduced *Lysiphlebus testaceipes* (Cress.) (Hym., Aphidiidae) in Mediterranean France. *Journal of Applied Entomology* 105, 74–87.

Stroyan, H.I.G. (1961) Identification of aphids living on *Citrus. FAO Plant Protection Bulletin* 9, 45–65.

Takanashi, M. (1990) Development and reproductive ability of *Lysiphlebus japonicus* Ashmead (Hymenoptera: Aphididae) parasitizing the brown citrus aphid, *Toxoptera citricidus* (Kirkaldy) (Homoptera: Aphididae). *Japanese Journal of Applied Entomology and Zoology* 34, 237–243.

Tang, Y.Q. and Yokomi, R.K. (1996) Effect of parasitism by *Aphelinus spiraecolae* (Hymenoptera: Aphelinidae) on development and reproduction of spiraea aphid (Homoptera: Aphididae). *Environmental Entomology* 25, 703–707.

Toda, S., Komazaki, S., Tomita, T. and Kono, Y. (2004) Two amino acid substitutions in acetylcholinesterase associated with pirimicarb and organophosphorous insecticide resistance in the cotton aphid, *Aphis gossypii* Glover (Homoptera: Aphididae). *Insect Molecular Biology* 13, 549–553.

Tremblay, E. and Barbagallo, S. (1983) *Lysiphlebus testaceipes* (Cr.), a special case of ecesis in Italy. In: Cavalloro, R. (ed.) *Aphid Antagonists*. Balkema, Rotterdam, pp. 65–68.

Tremblay, E., Barbagallo, S., Micieli De Biase, L., Monaco, R. and Ortu, S. (1980) Composizione dell'entomofauna parassitica vivente a carico degli afidi degli agrumi in Italia (Hymenoptera Ichneumonoidea, Homoptera Aphidoidea). *Bollettino Laboratorio Entomologia Agraria Portici* 37, 209–216.

Valdes-Estrada, M.E., Aldana-Llanos, L. and Evangelista-Lozano, S. (2005) Incidencia de afidos en el cultivo de *Carica papaya* L. *Proceedings of the International Society for Tropical Horticulture* 48, 108–110.

Verghese, A. and Tandon, P.L. (1987) Interspecific associations among *Aphis gossypii*, *Menochilus sexmaculatus* and *Camponotus compressus* in a guava ecosystem. *Phytoparasitica* 15, 289–297.

Wallace, J.M. (1975) Vein enation–woody gall. In: Bové, J.M. and Vogel, R. (eds) *Description and Illustration of Virus and Virus-Like Diseases of Citrus*. SETCO-IFRA, Paris, 6 pp.

Watterson, G.P. and Stone, J.D. (1982) Parasites of black-margined aphids and their effect on aphid population in Far-West Texas. *Environmental Entomology* 11, 667–669.

Yokomi, R.K. and DeBorde, R.L. (2005) Incidence, transmissibility, and genotype analysis of citrus tristeza virus (CTV) isolates from CTV eradicative and noneradicative districts in Central California. *Plant Disease* 89, 859–866.

31 Decision Support Systems

Jonathan D. Knight[1] and Deborah J. Thackray[2,3]

[1]*Centre for Environmental Policy, Imperial College London, Silwood Park Campus, Ascot, Berks, SL5 7PY, UK;* [2]*Centre for Legumes in Mediterranean Agriculture, MO80, Faculty of Natural and Agricultural Sciences, University of Western Australia, 35 Stirling Highway, Crawley, WA 6009, Australia;* [3]*Plant Pathology, Department of Agriculture and Food Western Australia, Locked Bag 4, Bentley Delivery Centre, WA 6983, Australia*

Introduction

Over the years, much research on aphids has been targeted at reducing the damage they inflict to crops or the viruses they transmit. This has been achieved by improving our understanding of aphid biology and behaviour and of virus epidemiology. While much useful information has reached end-users, usually farmers or advisers, and helped reduce yield and quality losses, a significant proportion of this knowledge and information has remained in the domain of the scientist. This may be through lack of funding, neglect, inadequate communication of the results to end-users, or because the results are too difficult to simplify for commercial use. Traditionally, information has passed from research institutes to end-users *via* subsidized government advisory services. However, generally in most developed countries, access to up-to-the-minute information and advice now involves some monetary cost being passed on to farmers or advisers. For a number of years, alternatives to the 'traditional' method of on-farm advice have included a rapid rise in non-government advisers and the use of decision support aids, which have been developed with a varying degree of success. These aids range from highly sophisticated 'Expert Systems' that seek to replace human advisers entirely, to relatively simple models or forecasts that growers and advisers can use in support of their decision making. Used either alone or in combination, they are addressed here under the collective term 'Decision Support Systems' (DSSs). Their common aim is to improve the decision making of farmers and advisers for crop management, thereby increasing profits and/or reducing environmental impact.

What is a Decision Support System?

The term 'Decision Support System' (DSS) has been used rather loosely to include a number of different methods of improving pest management and decision making over control measures. In this chapter, DSS is taken to mean one or more models, databases, rules, or pieces of information that together support the decision-making process for control of aphids causing direct feeding damage or vectoring damaging viruses. The DSS does this by providing information to users in a form that does not require specialist knowledge of the

©CAB International 2007. *Aphids as Crop Pests*
(eds H. van Emden and R. Harrington)

underlying science or computer program. In some cases, queries from users might be accommodated, for example to compare current catches of aphids with historical catches or to run a predictive model with a range of different weather or planting scenarios. Usually, a model or a database on its own would not allow untrained users to access such information easily. DSSs therefore wrap the functional core in an outer layer that allows a flexible and easy exploration of the data, model, expertise, or other information contained within. Hansen (1999) suggested that a model should be sufficiently reliable to serve as a decision-making tool. In reality, models are often developed for research purposes and are not suitable for use by untrained individuals. Models originating in this way rarely make suitable starting points for DSSs. The key to success is to develop DSSs from farmers' and advisers' points of view and with their close involvement.

Are Aphids Suitable Subjects for the Development of DSSs?

The development of a DSS is only justified where the effort required is repaid by the benefits that arise. Thus, any system has to address either a widespread problem or one that results in significant losses, or both. Also, the problem must be sufficiently difficult to solve, such that simple rules cannot be used to make decisions over management. Using these criteria, aphids and the viruses they transmit would appear to be worthy subjects for the development of these systems. Both cause widespread damage to crops and, in many cases, aphids have complex population dynamics. Including virus transmission in the system makes it even more complex and, importantly, both aphids and viruses have been the subjects of a great deal of research that can be used in developing DSSs. In addition, measuring some of the parameters that are useful in prediction, such as aphid flight, is not possible at the farm level and has to be done at a larger scale to provide useful data,

followed by their interpretation for use at the local level.

Examples of Aphid and Virus DSSs

Several DSSs have been developed that have aimed to improve decision making for aphid and virus control. One of the earliest of these was EPIPRE, a disease and pest management system developed in the Netherlands for winter wheat (Zadoks, 1981). This system relied on farmers supplying information about the crop and numbers of aphids to a central computer where the predictive model was run, with advice on actions to take delivered *via* advisory services. Since then, systems have tended to become more complex. However, with the development of superior and cheaper technology, it is now possible to have self-contained DSSs on farm computers or available *via* the Internet. The range of DSSs for aphids and viruses is quite large, with examples for many crops, the majority coming from areas where the problem is most difficult and the rewards greatest. Table 31.1 shows some examples of DSSs developed to improve aphid control and management in crops in different countries.

Many of these systems are not developed to function individually but are integrated with those for other agronomic problems that occur in the crop of interest. This undoubtedly increases the value of DSSs, since it enables growers to find advice on many management aspects of the crop, thereby facilitating an integrated approach, ensuring maximum benefit to end-users with minimum impact on the environment. This approach does present its own problems, since the process of specifying and developing an all-encompassing system is much more challenging and expensive than for one that addresses a single problem. An example of such a system is DESSAC (decision support system for arable crops), currently being developed in the UK, which will include modules for fungicide application to winter wheat, barley yellow dwarf virus (BYDV) vector control, spring barley production, oilseed rape pest control, soil nitrogen assessment, and environmental management (Brooks, 1998).

Table 31.1. A selection of DSSs for aphid and virus control, published from 1981–2004.

Crop	Target aphid	Country	Reference
Cereals	Various cereal aphids	Netherlands	Zadoks, 1981
Cotton	*Aphis gossypii*	USA	Helm *et al.*, 1987
Cereals	Various cereal aphids	Switzerland	Derron and Forrer, 1989
Cereals	Various cereal aphids	UK	Mann and Wratten, 1991
Cereals	Various cereal aphids	Spain	Gonzalez-Andujar *et al.*, 1992
Hops	*Phorodon humuli*	Czech Republic	Mozny *et al.*, 1993
Field beans	*Aphis fabae*	UK	Knight and Cammell, 1994
Cereals	Various cereal aphids	Denmark	Murali *et al.*, 1999
Cereals	Various cereal aphids	UK	Harrington *et al.*, 1999
Cotton	*Aphis gossypii*	Australia	Hearn and Bange, 2002
Lupins	Various lupin aphids and *Cucumber mosaic virus*	Australia	Thackray *et al.*, 2004
Wheat	*Schizaphis graminum*	USA	Elliott *et al.*, 2004

Development and Delivery of DSSs

Success in developing a DSS depends on many factors, including the availability of adequate research and other information that can be assimilated into a system, the availability of necessary skills and funds needed to develop the software or other means of delivery, end-users being able to access and use the product, and the existence of a long-term funding commitment to update the system regularly over time. Assembling the relevant components often requires different information sources, such as models and databases, to be brought together with expert interpretation and presented in a coherent fashion. There may be a need to re-work research models that are too complicated, or even to reject them and start again. Presentation and delivery of DSSs to end-users requires careful planning since the appearance, ease of use, and reliability of DSSs are very important to their adoption. The end-users' requirements must be considered in determining the shape and form of DSSs (Arinze, 1992), and potential users should always be involved as much as possible in their development.

Several methods have been used for distributing DSSs or their outputs to end-users, including hand-held cardboard calculators, computer programs installed via CD-ROMs on the end-users' computers, and web-based applications and forecasts that are accessed over the Internet (e.g. Mann and Wratten, 1991; Thackray *et al.*, 2004). The method chosen depends on various factors, including the number of end-users that have access to a particular technology. Many farmers and advisers use personal computers for the day-to-day running of their businesses, so a CD-ROM containing the DSS may be a good way to distribute it. However, if the DSS relies on the provision of up-to-date information that is not collected by the grower, the program will need to be updated on a regular basis from some other source, which is best achieved via an Internet site or through a modem. The current trend is towards systems that reside on a remote server accessible via the Internet. This has the advantage that the data in the program are always up-to-date and any changes can be made quickly and easily. However, problems arise if the server or service does not function correctly at all times.

Case Studies

Two DSSs illustrate contrasting approaches to assisting farmers and advisers in making decisions about the need for and timing of insecticides to prevent aphids from spreading BYDV in cereal crops. The first, developed in

New Zealand, uses the historical relationship between suction trap catches of cereal
aphids in autumn/winter and BYDV incidence in the following summer to infer
likely BYDV risk in two winter cereal-growing
regions. The second, developed in Western
Australia, is based on a computer simulation
model. This uses climate data, especially
summer/autumn rainfall before the start of
the growing season, to forecast arrival and
build-up of aphids within winter cereal crops,
subsequent spread of BYDV, and potential
yield losses in the 'grainbelt' region of Western Australia. With both the New Zealand
and Western Australian situations, BYDV
epidemiology is simpler than in Northern
Europe because the only significant BYDV
vectors are *Rhopalosiphum padi* (bird cherry–
oat aphid) and *Rhopalosiphum maidis* (corn
leaf aphid) (Farrell and Sward, 1989; Jones
et al., 1990; Farrell and Stufkens, 1992;
McKirdy and Jones, 1993), and in New Zealand and Western Australia they are exclusively anholocyclic, removing the need to
distinguish aphid forms that do not colonize cereals. Outbreaks of aphids and epidemics of BYDV vary between years and
cropping regions. Greatest yield loss results
from infection at early growth stages (up to
stem extension, GS30; Doodson and Saunders,
1970; Zadoks *et al.*, 1974; Thackray *et al.*,
2005).

In the Canterbury region of New Zealand,
R. padi is the main BYDV vector and has
distinct flight peaks in spring and autumn,
with periods of low flight activity during
winter and summer (Farrell and Stufkens,
1989, 1992). Primary BYDV infection of
autumn-sown cereal crops results from
aphids infected with BYDV flying from
other nearby cereal crops or pastures (e.g.
grass pastures, forage cereals). Both primary
infection in autumn and secondary spread
through winter into spring cause yield
losses. Severe BYDV outbreaks are usually
associated with substantial flight activity
from April until June (late autumn-early
winter). However, severe outbreaks have
also occurred in years with little early flight
activity but with mild winters, allowing late
aphid flights in June and July and considerable secondary virus spread over winter

(Lowe, 1963; Farrell and Stufkens, 1992;
Teulon *et al.*, 1999, 2001). Recent trends
towards earlier sowing have increased the
need for earlier forecasts of BYDV risk.
Increased virus incidence is associated
with greater primary infection as crops are
emerging during the greatest flight period,
and also more secondary infection due to an
extended warmer period before winter.

In the Western Australian 'grainbelt',
which has a Mediterranean-type climate with
a marked summer drought, potential aphid
vectors of BYDV survive the dry summer
period (from 3-5 months) between cropping
seasons in the parthenogenetic state, on
perennial grass weeds and crop volunteers
in isolated wet spots (Jones *et al.*, 1990;
McKirdy and Jones, 1993; Berlandier and
Cartwright, 1998; Hawkes and Jones, 2002,
2005). Some of these plants also act as BYDV
reservoirs from which aphids carry BYDV
into crops (Jones *et al.*, 1990; McKirdy and
Jones, 1993). Although the main aphid
flights are in autumn and spring, flights
continue at reduced levels throughout the
mild winter (Thackray *et al.*, 2005). Crops
sown early, from mid-April to late May, that
are exposed to early migrations of aphids,
suffer most damage in autumn and early
winter (April to mid-July). However, in
years of late migration, spread is mostly in
winter and early spring, which is more
damaging in late than early-sown crops.
Cereal crops are most prone to yield losses
when BYDV spread occurs during the first
8-10 weeks after crop emergence (McKirdy
and Jones, 1993, 1996, 1997; Thackray *et al.*,
1998, 2005). Historically, high rainfall zones
are most prone to BYDV because here aphids
more often arrive and start spreading BYDV
early in the life of the crop, with yield losses
of up to 50% in crops with high incidences
of early disease (McKirdy and Jones, 1996,
1997; Thackray *et al.*, 2005).

Strategically timed applications of
pyrethroid insecticides at early crop growth
stages give the most reliable form of control
of aphid vectors of BYDV and are relatively
cheap (McKirdy and Jones, 1996). Therefore,
although BYDV epidemics are sporadic,
many growers in New Zealand and Western
Australia use insecticides prophylactically

in cereal crops (Greer and Teulon, 2003). However, increased costs of production, the possibility of insecticide resistance developing in pest populations, lower wheat prices, the trend towards more environmentally responsible farming methods, and greater use of information technology have led to a need to plan insecticide use according to forecasts of aphid outbreaks and BYDV epidemics. By forecasting risk, both the New Zealand and Western Australian DSSs aim to optimize use of insecticides and improve timing of foliar sprays when they are necessary to control aphids spreading BYDV.

New Zealand Decision Support System
(www.AphidWatch.com)

A BYDV risk forecast for autumn-sown cereals was formally made available to users in Canterbury, New Zealand in 1997. Using a similar methodology to that used for the BYDV infectivity index in the UK (Northing *et al.*, 2004), it is based on 8 years of data relating aphid numbers caught in a 7.5 m high suction trap (Johnson and Taylor, 1955) at Lincoln, Canterbury, in June and July to the proportion of crops sown locally after mid-May with economically damaging infections (> 5%) of BYDV (Farrell and Stufkens, 1992) (Fig. 31.1). More recently, Teulon *et al.* (2004b) reported a significant positive relationship between numbers of aphids caught in suction traps and the numbers infesting wheat fields. Since 1997, the aphid monitoring/forecasting programme has been extended to four suction traps in the Canterbury plain and one in Southland. Due to similar flight patterns recorded in some traps on the Canterbury plain, it may be possible to reduce the number of suction traps in that area (Teulon *et al.*, 2004a). Predictions based on April and May suction trap catches for early-sown crops (sown before mid-May) and June and July catches for late-sown crops (sown after mid-May) are made in early June and August, respectively. This enables DSS users to make decisions about the need for applications of insecticides to prevent secondary spread of BYDV in spring (Farrell and Stufkens, 1989).

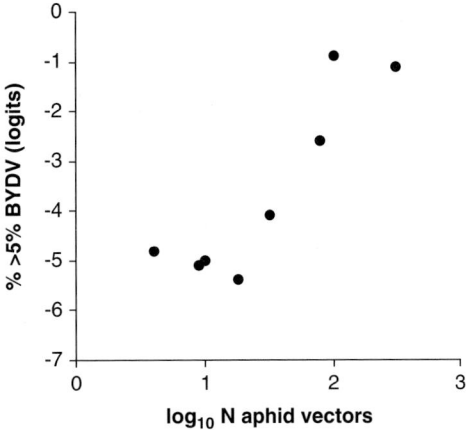

Fig. 31.1. Relationship between $\log_{10}$ N aphid vectors of BYDV trapped in June–July at Lincoln, New Zealand (x) and logit % crops with >5% BYDV infection in Canterbury in the following late November and early December (y):
$y = 2.311 \, x \, (\pm 0.523) - 6.91 \, (\pm 1.15)$; $r^2 = 0.82$ (from Farrell and Stufkens, 1992 with permission).

The forecast for early-sown crops is rather tentative because the model was not developed with data from such crops. In 1998 and 2000, although overall BYDV incidence was low, the forecast considerably underestimated incidence in crops emerging after 1 June in 1998 (Teulon *et al.*, 1999) and up until late-May in 2000 in early-sown crops and crops sown at low elevations (Teulon *et al.*, 2001). Extensive BYDV infection still occurred in these crops without large aphid flights in April–May (2000) or June–July (1998). Other factors, such as the influence of mild winter weather conditions on secondary spread, need to be considered in refining the forecast (Teulon *et al.*, 1999, 2001). In addition, the forecast for Canterbury relies on the aphid catches from four suction traps to represent the whole region.

To assist in decisions about BYDV control, growers can access the forecasts for aphid flights and virus risk by various means. These include an Internet site (www.AphidWatch. com), direct-dial telephone message, and reports to the Foundation for Arable Research, which forwards the information to growers by e-mail, and media articles. The DSS on the Internet site shows graphical representations of weekly aphid flights from April–July at each suction trap site, and indicates how the current year's flight patterns can be used to predict BYDV incidence by comparing with those in previous years of high or low virus incidence (Fig. 31.2). Each graph is updated on a weekly basis, usually within 2 days of the trap sample being received for counting. General information on BYDV, its aphid vectors and their control is included on the site. The site also contains information on other pest aphids in potato, lettuce, and squash crops. Due to low levels of funding in recent years, this system has been running at a subsistence level. However, severe outbreaks of BYDV in cereals in Canterbury in the 2005/06 season provided strong interest in underpinning the system with adequate funding.

Despite its limitations, AphidWatch is widely used by farmers, who also support it financially. The monitoring information was judged as a good or satisfactory use of levy funds by over 92% in a farmer survey (Greer and Teulon, 2003). In 2000, peak usage of the forecast occurred in May and June with, on average, two visitor sessions to the Internet and one telephone call per day for this period. Bicknell *et al.* (2000) reviewed its financial value. Results suggested an average annual benefit to the region of around NZ$450,000, but the benefit varied greatly depending on forecast accuracy and adoption rate. In years of severe BYDV outbreaks, the cost of failing to control spread is high, but the cost of spraying in a non-outbreak year is relatively low. Bicknell *et al.* (2000) calculated that at 2000 prices and by saving application costs by including insecticides with herbicide applications, a BYDV spray could be justified on economic grounds if it resulted in an overall yield increase of only 1.5%. The value of the forecast, assuming 100% accuracy, was shown to be highest when the probability of a BYDV epidemic was 14%. If the forecast is 80% accurate, the forecast strategy is optimal for outbreak probabilities between 3–39% only. Above 39%, a prophylactic spray regime is, on average, more profitable.

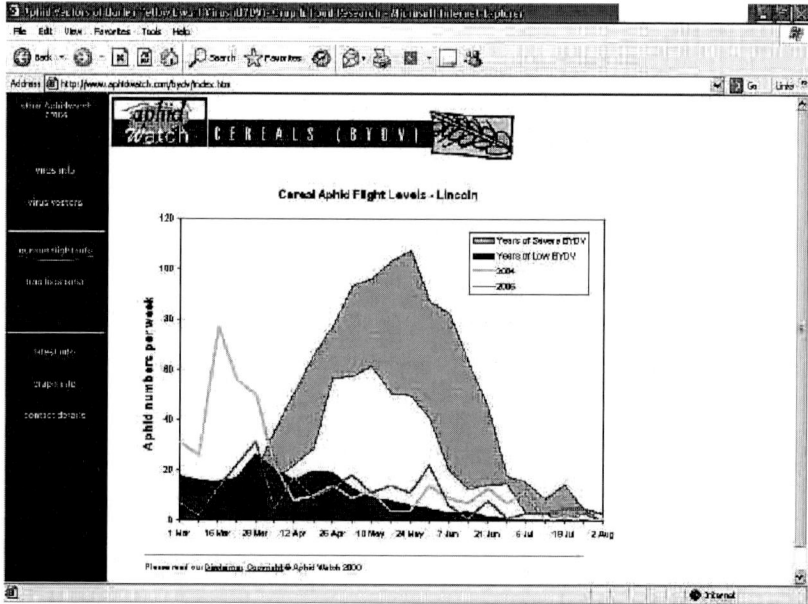

Fig. 31. 2.   Example output from the NZ Decision Support System (www.AphidWatch.com), with a graph illustrating the pattern of cereal aphid catches in a suction trap at Lincoln, New Zealand in 2004 and 2005, compared with the pattern in years of severe and low BYDV incidence.

Western Australian Decision Support System (www.agric.wa.gov.au/bydv)

A simulation model was developed to forecast aphid outbreaks and BYDV epidemics in cereals growing in the 'grainbelt' region of Western Australia. The design was based on a model for aphid and virus control in lupin (Thackray *et al.*, 2004). The major steps in the BYDV model calculations are shown in Fig. 31.3. Unlike the New Zealand DSS, which uses suction trap aphid data to infer likely BYDV risk, the Western Australian model uses temperature and rainfall data for a given locality to forecast first aphid arrival and subsequent BYDV spread. Analysis of historical data showed that the magnitude of late summer and early autumn (March–April) rainfall before sowing was related directly to the date of the first aphid arrival in cereal crops in a given locality (Thackray and Jones,

1999; Thackray *et al.*, 2000, 2001, 2002, 2005) (Fig. 31.4). Sufficient rainfall supports growth of grass weeds and cereal volunteers on which cereal aphids multiply before they fly to crops. Because of the Mediterranean-type climate, little or no rain at this time means that few or no plants are available to support aphids in the months before crops are sown, so first arrival of aphids occurs much later (Hawkes and Jones, 2002, 2005). An estimate of soil water content available to plants is used in calculating a 'background aphid' population in each locality, and the change in this population is used to predict the development and subsequent migration of alatae into cereal crops. A fixed proportion of these carry BYDV into the crop. Aphid build-up and movement within the crop and spread of BYDV during the growing season are then calculated. The proportion of plants that is infective and the length of time for

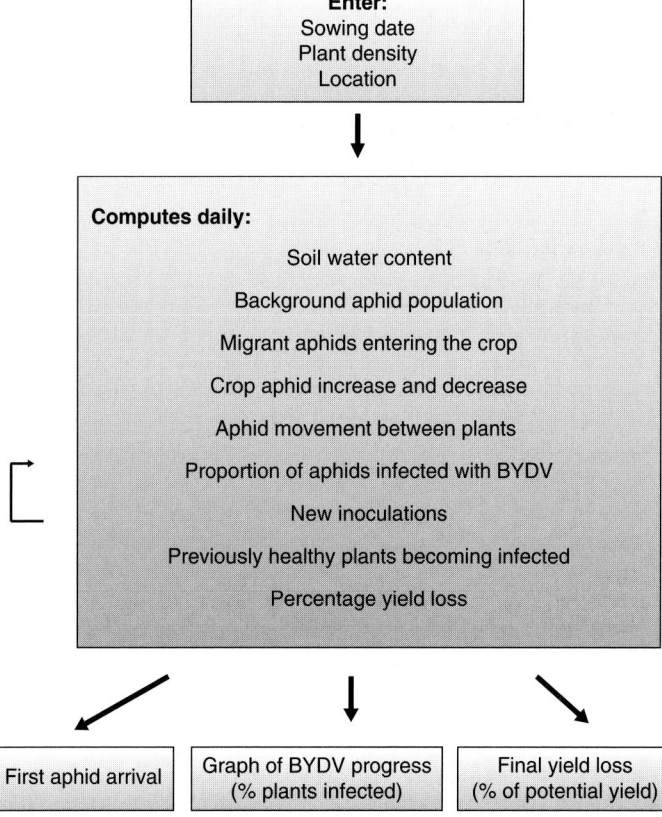

Fig. 31.3. Simplified flow diagram showing the five main steps in the Western Australian aphid and BYDV forecasting model.

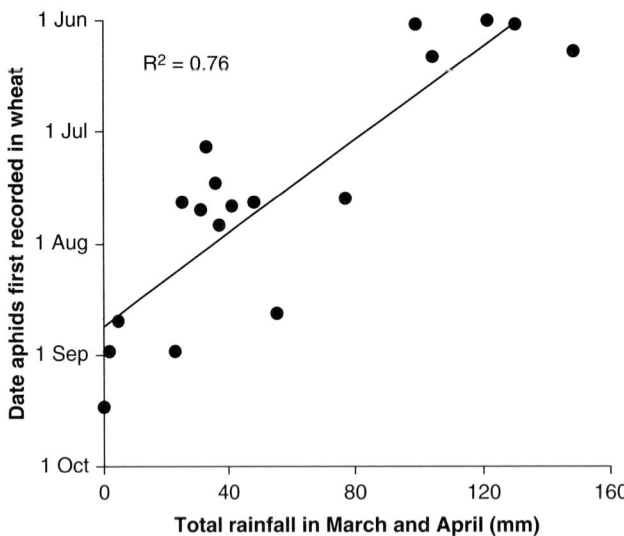

Fig. 31. 4. Effect of total rainfall in March and April (x) on the date aphids were first recorded on wheat (y) at five different sites in the Western Australian grainbelt from 1998–2001: $y = -0.6279\,x + 37,125$; $r^2 = 0.7587$ (data from Thackray and Jones, 1999; Thackray *et al.*, 2000, 2001, 2002).

which they have been infective at the time the crop begins to senesce are both used to calculate the percentage of potential yield lost to BYDV infection. Model simulations were verified and validated extensively against field data.

The aphid/BYDV model was incorporated into a personal computer (PC)-based DSS for use by farmers and advisers in targeting and timing insecticide sprays against BYDV aphid vectors in cereal crops in medium and high rainfall zones. Predictions for the date of first aphid arrival in crops, final BYDV incidence, grain-yield losses from BYDV, and need for and optimum spray timings for BYDV control are given. From 2002, users have been able to access DSS predictions through an Internet site (www.agric. wa.gov.au/bydv). Seasonal forecasts of risk of yield loss for different sowing dates in each shire in the medium and high rainfall zones are in the form of colour-coded maps of the grainbelt (Fig. 31.5). Photographs and background information on BYDV, aphids, control measures, and the DSS are also provided. A personalized forecast that depends on individual user inputs for location, temperature, rainfall, and plant density, is planned for the website. Automation of climate data retrieval and map generation is also being undertaken, so that the costs involved

in providing regular forecasts will be substantially reduced (Maling *et al.*, 2006).

Before completion of the Western Australian DSS, a market survey of agribusiness advisers was carried out. The majority of advisers were interested in using the DSS. In high-risk areas, most said they would use it to decide when not to apply insecticides rather than when to apply them. Several were keen to have access to the simulation model itself, which they would use to demonstrate to their grower clients the effect of different cropping and climate scenarios on BYDV spread and yield loss. Some wished to input information from their own observations of aphid numbers in crops. Many believed that the DSS would be a simple and effective method of bringing together the complex issues affecting aphid outbreaks and BYDV epidemics in cereal crops. Some requested that threshold BYDV levels and economics of yield loss and control be included. Almost all preferred access to the DSS to be *via* the Internet so that updates would be available immediately. A few suggested a CD-ROM of the simulation model be made available with links to the Internet site allowing current climate and risk information to be downloaded. Most advisers were prepared to pay for access, at around US$25.

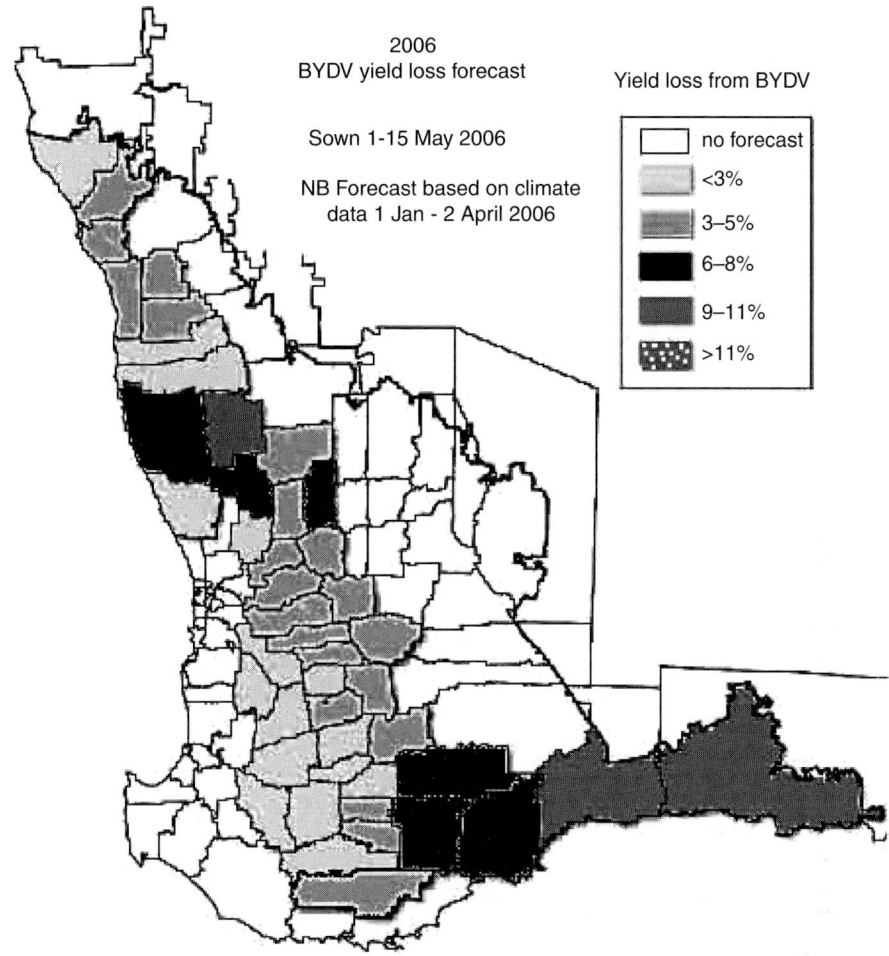

Fig. 31. 5. Example output from the WA Decision Support System (www.agric.wa.gov.au/bydv), with colour-coded (in the original) maps of the grainbelt illustrating different degrees of yield loss risk from BYDV for each locality in the medium and high rainfall zones.

Conclusions

Early attempts at providing decision support through the medium of computers had to be able not only to provide accurate information but also to overcome any resistance to a new technology and approach. Today, the use of computers in agriculture is probably the norm, so that particular obstacle has been largely removed. However, the need for relevant and accurate information for end-users remains, and systems have to be designed carefully to achieve this end. DSSs offer a valuable way of explaining the dynamics of aphids and transferring knowledge and information about them and the viruses they transmit in a user-friendly way to people who have to make management decisions. In the ever-changing circumstances of agriculture, they provide a valuable vehicle for delivering information and forecasts on which to base decisions, in a simple, timely, and cheap way. The systems rely on having up-to-date information but, in the past, providing this to end-users has been difficult. However, recent technological change has made PC-based systems easier and cheaper to maintain, and web-based implementation enables continuous updating. There are obvious overheads associated with keeping the DSSs current and operational, but these are being reduced all the time. Whilst DSSs can still be complex to construct, they offer an effective complement to information from agricultural advisers and can provide a flexible and rapid service to both farmers and advisers.

Acknowledgements

The Agricultural Market and Research Development Trust (Progressive Farming Grant), the Foundation for Arable Research, and the Foundation for Research, Science and Technology provided financial support for the New Zealand work summarized here. The Grains Research and Development Corporation of Australia, the Western Australian Department of Agriculture, and the Centre for Legumes in Mediterranean Agriculture provided financial support for the Western Australian work summarized here. We thank David Teulon for assistance in describing the New Zealand BYDV DSS, Roger Jones for guidance in developing the Western Australia aphid/BYDV forecasting model, and the research staff and farmers who provided much of the data used in development and validation of the DSSs.

References

Arinze, B. (1992) A user enquiry model for DSS requirements analysis: a framework and case study. *International Journal of Man-Machine Studies* 37, 241–264.

Berlandier, F.A. and Cartwright, L.M. (1998) Survival of aphids over summer in Western Australia. In: Zalucki, M.P., Drew, R.A.I. and White, G.C. (eds) *Pest Management – Future Challenges, Volume 2*. University of Queensland Press, Brisbane, p. 299.

Bicknell, K., Greer, G. and Teulon, D.A.J. (2000) The value of forecasting BYDV in autumn-sown cereals. *New Zealand Plant Protection* 53, 87–92.

Brooks, D.H. (1998) Decision Support System for Arable Crops (DESSAC): an integrated approach to decision support. *Proceedings of the Brighton Crop Protection Conference, Pests and Diseases, November 1998* 1, 239–246.

Derron, J.O. and Forrer, H.R. (1989) Timing treatments against cereal aphids: a decision-aiding program on Videotext. *Revue Suisse d'Agriculture* 21, 133–136.

Doodson, J.K. and Saunders, P.J.W. (1970) Some effects of barley yellow dwarf virus on spring and winter cereals in field trials. *Annals of Applied Biology* 66, 361–374.

Elliott, N.C., Royer, T.A., Giles, K.L., Kindler, S.D., Porter, D.R., Elliott, D.T. and Waits, D.A. (2004) A web-based decision support system for managing greenbugs in wheat. *Crop Management, 6th October 2004* (online journal of Pest Management Network, St Paul).

Farrell, J.A. and Stufkens, M.W. (1989) Flight activity and cereal host relationships of *Rhopalosiphum* spp. in Canterbury. *New Zealand Journal of Zoology* 15, 499–505.

Farrell, J.A. and Stufkens, M.W. (1992) Cereal aphid flights and barley yellow dwarf virus infection of cereals in Canterbury, New Zealand. *New Zealand Journal of Crop and Horticultural Science* 20, 407–412.

Farrell, J.A. and Sward, R.J. (1989) Barley yellow dwarf virus serotypes and their vectors in Canterbury, New Zealand. *Australian Plant Pathology* 18, 21–23.

Gonzalez-Andujar, J.L., Ceca, J.L.G. and Fereres, A. (1992) Expert System for identification and control of aphids in cereals. *AI Applications* 6, 60–61.

Greer, G. and Teulon, D.A.J. (2003) Farmer survey of yellow dwarf viruses in autumn-sown cereals in Canterbury. *New Zealand Plant Protection* 56, 257–261.

Hansen, L.M. (1999) Effect of weather during the spring on the time of arrival of bird cherry–oat aphid (*Rhopalosiphum padi* L.) in spring barley (*Hordeum vulgare* L.) fields. *Acta Agriculturae Scandinavica, Section B, Plant Soil Science* 49, 117–121.

Harrington, R., Mann, J.A., Burgess, A.J., Tones, S.J., Rogers, R., Foster, G.N., Blake, S., Morrison, S.F., Ward, L., Barker, I., Morgan, D. and Walters, K.F.A. (1999) Development and validation of decision support methodology for control of barley yellow dwarf virus. *HGCA Project Report No. 205*, 86 pp.

Hawkes, J. and Jones, R.A.C. (2002) Distribution and incidence of aphids and barley yellow dwarf virus in over-summering grasses in the Western Australian wheatbelt. In: Jettner, R. (ed.) *2002 Cereals Update*. Department of Agriculture, Western Australia, Perth, pp. 45–46.

Hawkes, J.R. and Jones, R.A.C. (2005) Incidence and distribution of *Barley yellow dwarf virus* and *Cereal yellow dwarf virus* in over-summering grasses in a Mediterranean-type environment. *Australian Journal of Agricultural Research* 56, 257–270.

Hearn, A.B. and Bange, M.P. (2002) SIRATAC and *Cotton OGIC*: persevering with DSSs in the Australian cotton industry. *Agricultural Systems* 74, 27–56.

Helm, G.L., Richardson, J.W., Rister, M.E., Stone, N.D. and Loh, D.K. (1987) Cotflex – a farm-level Expert System to aid farmers in making farm policy decisions. *American Journal of Agricultural Economics* 69, 1097–1098.

Johnson, C.G. and Taylor, L.R. (1955) The development of large suction traps for airborne insects. *Annals of Applied Biology* 43, 51–62.

Jones, R.A.C., McKirdy, S.J. and Shivas, R.G. (1990) Occurrence of barley yellow dwarf viruses in over-summering grasses and cereal crops in Western Australia. *Australian Plant Pathology* 19, 90–96.

Knight, J.D. and Cammell, M.E. (1994) A Decision Support System for forecasting infestations of the black bean aphid, *Aphis fabae* Scop., on spring sown field beans, *Vicia faba*. *Computers and Electronics in Agriculture* 10, 269–279.

Lowe, A.D. (1963) Warning services on aphid flights. *Proceedings of the 16th New Zealand Weed Control Conference, Wellington, July 1963*, pp. 108–109.

Maling, T., Diggle, A., Thackray, D., Siddique, K. and Jones, R. (2006) Development of a generic forecasting and decision support system for diseases in the Western Australian wheatbelt. In: McLarty, A. (ed.) *2006 Lupins and Pulses Updates*. Department of Agriculture, Western Australia, Perth, pp. 62–64.

Mann, B.P. and Wratten, S.D. (1991) A computer-based advisory system for cereal aphids – field testing the model. *Annals of Applied Biology* 118, 503–512.

McKirdy, S.J. and Jones, R.A.C. (1993) Occurrence of barley yellow dwarf virus serotypes MAV and RMV in over-summering grasses. *Australian Journal of Agricultural Research* 44, 1195–1209.

McKirdy, S.J. and Jones, R.A.C. (1996) Use of imidacloprid and newer generation synthetic pyrethroids to control the spread of barley yellow dwarf luteovirus in cereals. *Plant Disease* 80, 895–901.

McKirdy, S.J. and Jones, R.A.C. (1997) Effect of sowing time on barley yellow dwarf virus infection in wheat: virus incidence and grain yield losses. *Australian Journal of Agricultural Research* 48, 199–206.

Mozny, M., Krejci, J. and Kott, I. (1993) CORAC, hops protection management systems. *Computers and Electronics in Agriculture* 9, 103–110.

Murali, N.S., Secher, B.J.M., Rydahl, P. and Andreason, F.M. (1999) Application of information technology in plant protection in Denmark: from vision to reality. *Computers and Electronics in Agriculture* 22, 109–115.

Northing, P., Walters, K.F.A., Barker, I., Foster, G., Harrington, R., Taylor, M., Tones, S. and Morgan, D. (2004) Use of internet for provision of user specific support for decisions on the control of aphid-borne virus. In: Dedryver, C.A., Simon, J.C., Rispe, C. and Hullé, M. (eds) *Aphids in a New Millennium: Proceedings of the 6th International Aphid Symposium, Rennes, September 2001*, 331–336.

Teulon, D.A.J., Stufkens, M.A.W., Nicol, D. and Harcourt, S.J. (1999) Forecasting barley yellow dwarf virus in autumn-sown cereals in 1998. In: O'Callaghan, M. (ed.) *Proceedings of the 52nd Plant Protection Conference, Auckland, August 1999*. New Zealand Plant Protection Society, Auckland, pp. 187–191.

Teulon, D.A.J., Fletcher, J.D. and Cromey, M.G. (2001) Localised severe incidence of barley yellow dwarf virus in winter wheat. *New Zealand Plant Protection* 54, 253.

Teulon, D.A.J., Lankin, G.O., Stufkens, M.A.W., Lee, J. and Travis, G.R. (2004a) Local variation in cereal aphid flight activity in Canterbury. *New Zealand Plant Protection* 57, 221–226.

Teulon, D.A.J., Stufkens, M.A.W. and Fletcher, J.D. (2004b) Crop infestation by aphids is related to flight activity detected with 7.5 metre high suction traps. *New Zealand Plant Protection* 57, 227–232.

Thackray, D.J. and Jones, R.A.C. (1999) Forecasting barley yellow dwarf virus risk in cereals. In: Zaicou-Kunesch, C. and Kerr, N. (eds) *1999 Cereals Updates*. Department of Agriculture, Western Australia, Perth, pp. 40–42.

Thackray, D.J., Ward, L. and Jones, R.A.C. (1998) Aphid arrival and build up in relation to BYDV levels and yield loss. In: Anderson, W. and Blake, J. (eds) *Highlights of Cereal Research and Development in Western Australia, 1998*. Department of Agriculture, Western Australia, Perth, pp. 60–61.

Thackray, D.J., Hawkes, J. and Jones, R.A.C. (2000) Forecasting aphid and virus risk in cereals. In: Zaicou-Kunesch, C. and Kerr, N. (eds) *2000 Cereals Updates*. Department of Agriculture, Western Australia, Perth, pp. 57–59.

Thackray, D.J., Hawkes, J. and Jones, R.A.C. (2001) Further progress in forecasting aphid and virus risk in cereals. In: Jettner, R. and Johns, J. (eds) *2001 Cereals Update*. Department of Agriculture, Western Australia, Perth, pp. 67–69.

Thackray, D.J., Hawkes, J. and Jones, R.A.C. (2002) A decision support system for control of aphids and BYDV in cereal crops. In: Jettner, R. (ed.) *2002 Cereals Update*. Department of Agriculture, Western Australia, Perth, pp. 57–58.

Thackray, D.J., Diggle, A.J., Berlandier, F.A. and Jones, R.A.C. (2004) Forecasting aphid outbreaks and epidemics of cucumber mosaic virus in lupin crops in a Mediterranean-type environment. *Virus Research* 100, 67–82.

Thackray, D.J., Ward, L.T., Thomas-Carroll, M.L. and Jones, R.A.C. (2005) Role of winter-active aphids spreading *Barley yellow dwarf virus* in decreasing wheat yields in a Mediterranean-type environment. *Australian Journal of Agricultural Research* 56, 1089–1099.

Zadoks, J.C. (1981) EPIPRE: a disease and pest management system for winter wheat developed in the Netherlands. *EPPO Bulletin,* 11, 365–369.

Zadoks, J.C., Chang, T.T. and Konzak, C.F. (1974) A decimal code for the growth stages of cereals. *Eucarpia Bulletin 7*, 11 pp.

Taxonomic Glossary

with common names where appropriate

APHIDS AND THEIR NATURAL ENEMIES

Aphidoidea

Acyrthosiphon brevicorne Hille Ris Lambers

Acyrthosiphon kondoi Shinji – blue alfalfa aphid

Acyrthosiphon pisum (Harris) – pea aphid

Acyrthosiphon pisum destructor (Johnson)

Acyrthosiphon pisum ononis (Koch)

Acyrthosiphon svalbardicum Heikinheimo

Adelges (Sacchiphantes) abietis (L.) – spruce pineapple-gall adelgid

Adelges (Gilletteella) cooleyi (Gillette) – Cooley spruce gall adelgid

Adelges (Aphrastasia) funitecta Dreyfus = *A. tsugae* Annand – hemlock woolly adelgid

Adelges (Adelges) laricis Vallot – larch adelgid

Adelges (Dreyfusia) piceae Ratzeburg – balsam woolly adelgid

Adelges tsugae (see *A. (Aphrastasia) funitecta*)

Amphorophora agathonica Hottes

Amphorophora idaei (Börner)

Amphorophora rubi (Kaltenbach) – Rubus aphid

Anuraphis farfarae (Koch) – pear–coltsfoot aphid

Aphanostigma iaksuiense (Kishida)

Aphanostigma piri (Cholodkovsky) – pear phylloxera

Aphis barbarae Robinson

Aphis citricida (see *Toxoptera citricidus*)

Aphis craccivora Koch – cowpea aphid, groundnut aphid

Aphis eugeniae van der Goot

Aphis euonymi F.

Aphis fabae Scopoli – black bean aphid

Aphis fabae cirsiiacanthoides Scopoli

Aphis fabae mordvilkoi Börner & Janich

Aphis farinosa Gmelin

Aphis frangulae Kaltenbach

Aphis glycines Matsumura – soybean aphid

Aphis gossypii Glover – cotton or melon aphid

Aphis grossulariae Kaltenbach – gooseberry aphid

Aphis helianthi Monell

Aphis idaei van der Goot

Aphis illinoisensis Shimer – grape vine aphid

Aphis maidiradicis Fitch – corn root aphid

Aphis nasturtii Kaltenbach – buckthorn–potato aphid

Aphis pomi De Geer – green apple aphid

Aphis rubicola Oestlund

Aphis schneideri (Boerner)

Aphis solanella Theobald

Aphis spiraecola Patch = *A. citricola* van der Goot of many authors – green citrus aphid, spiraea aphid

Aphis vaccinii (Börner)

Aulacorthum (Neomyzus) circumflexum (Buckton) – mottled arum aphid

Aulacorthum magnoliae (Essig & Kuwana)

Aulacorthum solani (Kaltenbach) – glasshouse and potato aphid, foxglove aphid

Baizongia pistaciae (L.)

Brachycaudus amygdalinus (Schouteden) – leaf-curling almond aphid

©CAB International 2007. *Aphids as Crop Pests* (eds H. van Emden and R. Harrington)

Brachycaudus cardui (L.)

Brachycaudus helichrysi (Kaltenbach) – leaf-curling plum aphid

Brachycaudus persicae (Passerini) – black peach aphid

Brachycaudus prunicola (Kaltenbach) – brown plum aphid

Brachycaudus schwartzi (Börner) – brown peach aphid

Brevicoryne brassicae (L.) – cabbage aphid

Capitophorus elaeagni (del Guercio)

Capitophorus horni (Börner)

Cavahyalopterus graminearum Mimeur (see *Diuraphis noxia*)

Cavariella aegopodii Scopoli – willow–carrot aphid

Ceratovacuna lanigera Zehntner – sugar-cane woolly aphid

Ceratovacuna nekoashi (Sasaki)

Chaetosiphon fragaefolii (Cockerell) – strawberry aphid

Chaitophorus stevensis Sanborn

Chromaphis juglandicola (Kaltenbach) – walnut aphid

Cinara cronartii Tissot & Pepper – black pine aphid

Cinara cupressi (Buckton)

Cinara (*Cedrobium*) *laportei* (Remaudière) – cedar aphid

Cinara piceicola (Cholodkovsky)

Cinara pinea (Mordvilko)

Cinara todocola (Inouye)

Daktulosphaira vitifoliae (Fitch) (see *Viteus vitifoliae*)

Diuraphis muehlei (Börner)

Diuraphis noxia (Kurdjumov) = *Cava-hyalopterus graminearum* Mimeur – Russian wheat aphid

Diuraphis tritici (Gillette) – western wheat aphid

Drepanosiphum platanoidis Schrank – syc-amore aphid

Dysaphis anthrisci Börner

Dysaphis devecta (Walker) – rosy leaf-curling aphid

Dysaphis plantaginea (Passerini) – rosy apple aphid

Dysaphis pyri (Boyer de Fonscolombe) – pear–bedstraw aphid

Dysaphis reamuri (Mordvilko)

Elatobium abietinum (Walker) – spruce aphid

Ericaphis (= *Fimbriaphis*) *fimbriata* (Richards)

Ericaphis (= *Fimbriaphis*) *scammelli* (Mason)

Eriosoma lanigerum (Hausmann) – woolly apple aphid

Eriosoma lanuginosum (Hartig) – woolly pear aphid

Eriosoma pyricola Baker & Davidson

Eriosoma ulmi (L.) = *Schizoneura ulmi* (L.)

Eucallipterus tiliae (L.) – lime aphid

Greenidea anonae (Pergande)

Greenidea ficicola Takahashi

Greenidea formosana Maki

Greenidea mangiferae Takahashi

Hayhurstia atriplicis (L.)

Hormaphis hamamelidis (Fitch)

Hyadaphis tataricae (Aizenberg)

Hyalopterus amygdali (Blanchard) – mealy peach aphid

Hyalopterus pruni (Geoffroy) – mealy plum aphid

Hyperomyzus lactucae (L.) – black currant–sowthistle aphid

Hysteroneura setariae (Thomas)

Illinoia azaleae (Mason)

Illinoia borealis (Mason)

Illinoia liriodendri (Monell)

Illinoia pepperi (MacGillivray) – blueberry aphid

Lipaphis erysimi (Kaltenbach) (see *L. pseudobrassicae*)

Lipaphis pseudobrassicae (Davis) = *L. erysimi* (Kaltenbach) of many authors, especially pre-2000 – mustard aphid, turnip aphid, false cabbage aphid

Longistigma caryae (Harris)

Macrosiphoniella tanacetaria (Kaltenbach) – tansy aphid

Macrosiphum albifrons Essig – lupin aphid

Macrosiphum euphorbiae (Thomas) – potato aphid

Macrosiphum rosae (L.) – rose aphid

Macrosiphum stellariae Theobald

Macrosiphum tinctum (Walker)

Megoura viciae Buckton – vetch aphid

Melanaphis pyraria (Passerini) – brown pear aphid

Melanaphis sorghi (Theobald) – sugarcane aphid

Melaphis rhois (Fitch)

Metopeurum fuscoviride Stroyan

Metopolophium dirhodum (Walker) – rose–grain aphid

Metopolophium festucae (Theobald)

Microlophium carnosum (Buckton) – nettle aphid

Mindarus kinseyi Voegtlin

Monellia caryella (Fitch) – black-margined pecan aphid

Monelliopsis pecanis Bissell – yellow pecan aphid

Myzocallis boerneri Stroyan – Turkey oak aphid

Myzocallis coryli (Goetze) – hazel aphid

Myzus amygdalinus (Nevsky)

Myzus antirrhinii (Macchiati)

Myzus ascalonicus Doncaster – shallot aphid

Myzus cerasi (F.) – cherry blackfly

Myzus certus (Walker)

Myzus dianthicola Hille Ris Lambers

Myzus mumecola (Matsumura)

Myzus myosotidis (Börner)

Myzus persicae (Sulzer) – peach–potato aphid, green peach aphid

Myzus persicae nicotianae Blackman – tobacco aphid

Myzus siegesbeckiae Takahashi

Myzus varians Davidson – cigar-rolling peach aphid

Nasonovia ribisnigri (Mosley) – currant–lettuce aphid

Nippolachnus piri Matsumura

Pemphigus betae Doane – sugarbeet root aphid

Pemphigus bursarius (L.) – lettuce root aphid

Pemphigus fuscicornis (Koch)

Pemphigus obesinymphae Aoki & Moran

Pemphigus phenax (Börner & Blunck)

Pemphigus populitransversus Riley

Pemphigus spyrothecae Passerini

Pentalonia nigronervosa Coquerel – banana aphid

Periphyllus testudinaceus (Fernie)

Phorodon humuli (Schrank) – damson–hop aphid

Phyllaphis fagi (L.) – woolly beech aphid

Phylloxera caryaecaulis (Fitch) – hickory leaf stem gall aphid

Phylloxera devastatrix Pergande – pecan phylloxera

Phylloxera notabilis Pergande – pecan leaf phylloxera

Pterochloroides persicae (Cholodkovsky)

Pyrolachnus pyri (Buckton)

Rhopalomyzus lonicerae (Siebold)

Rhopalosiphoninus latysiphon (Davidson) – potato root aphid

Rhopalosiphum insertum (Walker) – apple–grass aphid

Rhopalosiphum maidis (Fitch) – corn leaf aphid, maize aphid

Rhopalosiphum nymphaeae (L.)

Rhopalosiphum padi (L.) – bird cherry–oat aphid

Rhopalosiphum rufiabdominale (Sasaki) – rice root aphid

Schizaphis graminum (Rondani) – greenbug

Schizaphis hypersiphonata Basu

Schizaphis piricola (Matsumura)

Schizaphis pyri Shaposhnikov

Schizaphis rotundiventris (Signoret)

Schizolachnus pineti (F.)

Schizolachnus piniradiatae (Davidson)

Schizoneura ulmi (L.) (see *Eriosoma ulmi*)

Sinomegoura citricola (van der Goot)

Sipha flava (Forbes) – yellow sugarcane aphid

Sitobion avenae (F.) – grain aphid

Sitobion fragariae (Walker) – blackberry–cereal aphid

Sitobion miscanthi (Takahashi)

Smynthurodes betae Westwood

Symydobius oblongus (von Heyden)

Taiwanaphis kalipadi (Raychaudhuri & Ghosh)

Tetraneura fusiformis Matsumura

Thelaxes dryophila (Schrank)

Therioaphis riehmi (Börner) – sweet clover aphid

Therioaphis trifolii (Monell) – yellow clover aphid, alfalfa aphid

Therioaphis trifolii maculata (Buckton) – spotted alfalfa aphid

Tinocallis caryaefoliae (Davis) – black pecan aphid

Tinocallis kahawaluokalani (Kirkaldy) – crape myrtle aphid

Tinocallis platani (Kaltenbach)

Toxoptera aurantii (Boyer de Fonscolombe) – black citrus aphid

Toxoptera citricidus (Kirkaldy) – tropical citrus aphid

Toxoptera odinae (van der Goot)

Tuberolachnus salignus (Gmelin) – willow aphid

Uroleucon jaceae (L.)
Uroleucon nigrotuberculatum (Olive)
Uroleucon tanaceti (L.)
Viteus (= *Daktulosphaira*) *vitifoliae* (Fitch)
 – grape phylloxera

Natural enemies of aphids

DERMAPTERA
Forficula auricularia L. – European earwig

HEMIPTERA – HETEROPTERA
Anthocoris nemoralis (F.)
Anthocoris nemorum (L.) – common flower
 bug
Deraeocoris brevis (Uhler)
Geocoris punctipes (Say) – big-eyed bug

DIPTERA – CECIDOMYIIDAE
Aphidoletes aphidimyza (Rondani)
Endaphis gregaria Gagné

DIPTERA – SYRPHIDAE
Epistrophe nitidicollis (Meigen)
Episyrphus balteatus (DeGeer)
Eupeodes (Metasyrphus) corollae (F.)
Eupeodes (Metasyrphus) nielseni (Dusek &
 Laska)
Megasyrphus erraticus (L.)
Platycheirus parmatus (Rondani)
Pseudodorus clavatus (F.)
Sphaerophoria scripta (L.)
Syrphus ribesii (L.)

NEUROPTERA
Chrysopa cognata McLachlan
Chrysopa formosa Brauer
Chrysopa oculata Say
Chrysopa pallens (Rambur)
Chrysopa phyllochroma Westwood
Chrysoperla carnea (Stephens) – green
 lacewing
Chrysoperla lucasina (Lacroix)
Chrysoperla plorabunda (Fitch)
Chrysoperla rufilabris (Burmeister)
Chrysoperla sinica (Tjeder)
Cunctochrysa jubigensis (Hölzel)
Hemerobius pacificus Banks
Micromus angulatus (Stephens)
Nineta vittata (Wesmael)
Peyerimhoffina gracilis (Scheider)

COLEOPTERA – COCCINELLIDAE
Adalia bipunctata (L.) – 2-spot ladybird
Adalia decempunctata (L.) – 10-spot
 ladybird
Cheilomenes sexmaculatus (F.)
Coccidula rufa (Herbst)
Coccinella magnifica Redtenbacher
Coccinella septempunctata L. – 7-spot
 ladybird
Coccinella septempunctata brucki Mulsant
*Coccinella undecimpunctata undecim-
 punctata* L.
Coelophora biplagiata (Swartz in
 Schoenherr)
Coleomegilla maculata (De Geer) – 12-spot
 ladybird
Coleomegilla maculata lengi Timberlake
Cycloneda sanguinea (L.)
Eriopis connexa (Germar)
Harmonia axyridis (Pallas) – harlequin
 ladybird
Hippodamia convergens Guerin – conver-
 gent ladybird
Hippodamia sinuata Mulsant
Hippodamia variegata (Goeze) – variegated
 ladybird
Platynaspis luteorubra (Goeze)
Propylea japonica (Thunberg)
Propylea quatuordecimpunctata (L.) – 14-
 spot ladybird
Rodolia cardinalis (Mulsant)
Scymnus interruptus (Goeze)
Scymnus nigrinus Kugelann
Synonycha grandis (Thunberg)

COLEOPTERA – CARABIDAE
Bembidion lampros (Herbst)
Harpalus rufipes (DeGeer)
Platynus dorsalis (Pontoppidan)
Poecilus cupreus (L.)
Pterostichus madidus (F.)
Pterostichus melanarius (Illiger)
Trechus quadristriatus (Schrank)

COLEOPTERA – STAPHYLINIDAE
Tachyporus chrysomelinus (L.)
Tachyporus hypnorum (F.)
Tachyporus obtusus (L.)

HYMENOPTERA – Parasitoids
Aphelinus abdominalis (Dalman)
Aphelinus albipodus Hayat & Fatima

Aphelinus asychis Walker
Aphelinus flavus Thomson
Aphelinus hordei Kurdjumov
Aphelinus mali (Haldeman)
Aphelinus perpallidus Gahan
Aphelinus semiflavus Howard
Aphelinus spiraecolae Evans & Schauff
Aphelinus varipes (Foerster)
Aphidius colemani Viereck
Aphidius eadyi Starý, Gonzales & Hall
Aphidius ervi Haliday
Aphidius funebris Mackauer
Aphidius hieraciorum Starý
Aphidius matricariae Haliday
Aphidius nigripes Ashmead
Aphidius picipes (Nees)
Aphidius pisivorus Smith
Aphidius rhopalosiphi DeStefani-Perez
Aphidius rosae Haliday
Aphidius salicis Haliday
Aphidius smithi Sharma & Subba Rao
Aphidius sonchi Marshall
Aphidius uzbekistanicus Luzhetzki
Binodoxys angelicae (Haliday)
Binodoxys indicus (Sharma & Subba Rao)
Diaeretiella rapae (M'Intosh)
Diaeretus leucopterus (Haliday)
Ephedrus californicus Baker
Ephedrus cerasicola Starý
Ephedrus nacheri Quilis
Ephedrus plagiator (Nees)
Euaphidius cingulatus (Ruthe)
Falciconus pseudoplatani (Marshall)
Lysiphlebia japonica Ashmead
Lysiphlebus cardui (Marshall)
Lysiphlebus confusus Tremblay & Eady
Lysiphlebus fabarum (Marshall)
Lysiphlebus hirticornis Mackauer
Lysiphlebus testaceipes (Cresson)
Monoctonia pistaciaecola Starý
Monoctonus crepidis (Haliday)
Monoctonus nervosus (Haliday)
Monoctonus paulensis (Ashmead)
Pauesia bicolor (Ashmead)
Pauesia californica (Ashmead)
Pauesia cedrobii Starý & Leclant
Pauesia cinaravora Marsh
Pauesia juniperorum (Starý)
Pauesia picta (Haliday)
Pauesia pini (Haliday)
Pauesia silvestris (Starý)

Pauesia unilachni (Gahan)
Praon barbatum Mackauer
Praon exsoletum (Nees)
Praon gallicum Starý
Praon palitans Muesebeck = *Praon exsoletum* (Nees)
Praon volucre (Haliday)
Pseudopauesia prunicola Halme
Trioxys angelicae (Haliday) = *Binodoxys angelicae* (Haliday)
Trioxys betulae Marshall
Trioxys cirsii (Curtis)
Trioxys complanatus Quilis
Trioxys curvicaudus Mackauer
Trioxys falcatus Mackauer
Trioxys pallidus (Haliday)
Trioxys tenuicaudus Starý
Trioxys utilis Muesebeck = *Trioxys complanatus* Quilis

ARANEAE – LINYPHYIIDAE
Erigone atra Blackwall
Oedothorax apicatus (Blackwall)

ARANEAE – TETRAGNATHIDAE
Pachygnatha degeer Sundevall

ENTOMOPATHOGENS
Beauveria bassiana (Balsamo-Crivelli) Vuillemin
Conidiobolus obscurus (Hall & Dunn) Remaudière & Keller
Entomophthora planchoniana Cornu
Lecanicillium lecanii (Zimmerman) Zare & Gams
Lecanicillium longisporum (Petch) Zare & Gams
Lecanicillium muscarium (Petch) Zare & Gams
Metarhizium anisopliae (Metschnikoff) Sorokin
Neozygites fresenii (Nowakowski) Remaudière & Keller
Paecilomyces fumosoroseus (Wize) Brown & Smith
Pandora neoaphidis (Remaudière & Hennebert) Humber
Verticillium lecanii (Zimmerman) Viegas (now *Lecanicillium longisporum* or *L. muscarium* above)
Zoophthora phalloides Batko
Zoophthora radicans (Brefeld) Batko

OTHER INSECTS

Aleyrodoidea

Bemisia tabaci (Gennadius) – sweet potato whitefly, tobacco whitefly

Thysanoptera

Frankliniella occidentalis (Pergande) – western flower thrips

Lepidoptera

Autographa gamma (L.) – silver Y moth
Ephestia kuehniella Zeller – Mediterranean flour moth
Malacosoma disstria Hübner – forest tent caterpillar
Spodoptera frugiperda (J.E. Smith) – fall armyworm

Diptera

Delia radicum (L.) – cabbage root fly
Drosophila melanogaster Meigen
Sitodiplosis mosellana (Géhin) – orange wheat blossom midge

Coleoptera

Anthonomus grandis Boheman – boll weevil
Leptinotarsa decemlineata (Say) – Colorado beetle
Phyllotreta cruciferae (Goeze) – flea beetle
Psylloides chrysocephala (L.)
Sitona lineatus (L.) – pea and bean weevil

Hymenoptera – Formicidae

Formica exsectoides Forel – Allgegheny mound ant
Formica polyctena Foerster
Linepithema humile (Mayr)

Hymenoptera – Braconidae

Dolichogenidea tasmanica (Cameron)

ACARINA

Aculus tetanothrix (Nalepa)
Oligonychus pratensis (Banks) – Banks grass mite

NEMATODA

Heterodera avenae Wollenweber – cereal cyst nematode
Meloidogyne incognita (Kofoid & White) Chitwood – root-knot nematode

OTHER MICROORGANISMS

Agrobacterium tumefaciens (Smith & Townsend) Conn – crown gall
Bacillus thuringiensis Berliner – Bt
Bacillus thuringiensis var. *tenebrionis* (Krieg) – Bt
Buchnera aphidicola Munson, Baumann & Kinsey
Capnodium citri Berkeley & Desmaszi
Erysiphe polygoni DC – powdery mildew
Escherichia coli (Migula) Castellani & Chalmers
Hamiltonella defensa Moran *et al.*
Neotyphodium coenophialum (Morgan-Jones & Gams) (Glenn, Bacon, Price & Hanlin)
Neotyphodium lolii Latch, Christensen & Samuels (Glenn, Bacon, Price & Hanlin)
Regiella insecticola Moran *et al.*
Rickettsia belli (da Rocha-Lima)
Serratia symbiotica Moran *et al.*
Verticillium albo-atrum Reinke & Beethold – lucerne wilt

PLANTS

Abelmoschus esculentus (L.) Moench – okra

Abies balsamea (L.) Miller – balsam fir

Abies concolor (Gordon & Glendenning) Lindley ex F.H. Hildebrand – white fir

Abies fraseri (Pursh) Poiret – Fraser fir

Abies sachalinensis (F. Schmidt) Masters

Acer pseudoplatanus L. – sycamore

Actinidia polygama (Siebold & Zuccagno) Maximowicz

Aegilops ventricosa Tausch – barbed goatgrass

Agropyron desertorum (Fischer ex Link) Schultes – crested wheat grass

Agropyron intermedius (see *Agropyron intermedium*)

Agropyron intermedium (Host) Beauvisage – intermediate wheat grass

Agropyron smithii Rydberg – western wheat grass

Allium cepa L. – onion

Allium cepa L. var. *aggregatum* G. Don – shallot

Amaranthus caudatus L. – grain amaranth

Anacardium occidentale L. – cashew nut

Annona cherimola Miller – custard apple

Annona muricata L. – soursop

Annona reticulata L. – bullock's heart

Annona squamosa L. – sugar apple

Arabidopsis thaliana (L.) Heynhold

Arachis hypogaea L. – groundnut

Arrhenatherum elatius (L.) P. Beauvois ex J. Presl & C. Presl

Asclepias curassivica L. – milkweed

Avena barbata Pott ex Link

Avena macrostachya Balansa & Durieu

Avena sativa L. – oat

Benincasa hispida (Thunberg) Cogniaux – ash gourd, wax gourd, winter melon

Beta vulgaris L. – sugarbeet, beet

Beta vulgaris L. subsp. *vulgaris* L. – mangold

Betula pendula Roth – silver birch

Brassica carinata A. Braun – Ethiopian mustard

Brassica fruticolosa Cirillo

Brassica juncea (L.) Czernajew – Indian mustard, mustard greens

Brassica napus L. – oilseed rape

Brassica napus L. var. *napobrassicae* (L.) Reichenbach – swede

Brassica nigra (L.) W.D.J. Koch

Brassica oleracea L.

Brassica oleracea L. var. *capitata* L. – cabbage

Brassica oleracea L. var. *viridis* L. – kale, collards

Brassica rapa L.

Brassica rapa L. subsp. *campestris* (L.) A.R. Clapham

Brassica spinescens Pomel

Bromus tectorum L. – cheat grass

Calendula officinalis L. – marigold

Capsella bursa-pastoris (L.) Medikus

Capsicum annuum L. – bell pepper, chilli

Capsicum frutescens L. – Tabasco pepper

Carica papaya L. – papaya, pawpaw (UK)

Carya glabra (Miller) Sweet – hickory

Carya illinoinensis (Wangenheim) K.H.E. Koch – pecan

Catalpa bignonioides Walter

Cedrus atlantica (Endlicher) Carrière

Celastrus orbiculatus Thunberg

Chenopodium album L. – fathen

Chloris verticillata Nuttall – humble windmill grass

Chrysanthemum morifolium Rematuelle – chrysanthemum

Cirsium arvense L. – thistle

Citrullus lanatus (Thunberg) Matsumura & Nakai – watermelon

Citrus × *paradisi* McFadden – grapefruit

Citrus reticulata Blanco – clementine, mandarin orange

Citrus sinensis (L.) Osbeck – orange

Citrus unshiu Markow

Colocasia esculenta (L.) Schott – taro

Corylus maxima Miller – filbert nut

Crataegus azarolus L. – azarole

Cucumis melo L. – melon, cantaloupe

Cucumis sativus L. – cucumber

Cucurbita maxima Duchesne ex Lamarck – pumpkin

Cucurbita moschata Duchesne ex Poiret – squash

Cucurbita pepo L. – courgette, summer squash, muskmelon, pumpkin, zuchini squash

Cydonia oblonga Miller – quince

Cynara scolymus L. – globe artichoke

Dactylis glomerata L. – cocksfoot

Datura stramonium L. – jimsonweed

Dendranthema × *grandiflorum* (Ramatuelle) Kitamura – chrysanthemum

Echinochloa crusgalli (L.) Beauvisage – barnyard grass

Elettaria cardamomum (L.) Maton – cardamom

Elymus canadensis L. – Canada wild rye

Elytrigia repens (L.) Desvaux ex Nevski – couch grass

Eryobotria japonica (Thunberg) Lindley – loquat

Euonymus europaeus L. – spindle

Fagopyrum esculentum Moench – buckwheat

Festuca arundinacea Schreber – tall fescue

Fragaria × *ananassa* Duchesne – strawberry

Frangula alnus Miller – buckthorn

Galanthus nivalis L. – snowdrop

Gevuina avellana Molina – hazelnut (Chile)

Glycine max (L.) Merrill – soybean, soya bean

Gossypium arboreum L.

Gossypium barbadense L. – sea island cotton

Gossypium hirsutum L. – cotton

Hamamelis virginiana L. – witch hazel

Helianthus annuus L. – sunflower

Hemizygia petiolata Ashby

Hibiscus syriacus L.

Holcus lanatus L. – Yorkshire fog

Holcus mollis L. – creeping soft grass

Hordeum spontaneum K.H.E. Koch – wild barley

Hordeum vulgare L. – barley

Humulus lupulus L. – hop

Hyacinthus orientalis L. – hyacinth

Ipomea batatas (L.) Lamarck – sweet potato

Lactuca sativa L. – lettuce

Lactuca virosa L. – great lettuce

Lagerstroemia indica L. – crape myrtle

Lens culinaris Medikus – lentil

Liriodendron tulipifera L. – tulip tree

Litchi chinensis Sonnerat – litchi

Lobularia maritima (L.) N.A. Desvaux – sweet alyssum

Lolium perenne L. – perennial rye grass

Lupinus angustifolius L. – narrow-leaved lupin

Lycopersicon esculentum Miller – tomato

Lycopersicon hirsutum Dunal f. *glabratum*

Lycopersicon peruvianum (L.) Miller

Macadamia integrifolia Maiden & Belche – macadamia nut

Malus domestica Borkhauser – apple

Malus sylvestris Miller

Mangifera indica L. – mango

Medicago sativa L. – lucerne, alfalfa (USA)

Melothria pendula L. – creeping cucumber

Mentha × *piperita* L. – peppermint

Mespilus germanica L. – medlar

Miscanthus sinensis Andersson – Japanese pampas grass

Momordica charantia L. – balsam apple

Musa paradisiaca L. – banana

Nepeta cataria L. – catmint

Nepeta racemosa Lamarck

Nerium oleander L. – oleander

Nicotiana tabacum L. – tobacco

Oryza sativa L. – rice

Oryzopis hymenoides (Roemer & Schultes) Ricker – Indian rice grass

Passiflora edulis Sims – passion fruit

Persea americana Miller – avocado

Petroselinum crispum (Miller) Nyman ex A.W. Hill – parsley

Phacelia tanacetifolia Bentham – phacelia

Phaseolus lunatus L. – lima bean

Phaseolus vulgaris L. – bean

Philadelphus coronarius L. – mock orange

Phleum pratense L. – timothy

Picea abies (L.) G.K.W. Karst – Norway spruce

Picea sitchesis (Bongaral) Carrière – Sitka spruce

Pinus banksiana Lambert – jack-pine

Pinus resinosa Aiton

Pinus sylvestris L. – Scots pine

Pistacia palaestina Boissier – pistachio (not the edible nut)

Pisum sativum L. – pea

Poa compressa L. – Canada blue grass

Polygonum aviculare L. – knotgrass

Populus deltoides Bartram ex Marshall – eastern cottonwood

Populus nigra L. var. *italica* Münchhausen – Lombardy poplar

Populus tremuloides Michaux – American aspen

Prunus armeniaca L. – apricot

Prunus avium (L.) L. – wild cherry

Prunus cerasifera Ehrhart – cherry-plum

Prunus cerasus L. – cherry

Prunus cornuta (Wallich ex Royle) Steudel

Prunus domestica L. – plum

Prunus dulcis (Miller) D.A.Webb – almond

Prunus insititia L. – damson

Prunus mume Siebold & Zuccagno – Japanese apricot

Prunus nigra Aiton

Prunus padus L. – bird cherry

Prunus persica (L.) Batsch – peach

Prunus persica (L.) Batsch var. *nectarina* (Aiton) Maximowicz – nectarine

Prunus persica var. *nucipersica* (Suckow) C.K. Schneid – nectarine

Prunus spinosa L. – sloe

Prunus virginiana L. – common chokecherry

Psidium guajava L. – guava

Pyrus communis L. – pear

Pyrus pyrifolia (N.L. Burman) Nakai – Japanese pear, 'Nashi'

Raphanus sativus L. – radish

Ribes nigrum L. – black currant

Ricinus communis L. – castor oil plant

Rubia cordifolia L.

Rubus fruticosus agg. – blackberry

Rubus idaeus L. subsp. *idaeus* – European red raspberry

Rubus idaeus L. subsp. *strigosus* (Michaux) (Maximowicz) – North American red raspberry

Rubus laciniatus Willdenow – cut-leaved blackberry

Rubus × *loganbaccus* L.H. Bailey hybrids – loganberry and boysenberry

Rubus occidentalis L. – black raspberry

Rubus ursinus Chemisso & Schlechtendal – California blackberry

Rumex obtusifolius L. – broad-leaved dock

Salix acutifolia Willdenow

Sambucus nigra L. – elder

Secale cereale L. – rye

Sinapis alba L. – white mustard, yellow mustard

Solanum berthaultii Hawkes

Solanum etuberosum Lindley

Solanum melongena L. – aubergine

Solanum nigrum L. – black nightshade

Solanum pennellii Corell

Solanum polyadenium Greenman

Solanum tuberosum L. – potato

Solidago virgaurea L. – goldenrod

Sorbus domestica L. – service tree

Sorghum bicolor (L.) Moench – sorghum

Sorghum × *drummondii* (Nees ex Steudel) Millspaugh & Chase – Sudan grass

Sorghum halapense (L.) Persoon – Johnson grass

Spiraea thunbergii Siebold & Blume

Stellaria holostea L. – greater stitchwort

Tanacetum vulgare L. – tansy

Tilia cordata Miller – small-leaved lime

Tilia europaea L. – European lime

Tilia platyphyllos Scopoli – large-leaved lime

Trifolium fragiferum L. – strawberry clover

Trifolium pratense L. – red clover

Trifolium subterraneum L. – subterranean clover

× *Triticosecale* sp. – triticale

Triticum aestivum L. – wheat

Triticum monococcum L. – einkorn

Triticum monococcum L. subsp. *aegilopoides* (Link) Thellung

Triticum turgidum L. – emmer

Triticum urartu Tumanian ex Gandilyan

Tropaeolum majus L.

Tsuga canadensis (L.) Carrière

Tsuga caroliniana Engelmann

Ulmus davidiana var. *japonica* (Sargent ex Rehdes) Nakai – Japanese elm

Ulmus glabra Hudson – wych elm

Vaccinium angustifolium Aiton – lowbush blueberries

Vaccinium corymbosum L. – highbush blueberries

Vaccinium macrocarpon Aiton – American cranberry

Viburnum opulus L. – guelder-rose

Viburnum trilobum Marshall

Vicia faba L. – broad bean, field bean

Vigna unguiculata (L.) Walpers – cowpea

Vitis arizonica Engelmann – canyon grape

Vitis berlandieri Planchon

Vitis riparia Michaux

Vitis rupestris Scheele

Vitis vinifera L. – grape vine

Zea mays L. – maize, corn (USA)

Index

Page numbers in **bold** refer to illustrations and tables